Embryology of Angiosperms

Edited by B. M. Johri

With 278 Figures

Springer-Verlag
Berlin Heidelberg New York Tokyo 1984

Professor Brij Mohan Johri
Department of Botany
University of Delhi
Delhi-110007, India

ISBN 3-540-12739-9 Springer-Verlag Berlin Heidelberg New York Tokyo
ISBN 0-387-12739-9 Springer-Verlag New York Heidelberg Berlin Tokyo

Library of Congress Cataloging in Publication Data. Main entry under title: Embryology of angiosperms. Includes bibliographies. 1. Angiosperms. 2. Botany—Embryology. I. Johri, B.M. QK495.A1E43 1984 582.13′0433 83-20430.

Typesetting, printing, and bookbinding: Brühlsche Universitätsdruckerei, Giessen
2131/3130-543210

Dedicated to
Professor John M. Coulter
and
Professor Charles J. Chamberlain

Foreword

Thirty-four years have elapsed since the publication of the late Professor P. Maheshwari's text, *An Introduction to the Embryology of Angiosperms*, a work which for many years served as an invaluable guide for students and a rich source book for research workers. Various texts dealing with sections of the broad spectrum of topics encompassed by Maheshwari in his book have appeared in the interim, but a compendious modern work dealing with the whole field has been lacking. This present volume splendidly meets the need, and it is altogether fitting that Professor B. M. Johri, long an associate and close colleague of Professor Maheshwari and himself a prolific contributor to the subject, should have undertaken the task of editing it.

When Maheshwari wrote, it was still feasible for one author to handle the subject, but today even someone with his fine breadth of vision and depth of understanding could not, alone, do it justice. So the effort has to be a collaborative one; and Professor Johri's achievement has been to bring together a team of authoritative collaborators, assign them their responsibilities, and put them to work to produce a text as integrated in its treatment as the diversity of the subject would allow. The product vividly illustrates the advances that have been made in the study of angiosperm reproductive systems in the last 30 years, and the book is surely destined to become the new standard for student and researcher alike.

Like many surveys of its kind, the text inevitably shows just how dependent progress during the last few decades has been on the advent of new technologies. The great observers of earlier generations reached the limits set by the resolution of the optical microscope in their investigations of the cytological and karyological phenomena associated with plant reproduction, and were restricted further by the preparation techniques they had available. Their modern successors have been blessed with new extensions of vision – on the one hand through the advent of electron microscopy, and on the other, through the further technical development of optical microscopy, marked, for example, by the introduction of phase and differential interference contrast, fluorescence techniques and image-processing systems. At the same time, specimen handling methods have improved apace, and the crumbling of hitherto stoutly maintained interdisciplinary barriers has meant that many workers today have gained a new competence to interpret what they observe through the support of physiological and biochemical experimentation. All this adequately accounts for the flood of new information. Yet, if one message emerges from the present text as a whole – urged by author after author – it is that we are now but on the threshold of a new wave of discovery. No living group shows the diversity of reproductive method to be seen in the angiosperms; and no task in the whole field of biological research is more important for mankind than working the systems out and finding how to manipulate them for practical ends, for after all the flowering plants provide the indispensable basis of human life on earth. This volume is not only an epitome of knowledge, but an invitation and a challenge – a challenge, particularly, for the next generation of workers, some of whom may well find in its pages good reason for ignoring the facile appeal of test-tube biology in favour of making their contribution towards finding out more about how real plants go about the job of reproducing.

July 1984

J. Heslop-Harrison, F.R.S.

Royal Society Research Professor
University College of Wales
Welsh Plant Breeding Station
Plas Gogerddan (near Aberystwyth)
Great Britain

Preface

Since the publication of *Recent Advances in the Embryology of Angiosperms* (ed. P. Maheshwari) in 1963, there has been phenomenal progress in almost all areas of reproduction in flowering plants. This progress has resulted from investigations based on optical, electron (TEM, SEM), fluorescence, phase contrast, and interference (Nomarski effect) microscopy, cytochemistry, histochemistry, physiology and biochemistry. However, all this knowledge is scattered in a large number of periodicals and it has become a formidable task for both teachers and students to collect the necessary information. Hence this volume *Embryology of Angiosperms* – an advanced treatise comprising 16 chapters written by specialists who have devoted years of study to the subject.

The significant advances of the last 20–30 years have been discussed. The authors have raised several questions to seek new information, and have made appropriate mention of many unsolved problems. The text is adequately illustrated with line drawings and half-tones, including electron micrographs.

Every effort has been made to present a comprehensive up-to-date account. I have earnestly endeavoured to achieve uniformity in the format. Suggestions for further improvement would be most welcome.

I have no doubt that post-graduate and research students will find *Embryology of Angiosperms* a good source material. To those teaching "reproduction in flowering plants", the volume offers a fount of ready-made material.

July 1984 B. M. Johri

Acknowledgements

Through Dr. G.A. Nogler, Dr. Konrad F. Springer invited me to prepare an English translation of Professor A. Rutishauser's book *Embryologie und Fortpflanzungsbiologie der Angiospermen.* My experience of teaching undergraduate and postgraduate students for over three decades at the Universities of Agra, Rajasthan and Delhi has provided a deep insight into the various aspects of plant embryology. Also, over a span of almost 50 years, I have had unique opportunities to discuss problems of angiosperm embryology with eminent embryologists during my numerous visits both in India and abroad, and at international meetings. The idea of editing a volume consisting of contributions on recent developments in the embryology of angiosperms emerged out of these interactions. Dr. Springer readily agreed to my suggestion, and I am deeply indebted to him.

Professor J. Heslop-Harrison has done me a great favour in agreeing to prepare the "Foreword" to this volume.

I am grateful to the authors who accepted my invitation to write the chapters. The manuscripts were read by my colleagues in the University of Delhi and I especially thank Professor N.S. Rangaswamy, Dr. N.N. Bhandari, Dr. M.R. Vijayaraghavan, Dr. K.R. Shivanna and Dr. S. Natesh for offering valuable suggestions. I greatly appreciate the help rendered by Dr. K. B. Ambegaokar, who redrew several diagrams, reassembled many plates of illustrations, and prepared the indices.

Professor Dr. G. Melchers (Tübingen) provided photographs of the tomato + potato somatic hybrid. Professor Dr. E. Battaglia (Rome) sent a reprint containing the original diagrams of Fritillaria type of embryo sac. Professor Dr. F. Pospisil (Prague) sent photographs of pollen embryo sacs. Professor Dr. O. Erdelská (Bratislava) prepared an illustrated write-up on the role of microcinematography in embryology. Dr. Scott D. Russell (Norman) sent me his unpublished manuscripts, and electron micrographs of male gametes and fertilization in *Plumbago*. Dr. R. Wunderlich (Vienna), Dr. M. Luxova (Bratislava), Dr. L. Ahlstrand (Göteborg) and Dr. B.A. Fineran (Christchurch) made available the literature needed. I am much obliged to all of them for their gracious help.

The facilities at the Department of Botany, University of Delhi, made it possible to prepare the *Embryology of Angiosperms,* and I am grateful to Professor S.C. Maheshwari, Professor H.Y. Mohan Ram, Professor R.N. Kapil and Professor R.N. Chopra (present Head of the Department) who have been continuously looking after my interest since I retired in 1974. I have received much inducement to complete the book expeditiously from my wife Raj, my daughter-in-law Meera, and my son Lalit.

My sincere appreciation is extended to Mr. R. K. Gupta for typing several chapters, and to Mr. Krishan Lal for preparing some illustrations.

From the house of Springer, I am especially thankful to Dr. Dieter Czeschlik, Life-Science Editor, for valuable advice, to the Copy-Editor who made much improvement in the text, to Mrs. Linda Teppert for processing the publication, and to Miss Claudia Grössl who looked after the production of this book.

July 1984 B. M. Johri

Contents

1 Embryology: Then and Now

B.M. Johri and K.B. Ambegaokar (With 22 Figures) 1

1.1 Introduction 1
1.2 Techniques to Study Embryology 2
1.3 Embryo Sacs in Anthers 5
1.3.1 *Leptomeria* 5
1.3.2 *Calotis* 6
1.3.3 *Hyacinthus* 6
1.4 Fritillaria Type of Embryo Sac 12
1.5 Unusual Embryological Features in Loranthaceae . . . 14
1.5.1 Ovary-Ovule Complex 14
1.5.2 Embryo Sac 16
1.5.3 Endosperm 18
1.5.4 Embryo 19
1.5.5 Reproductive Calendar in *Struthanthus vulgaris* 21
1.6 Nutrition of the Ovule and Seed 22
1.6.1 Vascular Supply 23
1.6.2 Haustoria 23
1.6.3 Transfer Cells 28
1.7 Role of Persistent Pollen Tubes 30
1.8 Embryos of Doubtful Origin 32
1.8.1 Antipodal Embryos 33
1.8.2 Endospermal Embryos 33
1.9 Gymnospermous Characters in Angiosperms 37
1.9.1 *Tambourissa* 37
1.9.2 *Butomopsis* 39
1.9.3 *Paeonia* 42
1.10 Future Research 43
References 47

2 The Microsporangium

N.N. Bhandari (With 21 Figures) 53

2.1 Introduction 53
2.2 Morphology and Structure 53
2.3 Ontogeny 55
2.4 Wall Layers 57

2.4.1 Epidermis 57
2.4.2 Endothecium 58
2.4.3 Middle Layers 60
2.4.4 Tapetum 60
2.5 Ubisch Bodies/Orbicules 77
2.5.1 Origin 79
2.5.2 Relation to Exine Deposition: By-Product 82
2.6 Tapetal Membrane: Structure, Origin, and Significance . 83
2.7 Pollenkitt and Tryphine 85
2.8 Sporogenous Tissue: Ultrastructure 88
2.8.1 Cytoplasmic Membranes and Ribosome Population . . 88
2.8.2 Nucleus 90
2.8.3 Nucleolar Cycle 92
2.8.4 Cytoplasmic Organelles During Meiosis 94
2.8.5 Plasmodesmata and Cytoplasmic Connections 95
2.9 Initiation and Control of Meiosis 97
2.9.1 Duration of Meiosis 99
2.9.2 Synthesis of Callose: Deposition and Significance . . . 100
2.9.3 Cytokinesis 106
2.9.4 Cellulosic Wall of Microspore Mother Cell 107
2.10 Some Unusual Features 109
2.11 Conclusions 110
References 111

3 The Ovule

F. Bouman (With 21 Figures) 123

3.1 Historical 123
3.2 Ovular Morphology 125
3.3 Ovule Initiation 126
3.4 Nucellus 128
3.5 Megasporogenesis 131
3.6 Integuments 138
3.7 Types of Ovules 144
3.8 Vascular Supply of the Ovule 147
3.9 Special Structures 148
3.10 Ovule Reduction 151
3.11 Concluding Remarks 153
References 153

4 The Female Gametophyte

M.T.M. Willemse and J.L. van Went (With 16 Figures) 159

4.1 Introduction 159
4.2 Types of Megagametophyte Development 159
4.3 Development of the Megagametophyte 164
4.3.1 The Megaspore 164

4.3.2 The Coenocytic Megagametophyte 165
4.4 The Relation with the Nucellar Tissue 168
4.5 Organization of the Embryo Sac 170
4.5.1 The Synergids 174
4.5.2 The Egg Cell 183
4.5.3 The Central Cell 185
4.5.4 The Antipodal Cells 188
4.6 Female Gametophyte Development 190
References 191

5 The Pollen Grain

R.B. Knox (With 24 Figures) 197

5.1 Introduction 197
5.2 Pollen Structure and Cytochemistry 198
5.2.1 Pollen Shape 199
5.2.2 Pollen Cytology 199
5.2.3 Pollen-wall Structure 204
5.2.4 The Pollen Tube 208
5.2.5 Structural Adaptations of Pollen for Dispersal 210
5.3 Formation of Pollen 225
5.3.1 Cytology of Development 225
5.3.2 The Primexine 232
5.3.3 Establishment of Apertures 237
5.3.4 Exine Differentiation and Maturation 238
5.3.5 Intine Synthesis and Deposition 242
5.3.6 Mechanism of Cohesion in Composite Pollen 244
5.3.7 Pollen Sterility 245
5.4 Pollen Germination and Pistil Interactions 245
5.4.1 Pollen Quality 245
5.4.2 Pollen Germination and Tube Development 247
5.4.3 Diagnostic Landmarks of Pollination 249
5.4.4 Role of Pollen in Fertilization and Seed-setting 252
5.5 Pollen-wall Proteins and Allergens 253
5.6 Conclusions and Future Developments 259
References 261

6 Fertilization

J.L. van Went and M.T.M. Willemse (With 13 Figures) 273

6.1 Introduction 273
6.2 Stigma and Style 273
6.2.1 Function and Structure of the Stigma 273
6.2.2 The Nature of the Stigma Covering 275
6.2.3 Function and Structure of the Style 277
6.3 Pollen Germination 279
6.3.1 Sticking and Hydration 279

6.3.2 Pollen Tube Formation and Growth 281
6.3.3 The Sperm Cell 283
6.4 Pollen-Pistil Interaction 284
6.4.1 Pollen Recognition, Acceptance, and Rejection 284
6.4.2 Pollen Incompatibility 286
6.4.3 Entry of Pollen Tube into Stigma and Style 289
6.4.4 Pollination Effects 291
6.5 Entrance and Discharge in Embryo Sac 292
6.5.1 Course of the Pollen Tube 292
6.5.2 Entry into the Embryo Sac 294
6.5.3 Growth Through the Filiform Apparatus 297
6.5.4 Entry into the Synergid 298
6.5.5 Transfer of Tube Content 301
6.6 Fusion of Gametes 302
6.6.1 Fusion of Nuclei 305
6.7 The Progamic Phase and Fertilization 308
References 309

7 The Endosperm

M.R. Vijayaraghavan and K. Prabhakar (With 30 Figures) . 319

7.1 Introduction 319
7.2 Nuclear Endosperm 319
7.2.1 Wall Formation 322
7.2.2 Cucurbitaceae 322
7.2.3 Leguminosae 323
7.2.4 Palmae 323
7.2.5 Proteaceae 324
7.2.6 Histochemistry and Ultrastructure 324
7.3 Cellular Endosperm 330
7.3.1 Acanthaceae 330
7.3.2 Cyrillaceae 330
7.3.3 Gesneriaceae 330
7.3.4 Icacinaceae 331
7.3.5 Santalaceae 331
7.3.6 Loasaceae 332
7.3.7 Loranthaceae 332
7.3.8 Scrophulariaceae 335
7.3.9 Histochemistry and Ultrastructure 337
7.4 Helobial Endosperm 338
7.4.1 Salient Features in Monocotyledons 339
7.4.2 Histochemistry and Ultrastructure 341
7.5 Ruminate Endosperm 342
7.6 Central Cell 345
7.6.1 Central Cell Cytoplasm 346
7.6.2 Central Cell Nucleus 349
7.6.3 Covering of the Central Cell 349

7.7 Wall Formation in Endosperm 350
7.8 Cytology of the Endosperm 355
7.9 Reserve Materials in Developing Endosperm 361
7.9.1 Protein Bodies 361
7.9.2 Starch 363
7.10 Embryo-Endosperm Relationship 364
7.11 Incorporation of Nucellus and Integument in Endosperm Formation 368
7.12 Conclusions and Prospects 369
References 370

8 The Embryo

S. Natesh and M.A. Rau (With 29 Figures) 377

8.1 Historical 377
8.2 Zygote 378
8.2.1 Structure and Composition 379
8.2.2 Size Adjustments 382
8.2.3 Polarity 383
8.3 Early Embryogenesis 384
8.3.1 Cell Patterns 384
8.3.2 Tetrad, Quadrant, and Octant Proembryos 385
8.3.3 Stages Leading to Mature Embryo 387
8.4 Classification Based on Early Development of the Embryo 388
8.4.1 Systems Suggested by Schnarf and Johansen 388
8.4.2 System Suggested by Souèges 392
8.4.3 Other Systems of Classification 393
8.5 Differentiation in Embryo 397
8.6 Dicot and Monocot Embryo 398
8.7 The Grass Embryo 400
8.8 The Embryo in Palms 403
8.9 Ultrastructural and Cytochemical Aspects 405
8.10 Suspensor: Structure and Function 414
8.11 Deviations from Usual Development 424
8.11.1 Embryogeny in *Paeonia* 424
8.11.2 Embryos Devoid of Organs 425
8.11.3 Chimeral Embryos 430
8.12 Concluding Remarks 433
References 434

9 Polyembryony

K.K. Lakshmanan and K.B. Ambegaokar (With 12 Figures) . . 445

9.1 Introduction 445
9.2 Classification 446

9.2.1 Simple Polyembryony 447
9.2.2 Multiple Polyembryony 447
9.3 Nucellar Polyembryony 447
9.3.1 Rutaceae 448
9.3.2 Anacardiaceae 452
9.3.3 Myrtaceae 452
9.3.4 Cactaceae 454
9.3.5 Orchidaceae 454
9.4 Integumentary Polyembryony 454
9.4.1 Endothelial Polyembryony 455
9.5 Zygotic and Suspensor Polyembryony 456
9.5.1 Zygotic Polyembryony 456
9.5.2 Suspensor Polyembryony 458
9.6 Synergid Polyembryony 458
9.6.1 Twins and Triplets 461
9.7 Polyembryony in Hybrids 463
9.8 Causes of Polyembryony 465
9.9 Induction of Polyembryony 466
9.10 Utilization of Plural Embryos 470
References 470

10 Gametophytic Apomixis

G.A. NOGLER (With 8 Figures) 475

10.1 Introduction 475
10.1.1 Definitions 475
10.1.2 Synonyms 476
10.1.3 Remarks on Terminology 477
10.2 Embryology of Gametophytic Apomicts 477
10.2.1 Development of Unreduced Embryo Sacs 477
10.2.2 Formation of Endosperm and Embryo 487
10.2.3 Male Meiosis and the Male Gametophyte 493
10.3 Apomixis and Sexuality 494
10.3.1 Sexual Potentials in Gametophytic Apomicts 494
10.3.2 Possibilities of Influencing the Degree of Apomixis . . . 495
10.4 Causes and Consequences of Apomixis 497
10.4.1 Heterozygosity and Variability 497
10.4.2 Causes of Gametophytic Apomixis; Components of Apomixis in Sexual Plants 498
10.4.3 Apomixis and Polyploidy 500
10.4.4 Agamic Species Complexes and Their Evolution 501
10.5 Genetics of Gametophytic Apomixis 503
10.5.1 Dominant or Recessive Inheritance? 503
10.5.2 Some Critical Remarks 504
10.5.3 Some Results, Conclusions, and Prospects 506
10.6 Apomixis and Breeding. Conclusions 508
References 510

11 Role of Polyploidy in Reproductive Organs and Tissues

F. D'Amato (With 16 Figures) 519

11.1 Introduction 519
11.2 Methods of Polyploidization and Ploidy Determination . 519
11.2.1 Acytokinetic Mitosis (AM) 520
11.2.2 Spindle Fusion (SF) 520
11.2.3 Restitutional Mitosis (RM) 520
11.2.4 Endomitosis (Em) 520
11.2.5 Chromosome Endoreduplication (Er) 521
11.2.6 Ploidy Determination 522
11.3 Anther 523
11.3.1 Anther Wall and Connective 523
11.3.2 Tapetum 523
11.3.3 Pollen Grains 529
11.4 Ovule 530
11.4.1 Nucellus 530
11.4.2 Integument(s) and Endothelium 530
11.5 Embryo Sac 531
11.5.1 Egg Cell 531
11.5.2 Synergids 532
11.5.3 Proendospermic Cell 533
11.5.4 Antipodal Cells 534
11.6 Endosperm 537
11.6.1 Origin and Types of Endosperm 537
11.6.2 Polyploidy in the Endosperm 537
11.7 Embryo 543
11.7.1 Embryo Proper 543
11.7.2 Suspensor 546
11.7.3 Nonzygotic Embryos 557
11.8 Physiological Role of Polyploidy 558
11.9 Future Research 560
References 561

12 The Seed: Structure

F.D. Boesewinkel and F. Bouman (With 16 Figures) 567

12.1 Introduction 567
12.1.1 Importance of Seeds 567
12.1.2 Historical Survey 568
12.2 Structure of Seeds 569
12.2.1 General Morphology 569
12.2.2 Vascular Supply of the Seed 571
12.2.3 Seed Types 572
12.2.4 Development and Structure of Seed-coats 575
12.2.5 Differentiation 579

12.2.6 The Seed-coat Surface 585
12.2.7 Stomata in Seed-coats 589
12.3 Special Structures, Adaptations, and Functions 590
12.3.1 Sarcotestal Seeds 590
12.3.2 Arillate Seeds 590
12.3.3 Jaculator 594
12.3.4 Winged Seeds 594
12.3.5 Haired Seeds 594
12.3.6 Dust and Balloon Seeds 595
12.3.7 Hydrochoric Seeds 596
12.3.8 Mucilage Seeds 596
12.3.9 Active Dispersal of Seeds (Autochoric Seeds) 598
12.3.10 Color of Seeds 598
12.4 Adaptations to Germination 599
12.4.1 Opercula 599
12.4.2 Imbibition Lids 602
12.4.3 Epistase and Hypostase 602
12.5 Rumination and Labyrinth Seeds 603
12.5.1 Labyrinth Seeds 605
12.6 Storage Tissue 605
12.7 Concluding Remarks 607
References 608

13 The Seed: Germination

J.V. JACOBSEN (With 7 Figures) 611

13.1 Introduction 611
13.2 General Considerations 612
13.2.1 Internal Morphology 612
13.2.2 Composition of Seeds and Reserve Foods 613
13.3 Germination: Structural, Compositional, and Physiological Changes 614
13.3.1 Dicotyledons 616
13.3.2 Monocotyledons 629
13.4 Storage and Aging of Seeds: Structure and Physiology . 637
References 639

14 Embryology and Taxonomy

J.M. HERR, JR. (With 11 Figures) 647

14.1 Introduction 647
14.2 Embryological Features of Taxonomic Importance . . . 649
14.3 Applications of Embryology to Problems of Family Relationships 651
14.3.1 Podostemaceae 651

14.3.2 Basellaceae 655
14.3.3 Stachyuraceae 657
14.3.4 Frankeniaceae 659
14.3.5 Tropaeolaceae 661
14.3.6 Salvadoraceae 662
14.3.7 Centrolepidaceae 664
14.4 Applications of Embryology to Problems of Taxonomic Structure Within Families 666
14.4.1 Aizoaceae 667
14.4.2 Vacciniaceae 667
14.4.3 *Zannichellia* 669
14.5 Applications of Embryology to Problems of Generic Affinity and Assignment 671
14.5.1 *Ploiarium* 671
14.5.2 *Filicium* 672
14.5.3 *Nyctanthes-Jasminum* 673
14.5.4 *Ophiopogon* 677
14.5.5 *Kyllinga* 678
14.5.6 *Burmannia pusilla* 679
14.6 Contributions of Embryology to Taxonomic Problems Still Unsolved 680
14.6.1 Erythroxylaceae-Linaceae 680
14.6.2 Theligonaceae 681
14.6.3 *Blandfordia* 682
14.6.4 Philydraceae 682
14.6.5 Eriocaulaceae 683
14.7 Confirmation or Alteration of Earlier Conclusions . . . 683
14.7.1 Cactaceae 684
14.7.2 Garryaceae 684
14.7.3 Orobanchaceae 684
14.7.4 *Pentaphragma* 685
14.8 Embryology in Relation to Taxonomy – the Prospect . . 685
14.8.1 *Cornus* 686
14.8.2 *Glycine* and *Phaseolus* 688
14.8.3 *Eustachys petraea* and *E. glauca* 690
References 691

15 Homologies and Phylogeny

M. Favre-Duchartre (With 17 Figures) 697

15.1 Introduction 697
15.1.1 Sexuality and Overall Phylogeny 699
15.2 Microsporangia 700
15.2.1 Microsporangium Wall 700
15.2.2 Microsporangium Content 701
15.2.3 Microsporangium Number and Position 702

15.3 Megasporangia 703
15.3.1 General Structure 703
15.4 Ontogeny of Nucellus 705
15.4.1 Tapetum/Endothelium 706
15.4.2 Megasporangium Content 706
15.4.3 A Proposed Macroevolution of Meiosporic Structures . 706
15.5 Male Gametophyte 707
15.5.1 Extrinsic to Angiosperms 707
15.5.2 Intrinsic to Angiosperms 709
15.6 Female Gametophyte 711
15.6.1 Female Sexual Structures in Chlorophyta Extrinsic to Angiosperms 711
15.6.2 Embryo Sac 713
15.6.3 Diversity in Embryo Sacs 715
15.6.4 Ontogeny of Embryo Sac as Compared to Female Gametophyte of *Taxus* 717
15.7 Pollen Reception by Female Organ 717
15.7.1 Fertilization 719
15.7.2 Double Fertilization, Extrinsic to Angiosperms 719
15.7.3 Intrinsic to Angiosperms 721
15.7.4 Cytology of Zygote Formation 721
15.8 Embryogenesis 723
15.8.1 Albuminogenesis (or Endosperm Formation) in Angiosperms 724
15.8.2 Embryo 726
15.9 Seed 728
15.9.1 Time Relations in the Development of Seed 728
15.9.2 Diversity in Seed Structure 730
15.10 Concluding Remarks 730
15.10.1 Questions Raised by Natural Diversities 730
15.10.2 Questions Raised by Suggested Experiments 731
References 731

16 Experimental Embryology

B.M. Johri and P.S. Rao (With 15 Figures) 735

16.1 Introduction 735
16.2 The Problems: Role of Experimental Embryology . . . 736
16.3 Techniques of Cell, Tissue, and Organ Culture 738
16.4 Floral Organs and Buds 739
16.4.1 In Vivo Studies 739
16.4.2 In Vitro Studies 740
16.5 Anther and Pollen Culture 743
16.5.1 Cultural Procedure and Nutritional Requirements . . . 743
16.5.2 Anther Culture 746
16.5.3 Pollen Culture 749

16.5.4 The Pollen Plants 751
16.5.5 Use of Haploids 751
16.6 Ovary, Ovule, and Nucellus Culture 752
16.6.1 Ovary Culture 753
16.6.2 Ovule Culture 754
16.6.3 Nucellus Culture 757
16.7 Control of Fertilization 761
16.7.1 Intra-ovarian Pollination and Fertilization 761
16.7.2 In Vitro Pollination of Ovules and Fertilization 762
16.7.3 Inter- and Intrageneric Hybrids 764
16.8 Endosperm Culture 765
16.8.1 Immature Endosperm 766
16.8.2 Mature Endosperm 766
16.8.3 Embryo-Endosperm Relationship 766
16.8.4 Organogenesis in Endosperm Cultures 767
16.9 Embryo Culture 769
16.10 Polyembryony 773
16.11 Control of Embryogenesis 777
16.11.1 Loss of Embryogenic Potency 778
16.11.2 Ontogeny of Embryos 779
16.12 Protoplast Culture 782
16.12.1 Embryogenesis and Regeneration 782
16.12.2 Protoplast Fusion and Somatic Hybridization 784
16.12.3 "Pomatoes" and "Topatoes" 788
16.13 Concluding Remarks 788
References 789

Plant Index 803

Subject Index 819

Contributors

AMBEGAOKAR, K.B., Department of Botany, University of Delhi, Delhi 110007, India

BHANDARI, N.N., Department of Botany, University of Delhi, Delhi 110007, India

BOESEWINKEL, F.D., Hugo de Vries Laboratory, University of Amsterdam, 1018 DD Amsterdam, The Netherlands

BOUMAN, F., Hugo de Vries Laboratory, University of Amsterdam, 1018 DD Amsterdam, The Netherlands

D'AMATO, F., Institute of Genetics, University of Pisa, 56100 Pisa, Italy

FAVRE-DUCHARTRE, M., Botanical Laboratory, Faculty of Science, University of Reims, 51062 Reims Cedex, France

HERR, J.M., JR., Department of Biology, University of South Carolina, Columbia, SC 29208, USA

JACOBSEN, JOHN V., Division of Plant Industry, Commonwealth Scientific and Research Organization, Canberra City, A.C.T. 2601, Australia

JOHRI, B.M., Department of Botany, University of Delhi, Delhi 110007, India

KNOX, R.B., Plant Cell Biology Research Centre, School of Botany, University of Melbourne, Parkville, Victoria 3052, Australia

LAKSHMANAN, K.K., Department of Botany, Bharathiar University, Coimbatore 641041, India

NATESH, S., Department of Botany, SGTB Khalsa College, University of Delhi, Delhi 110007, India. (Present address: Department of Science and Technology, Technology Bhavan, New Mehrauli Road, New Delhi 110016, India

NOGLER, G.A., Institute for General Botany, Federal University of Technology (ETH), 8092 Zurich, Switzerland

PRABHAKAR, KUMKUM, Department of Botany, Maitreyi College, University of Delhi, Netaji Nagar, New Delhi 110023, India

RAO, P.S., Plant Morphogenesis and Tissue Culture Section, Bio-Organic Division, Bhabha Atomic Research Centre, Trombay, Bombay 400085, India

RAU, M.A., Department of Applied Botany, University of Mysore, Manasa Gangotri, Mysore 570006, India (962 Lakshmipuram, Mysore 570004, India)

VAN WENT, J.L. Department of Plant Cytology and Morphology, Agricultural University, Botanical Laboratory, 6703 BD Wageningen, The Netherlands

VIJAYARAGHAVAN, M.R., Department of Botany, University of Delhi, Delhi 110007, India

WILLEMSE, M.T.M., Department of Plant Cytology and Morphology, Agricultural University, Botanical Laboratory, 6703 BD Wageningen, The Netherlands

1 Embryology: Then and Now

B. M. Johri and K. B. Ambegaokar

1.1 Introduction

Maheshwari (1950) has summarized the history of development of our knowledge of embryology of angiosperms. The highlights are presented here. Sexuality in angiosperms has been known since the third century B.C. when the Arabs and Assyrians used to perform the ritual of artificial pollination of the date palm. They were not, however, aware of sexuality as such, nor what followed after dusting the female flowers with male flowers.

Camerarius (1694), at the University of Tübingen, established sexuality in *Mercurialis annua*. Kölreuter (1761), also at Tübingen, demonstrated that seeds could be produced by artificial pollination in *Dianthus, Hyoscyamus, Matthiola,* and *Nicotiana*. However, he hardly visualized the great importance of the technique of artificial pollination in plant breeding and crop improvement.

A little later, Amici (1824), an Italian astronomer and also a microscope maker, while studying the stigma of *Portulaca oleracea,* observed that a pollen grain burst forth into a tube. He (Amici 1830) made a detailed examination of *Hibiscus syriacus* and *P. oleracea,* and pointed out that the pollen tubes had grown, bit by bit, and came in contact with the ovule, one for each ovule.

In a meeting held at Padua in 1842, Amici emphasized that the embryo did not originate from the tip of the tube, but from a portion of the ovule which had already existed before the entry of the pollen tube. Finally, in *Orchis,* Amici (1847) reported that in an embryo sac a "germinal vesicle" (egg cell) was present before the entry of the pollen tube, and the embryo originated not from the tip of the tube, but from the germinal vesicle.

Schleiden (1837, 1845) confirmed Amici's work about the growth and entry of the pollen tube into the ovule through the micropyle. However, he still maintained that the tip of the pollen tube becomes the "embryonal vesicle", and undergoes a number of divisions to form the embryo.

Hofmeister (1849) confirmed Amici's observations, and pointed out that in all the 38 species belonging to 19 genera of angiosperms examined by him the embryo originated from a pre-existing cell in the embryo sac, and not from the tip of the pollen tube. His observations and description were convincing, and were confirmed by other workers. Thus ended the fierce controversy between Schleiden and his opponents.

Hofmeister (1849) also studied the formation of tetrads during microsporogenesis in *Tradescantia;* development and organization of female gametophyte, and formation of Cellular type of endosperm in *Monotropa hypopitys*.

Strasburger (1879) worked out the development of female gametophyte in *Polygonum divaricatum,* and observed the formation of linear tetrad, establishment of polarity at binucleate stage, and organization of mature eight-nucleate embryo sac. In 1884 he reported the widespread occurrence of binucleate (bicelled) pollen grains in several taxa, the formation of vegetative and generative cell in a pollen grain, and the formation of male gametes. But a significant research was the discovery of syngamy in *Monotropa*. In *M. hypopitys* he (Strasburger 1884) observed the tip of the pollen tube with a vegetative nucleus and two male gametes, the entry of the pollen tube through the micropyle, and the process of fusion of one male gamete with the egg nucleus. He also mentioned that the cytoplasm of the pollen tube does not enter into the egg cell.

Strasburger (1878) traced the development of adventive embryos from the nucellus in *Citrus aurantium, Coelebogyne* (*Alchornea*) *ilicifolia, Funkia* (= *Hosta*) *ovata*, and *Nothoscordum fragrans*.

Nawaschin (1898) discovered double fertilization in *Fritillaria tenella* and *Lilium martagon*. He observed the union of one male gamete with the egg cell (syngamy) and the fusion of the other with the two polar nuclei (triple fusion). Guignard (1899) also observed double fertilization in *Lilium* and *Fritillaria*, and illustrated his findings with magnificent illustrations.

Hanstein (1870) investigated the development of embryo of *Capsella bursa-pastoris*. He could follow the formation of quadrants, octants, delimitations of three germinal layers, and even hypophysis. In monocotyledons he studied the development of embryo of *Alisma plantago*.

With all the basic facts about reproduction in flowering plants known by the end of the nineteenth century, embryology of angiosperms became an established discipline. To begin with, most investigators looked for specific stages of development. With stage-oriented approach, some of the more interesting features were missed, and this often led to misinterpretations.

By the 1920's a great deal of embryological information had accumulated, and it prompted Schnarf (1931) to publish the first comparative account under the title „Vergleichende Embryologie der Angiospermen".

With improvements in microscopy, and refinements in techniques, there have been significant advances in our knowledge of embryology. These are dealt with in different chapters.

Chapter 1 is devoted to techniques to study embryology, nutrition of ovule and seed, and some unusual and interesting features in the embryology of angiosperms.

1.2 Techniques to Study Embryology

The earlier investigations were carried out by selecting such taxa as had a more or less transparent ovary wall and ovules. The larger ovules were dissected to follow the later stages of development of endosperm and embryo. All stages of microsporogenesis and male gametophyte could be studied from squash preparations. The "rocking" microtome made it possible to cut sections, but serial sections were rather difficult; the "rotary" microtome made it much easier to cut the sections swiftly and serially. Paraffin wax is still used as an embedding medium, though plastic-embedded materials yield very thin sections. Ultratome is used for cutting even thinner sections (materials embedded in resins) for electron microscopy.

With advances in microscopy, and use of cyto- and histochemical techniques, it became possible to study major features of wall structure, nuclear number and behaviour during megasporogenesis and in the female gametophyte, development of endosperm, embryo, and seed, and distribution of metabolites at various stages of development. The electron microscope opened up an entirely new field of observation and interpretation of fine structure. Examples of such studies are detailed in several chapters.

Polarizing optics can be used for studying starch grains, cell walls and crystalline inclusions in cells. The distribution of crystals and starch grains in fresh and cleared material is easily determined.

Phase contrast microscopy provides an extremely sensitive technique for the examination of living cells, and weakly stained (IKI) specimens show a remarkable improvement. The spectacular results achieved by Molè-Bajer and Bajer (1968) on mitosis in living cells of endosperm owed their quality to the very thin layers of flattened cells.

Fluorescence microscopy has demonstrated the presence of callose during microsporogenesis (see Fig. 2.18 A, B), and interaction between pollen and stigma in compatible and incompatible reactions. Sporopollenin in the exine of pollen grains is intensively autofluorescent. In the absence of autofluorescence, fluorochromes are used to excite fluorescence (see O'Brien and McCully 1981).

Nomarski optics (differential interference contrast microscope) are used for thick specimens, and the image shows excellent relief (see Fig. 10.6). Herr (1982) has used such optics to study megasporogenesis in cleared wholemounts of ovules of *Planera aquatica* and *Cassia occidentalis*.

The scanning electron microscope has been used to great advantage in the study of surface structures (see Figs. 12.11 and 12.12). In *Vaccinium* (Brisson and Peterson 1975) the freeze-fractured endosperm haustorium (Fig. 7.12 A–C) revealed conspicuous wall ingrowths (characteristic of transfer cells) which provide a much larger surface for absorption of nutrients.

The transmission electron microscope has significantly advanced our knowledge of a wide range of structures. The wall ingrowths in synergids (Fig. 4.8), antipodal cells (Fig. 4.16 A, B), and suspensor (Fig. 8.24) are only a few examples.

Bajer (1955, 1957) worked with living endosperm for studying mitosis. The endosperm cells lack cellulose walls and, hence, the penetration of various chemicals is facilitated. The phases of mitosis are comparatively better seen in the endosperm cells. Some of the monocotyledonous taxa (*Clivia, Colchicum, Haemanthus, Iris*) have thin, partially transparent ovules and are suitable for squash studies.

After 3–6 weeks of pollination endosperm cells show suitable stages of mitotic division. On a large coverslip (1–2 mm thick), within a ring of vaseline of 2-cm diameter, hot agar with sugar solution (agar 0.4%–0.5%, glucose 3%–4.5%) is smeared. The endosperm (with liquid contents) is pressed out of the ovule on to the agar, and covered with another coverslip. The edges of the coverslips are sealed with vaseline or liquid paraffin. By carefully pressing the coverslip the living cells and mitotic spindles are flattened to chromosome thickness. This preparation is inverted and mounted on a suitable metal holder.

Bajer analyzed mitosis from the commencement to the end of the process. The preparations were examined under phase contrast microscope, and photographed.

Molè-Bajer (1955) studied the effect of chemicals, which have an inhibitory effect on respiration, on living endosperm. The treatment was given directly to the excised endosperm, or through agar-glucose medium, or by injection into the embryo sac before excising the endosperm.

Herr (1971) has suggested a clearing squash technique for the study of sequential development of anthers and ovules. The pistils are fixed for 24 h in formalin-propionic-50%-ethanol, or Randolph's modified Navashin fluid (see Johansen 1940), and stored in 70-% ethanol. The material is then treated with a clearing fluid (for 24 h) which contains 85%-lactic acid: chloral hydrate: phenol: clove oil: xylene (2:2:2:2:1 – by weight). (For the sake of convenience we designate this as "Herr's fluid" although the author designated it 4½-clearing fluid.) From these transparent pistils the ovules are dissected, transferred to absolute ethanol, and upgraded in mixtures of ethanol: xylene (3:1, 2:2, 1:2). Finally, the ovules are mounted in "Piccolyte", or "Permount", on specially prepared micro-well slides (designated Raj-slides). This is prepared by fixing, with permount or balsam, two thick cover-slips, 1 cm apart, on a microscopic slide. Originally, Muir et al. (1958) had improvised such a micro-well slide for growing single cells. It is placed on a warming plate at 50 °C for 3 days to harden the mounting medium of the two cover-slips. The ovules (with the clearing fluid) are placed in between the cover-slips, and covered by another cover-glass which rests on the two affixed cover-slips. The clearing fluid gives suffcient transparency to study the cellular structure of the ovules, under phase contrast optics.

Sometimes, a pre-treatment with lactic acid (85%) for 24 h is essential. Smith (1973) placed the ovules of *Cornus* in lactic acid for 4 days before treating them with the clearing fluid. Otherwise, treatment with 10%-potassium hydroxide solution for 2 min, followed by dehydration with 95%-ethanol, is necessary.

Herr (1973) further suggested the use of three modified mixtures for the study of ovules with interference (Nomarski) optics.

1. Iodine-potassium iodide (IKI) + Herr's fluid – prepared by adding 100 mg iodine, 500 mg potassium iodide, and 9.0 gm of Herr's fluid. The ovules pretreated with Herr's fluid are selected under phase contrast optics, and transferred to IKI + Herr's fluid. This mixture increases the contrast of cell structure, sharpens the features, and detects the presence of starch.
2. Benzyl-benzoate (BB) + Herr's fluid – composed of lactic acid: chloral hydrate: phenol: clove oil: xylene: benzyl benzoate (2:2:2:2:1:1 – by weight). The ovules are treated with this mixture for about 2 weeks. While the cytological features, such as nuclei, nucleoli, and vacuoles do not stand out, the cell outlines of the ovules become sharply defined.
3. In contrast to the previous two mixtures, potassium permanganate + Herr's fluid is unstable, and has to be prepared for immediate use – 3 mg potassium permanganate are dissolved in 1 gm Herr's fluid. The treatment with this fluid

sharpens the structure of ovule, particularly the cell layer topography under phase contrast or interference (Nomarski) optics.

Herr's (1982) procedure to make permanent preparations of ovules treated with BB + Herr's fluid is also worthy of attention. A satisfactory mountant is the epoxy resin combination which is used in electron microscopy (Spurr 1969). Ten drops of vinylcyclohexene dioxide are combined with 6 drops of diglycidyl ether of polypropyleneglycol, and 26 drops of nonenyl succinic anhydride. The mixture is freshly prepared and a graded series with clearing fluid: epoxy medium – 3:1, 2:2, 1:3 – is used in upgrading the treated ovules. The ovules cleared for 24 h are passed through this graded series with 15 min exposure to each mixture and, finally, mounted in epoxy medium. A drop of cure accelerator (dimethylamino-ethanol) is added to the medium, and the ovules are mounted on micro-well slides. The preparations are kept in an oven at 60 °C for 24 h, and then observed with phase contrast or interference (Nomarski) optics. The Spurr epoxy mountant avoids the shrinkage artifact.

1.3 Embryo Sacs in Anthers

1.3.1 *Leptomeria*

While the sporogenous cells in anthers normally develop into microspore mother cells, undergo meiosis, and give rise to tetrads, microspores, and pollen grains, in *Leptomeria billardierii* Ram (1959 a) observed a rather unusual feature. Some of the sporogenous cells enlarge, and develop into an uni- or binucleate embryo sac-like structure (Fig. 1.1 A). Two subsequent mitotic divisions result in the formation of an eight-nucleate "embryo sac". These nuclei even organize like a typical embryo sac with three-celled egg apparatus, two polar nuclei, and three antipodal cells (Fig. 1.1 B). This phenomenon has not been reported in any other angiosperm so far.

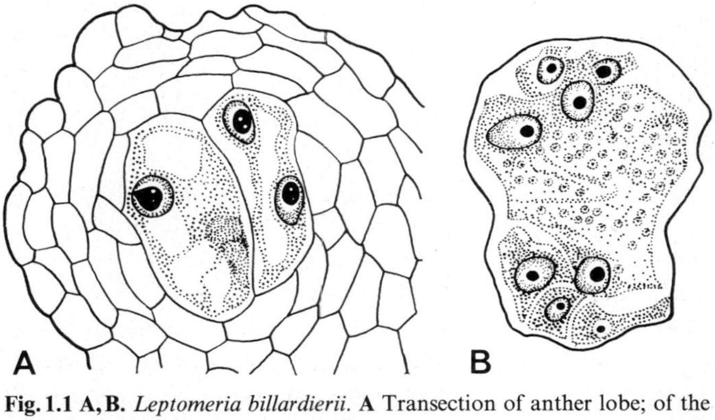

Fig. 1.1 A, B. *Leptomeria billardierii.* **A** Transection of anther lobe; of the two sporogenous cells, one is a binucleate "embryo sac". **B** Organized eight-nucleate "embryo sac". (After Ram 1959 a)

1.3.2 *Calotis*

Davis (1968) describes very interesting behaviour of the anther tapetum in *Calotis lappulacea*. The material was collected in Australia from the Armidale district, Narrabri (100 miles west of Armidale), and Cobar (western New South Wales).

During microsporogenesis the tapetal cells become multinucleate (8–16 nuclei). The release of microspores is associated with the formation of periplasmodium. The mature pollen grains are two- or three-celled.

In the material from all localities, during tetrad formation, one or more tapetal cells become vacuolate and increase in size (Fig. 1.2 A–C). These cells lose contact with the anther wall, some increase considerably in size, and simulate developing embryo sacs (Fig. 1.2 D–G).

In one anther in the Cobar material (Fig. 1.2 E, G), one embryo sac showed the egg apparatus, two antipodal cells (one two-nucleate), and an elongated polar cell (with three nuclei instead of the usual two). The rest of the locule contained only one- or two-nucleate plasmodial tapetal cells (protoplasts). In the corresponding locule of the same anther, only two embryo sac-like structures were present, otherwise it was filled with periplasmodium and uninucleate pollen grains (Fig. 1.2 E, F).

The abnormal behaviour of tapetal cells varies between anthers of the same floret, and floret to floret. In affected anthers there may be many such cells in a microsporangium, two or three degenerated tapetal cells in situ, and neither periplasmodium nor pollen grains are formed (Fig. 1.2 E).

In Fig. 1.2 G the resemblance to an embryo sac is very remote. If at all, it is inversely oriented. We consider that each component of the embryo sac is an independent tapetal cell (protoplast) – at the upper end one is binucleate, one uninucleate, the central elongated protoplast is three-nucleate, and there are three uninucleate protoplasts at the lower end. In Fig. 1.2 F the fate of the two tapetal protoplasts (one two- and the other three-nucleate) can hardly be predicted, and it would be erroneous to interpret them as tapetal embryo sacs.

Davis (1968) refers to Ram's report on the origin of embryo sac-like structures, in *Leptomeria billardierii,* from microspore mother cells (actually sporogenous cells). She considers that the condition in *Leptomeria* may be the same as in *Calotis*. The investigations on *Leptomeria* were conducted under the supervision of one of us (BMJ), and the microscopic preparations were also examined by Professor J. Heslop-Harrison who confirmed Ram's observations. We are convinced that in *Leptomeria* the embryo sac-like structures do develop from sporogenous cells. This is an obligate apomict, and pollen grains do not ever develop in the anthers.

1.3.3 *Hyacinthus*

Němec (1898), for the first time, discovered that the pollen grains in petaloid stamens of *Hyacinthus orientalis* either degenerate, or germinate and give rise to pollen embryo sacs. The pollen grains are of two types, small and large, and both types may develop from the same tetrad.

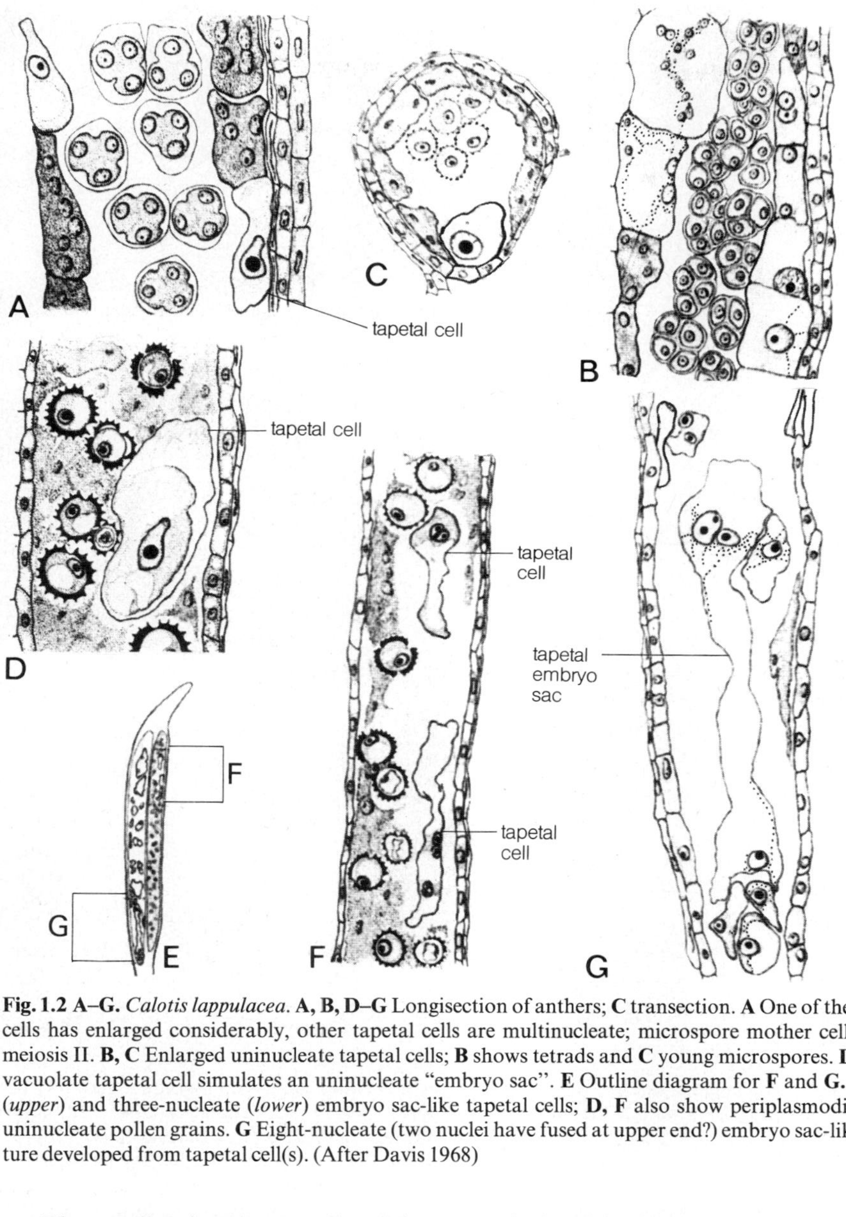

Fig. 1.2 A–G. *Calotis lappulacea*. **A, B, D–G** Longisection of anthers; **C** transection. **A** One of the tapetal cells has enlarged considerably, other tapetal cells are multinucleate; microspore mother cells are at meiosis II. **B, C** Enlarged uninucleate tapetal cells; **B** shows tetrads and **C** young microspores. **D** Large, vacuolate tapetal cell simulates an uninucleate "embryo sac". **E** Outline diagram for **F** and **G. F** Two- (*upper*) and three-nucleate (*lower*) embryo sac-like tapetal cells; **D, F** also show periplasmodium and uninucleate pollen grains. **G** Eight-nucleate (two nuclei have fused at upper end?) embryo sac-like structure developed from tapetal cell(s). (After Davis 1968)

The small uninucleate pollen (Fig. 1.3 A) is devoid of reserve food, and rarely develops into embryo sac. Its nucleus may divide to form two nuclei (Fig. 1.3 B), and they may be partitioned by a wall into two equal cells (Fig. 1.3 C). It may germinate and simulate an early stage of the embryo sac (Fig. 1.3 D).

The large pollen grains have abundant food reserve, may remain uninucleate, or become bicelled (Fig. 1.3 E). In the latter case the generative cell degenerates. As the pollen grain germinates, the protoplast enlarges and elongates, and the

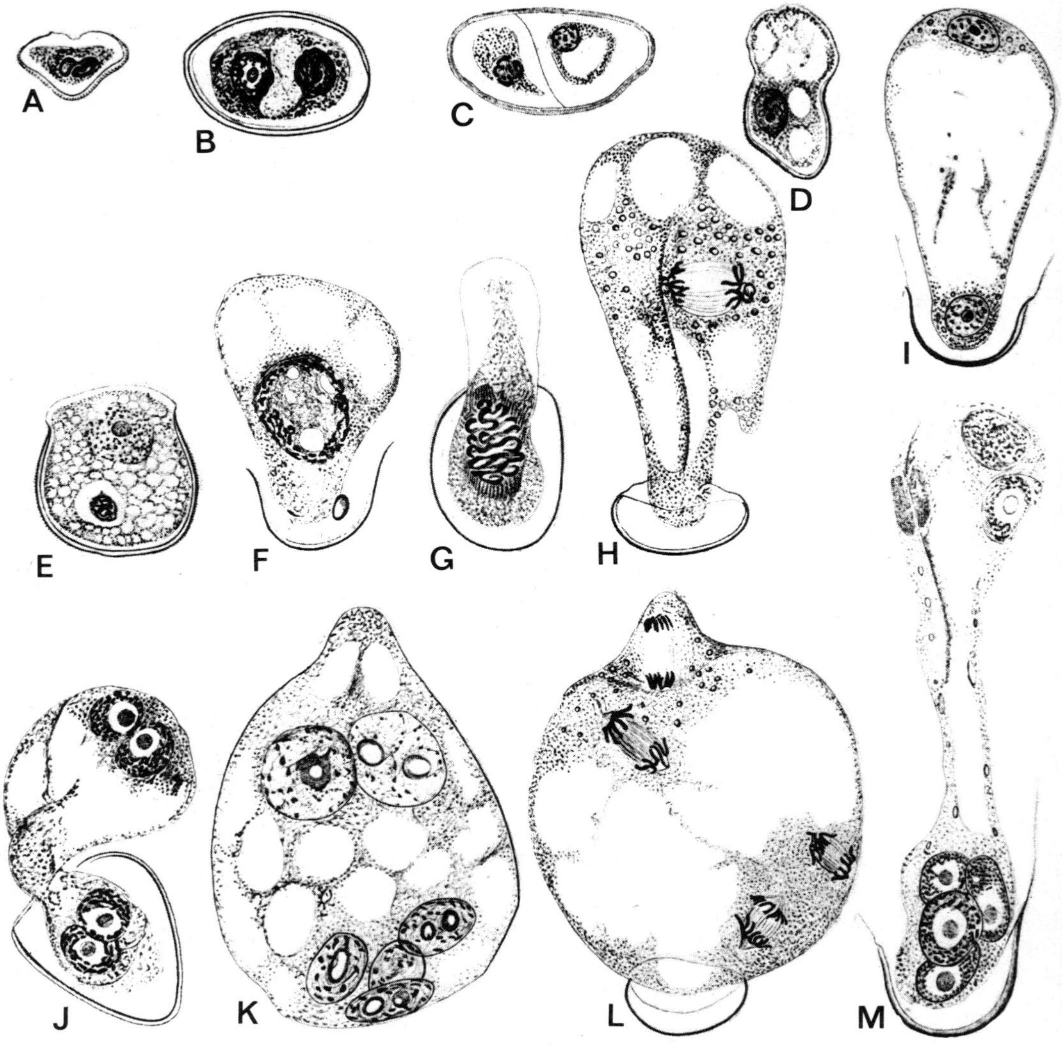

Fig. 1.3 A–P. *Hyacinthus orientalis*. **A** Small pollen grain. **B** Pollen grain with two equal nuclei. **C** The nuclei are separated by a wall. **D** Germinated small pollen grain. **E** Large binucleate (bicelled) pollen grain. **F–H** Germinated pollen grain with vegetative nucleus in division. **I–K** Binucleate (**I**), four-nucleate (**J**), and six-nucleate (**K**) pollen embryo sacs. **L** Simultaneous division of nuclei. **M** Eight-nucleate pollen embryo sac. Note the exine in **F–I** and **L, M. N** Tetrad with three "pollen grains" (p_1, p_2, p_3). The fourth has developed into a three-nucleate "embryo sac", P_3 is without cuticle (exine). **O** Tetrad with three pollen grains (p_1, p_2, p_3), p_2, p_3 without cuticle (exine). The fourth has given rise to an embryo sac which has degenerated. **P** Tetrad with three pollen grains (p_1, p_2, p_3). p_2, p_3 without cuticle (exine). The fourth has developed into an embryo sac, *e* endosperm nucleus, *ov* ovarian apparatus, the two antipodals (an_1, an_2) are in division phase. (After Němec 1898; courtesy Prof. RNDr. F. Pospisil, Praha, Czechoslovakia)

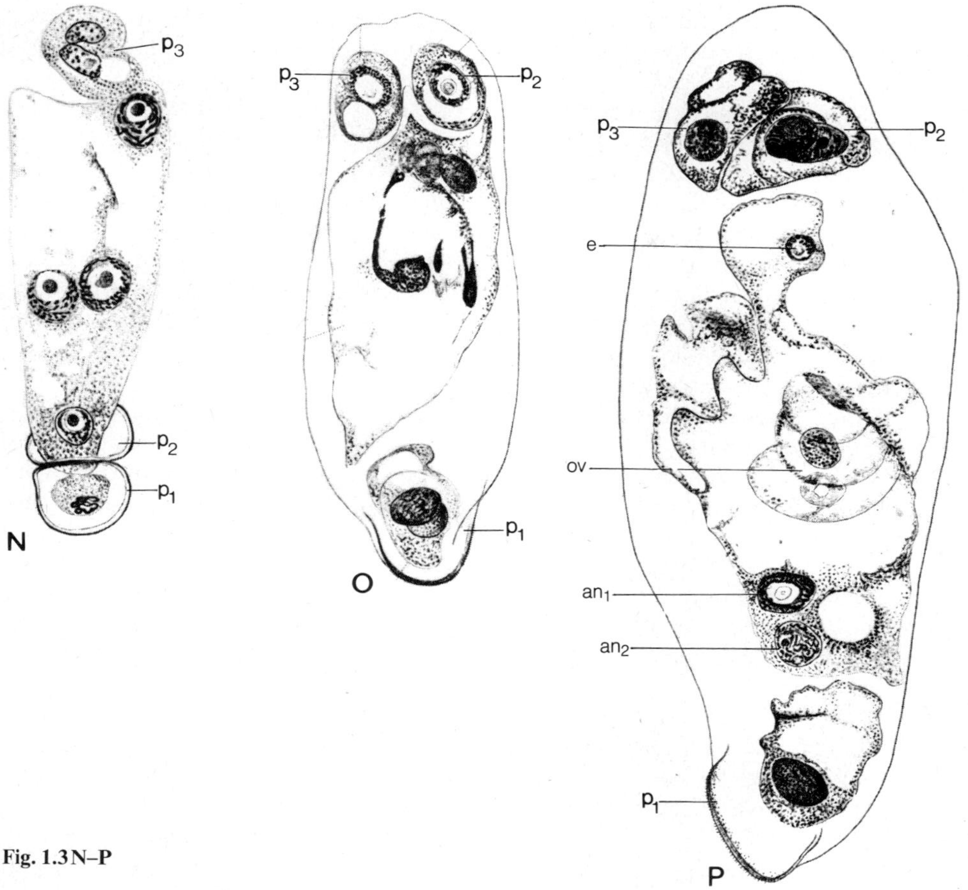

Fig. 1.3 N–P

vegetative nucleus divides (Fig. 1.3 F–H). The two- and four-nucleate pollen embryo sacs are polarized (Fig. 1.3 I, J). Sometimes, the lower nuclei divide earlier so that the "embryo sac" becomes six-nucleate (Fig. 1.3 K). With the division of all the four nuclei (Fig. 1.3 L), an eight-nucleate embryo sac is produced (Fig. 1.3 M). It may be noted that the exine remains attached throughout (Fig. 1.3 F–J, L, M).

Němec also mentions that one component of a tetrad may develop into a pollen embryo sac while the remaining three constituents may remain attached (Fig. 1.3 N–P). The pollen embryo sac is three-nucleate in Fig. 1.3 N, the embryo sac has degenerated in Fig. 1.3 O, and in Fig. 1.3 P Němec considers the smaller upper nucleus as endosperm nucleus, the middle region as ovarian (egg) apparatus, and the two lower nuclei as antipodals in division phase.

We suggest a reinterpretation of Fig. 1.3 N–P. Figure 1.3 N shows an uninucleate (nucleus in division) pollen grain attached to another pollen grain, which has developed into an "embryo sac". The upper two nuclei show cytoplasmic delimitation, and along with the adjoining third nucleus one can visualize the egg apparatus, the two nuclei in the centre may represent polars, and the nucleus at the lower pole the antipodal. Figure 1.3 O may also show an embryo sac developed from pollen grain p_1. The upper nuclei p_2, p_3 simulate synergids, the three nuclei lying below can represent the egg and two polars, then there are two antipodals.

In Fig. 1.3 P also the "embryo sac" appears to have originated from a single pollen. The three nuclei in the upper region, clearly demarcated into cells, can represent the egg apparatus, two nuclei in the centre may be polars, and the remaining three nuclei are the antipodals. The small nucleus (in the upper part of the embryo sac) cannot be the endosperm nucleus as interpreted by Němec.

Stow (1930) reported development of "embryo sac" from single pollen grain in *Hyacinthus orientalis* var. La Victor, and Naithani (1937) in *H. orientalis* var. Yellow Hammer. Němec's observations about the origin of a pollen embryo sac from one of the "pollen grains" in a tetrad, with the other three remaining attached, is not substantiated by these workers.

For early flowering, Naithani (1937) treated the bulbs of *H. orientalis* var. Yellow Hammer at about 70 °F for a few days. Then the temperature was raised to 80°–82 °F for 5 weeks. Later, it was lowered to 70 °F, and the bulbs were brought to room temperature (60 °F). The pollen grains of the treated bulbs showed embryo sac-like structures in the anthers.

According to Naithani, the temperature treatment given to immature bulbs resulted in the production of pollen embryo sacs. During their development a large quantity of reserve food is consumed and the remaining pollen grains collapse due to starvation. Naithani does not attribute any significance to the abortion of pollen grains.

It may be pointed out that in many bulbs the inflorescence is already well differentiated. Naithani (1937) explained the reversal of sex on the basis of Mendelian principles. In Yellow Hammer the genetic balance towards "maleness" and "femaleness" is affected due to temperature treatment.

Geitler (1941) observed pollen embryo sacs with eight to ten nuclei in *Ornithogalum nutans*. The eight-nucleate "embryo sac" simulated a three-celled egg apparatus, two polar nuclei, and three antipodals towards the exine.

Stow (1930) induced pollen embryo sacs in *Hyacinthus orientalis* var. La Victor. Embryo sac-like giant pollen grains are formed only when the bulbs are subjected to 17°–20 °C during the meiosis of pollen mother cells. The "giant" pollen grains are formed only in the normal anthers, and not in petaloid anthers. Stow emphasizes that pollen embryo sacs do not develop if the anthers obtained from the bulb are subjected to high temperature, and then placed in a cold place, or when the bulbs are not treated with temperature but kept in green house. The uninucleate pollen enlarges, undergoes mitotic divisions, and gives rise to pollen embryo sac. The pollen embryo sacs are mostly associated with 90% collapsed pollen grains.

Stow (1934) placed on agar medium pollen embryo sacs with the normal pollen of another variety. The pollen tube grew towards the pollen embryo sac and coiled around it. In one case a sperm was observed to enter the pollen embryo sac. In another, the pollen embryo sac showed 16 nuclei, and Stow presumed that these developed by the division of "primary endosperm nucleus". This is doubtful, and the supernumerary nuclei could have resulted from an additional mitotic division of eight nuclei.

"In the normal condition, the "male potency" is dominant over to the "female", so that the generative nucleus, which becomes sperm nuclei at last, is produced. But in some abnormal conditions, when the abundant quantity of necrohormone produced from dead pollen grains acts together with the suitable temperature, the dominancy of the "male potency" is weakened or disappears. At last the "female potency" becomes stronger, thus the embryo sac-like structure is formed" (Stow 1930, p. 436).

Němec (1898) had also pointed out that the large pollen grains acquire female potency, and develop into "embryo sac".

The question of the effect of "necrohormone" (see Haberlandt 1921) is debatable, since no such hormone has ever been discovered. The degenerated pollen grains may contribute some growth factors, but the acquisition of "female" potency in pollen grains is doubtful.

It is a horticultural practice to give temperature treatment to the bulbs for inducing early flowering. Florists keep the bulbs indoors in subdued light at a temperature of 50 °F for 2 weeks. Then the bulbs are subjected to 70 °F (see Bailey 1950). Only Stow (1930) and Naithani (1937) conducted suitable experiments to show that temperature treatment leads to the formation of pollen embryo sac. It is likely that such a condition is common when the bulbs are given temperature treatment.

The latest report on pollen embryo sacs is that of Vijayaraghavan and Ratnaparakhi (1977), in *Heuchera micrantha*. Some of the pollen grains enlarge, their cytoplasm becomes vacuolate, and they degenerate. In other pollen grains the nucleus of the microspore undergoes one or more mitotic divisions, the pollen grain elongates like an embryo sac, polarity is established, and the disposition of nuclei simulates 2+2 (Polygonum) type of gametophyte. Of the eight nuclei, three at each pole are comparable to egg apparatus and antipodals, and two nuclei in the centre represent polars. The delimitation of nuclei by walls is rare. Occasionally, the embryo sac may contain 8 or 16 nuclei recalling the Plumbago and Penaea type of embryo sac.

The pollen grains, which give rise to embryo sac-like structures, are devoid of starch, and are associated with degenerated microspores, or arrested meiocytes at different stages of development. The exine of the pollen grain ruptures as the pollen embryo sac elongates, but persists at one pole. During the development of pollen embryo sacs the tapetal cells remain healthy and may become palisade-like (Vijayaraghavan and Ratnaparakhi 1977).

The petaloid anther and normal anther in different cultivars of *Hyacinthus orientalis* need reinvestigation. Are pollen embryo sacs formed in both types of anthers, or any in normal anthers? At what stage of microspore mother cells is the temperature treatment to bulbs most effective? The precise temperature and the duration of treatment must also be determined.

The large-scale degeneration of pollen grains alone cannot be the main factor in the induction of pollen embryo sacs. To what extent temperature is responsible requires further critical studies. To presume that the "male potency" of pollen grains changes to "female potency", resulting in pollen embryo sacs, has also to be demonstrated, possibly by using cyto- and histochemical techniques.

1.4 Fritillaria Type of Embryo Sac

In *Lilium martagon* Sargant (1896), for the first time, noticed the fusion of three spindles, of the primary four-nucleate (1+3) embryo sac, at the chalazal end (Fig. 1.4 A). Thus, the secondary four-nucleate embryo sac had two haploid micropylar and two triploid chalazal nuclei (Fig. 1.4 B).

The work of Sargent remained largely unknown. For almost 30 years embryologists considered that the eight-nucleate embryo sac in *Lilium* developed directly from the megaspore mother cell through two meiotic and one mitotic division. Such a sequence of development was designated as the Lilium type (now called Adoxa type). There is no doubt that the investigators did observe the large size of nuclei at the chalazal pole, and attributed it to the excessive availability of nutrition.

Bambacioni-Mezzeti (1928) did not feel convinced that the large size of nuclei (at the chalazal pole of the secondary embryo sac) was really due to excessive supply of nutrition. After a painstaking and critical study of *Fritillaria persica,* she discovered that the hypodermal archesporial cell (Fig. 1.5 A) becomes the megaspore mother cell which undergoes two meiotic divisions, and the coenomegaspore shows 1+3 arrangement of nuclei (Fig. 1.5 B–E). During the first mitotic division the three spindles at the chalazal pole fuse (Fig. 1.5 F, F′). At the completion of the division there are two triploid chalazal nuclei, and two haploid micropylar nuclei. This is the secondary four-nucleate stage of the embryo sac (Fig. 1.5 G–I). It was precisely this behaviour which had been observed by Sargant.

One more mitotic division leads to eight-nucleate embryo sac with four micropylar haploid nuclei which organize into the egg apparatus and upper polar nucleus, and four chalazal triploid nuclei (Fig. 1.5 J–N) which organize into three antipodal cells and one triploid lower polar nucleus.

The pattern of development in *Lilium, Fritillaria* and several other taxa has been designated as the Fritillaria type.

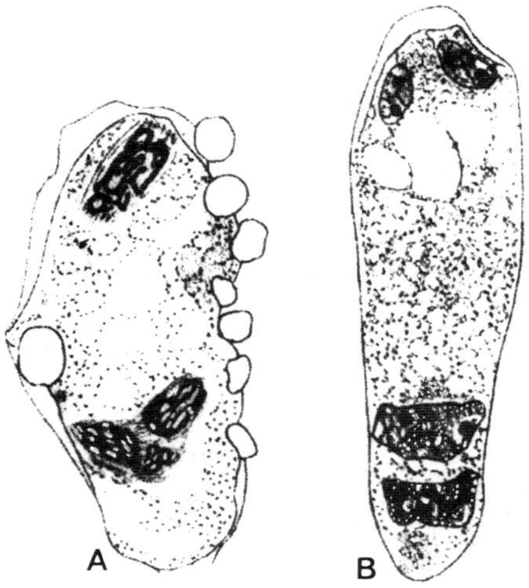

Fig. 1.4 A, B. *Lilium martagon.* **A** Coenomegaspore with 1+3 arrangement of nuclei, all the nuclei are in division phase, and chalazal spindles are in the process of fusion. **B** Secondary four-nucleate embryo sac with two haploid micropylar and two triploid chalazal nuclei. (After Sargant 1896, from Mazzolani 1978; courtesy Prof. E. Battaglia, Rome)

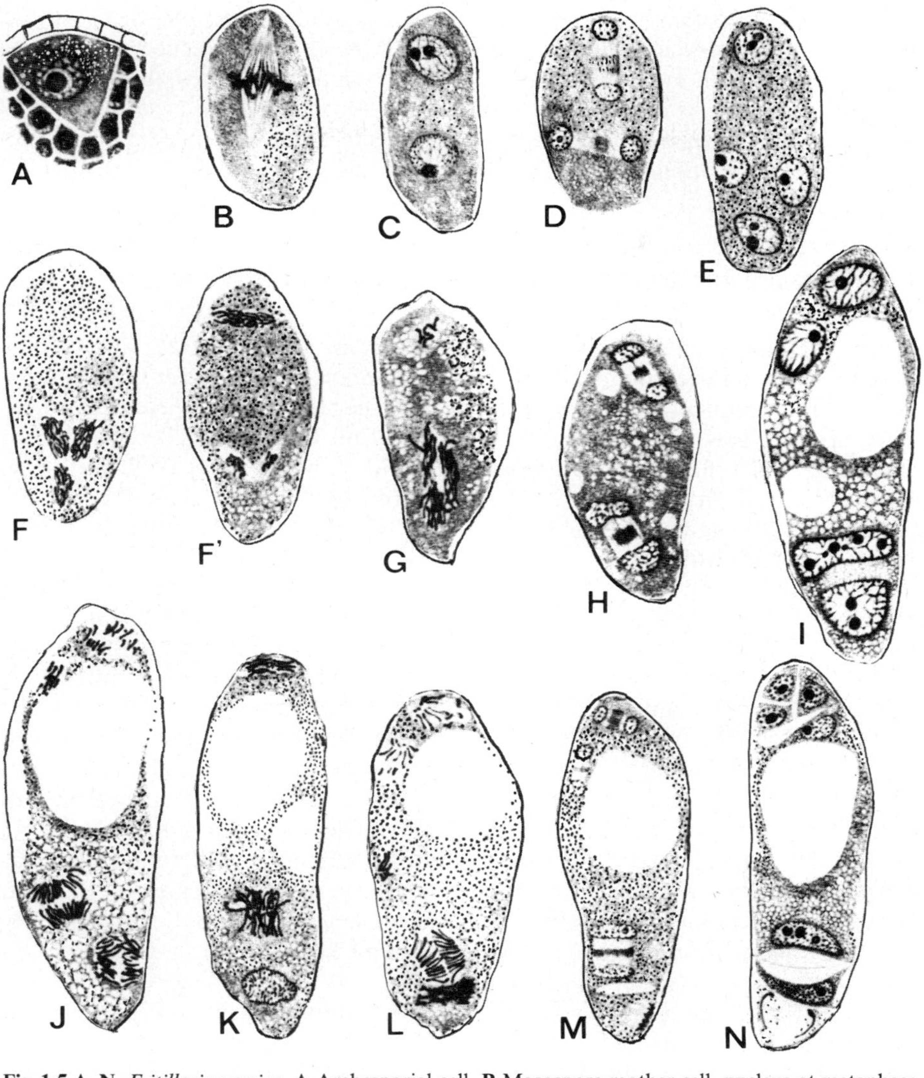

Fig. 1.5 A–N. *Fritillaria persica.* **A** Archesporial cell. **B** Megaspore mother cell, nucleus at metaphase of meiosis I. **C** Two megaspore nuclei. **D** Meiosis II, late telophase. **E** Coenomegaspore with 1 + 3 arrangement of nuclei. **F, F′** Consecutive sections with nuclei (1 + 3) in division. The three spindles at chalazal end fuse during mitosis. **G, H** Later stages. **I** Secondary four-nucleate embryo sac with two micropylar haploid and two chalazal triploid nuclei. **J–M** Mitotic divisions of nuclei. **N** Organized eight-nucleate embryo sac with haploid egg apparatus and upper polar nucleus, and triploid lower polar and three antipodals. (After Bambacioni-Mezzetti 1928, from Mazzolani 1978; courtesy Prof. E. Battaglia, Rome)

It is very important that in such studies one should not feel satisfied with the formal stages of development, but follow very closely the sequence of development. What is even more important is that any unusual features observed must be accounted for, even if they are difficult to interpret. If Bambacioni-Mezzetti had not probed deeper into the reasons for larger size of the chalazal nuclei, the discovery of the Fritillaria type of embryo sac would have been delayed further.

1.5 Unusual Embryological Features in Loranthaceae

The investigations on the embryology of Santalaceae, Loranthaceae, and Balanophoraceae have raised many controversies. Inaccurate observations, and misinterpretations frequently occur in the literature. The reproductive systems in all the three taxa are now well known (Santalaceae – Ram 1957, 1959 a, b, c, P. C. Joshi 1960, Agarwal 1961, 1962, Loranthaceae – B. M. Johri and Raj 1969, Raj 1970, B. M. Johri and Bhatnagar 1972, Bhatnagar and Johri 1983, Balanophoraceae – Arekal and Shivamurthy 1978, Engell 1979). Only the unusual features in Loranthaceae are dealt with here.

It may be of interest to relate that in the 1920's Winfield Dudgeon (Ewing Christian College, Allahabad, India) suggested to one of the research scholars to study the embryology of *Dendrophthoe falcata*. As was the custom in those days, the stigma with the style was usually trimmed. A very large number of microtome preparations were examined, but nothing worthwhile turned up. P. Maheshwari (a pupil of Dudgeon) joined Agra College, Agra, in 1929. He had also examined the preparations, but, in 1936, he passed on the slides (made at Allahabad) to Bahadur Singh. B. Singh could not make any progress either. He decided therefore that if the ovaries did not show the gametophyte, he might examine the style. He cleared the gynoecium and an examination immediately revealed the extensions of embryo sacs in the style. This solved the mystery. A detailed paper by B. Singh (1952) appeared in *Journal of the Linnean Society of Botany* and, later, another on *Macrosolon cochinchinensis* (Maheshwari and Singh 1952) in *Botanical Gazette*.

At that time one of us (BMJ) undertook the study of *Helixanthera ligustrina* (Maheshwari and Johri 1950 a). Since then B. M. Johri and co-workers have made a comparative study of 19 genera, and 27 species (B. M. Johri and Bhatnagar 1972). Several other interesting features have been reported, recently, which deserve attention.

1.5.1 Ovary-Ovule Complex

The Loranthaceae lack a normal ovule, but the ovary-ovule complex shows a gradation with the total absence of placenta (earlier referred to as mamelon) to a well-developed placenta:

1. Placenta absent: *Helixanthera ligustrina* (Fig. 1.6 A_1, A_2), *Dendrophthoe falcata* (Singh 1952), *Barathranthus axanthus* (Prakash 1963).

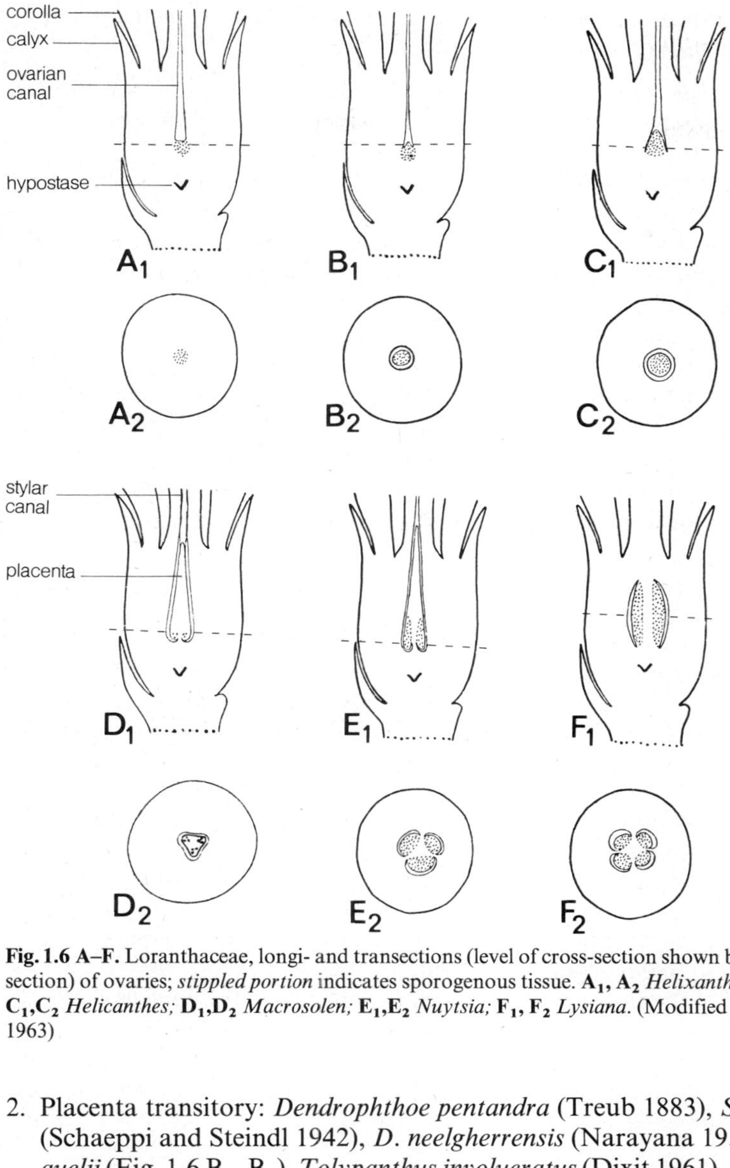

Fig. 1.6 A–F. Loranthaceae, longi- and transections (level of cross-section shown by *dotted line* in longisection) of ovaries; *stippled portion* indicates sporogenous tissue. **A_1, A_2** *Helixanthera;* **B_1, B_2** *Amyema;* **C_1,C_2** *Helicanthes;* **D_1,D_2** *Macrosolen;* **E_1,E_2** *Nuytsia;* **F_1, F_2** *Lysiana.* (Modified after Kapil and Vasil 1963)

2. Placenta transitory: *Dendrophthoe pentandra* (Treub 1883), *Scurrula montana* (Schaeppi and Steindl 1942), *D. neelgherrensis* (Narayana 1956), *Amyema miquelii* (Fig. 1.6 B_1, B_2), *Tolypanthus involucratus* (Dixit 1961), *Struthanthus vulgaris* (Venturelli 1981).
3. Placenta prominent and unlobed, ovary unilocular: *Amyema gravis* (Schaeppi and Steindl 1942), *Helicanthes elastica* (Fig. 1.6 C_1, C_2), *Tapinostemma acaciae* (Johri and Prakash 1965).
4. Placenta three-lobed, ovary unilocular: *Macrosolen cochinchinensis* (Fig. 1.6 D_1, D_2).
5. Placenta three-lobed, ovary trilocular: *Nuytsia floribunda* (Fig. 1.6 E_1, E_2), *Lepeostegeres gemmiflorus* (Dixit 1958 b), *Amylotheca dictyophleba* (Raj 1970).

6. Placenta four-lobed, ovary tetralocular: *Lysiana exocarpi* (Fig. 1.6 F_1, F_2), *Peraxilla tetrapetala* (Prakash 1960).

In contrast, *Phthirusa pyrifolia* (Kuijt and Weberling 1972) lacks the ovarian cavity which is characteristic of the family. The archesporium differentiates just above the collenchymatous pad (hypostase).

In the tribe Elytrantheae the placenta is four-lobed in a tetralocular ovary (*Lysiana, Peraxilla*), or three-lobed in a unilocular ovary (*Macrosolen*). In Nuytsieae the placenta is three-lobed in a trilocular ovary (*Nuytsia*). In Lorantheae the placenta may be absent, rudimentary (*Amyema, Dendrophthoe*), or well developed, unlobed, and free from the ovary wall (*Helicanthes, Lepeostegeres*). Thus, no taxonomic significance can be attached to the presence, or absence, and structure of the placenta.

Cocucci (1983) emphasises the importance of gynoecium, mamelon, and ovules (lobes of mamelon) in the taxonomy of Loranthaceae. Besides the three well recognised tribes (Elytrantheae, Nuytsieae, and Lorantheae), he erects two additional tribes (Tapinostemmae, and Lysianeae). He also questions the interpretation of the "mamelon" as the placenty.

1.5.2 Embryo Sac

The concurrent development of several embryo sacs and, in several taxa, the extension of the tip to various heights in the style, and up to the stigma, is a significant feature of Loranthaceae. In *Peraxilla tetrapetala* (Prakash 1960) and *Atkinsonia ligustrina* (Prakash 1961) the embryo sacs remain confined to the ovary (Fig. 1.7 A). The tip reaches up to the middle of the style in *Tapinanthus uhehensis* (Dixit 1956), *Helicanthes elastica* (Johri et al. 1957), and *Amyema miquelii* (Fig. 1.7 B), and up to the stigma in *Tupeia antarctica* (Smart 1952) and *Barathranthus axanthus* (Prakash 1963). In *Helixanthera ligustrina* (Maheshwari and Johri 1950 a) the tip reaches up to the stigmatic epidermis (Fig. 1.7 C).

In *Mouquiniella rubra* (Fig. 1.7 D) the tip extends up to the stigma and then curves backwards so that it appears 1-shaped.

The lower end of the embryo sac invariably extends up to the hypostase.

In *N. floribunda* (Narayana 1958 b) the upper end of the embryo sac develops a lateral caecum (which may sometimes branch) in the style (Fig. 1.7 E), and the lower end ruptures the tubular hypostase and grows beyond it. This is not reported in any other taxon.

Fig. 1.7. A *Atkinsonia ligustrina,* longisection of ovary showing mature embryo sacs. **B** *Amyema miquelii,* tip of embryo sac with egg apparatus and upper polar nucleus, at two-thirds the height of style. **C** *Helixanthera ligustrina,* longisection of stigma with five embryo sacs; one reaches as far as the stigmatic epidermis. **D** *Moquiniella rubra,* longisection of stigma with four embryo sacs; three reach up to the stigma and their tips curve backwards so that the egg apparatus/proembryo acquire reverse polarity. **E** *Nuytsia floribunda,* tip of embryo sac in the stylar tissue. Note the lateral caecum. **F** Longisection of carpel. The names of different taxa indicate the height to which the tip of the embryo sac reaches in the style and stigma. (**A** After Prakash 1961; **B, F** after B. M. Johri and Bhatnagar 1972; **C** after Maheshwari and Johri 1950 a; **D** after B. M. Johri and Raj 1965; **E** after Narayana 1958 b)

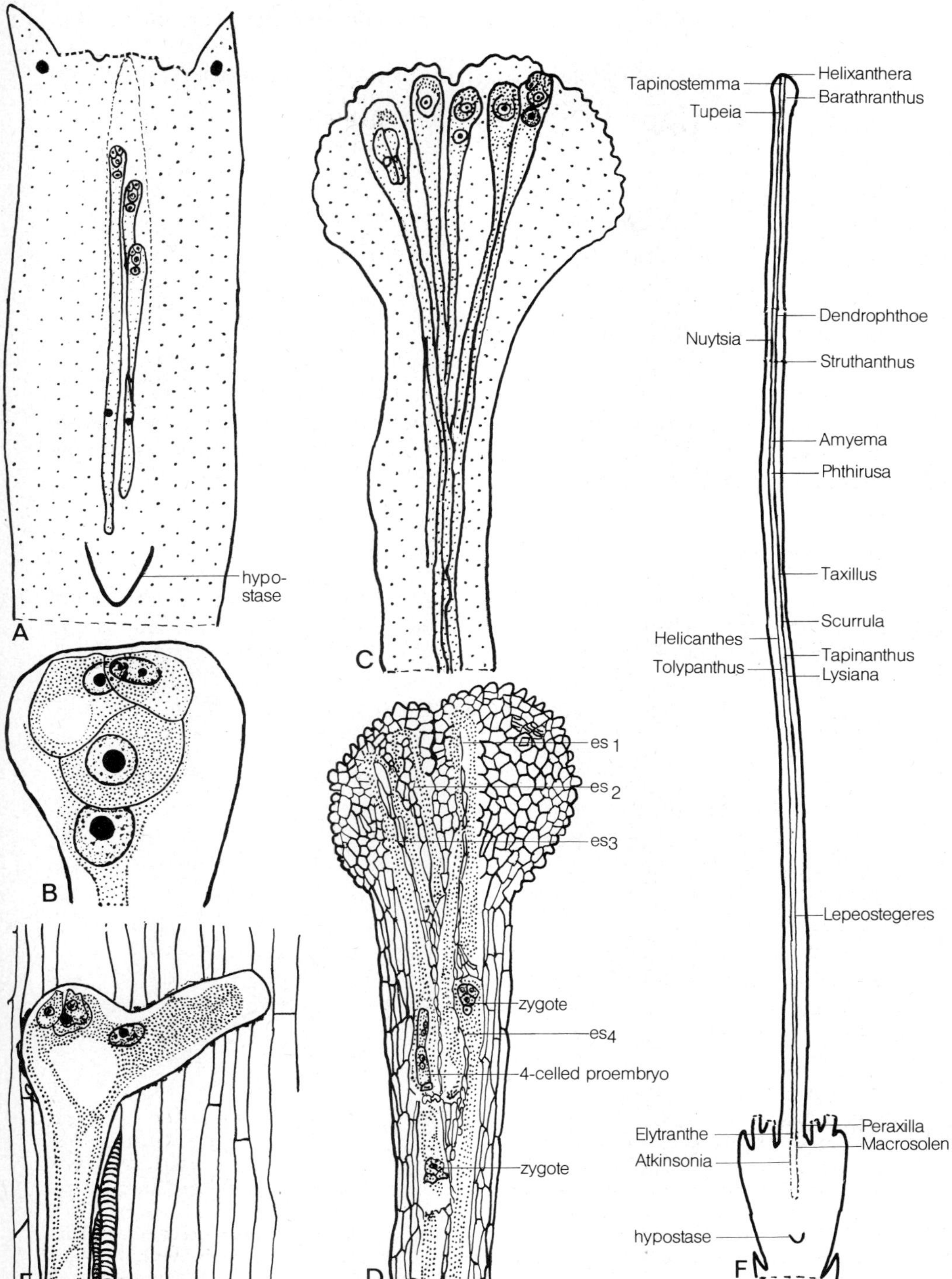
hypo-
stase
A
B
C
D
E
F
es 1
es 2
es3
zygote
es4
4-celled proembryo
zygote
Tapinostemma
Helixanthera
Barathranthus
Tupeia
Dendrophthoe
Nuytsia
Struthanthus
Amyema
Phthirusa
Taxillus
Scurrula
Helicanthes
Tapinanthus
Tolypanthus
Lysiana
Lepeostegeres
Elytranthe
Peraxilla
Macrosolen
Atkinsonia
hypostase

Fig. 1.7 A–F

The extension of the tip of the embryo sac in different taxa is represented in Fig. 1.7 F.

The length of the embryo sac, in some members of Loranthaceae, is as follows:

Plant/Investigator	Length of embryo sac ca. mm
Helicanthes elastica (B.M. Johri et al. 1957)	16
Amyema congener, A. miquelii, A. pendula and *A. preissii* (Dixit 1958a)	22
Tolypanthus involucratus and *T. lagenifer* (Dixit 1961)	25
Dendrophthoe falcata (B. Singh 1952)	28
Tapinanthus rubromarginatus (Pienaar 1952) and *T. uhehensis* (Dixit 1956)	35
Tapinostemma acaciae (B.M. Johri and Prakash 1965)	42
Moquiniella rubra (B.M. Johri and Raj 1965)	48

In contrast to several embryo sacs developing concurrently in a gynoecium, only a single embryo sac develops in *Phthirusa pyrifolia* (Kuijt and Weberling 1972) and *Struthanthus vulgaris* (Venturelli 1981).

1.5.3 Endosperm

The primary endosperm nucleus, in each embryo sac, moves to the lower part of the embryo sac, undergoes a division followed by a wall. Further divisions give rise to Cellular type of endosperm. In a cross-section of the ovary several embryo sacs show endosperm (Fig. 1.8 A). Due to the all-round growth of endosperm, the ovarian tissue between the individual endosperms is crushed. Thus, the endo-

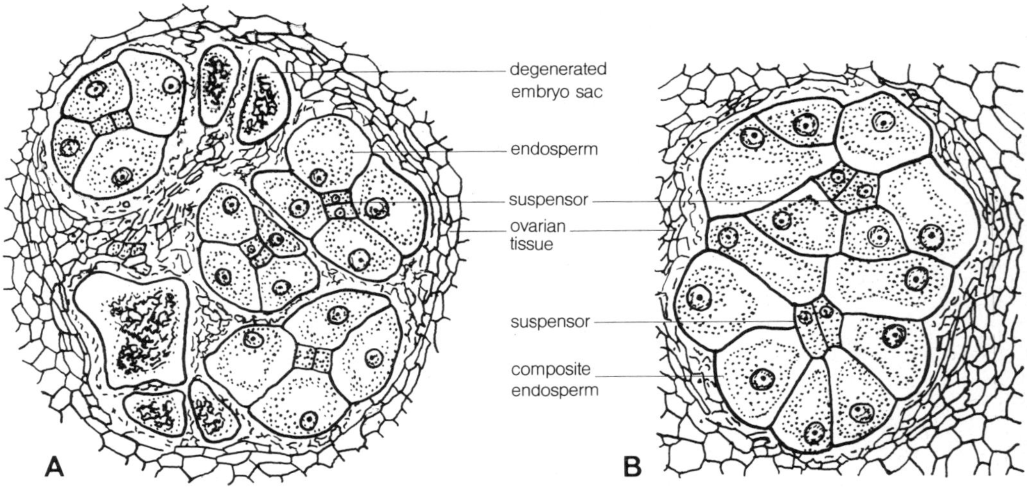

Fig. 1.8 A, B. *Tolypanthus involucratus*. **A** Transection of ovary showing four embryo sacs each containing four-seriate endosperm. Note the biseriate suspensor (t.s.) of proembryo in each. Five embryo sacs have degenerated. **B** Same, later stage. Note composite endosperm with two biseriate suspensors. (After Dixit 1961)

sperms in an ovary come close to each other and, finally, fuse to form a composite structure (Fig. 1.8 B). The development of composite endosperm was first observed in *Dendrophthoe pentandra* and *Scurrula atropurpurea* (Rauch 1936). The shape of the vase-shaped mature endosperm is determined by the vascular supply of the ovary.

In contrast to other members of Loranthaceae, in *Struthanthus vulgaris* (Venturelli 1981) the endosperm develops from a single embryo sac, and shows a lateral haustorium (see Fig. 1.10 E, F).

1.5.4 Embryo

The zygote elongates, becomes vacuolate in the upper region, and undergoes a vertical division followed by a transverse division. The four-celled proembryo elongates, and repeated transverse divisions result in a biseriate filamentous proembryo. The suspensor elongates until the apical embryonal cells reach up to the hypostase. If the elongation of the suspensor continues, it becomes coiled. Further coiling may occur due to the pushing up of the embryo. The remnants of the suspensor may persist in the mature embryo.

While the composite endosperm is being formed, the apical embryonal tiers divide and form a mass of cells which differentiates into a globular, heart-shaped, and dicotyledonous embryo.

The proembryo grows through the endosperm and, in a cross-section, the biseriate suspensor can be clearly seen (Fig. 1.8 A, B).

Several embryos develop concurrently in the same ovary as in *Dendrophthoe falcata* (B. Singh 1952). and *Helixanthera ligustrina* (B. M. Johri and Bhatnagar 1972). Finally, only one reaches maturity, and others are arrested at different stages of growth.

The embryo is typically dicotyledonous. The cotyledons are closely appressed to each other giving a pseudomonocotyledonous appearance as in *Amyema, Barathranthus, Dendrophthoe, Helicanthes, Helixanthera, Lysiana, Scurrula, Tapinostemma,* and *Tolypanthus.*

The cotyledons remain free in *Atkinsonia, Elytranthe, Lepeostegeres, Macrosolen, Nuytsia, Peraxilla, Struthanthus, Tapinanthus, Taxillus,* and *Tupeia.*

In *Lysiana exocarpi* (Narayana 1958 a) one of the cotyledons is larger. The branching or bifurcation of cotyledons is common.

In *Psittacanthus schiedeanus* (Kuijt 1967) eight to ten massive, awl-shaped cotyledons form a whorl around the shoot apex; in *P. cuneifolius* Bhatnagar and Chandra (1968) report only two cotyledons.

The embryo lacks a true radicle, and the swollen radicular end (which usually remains exposed outside the endosperm) is only a hypocotyledonary extension. It shows endarch, collateral bundles. In *Amyema* (Dixit 1958 a), *Dendrophthoe neelgherrensis* (Narayana 1954, 1956), *Helicanthes elastica* (B. M. Johri et al. 1957), and *Lysiana exocarpi* (Narayana 1958 a) the radicular end is studded with branched or unbranched outgrowths. B. M. Johri and Bajaj (1964) have demonstrated, through embryo culture, that in *A. pendula* the outgrowths represent suppressed shoot apices.

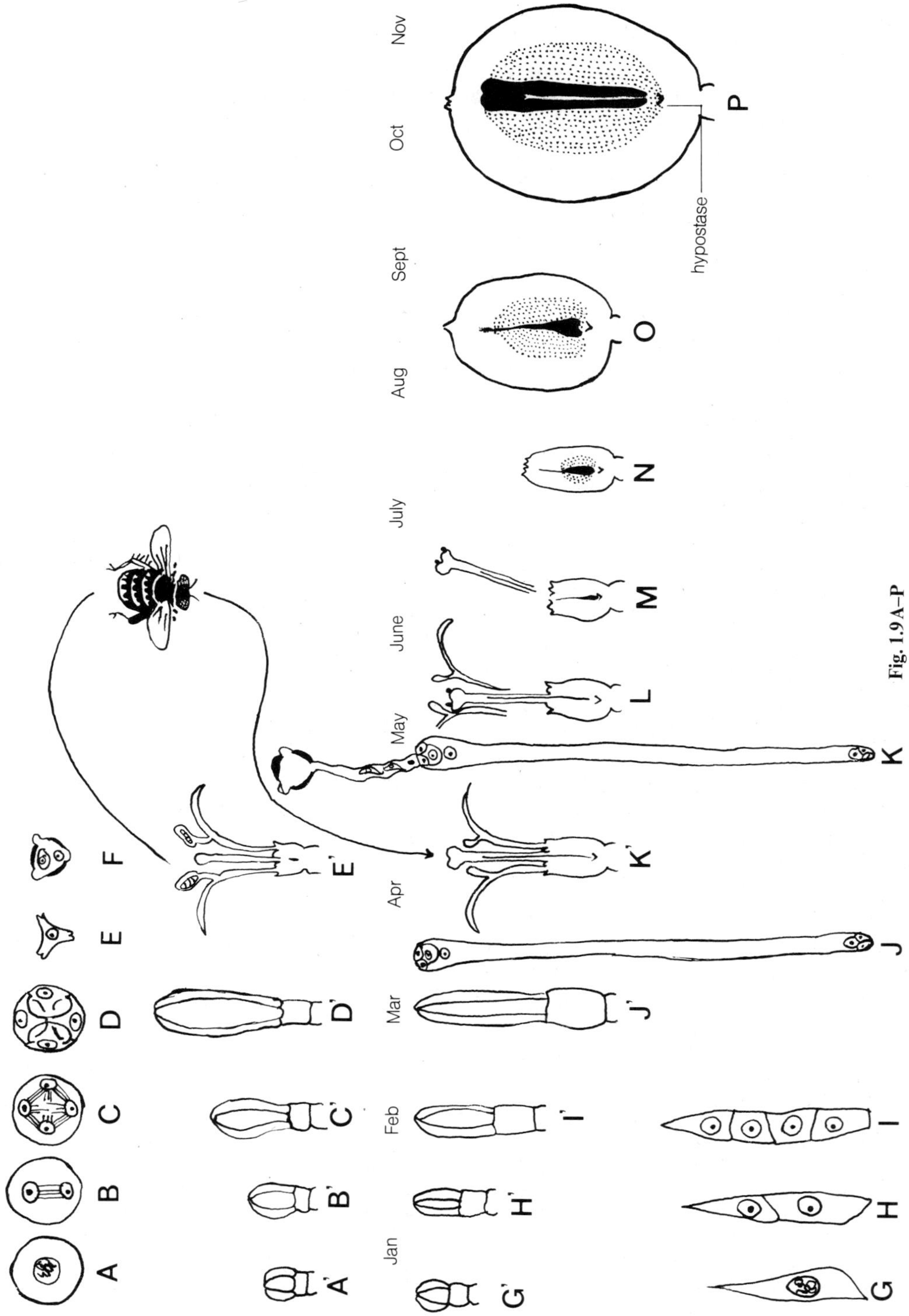
Jan
Feb
Mar
Apr
May
June
July
Aug
Sept
Oct
Nov
A
B
C
D
E
F
A'
B'
C'
D'
E'
G'
H'
I'
J'
K'
G
H
I
J
K
L
M
N
O
P
hypostase

Fig. 1.9A–P

1.5.5 Reproductive Calendar in *Struthanthus vulgaris*

Venturelli (1981) has given a reproductive calendar (as in Brazil) for *Struthanthus vulgaris* (Fig. 1.9 A–P). The male and female flower buds appear in January. The flowers bloom in April, followed by pollination. By June–July the floral parts wither, and the endosperm, embryo, and fruit mature in November.

The author emphasizes the correlation between the size of bud, flower, and ovary, and temporal development of microsporogenesis (Fig. 1.9 A–E), megasporogenesis (Fig. 1.9 G–I), male (Fig. 1.9 E, F), and female gametophytes (Fig. 1.9 J, K), pollination (Fig. 1.9 É, Ḱ), fertilization (Fig. 1.9 K, L), endosperm and embryo (Fig. 1.9 M–P).

The microspore mother cells (Fig. 1.9 A) undergo meiotic divisions (Fig. 1.9 B, C), cytokinesis is simultaneous (Fig. 1.9 C), and the tetrads are tetrahedral (Fig. 1.9 D). The uninucleate pollen grain is triradiate (Fig. 1.9 E). The pollen grain is two-celled at the dispersal stage (Fig. 1.9 F).

After two meiotic divisions, the megaspore mother cell gives rise to a linear tetrad (Fig. 1.9 G–I). Several tetrads are formed, but only one megaspore develops into an eight-nucleate, Polygonum type of embryo sac (Fig. 1.9 J, 1.10 A–D). The antipodals are ephemeral. The hypostase is lignified.

Pollination is entomophilous, and double fertilization occurs (Fig. 1.9 K, L).

The cellular endosperm develops a lateral haustorium (Fig. 1.10 E, F). The division of zygote is vertical, followed by several transverse divisions to form a biseriate proembryo (Fig. 1.10 G, H). The suspensor elongates and pushes the proembryonal apex to the basal portion of the embryo sac, where the endosperm is well developed (Fig. 1.10 I). The mature embryo is green and dicotyledonous (Fig. 1.9 O–P).

The reduction (or progression) of placenta, tip of embryo sacs limited to the the ovary or extension in the style (even up to stigma), composite endosperm, embryo without a true radicle, and endosperm (seed-coat absent) directly surrounded by the pericarp are highly specialized characters. It is difficult to visualize that this specialization is due to the semiparasitic habit of the loranthaceous taxa.

Fig. 1.9 A–P. *Struthanthus vulgaris*, reproductive calendar. **A–D** Sequential development of microspore mother cell up to tetrad; **A′–D′** corresponding stages of male bud. **E′** Open flower at pollination (with pollen as at **E, F**). **G–I** Megaspore mother cell (**G**), dyad (**H**), and tetrad (**I**); **G′–I′** corresponding stages of female bud. **J, J′** Mature embryo sac and corresponding stage of female bud. **K, K′** Embryo sac at fertilization, and corresponding stage of flower. **L** Post-pollinated flower. Perianth and androecium have fallen off. **M** Fertilized ovary with young endosperm and embryo. The style and stigma also fall off. **N–P** Successive stages in the development of fruit (pericarp *white*, endosperm *stippled*, embryo *black*; note the hypostase in **K′, L–P**). (After Venturelli 1981)

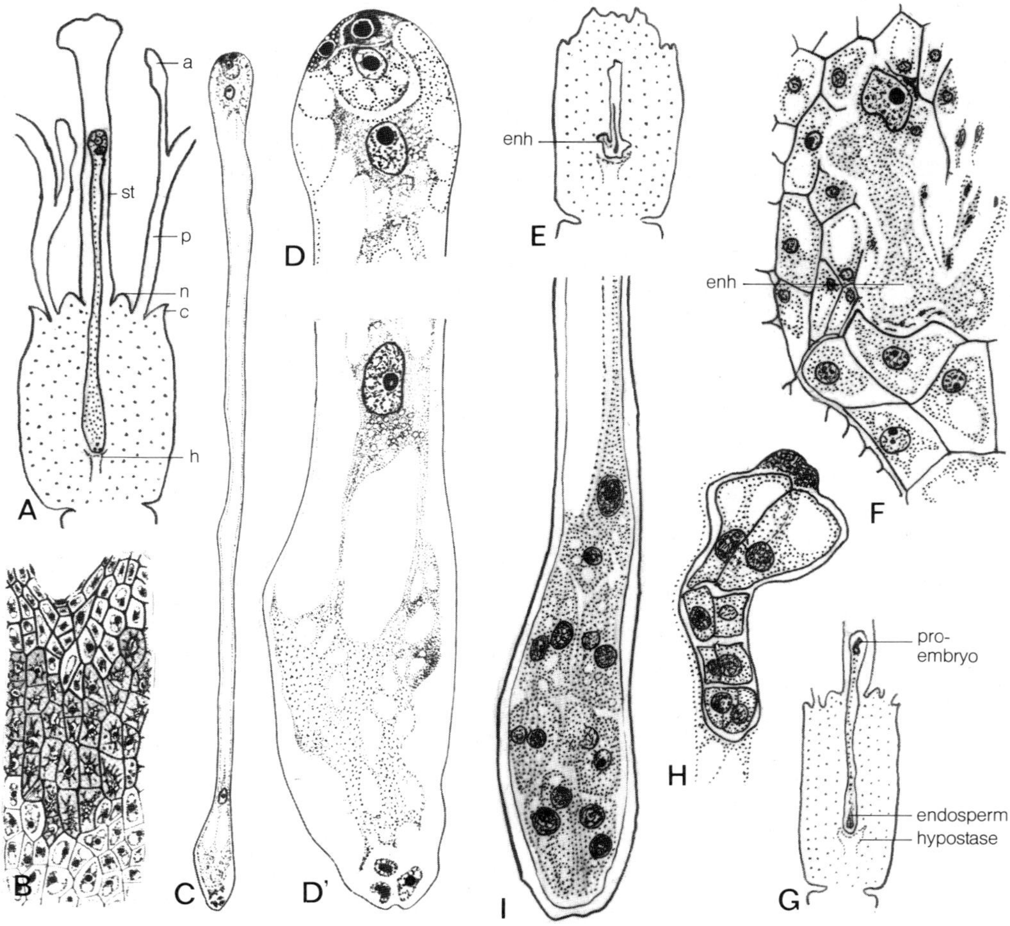

Fig. 1.10 A–I. *Struthanthus vulgaris*. **A** Longisection of flower showing the elongation of the embryo sac in the style (*a* anther, *c* calyculus, *h* hypostase, *n* nectary, *p* perianth, *st* style). **B** Hypostase with thick-walled cells, from **A. C** Eight-nucleate embryo sac from **A. D, D′** Micropylar (**D**) and chalazal (**D′**) regions from **C. E** Ovary with an embryo sac showing endosperm with lateral haustorium *(enh)*. **F** Endosperm haustorium from **E. G** Portion of ovary with style showing an embryo sac with biseriate proembryo, and cellular endosperm. **H** Proembryo from **G. I** Endosperm from **G.** (After Venturelli 1981)

1.6 Nutrition of the Ovule and Seed

The ovules have a well-developed funicular vascular supply (Figs. 12.2 A–E, 12.3 A–N) which, during the maturation of seed, becomes further differentiated. Besides the vasculature there are several types of haustoria which absorb nutrition from the adjoining tissues, and supply it to the endosperm and embryo. In later stages the embryo mostly grows at the expense of the endosperm.

1.6.1 Vascular Supply

The major supply of nutrients to the ovules is through the vascular strand. The vasculature in the ovules is variable, and is prominently developed in the primitive families. It may be represented by only elongated, thin-walled cells (pro-vascular strand) in some of the advanced families. In Orchidaceae, where the ovules are rather minute, there is no vascular supply.

In bitegmic ovules the vascular strand may branch, and both the integuments may be vascularized as in *Aleurites triloba* (Corner 1976); both the integuments as well as the aril are vascularized in *Myristica fragrans* (Mohana Rao 1974).

The vascular strand branches at the chalaza and the branches extend only into the outer integument up to the micropyle in Leguminosae (*Adenanthera*), Malvaceae (*Hibiscus*), Moringaceae (*Moringa*), Rhamnaceae (*Ventilago*), Rosaceae (*Prunus*). The branches extend only into the inner integument, as in Dipterocarpaceae, Elaeocarpaceae, Euphorbiaceae (*Ricinus*), Sterculiaceae (*Pentapetes*). In unitegmic ovules the vascular supply may reach up to the micropyle as in Boraginaceae, Compositae, Convolvulaceae, Cuscutaceae (see Corner 1976).

In larger ovules the vascular supply is extensive and, generally, terminates at the chalaza, or branches there. In *Trewia polycarpa* (R. P. Singh 1962) the chalazal vascular strand gives off a branch to the outer integument, and the rest of it forms a cup at the base of nucellus.

In *Sterculia colorata* (Venkata Rao 1953) six to nine prominent vascular strands extend into the testa up to the micropyle, and give off a few branches. But in *Pterospermum heyneanum* (Venkata Rao 1949) and *Pterygota alba* (Corner 1976), during megasporogenesis, the ovule develops a wing-like outgrowth at the chalaza. The funicular vascular supply curves into this outgrowth, appears as a stout rib, and gives out numerous oblique branches.

Tracheidal elements occur in the nucellus as in Asclepiadaceae, Capparidaceae, Liliaceae, Olacaceae, Resedaceae, Thymelaeaceae (see B. M. Johri and Kapil 1953), and Icacinaceae (*Nothapodytes foetida* – Swamy and Ganapathy 1957). In *Acalypha indica* (B. M. Johri and Kapil 1953) the nucellar tracheids persist in the seed.

Linostoma pauciflorum (Corner 1976) shows 10–12 vascular bundles along the periphery of the nucellus, up to the micropyle. They may even branch once or twice, and persist in the seed. Whether these tracheids are involved in the transport of nutrients has yet to be demonstrated.

Tracheids have also been observed in the sporogenous tissue, as in *Casuarina equisetifolia* (Swamy 1948), *Amyema miquelii* (Dixit 1958 a), *Dendrophthoe neelgherrensis* (Narayana 1956), and *Nuytsia floribunda* (Narayana 1958 b).

1.6.2 Haustoria

During pre- and post-fertilization development of ovule, in some taxa, the haustoria from the megaspores, embryo sac, synergids, antipodals, endosperm, embryo, and suspensor are also involved in the absorption and transport of nutrients. The haustoria show hypertrophied nuclei, and are richly cytoplasmic.

1.6.2.1 Megaspore Haustoria

Development of megaspore haustoria is a rare feature, and has been observed in the members of Crassulaceae. In *Rosularia pallida* (Fig. 3.10 A), *Sedum sempervivoides* (Mauritzon 1933) and *S. chrysanthum* (Subramanyam 1967) the four megaspores of the tetrad produce thin, tubular processes which intertwine with one another, and extend up to the micropyle. Some of the haustoria become extranucellar, and lie at the base of the micropyle (*Rosularia*).

1.6.2.2 Embryo Sac Haustoria

In *Exocarpos menziesii* (Fagerlind 1959) the haustoria develop at the four-nucleate stage of gametophyte, and reach maximal growth at the mature embryo sac stage. The finger-like branched outgrowths show extensive growth from the upper region of the embryo sac (Fig. 1.11 A, B); some haustoria also grow downwards (Fig. 1.11 A). In the fertilized embryo sac of *E. sparteus* (Ram 1959 b) the haustoria develop in the middle region of the embryo sac (Fig. 1.11 C).

1.6.2.3 Synergid Haustoria

Synergid haustoria are reported in Compositae, Crassulaceae, Gramineae, and Santalaceae.

The tips of the synergids grow beyond the micropyle in *Cortaderia selloana* (Fig. 4.12), and the haustoria extend in opposite direction between the outer integument and the ovary wall. Philipson (1978) refers to one case in *C. jubata* where the haustorium penetrated the nucellus laterally extending into the integuments. Later, it reached the ovary wall adjacent to the placenta. The haustorium persists up to the early globular stage of the proembryo.

The tips of synergids extend into the micropyle, and fill the whole of the micropylar canal, in *Osteospermum sinuatum* (Fig. 1.12 A). In *Ursinia anthemoides* subsp. *versicolor* (Ahlstrand 1978), and *Calendula suffruticosa* (Ahlstrand 1979), the tips grow beyond the micropyle, and one of the synergids develops into a conspicuous haustorium; the other synergid collapses during fertilization.

The most extensive synergid haustoria (Agarwal 1962) develop in *Quinchamalium chilense* (Santalaceae). The tubular outgrowths, from the tip of each synergid, grow along the vascular supply of the carpel, and reach up to one-third the length of the style (Fig. 1.12 B–D).

1.6.2.4 Antipodal Haustoria

The antipodals are often ephemeral, and degenerate either before or soon after the organization of the mature embryo sac, or after fertilization. In several taxa the antipodals persist until a late stage, and acquire various degrees of ploidy (see

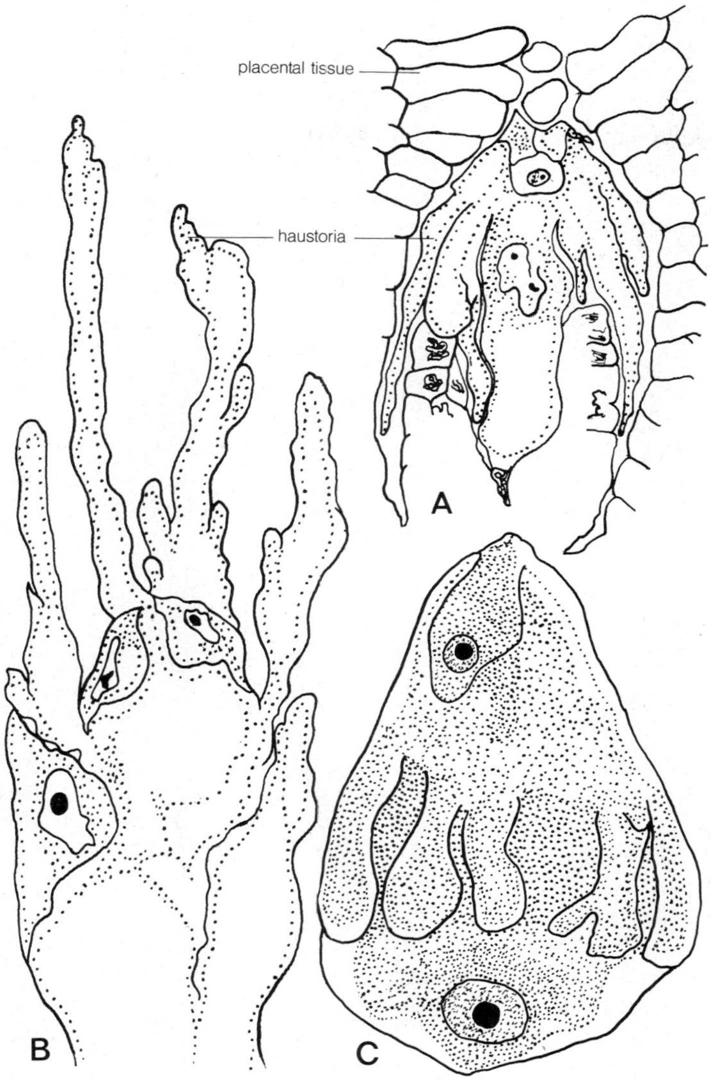

Fig. 1.11 A, B. *Exocarpos menziesii.* **A** Organized embryo sac with finger-like haustoria growing downwards. **B** Later stage. Note profuse development of haustoria. (After Fagerlind 1959). **C** *E. sparteus*, embryo sac with haustoria from the middle region. (After Ram 1959 b)

Chap. 11). One of the antipodal cells may elongate and develop into a conspicuous haustorium, as in members of Compositae, Crassulaceae, and Rubiaceae.

In *Galium mollugo* (Fig. 1.12 E) and *Putoria calabrica* (Fagerlind 1936) the basal antipodal cell forms an aggressive haustorium in the chalazal tissue, whereas in *Artemisia tridentata* (Diettert 1938) the haustorium penetrates through the chalazal tissue, and finally enters the ovarian chamber. In *Osteospermum sinuatum* (Fig. 1.12 A) the haustorium becomes trinucleate.

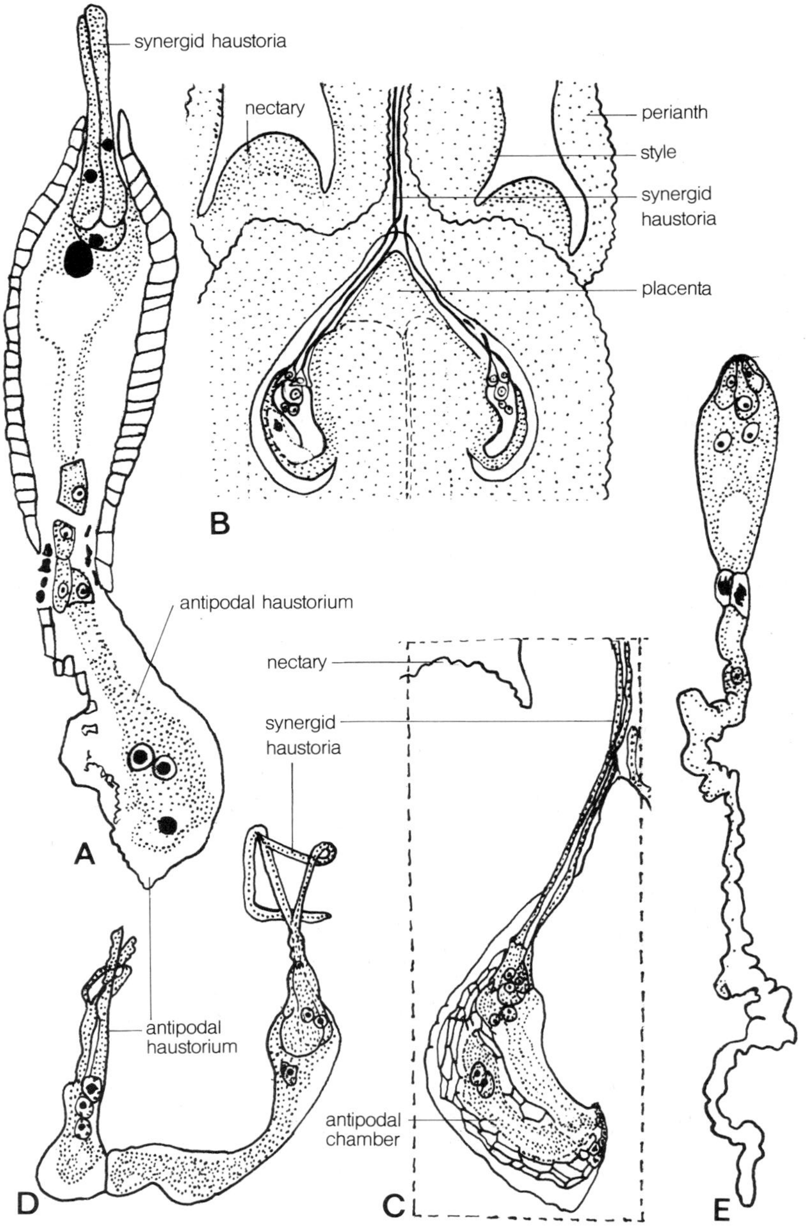

Fig. 1.12. A *Osteospermum sinuatum,* mature embryo sac with synergid haustoria. Note also antipodal haustorium. (After Ahlstrand 1979). **B–D** *Quinchamalium chilense;* **B** longisection of lower portion of gynoecium; **C** embryo sac from **B** (*left*); **D** mature embryo sac (dissection; note extensive synergid haustoria in **B–D**, and branched antipodal haustorium in **D**). (After Agarwal 1962). **E** *Galium mollugo,* mature embryo sac with much-elongated basal antipodal cell. (After Fagerlind 1937)

The antipodal haustorium, in *Quinchamalium chilense* (Agarwal 1962), deserves special mention. The antipodal nuclei do not organize into individual cells so that a three-nucleate chamber (Fig. 1.12 C, D) is formed; it may even become multinucleate. The tip of the chamber elongates and extends as far as the apex of the placenta. There it branches, and invades the placental tissue.

Both synergid and antipodal haustoria have been observed in *Quinchamalium chilense* (Agarwal 1962) and *Osteospermum sinuatum* (Ahlstrand 1979).

1.6.2.5 Endosperm Haustoria

An extensive development of chalazal haustorium has been reported in Cucurbitaceae (Fig. 7.3), Icacinaceae, Leguminosae, Santalaceae (Fig. 7.9 A, B), Opiliaceae (Fig. 7.9 C), and some other taxa.

In living material, streaming of cytoplasm has been observed in the coiled haustorium (vermiform appendage) of *Grevillea robusta* (Kaushik 1941), and the coenocytic haustorium of *Trichosanthes anguina* (Chopra 1955).

In Cucurbitaceae, the length of the chalazal haustorium is maximal (19,000 μm) in *Sechium edule* (D. Singh 1964).

The development of unicellular, finger-like appendages all over the surface of the endosperm in *Lomatia polymorpha* (Fig. 7.4) is unique, and increases the absorptive surface.

In *Nothapodytes foetida* (Swamy and Ganapathy 1957) the basal cell of the endosperm proper becomes haustorial, and produces a sac-like intercellular extension in the chalazal region. The haustorium establishes chain-like cytoplasmic connections with the adjoining cells, at the point of contact, by forming apertures in the intervening walls. The cytoplasm and nuclei of the affected cells are incorporated into the haustorium.

The nucleus of the haustorium becomes multinucleolate and amoeboid. As the protoplasts are absorbed by the haustorium, the nucleus of the haustorium extends into the affected cells, becomes lobed, and the lobes dilate at the tip.

The activity of the haustorium declines at the initiation of cotyledonary primordia in the embryo. The absorbed protoplasts swell and become gelatinous. Two or three tiers of endosperm cells, adjoining the haustorium, also develop tubular prolongations which grow irregularly in the surrounding chalazal tissue.

In most members of Sympetalae, endosperm haustoria develop terminally. Their morphology, duration of persistence and extensions are summarized in Chapter 7.

An extensive development of micropylar and chalazal haustorium has been reported in *Melampyrum lineare* (Arekal 1963). The mature embryo sac develops six to eight narrow lateral processes. After fertilization these processes form extensions of the micropylar haustorium. One of these branches extends downwards through the integument, and grows towards the chalaza. It ramifies in the funiculus (Fig. 7.10 C). The growth of the haustorial branches is intercellular.

The main binucleate chalazal haustorium enlarges a great deal, and its nuclei increase in size. It is conical and densely cytoplasmic, and establishes direct contact with the funicular vascular strand of immature seed (Fig. 7.10 C).

1.6.2.6 Suspensor Haustoria

The morphology, structure, and function of the suspensor is discussed in Chapter 8. The suspensor provides nourishment to the young proembryo.

The development of embryo, in Crassulaceae, is of the Caryophyllad type (Subramanyam 1962). The basal cell of the two-called proembryo does not divide, but becomes large and vesicular. It encloses a hypertrophied nucleus and, in the micropylar region, develops haustorial branches which ramify in the integument, in *Sedum ternatum* (Subramanyam 1963).

In the absence of endosperm, the suspensor is very conspicuous in Podostemaceae. In *Dicraea stylosa* (Mukkada 1962) the basal cell of the embryo shows hypha-like haustorial extensions which grow between the starch-containing cells of the integuments. The basal cell enlarges in *Indotristicha ramosissima* (Mukkada and Chopra 1973), and becomes vesicular and coenocytic. The size of the haustorium and the number of nuclei are maximal at the late heart-shaped stage of the embryo. The number gradually decreases at the cotyledonary stage of the embryo.

The basal cell divides vertically in *Pisum sativum* (Cooper 1938) and *Haloragis micrantha* (Nijalingappa 1975). In *P. sativum* both the cells elongate considerably, become haustorial, and contain as many as 64 nuclei at the globular stage of the proembryo; in *H. micrantha* the two juxtaposed cells enlarge, become haustorial, but remain uninucleate.

The development of suspensor haustoria is noteworthy in Orchidaceae; the endosperm is either absent or poorly developed. In *Cymbidium bicolor* (Fig. 8.20 F–H) the basal cell of the two-celled proembryo divides, and the daughter cells develop into four or five prominent tube-like, unicellular, unbranched haustoria.

In *Asperula* (Lloyd 1902; see Maheshwari 1950) six to seven apical cells of the suspensor are filamentous, while the basal cells form uninucleate, balloon-like structures which bulge into the endosperm tissue.

1.6.3 Transfer Cells

The transfer cells are meant for short-distance transport of solutes through wall ingrowths.

In a fully grown cell ingrowths develop from the wall and project into its lumen. These are specialized secondary structures on the inner surface of the primary wall; the middle lamella is not concerned.

The transfer cells exhibit three patterns: (1) The wall ingrowths develop all round the cell wall. (2) The ingrowths develop only in that portion of the wall which lines the adjoining cell. When the transfer cell is connected with more than one cell, the development corresponds to each cell. (3) The wall ingrowths may develop only along a strip girdling the cell.

The wall ingrowths are of two types: filiform or papillate; or branched and interconnected to form a labyrinth. The plasma membrane follows the contour of the wall ingrowths even if they are irregular or of the labyrinth type. The absorptive surface of the plasma membrane increases considerably, which facilitates

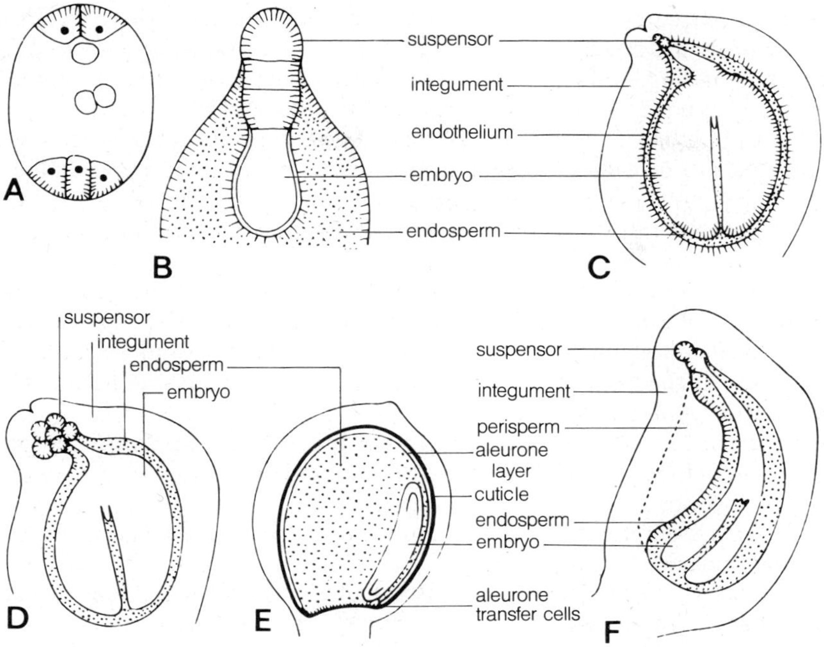

Fig. 1.13 A–F. Transfer cells (all figures diagrammatic). **A** Wall ingrowths in synergids and antipodals. **B** Wall ingrowths in suspensor cells (there are no ingrowths on the periclinal walls), and outer and inner epidermes of endosperm. **C** *Vicia,* longisection of seed. Note wall ingrowths in persistent suspensor cells, epidermis of cotyledons (endosperm wall ingrowths are not shown), and in endothelium. **D** *Phaseolus,* longisection of seed. **E** Longisection of grass seed. Note aleurone transfer cells at placenta. **F** *Mesembryanthemum,* longisection of seed. Wall ingrowths are present in persistent suspensor cells and epidermis of endosperm facing the perisperm. (After Gunning and Pate 1974)

transmembrane flux of solutes. The plasmodesmata, in transfer cells, are not restricted to any specific region. The transfer cells exhibit polarity, and are species-specific.

The transfer cells usually develop at the junction of the boundaries of the organs through which the nutrients are to be transported. The two adjoining organs form a donor–receptor system. The site of transfer cells in various reproductive tissues and organs is diagrammatically represented in Fig. 1.13 A–F.

The transfer cells occur in the anther tapetum of *Paeonia,* the pollen tube of compatible pistil in *Lilium,* the stylar tissue in contact with pollen tube of *Muscari,* and the transmitting tissue of style in *Fritillaria* and *Lilium* (see Pate and Gunning 1972).

According to Gunning and Pate (1974), the wall ingrowths are penetrated by a network of channels, probably larger than 2 mm in diameter, and are lined with fixed anionic charges. These charges have ion-exchange capacity as well as favourable conditions of electro-osmosis. The potential difference between the cytoplasm and the surrounding apoplast could drive the coupled transport of solutes and water within the ingrowth itself. Thus, the electro-osmotic phenomena "provides a mechanism for mass-flow within the wall ingrowths ...". The mass-flow could be the effect of geometry of the wall membrane apparatus according to the "standing gradient osmotic flow" hypothesis (developed by animal physiologists). Consequently, there are two mechanisms by which entry of solutes occurs in the wall ingrowths from the adjoining apoplast: (a) diffusion according to their own concentration gradient, and (b) the solutes are swept in with the flow of water.

The short-distance transport of nutrients takes place from the ovular tissues to the embryo sac.

At the time of fertilization the embryo sac in *Jasione montana* (Erdelská 1975) expands, and in the submicropylar region interruptions occur in the cuticle around the nucellus and integument. The wall of the embryo sac develops a "transfer region" at the site of interruptions of the cuticle. Thus, the uptake or exchange of nutrients takes place.

The synergid of *Spinacia* exhibits a well-developed filiform apparatus in the micropylar region (Figs. 4.7–4.9). The filiform apparatus consists of long wall ingrowths which branch, interconnect, and extend into the cytoplasm. The surface of the plasma membrane is increased, and the nutrients are transported to the egg cell and central cell.

In the embryo sac of *Plumbago,* in the absence of synergids, the wall ingrowths (= filiform apparatus) are present at the micropylar region of egg.

The walls of the antipodal cells, bordering the nucellus, also have conspicuous ingrowths through which the nutrients are transported.

The endosperm tissue has transfer cells in the outer and the inner epidermis. It draws nutrients from the nucellus, and provides it to the young embryo.

In *Vaccinium macrocarpron* (Brisson and Peterson 1975) the endosperm develops micropylar and chalazal haustoria which have highly complex finger-like wall ingrowths (Fig. 7.12 A, B).

The suspensor cells in contact with the endosperm tissue also develop ingrowths on their lateral walls. The wall ingrowths in the suspensor cells of *Phaseolus coccineus* (Fig. 8.21 A) are very prominent.

1.7 Role of Persistent Pollen Tubes

In a number of taxa, after fertilization, the pollen tube persists in the micropyle, as in Cucurbitaceae (Longo 1903, Kirkwood 1906, B. M. Johri and Roy Chowdhury 1957, D. Singh 1963), Passifloraceae (Cook 1909, Raju 1956), Nyctaginaceae (Cooper 1949, P. Maheshwari and Johri 1950 b), Hydrocharitaceae (P. Maheshwari and Johri 1950 b), Orobanchaceae (Tiagi 1951), Gesneriaceae (Tiagi 1952), and Malvaceae (Venkata Rao 1952).

In *Cucurbita* (Longo 1903) the embryo sac is unable to obtain adequate nutrients as the nucellar epidermis is cutinized, and the hypostase is suberized. In such a condition the pollen tube expands into a "bulla", develops branches into the nucellus and inner integument, and provides nutrition to the developing embryo. Longo presumed that this signifies the haustorial role of the persistent pollen tube. Kirkwood (1906) pointed out that the bulla formation is associated with starch grains in the integument. B. M. Johri and Roy Chowdhury (1957) also observed such a swelling of the pollen tube, after entering the embryo sac of *Citrullus colocynthis*. It persisted up to the eight-nucleate stage of endosperm, and globular proembryo. It often showed a nucleus, probably the vegetative nucleus.

Cook (1909) mentions the haustorial role of the persistent pollen tube in Passifloraceae (*Passiflora adenophylla*). The pollen tube enters the embryo sac, continues its growth before discharging the sperms, and coils around itself in the embryo sac. Sometimes it invades the constituents of the embryo sac (including the egg apparatus) and fills the entire cavity of the gametophyte.

Raju (1956) also observed the coiling of the pollen tube inside the embryo sac, at the time of fertilization, in *P. calcarata*, and its persistence during the maturation of seed. Raju emphasizes that the wall of the pollen tube is thick, and it does not get crushed. However, there is no indication of any activity, and it does not have any haustorial role.

Venkata Rao (1952) observed that after discharge pollen tubes persist up to a late stage of the embryo, in some Malvaceae. He points out that "Even after the discharge of the gametic nuclei, the tube does not collapse or show signs of degeneration". The membrane of the persistent pollen tube is "tough" lined by a thin layer of cytoplasm, and it crushes the adjoining nucellar tissue. Its long persistence and the presence of the ruptured end at the base of the embryo suggests, probably, a haustorial role.

D. Singh (1963) observed persistent pollen tube in 22 species belonging to 15 genera of Cucurbitaceae. He noticed, in five taxa, in some ovules, the formation of bulla in the apical, middle, and basal region in the nucellus. Due to such an expansion the adjoining nucellar cells are crushed. The wall of the pollen tube is thick and, as the pollen tube does not show any sign of degeneration, it stains like a healthy organ. Its dilation has no positive correlation with starch grains in the integuments. The persistent pollen tubes are considered to have a haustorial role.

Cooper (1949) observed the long persistence of the pollen tube in *Oxybaphus nyctagineus*, and presumed that it becomes haustorial and transports nutrients from the secretory cells of the nucellus to the embryo. Maheshwari and Johri (1950b) also reported persistent pollen tube in the fruits of *Boerhaavia diffusa* (Fig. 1.14A, B). It stains uniformly and densely.

Maheshwari and Johri (1950b) emphasized that merely the long persistence of the pollen tube cannot be interpreted to indicate a haustorial function.

Persistent pollen tubes have also been observed in *Hydrilla verticillata* (Fig. 1.14C–E), *Ottelia alismoides* (Fig. 1.14F), *Aeginetia indica*, and *Orobanche cernua* (Tiagi 1951, 1952). Tiagi confirms that the remnants of pollen tubes are merely dead structures without any haustorial role.

A great drawback in the above studies has been that living material has never been examined. During the development of seed, microtome sections are usually cut 10–15 μm thick, or even thicker, so that thin sections of the persistent pollen tube have not been examined either. The examination of living material under phase contrast microscopy, use of cyto- and histochemical techniques, combined with fine structure, would provide valuable information.

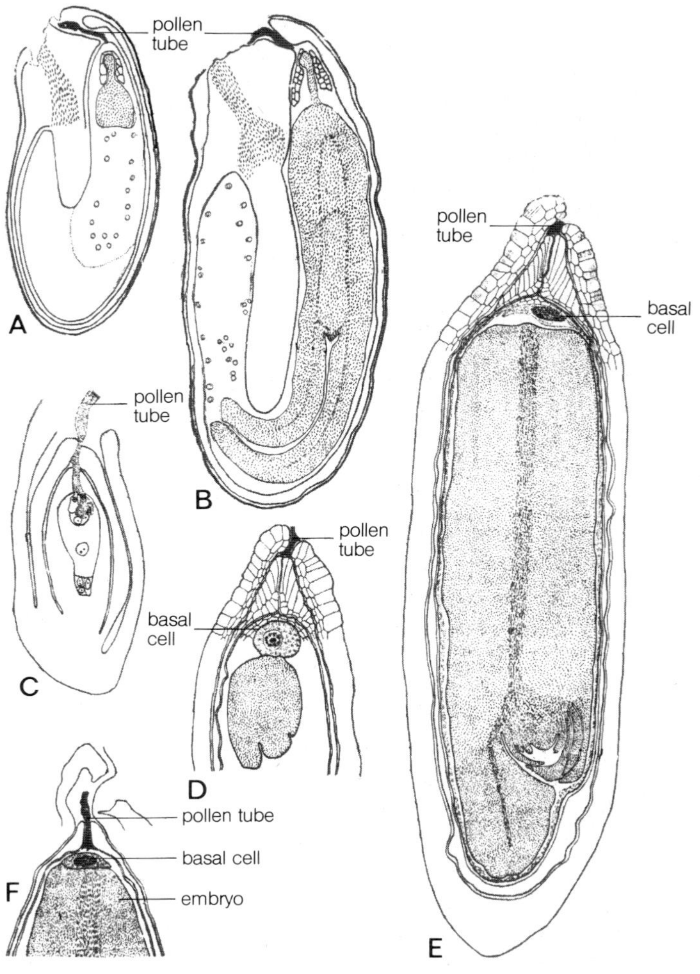

Fig. 1.14. A, B *Boerhaavia diffusa;* **A** longisection of ovule at heart-shaped embryo, **B** mature embryo. **C–E** *Hydrilla verticillata;* **C** eight-nucleate embryo sac at fertilization, **D** differentiated embryo, **E** mature seed (l.s.). **F** *Ottelia alismoides,* upper part of mature seed. Note remnants of pollen tube in the micropyle in **A, B, D–F,** and basal cell (of two-celled proembryo) in **D–F.** (After Maheshwari and Johri 1950b)

1.8 Embryos of Doubtful Origin

In the embryo sac the zygotic embryo is a normal feature. Sometimes, one or both synergids may also give rise to embryos (see Chap. 9). There are a few reports of embryos developing from the antipodal cells, and two reports of embryos of endospermal origin.

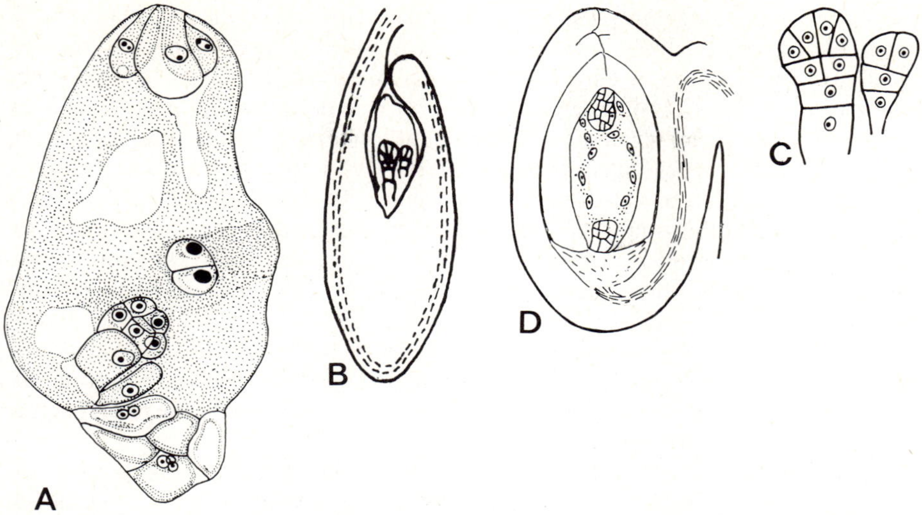

Fig. 1.15. A *Paspalum scrobiculatum*. **B, C** *Solidago canadensis*. **D** *Ulmus campestris*. **B, D** Longisections of ovules. **A** Unfertilized embryo sac. Several antipodal cells have enlarged, and one has produced a multicellular mass. **B** Outline diagram for **C. C** Two multicelled antipodal cells. **D** Zygotic proembryo, free nuclear endosperm, and a multicelled antipodal. The multicelled antipodals simulate proembryos. (**A** After Narayanaswami 1954; **B, C** after Pullaiah 1978; **D** after Guignard and Mestre 1966)

1.8.1 Antipodal Embryos

The antipodal cell(s) undergo irregular divisions which have been interpreted as embryos: *Allium odorum* (Modilewski 1931), *Elatostema eusinuatum* (Fagerlind 1944), *Paspalum scrobiculatum* (Fig. 1.15 A), *Solidago canadensis* (Fig. 1.15 B, C), *Ulmus campestris* (Fig. 1.15 D), and *U. glabra* (Ekdahl 1941).

Presumably, the stimulus of fertilization activates sometimes only one (Fig. 1.15 A, D), sometimes two (Fig. 1.15 B, C) antipodals. These undergo mitotic divisions forming a cellular mass which may simulate a young globular proembryo. Further growth usually remains arrested. Not a single instance is known of an organized and a partially or fully differentiated embryo.

Fagerlind (1944) reports that antipodal embryos develop in apomictic *Elatostema,* and one of his figures shows a fully differentiated dicotyledonous embryo at the chalazal end of the embryo sac. The polyembryony in this taxon deserves reinvestigation.

We consider the concept of antipodal embryo to be completely erroneous.

1.8.2 Endospermal Embryos

1.8.2.1 *Isomeris*

Billings (1937) made several erroneous observations in *Isomeris arborea* (see Sachar 1956). According to Billings: "An examination of much material in the sporogenous cell stage, and following (stages) – *parenthesis ours* – , plainly indicates that the cell typically develops directly into the embryo sac without the formation of megaspores. A single instance was observed, however, in which they were present ..." He further describes: "The typical definitive embryo sac in *I. arborea* has three nuclei which are organized into what appears to be an egg apparatus ... The two upper (cells), of pyriform shape with their associated cytoplasm, resemble synergids and are doubtless truly such. The lower nucleus may appear like an egg but perhaps more like an endosperm nucleus. There is a complete absence of polar nuclei and antipodals" ... "No true egg is formed". In a few instances where more than three nuclei were present, the nuclei were not organized into cells.

About the pollen tube Billings adds: "A pollen tube may enter the micropyle and form a bulbous terminus in the embryo sac. It does not discharge its contents, but instead develops a thickened wall and remains intact during the earlier stages of endosperm development".

Contrary to these observations, in the same taxon, Sachar (1956) observed the development of an eight-nucleate, Polygonum type of embryo sac. The pollen tubes were frequently present in the ovules, and in one case he noticed syngamy and triple fusion.

The development of endosperm is of the Nuclear type, and wall initiation occurs at the time of differentiation of cotyledons (Sachar 1956). Billings (1937) also observed similar development but, in addition, he noted the differentiation of dark "patches" of coarse granular cytoplasm with nuclei. Later on these nucleated patches get delimited by walls from the surrounding cytoplasm, and form endosperm nodules. The nodules do not extend into the short arm of the J-shaped embryo sac.

Billings (1937) presumed that "The embryo arises by a direct outgrowth of an endosperm nodule which is generally located near the concave base of the long arm of the embryo sac, though it may originate from a nodule located near the base of the convex curve of the sac" (Fig. 1.16 A, B).

The development of embryo from the endosperm has long since been doubted. Maheshwari and Sachar (1954) and Sachar (1956) made a detailed study of *I. arborea*. In the sections they observed, again and again, the so-called endosperm nodules described by Billings, but remained convinced that the embryo could not originate from the nodules. Finally, Sachar dissected a seed, and observed a highly coiled suspensor of uniseriate multinucleate cells with a few embryonal cells at the tip (Fig. 1.16 C–H′). What appeared to be multinucleate endosperm nodules in sections were portions of the suspensor cells (Fig. 1.16 E, F). Thus, the zygotic origin of embryo was established in *Isomeris*.

Fig. 1.16 A–H. *Isomeris arborea*. **A** Longisection of ovule at about the time embryo formation begins. The usual position of the so-called endosperm nodule that gives rise to the embryo is indicated at *e*. **B** Multinucleate endosperm nodules (as seen in section parallel to one side of embryo sac). (After Billings 1937). **C** Lower part of embryo sac (wholemount) with proembryo and free-nuclear endosperm. **D** Proembryo from **C**. **E** Embryo sac (wholemount). Note the proembryo with a long suspensor; endosperm omitted. **F** Proembryo with a highly coiled, uniseriate, suspensor with coenocytic cells, from **E**. **G** Older ovule, the proembryo has a long suspensor. **H, H′** Proembryo with suspensor from **G**. The uninucleate cells at the tip (**D, F, H′**) are progenitors of the embryo proper. (After Sachar 1956)

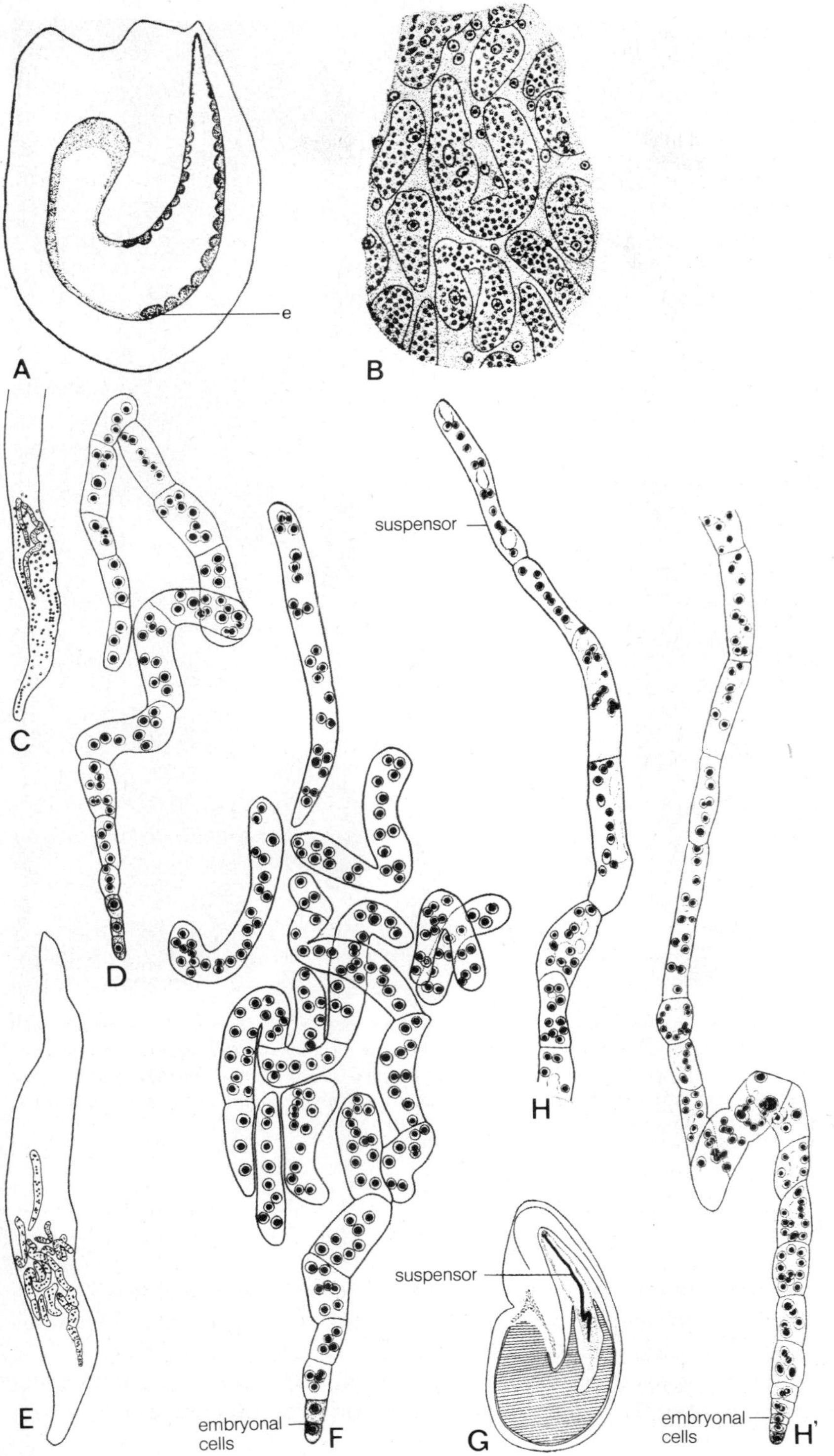
e
A
B
C
D
suspensor
H
E
embryonal
cells
F
suspensor
G
embryonal
cells
H'

Fig. 1.16 A–H

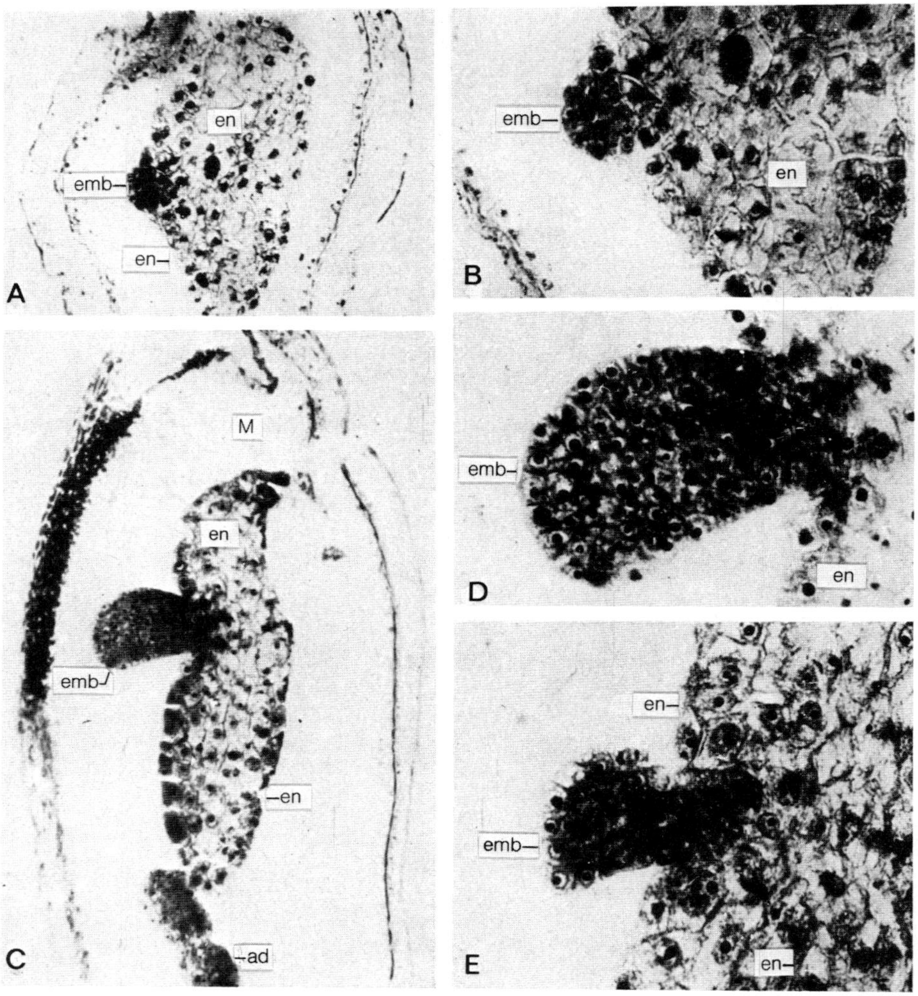

Fig. 1.17. A–E *Brachiaria setigera.* **A** Longisection of ovule. Note early initiation of "embryo" (*emb*) from endosperm (*en*). **B** "Embryo" (*emb*) from **A.** **C** Longisection of ovule to show a large embryo (*emb*) developed from endosperm (*en*). An adventitious embryo (*ad*) at the chalazal end has developed from an aposporous embryo sac; *M* micropyle. **D** Endospermal embryo (*emb*) from **C.** **E** Embryo (*emb*) from endosperm (*en*) with a suspensor-like attachment. (After Muniyamma 1977)

1.8.2.2 *Brachiaria*

Another recent report on the origin of embryo from the endosperm is by Muniyamma (1977) in *Brachiaria setigera* – tetraploid, facultative, pseudogamous apomict. Polyembryony, in this species, is said to be due to development of the embryo from the egg, synergid, and endosperm. Diploid embryos originate from unreduced egg or synergid, while triploid embryo originates from the endosperm.

Six aposporous embryo sacs (from 675 post-fertilized ovules) contained embryos which had originated from endosperm. Syngamy was not observed in 600 aposporous embryo sacs. In *B. setigera* triple fusion gives rise to a triploid endosperm (3n = 54). She states: "The presence of triploid endosperm coupled with the failure of fertilization of the egg in aposporous embryo sacs further suggested that the triploid embryos obtained were from endosperm". The site of embryo formation also suggests its origin from the endosperm. In addition, aceto-carmine root tip squashes of one of the triplet seedling substantiated its 3n composition (3n = 54).

The peripheral cells of the mature endosperm become meristematic, and develop into a globular or spindle-shaped mass of cells (Fig. 1.17 A–E). They are attached to the endosperm by a suspensor-like structure (Fig. 1.17 C–E). In one ovule (Fig. 1.17 C) the egg had not divided, and an embryo developed from the endosperm. In the lower part of the embryo sac an adventive embryo, of adjacent aposporous embryo sac, was present. In other ovules the embryos from egg cells were observed.

The cytological studies of root tips of twin and triplet seedlings showed combinations of diploid-diploid, diploid-triploid, and diploid-diploid-triploid. She obtained only two triploid seedlings out of 200 seeds germinated. Except the outgrowth of a cellular mass from the endosperm, Muniyamma did not observe any comparable development to that of a monocot embryo. It is very doubtful whether the triploid seedlings developed from endospermal embryo. In view of the fact that fertilization occurs and triploid endosperm develops, it is likely that, rarely, one of the synergids of an aposporous embryo sac may also be fertilized, and develop into an embryo which may account for a triploid seedling. In any case, a critical study is called for to determine if any embryo ever develops from the endosperm. Until then we consider that Muniyamma's claim is unjustified.

1.9 Gymnospermous Characters in Angiosperms

1.9.1 *Tambourissa*

The carpellary arrangement in *Tambourissa* (Monimiaceae) reminds of gymnosperms. Endress (1980) describes the receptive site of the pollen grains on the "hyperstigma" which is quite different from the condition in other angiosperms.

The unisexual flowers of *T. religiosa* are enclosed in an urceolate cup (Fig. 1.18 A). The perianth lobes are extremely reduced; the anther, or the stigma, is short, and never protrudes beyond the cup. The female floral cups develop hyperstigma. The tepals are small and their epidermes and the cells lining the canal, enclosed by the mouth of cup, secrete mucilage which fills the centre of the cup. During anthesis the mucilage oozes out (Fig. 1.18 B).

The carpels are embedded in the thick, fleshy wall of the cup, and are connected with the centre of the cup by narrow channels (Fig. 1.18 C, D). The carpellary surface is also secretory (Fig. 1.18 E). The continuity of mucilage from the hyperstigma to the micropyle is maintained (Fig. 1.18 F). Possibly, the mucilage consists of pectic substances.

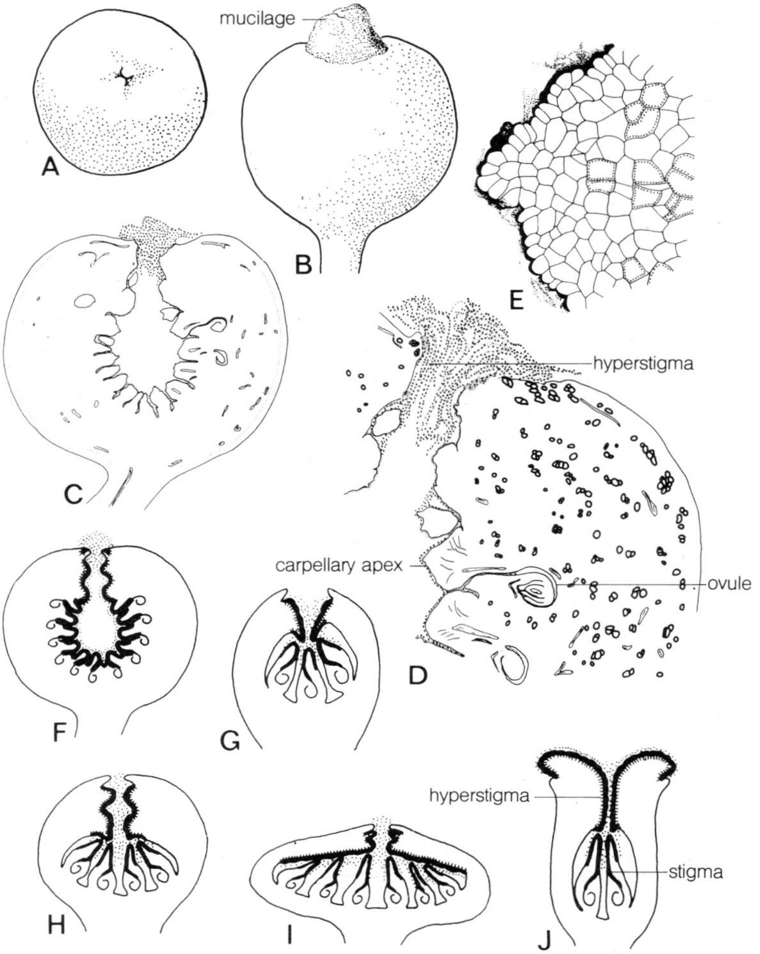

Fig. 1.18. A–F *Tambourissa religiosa.* **G** *Kibara.* **H** *Wilkiea macrophylla.* **I** *W. huegeliana.* **J** *Hennecartia.* **A, B** Female flower, before (**A**) and at anthesis (**B**). **A** As seen from above, mucilage removed from the pore. **B** Lateral view, with mucilage. **C** Entire female flower in longisection. **D** Upper portion (*right*) of **C**. Floral channel with hyperstigma, mucilage (*dotted*), and upper carpels. Note groups of stone cells. **E** Carpellary apex with secretory epidermis from **D**, mucilage *dotted*. **F–J** Schematic longisections of female flowers. *Dark lines studded with short stubs* represent hyperstigma (secretory), *plain dark lines* represent stigma and ventral slit, mucilage *dotted*. (After Endress 1980)

The hyperstigma in *T. religiosa* is highly evolved. In addition to the secretions of the tepal epidermes, the tissue around the hyperstigma decays, and produces a large quantity of mucilage.

Hyperstigma also develops in the floral cups of *Kibara* (Fig. 1.18 G), *Wilkiea macrophylla* (Fig. 1.18 H), *W. huegeliana* (Fig. 1.18 I), and *Hennecartia* (Fig. 1.18 J) – all members of the Monimiaceae. The hyperstigma in *W. huegeliana* extends up to the middle of the central cavity of the cup.

As far back as 1885, Solms-Laubach [1] imagined that wind-blown pollen grains of *Tambourissa* must have been caught in the funnel-shaped pore of the female floral cup; he did observe mucilage in the floral cup. Perkins and Gilg [1] (1901) and Perkins [1] (1925) presumed that in the entire tribe Mollinedieae when the stigma becomes receptive the floral cup falls off and the carpels are exposed. They also presumed that in *Tambourissa* the pollen grains reach the carpel through the pore. De Candolle [1] (1868) and Bentham and Mueller [1] (1870) had pointed out earlier that the female perianth falls off only after anthesis.

The investigations of Endress (1980) provide the first accurate account. The floral cup splits open and falls off only after anthesis, during the development of young fruits. The pollen grains brought by the insects, or some such agent, are caught up in the mucilage present at the apex of the female floral cup. The mucilage is continuous from the floral pore to the carpels and the ovules, and acts as a transmitting tissue. The pollen grains may germinate, and the tubes pass through the floral pore to the carpels, and then to the micropyle of the ovule. Finally, fertilization takes place.

Thus, there is a change in the site of reception and germination of pollen grains. Endress (1980) terms this feature as "angiocarpy" reflecting an intermediate behaviour between angiosperms and gymnosperms.

1.9.2 *Butomopsis*

In angiosperms the pollen grains usually land on the stigma, germinate, the tubes grow through the style, enter the ovule through micropyle, and bring about fertilization.

Butomopsis lanceolata (Alismataceae) presents an interesting variation. In a few instances B. M. Johri (1936) discovered groups of five or eight three-celled pollen grains lying in the stylar canal (Fig. 1.19 A–E), presumably containing mucilage. A germinated pollen grain and portions of pollen tubes were also observed in the stylar canal (Fig. 1.19 E). In an ovary a pollen grain had germinated directly on the surface of the ovule (Fig. 1.19 F, G).

According to Sahni (see Sahni and Johri 1936), the occurrence of pollen grains "in these positions, even though occasional, is a distinctly gymnospermous feature. It seems impossible to explain the entry of these spores, even into the upper part of the canal, except on the assumption that they were drawn in by some sort of suction mechanism like that of the "stigmatic drop" of gymnosperms, the stylar canal functioning like a micropyle".

Maheshwari (1960) compared the presence of pollen grains in the stylar canal of *B. lanceolata* with those of *Ephedra* and *Gnetum*. The wide-open hollow styles may secrete the "pollination drop". In *Gnetum* (Vasil 1959) the pollen grains germinate in the stylar canal some distance away from the nucellus, or on the nucellus.

B. M. Johri and Bhatnagar (1957) observed germinated pollen grains in the ovary of *Limnocharis emarginata* and *Trillium sessile* var. *giganteum*. Intracarpel-

1 For citations see Endress (1980)

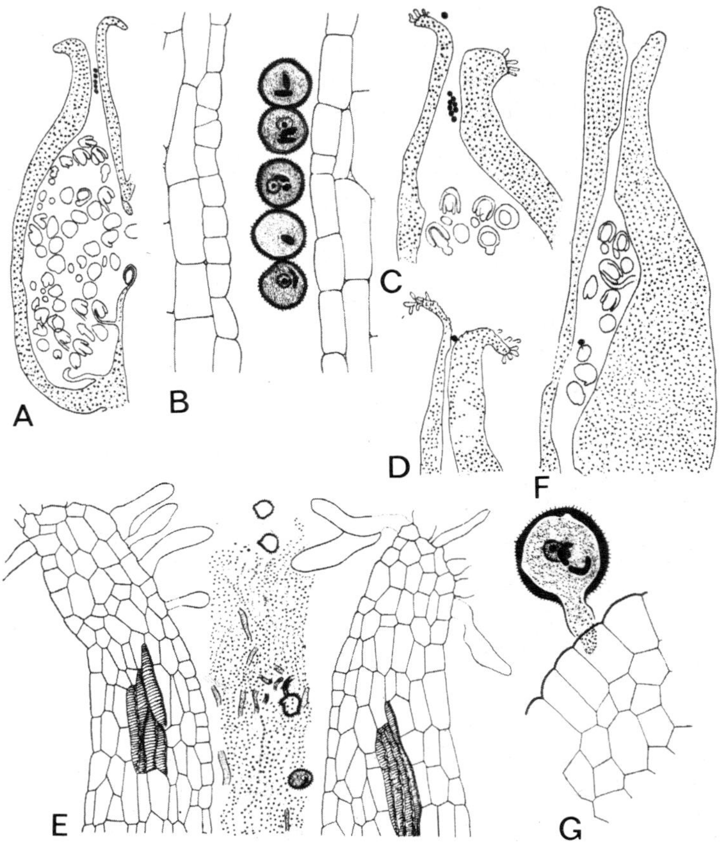

Fig. .1.19 A–G. *Butomopsis lanceolata.* **A** Longisection of a carpel with five pollen grains in the style. **B** Portion of style with pollen grains from **A.** **C** Upper region of style (longisection) with eight pollen grains. **D** Pollen grain in the upper region of style. **E** Upper part of style with two pollen grains in the apical region. The canal also shows two pollen grains (one germinated) and fragments of pollen tubes. Mucilage *dotted.* **F** Portion of carpel with a germinated pollen grain on the surface of ovule. **G** Portion of ovule with germinated pollen grain, from **F**. (After B. M. Johri 1936)

lary pollen grains are now reported in several other taxa (see B. M. Johri and Bhatnagar 1957, M. M. Johri 1962).

While discussing this condition in *Butomopsis,* Sahni (see Sahni and Johri 1936) recalls the discovery "of pollen grains inside the supposed "angiospermic" fruit of a new species of *Caytonia, C. thomasi* T. M. Harris". The Caytoniales have been treated as possible ancestors to angiosperms, since the seeds are enclosed in recurved cupules (carpel-like sheath). Andrews (1961) emphasizes: "It should be noted very clearly that the Caytoniales are not angiosperms but in the nature of their seed-bearing organs they come very close to being so."

The berry-like fruits (cupules) of *Caytonia nathorsti* (Harris 1951) are produced on a dorsiventral, ridged rachis in acropetal succession (Fig. 1.20 A). Each fruit contains about 8 seeds in *C. sewardi,* about 15 in *C. nathorsti,* and about 30

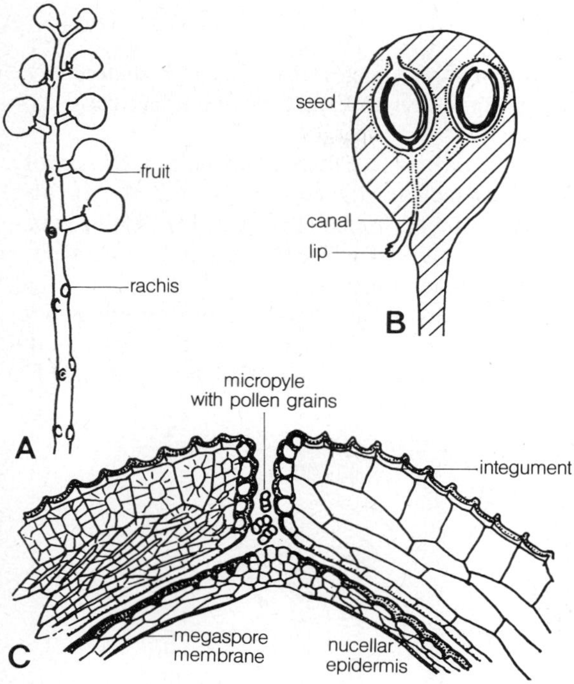

Fig. 1.20 A–C. *Caytonia nathorsti,* restorations. **A** Fruiting branch. Note the raceme type of arrangement, the lower fruits have been shed. **B** Longisection of young fruit with two orthotropous seeds. One seed is cut in vertical plane and shows a canal from the "lip" up to the micropyle. **C** Longisection of upper part of seed through the micropyle. The micropyle contains pollen grains. The integument has cutinized epidermis (*stippled*), and two to four layers of sclereids. The nucellus has a thick cutinized epidermis (*stippled*). The next two layers are imaginary, the innermost layer is the megaspore membrane. (After Harris 1951)

in *C. thomasi.* Near the stalk, the cupule shows a flap-like outgrowth, the "lip". From the lip ventral slits, organized canals, lead to the micropyle of the seeds (Fig. 1.20 B). Pollen grains have been observed in the micropyle (Fig. 1.20 C); wind-blown pollen grains caught in the pollination drop (secretions of ovules).

The formation of a continuous cavity from the hyperstigma of the cup to the micropyles of the carpels, and development of the transmitting medium in *Tambourissa religiosa* (Endress 1980) is somewhat comparable to the mucilage-containing slits in the cupule of Caytoniales. However, in Caytoniales, the pollen is deposited on the "mouth" of the cupule and is sucked in, presumably in a mucilage drop, and comes to lie in the micropyle on the surface of the nucellus (Harris 1951).

Sahni (see Sahni and Johri 1936) sums up: "... Harris's discovery on the one hand, and Johri's on the other, are enough to set one thinking, to say the least ... The whole question is very intriguing, and we obviously need more facts, covering a wider field; but Mr. Johri deserves thanks for bringing to notice what can only be regarded as a relic of gymnospermy in a confirmed and unquestionable angiosperm."

1.9.3 *Paeonia*

Yakovlev and Yoffe (1957) investigated three species of *Paeonia* (*P. anomala, P. moutan,* and *P. wittmanniana*), and observed an unusual type of embryogeny. The zygote forms a coenocyte as in *Ginkgo*.

According to Yakovlev (1969), in contrast to other angiosperms, the zygote enlarges, undergoes synchronous free-nuclear divisions (Fig. 1.21 A–D), and forms a coenocyte with peripherally arranged nuclei (Fig. 1.21 D). Wall formation is initiated centripetally, and several layers of cells enclose a large central cavity (Fig. 1.21 E). This is the first phase of development.

In the second phase, some groups of cells in the distal region become active, undergo mitotic divisions, and form meristematic zones. In the third phase, the meristematic groups continue to grow and produce five to six buds (Fig. 1.21 F).

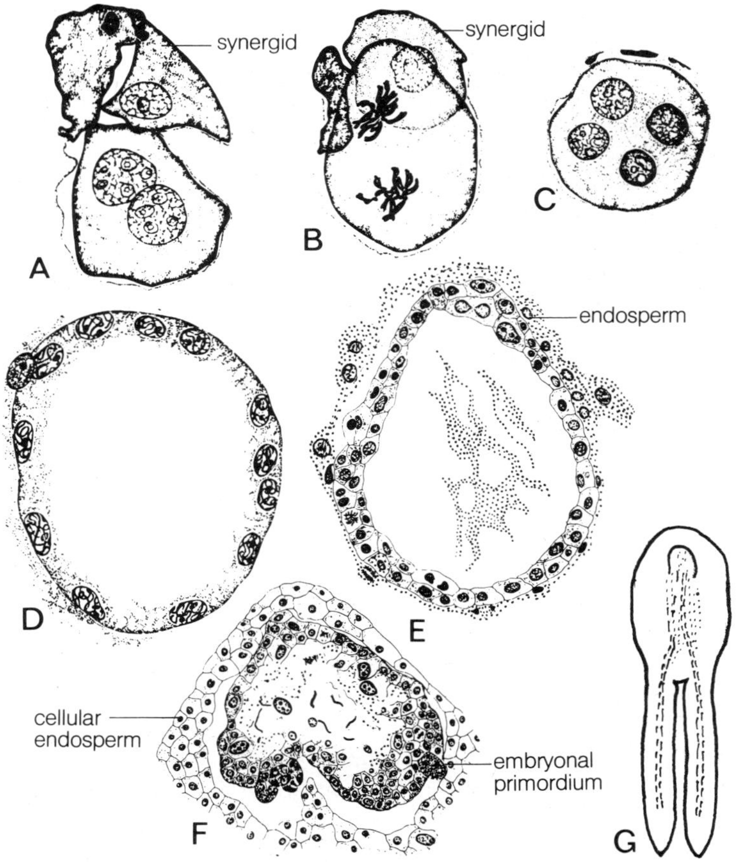

Fig. 1.21 A–G. *Paeonia lactiflora.* **A** Two-nucleate proembryo. **B** Both nuclei in division. **C** Four-nucleate proembryo. **D** Multinucleate proembryo with peripheral arrangement of nuclei. The dark nucleus in the upper region (*left*) is an endosperm (free-nuclear) nucleus. **E** Later stage. Wall formation has taken place in the proembryo, the endosperm is still free-nuclear. **F** The proembryo undergoes active cell division, and embryonal primordia differentiate. Note the cellular endosperm. **G** Dicotyledonous embryo. (After Yakovlev 1969)

As many as 25 such buds were counted. One of the buds, finally, grows and differentiates into a typical dicotyledonous embryo (Fig. 1.21 G).

Murgai (1959) reported that the embryo in *Paeonia* is like other angiosperms. The zygote divides transversely. The basal cell undergoes rapid growth and forms a massive, coenocytic suspensor. The nuclei get arranged peripherally with a single vacuole in the centre. Then centripetal growth occurs and the suspensor haustorium becomes cellular. Murgai (1959) failed to confirm the formation of a peripheral meristematic ring as mentioned by Yakovlev (Fig. 1.21 F); due to the paucity of material she could not trace the fate of the apical cell. She states that the derivatives of the apical cell become indistinguishable from those of the basal cell. Only in *P. anomala* did she (Murgai 1959) observe the degeneration of the apical cell. She presumed that in such cases the embryo must originate from the basal cell.

Murgai (1959, 1962) studied four species of *Paeonia* (*P. albiflora, P. actiflora, P. delavayi,* and *P. suffruticosa*) and contradicted the observations of Yakovlev and Yoffe. Cave et al. (1961), Matthiessen (1962), and Moscov (1964) have supported the findings of Yakovlev and Yoffe. Yakovlev (1969) investigated another species, *P. lactiflora,* and confirmed his earlier observations. There is no question that his findings are beyond doubt.

The presence of a coenocytic phase during embryogenesis in *Paeonia* also reminds of gymnosperms. In angiosperms the division of the zygote is invariably followed by a wall, whereas the coenocytic proembryo is a characteristic of gymnosperms, except *Gnetum* and *Sequoia* (Vasil 1959). Yakovlev (1969) also emphasizes that the coenocytic proembryo in *Paeonia* is comparable to the condition in gymnosperms (*Ginkgo*). Yakovlev and Yoffe (1957) consider that "Each group has its own specific characters and biological peculiarities".

While the ancestry of angiosperms is unknown, the carpellary structure and pollination in *Tambourissa,* pollination in *Butomopsis,* and proembryo in *Paeonia* are undoubtedly gymnospermous features. All these angiospermous taxa are so far removed from gymnosperms that no relationship of the two groups can be visualized.

1.10 Future Research

The earlier investigators on embryology based their studies on the examination of semitransparent ovary and ovule (orchids), cleared wholemounts, dissections, and hand and microtome sections. While these were fairly adequate, they were not always quite reliable. The emphasis was essentially on tracing various stages of development. With the stage-oriented outlook, the important features were often ignored. A perusal of publications of some of the European laboratories indicates that the sequence of reproductive stages was based on several species, or genera, without drawing attention to special features. Gradually, however, this trend changed in the 1920's and detailed studies were undertaken. Thus, comparative embryology came to the fore (see Schnarf 1931), and evolutionary tendencies and phylogenetic relationships were emphasized. Later, polarizing, phase contrast, fluorescence, electron (TEM, SEM), and interference (Nomarski) contrast optics, and cyto- and histochemistry, added several new dimensions to investigations on the structure and function of reproductive organs and tissues. This would be fully borne out from the chapters that follow.

Cass (1973) studied the morphology of pollen tubes and isolated living sperm cells in barley, in ovule and embryo sac "juice". Fifteen ovules were crushed in a hypodermic syringe, and the contents used as mounting medium. The recently opened anthers and pollinated stigma were teased in this medium, and examined under Nomarski-interference optics.

The spindle-shaped sperm cells showed compact cytoplasm, and a conspicuous nucleus with condensed chromatin. The male gametes remained in contact (with each other) for varying periods, and did not show any directional movement.

The sperm cells changed from spindle shape to spherical, and then again to spindle shape. These changes are not due to osmotic effect, but mainly to cytoplasmic microtubules present around the periphery of the sperm cell, and they regulate the shape. While entering or leaving the pollen tube, the shape of the sperm changes. The diameter of the subterminal pore of the tube is smaller than that of the sperm cells, and the latter undergo compression during exit. They also change shape as the course of pollen tube through the micropyle and synergid is contorted.

The cytoplasm of the sperm cells is active with small external bulges appearing and disappearing rapidly. In barley it lacks plastids or plastid-like bodies, but these are present in the cytoplasm of vegetative cell.

The cytoplasmic sheath of the sperm cell enters the egg apparatus at the time of discharge of the pollen tube (Cass and Jensen 1970). Cass (1973) questions whether the constituents, such as mitochondria which enter the egg cell or central cell, could alter the development. Cass suggests that further information concerning the motility of sperm cells may result if the isolated sperm cells are studied in the pollen tube "sap", or embryo sac "juice", or in a combination of both.

Russell and Cass (1983) have indicated that, in *Plumbago zeylanica,* it is possible to distinguish the two male gametes, in a pollen tube, on the basis of the number of mitochondria and plastids. Would it be possible to predict and demonstrate which of the male gametes will take part in syngamy, and which one in triple fusion, during double fertilization?

Russell also emphasizes that the male cytoplasm is involved in syngamy. Further information is awaited.

An even more fascinating and exciting area of research is the examination of living materials through cine films. A. Bajer and J. Molé-Bajer (1963) used microcinematographic techniques to study various aspects of mitosis in the endosperm of *Haemanthus* and some other taxa. Erdelská (pers. comm. 1982) studied pollen grain and pollen tube, female gametophyte, fertilization, and early stages of endosperm and embryo of *Galanthus, Jasione,* and *Torenia* (see Erdelská 1983).

Microcinematography clearly brings out the structure of thin-walled, transparent pollen grains; the agitation, circulation and rotation of contents of pollen grains, germination and growth of pollen tube, velocity and character of plasma streaming, movement of the vegetative nucleus and generative cell, division of the generative cell, and movement of male gametes.

Erdelská (pers. comm. 1982) states "... however only small attention was devoted to the application of the microcinematography in the developmental phase beginning with the release of the microspores from tetrad till the maturation of the pollen grain which would be interesting also with respect to the induction of androgenesis in vitro".

The ovules and the dissected embryo sac have to be mounted in a suitable medium to retain their transparency. Silicone fluids (inert media), sucrose solution, and liquid or semisolid agar medium provide suitable osmotic environment. By the use of Nomarski contrast optics the transparency of the partially dissected embryo sac can be intensified.

For microcinematographic investigation it is necessary to select some transparent thin-walled extramicropylar extension of the embryo sac, as in *Torenia fournieri* (Fig. 1.22 A), partly dissected embryo sac of *Galanthus nivalis* (Fig. 1.22 B), or embryo sacs enclosed by thin integuments (Campanulaceae, Monotropaceae, and Orchidaceae).

Microcinematography helps in understanding the behaviour and movement of different organelles in the central cell of the embryo sac, contraction cycle of nucleolar vacuoles in secondary nucleus, speed of movement of sperm cell in the central cell, and the time required for the fusion of the sperm cell with the secondary nucleus (Erdelská, pers. comm. 1982).

In *Jasione montana* (Erdelská 1968, 1973) the contraction cycle of the nucleolus of the secondary nucleus ranges from 7 min to several hours. The speed of movement of sperm in the central cell of the embryo sac of *Galanthus nivalis* is about 3 μm per min. Such a direct measurement facilitates the collection of data concerning the time required for the fusion of sperm with the secondary nucleus, during double fertilization. In *Galanthus nivalis* this process takes about 150 min at 25 °C in a silicone fluid medium, while the fusion of nucleoli takes about 10–15 min (Erdelská 1974).

Erdelská (1983) studied the development of endosperm, in situ, in *Galanthus nivalis* and *Jasione montana*. The type of endosperm, either Nuclear or Cellular, does not affect the duration of the mitotic cycle. The mitotic cycle during earlier development takes 5–8 h in *J. montana,* and 12.45 h in *G. nivalis*.

The examination of living material of *J. montana* revealed that the division of the primary endosperm nucleus (Fig. 1.22 C) is followed by a transverse wall (Fig. 1.22 D). The next division, in the chalazal chamber, is vertical (Fig. 1.22 E). Multicellular endosperm and linear growth of zygote is shown in Fig. 1.22 F. The elongated tubular zygote is 60–80 μm long, and the growth rate is 5–6 μm per hour (Erdelská and Klasová 1978). This phase lasts for about 12 h (see Erdelská 1983). The mitotic cycle of the endosperm nuclei is short, as compared to that in somatic cells.

The rhythm of development of endosperm and embryo can also be analyzed during the first few stages. Later, the thickness of endosperm masks the embryo (Erdelská 1983, see also Erdelská and Klasová 1978).

One can also conduct a qualitative and quantitative analysis of different organelles of the pollen grain, embryo sac, endosperm, and embryo.

Finally, we wish to draw attention to some unsolved problems.

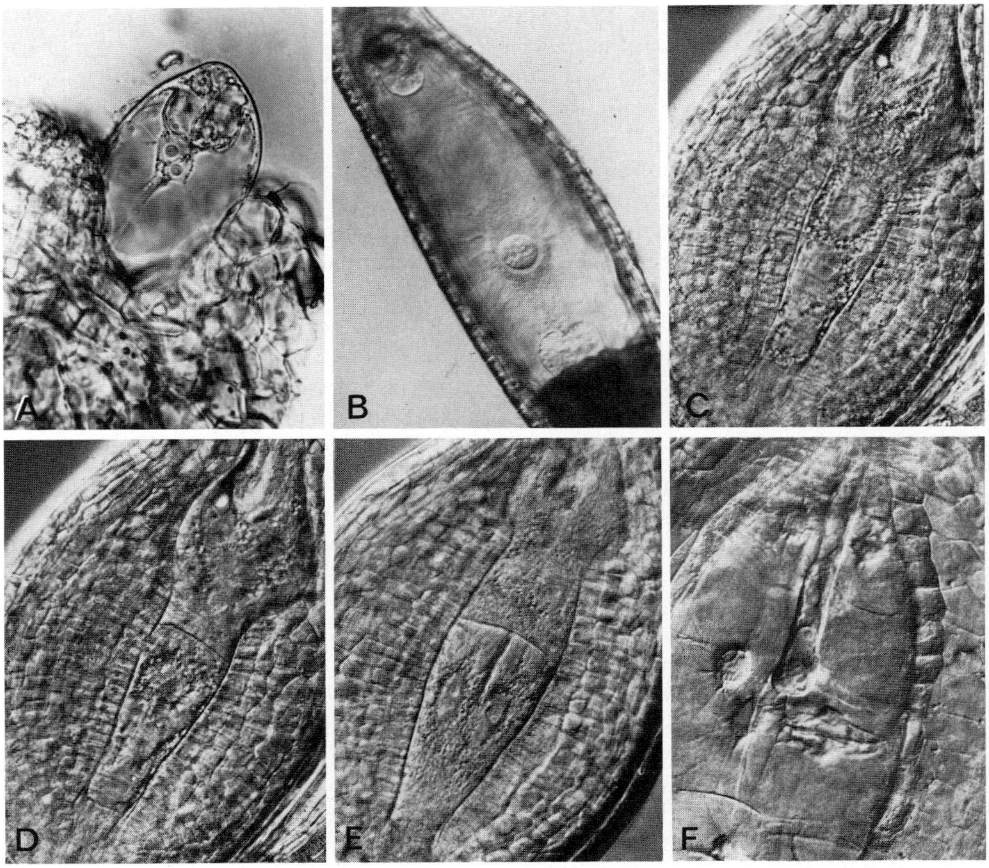

Fig. 1.22. A *Torenia fournieri*. **B** *Galanthus nivalis*. **C–F** *Jasione montana*. **A** Micropylar region of ovule. The embryo sac shows egg apparatus and polar nuclei. **B** Partially isolated organized embryo sac. **C** Embryo sac with primary endosperm nucleus. **D** Two-celled endosperm. **E** Four-celled endosperm. **F** Micropylar region of embryo sac with zygote and endosperm. Note the endothelium in **C–F.** All observations based on living material. (**A, B** from cine film C1463/1982; **C–E** cine film E1759/1971; **F** cine film E1759/1971 – Encyclopaedia cinematographica, IWF, Göttingen. Courtesy Prof. Dr. O. Erdelská, Bratislava, Czechoslovakia)

What factor(s) trigger the onset and regulation of meiosis in the anther and ovule? The synthesis of callose during microsporogenesis is under the control of the meiocytes themselves, but its digestion (by β 1,3-glucanase) is under the control of the tapetum. This is very interesting, indeed.

Synthesis of callose occurs during megasporogenesis of mono- and bisporic embryo sacs but, so far, callose formation in a coenomegaspore has not been reported. If one assumes that the callose is formed by the megaspore mother cell itself(?), how is the digestion brought about?

In a coenomegaspore the disposition of nuclei may be 2+2 (Adoxa type), 1+1+1+1 (Plumbago and Penaea types), 1+3 (Drusa type), or 1+(3) (Plumbagella and Fritillaria types). What are the causative factors which bring about this variation in the disposition of nuclei?

What is the fate of the "discharge" from the pollen tube when it bursts in the synergid? What is the precise mechanism of movement of gametes – one to the egg, the other to the polars (or their fusion product)? One would also like to have further information on the interaction between male gametes and egg and polars, at the time of double fertilization. How is actual fusion brought about? We also need additional information concerning the fate of the cytoplasmic sheath around the male gametes, in angiosperms in general.

During the development of endosperm, what is the mechanism to determine whether it would be of the Nuclear, Helobial, or Cellular type? What is the endosperm : embryo relationship? Sometimes, the embryo sac has only a single polar nucleus so that the egg and polar nucleus are genetically identical as in Onagraceae (Subramanyam and Govindu 1948) and *Butomopsis* (B. M. Johri 1936). The egg and the polar are fertilized by one male gamete each, which are also genetically identical. It follows that the zygote and primary endosperm nucleus are also genetically identical. In spite of this identical nature, the zygote produces the embryo, and the primary endosperm nucleus the Helobial type of endosperm, in *Butomopsis*. Is this due to the different biochemical milieu of the egg cell, and the central cell?

We do have some knowledge of the cytology and fine structure of the embryo sac and its constituents. What we need urgently is the biochemical information concerning the synergids, egg, polars, antipodals, zygote, primary endosperm nucleus, and central cell, at various stages of development.

Lastly, in several taxa, the mature seed contains only a globular "embryo" as in Orchidaceae, Podostemaceae, Orobanchaceae. These are referred to as "undifferentiated" and "unorganized" embryos, in comparison with normal monocot and dicot embryos. Such a concept is questionable, and Natesh and Rau discuss this aspect in Chapter 8. They emphasize that there is internal differentiation and organization at the cellular, subcellular, and molecular levels.

There are several other aspects about which we do not have adequate information, e.g. factors affecting polarity, and the role of hypostase, nucellus, and endothelium. What causes lead to the development of aposporous embryo sacs, and nucellar polyembryony? How does the stimulus of pollination, recognized and accepted, leads to the development of the unreduced egg into a diploid embryo? Similarly, how is the stimulus of fertilization recognized and accepted by the synergid when it develops into a haploid embryo?

The "embryology of angiosperms" is now emerging as a very dynamic area of investigations and, with all the recent approaches, significant advances are expected in the near future.

References

Agarwal S (1961) The synergid haustoria of *Quinchamalium chilense* Lam. Nature 192:1313–1314

Agarwal S (1962) Embryology of *Quinchamalium chilense* Lam. In: Plant embryology: A symposium. Council of Scientific and Industrial Research, New Dehli, pp 162–169

Ahlstrand L (1978) Embryology of *Ursinia* (Compositae). Bot Not 131:487–496

Ahlstrand L (1979) Embryology of Arctotideae and Calenduleae (Compositae). Ph.D. Thesis Univ Göteborg, Göteborg
Amici GB (1824) Observations microscopiques sur diverses espèces de plantes. Ann Sci Nat Bot Biol 2:41–70, 211–248
Amici GB (1830) Note sur le mode d'action du pollen sur le stigmate. Extrait d'une lettre d'Amici à Mirbel. Ann Sci Nat Bot Biol 21:329–332
Amici GB (1847) Sur la fécondation des Orchidées. Ann Sci Nat Bot Biol 718:193–205
Andrews Jr HN (1961) Studies in palaeobotany. Wiley, New York London
Arekal GD (1963) Embryological studies in Canadian representatives of the tribe Rhinantheae, Scrophulariaceae. Can J Bot 41:267–302
Arekal GD, Shivamurthy GR (1978) Female gametophyte in two species of *Balanophora* – A reinvestigation. Phytomorphology 28:7–13
Bailey LH (1950) The standard cyclopedia of horticulture, vol II. Macmillan, New York
Bajer A (1955) Living smears from endosperm. Experientia (Basel) 11:221–222
Bajer A (1957) Cine-micrographic studies on mitosis in endosperm III. The origin of mitotic spindle. Exp Cell Res 13:493–502
Bajer A, Molé-Bajer J (1963) Cine-analysis of some aspects of mitosis in endosperm. In: Rose GG (ed) Cinemicrography in cell biology. Academic Press, London, pp 357–409
Bambacioni-Mezzetti V (1928) Ricerche sulla ecologia e sulla embriologia di *Fritillaria persica* L. Ann Bot (Rome) 18:7–37
Bhatnagar SP, Chandra S (1968) Endosperm in *Psittacanthus*. Curr Sci 37:704–706
Bhatnagar SP, Johri BM (1983) Embryology of the Loranthaceae. In: Calder DM, Bernhardt P (eds) The biology of mistletoes. Academic Press, Sydney, pp 47–67
Billings FH (1937) Some new features in the reproductive cytology of angiosperms illustrated by *Isomeris arborea*. New Phytol 36:301–326
Brisson JD, Peterson RL (1975) SEM of fractured plant material embedded in glycol methacrylate. Microscop Soc Can 2:64–65
Camerarius RJ (1694) De sexu plantarum epistola. Tübingen
Cass DD (1973) An ultrastructural and Nomarski-interference study of the sperms of barley. Can J Bot 51:601–605
Cass DD, Jensen WA (1970) Fertilization in barley. Am J Bot 57:62–70
Cave MS, Arnott HJ, Cook SA (1961) Embryogeny in the California peonies with reference to their taxonomic position. Am J Bot 48:397–404
Chopra RN (1955) Some observations on endosperm development in Cucurbitaceae. Phytomorphology 5:219–230
Cocucci AE (1983) New evidence from embryology in angiosperm classification. Nord J Bot 3:67–73
Cook MT (1909) Notes on the embryo sac of *Passiflora adenophylla*. Bull Torrey Bot Club 36:373–374
Cooper DC (1938) Embryology of *Pisum sativum*. Bot Gaz 100:123–132
Cooper DC (1949) Flower and seed development in *Oxybaphus nyctagineus*. Am J Bot 36:348–355
Corner EJH (1976) The seeds of Dicotyledons, vol I and II. 1st edn. Cambridge Univ Press, Cambridge
Davis GL (1968) Apomixis and abnormal anther development in *Calotis lappulacea* Benth. (Compositae). Aust J Bot 16:1–17
Diettert RA (1938) The morphology of *Artemisia tridentata* Nutt. Lloydia 1:3–74
Dixit SN (1956) Morphological and embryological studies in the Loranthaceae-Loranthoideae. Ph D Thesis, Univ Delhi
Dixit SN (1958a) Morphological and embryological studies in the family Loranthaceae IV. *Amyema* van Tiegh. Phytomorphology 8:346–364
Dixit SN (1958b) Morphological and embryological studies in the family Loranthaceae V. *Lepeostegeres gemmiflorus* Bl. Phytomorphology 8:365–376
Dixit SN (1961) Morphological and embryological studies in the family Loranthaceae VIII. *Tolypanthus* Bl. Phytomorphology 11:335–345
Ekdahl I (1941) Die Entwicklung von Embryosack und Embryo bei *Ulmus glabra* Huds. Sven Bot Tidskr 35:143–156
Endress PK (1980) Ontogeny, function, and evolution of extreme floral construction in Monimiaceae. Plant Syst Evol 134:79–120
Engell K (1979) Morphology and embryology of Scybalioideae (Balanophoraceae) 1. *Corynaea crassa* Hook. f. var. *sprucei* (Eichl.) B. Hansen. Bot Tidsskr 73:155–166

Erdelská O (1968) Embryo sac of the species *Jasione montana* L. studied on living material. Biologia (Bratisl) 23:753–758

Erdelská O (1972) *Jasione montana* (Campanulaceae). Entwicklung des Endosperm. Encyclopaedia cinematographica E 1759/1971, Inst wiss Film, Göttingen

Erdelská O (1973) Nucleolarvakuolen in den Zellen des Embryosacks von *Jasione montana* L. Protoplasma 76:123–127

Erdelská O (1974) Contribution to the study of fertilization in the living embryo sac. In: Linskens HF (ed) Fertilization in higher plants. North-Holland, Amsterdam

Erdelská O (1975) Pre-fertilization development of ovule of *Jasione montana* L. Phytomorphology 25:76–81

Erdelská O (1983) Microcinematographical investigation of the female gametophyte, fertilization, and early embryo and endosperm development. In: "Fertilization and embryogenesis in ovulated plants." VEDA, Bratislava, Czechoslovakia, pp 49–54

Erdelská O, Klasová A (1978) La région micropylaire du sac embryonnaire de *Jasione montana* L. avant et après fécondation. Soc Bot Fr Actualités Bot, pp 249–252

Fagerlind F (1936) Die Embryologie von *Putoria*. Sven Bot Tidskr 30:362–372

Fagerlind F (1937) Embryologische, zytologische und bestäubungsexperimentelle Studien in der Familie Rubiaceae nebst Bemerkungen über einige Polyploiditätsprobleme. Acta Horti Bergiani 11:195–470

Fagerlind F (1944) Die Samenbildung und die Zytologie bei agamospermischen und sexuellen Arten von *Elatostema* und einigen nahestehenden Gattungen nebst Beleuchtung einiger damit zusammenhängender Probleme. K Sven Vetenskapsakad Handl III 21:1–130

Fagerlind F (1959) Development and structure of the flower and gametophytes in the genus *Exocarpos*. Sven Bot Tidskr 53:257–282

Geitler L (1941) Embryosäcke aus Pollenkörnern bei *Ornithogalum*. Ber Dtsch Bot Ges 59:419–423

Guignard L (1899) Sur les anthérozoides et la double copulation sexuelle chez le végétaux angiospermes. Rev Gén Bot 11:129–135

Guignard JL, Mestre JC (1966) Sur le développement d'embryons à partir des antipodes chez l'*Ulmus campestris* L. Bull Soc Bot Fr 113:227–228

Gunning BES, Pate JS (1974) Transfer cells. In: Robards AW (ed) Dynamic aspects of plant ultrastructures. McGraw Hill, New York, pp 441–480

Haberlandt G (1921) Wundhormone als Erreger von Zellteilungen. Beitr Allgem Bot 2:1–53

Hanstein J (1870) Die Entwicklung des Keimes der Monocotylen und Dicotylen. Bot Abhandl Bonn 1:1–112

Harris TM (1951) The relationships of the Caytoniales. Phytomorphology 1:29–38

Herr Jr JM (1971) A new clearing-squash technique for the study of ovule development in angiosperms. Am J Bot 58:785–790

Herr Jr JM (1973) The use of Nomarski-interference microscopy for the study of structural features in cleared ovules. Acta Bot Indica 1:35–40

Herr Jr JM (1982) An analysis of methods for permanently mounting ovules cleared in four-and-a-half type clearing fluids. Stain Technol 57:161–169

Hofmeister W (1849) Die Entstehung des Embryo der Phanerogamen. Leipzig

Johansen DA (1940) Plant microtechnique. McGraw-Hill, New York

Johri BM (1936) The life history of *Butomopsis lanceolata* Kunth. Proc Indian Acad Sci Sect B 4:139–162

Johri BM, Bajaj YPS (1964) Growth of embryos of *Amyema, Amylotheca,* and *Scurrula* on synthetic media. Nature 204:1220–1221

Johri BM, Bhatnagar SP (1957) Intra-carpellary pollen grains in angiosperms. Phytomorphology 7:292–296

Johri BM, Bhatnagar SP (1972) Loranthaceae. Botanical monograph No 8. Council of Scientific and Industrial Research, New Delhi

Johri BM, Kapil RN (1953) Contribution to the morphology and life history of *Acalypha indica* L. Phytomorphology 3:137–151

Johri BM, Prakash S (1965) Morphological and embryological studies in the family Loranthaceae XI. *Tapinostemma acaciae* (Zucc.) van Teigh. Phytomorphology 15:150–158

Johri BM, Raj B (1965) Embryo sac development in *Moquiniella*. Nature 205:415–416

Johri BM, Raj B (1969) Morphological and embryological studies in the family Loranthaceae XII. *Moquiniella rubra* (Spreng. f.) Balle. Oesterr Bot Z 116:475–485
Johri BM, Roy Chowdhury C (1957) A contribution to the embryology of *Citrullus colocynthis* Schrad. and *Melothria maderaspatana* Cogn. New Phytol 56:51–60
Johri BM, Agrawal JS, Garg S (1957) Morphological and embryological studies in the family Loranthaceae I. *Helicanthes elastica* (Desr.) Dans. Phytomorphology 7:336–354
Johri MM (1962) Intra-carpellary pollen grains in *Fritillaria* and *Lilium*. Curr Sci 31:255–256
Joshi PC (1960) Morphological and embryological studies in the family Santalaceae V. *Osyris wightiana* Wall. Phytomorphology 10:239–248
Kapil RN, Vasil IK (1963) Ovule. In: Maheshwari P (ed) Recent advances in the embryology of angiosperms. Intl Soc Plant Morphologists, Univ Delhi, Delhi, pp 41–68
Kaushik SB (1941) Development of the vermiform appendage in *Grevillea robusta* Cunn. Proc Indian Acad Sci Sect B 14:137–140
Kirkwood JE (1906) The pollen tube in some of the Cucurbitaceae. Bull Torrey Bot Club 33:327–342
Kölreuter JC (1761) Vorläufige Nachricht von einigen das Geschlecht der Pflanzen betreffenden Versuchen und Beobachtungen.
Kuijt J (1967) On the structure and origin of the seedling of *Psittacanthus schiedeanus* (Loranthaceae). Can J Bot 45:1497–1506
Kuijt J, Weberling F (1972) The flower of *Phthirusa pyrifolia* (Loranthaceae). Ber Dtsch Bot Ges 85:467–480
Longo B (1903) La nutrizione dell'embrione della *Cucurbita* operata per mezzo del tubetto pollinico. Ann Bot (Rome) 1:71–74
Maheshwari P (1950) An introduction to the embryology of angiosperms. McGraw-Hill, London New York
Maheshwari P (1960) Evolution of the ovule. 7th Sir Albert Charles Seward memorial lecture, Birbal Sahni Institute of Palaeobotany, Lucknow, pp 3–13
Maheshwari P, Johri BM (1950a) Development of embryo sac, embryo, and endosperm in *Helixanthera ligustrina* (Wall.) Dans. Nature 165:978–979
Maheshwari P, Johri BM (1950b) The occurrence of persistent pollen tubes in *Hydrilla, Ottelia,* and *Boerhaavia,* together with a discussion of the possible significance of this phenomenon in the life history of angiosperms. J Indian Bot Soc 29:1–16
Maheshwari P, Sachar RC (1954) A reinvestigation of the embryogeny of *Isomeris arborea* Nutt. Curr Sci 23:61–63
Maheshwari P, Singh B (1952) Embryology of *Macrosolen cochinchinensis*. Bot Gaz 114:20–32
Matthiessen A (1962) A contribution to the embryogeny of *Paeonia*. Acta Horti Bergiani 20:57–61
Mauritzon J (1933) Studien über die Embryologie der Familien Crassulaceae und Saxifragaceae. Diss Univ Lund
Mazzolani G (1978) Embriologia vegetale: Ricordo della prof.ssa Valeria Bambacioni Mezzeti. Ann Bot (Rome) 37:65–102
Modilewski J (1931) Die Embryobildung bei *Allium odorum* L. Bull Jard Bot Kiev 12/13:27–48
Mohana Rao PR (1974) Nutmeg seed: Its morphology and developmental anatomy. Phytomorphology 24:262–273
Molé-Bajer J (1955) A simple method for examining the action of chemicals on mitosis in living endosperm. Acta Soc Bot Pol 24:619–625
Molé-Bajer J, Bajer A (1968) Studies of selected endosperm cells with the light and electron microscope: The technique. Cellule 67:257–265
Moscov IV (1964) On the development of the embryo in several species of *Paeonia* (in Russian). Bot Zh 49:887–894
Muir WH, Hildebrandt AC, Riker AJ (1958) The preparation, isolation, and growth in culture of single cells from higher plants. Am J Bot 45:589
Mukkada AJ (1962) Some observations on the embryology of *Dicraea stylosa* Wight. In: Plant embryology: A symposium. Council of Scientific and Industrial Research, New Delhi, pp 139–145
Mukkada AJ, Chopra RN (1973) Post-fertilization development in *Indotristicha ramosissima* (Wight) van Royen. New Phytol 72:639–646
Muniyamma M (1977) Triploid embryos from endosperm in vivo. Ann Bot 41:1077–1079
Murgai P (1959) The development of the embryo in *Paeonia* – A reinvestigation. Phytomorphology 9:275–277

Murgai P (1962) Embryology of *Paeonia* together with a discussion on its systematic position. In: Plant embryology: A symposium. Council of Scientific and Industrial Research, New Delhi, pp 215–223
Naithani SP (1937) Chromosome studies in *Hyacinthus orientalis* L. III. Reversal of sexual state in the anthers of *H. orientalis* L. var. Yellow Hammer. Ann Bot 1:369–377
Narayana R (1954) Contribution to the embryology of *Dendrophthoe* Mart. Phytomorphology 4:173–179
Narayana R (1956) Morphological and embryological studies in the Loranthaceae-Loranthoideae III. *Dendrophthoe neelgherrensis* (W. and A.) van Tiegh. J Mysore Univ Sect B 16:185–205
Narayana R (1958a) Morphological and embryological studies in the family Loranthaceae II. *Lysiana exocarpi* (Behr.) van Tiegh. Phytomorphology 8:146–168
Narayana R (1958b) Morphological and embryological studies in the family Loranthaceae III. *Nuytsia floribunda* (Labill.) R. Br. Phytomorphology 8:306–323
Narayanaswami S (1954) The structure and development of the caryopsis in some Indian millets II. *Paspalum scrobiculatum* L. Bull Torrey Bot Club 81:288–299
Nawaschin SG (1898) Resultate einer Revision der Befruchtungsvorgänge bei *Lilium martagon* und *Fritillaria tenella*. Bull Acad Imp Sci St Petersburg 9:377–382
Němec B (1898) Über den Pollen der petaloiden Antheren von *Hyacinthus orientalis* L. Bull Intl Acad Sci Boh 5:17–23
Nijalingappa BHM (1975) Embryological studies in *Haloragis micrantha*. Phytomorphology 25:146–155
O'Brien TP, McCully ME (1981) The study of plant structure: Principles and selected methods. Termarcarphi, Melbourne
Pate JS, Gunning BES (1972) Transfer cells. Annu Rev Plant Physiol 23:173–196
Philipson MN (1978) Apomixis in *Cortaderia jubata* (Gramineae). N Z J Bot 16:45–59
Pienaar R de V (1952) The origin and development of the embryo sac and young embryo of *Loranthus rubromarginatus* Engl. Trans R Soc S Afr 33:223–237
Prakash S (1960) Morphological and embryological studies in the family Loranthaceae IV. *Peraxilla tetrapetala* (Linn. f.) van Tiegh. Phytomorphology 10:224–234
Prakash S (1961) Morphological and embryological studies in the family Loranthaceae VIII. *Atkinsonia ligustrina* (Cunningh.) F. V. Muell. Phytomorphology 11:325–334
Prakash S (1963) Morphological and embryological studies in the family Loranthaceae X. *Barathranthus axanthus* (Korth) Miq. Phytomorphology 13:97–103
Pullaiah T (1978) Studies in the embryology of Compositae III. The tribe Astereae. Bot Mag Tokyo 91:197–205
Raj B (1970) Morphological and embryological studies in the family Loranthaceae XIII. *Amylotheca dictyophleba* van Tiegh. Oesterr Bot Z 118:417–430
Raju MSV (1956) Embryology of Passifloraceae 1. Gametogenesis and seed development of *Passiflora calcarata* Mast. J Indian Bot Soc 35:126–137
Ram M (1957) Morphological and embryological studies in the family Santalaceae I. *Comandra umbellata* (L.) Nutt. Phytomorphology 7:24–35
Ram M (1959a) Occurrence of embryo sac-like structure in the microsporangia of *Leptomeria billardierii* R. Br. Nature 184:914–915
Ram M (1959b) Morphological and embryological studies in the family Santalaceae II. *Exocarpus*, with a discussion on its systematic position. Phytomorphology 9:4–19
Ram M (1959c) Morphological and embryological studies in the family Santalaceae III. *Leptomeria* R. Br. Phytomorphology 9:20–33
Rauch KV (1936) Cytologisch-embryologische Untersuchungen an *Scurrula atropurpurea* Dans. und *Dendrophthoe pentandra* Miq. Ber schweiz Bot Ges 45:5–61
Russell SD, Cass DD (1983) Unequal distribution of plastids and mitochondria during sperm cell formation in *Plumbago zeylanica*. In: Mulcahy DL, Ottaviano E (eds) Pollen: Biology and implications for plant breeding. Elsevier, Amsterdam, pp 135–140
Sachar RC (1956) The embryology of *Isomeris:* A reinvestigation. Phytomorphology 6:346–363
Sahni B, Johri BM (1936) Pollen grains in the stylar canal and in the ovary of an angiosperm. Curr Sci 4:587–589
Sargant E (1896) The formation of the sexual nuclei in *Lilium martagon*. Ann Bot 10:445–472
Schaeppi H, Steindl F (1942) Blütenmorphologische und embryologische Untersuchungen an Loranthoideen. Vierteljahrsschr Naturforsch Ges Zur 87:301–372

Schleiden MJ (1837) Einige Blicke auf die Entwicklungsgeschichte des vegetablischen Organismus bei den Phanerogamen. Arch Bwl Naturgeschichte III, 1:289–320

Schleiden MJ (1845) Über Amici's letzten Beitrag zur Lehre von der Befruchtung der Pflanzen. Flora: 593–600

Schnarf K (1931) Vergleichende Embryologie der Angiospermen. Bornträger, Berlin

Singh B (1952) The embryology of *Dendrophthoe falcata* (Linn. f.) Ettings. J Linn Soc Lon Bot 53:449–473

Singh D (1963) Studies on the persistent pollen tubes of the Cucurbitaceae. J Indian Bot Soc 42:208–213

Singh D (1964) A further contribution to the endosperm of the Cucurbitaceae. Proc Indian Acad Sci Sect B 60:399–413

Singh RP (1962) Forms of ovules in Euphorbiaceae. In: Plant embryology: A symposium. Council of Scientific and Industrial Research, New Delhi, pp 124–128

Smart C (1952) The life history of *Tupeia* Cham. et Schl. Trans R Soc N Z 79:459–466

Smith BB (1973) The use of a new clearing technique for the study of early ovule development, megasporogenesis, and megagametogenesis in five species of *Cornus* L. Am J Bot 60:322–338

Spurr AR (1969) A low-viscosity epoxy resin embedding medium for electron microscopy. J Ultrastruc Res 26:31–43

Stow I (1930) Experimental studies on the formation of the embryo sac-like giant pollen grain in the anther of *Hyacinthus orientalis*. Cytologia 1:417–439

Stow I (1934) On the female tendencies of the embryo sac-like giant pollen grain of *Hyacinthus orientalis*. Cytologia 5:88–108

Strasburger E (1878) Über Polyembryonie. Jenaische Ztschr Naturw 12:647–670

Strasburger E (1879) Die Anginospermen und die Gymnospermen. Jena

Strasburger E (1884) Neue Untersuchungen über den Befruchtungsvorgang bei den Phanerogamen. Jena

Subramanyam K (1962) Embryology in relation to systematic botany with particular reference to the Crassulaceae. In: Plant embryology: A symposium. Council of Scientific and Industrial Research, New Delhi, pp 94–112

Subramanyam K (1963) Embryology of *Sedum ternatum* Michx. J Indian Bot Soc (Maheshwari Comm Vol) 42A:259–275

Subramanyam K (1967) Some aspects of the embryology of *Sedum chrysanthum* (Boissier) Raymond-Hamlet with a discussion on its systematic position. Phytomorphology (Panchanan Maheshwari Mem Vol) 17:240–247

Subramanyam K, Govindu HC (1948) The development of the female gametophyte in *Oenothera odorata* Jacq. J Indian Bot Soc 27:25–30

Swamy BGL (1948) A contribution to the life history of *Casuarina*. Proc Am Acad Arts Sci 77:1–32

Swamy BGL, Ganapathy PM (1957) A new type of endosperm haustorium in *Nothapodytes foetida*. Phytomorphology 7:331–336

Tiagi B (1951) Studies in the family Orobanchaceae III. A contribution to the embryology of *Orobanche cernua* Loeffl. and *O. aegyptiaca* Pers. Phytomorphology 1:158–169

Tiagi B (1952) Studies in the family Orobanchaceae II. A contribution to the embryology of *Aeginetia indica* Linn. Bull Torrey Bot Club 79:63–68

Treub M (1883) Observations sur les Loranthacées IV. *Loranthus pentandrus* L. Ann Jard Bot Buitenz 3:184–189

Vasil V (1959) Morphology and embryology of *Gnetum ula* Brongn. Phytomorphology 9:167–214

Venkata Rao C (1949) Contributions to the embryology of Sterculiaceae I. J Indian Bot Soc 28:180–197

Venkata Rao C (1952) Occurrence of persistent pollen tubes in Malvaceae. Curr Sci 21:49–50

Venkata Rao C (1953) Contributions to the embryology of Sterculiaceae V. J Indian Bot Soc 32:208–238

Venturelli M (1981) Embriologia de *Struthanthus vulgaris* (Loranthaceae-Loranthoideae). Kurtziana 14:73–100

Vijayaraghavan MR, Ratnaparkhi S (1977) Pollen-embryo sacs in *Heuchera micrantha* Dougl. Caryologia 30:105–119

Yakovlev MS (1969) Embryogenesis and some problems of phylogenesis. Revue Cytol Biol Vég 32:325–330

Yakovlev MS, Yoffe MD (1957) On some peculiar features in the embryogeny of *Paeonia* L. Phytomorphology 7:74–82

2 The Microsporangium

N. N. BHANDARI

2.1 Introduction

Two treatises on the embryology of angiosperms by Karl Schnarf (1929, 1931) were followed by classical works of P. Maheshwari (1950, 1963), and these still serve as excellent reference volumes. Davis (1966) brought out her book on systematic embryology. The above compilations, however, are restricted to the then available light microscopic data. During the last two decades significant electron microscopic observations were made on the reproductive biology in general and on gametophytes in particular, and have yielded valuable ultrastructural details on the change-over of the sporophytic-gametophytic phase. The anther and pollen have received adequate attention at the hands of Prof. J. Heslop-Harrison and his school, who have discussed earlier notions, and suggested fundamental concepts. In this chapter microsporangium, an attempt has been made to interlink both light and electron microscope findings and present a coherent account on the structure and development of anther, microspore, and associated tissues. The histochemical and physiological aspects have also been considered.

2.2 Morphology and Structure

In most angiosperms the typical stamen comprises an anther inclusive of microsporangia and intervening connective, and a filament (Fig. 2.1 A). The primitive stamen, unlike the advanced type, lacks any distinction between filament and anther. In the putative primitive angiosperms – the woody Ranales – the stamen is laminar, i.e. a broad expanded structure with scanty or no differentiation in the sterile and fertile parts. The microsporangia are not protuberant, but are embedded abaxially (Degeneriaceae, Annonaceae, Himantandraceae), or adaxially (*Austrobaileya, Magnolia*). Such a primitive stamen has undergone reduction in the extensive sterile lamina, accompanied by retraction of margins. This course of evolution is amply substantiated by the woody Magnoliaceae (Canright 1952) and aquatic Nymphaeaceae (Moseley 1958). The connective tissue, which separates the microsporangia, is massive in the laminar stamen and undergoes progressive reduction. A similar reductional trend has been observed for the distal connective appendages, frequently met with in Magnoliaceae, Cercidiphyllaceae, Eupteleaceae, and Nymphaeaceae (see Eames 1961).

The angiospermous anther normally comprises four microsporangia (in 190 families), but it may be bisporangiate as in Adoxaceae, Circeasteraceae, Epacridaceae, Malvaceae, Philydraceae, and Restionaceae. Bixaceae have eight spo-

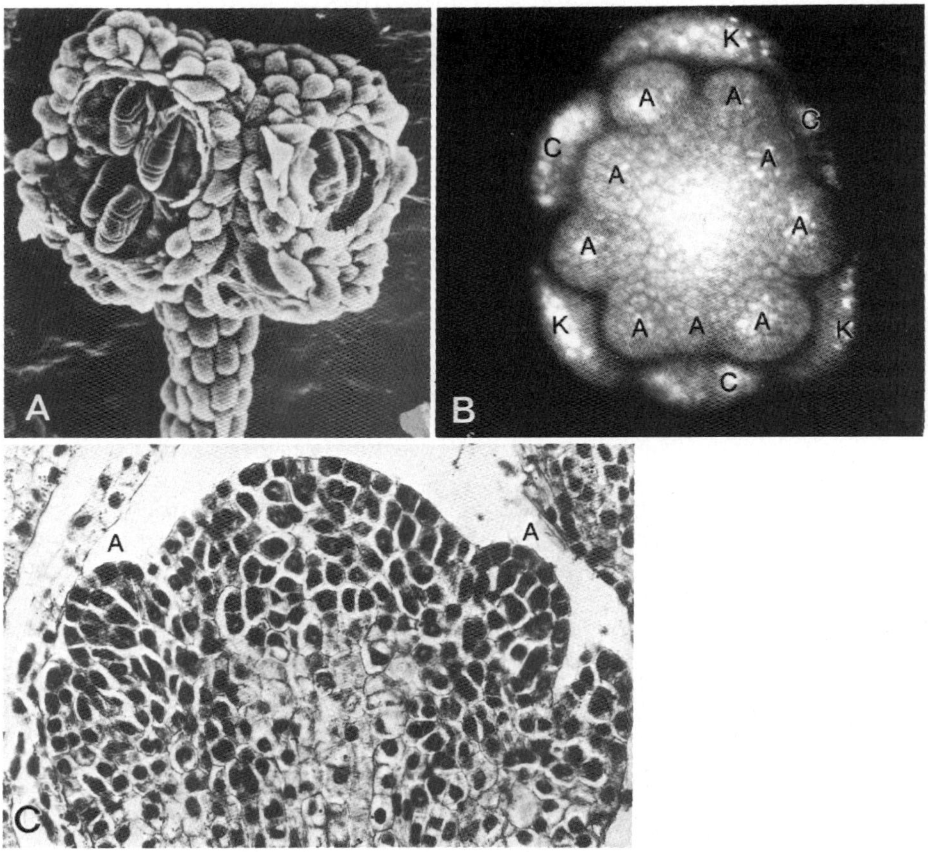

Fig. 2.1. A Scanning electron micrograph of dehisced anther of *Acacia,* showing polyads of pollen grains. **B** Top view of a floral bud at inception of staminal primordia (*A*), corolla (*C*), and calyx (*K*). **C** Longisection of floral bud; note differentiation of staminal primordia (*A*). (**A** Courtesy J. Kenrick, Malvern, Australia; **B, C** after Sattler 1973)

rangia. Multisporangiate anthers of Rhizophoraceae, Gentianaceae, and Loranthaceae (Johri and Bhatnagar 1971) result from apparent partitioning of the sporogenous tissue by sterile septae. Both bi- and tetrasporangiate anthers co-occur in about 4 monocotyledonous and 12 dicotyledonous families, such as Araceae, Asclepiadaceae, Celastraceae, Compositae, Cucurbitaceae, Lauraceae, Lemnaceae, Moringaceae, and Piperaceae (see Davis 1966). In *Cucumis sativus* and *Echinocystis lobata* (Cucurbitaceae) bi- and tetrasporangiate anthers are met with in the same flower. The morphology of the stamen in *Arceuthobium* (Viscaceae) has been variously interpreted as unisporangiate (Heinricher 1915, Dowding 1931), or multisporangiate (Städtler 1923). Pisek (1924) considered the partitions as extensions of the tapetum. Bhandari and Nanda (1968) have, however, confirmed the unisporangiate nature ontogenetically, and the horseshoe-shaped archesporium surrounds incompletely the sterile central columella.

2.3 Ontogeny

Sattler (1973), in a beautifully illustrated volume, describes floral organogenesis in 50 species of 44 angiospermous families. After the inception of calyx and corolla, the staminal primordia arise centripetally (centrifugal in Paeoniaceae; see also Cheung and Sattler 1967) as rounded or crescent-shaped projections on the floral receptacle (Fig. 2.1 B). These primordia comprise unilayered dermatogen (protoderm) which conforms to tunica-corpus concept, and covers the multicellular hump-like tissue (Fig. 2.1 C). They enlarge rapidly and assume the outline of the anther. In some grasses special cells – the specific cells – ensheath the vascular bundle. In *Zea mays* Korobova (1971) observed sequential nuclear divisions and fusions correlated with the differentiation of microsporangium.

Maheshwari (1950) described the hypodermal origin of uni- or multicelled archesporium in each lobe of a young four-lobed anther. In *Cananga odorata* (Periasamy and Swamy 1959) a row of single-celled archesporium extends throughout the entire length of a young anther. Single-celled archesporium (Fig. 2.2 A) is also met with in *Triticale* (Bhandari and Koshla 1982) Two groups of archesporium may result due to sterilization of middle cells in the massive archesporial tissue present in each half of a young anther in *Nigella damascena* (Bhandari et al. 1976). Critical studies concerning the early ontogeny of anther wall are rather meagre. In *Triticale* Bhandari and Khosla (1982) have shown that a single archesporial cell is formed by the "archesporial initial" after it has cut off two lateral cells. The lateral cells are largely responsible for organizing the wall layers, whereas the unicelled archesporium divides (Fig. 2.2 B) repeatedly to give rise to the central mass of sporogenous cells. The primary parietal layer divides and forms secondary parietal layers which give rise to endothecium, middle layer(s) and tapetum (Fig. 2.2 B–D). Brunkener (1975), however, suggested that the development of early anther need not be so specific, and may show certain flexibility – particularly in the differentiation of archesporium. Additional layers of sporogenous tissue are added by the "regenerated initials" – the primary parietal layer.

Based on the behaviour of secondary parietal layers, Davis (1966) recognized four types (see also Poddubnaya-Arnoldi 1976) of anther wall:

1. *Basic type*. The outer and inner secondary parietal layers divide periclinially, and form the endothecium, two middle layers, and tapetum.
2. *Dicotyledonous type*. The outer secondary parietal layer divides to give rise to endothecium and middle layer, while the inner layer directly functions as tapetum.
3. *Monocotyledonous type*. The inner secondary parietal layer divides and produces the middle layer and tapetum, whereas the outer one forms the endothecium.
4. *Reduced type*. The outer and inner secondary parietal layers are transformed into endothecium and tapetum, respectively, and a middle layer is absent.

These types are mostly family-specific. However, the first and third types occur concurrently in Thymelaeaceae, first and second types in Sterculiaceae, and second and third types in Euphorbiaceae.

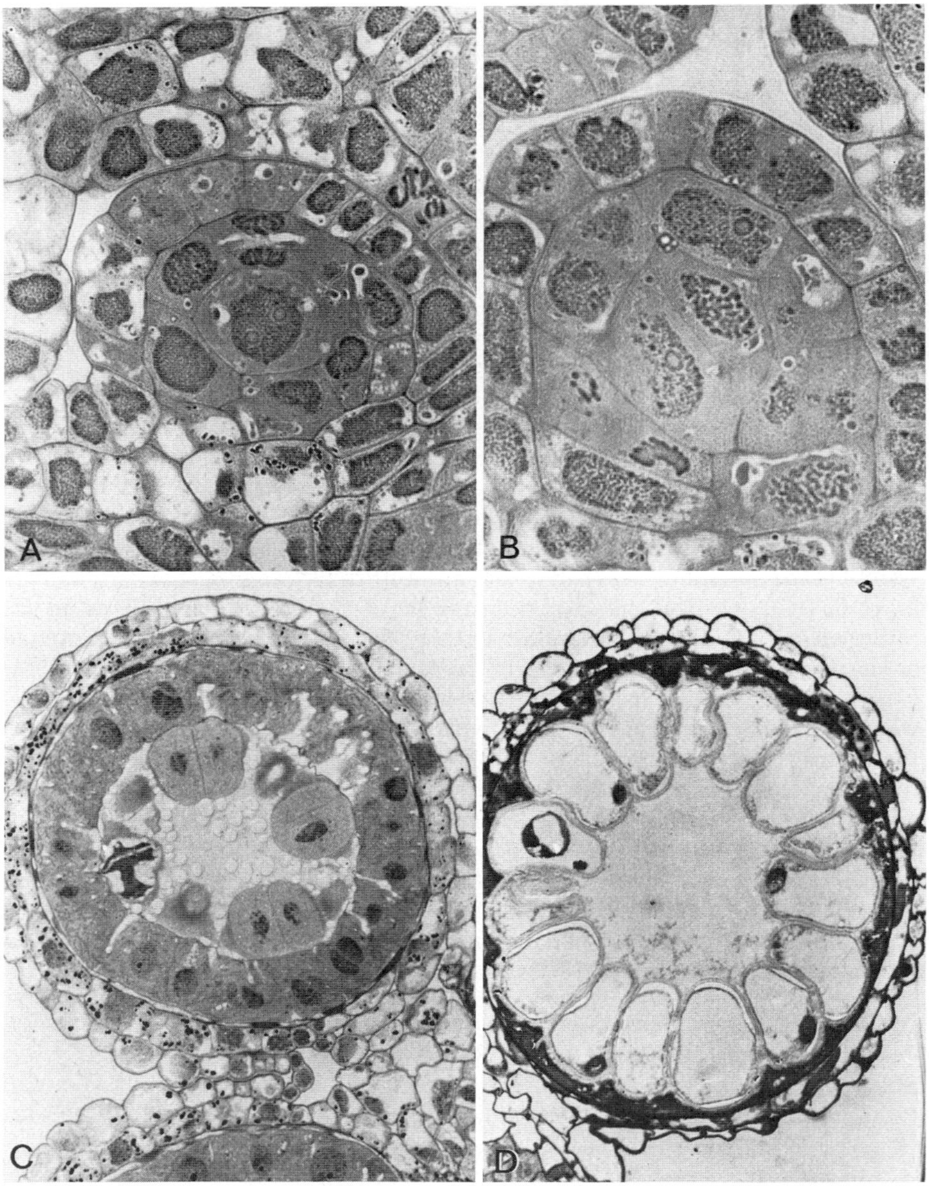

Fig. 2.2 A–D. Transections of anthers of *Triticale* to show differentiation of wall layers and development of pollen. **A–D** PAS-aniline blue staining. **A, B** One locule to show sporogenous cells surrounded by parietal layer; note periclinal division in one parietal cell. **C** Same, at dyad stage. The dyad cells are arranged at the periphery and cell plates are laid perpendicular to the tapetal surface; typical graminaceous pattern. The locular space is packed with lipid droplets. **D** A locule showing peripherally arranged uninucleate microspores. The tapetum has become compressed, while the middle layer is completely obliterated. (After Bhandari and Khosla 1982)

2.4 Wall Layers

In most angiosperms the anther wall consists of the epidermis, endothecium, middle layer(s), and tapetum. In primitive angiosperms the anther wall is massive (Bhandari 1971), which is primarily due to a larger number of middle layers (Magnoliaceae, Degeneriaceae, and Ranunculaceae) and, occasionally, bilayered tapetum as in Magnoliaceae, Trochodendraceae, Schisandraceae, and Illiciaceae. Contrarily, in advanced families the anther wall is thin, chiefly because there is only one middle layer. The wall layers, except the tapetum, are only casually referred to as regards the presence or absence of fibrous bands in endothecium, and the ephemeral nature of the middle layer(s) (Poddubnaya-Arnoldi 1976). Their role in the development of pollen grains is seldom discussed. Reznickova and Willemse (1980) consider this aspect in *Lilium*, and point out that the cells of the inner middle layer (next to tapetum) have few ribosomes, ER, some dictyosomes, plastids containing electron-dense granules, and lipid granules which increase in number and fuse during the maturation of young microspores. These cells disintegrate before mitosis in pollen grains. In the middle layer 2, the cells contain polysomes, polyvesicular bodies, some ER, amyloplasts, a few lipid granules, and a large vacuole. Starch is digested before vacuolation in the microspore, and accumulates again before pollen mitosis. The lipid granules increase after mitosis, and amyloplasts accumulate electron-dense granules. This middle layer 2 degenerates at the two-celled stage of pollen, and the starch and lipid granules are digested. Other middle layers (3–5) are comparable to the second layer but contain more starch. Before pollen mitosis, in the middle layers 1–5, some vesicles with protein have been observed; ER and dictyosomes produce vesicles and lipid granules accumulate. After mitosis, lipid is digested, amyloplasts lose starch, and some electron-dense vesicles appear. Dilations in mitochondrial cristae mark the setting in of disintegration of the middle layers. The endothecium is also comparable to middle layers 4 and 5. Reznickova and Willemse (1980) state that because of the increase and decrease of starch in endothecium and middle layers, the breakdown products of starch is one of the main sources of locular fluid used during the development of microspore and pollen grains (including its wall). Because of the lipid granules and carotene contents in middle layers, these are concerned in "Pollenkitt" formation. Such studies need to be augmented for elaboration and understanding of the functional aspects of the middle layers.

2.4.1 Epidermis

The single-layered external covering of the sporangial wall may remain intact as in Amaryllidaceae, Anacardiaceae, Araceae, Balanophoraceae, Begoniaceae, Canellaceae, Degeneriaceae, Gramineae, Lauraceae, Liliaceae, Magnoliaceae, Trochodendraceae (see Bhandari 1971, Davis 1966), and Winteraceae. The epidermal cells may become stretched, compressed, scattered, or completely sloughed off (the cells could also be missed when compressed) during the course of maturation of anther as in some members of Cannabinaceae, Moraceae, Ulmaceae, *Dorstenia,* and *Broussonetia* (see Eames 1961). In *Aristolochia* (Johri and Bhatnagar 1955), *Calycanthus* (Mathur 1968), and *Chelone glabra* (Arekal 1963) the anthers are covered all over by hairs of epidermal origin. In *Arceuthobium* (Bhandari and Nanda 1968) the epidermis develops fibrous bands like that of endothecium, and is termed exothecium. In *Zeuxine longilabris* (Karnath et al. 1979) the epidermal cells simulate tapetum, and even become binucleate. In *Trochodendron aralioides* (Yoffe 1965) and *Triticale* (Bhandari and Khosla 1982) the epidermal cells develop cuticular fibrillar projections. Cheng et al. (1981) investigated the structure of cuticle in corn, which comprises homogeneous and reticular layers composed of alkanes.

2.4.2 Endothecium

During the maturation of anther, the endothecial cells acquire thickenings which are present even in reduced aquatic plants such as *Utricularia* and *Wolffia* (Maheshwari 1950). The fibrous bands, which arise chiefly along the inner tangential walls, extend outward and upwards terminating near the outer tangential wall. In *Trochodendron aralioides* (Yoffe 1965), however, the fibrous thickenings appear to continue along the outer tangential walls as well resulting in complete annular or ring-like bands. In *Lens* (Biddle 1979) bulbous thickenings develop in the endothecial cells (Fig. 2.3 A). Such an interpretation can also perhaps result from sections having been cut at right angles to the fibrous bands which have thinner bases. In fact, the best way to study the thickenings is from the whole-mounts of endothecial cells. Eames (1961) has reviewed the structural variation in fibrous thickenings. The cytoplasm before it senesces, at the mature pollen grain stage (Fig. 2.3 B), contains RER in long rope-like strands, polysomes, and plastids, some even with starch (Biddle 1979). In *Magnolia stellata* Kapil and Bhandari (1964) observed reticulate thickenings. In *Eucommia ulmoides* (Eckardt 1963) the endothecium is bilayered, while in *Chelone glabra* (Arekal 1963) the endothecial cells undergo periclinal divisions and two to six layers are produced. Bhandari (1971) reports that in Magnoliaceae, Schisandraceae, and Winteraceae the endothecium becomes irregularly bilayered, due to periclinal divisions in some cells. Additionally, varying number of layers of the connective tissue develop fibrous bands and simulate endothecium (Maheswari Devi and Lakshminarayana 1977 b). Multilayered endothecium is common in relatively more primitive families which indicates that it is a primitive feature, and there is a phylogenetic trend towards reduction of endothecial tissue. The endothecium is limited to the protuberant part of microsporangium, and originates from the parietal layer. The endothecium-like tissue, whenever present, towards the inner side, is always contributed by the connective tissue (see Bhandari 1971). In *Triticale,* however, Bhandari and Khosla (1982) have shown a complete ring of endothecial layer surrounding the tapetum and sporogenous tissue. This originates exclusively from the parietal layer. De Fossard (1969) reported that in *Chenopodium rubrum* the endothecial thickenings contain mostly α-cellulose. He also corrected (through personal communication) Vasil (1967) that in most angiosperms these are made up of cellulose.

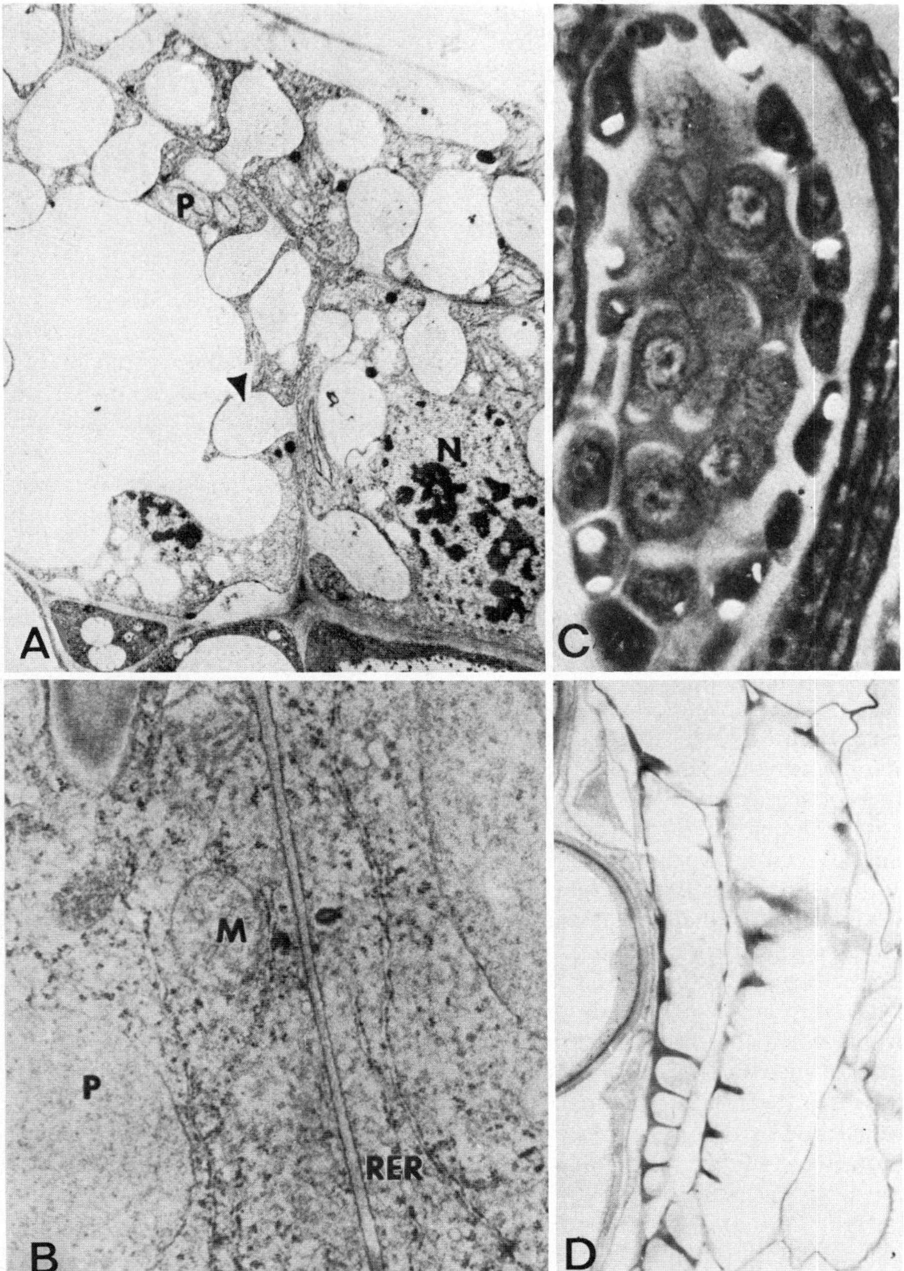

Fig. 2.3. A, B *Lens culinare*. **C** *Desmodium*. **D** *Triticale*. **A** Endothecial cells have bulbous wall thickenings (*arrow*), while the plastids (*P*) lack starch, *N* nucleus. **B** Cytoplasm of mature endothecial cell; note rough endoplasmic reticulum (*RER*), plastid (*P*), mitochondria (*M*), and ribosomes. **C** Tapetal cells of *Desmodium* showing crystals: some cells have even two crystals. **D** Endothecial cells showing PAS-positive fibrous thickenings. (**A, B** after Biddle 1979; **C** courtesy Reeta Khosla, Delhi; **D** after Buss et al. 1969)

In *Triticale* (Bhandari and Khosla 1982) the cellulosic nature of thickening has been confirmed by a strong PAS reaction (Fig. 2.3 C). Biddle (1979) mentions a small amount of pectin and lignin as well in *Pisum* and *Lens,* respectively. The endothecium brings about the dehiscence of anthers.

2.4.3 Middle Layers

Depending upon whether outer and inner primary parietal layers, or any one or none of them, contribute to the formation of middle layer(s), their number varies from none to two. The middle layers are transient or ephemeral, and become compressed, crushed, or obliterated even before the microsporangium is fully mature and ready to dehisce (Swamy and Krishnamurthy 1980). In *Nigella damascena* (Bhandari et al. 1976) and *Lilium* (Reznickova and Willemse 1980) two, or two to five, middle layers persist until the dehiscence of anther. In *Lilium* these even contribute towards the development of pollen. Sometimes, the middle layer lying immediately below the endothecium may also develop fibrous thickenings as in *Agave, Amyema, Argemone, Crinum, Olax, Oxychloe,* and Zingiberaceae.

2.4.4 Tapetum

The tapetum is the innermost layer of anther wall and surrounds the sporogenous tissue. Because of its strategic position, Maheshwari (1950) regarded it "of considerable physiological significance". Currently, it is generally assumed that the tapetum may be involved in three different aspects of pollen development, namely: nourishment of microspores, formation of exine, and synthesis and release of materials which take part in the deposition of tryphine and "Pollenkitt".

In angiosperms the tapetum is of two types: glandular or secretory in which the tapetal cells remain intact and persist in situ, and the periplasmodial or amoeboid type when the walls of tapetal cells break down, and the protoplasts protrude into the locule and fuse to form a coenocytic plasmodium. Of the latter, Clausen (1927) distinguished four subtypes depending on whether the periplasmodium is formed when the microspore mother cells are undergoing meiosis (Triglochin type), at the tetrad stage (Butomus, Sparganium types), or as soon as microspores have separated (Sagittaria type). Biddle (1979) suggested that the condition of the nucleus and cytoplasm, at the time of dissolution of wall, is a more reliable criterion to classify the tapetum as secretory or plasmodial. In the former the protoplast and the nuclei collapse, whereas in the latter the cells retain nuclei and normal cytoplasm at dissolution. The classification of Clausen (1927), based on the time of dissolution of tapetal walls and formation of plasmodium, is no longer valid since even in secretory tapetum this timing is extremely variable. The walls breakdown during meiosis I in *Avena* (Steer 1977), at tetrad stage in *Beta* (Hoefert 1971), *Lens* and *Pisum* (Biddle 1979), and during early pollen development in *Cannabis* and *Silene* (Heslop-Harrison 1963 a), while the tapetum does not degenerate till anthesis in *Helleborus* (Echlin and Godwin 1968 a).

2.4.4.1 Types. General Organization

Davis (1966) points out that of the 231 families of angiosperms where the information on tapetum is available, 81 families show glandular type, 29 amoeboid type, and 21 both glandular and amoeboid type. A survey of subsequent literature revealed that about 200 species spread over 68 dicotyledonous and 11 monocotyledonous families have been studied. Of these, 176 show glandular type, 7 amoeboid type, while in the remaining plants precise information is not available. In 17 families, subsequent to Davis' compilation, additional information is now available, and all of them, including three monocots, show glandular type.

According to Davis (1966), the tapetum in Pandanaceae is of the glandular type. Cheah and Stone (1975), on the other hand, report that in *Pandanus parvius* the tapetal cell walls disintegrate releasing the nuclei and protoplast into the loculus, and the tapetum is of the amoeboid type.

Generally, the tapetum comprises a single layer, exceptionally it may divide and become biseriate throughout as in *Pyrostegia, Tecoma,* some Bignoniaceae, *Magnolia* (Kapil and Bhandari 1964), and *Buckleya lanceolata* (Raj 1969). A multiseriate condition is known in *Combretum grandiflorum* (Combretaceae) and *Oxystelma esculentum* (Maheswari Devi and Lakshminarayana 1977a). The tapetum is single-layered towards the epidermis, and multilayered towards the connective in some Rubiaceae (Periasamy and Parameswaran 1965), Vacciniaceae, and Oleaceae (Maheswari Devi 1975). The reports of origin of the entire tapetum from the parietal layer was chiefly due to the fact that the investigators confined their attention to the protuberant (outer) region of microsporangium. Periasamy and Swamy (1966) suggest the dual origin of tapetum in angiosperms: contributed by the parietal layer towards the protuberant region, and connective tissue towards the inner half of loculus. Ontogenetic studies confirming the dual origin of tapetum are fewer because of the morphological similarity after alignment of tapetum derived from the parietal and connective tissues, and because of the rapid differentiation along the connective region. In *Anemone rivularis* (Bhandari 1968), however, because of slow development, it was possible to demonstrate the origin of tapetum towards the inner side from the connective tissue. Plants with dimorphic tapetum such as *Salvia mellifera* (Carlson and Stuart 1936) and some species of Acanthaceae, Scrophulariaceae, and Labiatae are ideal materials to confirm the dual origin. In *Alectra thomsoni* (Vijayaraghavan and Ratnaparkhi 1973) and *Celsia coromandelina* (Kapoor et al. 1978) the tapetal cells which differentiate from the connective tissue are not only structurally different, but also differentiate and develop precociously. That the tapetum is of dual origin in most angiosperms is a logical conclusion. In some instances, however, it could arise from the sporogenous tissue as well (Bhandari 1971). In *Triticale* Bhandari and Khosla (1982) report that the tapetum originates as a concentric layer around the sporogenous tissue, solely from the parietal tissue, and is not of dual origin. They presume a similar mode of origin of tapetum in the other graminaceous members, although early ontogenetic studies are wanting.

Concurrent with meiosis in pollen mother cells, the synthetic activity in the tapetum increases and certain food reserves, such as starch, lipids, tryphine, etc., are stored. These are degraded or released as such in the anther locule. Rarely, calcium oxalate raphides, or oval to spherical discs, or prismatic crystals have been reported in plasmodial (Mepham and Lane 1969b) and secretory tapetum (Fig. 2.3D) (Buss et al. 1969, Buss and Lersten 1972). Mepham and Lane (1969 a, b) report that the raphides lie free in the cytoplasm, and are not associated with any organelle. The role of such crystalline inclusions in the development of the tapetum or pollen is obscure.

2.4.4.2 Ultrastructure

Periplasmodial Tapetum. Since the recognition of tapetum as an important tissue, particularly in relation to pollen development, during the last 10 or 12 years it has been the subject of detailed study with EM. Echlin (1971) reviewed the structural and functional aspects of tapetum during microsporogenesis. Of the three studies on the plasmodial tapetum in *Tradescantia bracteata* (Mepham and Lane 1969a, b), *Helianthus annuus* (Horner 1977), and in *Mahonia aquifolium* (Roland-Heydacker 1980), the first one furnishes considerable information. In *Tradescantia* and *Helianthus* the tapetum is uniseriate, while in *Mahonia* it has three layers which release their cellular contents into the locule, one after the other (Fig. 2.4 E, F). In *Tradescantia bracteata* the tapetal cells, preceding meiosis, show prominent population of organelles. The plastids have dense matrix, pro-lamellar bodies in variable organizational state, and different inclusions like plastoglobuli, large polysaccharide granules, and bodies of moderate electron density and characteristic granular appearance. Raphides, without any connection to membranous structure, accumulate and grow abundantly. Styloid-like crystals have been observed in *Helianthus annuus* (Horner 1977). The peripheral tapetal cells become extremely vacuolate and large dictyosomes are formed which produce numerous small, single membrane-bound vacuoles. Many such vacuoles become enclosed by an element of ER, and result in multivesiculate bodies which also show double-membraned vacuoles (Fig. 2.4A). The outer membrane of such multivesiculate bodies becomes confluent with the plasmalemma, and discharge the contained vesicles between the plasmalemma and cell wall (Fig. 2.4A). Also, the dictyosome-derived vesicles are similarly extruded individually. In *Ipomoea purpurea* (Godwin et al. 1967) and *Helleborus foetidus* (Echlin and Godwin 1968a), which show secretory tapetum, similar polyvesicular bodies have been observed but have not been conclusively shown to bring about the lysis of tapetal cells. Concurrently, a wave of lysis ensues from the tapetal cells inward, bringing about dissolution of largely pectic cell walls and allowing the plasmodial tapetum to invade the locule. It affects the lysis of walls of microspore mother cell which are finally surrounded by the infiltrating tapetal protoplasts. In *Helianthus* (Horner 1977), however, no mention has been made of such multivesiculate bodies even though the dissolution of tapetal walls takes place at the late tetrad stage. The vesicles presumably release hydrolytic enzymes. No sooner do the tapetal plasmalemma breakdown along the radial surface, the cell contents become confluent carrying the mitochondria, plastids, nuclei, and raphides inwards and engulf the meiocytes.

The periplasmodial cytoplasm shows signs of reorganization by the reappearance of rough ER, dictyosomes and formation of microtubules. Prior to callase

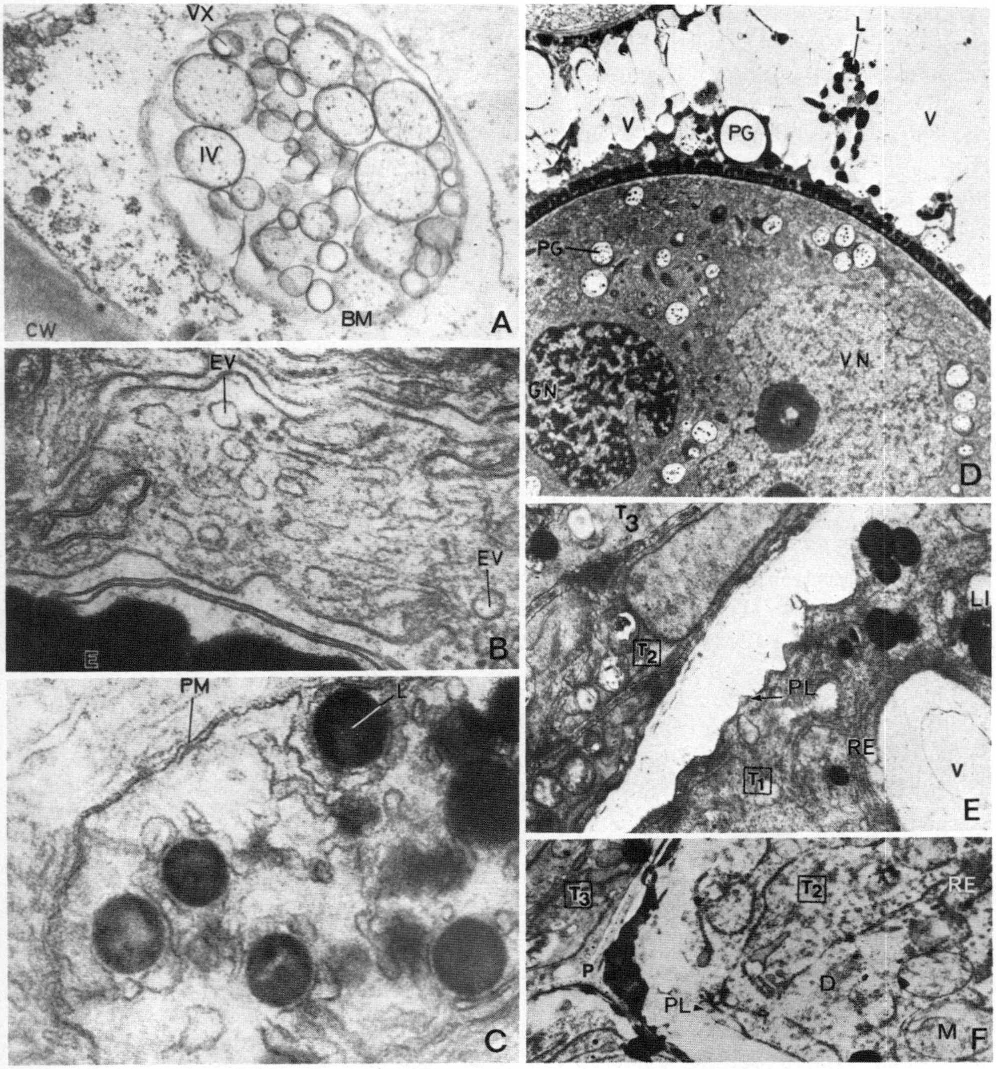

Fig. 2.4. A–D *Tradescantia bracteata*. **E, F** *Mahonia aquifolia*. **A** Multivesicular body in the cytoplasm of a tapetal cell just before lysis of cell wall (*CW*), showing included vesicles (*IV*); some vesicles (*VX*) are with double membrane. The bounding membrane (*BM*) has been cut obliquely and appears diffuse. **B** Region of tapetal cytoplasm at the end of tetrad stage of pollen development showing electron-lucent vesicles (*EV*) apparently involved in callose degradation; note pollen exine (*E*). **C** One of a series of sections taken through part of a tapetal plastid (*PM* plastid membrane) showing a stage in the extrusion of a lipid globule (*L*). **D** Part of tapetal cytoplasm, and a pollen grain with generative (*GN*) and vegetative (*VN*) nuclei. The tapetal cytoplasm between pollen grains shows large vacuoles (*V*), lipid globule (*L*) encrusting tapetal membranes, and unstable polysaccharide granules (*PG*). **E** Successive release of periplasmodial protoplast at the microspore stage in first layer (T_1), second layer (T_2), and third layer of tapetum (T_3). Lipid globules (*LI*) and plasmalemma (*PL*) are clearly visible in (T_1); *RE* endoplasmic reticulum. **F** General aspect of disorganization of tapetum two (T_2) and three (T_3); there is overall disruption and degeneration of organelles such as mitochondria (*M*), dictyosome (*D*) and endoplasmic reticulum (*RE*). Remains of pectocellulosic walls (*P*) between (T_2) and (T_3) can be seen; *PL* plasmalemma. (**A–D** after Mepham and Lane 1969 a; **E, F** after Roland-Heydacker 1980)

synthesis in the tapetum which degrades special callose wall, the dictyosome-derived electron-lucent vesicles (Fig. 2.4 B, E) fuse with the membranes surrounding the tetrads. In the plastids the number of plastoglobules increases, and the outer membrane of the pro-lamellar membranous system becomes incorporated within the plastid membrane, thus extruding many of these into the tapetal cytoplasm (Fig. 2.4 C). In *Helianthus*, at the tetrad stage, the tapetal cells have dense cytoplasm and organelles are difficult to distinguish, but mitochondria, plastids, and vacuoles are quite evident. There are numerous dictyosomes with associated vesicles.

Vacuolation of plasmodium increases owing to increased hydration which brings about a close contact between tapetal cytoplasm and pollen grains (Fig. 2.4 D). Such a high degree of vacuolation has also been observed in *Helianthus* and *Mahonia* (Fig. 2.4 E, F). Consequently, tryphine – polysaccharide granules, flattened lipid globuli, and other cytoplasmic debris – is deposited as a superficial layer on pollen exine (Fig. 2.4 D).

Secretory Tapetum. The glandular tapetum has been studied in a much larger number of taxa: *Poa annua* (Rowley 1963), *Silene* and *Cannabis* (Heslop-Harrison 1963 a, b), *Helleborus foetidus* (Echlin and Godwin 1968 a), *Lilium longiflorum* (Heslop-Harrison and Dickinson 1969), *Allium cepa* (Risueño et al. 1969), *Beta vulgaris* (Hoefert 1971), *Citrus limon* (Horner and Lersten 1971), *Sorghum bicolor* (Christensen et al. 1972), *Capsicum annuum* (Horner and Rogers 1974), *Antirrhinum majus, Gentiana acaulis* (Lombardo and Carraro 1976 a, b), *Pelargonium zonale, Kalanchoe obtusa* (Carraro and Lombardo 1976), *Avena sativa* (Steer 1977), *Lycopersicum peruvianum* (Pacini and Cresti 1978), *Olea europea* (Pacini and Juniper 1979 b), *Pisum sativum* and *Lens culinare* (Biddle 1979), and *Lilium* (Reznickova and Willemse 1980).

Tapetum Before and During Meiosis. In *Avena* (Steer 1977) the newly formed tapetal cells show plasmodesmatal connections between the adjacent cells, and also

Fig. 2.5. A, E, F *Helleborus foetidus*; **B, C** *Avena sativa*; **D** *Citrus limon*. **A** Polyvesicular body (*PVB*) at the junction of two tapetal cells. The cell wall (*CW*) is relatively thin and associated with microtubules (*MT*) in the peripheral cytoplasm. Plastids (*P*) and pro-Ubisch bodies (*PB*) are seen in the tapetal cytoplasm. **B** Subplasma-membrane microtubules (*MT*) lie on either side of the cellulose wall separating a tapetal cell (*right*) from a pollen mother cell. There is a noticeable halo around each microtubule in the tapetal cell, suggesting the presence of non-staining component. **C** Radial wall between two tapetal cells at metaphase I in the meiocytes. Only the middle lamella (*ML*) remains of the original cellulosic wall. Microtubules (*MT*) run both alongside the wall and into the cytoplasm. **D** Portion of cytoplasm in a tapetal cell at a stage when young microspores have been released. Endoplasmic reticulum (*E*) and dictyosomes (*D*) with associated vesicles (*Z*) are evident. **E** Periphery of tapetum at the time of primexine formation in the microspore. The cell wall is less electron-dense and begins to show a more open fibrous appearance containing irregularly-shaped electron-dense granules. The extruded pro-Ubisch bodies (*PB*) have a thin coating of sporopollenin. The endoplasmic reticulum is associated with the developing pro-Ubisch bodies. Note regions of unit membrane dimensions surrounding the extruded pro-Ubisch body (*asterisk*). **F** Details of electron-transparent vesicles on the outside of tapetal cells, at the stage of final dissolution of tapetal cell wall and special callose walls. The tapetum is in a phase of senescence. **G** Semi-diagrammatic representation of a part of anther locule showing cytoplasmic streaming (*arrows*) which is presumed to control the uptake and release of substances by the tapetal cells. (**A, E, F** after Echlin and Godwin 1968 a; **B, C** after Steer 1977; **D** after Horner and Lersten 1971; **G** adapted from Pacini and Juniper 1979 a)

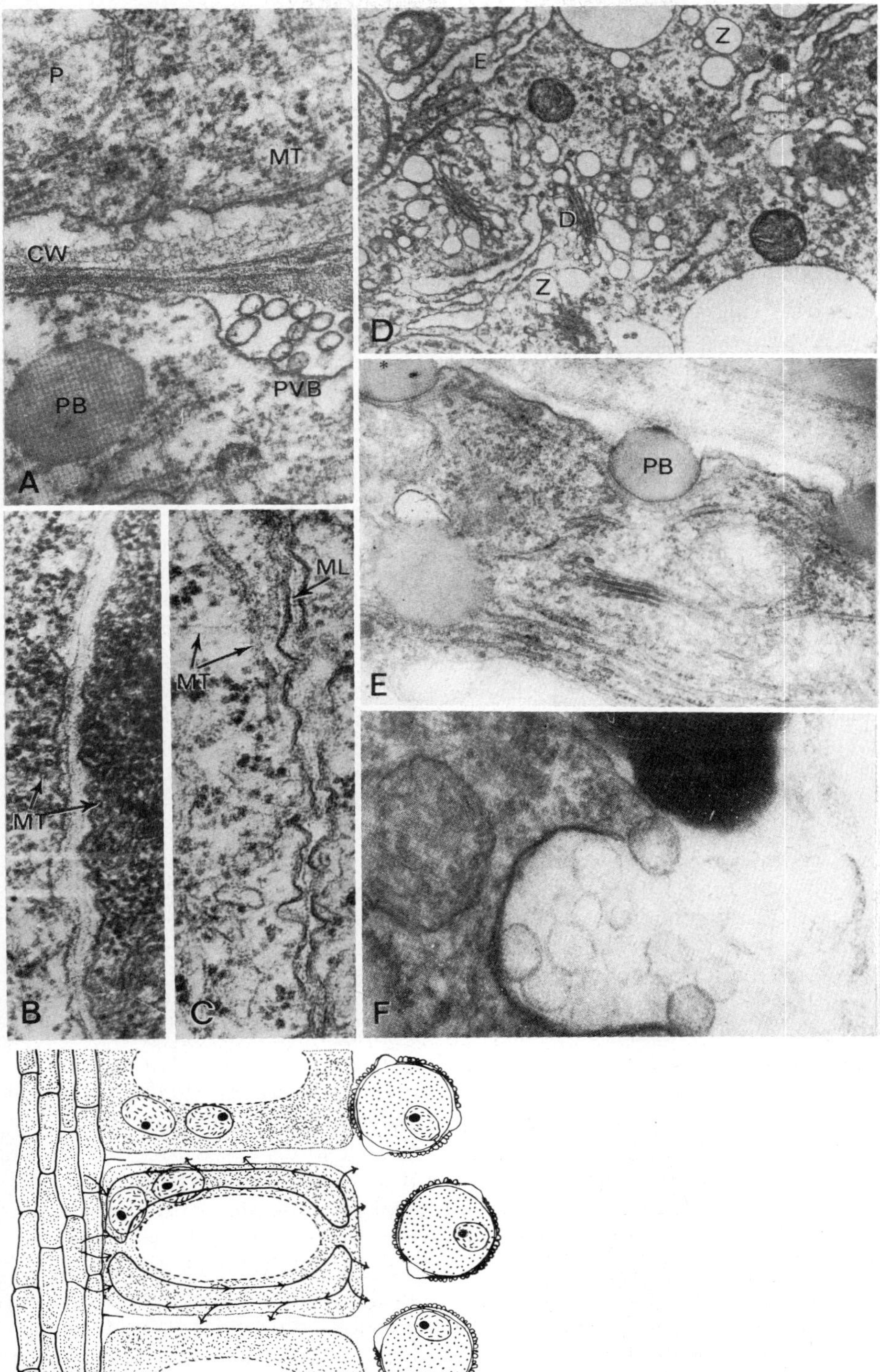
P
MT
CW
PVB
PB
A
E
Z
D
Z
D
PB
E
MT
B
ML
MT
C
F
G

between the sporogenous cells. Microtubules run parallel to the long axis of anther along the tangential walls (Fig. 2.5 A–C) and tangentially or radially along the radial walls. However, prior to nuclear division which results in binucleate tapetal cells, aggregates of such microtubules are formed along the radial and tangential walls. Even after the primary wall is completely dissolved, microtubules are still present beneath it (Fig. 2.5 A, C). Similar microtubules have also been reported (at this stage) in *Olea* (Pacini and Juniper 1979 b) and *Helleborus* (Echlin and Godwin 1968 a). The microtubules are normally concurrent with the orientation of cellulose fibrils during early stages of pollen-wall formation or its subsequent modification, owing to differential thickenings (Hoefert 1971). However, it is rather enigmatic that the microtubules persist beneath the plasmalemma of tapetal cells even when their walls have dissolved, and when no fresh synthesis of cellulose is expected. Steer (1977) presumes that either these microtubules represent the relics of earlier activity, or contribute to the maintenance of tapetal cell shape. Further, the absence of microtubules in the periplasmodial tapetum tends to support the latter possibility. In *Helleborus* (Echlin and Godwin 1968 b), at the sporogenous stage, organelles in the tapetal cells are recognized with difficulty, though mitochondria, plastids, and a number of "grey bodies" or pro-Ubisch bodies can be identified. Some dictyosomes with peripherally associated vesicles, ribosomes, and associated profiles of SER are also present (Fig. 2.5 D, E). In *Helleborus* (Echlin and Godwin 1968 a) and *Beta* (Hoefert 1971), however, a large number of vacuoles occur interspersed in the tapetal cytoplasm. Some protein inclusions have been observed in plastids in *Lycopersicum* (Pacini and Cresti 1978). In *Olea* (Pacini and Juniper 1979 b), because of similar ribosomal density, the tapetal cells are difficult to distinguish from the sporogenous cells. However, at the leptotene stage, the density of ribosomes in the tapetal cells is higher than that of the microspore mother cells. In *Lycopersicum peruvianum* (Pacini and Cresti 1978), *Avena* (Steer 1977), and *Helleborus* (Echlin and Godwin 1968 a) the ribosomes are fewer in the tapetal cells. In *Helleborus* and *Avena* the walls of tapetal cells are thin, and comprise the middle lamella, and a small amount of cellulosic primary wall composed of numerous fine, lightly stained fibrils. Plasmodesmata connect the adjoining tapetal cells and to the microspore mother cells in *Avena, Beta, Citrus, Helleborus, Lens, Lycopersicum, Olea, Pisum,* and *Sorghum*. In *Lens* and *Pisum* such connections were not observed between the tapetal and sporogenous cells.

In *Sorghum* (Christensen et al. 1972) the tapetal cytoplasm contains distinct dictyosomes, vesicles, mitochondria, and an extensive and enlarged membrane-bound tubular system of ER, as observed earlier in *Helleborus* (Echlin and Godwin 1968 a). The amplification of tapetal endoplasmic reticulum, both smooth and rough, during meiotic divisions in microspore mother cells appears to be of universal occurrence, as in *Avena, Beta, Citrus, Lens, Lycopersicum, Olea,* and *Pisum*. The cytoplasm presents a denser appearance because of the increase in ribosomal population and the formation of pro-orbicular bodies which are extruded. At the tetrad stage the dictyosomes, so far relatively inactive, produce a large number of vesicles which become included in larger vesicles, the polyvesicular bodies (Fig. 2.5 A, F). The membrane of such polyvesicular bodies becomes confluent with the plasma membrane and, thus, releases numerous vesicles into the space between plasmalemma and tapetal cell (Fig. 2.5 A, F). Subsequently, the cell wall undergoes autolysis, concurrent with the dissolution of callosic wall of the tetrad. In *Olea europaea,* Pacini and Juniper (1979 a) consider that the tapetal cells are polarized in structure and function. Such polarization is brought about by convective cyclosis (Fig. 2.5 G) all round the central vacuole, and imposes a significant alignment and disposition of ER, plastids, and mitochondria.

The structural changes in the tapetal cytoplasm and organelles, narrated briefly, should serve as an introduction since the origin of pro-orbicules and their extrusion, synthesis of sporopollenin and its deposition on Ubisch bodies, formation and release of callase, origin and development of orbicular wall (tapetal membrane), and the role of tapetum in the formation and deposition of sporophytic proteins, Pollenkitt and tryphine on the pollen surface have been dealt with separately.

2.4.4.3 Cytology of Tapetum

Concomitant with the onset of meiosis the tapetal cells exhibit significant cytological variations. The single nucleus may undergo one or two mitotic cycles resulting in two to four nuclei which may fuse so that a polyploid nucleus is formed. Or, through endomitosis and restitution, the tapetal cells become polyploid (see Carniel 1952). The cells may remain uninucleate throughout (see Davis 1966), or become multinucleate, and 13 nuclei are reported in *Hepatica acutiloba* (Schnarf 1931). They often fuse resulting in various levels of polyploidy. Maheshwari (1950) recognized three types of nuclear divisions: mitotic, "sticky type" or amitotic division, and endomitotic. Davis (1961) studied nuclear behaviour in *Podolepis jaceoides*. The nuclei may become up to 16-ploid, and she terms this phenomenon as "sympathetic" in view of the fact that the stimulus for initiation of meiosis extends beyond the sporogenous tissue to tapetum. Such a meiotic stimulus has not yet been demonstrated.

Bhandari et al. (1976) report that, in *Nigella damascena,* polyploidy up to 16n is achieved by both mitotic and endomitotic divisions of tapetal nuclei. Oksala and Therman (1977) also observed a similar formation of octoploid tapetal nuclei (Fig. 2.6 A–C) in *Eremurus,* and diagrammatically outlined the modes of their origin (Fig. 2.6 A). Fusion of approximate poles of two metaphase spindles results in polyploid restitution nuclei in *Nigella* and *Eremurus* (Fig. 2.6 D), and many other taxa. Ünal (1978) observed normal and "sticky" mitoses in *Anemone pavonia,* but polyploid nuclei resulted only after nuclear fusions. Carniel (1963) also reported such a polar fusion of spindles.

The exact mode of polyploidy is controversial. Two opposing views concern true endomitosis in tapetal cells. Witkus (1945) and D'Amato (1952, 1954) consider that typical endomitosis prevails in most species, while Geitler (1953) and Tschermak-Woess (1971) believe that stages resembling endomitosis are actually phases of "gehmto" (inhibited) mitosis. The latter is initiated typically as an intranuclear division but, subsequently, at endometaphase the nuclear membrane disappears and the chromosomes (are believed to) divide in the metaphase plane to form a restitution nucleus. In this respect it resembles C-mitosis in which the

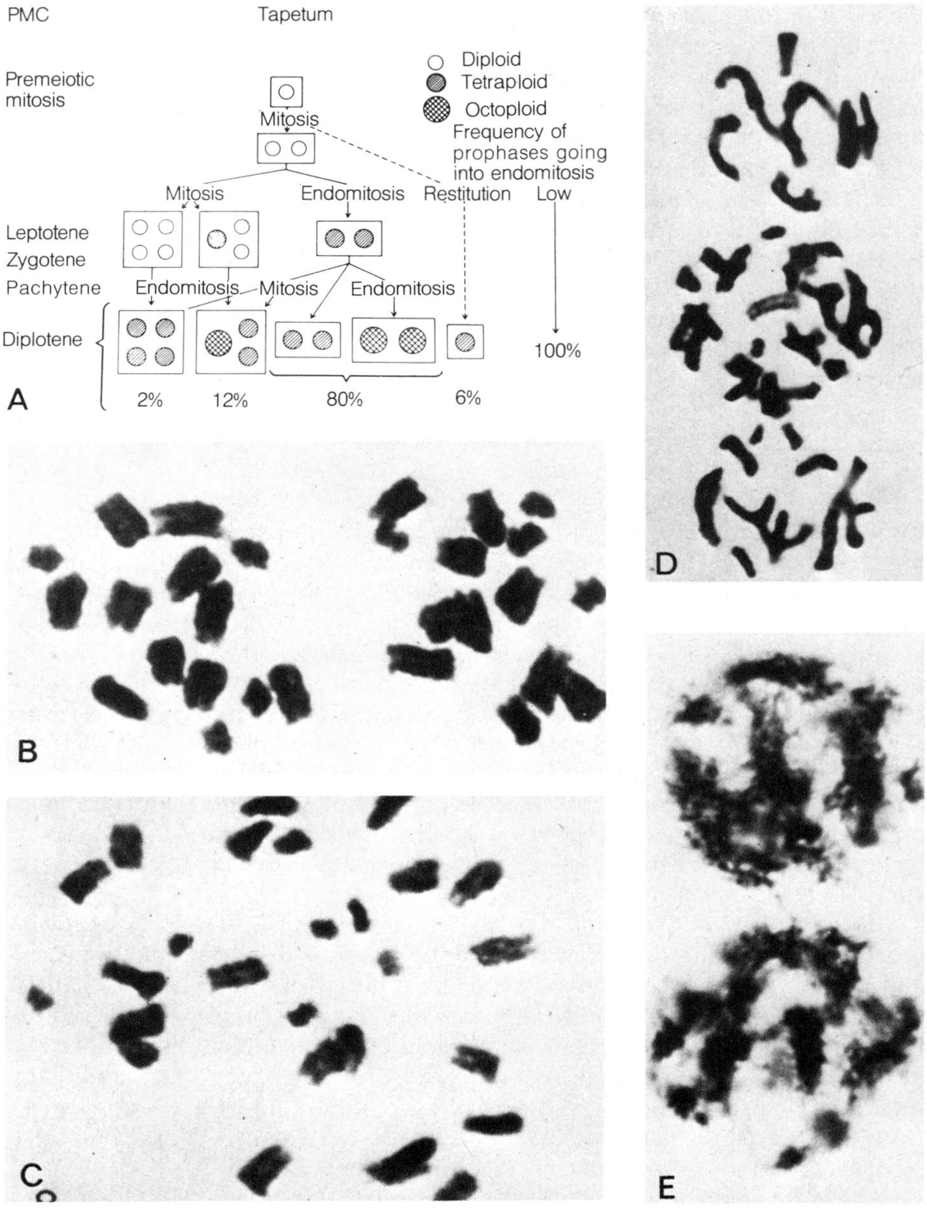

Fig. 2.6. A The main pathway in the development of tapetal cells of *Eremurus*. In the cells with three nuclei, the largest nucleus has been formed through restitution. The percentage of cells with different numbers of nuclei have been determined from a total of 283 cells. **B** Diploid endometaphase. **C** Tetraploid endometaphase (only one metaphase shown). **D** Cell with one tetraploid (*centre*) and two diploid nuclei in endomitosis. **E** Diploid endotelophase. (Adapted from Oksala and Therman 1977)

spindle is defective or absent, and the chromosomes divide, separate, and organize a restitution nucleus, or micronuclei including one or more chromosomes. Mechelke (1952) investigated tapetal cytology in cultivated var. and wild diploids, and in tetraploids of *Antirrhinum majus*. He pointed out that polyploid nuclei are formed through "inhibited" mitosis. Endomitosis has been reported by Witkus (1945) in *Spinacia oleracea,* Turula (1958) in *Cucurbita pepo,* and Greilhuber (1974) in some members of Scrophulariaceae. However, Oksala and Therman (1977) consider that the report of Mechelke (1952) in *Antirrhinum majus* is not so well documented as to be regarded as an example of "inhibited mitosis". In their own studies they observed only mitosis and endomitosis (Fig. 2.6 E), and "inhibited mitosis" did not occur in *Eremurus*. The reader is referred to Chapter 12 by D'Amato for further details.

2.4.4.4 Synthesis of Sporopollenin: Deposition

Sporopollenin is a highly resistant chemical substance(s) in the pollen exine and exosporium of a large number of spore walls (Shaw 1971). The reader is referred to a treatise on sporopollenin by Brooks et al. (1971), and Chapter 5 by Knox.

The synthesis of sporopollenin occurs both in the tapetum and in the cytoplasm of young spores. In the microspores it is deposited in an orderly fashion to form the delicate patterns on the wall of pollen grains. The capacity to synthesize sporopollenin is also present in the tapetum of some plants, the controlling mechanisms are either absent or non-functional (Echlin 1971). Mepham and Lane (1970) consider that, at least in *Tradescantia,* the entire pollen exine sporopollenin is secreted, controlled, and polymerized by the microspore itself, and the tapetum makes no contribution. This aspect requires further attention.

The deposition of sporopollenin (whether of tapetal or microspore origin) is invariably extracellular, and there are not many well-documented instances of sporopollenin being deposited within the cytoplasm. *Allium cepa* (Risueño et al. 1969) is an exception, and Echlin (1971) questions this investigation.

Site of Sporopollenin Precursors in Microspores and Tapetum. In anthers with a secretory tapetum electron-dense bodies surrounded by a limiting membrane, the pro-Ubisch bodies are primary sites of sporopollenin polymerization (see also Sect. 5).

It is most likely that, during early stages of pollen-wall development, the sporopollenin is formed from precursors located within the pollen grain cytoplasm. The medium electron-dense spherosome-like structures reported by Echlin and Godwin (1968 b) probably represent deposits or synthetic sites of such precursors.

The cellulosic primexine forms a template on which the sporopollenin is deposited. Waterkeyn and Bienfait (1971) also reported that the primexine in *Ipomoea* does not show any trace of fluorescence, which indicates the absence of sporopollenin. The control for this template of pattern determination may reside in (1) interaction between sporophytic tapetum and developing microspores, (2) sporophytic control from relict information possessed by diploid mother cells and passed on to the haploid microspores, (3) gametophytic control through information inherited and segregated at meiosis, and (4) haplophytic control (Wodehouse 1935). Current evidence favours sporophytic control of wall patterns, but the mechanism is unknown.

According to Jonker (1971), the synthesis of sporopollenin in mitochondria of tapetal protoplasts during metaphase is later transported to the exterior of tapetal cells. Precise mechanism of the transport is not fully understood.

Transfer of Sporopollenin Precursors from Tapetum to Microspores. It is presumed that any material passing from the tapetum to the microspores would be a precursor of unpolymerized sporopollenin (protosporopollenin?), and three pathways are postulated by which this material could find its way into the anther cavity: (1) solubilized precursor could be deposited on pre-existing sites such as thin lamellae or tapes on and within the maturing exine; (2) the precursor could be deposited or polymerized in the form of Ubisch bodies; (3) the precursor could be in the form of small granules or accretions laid down onto a pre-existing site. Ahokas (1975) has demonstrated the presence of late poral esterase under the pore. This can hydrolyse the prosporopollenin which is, subsequently, transferred to below the annulus. This enzyme is located in the membranes immediately beneath the pore.

Process of Polymerization. Based on Shaw's (1971) study on *Lilium henryi,* pro-orbicules formed in the tapetum consist of a solution of carotenoids and carotenoid esters in a fat and/or hydrocarbon solvent, stabilized presumably by an emulsifying agent, perhaps of a protein origin.

When the pro-orbicules leave the tapetal environment, they come in contact with an aqueous phase which contains a specific enzyme system and a source of molecular or chemical oxygen. A peroxidase enzyme system has been reported by Linskens (1967), and this enzyme reaches its maximum concentration in the early post-tetrad period during which deposition of sporopollenin is in progress. The enzyme operates as an ionic catalyst since an alternative free radical oxidation seems unlikely. Co-polymerization of the carotenoid mixture occurs in what appears to be a fine suspension polymerization process in *Lilium henryi.*

Site of Sporopollenin Deposition. The deposition of sporopollenin starts on membrane-like lamellae. These structures are described as an universal mode of sporopollenin deposition, and their origin is related to elements of RER or vesicles (Dickinson and Heslop-Harrison 1968, 1971, Dickinson 1971). The origin of the membrane-like lamellae outside the microspore is not known, but they may be elements of the plasmamembrane of tapetal cells.

Apart from the lamellae, an electron-transparent layer on the surface of the exine is present when it becomes more compact. This means that the lamellae are dependent on the presence of sporopollenin (Willemse and Reznickova 1980).

In *Lilium* (Reznickova and Willemse 1980) the lamellae are formed by the undulated plasma membrane (Fig. 2.7 A), and not by ER or vesicles (Fig. 2.7 B) as reported in other cases. Outside the plasma membrane a special surface is formed as a glycocalyx (Rowley and Skvarla 1975).

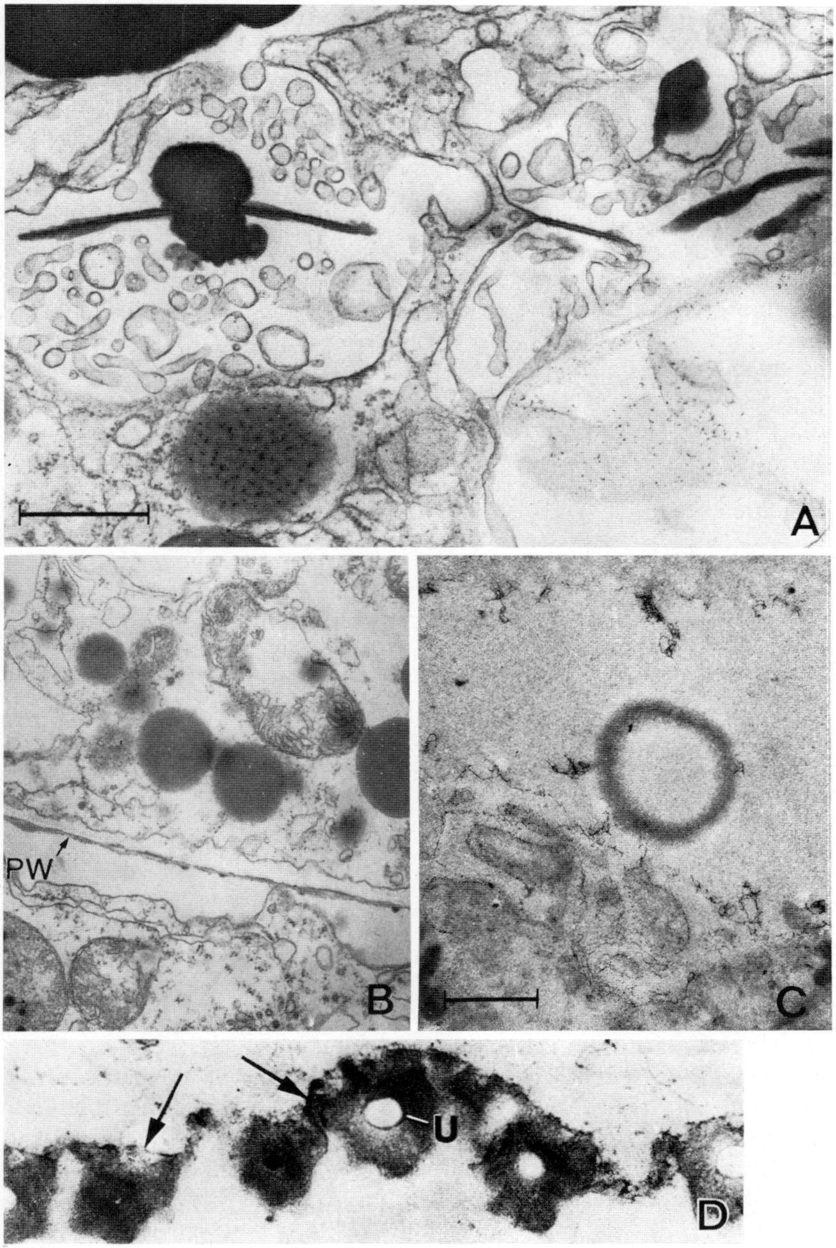

Fig. 2.7. A–C *Lilium longiflorum*. **D** *Phleum pratense*. **A** Tapetal cell during the vacuolate stage of microspore showing sporopollenin on the folds of plasma membrane. **B** Peritapetal wall (*PW*) with electron-dense globules at the stage just before microspore mitosis. **C** The outer layer of orbicule stains for polysaccharides. Between the undulating plasma membrane fibrils of polysaccharides are present. **D** Glycocalyx membrane (*arrows*) where initial deposition of sporopollenin is affected. *U* Ubisch body. (**A–C** courtesy MTM Willemse, Wageningen, The Netherlands; **D** after Rowley and Skvarla 1974)

In relation to the deposition of sporopollenin, the receptor surfaces (Fig. 2.7 C) are described as ordered polymerization surfaces (Dickinson 1976), and are related to the glycocalyx (Fig. 2.7 D) of the plasma membrane (Rowley and Prijanto 1977).

Mainly after release from plasma membrane, the membrane-like lamellae are sites of sporopollenin deposit. They are sometimes visible around the orbicules as well as around the sexine.

Role of Endothecium and Middle Layers in Formation of Sporopollenin. Because of the increase and decrease of starch in the middle layer(s), and endothecial cells, its breakdown products are considered as the main source of nutrition for the development of microspores and pollen. An intermediate product in the loculus could be active acetate as a common precursor of lipids, carotenoids, and probably sporopollenin (Atkinson et al. 1972). The breakdown of starch can be related to the formation of such a product (Reznickova and Willemse 1980).

2.4.4.5 Functions of Tapetum: Cause of Sterility

Irrespective of the origin and type of tapetum, together with the temporal variation displayed in its disintegration, it is evident that this tissue maintains a very delicate balance vis-a-vis the differentiation of sporogenous tissue. Any factor(s) which upsets this harmony between the two tissues results in asynchronous development and, finally, degeneration of pollen. Williams and Heslop-Harrison (1979) compared the incorporation of [^{3}H]-uridine in plasmodial tapetum of *Rhoeo spathacea* and the secretory tapetum of *Lilium longiflorum* during meiosis in the pollen mother cells. Both show substantial increase in nucleolar RNA synthesis followed by an increase in cytoplasm basophilia. They conclude that in spite of considerable differences in the behaviour, the plasmodium and secretory tapeta can gear the overall sequence of RNA and protein synthesis related to pollen development.

Numerous investigations on the cytoplasmic male sterile (CMS) lines indicate that tapetum, through altered physical or physiological factors, initiates the process of abortion of pollen. Excellent reviews (Harvey et al. 1972, Heslop-Harrison 1972, Laser and Lersten 1972) concern the time and site of initiation, mode of action, and factors responsible for the abortive processes in male sterile lines.

Apart from tapetum, endothecium or conducting strand of the filament may also influence the abortive process. Reddy and Reddi (1974) consider that even though the tapetal layer is persistent, other abnormalities are operative in some CMS lines. They observed a positive correlation between pollen sterility and thickness of endothecium at maturity, consequently, the anthers are non-dehiscent and prevent the release of pollen.

Rohrbach (1965) concluded that in *Beta* CMS nutritional imbalance of anthers is due to reduced activity of the conducting strand and, consequently, the pollen aborts. Joppa et al. (1966) in *Triticum,* and Alam and Sandal (1967) in *Sorghum,* observed poorly developed vasculature in the stamens of CMS lines. This impeded or blocked the transport of nutrients so that there was inadequate accumulation of starch in the anther tapetum of CMS plants. Such a failure of sufficient nutrients to reach the developing pollen grains could possibly initiate their abortion. Laser (1972) did not notice any appreciable structural difference in the vasculature of CMS and normal line in *Sorghum* and, thus, could not support the conclusions of Joppa et al. (1966).

Raj (1969) observed, in *Sorghum,* that the tapetal cells enlarge considerably and occlude the entire pollen sac. Consequently, the sporogenous tissue is squeezed and crushed (see also Pritchard and Hutton 1972). In *Capsicum annuum* Horner and Rogers (1974) emphasized that when tapetal cells malfunction, as evidenced by increased vacuolation, cell enlargement, and lack of dense cytoplasm, the development of microspore is arrested (these are subjected to mechanical pressure). Horner (1977) reported that, at the tetrad stage, the tapetal cells begin to enlarge and, finally, obliterate the locular space in sunflower.

Novak and Betlach (1970) and Novak (1971) argued that in *Capsicum* the secretory function of hypertrophied tapetum is highly disturbed, this resulting essentially in meiotic desynchronization in pollen mother cells. Novak (1971) supported Rick's (1945) suggestion that in tomato the primary influence of "sterile" hereditary factors is manifested through the mediation of disturbed function of tapetum. The premature degeneration of tapetal tissue results in the starvation of sporogenous tissue and, hence, its degeneration. Similar conclusions were reached by Chauhan and Singh (1966) in their study on hexaploid wheat (cv. Norin 26). Also, the persistence of tapetum blocks the trophic process necessary for the development of microspores into viable pollen in *Petunia* (Singh 1975). Nakashima and Hosokawa (1971) furnished experimental evidence of such a nutritional blockade in sugar beet. They observed that [^{14}C] compounds accumulated much more in the tapetum of cytoplasmic male sterile (CMS) line, at the tetrad stage, in comparison with fertile lines. Correspondingly, the parenchyma in the connective of sterile lines showed much less concentration of [^{14}C] at microspore stage than that of fertile lines. At the mature pollen grain stage, however, the reverse was true. Nakashima and Hosokawa (1971) rightly concluded that nutrients were not being transferred from the abnormal CMS tapetum to microspores, which caused pollen abortion.

Criswell et al. (1974) demonstrated that in maize CMS tassels accumulate [^{14}C]-labelled substances which are quantitatively much fewer than those of the fertile tassels. These observations have since been confirmed by Nakashima and Hosokawa (1974).

Cheng et al. (1979) did not observe any difference in tapetum in the male sterile (MS) and male fertile (MF) lines in *Zea mays* until the tetrad stage; subsequently, in MS lines vacuolar areas appeared in the tapetal cells which disorganized by the time young pollen grains were formed. The formation of pro-orbicules in MS lines, in contrast to MF lines, is completely suppressed. Cheng et al. suggested that such a failure to accumulate photosynthate could also be due to a physiological failure in the anther, particularly the "hormone-directed transport" of nutrients which can result in differential accumulation in various tissues.

In *Sorghum* and *Zea mays* the male sterile and fertile lines show a gradual increase in the reducing sugars in an identical way. The starch, however, persists in endothecium and parenchymatous cells of the connective in fertile line, but disappears in the CMS line by anthesis (Nakashima and Hosokawa 1970). Contrarily, in *Beta* Hosokawa et al. (1963) observed a gradual decline in starch content in normal anther until mature pollen grains were formed, whereas in sterile anthers starch was present throughout the development of anther. Fukasawa et al. (1957) described a decrease in sucrose content of anthers of male sterile plants of *Triticum*. In *Raphanus* (Ogura 1968) young anthers in sterile lines had less fructose than fertile anthers, while older stages showed more sucrose. Fructose, together with glucose, continued to be much less.

The activity of various enzymes in the male sterile lines is usually at a very low level. In *Cucumis melo* Chauhan and Singh (1968) pointed out low activity of acid phosphatase during microsporogenesis, and Fukasawa (1961) of succinic dehydrogenase. In male sterile lines of maize (Dmitrieva and Khavzhinskaia 1962), amylase activity was lower than in fertile lines, but polyphenoloxidase and peroxidase showed higher activity.

Frankel et al. (1969) discovered that premature dissolution of callose before the completion of meiosis, or even earlier, in male sterile lines of *Petunia* is correlated with precocious increase in the activity of callase synthesized by tapetum, and results in early pollen abortion. Subsequently, Izhar and Frankel (1971) established that pH of the anther locule played a very critical role in the activation of callase. A drop in the pH increased its activity at early stages of meiosis in anthers of male sterile lines, which brings about callose digestion and subsequent degeneration of microspore mother cells (MMCs) during meiotic prophase. In *Petunia,* therefore, untimely increase in callase activity results in pollen sterility. A relationship between increased callase activity in anther tapetum and callose dissolution has also been confirmed by Mepham and Lane (1969a) in *Tradescantia,* in which normal pollen is formed. Horokawa et al. (1963) had earlier observed differences in pH values of fertile and sterile anthers, at different stages of development. Abnormal behaviour of special callose wall in MMCs has often been related to CMS sterility in *Sorghum* (Damon 1961, Erichsen and Ross 1963a, b, Warmke and Overman 1972a), *Petunia* (Frankel et al. 1969) and *Allium cepa* (Nanda and Gupta 1974).

A comparison of free amino acid content has been reported in male sterile and male fertile lines (Khoo and Stinson 1957). Fukasawa (1954) in wheat (with *Aegilops* cytoplasm), and Tubý (1963) in apple, pointed out low proline content in male sterile lines. It was totally absent in *Zea* and *Sorghum* (Nakashima and Hosokawa 1970, Nakashima 1975), *Beta* (Horokawa et al. 1963), and *Raphanus* (Ogura 1968).

On the other hand, aspartic acid (Rohrbach 1965) and alanine in *Beta* (Horokawa et al. 1963, Rohrbach 1965), and glycine in *Sorghum vulgare* (Brooks 1962), are quantitatively more in sterile than in fertile lines. Alam and Sandal (1972), however, did not notice any significant difference in free amino acids at premeiotic stage. But, during pollen development, alanine, glutamic acid, proline, phenylalanine, and tyrosine were significantly lower in male sterile lines of *Sorghum vulgare* var. *sudanense*. Increase in asparagine content in *Petunia hybrida* effected abortion of PMC at prophase I in CMS line, while in genic male sterile lines degeneration occurred at microspore tetrad stage. This led Izhar and Frankel (1973) to suggest that amino acid imbalance could probably induce changes in pH in anthers of fertile and sterile lines.

Electron microscopic studies provide further information on tapetal abnormalities at the subcellular level, in CMS lines. De Vries and Ie (1970) did not observe any differences in the organelles in tapetal cells of normal and CMS line except that, in the latter, some small dark bodies are present together with some plastid-like structures filled with lipid droplets. In CMS of sugar beet the anther tapetum has abnormal and degenerated mitochondria and, in contrast to fertile lines, lacks the capacity to produce Ubisch bodies at the microspore stage (Nakashima 1975). In *Helianthus annuus* (Horner 1977), at the tetrad stage, in the enlarged tapetal cell of CMS anthers the walls degenerate partially but, contrary to the fertile line, a periplasmodium is never formed (see also Warmke and Overman 1972b). The mitochondria are pleomorphic, whereas the rest of the organelles appear to be intact. Some dense material is present in the fragmentary ER. The dictyosomes are missing, but the plastids are quite conspicuous. The presence of dense osmiophilic bodies and a relatively large heterogeneous membrane structure are additional features restricted to CMS tapetum. Contrarily, in normal (N) tapetum there is no cell enlargement, the cell walls expand and, soon after the release of microspores, they disappear and a plasmodium is formed. There are numerous mitochondria, plastids, dictyosomes associated with vesicles, vacuoles with fibrous contents, and ribosomes.

Young et al. (1979), in their study of male sterile hybrid between wheat and *Agropyron intermedium,* observed that pollen-wall malformations were very frequent in meiotically unstable CMS lines. Fewer tapetal abnormalities were observed and, contrary to other reports (Nakashima 1975, Horner 1977, Warmke and Lee 1977, 1978), early degeneration of mitochondria was not observed. An unusual feature reported by Young et al. is the formation of an orbicular wall completely surrounding each tapetal cell. This might have prevented the rupture of plasmalemma and the subsequent formation of plasmodium.

Occasionally, the inability of the microspores to absorb nutrients, and not the defective secretion of tapetum, has been stated to be the cause of male sterility as in *Allium cepa* (Virnich 1967). Similarly, Singh and Rhoades (1961) suggested that in squash the pollen abortion is due to the absence of meiosis in PMCs, rather than the defective or abnormal tapetum. In barley the PMCs degenerate even before the breakdown of tapetal cell walls. Structural differences have been reported in the pollen of normal (N) and cytoplasmic male sterile (CMS) lines in wheat (De Vries and Ie 1970) and, in fact, very few functional organelles are present in the pollen of CMS lines. Even in the PMCs fewer organelles have been reported (Turbin et al. 1974). The mitochondria appear in a disturbed state, and many more vacuoles are present in the cytoplasm. It is believed that the lack of functional organelles in the PMC or pollen is the prime cause of pollen degeneration in CMS lines.

In *Triticale* cv. *Tri*-I the plants produced from seeds treated with 0.2%, 0.4%, and 0.6% EMS revealed a direct correlation of dosage to the extent of abnormalities induced in the development of anther (Khosla 1982). EMS affected the ontogeny of anther and induced (1) enlargement of tapetal cells (Fig. 2.8 A) which squeezed out the developing PMCs, (2) transformation of glandular to plasmodial tapetum (Fig. 2.8 B), (3) coenocytic mass from sporogenous PMCs (Fig. 2.8 B), (4) nuclear fragmentation, (5) transfer of nuclear material in PMC and ta-

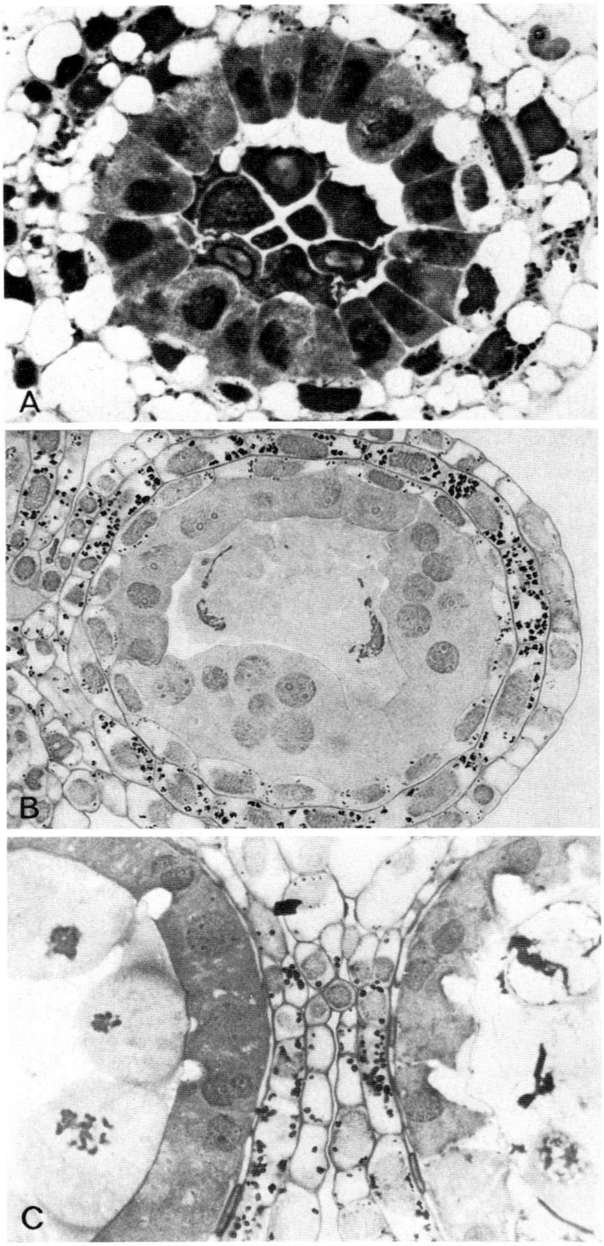

Fig. 2.8 A–C. *Triticale*. Transections of EMS-treated anthers showing abnormal development; PAS-aniline blue staining. **A** 0.6%, **B** 0.4% EMS treatment. **A** Tapetum becomes hypertrophied and, thus, brings about the degeneration of microspore mother cells. Note the enlargement of middle layer cells. **B** The central mass formed by sporogenous cells has degenerated nuclei. Intratapetal syncytium occludes the locular cavity. **C** Two adjacent locules showing normal and abnormal development of tapetum and microspore mother cells. (**A–C** courtesy Reeta Khosla, Delhi)

petum, and (6) disruption in the formation of pollen exine (Fig. 2.8 C). In Gramineae there exists a special spatial arrangement between the tapetum and microspore mother cells (Fig. 2.8 C), and it has been noted that the meiocytes in direct contact with the abnormal or degenerating tapetum abort, while those abutting the healthy tapetum develop normally within the same locule. It is suggestive of a cell-to-cell relationship of healthy tapetum to the development of normal, viable pollen grains.

According to Flavell's (1974) hypothesis, in CMS lines some substance/s similar to pathotoxin (Barratt and Flavell 1975, Petersen et al. 1975) is/are synthesized in the tissue of anther alone inhibiting the activity of organelles with altered structure. Flavell suggested that mitochondria, chloroplast and, perhaps, cell membrane furnish the site of action. According to Laser and Lersten (1972), however, more than one mechanism is operative, since pollen abortion occurs at almost every stage of development. Horner (1977) considers that the alteration of specific structure, cytoplasm or organelles of tapetum, and/or sporogenous cells during microsporogenesis disturbs the balance between these two tissues.

2.5 Ubisch Bodies/Orbicules

The secretory tapetum, in many angiosperms, in contrast to the periplasmodial, is characterized by the development of Ubisch bodies or orbicules [designated by Heslop-Harrison (1968 b) in preference to the term Ubisch bodies] along the tangential surface facing the anther locule. I personally prefer the term Ubisch bodies since it has been in usage in literature much longer, and is well understood both by LM and EM embryologists.

Rosanoff (1865), Mascré (1922), and Schnarf (1923) seem to have recorded the occurrence of granular bodies along the inner "walls" of tapetal cells in some members of Solanaceae, Boraginanceae, and Liliaceae. Ubisch (1927) and Kosmath (1927), simultaneously, prepared an up-to-date list of plants showing such granules. They pointed out their similarities to pollen exines, in staining properties. Chemically, these are composed of sporopollenin. Bhandari and Kishori (1971) demonstrated specific staining of these two depositions, with spirit-soluble aniline blue.

The Ubisch bodies may remain individually separate or, during the course of development, fuse to form larger aggregations or compound bodies (Echlin and Godwin 1968 a, Christensen et al. 1972, Bhandari and Kishori 1973).

Reznickova et al. (1980) followed the successive changes in size during the ontogeny of orbicules in *Lilium*. The pro-orbicules, at the time of extrusion, are 0.30–0.90 μm in diameter. The flattened orbicules undergo a change in shape and become rounded (Fig. 2.9 A–C), and vary from 1.25 μm to 3.91 μm in diameter; about 50%–70% have the average size of 1.91 μm at microspore stage with a distinct wall, 2.58 μm at microspore with small vacuole stage, and 3.5 μm at late vacuolate stage. Subsequently there is no change until one day before anthesis, when a second generation of orbicules can be distinguished and there is a perceptible increase in size. In addition to the orbicules they also report the presence of rod-like structures made up of sporopollenin (Fig. 2.9 A).

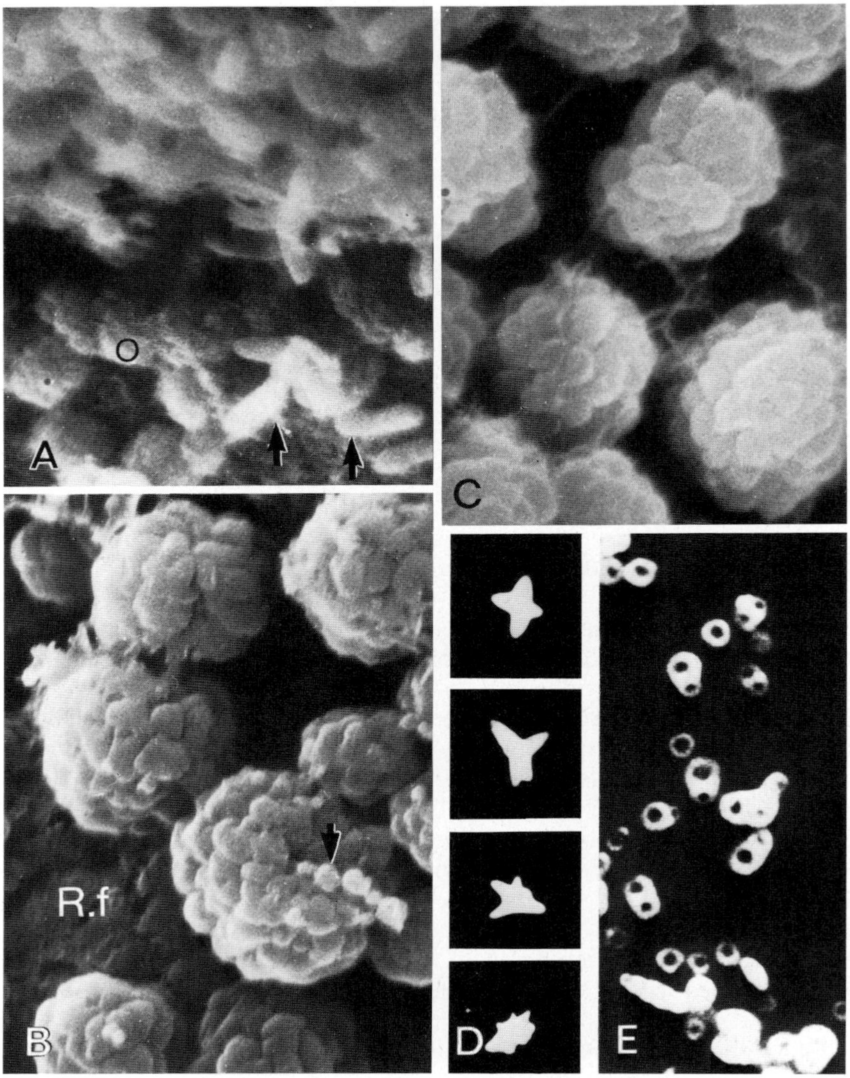

Fig. 2.9 A–E. SEM and fluorescence micrographs showing the inner surface of pollen sac, and orbicules. **A–C** *Lilium;* **D** *Ipomoea rubrocoeruler;* **E** *Oxalis acetosella.* **A** At the young microspore stage, orbicules (*O*) and rodlike bodies (*arrows*) are present on the locular face of tapetal cell. **B** The flattened form of orbicules subsequently changes to a more rounded shape from young microspore stage to generative cell stage in pollen; reticulate formation (*R.f*) is seen at the base of orbicule. **C** Locular surface of anther 2 days before anthesis; the orbicules are linked by fibrils. **D, E** Stellate (**D**) and single or compound (**E**) Ubisch bodies. (**A–C** after Reznickova et al. 1980; **D, E** after Waterkeyn and Bienfait 1970)

Waterkeyn and Bienfait (1971) observed primuline-induced fluorescence in Ubisch granules comparable to pollen exine, and reported stellate as well as small spheroidal bodies in *Ipomoea,* perforated circular discs (which appear polygonal when closely packed) in *Lilium,* and large multiperforate plates in addition to perforated circular discs in *Oxalis* (Fig. 2.9 D, E).

2.5.1 Origin

During the last two decades EM studies have illucidated the morphology of Ubisch bodies (Rowley et al. 1959, Chambers and Godwin 1961, Rowley 1962, 1964, Yamazaki and Takeoka 1962, Dahl and Rawley 1965). Ontogenetic studies concern *Cannabis sativa* and *Silene pendula* (Heslop-Harrison 1962, 1963 a), *Poa annua* (Rowley 1963), *Zea mays* (Skvarla and Larsen 1966), *Oxalis* (Carniel 1967), *Helleborus foetidus* (Echlin and Godwin 1968 a, b), *Citrus limon* (Horner and Lersten 1971), *Sorghum bicolar* (Christensen et al. 1972), *Phleum pratense* (Rowley and Skvarla 1974), *Antirrhinum majus, Gentiana acaulis* (Lombardo and Carraro 1976a, b), *Pelargonium zonale, Kalanchoe obtusa* (Carraro and Lombardo 1976), *Avena* (Steer 1977), *Soja hispida* (Madjd and Roland-Heydacker 1978), *Olea* (Pacini and Juniper 1979 a, b), and *Lilium* (Reznickova and Willemse 1980).

Krjatchenko (1925) suggested that in *Lilium croceum* Ubisch bodies were formed from lipid droplets, arising probably from mitochondria in the tapetal cells. Heslop-Harrison (1962) initially proposed mitochondrial origin but later withdrew this proposition (Heslop-Harrison 1964).

Echlin and Godwin (1968 a) traced the origin and development of Ubisch bodies in *Helleborus foetidus.* At the sporogenous stage numerous membrane-bound "grey bodies" or the pro-Ubisch bodies appear in the tapetal cytoplasm, prominently aggregated towards the anther locule (Fig. 2.5 A, E). ER cisternae and ribosomes are in close association with the developing pro-Ubisch bodies at the tetrad stage. The limiting membrane is discontinuous at sites of radiating out rays of ribosomes (Fig. 2.5 C). The similarity of material present in ER and pro-Ubisch bodies as well as their close association with ribosomes indicate that ER and ribosomes are the organelles involved in the synthesis of pro-Ubisch bodies. Echlin and Godwin observed a decline in the number of mitochondria and their activity, and the absence of electron-dense material, even though in association with ER. This excludes any speculation concerning their involvement in the formation of pro-Ubisch bodies. The dictyosomes with peripheral vesicles are also not associated with their origin; the source is, therefore, still obscure in *Helleborus.*

Risueño et al. (1969) observed dense bodies aligned along the plasmalemma in *Allium cepa.* The ER running parallel to the latter widens at places, and becomes filled with material of similar electron density as in pro-Ubisch bodies, revealing its positive role in their origin. No ribosomal association was observed. Heslop-Harrison and Dickinson (1969) were unable to find any relationship of ER and ribosomes with the origin of pro-Ubisch bodies, but commented upon their resemblance to spherosomes (see also Echlin 1971).

In *Citrus limon* (Horner and Lersten 1971) there is no evidence for the presence of pro-orbicular bodies in the tapetal cytoplasm (Fig. 2.5 B). However, Horner and Lersten observed vesicles of dimensions similar to the orbicules present in the cytoplasm, and their subsequent fusion with plasmalemma. According to them, ER or dictyosomes may produce these vesicles which may contain pro-orbicules and, possibly, also the precursors of sporopollenin. Similarly, initial and continued association of ER with the pro-Ubisch bodies has been observed by Christensen et al. (1972) in *Sorghum bicolor*. Some osmophilic bodies occur in *Soja hispida* (Madjd and Roland-Heydacker 1978) between the spaces along the inner tangential and radial walls of tapetal cells. The origin of these bodies could not be related to any organelle.

Pacini and Juniper (1979 a) observed the formation of orbicules far from the surface of plasmalemma in *Olea europea;* these bodies lack the pro-orbicular core. The organellar source of these orbicules is not clearly stated, and cannot be made out from illustrations.

During the extrusion of pro-orbicules, the surrounding membrane fuses with the plasmalemma releasing them into the space between plasma membrane and tapetal cell wall (Figs. 2.5 C, 2.10 A, B, 2.11 A). Immediately thereafter the pro-Ubisch bodies become covered with sporopollenin which is deposited on either side of a unit membrane (Figs. 2.5 C, 2.10 C). Similar observations have been made by Heslop-Harrison and Dickinson (1969) and Dickinson (1970) in *Lilium*. Echlin (1971) considered that the newly formed sporopollenin is in a low state of polymerization. Reznickova and Willemse (1980), however, point out that the pro-orbicules are surrounded by a layer of loose fibrillar electron-dense material staining positively for polysaccharides (Fig. 2.7 C). Such fibrillar material is also present between the undulations of plasmalemma and, probably, dictyosome-derived vesicles transport it across the latter. Rowley and Skvarla (1974, 1975) also reported acid polysaccharide [stained positively with PTA (phosphotungstic acid) in chromic acid] coating around the core of Ubisch bodies (pro-orbicules) which has been termed glycocalyx in *Phleum pratense* (Fig. 2.7 D) and *Canna*.

In *Lilium* Heslop-Harrison and Dickinson (1969) and Reznickova and Willemse (1980) occasionally observed a persistent stalk connecting the Ubisch bodies with the tapetal cytoplasm. The stalks are actually portions of plasmalemma entering the base of orbicules (Fig. 2.10 D). This contact continues until the formation of the tapetal wall.

Echlin (1971) expressed the opinion that the accretion of sporopollenin on the pro-Ubisch bodies is always extracellular. Pro-orbicules with deposited sporopollenin have not been observed in the tapetal cytoplasm. Risueño et al. (1969), however, observed, in *Allium cepa,* that deposition is effected in the ER cisternae. Even such small channels – leading to the exterior of plasmalemma – provided the passage for the extrusion of pro-orbicules. Reznickova and Willemse (1980) described the deposition of sporopollenin on the folds of plasma membrane, which forms a transport system; such membranous lamellae have been observed external to plasmalemma. At the same time, the ER cisternae produce some globules, smaller than orbicules. They stick to Ubisch bodies to increase its size and probably also represent the second generation of orbicules (Fig. 2.10 E).

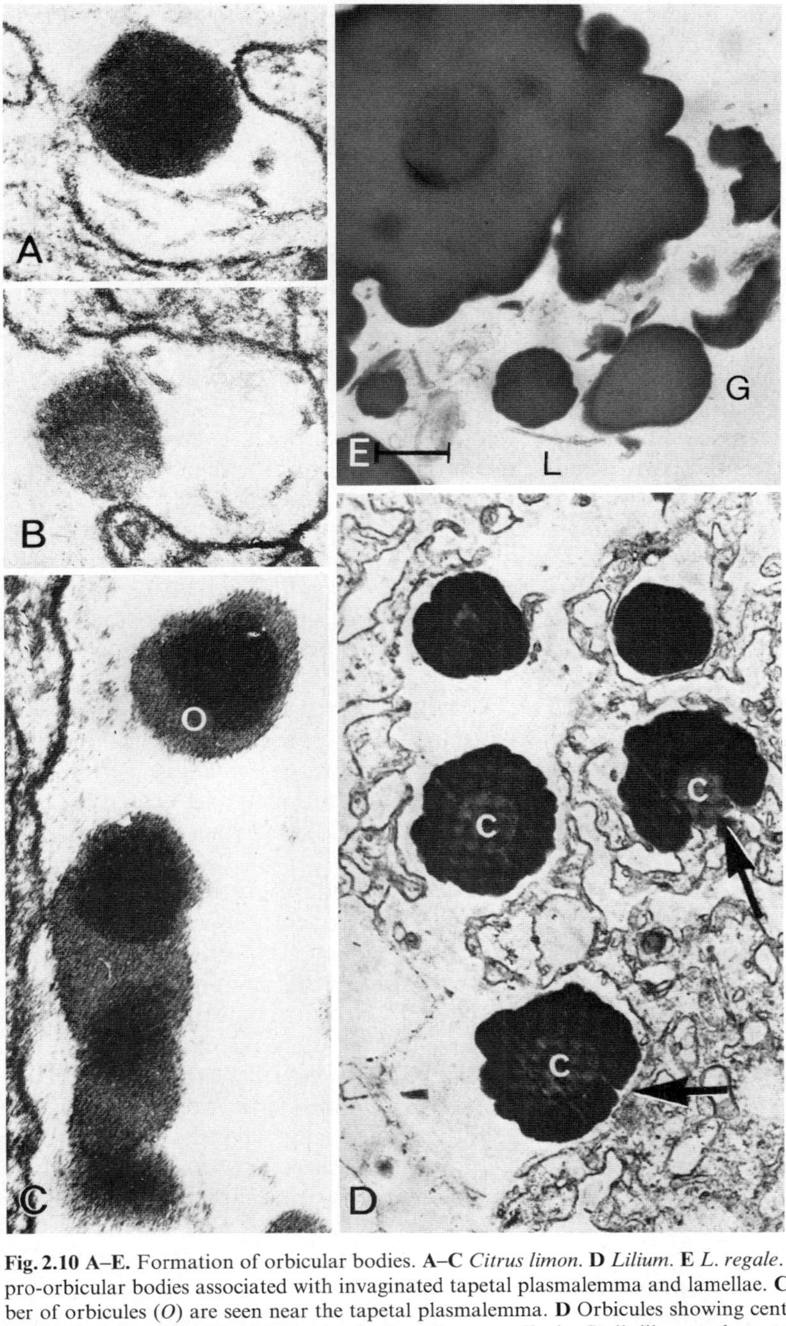

Fig. 2.10 A–E. Formation of orbicular bodies. **A–C** *Citrus limon*. **D** *Lilium*. **E** *L. regale*. **A, B** Extruded pro-orbicular bodies associated with invaginated tapetal plasmalemma and lamellae. **C** Same, a number of orbicules (*O*) are seen near the tapetal plasmalemma. **D** Orbicules showing central (*C*) pro-orbicular body coated with peripheral accretion of sporopollenin. Stalk-like attachment to the lamellae is clear (*arrows*). **E** Orbicule towards the end of development. Note the presence of many globules (*G*) (probably of sporopollenin) and membrane-like lamellae (*L*). (**A–C** adapted from Horner and Lersten 1971; **D** from Heslop-Harrison and Dickinson 1969; **E** courtesy MZM Willemse)

The presence of larger "grey bodies" in *Helleborus* (Echlin and Godwin 1968 a, 1969) in tapetal cytoplasm, subsequent to extrusion of pro-orbicules, more or less of similar dimensions, casts doubt on their being the actual progenitors of orbicules (see also Pacini and Juniper 1979 a).

2.5.2 Relation to Exine Deposition: By-Product

The functions of Ubisch bodies, particularly the role these might play in the deposition of pollen exine, are not yet clearly understood. Gorezynski (1934; cited in Maheshwari 1950) observed that in *Cardamine* the exine develops first on the side of the pollen grains in contact with tapetum bearing Ubisch granules. Maheshwari, therefore, speculated on these granules having some contribution towards exine formation. Carniel (1967), however, disregarded any connection between the orbicules and exine development in *Oxalis*. Christensen et al. (1972) refer to time-bound correlation in the development of orbicular wall and pollen exine, and a significant similarity in their structure. However, they regarded it as a vestigial capacity of tapetum since, phylogenetically, it might have arisen from the sporogenous tissue itself. Bhandari and Kishori (1973) investigated *Nigella damascena* and agree with this conclusion. Dickinson and Bell (1972) compared their results in *Pinus banksiana* with angiosperms, and considered the Ubisch bodies as a by-product of tapetal activity necessitated for the formation of peritapetal membrane.

Heslop-Harrison and Dickinson (1969) discarded the hypothesis that orbicules are concerned in the transport mechanism of sporopollenin, since the latter is a polymerized substance of tapetal metabolism. Also, Knox and Heslop-Harrison (1969) state that there is no evidence of any enzyme system capable of degrading sporopollenin, although many other hydrolytic enzymes have been localized in pollen wall.

Banerjee and Barghoorn (1971 a, b) point out that freshly deposited sporopollenin is not resistant, but becomes resistant subsequently. Some sporopollenin strands form a continuous bridge between the pollen grain spinules and Ubisch bodies, which possibly effect transfer of substrate from the tapetal cell to exine. Some depolymerization, possibly controlled by the centrum of Ubisch bodies, has also been reported. A similar conclusion, based purely on morphological observations, was reached by Gherardini and Healey (1969) in *Pharbitis*. Recent observations of Ahokas (1975) on the mutants of male sterile lines in barley suggested some esterases to be connected with sporopollenin degradation around the pore in pollen. It is very likely that, in the near future, a regular mechanism of sporopollenin degradation can be traced, originating from tapetum and being released in the locules. The lack of Ubisch bodies in the periplasmodial tapetum might also indicate the presence of some system of direct utilization of sporopollenin precursors during pollen-wall formation.

The evidence in the work of Risueño et al. (1969) does show that the orbicules bodily become part and parcel of the mature exines because of apparent continuity between the border surface of Ubisch body and the exine. Risueño et al., however, hesitated to speculate on this aspect. In *Soja hispida* Madjd and Roland-

Heydacker (1978) report that the orbicules are pushed very close to and in contact with the pollen surface by mucilaginous secretion resulting from the fusion of vesicles derived from endoplasmic reticulum. Some of the orbicules assume a finger-nail-like shape, become inserted between the tectum perforations, and enter the columella. Reznickova and Willemse (1980) observed a similar association between pollen exine and tapetum-derived sporopollenin globules in *Lilium*.

2.6 Tapetal Membrane: Structure, Origin, and Significance

Banerjee (1967) observed acetolysis-resistant, intricately patterned tapetal membrane in some grasses. Structurally, the tapetal membrane comprises three layers – fenestrated, reticular, and orbicular (Fig. 2.11 A). Bhandari and Kishori (1969, 1973) report that, in *Nigella damascena*, the original tapetal cell walls disintegrate and, instead, a new membrane resistant to acetolysis becomes organized along the thecal face of tapetal cells. The Ubisch bodies get attached to the tapetal membrane, and it reveals the imprint of the tapetal cells which act as a template (Fig. 2.11 B). Incidentally, *Dactylis glomerata, Phleum pratense, Zea mays, Spartina patens,* and *Nigella damascena* (see Banerjee 1967, Bhandari and Kishori 1973) show glandular tapetum, and Ubisch bodies. In *Helleborus foetidus* (Echlin and Godwin 1968 a), *Lilium longiflorum* (Heslop-Harrison 1968 a, b), *Citrus limon* (Horner and Lersten 1971), and *Olea europaea* (Pacini and Juniper 1979 a), with secretory tapetum and orbicules, there is no tapetal membrane (Fig. 2.11 C). Similarly, Rowley (1962), and Skvarla and Larson (1966) did not observe any reticulum interlinking orbicules, and were equally unaware of a tapetal membrane. In some members of Compositae, which have plasmodial tapetum, Heslop-Harrison (1969) reported an acetolysis-resistant extratapetal membrane, in the form of a complete sac surrounding mature pollen. De Vries and Ie (1970) reported, in *Triticum*, the presence of orbicular wall (tapetal membrane) with sporopollenin reticulum interconnecting the Ubisch bodies. Christensen et al. (1972) investigated the development of tapetal wall in *Sorghum bicolor*, which is closely linked with the origin and amplification of orbicules (Fig. 2.11 D–F). At the late vacuolate microspore stage, and subsequently, polymerization of additional sporopollenin in inter-orbicular areas produces a reticulum (Fig. 2.11 F), which together with Ubisch bodies constitutes the tapetal wall. Beneath the reticulum develops a layer of fibrillar material (Fig. 2.11 F). At the engorged pollen-grain stage, the tapetum degenerates but the distinct massive orbicular wall persists (Fig. 2.11 F). In the anthers of *Lilium* (hybrid "Enchantment") Reznickova and Willemse (1980) point out that after the dissolution of tapetal cell walls at the tetrad stage a layer of fine fibrillar material, staining weakly for polysaccharides, develops between the orbicules (Fig. 2.7 C) and tapetal cell to which the orbicules stick. This thin layer with frequent electron-dense spots persists until dehiscence without much change. In addition to tapetal wall, Reznickova and Willemse find a peritapetal wall too between the adjacent middle layer and tapetal cytoplasm, which also persists till anthesis.

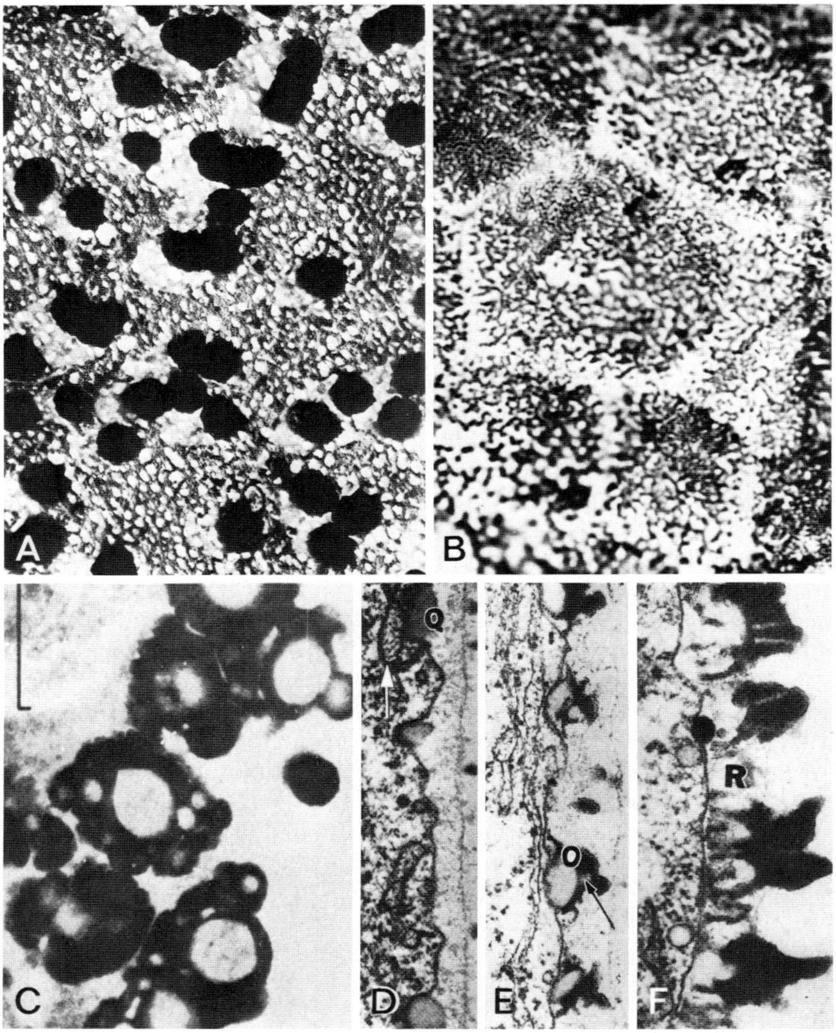

Fig. 2.11 A–F. Development of orbicular wall/tapetal membrane. **A** Electron micrograph of an isolated tapetal membrane of *Zea mays* obtained from a dehisced anther, showing the fenestrated first layer. The *light areas* at one side of the orbicules were produced by chromium shadowing. **B** Light microscope photomicrograph of tapetal membrane of *Nigella damascena* obtained from a dehisced anther, showing the configuration of several tapetal cells which acted as template for its formation. **C** The cytoplasm of the tapetum is completely disorganized, and aggregated Ubisch bodies line the outer region of the remains of tapetum in *Helleborus foetidus*. There is no tapetal membrane. **D** Late tetrad stage, pro-orbicules (*Q*) are evident between tapetal primary wall and plasmalemma; they are often associated with ER (*arrow*). **E** Sporopollenin deposition on orbicules (*O*) continues during early vacuolate microspore stage; incipient channels can be seen (*arrow*). **F** Late vacuolate microspore stage. Sporopollenin reticulum (*R*) between orbicules has completed orbicular wall and channels are slightly occluded; pro-orbicular material is less distinct. (**A** from Banerjee 1967, **B** after Bhandari and Kishori 1973, **C** after Echlin and Godwin 1968 a, **D–F** adapted from Christensen et al. 1972)

I prefer to use the terms tapetal and peritapetal (extra-tapetal) membranes which develop along the inner (towards the locule) or outer tangential wall (adjacent to middle layer). The tapetal and peritapetal membranes have been observed on both sides of tapetal cells in some members of Poaceae and Solanaceae (Banerjee 1967, Ahokas 1975, Ogorodnikova 1976). Dickinson and Bell (1972) observed peritapetal membrane in *Pinus banksiana* where the tapetum is glandular.

Gupta and Nanda (1972) assumed that plants with amoeboid tapetum show peritapetal membrane, and with glandular tapetum only tapetal membrane. However, in *Lilium,* which has glandular tapetum, both the membranes are present (Reznickova and Willemse 1980).

The precise role of tapetal membranes is rather enigmatic. One possible function, as suggested by Heslop-Harrison (1969), may be to act as a "culture sac" which surrounds the maturing pollen and labile periplasmodium, or the contents released by the disintegration of secretory tapetum. It might also help to prevent a quick loss of water from within the anther locule, which could indirectly affect Pollenkitt deposition on exine.

2.7 Pollenkitt and Tryphine

The general term Pollenkitt is applied to various substances responsible for imparting stickiness to the pollen (see Troll 1928, Pohl 1930, Knoll 1930, 1936). Heslop-Harrison (1968a, b, 1969) observed that in *Lilium longiflorum* the Pollenkitt material is of tapetal origin, and comprised mainly lipid and carotenoids, chiefly α-carotene-5 and 6-epoxide. He also suggested that these coloured carotenoids are not used up in the synthesis of sporopollenin since these appear much later after the pollen exine is well formed. Mepham and Lane (1969a, b) state that concomitant with the release of microspores lipid globuli are released from the periplasmodial plastids into the cytoplasm (Fig. 2.4C). A similar plastidial origin of Pollenkitt coating is reported in *Lilium* (Dickinson 1973). This, eventually, is deposited on the surface of pollen and, possibly, forms a part of Pollenkitt (Fig. 2.4 D, 2.12A–E). In *Artemisia mutellina* (Hesse 1979a), soon after degeneration of the tapetum, lumps of Pollenkitt material composed of lipid droplets of various shape and size, together with other cellular organelles, float in the anther locule (Fig. 2.12A, C). At the time of anthesis, underneath an electron-transparent Pollenkitt crust the lipid droplets become entrapped and adpressed to the pollen exine (Fig. 2.12B). Mepham and Lane, however, observed the interbacular spaces in pollen exine filled with lipids exuded from within the pollen grain itself. The functions of Pollenkitt are not very well understood (Pacini and Juniper 1979a, b). Its role in attracting insects, protecting against UV radiation damage, and adhering to insect body because of its sticky nature, has been suggested by Heslop-Harrison (1968b). Hesse (1977, 1978a, b, 1979a, b, 1980) has made an extensive correlative study of the nature of Pollenkitt and its function in closely related entomophilous and anemophilous species of *Acer,* and some members of Polygonaceae, Oleaceae, Scrophulariaceae, Plantaginaceae, and Asteraceae. He

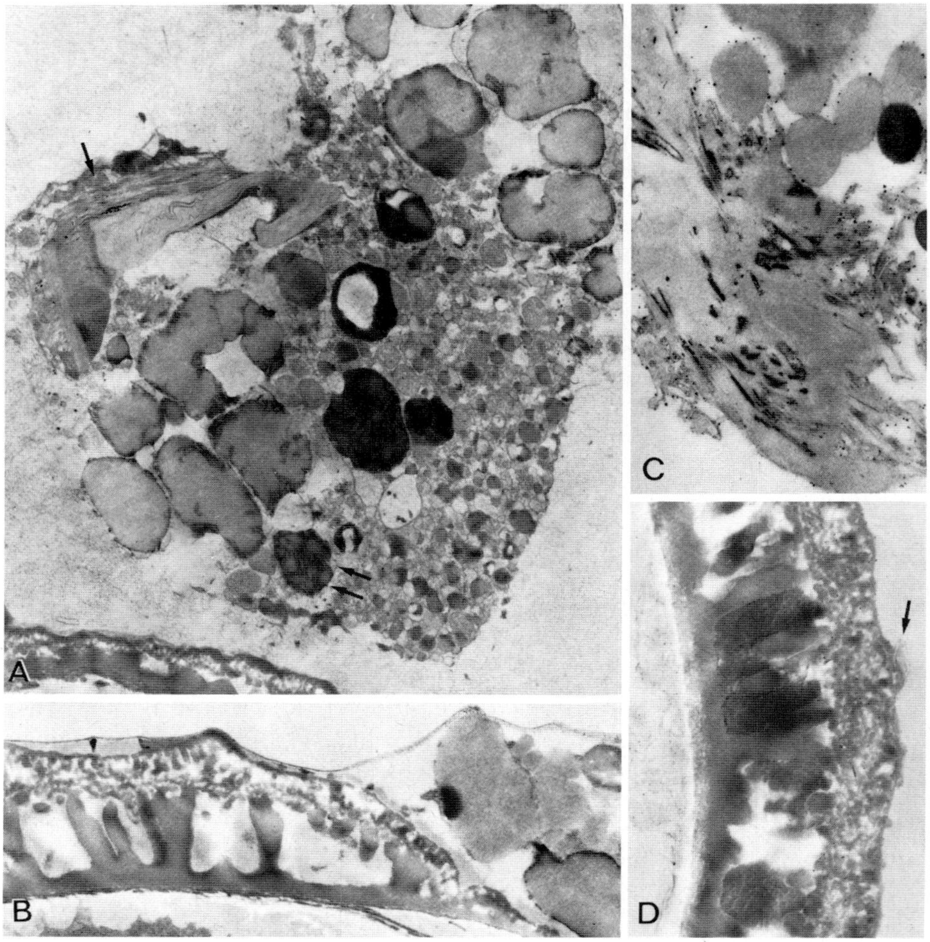

Fig. 2.12. A, B Pollenkitt in *Artemisia mutellina,* an entomophilous species. **A** After the degeneration of tapetum, lumps of Pollenkitt material float in the locule. It is filled with lipid droplets of various shape and size. **B** At the opening of pollen sac, on the tectum lies one electron transparent Pollenkitt crust which encloses lipid droplets. **C–E** Pollenkitt in entomophilous, amphiphilous, and anemophilous species of *Artemisia*. **C** *A. abisinthium,* after the degeneration of tapetum heterogenous large Pollenkitt lumps float in the locule. **D** *A. chamaemelifolia,* at the opening of pollen sac the empty spaces in the sexine are filled mostly with electron-thick sticky material, while on the tectum lie numerous electron-dense (*arrow*) Pollenkitt materials. **E** *A. vulgaris,* shortly before opening of the pollen sac a large amount of homogenous sticky material of medium electron density lies against the tectum. **F** Part of tapetal cytoplasm facing the loculus and exine sculptures during the first haploid mitosis. The tips of the exine sculptures are in contact with tapetal plasmalemma (*1*). When they detach from it, a thin electron-dense irregular fading layer, the tapetal printing (*2*), is present on their surfaces. **G** Two tapetal cells, tapetal intercellular space and a part of microspore during first haploid mitosis. Inside the space and loculus three different substances are evident: orbicules (*O*), granular electron-dense masses (*asterisks*) and fibrillar proteins (*arrows*). One elaioplast (*E*) is present in the cytoplasm. (**A–E** after Hesse 1979 b; **F, G** after Pacini and Juniper 1979 b)

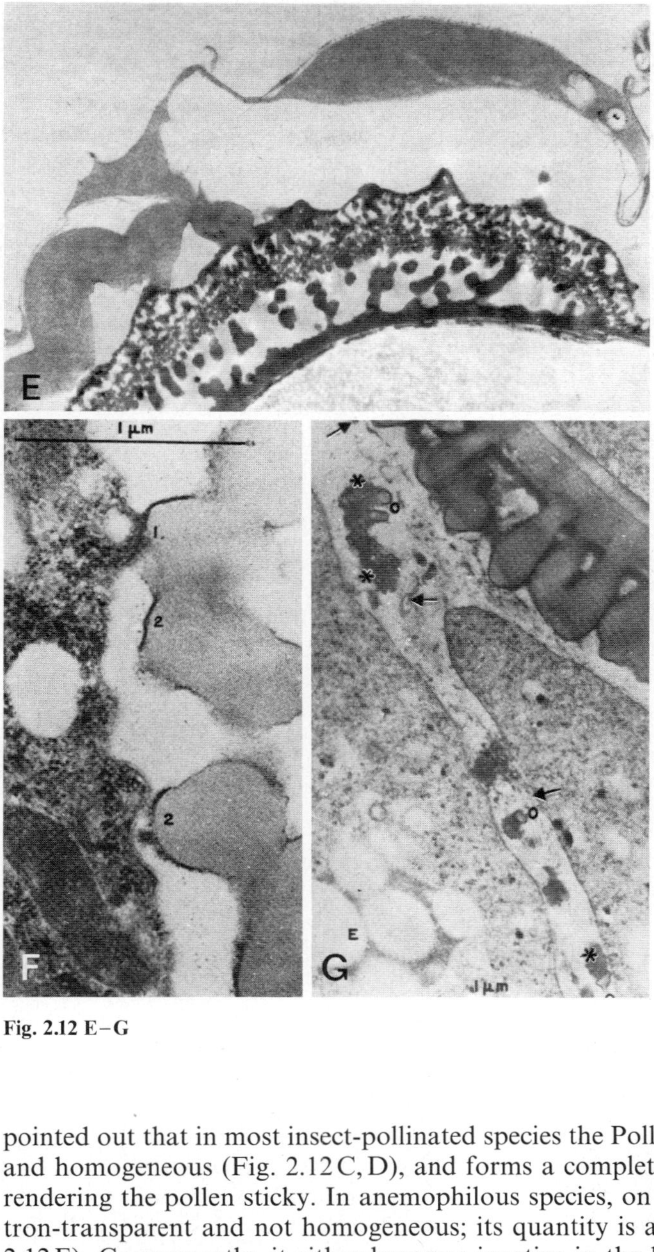

Fig. 2.12 E–G

pointed out that in most insect-pollinated species the Pollenkitt is electron-dense and homogeneous (Fig. 2.12 C, D), and forms a complete coating on the exine, rendering the pollen sticky. In anemophilous species, on the contrary, it is electron-transparent and not homogeneous; its quantity is also much smaller (Fig. 2.12 E). Consequently, it either becomes inactive in the loculus, or sinks to the bottom of the exine perforations. The pollen is, thus, non-sticky and powdery. Pacini and Juniper (1979 b) reported a new feature of "tapetal printing" on pollen exine in *Olea europaea* (Fig. 2.12 F, G).

In *Acer* (Hesse 1979 b) various species show differential depositions: (1) inside the exine cavities (*A. negundo*), (2) inside the exine cavities, with only small droplets adhering to the tectum (*A. campestre*), (3) average amount of granular material deposited both inside the cavities and outside as a slender film (*A. pseudoplatanus, A. opalus*), and (4) a good amount of granular material filled in the cavities of exine and deposited as a thick layer over the surface of tectum (*A. platanoides*). Thus, the pollen is (1) powdery, (2) moderately sticky, (3) sticky, and (4) very sticky which can be correlated to anemophilic, amphiphilic, and entomophilic modes of pollination.

Distinction is made betwen the Pollenkitt and tryphine, though these are similar in many respects (Heslop-Harrison 1968 b, Echlin 1971, Dickinson 1973). The former is a synthetic activity, while the latter is formed from the remains of degenerated tapetum. These two differ in containing hydrophobic or hydrophilic lipids, respectively. The latter also includes some cytoplasmic structures such as ER, and pro-Ubisch bodies in *Helleborus* which are tansitory. However, the existence of Pollenkitt in *H. foetidus* is not certain (Echlin and Godwin 1968 a). It should be interesting to find co-occurrence of both in any taxon.

2.8 Sporogenous Tissue: Ultrastructure

Concurrent with meiosis, conspicuous changes occur in the cytoplasm and organelles of microsporocyte (Heslop-Harrison 1971). The sporophytic-gametophytic transition, although ultimately an expression of genetic potential, must depend upon extranuclear factors. The continuity of plastids and mitochondria through somatic cell lineages and microsporogenesis, and other changes in the cytoplasm, are no doubt connected with the sporophytic-gametophytic transition.

2.8.1 Cytoplasmic Membranes and Ribosome Population

Pacini and Cresti (1978) have shown, in *Lycopersicum*, that before meiosis the ribosomes per unit area are more numerous in microspore mother cells than in tapetum, but the contrary is true after the beginning of meiotic prophase. In microspore mother cells of *Lilium* large amounts of cytoplasm are generally invested in double membranes; concentric multimembrane inclusions are also formed quite regularly (Dickinson 1971). Similar structures have been observed in *Pisum* and *Lens* (Biddle 1979). As the meiocytes enter prophase, there is an abundance of free ribosomes and an appreciable amount of ribosomal endoplasmic reticulum. The process of encapsulation of the cytoplasm commences in late leptotene to early zygotene, just prior to elimination of cytoplasmic ribosomes. The outer membranes lose the major part of their surface polysome population, form cup-shaped profiles and, finally, come to envelop a portion of the premeiotic cytoplasm (Fig. 2.14 A, B). This process appears completely non-selective, in that lipid droplets, organelles, or even other membranes may also become invested (Dickinson and Andrews 1977). A substantial fall in ribosome number occurs during the zygotene-pachytene interval (Fig. 2.13 A, B), and by mid-prophase about 15% of cell cytoplasm becomes included in double- or multimembrane sys-

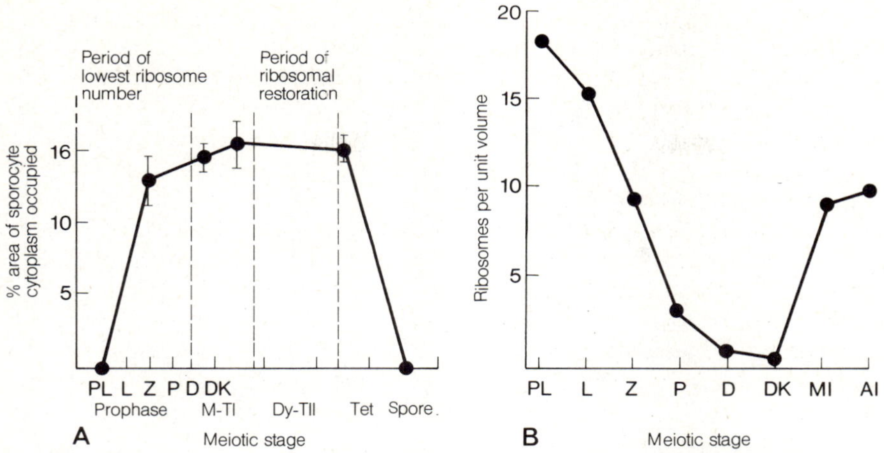

Fig. 2.13. A Relative proportion of cytoplasm encapsulated in a double membrane during prophase, metaphase to telophase I (*M-TI*), dyad to telophase II (*Dy-TII*), tetrad (*Tet*), or multi-membrane bounded inclusions in *Lilium*. **B** Changes in the mean number of ribosomes per unit volume of cytoplasm in *L. henryi* at different stages: *PL* preleptotene, *L* leptotene, *Z* zygotene, *P* pachytene, *D* diplotene, *DK* diakinesis, *MI* metaphase I, *AI* anaphase I. (**A** after Dickinson and Andrews 1977, **B** after Dickinson and Heslop-Harrison 1977)

tems (Fig. 2.13 A, B). During the process of degradation the ribosomes enclosed in these membrane-bound bodies are far less affected by agents effecting their eradication/disintegration in the general cytoplasm. The ribosomes are usually concentrated in the centre, because those on the surface of investing membranes are degraded. Granular deposits persist, however, suggesting that only one component of the ribosome is eliminated. By late prophase, the ribosome level is lowest (Fig. 2.13 A). There is practically no further change until the breakup of these membranes, which begins at the end of meiosis (Fig. 2.14 A–C). Some ribosomes become associated to form polysomes (Fig. 2.14 B). In *Lilium* they persist until early stage of microspore (Dickinson and Heslop-Harrison 1977).

Scheer and Franke (1972), in MMCs of *Canna generalis,* describe annulate lamellae of irregularly spaced interconnected cisternae bearing ribosomes, and continuous with rough endoplasmic reticulum (see also Kessel 1968). They consider annulate lamellae as a "degenerate form of rough endoplasmic reticulum" which suggests that these do not necessarily have any particular function, such as storage of ribosomal or messenger RNA.

The ribosome population is restored during or immediately after meiotic divisions. Dickinson and Heslop-Harrison (1970) suggest that the nucleoloids released into the cytoplasm at anaphase I and II may represent the main mechanism through which the ribosome population of the cytoplasm is restored. The nucleoloids are regarded as packeted ribosomes or ribosomal subunits, which are dispersed, or assembled and dispersed, after exclusion from the nucleus at each successive division. This phase of ribosome degradation is complemented by a decline of cytoplasmic rRNA during zygotene-pachytene gap in *Cosmos bipinnatus* (Knox et al. 1970), and in *Lilium* and *Trillium* (Mackenzie et al. 1967).

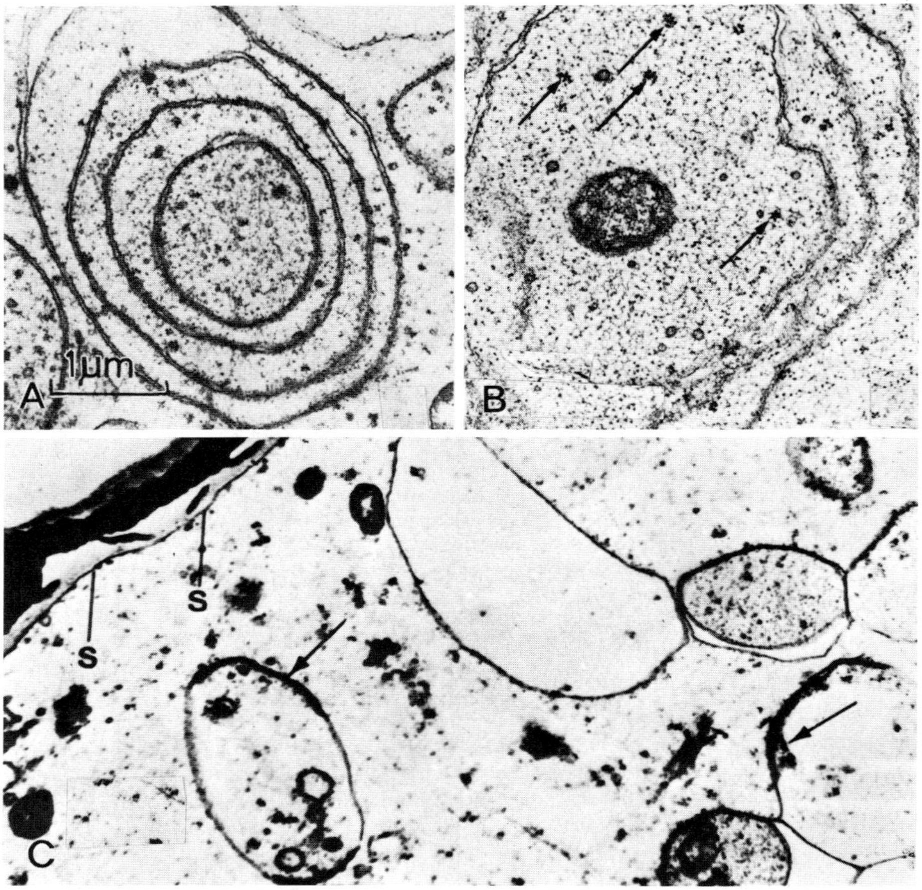

Fig. 2.14 A–C. *Lilium*. **A, B** Inclusion invested by four layers of double membrane. Note the increase in concentration of ribosomes towards the centre of inclusion, and the polysomal configuration (*arrows*) in **B. C** Young microspore cytoplasm showing the splitting of double membranes. The regions of membrane heavily encrusted with ribosomes (*arrow*) oriented towards the cell surface (*S*). (After Dickinson and Andrews 1977)

Evidently, the cytoplasm of MMCs receives morphogenetic information before meiotic cytokinesis, and it appears reasonable that this is conserved or carried over through the ribosome-elimination phase into the spores via the membrane-bound inclusions.

2.8.2 Nucleus

The post-pachytene nuclear envelope generates numerous double membrane-bound inclusions by sacculation of the inner nuclear membrane and complementary blebbing of the outer membrane of the nuclear envelope (Fig. 2.16A). It is

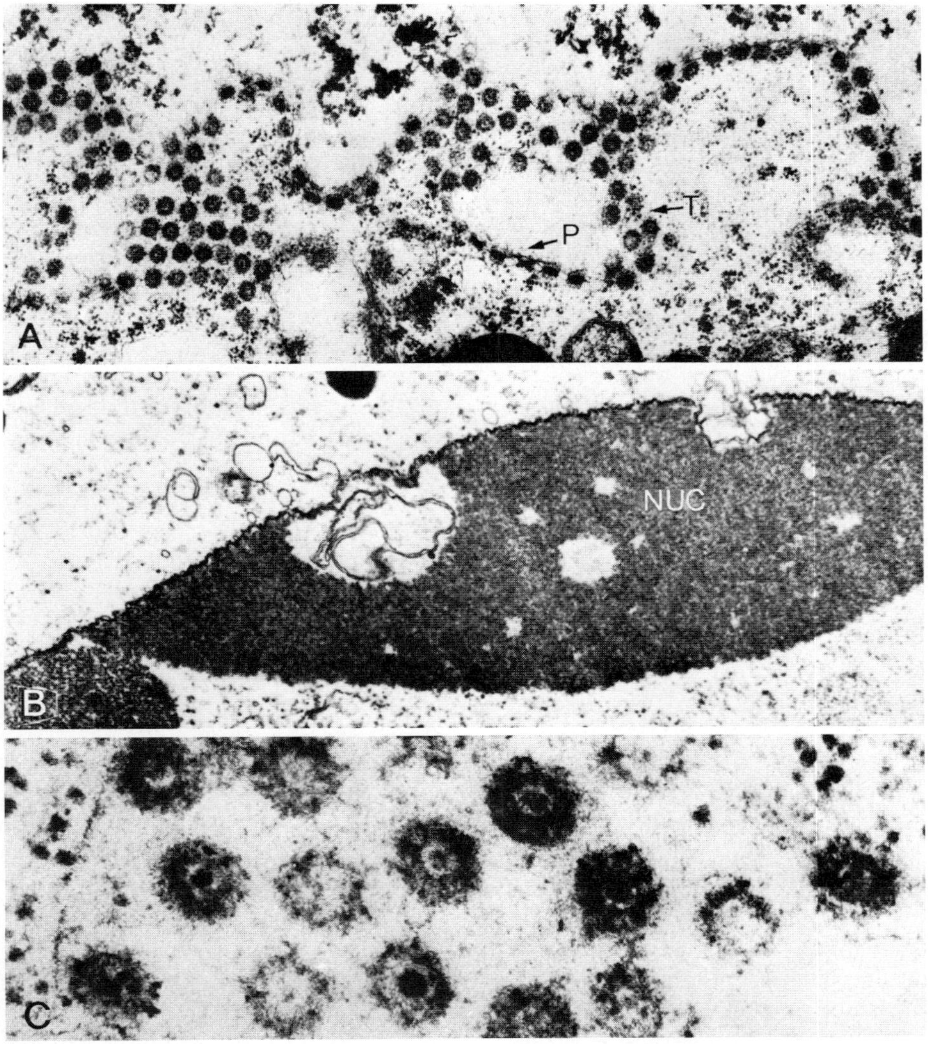

Fig. 2.15. A Electron micrograph from a pollen mother cell at late zygotene of *Fritillaria lanceolata*, showing arrays of annuli face-on, and pore complexes in perpendicular (*P*) and tangential (*T*) sections. All those seen face-on contain a central granule. **B** Pollen mother cells of lily "Formobel" showing absence of pores in the region of an envelope at which the nucleolus (*NUC*) is adpressed during leptotene. The envelope runs tight against the nucleolus, and at two places invaginates vacuoles. **C** Zygotene in *F. lanceolata* illustrating variability in the appearance of central plug. (From LaCour and Wells 1974)

also coupled with inwardly directed activity associated with the inner nuclear membrane. This development is a general one in plants (see Dickinson and Heslop-Harrison 1977, Sheffield et al. 1979). In *Lycopersicum* vacuole formation coincides with the completion of synaptonemal complex development. Rasmussen (1976) termed such dilations of the perinuclear space, caused by invagination of the inner membrane of the nuclear envelope, as nuclear vacuoles. La Cour and

Wells (1974) have studied the distribution and fine structure of nuclear pores (Fig. 2.15 A–C) at early to mid-meiotic prophase in the MMCs of *Fritillaria lanceolata, Phaedranassa virdiflora, Tulbaghia violacea,* an F_1 hybrid of *Allium fistulosum* × *A. cepa,* and lily var. "Formobel". Except in the region where nucleolus is adpressed to the nuclear membrane (Fig. 2.15 B), the pore clusters are distributed randomly over the nuclear membrane (Fig. 2.15 A, C). The number is variable, and directly proportional to the level of metabolic activity of the cell. The nuclear pores are evidently involved in nucleo-cytoplasmic interaction.

It is possible that the accumulation within the vacuoles represents nuclear material which is undergoing degradation or reorganization. It is pertinent to point out that such a reorganization in the nucleus assumes meaningful importance, since contemporary cytoplasmic reformation is being conducted by elimination of ribosomal population (Dickinson and Heslop-Harrison 1977). The restandardization of the non-genetic components of nucleus in preparation for the gametophytic generation might, therefore, involve elimination of nucleoplasmic structures associated specifically with sporophytic growth.

2.8.3 Nucleolar Cycle

Meiocytes enter prophase with one, two, or more nucleoli, depending upon genotype. At the time of synapsis, when there is more than one, the number is reduced; subsequently there is usually an increase in volume (Lin 1955). Often, there is a change in shape due to the flattening of nucleolus towards one pole (Latter 1926, Frankel 1937); Biddle (1979) reported a distinct nucleolar cap in *Pisum sativum* (Fig. 2.16 B), observed earlier in *Lilium* (Moens 1968). In fact, Biddle considered the polar nucleolar cap as a marker for initiation of meiosis in pollen mother cells.

In *Lilium* the spherical shape is restored at diplotene, and during pachytene supernumerary nucleoli are formed at the nucleolar organizing sites of the nucleolar-organizing chromosomes. When the microspore mother cell nucleolus is losing basiphilia and becoming vacuolate, they frequently lie in chains, the smallest showing the highest basiphilia. Late diplotene nuclei contain supernumerary nucleoli which are not associated with chromosomes, and probably these get detached from nucleoli organizing regions of chromosomes. The accessory nucleoli are released after dissolution of the nuclear membrane at the end of diakinesis. There are instances, however, where the nucleolus of mother cell does persist through the meiotic divisions (Frankel 1937). Luck and Jordan (1977) investigated the nucleolar cycle during microsporogenesis in *Endymion non-scriptus*. The zygotene nucleoli still in association with the nucleolar organizer part of chromosome show the two components: the granular and fibrillar regions (Fig. 2.16 C). The two segregated regions, quite distinct by diplotene, undergo fragmentation (Fig. 2.16 D) and, eventually, the fibrillar masses disappear. Concurrently, the granular masses become looser and increase in number (Fig. 2.16 E). Large areas of nucleolar envelope, however, remain intact (Fig. 2.16 C). During metaphase I numerous pre-nucleolar bodies coalesce on to the nucleolar organizer to form the main nucleolus. The cycle is repeated during meiosis II as well. Luck and Jordan conclude that the fibrillar region shows continuity through nuclear divisions.

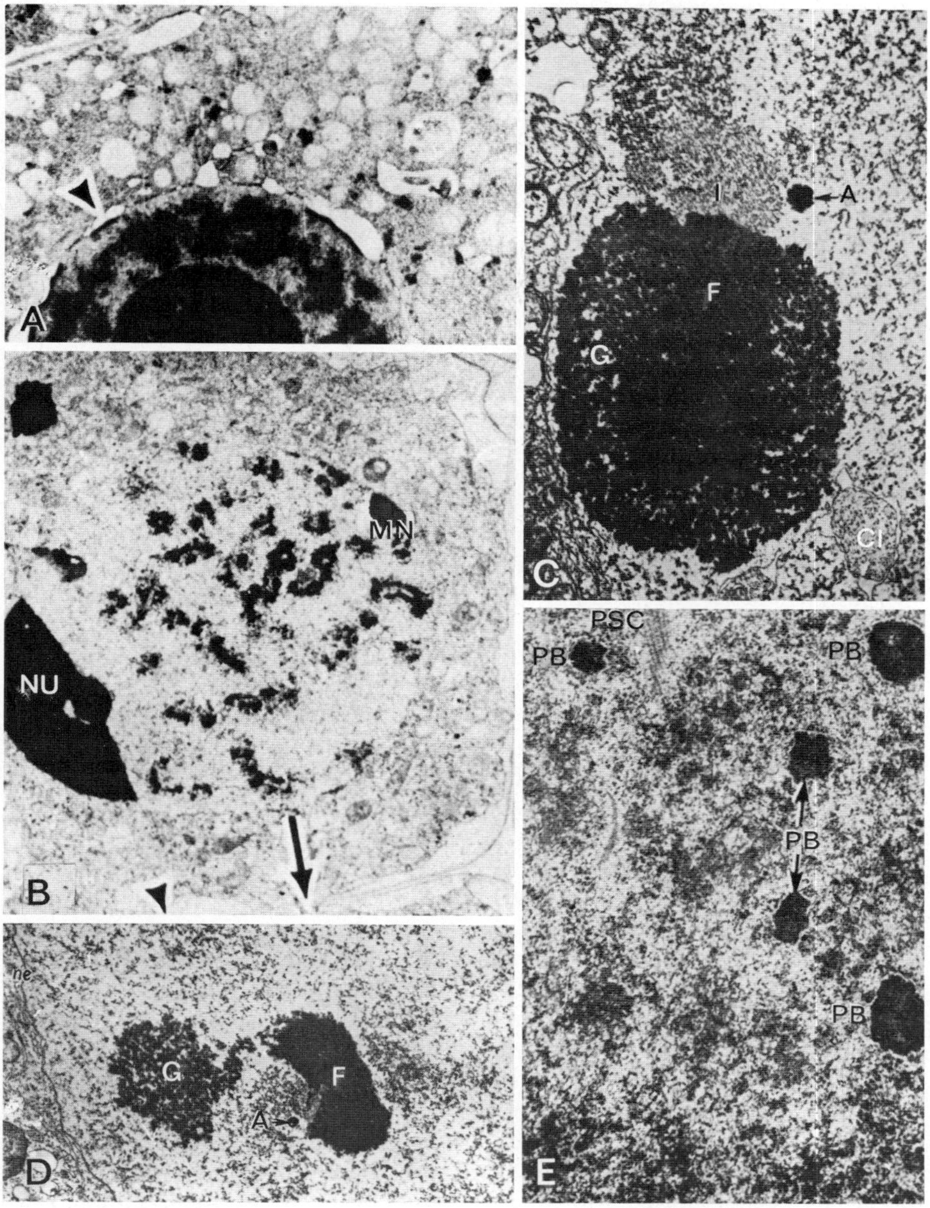

Fig. 2.16. A Pollen mother cell in *Pisum* at premeiotic interphase; large dilations (*arrow*) occur between outer and inner nuclear envelopes, primarily between pores. Dilated rough endoplasmic reticulum cisternae occur in cytoplasm. **B** Same, at diplotene of prophase I; an electron-dense crescentic nucleolus (*NU*) and two micronucleoli (*MN*) are present in the nucleus. Cytomictic channels (*arrow*) connect adjacent meiocytes. **C** Segregated zygotene nucleolus showing granular (*G*) and fibrillar (*F*) regions with nucleolus organizer. An accessory nucleolus (*A*) is on the surface of the organizer, and cytoplasmic invaginations (*CI*) are present in the nucleus. **D** The granular (*G*) and fibrillar (*F*) parts separate and fragment during pachytene-diplotene stage. Note an accessory nucleolus (*A*). **E** At the dyad stage a large number of pre-nucleolar bodies (*PB*) and polysynaptonemal complex (*PSC*) can be noticed. These coalesce on the nucleolar organizer and form nucleolus. (**A, B** after Biddle 1979; **C–E** after Luck and Jordan 1977)

Bodies cytochemically and structurally similar to nucleoli become visible in the spindle region during anaphase. They do not appear to be formed at the nucleolar-organizing regions of the chromosomes. Most of these are released into the cytoplasm as nucleoloids. The residue enclosed within the envelope of dyad nuclei is released into the cytoplasm during anaphase II. Similar structures have been observed by Latter (1926) in *Lathyrus odoratus,* Håkansson and Levan in *Pisum* (1942), Lindemann (1956) in *Bellevalia* and *Agapanthus,* and Dickinson and Heslop-Harrison (1970) in *Lilium.*

2.8.4 Cytoplasmic Organelles During Meiosis

Guillermond (1920, 1924) and Py (1932) presumed that organelles in the cytoplasm pass through a granular period in prophase before their restoration in the young spores. Bal and De (1961) reported that the organelle lineage is broken at this point, with its subsequent de novo origin. Later investigations, however, have shown that cycles of dedifferentiation and redifferentiation occur in both plastids and mitochondria (Maruyama 1968, Dickinson and Heslop-Harrison 1970, 1977).

During the premeiotic period profiles suggestive of the division of plastids are frequent. Biddle (1979) observed plastids in this state in *Pisum sativum,* unlike *Helleborus.* Normal amyloplasts in the tissue are 0.5–1.5 μm, and may contain one or two starchgrains, osmiophilic globuli, ribosomes, and DNA fibrils. All through prophase the number of divisional configurations decline. The starch grains disappear, while the lamellar system regresses and the ribosomes present in the centre of the plastid get degraded. Towards late zygotene, the plastids assume isodiametric form, dedifferentiate completely, and become almost indistinguishable from the double-membrane structures; only the osmiophilic globules remain intact. Plastids may also be ellipsoidal, with few or more discernible ribosomes, and a severely reduced lamellar system. Subsequently, and through metaphase I, the plastid population remains relatively unaltered.

An unusual, and possibly unique, structural feature appears at interphase consisting of an association between a membrane component, in the form of a flattened tubule or ribbon, and a cluster of particles of diverse size. The granules measure 15–60 μm and may be interspersed with a small number of osmiophilic bodies or droplets. Diffuse masses of fibrils are also present. This membrane–particle association is probably present in all plastids from the dyad to the late tetrad stages (Dickinson and Heslop-Harrison 1970, 1977). With rare exception, only one membrane–particle association is present in each plastid at the dyad stage. Its significance cannot be specified, but its consistent occurrence suggests that it may be concerned with the redifferentiation of plastids, which occurs from late tetrad onwards. During this stage, divisional figures of plastids are invariably accompanied by synchronous division of the membrane–particle association.

The mitochondria show regular changes which closely synchronize with nuclear and other cytoplasmic events. During the premeiotic period the cells entering meiotic prophase have normal mitochondria (1.0 μm) much like the somatic cells. In the leptotene stage, however, the mitochondria begin a phase of division followed by reduction in size (maximum diameter about 0.3 μm). They are spherical and retain a small number of cristae, but the matrix is rather electron-dense, or opaque, and rich in ribosomes. In the young spores, an enlargement and recovery of the normal structure is common.

This cycle of dedifferentiation and redifferentiation in plastids and mitochondria is presumably connected with the sporo-gametophytic transition.

De Vries and Ie (1970) and Turbin et al. (1974) noted significant differences in the cytoplasm of normal wheat pollen grains and cytoplasmic male-sterile (CMS) pollen mother cells. CMS microspores contained fewer organelles, presumably of low metabolic activity. Similarly, pollen mother cells had fewer organelles than the normal cells. The mitochondria have distorted matrix and irregular cristae. There are fewer dictyosomes, and cytosegnesomes are present isolated by three or four concentric agranular cisternae. Pollen mother cells are also more vacuolate. Unfortunately, the stage at which abortion occurrs could not be determined.

Skvarla and Kelly (1971) point out that in the premeiotic period the golgi constitute several inter-associated dictyosomes in *Canna generalis;* also present are mirror-image twins. The smooth endoplasmic reticulum is frequently associated with the golgi apparatus. The ribosomes near the dictyosomes are mostly situated in the vicinity of smaller face. Additional non-vesicular cisternae are added by convergent tubules of smooth endosplasmic reticulum. In meiotic prophase there is no clear morphological evidence of secretory activity by the dictyosomes but, with the beginning of the deposition of intine, intense vesiculation occurs which also reduces the number of cisternae (see also Sect. 8.1).

2.8.5 Plasmodesmata and Cytoplasmic Connections

The existence of plasmodesmata (Fig. 2.17 A–E) interlinking the meiocytes has been reported in a number of angiosperms (Heslop-Harrison 1964, 1966a, b). Biddle (1979) observed cytomictic channels in meiotic cells in *Pisum sativum,* but these were absent in *Lens*. Horner and Lersten (1971) also state that the cytoplasmic channels were absent in *Citrus limon*. Whelan (1974), however, observed, under SEM, cytomictic connections through the callose wall of adjacent meiocytes in *Allium* (Fig. 2.17 A, B). These were of variable size, and increased in diameter between pachytene and metaphase. In fractured material he has shown intermeiocytic channels through which cytoplasmic strands connect the two pollen mother cells (Fig. 2.17 E).

Pacini and Cresti (1978) have shown that in *Lycopersicum* some plasmodesmata close up, while from others originate cytomictic channels large enough to allow the passage of even plastids, in addition to other cytoplasmic contents (Fig. 2.17 C, D). With the deposition of callose wall several interconnections are severed though some may persist until metaphase or anaphase I (Whelan 1974). According to Heslop-Harrison (1964, 1966b), through these plasmodesmata may pass solutes and other factors from one meiocyte to another so that the entire mass has a synchrony in initial development.

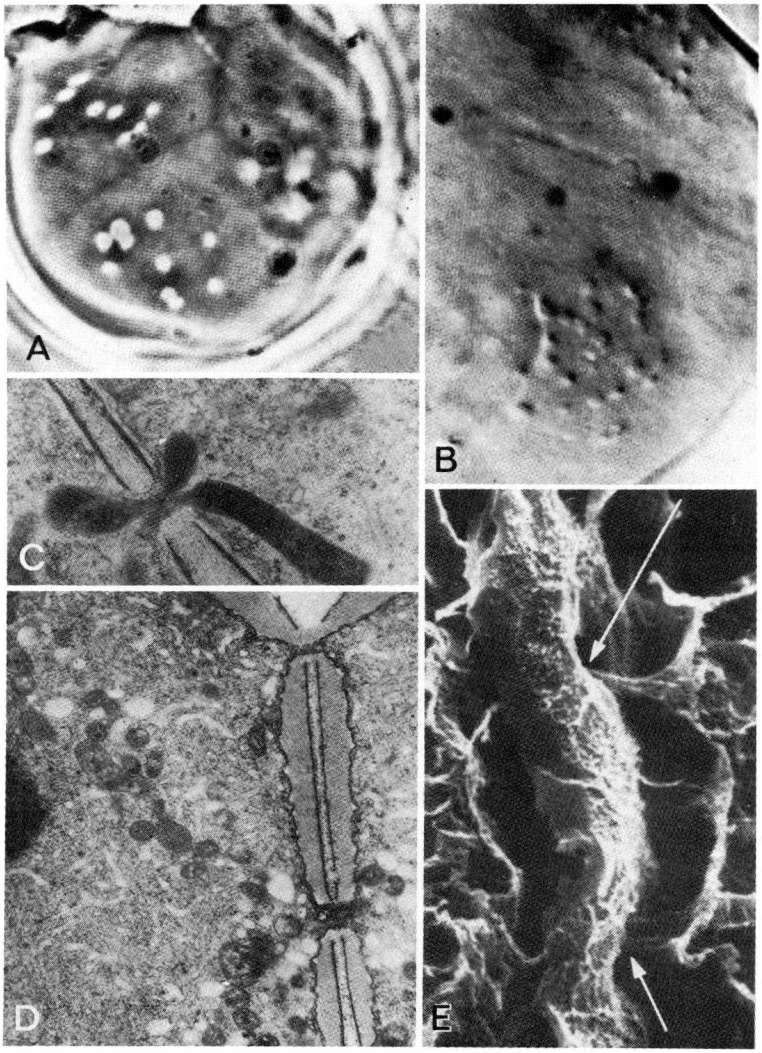

Fig. 2.17. A *Monadra,* callose wall in meiocytes in metaphase I. The holes appear to be relatively uniform in size, but increase in diameter between pachytene and metaphase I. **B** Enlargement of central area to illustrate the variation in size of holes in the developing meiocytes callose wall. They appear to be confined to specific areas of cell interfaces. **C, D** Electron micrographs through cytomictic channels, showing the passage of chloroplast in **C** and other organelle in **D** from one microspore mother cell to the other. **E** Fracture through two adjacent meiocytes showing holes in the callose walls and intermeiocyte connections (*arrow*). (**A, B, E** adapted from Whelan et al. 1974; **C, D** from Pacini and Cresti 1976)

It is evident that the cytoplasm and its organelles undergo a kind of reorganization and, as interpreted by Dickinson and Heslop-Harrison (1977), it is meant for the sporophytic cell to get rid of sporophytic messages and prepare itself for undertaking the functions of gametophytic generation. The synthesis and development of callose encasement furnishes compartments for autonomous development of haploid gametophytic cells. It seems unlikely that the cycles of de-, and redifferentiation are prerequistes of meiosis but rather a characteristic of cells undergoing major phase changes.

Albertson and Palmer (1979) studied microsporogenesis in fertile and male-sterile soya beans, and concluded that nuclear and cytoplasmic events must occur at very precise times for the successful development of haploid pollen from diploid pollen mother cell (see also Sect. 4.4.5).

2.9 Initiation and Control of Meiosis

Meiosis is the most important event in the life cycle of a plant, and its completion ends the sporophytic generation and heralds the gametophytic take-over. Nevertheless, this phenomenon is not yet fully understood, and attention is invited to the contributions of Darlington (1977), Riley and Flavell (1977), Bennett (1977), LaCour and Wells (1977), Stern and Hotta (1977), Dickinson and Heslop-Harrison (1977), and the excellent reviews by Vasil (1967, 1973 a, b), and Mascarenhas (1975).

The most fundamental question, yet unanswered, is what factor(s) precisely trigger(s) the mitotic to meiotic change-over? Riley and Flavell pointed out that much is not known about the cause of this switch-over except that "it is very well-protected from error and may consist of a number of integerated steps". Evidence could have been expected from studies on the culture of anther and PMCs. However, to the best of my knowledge (also personal discussion with my colleague Dr. S. S. Bhojwani), nothing concrete has emerged. According to Vasil (1967), two bottlenecks that must be surpassed are (1) to culture premeiotic anthers in order to obtain mature and normal pollen grain, and (2) to induce mitosis in microspore nucleus in vitro. Vasil emphasizes that the first barrier has never been overcome, and premeiotic anthers "continue to be stubbornly immune to all the nutrient media and treatments" given them under cultural conditions. Even though Pereira and Linskens (1963) claimed that, to some extent, it is possible to induce meiosis in premeiotic anthers cultured in a medium containing 250 mg/l of glutathione, and the second barrier has been removed by glutamic acid in some cases. The source of stimulus for the induction of meiosis is unknown. Gregory (1940) observed that in tomato branches bearing premeiotic anthers, when grafted on a healthy flowering stock, resulted in normally developed pollen grains, whereas the development was arrested when cultured as such. Similarly, Sussex and Steeves (1958) showed that in ferns when leaf primordia were cultured, these developed further and even differentiated sporangia. The development of sporangia was, however, arrested at the premeiotic stage of sporogenous cells. It is clear that the control of meiosis depends on the synthesis of factor(s) in some tissue other than the sporogenous cells, and its transport to the anther may trigger the meiotic process. Clutter and Sussex (1965) have shown that the stimulus for meiosis arises in the vegetative shoots of plants. It is highly specific in that it affects only the spore mother cells, and not even the tapetal cells through which it must pass. In a number of plants anthers at the early prophase and leptotene-zygotene stage have been successfully cultured until tetrad, young microspore, and even two-celled pollen grain stages (Vasil 1967, 1973 a, b).

Heslop-Harrison (1959) points out that "successful initiation of meiosis in the outgrowing stamen is dependent upon some stimulus passing into it over a period of time, so that it is not capable of self-differentiation to a degree comparable with the leaf primordium of *Osmunda* L." Bennett et al. (1973) suggest that, in wheat, before meiosis, three mitotic cell cycles are accomplished in 25, 35, and 55 h, respectively, in continuous light. There are practically no investigations to indicate the precise stage at which the meiotic division can be reversed; it is clear though that beyond leptotene the change is irreversible. Stern and Hotta (1969) were able to reverse the mitotic to meiotic switch in *Lilium longiflorum* only if the PMCs were at a stage preceding leptotene, despite the prolonged prerequiste DNA replication period. Riley and Flavell (1977) make a distinction between the two aspects, namely, the readiness and commitment to meiosis. The latter is irrevocable. Dover and Riley (1973) have, however, demonstrated that treatment of spindle inhibitors applied before premeiotic S-period can affect the meiotic process, which obviously must have triggered changes at the subcellular level.

DNA synthesis marks the first phase of mitotic or meiotic cell division, and differences do exist between mitotic and meiotic DNA synthesis. In wheat, premitotic S-phase lasts about 35 h as against the approximately 8 h long premeiotic S-period at 20 °C (Bennett et al. 1971). Callan (1972) concluded that the period of synthesis is due to wider distances between the initiation points during replication. The replication is predominantly unidirectional from the point of initiation in premeiotic S-phase. It is possible that the changed pattern of DNA replication is related to the preparation of meiotic division. High temperature and continuous light inhibit premeiotic synthesis while the premitotic remains unaffected, which indicates different controls for the two processes. In *Lilium* microsporocytes, 0.3%–0.4% of the total DNA content, and the Z-DNA (Stern and Hotta 1977) which has a distinct base composition, are replicated not during the premeiotic S-phase but during zygotene. Chromosomes do not pair if there is no replication (Hotta and Stern 1971, Stern and Hotta 1973).

Vasil (1959) has discussed the possible role of tapetal breakdown products in DNA synthesis, correlated with the localized activity of enzymes, degradation of DNA into nucleotide, and its dephosphorylation into deoxyribosides, in the anther wall. Sheridan and Stern (1967) reported meiotic histone in addition to the mitotic histones in *Lilium* and *Tulipa*. The meiotic histone appears at premeiotic DNA synthesis, persists through meiosis and, subsequently, disappears before the next nuclear division.

Riley and Flavell (1977) conclude that regulation of meiosis can be affected by environmental factors which can disturb many components of the process such as contraction and congression of chromosomes, pairing and disjunction, and spindle formation and chromosome movement, although these do not interfere with the course of division. However, inhibition of protein by cycloheximide, and DNA synthesis by deoxyadenosine, do interfere with the course of meiosis (Stern and Hotta 1967, Parchman and Stern 1969, Flavell and Walker 1973). Events preprogrammed before meiosis are not affected by internal anomalies. Riley and Flavell pressume that the developmental process depends critically on the induction of specific proteins, as for example spindle proteins, recombination enzymes and synaptonemal complex proteins, and RNA and DNA synthesis at a particular stage of development.

Studies on cytoplasmic male sterility (see Sect. 4.4.5) bring out very clearly that a delicately balanced functional and biochemical relationship must exist between the tapetum and pollen mother cells throughout, so that perfectly normal and viable pollen is formed. Any disturbance in this harmony results in abnormalities and, consequently, male sterility. Bennett and Smith (1979a) observed the formation of paracrystals in the tapetum of colchicine-treated anthers, in wheat up to 8% of the cytoplasmic volume of tapetal cells. Such a large quantity of substance composing such structures could be proteinaceous, notably tubulins, actomyosin, and several nuclear histones. The paracrystals have also been observed in PMCs (Bennett and Smith 1979b); their disappearance during early premeiotic interphase may indicate that the constituent material may pass into other adjacent tissue, presumably to PMCs. Bennett and Smith tentatively proposed that fibrillar material in wheat PMCs may comprise protein subunits synthesized in tapetum, but colchicine interferes with its movement to sporogenous cells where the formation of intranuclear fibrillar material is, consequently, delayed. Its precise role in initiation of meiotic process is, however, not clear.

2.9.1 Duration of Meiosis

Bennett (1977) listed 24 diploid, 1 triploid, 10 tetraploid, 3 hexaploid, and 1 octoploid species in which the time and duration of meiosis have been noted. Meiotic duration ranges from less than 24 h in *Petunia* (Izhar and Frankel 1973) to about 3 months in *Trillium* (Hotta and Stern 1963). Such a variation exists even in plants grown at a constant temperature of 20 °C, from 18 h in *Capsella bursa-pastoris* to about 10.5 days in *Lilium* hybrid cv. Black Beauty.

The duration of meiosis decreases with the increasing temperature. Dehydration and physical damage have either no effect, or may bring about complete cessation of the meiotic process (Bennett et al. 1971).

There is a positive correlation between nuclear DNA content and decrease in the duration of meiosis. There is no effect of the diploid chromosome number. A comparison of diploids and polyploids with more or less the same DNA content indicates that the polyploids have shorter meiotic duration (Bennett 1972). However, it increases in a series of related polyploids, e.g. an octoploid showed maximal increase over the related diploid than its tetra- and hexaploid. Bennett et al. (1974) pointed out that proportional loss of segments of chromosomes decreases the duration of meiosis. The changes in duration involve not only one or two but all stages of meiosis (see Bennett 1977).

2.9.2 Synthesis of Callose: Deposition and Significance

During meiotic divisions a distinct, refractile wall, "special callose wall" (Fig. 2.17 A, B) is secreted around the microspore mother cells. This wall is composed of callose-β-1, 3-glucans (Rowley 1959, Heslop-Harrison 1968 a, b). This is generally present in small quantities in structurally different plant tissues, and is a substance which has special physical and physiological properties, e.g. it is rapidly synthesized and degraded with equal ease (Eschrich 1961, Eschrich and Eschrich 1964, Waterkeyn 1964). The emittance of secondary fluorescence by the use of aniline blue (0.005%), a technique developed by Arens (1949) and standardized by Eschrich and Currier (1964), has been extensively used to elucidate callose deposition (Fig. 2.18 A–E) and its structure. Waterkeyn (1962), in his pioneer investigations, demonstrated temporal relationship of six layers composing the callose wall in *Helleborus foetidus*. A layered appearance of callose wall has also been observed in *Zea mays* (Skvarla and Larson 1966). *Eleocharis* (Dunbar 1973) seems to be the only example in which tetrads remain embedded in callose synthesized premeiotically, and microspores do not add to it. In contrast to most angiosperms, *Pergularia daemia* (Asclepiadaceae) is characterized by compound pollen or pollinia, and the callose wall does not develop around the microspore mother cells during or after meiosis (Vijayaraghavan and Shukla 1977).

During prophase 1, the callose deposition is initiated between the plasmalemma and the original cellulosic cell wall, particularly along the corners of the PMCs (Fig. 2.18 A), and extends laterally until a continuous layer is laid down. In some graminaceous members, soon after some space develops by separation of PMCs, the callose is first deposited along the inner tangential cell walls (see also Kapil and Tiwari 1978) towards the centre of the anther locule, as small pegs (Fig. 2.18 B, C) and, subsequently, extends towards the outer tangential walls. The callose wall is not of uniform thickness. According to Waterkeyn (1962), the density of callose in different layers can be correlated with meiotic stages, while Christensen et al. (1972) point out that these are artefacts on account of fixatives and the processes of fixation. Nevertheless, Biddle (1979) reports that in *Lens* and *Pisum* callose lying closest to the plasmalemma is more electron-dense.

In *Helleborus foetidus* (Echlin and Godwin 1968 a) the plasmalemma is convoluted and, at places, distinct gaps have been observed between the plasmalemma and wall of pollen mother cell which contains certain electron-transparent material (Fig. 2.21 B). A material of similar electron density is reported in discrete vesicles distributed within the cytoplasm of meiocyte (Fig. 2.19 A–C) and, according to Echlin and Godwin, it could be callose. However, it has not been ascertained if the callose is formed in the dictyosomes or ER. In maize Skvarla and Larsen (1966) noticed that, prior to callose deposition, the plasma membrane receded from the cell wall and a large number of vesicles of variable size occupied this space. The preponderance of dictyosomes and vesicles in the peripheral cytoplasm of the meiocyte at this stage, coupled with the inability to demonstrate ER in this region, indicate the possible role of dictyosomes in the production of callose (Heslop-Harrison 1966 a). Eschrich (1962) observed a thick lipoprotein layer in this zone, and considers that it is concerned in callose deposition. However, a similar, though narrower, clear zone referred to by Echlin and Godwin (1968 a, b) did not stain for either lipids or proteins.

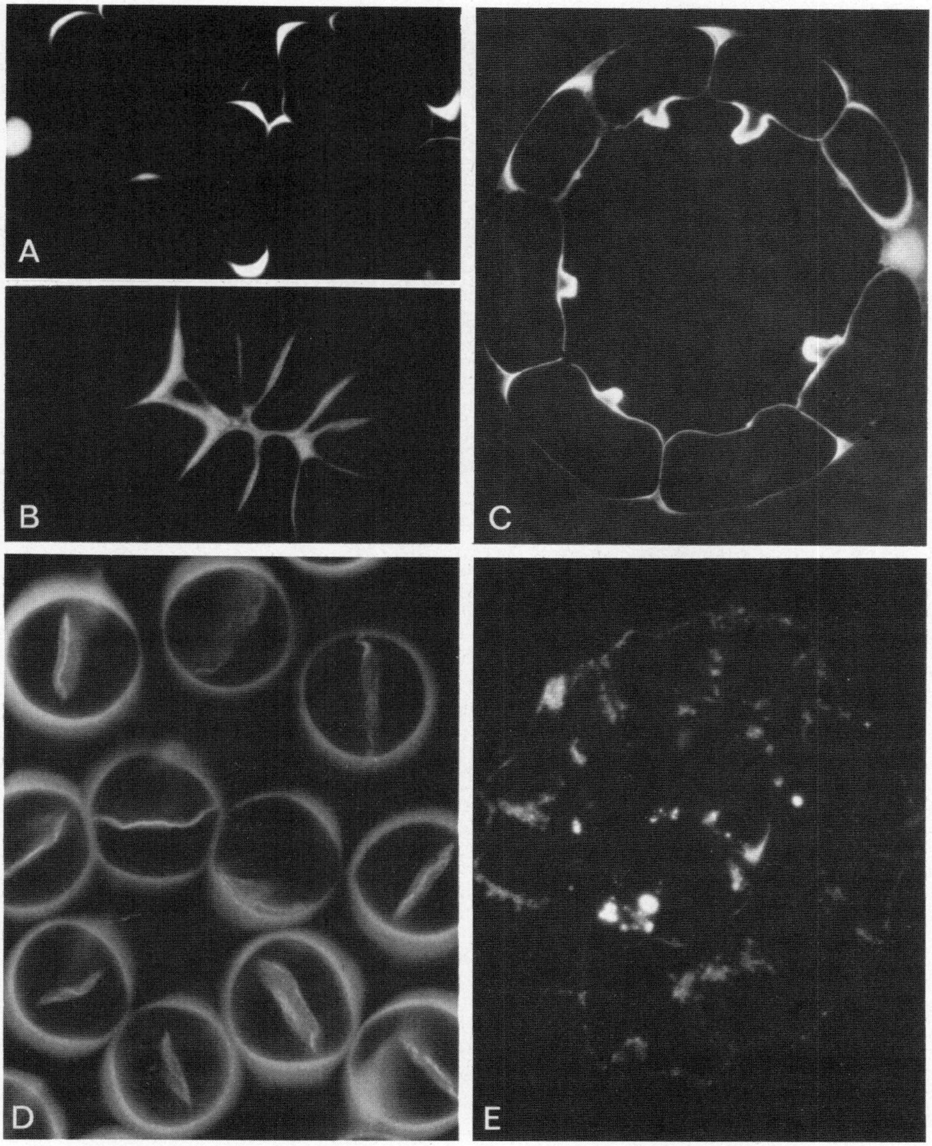

Fig. 2.18. A *Decaisnea fargesii*. **B–E** *Triticale*. **A** Synthesis and deposition of callose during microsporogenesis; aniline blue-induced fluorescence. Cross section of anther showing asymmetrical deposition of callose along one side of microspore mother cell (MMC). **B** Cross section of anther locule, at sporogenous cell stage. Special callose envelope extends along the radial walls to outside. **C** Same, at prophase I in MMC exhibiting asymmetrical callose deposition. The meiocytes are not completely isolated from the tapetum. **D** Smear of dyads showing the centripetal extension of cell plate during cytokinesis. **E** EMS-treated anther locule, showing hypertrophied tapetum and plasmodial mass of MMCs. Note poor and irregular callose deposition in MMCs; the tapetal cells also show callose deposition along the periclinal and radial walls. (**A** original, **B–E** courtesy Reeta Khosla)

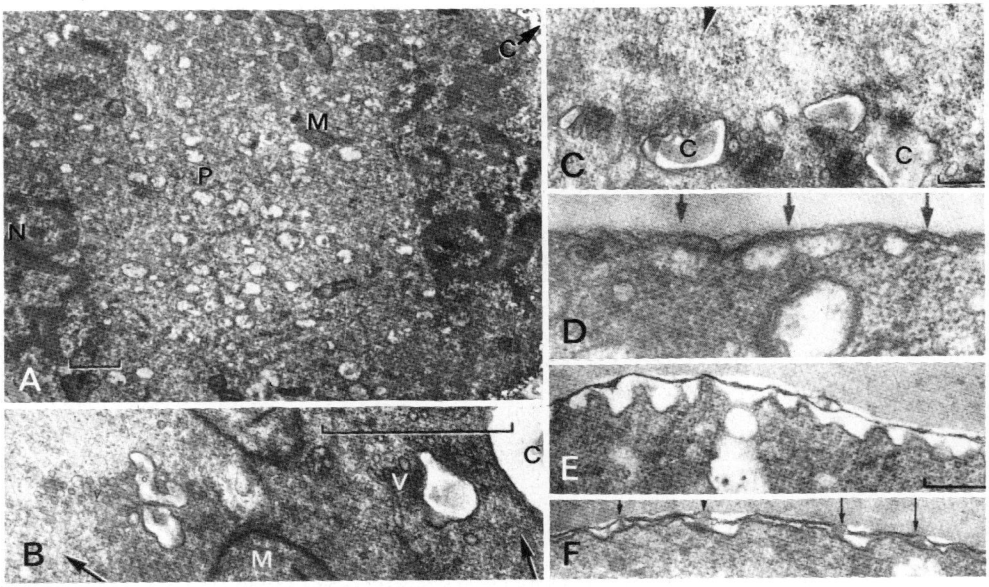

Fig. 2.19 A–F. *Helleborus foetidus.* **A** Meiocyte at the end of first meiotic division. Note the two nuclear masses (*N*), plastids (*P*), and mitochondria (*M*) in the central region of cell. There is some withdrawal between the surface of cell and callose (*C*). **B** Central region of tetrad at the end of second meiosis, showing gradual coalescence of dictyosome-derived vesicles (*V*). Microtubules (*arrow*) and mitochondria (*M*) are common in this region. Note close similarity between electron density of callose (*C*) and material within the vesicles. **C** Central region of tetrad at a slightly later stage than **B.** Note microtubules (*arrow*) between the vesicles. **D** Association of fragments of endoplasmic reticulum (*arrows*) at the periphery of microspore. **E** Convolutions in the microspore membrane; fibrous layer beginning to form between callose and tips of cytoplasmic projections. **F** Incipient loss in density in certain regions (*arrows*) of fibrous layer, coupled with the deposition of material of moderate electron density. (**A–F** after Echlin and Godwin 1968 a)

At the termination of meiosis, and at a time when the tapetal cells have become capable of producing sporopollenin, the callose wall is rapidly broken down and the microspores are released into the anther locule (Heslop-Harrison and Dickinson 1969). The dissolution of callose in *Lens* and *Pisum* is generally centripetal (Biddle 1979); in tetrahedral quartets, however, the dissolution begins at the corners. There is a positive relationship between pH, callase activity, and breakdown or degradation of callose (Izhar and Frankel 1971). In fertile *Petunia hybrida* anthers, pH during meiosis is 6.8 to 7.0, and the activity of callase cannot be detected. However, at the tetrad stage, the pH drops to 5.9–6.2, and there is a sharp increase in callase activity resulting in the digestion of callose around the tetrads and release of microspores in the anther locule. On the other hand, in the anthers of male-sterile *Petunia,* the drop of pH and callase activity are precocious and, consequently, the development of microspores is arrested. In other sterile genotypes, the pH of locule remains high; callase activity is not detected at the end of tetrad formation with the result that the callose wall remains intact until a very late stage. The experiments performed by Izhar and Frankel (1971) indicate that the timing of callase activity is critical for normal development of male gametophyte, and the drop in pH is a prerequisite for callase activity. Certainly, there is a precise time during microsporogenesis when callose dissolution should occur to ensure normal development of pollen.

In a fluorescence microscopic study of the behaviour of callose, along with the activity of β-1,3-glucanase, the enzyme which brings about the degradation of callose shows a sharp temporal correlation between callose deposition and degradation process with glucanase activity (Stieglitz and Stern 1973). The activity of the enzyme is negligible during the first meiotic division during which callose is being synthesized. But, during the second meiotic division, the activity of the enzyme rises and a peak is obtained at the time of release of spores. The fact that callose degradation is not observed in in vitro grown anthers, in the absence of somatic tissue, suggests that the enzyme is secreted from the somatic tissue (Stieglitz and Stern 1973). According to Mepham and Lane (1969a, b), the dissolution of callose is tapetum mediated. In *Tradescantia* they observed that the presence of tapetum is essential for dissolution of the callose special wall.

2.9.2.1 Functions of Callose

The development of callose at pollen mother-cell stage, and its degradation a little after the completion of meiosis suggests that this callose wall layer performs some special function(s). The callose wall not only isolates the sporogenous tissue from the somatic tissue, but it also isolates the individual microspores. Some well-established functions of callosic wall, and their direct role in pollen ontogeny, may be enumerated. It gives mechanical isolation to the developing microspores, thereby preventing cell coherence and, by their rapid and total dissolution, set the microspores free. According to Waterkeyn (1962), it is an important biological event.

The callose layer also functions as a kind of chemical isolation, establishes a selective barrier between genetically different haploid cells, which must pass through their developmental stages unexposed to the influence of their sister spore, or of the adjoining spores and somatic tissue (Heslop-Harrison 1964, 1966a, Heslop-Harrison and Mackenzie 1967). Heslop-Harrison and Mackenzie followed the incorporation of [^{14}C]-labelled thymidine in detached buds, and inflorescences of *Lilium henryi*. The radioactive thymidine moved freely into the anther locules from the archesporial phase until the pachytene. Subsequently, the tracer was excluded from the mother cell until the dissolution of tetrads. On their release the young spores take up the radioactivity readily. The incorporation of thymidine occurs mostly in premeiotic and early leptotene period initially, and at the disappearance of callose walls. This suggests that the callose wall is resistant to the passage of labelled thymidine derivatives. Therefore, these authors suggest that callose acts as a barrier or "molecular filter" to the exchange of at least some macromolecules, and also provides genetic autonomy to each developing sporocyte.

Southworth (1971) demonstrated that in *Gerbera jamesonii* glucose and sodium acetate pass through the callosic wall, while phenylalanine does not. Mascarenhas (1975), however, does not favour this hypothesis because (1) phenylalanine, a molecule smaller than glucose, should have entered. (2) Heslop-Harrison and Mackenzie (1967) studied the uptake of thymidine at periods of time when DNA synthesis did not occur. Possibly, it could be that the sporocytic cell was impermeable to thymidine during such periods, rather than that callose obstructed its movement. (3) Rowley and Dunbar (1970) report the easy passage of colloidal iron in *Populus* and *Salix*. (4) The callose wall does not stop the entry of even tetrazolium salts as shown from a study of succinic dehydrogenase and cytochrome oxidase activity in pollen tetrads of *Paeonia*. Khosla (unpublished) also observed the induced fluorescence in PMC at prophase and pollen tetrads of *Triticale* by FDA, normally used in the viability test of pollen.

It has also been suggested that the callose wall protects the developing sporocytes from the harmful hormonal and nutritional influence of the adjoining somatic cells (Godwin 1968, De Halac and Harte 1975). Barskaya and Balina (1971), while studying the effect of atmospheric drought on microsporogenesis, observed that in Sax beans moderate drought inflicts considerable damage on reproductive cells not protected by callose, but does not cause dehydration in the tetrads enclosed by the callose special wall. They point out that callose protects reproductive cells from dehydration. It has long been known that callose is hygroscopic in nature, and has a pronounced ability to absorb water.

Premature dissolution of callose around the tetrads may contribute to sterility. The published reports show that abortion can occur at almost any time during microsporogenesis, and that probably several phenomena and mechanisms are involved. Izhar and Frankel (1971) provide impressive evidence that faulty timing of callose digestion is a primary factor in microspore mother cell and microspore abortion in cytoplasmic male-sterile and genic male-sterile lines. Warmke and Overman (1972) studied the behaviour of callose in sterile and fertile plants. The behaviour of callose is more or less similar through the early meiotic stages. Later, however, the central callose mass in fertile anthers splits into sectors along the plane of the original microsporocyte walls, and forms a covering that isolates the microsporocytes and young microspores. This callose wall then dissolves to free the microspores from the tetrads. In sterile plants, on the other hand, callose starts separating from the developing microsporocytes, and begins to accumulate in the form of an amorphous mass in the centre of locule. This mass becomes fibrous and diffuse, and disappears during early meiosis. As the callose wall is dissolved early, the microspores do not get physical and physiological isolation. These studies offer a new insight into the role of callose in pollen abortion in male-sterile lines.

In *Triticale* treatment with 0.6% ethyl methyl sulfonate causes male sterility, and in such instances the callose is deposited in the form of granules rather than a layer (Fig. 2.18 E). The tapetum too shows the accumulation of callose, and in such instances the microspore mother cells abort (Bhandari and Khosla, unpublished).

Callose wall also seems to play an effective role in the establishment of the very first pattern of exine (Fig. 2.19 E, F). Various hypotheses have been proposed to explain the role of callose in laying down the pattern of primexine:

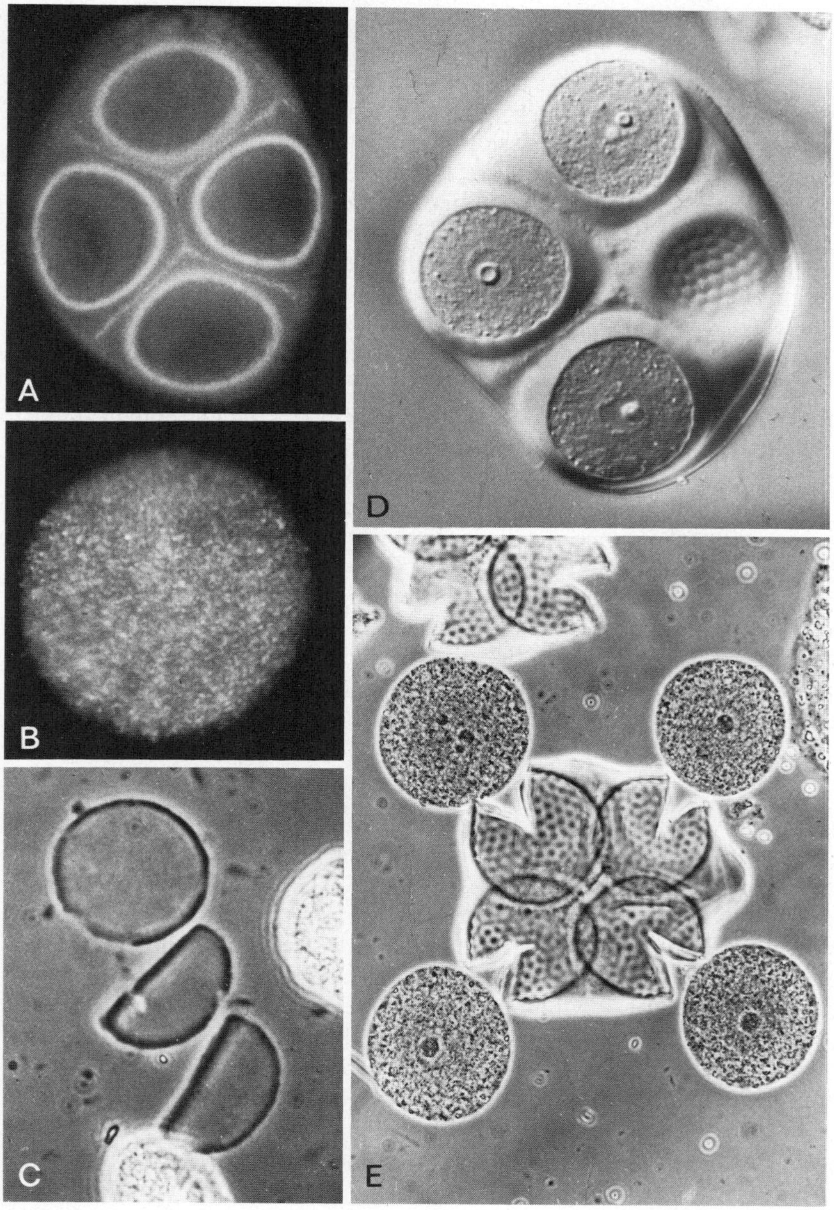

Fig. 2.20 A–E. Special callosic wall at tetrad stage, and at release of microspores. **A** Microspore tetrad of *Ipomoea purpurea* after aniline blue fluorescence; only the special callose wall shows fluorescence. Note bright fluorescence of spore chambers and embedded cell plates. **B** A naturally released heterotypic cell plate of microspore tetrad of *Lilium regale*. **C** Cell plates of microspore tetrad of *Tradescantia virginiana*. The circular cell plate corresponds to the heterotypic, and the semicircular to the homotypic cell plates (phase contrast). **D** Tetrad of microspores of *Lavatera trimestris* still embedded in callose wall. One spore escaped after gentle crushing leaving the empty spore chamber with characteristic inner markings (Nomarski Interference contrast). **E** Squashed and crushed microspore tetrad of *Althaea rosea*. The microspores have escaped leaving behind spore chambers which show characteristic relief (phase contrast; formaldehyde 10% + OSO_4, buffered at pH 7). (**A–E** courtesy L Waterkeyn, Louvain, Belgium)

1. The callose wall supplies carbon compounds, like glucose, for the development of cellulosic primexine which furnishes a basic framework of the future exine (Larson and Lewis 1962). Heslop-Harrison (1968 b) states that the early breakdown products of callose wall are 1,3-linked oligosaccharides, such as laminaribiose and laminaritriose. These oligosaccharides have a very transient existence in the anther fluid, and are soon broken down to glucose. A considerable increase of glucose has been noted in the anther extracts of *Cucurbita,* at the dissolution stage of tetrads (Eschrich 1961).

2. It acts like a template, or mould (Fig. 2.20 A–C) for the future pattern of exine. The inner face of an empty spore chamber shows delicate, negative replica of the primexine (Fig. 2.19 B). According to Waterkeyn and Bienfait (1970) "it is not a perfect negative template, it nevertheless plays an effective role in the definite exine patterning". This hypothesis is supported by the work of Ford (1971) in Epacridaceae; the part of tetrad with scanty callose has thin and irregular exine. Further evidence in support of the above hypothesis is in *Pergularia daemia.* In this plant there is total absence of callose during microsporogenesis, and the exine is thin, fragmentary, and lacks embellishment (Vijayaraghavan and Shukla 1977). Further, callose is formed as a normal and universal constituent of the special wall at the tetrad stage and important developmental stages occur when the callose wall is still present. This indicates that the special wall plays an essential role in the initial steps of pollen development.

2.9.3 Cytokinesis

The four products of meiosis become separated from each other through the process of cytokinesis accompanied by formation of cell plates. Two basic types have been recognized (Maheshwari 1950, Davis 1966): (1) the successive type in which the two cell plates are laid down in centrifugal manner immediately after the first and second meiotic division, and (2) the simultaneous type where the isolation occurs by concurrent centripetal furrows. Periasamy and Swamy (1959) emphasized that the terms should not only represent temporal relationship but two different cytological processes; the former is effected by cell plates, and the latter by furrows. Reviewing much contrary data, Sampson (1963) suggested that cytokinesis in pollen mother cell should be described as either "successive" or "simultaneous", by "cell plates" or "constriction furrows". Sampson (1969) subsequently summarized the literature and recognized three main categories of the process of cytokinesis:

1. Successive centrifugal cell plates at the end of meioses I and II, e.g. *Zea mays.*
2. Simultaneous centripetal constriction furrows at the end of meiosis II, e.g. *Nicotiana.* There are three variants:
   a) A heterotypic furrow appears at the end of meiosis I. A homeotypic furrow, however, initiated at the end of meiosis II reaches the centre and quartet formation occurs in a simultaneous manner, e.g. *Magnolia kobus* (Stoudt 1960).
   b) Heterotypic furrowing completes the division of PMC into dyad before homeotypic furrowing, e.g. *Magnolia liliflora* (Hayashi 1960).

c) There is a partially deposited cell plate (*Zygogynum*), or evanescent cell plate precursor (*Pseudowintera, Magnolia*).

3. Simultaneous centrifugal cell plate formation at the end of meiosis II, e.g. *Helleborus* (Waterkeyn 1962), *Illicium, Schisandra* and *Kadsura* (Hayashi 1960, Kapil and Jalan 1964). Apparent reconstruction of telophase I phragmoplast, and appearance of secondary spindles at the end of meiosis II, e.g. *Laurelia*.

Cytokinesis in PMCs of *Helleborus* (Waterkeyn 1962) involves almost simultaneous occurrence of granulose cell plates across the equators of six spindles. In *Laurelia* (Sampson 1969), however, cytokinesis occurs by formation of four cell plates. Periasamy and Swamy (1959) inferred that the six spindle-like regions, which appear when the spindles of the second meiotic division disappear, are formed only because of fine vacuolation of the cytoplasm in the thinned-out portions between the nuclei.

Longly and Waterkeyn (1979) report the morphology and structure of lamellae which they termed as embedded cell plates. According to them, the number and structure of these cell plates varies in plants with simultaneous or successive mode of cytokinesis. In the former (e.g. *Lavatera* and *Ipomoea*) four, five or six plates are formed and their borders are reinforced. In the successive type there are three plates (Fig. 2.20D, E) and their structure varies with the species (e.g. *Sansevieria, Canna, Lilium*). Longly and Waterkeyn (1979) admit that such differences in structure may be due to (1) velocity of embedding process, (2) direction of growth of the cell, and (3) stage at which the cell plates become embedded. They concluded that in *Lilium* the embedded cell plates present obvious chemical and structural analogy with the cell plate formed during somatic cytokinesis.

In spite of recent ultrastructural studies, the two basic patterns in which cytokinesis occurs in pollen mother cells in angiosperms remain unresolved both at the ontogenetic and phylogenetic level. To decide which of the two basic types is primitive is rather difficult. The presence of cell plate or cell plate precursor has been recorded in many members of Winteraceae such as *Drimys winteri* (Bhandari and Venkataraman 1968) and *Pseudowintera axillaris* (Sampson 1963). Hayashi (1960) observed such structures in *Magnolia liliflora* and Farr (1918) in *M. tripetala*. According to Davis (1966), simultaneous cytokinesis is derived from the successive type.

Sampson (1969, 1970) considers that cytokinesis in *Laurelia* is of an intermediary nature in the evolution of simultaneous from successive type. He argues that the first step in evolution is close to simultaneous formation of cell plates followed by strictly simultaneous centrifugal cell plates coupled with centripetal furrowing and, finally, disappearance of cell plate formation. The presence of temporary cell plates at the close of meioses I and II is definitely suggestive of an intermediate form, but the direction of evolution is by no means certain.

2.9.4 Cellulosic Wall of Microspore Mother Cell

A number of LM and TEM studies indicate that the pectocellulosic wall of the MMC disintegrates before the end of meiotic prophase and concurrent with the

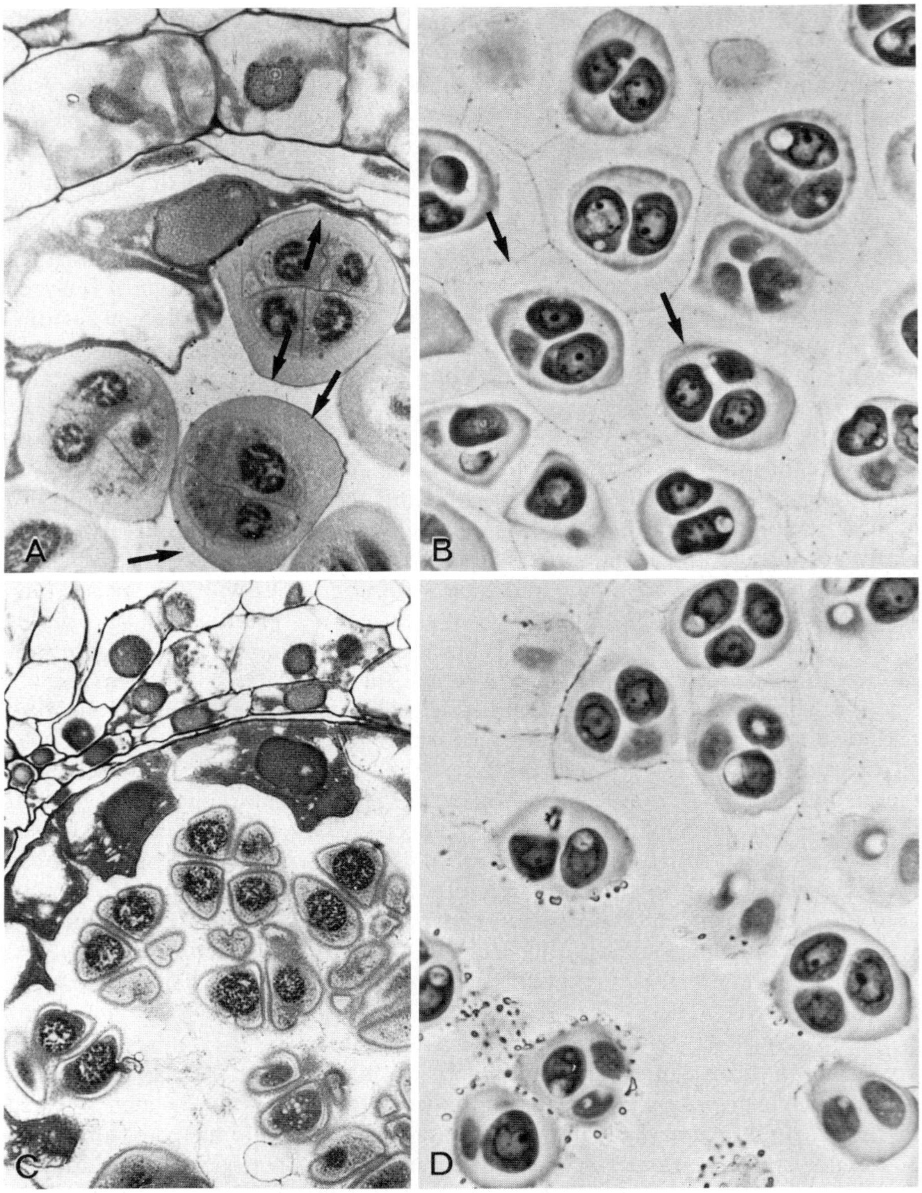

Fig. 2.21. A Portion of transection of anther lobe of *Allium tuberosum*. Note the presence of PAS-positive pecto-cellulosic wall around individual tetrads (*arrows*). **B** Same, in *Cyclamen persicum*, the tenuous wall (*arrow*) still persists; the tetrads of microspores are enclosed by callosis wall (*arrow*). **C** Simultaneous dissolution of both pecto-cellulosic and callosic walls to release the tetrads in *A. tuberosum*. **D** Same, showing flocculation and formation of PAS-positive globules around the callosic wall in *C. persicum*. (**A, B, D** after Bhandari et al. 1981; **C** original)

deposition of callosic wall (see Maheshwari 1950, Heslop-Harrison 1972). In *Helleborus foetidus* (Echlin and Godwin 1968a) and *Olea europaea* (Pacini and Juniper 1979a) it may be observed at places in a highly reduced form. In a recent histochemical study Bhandari et al. (1981) demonstrated the persistence of PAS-positive pectocellulosic wall until tetrad formation around individual PMC in *Allium tuberosum* (Fig. 2.21 A), or in situ in *Cyclamen persicum* (Fig. 2.21 B). In *A. tuberosum* both the callose layer and primary cellulosic wall dissolve simultaneously to release the young microspores (Fig. 2.21 C). In *C. persicum,* on the other hand, at the end of telophase II, the wall becomes tenuous, breaks down on account of flocculation and, as a result thereof, remnants of original wall adhere to the callosic wall in the form of numerous globules (Fig. 2.21 D). The purpose of persistence of the original cellulosic wall in addition to the callosic encasement is not yet understood. Perhaps, it acts as an added barrier augmenting the prevention of entry of macromolecules by callosic wall, as suggested by Heslop-Harrison (1972).

2.10 Some Unusual Features

A peculiar behaviour, the formation of embryo sac-like structures, has been observed in the pollen grains of some species of Liliaceae (Němec 1898). Němec, for the first time, observed this unique feature in *Hyacinthus orientalis,* generally referred to as Němec phenomenon. He observed the formation of eight-nucleate microspores resembling an embryo sac (see also Chap. 1 and Ram 1959). Vijayaraghavan and Ratnaparkhi (1977) report, in *Heuchera micrantha,* the development of such pollen grains along the lines of Polygonum, Allium, and Adoxa type of female gametophyte. Rarely, structures resembling Plumbago and Penaea types were noticed. Several variations occur in the number and position of the nuclei in the so-called pollen embryo sac. Divergent views have been expressed on the origin of these embryo sacs: by divisions of vegetative nucleus (Němec 1898), generative nucleus (De Mol 1923), or microspore nucleus (Stow 1930, Naithani 1937, Geitler 1941, Vijayaraghavan and Ratnaparkhi 1977).

In *Heuchera* Vijayaraghavan and Ratnaparkhi (1977) observed exine all around the pollen embryo sac, while Stow (1930), Naithani (1937), and Geitler (1941) observed exine only at one end. Stow considers the cells at the exine end to be equivalent to the egg and synergids, and the three nuclei at the other end to represent the antipodals, while Geitler offers an entirely reverse interpretation.

From the work of Stow (1930) and Vijayaraghavan and Ratnaparkhi (1977), it may be concluded that the presence or absence of starch does not affect the development of pollen embryo sacs. Persistent healthy tapetum at the microspore stage could provide nutrition to the developing pollen embryo sacs (Hayward and Manthriratna 1972, Vijayaraghavan and Ratnaparkhi 1977).

Another feature worth mentioning is the origin of androgenic haploids from microspores. Pollen dimorphism is instrumental in initiating the formation of androgenic embryos.

2.11 Conclusions

Davis (1966) remarked "Far from the descriptive phase of angiosperm embryology being over, it is hardly begun ..." She pointed out that in 80 of the 411 families no embryological information could be traced. Families in which no substantial description on anther exists exceed more than hundred. On a subsequent perusal of literature I could add 17 families to her list and, therefore, it must be emphasized that further investigations are essential. Early ontogeny of anther, in particular, needs to be studied in detail at the ultrastructural level (see Steer 1977) to understand the cytoplasmic changes which stimulate the differentiation into various anther wall layers. This has assumed more relevance because of the physiological attributes of the middle layer and endothecium as a resource pool during the development of pollen (Reznickova and Willemse 1980). Autoradiographic investigations, such as that of Williams and Heslop-Harrison (1979) comparing the efficiency of glandular and plasmodial tapetum, must be undertaken.

The role of tapetum in pollen abortion is another aspect which requires careful consideration. Whether structural changes in cytoplasm and organelles are the cause or the effect, is not certain. The enlargement of tapetal cells and adjoining tissues strangulating the sporogenous cells is only a physical and visual factor, and too simple an explanation which takes no account of the biochemical changes involved leading to nutritional imbalance. Is it because of the retention of nutrients by tapetum (Nakashima and Hosokawa 1971, 1974, Criswell et al. 1974), or because of qualitative and quantitative differences in enzyme profiles (Chauhan and Singh 1968, Fukasawa 1961, Dmitrieva and Khavzhinskaia 1962)? Frankel et al. (1969) and Izhar and Frankel (1971) have shown a direct correlation of callase activity and early pollen abortion; such studies in respect to other enzymes are desirable. A low content of free amino acid proline in apple, its complete absence in *Zea* and *Sorghum* and, contrarily, quantitatively more aspartic acid and alanine in *Beta* and glycine in *Sorghum vulgare,* have been reported. Izhar and Frankel (1973) suggested that amino acid imbalance caused pollen abortion because increase in aspargine content induced pH changes. Similar studies for other amino acids should prove fruitful.

The role of Ubisch bodies in exine formation is enigmatic. A few studies do indicate that they become bodily embedded in the tectum, but one wonders if they, at least during early development, could also form the reservoir of base material, the sporopollenin, required for amplification of the pattern laid down in the primexine. It could be so if an enzyme system(s) capable of degrading proto- or sporopollenin is met with outside the tapetal plasmalemma in the anther locule. Ahokas (1975) has demonstrated late poral esterase which can hydrolyse protosporopollenin. Efforts will have also to be made to demonstrate the enzyme system which can bring about polymerization of sporopollenin precursors outside cellular confines. Similarly, one would like to know more about the functional aspects of the tapetal and peritapetal wall/membrane which is a special structure formed after the dissolution of tapetal cell walls.

Initiation and control of meiosis is another important process in which our understanding is incomplete. The precise biochemical changes effecting or accompanying mitosis-meiosis shift are yet to be clarified. Anther culture, though it opened up new vistas in androgenic haploids has, unfortunately, not provided any insight into the above-mentioned problems. The synthesis of substance/s, if any, and the site have eluded scientific pursuit. Can tapetum be the tissue triggering such an activity in the sporogenous cells?

Through the investigations of Heslop-Harrison and his school, the cytoplasmic changes during the sporophytic-gametophytic shift are now well understood. However, the role of callose is still speculative. Does all the evidence point to the role of molecular filter or, as pointed out by Mascarenhas (1975), is it still doubtful? Early dissolution of callose has certainly something to do with pollen abortion, but the details of precise mechanism are not fully understood. Waterkeyn and Bienfait (1970) indicated the role of callose interior as a negative template for definite exine patterning. The ultrastructural investigations which could have provided fine details of the laying of primexine, vis-a-vis callose deposition, are not complementary.

References

Ahokas H (1975) Male sterile mutants of barley. 2. Cytochemistry of non-mutant of msglef microspores and pollen. Hereditas 81:33–46

Alam S, Sandal PC (1967) Cyto-histological investigations of pollen abortion in male sterile sudan grass. Crop Sci 7:587–589

Alam S, Sandal PC (1972) Relationships among free amino acids in male-sterile and male-fertile sudan grass, *Sorghum vulgare* var. *sudanense*. Bangladesh J Bot 1:199–204

Albertsen MC, Palmer RG (1979) Comparative light-microscopic and electron-microscopic study of microsporogenesis in male sterile (MS_1) and male fertile soybeans [*Glycine max* (L.) Merr.]. Am J Bot 66:253–265

Arekal GD (1963) Contribution to the embryology of *Chelone glabra*. Phytomorphology 13:376–388

Arens K (1949) Prova de calose por meio da microscopia a luz fluorescente e aplicasôes do metado. Lilloa 18:71

Atkinson AW, Gunning Jr BES, Gunning PCZJ (1972) Sporopollenin in the cell wall of *Chlorella* and other algae: Ultrastructure, chemistry, and incorporation of C^{14} acetate, studied in synchronous cultures. Planta 107:1–32

Bal AK, De DN (1961) Developmental changes in the submicroscopic morphology of cytoplasmic components during microsporogenesis in *Tradescantia*. Dev Biol 3:241–254

Banerjee UC (1967) Ultrastructure of tapetal membrane of grasses. Grana Palynol 7:365–377

Banerjee UC, Barghoorn ES (1971 a) How is pollen grain and spore exine pattern controlled? (Abstr) Am J Bot 58:457

Banerjee UC, Barghoorn ES (1971 b) The tapetal membranes in grasses and Ubisch body control of mature exine pattern. In: Heslop-Harrison J (ed) Pollen: Development and physiology. Butterworth, London, pp 126–127

Barratt DHP, Flavell RB (1975) Alterations in mitochondria associated with cytoplasmic and nuclear genes concerned with male sterility in maize. Theor Appl Genet 45:315–321

Barskaya EI, Balina NV (1971) The role of callose in plant anthers. Fiziol Rast (English Transl) 18:605–610

Bennett MD (1972) Nuclear DNA content and minimum generation time in herbaceous plants. Proc R Soc Lond B Biol Sci 181:109–135

Bennett MD (1977) The time and duration of meiosis. Philos Trans R Soc Lond B Biol Sci 277:201–226

Bennett MD, Smith JB (1979 a) Colchicine-induced paracrystals in the tapetum of wheat anthers. J Cell Sci 38:23–32

Bennett MD, Smith JB (1979b) The effect of colchicine on fibrillar material in wheat meiocytes. J Cell Sci 38:33–47

Bennett MD, Chapman V, Riley R (1971) The duration of meiosis in pollen mother cells of wheat, rye, and *Triticale*. Proc R Soc Lond B Biol Sci 178:259–275

Bennett MD, Rao MK, Smith JB, Bayliss MW (1973) Cell development in the anther, the ovule and the young seed of *Triticum aestivum* L. var. Chinese Spring. Philos Trans R Soc Lond B Biol Sci 266:39–81

Bennett MD, Stern H, Woodward M (1974) Chromatin attachment to the nuclear membrane of wheat pollen mother cells. Nature 252:395–396

Bhandari NN (1968) Studies in the family Ranunculaceae X. Embryology of *Anemone* L. Phytomorphology 18:487–497

Bhandari NN (1971) Embryology of the Magnoliales and comments on their relationships. J Arnold Arbor Harv Univ 52:1–39, 285–304

Bhandari NN, Khosla R (1982) Development and histochemistry of anther in *Triticale* cv. Tri-1. I. Some new aspects in early ontogeny. Phytomorphology 32:18–27

Bhandari NN, Kishori R (1969) Some observations on the ontogeny, histochemistry, and cytology of anther wall in *Nigella damascena*. (Abstr.). In: Johri BM, Kapil RN, Rashid A (eds) Seminar on morphology, anatomy, and embryology of land plants. Univ Delhi, pp 60–62

Bhandari NN, Kishori R (1971) Ubisch granules on tapetal membranes in anthers; rapid selective staining by spirit blue. Stain Technol 46:15–17

Bhandari NN, Kishori R (1973) Development of tapetal membrane and Ubisch-granules in *Nigella damascena* – A histochemical approach. Beitr Biol Pflanz 49:59–72

Bhandari NN, Nanda K (1968) Studies in the Viscaceae. I. Morphology and embryology of the Indian dwarf mistletoe – *Arceuthobium minutissimum*. Phytomorphology 18:435–450

Bhandari NN, Venkataraman R (1968) Embryology of *Drimys winteri*. J Arnold Arbor Harv Univ 49:509–525

Bhandari NN, Kishori R, Natesh S (1976) Ontogeny, cytology, and histochemistry of anther tapetum in relation to pollen development in *Nigella damascena*. Phytomorphology 26:46–59

Bhandari NN, Bhargava M, Geier T (1981) A persisting cellulosic wall of microspore mother cells during microsporogenesis in *Allium tuberosum* Rottl. and *Cyclamen persicum* Mill. Ann Bot 48: 425–431

Biddle JA (1979) Anther and pollen development in garden pea and cultivated lentil. Can J Bot 57:1883–1900

Brooks J, Grant PR, Muir MD, Gizel P van, Shaw G (1971) Sporopollenin. Academic Press, London

Brooks MH (1962) Comparative analyses of some of the free amino acids in anthers of fertile and genetic cytoplasmic male sterile *Sorghum*. Genetics 47:1629–1638

Brunkenner L (1975) Beiträge zur Kenntnis der frühen Mikrosporangienentwicklung der Angiospermen. Sven Bot Tidskr 69:1–27

Buss PA Jr, Lersten NR (1972) Crystals in tapetal cells of the Leguminosae. Bot J Linn Soc 65:81–85

Buss PA Jr, Galen FD, Lersten NR (1969) Pollen and tapetum development in *Desmodium glutinosum* and *D. illinoense* (Papilionoideae, Leguminoseae). Am J Bot 56:1203–1208

Callan HG (1972) Replication of DNA in the chromosomes of eukaryotes. Proc R Soc Lond B Biol Sci 181:19–41

Canright JE (1952) The comparative morphology and relationships of the Magnoliaceae I. Trends of specialization in the stamen. Am J Bot 39:484–497

Carlson EM, Stuart BC (1936) Development of spores and gametophytes in certain New World species of *Salvia*. New Phytol 35:68–91

Carniel K (1952) Das Verhalten der Kerne in Tapetum der Angiospermen mit besonderer Berücksichtigung von Endomitosen und sogenannten Endomitosen. Oesterr Bot Z 110:145–176

Carniel K (1963) Das Antherentapetum. Ein kritischer Überblick. Oesterr Bot Z 110:145–176

Carniel K (1967) Licht- und elektronenmikroskopische Untersuchungen der Ubisch Körperentwicklung in der Gattung *Oxalis*. Oesterr Bot Z 114:490–501

Carraro L, Lombardo G (1976) Tapetal ultrastructural changes during pollen development. II. Studies on *Pelargonium zonale* and *Kalanchoe obtusa*. Caryologia 29:339–344

Chambers TC, Godwin H (1961) The fine structure of the pollen wall of *Tilia platyphyllos*. New Phytol 60:393–399

Chauhan SVS, Singh SP (1966) Pollen abortion in male sterile haploid wheat (Norin) having *Aegilops ovata* L. cytoplasm. Crop Sci 6:532–535
Chauhan SVS, Singh SP (1968) Studies on pollen abortion in *Cucumis melo* L. Agra Univ J Res Sci 17:11–12
Cheah CH, Stone BC (1975) Embryo sac and microsporangium development in *Pandanus* (Pandanaceae). Phytomorphology 25:228–238
Cheng PG, Greyson RL, Welden DB (1979) Comparison of anther development in genic male sterile (ms 10) and in male fertile corn (*Zea mays*) from light microscopy and scanning electron microscopy. Can J Bot 57:578–596
Cheng PC, Greyson RI, Chow YS, Walden DB (1981) The cuticle of corn anther: Ultrastructure and waxy composition. Abstr XIII Intl Bot Congr Sydney, p 46
Cheung M, Sattler R (1967) Early floral development of *Lythrum salicaria*. Can J Bot 45:1609–1621
Christensen JE, Horner HT, Lersten NR (1972) Pollen wall and tapetal orbicular wall development in *Sorghum bicolor*. Am J Bot 59:43–52
Claussen P (1927) Über das Verhalten des Antherentapetums bei einigen Monocotylen und Ranales. Bot Arch 18:1–27
Clutter ME, Sussex IM (1965) Meiosis and sporogenesis in excised fern leaves grown in sterile culture. Bot Gaz 126:72–78
Criswell JG, Hume DJ, Tanner JW (1974) Effect of cytoplasmic male sterility on accumulation and translocation of ^{14}C-labelled assimilates in corn. Crop Sci 14:252–254
Dahl OA, Rowley JR (1965) Pollen of *Degeneria vitiensis*. J Arnold Arbor Harv Univ 46:308–323
D'Amato F (1952) Polyploidy in the differentiation and function of tissues and cells in plants. A critical review of literature. Caryologia 4:311–358
D'Amato F (1954) A brief discussion on "endomitosis". Caryologia 6:341–345
Damon EG (1961) Studies on the occurrence of multiploid sporocytes in three varieties of cytoplasmic male-sterile and the normal fertile variety, resistant wheat and sorghum. Phyton (Argentina) 17:193–203
Darlington CD (1977) Meiosis in perspective. Philos Trans R Soc Lond B Biol Sci 277:185–190
Davis GL (1961) The life history of *Podolepis jaceoides* (Sins) Voss. 1. Microsporogenesis and male gametogenesis. Phytomorphology 11:86–97
Davis GL (1966) Systematic embryology of the angiosperms. Wiley, New York
De Fossard RA (1969) Development and histochemistry of the endothecium in the anthers of in vitro grown *Chenopodium rubrum*. Bot Gaz 130:10–22
De Halac IN, Harte C (1975) Female gametophytic competence in relation to polarisation phenomena during megagametogenesis and development of the embryo sac in the genus *Oenothera*. In: Muscahy DL (ed) Gamete competition in plants and animals. North-Holland Publishing, Amsterdam, pp 43–56
De Mol WE (1923) Duplication of generative nuclei by means of physiological stimuli and its significance. Genetica 5:225–272
De Vries APh, Ie TS (1970) Electron microscopy of anther tissue and pollen of male sterile and fertile wheat *Triticum aestivum* L. Euphytica 19:103–120
Dickinson HG (1970) Ultrastructural aspects of primexine formation in the microspore tetrad of *Lilium longiflorum*. Cytobiologie 1:437–449
Dickinson HG (1971) Nucleo-cytoplasmic interaction following meiosis in the young microspores of *Lilium longiflorum*. Events at the nuclear envelope. Grana 2:107–127
Dickinson HG (1973) The role of plastids in the formation of pollen grain coatings. Cytobios 8:25–40
Dickinson HG (1976) The deposition of acetolysis resistant polymers during the formation of pollen. Pollen Spores 18:321–334
Dickinson HG, Andrews L (1977) The role of membrane-bound cytoplasmic inclusions during gametogenesis in *Lilium longiflorum* Thumb. Planta 134:229–240
Dickinson HG, Bell PR (1972) The role of the tapetum in the formation of sporopollenin containing structures during microsporogenesis in *Pinus banksiana*. Planta 107:205–215
Dickinson HG, Heslop-Harrison J (1968) Common mode of deposition for the sporopollenin of sexine and nexine. Nature 220:926–927
Dickinson HG, Heslop-Harrison J (1970) The ribosome cycle, nucleoli, and cytoplasmic nucleoids in the meiocytes of *Lilium*. Protoplasma 69:181–200

Dickinson HG, Heslop-Harrison J (1971) The mode of growth of the inner layer of the pollen grain exine in *Lilium*. Cytobios 4:233–243
Dickinson HG, Heslop-Harrison J (1977) Ribosomes, membranes, and organelles during meiosis in angiosperms. Philos Trans R Soc Lond B Biol Sci 277:327–342
Dmitrieva AN, Khavzhinskaia (1962) Physiological-biochemical differences between fertile forms of corn and their analogues with male cytoplasmic sterility. Tr Ukr Nauhno-IssLed Inst Rastenievool Sel Genet 7:147–153
Dover GA, Riley R (1973) The effect of spindle inhibitors applied before meiosis on meiotic chromosome pairing. J Cell Sci 12:143–161
Dowding ES (1931) Floral morphology of *Arcuethobium americanum*. Bot Gaz 91:42–54
Dunbar A (1973) Pollen development in the *Eleocharis palustris* group (Cyperaceae). I. Ultrastructure and ontogeny. Bot Not 126:197–254
Eames AJ (1961) Morphology of the angiosperms. McGraw Hill, New York
Echlin P (1971) The role of tapetum during microsporogenesis of anther. In: Heslop-Harrison J (ed) Pollen: Development and physiology. Butterworth, London, pp 41–61
Echlin P, Godwin H (1968 a) The ultrastructure and ontogeny of pollen in *Helleborus foetidus* L. I. The development of the tapetum and Ubisch bodies. J Cell Sci 3:161–174
Echlin P, Godwin H (1968 b) The ultrastructure and ontogeny of pollen of *Helleborus foetidus* L. II. Pollen grain development through the callose special wall stage. J Cell Sci 3:175–186
Eckardt T (1963) Some observations on the morphology and embryology of *Eucommia ulmoides*. J Indian Bot Soc (Maheshwari Comm Vol) 42A:27–34
Erichsen AW, Ross JG (1963 a) Inheritance of colchicine-induced male sterility in *Sorghum*. Crop Sci 3:335–338
Erichsen AW, Ross JG (1963 b) Irregularities at microsporogenesis in colchicine-induced male sterile mutants in *Sorghum vulgare* Pers. Crop Sci 3:481–483
Eschrich W (1961) Untersuchungen über den Ab- und Aufbau der Callose. Z Bot 49:153–218
Eschrich W (1962) Elektronenmikroskopische Untersuchungen an Pollenmutterzellen-Callose. IV. Mitteilung über Callose. Protoplasma 55:419–422
Eschrich W, Currier HB (1964) Identification of callose by its diachrome and fluorochrome reactions. Stain Technol 39:303–307
Eschrich W, Eschrich B (1964) Das Verhalten isolierter Callose gegenüber wäßrigen Lösungen. Ber Dtsch Bot Ges 77:329–331
Farr GH (1918) Cell division by furrowing in *Magnolia*. Am J Bot 5:379–395
Flavell R (1974) A model for the mechanism of cytoplasmic male sterility in plants with special reference to maize. Plant Sci Lett 3:259–263
Flavell RB, Walker GWR (1973) The occurrence and role of DNA synthesis during meiosis in wheat and rye. Exp Cell Res 77:15–24
Ford JH (1971) Ultrastructural and chemical studies of pollen wall development in the Epacridaceae. In: Brooks J, Grant PR, Muir M, Gijzel P van, Shaw G (eds) Sporopollenin. Academic Press, London, pp 686–707
Frankel O (1937) The nucleolar cycle in some species of *Fritillaria*. Cytologia 8:37–47
Frankel R, Izhar S, Nitsen J (1969) The timing of callose activity and cytoplasmic male sterility in *Petunia*. Biochem Genet 3:451–455
Fukasawa H (1954) On the free amino acids in anthers of male sterile wheat and maize. Jpn J Genet 29:135–137
Fukasawa H (1961) The reduction of triphenyl tetrazolium chloride by dehydrogenase in cytoplasmic male-sterile wheat. Nucleus Cytoplasm 2:32–35
Fukasawa H, Mito K, Fujiwara M (1957) Preventive effect of sugars on the pollen degeneration of wheat plant. Bot Mag Tokyo 70:251–257
Geitler L (1941) Embryosäcke aus Pollenkörnern bei *Ornithogalum*. Ber Dtsch Bot Ges 59:419–423
Geitler L (1953) Endomitose und endomitotische Polyploidisierung. Protoplasmatologia VI C. Springer, Wien
Gherardini GL, Healey PL (1969) Dissolution of outer wall of pollen grain during pollination. Nature 224:718–719
Godwin H (1968) The origin of exine. New Phytol 67:667–678
Godwin H, Echlin P, Chapman B (1967) The development of pollen grain wall in *Ipomoea purpurea* (L.). Roth. Rev Palaeobot Palynol 3:181–195

Gorczyriski T (1934) Zytologische Analyse einiger Pollenentwicklungsvorgänge bei der Apfelsorte „Schöner von Boskoop". Acta Soc Bot Pol 11:103–118
Gregory WC (1940) Experimental studies on the cultivation of excised anthers in nutrient solution. Am J Bot 27:687–692
Greilhuber J (1974) Uncommon karyological features in the anther tapetum of some Pedicularieae (Scrophulariaceae). Caryologia 27:169–182
Guillermond A (1920) Sur l'evolution du chondriome pendant la formation des grains de pollen de *Lilium candidum*. C R Acad Sci Paris 170:1003–1006
Guillermond A (1924) Recherches sur l'evolution du chondriome pendant le developement du sac embryonaire et des cellules-méres des grain de pollen dans les Liliacées et sur la significance des formations ergastoplasmiques. Ann Sci Nat Bot Biol Vég 6:1–52
Gupta SC, Nanda K (1972) Occurrence and histochemistry of anther tapetal membrane. Grana Palynol 12:99–104
Håkansson A, Levan A (1942) Nucleolar conditions in *Pisum*.Hereditas 28:436–440
Harvey PH, Levings CS, Wernsman EA (1972) The role of extrachromosomal inheritance in plant breeding. Adv Agron 24:127
Hayashi Y (1960) On the microsporogenesis and pollen morphology in the family Magnoliaceae. Sci Rep Tohuku Univ Fourth Ser (Biol) 27:45–52
Hayward MP, Manthriratna MAPP (1972) Pollen development and variation in the genus *Lolium*. 1. Pollen size and tapetal relationships in male fertiles and male steriles. Z Pflanzenzucht 67:131–144
Heinricher E (1915) Über Bau und Biologie der Blüten von *Arceuthobium oxycedri* (DC) MB Sber. Akad Wiss Wien 124:481–504
Heslop-Harrison J (1959) Growth substances and flower morphogenesis. J Linn Soc Bot 56:269–281
Heslop-Harrison J (1962) Origin of exine. Nature 195:1069–1071
Heslop-Harrison J (1963 a) An ultrastructural study of pollen wall ontogeny in *Silene pendula*. Grana Palynol 4:7–24
Heslop-Harrison J (1963 b) Ultrastructural aspects of differentiation in sporogenous tissue. Symp Soc Exp Biol 17:315–340
Heslop-Harrison J (1964) Cell walls, cell membranes, and protoplasmic connections during meiosis and pollen development. In: Linskens HF (ed) Pollen physiology and fertilization. North-Holland, Amsterdam, pp 39–47
Heslop-Harrison J (1966 a) Cytoplasmic continuities during spore formation in flowering plants. Endeavour 25:65–72
Heslop-Harrison J (1966b) Cytoplasmic connections between angiosperm meiocytes. Ann Bot 30:221–230
Heslop-Harrison J (1968 a) Wall development within the microspore tetrad of *Lilium longiflorum*. Can J Bot 46:1185–1191
Heslop-Harrison J (1968 b) Tapetal origin of pollen coat substances in *Lilium*. New Phytol 67:779–786
Heslop-Harrison J (1969) An acetolysis-resistant membrane investing tapetum and sporogenous tissue in the anthers of certain Compositae. Can J Bot 47:541–542
Heslop-Harrison J (1971) The cytoplasm and its organelles during meiosis. In: Heslop-Harrison J (ed) Pollen: Development and physiology. Butterworth, London, pp 16–31
Heslop-Harrison J (1972) Sexuality in angiosperms. In: Steward FC (ed) Plant physiology, vol 6C. Academic Press, New York, pp 133–289
Heslop-Harrison J, Dickinson HG (1969) Time relationships of sporopollenin synthesis associated with tapetum and microspores in *Lilium*. Planta 84:199–214
Heslop-Harrison J, Mackenzie A (1967) Autoradiography of (2-^{14}C)-thymidine derivative during meiosis and microsporogenesis in *Lilium* anthers. J Cell Sci 2:387–400
Hesse M (1977) Vergleichende Untersuchungen zur Entwicklungsgeschichte und Ultrastruktur von Pollenklebstoffen bei verschiedenen Angiospermen. Mitt Bot Linz 9:237–258
Hesse M (1978 a) Entwicklungsgeschichte und Ultrastruktur von Pollenkitt und Exine bei nahe verwandten entomophilen und anemophilen Angiospermensippen: Ranunculaceae, Hamamelidaceae, Platanaceae und Fagaceae. Plant Syst Evol 130:13–42
Hesse M (1978 b) Vergleichende Untersuchungen zur Entwicklungsgeschichte und Ultrastruktur von Pollenklebstoffen bei verschiedenen Angiospermen. Linz Biol Beitr 9:237–258
Hesse M (1979 a) Entwicklungsgeschichte und Ultrastruktur von Pollenkitt und Exine bei nahe verwandten entomophilen und anemophilen Sippen der Oleaceae, Scrophulariaceae, Plantaginaceae und Asteraceae. Plant Syst Evol 132:107–139

Hesse M (1979b) Ultrastruktur und Verteilung des Pollenkitts in der Insekten-Windblütigen Gattung *Acer* (Aceraceae). Plant Syst Evol 131:277–289

Hesse M (1980) Entwicklungsgeschichte und Ultrastruktur von Pollenkitt und Exine bei nahe verwandten entomophilen und anemophilen Angiospermensippen der Alismataceae, Liliaceae, Juncaceae, Cyperaceae, Poaceae und Araceae. Plant Syst Evol 134:229–267

Hoefert LL (1971) Pollen grain and sperm cell ultrastructure in *Beta*. In: Heslop-Harrison J (ed) Pollen: Development and physiology. Butterworth, London, pp 68–69

Horner HT Jr (1977) A comparative light and electron-microscopic study of microsporogenesis in male-fertile and cytoplasmic male-sterile sunflower (*Helianthus annuus*). Am J Bot 64:745–759

Horner HT Jr, Lersten NR (1971) Microsporogenesis in *Citrus limon* (Rutaceae). Am J Bot 58:72–79

Horner HT Jr, Rogers MA (1974) A comparative light and electron microscopic study of microsporogenesis in male-fertile and cytoplasmic male-sterile pepper (*Capsicum annuum*). Can J Bot 52:435–441

Horokawa S, Chikahiro T, Takeo T (1963) Histochemical studies on the cytoplasmic male-sterility of sugar-beets, with special reference to the hydrogen ion concentration, carbohydrates, and amino acids in anthers. Jpn J Breed 13:111–124

Hotta Y, Stern H (1963) Synthesis of messenger-like ribonucleic acid and protein during meiosis in isolated cells of *Trillium erectum*. J Biophys Biochem Cytol 199:45–58

Hotta Y, Stern H (1971) Analysis of DNA synthesis during meiotic prophase in *Lilium*. J Mol Biol 55:337–355

Izhar S, Frankel R (1971) Mechanism of male-sterility in *Petunia*. I. The relationship between pH, callase activity in the anthers and the breakdown of microsporogenesis. Theor Appl Genet 41:104–108

Izhar S, Frankel R (1973) Mechanism of male-sterility in *Petunia*. II. Free amino acids in male fertile and male sterile anthers during microsporogenesis. Theor Appl Genet 43:13–17

Johri BM, Bhatnagar SP (1955) A contribution to the morphology and life history of *Aristolochia*. Phytomorphology 5:123–137

Johri BM, Bhatnagar SP (1971) Loranthaceae. Botanical Monograph No 8. CSIR, New Delhi

Jonker FP (1971) Summary and general discussion. In: Brooks J, Grant PR, Muir M, Gijzel P van, Shaw G (eds) Sporopollenin. Academic Press, London, pp 686–707

Joppa LR, McNeal FH, Walsh JR (1966) Pollen and anther development in cytoplasmic male sterile wheat *Triticum aestivum*. Crop Sci 6:296–297

Kapil RN, Bhandari NN (1964) Morphology and embryology of *Magnolia* Dill. ex Linn. Proc Natl Inst Sci India Part B Biol Sci 30:245–262

Kapil RN, Jalan S (1964) *Schisandra:* Its embryology and systematic position. Bot Not 1964:285–306

Kapil RN, Tiwari SC (1978) Plant embryological investigations and fluorescence microscopy: An assessment of integration. Intl Rev Cytol 53:291–331

Kapoor T, Vijayaraghavan MR, Parulekar NK (1978) Ontogeny, structure, and differentiation of anther tapetum in *Celsia coromandeliana*. Phyton (Argentina) 18:209–216

Karnath KA, Bhat PK, Arekal DA (1979) Tapetum-like anther epidermis in *Zeuxine longilabris* (Lindl.) Benth. ex HK. (Orchidaceae). Curr Sci 48:542–553

Kessel RG (1968) Annulate lamellae. J Ultrastr Res Suppl 10:1–82

Khoo U, Stinson HT (1957) Free amino acid differences between cytoplasmic male sterile and normal fertile anthers. Proc Natl Acad Sci USA 43:603–607

Khosla R (1982) Developmental and histochemical studies on anther, ovule and endosperm in the control and EMS-treated plants of triticale. Ph D Thesis, Univ Delhi

Knoll F (1930) Über Pollenkitt und Bestäubungsart. Z Bot 23:610–675

Knoll F (1936) Eine Streuvorrichtung zur Untersuchung der Pollenverkittung. Oesterr Bot Z 85:161–182

Knox RB, Dickinson HG, Heslop-Harrison J (1970) Cytochemical observations on changes in RNA content and acid phosphatase activity during the meiotic prophase in the anther of *Cosmos bipinnatus* Cav. Acta Bot Neerl 19:1–6

Korobova SN (1971) On the development of the specific cells during micro- and macrosporangium formation in *Zea mays*. Ann Univ ARERS 9:58–61

Kosmath L (1927) Studies über das Antherentapetum. Oesterr Bot Z 76:235–241

Krjatchenko D (1925) De l'activité des chondriosomes pendant le développement des grains de pollen et des cellules nourricières du pollen dans *Lilium croceum* Chaix. Rev Gen Bot 37:193–211

Lacour LF, Wells B (1974) Nuclear pores at prophase of meiosis in plants. Philos Trans R Soc Lond B Biol Sci 268:95–100
Lacour LF, Wells B (1977) Some morphological aspects of the synaptonemal complex in higher plants. Philos Trans R Soc Lond B Biol Sci 277:259–266
Larson A, Lewis CW (1962) Pollen wall development in *Parkinsonia aculeata*. Grana Palynol 3:21–27
Laser KD (1972) A light and electron microscope study of the stamen vascular bundle in cytoplasmic male-sterile *Sorghum bicolor* (Abstr). Am J Bot 59:653
Laser KD, Lersten NR (1972) Anatomy and cytology of microsporogenesis in cytoplasmic male sterile angiosperms. Bot Rev 38:425–454
Latter J (1926) The pollen development of *Lathyrus odoratus*. Ann Bot 40:277–313
Lin M (1955) Chromosome control of nuclear composition. Chromosoma 7:340–370
Lindemann R (1956) Vergleichende cytologische Untersuchungen an den Liliaceen *Bellevalia romana, Agapanthus umbellatus* und *Lilium regale*. Planta 48:2–28
Linskens HF (1967) Isolation of ribosomes from pollen. Planta 73:194–200
Lombardo G, Carraro L (1976a) Tapetal ultrastructural changes during pollen development I. Studies on *Antirrhinum majus*. Caryologia 29:113–125
Lombardo G, Carraro L (1976b) Tapetal ultrastructural changes during pollen development III. Studies on *Gentiana acaulis*. Caryologia 29:345–349
Longly B, Waterkeyn L (1979) Étude de la cytokinèse. III. Les cloisonnements simultanés et successifs des microsporocytes. Cellule 73:66–80
Luck BT, Jordan EG (1977) The nucleolus and meiosis during microsporogenesis in *Endymion non-scriptus* (L.). J Cell Sci 25:111–123
Mackenzie A, Heslop-Harrison J, Dickinson HG (1967) Elimination of ribosomes during meiotic prophase. Nature 215:997–999
Madjd A, Roland-Heydacker F (1978) Sécrétions et dégénérescence des cellules du tapis dans l'anthere du *Soja hispida* Moench, Papilionaceae. Grana 17:167–174
Maheshwari P (1950) An introduction to the embryology of angiosperms. McGraw-Hill, New York
Maheshwari P (1963) Recent advances in the embryology of angiosperms. Intl Soc Plant Morphologists, Univ Delhi
Maheswari Devi H (1975) Embryology of jasminums and its bearing on the composition of Oleaceae. Acta Bot Indica 3:52–61
Maheswari Devi H, Lakshminarayana K (1977a) Embryology of *Oxystelma esculentum*. Phytomorphology 27:59–67
Maheswari Devi H, Lakshminarayana K (1977b) Embryological studies in Gentianaceae. J Indian Bot Soc 56:182–188
Maruyama L (1968) Electron microscopic observations of plastids and mitochondria during pollen development in *Tradescantia palludosa*. Cytologia 33:482–497
Mascarenhas JP (1975) The biochemistry of angiosperm pollen development. Bot Rev 41:259–314
Mascré M (1919) Sur le rôle de l'assise nourricière du pollen. C R Acad Sci Paris 168:1120–1122
Mascré M (1922) Sur l' étamine des Boraginées. C R Acad Sci Paris 175:987–988
Mathur SL (1968) Development of the male and female gametophytes of *Calycanthus fertilis* Walt. Proc Natl Inst Sci India Part B Biol Sci 34:323–329
Mechelke F (1952) Die Entstehung der polyploiden Zellkerne des Antherentapetums bei *Antirrhinum majus* L. Chromosoma 5:246–295
Mepham RH, Lane GR (1969a) Role of the tapetum in the development of *Tradescantia* pollen. Nature 221:282–284
Mepham RH, Lane GR (1969b) Formation and development of the tapetal periplasmodium in *Tradescantia bracteata*. Protoplasma 68:175–192
Mepham RH, Lane GR (1970) Observations on the fine structure and developing microspores of *Tradescantia bracteata*. Protoplasma 70:1–20
Moens PB (1968) The structure and function of the synaptinemal complex in *Lilium longiflorum* sporocytes. Chromosoma 23:418–451
Moseley MF Jr (1958) Morphological studies in the Nympheaceae. I. The nature of stamens. Phytomorphology 8:1–29
Naithani SP (1937) Chromosome studies in *Hyacinthus orientalis* L. III. Reversal of sexual state in the anthers of *H. orientalis* L. var. "Yellow hammer". Ann Bot 1:369–377

Nakashima H (1975) Histochemical studies on the cytoplasmic male sterility in some crops. IV. Electron microscopic observations in sugar-beet anther. Mem Fac Agric Hôkkaido Univ Ser B 9:247–252

Nakashima H, Hosokawa S (1970) Histochemical studies on the cytoplasmic male sterility of some crops. I. Changes of carbohydrates and amino acids in anthers (in Japanese). Mem Fac Agric Hôkkaido Univ Ser B 7:201–208

Nakashima H, Hosokawa S (1971) Histochemical studies on the cytoplasmic male-sterility of some crops. II. Changes of DNA, polysaccharides and proteins, and translocation of photosynthetic products of sugar-beets. Mem Fac Agric Hôkkaido Univ Ser B 8:1–4

Nakashima H, Hosokawa S (1974) Histochemical studies in the cytoplasmic male sterility of some crops. III. Distribution of ^{14}C assimilates in maize (in Japanese). Proc Crop Sci Soc Jpn 43:1–7

Nanda K, Gupta SC (1974) Malfunctioning tapetum and callose wall behaviour in *Allium cepa* microsporangia. Beitr Biol Pflanz 50:465–472

Němec B (1898) Über den Pollen der petaloiden Antheren von *Hyacinthus orientalis* L. Bull Intl Acad Sci Bohème 5:17–23

Novak F (1971) Cytoplasmatische männliche Sterilität bei Weidelgras (*Lolium* sp.). Z Pflanzenzucht 65:206–220

Novak F, Betlach J (1970) Development and karyology of the tapetal layer and anther in sweet pepper (*Capsicum annuum* L.). Biol Plant (Prague) 12:275–280

Ogorodnikova VF (1976) The ultrastructure of tapetal cells in the anthers of some cultivated plants (in Russian). Tr Prikl Bot Genet Sel 58:124–129

Ogura H (1968) Studies on the new male-sterility in Japanese radish, with special reference to the utilization of this sterility towards the practical raising of hybrid seeds. Mem Fac Agric Kagoshima Univ 6:39–78

Oksala T, Therman E (1977) Endomitosis in tapetal cells of *Eremurus* (Liliaceae). Am J Bot 64:866–872

Pacini E, Cresti M (1976) Close association between plastids and endoplasmic reticulum cisterns during pollen grain development in *Lycopersicum peruvianum*. Ultrastruct Res 57:260–265

Pacini E, Cresti M (1978) Ultrastructural characteristics of tapetum and microspore mother cells in *Lycopersicum peruvianum* during meiotic prophase. Soc Bot Fr Actualites Botaniques 125:121–128

Pacini E, Juniper BE (1979a) The ultrastructure of the pollen grain development in the olive (*Olea europaea*). I. Proteins in the pore. New Phytol 83:157–163

Pacini E, Juniper BE (1979b) The ultrastructure of pollen grain development in the olive (*Olea europaea*). II. Secretion by the tapetal cells. New Phytol 83:165–174

Parchman LG, Stern H (1969) The inhibition of protein synthesis in meiotic cells and its effect on chromosome behaviour. Chromosoma 26:298–311

Pereira ASR, Linskens HF (1963) The influence of glutathione and glutathione antagonists on meiosis in excised anthers of *Lilium henryi*. Acta Bot Neerl 12:302–314

Periasamy K, Parameswaran N (1965) A contribution to the floral morphology and embryology of *Tarenna asiatica*. Beitr Biol Pflanz 41:123–138

Periasamy K, Swamy BGL (1959) Studies in the Annonaceae. I. Microsporogenesis in *Cananga odorata* and *Miliusa wightiana*. Phytomorphology 9:251–263

Periasamy K, Swamy BGL (1966) Morphology of the anther tapetum in angiosperms. Curr Sci 35:427–430

Peterson PA, Flavell RB, Barratt DHP (1975) Altered mitochondrial membrane activities associated with cytoplasmically inherited disease sensitivity in maize. Theor Appl Genet 45:309–314

Pisek A (1924) Antherenentwicklung und meiotische Teilung bei der Wacheldermistel [*Arceuthobium oxycedri* (DC.)Bieb.], Antherenbau und Chromosomenzahlen von *Loranthus europaeus* Jacq. Sitzber Math Nat Kl Akad Wiss Wien 133:1–15

Poddubnaya-Arnoldi VA (ed) (1976) Cytoembryology of angiosperms. Muskva (in Russian)

Pohl F (1930) Kittstoffreste auf der Pollenoberfläche windblütiger Pflanzen. Untersuchungen zur Morphologie und Biologie des Pollens II. Beih Bot Zbl Abt A 46:286–305

Pritchard AJ, Hutton EM (1972) Anther and pollen development in male sterile *Phaseolus atropurpureus*. J Hered 63:280–282

Py G (1932) Recherches cytologiques sur l'assise nourricière des microspores et les microspores des plants vasculaires. Rev Gen Bot 44:316–413, 450–462, 484–512

Raj AY (1969) Histological studies in male-sterile and male-fertrile *Sorghum*. Indian J Genet Plant Breed 28:335–341

Ram M (1959 a) Occurrence of embryo sac-like structures in the microsporangia of *Leptomeria billardieri*. Nature 184:912–915
Ram M (1959 b) Morphological and embryological studies in the family Santalaceae. III. *Leptomeria*. Phytomorphology 9:20–33
Ramussen SW (1976) The synaptonemal complex in *Bombax mori* females. Philos Trans R Soc Lond B Biol Sci 277:343–350
Reddy GB, Reddi MV (1974) Cytohistological studies on certain male sterile lines of pearl millet (*Pennisetum typhoides* S. and H.). Cytologia 39:585–589
Reznickova SA, Willemse MTM (1980) Formation of pollen in the anther of *Lilium*. II. The function of the surrounding tissue in the formation of pollen and pollen wall. Acta Bot Neerl 29:141–156
Reznickova SA, Aelst AC van, Willemse MTM (1980) Investigation of exine and orbicule formation in the *Lilium* anther by scanning electron microscopy. Acta Bot Neerl 29:157–164
Rick CM (1945) Genetics and development of nine male-sterile tomato mutants. Hilgardia 18:599–633
Riley R, Flavell RB (1977) A first view of the meiotic process. Philos Trans R Soc Lond B Biol Sci 277:191–196
Riseuño MC, Gimenez-Martin G, Lopez-Saez JF, Garcia MIR (1969) Origin and development of sporopollenin bodies. Protoplasma 67:361–374
Rohrbach U (1965) Beiträge zum Problem der Pollensterilität bei *Beta vulgaris* L. I. Untersuchung über die Ontogenese des Phanotyps. Z Pflanzenzucht 53:105–124
Roland-Heydacker F (1980) Aspects ultrastructuraux de l'ontogénie du pollen et du tapis chez *Mahonia aquifolium* Nutt. Berberidaceae. Pollen Spores 21:259–278
Rosanoff S (1865) Zur Kenntnis des Baues und der Entwicklungsgeschichte des Pollens der Mimoseae. Jahrb Wiss Bot 4:441–450
Rowley JR (1959) The fine structure of the pollen wall in the Commelinaceae. Grana Palynol 2:3–31
Rowley JR (1962) Nonhomogeneous sporopollenin in microspores of *Poa annua*. Grana Palynol 3:3–19
Rowley JR (1963) Ubisch body development in *Poa annua*. Grana Palynol 4:25–36
Rowley JR (1964) Formation of the pore in pollen of *Poa annua*. In: Linskens HF (ed) Pollen physiology and fertilization. North-Holland, Amsterdam, pp 56–69
Rowley JR, Dunbar A (1970) Transfer of colloidal iron from sporophyte to gametophyte. Pollen Spores 12:305–328
Rowley JR, Prijanto B (1977) Selective destruction of the exine of pollen grains. Geophytology 7:1–23
Rowley JR, Skvarla JJ (1974) Plasma membrane glycocalyx origin of Ubisch body wall. Pollen Spores 16:441–448
Rowley JR, Skvarla JJ (1975) The glycocalyx and initiation of exine spinules on microspores of *Canna*. Am J Bot 62:479–485
Rowley JR, Muhlethaler K, Frey-Wyssling A (1959) A route for the transfer of materials through the pollen wall. J Biophys Biochem Cytol 6:537–539
Sampson FB (1963) The floral morphology of *Pseudowintera*, the New Zealand member of the vessel-less Winteraceae. Phytomorphology 13:403–423
Sampson FB (1969) Cytokinesis in pollen mother cells of angiosperms, with emphasis on *Laurelia novae-zelandiae* (Monimiaceae). Cytologia 34:511–634
Sampson FB (1970) Unusual features of cytokinesis in meiosis of pollen mother cells of *Pseudowintera traversii* (Buchen) Dandy (Winteraceae). Beitr Biol Pflanz 47:71–77
Sattler R (1973) Organogenesis of flowers. Univ Toronto
Scheer U, Franke WW (1972) Annulate lamellae in plant cells: Formation during microsporogenesis and pollen development in *Canna generalis* Bailey. Planta 107:131
Schnarf K (1923) Kleine Beiträge zur Entwicklungsgeschichte der Angiospermen. 4. Über Verhalten des Antherentapetums einiger Pflanzen. Oesterr Bot Z 1923:242–245
Schnarf K (1929) Embryologie der Angiospermen. Bornträger, Berlin
Schnarf K (1931) Vergleichende Embryologie der Angiospermen. Bornträger, Berlin
Shaw G (1971) The chemistry of sporopollenin. In: Brooks J, Grant PR, Muir M, Gijzel P van, Shaw G (eds) Sporopollenin. Academic Press, London, pp 305–350
Sheffield E, Cawood AH, Bell PR, Dickinson HG (1979) The development of nuclear vacuoles during meiosis in plants. Planta 146:597–601
Sheridan WF, Stern H (1967) Histones of meiosis. Expl Cell Res 45:323–335
Singh IS (1975) Induced pollen sterility in *Petunia*. Mode of inheritance and tapetal behaviour. Ann Amelior Plant (Paris) 25:303–319

Singh SP, Rhondes AM (1961) A morphological and cytological study of male sterility in *Cucurbita maxima*. Proc Am Soc Hortic Sci 78:375–378

Skvarla JJ, Kelly AG (1971) Dictyosome development during microsporogenesis in *Canna generalis*. In: Heslop-Harrison J (ed) Pollen: Development and physiology. Butterworth, London, pp 62–67

Skvarla JJ, Larson JA (1966) Fine structural studies of *Zea mays* pollen. I. Cell membranes and exine ontogeny. Am J Bot 53:1112–1125

Southworth D (1971) Incorporation of radioactive precursors into developing pollen walls. In: Heslop-Harrison J (ed) Pollen: Development and physiology. Butterworth, London, pp 115–120

Städtler G (1923) Über Reduktionserscheinungen im Bau der Antherenwand von Angiospermen. Flora 16:85–108

Steer MW (1977) Differentiation of the tapetum in *Avena*. The cell surface. J Cell Sci 25:125–138

Stern H, Hotta Y (1967) Chromosome behaviour during development of meiotic tissue. In: Goldstein L (ed) The control of nuclear activity. Prentice Hall, Englewood Cliffs, New York, pp 47–76

Stern H, Hotta Y (1969) DNA synthesis in relation to chromosome pairing and chiasma formation. Genetics 61:27–39

Stern H, Hotta Y (1973) Biochemical contents of meiosis. Annu Rev Genet 7:37–66

Stern H, Hotta Y (1977) Biochemistry of meiosis. Philos Trans R Soc Lond B Biol Sci 277:277–294

Stieglitz H, Stern H (1973) Regulation of β-1,3 glucanase activity in developing anthers of *Lilium* microsporocytes. Dev Biol 34:169–173

Stoudt HM (1960) Sporogenesis in *Magnolia kobus*. I. Microsporogenesis. Proc PA Acad Sci 34:29–33

Stow I (1930) Experimental studies on the formation of embryo sac-like giant pollen grain in the anthers of *Hyacinthus orientalis*. Cytologia 1:417–439

Sussex IM, Steeves TA (1958) Experiments on the control of fertility of fern leaves in sterile culture. Bot Gaz 119:203–208

Swamy BGL, Krishnamurthy KV (1980) From flower to fruit. Tata McGraw-Hill, New Delhi

Troll W (1928) Über Antherenbau, Pollen und Pollination von *Galatanthus nivalis*. Flora 23:321–343

Tscherrmak-Woess E (1971) Endomitose. In: Altman HW (Hrsg) Der Zellkern. Handb Allgem Pathol 2:569–625

Tupý J (1963) Free amino acids in apple pollen from the point of view of its fertility. Biol Plant (Prague) 5:154–160

Turala K (1958) Endomitosis in the tapetal cells of *Cucurbita pepo* L. Acta Biol Cracov Ser Viol 1:25–34

Turbin NV, Palilova AM, Alrashenok MV, Lyuikina (1974) Fine structure of cytoplasm of anthers of sterile wheat. Dokl Bot Sci 214:10–12

Ubisch GV (1927) Zur Entwicklungsgeschichte der Antheren. Planta 3:490–495

Ünal M (1978) Cytological and embryological studies on *Anemone pavonina* Lam. Istanbul Univ Fen Fak Med 43B:185–208

Vasil IK (1959) Cultivation of excised anthers in vitro. Effect of nucleic acids. J Expl Bot 10:399–408

Vasil IK (1967) Physiology and cytology of anther development. Biol Rev 42:327–373

Vasil IK (1973a) Fine structure of pollen ontogeny in seed plants. (Abstr). Am J Bot 59:657

Vasil IK (1973b) The new biology of pollen. Naturwissenschaften 60:247–253

Vijayaraghavan MR, Ratnaparkhi S (1973) Dual origin and dimorphism of the anther tapetum in *Alectra thomsoni* Hook. Ann Bot 37:355–359

Vijayaraghavan MR, Ratnaparkhi S (1977) Pollen embryo sacs in *Heuchera micrantha* Dougl. Caryologia 30:105–119

Vijayaraghavan MR, Shukla AK (1977) Absence of callose around the microspore tetrad and poorly developed exine in *Pergularia daemia*. Ann Bot 41:923–926

Virnich H (1967) Untersuchungen über das Verhalten der männlichen Sterilität und anderer Eigenschaften bei polyploiden Zwiebeln (*Allium cepa* L.) als Grundlage für eine Nutzung in der Hybrid-Züchtung. Z Pflanzenzucht 58:205–244

Warmke HE, Lee SLJ (1977) Mitochondrial degeneration in Teses cytoplasm male-sterile corn anthers. J Hered 68:213–222

Warmke HE, Lee SLJ (1978) Pollen abortion in T-cytoplasmic male-sterile corn (*Zea mays*): A suggested mechanism. Science 200:561–563

Warmke HE, Overman MA (1972a) Cytoplasmic male sterility in *Sorghum*. I. Callose behaviour in fertile and sterile anthers. J Hered 63:103–108

Warmke HE, Overman MA (1972b) Tapetal behaviour in fertile and sterile anthers. J Hered 63:227–233

Waterkeyn L (1962) Les parois microcytaires de nature callosique chez *Helleborus* et *Tradescantia*. Cellule 62:225–255
Waterkeyn L (1964) Callose microsporocytaire et callose pollinique. In: Linskens HF (ed) Pollen physiology and fertilization. North-Holland, Amsterdam, pp 52–58
Waterkeyn L, Bienfait A (1970) On a possible function of the callosic special wall in *Ipomoea purpurea* (L.) Roth. Grana 10:13–20
Waterkeyn L, Bienfait A (1971) A function of the sporocyte special wall. (Abstr). In: Heslop-Harrison J (ed) Pollen: Development and physiology. Butterworth, London, p 125
Whelan EDP (1974) Discontinuities in the callose wall intermeiocyte connections, and cytomixis in angiosperm meiocytes. Can J Bot 52:1219–1224
Whelan EDP, Haggis GH, Ford EJ (1974) Scanning electron microscopy of callose wall and intermeiocyte connections in angiosperms. Can J Bot 52:1215–1218
Willemse MTM, Reznickova SA (1980) Formation of pollen in the anther of *Lilium*. I. Development of the pollen wall. Acta Bot Neerl 29:127–140
Williams EG, Heslop-Harrison J (1979) A comparison of RNA synthetic activity in the plasmodial and secretory types of tapetum during the meiotic interval. Phytomorphology 29:370–380
Witkus ER (1945) Endomitotic tapetal cell divisions in *Spinacia*. Am J Bot 32:326–330
Wodehouse RP (1935) Pollen grains. McGraw-Hill, New York
Yamazaki T, Takeoka M (1962) Electron microscope investigations of the fine details of the pollen surface in Japanese gymnosperms. Grana Palynol 3:3–12
Yoffe MD (1965) On the embryology of *Trochodendron aralioides*. Embryo and endosperm development. In: Yakovlev MS (ed) Flower morphology and reproductive processes in angiosperms. Nauka, Leningrad, pp 177–185
Young BA, Schulzschaeffer J, Carol TW (1979) Anther and pollen development in male-sterile intermediate wheat grass plants derived from wheat × wheat grass hybrids. Can J Bot 57:602–618

3 The Ovule

F. Bouman

3.1 Historical

Of the first classical microscopists, Grew (1671) and Malpighi (1675), the latter was so impressed by the analogous processes of seed maturation and fetal development that he applied the terms "amnion" and "chorion" to the envelopes of the ovule. Grew and Malpighi attempted an ontogenetic approach, and studied ovules and developing and mature seeds. However, the resolving power of their microscopes was not sufficient to distinguish the boundaries between the nucellus and integuments. Their descriptions of the ovule and of seed coats are mainly stratigraphical, i.e., they emphasized the most conspicuous layers by differences in cell shape, cell wall thickness and consistency. This stratigraphical approach persisted till the first decade of the nineteenth century, and resulted in a profuse application of alternative terms, causing a terminological chaos (Bouman 1974, 1978 a). Grew described a designedly formed, small foramen (micropyle) in the outer coat of young leguminous seeds. He compared this foramen with the bung-hole of beer and wine barrels and assumed that it functions as a passage for "aiery particles to exite the fermentation". During germination the foramen would, in this train of thought, promote the eruption of the radicle. After an analysis of the seeds of some 1,200 species, Turpin (1806) concluded that the "micropyle" is of universal occurrence and forms an essential part of all mono- and dicotyledonous seeds. He rejected Grew's ideas and attributed to the micropyle a function associated with fertilization.

The generally held view, expressed by Gaertner (1788) and Mirbel (1815), that the unfertilized ovule is a homogeneous, parenchymatous structure in which only an epidermis could be distinguished, and which exhibits a clear stratification only on account of the "special forces of fertilization", was opposed by Treviranus (1815) who demonstrated that in early ontogenetic stages the outer ovular coats can already be discerned. Brown (1826) provided the first correct description of the "unpregnated" angiospermous ovule. He distinguished two concentric membranes enclosing a nucleus (nucellus) of a pulpy structure, an inner coat or membrana interna (= inner integument) unconnected with the outer one (membrana externa) or with the nucellus except at its origin. He reported that both coats have an opening corresponding with the apex of the nucellus which is never perforated. From the study of ovules of Nymphaeaceae, Piperaceae, and Scitamineae, Brown concluded that the storage tissue could be formed either inside the embryo sac, or from the cells of the nucellus. His findings were confirmed and augmented by Brongniart (1827) and Mirbel (1829, 1830). The former described unitegmic ovules in Asclepiadaceae, Scrophulariaceae, and *Peperomia,* and ategmic ovules in *Thesium*. Mirbel introduced the terms nucellus (nucelle) for the central unper-

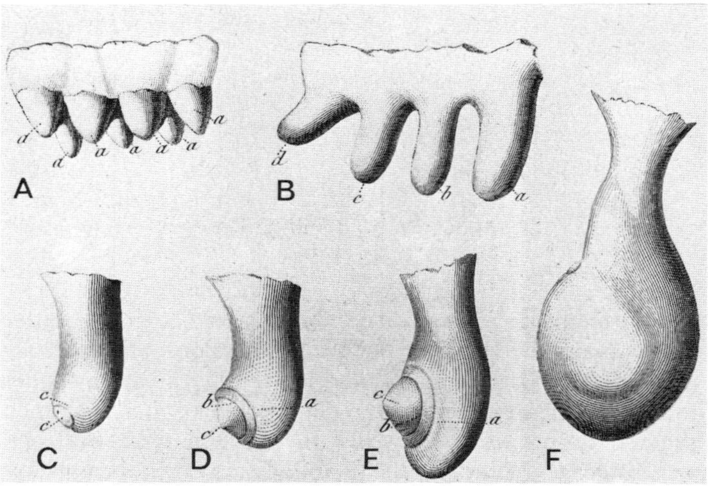

Fig. 3.1 A–F. *Cucumis anguria*, first published detailed figure of ovule ontogeny. **A, B** Ovular primordia. **C, D** Initiation of the outer and inner integument. **E** Later stage. **F** Mature ovule with closed exostome. (From Mirbel 1829)

forated core of the ovule, and exo- and endostome for the openings in the outer and inner integument, respectively (Fig. 3.1 A–F). He also made a more extensive survey of the types of ovules distinguished by Brongniart, and recognized orthotropous, campylotropous, anatropous, and amphitropous ovules. Fritsche (1835) described correctly that, as a rule, the outer integument is initiated before the inner one, and observed the integumentary primordia as lateral outgrowths which gradually envelop the nucellus and close almost completely at the apex. Schleiden (1837, 1839) made a detailed survey of ovule and seed development and introduced the terms integumentum simplex, integumentum externum, and integumentum internum (single, outer, and inner integument, respectively). He also recognized the hemianatropous condition as an intermediate type between atropy and anatropy. Hofmeister (1849), for the first time, gave a detailed account of the initiation of ovule and integuments in Orchidaceae. Hofmeister erroneously concluded that the entire ovule was derived from a single epidermal cell of the placenta. Hanstein's histogen theory provided the foundation for histogenetic studies of the ovule. Hanstein's pupil Schmitz (1872) showed that the ovule primordium in Piperaceae is built up from three components: dermatogen, periblem, and plerome. By studying ovular initiation in Orchidaceae and some other angiosperms, Strasburger (1872, 1879) disproved Hofmeister's incorrect conclusions and demonstrated that the ovules are of subdermal derivation, even in orchids. Warming (1878) confirmed that the angiospermous ovule is always initiated subdermally. He distinguished three types of initiation, viz., (1) in the cell layer immediately below the epidermis, (2) in the third cell layer, and (3) by simultaneous cell division in the second and third layers. He also generalized that the inner integument is exclusively epidermal and originally only two cell layers thick, whereas the outer integument was supposed to be sometimes of dermal, more often of subdermal, origin. In unitegmic ovules the single integument is of dermal origin, sometimes subdermal. Van Tieghem (1898) introduced the terms "tenuinucellate" (with a thin nucellus) and "crassinucellate" (with a bulky and more persistent nucellus).

Inspired by the acquired insights in animal reproduction, the process of megasporogenesis in ovules became better understood at the end of the nineteenth century. Strasburger (1888) and Guignard (1889) independently showed that a reduction of the number of chromosomes takes place in connection with the development of the reproductive cells. The recognition of megaspore tetrads was first hampered by the study of species with tetrasporic embryo sacs, but became gradually clear (Strasburger 1894).

3.2 Ovular Morphology

The ovule is the forerunner of the seed. Therefore, studies of the ovule are of vital importance for the understanding of the structure and functions of the seed. The ovule consists of nucellus, integument(s), chalaza, raphe, and funicle (Fig. 3.2).

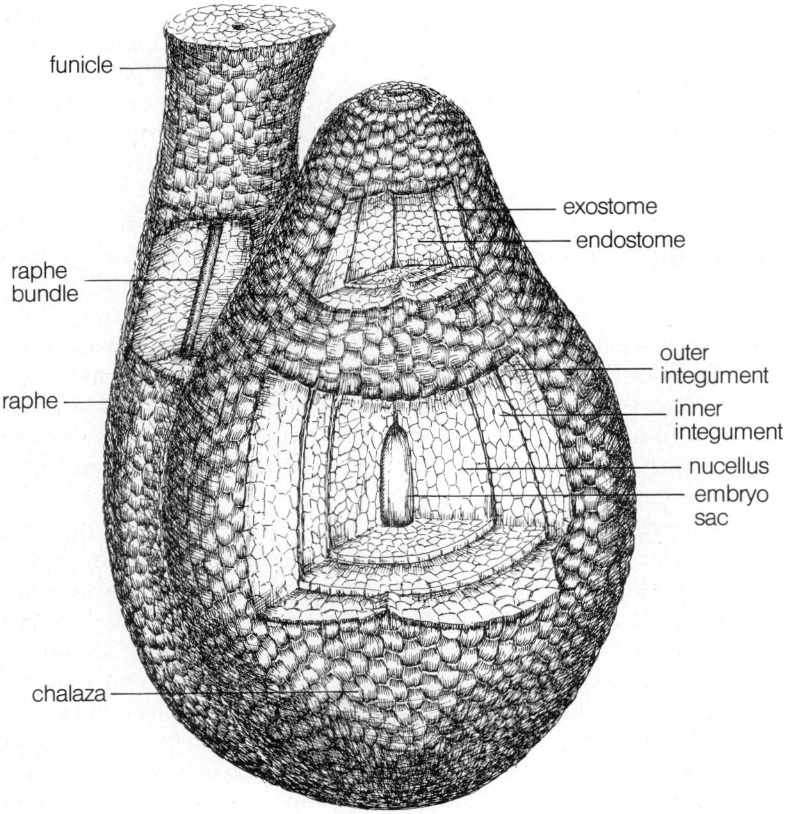

Fig. 3.2. Bitegmic, anatropous ovule with segments removed to show internal structure

The nucellus represents the homolog of the megasporangium of the progenitors of the seed plants, i.e., the organ in which the process of megasporogenesis takes place. Since, in the Spermatophytes, the megaspores are not liberated from the megasporangia any longer, it is also the site of megaspore development (megagametogenesis). Within the megagametophyte (embryo sac) endosperm and embryo are formed after fertilization.

The nucellus is enveloped by one or two integuments. They overgrow the nucellus and arch over its apex to form the micropyle which acts as a passage for the pollen tube. The integuments protect and nourish the nucellus and its contents. In the mature seed the integuments, together with the chalaza and raphe, form the seed-coat.

The most common type of ovule is the anatropous one. During its development this ovule undergoes a curvature of 180°, so that it becomes completely inverted and the micropyle comes to lie close to the place of attachment of the ovule.

The ovule is attached to the gynoecial placenta by a stalk, the funicle. In most ovules the funicle is provided with a vascular bundle. The part of the funicle that runs parallel with the nucellus is called the raphe. The chalaza is the region between raphe and nucellus, where the integument(s) are inserted. In fact, funicle, raphe, chalaza, and nucellus form a continuous tissue, the central axis of the ovule, and no sharp demarcation lines can be drawn between them. The structure of the ovule can become complicated by curvature as in campylotropous and amphitropous ovules.

Already before fertilization special structures, such as appendages (arils and arilloids), operculae, hypostases, and epistases, may start their development.

3.3 Ovule Initiation

Our knowledge about ovular initiation has not advanced much since Warming's publication of 1878. Floral morphologists usually limit their study to carpel initiation, and embryologists start with the ovule primordium without paying attention to its initiation. In most publications ovule primordia are described as mere bulges arising from the placenta, and no more precise information is given about its early histogenesis.

Ovule primordia are initiated by periclinal divisions in the second or third layer of the placenta, so that we can distinguish two- or three-zonate primordia, respectively (Fig. 3.3). In transverse sections they always exhibit a radial symmetry. Kordyum (1968; see also Fagerlind 1961, Brunkener 1977) states that the histogenetic organization of the ovule primordia is similar to that of the shoot apex (according to the concept of tunica and corpus). This may be true to some extent, but it is more appropriate to use the neutral terminology of two- and three-zonate (or di- and tri-zonate) ovule primordia, since we do not have a fair cognizance of the continuity of the independence of the tunica layers during the transition from the vegetative apex to the floral primordium, and the consequent formation of the carpellary, placental, and ovular meristems.

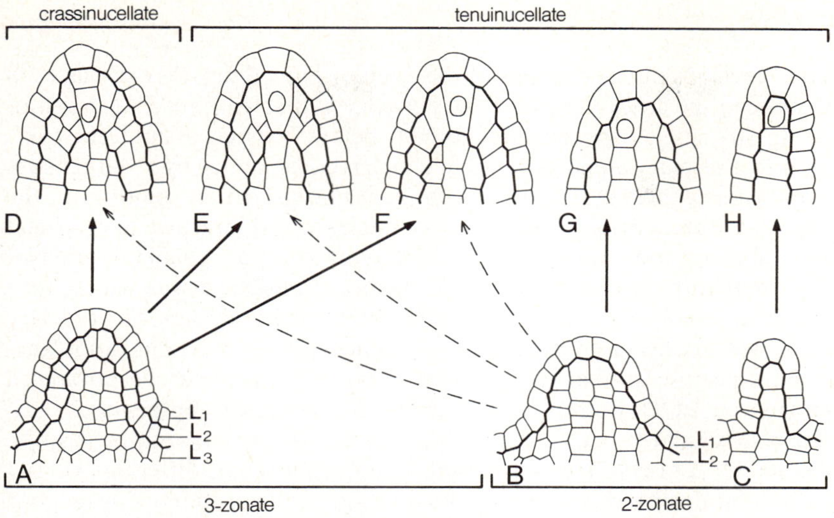

Fig. 3.3 A–H. Types of ovule initiation (*lower row*), and corresponding nucellar types (*upper row*). L_1, L_2, and L_3 dermal, subdermal and third layer of the ovular meristem, respectively

The tri-zonate ovular primordium is initiated by periclinal divisions in the third zone of the placental meristem, the two outer layers dividing only anticlinally in the beginning (Fig. 3.3 A). This three-zonate condition is found in many families: Berberidaceae, Betulaceae, Caricaceae, Casuarinaceae, Cistaceae, Cruciferae, Cucurbitaceae, Euphorbiaceae, Geraniaceae, Magnoliaceae, Papaveraceae, Ranunculaceae, Resedaceae, Rosaceae, Rutaceae, Sterculiaceae, Umbelliferae, Violaceae, Winteraceae, Convolvulaceae, Scrophulariaceae, Solanaceae, Amaryllidaceae, Liliaceae, Zingiberaceae and others. Highly convincing evidence of the existence of tri-zonate ovular primordia has been provided by Satina (1945) during her experimental work on cytochimeras of *Datura*.

The di-zonate ovular primordium is initiated by periclinal divisions in the second zone (subdermal layer) of the placental meristem (Fig. 3.3 B, C). Consequently, it is mostly smaller, and covered by the anticlinally dividing dermal layer only. Dizonate ovules occur in: Balsaminaceae, Begoniaceae, Droseraceae, Hypericaceae, Gesneriaceae, Lobeliaceae, Monotropaceae, and Orobanchaceae. In the extreme case of reduction the ovule is initiated in a single subdermal cell which, by periclinal divisions, gives rise to an axial row of cells covered by a dermal layer (Podostemonaceae, Orchidaceae).

The widespread occurrence of three-zonate ovular primordia in families of a supposedly primitive status and in derived ones suggests that the three-zonate ovular primordium is basic within the angiosperms.

Most probably, there is a correlation between the type of ovule initiation and the size of the ovules and seeds. The three-zonate ovular primordium possesses the greatest ontogenetic potential. Generally, tri-zonate ovular primordia give rise to larger ovules, with well-developed funicles, raphes, and nucelli, than two-zonate primordia.

3.4 Nucellus

The nucellus develops out of the apex of the ovular primordium. As early as 1878 Warming noticed the striking similarities between anther and nucellar development. During early stages one cell, lying directly below the nucellar epidermis, differentiates into a primary archesporial cell, which is conspicuous by its larger size, denser cytoplasmic contents and more prominent nucleus. In the majority of the angiosperms the archesporial cell divides periclinally to form the primary parietal and primary sporogenous cells. The former may undergo a variable number of further periclinal and anticlinal divisions to form the parietal tissue which, ultimately, is one to ten or more cells thick (up to 40 in Vochysiaceae). The primary sporogenous cell usually does not undergo further mitotic divisions, and functions as the megaspore mother cell. In more derived forms the archesporial cell does not form a parietal layer but directly acts as the megaspore mother cell.

In some taxa regularly, in others more occasionally, more than one megaspore mother cell may be present. This is brought about either by the differentiation of several archesporial cells, or by the mitotic division of the primary sporogenous cell. Multicellular archesporia occur in many taxa, they are common in Casuarinaceae, Rosaceae, Compositae, Rubiaceae, Umbelliferae, and Liliaceae. They play a role in polyembryony and apomixis.

The current nucellar classification is based upon the development of the archespore. In crassinucellate ovules the megaspore mother cell is separated from the nucellar epidermis by one or several parietal layers (Fig. 3.3 D) whereas in tenuinucellate ovules, by the absence of parietal cells (Fig. 3.3 E–H), the megaspore mother cell lies directly below the nucellar epidermis. This terminology dates from van Tieghem (1898) and was redefined by Asplund (1920). However, this classification has been much debated (van Tieghem 1901, Warming 1913, K.V.O. Dahlgren 1927, Mauritzon 1939, Hamann 1966, Rutishauser 1969, Bhatnagar and Johri 1972). The differences in nucellar morphology and size are caused by a number of characters, viz., the structure of the apex of the ovule primordium, the number of archesporial cells and whether or not the latter form parietal cells, the amount of parietal tissue, the presence of cells adjoining the archesporium, the mitotic activity of the adjacent and underlying cells of the embryo sac, the period, manner, and amount of resorption of the nucellar tissue by the growing embryo sac, the function of the nucellus as food-storing tissue, the mitotic activity of the dermal layer, the time and level of the initiation of the integument, and others.

Since the nucellar morphology is determined by so many characters, exhibiting reduction as well as specialization trends, no classification can ever adequately reflect the great diversity in nucellar structure and dimensions. In spite of the restrictions of the customary classification in crassi- and tenuinucellate ovules, it appears to be the best available and has proved to be of much practical value (Young and Watson 1970, R. Dahlgren 1975, 1980). Ontogenetically, the nucellus develops from the apex of the ovule primordium. Accordingly, the structure of this apex is the point of departure, and also the basis for description of the nucellus (Kordyum 1968). Figure 3.3 shows a simplification of Kordyum's scheme, which demonstrates that the tenuinucellate ovules form a rather heterogenous

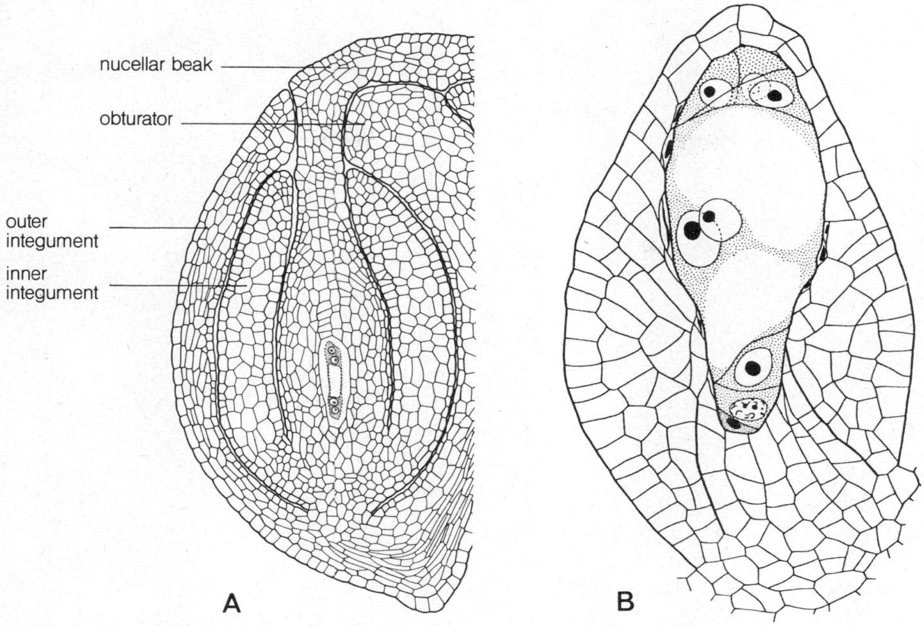

Fig. 3.4. A Ovule of *Chrozophora obliqua* with nucellar beak protuding from the micropyle. (From Bor and Kapil 1976). **B** Nucellus of *Medeola virginiana,* showing multilayered epidermis laterally. (From Berg 1962)

category and explains why so often tenui- and crassinucellate taxa are found within the same angiospermous family (see Davis 1966) or, occasionally, even within the same species. One of the questions to be solved in the near future is whether crassinucellate ovules always arise from three-zonate ovule primordia, or whether they can also develop from two-zonate primordia.

Generally, the dermal layer of the nucellus undergoes only anticlinal divisions, thus remaining one layer thick. However, in many cases the dermal cells of the nucellar apex divide periclinally to form a plurilayered epidermis or nucellar cap. Massive nucellar caps are known to occur in Aceraceae, Bombacaceae, Burseraceae, Combretaceae, Meliaceae, Petiveriaceae, Rhamnaceae, Rosaceae, Simaroubaceae, Vitaceae, and Agavaceae. In the extreme case the nucellus becomes beak-shaped. The nucellar beak protrudes from the micropyle or even extends beyond both integuments finally to reach the obturator outside the ovule (Fig. 3.4 A). Nucellar beaks are known in Aextoxicaceae, Cucurbitaceae, Euphorbiaceae, Malpighiaceae, Polygonaceae, Rhamnaceae, and Trapaceae.

In Liliaceae (Berg 1962) and Gramineae (Maze and Bohm 1973) the nucellar epidermis remains one- or few-layered at the apex, but becomes plurilayered around and below the middle of the embryo sac, thus contributing toward the bulk of the nucellus (Fig. 3.4 B). In Zingiberaceae, Nymphaeaceae, Centrolepidaceae, and some Liliaceae the cells of the nucellar epidermis undergo a pronounced radial stretching, sometimes followed by cell wall thickening (the nucellar pad).

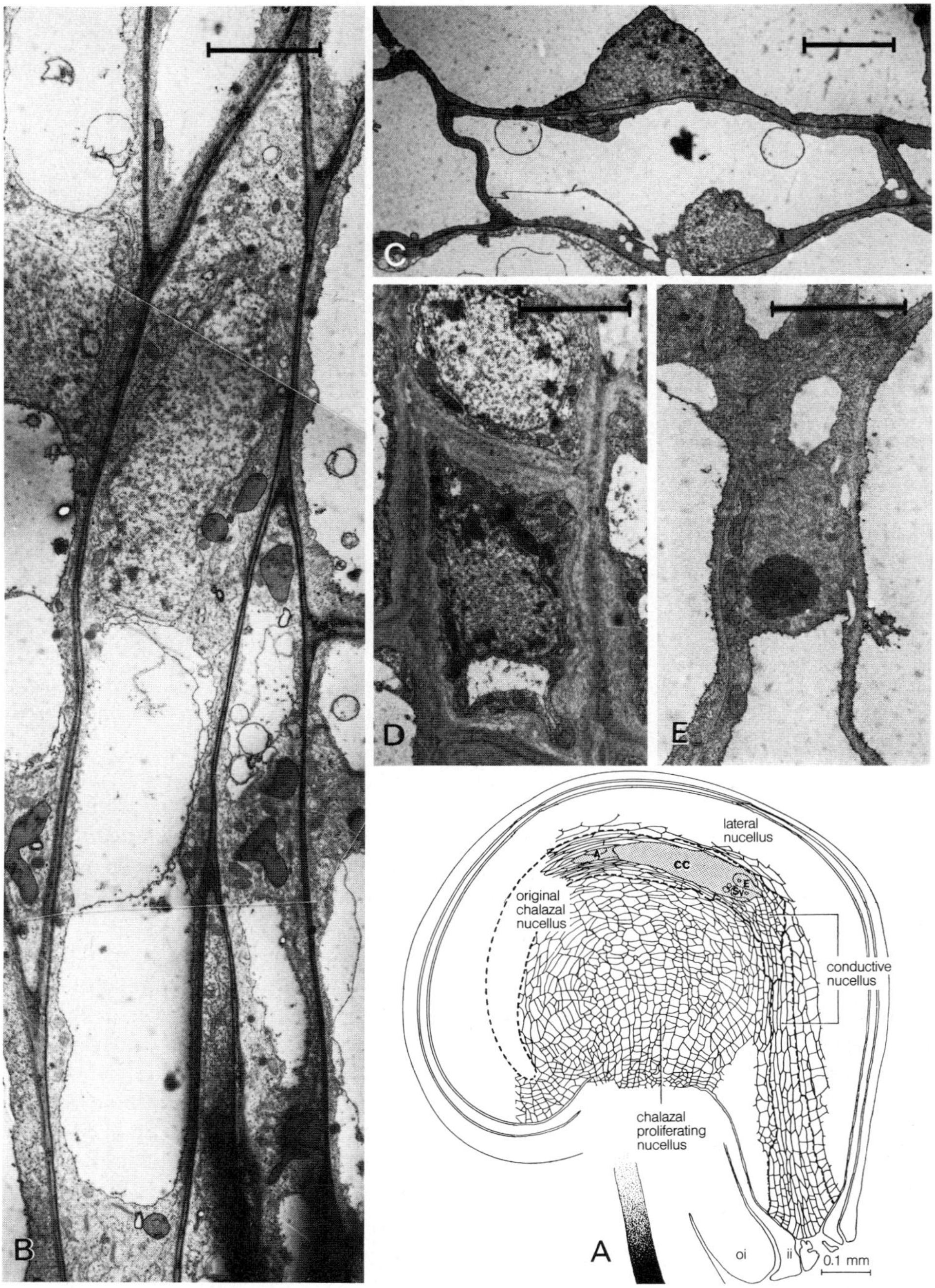

Fig. 3.5 A–E. Different cell types of the nucellus in the ovule of *Spinacia oleracea*. **A** Median section of ovule showing location of the cell types. Note outer (*oi*) and inner (*ii*) integuments. *E* egg; *Sy* synergid; *cc* central cell; *A* antipodes. **B** Original chalazal nucellar tissue. **C** Lateral nucellar tissue. **D** Transmitting (transfer) cells near the micropylar part of embryo sac; this tissue belongs to the conductive nucellus **E** Chalazal, proliferated nucellar tissue. The scales represent a length of 5 μm in **B** to **E.** (Courtesy HJ Wilms, Wageningen, The Netherlands)

In an attempt to construct a more detailed classification of the nucellus, Davis (1966) introduced the term pseudocrassinucellate for tenuinucellate ovules provided with a plurilayered epidermis. Although this term may be useful for embryological descriptions, it seems incorrect to consider it as equivalent to crassi- and tenuinucellate. The incidence of periclinal divisions in the nucellar epidermis is far less essential than the presence of parietal cells. If the term is to be used at all, it must be employed for a subdivision of both the tenui- and crassinucellate categories.

The histochemistry and the fine structure of nucellar cells have been scarcely studied. Noher de Halac (1980) described the differentiation of nucellar cells in *Oenothera biennis*. In the beginning the nucellar cells are of a meristematic nature. Several zones in the nucellus can be distinguished during differentiation, when cells undergo different kinds of cell degeneration. In the cells surrounding the megaspores, cytolysomes, and smooth ER vesicles enveloping cytoplasmic compounds indicate autophagic activity. Concomitant with the differentiation of the embryo sac these cells undergo degeneration. Toward the chalaza three other types are present, which exert a secretory function, a selective action on the growing embryo sac, and a function as a reservoir, respectively. Wilms (1980) examined the ultrastructure and histochemistry of the amphitropous ovule of *Spinacia oleracea*. In the nucellus four specialized tissues can be distinguished, whose cells have common features as well as distinguishing characteristics related to their position and function (Fig. 3.5 A–E). The conductive tissue plays a role in the penetration and growth of the pollen tube. A gradual dissolution of the longitudinal middle lamellae of these cells starts at the embryo sac and extends to the micropylar region. Ultimately, the nucellar cuticle is attacked and disintegrates. The original chalazal tissue most probably serves as a direct transporting pathway from the chalaza to the developing embryo sac, their cells have the characteristics of transfer cells. The chalazal proliferating tissue and the lateral tissue store reserve metabolites.

The destiny of the nucellus in the mature seed may be quite different. In small ovules the nucellar tissue can already be resorbed in the mature ovule, in most bigger ovules it is completely or almost completely consumed during embryo and endosperm development, whereas in, e.g., Piperaceae, Nymphaeaceae, and Zingiberaceae it undergoes a pronounced development and forms a persistent storage tissue, the perisperm.

3.5 Megasporogenesis

The megaspore mother cell (megasporocyte or meiocyte) (Fig. 3.6 A) divides meiotically. This process consists of two successive divisions, and gives rise to four haploid megaspores whose mutual arrangement varies (Rembert 1971). The first or heterotypic division is always transverse and results in two dyad cells (Fig. 3.6 B). Usually, the second or homotypic division is also transverse, which results in a linear tetrad (Fig. 3.6 C, D). Frequently, the micropylar dyad can also divide longitudinally to form a T-shaped tetrad (Fig. 3.6 E, F). According to the data

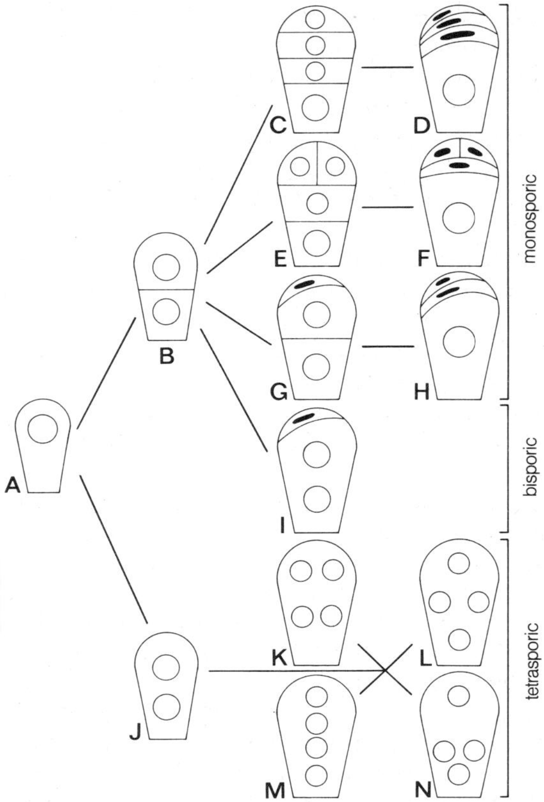

Fig. 3.6 A–N. Different types of megasporogenenis. **A** Meiocyte. **B, J** Dyads. **C, D** Linear tetrads. **E, F** T-shaped tetrads. **G, H** Reduction of the micropylar dyad. **I** Bisporic development. **K–N** Tetrasporic development

collected by Davis (1966), 213 angiospermous families are characterized by having exclusively linear tetrads, while in 56 families T-shaped tetrads occur next to linear ones. Especially within this group also tetrads of an intermediate type occur, in which the wall between the two micropylar megaspores lies at an angle of 45°. Other configurations as ⊥-shaped (Musaceae), isobilateral (Crassulaceae), and tetrahedral (Hydrocharitaceae) ones are rare, and mostly found in combination with linear and T-shaped tetrads. Meiosis II may be synchronous or heterochronous (George et al. 1979). The division of one of the dyad cells can be further suppressed in such a way that no separating cell wall is formed, and the dyad becomes binucleate (Hakki 1972), or that the division is abortive. Complete suppression results in a row of three cells, of which only two are megaspores (Fig. 3.6G, H). Such triads have been described in Amaranthaceae, Araliaceae, Cactaceae, Caryophyllaceae, Orchidaceae, Petiveriaceae, and Phytolaccaceae.

The above-mentioned suppression of the development of the nonfunctional dyad may be interpreted as a trend toward further reduction. However, failure of wall formation in the functional dyad, or even already in the meiocyte, represents a trend to specialization. By this elimination of cytokinesis, binucleate, or tetranucleate coenomegaspore(s) are formed, respectively (Fig. 3.6 I–N). These conditions will lead to a bisporic and tetrasporic development of the embryo sac, which is more extensively discussed in Chapter 4.

In the majority of angiosperms (268 families according to Davis 1966) the three micropylar megaspores degenerate and only the chalazal one develops into the megagametophyte (embryo sac). Only in exceptional cases does the subchalazal (Caesalpiniaceae, Cunoniaceae, Symplocaceae), the micropylar (Balanophoraceae, Calenduleae, Onagraceae, *Rosa, Furcraea*) or any given megaspore (*Casuarina, Gloriosa, Ostrya, Poa,* and *Rubus*) becomes functional.

The degeneration of the nonfunctional megaspores is accompanied with a high density of ribosomes, appearance of dictyosomes and lipid granules, vesicles and autophagic vacuoles, and results in a loss of turgescence and a low cellular activity (Willemse and de Boer-de Jeu 1981). The products of degeneration of the megaspores and surrounding nucellar cells can be used by the functional megaspore.

The introduction of advanced techniques, such as the use of fluorescence microscopy, Normarski interference microscopy, and electron microscopy, has opened new perspectives for a more detailed study of megasporogenesis.

By applying the aniline-blue fluorescence method to ovular squashes of *Orchis maculata,* Rodkiewicz (1967) was the first to demonstrate callose formation during megasporogenesis in angiosperms. Subsequently, a fair number of observations have been made (see for reviews Rodkiewicz 1970, Kuran 1972, Kapil and Tiwari 1978 a). In the course of megasporogenesis callose appears transiently in the cell walls of plants with a mono- or bisporic type of embryo sac, but the polysaccharide has till now not been observed in species with a tetrasporic type (Rodkiewicz 1970). The formation of callose in the walls during megasporogenesis is similar to that in microsporogenesis. Callose always appears in the early meiotic prophase. In the Polygonum type of embryo sac callose is first deposited at the chalazal pole of the meiocyte. In the late prophase, and especially in the first metaphase, the whole meiocyte is surrounded by a callose-containing wall. In the dyad and tetrad stage callose is especially abundant in the opposing walls of the megaspores, sometimes less so in their sidewalls. Ultimately, the callose fluorescence decreases, starting from the chalazal end of the dyad or the functional megaspore (Fig. 3.7 A–J). In orchids the callose persists longer in the chalazal wall of the functional megaspore. However, in this case the wall appears as an elaborate, sievelike structure with numerous nonfluorescent areas. During later stages these areas become transformed into larger fields. Electron microscopic photomicrographs show a noncallosic outer wall with fingerlike protrusions, and a sievelike callosic inner layer spread in between (Fig. 3.7 K–M). The wall and callose layer are crossed by plasmodesmata (Rodkiewicz and Bednara 1976, Rodkiewicz 1978). Such sievelike cross walls have also been observed between the nonfunctional megaspores of *Epilobium.*

The presence of callose seems to be more than just an archaic feature. It is assumed that the callose wall forms a molecular filter decreasing the permeability of the cell wall. In this way the generative cells become temporarily isolated, which enables the cells to embark upon an independent course of development.

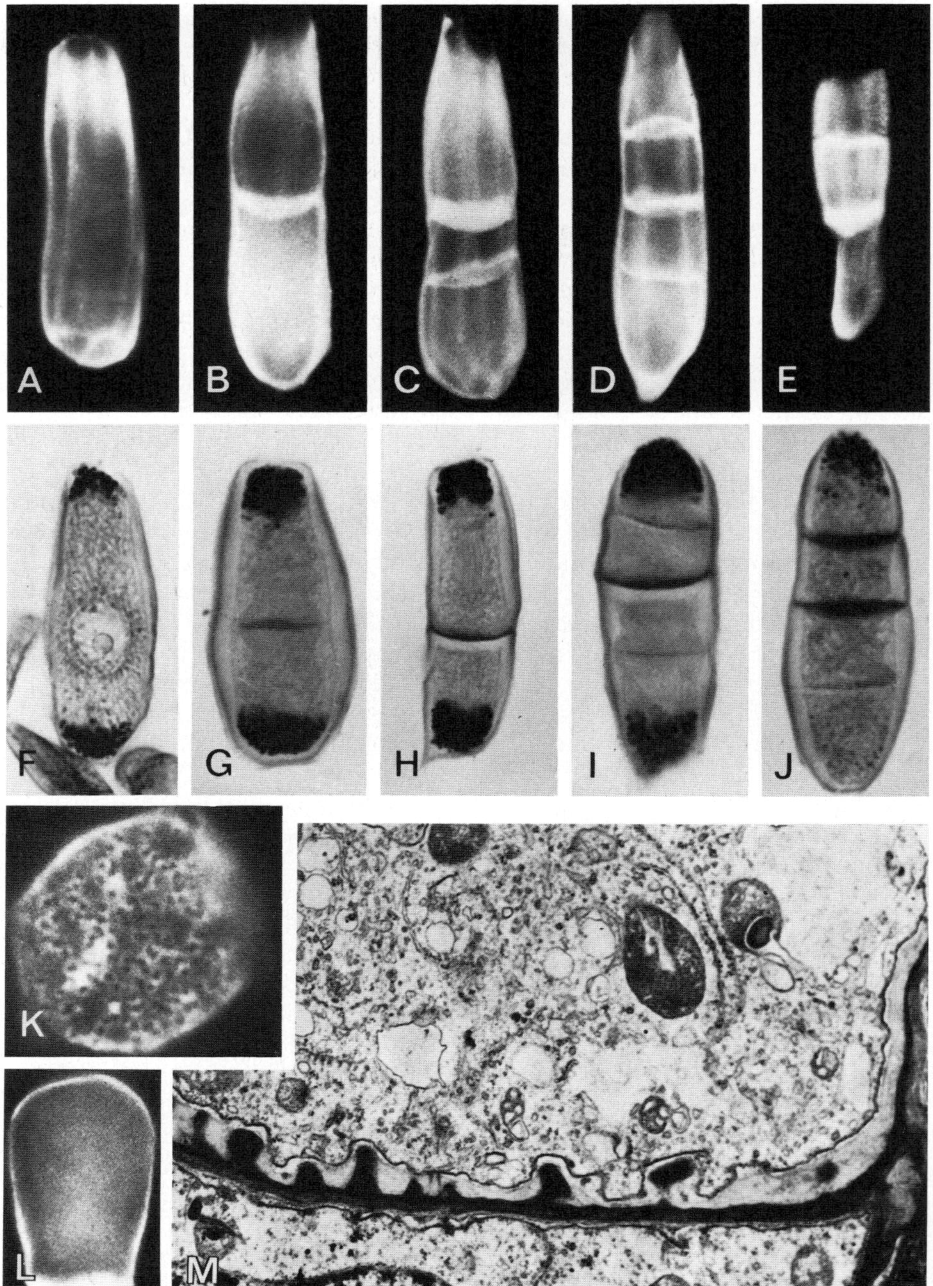

Fig. 3.7 A–M. Callose accumulation during megasporogenesis. **A–E** *Fuchsia hybrida,* fluorescence micrographs of meiocyte, dyad, triad, and tetrad. **F–J** *Epilobium palustre,* megasporogenesis, PAS-staining. **K–M** *Epipactis sp.* **K** Chalazal wall of the meiocyte with sieve-like distribution of callose. **L** Meiocyte. **M** Electron micrograph of chalazal wall. (Original photographs by courtesy of Rodkiewicz, Sniezko, and Bednara, Lublin, Poland)

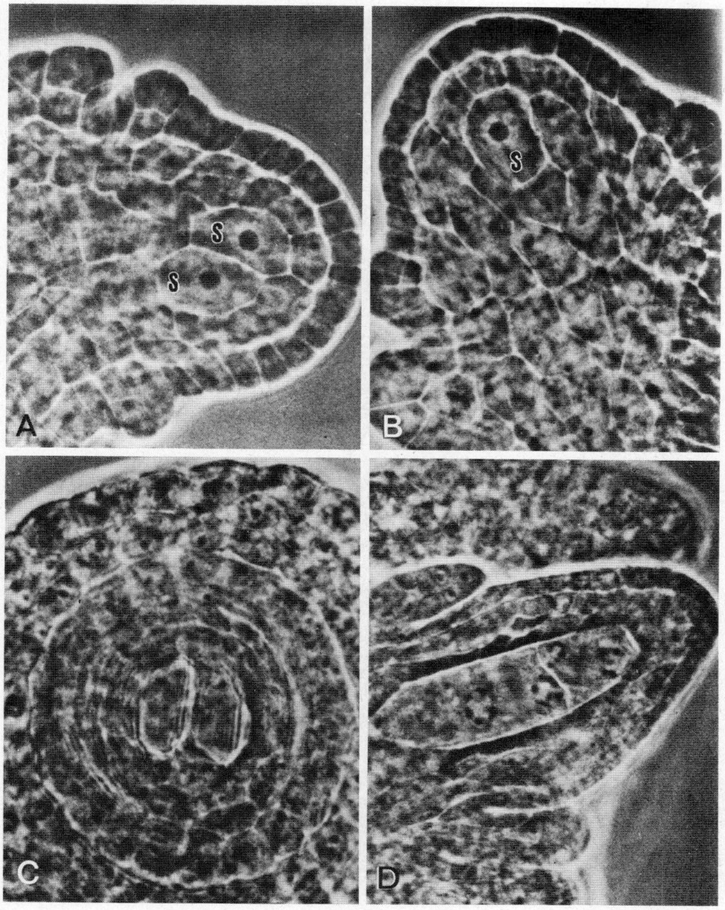

Fig. 3.8 A–D. Optical sections under phase-contrast optics. **A, B** *Glycine max,* two sections of the same ovule showing multicellular sporogenous (*S*) tissue. **C, D** *Phaseolus aureus,* cross section of an ovule containing two dyads (**C**), and ovule during the second meiotic division (**D**). (From George et al. 1979)

Besides, the uneven distribution of callose in megasporogenesis is clearly connected with the strong polarization of the cells. The beginning of callose deposition, and its disappearance, correspond with the localization of the active megaspore in the tetrad. This is clearly endorsed by the fact that in the Oenothera type of embryo sac both the formation of callose and the differentiation of the functional megaspore are reversed in respect of the micropylar side (Fig. 3.7 A–J).

The study of cleared, unsectioned ovules and of squashed ovules with phase contrast or Nomarski interference optics (Herr 1971, 1973) facilitates the study of archesporial differentiation, megasporogenesis, and megagametogenesis by circumventing time-consuming sectioning and allows more quantitative research of these reproductive processes within or between closely related species (Fig. 3.8 A–D) (George et al. 1979, Greene 1978). By means of these techniques Rembert

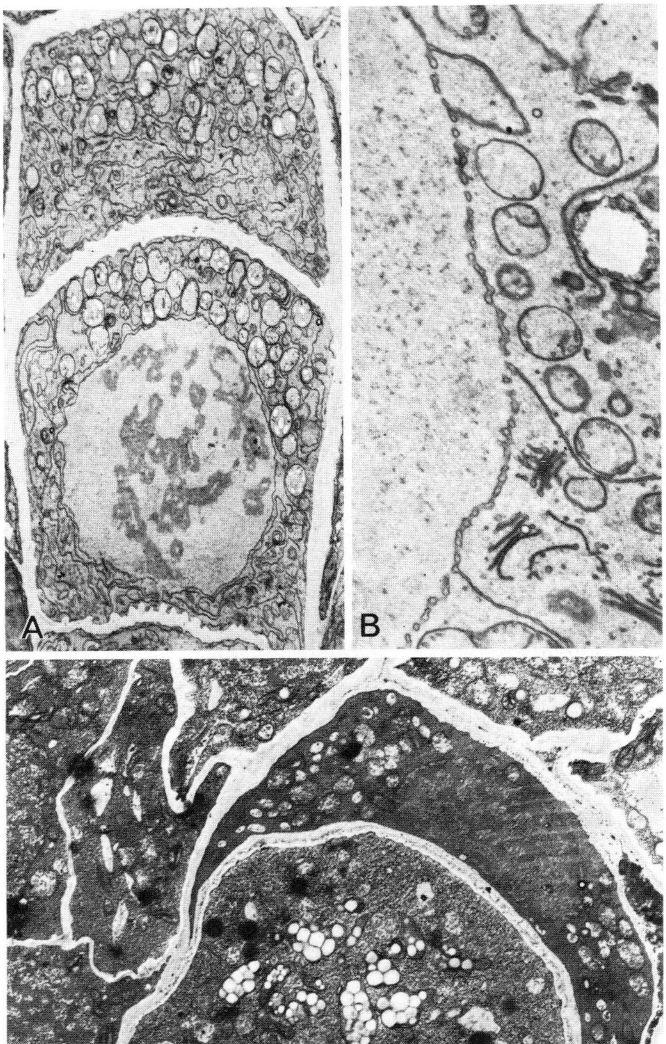

Fig. 3.9 A–C

(1977) demonstrated that in *Trifolium repens* the early embryo sac develops faster than it is allowed to expand in the nucellar mass. Released embryo sacs unfurl to reveal a contour at least 50% expanded over its previous situation.

One of the topics in the ultrastructural research of ovules is the polarity of the sporogenic and gametophytic tissue. This polarity is expressed by the selection of the functional megaspore and the degeneration of the nonfunctional ones, the position of the nucleus or nuclei, the orientation of the spindles, nuclear migration, and in some cases fusion and degeneration, cell elongation along the micropylar-chalazal axis in combination with vacuolation, and the rate of differentiation between the micropylar and chalazal cells of the embryo sac (Willemse 1981). Possible polarizing factors are of mechanical, nutritive, and organizing (hormonal control) nature. Tectonic stress imposed by the surrounding tissue determines the orientation of spindles and phragmoplasts (Lintilhac 1974). The presence of a polar nutritive flow (Mogenson 1981) can change by the differentiation of hypostase-like structure. At least, polarity seems to be an expression of a genetically regulated development of the gametophyte in relation to the developing ovular tissue. Electron microscopic studies on polarization of cell organelles during megasporogenesis have been too few to permit generalizations. Willemse and Franssen-Verheijen (1978) found many similarities in the development of plastids, mitochondria, ribosomes, RER, SER, and microtubular pattern during megasporogenesis and microsporogenesis in *Gasteria*. The time of initiation, and the extent to which the meiocyte shows polarization by an unequal distribution of the organelles in its cytoplasm, seem to vary in different species (Fig. 3.9 A–C). Bednara and Rodkiewicz (1974) described a distinct polarization in the meiocyte in *Epilobium*. As a consequence, the active megaspore has the maximal number of plastids, and the minimal number of dictyosomes per unit area, whereas the mitochondrial population is uniformly dense in all the megaspores. Godineau (1968) could not detect a polarity in the meiocyte of *Crepis*. Some of the earlier ideas concerning the polarity and the number and disposition of the nuclei during megasporogenesis and gametogenesis are discussed by Johri (1963).

An uncommon but notable phenomenon is the formation of aggressive haustoria by the megaspores in some species of Crassulaceae, Rubiaceae, Lauraceae, and Rosaceae (Fig. 3.10 A–D). In *Sedum chrysanthum* the megaspores produce haustoria, grow toward the apex of the nucellus, and form an intertwined mass. As a result of the active growth of the megaspore tubes, the nucellar cells are consumed. One of the tubes often penetrates through the nucellar epidermis and swells to form a vesicle at the base of the micropyle (Subramanyam 1967).

Fig. 3.9 A–C. Megasporogenesis, ultrastructure. **A** Dyad of *Epipactis palustris*. **B** Same, detail of the chalazal part of nucleus with many pores. **C** Megaspore of *Gasteria verrucosa* with starch and lipid granules. (**A, B** courtesy J Bednara, Lublin; **C** courtesy MTM Willemse, Wageningen)

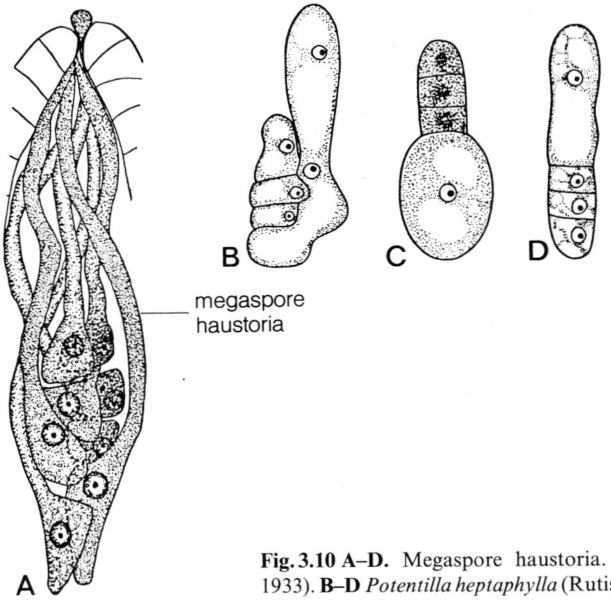

Fig. 3.10 A–D. Megaspore haustoria. **A** *Rosularia pallida* (Mauritzon 1933). **B–D** *Potentilla heptaphylla* (Rutishauser 1969)

3.6 Integuments

Angiospermous ovules may have two, one, or no integument(s), and are consequently called bitegmic, unitegmic, and ategmic. The initiation and early histogeny of the integuments have been studied inadequately. The inner integument is usually initiated before, sometimes simultaneously, with the outer integument. In Podostemonaceae and some Malvaceae the outer integument is initiated first. Other reports in which the outer integument is said to appear first need confirmation. The inner integument is almost always of fully dermal derivation. It is initiated by periclinal or oblique divisions in a circular zone around the ovular primordium mostly two to three cells high. Successive longitudinal divisions give rise to an integument two to three cells thick. The family Euphorbiaceae is the only one in which a subdermally initiated inner integument is reported (Bor and Bouman 1974).

The outer integument is initiated in either the dermal or subdermal layer of the ovular primordium. If dermal, the development is comparable to that described for the inner integument. According to Roth (1957), in *Capsella bursa-pastoris,* both the integuments grow by means of a rim of apical initials exhibiting a bifacial rhythm of cell division. Even if these observations are correct, this type of histogenesis does not hold good for the majority of ovules. The subdermal outer integument consists of a middle layer one to several cells in thickness, covered with a true outer and inner epidermis (Fig. 3.11 A–F). At the apex the dermatogen mostly divides periclinally to form the micropylar part of the outer integument. The type of initiation of the outer integument is generally a taxonomic character

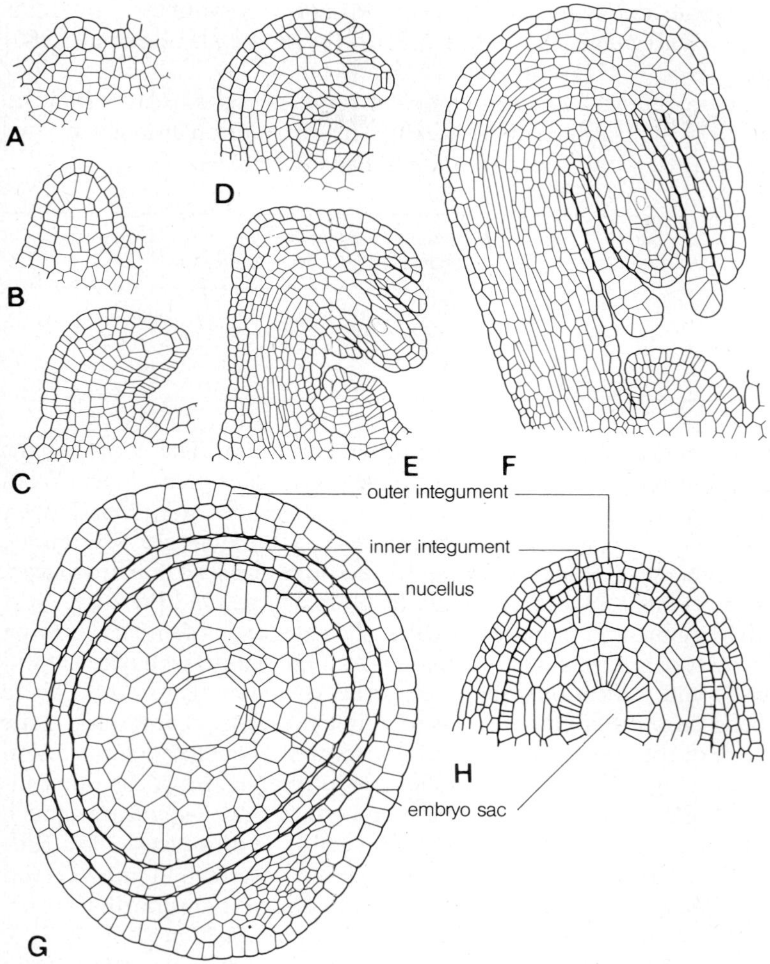

Fig. 3.11 A–H. Integument ontogenesis. **A–F** *Nigella damascena,* ovule and integument initiation (Bouman 1978 b). **G** *Polygala vulgaris,* cross section of an ovule showing partly subdermal outer integument (Verkerke and Bouman 1980). **H** *Lunaria annua,* cross section with subdermal part of the outer integuments laterally, inner integument multiplicative, nucellus resorbed; the layer of radially-elongated cells around the embryo sac is endothelium (Bouman 1975)

at the family level. Recent inquiry (Bouman 1974) has revived the opinion, held by Warming (1878) and Péchoutre (1902), in contrast to Netolitzky (1926), that the outer integument in angiosperms is originally of subdermal derivation. The outer integument in Magnoliaceae, Ranunculaceae, Rosaceae, and Papilionaceae is subdermal, that of most Monocotyledonae, Saxifragales, Parietales, Centrospermae, and Geraniales dermal. Dermalization of the outer integument must have taken place several times during the phylogeny of the angiosperms. Both dermal and subdermal outer integuments are present in a number of families:

Cruciferae, Malvaceae, and Rutaceae. A partly dermalized, asymmetric, outer integument occurs in *Polygala vulgaris* (Fig. 3.11 G) (Verkerke and Bouman 1980) and *Lunaria annua* (Fig. 3.11 H) (Bouman 1975).

The number of integuments is a very important embryological feature (Sporne 1969, Philipson 1974, R. Dahlgren 1975, 1980). The distribution among the families of the angiosperms, according to the data of Davis (1966), is:

	No. of families	Bi-tegmic	Uni-tegmic	Ategmic	Bi- or unitegmic	Unknown
Dicotyledons	341	152	91	4	16	74
Monocotyledons	69	52	0	0	3	14
Total	410	204	91	4	19	88

Bitegmy is the more common, and primitive condition. The evolutionary changeover from bitegmy to unitegmy must have taken place many times. Well-known families with both bitegmic and unitegmic species are: Corylaceae, Cytinaceae, Fabaceae, Fagaceae, Menispermaceae, Monimiaceae, Myrtaceae, Nyctaginaceae, Piperaceae, Ranunculaceae, Rosaceae, Salicaceae, Saxifragaceae, and, among Monocotyledons, Amaryllidaceae, Orchidaceae, and Poaceae. Next to families with bitegmic and unitegmic ovules, the category of families showing ovules with a structure intermediate between bi- and unitegmy (by having a bifid integument) can yield indications about the process of unitegmization. There seem to have been three different pathways to unitegmization. The simplest is the elimination of one of the two integuments by retardation or complete suppression of its development (Fig. 3.12 A, B). Examples of a probable loss of the outer integument are some Piperaceae (the single integument of *Peperomia* shows much resemblance to the inner one of *Piper;* Warming 1878), Hydnoraceae, and Rafflesiaceae. A very weakly developed inner integument is described in *Populus* (van Tieghem 1900). A second pathway, the fusion of integumentary primordia, is only possible if the two integuments are initiated in the same fashion (Fig. 3.12 C). Such a fusion between a dermal inner and a dermal outer integument gives rise to a single, massive, dermal integument as in Anacardiaceae and Balsaminaceae. A third pathway, integumentary shifting (Fig. 3.12 D), has recently been described in Ranunculaceae (Bouman and Calis 1977). It takes place in ovules with a dermal inner, and a subdermal outer integument. Integumentary shifting is a complicated ontogenetic process involving (1) a fusion of primordia in the sense that the initials of the two integuments give rise to a common structure, (2) a

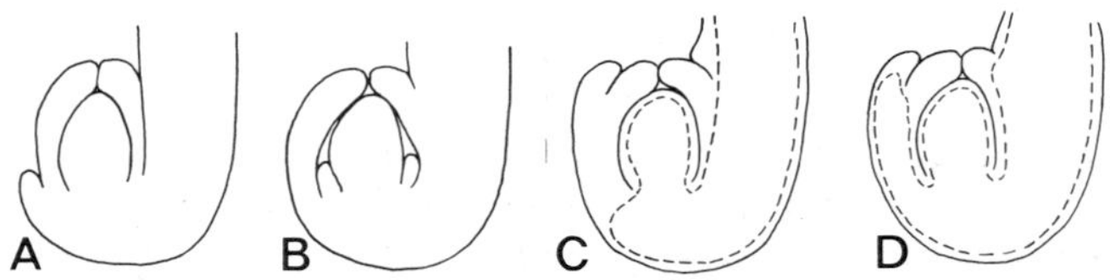

Fig. 3.12 A–D. Intermediate stages in unitegmization. **A** Reduction of the outer integument. **B** Reduction of the inner integument. **C** Integumentary fusion. **D** Integumentary shifting

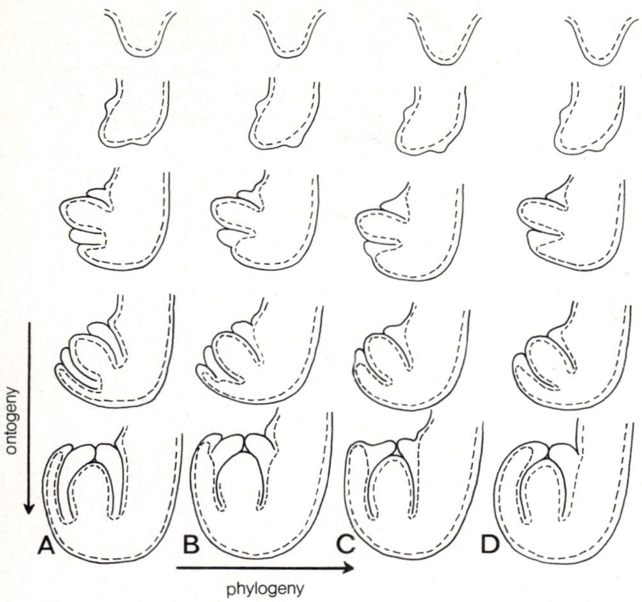

Fig. 3.13 A–D. Integumentary shifting. **A** A bitegmic ovule with subdermal outer and dermal inner integument. **B, C** Intermediate stages in integumentary shifting. **D** Unitegmic ovule

shifting of the inner integument, and (3) an arrested growth of the latter (Fig. 3.13 A–D).

The unitegmic condition must have been achieved several times, and along different ontogenetic pathways during the evolution of the angiosperms. This means that the condition of unitegmy is not necessarily homologous throughout the angiosperms. Detailed studies on the initiation of the integument in unitegmic families are scanty. Most data still date from Warming (1878) who concluded that the integument of sympetalous taxa, in spite of its thickness, is mainly of dermal derivation. Independent evidence of the dermal origin of the single integument was obtained by Satina (1945) from observations on cytochimeras of *Datura,* and by Bhandari et al. (1976) in *Scrophularia*. However, Bouman and Schrier (1979) showed that the single integument can also be of subdermal derivation (e.g., Convolvulaceae and Cuscutaceae), and that in *Gentiana* the sequence and position of the dermal and subdermal initials render it plausible that this single integument has been derived from a dermal inner and a subdermal outer integument by integumentary shifting.

Integuments often increase in thickness during ovule and seed development. Such integuments are called multiplicative (Corner 1976). The mode of multiplication depends on the type of integument. A subdermal (outer) integument can become plurilayered by periclinal divisions of (1) the outer dermal layer (Cucurbitaceae), (2) the subdermal, middle layer (most common), or (3) the inner dermal layer (Magnoliaceae, Vitaceae). Two-layered (i.e., dermal) integuments can become three-layered by periclinal divisions of the inner layer. In many cases only

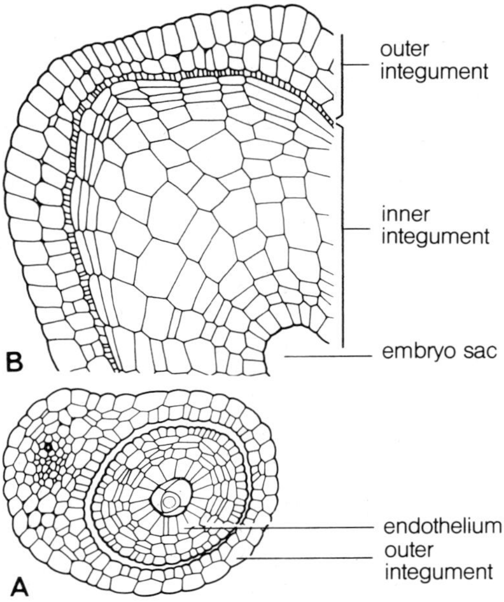

Fig. 3.14 A, B. *Linum usitatissimum*, development of the multiplicative inner integument. **A** Cross section of young ovule. **B** Cross section of immature seed. (From Boesewinkel 1980)

the middle layer divides further periclinally, thus giving rise to a plurilayered integument: e.g., the inner integument of Cruciferae (Fig. 3.11), Geraniaceae, Linaceae (Fig. 3.14 A, B), the outer integument of Oxalidaceae and Liliaceae. However, in primary or secondary three- or more-layered, dermal integuments both the inner and middle layers can participate (Dichapetalaceae). The growth in length and thickness of the integuments can be successive, or proceeds simultaneously. In the latter case the analysis of the precise course of ontogeny is rendered much more difficult.

One of the more remarkable phenomena in the histogenesis of integuments is the differentiation of an integumentary tapetum or endothelium (Fig. 3.15 A, B). This has been recorded in about 65 families of dicotyledons and seven of monocotyledons. Both the terms and criteria, and also the putative function, have long been much debated. An extensive review has recently been presented by Kapil and Tiwari (1978 b). The endothelium is difficult to define, because its morphological, cytological, and functional characteristics show much variation. It forms a new limiting layer of the embryo sac in many ovules, where the nucellus disorganizes at an early developmental stage. As such it is especially present in tenuinucellate or weakly crassinucellate ovules. The endothelium originates from either the inner layer of the single integument as in unitegmic ovules of sympetalous families, in which group it is best known, or from the inner layer of the inner integument as in bitegmic ovules of Celastraceae, Elaeocarpaceae, Erythroxylaceae, Lecythidaceae, Linaceae, Primulaceae, and Zygophyllaceae. It often surrounds the embryo sac almost completely, but may also be confined to a part of it as in Ericaceae, Gesneriaceae, and Scrophulariaceae. In most families it remains uniseriate but in Compositae it becomes bi- or multilayered. The cells become radially stretched and they contain prominent nuclei which often behave aber-

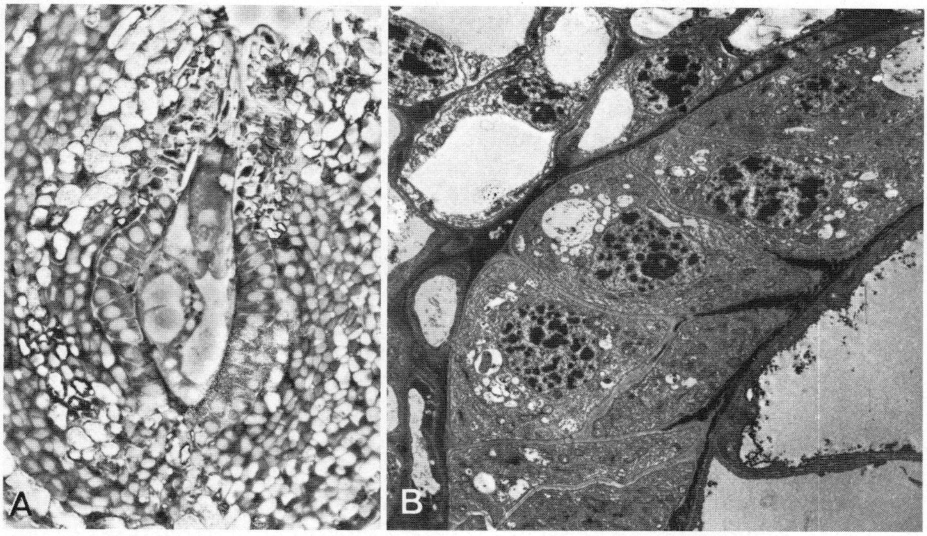

Fig. 3.15 A, B. *Impatiens walleriana.* **A** Ovule with embryo sac invested with endothelial layer. **B** Detail of endothelium, the wall near the embryo sac shows electron-dense and acetolysis-resistant material. (Courtesy J. L. van Went, Wageningen)

rantly. The endothelial cells can become bi- or multinucleate, or polyploid by endomitosis and nuclear fusion.

Ultrastructural studies (Godineau 1971, Newcomb 1973 a, b) are too few to allow generalization. During their development the character of the endothelial cells changes from meristematic to secretory (Engels and Petersen 1977). The endothelial cells are separated from the embryo sac by a prominent cuticle. The thickness of this cuticle varies at different sites and alters during ovular development (Erdelska 1975). Besides, the adjacent cell walls may become thick-walled. Several, not mutually exclusive, functions have been ascribed to the endothelium. The most important is the supply of nutrition to the embryo sac. Temporarily, this may take place simply by a transfer of nutrients supplied to the integument through the conducting tissue in funicle, raphe, and chalaza, or by the secretion of hydrolyzing enzymes such as amylases, proteases and others into the adjacent integumentary tissue.

The dissolution products find their way into the endothelium where the solutes may be metabolized before entering the embryo sac (Newcomb 1973 a). The endothelium can also act as a barrier tissue to resist, or to localize, the aggressive action of the growing embryo sac or endosperm. In many families the endothelium disintegrates and, subsequently, disappears during the later phases of seed development. Only in some families does the endothelial layer remain distinct in the mature seed, for instance in Polemoniaceae, Plantaginaceae, and Linaceae where it forms a pigment layer, and in Scrophulariaceae and Solanaceae.

According to Swamy and Krishnamurthy (1970), a typical endothelium occurs only in the dicotyledons. The reports of the presence of an endothelium in monocotyledonous taxa are said to be based on the employment of rather gross and unreliable criteria. A false endothelium is present in Begoniaceae and Droseraceae, where the cells of the persistent nucellar epidermis elongate radially and form a jacket around the embryo sac.

The great diversity in distribution, morphology, cytology, and functions renders it likely that the endothelium is the product of analogous developments, and that it has originated independently in several taxa during the evolution of angiosperms.

Lobed integuments have been observed in several families, Juglandaceae, Berberidaceae, Rosaceae, and Flacourtiaceae. Distal lobing may involve either the inner or the outer integument, or both. The phenomenon is supposed to be significant for the understanding of the evolution of the ovule. Lobing occurs in many extinct and recent gymnosperms. The lobing suggests that the integuments are compound organs. Unfortunately, none of the cases of lobed integuments in angiosperms can as yet be related to the location of vascular bundles (van Heel 1976).

In several centrospermalean families (Amaranthaceae, Cactaceae, Chenopodiaceae, Nyctaginaceae, and Phytolaccaceae), and also in Bombacaceae and Simaroubaceae (Narayana 1957) a temporary air space is formed between the integuments due to a more rapid growth of the basal region of the outer integument. Its functional significance, if any, is not known (Kapil and Prakash 1969).

3.7 Types of Ovules

The ovule may have various forms, depending on the shape and position of the different ovular parts. We can distinguish between ovules with straight nucelli (anatropous and atropous ones) and those with curved nucelli (campylotropous and amphitropous ones).

Anatropy is of most common occurrence among the angiosperms. According to Davis (1966), 204 families are exclusively anatropous. The anatropous curvature starts shortly after the initiation of the ovule. It is, in fact, an intercalary growth of the funicle. This growth takes place on the convex side of the ovular primordium in the zone below the attachment of the integument(s), and on the concave side in the zone between the outer and inner integument (in bitegmic ovules), or between the integument and nucellus (in unitegmic ovules). By this process the nucellus turns over 180° and the micropyle shifts toward the future hilum (Fig. 3.16 A–F). In anatropous ovules the outer (or single) integument is less strongly developed at the raphal side. Occasionally, the indication of its presence is restricted to an enlargement of a few epidermal cells; in extreme cases there is not a single trace of an integument in that region.

Fig. 3.16 A–F. SEM-photographs of developing ovules. **A** *Passiflora racemosa,* ovular primordia with beginning of anatropous curvature, and initiation of the inner and outer integument. **B, C** *P. verspitilio,* showing overgrowth of the nucellus beyond the integuments (**B**) and mature (**C**), anatropous ovule. **D–F** *Agrostemma gracile,* ovular primordia with initiation and development of the integuments (**D, E**), and mature (**F**) ovules. (Original photographs, courtesy of W. Verkerke, Amsterdam)

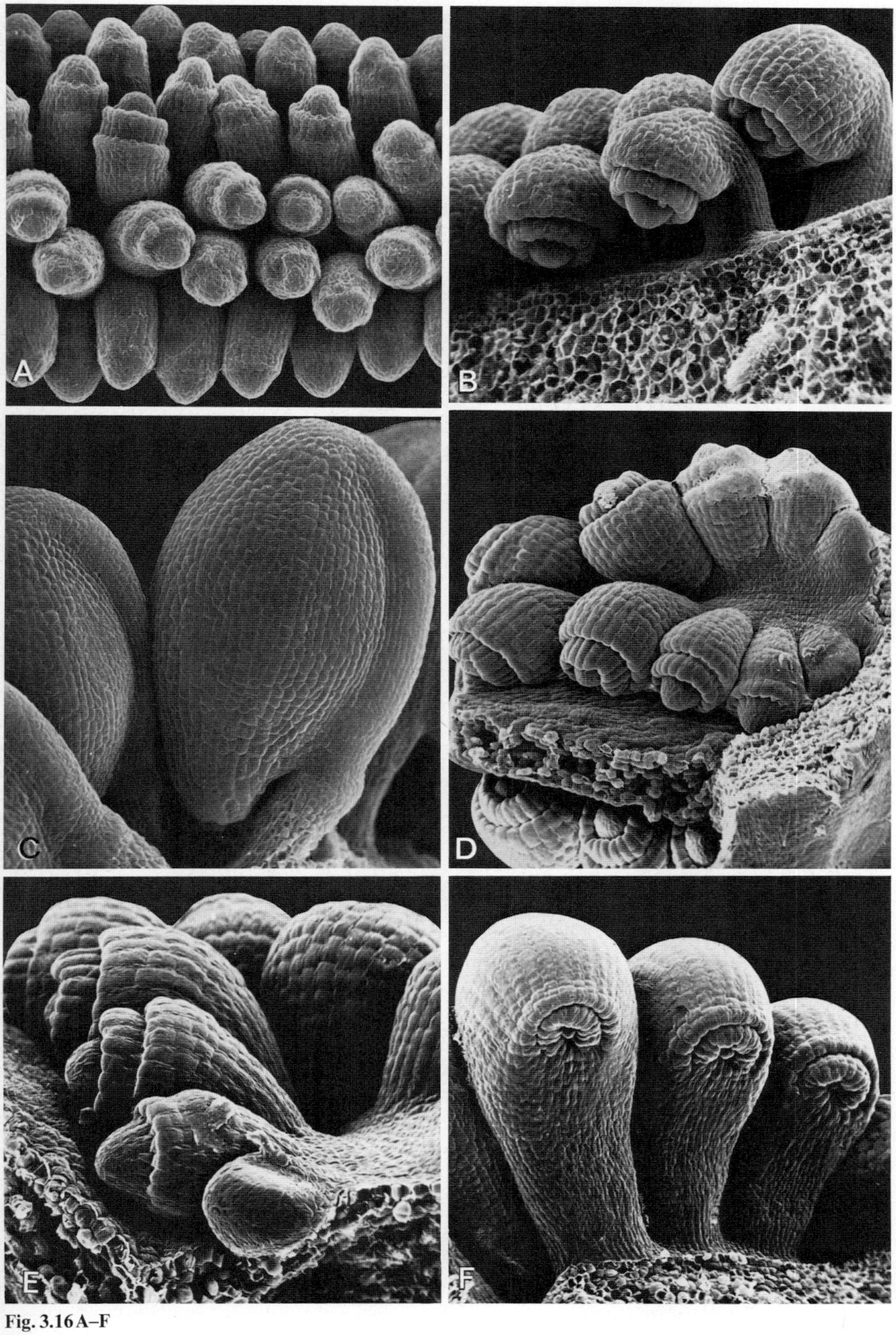

Fig. 3.16 A–F

After a study on the anatropous curvature, Bor (1978) concluded that the ovule consists of two separate developmental "units": (1) the funicular unit which gives rise to raphe, chalaza, and outer integument, and (2) the nucellar unit from which the inner integument is derived. This hypothesis does not seem to be in accordance with the ontogenetic facts.

Orthotropous or atropous ovules occur in 20 families (e.g., in Juglandaceae, Polygonaceae, Urticaceae, Najadaceae). Funicle, chalaza, nucellus, and micropyle lie in a direct line. Owing to its common presence in the gymnosperms, several authors consider the orthotropous ovule to be primitive. It is probable, however, that in at least several angiospermous families the orthotropous condition has been derived from the anatropous one, and may be correlated with the trend to mono-ovuly. The changeover to orthotropy has consequences for the pathway of the pollen tube. Orthotropous and anatropous ovules are connected by a complete range of transitional stages. If the nucellus forms an angle of about 90° with the funicle, the ovule is called hemi(ana)tropous. On the other hand, in the circinotropous ovule the curvature of the funicle exceeds 180° and it encircles the ovule more or less completely (Cactaceae, Plumbaginaceae).

Ovules with curved embryo sacs occur in many families of the Centrospermae, Cruciferae, Leguminosae, Alismataceae, Butomaceae and others. In most textbooks a distinction is made between campylotropous ovules characterized by a slightly curved, kidney-shaped embryo sac, and amphitropous ovules with a more strongly curved, horseshoe-shaped embryo sac. Since these do not provide clear-cut criteria and do not reflect ontogenetic aspects, other workers use the term campylotropous for all ovules with curved embryo sacs (Corner 1976). According

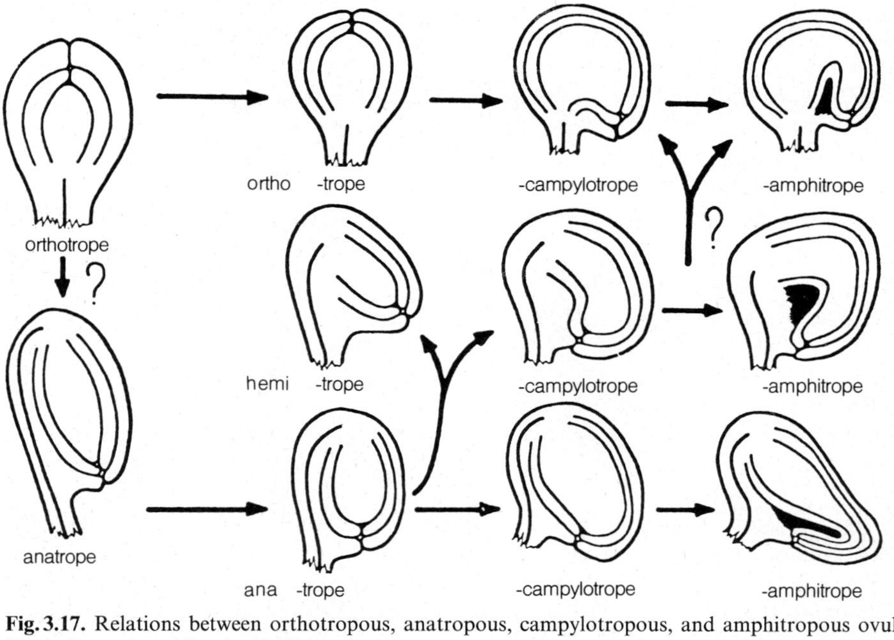

Fig. 3.17. Relations between orthotropous, anatropous, campylotropous, and amphitropous ovules. (From Bocquet and Bersier 1960)

to Bocquet (1959), and Bocquet and Bersier (1960), the bending of amphitropous ovules is accompanied by the formation of a so-called basal body which extends into the arch of the nucellus. They clearly demonstrated that campylotropy and amphitropy are independent processes in respect of anatropy. Combining the basic series of ortho-, hemi-, and anatropy with campylotropy and amphitropy, they distinguish nine different types of ovules (Fig. 3.17). This classification has not come into general use, and does not seem adequate to cover the great diversity. For instance, the basal body may be of different ontogenetic origin. It is sometimes formed by the nucellus or raphe, sometimes by the inner or outer integument, or by two or more in combination. Moreover, a curved embryo sac may also develop secondarily from a straight nucellus as a result of the resorbing activity of embryo sac or endosperm. Curvature of the embryo sac is clearly a derived character, and has the advantage that the embryo may become twice as long as the seed.

The great diversity in the distribution and structure of ovules with curved embryo sacs renders it plausible that they have developed more than once, and along several developmental pathways during the evolution of angiosperms.

3.8 Vascular Supply of the Ovule

Most angiospermous ovules have a very simple type of vascularization: the funicle and raphe contain a single bundle which ends blindly or fans out in the chalaza. But little attention has been paid to the structure of the funicular and raphal bundle. Collateral bundles seem to be most common but, probably, amphicribral ones are not rare (e.g., Cruciferae, Malvaceae, Pittosporaceae, Rutaceae, Styracaceae, Zingiberaceae). Multiple raphal bundles occur in Erythroxylaceae, Juglandaceae, Myricaceae, Myristicaceae, Papilionaceae, and Polygonaceae. Orthotropous ovules show a tendency toward a more concentric arrangement of vascular elements.

In a fairly large number of families the outer or single integument may also become vascularized and, in rarer cases, even the inner integument or the nucellus. By analogy with floral morphology, the importance of the ovular vascular supply is not consistently appraised in current literature. Some authors consider integumentary bundles as original and as representing vestiges of more extensive vascular patterns in the progenitors of the angiosperms, whereas others suppose, from a more functional point of view, that vascular bundles simply differentiate "where they are needed", so that their presence does not have any phylogenetic implications. There seems to be a general relation between the size of ovules and seeds and the degree of vascularization. Large ovules and seeds mostly have a more extensive vascular supply, whereas small ones show reductions. Xylem elements may become differentiated only in later developmental stages, or not at all, and the raphal bundle remains procambium-like (Begoniaceae, Gentianaceae). In the smallest ovules (Orchidaceae) there is not even a trace of any provascular tissue in the raphe.

Vascular bundles in the outer integument are present in at least 80 families (e.g., Annonaceae, Cucurbitaceae, Euphorbiaceae, Leguminosae, Magnoliaceae, Myristicaceae, Rosaceae, Sterculiaceae, and Vitaceae). The vascular pattern varies from a single bundle in the antiraphe to an extensive network of bundles (le Monnier 1872, Kühn 1928). In monocotyledons integumentary vascularization is more incidental.

Vascularization of the single integument occurs in, e.g., Alangiaceae, Boraginaceae, Compositae, Convolvulaceae, Juglandaceae, Myricaceae, and Oleaceae.

Vascularized inner integuments are far less common; and have only been reported in at least some species of Dipterocarpaceae, Elaeocarpaceae, Euphorbiaceae, and Myristicaceae (Corner 1976).

The problem of nucellar vascularization of the angiosperms has received much attention, especially in connection with the presence of vascular bundles in the nucellus of several extinct gymnosperms (Trigonocarpales and Cardiospermales). Vascular elements in the nucellus have been reported in taxa belonging to about 14 angiospermous families. The records vary from isolated tracheidal cells around the embryo sac (Asclepiadaceae, Casuarinaceae), or a subepidermal sheath of tracheid-like cells in *Luffa* (Gupta and Rajeswari 1977), to distinct vascular bundles connected with the vascular supply of the raphe as in Thymelaeaceae (Corner 1976) and Euphorbiaceae (Landes 1946). Except for the monogeneric Casuarinaceae, no other angiospermous family shows nucellar vascularization as a general character. Many recorded cases are poorly documented or have been contradicted. The whole subject needs a fresh reappraisal, especially since it is doubtful whether the reported structures are indeed true xylem elements or simply cells with reticulate cell thickenings, comparable with those sometimes found in integuments ("tracheidal seeds", Corner 1976).

3.9 Special Structures

The nucellar and chalazal tissue facing the antipodal end of the embryo sac may differentiate into a "hypostase" (van Tieghem 1901). Both structure and function of the hypostase are poorly known and, in recent embryological literature, the term hypostase is used in a rather loose sense (Singh and Gupta 1967). The features of the hypostase are rather diverse in different taxa: dense cytoplasm, accumulation of tannin-like substances, impregnation, and thickening of cell walls with cutin, suberin, lignin, or callose. The hypostase is situated above the chalazal vasculature and, depending on the form of the nucellar base, it may be present as a cluster of cells, or a disk- or cuplike cell plate. It may be extant only in the mature ovule, and in certain stages of seed development, but may also persist in the mature seed. The hypostase is known in many families, e.g., Amaryllidaceae, Cyperaceae, Liliaceae, Zingiberaceae, Anacardiaceae, Bixaceae, Euphorbiaceae, Lauraceae, Sapindaceae, Theaceae. Many functions have been attributed to the hypostase: it may act as a barrier tissue for stopping the encroachment of the embryo sac (see Venkateswarlu and Prakasa Rao 1972); it may connect the vascular supply with the embryo sac and, thus, facilitate transport of nutritional material (Tilton 1980 a); it may be responsible for the production of certain enzymes or hormones; play a role in the water balance of dormant seeds, or have a protective function in mature seeds. In the mature seeds of Celastraceae, Chloranthaceae, Clusiaceae, Flacourtiaceae, Staphyleaceae, Violaceae, and Vitaceae the chalazal cells become clearly sclerified, and plug the opening in the endotestal or exotegmic layer (Corner 1976). Undoubtedly, the hypostase has a variable structure and is multifunctional. Much detailed research is needed to elucidate its precise characters in different taxa.

The "podium" (K.V.O. Dahlgren 1940) and "postament" (Westermaier 1890) are structures more or less comparable to the hypostase in that they consist of nucellar tissue resistant to the absorbing activity of the embryo sac. A podium is a cuplike remnant of the nucellar base, which looks like a pedestal for the embryo sac. When a postament is present, only the axial part of the nucellar base persists, surrounded by the embryo sac and, often, having at its apex the antipodals (Ranunculaceae), or the chalazal endosperm chamber (several monocotyledonous families).

Also, at the micropylar end of the embryo sac nucellar remnants may persist. This is called "epistase," which plugs the micropyle, and is mostly formed by the nucellar cap or pad. It is known only in a few families: Zingiberaceae, Nymphaeaceae, and in some representatives of the Myrtales.

In *Ornithogalum* the cells of the nucellar cap and inner integument secrete an exudate, which fills the micropyle. After fertilization the micropyle is occluded by a plug of flocculent material, which probably serves to preclude pathogen invasion and prevent desiccation of embryo and endosperm (Tilton 1980 b).

In the direct vicinity of the micropyle often a special structure, the "obturator", is present, which has a function in the guidance and growth of the pollen tube. This obturator varies appreciably in structure and origin. In ovules with a distinct funicle the obturator is generally of funicular origin, while in more sessile ovules it is a derivative of the placenta or of a combination of placenta and funicle. It can be a prominent, often subdermal, bulge covered by secretory hairs, or simply a region with a papillate epidermal layer (Tilton and Horner 1980). Peterson et al. (1979) made a detailed ultrastructural study of this tissue in *Caltha*. On the basis of the presence of deeply staining cytoplasm, thick outer cell walls with electron-dense wall ingrowths, secreted material outside the cell walls, and numerous dictyosomes, they characterize the cells as transfer cells (Fig. 3.18 A, B). The transfer cells are supposed to be involved in the short-distance transport of metabolites and/or the secretion of chemotropic substances for the growth of the pollen tube. The surface of the obturator opens on to the micropyle, or may fill the micropyle, and is mostly continuous with the pollen-guiding tissue which covers the stigma, style, and locule cavity of the ovary. After pollination, the obturator degenerates.

In mature ovules the initials of "arils," "aril-like" appendages, micropylar collar, and operculum can already be distinguished (Fig. 3.19 A–C). Since these structures mainly function in seed dispersal, or seed germination, they will be discussed in Chapter 12 on seeds.

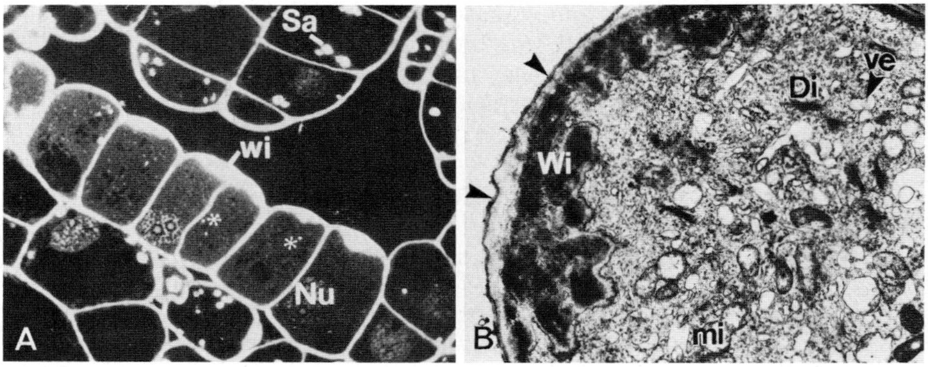

Fig. 3.18 A, B. *Caltha palustris*, obturator. **A** Carpel cross section treated with pseudo-Schiff's reagent for fluorescence microscopy, showing transfer cells (with wall ingrowths) lining locular cavity immediately beneath the micropylar end of ovule. **B** Electron micrograph of transfer cell with prominent wall ingrowth and external osmiophilic layer (*arrows*). Cytoplasm with dictyosomes produce a large number of vesicles or cisternae containing flocculent material. *Di* dictyosome, *mi* mitochondria, *Nu* nucellus, *Sa* starch, *ve* vesicles, *Wi* wall ingrowth. (From Peterson et al. 1979)

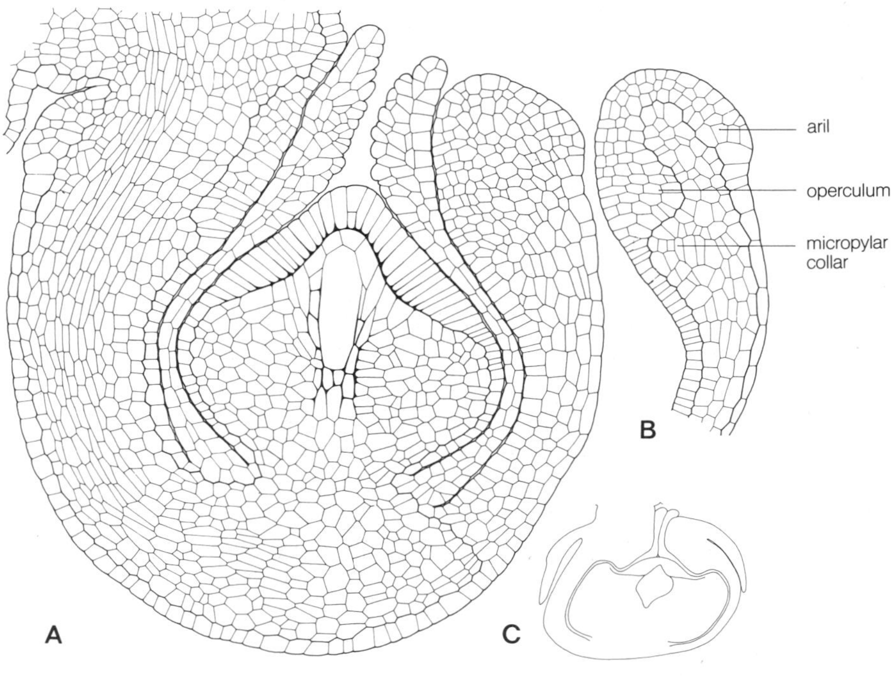

Fig. 3.19 A–C. *Costus cuspidatus*. **A** Longitudinal section of a fully developed ovule. **B** Detail of the apex of the outer integument showing initiation of aril, operculum, and micropylar collar. **C** Outline of an immature seed. (From Grootjen and Bouman 1981)

3.10 Ovule Reduction

The complexity of the process of ovule reduction in combination with superficial or incomplete studies has resulted in many misinterpretations and confusion in terminology. Fundamentally, the reduction of the ovule can be analyzed as the reduction of the nucellus, the integument, or of the entire ovule.

The process of nucellar reduction, to a smaller or greater extent, can be observed in a number of sympetalous families. The nucellus tends to become smaller and, ultimately, consists of only an archesporial cell partly immersed in the tissue lying between the attachment of the integument, and partly extending – covered by some epidermal cells – between the integument proper (Bhandari et al. 1976). A convincing series leading to complete reduction has been suggested by Fagerlind (1937) within the family Rubiaceae (Fig. 3.20 A–E), starting from the Phyllis type with archesporial cells partly extending between the free part of the integument, via the Bouvardia type with archesporial cells fully merged in the "chalazal" tissue but covered by several nucellar epidermal cells, and ending in the Oldenlandia type with mostly only one epidermal cell remaining. In the Rubia olivieri type the morphology of the ovule is obscured by a prominent stretching and periclinal division of the nucellar epidermis. Other instances of extreme nucellar reduction occur in the Opiliaceae (*Cansjera, Opilia*) and Santalaceae. The situation of the integumentary initials at the apex of the ovular primordium determines the size of the young nucellus and, therefore, nucellar reduction cannot only be described as an inhibition of the growth of the nucellar primordium, but also in terms of a more apical initiation of the integument.

The reduction of the integument, ultimately, results in ategmic ovules which lack a micropyle. Taxa showing a progressive range of integumentary reduction, not preceded by or concomitant with nucellar reduction, have so far not been recorded. The amaryllidaceous genus *Crinum* may provide an example.

Several allegedly ategmic ovules show rudiments of the integument in the form of a locally dividing epidermal layer, and had better be defined as "undifferentiated ovules". In most cases the ovular reduction is a complex process involving nucellar reduction, integumentary reduction, and reductions of the funicular and raphal tissues combined with the loss of anatropous curvature. The distinct reduction processes may start at different developmental stages, and operate at different rates. This is convincingly demonstrable within the Santalales. Indian embryologists have substantially contributed to our knowledge of the families of this

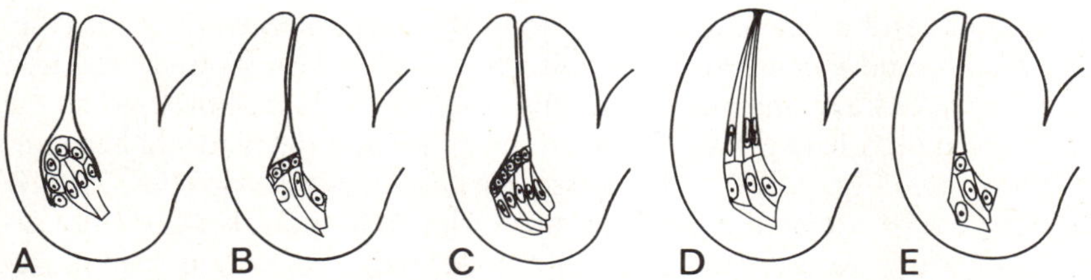

Fig. 3.20 A–E. Diagrams illustrating progressive reduction of the nucellus in Rubiaceae. **A** *Phyllis*. **B** *Bouvardia*. **C** *Vallantia*. **D** *Rubia*. **E** *Oldenlandia*. (From Fagerlind 1937)

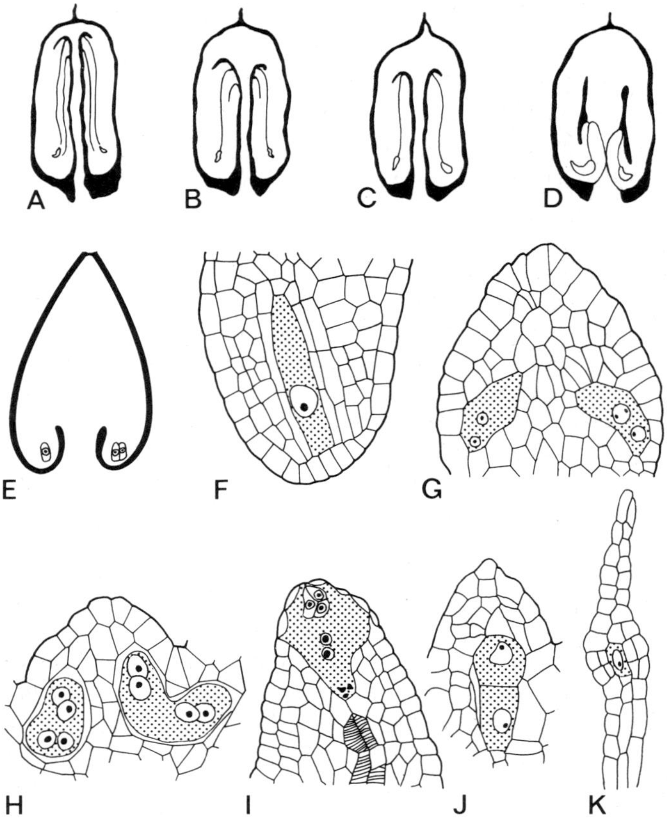

Fig. 3.21 A–K. Progressive reduction of ovules and placental-ovular complex in Santalales and *Balanophora*. **A** *Octoknema orientalis*, bitegmic ovules. **B** *Heisteria cyanocarpa*, fusion of outer and inner integuments. **C** *Tetrastylidium janeirense*, unitegmic ovules. (**A–C** From Fagerlind 1948). **D** *Thesium wightianum*, unitegmic hemianatropous ovules. (From Bhatnagar and Agarwal 1961). **E, F** *Santalum album*, ategmic ovules. (From Paliwal 1956). **G–J** Placental-ovular complexes showing reorientation and reduction of the sites of megaspore mother cells: **G** *Arceuthobium douglasii*. (From Bhandari and Nanda 1968). **H** *Korthalsella dacrydii*. (From Rutishauser 1935). **I** *Exocarpus bidwillii*. (From Bhatnagar and Joshi 1965). **J** *Eubrachion ambiguum*. (From Bhandari and Indira 1969). **K** *Balanophora elongata*, female flower. (From Fagerlind 1945)

order. The Olacaceae (Fig. 3.21 A–C) are the most primitive, and have bitegmic next to unitegmic and ategmic species. Among the Santalaceae, *Thesium* (Fig. 3.21 D–F), *Iodina*, and *Comandra* have unitegmic ovules with strongly reduced nucelli. Most genera have ategmic ovules, and the reduction can proceed so far that even the boundary between the rudimentary ovule and the placenta becomes obscure (*Exocarpus*, Fig. 3.21 I). This is also the case in Viscaceae (Fig. 3.21 G, H, J) where the ovules are no longer discernible. The archesporial cells differentiate directly along the placental surface. Within this family there is a gradual reduction of this placental-ovular complex and, in *Helixanthera* (Loranthaceae), a placenta is absent; the archesporial cell differentiates at the base of the ovarian cavity. Also, in the doubtfully related Balanophoraceae, a placental-ovular complex is present. In *Balanophora* (Fig. 3.21 K) itself the reduction has gone to extremes, and even involves the ovarian tissue (Engell 1979).

3.11 Concluding Remarks

The ovule is the forerunner of the seed. As such any information on the development, structure, and functions of the ovule is of paramount importance for the understanding of the reproductive processes in seed plants. The structure and functions of the seeds cannot be properly understood without studying the ovule in all its aspects. The embryological publications of the last decades have contributed much to the improvement of our cognizance of ovules. Nevertheless, there is still a large amount of research to be undertaken.

Detailed ontogenetic research into the initiation of ovules and integuments has stagnated since the end of the nineteenth century and restarted only recently. Certain problems that have to be solved include the relations between the different types of ovular primordia and the nucellar structure, such processes as unitegmization and dermalization of integuments, and the different types of campylotropy. In view of the existing diversity in the development and structure of the ovules, the ontogenetic potencies of the ovule are impressive. The comparative morphology of ovules is in its infancy, and may prove to be helpful in establishing evolutionary relationships at higher taxonomic levels.

The evolution of the seed plants is still one of the leading topics in botanical research. It is self-evident that the evolution of the seed and the ovule must undisputably be a part of this problem. There is not much clarity as regards the similarities and differences between gymnospermous and angiospermous types of ovules and the structure of the primitive ovule. Main evolutionary trends in ovule characters such as anatropy-atropy, crassinucellate-tenuinucellate, bitegmy-unitegmy-ategmy, subdermal-dermal integuments must be fitted into our overall evolutionary approach.

The ultrastructural studies of ovules have more or less become a routine. Electron microscopic studies of the cellular organization of the meiocyte, megaspores, embryo sac, and their surrounding sporophytic tissues, in combination with histochemical studies, will eventually solve such problems as the selection of the functional megaspore, the different pathways of the nutrient flow, the presence of specialized transfer cells or blocking tissues, etc.

References

Asplund E (1920) Studien über die Entwicklungsgeschichte der Blüten einiger Valerianaceen. K Sven Vetenskapakad Handl 61:1–66

Bednara J, Rodkiewicz B (1974) Megasporocyte and megaspore ultrastructure in *Epilobium*. Bull Acad Pol Sci Ser Sci Biol (2)22:847–850

Berg RY (1962) Contribution to the comparative embryology of the Liliaceae: *Scoliopus, Trillium, Paris*, and *Medeola*. Nor Vidensk Akad Oslo I NS 4:1–64
Bhandari NN, Indira K (1969) Studies on the Viscaceae IV. Embryology of *Eubrachion* (Hook. et Arn.) Engl. Bot Not 122:183–203
Bhandari NN, Nanda K (1968) Studies in the Viscaceae II. A reinvestigation of the female gametophyte of *Arceuthobium douglasii*. Am J Bot 55:1028–1030
Bhandari NN, Bouman F, Natesh S (1976) Ovule ontogeny and seed coat structure of *Scrophularia himalensis* Royle. Bot Jahrb Syst 95:535–548
Bhatnagar SP, Agarwal S (1961) Morphological and embryological studies in the family Santalaceae VI. *Thesium* L. Phytomorphology 11:273–282
Bhatnagar SP, Johri BM (1972) Development of angiosperm seeds. In: Kozlowski TT (ed) Seed biology, vol I. Academic Press, New York London, pp 77–149
Bhatnagar SP, Joshi PC (1965) Morphological and embryological studies in the family Santalaceae VII. *Exocarpus bidwillii* Hook. f. Proc Natl Inst Sci India Part B Biol Sci 31:34–44
Bocquet G (1959) The campylotropous ovule. Phytomorphology 9:222–227
Bocquet G, Bersier JD (1960) La valeur systématique de l'ovule: Développements tératalogiques. Arch Sci (Geneva) 13:475–496
Boesewinkel FD (1980) Development of ovule and testa of *Linum usitatissimum* L. Acta Bot Neerl 29:17–32
Bor J (1978) A note on anatropy versus orthotropy. Phytomorphology 28:219–224
Bor J, Bouman F (1974) Development of ovule and integuments in *Euphorbia milii* and *Codiaeum variegatum*. Phytomorphology 24:280–296
Bor J, Kapil RN (1976) Anatropy and ontogeny of the bitegmic ovule in *Chrozophora* A.H.L. Jussieu (Euphorbiaceae). Acta Bot Neerl 25:385–400
Bouman F (1974) Developmental studies of the ovule, integuments, and seed in some angiosperms. Thesis Univ Amsterdam
Bouman F (1975) Integument initiation and testa development in some Cruciferae. Bot J Linn Soc 70:213–229
Bouman F (1978a) Ovule initiation, ovule development, and seed coat structure in angiosperms. In: Varghese TM (ed) Vistas in plant sciences, vol 5. Intl Bio Sci, Hissar
Bouman F (1978b) Integumentary studies in the Polycarpicae V. *Nigella damascena* L. Acta Bot Neerl 27:175–182
Bouman F, Calis JIM (1977) Integumentary shifting – A third way to unitegmy. Ber Dtsch Bot Ges 90:15–28
Bouman F, Schrier S (1979) Ovule ontogeny and seed coat development in *Gentiana* with a discussion on the evolutionary origin of the single integument. Acta Bot Neerl 28:467–478
Brongniart A (1827) Mémoire sur la génération et développement de l'embryon dans les végétaux phanerogames. Ann Sci Nat 12:14–53, 145–172, 225–296
Brown R (1826) Character and description of *Kingia*. In: King PP (ed) Voyages of discovery, vol 2. London, pp 534–565
Brunkener L (1977) Spore-producing and apical meristems in vascular plants – A comparison. Bot Not 130:189–202
Corner EJH (1976) The seeds of dicotyledons, 2 vols. Cambridge Univ Press, Cambridge London New York Melbourne
Dahlgren KVO (1927) Die Morphologie des Nuzellus mit besonderer Berücksichtigung der deckzellosen Typen. Jahrb Wiss Bot 67:347–426
Dahlgren KVO (1940) Postamentbildungen in den Embryosäcken der Angiospermen. Bot Not 1940:347–369
Dahlgren R (1975) The distribution of characters within an angiosperm system. 1. Some embryological characters. Bot Not 128:181–197
Dahlgren R (1980) A revised system of classification of the angiosperms. Bot J Linn Soc 80:91–124
Davis GL (1966) Systematic embryology of the angiosperms. Wiley, New York London Sydney
Engell K (1979) Morphology and embryology of Scybalioideae (Balanophoraceae). I. Bot Tidsskr 73:155–166
Engell K, Petersen GB (1977) Integumentary and endothelial cells of *Bellis perennis*. Bot Tidsskr 71:237–244

Erdelska O (1975) Pre-fertilization development of ovule of *Jasione montana* L. Phytomorphology 25:76–81

Fagerlind F (1937) Embryologische, zytologische und bestäubungsexperimentelle Studien in der Familie Rubiaceae nebst Bemerkungen über einige Polyploiditätsprobleme. Acta Horti Bergiani 11:195–470

Fagerlind F (1945) Blüte und Blütenstand der Gattung *Balanophora*. Bot Not 1945:330–350

Fagerlind F (1948) Beiträge zur Kenntnis der Gynäceummorphologie und Phylogenie der Santalales-Familien. Sven Bot Tidskr 42:195–229

Fagerlind F (1961) The initiation and early development of the sporangium in vascular plants. Sven Bot Tidskr 55:299–312

Fritsche J (1835) Über die Entwicklung des Pflanzeneies in seinen frühesten Zuständen und über die Bildung der Häute derselben. Arch Naturgesch (Wiegmann) 1 Band 2:229–332

Gaertner J (1788–1798) De fructibus et seminibus plantarum, 3 vols. Stuttgart

George GP, George RA, Herr JM Jr (1979) A comparative study of ovule and megagametophyte development in field-grown and greenhouse-grown plants of *Glycine max* and *Phaseolus aureus* (Papilionaceae). Am J Bot 66:1033–1043

Godineau JC (1968) Ultrastructure des différents tissus de l'ovule du *Crepis tectorum* L. au moment de la prophase méiotique. CR Acad Sci Paris 266:1008–1010

Godineau JC (1971) Ultrastructure of the embryo sac of *Crepis tectorum* L. after cell formation and fusion of polar nuclei. Ann Univ ARERS 9:78–88

Greene CW (1978) A Nomarski-interference study of megasporogenesis and megagametogenesis in *Smelowskia calycina* (Cruciferae). Am J Bot 65:353–358

Grew N (1671) The anatomy of plants begun. London. Also in: The anatomy of plants (1682). London

Grootjen CJ, Bouman F (1981) Development of ovule and seed in *Costus cuspidatus* (Zingiberaceae), with special reference to the operculum. Bot J Linn Soc 83:27–39

Guignard L (1889) Etude sur les phénomènes morphologiques de la fécondation. Bull Soc Bot Fr 36:100–146

Gupta SC, Rajeswari VM (1977) Nucellar tracheids in ovules of *Luffa cylindrica*. Ann Bot 41:1005–1007

Hakki MI (1972) Blütenmorphologische und embryologische Untersuchungen an *Chenopodium capitatum* und *Chenopodium foliosum* sowie weiteren Chenopodiaceae. Bot Jahrb Syst 93:178–330

Hamann U (1966) Embryologische, morphologisch-anatomische und systematische Untersuchungen an Philydraceen. Willdenowia Beih 4:1–178

Heel WA van (1976) Distally-lobed integuments in *Exochorda, Juglans, Leontice* and *Bongardia*. Phytomorphology 26:1–4

Herr JM Jr (1971) A new clearing-squash technique for the study of ovule development in angiosperms. Am J Bot 58:785–790

Herr JM Jr (1973) The use of Nomarski-interference microscopy for the study of structural features in cleared ovules. Acta Bot India 1:35–40

Hofmeister W (1849) Die Entstehung des Embryo der Phanerogamen, eine Reihe mikroskopischer Untersuchungen. Leipzig

Johri BM (1963) Female gametophyte. In: Maheshwari P (ed) Recent advances in the embryology of angiosperms. Intl Soc Plant Morphologists, Univ Delhi, pp 69–103

Kapil RN, Prakash N (1969) Embryology of *Cereus jamacaru* and *Ferocactus wislizenii* and comments on the systematic position of the Cactaceae. Bot Not 122:409–426

Kapil RN, Tiwari SC (1978 a) Plant embryological investigations and fluorescence microscopy: An assessment of integration. Intl Rev Cytol 53:291–331

Kapil RN, Tiwari SC (1978 b) The integumentary tapetum. Bot Rev 44:457–490

Kordyum EL (1968) Peculiarities on early ontogeny in ovules with different type of archesporium in some representatives of angiosperms (in Russian with English summary). Tsitol Genet 2:415–424

Kühn G (1928) Beiträge zur Kenntnis der intraseminalen Leitbündel bei den Angiospermen. Bot Jahrb Syst 61:325–385

Kuran H (1972) Callose localization in the walls of megasporocytes and megaspores in the course of development of monospore embryo sacs. Acta Soc Bot Pol 41:520–534

Landes M (1946) Seed development in *Acalypha rhomboidea* and some other Euphorbiaceae. Am J Bot 33:562–568

Lintilhac PM (1974) Differentiation, organogenesis, and the tectonics of cell wall orientation. I. Preliminary observations on the development of the ovule in cotton. Am J Bot 61:129–140
Malpighi M (1675) Anatome plantarum. London
Mauritzon J (1939) Contribution to the embryology of the orders Rosales and Myrtales. Lunds Univ Arsskr (2)35:1–121
Maze J, Bohm LR (1973) Comparative embryology of *Stipa elmeri* (Gramineae). Can J Bot 51:235–247
Mirbel CFB (1815) Eléments de physiologie végétale et de botanique, 2 vols. Paris
Mirbel CFB (1829) Nouvelles recherches sur la structure et les développements de l'ovule végétale. Ann Sci Nat 17:302–318
Mirbel CFB (1830) Additions aux nouvelles recherches sur la structure et les développements de l'ovule. Mem Acad Sci Inst France 9:629–684
Mogenson HL (1981) Translocation of uranin within the living ovules of selected species. Am J Bot 68:195–199
Monnier G le (1872) Recherches sur la nervation de la graine. Ann Sci Nat Bot (5)16:233–305
Narayana LL (1957) Embryology of two Simaroubaceae. Curr Sci (Bangalore) 26:323–324
Netolitzky F (1926) Anatomie der Angiospermen-Samen. In: Linsbauer K (ed) Handbuch der Pflanzenanatomie. Borntraeger, Berlin
Newcomb W (1973a) The development of the embryo sac of sunflower *Helianthus annuus* before fertilization. Can J Bot 51:863–878
Newcomb W (1973b) The development of the embryo sac of sunflower *Helianthus annuus* after fertilization. Can J Bot 51:879–890
Noher de Halac I (1980) Fine structure of the nucellar cells during development of the embryo sac in *Oenothera biennis* L. Ann Bot 45:515–521
Paliwal RL (1956) Morphological and embryological studies in some Santalaceae. Agra Univ J Res Sci 5:193–284
Péchoutre F (1902) Contribution à l'étude du développement de l'ovule et de la graine des Rosacées. Ann Sci Nat Bot (8)16:1–158
Peterson RL, Scott MG, Miller SL (1979) Some aspects of carpel structure in *Caltha palustris* L. (Ranunculaceae). Am J Bot 66:334–342
Philipson WR (1974) Ovular morphology and the major classification of the dicotyledons. Bot J Linn Soc 68:89–108
Rembert DH Jr (1971) Phylogenetic significance of megaspore tetrad patterns in Leguminales. Phytomorphology 21:1–9
Rembert DH Jr (1977) Ovule ontogeny, megasporogenesis, and early gametogenesis in *Trifolium repens* (Papilionaceae). Am J Bot 64:483–488
Rodkiewicz B (1967) Walls with callose in the megaspores and hypostase of ovules of *Antirrhinum majus* observed in a fluorescence microscope. Bull Acad Pol Sci Ser Sci Biol (5)15:493–495
Rodkiewicz B (1970) Callose in cell walls during megasporogenesis in angiosperms. Planta 93:39–47
Rodkiewicz B (1978) Cytoplasm restructuring isolation and polarization in meiocytes during megasporogenesis in angiosperms (in Polish with English summary). Postepy Biol Komorki 5:135–154
Rodkiewicz B, Bednara J (1976) Cell wall ingrowth and callose distribution in megasporogenesis in some Orchidaceae. Phytomorphology 26:276–281
Roth I (1957) Die Histogenese der Integumente von *Capsella bursa-pastoris* und ihre morphologische Bedeutung. Flora 212–233
Rutishauser A (1935) Entwicklungsgeschichtliche und zytologische Untersuchungen an *Korthalsella dacrydii* (Ridl.) Danser. Ber Schweiz Bot Ges 44:389–436
Rutishauser A (1969) Embryologie und Fortpflanzungsbiologie der Angiospermen. Springer, Berlin Heidelberg New York
Satina S (1945) Periclinal chimeras in *Datura* in relation to the development and structure of the ovule. Am J Bot 32:72–81
Schleiden MJ (1837) Einige Blicke auf die Entwicklungsgeschichte des vegetablischen Organismus bei den Phanerogamen. Arch Naturgesch (Wiegmann) 1 Band 3:289–320
Schleiden MJ (1839) Über die Bildung des Eichen und Entstehung des Embryo bei den Phanerogamen. Verh Kais Leop Carol Akad Naturf 11:27–58
Schmitz F (1872) Die Blüthen-Entwicklung der Piperaceen. Hanstein's Bot Abh 2:1–74
Singh D, Gupta S (1967) The seeds of the Violaceae and Resedaceae – A comparison. J Indian Bot Soc 46:248–256

Sporne KR (1969) The ovule as an indicator of evolutionary status in angiosperms. New Phytol 68:555–566
Strasburger E (1872) Die Coniferen und die Gnetaceen. Fischer, Jena
Strasburger E (1879) Die Angiospermen und die Gymnospermen. Fischer, Jena
Strasburger E (1888) Über Kern- und Zelltheilung im Pflanzenreich, Fischer, Jena
Strasburger E (1894) The periodic reduction of the number of the chromosomes in the life-history of living organs. Ann Bot 8:218–316
Subramanyam K (1967) Some aspects of the embryology of *Sedum chrysanthum* (Boiss.) Raymond-Hamlet with a discussion on its systematic position. Phytomorphology 17:240–247
Swamy BGL, Krishnamurthy KV (1970) On the so-called endothelium in the monocotyledons. Phytomorphology 20:262–269
Tieghem P van (1898) Structure de quelques ovules et parti qu'on en peut tirer pour améliorer la classification. J Bot (Paris) 12:197–220
Tieghem P van (1900) Sur la structure de l'ovule et de la graine et sur les affinités des Salicacées. Bull Mus Natl Hist Nat 6:197–201
Tieghem P van (1901) L'hypostase, sa structure et son rôle constantes, sa position et sa forme variables. Bull Mus Natl Hist Nat 7:412–418
Tilton VR (1980a) Hypostase development in *Ornithogalum caudatum* (Liliaceae) and notes on other types of modifications in the chalaza of angiospermous ovules. Can J Bot 58:2059–2066
Tilton VR (1980b) The nucellar epidermis and micropyle of *Ornithogalum caudatum* (Liliaceae) with a review of these structures in other taxa. Can J Bot 58:1872–1884
Tilton VR, Horner HT Jr (1980) Stigma, Style, and obturator of *Ornithogalum caudatum* (Liliaceae) and their function in the reproductive process. Am J Bot 67:1113–1131
Treviranus LC (1815) Von der Entwicklung des Embryo und seiner Umhüllungen im Pflanzen-Ey. Berlin
Turpin P (1806) Sur l'organe par le quel le fluide fécondant peut s'introduire dans l'ovule des végétaux. Ann Mus Hist Nat 7:199–211
Venkateswarlu J, Prakasa Rao PS (1972) Embryological studies in some Combretaceae. Bot Not 125:161–179
Verkerke W, Bouman F (1980) Ovule ontogeny and its relation to seed-coat structure in some species of *Polygala* (Polygalaceae). Bot Gaz 141:277–282
Warming E (1878) De l'ovule. Ann Sci Nat Bot (6)5:177–266
Warming E (1913) Observations sur la valeur systématique de l'ovule. Mind Jap Steenstrup 24:1–45
Westermaier M (1890) Zur Embryologie der Phanerogamen. Nova Acta Leopold 57:1–30
Willemse MTM (1981) Polarity during megasporogenesis and megagametogenesis. Phytomorphology 31:124–134
Willemse MTM, de Boer-de Jeu MJ (1981) Megasporogenesis and early megagametogenesis. Acta Soc Bot Pol 50:105–114
Willemse MTM, Franssen-Verheijen MAW (1978) Cell organelle changes during megasporogenesis and megagametogenesis in *Gasteria verrucosa* (Mill.) Haw. Bull Soc Bot Fr 125:187–191
Wilms HJ (1980) Development and composition of the spinach ovule. Acta Bot Neerl 29:243–260
Young DL, Watson L (1970) The classification of dicotyledons, a study of the upper levels of the hierarchy. Aust J Bot 18:387–433

4 The Female Gametophyte

M. T. M. WILLEMSE and J. L. VAN WENT

4.1 Introduction

The female gametophyte develops from the megaspore formed in the nucellar tissue of the ovule. The megaspore enlarges and after some mitotic divisions a small coenocyte differentiates. In general, a megagametophyte consists, after cytokinesis, of an egg apparatus, a central cell, and antipodal cells.

The development of this highly specialized structure is characterized by a clear polarity from micropyle to the chalaza. At the micropylar side an egg cell and normally two synergids develop, while at the chalazal part three antipodal cells are formed. The largest portion is occupied by the central cell with two polar nuclei which, after fertilization, differentiates into endosperm.

During the development of the megagametophyte, nutrients are supplied by and transferred through the ovular tissue. When the megagametophyte is fully differentiated, its development becomes arrested. The duration of this resting stage depends on the moment of pollination and arrival of the pollen tube. The arrival of the pollen on the stigma activates the ovule which permits fertilization. Fertilization, in turn, triggers further development of the ovule into a seed.

4.2 Types of Megagametophyte Development

During megasporogenesis the formation of a cell wall after the first and second meiotic division usually results in a linear row of four megaspores. As a rule, three megaspores degenerate and only one remains functional. This megaspore is the onset of the megagametophyte. This type of development occurs in most of the angiospermous plants, and is called the monosporic type of megasporogenesis. In a number of plant species the cell-wall formation after the second meiotic division is absent, leading to the formation of a dyad of which each cell contains two nuclei. Degeneration of one cell of the dyad results in the formation of one functional dyad with two nuclei as starting point of the megagametophyte. This type is called the bisporic type. Finally, there are species in which cell wall formation does not occur after the meiotic divisions. Then a coenomegaspore with four nuclei is the onset of the megagemetophyte, called the tetrasporic type (see Fig. 3.6). These three types of megasporogenesis can be followed by various types of megagametogenesis, characterized by a variable number of nuclear divisions. Based on these criteria, several types of embryo sac development are distinguished (Fig. 4.1) (Schnarf 1929, Maheshwari 1950, Johri 1963). Because of fur-

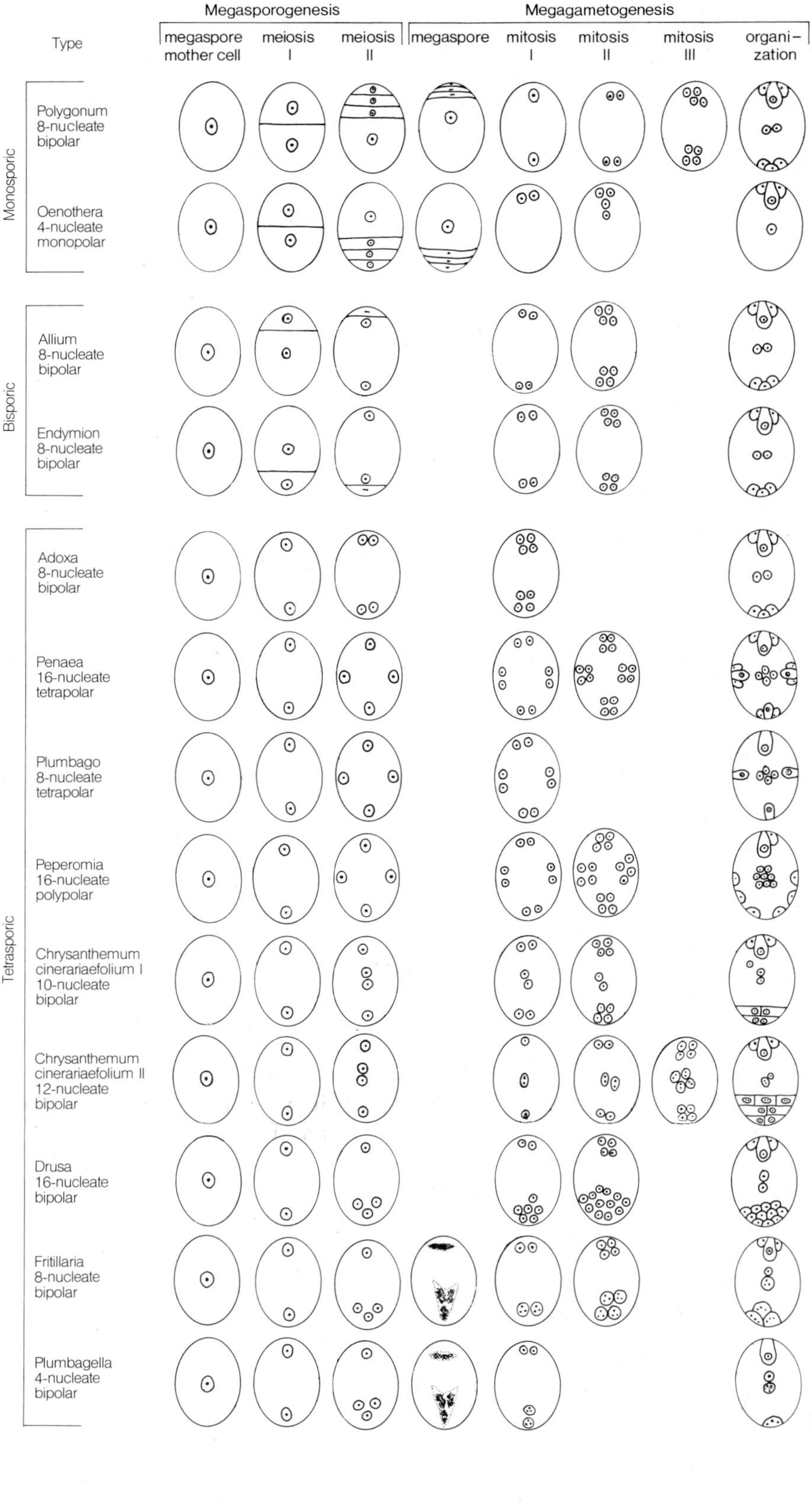

Megasporogenesis
Megagametogenesis
Type
megaspore mother cell
meiosis I
meiosis II
megaspore
mitosis I
mitosis II
mitosis III
organi-zation
Monosporic
Polygonum 8-nucleate bipolar
Oenothera 4-nucleate monopolar
Bisporic
Allium 8-nucleate bipolar
Endymion 8-nucleate bipolar
Tetrasporic
Adoxa 8-nucleate bipolar
Penaea 16-nucleate tetrapolar
Plumbago 8-nucleate tetrapolar
Peperomia 16-nucleate polypolar
Chrysanthemum cinerariaefolium I 10-nucleate bipolar
Chrysanthemum cinerariaefolium II 12-nucleate bipolar
Drusa 16-nucleate bipolar
Fritillaria 8-nucleate bipolar
Plumbagella 4-nucleate bipolar

Fig. 4.1

ther criteria, even more types have been proposed (Rutishauser 1969, Romanov 1971). Other criteria for the differences in embryo sac typology may be a sequence of cellular events. This results in a scheme represented in Fig. 4.2, based on the data of Rutishauser (1969). It becomes clear that the position of the nucleus is a determining factor. This position is influenced by the direction of the spindle, by its migration, fusion or degeneration.

So changes in cytological steps in megasporo- and megagametogenesis give rise to a different cell pattern in the embryo sac. Based on the high number of cytological configurations leading to the Polygonum type, and the high frequency of this type in plants, this process can be considered as basal program in the development. The use of typology is still of great value for comparative embryology, as is shown by Maheshwari (1963) and Johri (1970).

Many alternatives in development can be observed within a distinct type of development. For instance, in *Oenothera* (Noher de Halac and Harte 1975) different pathways have been observed in relation to the selection of the megaspore after meiosis. In *Allium* (Naumova 1978) a delay of postmeiotic mitoses and variation in number can take place, and in Ranunculaceae many types and variations during all developmental stages are present (Vijayaraghavan 1970).

Aberrations are very common during the development of the megagametophyte. The causes for such events in *Ophris* are the variability in megasporogenesis, the behavior of the nuclei of the embryo sac, and insufficient nutrient supply (Savina 1971). The formation of twin embryo sacs originating from two megaspore mother cells (Agarwal 1960, Tilton 1980, Willemse and Kapil 1981) indicates that during the meiotic induction also an aberration can start. All these examples of variation illustrate the variability in megasporo- and megagametogenesis.

Swamy and Krishnamurthy (1975) have simplified the typology of embryo sac development by focusing attention on the exclusive features of the megagametophyte. The attainment of haploidy, the establishment of polarity, an elaboration phase to the formation of the egg apparatus, and a mature organization are recognizable in all types. The micropylar pole of the embryo sac is very regular in structure, whereas the chalazal pole shows a great variability. Of causal nature are factors like the wall formation during meiosis, the number of mitotic divisions after formation of the megaspore, cell-wall formation after mitotic divisions, and the fusion of nuclei. With the time of establishment of polarity as the guiding factor, four types of embryo sacs are distinguished. The Polygonum type as representative for the "Supra-Homeotypic" category, the Allium type as representative of the "Homeotypic" category I, and the Penaea type and Fritillaria type as representatives of the "Homeotypic" category II.

Compared with the other more elaborate typologies, no new characteristics are added, but several fixed points and variable points in the development of the megagametophyte are taken into account. Primarily, the establishment of polarity is chosen as a guiding factor. Indeed, polarity is a factor which acts from the

Fig. 4.1. Schematic representation to show the origin and development of different types of embryo sacs. [Modified after Johri (1963) and Erdelská (1981), courtesy Dr. K. B. Ambegaokar, Delhi]

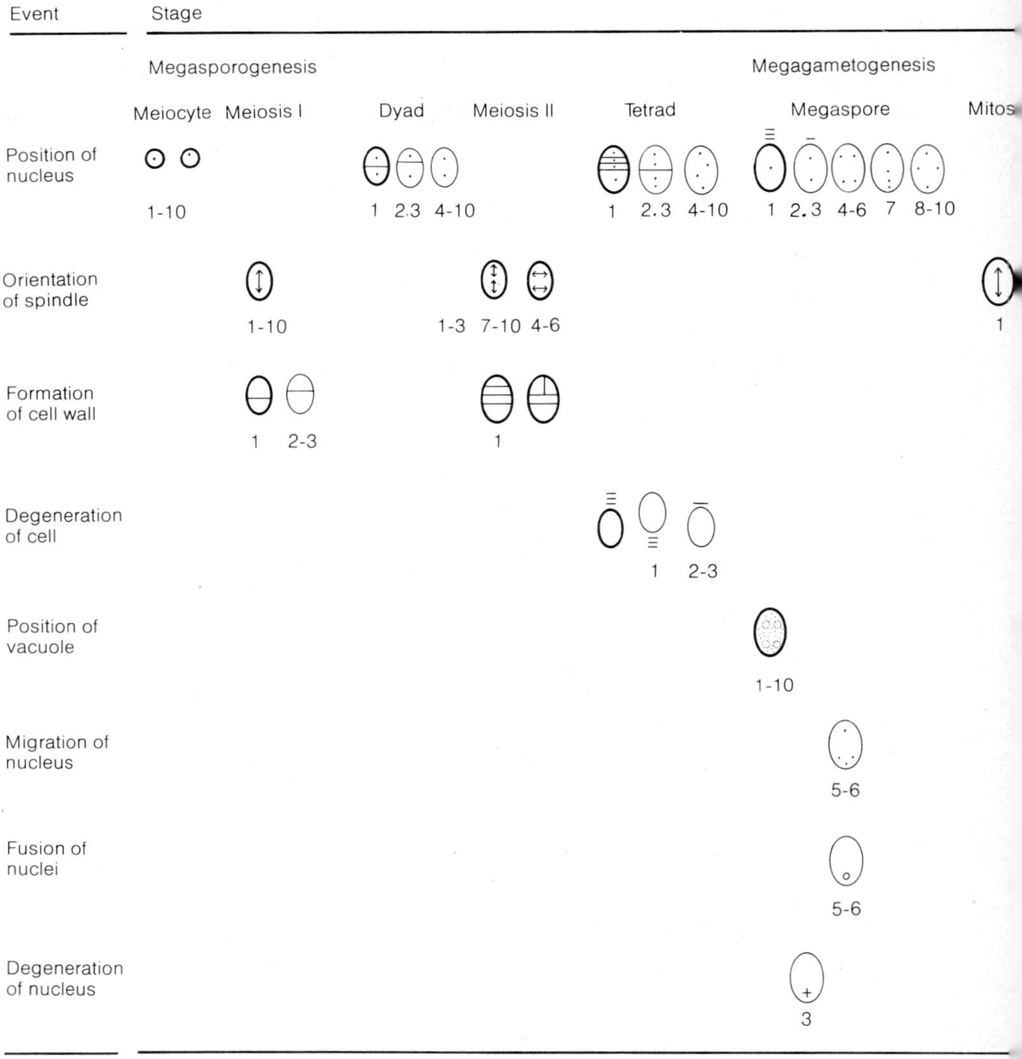

Fig. 4.2. The steps during megasporo- and megagametogenesis are related to the cellular events, presented in the vertical axis. The different types of embryo sac development are related to these events, and arranged according to the Polygonum type which is represented by a *darker line*. The other types are: *1* Oenothera, *2* Allium, *3* Podostemon, *4* Adoxa, *5* Fritillaria, *6* Plumbagella, *7* Drusa I, *8* Penaea, *9* Peperomia II, *10* Plumbago

start of megasporogenesis and determines the existence of the micropylar-chalazal axis.

Meiosis, polarity, and the number of postmeiotic mitoses mainly characterize the typology, but there are other factors also which determine the development of the embryo sac. The developing embryo sac is dependent on the surrounding tissue for the supply of nutrients. In a sterile ovule an embryo sac is formed, but

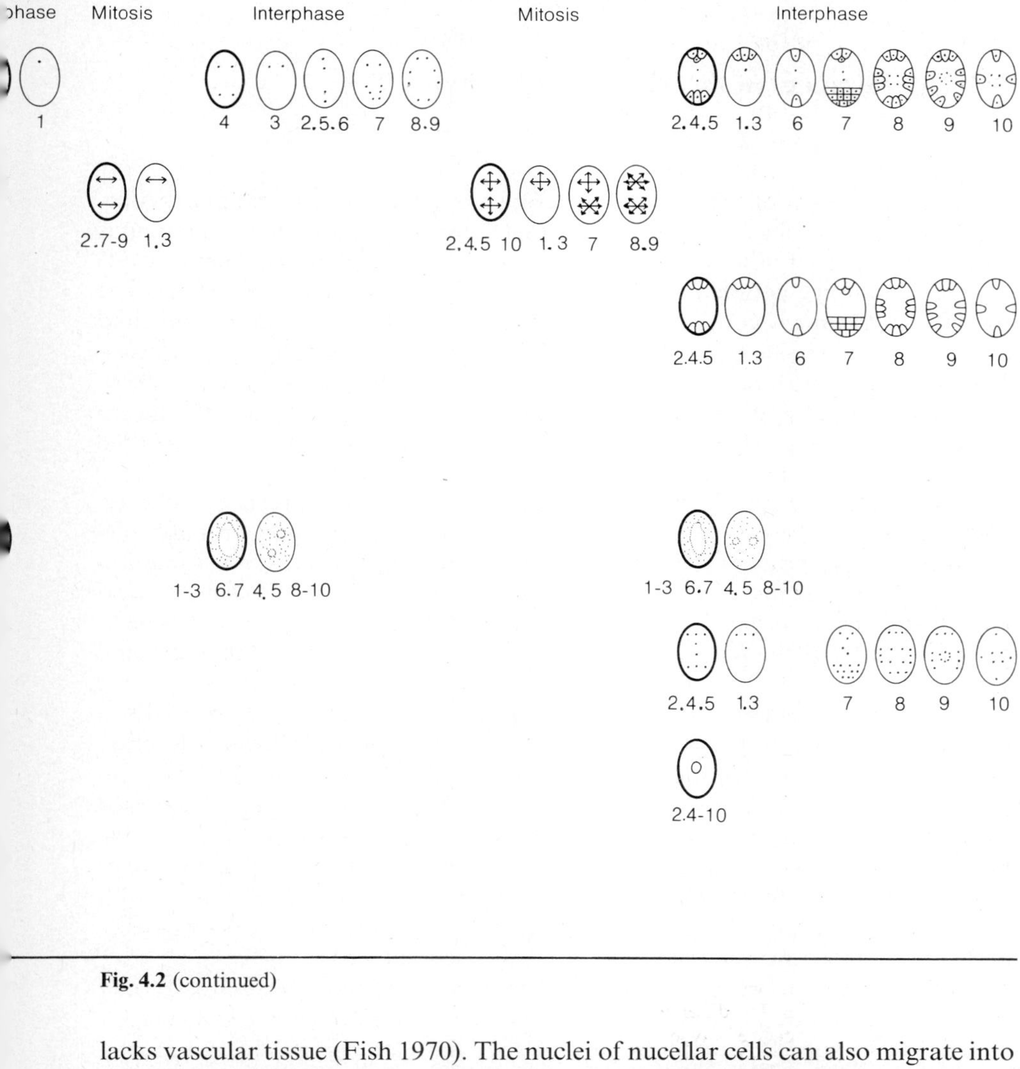

Fig. 4.2 (continued)

lacks vascular tissue (Fish 1970). The nuclei of nucellar cells can also migrate into the developing embryo sac as observed in *Pandanus* (Cheah and Stone 1975), and suggested in *Costus* (Panchaksharappa 1962). Seasonal influences are reported in the ovule of *Capsicum* (Dharamadhaj and Prakash 1978). In winter, a preference of the bisporic type of megasporogenesis has been observed, whereas in summer the monosporic type dominates.

The process of megasporogenesis and the formation of the megagametophyte show distinct phases of development. Such a sequence is genetically determined. The aberrations visible in the nucellar tissue and the effect of an environmental factor show that the process is a result of genetic as well as environmental influences. The induction, and probably the duration of the process, show the genetic base; nutrition and probably temperature are influences from outside, acting on the sequence of embryo sac development.

4.3 Development of the Megagametophyte

4.3.1 The Megaspore

The cellular content of the mature functional megaspore is prepared mostly during the meiotic prophase. Because of unequal divisions, the cell generally contains more than a quarter of the cell contents present just before division. There are differences in storage between the functional megaspores of different species. The plastids of *Impatiens* contain starch, in *Lilium* and *Allium* no starch but lipid bodies are present (De Boer-De Jeu 1978). In the developing megaspore of *Gasteria* both lipid and starch are present (Willemse and Franssen-Verheijen 1978). In *Conium* (Dumas 1971) the production of starch takes place when the megaspore starts growing. There is also a difference in the degree of vacuolation at the onset of the development of the megagametophyte.

The selection of one functional megaspore is based on a prepared polarity. This is visible, for instance, in the position of the nucleus, plastids, probably elements of the membrane system, the contact with the nucellar cells by plasmodesmata, and the position of a callose wall. But polarity is the result of a genetical plan modified by external factors like nutrition and ovular development (Willemse and Bednara 1979, Willemse and De Boer-De Jeu 1981, Willemse and Kapil 1981).

In general, this onset of polarity along the micropylar-chalazal axis results in the selection of one megaspore, but is expressed again in the following development. In *Lilium,* a tetrasporic type, the position of the four nuclei is initially related to the shape of the cell. Polarization at the four-nucleate stage is achieved by the movement of three nuclei to the chalazal pole. Polarity exists also in the position of plastids, and the endomembrane system. In *Allium* the plastids prefer a chalazal position. In *Impatiens* a bisporic type has plasmodesmatal contact with the chalazal nucellar cells; there is no clear polarized position of cell organelles at the onset of the four-nucleate stage (De Boer-De Jeu 1978).

In the central part of the functional megaspore the vacuolation will start only after an augmentation of cell organelles has taken place in the preceding stages.

The just-formed functional megaspore may have storage products and may get nutrients from its dissolving callose wall, if present, through open plasmodesmata from the chalazal nucellar cells. During this stage the nucellar tissue is still differentiating, a structure like S-cells (Korobova 1971), a hypostase, is still developing or still absent. Obviously, in this situation of early differentiation, the functional megaspore partly depends on the nucellar tissue for nutrients. Therefore, the start of the megagametophyte development might be partly autonomous.

Till the stage of the functional megaspore, the subcellular and functional activities during mega- and microsporogenesis are comparable. When growth of the

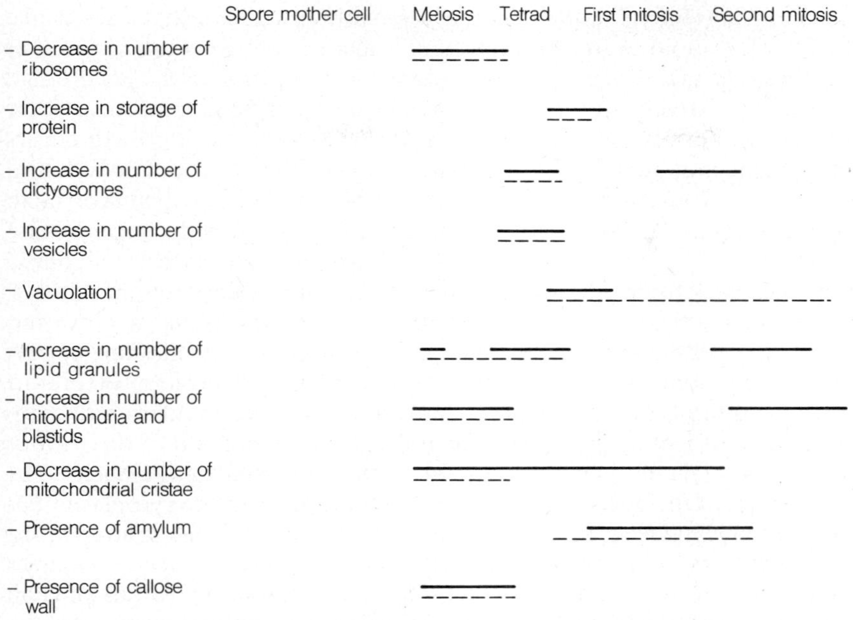

Fig. 4.3. A comparison of some events during the development of micro- (——) and megaspores (– – –) and the micro- (——) and megagametogenesis (– – –)

functional megaspore starts, less similarities remain. Based mainly on the data of ultrastructural studies, some events during sporogenesis are compared in Fig. 4.3 (Willemse and De Boer-De Jeu 1981).

4.3.2 The Coenocytic Megagametophyte

In the Polygonum type from the formation of the functional megaspore to the cellular stage of the embryo sac, three mitoses take place. The duration of this differentiation is very variable, and depends on different conditions. In *Lycopersicum* (Pacini and Sarfatti 1978) it takes about 50 h, and in *Gasteria* about 80 h.

The functional megaspore has a higher osmolarity than the degenerating megaspore(s), and enlarges (Rembert 1977). The maturing megaspore of *Pisum* (Medina 1980) has small vacuoles which coalesce during the formation of the large vacuole of the growing megagametophyte. This vacuole may be involved in the unequal partition of the cytoplasm.

Normally, the vacuole is situated in the micropylar half of the cell. The nucleus of the enlarging megaspore lies near the cell centre at the chalazal end. So a larger micropylar vacuole is present. With the formation of the two-nucleate stages in *Pelargonium,* the two vacuoles fuse and form one central vacuole (Tsai et al. 1973). In *Crepis* (Godineau 1973) and *Brodiaea* (Berg 1978) the micropylar vacuole increases and forms a large central vacuole. In *Stipa* (Maze and Bohm

1973) and *Triticum* (Batygina 1971) a central vacuole as well as a chalazal vacuole are present. This variation in the process of vacuolation can result in another composition of the cytoplasm. But it may also be an indication of the preferential place of cellular enlargement. In relation to the nucellar structure, the micropylar part permits an enlargement because of a lower resistance. Such growth occurs in *Oryzopsis* (Maze et al. 1970) and *Jasione* (Erdelská 1975) but can be followed as in *Dipcadi* by an enlargement in the direction of the chalaza (Chennaveeraiah and Mahabale 1962). A sign of the direction of enlargement is the degeneration of nucellar cells. In *Juglans* (Bouillot 1969) the position of the vacuole and nuclei is variable. The most frequent situation is a central vacuole with, after the mitoses, the nuclei located at the micropylar and chalazal part of the cytoplasm. However, it is possible that all the nuclei are formed and positioned in the micropylar part of the cytoplasm. Some of these nuclei will move to another position before the ultimate formation of the cellular megagametophyte. An accumulation of cytoplasm is always present at the micropylar pole. Before formation of the various cells, the cytoplasm is reorganized followed by the movement of the nuclei. The ultimate position of the nuclei depends on the organization of the cytoplasm. Polarity in the coenocytic megagametophyte is determined by the distribution of the cytoplasm as well as in its organization. The position of the vacuole or vacuoles is an indication for the cell polarity. In some Commelinaceae (Chikkannaiah 1962) and Zingiberaceae (Panchaksharappa 1962) the functional megaspore has only small vacuoles, and an increase of cytoplasm takes place during its enlargement.

In *Gasteria* an increase of the number of plastids, mitochondria, lipid granules, dictyosomes, and ribosomes takes place in the megaspore preceding the nuclear divisions. In the vacuolate four-nucleate stage only polysomes increase in number. So, in general, the main production of cell organelles occurs before vacuolation (Willemse and Franssen-Verheijen 1978).

In the four-nucleate stage, in the micropylar part many dictyosomes are present (Fig. 4.4). In *Allium, Lilium,* and *Impatiens* (De Boer-De Jeu 1978) the number of organelles remains unaltered up to the four-nucleate stage.

Only changes in ER and lipid bodies have been noted. In *Zea* (Russell 1979) the megaspore shows an increase in the number of organelles, while the two-nucleate stage has the maximal amount of organelles and during this stage the vacuole starts to increase. In *Crepis* (Godineau 1973) the number of organelles increases during the two- and four-nucleate stages. The ER forms many vesicles, and the plasma becomes rich in ribosomes. Based on the few quantitative data available, apparently only two different times of augmentation can be distinguished, in the functional megaspore and near the two-nucleate stage.

Of changes in organelles during the coenocytic phase, the functional meaning is obvious. In *Allium, Lilium*, and *Impatiens* (De Boer-De Jeu 1978) the number of mitochondrial cristae increase. This indicates a higher metabolic activity of the coenocyte. Storage products like starch and lipid droplets disappear in *Gasteria;* in *Allium* there is an increase in lipid droplets, and in *Impatiens* the starch increases. These changes can be related to, respectively, use and uptake of nutrients.

In *Zea* Russell (1979) observed a lytic complex formed in the megaspore which increases till the four-nucleate stage. It is situated partly around the vacuole, and

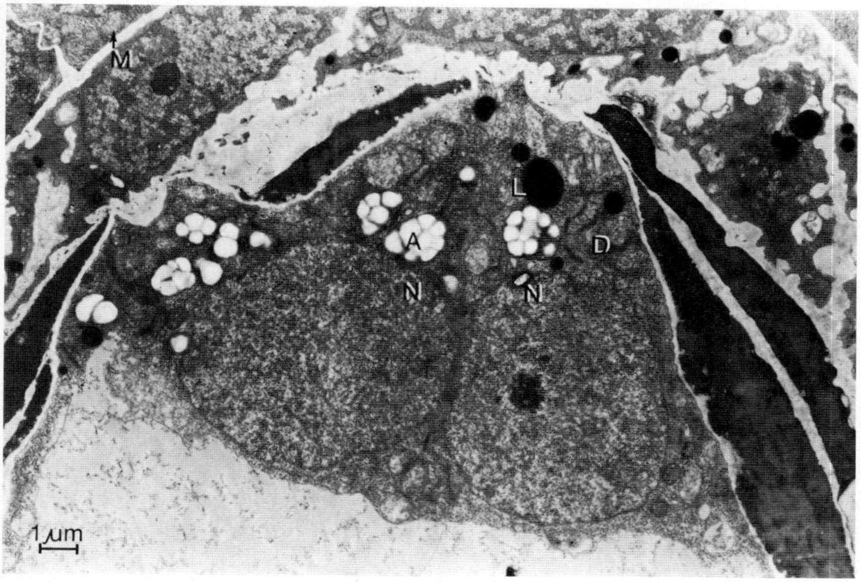

Fig. 4.4. Micropylar side of the four-nucleate megagametophyte of *Gasteria*, two nuclei with the surrounding organelles, *N* nucleus, *A* amyloplast, *L* lipid globule. Note the position of the dictyosomes (*D*) and micropylar side (*M*)

involved in a possible lytic activity which stops at the eight-nucleate stage. The function of this lytic activity is not clear.

In *Gasteria* (Willemse and Franssen-Verheijen 1978) and *Allium* (De Boer-De Jeu 1978) the nuclei enlarge in the coenocytic megagametophyte. This also occurs in *Lilium*. Besides, in *Allium*, in the cytoplasm nucleolar-like bodies have also been observed. Simultaneously the amount of ribosomes, which is high, increases, probably as restoration of the ribosome population (De Boer-De Jeu 1978). In light-microscopic observations the nucleoli show a core.

Histochemical tests on RNA and DNA in *Nicotiana* (Bannikova 1971) reveal a changing chromatin configuration in the nuclei of the two-nucleate stage. During vacuolation the RNA content is high, and at the eight-nucleate stage the micropylar cytoplasm has more RNA than the chalazal. A low RNA reaction has been observed in the area of the future central cell. The nuclei in the micropylar part are larger than the chalazal nuclei, and the largest nucleus in the micropylar zone is reserved for the future egg cell. For *Lilium* (Miki-Hirosige 1964) and *Stellaria* (Pritchard 1964), a higher RNA-level is stated in the young coenocyte, and this high level of RNA corresponds to an increase of protein content of the coenocyte. In *Dipcadi* (Panchaksharappa and Syamasundar 1975) the high level of RNA and proteins in the functional megaspore is very much reduced in the two- and four-nucleate megagametophyte stages. In these species the high content of RNA and protein in the degenerating megaspores perhaps plays a role as a precursor source for the developing megagametophyte. Staining of proteins in *Lilium* (Miki-Hirosige 1964), *Stellaria* (Pritchard 1964), *Zephyranthes* and *Lagenaria*

(Malik and Vermani 1975) shows a high response in the cytoplasm during the coenocytic development. Enzyme staining (Poddubnaya-Arnoldi et al. 1964, Malik and Vermani 1975, Willemse and De Boer-De Jeu 1981) revealed differences in distribution pattern in the coenocyte and embryo sac and surrounding ovular tissue, during ovule ontogeny. The function of these enzymes, in relation to their presence and location is still a topic of discussion (Malik and Vermani 1975, Willemse and Bednara 1979). Carbohydrate staining, as for starch, reducing sugars, and insoluble carbohydrates is, in general, positive but weak. These stainings permit to follow the possible quantitative changes. In *Dianthus* (Buell 1952) the starch content of the coenocyte increases from megaspore to the two- and four-nucleate coenocyte, and decreases till the four- and eight-nucleate coenocyte. A strong increase again takes place just before the stage of the mature embryo sac. Staining with periodic acid–Schiff reagent becomes positive during the coenocytic stage in *Zephyranthes* and *Lagenaria* (Malik and Vermani 1975). Remarkable is the positive reaction in the synergids of the mature embryo sac, which is also observed in *Paspalum* (Chao 1977). In *Stellaria* (Pritchard 1964) polysaccharides are absent. Only scanty data of the lipids during the coenocytic stages are available. In *Lilium* (Miki-Hirosige 1964) and in *Quercus* (Mogensen 1973) lipids seem to be absent.

Because of the variation in the developmental process, the data about the various cytological characters during the coenocytic stage of the megagametophyte cannot be generalized. The specificity of the cellular structure of various types is very clear, and the number of investigations about the cytological changes during the development is insignificant. The developing coenocyte is an active cell, and the rate of this activity is also determined by the surrounding tissue.

Before the formation of the cells in the embryo sac, the enlargement of the megagametophyte is completed. Studies of *Ammannia* (Smith and Herr 1971) show that the maximal size is reached at the eight-nucleate stage, while the greatest increase in growth along the micropylar-chalazal axis mainly occurs at the four-nucleate stage. At the eight-nucleate stage, there is a small reduction in the width. In such studies clearing techniques are involved (Smith 1973, 1974, Greene 1978).

The division of the nuclei of the coenocyte is not always completely synchronous. The nucleus in the chalazal part divides somewhat earlier (Mogensen 1970). After the eight-nucleate stage, the coenocytic megagametophyte becomes partly cellular. This process can start at first at the chalazal part, or it is a simultaneous process (Newcomb 1973).

4.4 The Relation with the Nucellar Tissue

The formation of the coenocytic megagametophyte depends on its surrounding ovular tissues, especially the nucellar tissue. The megagametophyte is surrounded by a thin cell wall containing PAS-positive material and, in combination with the cell walls of the nucellar cells, it becomes a thick covering composed of thin, appressed lamellae (Diboll and Larson 1966). Plasmodesmata are not frequent (Rifot 1971), or can be closed (Kapil 1974).

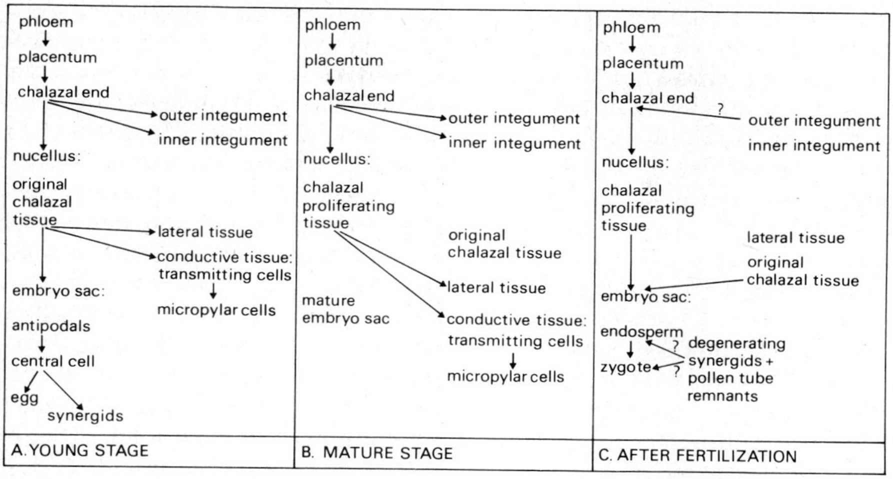

Fig. 4.5. Possible changes in the nutritive pathway in the ovule of spinach. (After Wilms 1980)

In *Jasione* (Erdelská 1975), however, the submicropylar part of the megagametophyte forms an extension, and the nucellar and integumentary cuticles are interrupted here. Nutritional pathways are possibly present in the chalazal region as plasmodesmata, and in the micropylar part by cuticular interruptions resulting from the embryo sac extension in the micropyle. However, our knowledge about the function of the pathways through plasmodesmata, or via the interruptions, is inadequate. The cytoplasmic contact by plasmodesmata is mainly in the chalazal region; in this part the nuclear divisions in the coenocyte can take place somewhat earlier and this might be an indication for a preference in chalazal nutritive transport. It is likely that the uptake of nutrients which pass the megagametophyte wall is the main pathway of transport. In the micropylar part of the embryo sac wall projections can be present, which can help in transference of metabolites (Johri 1981).

In the chalazal region nucellar tissue forms the intermediate link between the embryo sac and the transport tissue. In some cases enlarged cells with specific structures are present in the nucellar tissue, forming an epistase or hypostase. The hypostase is more commonly situated at the chalazal part of the megagametophyte (Masand and Kapil 1966). These morphological structures can be interpreted as a possible pathway of nutrients. Also, haustoria or extensions of the megagametophyte indicate the presence and location of a possible way of transport (Johri 1962, Maheshwari and Kapil 1966, Kapil 1974).

In autoradiographic studies of Coe (1954) on *Zephyranthes,* the chalazal portion of the ovule and the vascular tissue is quickly labelled at the stage of the mature embryo sac, while accumulation of labels is formed around the embryo sac. This indicates a preferential transport toward the chalazal part of the embryo sac, and the tissue around the embryo sac. The presence of products around the embryo sac has also been observed in histochemical studies. This accumulation can be related to the development after fertilization (Buell 1952), and is considered as a reserve (Malik and Vermani 1975).

Considering the various changes in the nucellar tissue, the degeneration of nucellar cells by enlargement of the coenocytic megagametophyte, the storage of reserves in the nucellus (Kapil 1974), as well as the continuous differentiation of the nucellar tissue during the coenocytic stage of the embryo sac (Willemse and Bednara 1979, Willemse and De Boer-De Jeu 1981), a special nutritive function of the nucellus is not probable. Moreover, the changes in the developing megagametophyte indicate more or less an uptake of water and, in some cases, a use of its storage products. There is a minor uptake of nutrients. The development of the coenocytic megagametophyte does not show, in this stage, a strong dependence on the surrounding nucellar and other ovular tissues.

In models of a possible pathway in the ovule of spinach (see Fig. 4.5), Wilms (1980) stated a shift from young stage to mature and postfertilization stage. After the young stage, the chalazal tissue proliferates and the embryo sac does not receive any more nutrition. After fertilization, however, this tissue and the original chalazal tissue, as well as degeneration products, supply nutrients to the now activated embryo sac.

4.5 Organization of the Embryo Sac

In most angiosperms the embryo sac consists of seven cells: the egg cell, two synergids, the central cell, and three antipodal cells. This type of embryo sac is referred to as "Normal" or "Polygonum" type. In the remaining angiosperm species other embryo sac types are reported, characterized by being composed of more or fewer cells, and by aberrant ploidy-level of some of their nuclei. Different cell numbers can result from the absence of one or both synergids, and a reduced or increased number of antipodal cells (Maheshwari 1950).

In the normal embryo sac type all nuclei, except that of the central cell, are haploid. The central cell contains two haploid nuclei, called the polar nuclei which, eventually, fuse and form the diploid central cell nucleus. Haploidy is a common feature for the nuclei of both egg cells and synergids, if present, in all types of embryo sacs. This is in contrast to the ploidy level of the nuclei of the central cell and antipodal cells, which varies greatly among the different embryo sac types.

Since the first description of the female gametophyte by Strasburger (1877), a large number of species has been studied, mainly by light microscopic methods. The results of these studies are well reviewed (Maheshwari 1950, 1963, Rutishauser 1969, Johri 1963, 1970, Coccuci 1973, Kapil and Tiwari 1978, Kapil and Bhatnagar 1981).

The egg cell and the synergids are large, elongated or pear-shaped cells, which are located at the micropylar pole of the embryo sac. These three cells show a triangular arrangement and share common surfaces, forming in this way the structure known as egg apparatus. The egg apparatus is attached to the embryo sac wall bordering only at the extreme micropylar pole, and the major portion of the cells is surrounded by the central cell. Usually, only the synergids can be in direct contact with the micropyle. The apex of the egg cell lies some microns below the

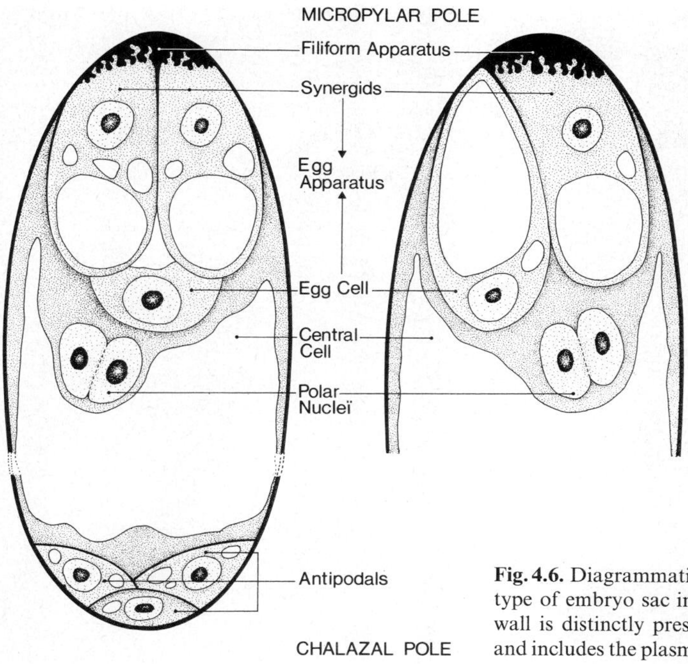

Fig. 4.6. Diagrammatic survey of the "normal" type of embryo sac in angiosperms. If the cell wall is distinctly present, the line is *thickened* and includes the plasma membrane

apices of the synergids at one side of the embryo sac. Since all cells of the egg apparatus have approximately the same length, the egg cell extends somewhat further into the central cell than the two synergids. Around the cells of the egg apparatus walls can be formed. The thickness of these walls diminishes strongly when vacuolation begins and the cells elongate. The remaining wall can be irregular or disappear locally (Fougère-Rifot 1981). For some species the synergids are reported to develop haustoria into and beyond the micropyle, as in *Ursinia, Calendula* (Dahlgren 1924), and *Cotula* (Davis 1971) (Compositae), *Quinchamalium* (Santalaceae) (Agarwal 1960), and *Cortaderia* (Poaceae) (Philipson 1977, 1981).

The central cell occupies the largest portion of the embryo sac, and ranges from the micropylar pole to the chalazal pole, where it borders the antipodal cells (Fig. 4.6). Usually, the antipodal cells are the smallest cells of the embryo sac, and in many species of the dicotyledons they are ephemeral. In monocotyledons, however, especially in grasses, the antipodal cells undergo a number of divisions which results in the formation of a mass of antipodal cells in the mature embryo sac.

All cells of the embryo sac show a specific morphology and structural organization (Figs. 4.7, 4.8). One of the main characteristics of the cells is polarity, expressed by polar distribution of the cytoplasm, and specific position of the nuclei and some cell organelles. Since the early 1960's, considerable new information has been obtained by the use of the transmission electron microscope and histochemistry. The better knowledge of the ultrastructure and composition of the

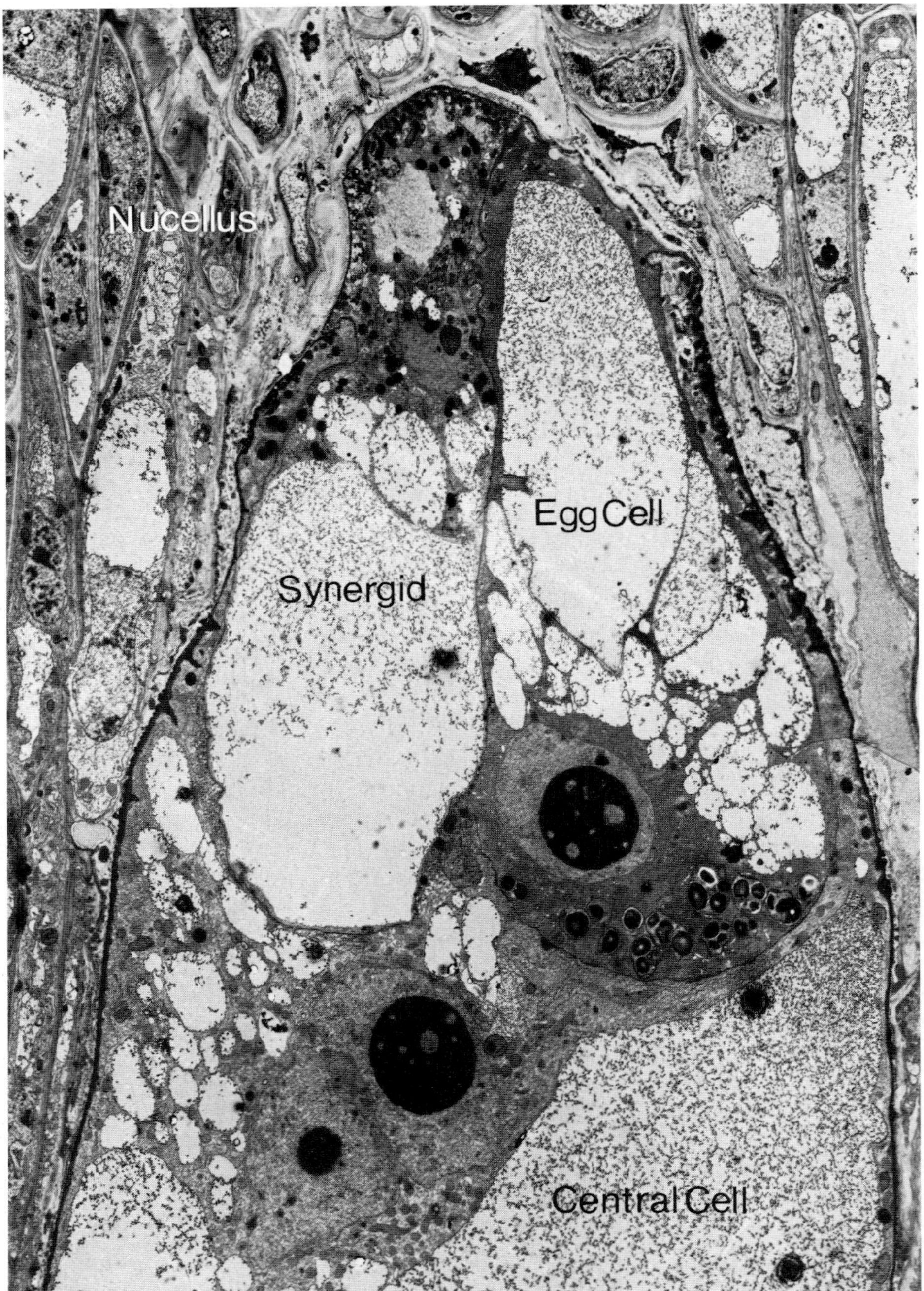

Fig. 4.7. Longitudinal section of the micropylar part of embryo sac of *Spinacia*. (Courtesy Dr. H. J. Wilms, Wageningen)

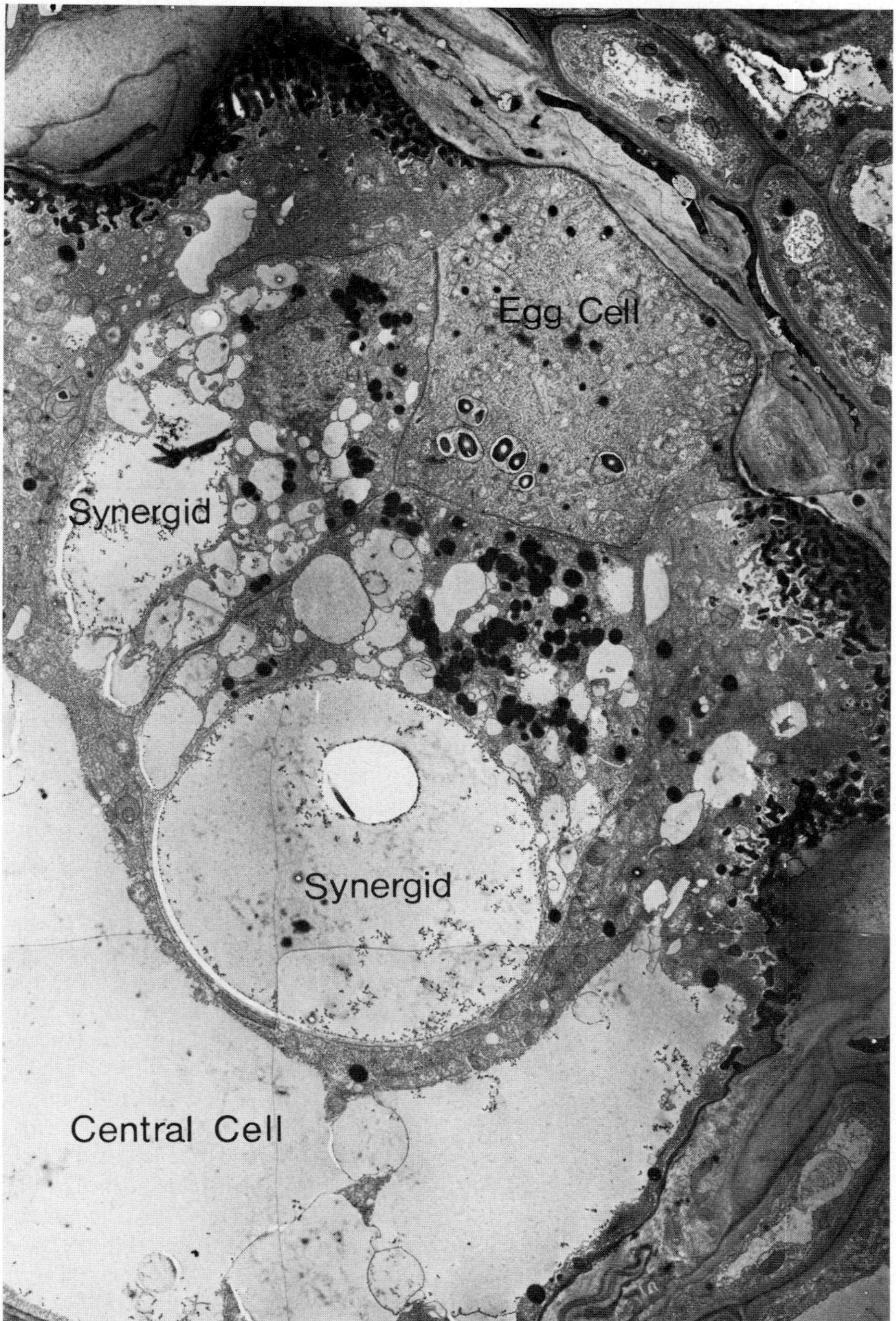

Fig. 4.8. Transverse section of the micropylar part of embryo sac of *Spinacia*. (Courtesy Dr. H.J. Wilms)

cells of the embryo sac, in turn, has resulted in a better understanding of the functioning of the embryo sac.

Thus far, extensive ultrastructural data on the mature embryo sacs of the following genera have been published: *Gossypium* (Jensen 1963, 1964, 1965a–c), *Torenia* (van der Pluym 1964), *Zea* (Diboll 1964, Diboll and Larson 1966, Chebotaru 1970, Russell 1979, van Lammeren 1981), *Lilium* (Eymé 1965, Mikulska and Rodkiewics 1967), *Linum* (B. Vazart and J. Vazart 1965, 1966, J. Vazart 1969, D'Alascio-Deschamps 1973), *Crepis* (Godineau 1966, 1969, 1971, 1973), *Petunia* (van Went and Linskens 1967, van Went 1970a, b), *Capsella* (Schulz and Jensen 1968a, b, 1973), *Epidendrum* (Cocucci and Jensen 1969), *Nerium* (Gori 1971), *Calendula* (Plisko 1971, 1974), *Helianthus* (Newcomb and Steeves 1971, Newcomb 1972, 1973), *Aquilegia* (Rifot 1971, Fougère-Rifot 1975, 1978, Vijayaraghavan et al. 1972), *Quercus* (Mogensen 1972, 1973, 1975), *Eschscholtzia* (Negi 1972), *Oenothera* (Jalouzot 1975), *Stipa* (Maze and Lin 1975), *Festuca* (Maze and Bohm 1977), *Paspalum* (Chao 1977), *Cortaderia* (Philipson 1977, 1981), *Plumbago* (Cass 1972a, b, Cass and Karas 1974, Russell and Cass 1981), *Agave* (Tilton 1974), *Nicotiana* (Mogensen and Suthar 1979), *Spinacia* (Wilms 1981a, b), *Euphorbia* (Gori 1977), *Proboscidea* (Mogensen 1978), *Plantago* (Vannereau 1978), *Conium* (Dumas 1978), *Jasione* (Berger and Erdelská 1973, Erdelská and Klasová 1978), and *Pandanus* (Cheah and Stone 1975).

4.5.1 The Synergids

In most species the synergids are only partly surrounded by a cell wall. Besides, this cell wall shows considerable variation in thickness. At the extreme micropylar pole the wall is strongly thickened, forming the structure known as "Filiform Apparatus" (FA) (Fig. 4.9). From the FA toward the base of the synergids the wall gradually thins. The chalazal half of the synergid is surrounded by the plasma membrane only. The plasma membrane lies close to the plasma membrane of the neighboring cells although, locally, the two membranes can be separated by electron-translucent spaces. There are only a few exceptions to this rule. Schulz and Jensen (1968a) showed the local presence of PAS-positive material at the chalazal region of the synergids in *Capsella*. The partial cell wall in this region appears to have a honeycomb structure. In *Epidendrum* (Cocucci and Jensen 1969) the synergid cell wall is even more complete. Several authors (Linskens 1968, van Went 1970c, Jensen 1972) proposed that the absence of a cell wall at the chalazal region of the synergid plays an essential role in the fusion process of male and female gametes. After the penetration and discharge of the pollen tube into one of the synergids, the plasma membrane of that synergid disintegrates and disappears. The absence of the plasma membrane and the absence of a cell wall in the chalazal portion of the synergid enable the male gamete, transferred through the synergid, to come in direct contact with the plasma membrane of the adjacent female gamete. In *Capsella,* where cell wall material is present in between the synergid and the female gamete, frequently a pore has been observed in this cell wall after successful fertilization (Schulz and Jensen 1968a). This indicates that the presence of a cell wall is not necessarily in contradiction with the proposed fusion mechanism.

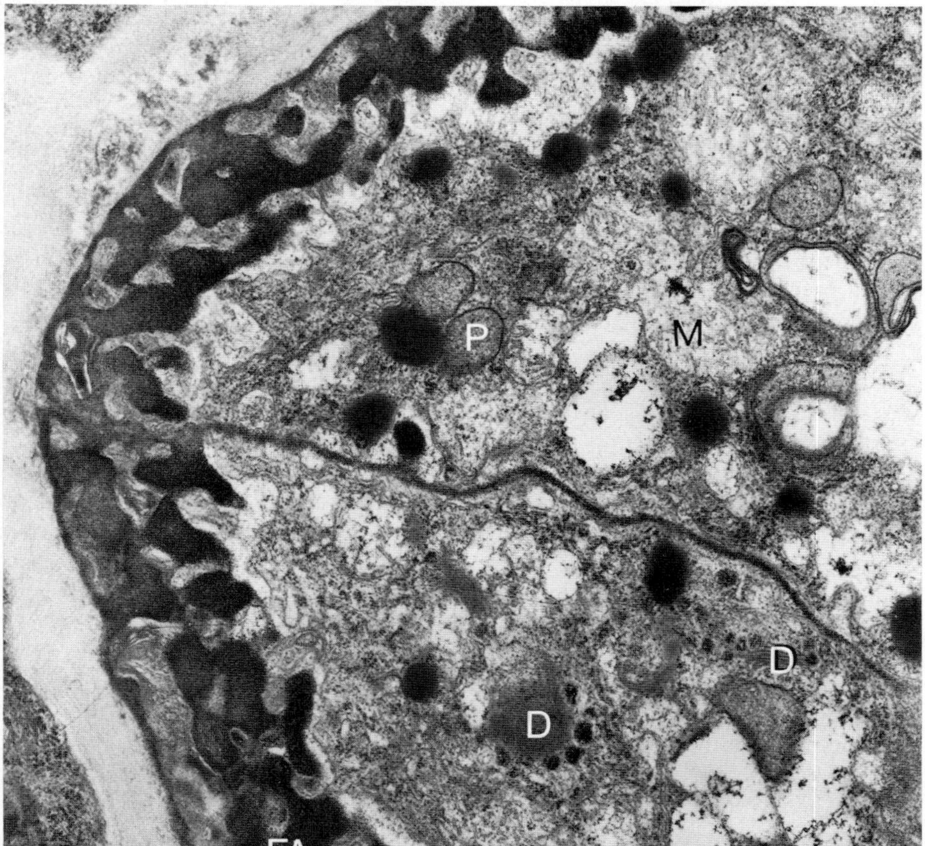

Fig. 4.9. *Spinacia*, synergids during the formation of filiform apparatus (*FA*); note numerous dictyosomes (*D*) with electron-dense Golgi vesicles; *M* mitochondria, *P* plastids

The morphology of the FA is highly variable among species. In *Torenia* (van der Pluym 1964) it is spherical, in *Petunia* (van Went 1970a) and *Helianthus* (Newcomb and Steeves 1971) it is wedge-shaped and located between the tips of the synergids, while in most other species it forms a broad layer of long, irregular cell-wall projections on top of the synergids. In *Petunia* and *Torenia* the FA exhibits a relatively homogeneous structure and smooth surface. The FA of *Petunia* contains a loosely organized network of cellulosic microfibrils, embedded in a matrix of pectic nature. At those places where the FA is connected to the lateral wall of the synergid, it is impregnated with an electron-dense material. A similar situation has been observed in *Jasione* (Berger and Erdelská 1973) where the lateral walls of the synergids are cutinized.

However, in most species studied, thus far, with the electron microscope, the FA shows a very complex organization and heterogeneous structure (Fig. 4.10A–D). It usually consists of a mass of wall projections extending deep into the cytoplasm, and vice versa cytoplasmic tunnels penetrating deep into the FA. As a result of this organization there is a tremendous increase of the plasma membrane surface in this region. The wall projections themselves show two structural phases. They consist of a central electron-dense core surrounded by an electron-translucent layer. According to Schulz and Jensen (1968a) the dense core consists of a tightly packed mass of microfibrils, while the peripheral region contains fewer loosely organized microfibrils set in a translucent matrix (Fig. 4.10A–D). In *Capsella* (Schulz and Jensen 1968a) the FA gives a strong PAS-positive reaction for insoluble carbohydrates, indicating its polysaccharide nature, whereas its reaction is negative for nucleic acids and proteins. Chao (1971) showed that the main component of the FA of *Paspalum* is hemicellulose.

In a few species the presence of plasmodesmata has been reported in the cell walls of the synergids, except for those parts which are in contact with the surrounding sporophytic tissue (Schulz and Jensen 1968a, Mogensen and Suthar 1979, Wilms 1981a).

Most of the cytoplasm of the synergids is located in the micropylar half of the cell. Invariably, the chalazal part of the synergid is occupied by one large or several smaller vacuoles, surrounded by a thin layer of cytoplasm. In *Gossypium* (Jensen 1965a) the vacuoles of the synergids are rich in ash (after microincineration), in contrast to the vacuole of the egg cell which is negative in ash content. The synergid vacuoles give a negative reaction after staining for carbohydrates, nucleic acids, and proteins. Although the chemical nature of the ash could not be established, Jensen assumed that it may be calcium, which has been shown to be important in the directional growth of pollen tubes (Mascarenhas and Machlis 1962a–c).

The cytoplasm of the synergids shows a complex organization, and is rich in organelles. Common features of synergid plasma are the presence of a large number of mitochondria and dictyosomes, extensive endoplasmic reticulum, and abundant ribosomes. In some genera, as *Gossypium* and *Capsella,* there is a tendency for mitochondria to be concentrated near the FA, although mitochondria are present throughout the cytoplasm. In other genera, as *Petunia, Spinacia,* and *Quercus,* the mitochondria are uniformly distributed. The ribosomes are both free and bound to membranes. Frequently, long ER profiles are found running paral-

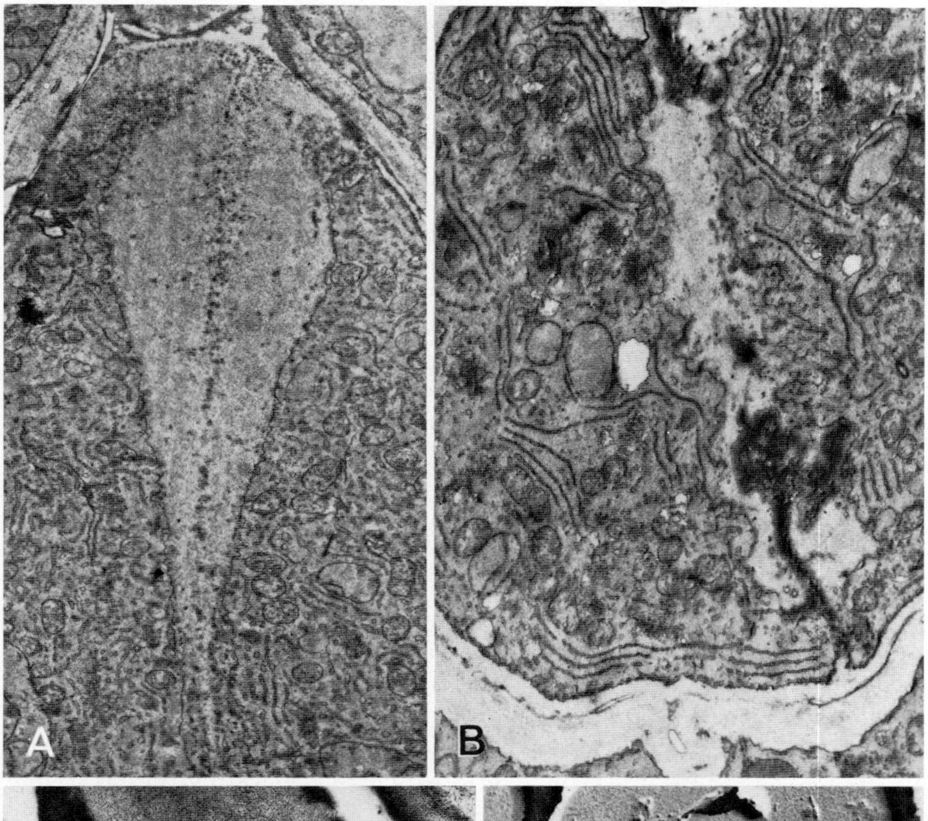

Fig. 4.10. A *Petunia*, synergids and filiform apparatus. **B** Transection. **C** At maturity the micropyle near FA is filled with electron-dense material. **D** The FA contains a loosely organized network of cellulosic fibrils

lel to each other, and are oriented parallel to the long axis of the cell. Although dictyosomes are reported to be numerous in the mature synergid, they are thought to be more active in the premature stage. Especially during the formation of the FA, the presence of active dictyosomes associated with numerous Golgi vesicles and dilated ER has been described for the synergids of *Helianthus* (Newcomb 1973) and *Gasteria* (Willemse and Franssen-Verheijen 1978). After the formation of FA, the numerous dictyosomes, their associated vesicles, and the dilated ER are no longer present. In other taxa, as *Aquilegia* (Vijayaraghavan et al. 1972), *Capsella* (Schulz and Jensen 1968 a), *Nicotiana* (Mogensen and Suthar 1979), and *Spinacia* (Wilms 1981 a), there are numerous dictyosomes and they actively produce vesicles.

Gossypium and *Nicotiana* have numerous plastids and contain starch. However, in most other taxa mentioned before plastids are less frequent and are devoid of starch grains or contain only a few small ones. In *Paspalum* starch is present in the immature synergid, but disappears in the mature cells. The main reserve storage in this taxon is lipid, which is present as numerous small bodies near the FA. Concrete lipid bodies are also found in the cytoplasm of the synergids of *Stipa* (Maze and Lin 1975) and *Spinacia* (Wilms 1981 a).

In a number of species degeneration of one of the synergids prior to the arrival of the pollen tube has been observed (Fig. 4.11): *Gossypium* (Jensen 1965 a), *Hordeum vulgare* (Cass and Jensen 1970), *Linum* (J. Vazart 1971), *Paspalum* (Chao 1971), *Stipa elmeri* (Maze and Lin 1975), *Nicotiana* (Mogensen and Suthar 1979), and *Spinacia* (Wilms 1981 a, b). The second synergid retains its original constitution. The degeneration results in a decreased volume of the synergid, disappearance of the vacuole, and an increased stainability and density of the cytoplasm. The start of the degeneration process is usually marked by the deposition of osmiophilic material on the membranes of the mitochondria (Mogensen 1972). Subsequently, the nucleus, dictyosomes, plastids, and mitochondria lose their identity by the disintegration of their membranes. Simultaneously, the plasma membrane disappears. Frequently, the ER remains intact until the final stage of degeneration. There are only few exact data concerning the possible relation between the moment of pollination, and synergid degeneration. Usually, data are collected as part of a study of the process of fertilization, and degeneration is observed after previous pollination. In cotton, however, Jensen and Schulz (1976) and Jensen et al. (1977) observed degeneration in cultured ovules which were not pollinated. This indicates that the process of degeneration is autonomous, although it probably can be influenced by pollination.

In other taxa, as *Petunia* (van Went 1970 a), *Helianthus* (Newcomb 1973), *Capsella* (Schulz and Jensen 1968 a), and possibly *Quercus* (Mogensen 1972), both synergids remain healthy until the moment of pollen tube penetration into the embryo sac. However, when the pollen tube has penetrated and ejected its content, the penetrated synergid shows degenerative changes. These postpenetration changes are strikingly similar to those during prepenetration degeneration as described above.

Generally, the synergids are considered as highly active cells. The presence of numerous organelles, especially mitochondria and dictyosomes, abundant ribosomes, and extensive RER indicate an intense metabolism. Histochemical studies

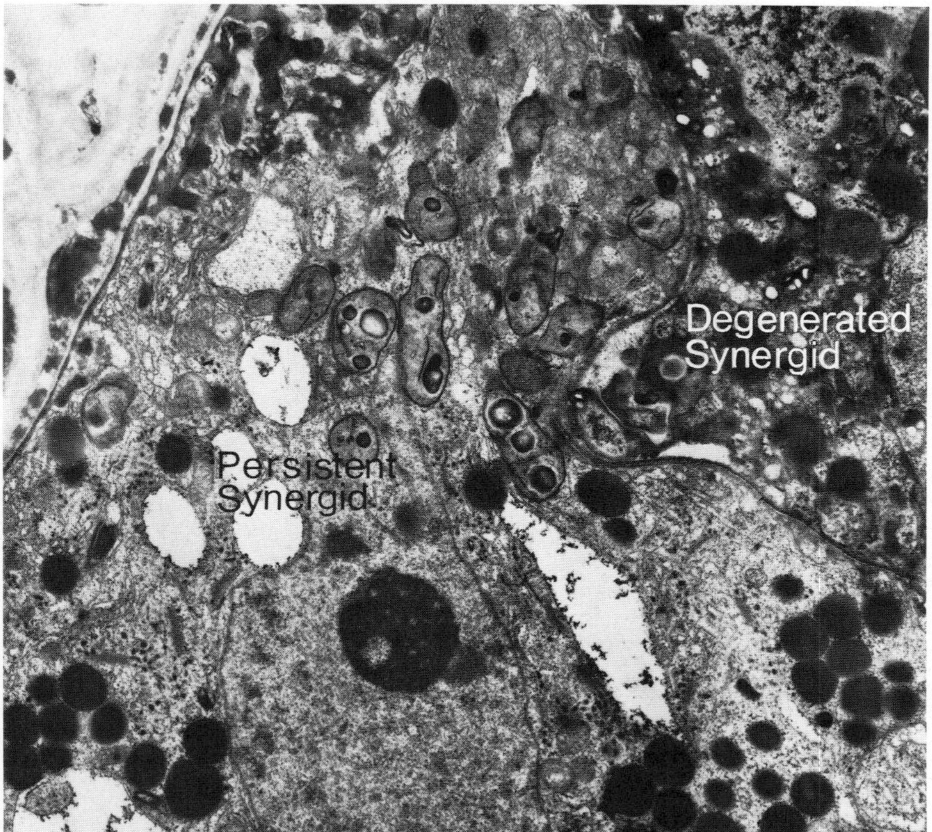

Fig. 4.11. At maturity one of the synergids of *Spinacia* degenerates, whereas the second one retains its original ultrastructure. (Courtesy Dr. H. J. Wilms)

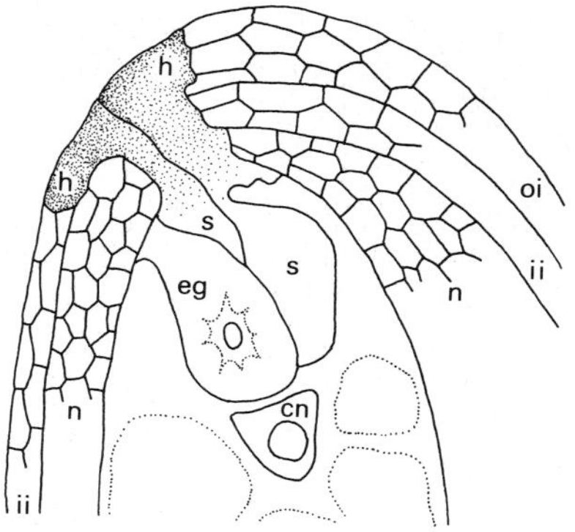

Fig. 4.12. Synergid haustoria in *Cortaderia* projecting through the micropyle. *cn* central cell nucleus, *eg* egg cell, *h* haustorium, *ii* inner integument, *n* nucellus, *oi* outer integument, *s* synergid. (After Philipson 1977)

show that the synergid cytoplasm is rich in protein, and RNA. Malik and Vermani (1975) observed activity of a number of enzymes in the mature synergids of *Zephyranthes* and *Lagenaria*. There was intense cytochrome oxidase and succinate dehydrogenase activity, together with acid phosphatase, ATPase, alkaline phosphatase, phosphorylase, and lipase activity. Shortly before fertilization they observed an increasing peroxidase activity. These data also suggest the synergids to be a seat of highly active metabolism.

Because of the polar distribution of the cytoplasm, most of this metabolic activity is confined to the micropylar region of the synergid. In the same region we find the filiform apparatus. The FA shows all characteristics of a transfer cell wall (Gunning and Pate 1969), and because of its convoluted nature it represents a strongly increased area of plasma membrane. Gunning and Pate related this type of cell wall to places of intense exchange, absorption, and secretion of substances. Since the FA borders the micropylar nucellar tissue in crassinucellate ovules, and the integuments and micropyle in tenuinucellate ovules, a transfer function of the FA and synergid must involve these tissues as well.

According to Jensen (1965a), the synergids in cotton play an important role in the nutrition of the embryo sac. Through their FA they are supposed to take up nutrients from the metabolite-laden nucellus. The observed close association of mitochondria with the plasma membrane of the FA strongly suggests that the energy necessary for absorption is being supplied by the mitochondria. The presence of many plastids, containing large amounts of starch in the synergid near the FA, also points to an absorptive function of the FA and endocytosis of the synergids. The extensive ER of the synergid, with its characteristic distribution and orientation, parallel to the long axis of the cell, is thought to function as an internal transport system, through which nutrients are transported toward the egg cell and central cell. The transfer of material to the latter cells may be facilitated by the absence of a cellulosic cell wall at the chalazal end of the synergid. A nutritional function is also proposed for the synergids in the taxon *Cortaderia* (Philipson 1977, 1981). In *Cortaderia* the tip of the synergids elongate and grow along and out of the micropyle (Fig. 4.12). Ultrastructural studies show the presence of strongly developed transfer walls in the haustorium, especially at the tip, and a dense ER parallel to its long axis.

Some authors doubt a nutritional function of the synergids. In many species the FA of the synergids is in direct contact with the micropyle only, whereas the integuments are covered with a cuticle. Only in a few species the plastids of the synergids contain starch (Fig. 4.13).

Most of the features usually mentioned as indicative of absorption can also be interpreted as indicative of a secretive function. In this respect it is interesting to mention the close resemblance between synergids and the secretory cells lining the stylar canal of the lily pistil (Rosen and Thomas 1970).

The secretory function of the canal cells is well established. The stylar canal cells show a similar polarity both in structure and function, as do the synergids. At their secretive surface they have a highly convoluted and thickened cell wall; this structure closely resembles that of the FA. They also contain numerous mitochondria, abundant ribosomes, and extensive rough ER, as do the synergids. Most of the canal cell cytoplasm is located at the secretory pole of the cell, while the opposite pole is usually filled with vacuoles.

In *Petunia* the synergids are in direct contact with the micropyle (van Went 1970a). The synergids become mature and develop their specific ultrastructure within a very short period, just prior to anthesis (Fig. 4.13). Only after the synergids have reached maturity the micropyle becomes filled with densely stained material, suggesting its synergidal origin. In *Spinacia* Wilms (1981a, b) observed that the middle lamellae of the nucellar tissue between the embryo sac and the micropyle is altered when the ovule becomes receptive. Dissolution of the middle lamellae starts near the FA, suggesting that the enzymes involved are secreted by the synergids.

Enzyme production and secretion have also been postulated by Chao (1971) for the synergids of *Paspalum*. The secreted enzymes are presumed to be responsible for the degradation of integumentary cells leading to the release of PAS-positive substances into the micropyle.

A secretory activity of the synergids is usually considered to be related to a chemotropically directed growth of the pollen tube in the ovary. Although, for a number of species, chemotropic activity of ovules has been established, there is no direct evidence that synergids are involved. On the contrary, according to several authors (see Steffen 1951), the absence of synergids in a number of taxa, as *Plumbago,* indicates that they are not involved in the directional growth of the pollen tube. It was, however, proved by Cass (1972a, b) and Cass and Karas (1974) that at least in *Plumbago* the absence of synergids is compensated by the presence of a filiform apparatus and synergid-like cytoplasm in the egg cell.

Well-established and documented is the role of the synergids in the entrance and discharge of the pollen tube into the embryo sac, and the subsequent transfer of the sperm cells to the egg cell and the central cell. In all species examined thus far with the electron microscope, the pollen tube enters one of the synergids through the FA. After its arrival in the cytoplasm of the synergid, the growth of the tube ceases and an opening is formed in the tube wall. Opening of the pollen tube can occur by simple rupturing at the tube tip, as proposed for *Petunia* (van Went 1970c). In other species, as *Gossypium* (Jensen and Fisher 1968), a well-defined pore is formed; the shape and position indicate that its formation is well regulated and active.

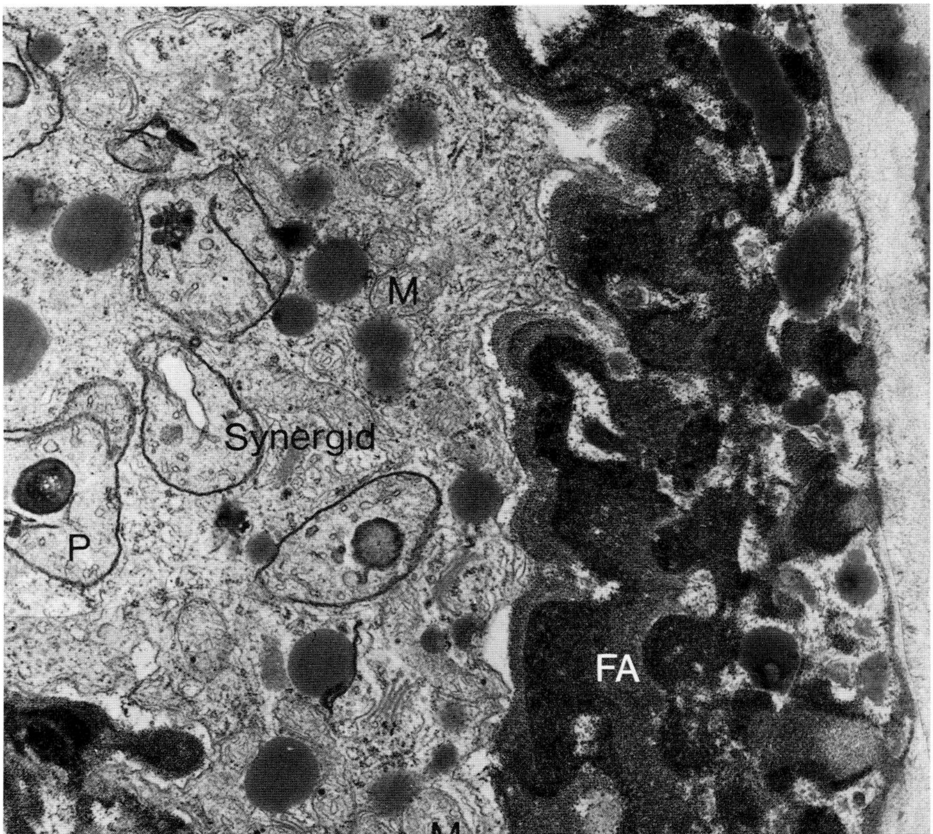

Fig. 4.13. The ultrastructure of persistent synergid and egg cell of *Spinacia* at maturity. *P* plastids, *M* mitochondria, *FA* filiform apparatus. (Courtesy Dr. H. J. Wilms)

Opening of the pollen tube leads to the effusion of an amount of tube cytoplasm into the synergid, including the vegetative nucleus and the sperm cells. The vegetative tube cytoplasm can be usually recognized by the presence of specifically structured organelles, and the presence of numerous small PAS-positive spheres. The amount of tube cytoplasm and its position in the synergid indicate that the ejection happens with force, and that there must be considerable differences in internal pressure between pollen tube and synergid. Apparently, the pollen tube and synergid cytoplasm do not merge. The tube cytoplasm forms a distinct mass in the center of the synergid, reaching from the FA to the base of the cell, whereas the synergid cytoplasm is forced to a peripheral position. Since the plasma membrane of the penetrated synergid has already degenerated, and the synergid cell wall is only partial, the sperm cells probably come in direct contact with the plasma membranes of the egg cell and central cell. The subsequent fusion of the plasma membrane of one sperm cell with the plasma membrane of the egg cell results in the formation of a bridge through which the sperm nucleus and, possibly, sperm cytoplasm can enter the egg cell. In the same way the second sperm cell is believed to fuse with the central cell.

Although our knowledge concerning the structure and function of the synergids has greatly increased during the past decade, many questions still remain to be answered. What is the exact nature of the metabolic activity of the synergids? Do they absorb and transfer, or do they synthesize and secrete? Why do both synergids develop and behave differently? What makes the pollen tube grow through the FA and enter the synergid cytoplasm? How is cessation and opening of the tube achieved?

4.5.2 The Egg Cell

In most species and embryo sac types the egg cell shows a similar organization, characterized by distinct polarity. Only a limited region of the egg cell is in contact with the embryo sac boundary, most of the cell is surrounded by the synergids and the central cell. Usually, the egg cell is only partially surrounded by a cell wall. Its chalazal one-half to one-third portion is limited by a plasma membrane only. Exceptional, in this respect, is the egg cell of *Capsella* (Schulz and Jensen 1968 b) which in its chalazal region is surrounded by a thin and irregular PAS-positive cell wall.

The egg cell wall is thickest at the micropylar region of the cell, and gradually thins toward the chalazal region. Cell-wall thickenings and protuberances, as usually reported in the other cells of the embryo sac, are completely absent in the egg cell. Different in this respect is the egg cell of *Plumbago* (Cass 1972 b, Cass and Karas 1974) which has a filiform apparatus-like structure in the micropylar region. In *Plumbago* the synergids are absent, and it is concluded that the egg cell in the Plumbago type of embryo sac also partly functions as a synergid.

In most species the egg cell is highly vacuolate. The micropylar two-thirds of the cell is almost completely filled by a single large vacuole, while the cytoplasm is restricted to a thin layer along the plasma membrane. The chalazal one-third is filled with cytoplasm containing the egg nucleus and, sometimes, some small additional vacuoles. In *Stipa elmeri* (Maze and Lin 1975) the egg cell is also

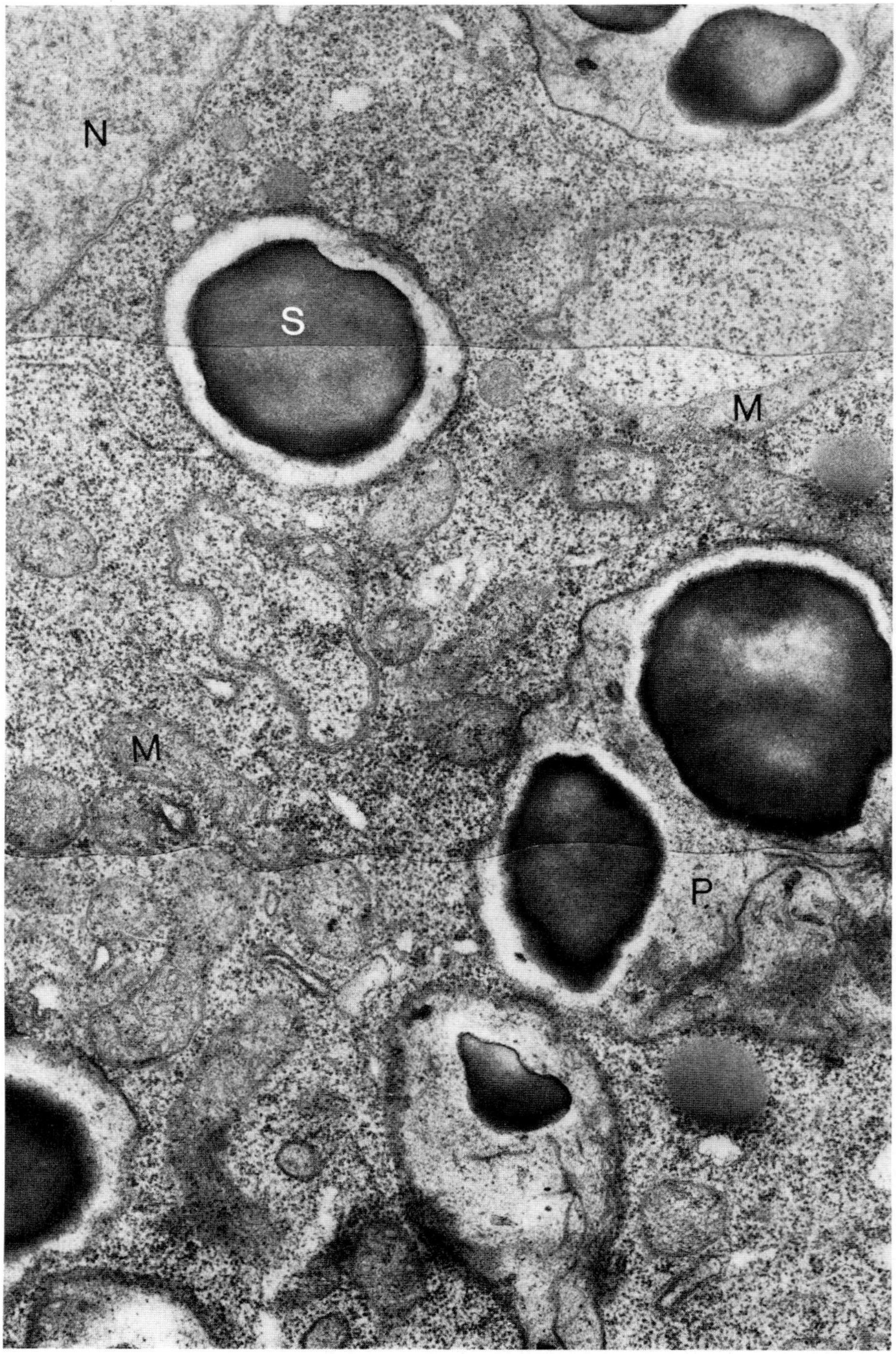

Fig. 4.14. The ultrastructure of the egg cell cytoplasm near the nucleus (*N*). *M* mitochondria, *P* plastids, *S* starch. (Courtesy Dr. H. J. Wilms)

vacuolate, but there are many small vacuoles randomly distributed in the cytoplasm. According to Jensen (1965 b), the vacuole of the *Gossypium* egg cell shows no ash residue after microincineration. In this respect the egg cell vacuole is completely different from the synergid vacuole which leaves a considerable amount of ash.

The mature egg cell nucleus gives only a faint reaction after Feulgen staining. The reduction in stainability during maturation of the nucleus is ascribed to the large increase in nuclear size and consequent dilution of the DNA (Pritchard 1964). Frequently, a large vacuolate nucleolus is present in the egg nucleus. According to Schulz and Jensen (1968 b) the structure of the nucleolus in *Capsella* suggests that ribosomes are not being produced. However, in *Helianthus* the presence of a large nucleolus is considered to be indicative of a high rate of ribosome synthesis (Newcomb 1973). The ultrastructure of the egg cell cytoplasm varies considerably among the various species studied (Fig. 4.14). In *Gossypium* the egg cell contains a large amount of ER, numerous ribosomes, numerous mitochondria with only few cristae, and only few dictyosomes and plastids (Jensen 1965 b). In *Petunia* ER is scarce, and polysomes, mitochondria, plastids, and dictyosomes are few (van Went 1970 b). The egg cell of *Zea mays* (Diboll and Larson 1966) contains many mitochondria and plastids, whereas there are only few dictyosomes, and a small amount of ER. A similar condition has been observed in *Helianthus* (Newcomb 1973). In *Spinacia* the number of mitochondria increases considerably and they mainly become clustered in the micropylar part of the egg cell, whereas large plastids accumulate starch in the chalazal part (Wilms 1981 a).

A common feature of egg cell cytoplasm is the presence of only a limited number of relatively inactive dictyosomes. In all other cytoplasmic aspects egg cells are very specific. Egg cells seem to be rich in protein (Jensen 1965 b, Schulz and Jensen 1968 b, Yu and Chao 1979, Wilms 1980); whereas some react strongly after RNA-staining while others give only a faint reaction. Most authors assume that the egg cell is in a relatively inactive state, on the basis of the cytoplasmic constitution. The ultrastructural changes that have been observed in the cytoplasm after fertilization underline this assumption. However, substantial biochemical or physiological evidence for this assumption is lacking.

As already indicated previously, a specific situation has been observed in *Plumbago* (Cass 1972 a, b, Cass and Karas 1974), where synergids are missing. Nearly all of the *Plumbago* egg cell is surrounded by a cell wall. At the micropylar pole where the egg cell is in contact with the embryo sac wall, a filiform apparatus is present. The egg cell contains one large vacuole which occupies most of the cell volume. There are two accumulations of cytoplasm: one at the micropylar pole associated with the FA, and the other at the chalazal pole containing the egg nucleus. The micropylar cytoplasm has many mitochondria and dictyosomes. The chalazal cytoplasm contains many plastids, mitochondria and polysomes, and conspicious RER. The *Plumbago* egg cell is considered as a highly metabolic active cell which combines both gametic and synergid functions.

4.5.3 The Central Cell

The central cell usually occupies the largest portion of the embryo sac, bordering and surrounding the major portion of the egg apparatus at the micropylar pole, and reaching to the antipodal cells or nucellus at the chalazal pole. The lateral cell wall of the central cell, therefore, forms the main part of the embryo sac wall in contact with the surrounding tissue.

The central cell is highly vacuolate, and its cytoplasm is confined to a thin layer along the embryo sac wall and accumulates near the egg apparatus and the antipodal cells. The micropylar accumulation contains the two polar nuclei, or the fusion nucleus. The time of fusion of the polar nuclei varies from species to species. In *Capsella* (Schulz and Jensen 1973) and *Helianthus* (Newcomb 1973) the fusion is completed before fertilization. In *Gossypium* (Jensen 1964) and *Spinacia* (Wilms 1981 a) fusion starts in the mature embryo sac, but is only completed after the arrival of the sperm nucleus. In *Spinacia* the polar nuclei form large protrusions before the fusion. The polar nuclei and fusion nucleus are the largest nuclei observed in the embryo sac. They generally show very little heterochromatin, and contain a large nucleolus. In *Helianthus* (Newcomb 1973) the nucleolus has a large vacuole, while in *Gossypium* (Jensen 1965 b) the nucleoli are relatively massive. Ultrastructural details of nuclear fusion in plants were first given by Jensen (1964) in his description of the fusion process of polar nuclei in *Gossypium*. In this taxon the nuclear envelopes are continuous with and connected by cisterns of the ER. Fusion of the nuclei starts with the contact and fusion of the outer nuclear membranes, followed by the contact and fusion of the inner membranes. In this way a number of small bridges are formed which, gradually enlarge and coalesce. After the nuclear fusion the nucleoli also fuse, indicating a completion on chromatin level.

Like the cells of the egg apparatus, the central cell is surrounded only by a partial cell wall. The cell wall is absent at the micropylar pole near the base of the egg apparatus at those places where the neighboring cells also lack cell walls. The wall against the nucellus or integuments is usually thick and regular. In *Helianthus* (Newcomb and Steeves 1971, Newcomb 1973), *Euphorbia* (Gori 1977), *Jasione* (Berger and Erdelská 1973), *Capsella* (Schulz and Jensen 1973), and *Spinacia* (Wilms 1981 a) the micropylar part of the lateral wall shows many fingerlike projections. In *Capsella* (Schulz and Jensen 1971), *Helianthus*, and *Euphorbia* similar projections also occur along the chalazal part. The wall projections strongly resemble the FA and, similarly, greatly increase the plasma membrane surface in that area. According to Newcomb and Steeves (1971), the presence of embryo sac wall projections suggest that the central cell plays an important role in the absorption of metabolites from the surrounding tissue. In *Jasione* (Berger and Erdelská 1973) the embryo sac is surrounded by a thick cuticle. However, in the region with the central cell-wall projections, the cuticle consists of small plaques with discontinuities. This configuration also suggests an absorptive function of this part of the central cell.

Most central cells contain complex organized cytoplasm which is rich in organelles (Fig. 4.15). There are numerous mitochondria with well-developed cristae; frequently there is extensive RER; there is a large number of dictyosomes appearing active in the production of vesicles; there are numerous ribosomes, both as monosomes and polysomes. The number and ultrastructure of plastids is variable. In *Helianthus* and *Linum* the plastids do not contain starch, whereas in *Gossypium, Capsella,* and *Spinacia* some starch is present, but less than in the

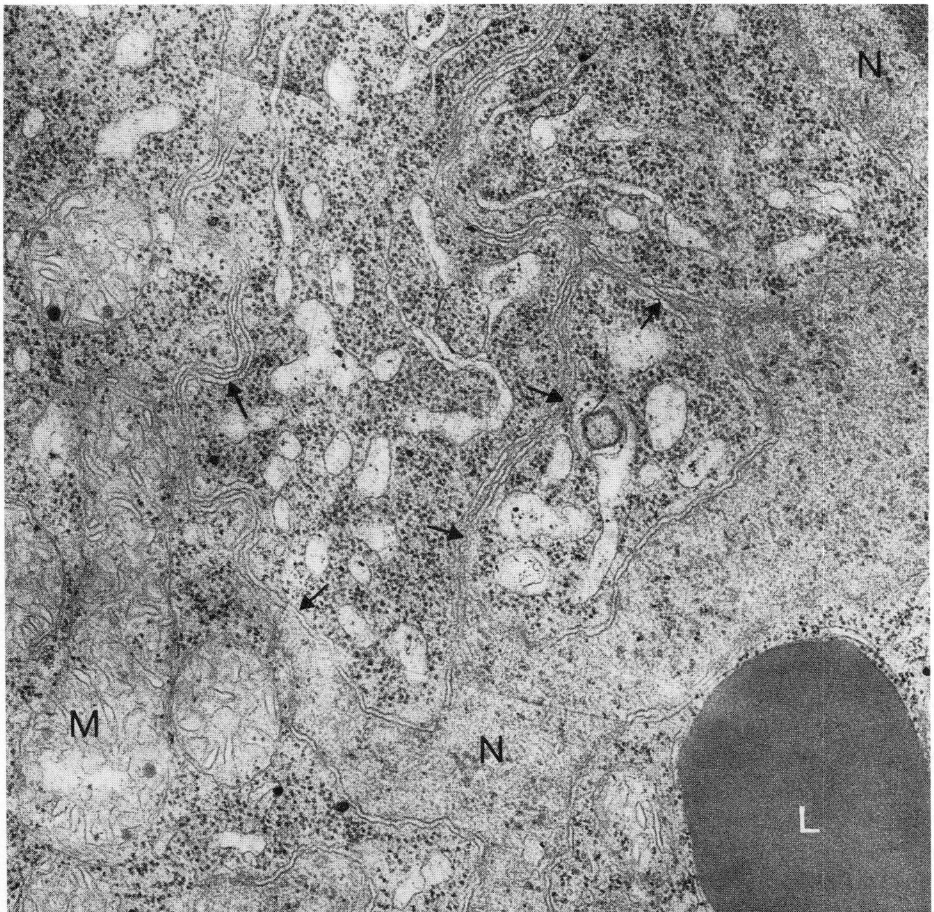

Fig. 4.15. The central cell of *Spinacia*. At maturity, the polar nuclei (*N*) form long protrusions (*arrows*). *M* mitochondria, *L* lipid. (Courtesy Dr. H. J. Wilms)

egg cell. There is abundant starch in the central cell of *Nicotiana rustica* before fertilization (Sehgal and Gifford 1979). In *Capsella* the plastids contain a well-developed thylakoid system, including grana, indicating that the cell is partly autotrophic. The plastids in the central cell of *Gossypium* are reported to contain phytoferretin. Besides starch, usually lipids are present in the central cell as reserve substances. The presence of lipids is accompanied by the presence of numerous microbodies. The microbodies possibly represent glyoxisomes containing the enzymes for β-oxidation of fatty acids, and the enzymes of the glyoxylate cycle for the conversion of acetyl-CoA to succinic acid.

In *Petunia* the central cell has a poor cytoplasmic constitution (van Went 1970 b). The number of mitochondria and dictyosomes is low, and the ER is only poorly developed. The plastids are large and contain high amounts of starch. Free ribosomes are very abundant and, likely, they are present as monosomes. The mature central cell is probably in a state of low metabolic activity. In this species triple fusion causes a dramatic change in the cytoplasmic ultrastructure. The number of mitochondria and dictyosomes, and the amount of RER rapidly increase, whereas the starch quickly disappears. Simultaneously, the ribosomes become organized as polysomes. Apparently, the newly formed endosperm cell becomes metabolically very active.

4.5.4 The Antipodal Cells

The antipodal cells show much variation among angiosperms. In many taxa of the dicotyledons they degenerate before or during the maturation of the embryo sac. In other species they persist even during embryo and endosperm formation, although their condition can be highly variable. Finally, in monocotyledons, especially in grasses, they can proliferate into a multicellular tissue consisting of up to 100 cells. The great variability makes it impossible to draw general conclusions on the structure and functions of antipodal cells.

In *Capsella* (Schulz and Jensen 1971) the antipodals are small, inactive cells. They are surrounded by cell walls of uniform thickness, invariably containing plasmodesmata. So the antipodals are interconnected as well as connected to the nucellus and central cell. This indicates the presence of a pathway for metabolites from the nucellus through the antipodals to the central cell, and vice versa. A similar condition is reported for *Helianthus* (Newcomb 1973) and *Jasione* (Berger and Erdelská 1973).

Frequently, the antipodal cell walls bordering the nucellus have finger-like wall projections, similar to those observed in the central cell and the filiform apparatus. In *Stipa* (Maze and Lin 1975), *Helianthus* (Newcomb 1973), *Zea* (Diboll 1968), and *Spinacia* (Wilms 1981 a) the presence of these transfer wall-like projections is accompanied by the presence of very active cytoplasm (Fig. 4.16 A, B). The cytoplasm contains abundant organelles as mitochondria, dictyosomes, plastids, and ribosomes. The ER is extensive, and consists of parallel and partly concentric cisterns.

Histochemical tests also indicate the presence of high metabolic activity. Malik and Vermani (1975) established enzyme activity, as peroxidase, succinic dehy-

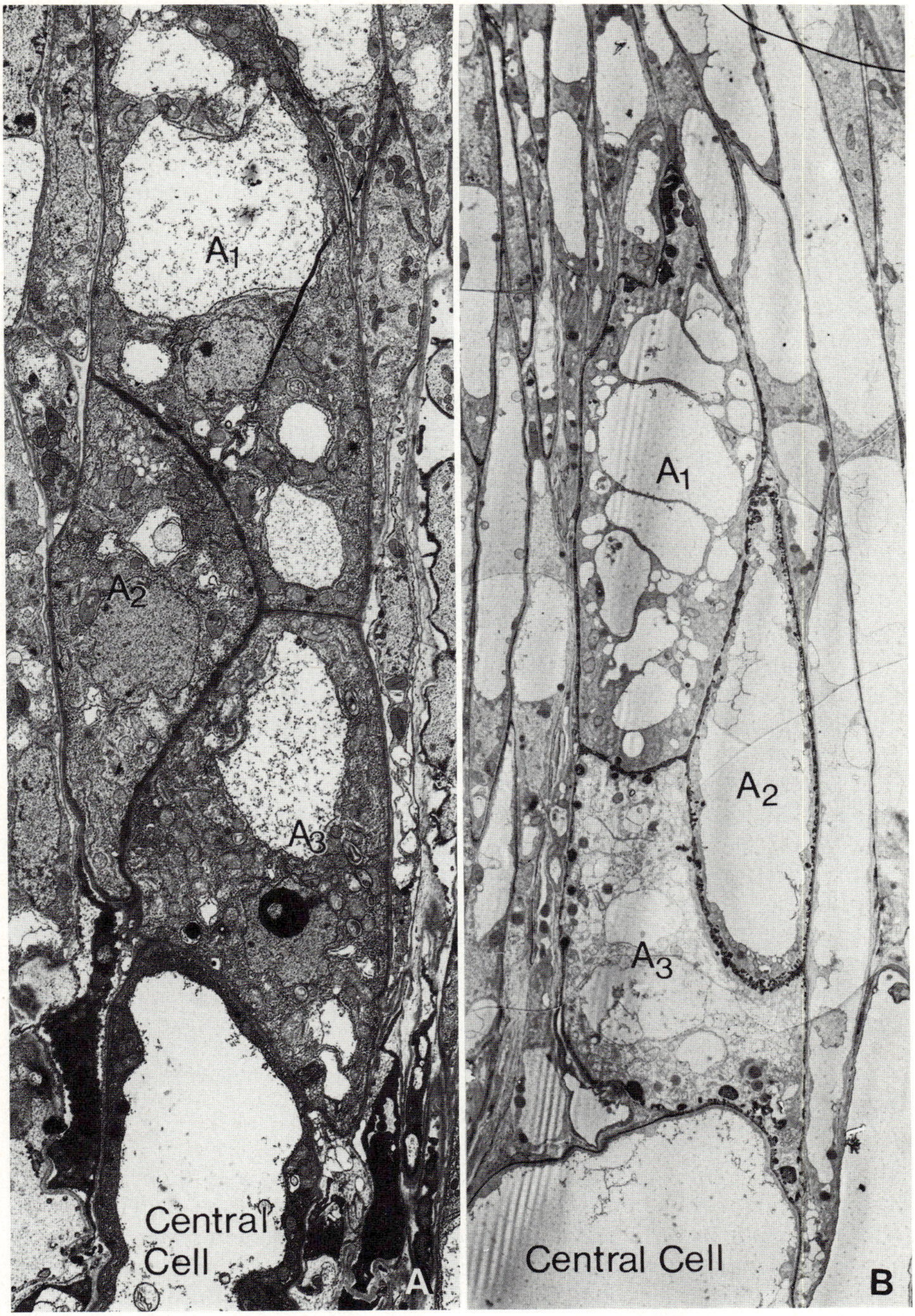

Fig. 4.16 A, B. The antipodals of *Spinacia*. **A** shows the young antipodals A_1, A_2, A_3, and **B** the antipodals of mature embryo sac. (Courtesy Dr. H. J. Wilms)

drogenase, phosphorylase, cytochrome oxidase, and lipase in the antipodal cells of *Zephyranthes* and *Lagenaria*. These cells also contain high amounts of protein, polysaccharides, lipids, and RNA. During maturation of the embryo sac there is an increase of RNA and protein. Similar results have been obtained for the antipodal cells of *Paspalum* (Yu and Chao 1979). In this taxon lipids are transferred from a solid to a liquid state during maturation. In many species the antipodal cells contain starch, which is used up during fertilization and subsequent development of endosperm, and embryo.

Variability has also been observed in the cellular organization of the antipodal tissue. In *Helianthus,* instead of three antipodals conforming to the Polygonum type of embryo sac, only two antipodal cells are present (Newcomb 1973). One of the antipodal cells is mononucleate, the second antipodal cell contains two nuclei. In *Linum* (J. Vazart 1969) the antipodal cells are only partly separated by walls. These incomplete walls subsequently dissolve, and a syncytium is formed. Especially in species with proliferating antipodals, cells can develop with more than one nucleus as in *Oryzopsis* and *Stipa* (Maze et al. 1970).

Invariably, the antipodal nuclei stain intensely for DNA. This might be partly due to the largely heterochromatic condition of the nucleoproteins. However, for many species a strong increase in the amount of DNA per nucleus has been reported. The enhancement in DNA content can be due to karyokinesis followed by nuclear fusion, endomitosis, and polyteny. In *Scilla* the level of endoploidy reaches as much as 1,024 C (Frisch and Nagl 1979).

Generally, three main functions can be attributed to antipodal cells. Firstly, they may be involved in the transfer of nutrients and serve as a pathway for metabolites from the nucellus to the central cell. This conclusion is based on the presence and location of plasmodesmata, and the presence of transfer cell-like walls, in combination with cytoplasm which appears active.

Secondly, they may serve as a storage for the developing endosperm and embryo. Many antipodals contain abundant starch, lipid, and proteins which are used after fertilization. Degeneration and breakdown of antipodal cells after fertilization can serve as an additional food supply. Thirdly, the antipodal cells may have a secretory function. This is indicated by the presence of extensive rough ER and abundant ribosomes. The antipodals possibly produce and secrete growth-controlling substances which regulate the development of the adjacent endosperm.

4.6 Female Gametophyte Development

The ovule has, from its onset, an axis of polarity from micropyle to chalaza. Along this axis evolves the development of the megagametophyte. This process is programmed by the genome present in the cell, and later the cells of the megagametophyte. But, considering the shape of the nucellus, the megaspore selection, the cellular contact and type of vacuolation, also influences of the surrounding tissue, are very likely.

Megagametophyte development is the result of an interaction between sporophytic tissue and gametophytic cells. The developing megagametophyte gets its nutrition from the surrounding tissue. The flow of these nutrients comes from the chalazal part, but after formation of the embryo sac the flow passes possibly the basal part and penetrates along the side of the embryo sac. Depending on the structural elaboration of the nucellar tissues, more pathways are possible. Nutrition flow and megagametophyte development are still not fully understood. Ultrastructural, histochemical, quantitative, and physiological approaches can elucidate, also, the relation of the megagametophyte development and its surrounding tissue. The function and appearance of special structures like the hypostase and integumentary tapetum, and the different outgrowths of embryo sac cells, are to be included in this process of interaction.

The differentiation patterns of the embryo sac result in a polarity with respect to the programmed receptivity of the micropylar part for the pollen tube. In this receptivity and acceptance the function of both synergids needs more attention together with the possible resting period of the embryo sac. About the function of the antipodals different opinions still exist.

Elucidation of the relation of megagametophyte development to pollination and seed formation will further add to our understanding of the early stages of ovular development. Starting with this background, abortion or deviation in development and polyembryony are linked to the process of ovular development.

References

Agarwal S (1960) Embryology of *Quinchamalium chilense* Lam. In: Plant Embryology: A symposium. Council of Scientific and Industrial Research, New Delhi, pp 162–169

Bannikova VP (1971) Developmental features of female gametophyte of *Nicotiana*. Ann Univ ARERS 9:73–77

Batygina TB (1971) Some peculiar features of macrosporogenesis and development of the female gametophyte in wheat. Ann Univ ARERS 9:46–50

Berg RY (1978) Development of ovule, embryo sac and endosperm in *Brodiaea* (Liliales). Norw J Bot 25:1–7

Berger Ch, Erdelská O (1973) Ultrastructural aspects of the embryo sac of *Jasione montana* L.: Cell walls. Caryologia 25:109–120

Bouillot J (1969) Individualisation progressive des cellules du sac embryonnaire chez le *Juglans regia* L. Rev Cytol Biol Vég 32:203–208

Buell KM (1952) Developmental morphology in *Dianthus*. II. Starch accumulation in ovule and seed. Am J Bot 39:458–467

Cass DD (1972a) Occurrence and development of a filiform apparatus in the egg of *Plumbago capensis*. Am J Bot 59:279–283

Cass DD (1972b) Ultrastructure of the egg of *Plumbago zeylanica*. (Abstr). Am J Bot 59:648

Cass DD, Jensen WA (1970) Fertilization in barley. Am J Bot 57:62–70

Cass DD, Karas I (1974) Ultrastructural organization of the egg of *Plumbago zeylanica*. Protoplasma 81:49–62

Chao CY (1971) A periodic acid-schiff's substance related to the directional growth of pollen tube into embryo sac in *Paspalum* ovules. Am J Bot 58:649–654

Chao CY (1977) Further cytological studies of a periodic acid-schiff's substance in the ovules of *Paspalum orbiculare* and *P. longifolium*. Am J Bot 64:922–930

Cheah CH, Stone BC (1975) Embryo sac and microsporangium development in *Pandanus* (Pandanaceae). Phytomorphology 25:228–238

Chebotaru AA (1970) Maize embryology: Ultrastructure of embryo sac before and at the moment of fertilization. 7° Congrès Intl Micr électr 3:443–444

Chennaveeraiah MS, Mahabale TS (1962) Morphological and embryological studies in *Dipcadi*. In: Plant Embryology: A symposium. Council of Scientific and Industrial Research, New Delhi, pp 12–22

Chikkannaiah PS (1962) Morphological and embryological studies in the Commelinaceae. In: Plant embryology: A symposium. Council of Scientific and Industrial Research, New Delhi, pp 23–36

Cocucci AE (1973) Some suggestions on the evolution of gametophytes of higher plants. Phytomorphology 23:109–124

Cocucci AE, Jensen WA (1969) Orchid embryology: The mature megagametophyte of *Epidendrum scutella*. Kurtziana 5:23–38

Coe GE (1954) Distribution of carbon-14 in ovules of *Zephyranthes drummondi*. Bot Gaz 115:342–346

Dahlgren KVO (1924) Studien über die Endospermbildung der Kompositen. Svensk Bot Tidskr 18:177–203

D'Alascio-Deschamps R (1973) Modifications ultrastructurales liées à la fécondation chez le Lin. Bull Soc Bot Fr 122:137–146

Davis GL (1971) The occurrence of synergid haustoria in *Cotula australis* (Less.) Hook. f. Aust J Sci 24:296

De Boer-De Jeu M (1978) Megasporogenesis, a comparative study of the ultrastructural aspects of megasporogenesis in *Lilium, Allium,* and *Impatiens*. Meded Landbouwhogesch Wageningen 16:1–128

Dharamadhaj P, Prakash N (1978) Development of the anther and ovule in *Capsicum*. Aust J Bot 26:433–439

Diboll AG (1964) Electron microscopy of female gametophyte and early embryo development in *Zea mays*. Ph D dissertation, Univ Texas

Diboll AG (1968) Fine structure development of the megagametophyte of *Zea mays* following fertilization. Am J Bot 55:787–806

Diboll AG, Larson DA (1966) An electron microscopic study of the mature megagametophyte in *Zea mays*. Am J Bot 53:391–402

Dumas A (1971) Au sujet des charactères ultrastructuraux de l'archespore et des très jeunes stades du sac embryonnaire de *Conium maculatum*. Ann Univ ARERS 9:51–57

Dumas A (1978) Le développement du sac embryonnaire de *Conium maculatum* L. et les phénomènes sécrétoires au moment de la fécondation. Bull Soc Bot Fr 125:193–199

Erdelská O (1975) Pre-fertilization development of ovule of *Jasione montana* L. Phytomorphology 25:76–81

Erdelská O (1981) Embryológia Krytosemenných Rastlín. VEDA Vydavatel'stvo Slovenskej Akadémie Vied. Bratislava

Erdelská O, Klasová A (1978) La region micropylaire du sac embryonnaire de *Jasione montana* L. avant et après la fécondation. Bull Soc Bot Fr 125:249–252

Eymé J (1965) Recherches sur la constitution cytoplasmique de l'archespore et du sac embryonnaire de *Lilium canditum* L. Extrait Botanîste 48:99–155

Fish RK (1970) Megagametogenesis in *Clematis* and its taxonomic and phytogenetic implications. Phytomorphology 20:317–327

Fougère-Rifot M (1975) L'édification de l'appareil filiforme et l'évolution cytoplasmique des synergides du sac embryonnaire d'*Aquilegia vulgaris*. C R Acad Sci Paris 280(D):2445–2447

Fougère-Rifot M (1978) La cellule centrale du sac embryonnaire d'*Aquilegia vulgaris* L.: Des noyaux polaires aux noyaux d'albumen laiteux. Bull Soc Bot Fr 125:207–213

Fougère-Rifot M (1981) Intercellular relations and wall structures in mature embryo sacs of four species of angiosperms. Acta Soc Bot Pol 50:131–138

Frisch B, Nagl W (1979) Patterns of endopolyploidy and 2C nuclear DNA content (Feulgen) in *Scilla* (Liliaceae). Plant Syst Evol 131:261–276

Godineau JC (1966) Ultrastructure du sac embryonnaire du *Crepis tectorum* L.: Les cellules du pôle micropylaire. C R Acad Sci Paris 263:852–855

Godineau JC (1969) Ultrastructure des synergides chez quelques composées. Rev Cytol Biol Vég 32:209–226

Godineau JC (1971) Ultrastructure of the embryo sac of *Crepis tectorum* L. after cell formation and fusion of polar nuclei. Ann Univ ARERS 9:78–88

Godineau JC (1973) Les sac embryonnaire des Angiospermes. Morphogénèse et infrastructure. Bull Soc Bot Fr 122:25–54

Gori P (1971) Un aspect original des parois et du cytoplasme des cellules de l'ovule chez le *Nerium oleander* L. C R Acad Sci Paris 273:328–331

Gori P (1977) Wall ingrowths in the embryo sac of *Euphorbia helioscopia*. Isr J Bot 26:202–208

Greene CW (1978) A Nomarski interference study of megasporogenesis and megagametogenesis in *Smelowskia calycine* (Cruciferae). Am J Bot 65:353–358

Gunning BES, Pate JS (1969) "Transfer cells" – plant cells with wall ingrowth, specialized in relation to short distance transport of solutes: Their occurrence, structure, and development. Protoplasma 68:107–133

Jalouzot MF (1975) Aspects ultrastructuraux du sac embryonnaire d'*Oenothera lamarckiana*. C R Acad Sci Paris 281:1305–1308

Jensen WA (1963) Cell development during plant embryogenesis. In: Meristems and differentiation. Brookhaven Symp Biol 16:179–202

Jensen WA (1964) Observations on the fusion of nuclei in plants. J Cell Biol 23:669–672

JensenWA (1965a) The ultrastructure and histochemistry of the synergids of cotton. Am J Bot 52:238–256

Jensen WA (1965b) The ultrastructure and composition of the egg and central cell of cotton. Am J Bot 52:781–797

Jensen WA (1965c) The composition and ultrastructure of the nucellus in cotton. J Ultrastruct Res 13:112–128

Jensen WA (1972) The embryo sac and fertilization in angiosperms. Harold L Lyon Arbor Lect 3:1–32

Jensen WA, Fisher DB (1968) Cotton embryogenesis: The entrance and discharge of the pollen tube in the embryo sac. Planta 78:158–183

Jensen WA, Schulz P (1976) An ultrastructural study of early endosperm development and synergid changes in unfertilized ovules of cotton. J Cell Biol 70:87A

Jensen WA, Schulz P, Ashton ME (1977) An ultrastructural study of early endosperm development and synergid changes in unfertilized cotton ovules. Planta 133:179–189

Johri BM (1962) Nutrition of the embryo sac. In: Maheshwari P, Johri BM, Vasil IK (eds) Proc Summer School Botany, Darjeeling, pp 106–118

Johri BM (1963) Female gametophyte. In: Maheshwari P (ed) Recent advances in the embryology of angiosperms. Intl Soc Plant Mophologists, Univ Delhi, pp 69–103

Johri BM (Convener) (1970) Symposium on "Comparative embryology of angiosperms". Indian Natl Sci Acad, New Delhi, Bull No. 41

Johri BM (1981) Transfer cells, their role in reproductive structures of angiosperms. Abstr No. 042014 XIII Intl Bot Congr, Sydney, p 61

Kapil RN (1974) Contemporary spectacle in angiosperm embryology. Acta Bot Indica 2:79–106

Kapil RN, Bhatnagar AK (1981) Ultrastructure and biology of female gametophyte in flowering plants. Int Rev Cytol 70:291–341

Kapil RN, Tiwari SC (1978) Plant embryological investigations and fluorescence microscopy: An assessment of integration. Int Rev Cytol 53:291–331

Korobova SN (1971) On the development of the "specific" cells during micro- and macrosporangium formation in *Zea mays*. Ann Univ ARERS 9:58–61

Lammeren AAM van (1981) Embryogenesis of *Zea mays*, an ultrastructural approach. Acta Soc Bot Pol 50:283–284

Linskens HF (1968) Egg-sperm interactions in higher plants. Acc Naz Lincei 104:47–60

Maheshwari P (1950) An introduction to the embryology of angiosperms. MacGraw Hill, New York

Maheshwari P (ed) (1963) Recent advances in the embryology of angiosperms. Intl Soc Plant Morphologists, Univ Delhi

Maheshwari P, Kapil RN (1966) Some Indian contributions to the embryology of angiosperms. Phytomorphology 16:239–291

Malik CP, Vermani S (1975) Physiology of sexual reproduction. I. A histochemical study of the embryo sac development in *Zephyranthes rosea* and *Lagenaria vulgaris*. Acta Histochem 53:244–280

Masand P, Kapil RN (1966) Nutrition of the embryo sac and embryo – A morphological approach. Phytomorphology 16:158–175

Mascarenhas JP, Machlis L (1962a) Chemotropic response of *Antirrhinum majus* pollen to calcium. Nature 196:292–293

Mascarenhas JP, Machlis L (1962b) The pollen tube chemotropic factor from *Antirrhinum majus*: Bioassay, extraction, and partial purification. Am J Bot 49:482–489

Mascarenhas JP, Machlis L (1962c) The hormonal control of directional growth of pollen tubes. Vitam Horm XX:347–371

Maze J, Bohm LR (1973) Comparative embryology of *Stipa elmeri* (Gramineae). Can J Bot 51:235–247

Maze J, Bohm LR (1977) Embryology of *Festuca microstachys*. Can J Bot 55:1768–1782

Maze J, Lin SCh (1975) A study of the mature megagametophyte of *Stipa elmeri*. Can J Bot 53:2958–2977

Maze J, Bohm LR, Mehlenbacher LE Jr (1970) Embryo sac and early ovule development in *Oryzopsis miliacea* and *Stipa tortilis*. Can J Bot 48:27–41

Medina FJ (1980) The occurrence of two different processes of cytoplasmic vacuolation in the megaspore formation in peas. Electron Microscopy 2:238–239

Miki-Hirosige H (1964) Metabolism in the ovary tissue of *Lilium longiflorum*. In: Linskens HF (ed) Pollen physiology and fertilization. North-Holland Publ, Amsterdam

Mikulska E, Rodkiewicz B (1967) Ultrastructure of the maturating embryo sac of *Lilium regale*. Acta Soc Bot Pol 36:555–566

Mogensen HL (1970) Megagametophyte development in *Agave parryi*. Phytomorphology 20:16–22

Mogensen HL (1972) Fine structure and composition of the egg apparatus before and after fertilization in *Quercus gambelii:* The functional ovule. Am J Bot 59:931–941

Mogensen HL (1973) Some histochemical, ultrastructural, and nutritional aspects of the ovule of *Quercus gambelii*. Am J Bot 60:48–54

Mogensen HL (1975) Fine structure of the unfertilized abortive egg apparatus in *Quercus gambelii*. Phytomorphology 25:19–30

Mogensen HL (1978) The effect of the pollen tube on synergid ultrastructure before and after penetration in *Proboscidea louisianica* (Mill.) Thell. Bull Soc Bot Fr 125:237–241

Mogensen HL, Suthar HK (1979) Ultrastructure of the egg apparatus of *Nicotiana tabacum* (Solanaceae) before and after fertilization. Bot Gaz 140:168–179

Naumova TN (1978) Peculiarity of macrogametogenesis and post-fertilization development of *Trillium camschatcense* Kev-Gawl. Bull Soc Bot Fr 125:183–186

Negi D (1972) Ultrastructure of the female gametophyte of *Eschcholtzia californica* before and after fertilization. Ph D Dissertation, Univ Calif Berkeley

Newcomb W (1972) The development of the embryo sac of sunflower *Helianthus annuus* before and after fertilization. Ph D Dissertation, Univ Saskatchewan, Saskatoon Sas

Newcomb W (1973) The development of the embryo sac of sunflower *Helianthus annuus* before fertilization. Can J Bot 51:863–878

Newcomb W, Steeves TA (1971) *Helianthus annuus* embryogenesis: Embryo sac wall projections before and after fertilization. Bot Gaz 132:367–371

Noher de Halac I, Harte C (1975) Female gametophyte competence in relation to polarisation phenomena during the megagametogenesis and development of the embryo sac in the genus *Oenothera*. In: Mulcahy DL (ed) Gamete competition in plants and animals. North-Holland Publ Amsterdam, pp 43–56

Pacini E, Sarfatti G (1978) The reproductive calendar of *Lycopersicum peruvianum* Mill. Bull Soc Bot Fr 125:295–299

Panchaksharappa MG (1962) Embryological studies in the family Zingiberaceae. I. *Costus speciosus* Smith. Phytomorphology 12:418–430

Panchaksharappa MG, Syamasundar J (1975) A cytochemical study of ovule development in *Dipcadi montanum* Dalz. Cytologia 40:141–149

Philipson MN (1977) Haustorial synergids in *Cortaderia* (Gramineae). N Z J Bot 15:777–778

Philipson MN (1981) The haustorial synergids of *Cortaderia* (Gramineae). Acta Soc Bot Pol 50:151–

Plisko MA (1971) An electron microscopic investigation of the characteristic features of megagametogenesis in *Calendula officinalis* L. Bot Zh (Leningr) 56:582–597

Plisko MA (1974) Ultrastructure of the integument in *Calendula officinalis* L. in the early period of embryogenesis. Bot Zh (Leningr) 59:246–251

Plijm JE van der (1964) An electron microscopic investigation of the filiform apparatus in the embryo sac of *Torenia fournieri*. In: Linskens HF (ed) Pollen physiology and fertilization. North-Holland Publ, Amsterdam, pp 8–16

Poddubnaya-Arnoldi VA, Zinger NV, Petrovskaya-Baranova TP (1964) A histochemical investigation of the ovules, embryo sacs, and seeds in some angiosperms. In: Linskens HF (ed) Pollen physiology and fertilization. North-Holland Publ, Amsterdam, pp 3–7

Pritchard HA (1964) A cytochemical study of embryo sac development in *Stellaria media*. Am J Bot 51:371–378

Rembert DH (1977) Ovule ontogeny, megasporogenesis, and early gametogenesis in *Trifolium repens* (Papilionaceae). Am J Bot 64:438–488

Rifot M (1971) L'évolution ultrastructurale des parois en rapport avec les transformations ontogéniques du gametophyte femelle chez *Aquilegia vulgaris* L. Ann Univ ARERS 9:66–72

Romanov JD (1971) Types of development of the embryo sac of angiosperms. (see Embryological problems. Scientific thoughts. Kiev) In: Poddubnaya-Arnoldi VA (ed) (1976) Cyto-embryology of angiosperm. Nauka, Moscov, p 174

Rosen WG, Thomas HR (1970) Secretory cells of Lily pistils. 1. Fine structure and function. Am J Bot 57:1108–1114

Russell SD (1979) Fine structure of megagametophyte development in *Zea mays*. Can J Bot 57:1093–1110

Russell SD, Cass DD (1981) Ultrastructure of fertilization in *Plumbago zeylanica*. Acta Soc Bot Pol 50:185–190

Rutishauser A (1969) Embryologie und Fortpflanzungsbiologie der Angiospermen. Eine Einführung. Springer, Berlin

Savina GI (1971) Development of the embryo sac in *Ophrys insectifera* L. Ann Univ ARERS 9:62–65

Schnarf K (1929) Embryologie der Angiospermen. In: Linsbauer K (ed) Handbuch der Pflanzenanatomie. Abt II, Teil 2. Bornträger, Berlin

Schulz P, Jensen WA (1968a) *Capsella* embryogenesis: The synergids before and after fertilization. Am J Bot 55:541–552

Schulz P, Jensen WA (1968b) *Capsella* embryogenesis: The egg, zygote, and young embryo. Am J Bot 55:807–819

Schulz P, Jensen WA (1971) *Capsella* embryogenesis: The chalazal proliferating tissue. J Cell Sci 8:201–227

Schulz P, Jensen WA (1973) *Capsella* embryogenesis: The central cell. J Cell Sci 12:741–763

Sehgal CB, Gifford EM Jr (1979) Developmental and histochemical studies of the ovules of *Nicotiana rustica* L. Bot Gaz 140:180–188

Smith BB (1973) Quantification in morphological studies with reference to the megagametophyte of angiosperms. ASB Bull 20:82

Smith BB (1974) Floral ontogeny of *Eupatorium perfoliatum* L. (Compositae) with special reference to why angiosperm embryologists should employ the clearing fluid 4-1/2 ("Herr Fluid"). ASB Bull 21:84

Smith BB, Herr JM Jr (1971) Ovule development, megagametogenesis, and early embryogeny in *Ammannia coccinea* Rothb. J Elisha Mitchell Sci Soc 87:192–199

Steffen K (1951) Zur Kenntnis des Befruchtungsvorganges bei *Impatiens glanduligera* Lindl. Cytologische Studien am Embryosack der Balsamineen. Planta 39:175–244

Strasburger E (1877) Über Befruchtung und Zellteilung. Fischer, Jena

Swamy BGL, Krishnamurty KV (1975) Embryo sac ontogenies in angiosperms. An elucidation. Phytomorphology 25:12–18

Tilton VR (1974) Ultrastructure, histochemistry, and function of the ovule of *Agave parryi*. Ph D Dissertation, Northern Arizona Univ

Tilton VR (1980) Aberrant ovules in angiosperms: A review of selected examples and new observations on *Ornithogalum caudatum* (Liliaceae). Proc Iowa Acad Sci 87:23–28

Tsai AH, Harney PM, Peterson RL (1973) Megasporogenesis and megagametogenesis in *Pelargonium hortorum*. Can J Bot 51:607–617

Vannereau A (1978) Etude ultrastructural du complex micropylaire du sac embryonnaire adulte de *Plantago lanceolata* L. Bull Soc Bot Fr 125:201–205

Vazart B, Vazart J (1965) Infrastructure de l'ovule de Lin, *Linum usitatissimum* L.: Les cellules du sac embryonnaire. C R Acad Sci Paris 261:3447–3450

Vazart B, Vazart J (1966) Infrastructure du sac embryonnaire du Lin (*Linum usitatissimum* L.). Rev Cytol Biol Vég 24:251–266

Vazart J (1969) Organisation et ultrastructure du sac embryonnaire du Lin (*Linum usitatissimum* L.). Rev Cytol Biol Vég 32:227–240

Vazart J (1971) Dégenerescence d'une synergide et pénetration du tube pollinique dans le sac embryonnaire de *Linum usitatissimum* L. Ann Univ ARERS 9:89–97

Vijayaraghavan MR (1970) Ranunculaceae. In: Johri BM (Convener) Symposium on "Comparative embryology of angiosperms". Indian Natl Sci Acad, New Delhi, Bull No. 41:45–52

Vijayaraghavan MR, Jensen WA, Ashton ME (1972) Synergids of *Aquilegia formosa:* Their histochemistry and ultrastructure. Phytomorphology 22:144–159

Went JL van (1970a) The ultrastructure of the synergids of *Petunia*. Acta Bot Neerl 19:121–132

Went JL van (1970b) The ultrastructure of the egg and central cell of *Petunia*. Acta Bot Neerl 19:313–322

Went JL van (1970c) The ultrastructure of the fertilized embryo sac of *Petunia*. Acta Bot Neerl 19:468–480

Went JL van, Linskens HF (1967) Die Entwicklung des sogenannten „Fadenapparates" im Embryosack von *Petunia hybrida*. Gen Breed Res 37:51–56

Willemse MTM, Bednara J (1979) Polarity during megasporogenesis in *Gasteria verrucosa*. Phytomorphology 29:156–165

Willemse MTM, De Boer-De Jeu MJ (1981) Megasporogenesis and early megagametogenesis. Acta Soc Bot Pol 50:105–114

Willemse MTM, Franssen-Verheijen MAW (1978) Cell organelle changes during megasporogenesis and megagametogenesis in *Gasteria verrucosa* (Mill.) Haw. Bull Soc Bot Fr 125:187–191

Willemse MTM, Kapil RN (1981) On some anomalies during megasporogenesis in *Gasteria*. Acta Bot Neerl 30:439–447

Wilms HJ (1980) Development and composition of the spinach ovule. Acta Bot Neerl 29:243–260

Wilms HJ (1981a) Ultrastructure of the developing embryo sac of spinach. Acta Bot Neerl 30:75–99

Wilms HJ (1981b) Pollen tube penetration and fertilization in spinach. Acta Bot Neerl 30:101–122

Yu SH, Chao ChY (1979) Histochemical studies of ovary tissues during the embryo sac development in *Paspalum longifolium* Roxb. Caryologia 32:147–160

5 The Pollen Grain

R. B. KNOX

5.1 Introduction

The pollen grain is the carrier of the male gametes or their progenitor cell, in higher plants. In a single unit, each grain contains all the genetic information required to specify an entire haploid plant organism (for example, pollen embryoids in tissue culture), or to unite with the female gamete at fertilization and form a diploid zygote and, hence, a new sporophyte. The male gametes, the sperm cells (or their progenitor, the generative cell) are housed entirely within the cytoplasm of the vegetative cell. This is dehydrated like a seed at maturity, and is filled with storage reserves. The vegetative cell is surrounded by a complex, intricately patterned outer wall, and its nucleus controls at least the initial growth and metabolism of the pollen tube following germination.

Such a structure represents one of the simplest, isolated cellular systems in flowering plants. Each grain is either bi- or tricellular at maturity (see Fig. 5.1 A–C), and is the product of only two rounds of cell division following microspore formation (as opposed to five rounds of mitosis necessary for pollen formation in gymnosperms). Thirty years ago, Maheshwari (1950) declared that pollen grain development in angiosperms is remarkably uniform, referring to the cytological processes of nuclear division. During the last 10 years it has become evident that while pollen grains at dispersal may show superficial similarities, nevertheless pollen grain development and wall structure proceeds along different routes in different groups of flowering plants (see review by Knox 1979). Each is distinctive, presumably an adaptation to its cellular environment and mode of pollen dispersal by wind, water or animal vectors.

The adaptations include:

1. Reduction in chromosome number, and genetic recombination essential to microspore formation (meiosis).
2. Development of cellular barriers to protect the integrity of the microspores from the genetically distinct parent cells (the callose special wall).
3. Protection of the male gametes from mechanical and environmental damage during dispersal (the pollen grain wall).
4. Vector transport, and adhesion to the stigma surface (pollen shape, sculpturing, and surface coatings).
5. Pollen-stigma recognition, and allergic response in man (pollen-wall proteins and glycoproteins).
6. Germination and pollen tube growth through the style to the ovary for fertilization (specific enzymes, heterotrophic nutrition).

It is the purpose of this chapter to outline the structure and developmental processes of pollen grains of flowering plants; particularly as they relate to the pro-

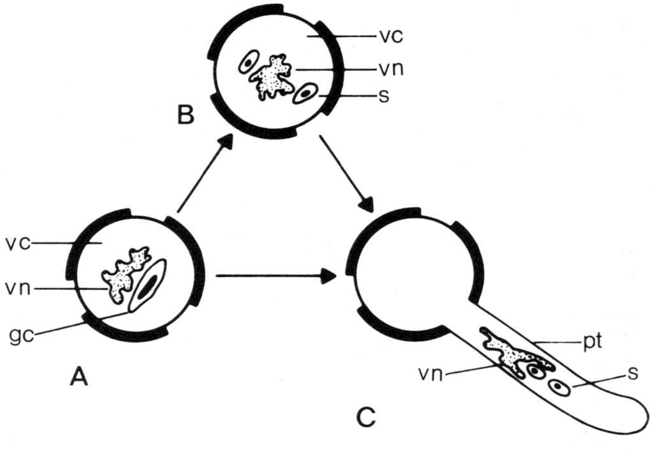

Fig. 5.1. **A** Mature bicellular. **B** Tricellular pollen grains. **C** Bicellular grain becomes tricellular after germination. Vegetative cell *vc*, generative cell *gc*, vegetative nucleus *vn*, sperm cells *s*, pollen tube *pt*. (Adapted from Dexheimer 1970)

cesses of pollination. The diversity in form and structure of pollen grains suggest they have arisen by remarkable processes of adaptation. On the one hand, to the environment during pollen dispersal whether it be wind or water currents, and, on the other, mutual co-adaptation with specific animal vectors. The latter present some of the most convincing examples of mutual co-evolution known in biology today. In all these cases, pollen structure has been influenced to a remarkable degree, which accounts for the species and group-specific patterns of sculpturing, and the composite nature of certain types of pollen.

In addition, the pollen grain and its pollen tube is an essential partner in fertilization and seed setting. In order to produce new types of crop plants through hybridization and manipulation of the reproductive system, it is essential to have an understanding of the structure and function of pollen. An attempt has been made to interpret research on pollen grains with this aim in view, and to point out deficiencies where they may exist.

5.2 Pollen Structure and Cytochemistry

The pollens of most species of flowering plants are free, cellular structures, each developing from a single microspore. In some families variation on this theme occurs, with the formation of composite grains. These may be tetrads, containing all four products of a single microporocyte linked together, or multiples of four forming polyads of 8, 16, 32, or 64 grains, or more complex masses, or pollinia. Among tetrad types, the grains may have 0, 1, 2, 3, or all 4 grains fertile, with the sterile aborted grains remaining attached, i.e., they may be nullads, monads, dyads, triads, or tetrads. Each microspore is walled off, and regulates its own destiny. However, in one family, the Cyperaceae, each grain forms from one of the four microspores, the others aborting with or without wall formation (see Maheshwari 1950).

5.2.1 Pollen Shape

The form of the pollen of angiosperms is a genetically determined character, and is established in three ways:

1. By siting of the microspores within the meiotic tetrad.
2. By the number and disposition of germinal apertures, determined within the meiotic tetrad.
3. By differential cell expansion during microspore and pollen development.

The first factor may control shape mechanically. The shape of the grains is established within the thick, rigid callose special wall, which shows little or no expansion during tetrad period, acting like a mould or template for initial pollen wall growth (see Sect. 5.3.1). The second factor is also determined within the callose wall since the sites of apertures are pre-determined to specific areas within the tetrad. For example, in many monocot pollen grains the single aperture is sited on the outer tangential surface of each microspore (see Sect. 5.3.3).

The third factor is manifested in grains with an unusual shape. Most pollen grains are spherical, and essentially do not change their gross shape after spore release period. However, in filiform grains of some sea-grasses, as in *Amphibolis* and *Halodule,* the tetrad of microspores shows little elongation until release when each grain shows characteristic tip growth – resulting in the long filiform grains that may reach 5 mm in length in *Amphibolis* (Ducker et al. 1978).

5.2.2 Pollen Cytology

In most families of flowering plants, the pollen grain is bicellular. The largest cell, the vegetative cell, contains a lobed central nucleus with pores in the nuclear membrane communicating with a cytoplasmic endomembrane system (Jensen et al. 1968 b; and review by Jensen 1974). The cytoplasm is also packed with organelles: mitochondria, plastids, golgi, and ribosomes; and storage organelles: starch grains, P (polysaccharide) particles (J. Heslop-Harrison 1979 a, b), lipid droplets and proteins stored within membrane vesicles of the endoplasmic reticulum. Stacked endoplasmic reticulum is an unusual ultrastructural feature of the cytoplasm of the vegetative cell of pollen grains (Rosen 1971, Jensen et al. 1974).

The generative cell is usually bounded by a plasma membrane and distinct cell wall, often containing callose, e.g., in orchids (J. Heslop-Harrison 1968 a). The cell is located entirely within the cytoplasm of the vegetative cell, is usually markedly elongate (Fig. 5.2 A), and contains a limited range of cytoplasmic organelles (Larson 1965, Jensen and Fisher 1968 a, Coccuci and Jensen 1969, Jensen 1974).

Division of the generative cell to form the pair of sperm cells most commonly occurs within the pollen tube shortly after its emergence and penetration into the stigma (Cresti et al. 1980). In tricellular types of pollen it occurs precociously in the maturing grains. In dicots, such as the sunflower, *Helianthus annuus,* the filiform sperm cells are approx. 40 μm in length (Fig. 5.2 B, C) but are smaller in *Brassica* (Fig. 5.2 D); their structure is known in cotton (Jensen and Fisher 1968 a), sugar beet (Hoefert 1969), and *Plumbago* (Russell and Cass 1981 a, b). The

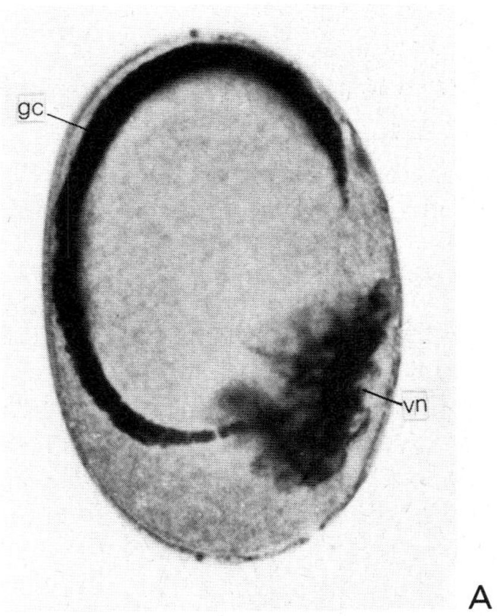

Fig. 5.2 A–D. Cytology of the generative cell in bicellular pollen of *Tradescantia virginiana*, and sperm cells in tricellular pollen of sunflower (*Helianthus annuus*) and kale (*Brassica oleracea*). **A** Mature grain of *Tradescantia*, stained with aceto-orcein, showing horseshoe-shaped generative cell *gc*, and vegetative nucleus *vn*; they appear connected. **B**, **C** Sunflower sperm cells *sc* and vegetative nucleus *v*. **B** Light microscopy using bright-field illumination. **C** Nomarski differential interference contrast optics. *e* spines of exine, *i* intine. (From Vithanage and Knox 1979). **D** Cytology of the sperm cells of *Brassica*, transmission electron micrograph of mature pollen grain. The sperm cells are enveloped by double unit membranes *pm* which are deposited in pockets or blebs, and contain an elongate nucleus *sn* and a thin layer of cytoplasm. The convoluted vegetative nucleus *vn* is at centre of the vegetative cell. *a* aperture, *p* pollen-coat. (From Dumas and Gaude 1982)

structure of the sperm cells of monocots has been studied in maize, *Zea mays*, where each cell is approx. 14 × 8 μm (Jensen 1974), in grasses (Larson 1965), orchids (Coccuci and Jensen 1969), barley (Cass and Karas 1975), rye (Karas and Cass 1976), and wheat (Chu and Hu 1981). Such studies have provided some answers to questions concerning the differentiation of sperm cells, and the role of their cellular organelles.

First, what specific processes are involved in sperm cell differentiation? Chu and co-workers at Peking University, Peking (Beijing), China, have demonstrated that there are four developmental periods in the tricellular pollen of wheat (Fig. 5.3):

1. Naked cell stage: After mitosis the sperm cell is enveloped in a discontinuous plasma membrane.
2. Walled cell stage: The cell is enclosed within a plasma membrane – surrounded by a callose wall and the plasma membrane of the host vegetative cell.
3. Cytoplasm-increasing stage: Period of organelle and cytoplasmic-differentiation.
4. Mature sperm cell: Wall becomes discontinuous, surrounding the sperm with an elliptical nucleus at one end and organelles (especially mitochondria) concentrated in the remainder of the filiform cell (Zhu et al. 1980, Chu and Hu 1981).

A range of organelles occurs in the sperm cytoplasm – including mitochondria, ER, Golgi bodies, ribosomes, small vacuoles, and plastids with reduced lamellar structure. In addition, microtubules, and numerous microfilaments are arranged parallel to the longitudinal axis of the cell.

Is there any evidence for sperm cell motility? Zhu et al. (1980) have approached this question in their ultrastructural studies of mature wheat pollen by

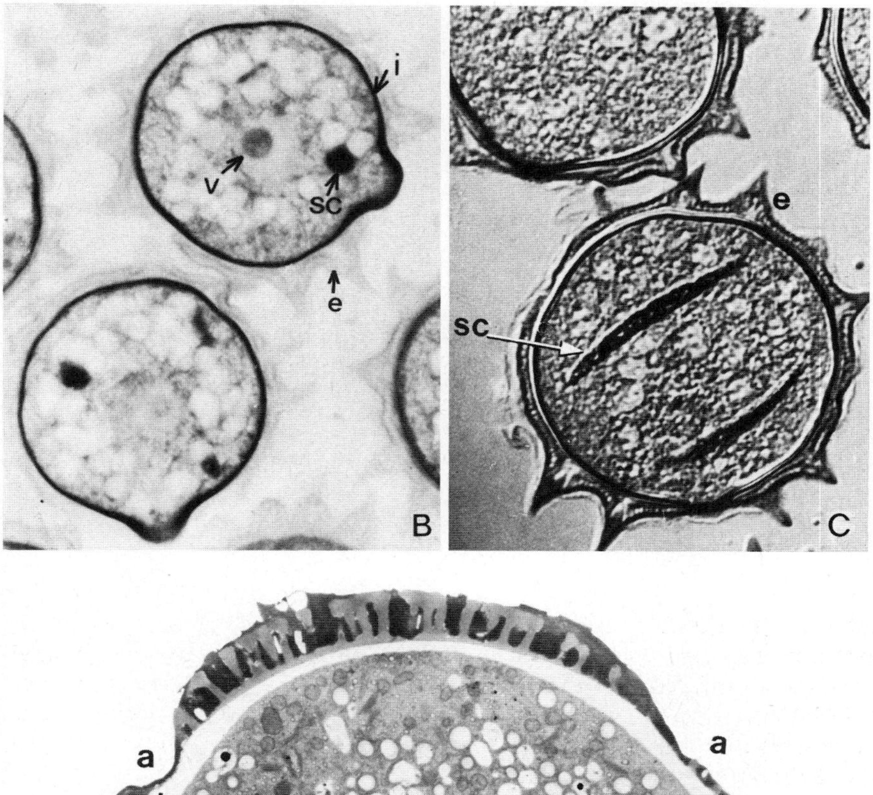

Fig. 5.2 B–D

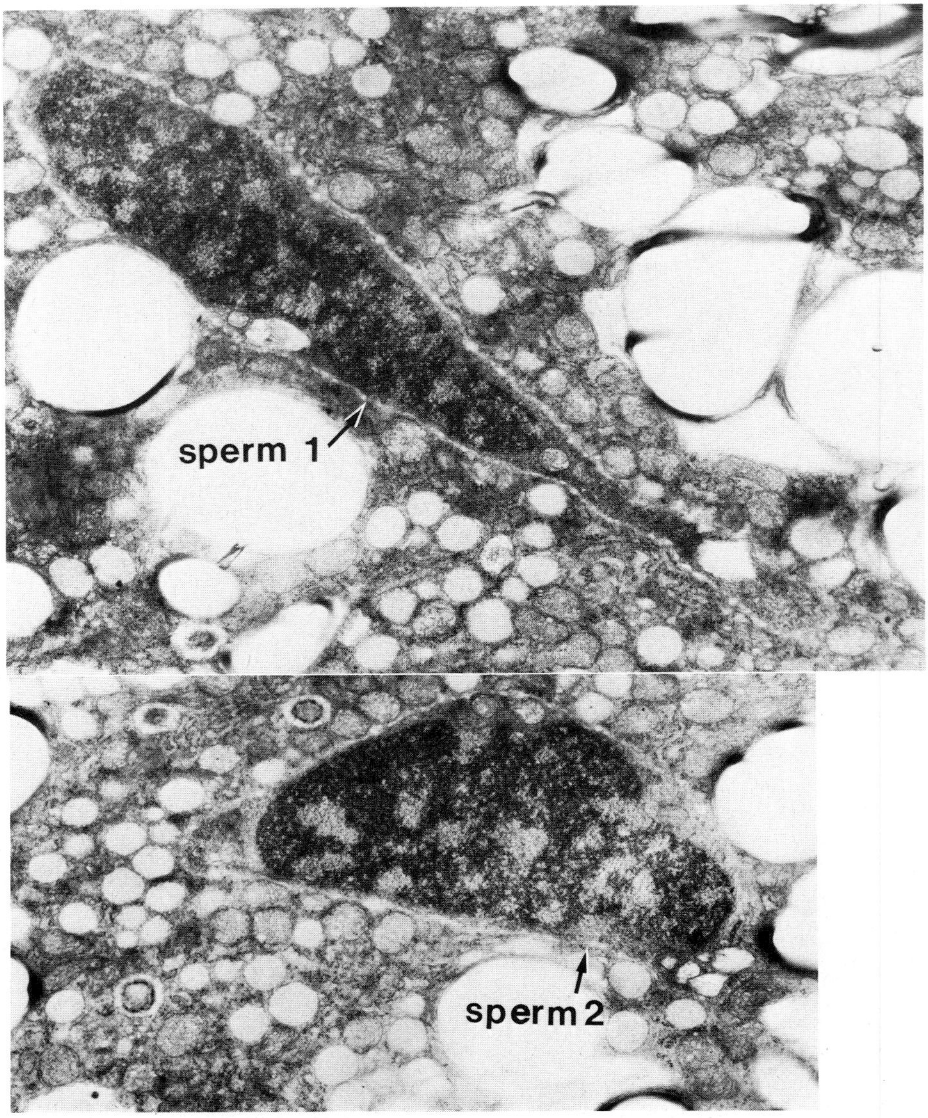

Fig. 5.3. Ultrastructure of the elongate sperm cell in a mature tricellular pollen grain of wheat, *Triticum aestivum*. Transmission electron micrograph showing the large elliptical sperm nucleus occupying the "head", while small vacuoles, mitochondria, microtubules, starch grains, and fibrillar structures are held within the plasma membrane. The double membrane of the sperm and vegetative cell has a characteristic blebbed appearance. (From Chu and Hu 1981)

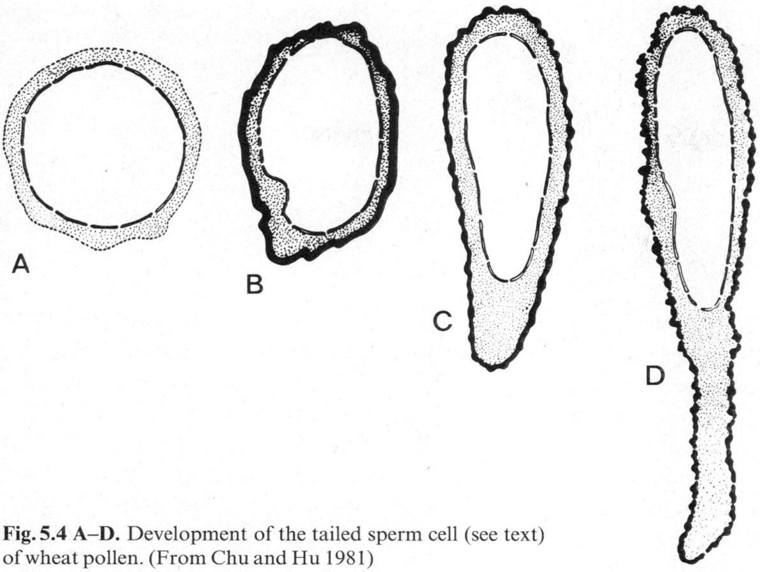

Fig. 5.4 A–D. Development of the tailed sperm cell (see text) of wheat pollen. (From Chu and Hu 1981)

using both glutaraldehyde-osmium fixation and potassium permanganate fixation to detect the sperm cells within the densely packed storage organelles. They first noted a tailed appearance of the sperm cell in mature pollen (Fig. 5.4 A–D), with nucleus in the head region, and mitochondria and ER extending along the tail, accompanied by longitudinally oriented microtubules (270 Å diameter), and numerous microfilaments (60–70 Å diameter) occurring throughout the cytoplasm, closely packed and parallel to the plasma membrane. These are similar to microfilaments found in pollen tubes of other angiosperms (Franke et al. 1972, Condeelis 1974, Cresti et al. 1975), and Chu and co-workers liken their appearance to microfilaments reported in the sperm tail of insects, and consider that the tailed structure is adapted for sperm movement. Chu and Hu (1981) cite recent work in which the sperm cells of pollen tubes of *Amaryllis* and *Clivia* are shown to possess ATPase activity, which is related to mechanical movement in other systems. In tricellular pollen the sperm cells and vegetative nucleus have to move, or be moved, passively into the pollen tube at germination; likewise in bicellular grains migration of the generative cell and vegetative nucleus occurs.

Finally, are there any detectable differences between the pair of sperms in the pollen tube? These have to fulfil quite different roles during double fertilization. In recent ultrastructural studies of pollen of *Plumbago zeylanica,* differences between the pair of sperm cells have been observed (Russell 1980, Russell and Cass 1981 a, 1983). One sperm cell consistently is attached to the vegetative cell by a projection, while the other is linked to the first by plasmodesmatal connections. Furthermore, the sperm attached to the vegetative nucleus contained significantly higher numbers of mitochondria, while the other had higher numbers of plastids. These data suggest that the two sperm cells may be dimorphic, and have quite different patterns of male cytoplasmic transmission (Hagemann 1976, 1981, Russell and Cass 1981 a, b, 1983).

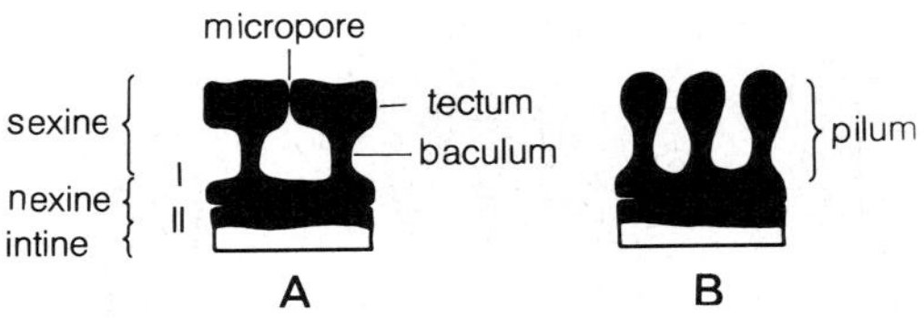

Fig. 5.5 A, B. Stratification of pollen walls. **A** Tectate exine, **B** pilate exine. (Modified from J. Heslop-Harrison 1968 b)

5.2.3 Pollen-wall Structure

There are two domains in the pollen wall (in the sense of J. Heslop-Harrison 1975 a): the exine, the outer patterned layer made of sporopollenin, and the intine, the inner smooth polysaccharide layer. These layers are chemically, morphologically, developmentally, and genetically distinct. The terminology applied to the wall of pollen grains is daunting, especially as it has been developed from early light microscope work, and then transposed to the images seen in the transmission and scanning electron microscopes (Erdtman 1952, Faegri and Iversen 1975).

The exine is made of sporopollenin – a wall polymer remarkable for its resistance to biodegradation, and considered to be formed by the oxidative polymerization of carotenoids and carotenoid esters (Shaw 1971, Brooks and Shaw 1978). The exine has many properties in common with lignin and cutin (see Kolattukudy 1980). The pollen wall may contain about 1% elemental silicon, as determined by X-ray microanalysis (Crang and May 1974). In morphological terms, there are two types of exine (Figs. 5.5, 5.6):

1. Where the patterned surface layer contains arcades covered by a roof or tectum, supported by rod-like bacula (or columellae) – the "tectate" type (Fig. 5.5 A, 5.6 A, B).
2. Where the patterned surface layer is open, the pattern comprising the bacula (pili) which may be ornamented or fused together at their tips – the "pilate" type (Fig. 5.5).

The exine layer is complex and comprises several substrata: (1) the outer tectum, when present, (2) the bacula (singular: baculum; may be called bacules), (3) the foot layer or nexine, and (4) the innermost nexine-2. Layers 1 and 2 are often termed the sexine, while 1, 2, and 3 are called the ektexine (see Fig. 5.7); layer 4 has been termed the endexine. In the Compositae the two nexine layers may be separated by spaces, the cavea. The terminology applied to grass pollen is compared in Fig. 5.7 as this type of grain has been the object of much recent research.

In tectate exines, the tectum is perforated by micropores, communication channels which lead to the crypt-like arcades within; in some specific pollen types, microchannels traverse the nexine layer linking with the nexine-2 or intine, e.g., in grass pollen (see review by J. Heslop-Harrison 1979 b). In pilate exines the spaces between the pili are open to the exterior. In both these types of grains it is usual for the surface cavities and exine arcades to be filled with pollen coat substances (see Sect. 5.2.5.1).

The intine forms the surface of the grain at the germinal apertures. Here, the exine layer is absent, greatly reduced or forms a cap-like operculum. In many pol-

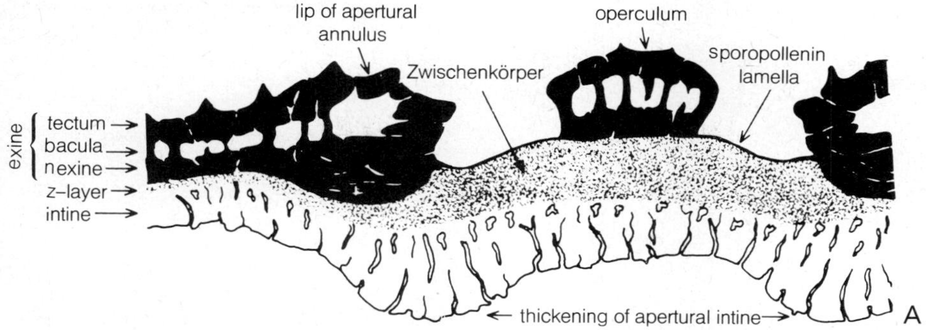

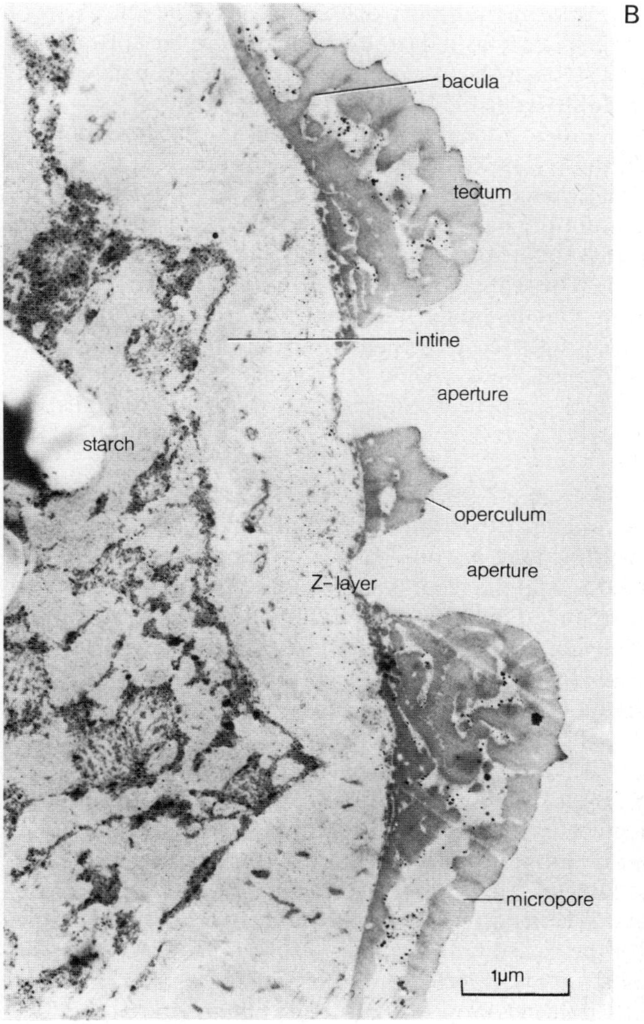

Fig. 5.6. A Aperture region of a grass pollen grain to identify various structural features of the wall. **B** Transmission electron micrograph of pore region of the pollen grain of rye grass, *Lolium perenne,* for comparison with **A**. (From Vithanage et al. 1982)

Heslop-Harrison (1979)	Erdtman (1952)	Rowley (1960)	Skvarla and Larson (1966)	Christensen, Horner and Lersten (1974)
exine [tectum, bacula, nexine]	exine [sexine [tectum, bacula], nexine]	exine [ektexine [tegillum, columellae], endexine]	exine [ektexine [tectum, columellae, foot layer]]	exine [tectum, bacula, nexine]
Z-layer	—	—	endexine	granular layer
intine	intine	intine	intine	intine

Fig. 5.7. Terminology applied by various authors to the stratigraphy of grass pollen wall. (From J. Heslop-Harrison 1979 b)

len types it comprises a thick polysaccharide layer, often with radially or tangentially oriented tubules, and is considerably thickened at the germinal apertures. In other types it may consist of more than one layer, differentiated by their characteristic appearance by transmission electron microscopy. The inner part of the intine is predominantly cellulosic, as judged by its microfibrillar appearance, while the outer tubular zone stains with ruthenium red, indicating a high content of acidic polysaccharides (Mühlethaler 1953, Sitte 1953, Bailey 1960).

The chemical and morphological identity of the two strata has been defined by chemical tests in which the layers can be differentially solubilized in organic solvents, and acidic (especially acetolysis – Erdtman 1960, Dickinson and Bell 1972) or alkaline solutions (Southworth 1974, J. Heslop-Harrison 1979 b, Kress and Stone 1982). The stratigraphy can be most readily demonstrated by cytochemical tests (see Table 5.1). These are based on the affinity of the exine for dyes binding lignin, phenolic compounds, lipids and negatively charged groups, and of the intine for cytochemical tests for polysaccharides, acidic polyanions, and proteins. The chemical differences between the exine sublayers – tectum (~ektexine) and nexine (~endexine) – may be distinguished by their differential affinity for negatively charged dyes (e.g., basic fuchsin) and for lipids (auramine O fluorescence).

The inner exine layer, the nexine-2 (or endexine) appears to be very variable in its chemical composition and ultrastructural appearance in different pollen systems. It is frequently present as an electron-dense lamellated layer, e.g., in the monocot *Clivia* (Afzelius 1955), and the dicot *Saintpaulia*. This intermediate layer between exine and intine frequently shows differential solubility by acetolysis, while its ultrastructural appearance suggests it may be composed of both sporopollenin and polysaccharides. Saad (1963) described several variants of this layer which he termed the medine. In several of Saad's descriptions, the medine forms a lens-shaped structure at the apertures, which appear similar to structures termed Zwischenkörper by other authors (see Rowley 1962), and more recently defined as the Z-layer by J. Heslop-Harrison and Y. Heslop-Harrison (1980) in grass pollen. The cytochemical properties of the nexine-2 in sunflower pollen (Vithanage and Knox 1979) certainly suggest it may comprise polysaccharides (affinity for decolorized aniline blue) and sporopollenin (shows auramine O fluorescence). This is an active area in current research, and further work with other pollen systems may resolve the varied nature of this transition zone of the exine.

The complexity of wall structure revealed in these free pollen types (monads) is also reported in composite pollen. Composite grains are of two types:

Table 5.1. Cytochemical reactions of the pollen grain wall to various cell-wall stains. + indicates a positive staining reaction, − a negative reaction; numbers in superscript refer to the references cited below. Use of the stains is described by Jensen (1962), Southworth (1973), J. Heslop-Harrison (1979b), or in the cited references

Cytochemical technique	Specificity	Pollen wall layers		
		Tectum (sexine, ektexine)	Nexine (endexine)	Intine
1. Lipid-containing components				
Auramine 0	Lipids, especially of cuticles	+[1,6,7,8,9,10,13]	+[6] −[7,13] absent [1,8,13]	−
Nile blue after pyridine extraction	Specific reactive groups of lipids; anionic groups	+[2,4] (blue)	+[2,4]	−
Osmium tetroxide	Specific reactive groups of lipids	+[2,3,4]	+[2,3,4]	−
Scarlet R (Sudan IV)	Lipids	+[1,8,10]	−	−
2. Lignin and phenolic components				
Autofluorescence	Phenolic compounds or lignin-like components	+[1,6,7,8,9,10]	+[6] −[7] absent [1,8]	−
Fast blue B salt	Non-specific binding to phenolic compounds	+[11]	absent [11]	−[11]
Maule reaction	Lignin components	+[2] (brown)	+[2] (brown)	
Toluidine blue	Acidic polyanionic groups stain blue, purple lignin and phenolic stain green; acidic polysaccharides (pectic acid) stain red or purple	+[1,4,6,7,8] (green) +[9] (blue)	+[4,6,7,10] (green)	+[1,4,5,6,7,8] (red) +[9,10] (Z-layer) (red)
3. Wall polysaccharides				
Alcian blue	Acidic polysaccharides	−[9,13]	−[9,13]	+[9] (Z-layer) +, −[13]
Calcofluor	*β*-glucans	−[8,9]	−[9] absent [8]	+[8,9] +, −[13]
Decol. aniline blue	Callose	−[2,6,8]	+[6] −[2] absent [8]	+[2,8,12] −[6]
Hydroxylamine Iron after methylation	Pectic acids	−	−	+[2]
PAS	Vicinal glycol groups of carbohydrates	−[1,3,4,6,7.8,9,10] absent [5]	+[6] −[3,4] absent [1,5,8]	+[1,3,4,5,6,7,8,12] +[9,10] (Z-layer) +, −[13]

Table 5.1 (continued)

Cytochemical technique	Specificity	Pollen wall layers		
		Tectum (sexine, ektexine)	Nexine (endexine)	Intine
Ruthenium red	Acidic polysaccharides (pectic acids)	– [1,2,3,9]	– [2,3,9] absent [1]	+ [1,2,3] + [9] (Z-layer)
Zinc chlor-iodide	Cellulose or hemicellulose	+ [4] (orange)	+ [4] (orange)	+ [2] (blue)
4. Charged groups				
Azure B	Negatively charged groups	+ [3,4]	+ [3,4]	–
Basic Fuchsin	Negatively charged groups	+ [1,6,7,8,10,13]	+ [6] – [7,13] absent [1,8,13]	–
5. Wall proteins				
Aniline blue black	Proteins	+ [3,4]	+ [3] – [4]	+ [3,4]
Coomassie blue	Proteins	+ [4] (absent after pronase)	– [4]	+ [4] (absent after pronase)

[1] *Gladiolus gandavensis* (Knox 1971), [2] *Ambrosia trifida* (Southworth 1973), [3] *Artemisia pynocephala* (Southworth 1973), [4] *Gerbera jamesonii* (Southworth 1973), [5] *Amphibolis antarctica* (Ducker et al. 1978), [6] *Helianthus annuus* (Vithanage and Knox 1979), [7] *Acacia subulata* (Kenrick and Knox 1979), [8] *Populus deltoides* (Ashford and Knox 1980), [9] *Secale cereale* (J. Heslop-Harrison 1979b), [10] *Lolium perenne* (Howlett et al. 1981), [11] *Hordeum vulgare* (Ahokas 1975), [12] *Vinca rosea* (Cousin 1980), [13] *Musa* (Kress and Stone 1982)

1. "Calymmate" grains united together by a continuous layer of sexine (i.e., tectum, bacula, and foot layer).
2. "Acalymmate" grains not united by continuous layer of sexine, but by other means (van Campo and Guinet 1961).

In calymmate tetrads, for example in Asclepiadaceae and Goodeniaceae, a common tectum links the grains together. In acalymmate complexes the tectum is not united between grains and the cohesion of the polyad or pollinium has to be assured by other means (Table 5.2); for example wall bridges or by a special adhesive layer. The development of composite pollen is considered later (see Sect. 5.3.6). Its ultimate expression is seen in the large pollinia and massulae found in the higher Orchidaceae (see Sect. 5.3.6).

5.2.4 The Pollen Tube

The pollen tube is produced after germination in vivo on the stigma, or during suitable incubation in vitro. For most angiosperms it is the vehicle for the male

Table 5.2. Mechanisms of cohesion between the grains of acalymmate polyads. (From Kenrick and Knox 1979)

Nature of wall link	Genus (family)	Reference
Free elements of sexine	*Dinizia* (Mimosaceae) *Typha* (Typhaceae) *Leschenaultia* (Goodeniaceae)	Guinet and LeThomas (1973) Skvarla and Larsen (1963) Knox and Friedrich (1973)
By bacula of the sexine only	*Drimys* (Winteraceae)	Smith (1967), Roland (1971), Agababyan (1972)
By layers of nexine-2	*Asteranthe*, *Hexalobus* (Annonaceae)	LeThomas and Lugardon (1972)
By bridges of nexine-2	*Epilobium*, *Ludwigia*, *Camissonia* (Onagraceae)	Skvarla et al. (1975)
	Inga, *Acacia* (Mimosaceae)	Barth (1965)
By bridges of nexine-2 and intine	*Acacia* (Mimosaceae) *Hedycarea* (Monimiaceae)	Guinet and Lugardon (1976) Sampson (1977)
By „Kittsubstanz" or cement	*Acacia* (Mimosaceae)	Barth (1965)

gametes, since it is within the tube that the generative cell divides to form the sperm cells. Both the generative cell and vegetative nucleus migrate into the tube soon after its formation, and move towards the tip of the tube. In *Lycopersicon peruvianum* they remain in the nuclear zone, the largest of the four regions of the growing pollen tube (Cresti et al. 1980), and division to form the sperm cells occurs immediately. Unfortunately, there is little information on the early events of pollen-tube formation and extension. Within the pollen tube the sperm cells are closely associated with the vegetative nucleus, and in some recent studies there is electron microscopic evidence that they may indeed be structurally linked, e.g., in *Gossypium* (Jensen and Fisher 1968 a, 1970) and *Plumbago* (Russell and Cass 1981 a, b, Russell 1982, 1983). Apart from *Lycopersicon,* scanty data are available on the events within the pollen tube cytoplasm, although ultrastructural studies have been carried out on the monocots *Lilium* (Rosen et al. 1964, Reiss and Herth 1979), *Clivia, Nymphea,* and *Ornithogalum* (Pargney 1982), and the dicots *Petunia* (Sassen 1964) and *Lobelia* (Pargney 1982).

The nature of the pollen-tube wall, and its formation at pollen germination is described in Section 5.4.2. In cotton, *Gossypium,* Jensen and Fisher (1969, 1970) have shown that the tube wall is thick and multilayered. The developing tube wall of *Lycopersicon* is bi-layered, comprising an outer pecto-cellulosic wall continuous with the intine, and an inner layer of callose, detected by its fluorescence with decolorized aniline blue (Cresti et al. 1977). A similar organization has been observed in *Nicotiana* pollen tubes (Kroh and Knuiman 1982). Four zones, reflecting functional specialization, have been detected by electron microscopy in *Lycopersicon:* (1) and (2) Apical and subapical zones: cytoplasm with abundant Golgi bodies and vesicles; thin fibrillar cell wall. (3) Nuclear zone: houses vegetative nucleus and elongate generative cell surrounded by two membranes, often with fibrillar material betwen them; callose layer in inner tube wall. (4) Zone of vacuolation and callose plug formation: very thick callosic wall ending with callose plug adjacent to pollen aperture (Cresti et al. 1977). The callose wall deposition in *Lycopersicon* tubes is similar to that described for *Petunia* (Cresti and van Went 1976) and *Nicotiana* (Kroh and Knuiman 1982). The participation of golgi vesicles in secretion of the tube wall polysaccharides, especially callose, has been suggested in several studies (Morre et al. 1971, Franke et al. 1972, Helsper et al. 1977). Indeed, Woude et al. (1971) have not only described the ultrastructure of the vesicles, but have also isolated a fraction containing them and carried out a chemical analysis.

Plugs of callose and related polysaccharides contain the living protoplast at the pollen tube tip. Ultrastructural studies of the plugs have been carried out with several dicots, e.g., *Lychnis* (Crang and Hein 1971), *Lycopersicon* (Cresti et al. 1980), *Oenothera* (Dickinson and Lawson 1975), *Petunia* (Cresti and van Went 1976, Herrero and Dickinson 1980), and *Prunus* (Cresti et al. 1979), and the monocots, e.g., *Lilium* (Nakamura et al. 1979), and *Secale* (J. Heslop-Harrison 1979 b). Presumably, the plugs seal the protoplast, permitting renewed tip growth, although considerable controversy exists concerning their chemical nature (Kroh and Knuiman 1982).

During its subsequent development, the pollen tube has its unique characteristics expressed both in its chemistry and structure. Structural differences occur in pollen tubes of *Rhododendron* during interspecific matings. Pollen tubes of one species exhibited markedly different phenotypes during growth through the styles of different species (Williams et al. 1982 a). At the same time, pollen tube tips showed a range of characteristic cytological abnormalities associated with interspecific incompatibility during in vivo growth on the stigma, and in the style and ovary (Williams et al. 1982 a, b). Studies need to be directed towards determining the nature of gene expression in the pollen tube tip – as this structure provides the cellular interface with much of the pistil during the interactions that determine the fate of the sperm cells.

5.2.5 Structural Adaptations of Pollen for Dispersal

The pollen grain forms only a small part of the flower whose entire structure is subject to selection pressures for adaptation to mode of pollination, especially potential interactions with animal pollinators. Nevertheless, it is the haploid gametophyte generation, and therefore might be expected to reflect genetic changes more rapidly and directly than the surrounding diploid parental tissue. Until recently, the evidence from pollen morphology has not supported this concept. The pollen grain has been seen as a conserved structure: it has species and group-specific exine patterning, and intraspecific variation has seldom been demonstrated. Together with the evidence of genetic experiments, this conservatism suggested that exine form and structure is under sporophytic control (Godwin 1968, J. Heslop-Harrison 1968 b, Knox 1976, 1979).

The adaptations that were first observed were correlations between exine structure, and mode of pollination in families with both wind-pollinated and ani-

Table 5.3. Sealing devices of pollen grains, classified according to type of exine. (From J. Heslop-Harrison 1979c)

Exine type	Sealing devices
Colpate; sulcate	(a) Infolding, with buckling, folding or interleaving shutters; lipid seal (e.g., *Lilium*)
Porate	
Single pore	Operculate; no lipid seal (Gramineae)
Multiple pores	(a) Operculate; with or without lipid seal (e.g., Caryophyllaceae) (b) Occlusion by sporopollenin granules; lipid seal (e.g., Malvaceae)
Colporate; triaperturate	Infolding, with simple folding or buckling shutters or interleaving sporopollenin plates; with or without lipid seals (commonest dicotyledonous types)
Inaperturate	Granulate exine; lipid seal (e.g., many Iridaceae)

mal-pollinated genera. Characters such as exine thickness and extent of sculpturing, type of aperture, and the presence of different adhesive components in the surface pollenkitt have also proved of adaptive significance. These features were highlighted by J. Heslop-Harrison (1979c) in a recent review of pollen walls as adaptive systems. He has defined the routes for water loss during anther dehydration and pollen transfer, and then water uptake at hydration. His findings are summarized in Table 5.3. In colpate and sulcate types (see Sect. 5.3.3) the exine has specific mechanisms for dehydration, including unfolding, e.g., in *Gladiolus* (Fig. 5.8). Contrast the dehydrated mature pollen in Fig. 5.8A with the germinated grain in Fig. 5.8B. In porate types, e.g., grass pollen (Fig. 5.8C), the whole grain can act in water-flow control as this is regulated both through the operculum and through micropores which traverse both sexine and nexine wall layers, and there is no extensive lipid seal covering the grain surface. The dry grains are wrinkled (Fig. 5.8C) while the germinated grains are spherical (Fig. 5.8D).

In polyporate grains the exine is usually massive, e.g., in *Malvaviscus* and *Ipomoea* pollen (Fig. 5.8E, F), and a thick lipid seal is present in mature pollen (removed in SEM preparation, Fig. 5.8E; present in living pollen, Fig. 5.8F). In colporate or triaperturate types, e.g., pollen of *Eucalyptus* (Fig. 5.8G), *Amyema* (Fig. 5.8H, I), *Ricinus* (Fig. 5.8J), *Datura* (Fig. 5.8K), *Cosmos* (Fig. 5.8L), and *Crossandra* (Fig. 5.8M), there is a combination of the mechanism shown by colpate and porate pollen types. The grains may be infolded or have shutters or plates (e.g., in Fig. 5.8K). The grains of *Crossandra stenostachya,* an African species of Acanthaceae, have a most unusual shape (Fig. 5.8M), long and narrow, 520×19 μm. The grains have a thick exine covered by a tectum, and a thick underlying nexine layer (Brummitt et al. 1980).

Finally, there are pollen types without preformed apertures, the inaperturate class in Table 5.3. Here, the exine may be fragile and much reduced, forming a surface reticulum in *Populus* (Fig. 5.8N) or isolated spinules in *Cinnamomum* (Fig. 5.8O). The intine is usually thick and prominent in these types. The surface of the grains may have a prominent lipid seal or be covered with mucilage (in aquatic plants with hydrophilous pollination, see Sect. 5.2.5.3). The unique filiform

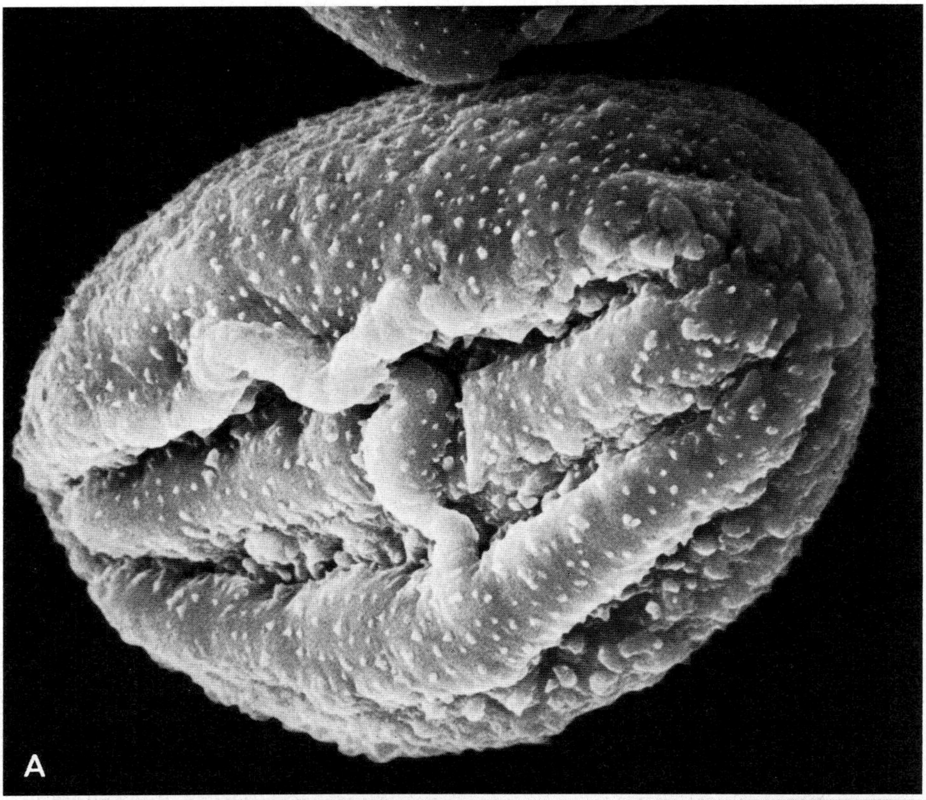

Fig. 5.8 A–C (legend see p. 215)

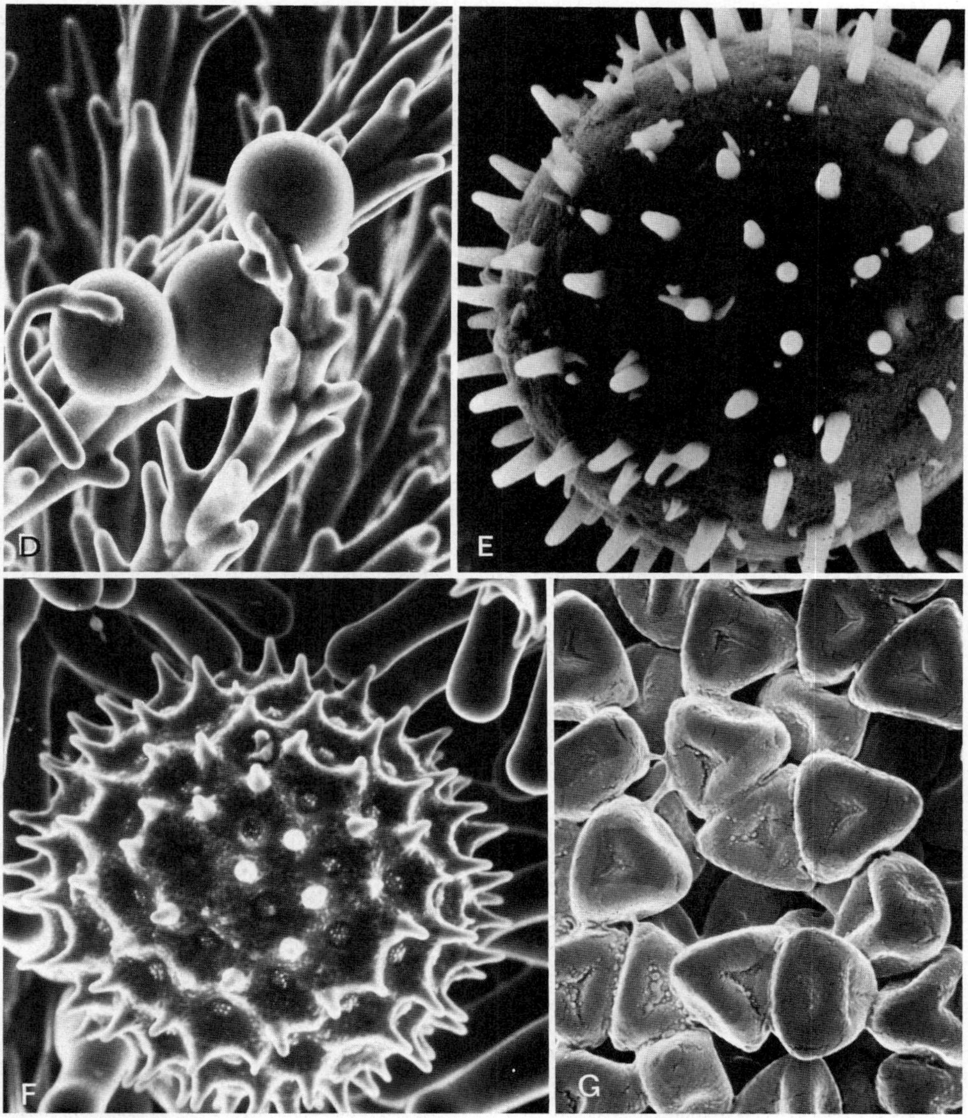

Fig. 5.8 D–G (legend see p. 215)

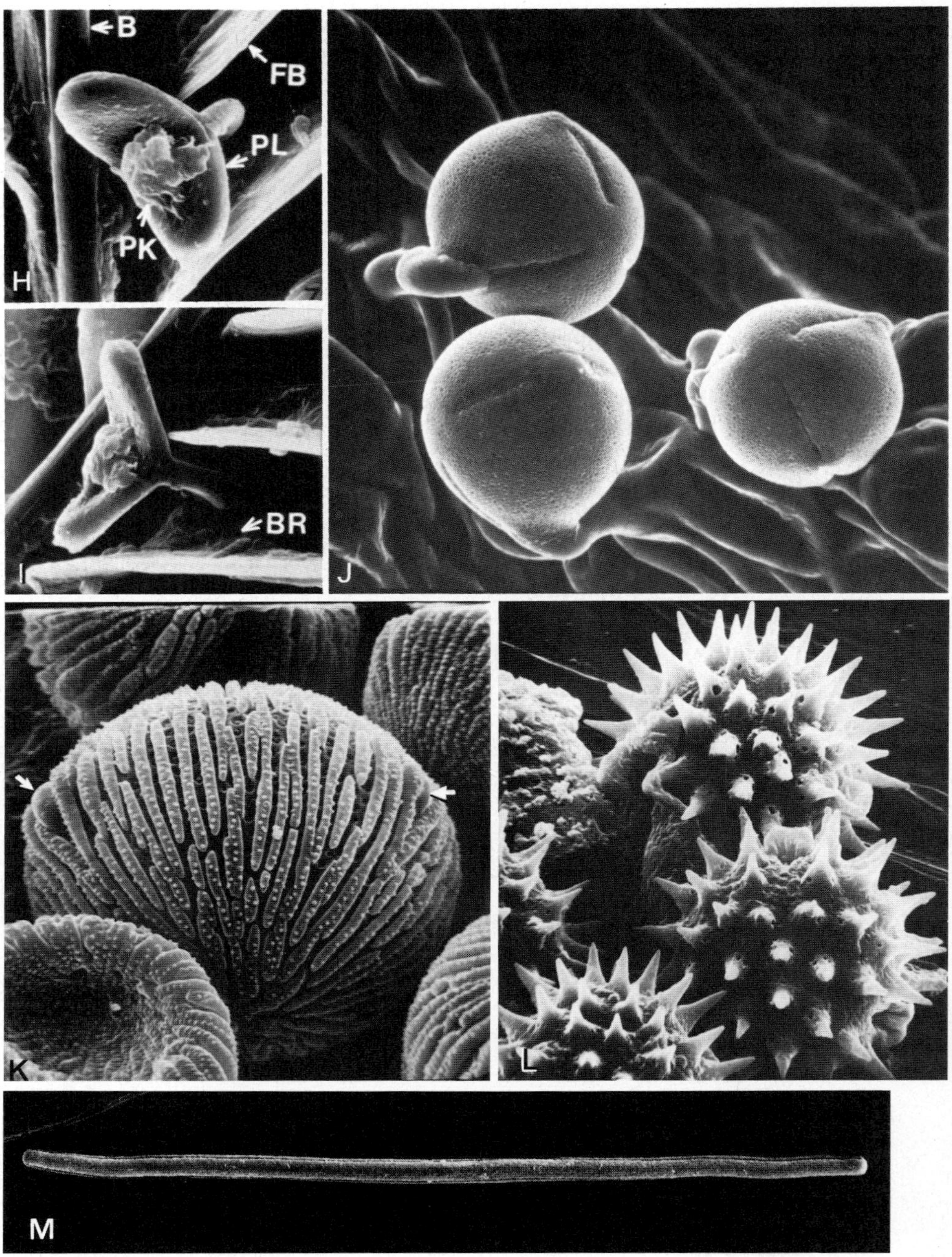

Fig. 5.8 H–M (legend see p. 215)

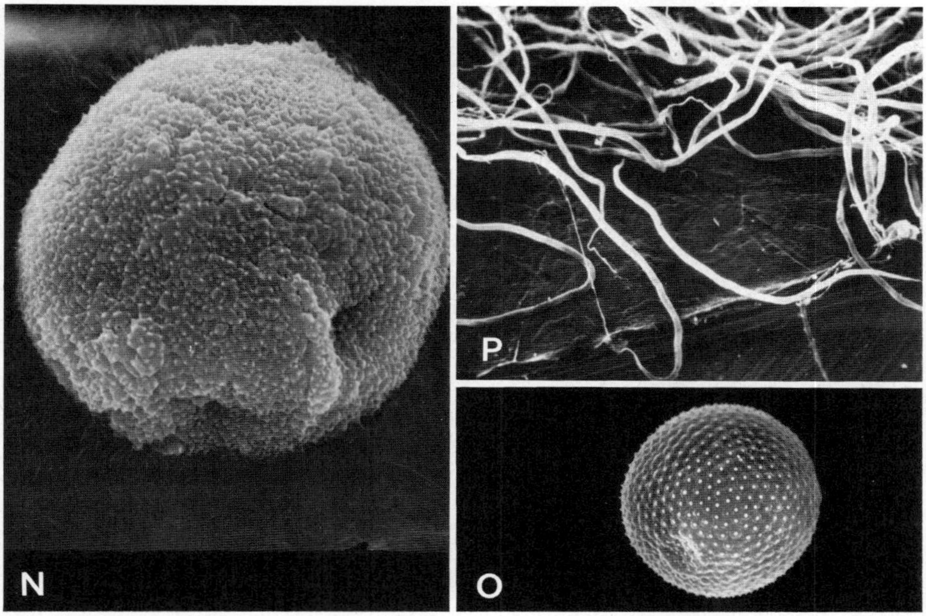

Fig. 5.8 A–P. Scanning electron micrographs of pollen grains, to show different shapes, exine pattern, and apertural mechanisms. (According to the classification of J. Heslop-Harrison 1979c). **A, B** Colpate or sulcate types: *Gladiolus*. The grain is spherical, and the exine covered with minute spines. The aperture is a longitudinal slit, or colpus, comprising infolding or, in other types, interleaving shutters; the grain surface is sealed by lipids. **A** Dry grain. **B** Germinated grain showing the pollen tube protruding from apertural intine. **C, D** Porate types: single pore, grasses (e.g., *Lolium perenne*). The grain is spherical, and the exine is covered with minute protuberances. The grains have a single pore (*a*) covered by an operculum, or cap of exine. There is no lipid seal. **C** Mature grain. **D** Germinated grains showing a single pollen tube each. **E, F** Porate types: multiple pores. **E** Grain of "Turk's cap", *Malvaviscus arboreus*, in which the exine is covered by spines, surrounding the apertures which are occluded by sporopollenin granules. (From J. Heslop-Harrison et al. 1973). **F** Grain of "morning glory", *Ipomoea coerulea*, in which the exine is covered by spines, surrounding the apertures which are covered by an operculum. **G–L** Colporate, triaperturate types: grains spherical or three-lobed, usually with surface ornamentation, the apertures are infolded sometimes with buckling shutters or interleaving sporopollenin plates. **G** Triangular grains of *Eucalyptus ficifolia*, with apertures sited at corners, and heavily sealed with lipids. **H–I** Three-lobed grains of the loranthoid mistletoe, *Amyema pendula*, lodged in the feathers of an Australian passerine bird. *B* Feather barb, *BR* barbicel, *FB* barbule, *PK* pollenkitt, *PL* grain lobe. (From Bernhardt and Calder 1981). **J** Grains of castor bean, *Ricinus communis*, with smooth reticulate exine, showing conspicuous tripartite infoldings, and emergent pollen tubes at stigmatic surface. **K** Grains of "trumpet lily", *Datura arborea*, showing striated exine surface composed of fused heads of bacula and microspores in foot layer beneath; two apertures can be seen by protrusion of the intine (*arrows*). **L** Grains of *Cosmos bipinnatus*. The three pores are evident by the tip of the intine, protruding between the spines which have conspicuous microspores around their bases. (From J. Heslop-Harrison 1968b). **M** Tricolpate grain of *Crossandra stenostachya* in equatorial view, showing two of the long colpi; the unusual elongate grains are 500 μm in length. (From Brummitt et al. 1980). **N, O** Inaperturate types: granulate exine. **N** *Populus deltoides*, pollen has a fragile, reticulate exine covering a thick intine. (From Ashford and Knox 1980). **O** Grain of *Persea borbonia* (Lauraceae) is covered with fine spines of exine overlying a very thick intine. (From Lieux 1978). **P** Inaperturate type: exine absent. The grains of the sea-grass, *Amphibolis antarctica*, are filiform, each up to 5 mm long, terminating in curious hook-like protuberances, and have no exine. (From Ducker et al. 1978)

Table 5.4. Summary of principal adaptations of pollen grains to various modes of pollination

Mode of pollination	Pollen form	Pollen wall	Pollen surface	Dispersal distance (kilometer)
Animal	Spherical $<300\ \mu m$, or aggregates	Thick, ornamented	Thick, oily, sticky-coating	0.1–5
Air	Spherical $>50\ \mu m$, aerodynamic	Thin, 2-layered	Thin, adhesive-coating	99% <1 km 1% <1 to 3 days in atmosphere
Water	Elongate <5 mm, or spherical in strings	Thin, 1-layered	Slime in submerged types	

pollen of many sea-grasses, e.g., *Amphibolis* (Fig. 5.8 P), has no exine. The term omniaperturate has been applied by Rowley and Vasanthy (1980) referring to the ability of these pollen types for de novo synthesis of an aperture at pollen germination, usually at a site adjacent to the stigma surface (see Sect. 5.3.3).

The structure and components of the pollen wall will now be examined in relation to their function in different modes of pollination. The broad features are summarized in Table 5.4 in relation to changes in the form of pollen, wall structure, the surface coatings, and the distance the pollen has to be transported for efficient dispersal.

5.2.5.1 Pollen Transport by Animals

Animal vectors of pollen range from insects to mammals, and a characteristic feature is that a reward or attraction is provided. This may be in the form of nectar, a sugary secretion produced by special floral glands, the nectaries, usually containing up to 15% sugars, especially sucrose, fructose, and glucose with traces of amino acids (Stanley and Linskens 1974). Nectaries are usually, but not always, sited within the flower and are structurally distinct from the anthers. Sugary exudates are a feature of some types of stigmas, and may serve as an attractant to licking insects, as well as an adhesive and germination medium for the pollen. In addition, pollen itself may be employed as an attractant – its high carbohydrate and protein content making a suitable food for animal pollinators. In this case special food anthers filled with sterile cells resembling pollen, or simply extra anthers, attract the pollinator. These lures are often supplemented by the presence of a scent or odour attractive to the pollinator. While the scent-gland cells are usually located in the petals, there is evidence that the anthers may also participate as in Winteraceae (Gottesberger et al. 1980).

Pollen grains of entomophilous plants usually have a greatly thickened exine, in which the inner nexine layer provides solid foundation for the long bacula and elaborately sculptured and ornamented tectum or pili, for example, the pollen of lily, *Lilium,* in monocots, and of *Cosmos* (Fig. 5.8 L), *Hibiscus* or *Malvaviscus* (Fig. 5.8 E) among dicots. The surface ornamentation apparently assists grain adhesion both to the pollinating insect and to the stigma surface. The grains may be large (> 300 μm) and are often spherical in shape.

A most conspicuous feature is that the grains are usually covered by a sticky adhesive pollen-coat (lipid seal) so that they adhere together in masses – almost an agglutination reaction, e.g., in *Eucalyptus* (Fig. 5.8 G). This is achieved by means of a surface pollen cement ("Pollenkitt", Pankow 1957) that forms a thick viscous coating over the grain surface, and may confer a characteristic stickiness, odour, and colour to the grains. This is termed pollen-coat materials (J. Heslop-Harrison 1968 b). The pollen-coat includes carotenoids or flavonoid pigments (see review by Stanley 1971) which are common in 80% of pollen grains. The sticky nature of the coating is due to the presence of glycoproteins, lipids, glycolipids, and monosaccharides which all contribute to the adhesive properties (Clarke et al. 1979).

These surface coatings are produced in the tapetal cells that line the anther cavity, and transferred to the surface of the pollen grains late in pollen development just before tapetal dissolution. This was first shown for *Lilium* pollen by J. Heslop-Harrison (1968 b), and has since been confirmed for pollen of many genera of dicots (see Knox et al. 1975, Hesse 1980 b).

The exine transfer is a complex process, involving the insertion into the exine arcades of proteins and glycoproteins, lipids, and carbohydrates synthesized in the tapetum. Most of the evidence is ultrastructural, and subject to artefacts of fixation and tissue processing. However, quite remarkable correlations have been demonstrated by M. Hesse at the University of Vienna. He used traditional ultrastructural techniques, and observed consistent differences between the ultrastructural appearance and distribution of pollen surface coatings in entomophilous compared with anemophilous pollen types in Oleaceae and Asteraceae (Hesse 1979 a), and Aceraceae (Hesse 1980 b). Pollen is fixed in glutaraldehyde, dehydrated in ethanol, and embedded in Spurr's plastic resin for ultrathin sectioning. Hesse distinguishes between the abundance, consistency and stickiness, and distribution of pollen-coat elements as judged by their electron density and homogeneity, the formation of connections between grains and loculi, and siting within the exine arcades, or coating the exine surface (Hesse 1978 a–c, 1980 a, b). When we consider that most of the soluble proteins, glycoproteins, and carbohydrates are likely to be lost by this method of processing (see Knox et al. 1980), the results obtained are remarkably consistent, and are valid since comparisons are only made between material treated in a similar way.

Hesse examined pollen of genera with both entomophilous and anemophilous species. In the entomophilous species, the pollen-coat is homogeneous and electron-dense and fills both the exine arcades, and provides a thick surface coating. This corresponds to pollen types that are exceedingly sticky. In contrast, in anemophilous species the pollen-coat is mainly confined to the exine arcades, and is markedly non-homogeneous and electron-lucent. This corresponds to pollen types that are exceedingly sticky. In contrast, in anemophilous species, the pollen-coat is mainly confined to the exine arcades, and is markedly non-homogeneous and electron-lucent. This corresponds to pollen types that are powdery and non-sticky. Hesse (1979 a, b) has related these types of pollen-coat to their role in pollen adhesion: "active" pollen-coat is homogeneous and located on the pollen sur-

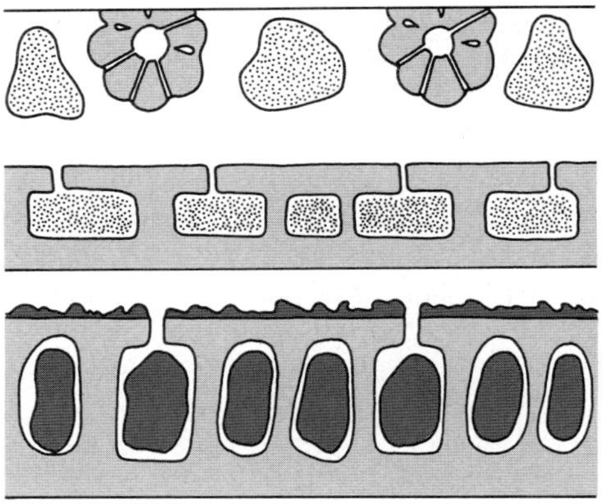

Fig. 5.9 A, B. Structural adaptations of the outer pollen wall, the exine, and surface pollen-coat associated with different modes of pollination. **A** Wind pollination, exine arcades contain electron-lucent material (*dottered areas*), which is also present attached to the wall and orbicules of the anther loculus (inactive pollen-coat; Hesse 1979). **B** Insect pollination, electron-dense pollen-coat present within exine arcades (*black material*) and on exine surface. The exine is markedly thicker. (Adapted from Hesse 1979a)

face; "inactive" pollen-coat is non-homogeneous and remains in the anther loculus (Fig. 5.9 A, B).

The type of pollen-coat present in various genera of Euphorbiaceae has been correlated with the development of "secondary" entomophily in *Euphorbia* by its ultrastructural appearance (Hesse 1980a). The effectiveness of the pollen-coat as an adhesive for insect pollination has been compared with viscin threads in several angiosperms. The pollen-coat agglutinates the grains together because it is a sticky surface component. Viscin threads, in contrast, are themselves non-sticky, thin, and non-elastic but flexible fibres of sporopollenin which link the grains together. In the Onagraceae the viscin threads may originate from within the composite grains of *Epilobium* (Fig. 5.10 A–C), as also in certain Caesalpinioideae (Cruden and Jensen 1979). In the Ericaceae the viscin threads are attached to the surface of the grains and may tie them like ropes to insect hairs and bristles (Fig. 5.10 D).

The surface appearance of these composite pollen grains which possess viscin threads is comparable to other composite types that have other means of ensuring adhesion of the pollen to visiting insects. The tetrad pollen of *Leschenaultia* (Fig. 5.10 E) and the polyads of *Acacia* (Fig. 5.10 F, G) have sticky pollen-coat materials. In orchids various configurations of grains occur, from tetrads to complex masses of grains (massulae) and pollinia (Fig. 5.10 H–L). Pollen-coat materials are present in and on the exine that surrounds the massulae of *Dactylorchis* (J. Heslop-Harrison 1968b).

The structural adaptations of pollen grains for pollinations by birds and mammals are less well known, although some principles are emerging from current research. There are two main sites of deposition of pollen on birds: on the beak, and on the head feathers (the cap). As with insect pollination, the site is determined in the placement of the anthers in the flower, deposition of pollen on the cap feathers requiring a much longer distance between the nectaries and the anthers. This is often achieved by means of a pollen presentation organ, usually the pistil which, as it uncoils during flower opening, squeezes the pollen onto its tip, a feature seen in many Proteaceae. The pollen is usually light in colour (white or pale yellow) and powdery. In Loranthaceae and Proteaceae the grains are often markedly three-lobed, or may be covered with spines which enable the grains to fit into the barbules of the head feathers (Fig. 5.8 H, I).

In contrast, in beak-pollinated pollen as in Myrtaceae, the pollen is usually smooth and spherical or triangular in shape, and is copiously covered with a thick viscous layer of pollen-coat (Fig. 5.8 G). Occasionally, it may be dark-brown, black, or blue in colour (see Inouye 1975). There is a conspicuous absence in the literature of structural information concerning the pollen of species pollinated by mammals, including bats, rats, mice, and marsupials. The pollen is carried on scales or bristles on the snout, and we may expect adaptations of shape, ornamentation, and pollen-coat materials. This gap in our knowledge is maximal in tropical species.

No account of pollen structure would be complete without mentioning the unusual pollen wall found in certain families of monocots, in the Scitamineae – where the intine is greatly thickened and the exine reduced sometimes to a few spinules, presumably an adaptation to the mode of pollination in the tropical rain forest. Recent work by Stone, Sellers, and Kress at Duke University, North Carolina, revealed that the exine is greatly reduced in seven out of the eight families of Zingiberales. These workers used acetolysis to dissolve the cytoplasmic and polysaccharide wall components, and only pollen of genera of the Costaceae had an intact exine after this treatment (Stone et al. 1980). In *Tapeinochilos* (Costaceae) the exine consists of a single outer layer formed by fusion of radially aligned rod-like units, so that there are no arcades. In the Heliconiaceae, *Heliconia* pollen has an even more remarkable exine which is confined to surface spinules linked by a thin zone of sporopollenin lamellae at the surface of the intine, which is greatly thickened and differentiated into four layers (Kress et al. 1978, Stone et al. 1979). These authors consider that this unusual pollen-wall structure may be related to pollination by humming birds.

A greatly reduced exine is also present in several tropical Lauraceae (Erdtman 1971, Lieux 1978). Ultrastructural studies of pollen of avocado *Persea americana* (Sedgley 1979) show that the exine is reduced to scattered surface spinules overlying the thick intine. Studies of these types of pollen have been important in understanding development of the exine (see Sect. 5.3.4), and the great reduction of exine must presumably be viewed as an adaptation to pollen transfer in the humid tropics. Such pollen adaptations to the environment may now be added to the list compiled by Tomlinson (1962) for the Scitamineae. It may also be analogous to the reduction in the exine in aquatic angiosperms such as sea-grasses (see Sect. 5.2.3).

5.2.5.2 Pollen in Air Currents

Grass pollen is one of the commonest types found in air currents. Each grain is very light ca. 22×10^{-9}g mass, and pollen grains are spherical (Fig. 5.8 C, D), ca. 50 μm in diameter, with a relatively thin wall and a powdery, non-sticky surface. The grains have a single aperture. The exine is tectate with the tectum raised upon short bacula from the foot layer, forming an intricate pattern of arcades (Fig. 5.7 A). The tectum surface carries many short spinules giving a roughened surface.

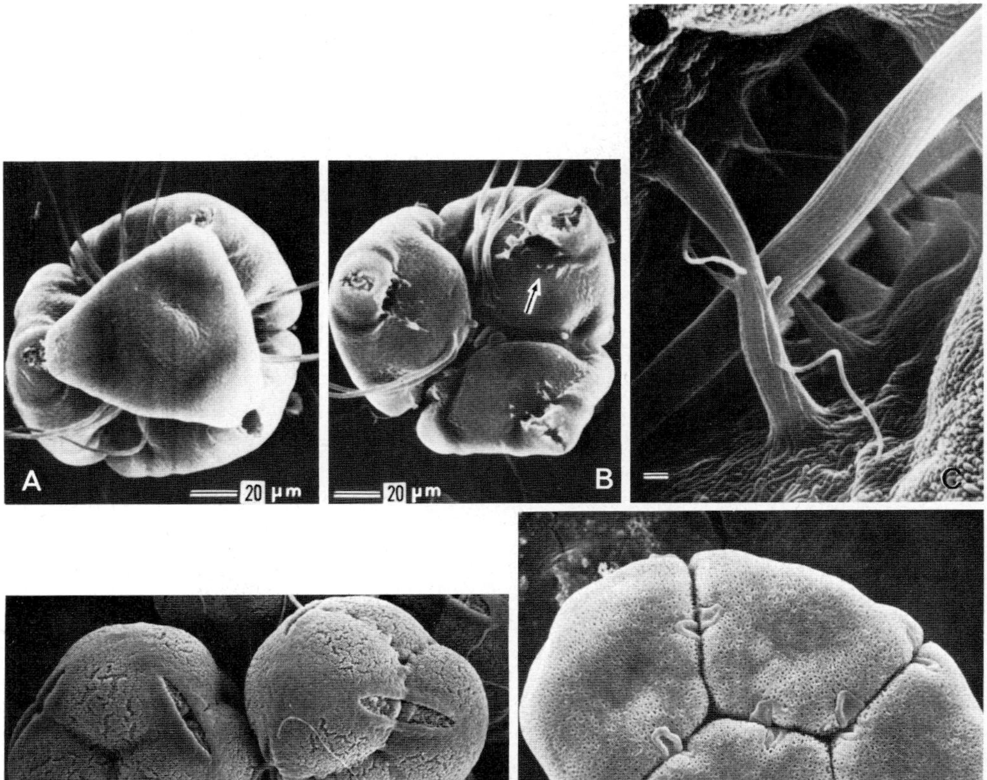

Fig. 5.10 A–G

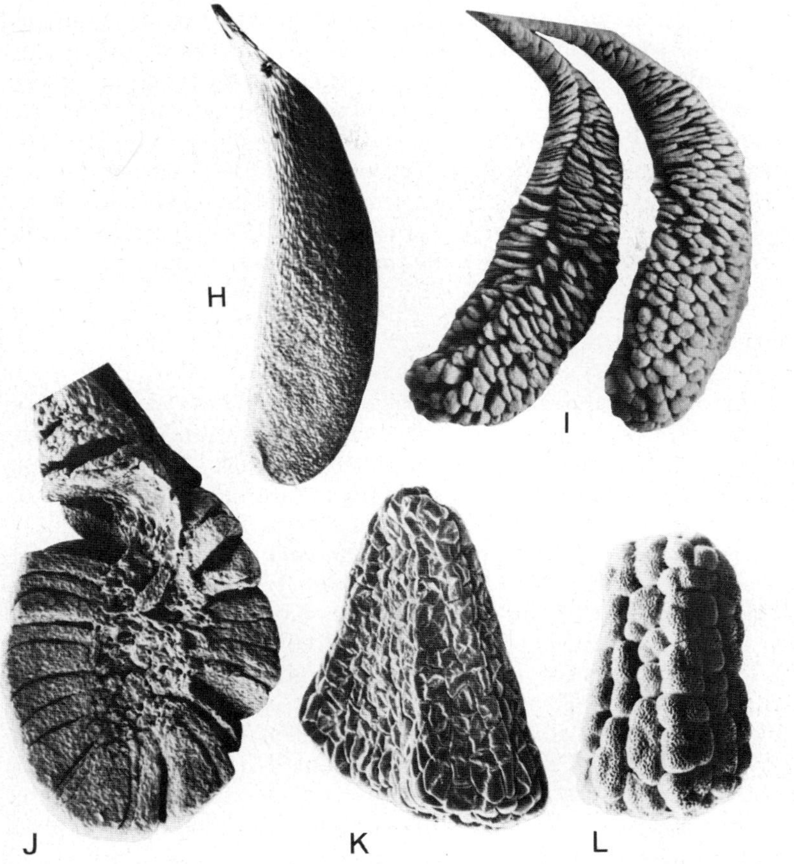

Fig. 5.10 A–L. Scanning electron micrographs of various types of composite grains. **A–C** Tetrads of *Epilobium rigidum* (Onagraceae). **A** Tetrad, showing viscin threads emerging from the proximal surface of each component grain. **B** Upper grain has been removed to show attachment points of exine bridges (*arrow*) beneath each aperture to indicate how the tetrads cohere. **C** Detail of viscin thread and its tight compound nature; the threads pass between exine bridges. (From Skvarla et al. 1978). **D** Tetrads of *Rhododendron schlippenbachii* (Ericaceae), showing linked common pores of sexine sporopollenin, held by slender viscin threads attached to the distal surface of each grain, and to the hairs of the hind legs of a Dipteran insect. (From Hesse 1980 b). **E** Tetrad of *Leschenaultia formosa* (Goodeniaceae), to show individual grains covered by a finely reticulate sexine, with the intine protruding at the apertures. (From Knox and Friederich 1974). **F, G** Polyad of 16 grains of *Acacia subulata*. **F** Entire polyad. **G** Detail of individual grain surface, with micropores (*arrow*) in the furrows. (From Kenrick and Knox 1979). **H–L** Pollinia of various orchids. **H** *Spiranthes*, **I** *Haemaria*, **J** *Ophrys* form discrete pollinia, **K** *Orchis*, **L** *Gymnadenia* form massulae, i.e., interconnected pollen sacs. (From Schill and Pfeiffer 1977)

Ultrastructural investigations by Rowley et al. (1959) and Rowley (1960, 1964) show that the tectum and foot layer are perforated by numerous channels 14 to 25 nm in diameter. The intine is well-developed, and markedly thickened at the apertures. Skvarla and Larson (1966) first noted the presence of an intermediate layer (which they termed endexine). This layer has recently been termed the Z-layer (J. Heslop-Harrison and Y. Heslop-Harrison 1980) and plays an important role in pollen germination.

In grasses each species has a similar appearance by light microscopy, and only slight differences in surface pattern are evident by scanning electron microscopy (Watson and Bell 1975). Most other windborne pollen has low sculpturing, e.g., ragweed (*Ambrosia* spp.) which has low spines (Skvarla and Larson 1975). Other features of windborne pollen include their seasonal periodicity and often high concentration in the atmosphere. While 99% of pollen grains are deposited within 1 km of their source, it is possible for pollen grains to remain up to 3 days in the upper atmosphere before deposition (see review by Knox 1979). It is also possible for pollen to be deposited and redeposited by air currents several times.

5.2.5.3 Pollen in Water Currents

Most aquatic plants flower above water in the air, and use animal vectors for pollination (see review by Cook 1982). A few are completely hydrophilous, flowering when submerged, and use water currents for pollen dispersal. The largest group are probably the sea-grasses which have some remarkable adaptations for submarine pollination.

The extraordinary pollen of the sea-grass, *Posidonia oceanica,* was first described and illustrated by F. Cavolini at Naples in 1806. He observed that the pollen was "different from that of other plants, being oblong like little eels, which with a sudden and brisk motion exploded, and scattered their sperm in the twinkling of an eye". He noted its cottonwool-like appearance when the anthers dehisced in the sea, forming a cloud of pollen in the water. Later, in 1826, the French botanist C. Gaudichaud reported on the thread-like pollen of the sea nymph, *Amphilobis antarctica,* from Shark Bay, Western Australia. He noticed its extrusion in rope-like masses from the anthers. He also observed that the pollen of the sea wrack (*Halophila*) had spherical grains released into the sea in long, sticky threads.

Later, Fritsche (1837) noted the exceptional appearance of the pollen of eel-grass, *Zostera,* which also formed long thread-like tubes tightly filled with granular cytoplasm resembling the pollen tubes developed by other pollen types at fertilization. Fritsche also recorded the presence of slime at the surface of pollen of *Najas, Caulinia, Zannichellia,* and *Zostera*. Bornet, in 1864, extended earlier observations by pointing out that the pollen grains of the Mediterranean eel-grass were coiled in a spiral arrangement within the anther.

The pollen grains of sea-grasses display three different adaptations to the marine environment:

1. Spherical grains, 100 × 150 μm in diameter, are embedded in spherical droplets of slime in the family Hydrocharitaceae, [turtle grass *Thalassia,* and *Enhalus* (Fig. 5.11 A)]. The filamentous shape is presumably an adaptation to transport in sea currents, as is the surface coating of slime.
2. Ellipsoidal grains (40 × 80 μm) released in linear tetrads within a tube of slime, in the family Hydrocharitaceae, sea wrack, *Halophila* (Fig. 5.11 B).
3. Filiform, thread-like grains occur in the families Zosteraceae (*Zostera, Heterozostera*), Posidoniaceae (*Posidonia*), Cymodoceaceae [*Amphibolis* where each grain may be up to 5 mm long, 30 μm in diameter (Fig. 5.11 C)].

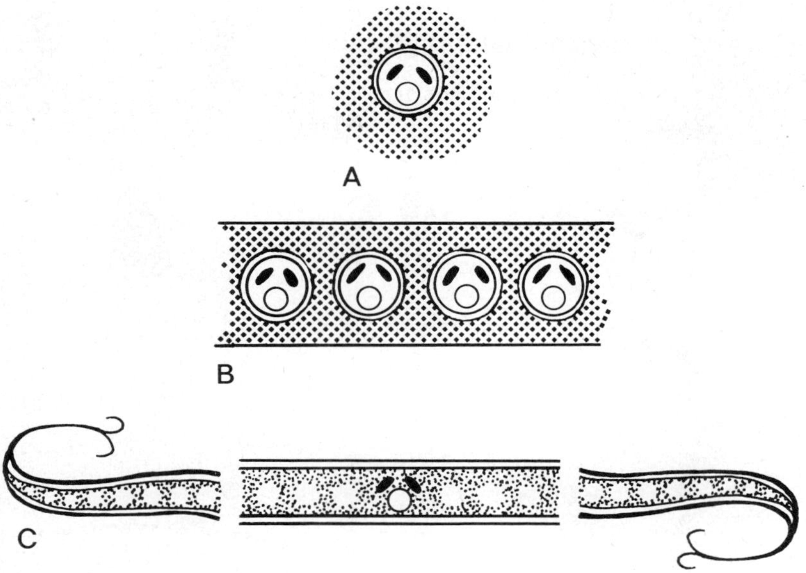

Fig. 5.11 A–C. Three different adaptations in morphology of sea-grass pollen. **A** Spherical grains of *Thalassia,* or *Enhalus,* embedded in a droplet of slime. **B** Ellipsoidal grains of *Halophila* held within a mucilaginous tube. **C** Filiform grains of *Amphibolis.* (Redrawn from Ducker and Knox 1976, Pettitt 1976, 1980)

A further remarkable adaptation for hydrophilous pollination occurs in the cell wall that encases the pollen of sea-grasses. Extreme differences exist in sea-grasses that have filiform pollen, and these are reflected, to some extent, in the other species with spherical grains. The wall in filiform types is essentially reduced in structural complexity, when compared with terrestrial pollen. The outer exine layer is absent, so that the pollen wall resembles the intine of terrestrial pollen. This structural difference occurs in filiform pollen of *Amphibolis* and *Zostera,* and in the string of spherical grains of *Halophila* (Ducker et al. 1978, Pettitt 1980). The pollen wall of these two kinds of sea-grasses is, therefore, unique among flowering plants. The wall resembles the intine of terrestrial pollen in two characteristic respects: (1) Its microfibrillar, electron-lucent appearance in thin sections viewed by transmission electron microscopy; this is consistent with its likely chemical composition of cellulose microfibrils embedded in a matrix containing acidic polysaccharides, suggested by the cytochemical evidence (Ducker et al. 1978, Pettitt 1980, McConchie et al. 1982a, b). (2) If fresh mature grains are freeze-sectioned, the pollen wall reacts positively for a marker enzyme, acid phosphatase, that is characteristically stored within the polysaccharide matrix of the intine of terrestrial pollen. In the ellipsoidal grains of *Thalassia* and *Enhalus* a rudimentary exine is, however, present (Pettitt and Jermy 1975, Pettitt 1980).

Finally, we may view these dramatic changes in pollen walls from a structural perspective. The advent of transmission electron microscopy has enabled the most intricate patterns and foldings of the wall to be detected. The surface

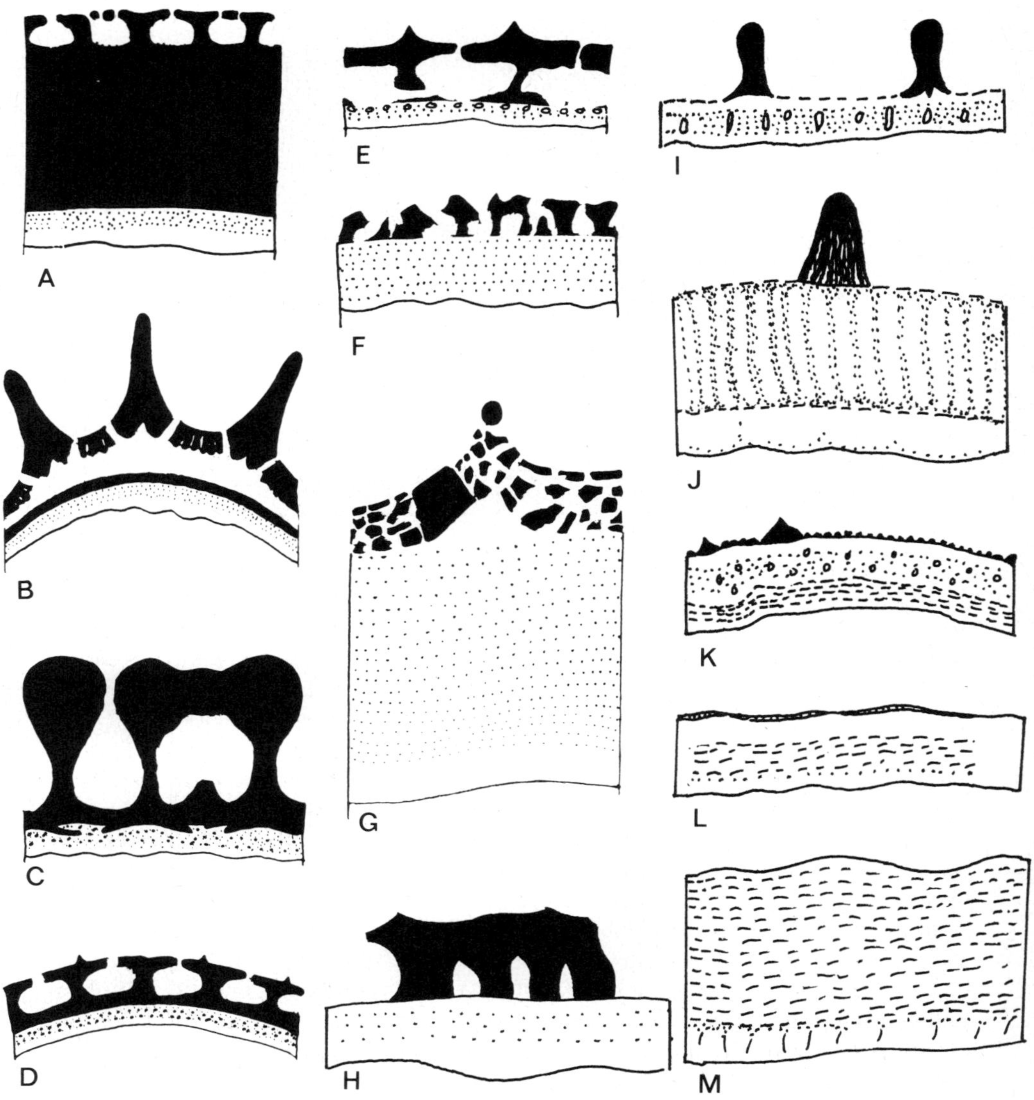

Fig. 5.12 A–M. Interpretations of the ultrastructure of pollen wall in selected angiosperms. Diagrams are based on transmission electron micrographs of the radial sections of wall. The exine is indicated in *black,* intine *white* or *stippled.* **A–D** Pollen grains in which the wall architecture prevents emission of components from the intine through the exine, by means of a solid nexine layer. **A** *Malvaviscus arboreus* (Dicot, Malvaceae). **B** *Cosmos bipinnatus* (Dicot, Asteraceae). **C** *Lilium longiflorum* (Monocot, Liliaceae). **D** *Silene vulgaris* (Dicot, Caryophyllaceae). **E–M** Pollen grains in which the wall architecture does not apparently prevent emission of internal components; all are monocots, except where indicated. **E** *Gladiolus gandavensis* (Iridaceae). **F** *Populus alba* (Dicot, Salicaceae). **G** *Crocus vernus* (Iridaceae) Modified from Knox et al. 1975). **H** *Potamogeton pectinatus* (Potamogetonaceae). (Redrawn from Pettitt and Jermy 1975). **I** *Thalassia hemprichii* (Hydrocharitaceae). (Redrawn from Pettitt 1976). **J** *Canna generalis* (Cannaceae). (Redrawn from Skvarla and Rowley 1970). **K** *Persea americana* (Dicot, Lauraceae). (Redrawn from Sedgley 1979). **L** *Najas flexilis* (Najadaceae). (Redrawn from Pettitt and Jermy 1975). **M** *Amphibolis* (Monocot, Cymodoceaceae)

changes we have discussed are summarized in Fig. 5.12 in terms of the form and thickness of the exine and intine layers. First, there are grains in which the exine is massive and ornamented, an adaptation for insect pollination (Fig. 5.12A–D). Second, there are many pollen types where channels pass through the exine, allowing entry and egress of molecules limited only by the presence of a lipid seal on the surface (Fig. 5.12E–G). Finally, we observe the remarkable trend for exine reduction in aquatic monocotyledons (Fig. 5.12H, I) which are, in many respects, similar to certain dicots and monocots of humid tropical climates (Fig. 5.12J, K). Extreme reduction has been observed in the hydrophilous monocots, with only a thin electron-dense coating in *Najas* (Fig. 5.12L), and complete absence of exine in *Amphibolis* (Fig. 5.12M). It is noteworthy that the structure of the exine layer itself, as well as its coatings, are important in the control of water movement, and cellular protection.

5.3 Formation of Pollen

5.3.1 Cytology of Development

The microspore, the first cell of the gametophyte generation, is the determinative cell of pollen development. After meiosis the tetrad of microspores is formed within the callose special wall. Microspore determinants are responsible for the design and initial construction of the complex outer pollen wall, the exine, and for the siting of the apertures, when present. The life of the microspore is terminated by the first mitotic division, an asymmetric division forming the vegetative and generative cells of the pollen grain.

Maheshwari (1949, 1950) considered the longevity of the microspore period, stating that it is shortest in tropical plants: *Tradescantia reflexa* 4 days, *Styrax obassia* 7 days, *Himantoglossum nircinum* 14–21 days. In the arctic species, *Uvularia sessilifolia, Empetrum nigrum,* and *Betula odorata,* he cited references showing it may extend throughout the winter period. In the canary grass, *Phalaris tuberosa* (Vithanage and Knox 1980), the microspore period lasts for 6 days out of 13 days required for pollen development (Fig. 5.13A).

The microspore cytoplasm is non-vacuolate when the spores are first released from the callose special wall at the end of tetrad period (spore release period), but vacuolation begins soon after. In grasses the vacuole expands as the volume of the pollen grain increases. This young spore period is the main focus of cell expansion in the entire pollen development (Fig. 5.13B), termed the vacuolate period. During this period, the diameter of the pollen grain of grasses more than doubles to about three-quarters of its final size. Towards the end of the vacuolate period, the amount of cytoplasm increases and the vacuole becomes resorbed about the time of the microspore division (Fig. 5.13B).

The first pollen grain mitosis is the division that heralds the end of the microspore period. In the grasses it occurs about half-way during pollen development (Fig. 5.13A). In the sea-grasses *Halodule* (Yamashita 1976) and *Amphibolis antarctica* (Ducker et al. 1978), the first pollen mitosis occurs almost immediately af-

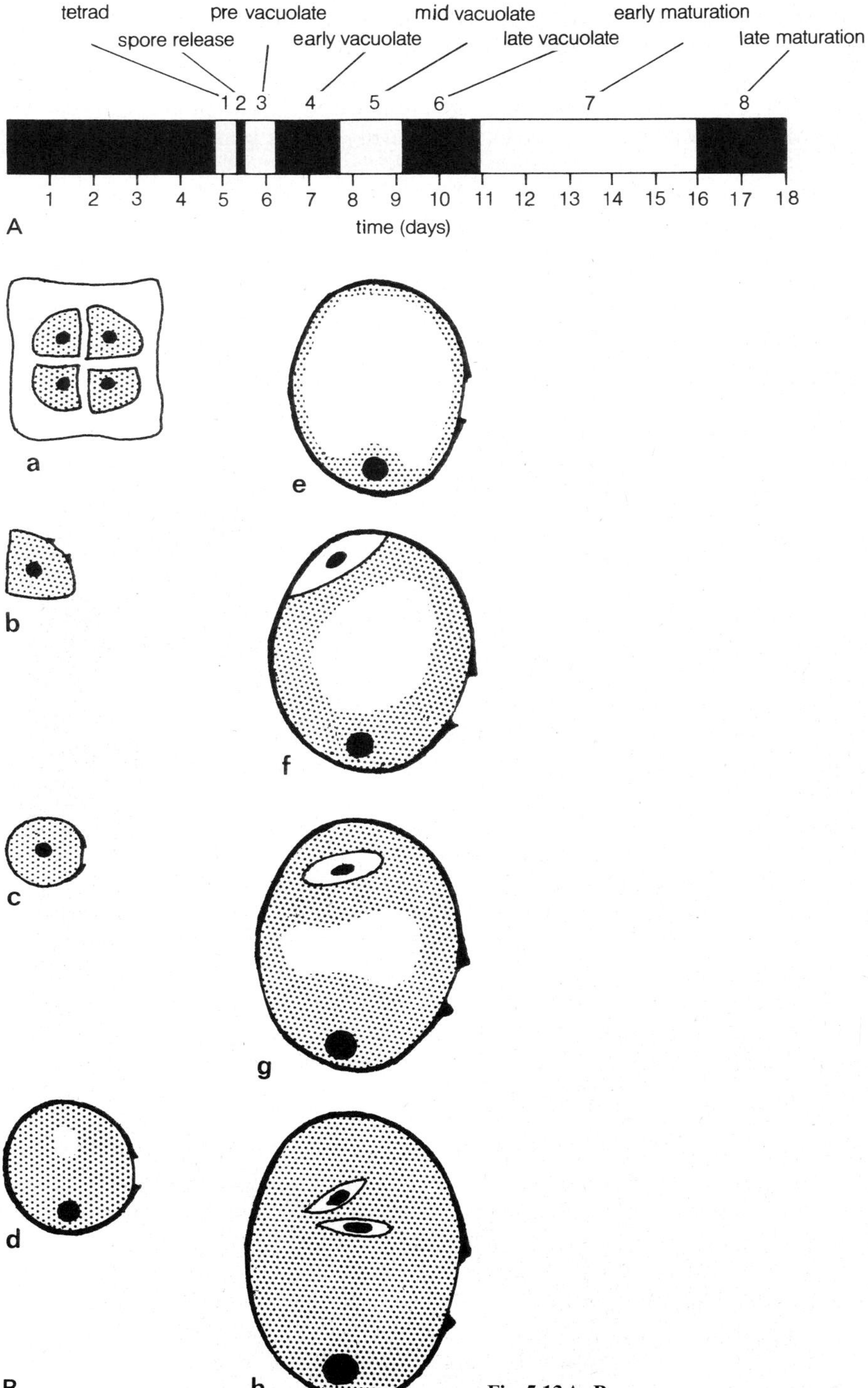
tetrad
spore release
pre vacuolate
early vacuolate
mid vacuolate
late vacuolate
early maturation
late maturation
1 2 3 4 5 6 7 8
1 2 3 4 5 6 7 8 9 10 11 12 13 14 15 16 17 18
A
time (days)
a
b
c
d
B
e
f
g
h

Fig. 5.13 A, B

Fig. 5.13. A Estimation of grass pollen development and periodicity. Inflorescences at a similar stage of development, after just emerging from the boot, were labelled and each day thereafter until anthesis, two of the labelled inflorescences were selected at random. The florets were excised, mixed, and a random sample of 50 florets taken. The pollen was treated for the FCR test (see text) to determine the developmental period. The frequency in which each stage occurs gives an estimate of its timing. **B** (*a–h*) and **C** (*a′–h′*) Morphological features of grass pollen development; diagrammatic (*a–h*) and fluorochromatic (*a′–h′*) reaction. *a* Period 1, tetrad. Developing spores do not show fluorescence as the thick callose layer prevents the FDA from entering; only those with damaged walls show fluorescence. *b* Period 2, spore release. *c* Period 3, pre-vacuolate. *d* Period 4, early vacuolate. *e* Period 5, mid-vacuolate showing large central vacuole and the peripheral nucleus, giving it a "signet ring" appearance. *f* Period 6, late-vacuolate showing the ovoid shape of grain, *g* Period 7, later stage. *h* Period 7, late-maturation showing absence of vacuole. (From Vithanage and Knox 1980)

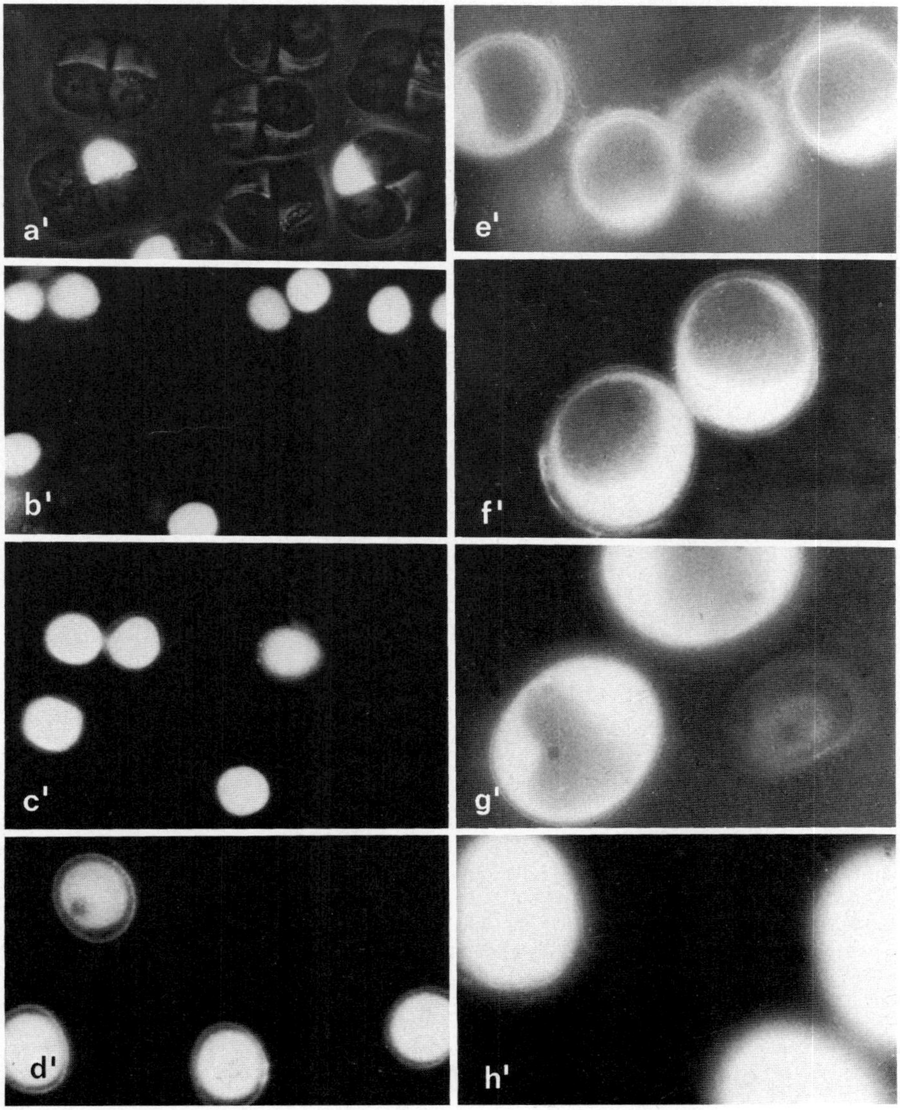

C

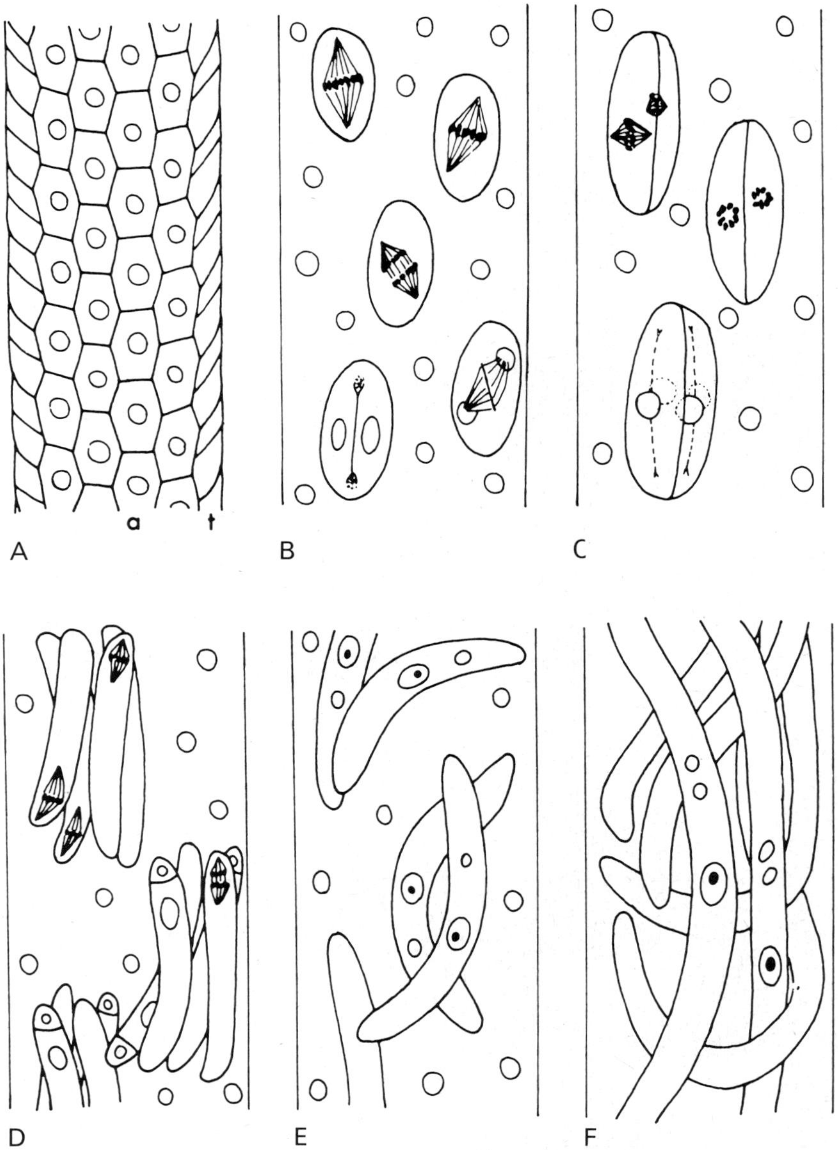

Fig. 5.14 A–H. Pollen grain mitosis and pollen development in two contrasting plant systems. **A–F** Filiform pollen of the sea-grass *Halodule*. **G, H** Composite grains (tetrads) of the Winteraceae. **A** Archesporial cells (*a*) are surrounded by a single file of tapetal cells (*t*). **B** Meiosis I in pollen mother cells, surrounded by plasmodial tapetum (nuclei indicated by *open circles*). **C** Meiosis II – tetrad formation. **D** Young spore period; first mitotic division in pollen grains and early two-celled stage. **E** Bicellular grains. **F** Tricellular filiform grains in mature anther. (From Yamashita 1976). **G** Pollen tetrad of *Belliolum* sp., showing synchronous division at first pollen mitosis, nuclei at metaphase. **H** Pollen tetrad of *Drimys winteri,* showing asynchronous division at first pollen mitosis. Division into the larger vegetative (tube) and smaller generative cell has been completed in the grain at *lower left*. The nucleus is at metaphase in the grain at *lower right*. Division has not begun in the grain at *upper centre*. (From Sampson 1980)

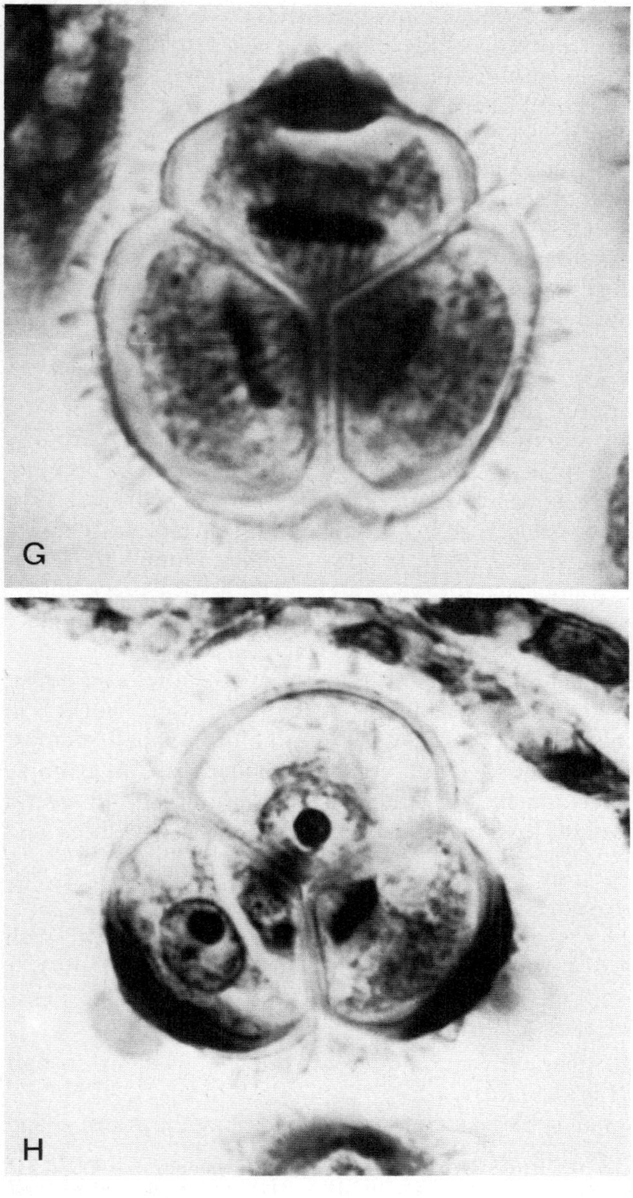

Fig. 5.14 G–H

ter microspore release from the tetrad, just as the grains begin to elongate (Fig. 5.14 A–F). In *Najas,* the division is even more precocious, occurring within the tetrad period. A pattern is manifest in this first mitosis. It is an unequal cell division, and is important for two reasons:

1. Partitioning of the cellular organelles into the vegetative and generative cells.
2. The geometric precision of its siting.

A relationship has been demonstrated in many pollen types between the siting of the aperture and the future position of the generative cell. Huynh (1976) embodied this in his "law of the longest distance", in which he showed that the generative cell will be cut off at a point furthest from the aperture. He considers that this siting might ensure that the vegetative nucleus would enter the pollen tube first. Some exceptions to this relationship have been noted by Sampson (1980), but he considers that the law is of general validity.

Pollen-grain mitosis is often synchronous within the anther, e.g., in orchids (J. Heslop-Harrison 1968 b). An interesting observation of synchronous pollen mitosis in the permanent tetrad grains of the Winteraceae has led Sampson (1980) to associate synchrony with the presence of cytoplasmic connections in the proximal walls of the tetrad of microspores (Fig. 5.14 G, H). In tetrads with asynchronous division connections are absent.

After the microspore mitoses, pollen development (sensu stricto) is under way, and is characterized by the accumulation of storage reserves – protein, carbohydrate, and lipid within the cytoplasm of the vegetative cell. This seems to be a major function of the vegetative cell, which becomes progressively dehydrated as the grain matures, and is ready for dehiscence. At this time, the pollen of most angiosperms remains bicellular (Brewbaker 1967). In a few families, notably the Compositae, Cruciferae, and Gramineae, the generative cell divides precociously forming a pair of elongate sperm cells which are present in the mature grain (see Figs. 5.2, 5.3, 5.4) – so that the grains are tricellular when released.

In order that the development and differentiation of microspores and pollen grains can be meaningfully described, it is important that the various workers in the field use a similar scoring system. At present the terminology applied is almost as complex as the names given to the various layers of the pollen wall, for example see review of grass pollen development by Vithanage and Knox (1980). In several recent studies of grass pollen, at least eight development periods have been recognized (Fig. 5.13 B):

1. Tetrad period: four microspores within callose special wall (Fig. 5.13 Ba).
2. Spore-release period: microspores are released from callose special wall with thin exine. Considerable distortion may occur during fixation (Fig. 5.13 Bb).
3. Pre-vacuolate period: grain spherical, with exine and pore clearly visible; dense non-vacuolate cytoplasm (Fig. 5.13 Bc).
4. Early vacuolate period: spherical vacuole present, with diameter up to half that of the pollen grain (Fig. 5.13 Bd).
5. Mid-vacuolate period: grain spherical with the vacuole filling the grain, giving a "signet-ring" appearance (Fig. 5.13 Be,f).
6. Late-vacuolate period: grain distinctly ovoid in shape, showing buildup of cytoplasm so that vacuole is reduced to one-half to one-third of grain volume (Fig. 5.13 Bg).
7. Early maturation period: first pollen mitosis has occurred, cytoplasm fills the grain, further reducing the size of vacuole (Fig. 5.13 Bg).
8. Late maturation period: second pollen mitosis has occurred, grains are tricellular, cytoplasm with starch grains completely fills the grain (Fig. 5.13 B,h).

The complex development schemes of earlier workers are based on the use of chromic or alcoholic fixatives (see review by Satake and Hayase 1970). The use of buffered aldehyde fixatives, whose osmolarity can be adjusted to that of the pollen grains, should prevent many of these artifacts. More recent schemes are all similar to that used in this chapter. This is based on the use of fresh pollen, with the fluorochromatic reaction (J. Heslop-Harrison and Y. Heslop-Harrison 1970) which provides a rapid evaluation of pollen development, based on the appearance by fluorescence microscopy of the characteristic vacuoles and cytoplasm of the microspore and vegetative cell (Fig. 5.13 Ba′–h′).

Having established the criteria for the development periods, it is practicable to compare these with anther length, and then to follow the sequential changes in pollen development with time. This has been done for canary grass, *Phalaris tuberosa* (Vithanage and Knox 1980). As might be expected, the major periods – vacuolate period and maturation period – are the longest, although it is surprising to find the maturation period of 7 days (Fig. 5.13 A). This period, however, is important since the storage reserves are synthesized, and also the pollen becomes partly desiccated. Schemes such as that used here are of great value in assessing the developmental origins of pollen sterility in cereals and grasses (see Satake 1976).

The periods of pollen development represent a topic that has been ignored by most workers. For example, in their major work on pollen biology, Stanley and Linskens (1974) divide wall development into two basic stages: tetrad and free spore. Each is further subdivided into ill-defined sub-periods according to deposition of wall layers. In considering the development, primarily in *Lilium* pollen, Dickinson (1982) has defined the events in terms of exine formation, with the addition of pollen mitosis and generative cell formation. Willemse and Reznickova (1980), in contrast, have conducted a detailed and elegant quantitative study of the changes in *Lilium* microspore and pollen development, based largely on estimates of cell and wall volume. The cell volume was estimated by application of a formula $V = \frac{3}{4} - LK\frac{L+K}{2}$ where L and K are the semi-major and semi-minor pollen axes measured from photographs of thin frozen sections. Twelve development periods are recognized. (Fig. 5.15):

MS2: microspore with noticeable wall,
MS3: microspore acquires vacuoles,
MS4: vacuolated microspore,
MS5: late-vacuolate microspore,
MS6: late interphase microspore,
P: pollen just after mitosis,
P2: young pollen grain,
P3: pollen with generative cell near its wall,
P4: pollen with lens-like generative cell, and
P6, P7, P8: pollen 4, 2, and 1 days, respectively, before dehiscence.

It is not clear how these arbitrary stages are actually defined in the experimental plants, although the data obtained are convincing. The pollen grains show an apparent two-phase growth curve when plotted against bud length. It is interesting that microspore mitosis ends at the transition to increased cell enlargement, and that there is a period of actual decrease in size just prior to dehiscence, presumably reflecting dehydration of the grains (Banerjee et al. 1965). An advantage

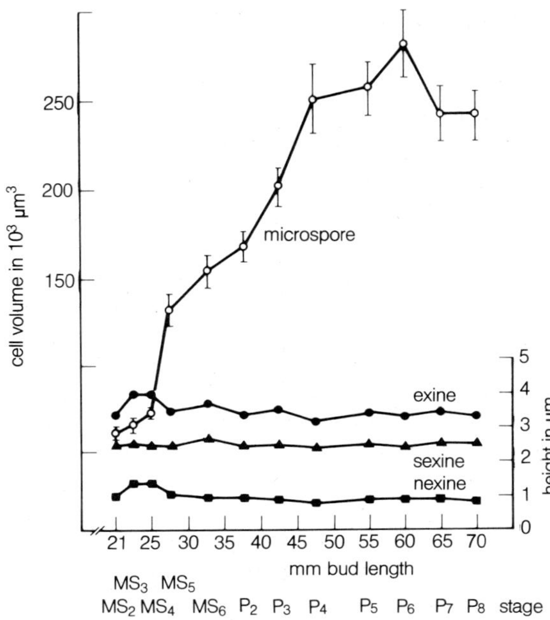

Fig. 5.15. Quantitative analysis of various parameters of microspore and pollen development in *Lilium*. Abscissa on LHS gives changes in cell volume (*open circles*), while abscissa on RHS gives changes in wall stratification, expressed as height in µm (*closed symbols*). Symbols used for developmental periods are described in text. (From Willemse and Reznickova 1980)

of this classification of microspore and pollen development periods is that finer tuning of the maturation period is ensured by the use of the differences in the cytology of the developing generative cell, when compared with the scheme adopted by Vithanage and Knox (1980). The latter scheme has proved useful in comparing development in terrestrial grasses with that in sea-grasses (Pettitt et al. 1981). It is evident that there is no ideal system, and that further progress needs to be made in the classification of developmental periods.

5.3.2 The Primexine

The pioneering studies of J. Heslop-Harrison, now at the Welsh Plant Breeding Station, Aberystwyth, and J. R. Rowley of the University of Stockholm, have provided an understanding of the origin, development and structural basis of wall patterning in angiosperm pollen. In 1963, J. Heslop-Harrison showed that the pattern elements of the pollen grain wall were laid down at tetrad period by the synthesis of the primexine. This layer is composed of microfibrillar cellulosic material, and is located between the plasma membrane of the microspore and the callose special wall. The primexine was regarded as a mould or template within which deposition of the protosporopollenin pattern elements occurs. Today, it is viewed as a matrix containing receptors for the exine wall polymer, protosporopollenin or sporopollenin.

These primexine elements are synthesized by the protoplast of the microspore while enclosed within the callose special wall. Rodlike probacula or flat trilaminar plaques are initiated at the outer face of the microspore plasma membrane, in the future wall zone. Structures, termed lamellations or white lines (Dickinson and

Heslop-Harrison 1968, Rowley and Dahl 1977), become inserted below the probacula providing an additional site for accumulation of protosporopollenin to complete the basic elements of microspore wall pattern. The models of sporopollenin polymerization have been reviewed by Buchen and Sievers (1981). When the callose wall is broken down enzymically, the microspores are released, and come into contact with sporopollenin derived from the tapetal cells, and further polymerization and formation of the mature acetolysis-resistant exine wall takes place.

Sporopollenin also occurs in the spore walls of certain algae and fungi (Atkinson et al. 1972, Gooday et al. 1973). In the green alga, *Chlorella*, the coating of the spores with sporopollenin is initiated in a manner similar to angiosperm microspores, via trilaminar plaques in the spore wall site external to the plasma membrane (Atkinson et al. 1972). It is noteworthy that *Chlorella* has one advantage not possessed as yet by any pollen system investigated, the occurrence of strains in which sporopollenin is absent. The existence of such mutant types would make it possible to explore the events and routes of sporopollenin synthesis, about which so little is known in chemical terms (see review by Brooks and Shaw 1978).

The development of the exine from the primexine matrix that has been described in *Silene* (J. Heslop-Harrison 1963) is typical for the pollen of many other angiosperms. Among monocotyledons, this differentiation system has been observed in *Zea* (Skvarla and Larson 1966), *Endymion* (Angold 1967), *Sorghum* (Christensen and Lersten 1972, Christensen and Horner 1974), *Lilium* (J. Heslop-Harrison 1968b, Dickinson 1970), and *Tradescantia* (Mepham and Lane 1969). Among dicotyledons it is consistent with the work of Larson and Lewis (1962) on *Parkinsonia*, Godwin et al. (1967) on *Helleborus*, Hoefert (1969) on *Beta*, Horner and Lersten (1971) on *Citrus*, Horner and Rogers (1974) on *Capsicum*, Nabli (1975, 1976) on *Teucrium*, *Marrubrium*, *Lavandula*, *Coleus*, and *Salvia*, Dickinson (1976a, b) on *Cosmos*, and Horner and Pearson (1978) on *Helianthus*.

In *Artemisia vulgaris* Rowley and Dahl (1977) described a variation on this theme: the formation of a mucopolysaccharide glycocalyx on the microspore surface during tetrad period. This is the precursor layer for the outer tectum. The layer is derived from a mat of tufts around which the polymer is deposited forming a pre-tectum (Fig. 5.16A). The probacula of *Artemisia* form on white line centred lamellations spatially separated from the plasma membrane, at the time of dissolution of the callose wall (see Rowley 1978, Rowley and Dahl 1982). Further differentiation results in massive deposition of sporopollenin to form the complex exine seen in Fig. 5.16B (see Sect. 3.4). In microspore development of monad pollen of Epacridaceae, Ford (1971) described a system in which even more extensive wall formation occurred within the callose wall at tetrad period.

The basic features of microspore exine formation are shown in Table 5.5 and involve sporopollenin synthesis at two periods in development: (1) at tetrad period within the callose special wall; and (2) during the young spore period, with deposition of sporopollenin synthesized by the tapetal cells (elaborating the foot layer, bacula, and tectum), and by the microspore protoplast (forming the nexine). Significant variations from this theme have been discovered in two groups of monocotyledons: in one only tetrad period synthesis occurs and, in the other, sporopollenin synthesis does not (apparently) occur at all (Table 5.5).

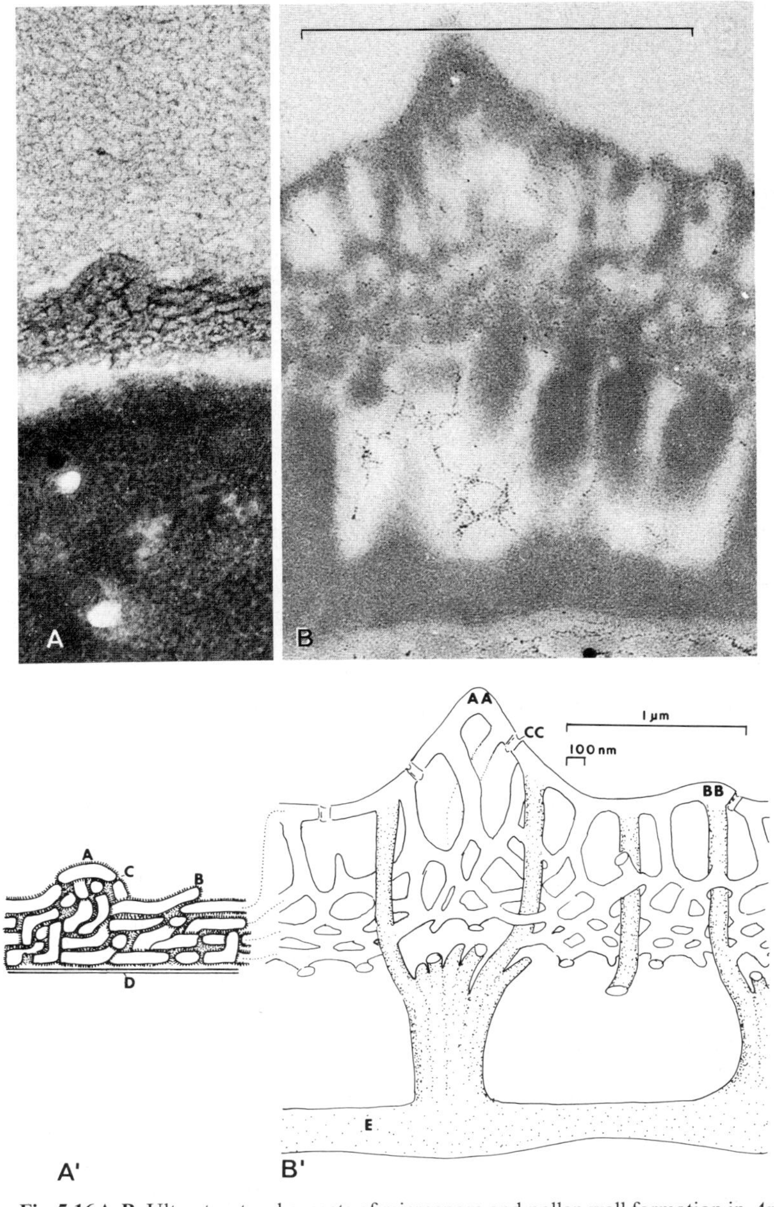

Fig. 5.16 A, B. Ultrastructural aspects of microspore and pollen wall formation in *Artemisia*. **A** Section of late-tetrad pretectal microspore surface coating. **B** Exine from the early pollen grain period. The interpretative drawing (**A'**, **B'**) while diagrammatic, is at the same scale as micrographs **A** and **B.** The tetrad tuft mat (pretectum) *at the left* is elevated and laterally expanded to form the tectal parts *at the right*, indicated by the *unstippled* elements. Elements of the major bacular system and the footlayer are *stippled*. Prespinule *A*, spinule *AA*, presumed prepunctuation *B*, presumed punctuation *BB*, presumed premicrochannel *C*, microchannel *CC*, plasma membrane *D*, foot layer of extexine (sexine) *E*. (From Rowley and Dahl 1977)

Table 5.5. A scheme showing the formation of the exine of pollen grains through two stages: (1) the transfer of wall information from the pollen mother cell to the young spores; (2) the transfer of wall polymer from the diploid tapetal cells and its role in exine thickening. The first scheme is based upon that given by J. Heslop-Harrison (1980); double-membrane inclusion (DMI) form an intracellular compartment in which determinants of the mother cell cytoplasm are conserved during meiosis. The options for the production of virtually exineless pollen types are indicated on the right, and refer to specific cases discussed in the text

1. Sporophytic diploid pollen mother cells	„Pattern“ genes transcribed. Products transferred to cytoplasm and plasma membrane	
	↓	
2. Meiosis	Organelles undergo cycle of de- and re-differentiation. ER membranes enclosed in DMI's	
	↓	→ No primexine formation. Exine absent in certain sea-grass pollen
3. Tetrad of microspores	DMI's release contents. Membranes participate in wall morphogenesis through initiation of primexine	
	↓	
	Synthesis and deposition of sporopolleninlike polymer in primexine	
	↓	→ No further accretion of exine polymer. Exine reduced in certain tropical monocots and dicots, and in many aquatic monocots
4. Young spore period	Tapetal cells synthesize sporopollenin, and transfer it to surface of microspores to form exine	

Such systems provide the means for elucidating the mechanism of exine formation. In the tropical Scitamineae the exine is reduced to scattered surface spinules and thin surface lamellae (Kress et al. 1978). Ultrastructural studies of *Heliconia* pollen have shown that the exine spinules and lamellae are laid down during tetrad period within the primexine (Fig. 5.17A–C). Subsequent elaboration of the exine does not occur until after release from the tetrad so that the mature wall is composed largely of intine. Stone et al. (1979) have evidence that this may be true also for other families in the Zingiberales, while Skvarla and Rowley (1970) showed a similar reduction of the exine in *Canna*. Subsequent work on another family of Scitamineae, the Costaceae, showed that in one genus, *Tapeinochilos,* the continuous exine layer is also laid down in the primexine, and no further accretions of sporopollenin are added after the tetrad period (Stone et al. 1980).

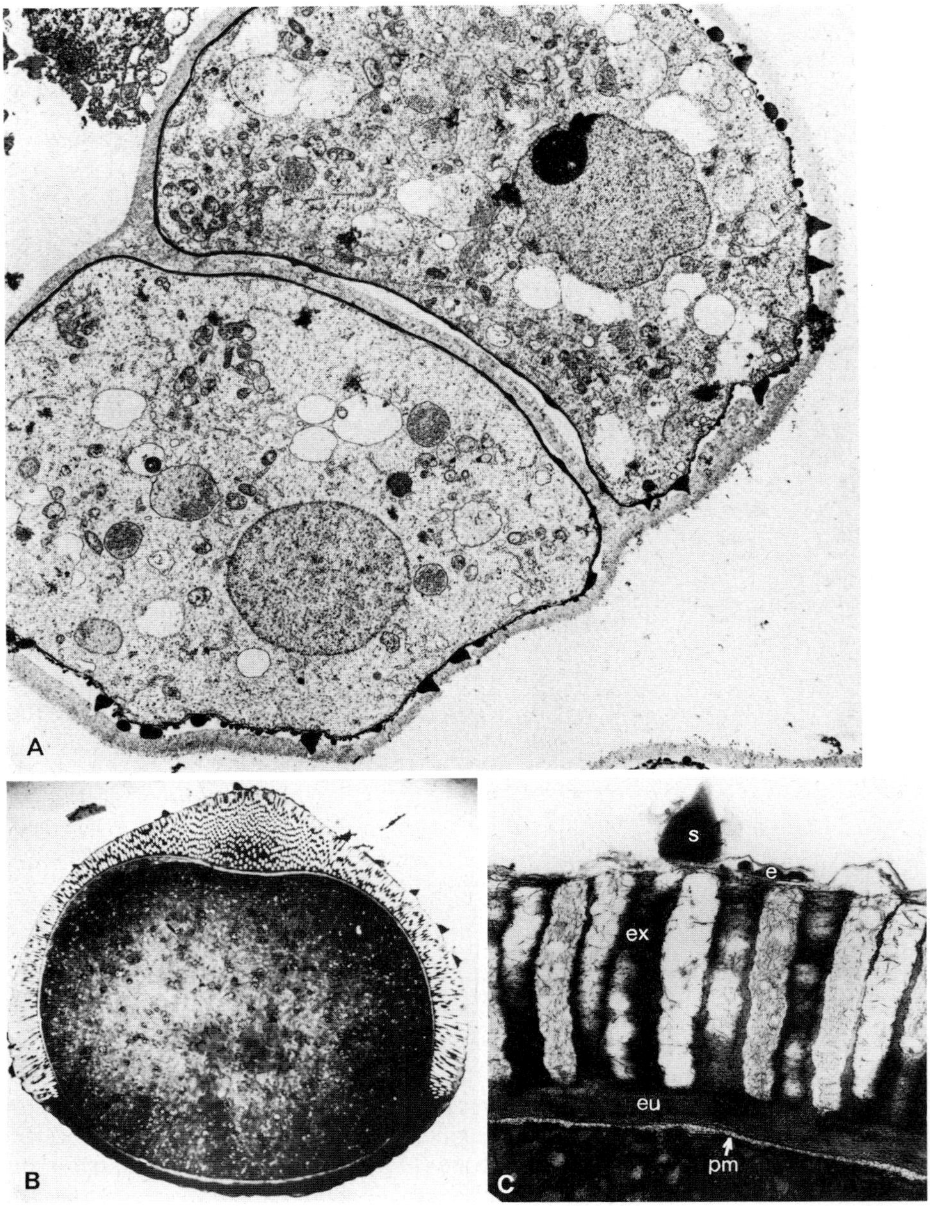

Fig. 5.17 A–C. Development of the pollen wall of *Heliconia nutans,* a tropical monocot with greatly reduced exine, shown by transmission electron micrographs. **A** Maturation of primexine, within the tetrad. Mid-probaculum stage with callose intact and primexine layer thin but continuous between microspores; nascent spinules localized on the outer, distal hemispheres. **B** Section of pollen grain in equatorial view, showing intine channels located on distal hemisphere. **C** Mature pollen wall, distal hemisphere, tenuously seated spinule (*s*) on fragmented exinous base (*e*), well developed channels in outer intine (*ex*) and radially and tangentially oriented microfibrils in inner intine (*eu*) plasmamembrane (*pm*). (From Stone et al. 1979)

Finally, the pollen of certain sea-grasses, especially the Australian sea nymph, *Amphibolis antarctica,* has been shown by cytochemical and ultrastructural studies to completely lack an exine layer at maturity (Ducker et al. 1978). In recent ultrastructural studies of microspores at tetrad period, Pettitt et al. (1981) observed that a primexine layer cannot be detected, implicating its absence for the lack of sporopollenin synthesis, and so for the loss of the exine layer. In related studies, Pettitt and Jermy (1975) and Pettitt (1976, 1980) show that other sea-grasses and aquatic plants have a greatly reduced exine, confined to isolated rodlike bacula, and it will be of interest to determine the nature of the primexine in these genera.

We can conclude from this evidence that the presence of a primexine is essential for exine differentiation. The pollens of angiosperms appear to have utilized many variations in producing their exine structure.

5.3.3 Establishment of Apertures

While some pollen types in both monocotyledons and dicotyledons do not have germinal apertures (inaperturate or omniaperturate types), most possess an area of wall devoid of exine, or covered by a cap or operculum at which pollen tube emergence may occur in a successful pollination. Wodehouse (1935) demonstrated that the type of microspore arrangement in the tetrad, and the points of microspore contact, served to delineate the sites of germinal apertures on the basis of contact symmetry and compensation for asymmetry. In the most common tetrad type, tetrahedral, each microspore contacts adjacent microspores at three points, corresponding to the sites of apertures in triporate pollen types. In square tetrads (termed planar by Walker and Doyle 1975) four apertures are formed, one at each of the designated contact sites, and a further pair opposite them. This explanation has been extended to higher aperture numbers (Walker 1974), and also to the formation of a spiraperture in pollen of *Tapeinochilos* (Stone et al. 1980).

There are two main types of aperture: simple furrows or pores; and complex types comprising an outer and inner aperture superimposed on each other. Huynh (1976) investigated the structure and arrangement of the furrow types of aperture, and considers there are two different kinds: (1) "sulcate" pollen in which the tangents to the furrows, each lying along one furrow on its mid-point, are parallel with the equatorial plane (Fig. 5.18 A, B); (2) "colpate" pollen in which the tangents to the furrows are borne perpendicularly on the equator (Fig. 5.18 C, D). The arrangement of the apertures is determined by their siting within the tetrad, which can be observed by staining with a dye specific for the exine, for example, congo red and maleic hydrazide (Stainier et al. 1967), or Calberla's stain (basic fuchsin; Ogden et al. 1974). The apertures are distinguished by the absence of staining. Huynh (1976) has identified four main patterns of aperture arrangement, based on the siting of the furrow in relation to the equator of the microspore in the tetrad (Fig. 5.18 A–D).

In all cases the aperture is defined during tetrad period, and ultrastructural studies have revealed that the site is frequently designated by the endomembrane system, ER being adpressed to the plasma membrane (J. Heslop-Harrison 1963).

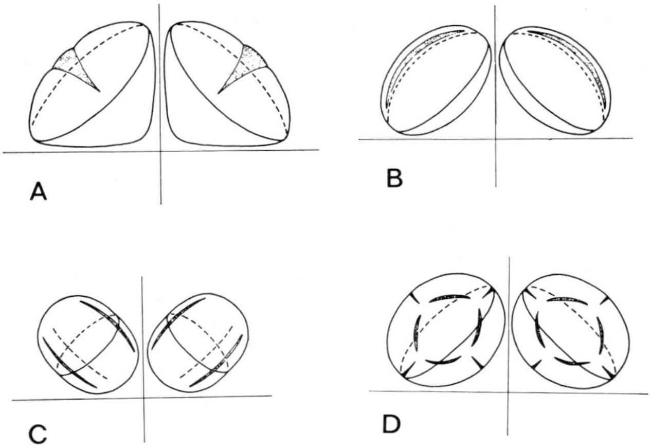

Fig. 5.18 A–D. Four types of microspore arrangement at tetrad period. Each diagram shows the upper pair of sister microspores of a tetrad. **A** Monosulcate microspores. **B** Monocolpate microspores. **C** Tricolpate microspores. **D** Pentacolpate microspores. (From Huynh 1976). Explanation of *perspective lines:* The more or less *ellipitc lines,* each of which encircles one microspore, represent the equators. The *horizontal line* and the *vertical line* represent, in each figure, the two planes, perpendicular to the plane of the figure and to each other, that run through the tetrad and touch both the microspores. All of the apertures are *dotted,* except for **C** where the third aperture is represented by an *interrupted line;* in **D** only the apertures on the upper faces of the microspores are represented

An inventory of the families which possess inaperturate pollen is given by Erdtman (1952). Among monocotyledons this type of pollen is frequent among the Helobiae, and in the Scitaminae, including the Musaceae, Zingiberaceae, Cannaceae, and Marantaceae. Among dicotyledons, the condition is frequent in the Polycarpiae, especially in the following families: Nymphaeaceae, Ceratophyllaceae, Annonaceae, Gomortegaceae, Monimiaceae, Lauraceae, Hernandiaceae, and Aristolochiaceae. These are mostly regarded as primitive groups, and generally possess monosulcate pollen (Muller-Stoll 1956). Exceptions are known in the Linaceae, Salicaceae, Icacinaceae, and Euphorbiaceae-Crotonoideae. These families possess some genera with inaperturate pollen, but accompanying related genera with normal three-colporate structure.

5.3.4 Exine Differentiation and Maturation

It is now appropriate to review both the ontogeny of the exine, and its structural characteristics.

The tectum, bacula, and foot layer (nexine) are laid down by the deposition of sporopollenin on the surface of membrane-like lamellae (Rowley and Southworth 1967), a lipoid substrate, or polymer already deposited (Dickinson 1976a, b). The major and minor spinules that may decorate the tectum are also determined. These events occur within the tetrad. Thereafter, at microspore release, subsequent development is dependent on the special circumstances peculiar to each pollen system. It usually involves the development of the inner nexine-2 or endexine layer (made of sporopollenin), and of the intine (made of polysaccharides).

Both these layers develop during the vacuolate period, e.g., in *Vicia* (Audran and Willemse 1982). While these have been studied in several systems, two cases are especially interesting: *Lilium,* in which both the wall layers develop simultaneously, and in certain Compositae, *Artemisia* and *Cosmos,* where there is a degree of separation.

J. Heslop-Harrison and Dickinson (1969), working with *Lilium,* obtained ultrastructural evidence to demonstrate that the formation of sexine and nexine is similar, and that the lamellae forming the sporopollenin-depositing surfaces are formed at the plasma membrane, so that the siting, orientation, and temporal control of exine deposition can be achieved. Dickinson (1971), subsequently observed lamellae forming the nexine-2 layer passing across the intine, apparently contributing new sporopollenin to the nexine layer. Dickinson considered that the evidence for passage of these plates across the polysaccharide layer was hardly convincing, and that the most likely explanation is that these images are of streams of plastic wall components (protosporopollenin) flowing from the plasma membrane, across the intine to the nexine. This may explain how both the nexine-2 and intine can thicken simultaneously. A similar mechanism may operate when the bacula and tectum are synthesized from the plasma membrane early in development.

Within the tapetal cells, sporopollenin is produced which will coat the young developing sexine. Sporopollenin accretion takes place upon a lipoid former or substrate (Dickinson 1976b). This lipoid material either coats the walls of the loculus or appears as droplets in the inward facing or radial walls of the tapetum. Sporopollenin is deposited on these structures, forming the peritapetal membrane lining the loculus and the orbicules present in many anther types. Sporopollenin accretion ceases about the time of the first mitotic division of the microspore. Dickinson (1976a, b) associated sporopollenin production with vesicles of the endoplasmic reticulum which are conspicuous in the tapetum. The amount of polymer produced in the microspore at tetrad period is considered to be too small to produce a conspicuous change in the microspore cytoplasm so that the organelles involved there are unknown. The tapetally synthesized exine polymer is, therefore, characteristically deposited on a lipoid substrate although the mature orbicules and peritapetal membrane are said to have no remaining lipophilic material (Dickinson 1976b). This seems unusual in view of the affinity of the exine, and orbicules of many pollen grains for lipophilic dyes such as Scarlet R, Auramine O, etc. (Southworth 1974, J. Heslop-Harrison 1979b, Kenrick and Knox 1979). The affinity of the polymer for various staining reactions also changes, being initially positive (when plastic at formation) and later negative (in mature pollen, e.g., electron stains, Rowley 1959, 1962), lignin dyes (J. Heslop-Harrison 1968a, 1969), periodic acid-Schiff reaction (Vithanage and Knox 1979), and autofluorescence characteristics (see Willemse and Audran 1982).

The layers of the exine can be distinguished from one another cytochemically by the use of certain dyes (Table 5.1, see Sect. 5.2.3). Especially useful are toluidine blue which generally stains the exine layers blue or green, and auramine O gives a brilliant yellow fluorescence, while the nexine-2 may not be fluorescent, for example, in *Acacia* polyads (Kenrick and Knox 1980). The nexine-2 in some pollen types is stained by decolorized aniline blue, giving a bright yellow fluorescence, for example, in sunflower, *Helianthus annuus* (Vithanage and Knox 1979).

Interesting and elegant developmental studies of the exine in *Artemisia* and other pollen types carried out by J. Rowley, O. Dahl and associates at the University of Stockholm, have revealed the existence of an exine glycocalyx network. This network is attached to the microspore plasma membrane and passes through the endexine and into the bacula, arcades, and tectum to the surface coating of the exine (Fig. 5.16 B). A novel feature of these studies is the use of electron stains claimed to be specific for proteins, carbohydrates or lipids (Rowley 1975, 1978, Rowley and Prijanto 1977). The exine network demonstrated is fibrillar in nature, and stains positively for proteins and acidic polysaccharides. These authors consider that the exine is not a static and inert wall layer, but is important as a cell surface zone. It permits the entry of small molecules, including colloidal iron, lanthanum, alcian blue, and ruthenium red (see references cited in Rowley and Dahl 1977). The permeability of the exine to small molecules is shown by its vital staining by safranin and basic fuchsin.

Rowley and co-workers employed chemical etching techniques to dissect away the sporopollenin, revealing the existence of the underlying glycocalyx network. Results have been interpreted in terms of a model of exine structure (Rowley et al. 1980, 1981 a, b), a development of an early model proposed by Rowley and Flynn (1968). The new exine model is based on filamentous subunits, 15–40 µm in diameter, each comprising an axially oriented branched tubule, 10–15 µm in diameter (Fig. 5.19 A–D). The complete subunit is almost entirely embedded in the sporopollenin. Most of the polysaccharide, protein, or lipid components of the subunit are thus not available to stains. Rowley et al. (1981 a) propose that these non-sporopollenin macromolecules are glycocalyx units, occur throughout the exine, and function as sporopollenin receptors.

An interesting feature of the exine is that it may provide what Rowley et al. (1981 a) term "built-in long term stability" for the macromolecules within the exine. If, as Rowley et al. (1981 a) suppose, these macromolecules originate like the glycocalyx of animal cells from the plasma membrane, then there is the possibility that they will be preserved in fossil pollen, accessible to analysis by all the techniques of macromolecular chemistry, after removal of the sporopollenin.

Finally, the surface of the exine must be considered as it is the site which first makes contact with the stigma at pollination. Here, signals may be generated and exchanged during the biocommunication that is an essential prelude to pollen–stigma interactions (Clarke and Knox 1978, J. Heslop-Harrison 1978). Exine surface coatings have been observed in many pollen types: *Clivia* (Afzelius 1955), *Populus* (Rowley and Erdtman 1967), *Aegiceras* (Dunbar and Erdtman 1969), and *Gladiolus* (Knox 1971). The most convincing demonstrations have resulted from the work of Rowley and co-workers, who have used cations (such as ferric iron, thorium, and lanthanum) or cationic dyes (such as ruthenium red or alcian blue) to stabilize polysaccharides. In *Populus* the exine coating after traditional buffered glutaraldehyde fixation, is a simple trilamellate structure (Rowley and Erdtman 1967). However, after addition of alcian blue or ruthenium red to the glutaraldehyde fixative, the exine surface coating consisted of a fine reticulum

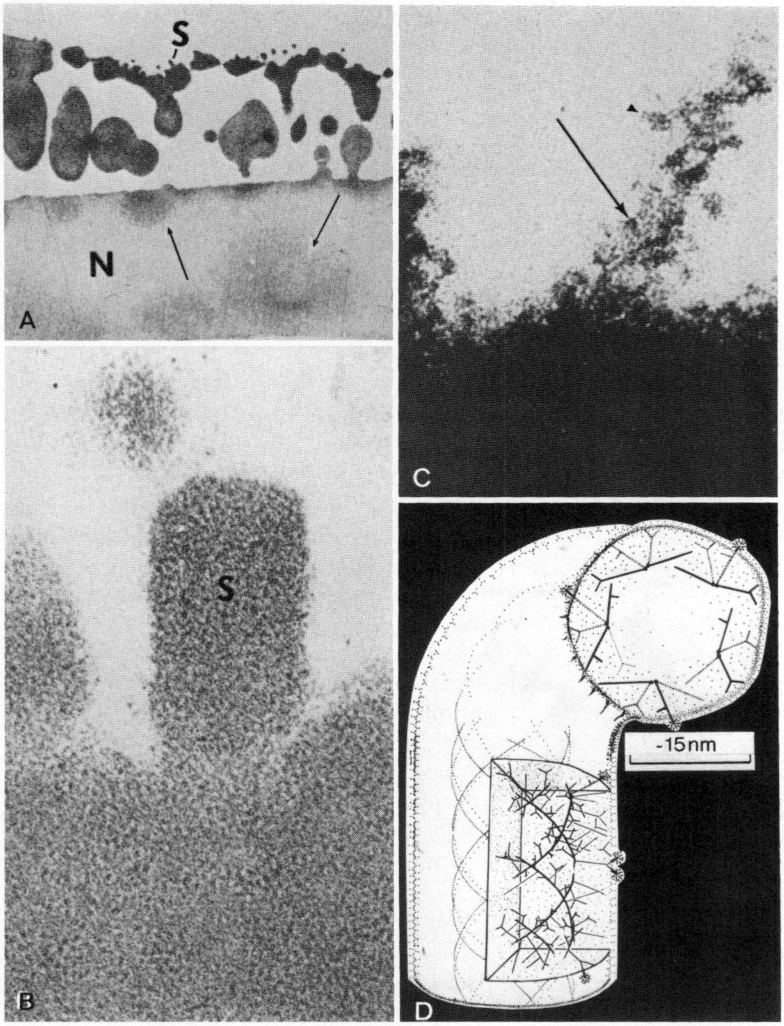

Fig. 5.19 A–D. Chemical dissection of the exine of *Malva alcea,* and the model of exine substructure proposed by Rowley et al. (1981 b). **A–C** Transmission electron micrographs of fresh pollen exines. **A** After acetolysis treatment nexine *N,* sexine *S.* **B** Similar exine sample, after various cleaning procedures; refluxed in acetone, boiled in water, refluxed in ethanol, and exposed to chromic acid. **C** as **B** but after exposure to boiling in 10% KOH for 10 min. This treatment has resulted in etching of exinous subunits (*arrow*) embedded in sporopollenin. **D** proposed model of rod-shaped exinous subunits. The *top portion* of the drawing shows the subunit in end-view, while the *lower part* shows the ten elements of the tubular glycocalyx complex, each with two branches exposed to the surface. (From Rowley et al. 1981 b)

(Rowley 1971). A structurally distinct zone at the surface of the tectum of *Brassica* pollen has been described by Dumas and Gaude (1982) using electron staining techniques.

The differentiation of the exine has also been studied by observing characteristic changes in the autofluorescence of the grains (Willemse 1972). Of especial interest are recent experiments where pollen of foreign species has been transplanted into the living anthers of *Lilium*, and observations of their characteristic exine autofluorescence made. Changes characteristic of *Lilium* pollen were found to occur also in transplanted pollen of *Gasteria* (Willemse 1981) and *Vicia* (Audran and Willemse 1982, Willemse and Audran 1982). These results indicate, once again, the important role of the sporophyte, mediated through the tapetal cells, in pollen wall development and maturation.

5.3.5 Intine Synthesis and Deposition

The intine is the final wall to be synthesized in the microspore (J. Heslop-Harrison 1968 b). Often it does not reach its final thickness until the maturation period – after the microspore mitosis. The intine is the essential wall layer of pollen grains, since none has been observed in which this layer is absent. The intine is a smooth layer, with a microfibrillar ultrastructural appearance, although it may possess several distinctive zones differentiated by their granularity, electron density, or other features including the presence of tubular or lamellar structures. Chemically, the intine is readily solubilized by acetolysis, by treatment with EDTA, ammonium oxalate, and monoethanolamine (Southworth 1973 and references cited therein, J. Heslop-Harrison 1979 b). Cytochemically, it shows a positive reaction with carbohydrate and protein stains (see Table 5.1).

The intine is synthesized from the haploid microspore or pollen protoplast. In some pollen types, Golgi bodies are frequent during intine synthesis, for example, Ranunculaceae (Roland 1971), *Olea europaea* (Pacini and Juniper 1979); while in others, endoplasmic reticulum and polyribosomes are abundant, for example, *Cosmos* (Knox and Heslop-Harrison 1970). The intine may contain proteins and glycoproteins stored in its polysaccharide matrix, either through microvillus-like extensions of the plasma membrane or by a process of exfoliation in which ribbons or leaflets or tubules containing the proteins appear within the wall. In *Heliconia* radially oriented vesicles become associated with the internal surface of the intine, and radially aligned channels are apparently dissolved in the intine (Stone et al. 1979). In the mature *Heliconia* intine these channels are sealed by a thin coating of exine on the external side, and by a thin layer of radially oriented microfibrils on the inner surface (Fig. 5.17 B). Skvarla and Rowley (1970) observed a similar tubular layer in the intine of *Canna* (Fig. 5.12 J) which developed soon after microspore release, while growth of the remaining innermost intine layer did not commence until after microspore mitosis. The outer (tubular) zone of the intine of *Cinnamomum* has been termed onciform (Rowley and Vasanthy 1980). An elaborate multilayered intine has been reported in several other pollen types, including the Caryocaraceae (Barth 1966), *Crocus* (Fig. 5.12 G), and the sea-grass *Amphibolis* (Fig. 5.12 M).

An important function of the intine, in its broadest sense, is to form the wall of the germinal apertures. It is at these sites that the intine reaches its greatest complexity, for example in *Malvaviscus* (J. Heslop-Harrison et al. 1973). In the Compositae the aperture development is of especial interest, as the oncus develops at late tetrad period, while the microspore is within the callose special wall.

This has been noted in several Compositae, including *Artemisia* (Rowley and Dahl 1977) and *Helianthus* (Horner and Pearson 1978, Vithanage and Knox 1979).

Indeed, aperture development in *Artemisia* occurs from white line centred membraneous lamellae which form at the site at which the lens-shaped oncus develops. At spore release period, a biconvex orate lens differentiates under the pore. Ultrathin sections show that the oncus is formed from the nexine (endexine) and is a complex lamellar structure, 2–3 μm in diameter. Later in pollen development it regresses, and is absent by the maturation period. Rowley and Dahl (1977) show that concurrently with oncus regression, an ellipsoidal disc of intine forms under the apertures, providing the tip zone of the putative pollen tube. A somewhat different interpretation of aperture development in pollen of the olive, *Olea europea,* is given by Pacini and Juniper (1979).

In grass pollen grains the aperture structure is also remarkable. The first ultrastructural studies, for example, of *Poa annua* pollen, revealed the existence of a lens-shaped oncus beneath the operculum, and above the intine layer (Rowley 1964). The layer had a distinctive ultrastructural appearance – homogeneous without the tubules that traversed the intine proper – and had a higher electron density. The name „Zwischenkörper" (intermediate body) had been applied to this layer by Rowley (1964) and subsequent authors followed his lead, or termed it endexine (see Fig. 5.7) (Skvarla and Larson 1966). Its nature and structure has only recently been resolved. J. Heslop-Harrison and Y. Heslop-Harrison (1980) showed cytochemically that it is composed of acidic polysaccharides (especially pectic polymers), becomes gel-like when pollen is hydrated, causes the operculum to lift off opening the aperture, and dissolves in front of the pollen tube at germination. A similar mechanism is considered to operate in the pollen of Cucurbitaceae.

Cytochemical studies of the intine have provided further evidence for the presence of several classes of wall polysaccharides (Table 5.1), and these have led to propositions that the intine is composed of several different types of polysaccharide: an outer zone that stains for pectic polysaccharides and appears to be homologous with the Z-layer of grass pollen; and inner zone/s that stain for β-glucans (calcofluor) and neutral polysaccharides (PAS reaction) (see Y. Heslop-Harrison 1982, Kress and Stone 1982). The entire intine may stain red with toluidine blue; in *Populus* pollen this metachromatic staining is absent if the section is digested with pectinase, indicating the presence of pectic polymers (Ashford and Knox 1980). In pollen of sea-grasses, the intine-like wall shows cytochemical reactions characteristic of both acidic and neutral polysaccharides (McConchie et al. 1982a, b).

The intine may also stain with a variety of protein stains (Tsinger and Petrovskaya-Baranova 1961, Knox and Heslop-Harrison 1969, 1970, Southworth 1973). The intine proteins have been shown by both light and electron microscope cytochemistry to be located in tubules or lamellar structures within the intine polysaccharide matrix, for example acid phosphatase in the intine of *Crocus* pollen (Knox and Heslop-Harrison 1971b). Intine proteins have been detected by their antigenic activity, and immunocytochemical methods have been used to investigate their sites of storage and routes of emission from moistened mature pollen (see Sect. 5.5).

5.3.6 Mechanism of Cohesion in Composite Pollen

The nature of the connections that link composite types of pollen was first explored by Roland (1965), and van Campo and Lugardon (1973). Most detailed work has been done on one system, the polyads of *Acacia* (Leguminosae, Mimosoideae). The polyads of this genus have intrigued botanists from early times, for example the botanical artist to George III, Francis Bauer, sketched the 8-, 12-, and 16-grain polyads during his stay at Kew, 1790–1840. Today, the scanning electron micrographs of *Acacia* polyads are spectacular, and show the flat disc form of the composite unit (Fig. 5.10 F, G). The precise mechanism for the origin of such diverse numbers of grains in each polyad was studied by Newman (1933, 1934). He showed that the 16-grained polyads of *Acacia baileyana* formed from a single sporogenous cell which divided twice by mitosis, these gave rise to four tetrahedrally arranged sets of tetrads of microspores which remained together within the anther loculus to form a polyad. The number of grains in each polyad is, therefore, determined by the number of pollen mother cells (microsporocytes) in each loculus.

Barth (1965), and more recently Guinet and Lugardon (1976), have demonstrated that the mature polyads of *Acacia* cohere by two mechanisms: (1) by the presence between the inner faces of the grains of „Kittsubstanz" or cement, and (2) by the presence of bridges of nexine linking adjacent grains (see Table 5.2). The most convincing evidence has been obtained for wall bridges linking the grains, following the pattern established for tetrads of three genera of Onagraceae by Skvarla et al. (1975) in which bridges of nexine-2 link the inner faces of the polyads. In transmission electron microscope studies of pollen-wall structure in mature polyads of *Acacia,* nexine-2 bridges have been demonstrated in three Brazilian species (Barth 1965) and in *A. farnesiana* (Guinet and Lugardon 1976). In the latter species the bridges comprise both the sporopollenin nexine layer and the more electron-dense polysaccharide intine layer. Guinet and Lugardon (1976) also detected these bridges by scanning electron microscopy in the Australian species *A. laccata*.

Such wall bridges may have originated from cytoplasmic connections that existed during development at the time when the intine was laid down. Cytoplasmic connections, almost at the limit of resolution of the light microscope, detected by their affinity for protein stains, linked adjacent microspores of *Acacia conferta* at tetrad period (Kenrick and Knox 1979). Their presence provides a visible means for the establishment of cohesion between adjacent microspores within the anther loculus. At tetrad period the microspores are held within the callose special wall, which provides a barrier of potentially reduced permeability between the microspores (Knox and Heslop-Harrison 1970). It is within this wall layer that the primexine is laid down. The initiation of wall links would ensure microspore cohesion after dissolution of the callose special wall. In *Acacia* the linkages were evident as gaps in the callose special wall.

Cytomictic channels (ca. 2 µm in diameter) have been recorded linking adjacent meiocytes at early prophase of meiosis in *Lilium* (J. Heslop-Harrison 1966), but these become closed by metaphase I. The cytoplasmic connections of *Acacia subulata* could possibly have their origin by the persistence of the prophase cytomictic channels. Continuity between these cytomictic channels and those linking pollen tetrads of *Hedycarea* (Monimiaceae) and *Exospermum* (Winteraceae) at maturity has been claimed by Sampson (1977, 1980).

5.3.7 Pollen Sterility

In many cases of genetic or cytoplasmic sterility, the block in development operates in the microsporocytes; in other cases, where pollen development is initiated, the size reached by the grains and the structure of the pollen wall provide clues to the period in development when sterility is expressed, if comparative studies of normal pollen development are made (see Knox and Heslop-Harrison 1966, Rowley and Flynn 1969, Horner 1977, Albertsen and Palmer 1979). In barley the staining reaction of the exine is altered in a cytoplasmic male sterile type (Ahokas 1978), suggesting modification of the chemical components of the exine. A cytoplasmic male-sterile mutant of wheat caused disorganized proliferation of bacula and tectum (Barlow 1978), while a cytoplasmic male-sterile mutant of broad bean caused reduced exine sculpture (Audran and Bouillot 1981). The role of glucanase enzymes in pollen sterility has been explored by Izhar and Frankel (1971). Alterations to the polysaccharides of the intine, including the deposition of callose, have been reported by Cousin (1980) in pollen sterility in *Vicia* following mycoplasma infection.

Viruses are known to be transmitted in pollen, e.g., barley stripe mosaic virus (Carroll and Mayhew 1976), and nepa virus have been observed by transmission electron microscopy in pollen of the olive, *Olea* (Pacini and Cresti 1977). Viruses have been detected in the pollen-coat materials (Hamilton et al. 1977). Fungal infection of *Rapanea* pollen is initiated by hyphal penetration through the apertures (Vasanthy and Pocock 1981).

5.4 Pollen Germination and Pistil Interactions

Pollen germination is triggered by hydration, either in vitro or in vivo, on the moist surface of the stigma. The pollen tubes may grow extremely rapidly and can be observed within 1 min in wheat (Chandra and Bhatnagar 1974). Successful in vitro conditions were established by Brewbaker and Kwack (1964), and have since been adapted for germination of a wide variety of pollen types by varying the carbohydrate or boron content. Maximal difficulties have been experienced in germinating tricellular pollen types. In these grains the mode of hydration appears to be the problem, and Bar-Shalom and Mattson (1977) have shown that semi-solid media, simulating the "dry" stigma surface, give highest percentage germination. After low-humidity storage of pollen of grasses, Shivanna and Heslop-Harrison (1981) have demonstrated that a period of conditioning in a humid atmosphere enhances germination.

5.4.1 Pollen Quality

Pollen quality is dependent on many factors, including the genetic background of the material, the environment in which the plants were grown to flowering, and the methods of pollen collection and storage in breeding experiments (see review by Stanley and Linskens 1974). Pollen of high viability is needed for breeding purposes and, often, scanty information is published on this topic, in many crop plants, not to mention specific experimental systems. The methods available will now be described and their advantages and disadvantages outlined, where known.

The collection and storage of pollen has been reviewed by Stanley and Linskens (1974). For storage, pollen must generally be dried, at least partially, over a desiccant before storage in vials at low temperatures. Bicellular types of pollen retain their viability longer in storage than tricellular types. Temperature and relative humidity are considered to be the major factors controlling successful storage. In a recent study of the storage of avocado, *Persea* pollen, Sedgley (1981) reported that the pollen quality, tested by observation of the ability of pollen to penetrate the ovules, was highest if pollen was stored at −196 °C. At other temperatures, and at times of storage longer than 1 month, while the pollen might germinate, it usually failed to penetrate the stigmas. Storage had the effect of greatly weakening pollen tube growth. Thawing of the pollen, even once, during storage, resulted in complete inviability.

Accordingly, what methods are available to determine pollen quality? Three classes of techniques are currently available, but none are completely satisfactory.

1. Direct Staining of Pollen. This method gives the percentage of filled normal grains produced by the anthers. It is assessed after staining pollen with, e.g., cotton blue in lactophenol (Hayman 1956), or 0.5% lissamine green (Willemse and Audran 1982), and viewed by bright-field microscopy, and is suitable for field conditions. It distinguishes between sterile or partially sterile plants, but does not give indication of loss of pollen viability in storage.

2. Enzyme-based Methods. Fluorochromatic reaction, FCR test. This microscopic test is applied to samples of pollen, and gives an estimate of the percentage of viable grains in the sample (J. Heslop-Harrison and Y. Heslop-Harrison 1970). The principle of the test is that a non-fluorescent ester enters the grains through the plasma membrane, is broken down by enzymic action to fluorescein, a highly fluorescent compound which is retained within the protoplast in viable grains, but lost from inviable grains. Its advantage is that it gives results within a few minutes; but its disadvantages are that it must be assessed at the time of preparation (no storage of slides) and requires fluorescence microscopy. A comparison of results of this test with pollen germination and other methods has been given by Shivanna and Heslop-Harrison (1981) and J. Heslop-Harrison et al. (1983).

Other enzyme-based methods are available which produce coloured reactions visible in the light microscope. The most widely used is the TTC topographical enzyme test (Stanley and Linskens 1974), based on the use of triphenyl tetrazolium chloride (TTC). The principle of the test is the acceptance of H^+ ions from enzymes in the pollen by TTC through flavoprotein nucleotides, which are intermediate H^+ ion acceptors for dehydrogenase enzymes. Unfortunately, not all pollen is equally sensitive to this test, giving false positive results. The reagents in the TTC test are stable at 5 °C for 3 months if stored in the dark.

3. Direct Pollen Germination. There are two methods available (see review by Stanley and Linskens 1974). Hanging Drop Test: pollen germinated in a small droplet of sterile medium (sucrose and salts solution) adhering to the lower side of microscope cover-slip raised above the surface of microscope slide by plastic O ring. Agar Slide Test: pollen germinated on surface of a thin film of solidified agar (containing a carbon source such as sucrose, together with boron and Ca^{2+}) on a microscope slide.

In both tests, the slide is sealed within a moist Petri dish, and the percentage pollen germination and tube length are assessed after a standard period (e.g. 2 to 6 h). Incubation at standard temperature is required. An interesting modification of the agar method is "David's Bread Loaf", a solid agar medium which is employed as 2 mm thick slices (Freeling and Cheng 1978). For use, pollen is dusted onto the newly cut surface with a camel hair brush. Pollen that is readily germinated in liquid media, e.g. of *Lilium,* or *Petunia,* may be sprayed on TLC plates for detection of growth inhibitors (Hodgkin 1982). A modification of the agar slide test has been used for similar purposes by Williams et al. (1982c).

These germination tests can also be used to assess the vigour of pollen tube growth from pollen of different genotypes. There is some evidence that vigour of tube growth varies, and that the most rapidly growing pollen tubes produce the most vigorous progeny (Ottaviano et al. 1980). A general principle of all these in vitro germination tests is that a viability rating of 40% or higher is generally satisfactory for field pollinations (Stanley and Linskens 1974).

A fourth test, involving NMR spectroscopy has recently been developed (Dumas et al. 1983), and is non-destructive of the pollen sample.

5.4.2 Pollen Germination and Tube Development

When pollen hydrates, the pollen tube emerges from one of the previously differentiated germinal apertures in the wall of aperturate grains. In *Lycopersicon peruvianum* detailed studies of pollen germination on the stigma, and in vitro have been made by Sarfatti, Cresti, Pacini and co-workers at the University of Siena, Italy. The grains have three semi-spherical apertures which are extruded at hydration. On germination "the pore opens like a port-hole door; ... the thicker portion of the pore, protruding through the corresponding part of the exine, breaks off along a large part of its edge forming a kind of door that opens towards the outside. The portion of the edge that remains attached to the intine functions as a hinge" (Cresti et al. 1977).

In many types of pollen, the pollen tube tip is preformed in the aperture, which may account for the structural complexity of the apertural intine, and the enzymes laid down within its walls. The emergence of the plug-like tip or oncus of pollen tubes has been reported in *Lychnis* (Crang and Miles 1969) and in various Compositae: *Cosmos* (Knox and Heslop-Harrison 1970), sunflower, *Helianthus* (Vithanage and Knox 1977), and *Artemisia* (Rowley and Dahl 1977). In *Artemisia* ultrastructural studies have demonstrated that the emergent tube is covered by a bubble of coarse fibrillar material through which the tube grows, surrounded by a modified intine-like wall. In germinating pollen of rye, *Secale cereale,* the oncus-

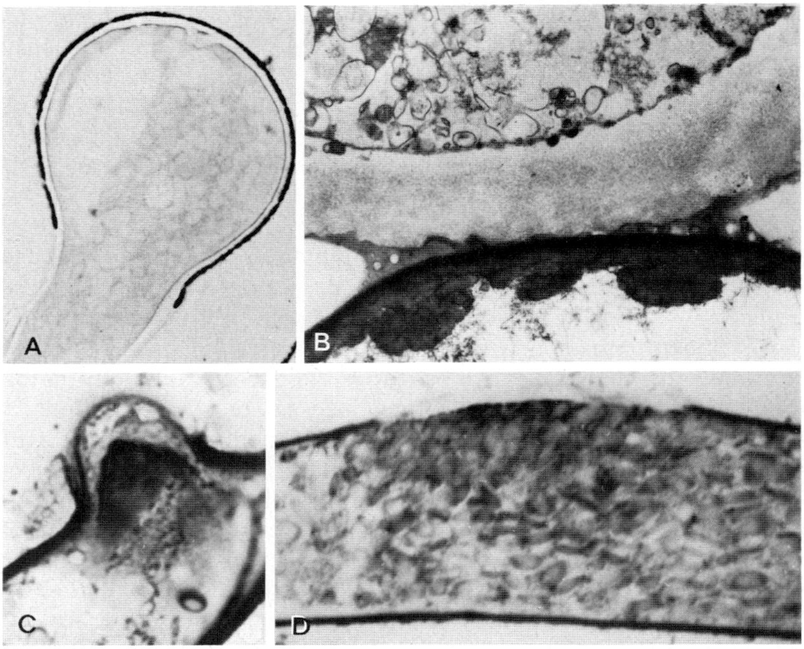

Fig. 5.20 A–D. Pollen grain germination in inaperturate types of pollen. **A** Spherical grains of *Populus*, the pollen tube appears continuous with the thick intine. (From Ashford and Knox 1980). **B–D** Filiform pollen of the sea-grass *Amphibolis*. **B** Transmission electron micrograph showing cohesion of the pollen wall (*upper*) to the stigmatic surface (*lower*); the microfibrillar pollen wall attaches by a meniscus of mucilage to the surface cuticle of stigma. **C** Light micrograph, showing the pollen tube emerging within a bubble of mucilage. **D** Light micrograph showing initial stage in tube formation, a portion of pollen wall loses its affinity for the dye toluidine blue, suggesting it may be undergoing localized depolymerization and breakdown. (From Pettitt et al. 1980)

like „Zwischenkörper" layer at the aperture forms a gel that swells displacing the sporopollenin cap, and the tube emerges within the dispersing gel (J. Heslop-Harrison 1979 b).

In a study of pollen germination in 12 genera with inaperturate grains, Muller-Stoll (1956) has outlined three common conditions: (1) in *Canna* and *Hedychium* the exine remains intact on hydration and the aperture, on germination, virtually punches the exit pore through it; (2) in *Arum* and *Aristolochia* slits appear in the exine at hydration, and the pollen tube exits through the slits; (3) in *Tulipa* and *Populus* the exine is brittle and tears into small portions at hydration, so that the germinating grains are almost naked, covered by the intine.

Muller-Stoll's (1956) work addressed questions concerning the evolution of the apertures. Today, a relevant question is their cell biology – how are these exits "punched" in the pollen wall?

In inaperturate pollen types, the tube is formed by a wall that appears continuous with the intine, for example in *Populus* (Fig. 5.20 A). There are differences in two cases. In avocado, *Persea americana*, the pollen tube at germination may be surrounded only by the plasma membrane, wall deposition does not occur until the pollen tubes have entered the stigma (Sedgley 1979). In the filiform pollen of the sea-grass, *Amphibolis antarctica* (Fig. 5.8 P), pollen tubes are formed by dissolution of the intine-like pollen wall by a process of localized digestion, focal autolysis (Fig. 5.20 B, C). The position of the aperture may be detected cytologically by a marked loss of staining affinity of the wall, indicating that the wall has apparently undergone dissolution. This may occur at one or more sites adjacent to the stigma. A tube-like outgrowth of the plasma membrane subsequently emerges within a mucilaginous bubble (Fig. 5.20 D). The tube wall develops soon after emergence, and is cytologically distinct from most of the pollen grain wall. In these cases extracellular secretions of slime, containing carbohydrates, are involved in pollen germination, especially the early stages of tube emergence and growth.

Growth of the pollen tube occurs at the tip, as detected in *Lilium* by the charcoal powder technique used by Rosen and co-workers (Rosen et al. 1964). Pollen-tube cytology and physiology may differ between artificial and natural germination. Tubes growing in vivo on the pistil developed transfer cell wall ingrowths at the tube tip (Rosen and Gawlik 1966), while those germinating in vitro in growth medium had only a thin wall at the tip. On the stigmas, pollen of *Lycopersicon* hydrates within 15 min, and germination begins after 3 h 30 min (Pacini and Sarfatti 1976). In contrast, in germination medium hydration occurred in seconds, and germination followed in 45 min (Cresti et al. 1977). Bicellular types of pollen may show two phases of growth: an initial period of growth, followed by a second period of more rapid growth (see Raff 1981, D. L. Mulcahy and G. B. Mulcahy 1983). In contrast, tricellular types of pollen show only a single growth period (D. L. Mulcahy and G. B. Mulcahy 1983). Questions that need to be answered concern the relationship between the vegetative cell and the germinating tube. Is the tube an extension of the vegetative cell, capable only of its limited gene expression? Is the tube wall simply an extension of the intine or one of its layers? It is well known that a callose wall is laid down secondarily in the pollen tube in many systems, but is the pollen-tube wall unique in its physical and chemical composition?

5.4.3 Diagnostic Landmarks of Pollination

Fertilization in flowering plants involves interactions between the pollen grain or its pollen tube, and the pistil. The pistil comprises several diploid sporophytic tissues which surround and protect the embryo sac, the female gametophyte. The pollen grain alights on the receptive cells of the stigma, where germination may occur. This involves hydration and swelling of the dry grains, so that the pollen tube may emerge, penetrate the stigmatic surface, and grow through the apoplast system of the stigmatic zone (see review by J. Heslop-Harrison 1975 a, 1978). Pollen-tube growth is always intercellular, through the gel matrix of the transmitting tissue of the style, or the stylar canal mucilage. Ultimately, the tubes pass into the ovary, usually growing through a track of mucilage leading them directly to the micropyle of an ovule.

The growth of the pollen tube, or even pollen germination itself, may be halted in incompatible pollinations at several points (Knox et al. 1976, Clarke and Knox 1978, Knox and Clarke 1980). Such incompatible matings are of two types: foreign pollinations where there are genetic or physiological barriers preventing fertilization; self-incompatibility where a genetically determined system operates to prevent selfing and, thus, inbreeding. The system is commonly based on a major gene S, which has many alleles (usually 30–40); although as many as four major genes controlling self-incompatibility have been recorded (see de Nettancourt 1977). Pollen grains, that are otherwise quite fertile, are unable to effect fertilization on their own pistil. The system operates quite simply; when pollen and pistil carry the identical S-allele, incompatibility ensues; where the S-alleles differ, compatible pollination occurs. The incompatibility reactions of pollen grains, therefore, depend on mutual interactions with the pistil, so that the S-gene can be considered as a supergene controlling mutual recognition.

It is hardly surprising to find that self-incompatibility responses are of two different types: (1) Sporophytic systems where the pollen grain response is governed by the phenotype of the parent sporophyte, when dominance of one allele over another is exhibited, and as a consequence all the pollen grains within any one anther behave alike in their incompatibility response. (2) Gametophytic systems in which the behaviour of individual pollen grains follows that of their own S-genotype, and there is no dominance and, consequently, in heterozygous situations the pollen grains in any one anther will be equally divided in their S-allele type (see de Nettancourt 1977).

In sporophytic self-incompatibility systems, such as those in the Cruciferae and Compositae, the site of pollen arrest is usually on the surface of the stigma (see reviews by J. Heslop-Harrison 1975 b, 1978), and experimental tests with proteins extracted from the surface of the pollen grains have demonstrated that pollen-held determinants control sporophytic S-gene action (Dickinson and Lewis 1973 b, J. Heslop-Harrison et al. 1973 b, 1974, Howlett et al. 1975). These proteins are laid down in the exine along with the pollen-coat after transfer from the diploid sporophytic tapetal cells (Fig. 5.21 A, B) (Dickinson and Lewis 1973 a, J. Heslop-Harrison et al. 1974).

In contrast, the gametophytically derived fraction in the pollen wall is held within the intine and in the protoplast itself, and the site of pollen-tube arrest is usually within the style. To date, it has not been possible to test experimentally whether the intine proteins are implicated in the control of gametophytic self-incompatibility because the interaction occurs deep within the style. Immunobiological tests have revealed the existence of an antigen in one pollen system – *Oenothera organensis* (Lewis 1952, Lewis et al. 1967), whose presence correlates precisely with the S-allele type of the pollen (see review by Knox and Clarke 1980). Unfortunately, this antigen has not been characterized, nor has its presence been investigated in the stigma of *Oenothera*. In other stigma systems, while S-specific antigens have been detected, they have shown no immunological identity with pollen fractions from the same system (see review by Clarke and Gleeson 1981).

The inhibition of pollen-tube growth following self-incompatible matings has recently been investigated in *Primula* spp. which have a different control system, known as heterostyly. Here both pollen and pistil are morphologically different

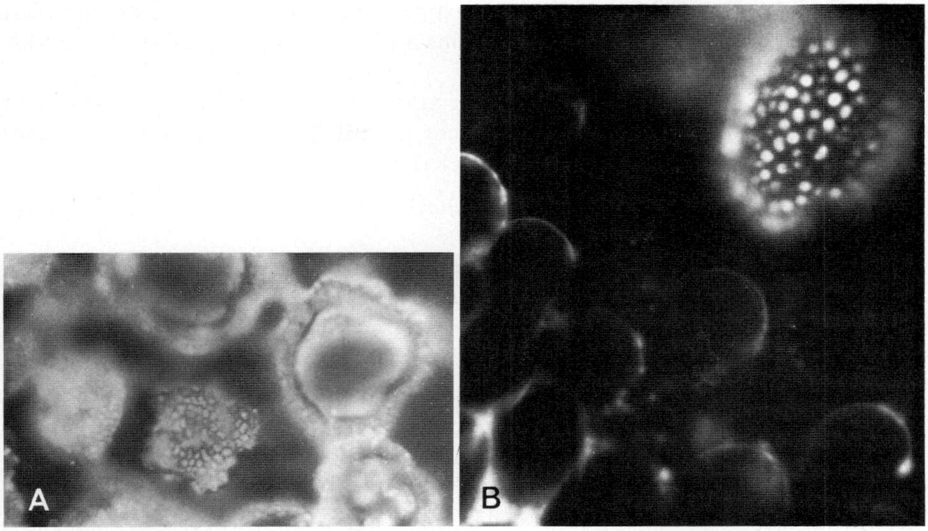

Fig. 5.21 A, B. Wall proteins in pollen of Cruciferae. In mature pollen the exine proteins may be visualized in the fluorescence microscope after staining with 1-ANS. **A** Frozen section of *Iberis* pollen. **B** Pollinated stigma in which a pollen grain has been stained. The protein-filled surface cavities appear *white* in these photographs. (From J. Heslop-Harrison et al. 1974)

in the two forms of pin and thrum flowers. The system is di-allelic, and pin pollen is inhibited on pin stigmas, and the reciprocal. Compatibility follows outcrossing between the two forms. Golynskaya et al. (1976) showed that pin extracts inhibited pin pollen germination in vitro, but not thrum pollen, and the converse, in *Primula obconica*. In recent studies of the system in *P. vulgaris*, Shivanna et al. (1981) have demonstrated that dialysates of high molecular weight fractions from stigma and style exert some differential effect on pollen tube growth. Those from thrum pistils retarding thrum tubes while having a lesser effect on pin tube, and the converse. Ferrari and Wallace (1977) have shown a similar effect in sporophytic self-incompatible *Brassica* pollen tube growth. Williams et al. (1982b) observed that isolated stylar glycoproteins of cherry, *Prunus avium*, may inhibit self pollen tube growth in this gametophytic system. These experiments suggest that control is exerted by the stylar S-gene product, but give no indication of the nature of the interaction with the pollen, pollen tube, or their putative receptors.

An interesting response to pollination is the production of callose in the wall of the pollen grain and pollen tube as a result of incompatible matings. The "rejection" callose deposits in the wall are detected cytochemically by fluorescence after staining with decolorized aniline blue, a technique first developed for pollen tubes by Linskens and Esser (1957). Callose contains a 1,3 β-glucan, and may be considered as an active rejection response (Dickinson and Lewis 1973a, b, Heslop-Harrison et al. 1974). These authors have shown experimentally that extracts of the exine proteins will induce a callose response in stigma cells of Cruciferae in self-incompatible but not compatible systems. In other cases, as in the grasses, which have a gametophytic self-incompatibility system, callose is induced only in the inhibited pollen grain and tube (J. Heslop-Harrison et al. 1978). Evidence concerning these sites of origin of the proteins and glycoproteins active in these responses will be given later (see Sect. 5).

In order to explore the role of pollen grains, and their tubes, in pollination, specific mutants of structural or regulatory genes are needed. Regrettably, only a few such pollen mutants have been detected:

1. Wall-shape mutant in sweet pea, *Lathyrus:* sporophytic inheritance (Bateson and Punnett 1909).
2. Waxy mutant in maize, *Zea;* gametophytic inheritance (Nelson 1962).
3. Alcohol dehydrogenase in maize, *Zea mays;* gametophytic inheritance (Freeling and Cheng 1978).
4. β-galactosidase in *Brassica;* gametophytic inheritance (Singh and Knox 1983).

The use of such mutants in controlled pollination experiments will provide a precise tool to explore the role of the specific gene products in the events of reproduction.

Many basic questions remain to be answered. What is the timing and site of generative cell division within the pollen tubes? An answer to this question is today known for *Lycopersicon,* and a few other systems. What is the route of tube growth through the ovary, after exiting from the style? Few advances in knowledge have followed the pioneering work of F. E. Knight (1910) with the date palm, *Phoenix.* He showed that the tubes followed a groove of mucilage to the micropyle.

5.4.4 Role of Pollen in Fertilization and Seed-setting

The essential role of the pollen grain is in sexual reproduction; it is crucial to the survival of flowering plants. The structural adaptations of the pollen wall, especially for pollinators, wind and water currents, have been stressed in this chapter – but are there any direct co-adaptations between pollen and stigma?

The remarkable co-evolution between the pollinia of orchids, containing several thousand pollen grains, and the development of adhesive mechanisms for attachment of the pollinia in specific parts of the body of pollinating insects, and later on the stigma were highlighted by Charles Darwin (1882). The cell biology of the adhesive mechanism for attaching the pollinia of asclepiads to visiting pollinators has been investigated by Schnepf et al. (1979). An unique translator organ, composed from lipids and mucilage, carries out this function.

Morphological adaptations have recently been noted in the family Leguminosae, subfamily Mimosoideae, whose species may have compound pollen grains or free grains (monads). In the Mimoseae, the different genera possess an unusual stigma morphology. The stigma cup varies from funnel-shaped in *Dinizia,* tubular in *Entada* and *Prosopis,* to narrow porate in *Adenanthera, Mimosa,* and *Piptadenia* (Lewis and Elias 1981). Polyads are recorded in genera with the two extremes – broader funnel-shaped stigmas, and very narrow porate types. Indeed, *Mimosa* has probably the smallest stigma, together with the smallest polyad recorded in the angiosperms; shown by Guinet (1981) to be only 6.5 μm in diameter. A similar correlation between stigma size and polyad size appears to operate in the Acacieae (Kenrick and Knox 1982).

In different species of *Acacia* there is a polymorphism in polyad grain number – 4, 8, 12, 16 grains in Australian species, and 16, 32, 48, and 64 in African species. The significance of these variations is that there is a parallel matching in seed number per pod. In species with 4-grain polyads the maximum pod seed number is 3; with 8-grain polyads, 8; with 16-grain polyads, 12–16; with 32-grain polyads, 21–24 seeds per pod (Kenrick and Knox 1982, Knox and Kenrick 1983). The relationship is further supported by data from two genera of Mimoseae with free grains, which have 20–25 seeds per pod (Kenrick and Knox 1981).

The conclusion is inescapable that there is a mutual co-adaptation of both pollen and ovary, with the aim of ensuring that a single polyad alighting on a stigma has the potential to set a whole pod of seeds. The system also ensures a highly efficient use of resources.

5.5 Pollen-wall Proteins and Allergens

One of the most remarkable features of the angiosperm pollen grain is the complexity of its wall which, as we have seen in this chapter, is usually a product of gene expression of parental determinants, the microspore, and tapetal cells. The role of this wall in protecting the male gametophyte during pollen dispersal has been emphasized, and the pollination strategies that might influence the evolution of pollen shape have been featured. However, the existence of the intricate, crypt-like arcades with their microchannels within the pollen wall and the complex and specific patterning all suggest that the exine plays a crucial role in recognition and other early events of pollen germination on the stigma surface. The exine arcades are filled with proteins (for example, in *Iberis,* Fig. 5.21 A, B), glycoproteins, carbohydrates, lipids, and pigments, to ensure both efficient pollen adhesion to pollinator or stigma surface, and also to initiate macromolecular contact with the stigma leading to pollen hydration and germination. Evidence for the presence of wall proteins will now be reviewed, together with the available experimental evidence concerning their functions.

Cytochemical methods, using freeze-sectioned anthers and pollen grains, have established that both the pollen wall and cytoplasm contain a wide range of enzymes. These tests depend on the production of coloured products on the hydrolysis of insoluble artificial substrates. Using such methods, the presence in the pollen wall of hydrolases, transferases, dehydrogenases, oxidases, ligases, and lyases has been demonstrated (Tsinger and Petrovskaya-Baranova 1961, Knox and Heslop-Harrison 1969, 1970). Many of these enzymes are apparently located within the polysaccharide matrix of the intine, especially the hydrolases. Others are present in the arcades of the exine especially the dehydrogenases; some of these enzymes are present in both sites (see reviews by Knox et al. 1975, Howlett et al. 1979). Wall-held enzymes occur in the spores of several lower plants (Pettitt 1979).

The pollen-wall proteins may play a role in pollen development, since they show quite characteristic fluctuations in activity. In pollen of *Brassica* the levels

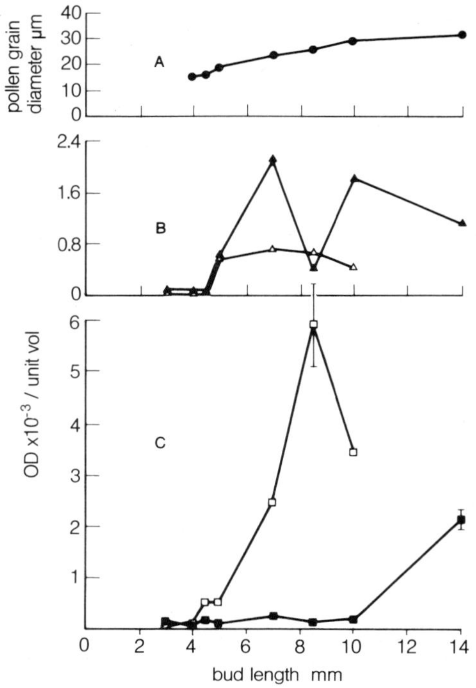

Fig. 5.22. Quantitative estimation of two pollen-wall enzymes during pollen development in *Brassica*, using two-wavelength microspectrophotometry. *A* Increase in pollen grain size. *B* Gametophytic marker enzyme: acid phosphatase activity in pollen (▲), and tapetal cells (△). *C* Sporophytic marker enzyme: esterase activity in parallel sections to those of *B* in pollen (■), and tapetal cells (□). (From Vithanage and Knox 1979)

of acid phosphatase activity show two peaks as detected by quantitative cytochemical methods (Fig. 5.22) (Vithanage and Knox 1976, 1979). The first peak at the vacuolate period of microspore development corresponds with the main period of pollen growth. The enzyme activity is located in the intine, where it is laid down with the polysaccharide matrix. Thereafter, activity declines until near-maturity when there is a massive increase in activity in the pollen protoplast (Fig. 5.22). Acid phosphatase is an intine enzyme, synthesized by the protoplast of the vegetative cell. In contrast, the enzyme esterase is of low activity until the maturation period of pollen development when there is a dramatic increase. All the activity is located in the exine cavities at the grain surface. At the same time, there has been a sudden decline in activity of the enzyme in the tapetal cells, suggesting this may be associated with transfer to the exine sites (Fig. 5.22).

The presence of allergenic proteins and glycoproteins in wind-borne pollen types led to the first immunocytochemical studies of flowering plant antigens being carried out on pollen (Hagman 1964, Knox et al. 1970), with the aim of locating the sites of allergens in the grains. In pollen that is dispersed by air currents, and so is able to make contact with the upper respiratory tract of man, allergens are released causing seasonal asthma and hay fever (see review by Howlett and Knox 1983). Immunocytochemical studies have revealed that the allergens of ragweed responsible for autumnal hay fever in North America are located in the outer wall of the pollen grain. Immunofluorescence studies of pollen, freeze-sectioned within the anther to minimize diffusion of antigens, showed that Antigen E (the principal allergen, a protein of MW 38,000; King et al. 1964, 1974) is concen-

Table 5.6. A scheme showing the origins of the proteins held within the two principal compartments of pollen wall – exine and intine and their modes of emission when the pollen is moistened

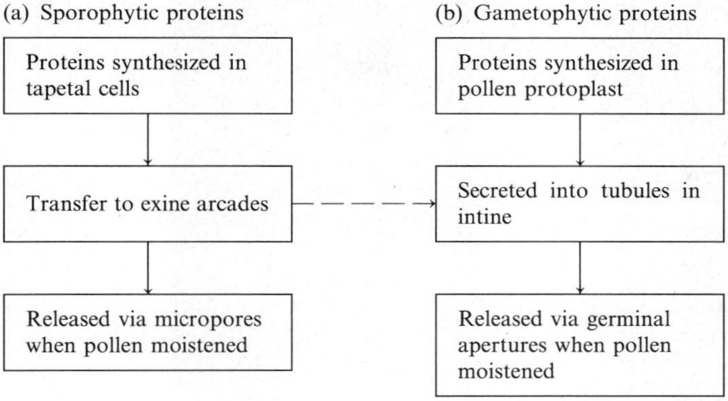

trated in the intine, and also in the exine, where it occurs in both the arcades and at the surface (Knox and Heslop-Harrison 1971 a). Birch pollen allergen is also present at the exine surface (Belin and Rowley 1971).

These immunofluorescence studies revealed that when the pollen is moistened, even only momentarily, the allergens will be released to coat the surface of the grains. The two principal routes for protein emission are shown in Table 5.6. When ragweed pollen was freeze-sectioned by preparing a slurry in 15% gelatine, and frozen within 30 s of pollen contact, all the antigens were found in the gelatine surrounding the grains. This extreme mobility of the antigens has been confirmed by making pollen prints of fresh pollen, to observe the routes and time of antigen emission. The pollen was sprinkled on adhesive tape, placed in contact with a thin agarose gel on a microscope slide, removed after varying intervals of time, and the film dried rapidly and processed for Antigen E activity by indirect immunofluorescence (Howlett et al. 1973). The pollen prints revealed that Antigen E is present in both the exine, from which it is immediately released, and intine at the apertures from which it is released more slowly during germination (Fig. 5.23 A–C). Freeze-substitution followed by paraffin-embedding with minimum exposure to high temperatures produced sections which gave the same localization in mature dry pollen as had been obtained with freeze-sectioned pollen within the anther (Howlett et al. 1973), but without the hazard of diffusion artifacts. The structure of the pollen wall of ragweed is adapted for rapid release of its wall-held proteins, and for wind-pollination (see Skvarla and Larson 1975).

These observations on the kinetics of allergen release suggest that the antigens play a role in pollination: in male–female recognition at the stigma surface. An interesting adaptation of immunofluorescence has provided circumstantial evidence for this in both ragweed and grasses: pollinated stigmas were placed into specific antipollen diffusate sera in order to fix the antigens in their sites during the pollination process. The pollen antigens were shown to be present both on the surface of pollen grains and tubes, but also bound to the stigma surface (Knox and Heslop-Harrison 1971 b, Knox 1973).

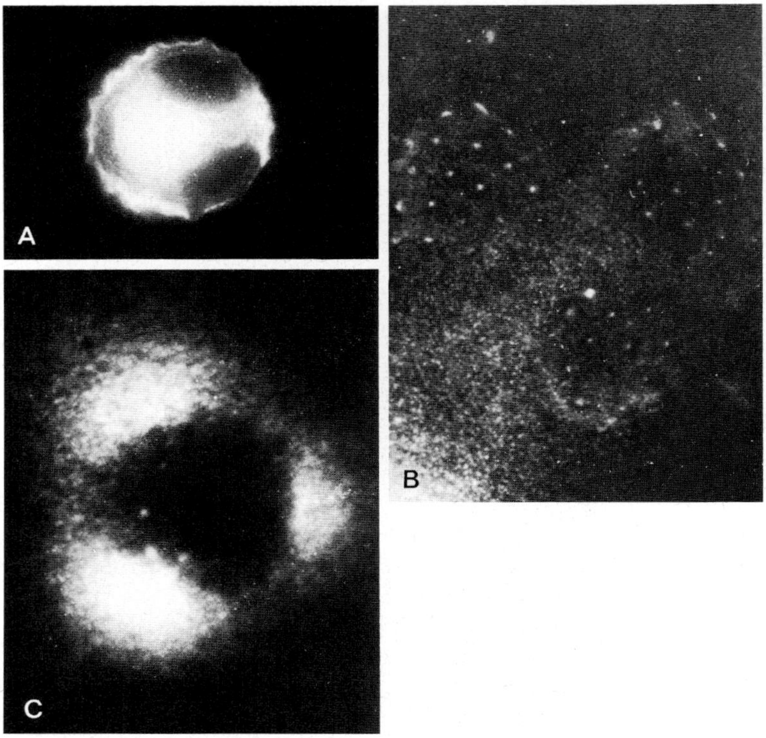

Fig. 5.23 A–C. Demonstration of the allergen Antigen E in pollen grain and pollen prints of ragweed, *Ambrosia elatior*. **A** Entire grain showing Antigen E located at surface of the three germinal apertures (*white*). **B** Rapid pollen-print on agarose film, showing imprint of grains for only 5 s and release of Antigen E from microspinules. **C** Longer term (3 min) pollen-print showing diffusion of Antigen E from the three colpi, with grain site indicated by *dark area in centre* of photograph. (From Howlett et al. 1973)

Grass pollen is dispersed in air currents, and is the major source of seasonal asthma and hay fever in spring and early summer in temperate climates. The allergens of grass pollen have been recently located by immunocytochemistry using antiserum to purified pollen components. The pollen of rye-grass, *Lolium perenne,* contains four groups of allergens, named Groups I–IV (Johnson and Marsh 1965, 1966, Marsh 1975). Group I allergen is the major allergenic fraction: it is a glycoprotein, MW 32,000, existing in multiple electrophoretic forms which differ only slightly in amino acid composition. A similar fraction has been isolated using ion exchange chromatography and gel filtration, and has all the properties of Group I allergen including similar behaviour by starch gel electrophoresis, antigenicity, and allergenic activity (Smart and Knox 1979).

A post-embedding staining technique has been developed to overcome the difficulties of antigen mobility in aqueous media (Knox et al. 1980). Anthers of rye-grass just prior to anthesis are either briefly fixed in cold methanol or freeze-dried, and placed directly in a water-soluble methacrylate resin, which has the advan-

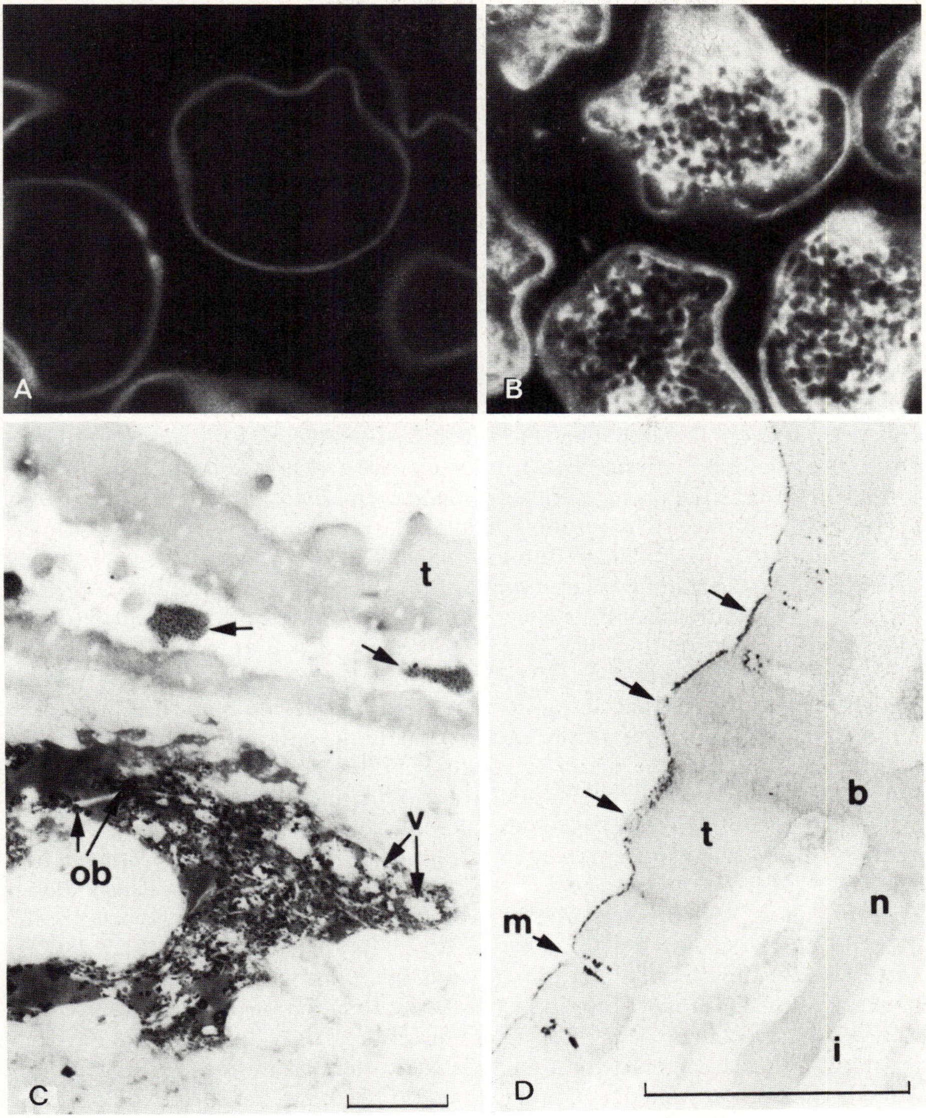

Fig. 5.24 A–D. Immunocytochemical localization of Group I allergen in mature pollen of ryegrass, *Lolium perenne*. **A, B** Immunofluorescence with FITC-labelled antibodies. **C, D** Immuno-electron microscopy with ferritin-labelled antibodies. **A** Normal (pre-immune) antibody control. **B, C** Anti-Group I allergen antibodies. In **B** the allergen is present in both wall and cytoplasmic sites. In **C** the allergen is present within the pollen-coat in exine arcades (*arrows*), vesicles *v*, and osmiophilic bodies *ob*. In **D** the allergen is located in the outer micropores *m* of the tectum and on the exine surface (*arrows*). In **D** tectum *t*, bacula *b*, nexine *n*, and intine *i* are also seen. (**A, B** From Howlett et al. 1981; **C, D** from Vithange et al. 1982)

tage of short infiltration and embedding time at 2 °C, and of being anhydrous so that there should be little possibility of the antigens showing diffusion artifacts. In immunofluorescence studies, the specific fluorescence is observed in the outer wall layer, and also in the cytoplasm of the sectioned pollen grains (Fig. 5.24A–D). In order to precisely determine the sites of these antigens within the grains, immuno-electron microscopic methods have been developed. The indirect method is used, with the anti-antibody labelled with ferritin (Knox et al. 1980). The results are spectacular: using specific antisera, ferritin molecules are observed bound to spherical material within the exine cavities, to wisps of material at the exine surface, to vesicles within the intine, and at the periphery of the cytoplasm (Fig. 5.24).

It appears that Group I allergen is present at the surface of mature rye-grass pollen. Vithanage et al. (1982), however, in their immunocytochemical studies, observed ferritin-labelling, indicating the presence of the allergen on the orbicules – spheres of sporopollenin that line the anther cavity (see Fig. 5.9A). Perhaps the allergen provides lubrication to ensure release of the grains at anthesis. The allergen is sited for a role in the earlier events of pollination, perhaps in recognition reactions at the stigma surface. It is certainly available to induce allergic reactions in the upper respiratory tract of susceptible humans. The latter response is presumably an unfortunate accident of evolution. It seems likely that most of the allergen is released very rapidly when the pollen is moistened – from both wall and cytoplasmic sites. Support for this view comes from recent work on rye (*Secale*) pollen by J. Heslop-Harrison (1979a, b). From both hydrodynamic and ultrastructural evidence, he concludes that the mature and partly desiccated rye pollen grain does not have an intact plasma membrane to regulate the entry of water. Instead, the plasma membrane in the partly dry state of pollen dormancy is represented by lipid micelles, and reforms soon after hydration. Under these conditions rapid release of allergens from cytoplasmic sites could be expected.

A further mechanism for allergen release has been suggested by Franchi and Pacini (1980) from their ultrastructural studies of the allergenic pollen of *Parietaria officinalis*. Wall projections, resembling the transfer cell ingrowths (Gunning and Pate 1969) occur in the intine, especially at the apertures, and link up directly with the wall tubules which are microvillus-like ingrowths. Franchi and Pacini have advanced four hypotheses concerning the function of these intine wall projections: (1) To take up substances from the anther cavity during pollen development. (2) To facilitate dehydration of grains during maturation. (3) To facilitate rapid hydration on stigma. (4) To ensure rapid release of gametophytic proteins.

The latter role bears directly on the release of potential allergens. The allergens of *Parietaria* pollen constitute 0.5–2% of its total protein, and are released into aqueous solution in 10–32 s (Falagiani 1976, cited by Franchi and Pacini 1980). The total amount of released allergen is far too much to be stored in the walls, and Franchi and Pacini (1980) postulate a cytoplasmic origin. However, the very rapid release time suggests that the allergens might be stored in the cavities of the tectate exine, and would not be detected in glutaraldehyde-fixed grains. However, the possibility that the pollen-wall ingrowths may be involved in active transport of cytoplasmic protein allergens is certainly attractive, and deserves further investigation.

Finally, the question must be asked whether there is any firm evidence supporting the role of the pollen-wall proteins in recognition reactions at the stigmatic surface? The first evidence came in 1972 from the work on interspecific incompatibility between two species of poplar. The "mentor" effect of compatible pollen in the pollen mixes used to overcome incompatible responses, was shown to be due to compatible pollen-wall proteins (Knox et al. 1972). In subsequent experiments, self-incompatibility in sporophytic *Cosmos bipinnatus* was overcome using extracts of compatible pollen-wall proteins (Howlett et al. 1975). At the same time, a bioassay for the rejection response of sporophytic self-incompatibility in various Cruciferae (J. Heslop-Harrison et al. 1973, Dickinson and Lewis 1973 b, J. Heslop-Harrison et al. 1974) had implicated exine-held proteins in initiating the incompatibility response.

The experiments have been extensively reviewed (Clarke and Knox 1978, J. Heslop-Harrison 1975 a, b, 1978, Knox 1976, Knox and Clarke 1980, Shivanna 1982). They provide the experimental basis for concluding that the pollen-wall proteins act in the biocommunication between pollen and stigma. Unfortunately, the active proteins or glycoproteins have yet to be characterized and isolated. The system remains one of the challenges for the future.

5.6 Conclusions and Future Developments

Thirty years ago, when Maheshwari (1950) reviewed the state of knowledge of pollen grains, the cytological features of the grains had been established. Since then, with the advent of transmission and scanning electron microscopy, a whole new dimension in our understanding of pollen structure and function has been discovered – the nature and diversity of the pollen grain wall, and the sperm cells which it encloses, aptly termed the "new biology" of pollen (Vasil 1973).

In 1837, Fritsche first named the twin outer envelopes the intine and exine. It has taken the intervening years to chart pollen diversity in terms of wall structure. Today, we are aware that all pollen grains are surrounded by a polysaccharide intine layer, but the superficial coat, the exine, may be absent in some aquatic pollen types and greatly reduced in quite unrelated types from humid tropical climates. In contrast, the exine may be massively developed in insect-pollinated types. The exine is the specific layer; its specificity lies not only in the extent of its thickness but in its patterning. The outer surface of the exine carries group-specific pattern attributes that form the basis of pollen classification and of palynology. Adaptations of the exine are under strict genetic control, but their expression, which is apparently under sporophytic control, has been remarkably responsive to environmental influences, especially the agents of pollen dispersal.

In the foregoing description we have seen how both form and structure of the pollen grain wall, especially of the exine, are influenced by the mode of dispersal. In air-current systems, the thickness and outer surface patterning of the exine are reduced, as is the amount and stickiness of pollen-coat materials; and powdery pollen grains have aerodynamic attributes aiding their dispersal in the atmosphere. Most types are tricellular, ready for rapid germination and pollen tube growth. In a similar way, water-dispersed pollen grains are adapted for hydrophilous pollination; the intine is thick, the exine usually reduced and fragile, or totally absent. Slime is secreted at the surface, instead of adhesive pollen-coat materials. In the extreme case of certain sea-grass pollen, the grains are filiform, long, and filamentous (<5 mm in length), and may flex and coil in water currents and around the stigma. Again, nearly all pollen types are tricellular at maturity.

In contrast, in animal-pollinated systems pollen grains are mostly bicellular, but there is an incredible and endless variety of adaptations reflected in both pollen form and wall structure. In free grains the shape may be an adaptation to vector transport, or in some cases, to the form and arrangement of the receptive stigma cells. This is seen in bird pollination, where three-lobed, almost winged grains are a feature of pollination via head feathers; and in insect pollination, in the array of spines and spinules that cover the grains. The pollen is usually heavily coated with coloured adhesive coat materials which enable the grains to stick to each other, to the anther wall, the vector, and the stigmatic surface. In insect-pollinated systems, the grains may cohere into tetrads, polyads, or pollinia, according to the requirements of the pollination system. Such systems have the apparent advantage that the pollinator, often travelling considerable distances between flowers, transports a package of pollen usually sufficient for complete seed-setting in the ovary. The remarkable mutual evolutionary responses of pollen grain and animal pollinator represent a particularly exciting field for exploration.

At the same time, progress is being made in understanding how the complex wall of the pollen grain is assembled, how its various strata are laid down at specific periods of development. We can anticipate that the development, and likely improvements in cytochemical specificity of electron stains, may greatly aid the search for knowledge of the mechanism of synthesis and polymerization of protosporopollenin and sporopollenin – the polymers that form the exine. Chemical dissection of exines has provided a new basis for understanding exine structure and formation. Coiled subunits composed of protein, polysaccharide and lipid components, act as sporopollenin receptors in the model of Rowley and associates. Evidence now needs to be sought for chemical specificity in addition to the unique structural patterns of the exine.

Finally, a great deal remains to be done to begin to understand the early events of pollination – pollen–stigma recognition, and pollen tube growth through the style to the ovary and the role of the sperm cells in double fertilization. The demonstration of links between the sperm cells and vegetative nucleus in some pollen systems highlights the need for more research into pollination.

Pollination leads to fertilization and seed-setting, and is essential for crops and the survival of mankind. The existence of modern high-resolution light and electron microscope techniques together with computer-aided stereological methods for quantitative analysis make it possible to monitor the progress of pollination, to detect sites where the interaction is blocked or errors occur. These sites are not available to routine biochemical analysis because of the complexity of the tissues involved. However, histochemical techniques make it feasible to characterize the macromolecules, and in some cases the molecules, at the sites of their interactions in the pollen grain, pollen tube, and the pistil. The way is open for the manipulation of reproduction in flowering plants in a way that has not previously been possible.

Germ banks for storage and supply of pollen in important agricultural crops are finally being established. The factors controlling pollen viability before and during storage are being investigated. In order to make further progress in understanding pollen biology and fertilization, we need to ensure three developments:

1. A search for further mutants of structural or regulatory genes that are expressed in pollen; these are needed to explore the molecular control mechanisms of pollination and fertilization.
2. Development of a simple microscopic test for monitoring pollen quality under field conditions.
3. A means for continued regular contact and exchange of ideas between scientists working on pollen biology, and those involved in the practice of plant genetics and breeding. There is no journal exclusively devoted to plant reproduction; conferences are infrequent. Only by improved contact can we be certain that new developments in understanding pollen and pollen tube biology will be expressed in our new crop plants.

Perhaps, the fascination of pollen is its ready availability as an isolated intact cellular system, in flowering plants which are otherwise massive, multicellular structures. While they are relatively simple in structural terms, they are as complex as other plant and animal cells in their functional aspects. However, they provide an unparalleled experimental system for exploration with the probes of modern cell biology, utilizing specific genetic systems and the tools of micro-electrophoresis and immunobiology, on the one hand, and the resolution of transmission electron microscopy and specific electron-staining methods on the other. With these approaches, we may expect the new biology of pollen to become a reality.

References

Afzelius BM (1955) On the fine structure of the pollen wall in *Clivia miniata*. Bot Not 108:141–143

Agababyan VS (1972) Ultrastructure of sporoderm of some primitive Angiospermae (in Russian). Bot Zh (Leningr) 57:955–959

Ahokas H (1975) Male-sterile mutants of Barley II. Cytochemistry of non-mutant and *msg6cf* microspores and pollen. Hereditas 81:33–46

Ahokas H (1978) Cytoplasmic male sterility in Barley II. Physiology and anther cytology of *msml*. Herediatas 89:7–21

Albertsen MC, Palmer RC (1979) A comparative light and electron microscopic study of microsporogenesis in male sterile (MS_1) and male fertile soybeans (*Glycine max* L.). Am J Bot 66:253–265

Angold RE (1968) The formation of the generative cell in the pollen grain of *Endymion nonscriptus*. J Cell Sci 3:573–578

Ashford AE, Knox RB (1980) Characteristics of pollen diffusates and pollen-wall cytochemistry in poplars. J Cell Sci 44:1–17

Atkinson AW, Gunning BES, John PCL (1972) Sporopollenin in the cell wall of *Chlorella* and other algae: Ultrastructure, chemistry, and incorporation of ^{14}C-Acetate studied in synchronous culture. Planta 107:1–32

Audran JC, Bouillot J (1981) Sporoderm infrastructural and cytochemical modifications in cytoplasmic male sterile broad bean (*Vicia faba* L.). Acta Soc Bot Pol 50:223–226

Audran JC, Willemse MTM (1982) Wall development and its autofluorescence of sterile and fertile *Vicia faba* L. pollen. Protoplasma 110:106–111

Bailey IW (1960) Some useful techniques in the study and interpretation of pollen morphology. J Arnold Arbor 41:141–148

Banerjee UC, Rowley JR, Alessio ML (1965) Exine plasticity during pollen grain maturation. J Palynol 1:70–89

Barlow K (1978) A study of chromosomal male sterility in wheat. Ph D Thesis Univ N. S. W., Sydney

Bar-Shalom D, Mattsson O (1977) Mode of hydration as an important factor in the germination of trinucleate pollen grains. Bot Tiddskr 71:245–251

Barth OM (1965) Feinstruktur des Sporoderms einiger brasilianischer Mimosoiden-Polyaden. Pollen Spores 7:429–441

Barth OM (1965) Feinstruktur des Sporoderms einiger brasilianischer Mimosoiden-Polyaden. Pollen Spores 35:651–655

Bateson W, Punnett RC (1911) On the interrelation of genetic factors. J Genet 1:297–305

Belin L, Rowley JR (1971) Demonstration of birch pollen allergen from isolated pollen grains using immunofluorescence and a single radial immunodiffusion technique. Intl Arch Allergy Appl Immunol 40:754–769

Bernhardt P, Calder DM (1981) The floral ecology of *Amyema pendula* and *A. quandang* (Loranthaceae). Bull Torrey Bot Club 108:213–230

Bornet E (1864) Recherches sur le *Phucagrostis major* Cavol. Ann Sci Nat Bot Biol Vég 5 1:5–51

Brewbaker JL (1967) The distribution and phylogenetic significance of binucleate and trinucleate pollen grains in the angiosperms. Am J Bot 54:1069–1083

Brewbaker JL, Kwack BH (1974) The calcium-ion and substances influencing pollen growth. In: Linskens HF (ed) Pollen: Physiology and fertilization. North-Holland, Amsterdam, pp 143–151

Brooks J, Shaw G (1978) Sporopollenin: A review of its chemistry, palaeochemistry, and geochemistry. Grana 17:91–97

Brummitt RK, Ferguson IK, Poole MM (1980) A unique and extraordinary pollen tube in the genus *Crossandra* (Acanthaceae). Pollen Spores 22:11–16

Buchen B, Sievers A (1981) Sporogenesis and pollen grain formation. In: Kiermayer O (ed) Cytomorphogenesis. Cell Biology Monographs 8:349–376. Springer, Vienna New York

Carroll TW, Mayhew DE (1976) Anther and pollen infection in relation to pollen and seed transmissibility of two strains of barley stripe mosaic virus in barley. Can J Bot 54:1604–1621

Cass DD, Karas I (1975) Development of sperm cells in barley. Can J Bot 43:1051–1062

Cavolini (1806) On the flowering of *Zostera oceanica* Linn. Ann Bot 2:77–91

Chandra S, Bhatnagar SP (1974) Reproductive biology of *Triticum*. II. Pollen germination, pollen tube growth, and its entry into the ovule. Phytomorphology 24:211–217

Christensen JE, Horner HT (1974) Pollen pore development and its spatial orientation during microsporogenesis in the grass *Sorghum bicolor*. Am J Bot 61:604–623

Christensen JE, Lersten NR (1972) Pollen wall and tapetal orbicular wall development in *Sorghum bicolor* (Gramineae). Am J Bot 59:43–58

Chu C, Hu S (1981) The development and ultrastructure of wheat sperm cell. Sci Sin (in press)

Clarke AE, Gleeson PA (1981) Molecular aspects of recognition and response in the pollen-stigma interaction. In: Loewus F, Ryan CA (eds) The phytochemistry of cell recognition and cell surface interactions. Rec Adv Phytochem 15:161–211

Clarke AE, Knox RB (1978) Cell recognition in flowering plants. Q Rev Biol 53:3–28

Clarke AE, Gleeson P, Harrison S, Knox RB (1979) Pollen-stigma interactions: Identification and characterization of surface components with recognition potential. Proc Natl Acad Sci USA 76:3358–3362

Coccuci A, Jensen WA (1969) Orchid embryology: Pollen tetrads of *Epidendrum scutillar* in the anther and on the stigma. Planta 84:215–229

Condeelis JJ (1974) The identification of F-actin in the pollen tube and protoplast of *Amaryllis belladonna*. Exp Cell Res 88:435–439

Cook CDK (1982) Pollination mechanisms in the Hydrocharitaceae. In: Symoens JJ, Hooper SS, Compere P (eds) Studies on aquatic vascular plants. Royal Bot Soc Belgium, Brussels, pp 1–15

Cousin MT (1980) Changes induced by mycoplasma-like organisms (MLO), etiologic agents of the Stolbur disease in the different tissues of the anther of *Vinca rosea* L. (Apocynaceae). Grana 19:99–125

Crang RE, Hein NB (1971) Enzyme degradation of *Lychnis alba* pollen tube walls. Cytologia 36:449–454

Crang RE, May G (1974) Evidence for silicon as a prevalent elemental component in pollen wall structure. Can J Bot 52:2171–2174

Crang RE, Miles GB (1969) An electron microscope study of germinating *Lychnis alba* pollen. Am J Bot 56:398–405

Cresti M, Went JL van (1976) Callose deposition and plug formation in *Petunia* pollen tubes in situ. Planta 133:35–40

Cresti M, Pacini E, Sarfatti G, Simoncioli C (1975) Ultrastructural features and storage function of *Lycopersicon peruvianum* pollen. In: Mulcahy DL (ed) Gamete competition in plants and animals. North-Holland, Amsterdam, pp 19–28

Cresti M, Pacini E, Ciampolini F, Sarfatti G (1977) Germination and pollen tube development in situ of *Lycopersicum peruvianum* pollen: Ultrastructural features. Planta 136:239–247

Cresti M, Ciampolini F, Pacini E, Sarfatti G, Conini B (1979) Ultrastructural features of *Prunus avium* L. pollen tubes in vivo. 1. The compatible pollen tube. Caryologia 32:433–440

Cresti M, Ciampolini F, Sarfatti G (1980) Ultrastructural investigations in *Lycopersicon peruvianum* pollen activation and pollen tube organization after self- and cross-pollination. Planta 150:211–217

Cruden RW, Jensen WA (1979) Viscin threads, pollination efficiency and low pollen-ovule ratios. Am J Bot 66:876–879

Darwin C (1882) The variation of animals and plants under domestication. Murray, London

Dexheimer J (1970) Recherches cytophysiologiques sur les grains de pollen. Rev Cytol Biol Vég Bot 33:169–234

Dickinson HG (1970) Ultrastructural aspects of primexine formation in the microspore tetrad of *Lilium longiflorum*. Cytobiologie 1:437–449

Dickinson HG (1976a) Common factors in exine desposition In: Ferguson IK, Muller J (eds) The evolutionary significance of the exine. Academic Press, London, pp 67–90

Dickinson HG (1976b) The deposition of acetolysis-resistant polymers during the formation of pollen. Pollen Spores 18:321–334

Dickinson HG (1982) The development of pollen. Rev Cytol Biol Vég Bot 5:5–19

Dickinson HG, Bell PR (1972) The identification of sporopollenin in sections of resin-embedded tissues by controlled acetolysis. Stain Technol 48:17–22

Dickinson HG, Heslop-Harrison J (1968) Common mode of deposition for the sporopollenin of sexine and nexine. Nature 220:926–927

Dickinson HG, Heslop-Harrison J (1971) The mode of growth of the inner layer of the pollen grain exine in *Lilium*. Cytobios 4:233–243

Dickinson HG, Lawson J (1975) The growth of the pollen tube wall in *Oenothera organensis*. J Cell Sci 18:519–532

Dickinson HG, Lewis D (1973a) Cytochemical and ultrastructural differences between intraspecific compatible and incompatible pollinations in *Raphanus*. Proc R Soc Lond B Biol Sci 183:21–35

Dickinson HG, Lewis D (1973b) The formation of the "tryphine" coating the pollen grains of *Raphanus* and its properties relating to self-incompatibility. Proc R Soc Lond B Biol Sci 184:149–165

Ducker SC, Knox RB (1976) Submarine pollination in sea grasses. Nature 263:705–706

Ducker SC, Pettitt JM, Knox RB (1978) Biology of Australian seagrasses: Pollen development and submarine pollination in *Amphibolis antarctica* and *Thalassodendron ciliatum* (Cymodoceaceae). Aust J Bot 26:265–285

Dumas C, Gaude T (1982) Secretions et biologie florale II. Leurs rôles dans l'adhesion et la reconnaissance pollen-stigmate. Données recentes, hypothèses et notion d'immunité végetable. Soc bot Fr Actualités Botaniques (in press)

Dunbar A, Erdtman G (1969) On the pollen grains of *Aegiceras corniculatum* (Myrsinaceae). Grana 9:63–71

Erdtman G (1960) The acetolysis method. A revised description. Sven Bot Tidskr 54:531–564

Erdtman G (1971) Pollen morphology and plant taxonomy. I. Angiosperms, 2nd ed. Almqvist and Wiksell, Stockholm

Faegri K, Iversen J (1975) Textbook of pollen analysis, 3rd edn. Hafner, New York

Ferrari TE, Wallace DH (1977) Incompatibility on *Brassica* stigmas is overcome by treating pollen with cycloheximide. Science 196:436–438

Ford JH (1971) Ultrastructural and chemical studies of pollen wall development in the Epacridaceae. In: Brooks J, Grant PR, Muir M, Gijzel P van, Shaw G (eds) Sporopollenin. Academic Press, London New York, pp 130–173

Franchi GG, Pacini E (1980) Wall projections in the vegetative cell of *Parietaria officinalis* L. pollen. Protoplasma 104:67–74

Franke WW, Herth W, Woude JW van der, Morre JD (1972) Tubular and filamentous structures in pollen tubes: Possible role as guide elements in protoplasmic streaming and vectorial migration of secretory vesicles. Planta 105:317–341
Freeling M, Cheng DSK (1978) Radiation-induced alcohol dehydrogenase mutants in maize following allyl alcohol selection of pollen. Genet Res 31:107–129
Fritsche CJ (1837) Über den Pollen. Mem Sav Etrang Acad St Petersburg 3:649–672
Gaudichaud-Beauprie C (1826) In: Freycinet L de „Voyage autour du Monde". Botanique Paris
Godwin H (1968) Pollen exine formation. Nature 220:389
Godwin H, Echlin P, Chapman B (1967) The development of the pollen grain wall in *Ipomoea purpurea* (L.) Roth. Rev Palaeobot Palynol 3:181–195
Golynskaya EL, Bashkirova NV, Tomchuk NN (1976) Phytohemaglutinins of the pistil in *Primula* as possible proteins of generative incompatibility. Sov Plant Physiol 23:69–76
Gooday GW, Fawcett P, Green P, Shaw G (1973) The formation of fungal sporopollenin in the zygospore of *Mucor mucedo:* A role for the sexual carotenogenesis in the Mucorales. J Gen Microbiol 74:233–239
Gottsberger G, Silberbauer-Gottsberger I, Ehrendorfer F (1980) Reproductive biology in the primitive relict angiosperm *Drimys brasiliensis* (Winteraceae). Plant Syst Evol 135:11–39
Guinet Ph (1981) Mimosoideae: The characters of the pollen grains. Part 1. In: Polhill RM, Raven PH (eds) Advances in legume systematics. Royal Bot Gard Kew, pp 835–857
Guinet Ph, Le Thomas A (1973) Interpretation de la repartition dissymetrique des couches de l'exine dans les pollens composés: Consequences relatives à la notion d'aperture. C R Acad Sci Paris Ser D:276:1545–1548
Guinet Ph, Lugardon B (1976) Diversité des structures de l'exine dans le genre *Acacia* (Mimosaceae). Pollen Spores 18:483–511
Gunning BES, Pate JS (1969) Transfer cells: Plant cells with wall ingrowth, specialized in relation to short distance transport of solutes – Their occurrence, structure, and development. Protoplasma 68:107–113
Hagemann R (1976) Plastid distribution and plastid competition in higher plants and the induction of plastome mutations by nitroso-urea compounds. In: Bucher T, Neupert W, Sebald W, Werner S (eds) Genetics and biogenesis of chloroplasts and mitochondria. North-Holland Publishing, Amsterdam, pp 331–338
Hagemann R (1981) Unequal plastid distribution during the development of the male gametophyte of angiosperms. Acta Bot Soc Pol 50:321–327
Hagman M (1964) The use of disc electrophoresis and serological reactions in the study of pollen and style relationships. In: Linskens HF (ed) Pollen: Physiology and fertilization. North-Holland, Amsterdam, pp 244–250
Hamilton RI, Leung E, Nichols C (1977) Surface contamination of pollen by plant viruses. Phytopathology 67: 395–399
Hayman DL (1956) The genetical control of incompatibility in *Phalaris coerulescens* Desf. Aust J Biol Sci 9:321–331
Helsper JPFG, Veerkamp JH, Sassen MDA (1977) β-Glucan synthetase activity in golgi vesicles of *Petunia hybrida*. Planta 133:303–308
Herrero M, Dickinson HG (1980) Pollen tube growth following compatible and incompatible intraspecific pollination in *Petunia hybrida*. Planta 148:217–221
Heslop-Harrison J (1963) An ultrastructural study of pollen wall ontogeny in *Silene pendula*. Grana 4:7–24
Heslop-Harrison J (1966) Cytoplasmic connections between angiosperm meiocytes. Ann Bot 30:221–230
Heslop-Harrison J (1968a) Synchronous pollen mitosis and the formation of the generative cell in massulate orchids. J Cell Sci 3:457–466
Heslop-Harrison J (1968b) The pollen grain wall. Science 161:230–237
Heslop-Harrison J (1975a) Physiology of the pollen-grain surface. Proc R Soc Lond B Biol Sci 190:275–300
Heslop-Harrison J (1975b) Incompatibility and the pollen-stigma interaction. Annu Rev Plant Physiol 26:403–425
Heslop-Harrison J (1978) Cellular recognition systems in plants. Studies in Biology No. 100. Arnold, London
Heslop-Harrison J (1979a) An interpretation of the hydrodynamics of pollen. Am J Bot 66:737–743

Heslop-Harrison J (1979 b) Aspects of the structure, cytochemistry, and germination of the pollen of rye (*Secale cereale* L.). Ann Bot Suppl 1:1–47

Heslop-Harrison J (1979 c) Pollen walls as adaptive systems. Ann MO Bot Gard 66:813–829

Heslop-Harrison J (1980) Compartmentation in anther development and pollen wall morphogenesis. In: Nover L, Lynen F, Mothes K (eds) Cell compartmentation and metabolic channelling. Fischer, Jena and Elsevier/North-Holland Biomedical Press, Amsterdam

Heslop-Harrison J, Dickinson HG (1969) Time relationships of sporopollenin synthesis associated with tapetum and microspores in *Lilium*. Planta 84:199–214

Heslop-Harrison J, Heslop-Harrison Y (1970) Evaluation of pollen viability by enzymatically induced fluorescence: Intracellular hydrolysis of fluorescein diacetate. Stain Technol 45:115–120

Heslop-Harrison J, Heslop-Harrison Y (1980) Cytochemistry and function of the Zwischenkörper in grass pollens. Pollen Spores 22:5–10

Heslop-Harrison J, Heslop-Harrison Y, Shivanna KR (1983) The evaluation of pollen quality and a further appraisal of the fluorochromatic (FCR) test procedure. Theoret Appl Genet 67:367–375

Heslop-Harrison J, Heslop-Harrison Y, Knox RB, Howlett B (1973) Pollen-wall proteins: "Gametophytic" and "Sporophytic" fractions in pollen walls of Malvaceae. Ann Bot 37:402–412

Heslop-Harrison J, Knox RB, Heslop-Harrison Y (1974) Pollen wall proteins: Exine-held fractions associated with the incompatibility response in Cruciferae. Theor Appl Genet 44:133–137

Heslop-Harrison Y, Heslop-Harrison J (1982) The microfibrillar component of pollen intine: Some structural features. Ann Bot 50:831–842

Hesse M (1978 a) Entwicklungsgeschichte und Ultrastruktur des Pollenkitts bei *Tilia* (Tiliaceae). Plant Syst Evol 129:13–30

Hesse M (1978 b) Entwicklungsgeschichte und Ultrastruktur von Pollenkitt und Exine bei nahe verwandten entomophilen, anemophilen Angiospermensippen, Ranunculaceae, Hamamelidaceae, Platanaceae und Fagaceae. Plant Syst Evol 130:13–42

Hesse M (1978 c) Vergleichende Untersuchungen zur Entwicklungsgeschichte und Ultrastruktur von Pollenklebstoffen verschiedener Angiospermen. Linzer Biol Beitr 9:237–258

Hesse M (1979 a) Entwicklungsgeschichte und Ultrastruktur von Pollenkitt und Exine bei nahe verwandten entomophilen und anemophilen Sippen der Oleaceae, Scrophulariaceae, Plantaginaceae und Asteraceae. Plant Syst Evol 132:107–139

Hesse M (1979 b) Ultrastruktur und Verteilung des Pollenkitts in der insekten- und windblütigen Gattung *Acer* (Aceraceae). Plant Syst Evol 131:277–289

Hesse M (1980 a) Ultrastruktur und Entwicklungsgeschichte des Pollenkitts von *Euphorbia cyporissias, E. palustris* und *Mercurialis perennis* (Euphorbiaceae). Plant Syst Evol 135:253–263

Hesse M (1980 b) Zur Frage der Anheftung des Pollens an Blüten besuchende Insekten mittels Pollenkitt and Viscinfaden. Plant Syst Evol 133:135–148

Hodgkin T, Lyon GD (1982) Germination of *Lilium* and *Petunia* pollen on TLC plates and their inhibition by extracts from *Brassica* tissues. In: Mulcahy DL, Ottaviano E (eds) Pollen: Biology and implications for plant breeding. Elsevier/North-Holland, Amsterdam, pp 343–349

Hoefert LL (1969) Fine structure of sperm cells in pollen grain of *Beta*. Protoplasma 68:237–240

Horner HT Jr (1977) A comparative light- and electron-microscopic study of microsporogenesis in male-fertile and cytoplasmic male-sterile sunflower (*Helianthus annuus*). Am J Bot 64:745–759

Horner HT Jr, Lersten NR (1971) Microsporogenesis in *Citrus limon* (Rutaceae). Am J Bot 58:72–79

Horner HT Jr, Pearson C (1978) Pollen wall and aperture development in *Helianthus annuus* (Compositae: Heliantheae). Am J Bot 65:293–309

Horner HT Jr, Rogers MA (1974) A comparative light and electron microscopic study of microsporogenesis in male-fertile and male-sterile pepper (*Capsicum annuum*). Can J Bot 52:435–441

Howlett BJ, Knox RB (1983) Allergic interactions. In: Heslop-Harrison J, Linskens HF (eds) Cellular interactions. Encycl Plant Physiol 14: Springer, Berlin Heidelberg New York (in press)

Howlett BJ, Knox RB, Heslop-Harrison J (1973) Pollen-wall proteins: Release of the allergen Antigen E from intine and exine sites in the pollen grains of ragweed and *Cosmos*. J Cell Sci 13:603–619

Howlett BJ, Knox RB, Paxton JD, Heslop-Harrison J (1975) Pollen-wall proteins: Characterization and role in self-incompatibility in *Cosmos bipinnatus*. Proc R Soc Lond B Biol Sci 188:167–182

Howlett BJ, Vithanage HIMV, Knox RB (1979) Pollen antigens, allergens, and enzymes. Curr Adv Plant Sci 35:1–17

Howlett BJ, Vithanage HIMV, Knox RB (1981) Immunofluorescence localization of two water-soluble glycoproteins, including the major allergen from pollen of ryegrass, *Lolium perenne*. Histochem J 13:461–480

Huynh KL (1976) Arrangement of some monosulcate, disulcate, trisulcate, dicolpate, and tricolpate pollen types in the tetrads, and some aspects of evolution in the angiosperms. In: Ferguson IK, Muller J (eds) The evolutionary significance of the exine. Academic Press, London, pp 101–124

Inouye DW (1975) Why don't more humming-bird pollinated flowers have dark-colored pollen? Am Nat 109:377–378

Izhar S, Frankel R (1971) Mechanism of male sterility in *Petunia:* The relationship of pH, callose activity in the anthers and the breakdown of microsporogenesis. Theor Appl Genet 41:104–108

Jensen WA (1962) Botanical histochemistry. Freeman, San Francisco

Jensen WA (1974) Reproduction in flowering plants. In: Robards AW (ed) Dynamic aspects of plant ultrastructure. Academic Press, New York London, pp 481–493

Jensen WA, Fisher DB (1968) Cotton embryogenesis: The sperm. Protoplasma 65:277–286

Jensen WA, Fisher DB (1970) Cotton embryogenesis: The pollen tube in the stigma and style. Protoplasma 69:215–235

Jensen WA, Fisher DB, Ashton M (1968) Cotton embryogenesis: The pollen cytoplasm. Planta 81:206–228

Jensen WA, Ashton M, Heckard LR (1974) Pollen of Scrophulariaceae contains stacked ER in cytoplasm. Bot Gaz 135:210–218

Johnson P, Marsh DG (1965) The isolation and characterization of allergens from the pollen of ryegrass (*Lolium perenne*). Europe Polym J 1:63

Johnson P, Marsh DG (1966) Allergens from common ryegrass pollen. I. Chemical composition and structure. Immunochemistry 3:91

Karas I, Cass DD (1976) Ultrastructural aspects of sperm cell formation in rye: Evidence for cell plate involvement in generative cell division. Phytomorphology 26:36–45

Kenrick J, Knox RB (1979) Pollen development and cytochemistry in some Australian species of *Acacia*. Aust J Bot 27:413–427

Kenrick J, Knox RB (1981) Post-pollination exudate from stigmas of *Acacia* (Mimosaceae). Ann Bot 48:103–106

Kenrick J, Knox RB (1982) Function of the polyad in reproduction of *Acacia* (Leguminosae, Mimosoideae). Ann Bot 50:721–727

King TP, Norman PS, Connell JT (1964) Isolation and characterization of allergens from ragweed pollen. II. Biochemistry 3:458–472

King TP, Norman PS, Tao N (1974) Chemical modifications of the major allergen of ragweed pollen, antigen E. Immunochemistry 11:83–92

Knox RB (1971) Pollen-wall proteins: Localization, enzymic, and antigenic activity during development in *Gladiolus* (Iridaceae). J Cell Sci 9:209–237

Knox RB (1973) Pollen-wall proteins: Pollen-stigma interactions in ragweed and *Cosmos* (Compositae). J Cell Sci 12:421–443

Knox RB (1976) Cell recognition and pattern formation in plants. In: Graham CF, Wareing PF (eds) The developmental biology of plants and animals. Saunders, Philadelphia, pp 141–168

Knox RB (1979) Pollen and allergy. Studies in Biology No 107. Arnold, London

Knox RB, Clarke AE (1980) Discrimination of self- and non-self in plants. In: Marchalonis JJ, Cohen N (ed) Contemp Topics Immunobiol. Plenum, New York, 9:1–36

Knox RB, Friederich E (1974) Tetrad pollen grain development and sterility in *Leschenaultia formosa*. New Phytol 73:251–258

Knox RB, Heslop-Harrison J (1966) Control of pollen fertility through the agency of the light regime in the grass *Dichanthium aristattum*. Phyton (Austria) 11:256–267

Knox RB, Heslop-Harrison J (1969) Cytochemical localization of enzymes in the wall of the pollen grain. Nature 223:92–94

Knox RB, Heslop-Harrison J (1970) Pollen-wall proteins: Localization and enzymic activity. J Cell Sci 6:1–27

Knox RB, Heslop-Harrison J (1971 a) Pollen-wall proteins: Electron-microscopic localization of acid phosphatase in the intine of *Crocus vernus*. J Cell Sci 8:727–733

Knox RB, Heslop-Harrison J (1971 b) Pollen-wall proteins: Fate of intine-held antigens in compatible and incompatible pollinations of *Phalaris tuberosa* L. J Cell Sci 9:239–251

Knox RB, Kenrick J (1983) Function of the polyad in the breeding system of *Acacia*. In: Mulcahy D, Ottaviano E (eds) Pollen: Biology and implications for plant breeding. Elsevier/North-Holland, Amsterdam, pp 411–418

Knox RB, Heslop-Harrison J, Reed CE (1970) Localization of antigens associated with the pollen grain wall by immunofluorescence. Nature 225:1066–1068

Knox RB, Willing RR, Ashford AE (1972) Role of pollen-wall proteins as recognition substances in interspecific hybridization in poplars. Nature 237:381–383

Knox RB, Heslop-Harrison J, Heslop-Harrison Y (1975) Pollen-wall proteins. In: Duckett JG, Racey PA (eds) The biology of the male gamete. Biol J Linn Soc, vol 7, Suppl. Academic Press, London, pp 177–187

Knox RB, Clarke AE, Harrison S, Smith P, Marchalonis JJ (1976) Cell recognition in plants: Determinants of the stigma surface and their pollen interactions. Proc Natl Acad Sci USA 73:2788–2792

Knox RB, Vithanage HIMV, Howlett BJ (1980) Botanical immunocytochemistry: A review with special reference to pollen antigens and allergens. Histochem J 12:247–272

Kolattukudy PE (1980) Cutin, suberin, and waxes. In: Stumpf PK, Conn EE (eds) The biochemistry of plants, vol 4. Stumpf PK (ed) Lipids: Structure and function. Academic Press, New York, pp 571–645

Kress J, Stone DE (1982) Pollen intine structure, cytochemistry and function in monocots. In: Mulcahy GB, Mulcahy DL, Ottaviano E (eds) Pollen: Biology and implications for plant breeding. Elsevier/North-Holland, Amsterdam, pp 159–164

Kress J, Stone DE, Sellars SC (1978) Ultrastructure of exineless pollen: *Heliconia* (Heliconiaceae). Am J Bot 65:1064–1076

Kroh M, Knuiman B (1982) Ultrastructure of cell wall and plugs of tobacco pollen tubes after chemical extraction of polysaccharides. Planta 154:241–250

Larson DA (1965) Fine structural changes in the cytoplasm of germinating pollen. Am J Bot 52:139–154

Larson DA, Lewis CW (1962) Pollen wall development in *Parkinsonia aculeata*. Grana Palynol 3:21–27

Lewis D (1952) Serological reactions of pollen incompatibility substances. Proc R Soc Lond B Biol Sci 140:127–135

Lewis D, Burrage S, Walls D (1967) Immunological reactions of single pollen grains, electrophoresis, and enzymology of pollen protein exudates. J Exp Bot 18:371–378

Lewis GP, Elias TS (1981) Tribe 3. Mimoseae Bronn. (1822). Part 1. In: Polhill RM, Raven PH (eds) Advances in legume systematics. Royal Bot Gard Kew, pp 155–168

Lieux M (1980) An atlas of pollen of trees, shrubs, and woody vines of Lousiana and other Southeastern states. Part 1. Ginkgoaceae to Lauraceae. Pollen Spores 22:17–57

Linskens HF, Esser K (1957) Über eine spezifische Anfärbung der Pollenschläuche und die Zahl Kallosepropfen nach Selbstung und Fremdung. Naturwissenschaften 44:16

Lloyd FE (1910) Development and nutrition of the embryo, seed, and kernel in the date, *Phoenix dactylifera*. A Rep Miss Bot Gard 21:103–164

Maheshwari P (1949) The male gametophyte in angiosperms. Bot Rev 15:1–75

Maheshwari P (1950) An introduction to the embryology of angiosperms. McGraw-Hill, New York

Marsh DG (1975) Allergens and the genetics of allergy. In: Sela M (ed) The antigens. 3:271–359

McConchie CA, Knox RB, Ducker SC (1982a) Ultrastructure and cytochemistry of the hydrophilous pollen of *Lepilaena* (Zannichelliaceae). Micron 13:339–340

McConchie CA, Knox RB, Ducker SC, Pettitt JM (1982b) Pollen wall structure and cytochemistry in the seagrass *Amphibolis griffithii* (Cymodoceaceae). Ann Bot 50:729–732

Mepham RH, Lane GR (1969) Formation and development of the tapetal periplasmodium in *Tradescantia bracteata*. Protoplasma 68:175–192

Moore DJ, Mollenhauer HH, Bracker CE (1971) Origin and continuity of Golgi apparatus. In: Beermann W, Reinert H, Ursprung H (eds) Origin and continuity of cell organelles. Springer, Berlin Heidelberg New York, pp 82–118

Mühlethaler K (1953) Untersuchungen über die Struktur der Pollenmembran. Mikroskopie (Wien) 8:103–112

Mulcahy DL (1983) A comparison of pollen tube growth in bi- and tri-nucleate pollen. In: Mulcahy GB, Mulcahy DL, Ottaviano E (eds) Pollen: Biology and implications for plant breeding. Elsevier/North-Holland, Amsterdam, pp 29–34

Muller-Stoll WR (1956) Über das Verhalten der Exine Nonaperturater Angiosperm-pollen bei Quellung und Keimung. Grana Palynol 1:38–46

Nabli MA (1975) Mise en evidence de deux lamelles primordiales, ectexinique et endexinique, dans l'exine de quelques Labiatae. C R Acad Sci Paris Ser D 281:251–254

Nakamura N, Miki-Hirosige H, Iwanami Y (1979) On the mechanism of callose wall and callose plug formation in germinating pollen. Jap J Palynol 24:33–44
Nelson OE (1962) The waxy locus in maize. I. Intralocus recombination frequency estimates by pollen and conventional analyses. Genetics 47:737–742
Nettancourt D de (1977) Incompatibility in angiosperms. Springer, Berlin Heidelberg New York
Newman IV (1933) Studies in the Australian Acacias. II. The life history of *Acacia baileyana* (F.v.M.). Part I. Some ecological and vegetative features, spore production and chromosome number. J Linn Soc Lond Bot 49:145–171
Newman IV (1934) Studies in the Australian Acacias. IV. The life history of *Acacia baileyana* (F.v.M.) Part II. Gametophytes, fertilization, seed production and germination, and general conclusion. Proc Linn Soc NSW 109:277–313
Ogden EC, Raynor GS, Hayes JV, Lewis DM, Haines JH (1974) Manual for sampling airborne pollen. Hafner, New York
Ottaviano E, Sari-Gorla M, Mulcahy D (1980) Pollen tube growth rates in *Zea mays:* Implications for genetic improvements of crops. Science 210:437–438
Pacini E, Cresti M (1977) Viral particles in developing pollen grains of *Olea europaea*. Planta 137:1–4
Pacini E, Juniper BE (1979) The ultrastructure of pollen grain development in the olive, *Olea europaea*. 1. Proteins in the pore. New Phytol 83:157–163
Pankow H (1957) Über den Pollenkitt bei *Galanthus nivalis*. Flora 146:240–253
Pargney JC (1982) Etude ultrastructural des tubes polliniques angiospermiens: Application de quelques techniques cytochimiques. Can J Bot 60:1167–1176
Pettitt JM (1976) Pollen wall and stigma surface in the marine angiosperms *Thalassia* and *Thalassodendron*. Micron 7:21–32
Pettitt JM (1979) Developmental mechanisms in heterospory: Cytochemical demonstration of spore-wall enzymes associated with B-lectins, polysaccharides and lipids in water ferns. J Cell Sci 38:61–82
Pettitt JM (1980) Reproduction in seagrasses: Nature of the pollen and receptive surface of the stigma in the Hydrocharitaceae. Ann Bot 45:257–271
Pettitt JM, Jermy AC (1975) Pollen in hydrophilous angiosperms. Micron 5:377–405
Pettitt JM, McConchie CA, Ducker SC, Knox RB (1980) Unique adaptations for submarine pollination in seagrasses. Nature 286:487–489
Pettitt JM, Ducker SC, Knox RB (1981) Submarine pollination in seagrasses. Sci Am 244:135–143
Raff J (1981) Cell recognition in *Prunus avium* L. Ph D Thesis, Univ Melbourne
Reiss HD, Herth W (1979) Calcium ionophore A23187 affects localized wall secretion in the tip region of pollen tubes of *Lilium longiflorum*. Planta 145:225–232
Roland F (1965) Precisions sur la structure et l'ultrastructure d'une tetrade calymée. Pollen Spores 7: 5–8
Roland F (1971) Characterization and extraction of the polysaccharides of the intine and of the generative cell wall in the pollen grains of some Ranunculaceae. Grana 11:101–106
Rosen WG (1971) Pistil-pollen interactions in *Lilium*. In: Heslop-Harrison J (ed) Pollen: Physiology and development. Butterworths, London, pp 239–254
Rosen WG, Gawlick SR (1966) Relation of lily pollen tube structure to pistil incompatibility and mode of nutrition. In: Proc VI Intl Congr Electron Microscopy, Kyoto. Mazuren, Tokyo, pp 313–314
Rosen WG, Gawlick SR, Dashek WV, Siegesmund KA (1964) Fine structure and cytochemistry of *Lilium* pollen tubes. Am J Bot 51:61–71
Rowley JR (1959) The fine structure of the pollen wall in the Commelinaceae. Grana Palynol 2:3–31
Rowley JR (1960) The exine structure of "cereal" and "wild"-type grass pollen. Grana Palynol 2:9–15
Rowley JR (1962) Stranded arrangement of sporopollenin in the exine of microspores of *Poa annua*. Science 137:526–528
Rowley JR (1964) Formation of the pore in pollen of *Poa annua*. In: Linskens HF (ed) Pollen: Physiology and fertilization. North-Holland, Amsterdam, pp 56–69
Rowley JR (1971) Implications of the nature of sporopollenin based upon pollen development. In: Brooks J, Grant PR, Muir M, Gijzel P van, Shaw G (eds) Sporopollenin. Academic Press, London, pp 174–219
Rowley JR (1975) Lipopolysaccharide embedded within the exine of pollen grain. In: Bailey GW (ed) 33 rd A. Proc EM Soc Am. Claitor's, Baton Rouge USA, pp 572–573
Rowley JR (1978) The origin, ontogeny, and evolution of the exine. Proc IV Inl Palynol Conf Lucknow 1:126–136

Rowley JR, Dahl AO (1977) Pollen development in *Artemisia vulgaris* with special reference to glycocalyx material. Pollen Spores 19:169–284
Rowley JR, Erdtman G (1967) Sporoderm in *Populus* and *Salix*. Grana Palynol 7:518–567
Rowley JR, Flynn JJ (1968) Tubular fibrils and the ontogeny of the yellow water lily pollen grain. J Cell Biol 39:159 a
Rowley JR, Flynn JJ (1969) Membranes, pressure, and exine form. A reinterpretation of exine formation on aborted pollen grains. Pollen Spores 11:169–180
Rowley JR, Prijanto B (1977) Selective destruction of the exine of pollen grains. Geophytology 7:1–23
Rowley JR, Southworth D (1967) Deposition of sporopollenin on lamellae of unit membrane dimensions. Nature 213:703–704
Rowley JR, Vasanthy G (1980) Sporopollenin accumulation on the surface coating of *Cinnamomum* pollen and its influence on exine resistance. Abstr Proc 5 th Intl Palynol Conf Cambridge, p 346
Rowley JR, Mühlethaler K, Frey-Wyssling A (1959) A route for the transfer of materials through the pollen grain wall. J Biophys Biochem Cytol 6:537–538
Rowley JR, Dahl AO, Rowley JS (1980) Coiled construction of exinous units in pollen of *Artemisia*. In: Bailey GW (ed) 38 th A. Proc EM Soc Am. Claitor's Baton Rouge USA, pp 252–253
Rowley JR, Dahl AO, Rowley JS (1981 a) Substructure in exines of *Artemisia vulgaris* (Asteraceae). Rev Palaeobot Palynol 35:1–38
Rowley JR, Dahl AO, Sengupta S, Rowley JS (1981 b) A model of exine substructure based on dissection of pollen and spore exines. Palynology 5:107–152
Russell SD (1980) Participation of male cytoplasm during gamete fusion in an angiosperm, *Plumbago zeylanica*. Science 210:200–201
Russell SD (1982) Fertilization in *Plumbago zeylanica:* Entry and discharge of the pollen tube in the embryo sac. Can J Bot 60:2219–2230
Russell SD (1983) Fertilization in *Plumbago zeylanica:* Gametic fusion and fate of the male cytoplasm. Am J Bot 70:416–434
Russell SD, Cass DD (1981 a) Ultrastructure of the sperms of *Plumbago zeylanica*. 1. Cytology and association with the vegetative nucleus. Protoplasma 107:85–107
Russell SD, Cass DD (1981 b) Ultrastructure and fertilization in *Plumbago zeylanica*. Acta Bot Soc Pol 50:185–189
Russell SD, Cass DD (1983) Unequal distribution of plastids and mitochondria during sperm cell formation in *Plumbago zeylanica*. In: Mulcahy DL, Ottaviano E (eds) Pollen: Biology and implications for plant breeding. Elsevier, Amsterdam, pp 135–140
Saad SI (1963) Sporoderm stratification: The "medine", a distinct third layer in the pollen wall. Pollen Spores 5:2–39
Sampson FB (1977) Pollen tetrads of *Hedycarya arborea* J.R. et G. Forst. (Monimiaceae) Grana 16:61–73
Sampson FB (1981) Synchronous versus asynchronous mitosis within permanent pollen tetrads of the Winteraceae. Grana 20:19–23
Sassen MM (1964) Fine structure of *Petunia* pollen grain and pollen tube. Acta Bot Neerl 13:175–181
Satake T (1976) Determination of the most sensitive stage to sterile-type cool injury in rice plants. Res Biol Hokkaido Natl Agric Exp Stn 113:1–35
Satake T, Hayase H (1970) Male sterility caused by cooling treatment at the young microspore stage in rice plants. V. Estimation of pollen developmental stage and the most sensitive stage to coolness. Proc Crop Sci Soc Jpn 39:468–473
Schill R, Pfeiffer W (1977) Untersuchungen an Orchideenpollinien unter besonderer Berücksichtigung ihrer Feinskulpturen. Pollen Spores 19:5–118
Schnepf E, Witzig F, Schill R (1979) Über Bildung und Feinstruktur des Translators der Pollinarien von *Asclepias curassavica* und *Gomphocarpus fructicosus* (Asclepiadaceae). Akad Wiss Lit Mainz Math-Naturwiss Kl Tropische Subtrop Pflanzenwelt 25
Sedgley M (1979) Structural changes in the pollinated and unpollinated avocado stigma and style. J Cell Sci 38:49–60
Sedgley M (1981) Storage of avocado pollen. Euphytica 30:595–599
Shaw G (1971) The chemistry of sporopollenin. In: Brooks J, Grant PR, Muir M, Gijsel P van, Shaw G (eds) Sporopollenin. Academic Press, London New York, pp 305–350
Shivanna KR (1982) Pollen-pistil interaction and control of fertilization. In: Johri BM (ed) Experimental embryology of vascular plants. Springer, Berlin Heidelberg New York, pp 131–174

Shivanna KR, Heslop-Harrison J (1978) Inhibition of the pollen tube in the self-incompatibility response of grasses. Incompatibility Newslett 10:5–7
Shivanna KR, Heslop-Harrison J (1981) Membrane state and pollen viability. Ann Bot 47:759–770
Shivanna KR, Heslop-Harrison J, Heslop-Harrison Y (1981) Heterostyly in *Primula*. 2. Sites of pollen inhibition, and effects of pistil constituents on compatible and incompatible pollen tube growth. Protoplasma 107:319–337
Singh MB, Marginson R, Knox RB (1983) Evidence for enzyme-substrate involvement in pollen-stigma recognition and adhesion. Nature
Sitte P (1953) Untersuchungen zur submikroskopischen Morphologie der Pollen und Sporenmembranen. Mikroskopie 8:290–299
Skvarla JJ, Larson DA (1963) Nature of cohesion within pollen tetrads of *Typha latifolia*. Science 140:173–175
Skvarla JJ, Larson D (1966) Fine structural studies of *Zea mays* pollen. 1. Cell membranes and exine ontogeny. Am J Bot 53:1112–1125
Skvarla JJ, Larson DA (1975) An electron microscopic study of pollen morphology in the Compositae with special reference to the Ambrosiinae. Grana Palynol 6:210–269
Skvarla JJ, Rowley JR (1970) The pollen wall of *Canna* and its similarity to the germinal aperture of other pollen. Am J Bot 57:519–529
Skvarla JJ, Raven PH, Praglowski J (1975) The evolution of pollen tetrads in Onagraceae. Am J Bot 62:6–35
Skvarla JJ, Raven PH, Chissoe WF, Sharp M (1978) An ultrastructural study of viscin threads in Onagraceae pollen. Pollen Spores 20:51–143
Smart IJ, Knox RB (1979) Rapid batch fractionation of ryegrass pollen allergens. Intl Archs Allergy Appl Immun 62:173–187
Smith FG (1968) "Dyads" in the Proteaceae. Grana Palynol 8:86–87
Southworth D (1973) Cytochemical reactivity of pollen walls. J Histochem Cytochem 21:73–80
Southworth D (1974) Solubility of pollen exines. Am J Bot 61:36–44
Stainier F, Huard D, Bronckers F (1967) Technique de coloration spécifique de l'exine des microspore jeunes groupées en tetrades. Pollen Spores 9:367–370
Stanley RG (1971) Pollen metabolism. In: Heslop-Harrison J (ed) Pollen: Development and physiology. Butterworths, London, pp 131–143
Stanley RG, Linskens HF (1974) Pollen. Springer, Berlin Heidelberg New York
Stone DE, Sellers SC, Kress WJ (1979) Ontogeny of exineless pollen in *Heliconia,* a banana relative. Ann MO Bot Gard 66:701–730
Stone DE, Sellers SC, Kress WJ (1980) Ontogenetic and evolutionary implications of a neotenous exine in *Tapeinochilos* (Zingiberales: Costaceae) pollen. Am J Bot 68:49–63
Thomas A le, Lugardon B (1972) Sur la structure des tétrades de deux Annonacées (*Asteranthe asterias* et *Hexalobus monopetalus*). CR Acad Sci Paris D275:1749–1752
Tomlinson PB (1962) Phylogeny of the Scitamineae – Morphological and anatomical considerations. Evolution 16:192–213
Tsinger VN, Petrovskaya-Baranova TP (1961) The pollen grain-wall a living physiologically active structure. Dokl Akad Nauk SSSR 138:466–469
Van Campo M, Guinet Ph (1961) Les pollens composés: l'exemple des Mimosacees. Pollen Spores 3:201–218
Van Campo M, Lugardon B (1973) Structure grenue infratectale de l'ectexine des pollens de quelques gymnospermes et angiospermes. Pollen Spores 15:171–187
Vasanthy G, Pocock SAJ (1981) A comparative study of anomalous and normal pollen of *Rapanea:* Morphology, elemental analyses, sterility, and fungal parasitism. Pollen Spores 23:349–379
Vasil IK (1973) The new biology of pollen. Naturwissenschaften 60:247–253
Vithanage HIMV, Knox RB (1976) Pollen-wall proteins: Quantitative cytochemistry of the origins of intine and exine enzymes in *Brassica oleracea*. J Cell Sci 21:423–435
Vithanage HIMV, Knox RB (1977) Development and cytochemistry of stigma surface and response to self- and foreign-pollination in *Helianthus annuus*. Phytomorphology 27:168–179
Vithanage HIMV, Knox RB (1979) Pollen development and quantitative cytochemistry of exine and intine enzymes in sunflower, *Helianthus annuus*. Ann Bot 44:95–106
Vithanage HIMV, Knox RB (1980) Periodicity of pollen development and quantitative cytochemistry of exine and intine enzymes in the grasses *Lolium perenne* and *Phalaris tuberosa*. Ann Bot 45:131–142

Vithanage HIMV, Howlett BJ, Jobson S, Knox RB (1982) Immunocytochemical localization of water-soluble glycoproteins, including Group I allergen, in pollen of ryegrass, *Lolium perenne*, using ferritin-labelled antibody. Histochem J 14:949–966

Walker J (1974) Aperture evolution in the pollen of primitive angiosperms. Am J Bot 61:1112–1137

Walker JW, Doyle JA (1975) The bases of angiosperm phylogeny: Palynology. Ann MO Bot Gard 62:664–723

Watson L, Bell EM (1975) Surface and structural survey of some taxonomically diverse grass pollens. Austr J Bot 23:981–990

Willemse MTM (1972) Changes in the autofluorescence of the pollen wall during microsporogenesis and chemical treatments. Acta Bot Neerl 21:1–16

Willemse MTM (1981) Microsporogenesis in vivo and in vitro. Autofluorescence of pollen wall of *Lilium* and changes in pollen wall of *Gasteria* in *Lilium* anther. Acta Bot Soc Pol 50:103–110

Willemse MTM, Audran JC (1982) Transfer of developing bean pollen in lily anthers. Acta Bot Neerl 31:221–226

Willemse MTM, Reznickova SA (1980) Formation of pollen in the anther of *Lilium*. Development of the pollen wall. Acta Bot Neerl 29:127–140

Williams EG, Knox RB, Rouse JL (1982a) Pollination sub-systems distinguished by pollen tube arrest after incompatible interspecific crosses in *Rhododendron* (Ericaceae). J Cell Sci 53:255–277

Williams EG, Knox RB, Rouse JL (1982b) Pollen-pistil interactions and the control of pollination. Phytomorphology 31:148–157

Williams EG, Ramm-Anderson S, Dumas C, Mau S-L, Clarke AE (1982c) The effect of isolated components of *Prunus avium* L. styles on in vitro growth of pollen tubes. Planta 156:517–519

Wodehouse RP (1935) Pollen grains. McGraw-Hill, New York

Woude WJ van der, Morre DJ, Bracker CE (1971) Isolation and characterization of secretory vesicles in germinated pollen of *Lilium longiflorum*. J Cell Sci 8:331–351

Yamashita T (1976) Über die Pollenbildung bei *Halodule pinifolia* und *H. unninervis*. Beitr Biol Pflanz 52:217–226

Zhu C, Hu SY, Xu LY, Li XR, Shen JH (1980) Ultrastructure of sperm cell in mature pollen grain of wheat. Sci Sin 23:371–376

6 Fertilization

J. L. VAN WENT and M. T. M. WILLEMSE

6.1 Introduction

In angiosperms a reduction of the gametophyte is manifest. After meiosis small gametophytes are formed with a low number of cells on a high level of differentiation. The microgametophyte produces two sperm cells and, in the macrogametophyte, one egg cell is formed.

The development and differentiation of the gametophytes are influenced by the sporophyte. The nutrition of the gametophytes and the duration of their development is governed by the sporophyte. The same holds good for the seed development, and the initial development of the new sporophyte. The reduction of the gametophyte, and the process of seed formation are strictly related to the higher differentiation and development of the sporophyte. Due to the dominance in the stigma, style, and ovule during the interaction with the pollen and pollen tube, as well as in the process of fertilization, the sporophyte is established. Only the specific interaction between gametophyte and sporophyte in a sequence of distinct stages results in a successful fusion of the gametes. This interaction is expressed in structural and functional changes in pollen, stigma, style, ovary, ovule, and embryo sac during the progamic phase.

Because of the limited number of species examined, the recent investigations can only give an incomplete view of data, and generalization is disputable.

6.2 Stigma and Style

6.2.1 Function and Structure of the Stigma

The stigma is the receptive surface for pollen, and consists of specialized surface cells which are in connection with the stylar tissue. The morphology of the stigma and style shows great diversity and depends on the composition of the flower and, in some cases, is related to the type of pollination.

The function of the stigma is to accept the pollen and to permit germination. The stigma can provide nutrients to the pollen and direct the pollen tube growth; it is also involved in the regulation of the flower metabolism.

A stigmatic surface, wet or dry, needs a correct physiological condition for the pollen, i.e., a balanced osmolarity and a sufficient water supply.

Because of the changing environmental factors, pollination can be delayed. Stigma and style, therefore, need good vitality to retain their function for some time. In recognition and receiving signals or stimuli both the pollen coating and the stigma coating are involved.

For the further development of flower, the presence of a stigma and changes in metabolism after pollination seem to be necessary. Also, without pollination in wilting plants the stigma functions as a regulating component.

The morphology of the stigma shows great diversity. Some comparative studies (Vasil and Johri 1964), mainly several ultrastructural ones, present detailed features of some plants (Konar and Linskens 1966a). Based on the morphology of the stigmatic surface, on the amount of secretion, and the nature of the surface cells of almost 1,000 species of about 900 genera of 250 families, a general classification (J. Heslop-Harrison et al. 1975a) is suggested by Y. Heslop-Harrison and Shivanna (1977). The stigmas are divided into two categories, those with copious fluid secretion – the wet stigmas – and those with limited surface secretion – the dry stigmas.

The dry stigmas have dispersed receptive cells on multiseriate branches (group 1, e.g., most Gramineae, as *Zea, Sorghum,* and *Hordeum*), or concentrated in distinct ridges, zones or heads (group 2). The second group can be subdivided into stigmas with nonpapillate surfaces, as in *Potamogeton* sp., *Cyperus* sp., and stigmas with papilla. The last type bears unicellular (mainly Liliaceae as *Allium, Tulipa,* and Iridiaceae, as *Crocus* and *Iris,* and Juncaceae, as *Luzula*), or multicellular papillae. The multicellular papillae can be uniseriate, as in *Caliphruria,* or multiseriate, as in *Limnobium.*

The wet stigmas have a receptive surface with small-to-medium size papillae (group 3), as in some Liliaceae (*Aloë, Gasteria*), Orchidaceae (*Epipactis, Ophrys*), and other plants, as *Canna;* or a nonpapillate surface, as in *Dracaena, Alpinia,* and the cells are often necrotic at maturity (group 4).

With these parameters, a relative taxonomic uniformity in some families can be estimated. Additionally, binucleate pollen seems to be correlated to wet and dry stigmas, whereas the trinucleate pollen prefers a dry stigma.

Dumas (1978), basing it on the position of the papillae, gives a better morphological classification in terminal and lateral stigmas. A further classification can be based on the function and nature of the secreted products. Stigmas may have an abundant secretion of lipophilic or polysaccharide substances, or a less abundant secretion with the same products but also with polyphenols, tannins, or flavonoids; they may have no liquid secretion at all. These stigmas are covered with waxes or proteins. This classification can be combined with the stylar structure. Hollow styles of monocotyledons produce an exudate of mainly polysaccharides. Dicotyledons with a solid style mostly produce a lipophylic exudate. Orchidaceae, Scrophulariaceae, and Solanaceae excrete a lipopolysaccharide substance.

This last classification includes the abundancy and nature of the stigmatic exudate. It should be of interest to involve the nature and morphology of the pollen covering in its function of sticking pollen onto the stigma.

In Commelinaceae (Owens and Kimmins 1981) there is extreme variability in stigma morphology. Not all surface ornamentations, as blisters, or in case the style is splittering, as in *Cyanotis,* can be correlated with pollination. In Boraginaceae a correlation exists between pollen and stigma size, preventing some types of pollination (Y. Heslop-Harrison 1981).

A particular morphological aspect of the stigma (Linskens 1976a) is the recurved lips, branches or lobes, which are sensitive to tactile stimulation and can close. In pollinated stigmas, after the primary closure of the branches, a reopening takes place followed by permanent closure. This occurs only when pollen is present, and depends on the amount and the species of pollen. The agent is heat-sensitive and present in the pollen. The mechanism of movement depends on changes in water permeability of the plasma membrane.

6.2.2 The Nature of the Stigma Covering

The receptive surface of the stigma is formed by surface cells with a more or less glandular nature. This layer is bordered by a transition zone of "stigmatoid" tissue without glandular cells (Dumas 1978). The stigmatoid tissue, also called the "neck" of the transmitting tissue (Herrero and Dickinson 1979), in turn borders on the transmitting and cortical tissue of the style. Earlier, the term stigmatoid tissue was reserved for the whole tissue into which the pollen tube penetrates. The term transmitting tissue is now more commonly used (Vasil 1974).

The dry stigma has a cuticle covered with a pellicle. The cuticle of *Silene* has discontinuities. Obviously, a route for water and a relation to the pellicle exist. Both the pellicle and the surface of the stigma remain hydrated. When the cutinized pellicle shrinks, the cuticle is closed by the rodlets (J. Heslop-Harrison et al. 1975a).

The pellicle is an extracellular product originating from the stigmatic cells of a hydrated protein film. The pellicle is a product of the outer cytoplasmic zone of the developing papillate cells. Here microbodies are present which release their granular and fibrillar content. In older papillate cells numerous esterase-containing bodies are present in the cytoplasm. The pellicle also shows an esterase activity (J. Heslop-Harrison et al. 1975a). A wet stigma surface can also show an esterase activity (Shivanna and Sastri 1981). By gel electrophoresis the pellicle protein of *Hibiscus* can be separated into 7 bands, of which two show an esterase activity (J. Heslop-Harrison et al. 1975b).

In young papillate cells of *Raphanus* (Dickinson and Lewis 1973), dictyosome vesicles are involved in the secretion which fills the space between the cell wall and plasma membrane. In *Silene* between the cuticle and the plasma membrane a pectocellulosic wall is present (J. Heslop-Harrison et al. 1975a). According to Dickinson and Lewis (1975), the papillate wall has a layered structure. The plasma membrane borders a thick cellulosic wall, which is covered by a thick electron-dense layer. A thin electron-transparent layer follows which is bordered by another thin electron-dense layer. This thin layer bears an irregular, incomplete waxy layer. In *Secale* the papillate wall shows six layers, a pellicle, a mucilage layer containing mucopolysaccharides, a cuticle, and a thick layer with a microfibrilar content containing pectine. The last layer adjoins a layer with one- to four-linked glucans. Vesicles with proteins are present in the wall (J. Heslop-Harrison and Y. Heslop-Harrison 1980). The nature of the papillate walls shows great diversity.

The exudate of wet stigmas has been studied in some species. The main components are amino acids, lipids and antioxidants, and proteins, whereas alkaloids show a variation. In relation to pollination, the exudate may serve as a nutritive for birds, bees, beetles, bats, and moths (Baker 1973), as well as a nutrient for providing boron for some pollen (Linskens 1976 a).

The exudate of wet stigmas differs in quantity. Lipophilic substances and polysaccharides are present in general. Additionally, polyphenols, tannins or flavonoids can accumulate in the exudate (Konar and Linskens 1966 a, b, Linskens and Kroh 1970, Dumas 1978). The composition of the exudate shows many differences, and can be considered as a complex liquid. These products are formed by the papillate cells, and transported through the cell wall and pass the cuticle by ruptures of interstices. The glandular cells usually function till, or long after, pollination, but in the wet stigma type of group 4, some papillate cells are necrotic before pollination.

In the hollow style of *Lilium* the stigmatic exudate has the same composition as the product in the canal. It is an aqueous solution with a mixture of acidic polysaccharides, glycoproteins and oligo-saccharides as galactose, arabinose, rhamnose, glucuronic acid and galaturonic acid. The exudate resembles in its composition the gum exudate of plant origin (Labarca et al. 1970). In *Trifolium* (J. Heslop-Harrison and Y. Heslop-Harrison 1982) the composition of the stigmatic fluid is comparable with the fluid in the hollow style and contains glucose, sucrose, and traces of galactose and arabinose, some proteins and esterases. However, the stigma fluid contains two more glycoproteins, and the stylar fluid has one specific glycoprotein.

Two phases of secretion are reported in *Petunia* (Konar and Linskens 1966 b); this secretion contains fats. A relation between the formation of the lipophilic secretion and the endoplasmic reticulum (ER) is stated; it permeates through the cell wall by diffusion. The cuticle bursts under the increasing pressure of the exudate (Kroh 1967). Before pollination the papillate cells of *Petunia* (Herrero and Dickinson 1979) start to degenerate, and large droplets of lipids are set free in the exudate and coalesce with the other lipid droplets. So the two phases of secretion include a continuous secretion and an addition of degradation products of stigmatoid cells. The secretion of granulocrine is claimed for *Forsythia* too (Dumas 1973).

In *Forsythia* the aggregation of ER is correlated with a large number of secretory vesicles and, therefore, the ER is related to the formation of these secretory vesicles. The lipophilic glandular fluid consists of neutral lipids containing hydrocarbons with terpenes, and some methylic esters of fatty acids. Histochemical tests show the presence of flavonoidic aglycons. The exudate can function in hydration of the pollen grain, stimulation or inhibition in pollen germination, as well as in nutrition. The phenolic compounds act as antioxidant. Interaction with chemical compounds, as auxins, is possible (Dumas 1974).

Fresh exudate of *Phaseolus* (Lord and Webster 1979) contains lipids (25 mg/ml), proteins (6–12 mg/ml), amino acids, as tyrosine, proline, alanine, and cysteine; sugars, mainly glucose, reducing acids, and phenols. Alkaloids are absent. The presence of sugars can be related to the stimulation of pollen germination. The lipids as a waxy component prevent desiccation and can be an attractive element for insects. The reducing acids prevent oxidation of the lipoidal substances. The protein can function in recognition, and the amino acids as nutrient for insects. Phenols are considered as chemical protection against chewing insects and pathogens, but they inhibit pollen germination too and might be involved in the barrier of self-pollination. Primarily, the exudate regulates the water supply to the pollen.

Shivanna and Sastri (1981) draw attention to the relation that in taxa with a wet stigma and solid style the factors for pollen germination are present from an early stage of pistil development, and the exudate is not per se involved in pollen germination. In taxa with a wet stigma and hollow style the exudate is involved in pollen germination.

The nature of the stigma coverings seems very complex, and till now no generalizations can be made. In wet stigmas the composition of the exudate has drawn much attention. In dry stigmas the composition of the pellicle has been studied, but also the bordering structures of the layered papillate wall. Both should be related to the pollen – stigma interaction.

6.2.3 Function and Structure of the Style

The morphology and anatomy of the style are very variable. During the development many changes take place. In relation to its transmitting function the style has either a canal lined with a glandular epidermis, the open type, which is mostly present in monocotyledons; or it has a solid core of transmitting tissue, the solid type, mostly present in dicotyledons. In some plants there is a half-closed type in which the transmitting tissue can be limited to one side of the stylar canal (Vasil and Johri 1964, Vasil 1974). In *Petunia* Linskens (1974b, c) showed three phases in stylar development: a young style (5 mm) with strong protein and RNA increase and cell divisions in the basal meristematic zone. The next stage is 5–15 mm where cell elongation is induced by the anther. The third phase leads to the mature style, and is triggered by the vegetative system and/or floral parts, such as calyx and corolla. This classification is based on the study of stylar development.

The transmitting tissue in a solid style is either loosely arranged or has a more compact structure. This depends on the amount and quality of the intercellular substance. In ultrastructural studies of stylar transmitting tissue the composition of the intercellular substance has been studied. In a short style, as in *Spinacia* (Wilms and van Aelst 1978, Wilms 1980b), it is scanty. In other types there is a large amount. In general, the long cells of the transmitting tissue show a thick cell wall with a small lumen, or a thin cell wall with a large amount of intercellular substance, secreted by the stylar cells and present in the intercellular spaces. The composition of these intercellular substances in *Diplotaxis* is of pectine-containing substances (Kroh and van Brakel 1973), and should be considered as a wall-like structure. But in *Petunia* (Sassen 1974), in between the cells of the transmitting tissue, a secretion product is present comparable with the mucilage of the canal cells in the open style of *Lilium*. The intercellular substance in the stigmatic zone of *Petunia* (Herrero and Dickinson 1980) is also sensitive to protease and contains esterase. Along the transmitting tissue only carbohydrates, peroxidase, and acid phosphatase have been demonstrated. The carbohydrate consists of pectic substances and low molecular acid carbohydrates, as galacturon and gluc-

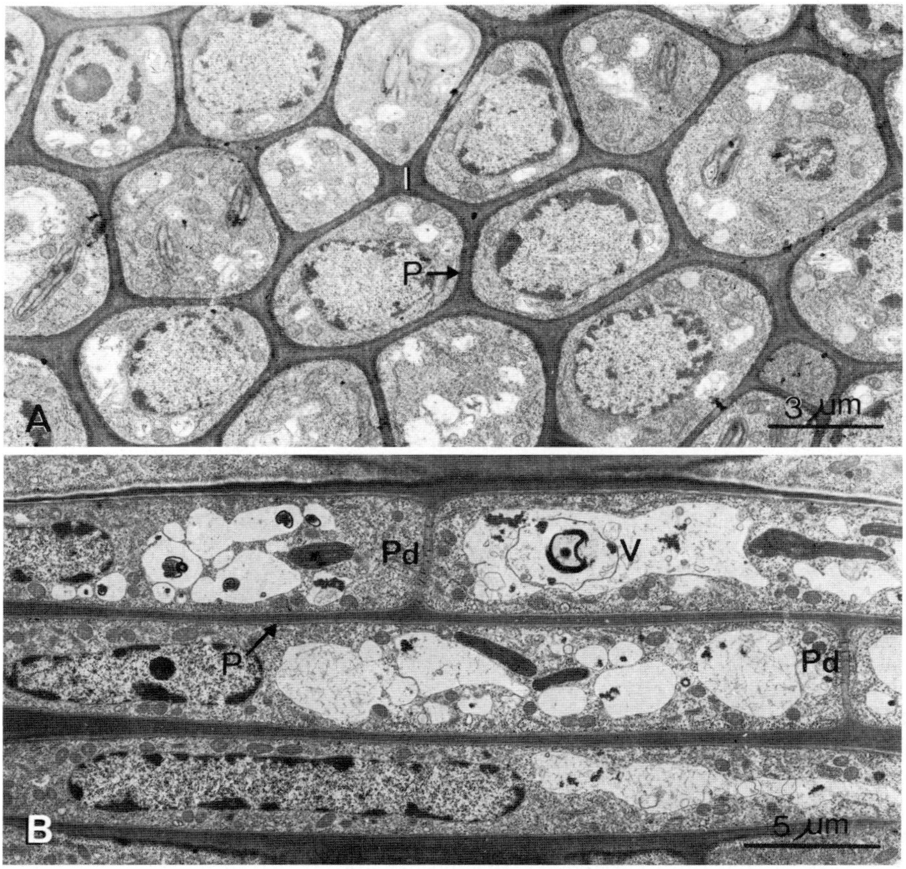

Fig. 6.1 A, B. Cross section (**A**) and longitudinal section (**B**) through an unpollinated style of *Lycopersicum peruvianum*. The cells of the transmitting tissue are elongated and surrounded by a primary wall (*P*) and intercellular substance (*I*). The cells are vacuolate (*V*) and connected by plasmodesmata (*Pd*) which results in cell rows

uronic acids. In *Gossypium* (Jensen and Fisher 1969) all cell walls of the transmitting tissue have a thick-layered structure. The innermost part consists of pectin and hemicellulose, in the second layer hemicellulose dominates, and the third layer is again rich in pectic substances and poor in hemicellulose and cellulose. The middle lamella, the fourth layer, is fibrous and the pectic substance is very dominant while hemicellulose is also present. The latter two layers react positively on staining for proteins. The pollen tube grows through these walls.

In *Lycopersicon* (Cresti et al. 1976a) the cell wall is thin and fibrillar (Fig. 6.1 A, B). The intercellular substance contains, in an early stage of development, polysaccharides of pectic nature; in the mature stage proteins are added. Comparable is also the situation in tobacco (Bell and Hicks 1976), *Malus* (Ciampolini et al. 1978), and *Prunus* (Cresti et al. 1978a). After irradiation of the style of *Prunus*, empty areas in the intercellular substance and necrotic cells indicate an inhibited or altered secretion during the development. The cells of the transmitting tissue are different from the stylar cortical cells. They are elongated and have plasmodesmatal contact in the transverse walls. The cell has a large vacuole, plastids, well-developed mitochondria, and an elaborate membrane system. Depending on the state of development, rough ER and polysomes are present. The rough ER is related to the formation of proteins in the intercellular substance during a later stage of development, as demonstrated in *Lycopersicon* (Cresti et al. 1976a).

Some studies report presence of special cellular structures or compounds, for instance phytoferritin related to senescence (Cresti et al. 1978b), calcium (Jensen and Fisher 1969) as mineral supply, and crystal-containing bodies (Bell and Hicks 1976) perhaps related to abscission. In the ontogeny of the style of *Petunia,* a change in protein pattern has been reported but there is no qualitative change in isoenzyme pattern of phosphatases, esterases, and peroxidases (Sastri and Shivanna 1980). During cell degeneration a callose deposition in the pit fields or, in extreme situation, along the whole surface of the wall takes place. In *Persea* (Sedgley 1977) it is suggested that the reduced pollen tube growth in the pistil is associated with the presence of callose.

In an open type style as *Lilium* (Dashek et al. 1971), secretion appears in the stylar canal. The canal cells have dense cytoplasm and are rich in organelles. The secretion product is a layered wall structure of pectin, a nonesterified pectin complex with protein and cellulose. This wall shows some similarities with the products in the canal. The cell membrane of the canal cell contains proteins and glycoproteins in paramural bodies. The cell can have more nuclei, and has a polar localization of organelles. A relation between a specific cell organelle and the secretion product is evident. Probably the bordering parenchyma cell produces the secretion, and the canal cell serves as transfer cell (Vasil 1974, Rosen 1975). In experiments with labelled glucose and myoinositol, an uptake in the canal cells and transport to the pollen tube wall has been demonstrated (Loewus and Labarca 1973). The uptake of the myoinositol in the intercellular substance and pollen has also been reported by Kroh (1973), and Kroh and Helsper (1974) for *Petunia.*

In *Gladiolus* (Clarke et al. 1977) the canal is covered by a cuticle which prevents the dehydration of the mucilaginous excretion. This cuticle remains intact, and the pollen penetrates the cuticle and grows through the secretion. In *Lilium,* at mature stage, the cuticle collapses.

The open style canal cells provide elements for pollen tube growth and, in this function, their secretion products can be compared with the intercellular substances of a solid style. However, the few but detailed data and differences suggest many more implications. To these different situations the condition for pollen tube growth should be worked out and compared. In stylar structure and function, an interesting field of research is the study of stylar development and growth.

6.3 Pollen Germination

6.3.1 Sticking and Hydration

Pollens are transported to the stigma by wind, animals, water or directly by the contact between open anther and stigma. The pollen dimension, mass, and sculpture reflect the mode of transport. The species-specific exine pattern can be covered by tryphine consisting mainly of the Pollenkitt and remnants of the locule fluid.

The first interaction between pollen and stigma is sticking. Adhesion results from the high viscosity and sticky nature of the exudate (Konar and Linskens 1966b). Factors in sticking in sequence of priority are: the surface tension, wind force, electrostatic force, electrodynamic force, gravity, and inertial force. Mainly, the extent of wetness of pollen, or stigma, determines the sticking. In such a situation the rate of sticking hardly depends on the composition of the fluid, the volume, mass or morphology of the pollen. Drying out of stigmas during aging causes the fall of pollen because of loss of surface tension and action of, for instance, a wind force (Woittiez and Willemse 1979).

In plants with two different pollen and stigma types, as in *Armeria maritima,* the shape of both determines successful pollination.

In *Brassica* incompatible pollen shows a less firm bond in sticking than compatible pollen. The duration of stay of incompatible pollen is shortened by a change in mobility of the proteinaceous component of the pollen coating (Stead et al. 1979).

Connected with sticking is the uptake of water by colloidal inhibition and endosmosis. All pollens swell in the presence of water indicating its rapid uptake. The water passes through the exine, and is taken up by the intine and the cell. Because of the desiccated condition of mature pollen, this rehydration takes place mainly along the contact site with preference of the colporal or poral zone, or through the canals in the exine. A hydrodynamic phenomenon based on the study of pollination in rye, and an early exudation of liquid by the pollen, have been observed. This outward movement is important for the release of products present in the intine. This exudation has a short duration and is localized in the apertures or pores. This type of water transport is not in accordance with normal membrane transport, or diffusion of water. It is suggested that a special plasma membrane condition as a nonstable bilayer is present. Immediately after rehydration, the plasma membrane is rebuilt. In the meantime, the exudation takes place because of the loss of an effective control by a normal plasma membrane, and a constant presence of stigmatic fluid. In this condition the pollen, without a normally functioning plasma membrane, swells quickly to a maximum water capacity and a further passage of water from the stigma to the grain than results in exudation. This hydrodynamical process strongly depends on the condition of the cytoplasm of the vegetative cell, and the thickness of the intine because of its imbibition capacity. However, the full hydration of pollen is not necessarily a prerequisite for germination. The uptake depends on the relative humidity of the surrounding atmosphere (J. Heslop-Harrison 1979). The uptake of fluoresceine diacetate in pollen affirms a dissociation of the membrane of the vegetative cell (Shivanna and Heslop-Harrison 1981).

When pollens stick and are rehydrated, the fluids of the pollen and stigma surface get mixed. Depending on the nature of these substances, the possibility of a separation of different products exists.

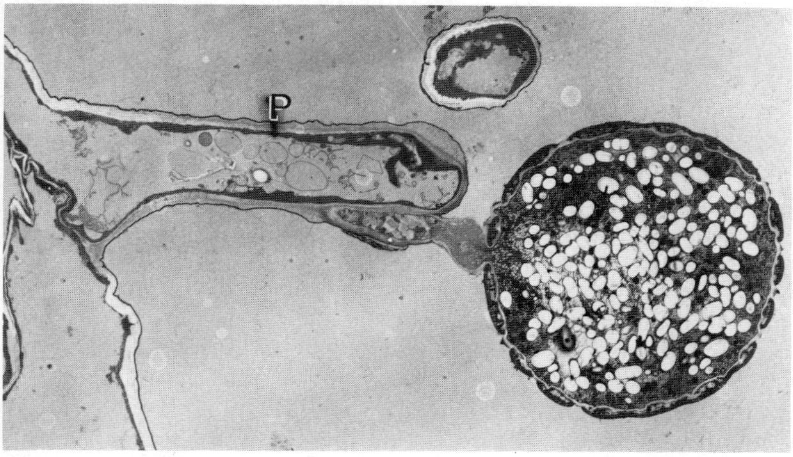

Fig. 6.2. Germinated spinach pollen on a papillar cell (*P*). Note the penetration of the pellicle, cuticle and papillar cell wall. The pollen loses contact with the pellicle. (After Wilms 1980 a)

6.3.2 Pollen Tube Formation and Growth

In vivo as well as in vitro pollen germination depends on several internal and external factors. Mature pollen shows significant differences in structure and function. From comparative studies it is known that the pollen development can reach various end stages. The tri- or binucleate pollen condition of mature pollen represents such a variation. But there are also the differences in the rate of vacuolation, and the kind and quantity of storage, such as starch or lipids, which lead to a classification of mature pollen in several groups (Baker and Baker 1979). Important is the state of the mitochondria, especially their number of cristae. Few cristae are indicative of a low metabolic activity (Hoekstra 1979). Even more important is the condition of the protein-synthesizing system, and the presence of the mechanism for pollen tube formation which includes production of plasma membrane and wall.

On this condition of the mature pollen depends whether a quick or slow germination on the stigma will follow. If the pollen is fully developed, as in the case of most trinucleate pollen, germination will start quickly. In less-developed pollen the stigma can provide an additive condition for germination. In all cases water or a high relative humidity is necessary; in some cases even a food supply, mostly carbohydrates, and minerals should also be present (Fig. 6.2).

After sticking, the pollen germination starts with the imbibition phase, which is usually followed by a lag phase, tube initiation and elongation (Linskens and Kroh 1970). From pollination experiments on young stigmas it is concluded that germination depends only to a low extent on the condition of the stigma.

Compiled from in vitro studies (Vasil 1974, Mascarenhas 1975, Stanley and Linskens 1974, Shivanna et al. 1979), pollen activation precedes tube formation. This includes an initial high respiration, the development of mitochondria to an active state, the utilization of the pollen storage products, and an enlargement of the membrane system as dictyosomes, for the formation of the plasma membrane and pollen tube wall.

In vivo, the lag-time period before germination in *Lycopersicon* (Cresti et al. 1977 a) involves activation of dictyosomes, development of RER cisterns and formation of polysomes, both related to the proteins. The activation of dictyosomes leads to production of small and large vesicles. The large vesicles are related to callose formation and deposition at the nonfunctional pores and, subsequently, against the new cell wall. Near the germinal pore both large and small vesicles are present. The small ones are related to the first formation of the pollen tube wall which opens the pore as a door. This process depends on the developmental state of the pollen, since differences in developmental state are related to lag-time phase, as observed in *Oenothera* (Dickinson and Lawson 1975 a).

The formation of the pollen tube involves a growth zone, vacuolation, and plasma streaming. The growth zone is the tip where new plasma membrane is formed, and a wall with the characteristics of a primary wall appears. This tip contains many organelles, mainly dictyosome vesicles which fuse with the plasma membrane excreting wall precursors and proteins (Engels 1974). In some species the ER is involved in this process. This zone is also characterized by the presence of ribosomes and polysomes. Microtubules are absent (Dickinson and Lawson 1975 b). This type of growth is comparable with that of the fungal hyphae (Chang and Trevithick 1974). Here the tip zone permits the excretion of proteins and enzymes. Germinating pollen also excretes enzymes (Stanley 1971, Vasil 1973, Stanley and Linskens 1974), among other products, probably at the tip zone mainly. The pollen tube wall consists of one outer pectic layer and an inner callosic layer, which originates in *Lilium* from the polysaccharide particles in the pollen tube (Miki-Hirosige and Nakamura 1982).

The cytoplasm in the remaining part of the pollen tube can be divided into different zones as observed in cotton (Jensen and Fisher 1970), *Impatiens* (van Went 1974), *Lilium* (Rosen 1971 a), *Lycopersicon* (Cresti et al. 1977 a), and *Prunus* (Cresti et al. 1979 b). Directly behind the tip zone dictyosomes and, in some pollen tubes, SER are present. Upon this zone mitochondria and, sometimes, a second type of vesicles become visible. The presence of vacuoles marks the following zone; here the vegetative nucleus and the generative cell or sperm cells are also present. The vacuole is formed by the second type of vesicles, tonoplast is produced (Coccuci 1973). The rest of the pollen tube contains a large vacuole and a thin layer of cytoplasm. In this part the streaming is usually manifest.

The transport of the vegetative nucleus and the generative cell is related to the growth of the tube and the cytoplasmic streaming. This streaming is caused by microfilaments (Franke et al. 1972, Herth 1978). The direction of the streaming is the result of interconnection between microfilaments, microtubules, and endoplasmic reticulum. In transport of the vegetative nucleus and generative cell fibrillar material near these structures is probably responsible for the movement (Cresti et al. 1976 b).

The tube growth can start after imbibition and activation within a very short period. This includes the possibility to form a new plasma membrane and a cell wall, and to start vacuolation. In quickly germinating pollen these processes are due to the presence of vesicles for the tip growth, and another type of vesicles for the formation of the vacuoles by fusion and uptake of water (van Went 1974, Dickinson and Lawson 1975 b).

In wheat Chandra and Bhatnagar (1974) observed the beginning of germination and initiation of the pollen tube. The sperm cells lay near the germinal pore and emerged before the vegetative nucleus. In growing pollen tubes the vegetative nucleus normally lies near the tip, and the generative cell or sperm cells follow. This means that in the fist stage of tube formation the position of the vegetative nucleus is reversed and, probably, the stratification of the tube is not yet complete.

Closely behind the tip zone an additional cell wall layer is formed, containing callose. Depending on the length of the pollen tube and the rhythmic growth (Korobova 1974), callose plug formation takes place (Cresti and van Went 1976). The region of callose plug formation is marked by electron density of the ground plasm, absence of plastids and dictyosomes, and presence of lipid bodies. The RER is involved in the synthesis of callose. Vesicles or spheres containing callose fuse with the plasma membrane. The plug formation shuts off part of the pollen tube. During pollen tube growth the cytoplasmic position and the number of cell organelles remain fairly constant. Pollen germination and pollen tube growth are processes of a free gametophytic plant. As can be concluded from in vitro experiments, the metabolism involved is complex and species-specific. The order in the pollen tube represents the zones of functions which seem independent of the surrounding tissues.

6.3.3 The Sperm Cell

The generative cell forms two sperm cells. The generative cell, as the result of an unequal division of the microspore, has few organelles. In some plants plastids are absent in the sperm cells (Hagemann 1979). The generative cell wall is initially continuous with the intine, but in most pollens the cell is set free in the cytoplasm of the vegetative cell.

In *Secale* (Karas and Cass 1976), during the cytokinesis, the generative cell is attached to the pollen wall and cell plate formation is clearly visible. The two sperm cells have large nuclei. The nucleus is surrounded by a layer of cytoplasm with mitochondria, dictyosomes, endoplasmic reticulum, and parallel arrays of microtubules. A cell wall around the round cell is absent. In *Heleocharis* (Carniel 1972) the sperm nuclei are fusiform and, after cell division, two fusiform cells appear but with a flat ending in the zone of separation. In *Gossypium* (Jensen and Fisher 1967) the nuclei are small with a small nucleolus. In the cytoplasm are conspicuous dictyosomes which produce vesicles and numerous polysomes. In *Plumbago* an unequal distribution of mitochondria and plastids is present in the generative cell. After division, the association with the nucleus of the vegetative cell persists. The sperm cell connected with the nucleus of the vegetative cell contains majority of mitochondria, the other the plastids (Russell 1981). In studies of the movement of sperm cells, the cellular connections of the sperm cells and with the nucleus of the vegetative cell should be included.

Only a few reports deal with the division of the generative cell in the pollen tube. The division is slow and, in some plants, a spindle seems to be absent (Stanley and Linskens 1974). The sperm cell can change its shape in response to conditions of squash preparations (Cass 1973). In *Beta* (Hoefert 1969a, b, 1971) microtubules are present oriented parallel to the long axis of the sperm cells. In *Hordeum* (Cass 1973) these microtubules are considered as elements of a cytoskeleton which preserve the cell shape.

The more or less fusiform shape of the cell permits them to move into the pollen tube. The movement of the generative cell or sperm cell is a result of cytoplasmic streaming (Navashin 1969). The direction of the movement depends on the existence of two opposite cytoplasmic flows. The streamlined shape of the generative cell permits to move toward the tip with the central streaming because of its position in the tube. The small end of the cells is directed toward the pollen grain, and the other broader end toward the tip.

Based on observation of living pollen of *Zea,* which germinate and form a tube in vitro, Korobova (1974) suggests that the independent movements of the vegetative nucleus and the sperm cells are related to a shift of the dynamic center of the cell. Because of the cytoplasmic movements, the forces which fix the nucleus in the cell are continuously changing. The position of the nuclei and sperm cells should be considered as dependent on, or a result of, various cellular forces.

The movement of the sperm cells is a cytological problem, on the same level as the more or less fixed position of a nucleus in a cell. In this respect the microtubules and filaments (Franke et al. 1972), and the band of fibrils in the cytoplasm (Cresti et al. 1976b) should be considered in a model of movement. But the influence of the current from base to the tip, as is present in the pollen tube of *Lilium,* should not be excluded (Weisenseel et al. 1975). The mechanism of movement is still not clear.

6.4 Pollen-Pistil Interaction

6.4.1 Pollen Recognition, Acceptance, and Rejection

After pollination of the wet stigma of *Cosmos,* the pollen is covered by a protein-lipid and carbohydrate coating (Knox 1973). In this situation products held by the exine of the pollen mix with the stigmatic exudate. In dry stigmas the pollen sticks on the pellicle which makes contact with the pollen covering. As has been demonstrated in *Silene* (J. Heslop-Harrison et al. 1975b), the papillate cells react on pollination with a change of permeability, plasma streaming, and plasma membrane.

In rye, after the initial contact, a penetration of pollen wall proteins could be proved by fluorescence microscopy. After 10 min the tips of the papillate cells are reached, after 24 h the cytoplasm of these cells also shows an intensive fluorescence (Vithanage and Heslop-Harrison 1979). In response to pollination of *Brassica* (Stead et al. 1980) the proteins in the exudate show a high turnover. Some proteins that are released from the pollen wall can be involved in immunological reactions. These types of proteins, the antigens, are considered as functional in recognition and may also be present in other related species (Knox 1973, Knox et al. 1975).

Table 6.1. Possible steps and events in pollen-pistil interaction after pollination

	Steps	Events
Pollen-stigma interaction	Contact pollen coating with stigma coating	Hydrophylic tryphine and lipophylic Pollenkitt in contact with proteinaceous pellicle or exudate result in sticking reactions. Outflow of exine components
	Attachment of pollen to stigma	Signals of recognition, such as by glycoproteins, by electrochemical way, or by membrane components as lipoproteins. Changes of stigma coating, as activation of esterase
	Acceptance of pollen by stigma	Imbibition and hydration of pollen. Outflow of pollen-wall components Signals for acceptance, such as by polypeptides or by membrane components as lipoproteins
	Rejection of pollen	No hydration of pollen. Callose formation in stigma surface cells
Germination	Activation of pollen	Activation of pollen metabolism, excretion of enzymes affecting the stigmatic surface, as cutinase
	Germination of pollen	Formation of the pollen tube, closure of non-functional pores. Production and excretion of enzymes
	Rejection of pollen tube	Possible growth of the tube stops after contact with the stigmatic surface
Pollen tube growth	Penetration in the stigmatic tissue	Zonal organization of pollen tube cytoplasm and vacuolation. Dissolution of intercellular substance and cell walls. Conversion and use of dissolved materials. Use of products of stigmatic tissue
	Directed growth of the pollen tube	Changes in storage products of the pollen and stylar cells in the transmitting tissue. Stylar cell degradation. Attraction of pollen tube, callose plug formation. Signal of acceptance probably by proteins
	Growth of pollen tube stops	Pollen tube growth stops after rejection, and lacks nutrition

In the recognition system the presence of some complementary products is usually involved. Enzymatic removal of the pellicle proteins in *Agrostemma* (J. Heslop-Harrison and Y. Heslop-Harrison 1975), *Crocus* (Y. Heslop-Harrison 1977), and *Raphanus* (Dickinson and Lewis 1975) prevents pollen tube entry. Also, concanavalin A binding compounds in the pellicle function as an adhesive for pollen (Linskens 1980a). These results support the idea of Christ (1959) that an inactive precursor present in the pollen is activated by a factor in the stigma. Another quick reaction after pollination is the enhancement of the esterase activity of the stigma.

The recognition is a stage necessary for pollen acceptance and germination. The recognition of an acceptable pollen is followed by pollen activation. The pollen starts to germinate and produces enzymes for penetration, as cutinase (Linskens and Heinen 1962), and for nutrition. The presence of such enzymatic activity is demonstrated by the prints made by the pollen on the papillate surface (J. Heslop-Harrison 1975a). In rejection the pollen germination or tube growth is blocked. Several factors are involved in the recognition reaction (Shivanna 1982). Table 6.1 summarizes the possible different steps and events in stigma and style.

6.4.2 Pollen Incompatibility

Pollen incompatibility reactions after pollination in angiosperms are manifest either in stigma, style, or ovule. The pollen does not germinate, tube growth is blocked on the stigma, in the style, in the ovary. Or, the pollen tube stops near the embryo sac, and fusion of sperm cell does not take place.

These phenomena and the suppression of the incompatibility are extensively described by Townsend (1971), Frankel and Galun (1977), de Nettancourt (1977), Shivanna et al. (1979), Linskens (1975a, b, 1976b), and Linskens et al. (1970), and is the main theme of the *Incompatibility News Letter*. The incompatibility is genetically based on one S-locus with many alleles, or has two (S and Z) or more genes. The expression of the gene activity can be manifest in the sporophyte or the gametophyte.

In the sporophytic incompatibility the products, probably of proteinaceous nature, are formed in the tapetum and added to the exine of the pollen grain. These products are set free on the stigma, and are involved in a complementary reaction (J. Heslop-Harrison et al. 1973).

In the gametophytic incompatibility the products involved are formed in the pollen during their development. These products are stored in the intine and are released during pollination on the stigma and act during or shortly after pollination. The incompatibility reaction is still under study, which means that it cannot be excluded that other sporophytic tissues such as stigma, style or ovule, as well as the embryo sac, are involved in the production of incompatibility substances.

In many families with gametophytic incompatibility the wet stigma dominates, and pollens are mostly binucleate. In sporophytic incompatibility the dry and papillate stigmas are more common but this is not true for Gramineae. The pollens are mostly trinucleate (Y. Heslop-Harrison and Shivanna 1977).

In plants with a wet stigma and solid style the recognition of pollen takes place in the stigma. In species with a wet stigma and hollow style the recognition is confined to the style (Shivanna 1980). In Poaceae with a dry stigma the contact with the pollen tube tip and stigma seems necessary, and there the recognition takes place (Thomas and Murray 1975, Shivanna et al. 1978b). In grasses (J. Heslop-Harrison 1982) the tip of the pollen tube is a very important place in the recognition reaction. The inhibition of the pollen tube growth occurs with a strong or weak reaction after contact with the pollen tube tip and the papillar cell. After the strong reaction the primary effect is a dislocation of the wall materials on the tip. The weak reaction is comparable with those in gametophytic systems where inhibition occurs in the transmissing tissue of the style (Shivanna et al. 1982).

The role of the pollen-coat is not clear. Washing out of the pollen-coat proteins of *Lilium, Pyrus,* and *Vicia* does not affect pollen germination (Shivanna 1978). But in *Brassica* different types of washing have a clear influence (Roggen 1974). A function of the *Brassica* tryphine, which contains many lipids, in recognition can be restricted to a preparation of a nonpolar environment for an interaction process (Roberts et al. 1979a). In *Lilium* Dickinson et al. (1982) stated that the recognition and response occurs in the style, but this is not a result of the inability to take up materials from the style. The suppressed growth of the tubes may be caused by the loss of the turgor pressure within the tube. Glycoproteins in the stylar fluid are possibly the major source to offer osmotically active molecules, which is inhibited in incompatible tubes.

In dry stigmas the pellicle in contact with the pollen coating creates the condition for the recognition reaction. In pollen-wall extracts of *Iberis* (J. Heslop-Harrison et al. 1974, 1975a) the proteins or glycoproteins are supposed to be involved in the incompatibility reaction. In *Brassica* Roberts et al. (1979b) report a glycoprotein in an extract of papillae of mature stigmas. This glycoprotein is absent in the immature stigma, which does not show an incompatibility reaction. The pollen-coat factor should react with the stigmatic glycoprotein, and a suppression of the pollen development takes place. The nature of the pollen-wall factor in *Brassica* is unknown, and there are no differences in the pattern between compatible and incompatible pollen-wall proteins. The presence of a special stigmatic glycoprotein has been reported earlier in studies on *Brassica* (Ockendon 1972, J. Heslop-Harrison 1975b, Nishio and Hinata 1977). The production of the stigma factors concerned with incompatibility is not always synchronous with the factors concerned in germination (Shivanna et al. 1978a).

In dry stigmas the rejection reaction results either in failure of pollen to germinate, or the occlusion of short pollen tubes with callose, or the failure to penetrate into stigma and style. The stigmatic papillae react with cell activation and production of callose in the papillate cells near the pollen or pollen tube.

The deposition of callose in *Raphanus* is possibly related to the perforation of the cuticle by the pollen tube, and can be compared with a wounding effect (Dickinson and Lewis 1973). The protein fraction of the tryphine causing the callose reaction consists of a protein with low molecular weight (Dickinson and Lewis 1975).

In wet stigmas with gametophytic incompatibility the pollen tube growth is blocked in the stigmatoid, or in the transmitting tissue. In the transmitting tissue the pollen tube growth ceases and callose is deposited at the tip, or the tip swells and bursts. The place of this reaction can be near the stigmatic tissue, as in *Oenothera* and *Alopecurus,* or on the upper half of the styles, as in *Petunia* or *Lycopersicum* (J. Heslop-Harrison 1978). During hydration of the pollen on the stigma proteins are released. The proteins present in the intine are considered to function in the recognition system of gametophytic incompatibility.

On ultrastructural level, the cytoplasm of the pollen tube of *Petunia* shows concentric ER, indicative of a block in protein synthesis, few polysomes, and numerous amyloplasts with phytoferritin (Cresti et al. 1979a). After incompatible pollination the stylar tissue shows more or less no change in reserve substances, and is comparable with stylar tissue after pollination prior to passage of the pollen tubes (Herrero and Dickinson 1979). In *Lycopersicum* the tube tip bursts reaching the zone of blocking, and an accumulation of particles appears in the tube, which consists of a mixture of proteins and substances for wall production. In these tubes concentric ER has also been observed (de Nettancourt et al. 1973 a, b, 1974). The ultrastructure of gamma-irradiated compatible pollen tubes shows a great similarity (Cresti et al. 1977 b).

In physiological studies, changes in enzymes (Bredemeijer 1974, 1979), glycoproteins (Linskens et al. 1970), amino acids (Linder and Linskens 1972), and glycosphingolipids (Coustaut et al. 1978, Delbart et al. 1980 a, b) are related to the process of incompatibility. The ideas on pollen–pistil interaction are mainly focussed on RNA and protein metabolism. Van der Donk (1975) proved a reaction of the RNA protein metabolism in the style of *Petunia* on pollination, and postulated a recognition reaction based on the polypeptides in the style and pollen.

As components acting in the signal polypeptides are involved, the question of the pathway of the signal still remains open. It is known that glycosphingolipids are relatively tissue-specific substances, and receptors for a variety of molecules (Delbart et al. 1980 a–c). It may be that fatty acids react with molecules in membranes, and play a role in recognition signals. In *Lilium* styles another mode of signalling was observed by means of an electrophysiological signal (Spanjers 1978). There are differences in the type of this signal in relation to the type of pollination. It seems that special vessel-like cells situated opposite to the xylem of the vascular bundle are involved.

For the stigma–pollen interaction several hypotheses have been proposed to specify the mechanism of recognition. The products of the S-genes are considered as complementary or opposed to each other. Dumas and Gaude (1981) compared the different models, and included some data on the presence of lipids to propose a hypothesis which is focused on both pollen-stigma adhesion receptors as lipids, and proteins which are complementary between pollen and stigma. The attachment of the pollen-coat is followed by hydration of the stigma pellicle and tryphine through the stigmatic cells. After acceptance the pollen is hydrated and a second recognition reaction of the pellicle-tryphine contact follows. After pollen hydration the second recognition reaction determines germination and when a tube is formed the penetration of the pollen tube follows. After rejection, hydration of pollen does not occur. These models are mainly restricted to the first step in pollen–stigma interaction. The stigmatic surface is involved in different functions including sticking, recognition, acceptance or rejection, and serves in pollen nutrition. The different steps in pollen and stigma interaction show a complex and programmed process. Only a few compounds, as proteins and glycoproteins in both the pollen and stigma covering are, in particular, involved in present studies; the function of other elements as lipids and carbohydrates need more attention. In the complex process of interaction the knowledge of pollen as well as stigma and style development should be involved, and will give a more complete view, too. During the rejection reaction in the style many changes in pollen and stylar metabolism are stated. The ordered connection of all these phenomena in relation to the incompatibility as well as the normal stylar metabolism has not yet been discussed. Also, the incongruency barrier, based on a lack of genetic information and the absence of coordination (de Nettancourt 1977, Linskens 1980 b, Shivanna

1982) should be involved. Barrier mechanisms without a relation to S-alleles, which function in incompatibility reactions and are considered to be based on other factors, are called incongruence. In plants resulting from an interspecific crossing as *Brassicoraphanus* (Dolstra 1982), the events and pollen–stigma interaction are similar to those of the parents.

6.4.3 Entry of Pollen Tube into Stigma and Style

After recognition and acceptance, the pollen tube is directed to the stigmatic surface and penetrates the stigma and stigmatic tissue. This ingrowth involves the excretion of cuticle and cell wall-affecting enzymes. In contact with the stylar tissue a nutritive relation develops. In open styles the pollen tube does not penetrate the stigmatic tissue.

Before pollination of *Petunia* (Herrero and Dickinson 1979) the stigmatoid or neck cells show a special type of wall contact, formed by ridges of walls which fit in a complementary furrow. These cells contain abundant ER and ribosomes. Between the stigmatoid cells the presence of carbohydrates and proteins, as demonstrated by protease digestion, and also enzymes as acid phosphatase, peroxidase and especially esterase, could be established. After pollination, not dependent on the incompatibility reaction, the stigmatoid cells contain an increasing number of polysomes. For a short period cytoplasmic inclusions, the embayments, associated with the plasma membrane, are added to the intercellular substance. Near the pollen tubes the cells can become necrotic, and start to degenerate after the passage of tubes. In this region, initially, the amount of starch is high, but one day after pollination it decreases, especially after a compatible pollination caused by the passage of the pollen tube. The same holds good for the lipid droplets. Before pollen tube arrival a synthesis probably of polypeptides and carbohydrate starts in the stigmatoid cells, and these products are excreted in the intercellular substance. The starch in these cells is used up during the passage of the pollen tube. Starch degradation provides free sugars for the pollen-wall formation and respiration. In incompatible crossing the latter process is blocked.

During pollination and pollen tube growth three phases of synthesis are distinct in the stigmatoid zone, the first before pollination, a second stimulated by pollination involving cellular excretion and, finally, a phase of transfer of nutritive substances to the pollen tube. During pollen germination the nutrition comes from internal as well as external sources. It should be emphasized that the pollen storage products can also function in animal transport, because of their use as nutrition by insects (Baker and Baker 1979).

During pollen germination in *Oenothera* (Dickinson and Lawson 1975a) the pollens are bound with threads of viscin to the stigmatic surface. This surface has a saturated lipid and droplets of proteins and carbohydrates in a layer of viscous fluid on papillate secretory cells. A region of alveolar parenchyma lies beneath the secretory cells, and the subjacent layer has loosely packed parenchyma cells coated with a fine film of carbohydrates and protein. These cells have very little starch content. The pollens contain carbohydrates as well as different types of membrane-bound bodies and vesicles. During tube formation a change in the membrane-bound bodies and the vesicles, in association with the mitochondria, takes place and vacuolation starts. During pollen tube growth the carbohydrate content remains constant. Compared with the incompatible pollen tube growth less carbohydrates are present, a difference in wall formation and change in membrane-bound bodies and vesicles occurs, callose plugs are formed and the growth is slow and finally stops within 24 h. The lack of free carbohydrate in the compatible pollen inhibits its growth. The role of surrounding tissue and the exudate is not clear. In *Persea* (Sedgley and Buttrose 1978), before pollination, the stigma and stigmatoid cells begin to become necrotic.

In *Petunia* (Herrero and Dickinson 1979), *Crocus* (Y. Heslop-Harrison 1977), *Oenothera* (Dickinson and Lawson 1975 a) and many other plants, the pollen tube grows intercellularly.

In barley (Cass and Peteya 1979) and *Spinacia* (Wilms and van Aelst 1978), on the dry stigmas the pollen tubes grow at first along the surface of the hairlike cells, and then penetrate its wall. In barley the pollen tube utilizes its own reserve substances for penetration. Following penetration of the cell wall, a thickened and looser texture of the wall is visible after the excretion of enzymes by the tube tip. In *Spinacia* the sticking of the pollen-coat to the pellicle is stable till germination elevates the pollen grain. The pollen tube penetrates the pellicle and cuticle, and grows either in the outer cell wall layer or along the inner side of the cell wall. In *Zea mays* (Kroh et al. 1979) the penetration of the silk via the multicellular hairs is restricted to two opposite sides of the silk, near the place of the vascular bundles and the transmitting tissue.

Pollen tube growth is a directed growth. On the stigmatic surface the tube grows to the transmitting tissue, and in this tissue to the ovary and then to the ovule.

In experimental studies on attraction, the major techniques used are the surface test, the angle test, and the depression test. All these tests present some difficulty in interpretation (Shivanna et al. 1979). Based on different experiments, an attraction can be demonstrated, but not in all species and not to the same extent (Miki-Hirosige 1964). In the studies on chemotropism a variety of data are presented in the reviews by Rosen (1975), Mascarenhas (1973, 1975), Kapil and Bhatnagar (1975), and Linskens (1976 a). In *Lilium* a chemotropic activity of the papillate cell to the pollen tube has been observed (Rosen 1971 b). In *Oenothera* mixtures of organic components are involved (Schildknecht and Benoni 1963). In *Antirrhinum* (Mascarenhas 1973) the presence of calcium is presumed to be a factor in pollen tube attraction, although this mechanism is not active in some other plants. Besides, a gradient of calcium also exists in the pollen tube, with the highest concentration at the tip (Reiss and Herth 1978). The tip zone should be altered or corrected during the growth. This part of the pollen tube is also a major zone of excretion, and probably of uptake of nutrition. Mascarenhas (1975) suggests that a missing component can result in a chemotropic reaction. A shift in the direction of growth of the tip center should be related to an uneven distribution factor. This influences the tip metabolism. In *Begonia* (Preil and Keyser 1975) the change in surface influences the pathway of the pollen tube. The cells around the micropyle have a smooth surface in contrast to the other cells which are covered with deposits of waxy substances.

Also, the situation is complex in *Spinacia,* near the nucellar tissue. Here the mass of pollen tubes branches and ceases to grow further; only one, or a few tubes penetrate the nucellar tissue and a selective attraction of one tube takes place (Wilms 1974). Such a reduction in the number of penetration of tubes also occurs in *Persea* (Sedgley 1976). The ovary has one ovule which is reached by only a few pollen tubes. The rest remain arrested, and the tips of the tubes swell.

In the chemotropic attraction of the pollen tube a significant role depends on the genetical background, the nutrient level in pollen and the transmitting tissues of the pistil, as well as gradients in pollen and pistil. Before and during entrance of pollen tubes, the interaction of pollen and stigma is determined by processes of nutrition and penetration. Both are mainly a feature of the pollen. The stylar and stigmatic nutrition, however, can be present in the exudate and stigmatoid and stylar tissue with its starch content. Carbohydrate nutrients seem to be very common. A lack of regulation in nutrition is observed in incompatibility crossings. So the redistribution of metabolites is a very important reaction on pollination.

6.4.4 Pollination Effects

Wilting of flowers of *Cucubalus* (de Vries 1900) can be related with inter- and intraspecific pollinations. In *Petunia* the process of wilting in relation to the stylar development and pollination has been studied by Gilissen (1976, 1977). Wilting of the corolla was accelerated by cutting off the stigma or the style halfway down. In young plants the rate of wilting is faster. The wilting takes place in early morning, except for cross-pollination which seems to be independent of a day rhythm.

Pollination with foreign pollen or killed *Petunia* pollen does not influence the wilting pattern of unpollinated flowers. These experiments indicate the presence of a sense organ. The process of wilting is a factor in the communication of the plant with its environment involved in pollination.

In the style the pollination leads to a signal for the regulation of floral development. Also, an activation of the ovary is induced (Linskens 1973). In translocation experiments with labelled glucose in *Petunia,* during cross-pollination (Linskens 1974a), the amount of labelled carbohydrates decreased in corolla and ovary. After self-pollination, this amount remains at a higher level. After cross-pollination an accumulation of carbohydrates takes place in the ovary when the pollen tubes have reached the base of the style. For labelled amino acids compatible pollination is marked by a more direct increase of label in calyx and corolla compared to an incompatible crossing. The more intense label in anther and style after self-pollination is remarkable. These data are indicative of a redistribution process, related to pollination. The style becomes an attraction center, and the ovary a sink after cross-pollination. A reaction on pollination is also the increase in metabolic activity after about 12 h, expressed in changes of the ribosomal protein and incorporation of C-14 leucine in the polysomes of the ovary of *Petunia.* Depending on different types of pollination, the pattern of protein synthesis changes. These reactions in the ovary show the existence of a long-distance transmission of information. This information communicates the fact of pollination and selection in the type of pollen (Deurenberg 1976, 1977).

Pollination changes the metabolic constitution of the flower, mainly in stigma, style, and ovary. The complexity and sequence of reactions should be related to the process of flowering, fertilization, and seed formation.

6.5 Entrance and Discharge in Embryo Sac

In angiosperms the pollen tube has to enter the embryo sac for successful fertilization. The pathway of the pollen tube from the placenta to the embryo sac is highly variable, related to the structure and organization of the ovular tissues. In most species the embryo sac is surrounded by one or two integuments, leaving a narrow opening called the micropyle. In between the integuments and the embryo sac there can be a varying amount of nucellar tissue, which has to be passed by the pollen tube.

6.5.1 Course of the Pollen Tube

Usually, the pollen tube enters the ovule through the micropyle, called porogamy. Exceptionally, the ovule is entered at the chalaza (chalazogamy), or laterally (mesogamy) (Maheshwari 1950). In a number of species the embryo sac protrudes through the micropyle prior to fertilization, and can be reached by the pollen tube directly, as in *Torenia fournieri* (van der Pluym 1964). For the passage of the pollen tube from the placenta to the micropyle several situations have been described. In a number of species special structures, such as obturator and papillae, formed by either the placenta or integuments, are present and serve as a pathway for the growing pollen tube and facilitate its entry into the micropyle (Kapil and Vasil 1963). Usually, the obturator is a swelling of the placenta which grows toward the micropyle. In other cases cells of the conducting tissue of the style, or cells of the inner integument, elongate and form a well-defined pathway for the pollen tube to reach the apex of the ovule.

However, in most species, the pollen tubes grow along the surface of the placenta, funiculus, and ovules in order to reach the micropyle. It is assumed that the growth of the pollen tube toward and into the micropyle is directed, although the nature of this directed growth is poorly understood. In *Impatiens glandulifera* Steffen (1951) observed ectotrophic growth on the funiculus, and directed growth through a slime canal on the margin of the funicle and ovule toward the micropyle. Numerous authors assume that the growth of the pollen tube is directed chemotropically by substances produced by the ovule, and secreted through the micropyle (Fig. 6.3).

Chemotropic activity of ovules, especially in their micropylar-halves, has been established in bioassays for a number of species. According to Mascarenhas and Machlis (1962a, b, 1964), in *Antirrhinum* directional growth is achieved by the presence of calcium gradients. This view is supported by the observation of Jensen (1965) that the synergids of cotton are rich in calcium. Brewbaker and Kwack (1964) established that calcium promotes pollen tube elongation, and Herth

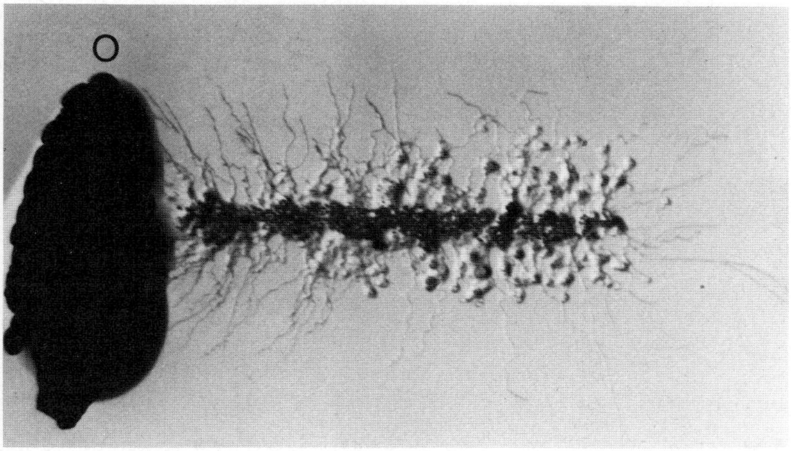

Fig. 6.3. Influence of ovules (*O*) on pollen tube growth in *Impatiens* in vitro. The pollen tubes show several reactions in relation to their position with respect to ovules. The nearest grow toward the ovules. The farthest pollen tubes grow straight as in the control. The remaining pollen tubes show an extreme curvature and inroll

(1978) stated that in *Lilium longiflorum* Ca^{2+} was needed to sustain oriented exocytosis at the pollen tube tip.

However, calcium alone has not been considered as the chemotropically active agent. Tsao (1949) isolated from lily styles a heat-stable, water-soluble substance of considerable molecular size with chemotropic activity. Noack (1960) found a heat-sensitive chemotropic factor with a molecular weight of approximately 900 in *Begonia*. Rosen (1961, 1964) was able to obtain a water- and alcohol-soluble chemotropic active extract from lily pistils. According to Vasilev (1970), the chemotropically active substance in lily belongs to the terpenoid group. This conclusion was based on the ultrastructural resemblance of the guiding stigmatoid cells in the stylar canal to terpenoidogenic cells in plants.

The present information on chemotropically directed growth of pollen tubes, and the chemical nature of the supposed chemotropic substances, is still inadequate. The comparability and interpretation of the various bioassays used are not uniformly accepted as reliable (Rosen 1964).

Most authors assume that the synergids are the origin of chemotropic substances involved in the directed growth of the pollen tube toward the embryo sac. The synergids contain an abundant number of mitochondria and dictyosomes, a well-developed endoplasmic reticulum, with partly parallel-arranged cisterns and numerous ribosomes indicating a high metabolic activity (see Jensen 1965, Schulz and Jensen 1968a, van Went 1970a, Vazart 1969a, Diboll and Larson 1966, Wilms 1980b). The synergids are supposed to produce the chemotropic substances which are secreted into the filiform apparatus. From the filiform apparatus they probably leach out into the micropyle, or the neighboring nucellar cells.

Coe (1954) reached a similar conclusion based on autoradiographic studies. He observed a maximal concentration of C-14-labelled compounds in the syner-

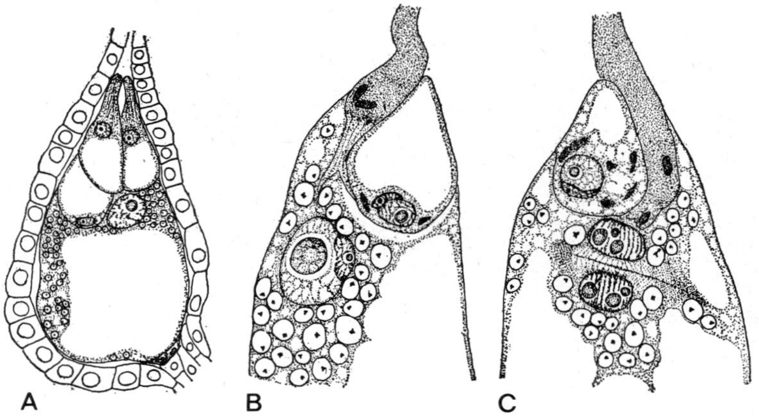

Fig. 6.4 A–C. Stages in fertilization in *Petunia*. **A** Mature embryo sac showing two synergids, egg cell, and central cell, surrounded by integumentary tapetum. **B** Upper portion of embryo sac after entrance of pollen tube. The pollen tube is presumed to form two branches through which the two male gametes were discharged. **C** Upper portion of embryo sac showing the zygote, and division in the primary endosperm cell. Note abundant starch grains in **A–C.** (After Cooper 1946)

gids and considerable amounts in the micropyle and neighboring tissue. In *Paspalum orbiculare* Chao (1971, 1977) observed the presence of a PAS-positive substance in the micropyle and other regions involved in guidance of the pollen tube. The substance contains considerable amounts of mucopolysaccharides (Chao 1979). He concluded that enzymes, secreted by the synergids or the filiform apparatus, were involved in the dissolution of cells of the integuments and conversion of dissolved substances.

A major objection to the role of synergids in chemotropic activity is their absence in a number of species. It was, however, proved by Cass (1972) and Cass and Karas (1974) that at least in *Plumbago* the absence of synergids is compensated by the presence of a filiform apparatus in the egg cell. Also, the egg cell shows a complex organization of cytoplasm, indicating a high metabolic level, similar to that of synergids in other species.

6.5.2 Entry into the Embryo Sac

Based on a large number of light microscope observations (Maheshwari 1950, 1963, Kapil and Bhatnagar 1975), it was presumed that the pollen tube could enter the embryo sac by one of the following main routes:

1. In between the cells of the egg apparatus, i.e., the synergids and egg cell.
2. Between the cells of the egg apparatus and the embryo sac wall.
3. Into one of the synergids, after previous degeneration or accompanied by the destruction of that synergid.

Since the light microscopic studies show so many different routes, it is even more noticeable that all electron microscope studies, so far, show the same out-

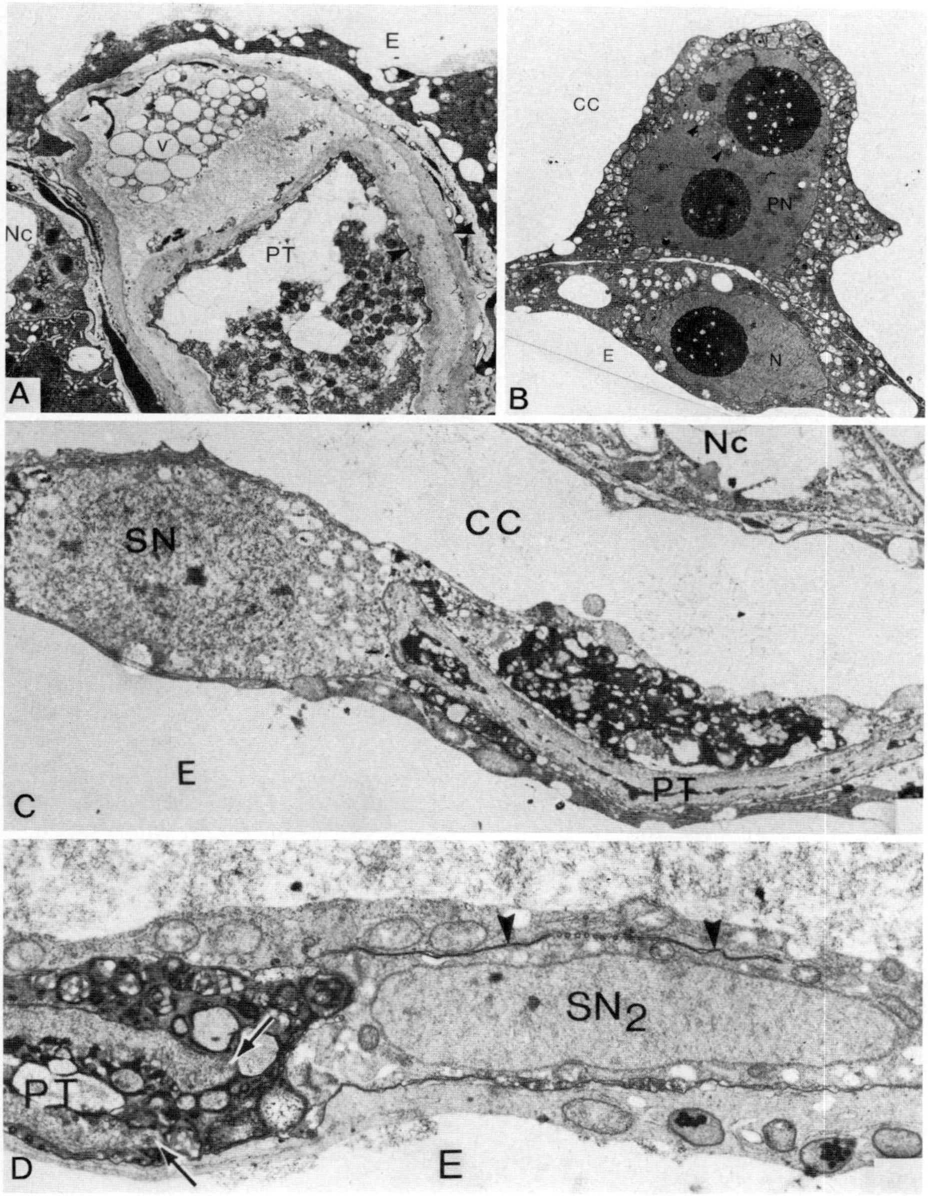

Fig. 6.5 A–D. *Plumbago zeylanica.* **A** Pollen tube (*PT*) at the entrance of the megagametophyte; *V* vacuoles. Egg (*E*) and nucellus (*NC*) are evident to the exterior of the pollen tube. The bi-lamellate nature of the pollen tube is indicated by *unlabelled arrowheads.* **B** Longisection of the fused central cell nucleus (*PN*) and the egg nucleus (*N*) in a mature megagametophyte. Cytoplasmic bodies within the central cell nucleus (*arrowheads*) presumably originate from prior fusion of the polar nuclei. **C** Unfused sperm cell located between the egg (*E*) and central cell (*CC*) soon after discharge of the pollen tube (*PT*). Sperm nucleus (*SN*) is evident within the sperm cell. **D** Spatial relationship of transmitted sperm nucleus (SN_2) with the pollen tube aperture (*unlabeled arrows*) and pollen tube cytoplasm (*PT*) soon after gametic fusion within the central cell. Membranes associated with the sperm nucleus (*unlabeled arrowheads*) appear to have arisen during sperm cell fusion with the central cell. Adjacent egg cell (*E*) is evident below. (**A, B** after Russell 1982; **C, D** after Russell 1983; courtesy Dr. S. D. Russell, Norman, USA)

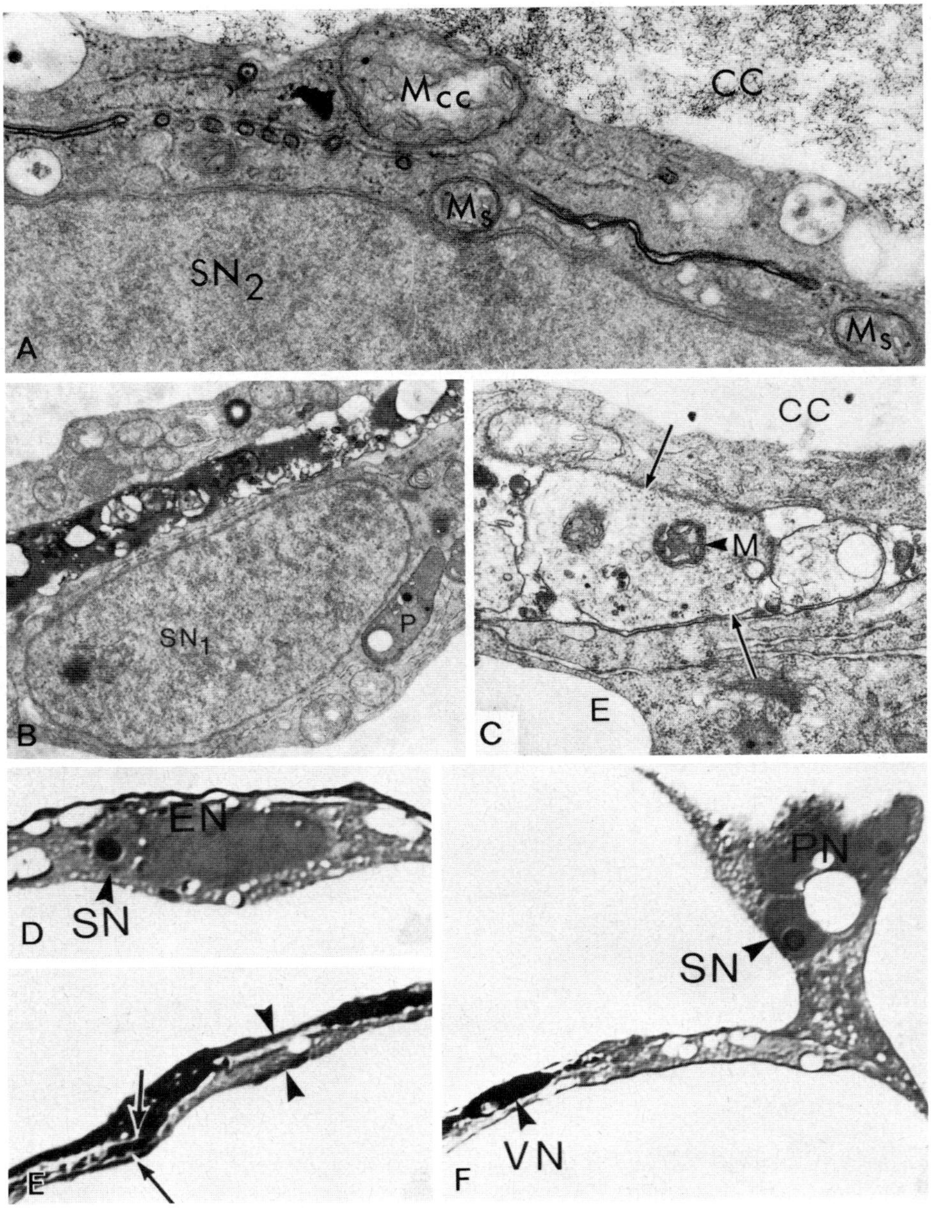

Fig. 6.6 A–F. *Plumbago zeylanica.* **A** Newly transmitted sperm nucleus (SN_2) and surrounding central cell cytoplasm (*CC*) are delimited by a densely staining vesiculated membrane segment. Mitochondria of paternal origin (M_s) are visible immediately next to the sperm nucleus, whereas mitochondria of maternal origin (M_{cc}) are located on the opposite side of the membrane segment. **B** Newly transmitted sperm nucleus (SN_1) within the egg cell. Plastids (*P*) flanking the sperm nucleus are of paternal origin. Degenerate, discharged pollen cytoplasm is evident between the egg and central cell. **C** Small, unfused cytoplasmic bodies (*unlabeled arrows*) of presumed sperm origin are present between the egg (*E*) and central cell (*CC*). Mitochondria (*M*) present within these cytoplasmic bodies are similar in size and mor-

standing pattern of penetration and discharge of pollen tubes in embryo sacs with three-celled egg apparatus (Fig. 6.4 A–C). Only for species which lack synergids, as *Plumbago* (Russell and Cass 1981, Russell 1982), a different pattern of pollen tube entry and discharge has been described (Figs. 6.5 A–D and 6.6 A–F). The mature megagametophyte consists of five cells: the egg, the central cell, and three antipodal cells. The mature egg is a highly polarized cell. The nucleus, numerous mitochondria and the majority of plastids are located at the chalazal end, while the micropylar end of the egg shows wall ingrowths, strongly resembling a filiform apparatus (Cass 1972, Cass and Karas 1974). The pollen tube penetrates the female gametophyte through the "filiform apparatus" and grows for another 80 μm between egg and central cell. The pollen tube discharges through a terminal aperture, resulting in release of male gametes, vegetative nucleus, and a limited amount of tube cytoplasm. The discharged male gametes remain cellular during their presence between egg and central cell.

6.5.3 Growth Through the Filiform Apparatus

The ultrastructural studies clearly show that the synergids play a vital role in the pollen tube entry. Penetration of the pollen tube in the embryo sac starts with the growth of the pollen tube into and through the filiform apparatus (FA). Essentially, the FA is an extension or elaboration of the micropylar synergid wall. Observations on living material (Erdelská 1974) have shown that the entry of the pollen tube and the release of the tube content cause considerable and very sudden changes in the organization of the embryo sac.

The changes in organization and the interaction of male and female material also result in drastic changes in staining properties of the cytoplasm, and these, in turn, interfere with the light microscopic observations and interpretation of stained sections. Furthermore, some of the structures involved are very near the limits of the resolving power of the light microscope, and it is possible that some conclusions based on light optical research are based on misinterpretations.

A well-established example is the conclusion of Cooper (1946) that in *Petunia* the pollen tube enters the embryo sac in between the cells of the egg apparatus (Fig. 6.4 A–C). Later electron microscopic studies have revealed that in *Petunia* the pollen tube actually penetrates into the synergid.

For this reason, in this chapter, we focus attention on observations based on electron microscopy and their interpretation (see van der Pluym 1964, Jensen

phology to paternally originating mitochondria. **D–F** Median longisections of a single embryo sac showing fusion of egg and sperm nucleus (**D**), pollen tube aperture (**E**), vegetative nucleus and fusion of sperm and central cell nuclei (**F**). Stained with aniline blue-black (for proteins); Nomarksi interference contrast microscopy. **D** Fusion of egg nucleus (*EN*) and sperm nucleus (*SN*) illustrating multiple fusion sites. Densely stained sperm nucleolus is evident near the center of the sperm nucleus. **E** Pollen tube aperture (*unlabeled arrows*) is evident within the densely staining discharged pollen cytoplasm. An area of modified egg central cell wall (*arrowheads*) apparently influences the distribution of discharged pollen cytoplasm. **F** Fusion of sperm nucleus (*SN*) and fused central cell nucleus (*PN*) within the central cell. The vegetative nucleus (*VN*) is located chalazal to the constricted egg – central cell junction indicated in the previous figure. (After Russell 1983; courtesy Dr. S. D. Russell)

1965, van Went and Linskens 1967, van Went 1970a, Chao 1971, Vijayaraghavan et al. 1972, Schulz and Jensen 1968a, Maze and Lin 1975, Newcomb 1973a, Wilms 1980b). In most species the cytoplasmic surface of synergid is highly convoluted, resulting in a strongly increased plasma membrane area. Histochemical studies reveal that the FA is PAS-positive (Jensen 1965, Chao 1971, Vijayaraghavan et al. 1972) indicating the presence of polysaccharides, merely of pectic and hemicellulosic nature. In *Petunia* (van Went 1970a) the presence of a very loosely organized cellulosic fibrillar network has been shown. Although the chemical composition and ultrastructure indicate that FA does not represent a solid barrier to the pollen tube, it is clear that it offers some resistance to tube growth which must be overcome. In *Epidendrum* Cocucci and Jensen (1969) observed that the pollen tube forms a cap over the micropylar part of the embryo sac, then an extension develops through the FA. Especially, the latter observation indicates that the final part of pollen tube growth is invariably directed, rather than being merely accidental. The growth pattern suggests that the pollen tube is guided toward the synergid.

6.5.4 Entry into the Synergid

After the passage of the filiform apparatus the pollen tube grows into one of the two synergids (Fig. 6.7A–D). In several species, as *Gossypium* (Jensen 1965, Jensen and Fisher 1968), *Zea* (Diboll 1968), *Epidendrum* (Cocucci and Jensen 1969), *Hordeum* (Cass and Jensen 1970), *Linum* (Vazart 1969b, 1970, D'Alascio-Deschamps 1973), *Lycopersicum* (Nettancourt et al. 1973a), *Spinacia* (Wilms 1980a), and *Quercus* (Mogensen 1972) one of the synergids starts degenerating before the arrival of the pollen tube. It has been stated that, in fact, pollination triggers the beginning of the degeneration process. However, Jensen et al. (1977) observed in cultured, nonpollinated and unfertilized ovules of cotton synergid changes strikingly similar to those after pollination in vivo. Apparently, in this case synergid degeneration is not triggered by pollination, but as a part of the common embryo sac development. In *Torenia* (van der Pluym 1964), *Petunia* (van Went 1970a), *Helianthus* (Newcomb 1973a, b), and *Capsella* (Schulz and Jensen 1968a) both synergids remain unchanged, until the pollen tube has penetrated one of them. The penetrated synergid then undergoes cytoplasmic changes that closely resemble those of the preentrance degeneration of synergid. Both degeneration processes are marked by decrease of cellular volume, collapse of the vacuoles, disintegration of the plasma membrane, and disorganization of the cell organelles.

In those species which do not show structural differences between the two synergids before the arrival of the pollen tube, it is not clear whether any of the synergids can be penetrated or not. Chemical or physiological differences between the two synergids, regulating the penetration event can not be excluded.

Shortly after the entrance of the pollen tube in the cytoplasm of the synergid, pollen tube growth ceases. Cessation can take place immediately after the passage through the filiform apparatus, as in *Petunia* (van Went 1970b), or the pollen tube can extend into the synergid for some distance, as in *Gossypium* (Jensen and Fisher 1968) and *Linum* (Vazart 1970).

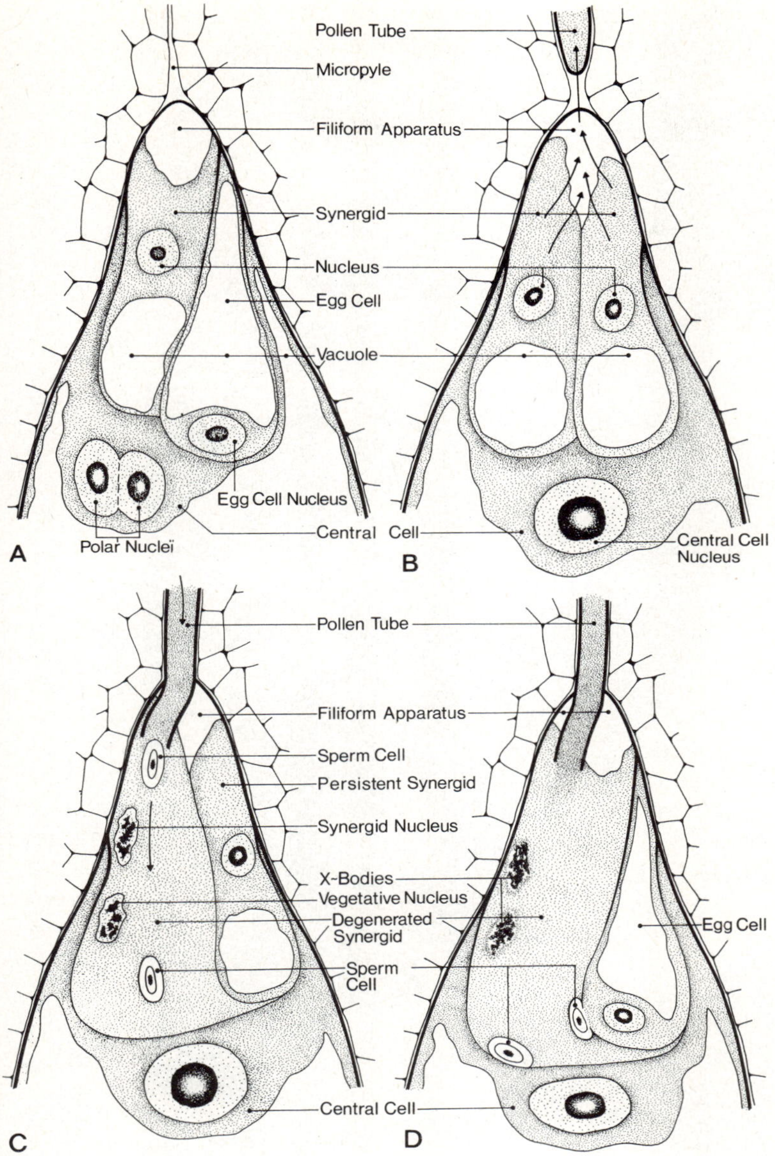

Fig. 6.7 A–D. Diagrammatic summary of pollen tube entrance, tube discharge, and gametic transfer in angiosperms. **A** Micropylar part of mature embryo sac. **B** Proposed chemotropic activity of synergids, and directed growth of pollen tube. **C** Discharge of pollen tube. **D** Proposed way of gametic transfer and fusion

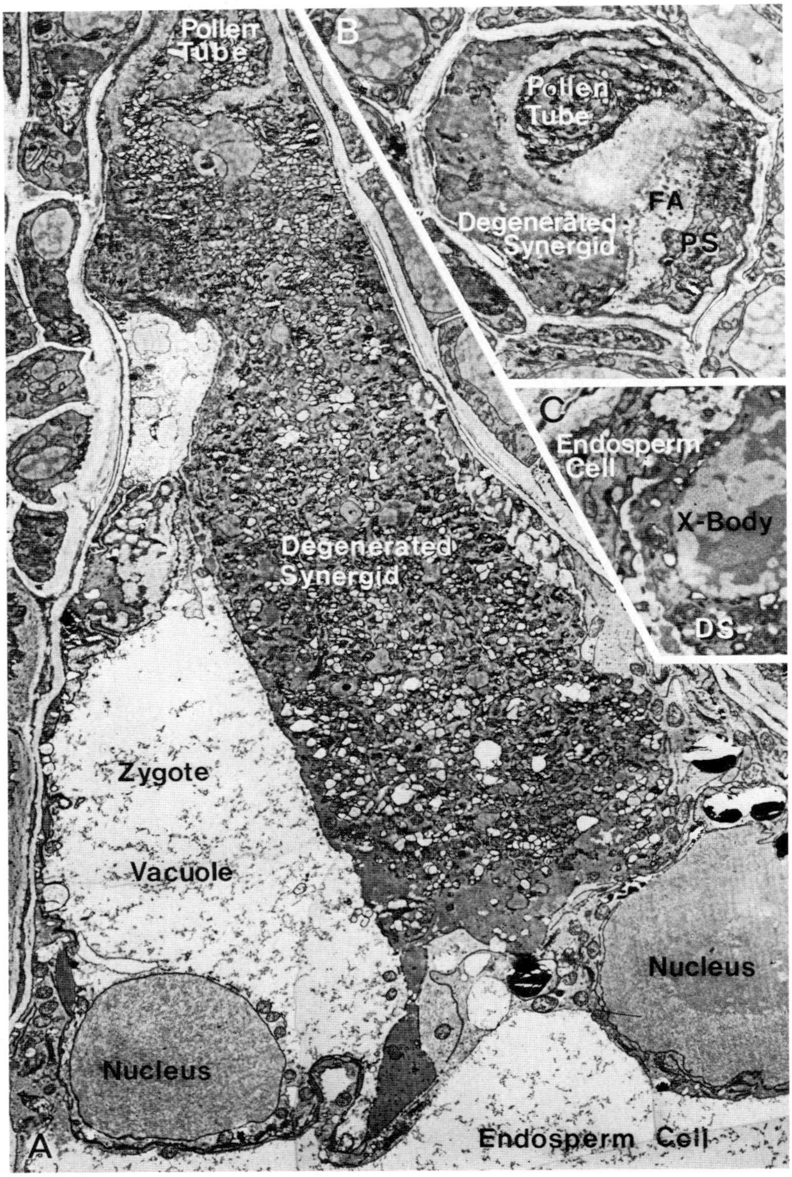

Fig. 6.8. A Longitudinal section of fertilized embryo sac of *Petunia*. **B** Transverse section through degenerated synergid, persistent synergid (*PS*), filiform apparatus (*FA*), and pollen tube. **C** X-body in the degenerated synergid (*DS*) of *Petunia*

It appears that the degeneration of the synergid, the entrance of the pollen tube, and cessation of tube growth are closely related. In a few cases entrance of the pollen tube in the persistent synergid, or the entrance of additional pollen tubes in that synergid, has been reported (Jensen and Fisher 1968). In these cases pollen tube growth does not cease in the normal way. Growth is continued until the degenerated synergid is reached and then, eventually, it stops. This indicates that the degenerated condition of the synergid cytoplasm is needed, or is responsible for tube growth cessation. It also indicates that in species in which both synergids apparently remain healthy until fertilization, the degeneration of cytoplasm of the penetrated synergid observed afterward might have actually taken place shortly before the arrival of the pollen tube and been triggered by the latter. However, then the question that remains to be answered is: why only one synergid degenerates, whether independently or triggered by the pollen tube, and why the other synergid does not, even when penetrated by a pollen tube?

After arrival inside the synergid, discharge of the pollen tube content takes place (Fig. 6.8 A–C). In *Gossypium* Jensen and Fisher (1968) observed that a subterminal pore develops on the side of the pollen tube toward the egg cell. The position of the pore and its regular diameter indicate the presence of a highly ordered process rather than a simple rupturing of the pollen tube. In *Petunia* (van Went 1970 b) the pollen tube opens at its tip, and it was concluded from the morphology of the opened tube that the opening merely occurs by bursting of the pollen tube.

6.5.5 Transfer of Tube Content

After the opening of the pollen tube a fair portion of the tube content, including the two sperm cells and the vegetative nucleus, is released into the synergid. The tube cytoplasm can easily be recognized, in spite of the degenerated appearance of both tube and synergid cytoplasm (Fig. 6.8 A–C). The tube cytoplasm contains numerous small PAS-positive spheres (Jensen and Fisher 1968, Schulz and Jensen 1968 a, van Went 1970 b). The pollen tube discharge forms a roughly cone-shaped mass spreading from the pore toward the chalazal end of the synergid. Not all of the tube cytoplasm is injected into the synergid. In cotton, soon after the release of the pollen tube cytoplasm, a plug is formed by fusion of PAS-positive spheres in the pollen tube near the FA. In other species, as *Petunia* (van Went 1970 b), the pollen tube collapses or is squeezed by the surrounding tissue.

In *Petunia* the high amount of released tube cytoplasm causes the bursting of the synergid and penetration of its cytoplasm in between the egg cell, central cell, and persistent synergid. However, in no case could tube cytoplasm be observed inside other cells than the penetrated synergid, and the neighboring cells were not affected by the close contact with this degenerated synergid cytoplasm. Since in *Petunia* walls are absent in the chalazal part of the egg apparatus, and the cells are surrounded only by plasma membranes, one might ask in what way the egg cell and central cell are protected, considering the dramatic cytoplasmic changes that occur in the penetrated synergid.

It is obvious that the penetrated synergid is involved in the transfer of the male gamete material to the egg cell and central cell. The first question to consider is in what condition the male material is released from the pollen tube: as naked nuclei or as complete cells?

6.6 Fusion of Gametes

There is abundant evidence that the male gametes of angiosperms are cellular and complete descriptions of their ultrastructure are available (Hoefert 1969 b, Jensen and Fisher 1968, Sanger and Jackson 1971 a, Larson 1965, Vazart 1970, Cass 1973, Cass and Karas 1975, Russell 1981). However, most studies focused their attention on sperm cells in mature trinucleate pollen grains or pollen tubes. The sperm cells are believed to be surrounded by a plasma membrane only, in contrast to generative cells which mostly show a cell wall. Their cytoplasm contains mitochondria, ER, ribosomes, dictyosomes, and microtubules. Plastids are frequently reported to be absent. Unfortunately, there are only a few reports dealing with the sperm material inside the synergid. Attempts at observing and recording the initial stages of the fertilization process in living material were not very successful (Erdelská 1974). The major problem is the low transparency of the penetrated synergid in optical microscopy.

There is, however, some evidence that the male gametes are deposited in the embryo sac as complete cells. With phase-contrast microscopy Cass and Jensen (1970) observed sperm nuclei surrounded by a distinct clear zone in the penetrated synergid in barley. This clear zone closely resembles the zone observed between the nucleus and boundary of the sperm cell in the pollen grain, which represents the male cytoplasm. Cass and Jensen (1970) also, rarely, observed penetration and discharge of a pollen tube into the persistent synergid. In these cases the discharge of complete sperm cells could be established. The movement or transport of the sperm cells after their discharge into the synergid remains an unresolved question.

According to Steffen (1963), isolated sperm cells show autonomous, amoeboid movement, which could be sufficient to reach the female egg cell and nucleus. However, Cass (1973) showed that isolated living sperm cells of barley did not exhibit any directional motility. Ultrastructurally, the presence of microtubules in these sperm cells could be established. The position of the tubules and the shape changes observed indicate that their chief role is regulation of cell shape. The suggestion of a microtubular role in regulation of cell shape is in agreement with the findings of Sanger and Jackson (1971 a, b). They induced generative cell-shape changes by treating pollen with isopropyl N-phenyl carbamate or colchicine; both compounds destroy cytoplasmic microtubules.

Our knowledge of the actual transfer of the sperm cells from the penetrated synergid to and into the egg cell and central cell is very limited. In cotton (Jensen 1964, 1968 b, 1972, Jensen and Fisher 1967), barley (Cass and Jensen 1970), and *Petunia* (van Went 1970 b) only the sperm nuclei have been identified in either the egg cell or the central cell. There is no structural evidence of the presence of male cytoplasm inside the female gametes. However, since the sperm cells only contain a highly reduced cytoplasm, it is likely that it is not easily recognized among the high amount of complex egg cell cytoplasm.

On the other hand, remnants of male gamete cytoplasm could not be found in the penetrated synergid either. The penetrated synergid contains two darkly stained bodies after the discharge of the pollen tube. These so-called X-bodies have been considered as the cytoplasmic remnants of the sperm cells (Cooper

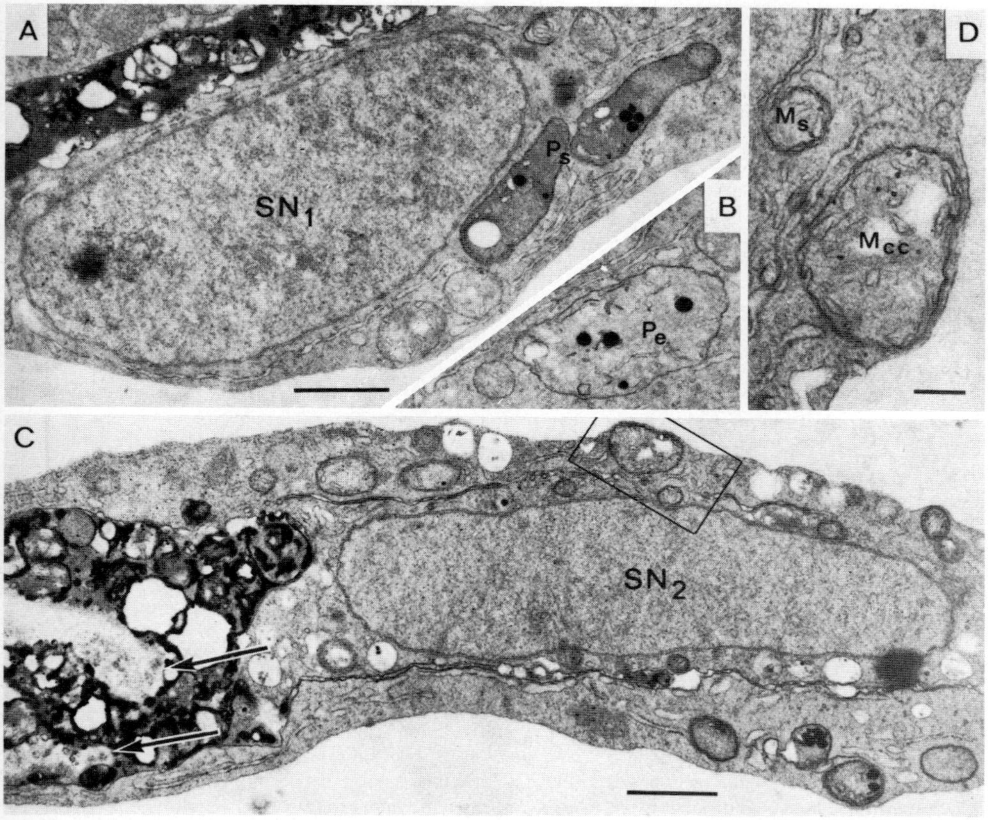

Fig. 6.9 A–D. Enlarged portions of the embryo sac of *Plumbago zeylanica* shortly after gamete fusion. **A** Sperm nucleus (SN_1) and two sperm plastids (P_s) in the egg cytoplasm (scale bar 1 μm). **B** Typical egg plastids (P_e) at the same magnification as **A**. **C** Second sperm nucleus (SN_2) shortly after fusion with the central cell nucleus. The *arrows* indicate both halves of the ruptured pollen tube wall (scale bar 1 μm). **D** Enlargement of marked region in **C**, showing sperm mitochondrion (M_s) at the periphery of the sperm nucleus, and a nearby central cell mitochondrion (M_{cc}) (scale bar 0.2 μm). (Courtesy Dr. S. D. Russell)

1946, Jensen and Fisher 1968). By specific staining and labelling it could, however, be proved that they contain DNA (Fisher and Jensen 1969, Cocucci and di Fulvio 1969), and they were identified as degenerated nuclei. From their morphology and position it was concluded that one of them represents the degenerated synergid nucleus.

The microscopic information about the fate of the sperm cell cytoplasm is limited. For *Plumbago zeylanica* Russell (1980, 1983) could establish clearly the transfer of male cytoplasm into both the egg cell and the central cell. It was possible to identify male mitochondria and plastids by their specific ultrastructure inside the egg cell and the central cell, directly after the cell fusion process (Fig. 6.9 A–D). Also, there is considerable genetical evidence that in a number of plant species, together with the male nucleus the cytoplasm is also transferred to the female

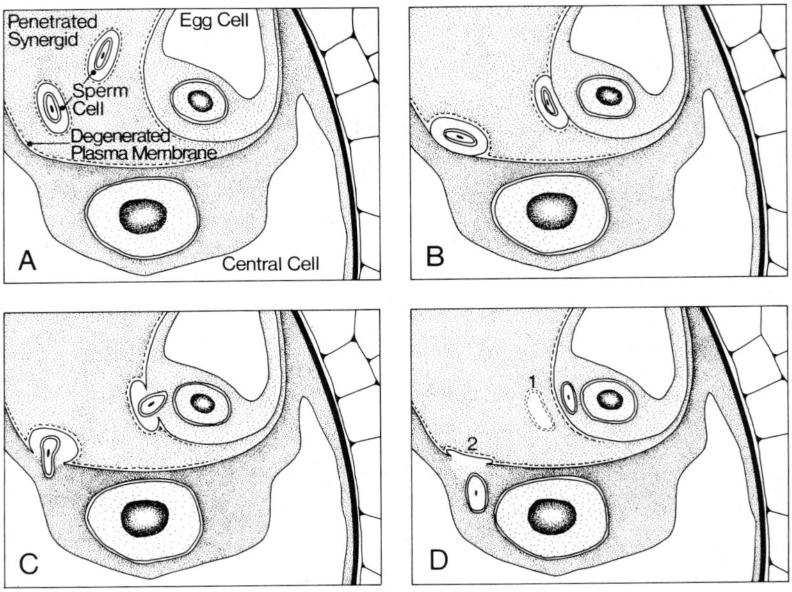

Fig. 6.10 A–D. Diagrammatic representation showing the proposed way of gametic fusion. **A** Basal portion of the penetrated synergid containing the two sperm cells, and the adjacent portions of egg cell and central cell (cf. Fig. 6.5). The synergid plasma membrane and the outer of the two membranes surrounding the sperm cell (i.e., the membrane belonging to the vegetative cell) are supposed to be degenerating, a result of the interaction of synergid cytoplasm and vegetative pollen tube cytoplasm. **B** As a result of the degeneration of the synergid and vegetative cell plasma membranes, the plasma membranes of the gametes (egg cell, central cell, and sperm cells) become exposed to each other and can fuse. **C** Fusion of the plasma membranes result in cytoplasmic bridges between the gametes through which sperm cell nuclei, and possibly cytoplasm, can enter the female gametes. **D** Possibly, only the male nucleus is transferred into the female gamete, and the male cytoplasm remains in the synergid (*1*), or both male nucleus and cytoplasm enter the female gamete (*2*)

gamete. In this respect the data on cytoplasmic male sterility and plastidal inheritance are indicative. In the past decade an attractive hypothesis has been proposed by several authors simultaneously (Linskens 1968, van Went 1970 b, Jensen 1972), which explains most of the known facts. It is stated that fertilization in angiospermous plants includes a real cell-fusion process (Fig. 6.10). It is proposed that the plasma membrane of one sperm cell comes in contact and fuses with the plasma membrane of the egg cell, while the plasma membrane of the second sperm cell fuses with the plasma membrane of the central cell. The fusion process leads to the formation of bridges between the two gametes, through which the sperm nucleus and possibly sperm cytoplasm can pass. In the proposed fusion process there is no need for pores or ruptures in the female gamete membrane to facilitate entrance of male material. It also provides a mechanism to prevent entry of synergid or vegetative pollen cytoplasm into the female gametes. One also speculates on the existence of a specialized fertilization membrane surrounding the female gamete. Such a membrane could be altered by the fusion with one of the sperm cells, making the receptor cell incapable to fuse with the second sperm

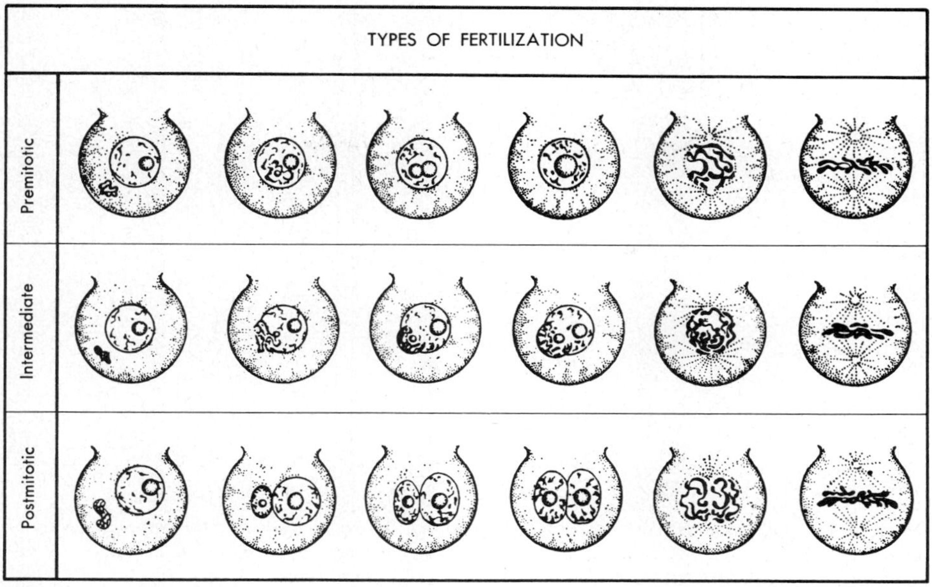

Fig. 6.11. Types of nuclear fusion in angiosperms, as based on the moment of first nuclear contact and completion of mixing of chromatin. In the "premitotic" type the nuclear fusion is completed before the zygotic division. In the "intermediate" type the chromatin of male and female nuclei remain partly separated until the first mitotic division. In the "postmitotic" type the chromatin of male and female nuclei remain separated completely until the mitotic division of zygote. (After Gerrasimova-Navashina 1960)

cell. The second sperm cell then fuses with the remaining receptive membrane of the central cell, and double fertilization is achieved. The fusion hypothesis is attractive, especially since in most species no ruptures or pores are found in the synergid–egg and synergid–central cell boundaries. However, in *Capsella,* which species is unique in having real walls in between synergid–egg and synergid–central cell, Schulz and Jensen (1968 a) observed the presence of a pore in the common synergid–egg wall near the opened tip of the pollen tube. It is believed that one sperm cell had entered the egg cell through this pore.

6.6.1 Fusion of Nuclei

Next in the process of double fertilization is the fusion of male and female nuclei. Based on light microscopic observations, various modes of transfer of genetic material and nuclear fusion have been described (Gerassimova-Navashina 1960). Three types of karyogamy are recognized, depending on differences in the condition of the sexual nuclei at the moment of fertilization: Premitotic, Postmitotic, and Intermediate. In the first type fusion is completed before the zygotic mitosis. This type is usually common in Gramineae and Compositae. In the second type the fusion is completed after the initiation of the zygotic mitosis during prophase-metaphase. It occurs in a number of Liliaceae (Fig. 6.11). The third type is char-

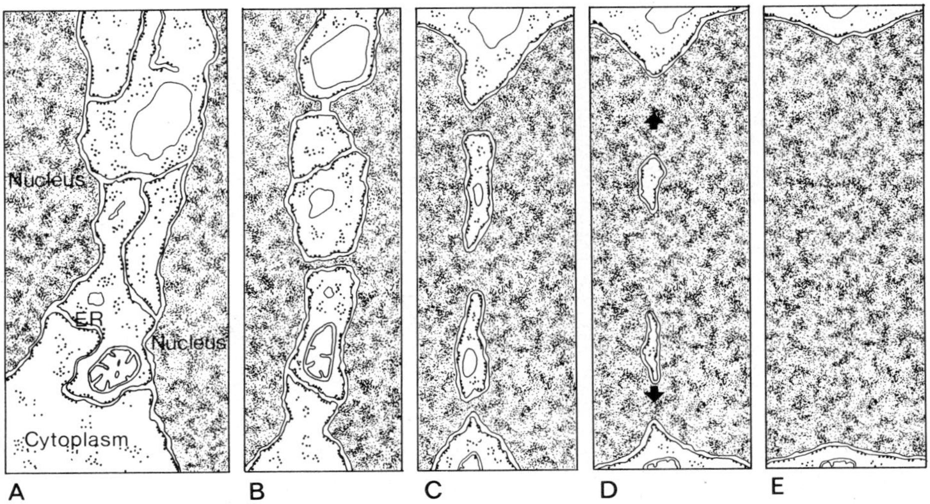

Fig. 6.12 A–E. Diagrammatic presentation of the cytoplasmic events leading to the complete fusion of gametic nuclei in plants. **A** Contact of *ER*. **B** Fusion of outer nuclear membrane. **C** Formation of nuclear bridges. **D** Enlargement and coalescence of bridges. **E** Fusion completed

acterized by an incomplete mixing of the sexual nuclei. The chromosomes remain apart for varying lengths of time and can still be seen in the prophase of the zygotic nucleus. Similar observations have been described by Hu and Zhu (1979). In the classical light microscopic literature it is usually stated that the male nucleus sinks into the female nucleus or becomes immersed into it. According to Luxova (1967), in barley the membrane of the egg nucleus apears locally interrupted temporarily to allow the entry of the sperm nucleus. Also, in *Caltha palustris* and *Ranunculus acris* (Brouland 1969) the disappearance of the membrane of the egg nucleus prior to nuclear fusion has been observed.

However, ultrastructural studies (Jensen 1964, van Went 1970b, Schulz and Jensen 1973) have shown that no interruptions or breakdown of nuclear membranes are involved in the nuclear fusion. The observed mechanism is shown in Figs. 6.12A–E and 6.13. The nuclear fusion starts with local contact and fusion of the outer membranes of the two nuclear envelopes which then become continuous. In this process elements of the neighboring and attached ER can be involved (Fig. 6.12A). Next sequence in the fusion process is the contact and fusion of the inner membranes of the nuclear envelopes. In this way bridges are formed between the two nuclei and their nucleoplasm become continuous (Fig. 6.12B). Subsequently, the nuclear bridges enlarge and coalesce until no separating membranes are left between the two nuclei (Fig. 6.12C–E). During the nuclear fusion process, cytoplasmic portions containing organelles become trapped between the nuclei, but this cytoplasm is squeezed out by the enlarging nuclear bridges. The observed sequence indicates that the membranes of both nuclei contribute to the envelope of the zygote nucleus. Soon after fertilization is achieved, division of the primary endosperm nucleus occurs, leading to the formation of endosperm. The

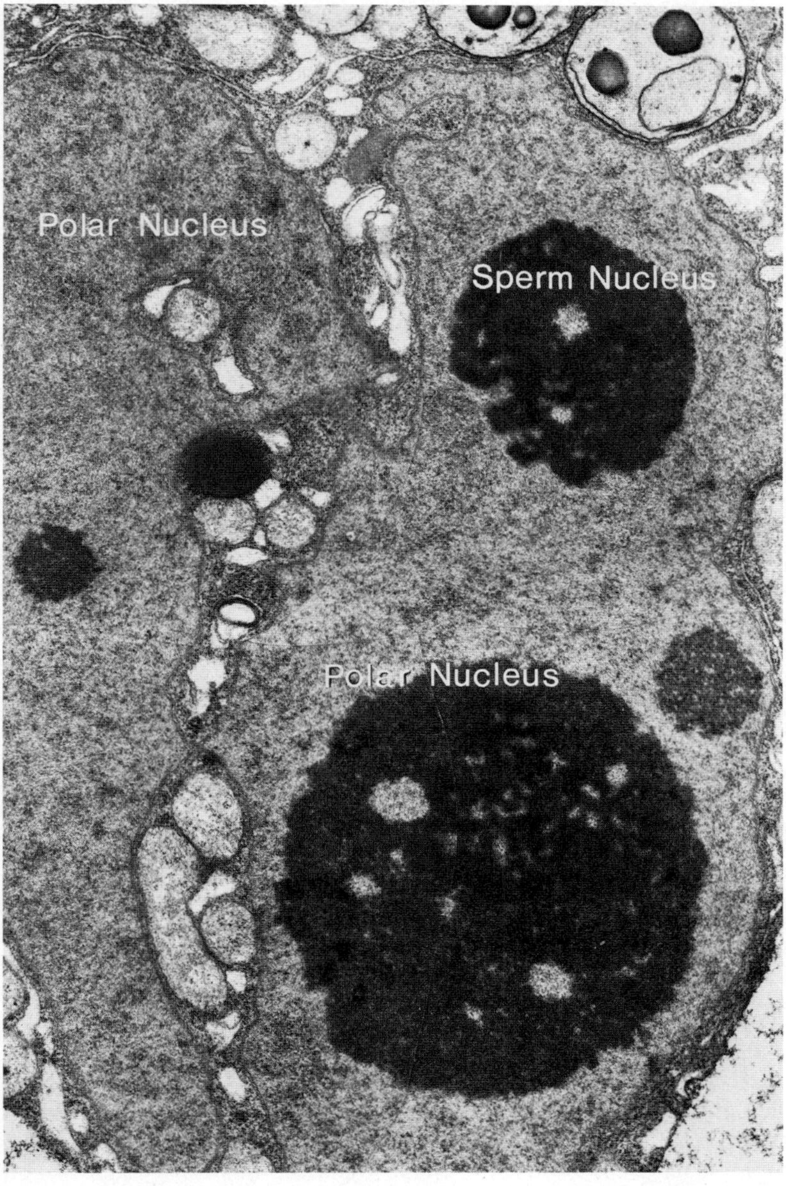

Fig. 6.13. Sperm nucleus fusing with the polar nuclei in *Spinacia*. (Courtesy Dr. H. J. Wilms, Wageningen, The Netherlands)

zygote remains undivided for some time, although considerable cellular and cytoplasmic changes can take place.

In cotton (Jensen 1963, 1968 a, b) and *Hibiscus* (Ashley 1972) the zygote shows drastic shrinkage after its formation (see Fig. 8.1 A, B). Shrinkage is accompanied by increasing density of the cytoplasm and loss of vacuolar volume. In cotton zygote starch is formed, the amount of ER increases, and the cytoplasm contains abundant RNA. In both *Epidendrum* (Cocucci and Jensen 1969) and cotton (Jensen 1968 b) marked changes in the number and aggregation of the ribosomes, leading to the formation of large helical polysomes, have been observed. In some species, as *Capsella* (Schulz and Jensen 1968 b), *Epidendrum* (Cocucci and Jensen 1969), *Hordeum* (Norstog 1972), and *Quercus* (Singh and Mogensen 1975), the zygote does not shrink after its formation. Usually, a reorganization of the cytoplasm takes place as in *Hordeum* where a peripheral localization of vacuoles is achieved. In *Capsella,* directly after fertilization, the dictyosomes are activated in relation to the formation of zygote cell wall. It is noticeable that the zygote wall lacks plasmodesmata. The zygote divides after a period of varying length following fertilization.

In the fertilized central cell the nucleus divides very rapidly following fertilization. In cotton, initially, little or no cytoplasmic reorganization has been observed (Jensen 1963). The early free-nuclear endosperm is a highly active, rapidly growing tissue (Schulz and Jensen 1974, 1977). As in the zygote, large helical polysomes are formed. In *Petunia,* however, marked cytoplasmic changes occur in the central cell after fertilization and prior to the first nuclear division (van Went 1970 b). There is formation of extensive RER and, simultaneously, with an increasing density of the cytoplasm, there is a decrease in the starch content of the plastids.

6.7 The Progamic Phase and Fertilization

Summarizing the previous notes, it becomes clear that the whole process from pollination till fertilization is a very complex interaction between gametophyte and sporophyte, and also between the two gametophytes with their gametes.

This complexity involves also the relation to animals as pollen vectors or visitors to flowers which results in pollination. The position and morphology of the stigma is clearly related to the mode of pollen transport. Most of these characters are all built up by the sporophyte, and result in the interaction of pollen and stigma. Thus, some factors in pollination, as sticking, exchange, the moment of pollen transport, and the intensity of the visit or duration of flowering, are prerequisites for a successful landing of the pollen.

Physical and chemical forces are involved in sticking of the pollen to the stigma. The pollen and stigma surface coatings, genetically determined, function in the first steps of recognition or rejection. In this context the preexisting sporophytic or gametophytic nature is important. The recognition signals may function on electric, membrane-bound or on molecular level, but are still not clear. Also, the recognition of interspecific pollen resulting in incongruency is unknown. The stage of pollen development and its vitality determine the moment of germination too. In this process of interaction, comparable with host–parasite relations, many steps are discerned. The sequence of these steps are not yet all known and need further investigation.

The pathway of the pollen tube passes through different zones. It penetrates tissues, or runs along wet phases, or gas-filled areas. Besides, the tube growth is a directed process and needs nutrition. The influence of the different tissues on each other, and the direction of growth, include a mechanism of control and attraction or regulation.

The stigma, style, and ovary are all involved in relation to the pollen tube growth. The study of the ontogeny of these organs, on structural and physiological level including hormonal control, would help in a further elaboration of this interaction process.

The factors for selective degeneration in the egg apparatus, and acceptance of the pollen tube have already been under study for a long time. Probably these phenomena depend not only on the normal developmental process, but are influenced by pollination too. The most intriguing pathway of the sperm cells in the pollen tube, the partition and directed movement in the egg apparatus still await a detailed explanation.

In summary, the progamic phase is a developmental process, expressed in the growth and function of the stamen and the germination of pollen. It includes a genetic programmed process sequence, influenced by the surrounding tissues which are involved in nutrition and can act on the direction of the growth. This complicated mechanism, partly to be compared with the fertilization in lower plants and lower tracheophytes, needs a structural and functional approach with a good genetically defined plant material under controlled conditions. The progamic phase and fertilization will be followed by seed formation. It is, therefore, necessary to include in these preceding phases even the preparation to subsequent seed development.

References

Ashley T (1972) Zygote shrinkage and subsequent development in some *Hibiscus* hybrids. Planta 108:303–317

Baker HG (1973) Stigmatic exudates and pollination. In: Brandjes NMB, Linskens HF (eds) Pollination and disposal. Univ Nijmegen, pp 47–60

Baker HG, Baker I (1979) Starch in angiosperm pollen grains and its evolutionary significance. Am J Bot 66:591–600

Bell J, Hicks G (1976) Transmitting tissue in the pistal of tobacco: Light and electron microscopic observations. Planta 131:187–200

Bredemeijer GMM (1974) Peroxidase activity and peroxidase isoenzyme composition in self-pollinated, cross-pollinated, and unpollinated styles of *Nicotiana alata*. Acta Bot Neerl 23:149–157

Bredemeijer GMM (1979) The distribution of peroxidase isoenzymes, chlorogenic acid oxidase, and glucose-6-phosphate dehydrogenase in transmitting tissue and cortex of *Nicotiana alata* styles. Acta Bot Neerl 28:197–203

Brewbaker JL, Kwack BH (1964) The calcium-ion and substances influencing pollen growth. In: Linskens HF (ed) Pollen physiology and fertilization. North-Holland, Amsterdam, pp 143–151

Brouland M (1969) Recherches cytologiques sur la fécondation chez quelques Renonculacées. Rev Cytol Biol Vég 32:309–317

Carniel K (1972) Elektronenmikroskopische Analyse der Pollenentwicklung von *Heleocharis palustris*. Oesterr Bot Z 120:223–234

Cass DD (1972) Occurrence and development of a filiform apparatus in the egg of *Plumbago capensis*. Am J Bot 59:279–283

Cass DD (1973) An ultrastructural and nomarski-interference study of sperms of barley. Can J Bot 51:601–605

Cass DD, Jensen WA (1970) Fertilization in barley. Am J Bot 57:62–70
Cass DD, Karas I (1974) Ultrastructural organization of the egg of *Plumbago zeylanica*. Protoplasma 81:49–62
Cass DD, Karas I (1975) Development of sperm cells in barley. Can J Bot 53:1051–1062
Cass DD, Peteya DJ (1979) Growth of barley pollen tubes in vivo. Ultrastructural aspects of early tube growth in the stigmatic hair. Can J Bot 57:386–396
Chandra S, Bhatnagar SP (1974) Reproductive biology of *Triticum* II. Pollen germination, pollen tube growth, and its entry into the ovule. Phytomorphology 24:211–217
Chang PLY, Trevithick JR (1974) How important is secretion of exoenzymes through apical cell walls of fungi? Arch Microbiol 101:281–293
Chao ChY (1971) A periodic acid-Shiff's substance related to the directional growth of pollen tube into embryo sac in *Paspalum* ovules. Am J Bot 58:649–654
Chao ChY (1977) Further cytological studies of a periodic acid-Shiff's substance in the ovules of *Paspalum orbiculare* and *Paspalum congiflorum*. Am J Bot 64:921–930
Chao ChY (1979) Histochemical study of a PAS-substance in the ovules of *Paspalum orbiculare* and *Paspalum longifolium*. Phytomorphology 29:381–386
Christ B (1959) Entwicklungsgeschichtliche und physiologische Untersuchungen über die Selbsterilität von *Cardamine pratensis* L. Z Bot 47:88–112
Ciampolini F, Cresti M, Pacini E (1978) Caratteristiche ultrastrutturali ed istochimiche del tessuto trasmittente stilare di melo. Proc Symp „Fertilità delle piante da frutto". Bologna, pp 544–551
Clarke AE, Considine JA, Ward R, Knox RB (1977) Mechanism of pollination in *Gladiolus:* Roles of the stigma and pollen tube guide. Ann Bot 41:15–20
Coccuci AE (1973) Orchid embryology: The membrane systems and the pollen tube growth. Caryologia 25:201–206
Cocucci A, Fulvio TE di (1969) Sobre la naturaleza nuclear de los „Cuerpos X". Kurtziana 5:317–323
Cocucci A, Jensen WA (1969) Orchid embryology: Megagametophyte of *Epidendrum scutella* following fertilization. Am J Bot 56:629–640
Coe GE (1954) Distribution of carbon-14 in ovules of *Zephyranthes drummondii*. Bot Gaz 115:342–346
Cooper DC (1946) Double fertilization in *Petunia*. Am J Bot 33:54–57
Coustaut D, Linskens HF, Moschetto Y, Delbart L (1978) Quelques aspects du métabolisme des glycosphingolipides des styles et du pollen chez *Petunia hybrida,* espicée auto-incompatible. Bull Soc Bot Fr 125:69–78
Cresti M, Went JL van (1976) Callose deposition and plug formation in *Petunia* pollen tubes in situ. Planta 133:35–40
Cresti M, Went JL van, Pacini E, Willemse MTM (1976a) Ultrastructure of transmitting tissue of *Lycopersicon peruvianum* style: Development and histochemistry. Planta 132:305–312
Cresti M, Went JL van, Willemse MTM, Pacini E (1976b) Fibrous masses and cell and nucleus movement in the pollen tube of *Petunia hybrida*. Acta Bot Neerl 25:381–383
Cresti M, Pacini E, Ciampolini F, Sarfatti G (1977a) Germination and early tube development in vitro of *Lycopersicon peruvianum* pollen: Ultrastructural features. Planta 136:239–247
Cresti M, Ciampolini F, Pacini E (1977b) Ultrastructural aspects of pollen tube growth inhibition after gamma irradiation in *Lycopersicum peruvianum*. Theor Appl Genet 49:297–303
Cresti M, Ciampolini F, Pacini E (1978a) Gamma irradiation of *Prunus avium* L. flower buds: Effects on stylar development – An ultrastructural study. Acta Bot Neerl 27:97–106
Cresti M, Ciampolini F, Pacini E, Sarfatti G (1978b) Phytoferritin in plastids of the style of *Olea europaea* L. Acta Bot Neerl 27:417–423
Cresti M, Ciampolini F, Pacini E, Sarfatti G, Went JL van, Willemse MTM (1979a) Ultrastructural differences between compatible and incompatible pollen tubes in the stylar transmitting tissue of *Petunia hybrida*. J Submicrosc Cytol 11:209–219
Cresti M, Ciampolini F, Pacini E, Sarfatti G, Donini B (1979b) Ultrastructural features of *Prunus avium* L. pollen tube in vivo. I. The compatible pollen tube. Caryologia 32:433–440
D'Alascio-Deschamps R (1973) Modifications ultrastructurales liées à la fécondation chez le lin. Bull Soc Bot Fr 122:137–146
Dashek WV, Thomas HR, Rosen WG (1971) Secretory cells of lily pistils. II. Electron microscope cytochemistry of canal cells. Am J Bot 58:909–920
Delbart C, Linskens HF, Bris B, Moschetto Y, Coustaut D (1980a) Analysis of glycosphingolipids of *Petunia hybrida,* a self-incompatible species. I. Composition in fatty acids and in long chain bases of pollen and unpollinated style. Dev Biol Proc 83:1–11

Delbart C, Bris B, Linskens HF, Linder R, Coustaut D (1980 b) Analysis of glycosphingolipids of *Petunia hybrida,* a self-incompatible species. II. Evolution of the fatty acids composition after cross- and self-pollination. Dev Biol Proc 83:241–254

Delbart C, Bris B, Linskens HF, Coustaut D (1980 c) Analysis of glycosphingolipids of *Petunia hybrida,* a self-incompatible species. III. Evolution of the chained bases after self- and cross-pollination. Dev Biol Proc 83:255–269

Deurenberg JJM (1976) In vitro protein synthesis with polysomes from unpollinated, cross- and self-pollinated *Petunia* ovaries. Planta 128:29–33

Deurenberg JJM (1977) Differentiated protein synthesis with polysomes from *Petunia* ovaries before fertilization. Planta 133:201–206

Diboll AG (1968) Fine structure development of the megagametophyte of *Zea mays* following fertilization. Am J Bot 55:787–806

Diboll AG, Larson DA (1966) An electron microscopic study of the mature megagametophyte in *Zea mays.* Am J Bot 53:391–402

Dickinson HG, Lawson J (1975 a) Pollen tube growth in the stigma of *Oenothera organensis* following compatible and incompatible intraspecific pollinations. Proc R Soc Lond B Biol Sci 188:327–344

Dickinson HG, Lawson J (1975 b) The growth of the pollen tube wall in *Oenothera organensis.* J Cell Sci 18:519–532

Dickinson HG, Lewis D (1973) Cytochemical and ultrastructural differences between intraspecific compatible and incompatible pollination in *Raphanus.* Proc R Soc Lond B Biol Sci 183:21–38

Dickinson HG, Lewis D (1975) Interaction between the pollen grain coating and the stigmatic surface during compatible and incompatible intraspecific pollinations in *Raphanus.* In: Duckett JG, Racey PA (eds) The biology of the male gamete. Biol J Linn Soc, vol 7, Suppl I. Academic Press, London, pp 165–175

Dickinson HG, Moriarty J, Lawson J (1982) Pollen-pistil interaction in *Lilium longiflorum:* The role of the pistil in controlling pollen tube growth following cross- and self-pollinations. Proc R Soc Lond Biol Sci 215:45–62

Dolstra O (1982) Synthesis and fertility of *Brassicoraphanus* and ways of transferring *Raphanus* characters to *Brassica.* Ph D Thesis, Agricult Univ, Wageningen PUDOC

Donk JAWM van der (1975) Molecular biological aspects of the incompatible reaction in *Petunia.* Ph D Thesis, Univ Nijmegen

Dumas C (1973) Contribution à l'étude cyto-physiologique du stigmate. III. Evolution et rôle du réticulum endoplasmique au cours de la sécrétion chez *Forsythia intermedia*; étude cytochimique. Z Pflanzenphysiol 70:119–130

Dumas C (1974) Some aspects of stigmatic secretion in *Forsythia.* In: Linskens HF (ed) Fertilization in higher plants. North-Holland, Amsterdam, pp 119–126

Dumas C (1978) Stigmates sécréteurs et lipides neutres sécretés. Bull Soc Bot Fr 125:61–68

Dumas C, Gaude T (1981) Stigma-pollen recognition: A new look. Act Soc Bot Pol 50:235–247

Engels FM (1974) Function of golgi vesicles in relation to cell wall synthesis in germinating *Petunia* pollen. II. Chemical composition of golgi vesicles and pollen tube wall. Acta Bot Neerl 23:81–89

Erdelská O (1974) Contribution to the study of fertilization in the living embryo sac. In: Linskens HF (ed) Fertilization in higher plants. North-Holland, Amsterdam, pp 191–195

Fisher DB, Jensen WA (1969) Cotton embryogenesis: Identification as nuclei, of the X-bodies in the degenerated synergid. Planta 84:122–133

Franke WW, Herth W, Woude WJ van der, Morré DJ (1972) Tubular and filamentous structures in pollen tubes: Possible involvement as guide elements in protoplasmic streaming and vectorial migration of secretory vesicles. Planta 105:317–341

Frankel R, Galun E (1977) Pollination mechanisms, reproduction, and plant breeding. Springer, Berlin Heidelberg New York

Gerassimova-Navashina H (1960) A contribution to the cytology of fertilization in flowering plants. Nucleus (Calcutta) 3:111–120

Gilissen LJW (1976) The role of the style as a sense-organ in relation to wilting of the flower. Planta 131:201–202

Gilissen LJW (1977) Style-controlled wilting of the flower. Planta 133:275–280

Hagemann R (1979) Genetics and molecular biology of plastids of higher plants. Stadler Genet Symp 11:91–116

Herrero M, Dickinson HG (1979) Pollen-pistil incompatibility in *Petunia hybrida:* Changes in the pistil following compatible and incompatible intraspecific crosses. J Cell Sci 36:1–18

Herrero M, Dickinson HG (1980) Ultrastructural and physiological differences between buds and mature flowers of *Petunia hybrida* prior to and following pollination. Planta 148:138–145
Herth W (1978) Ionophore A 23187 stops tip growth, but not cytoplasmic streaming, in pollen tubes of *Lilium longiflorum*. Protoplasma 96:275–282
Heslop-Harrison J (1975 a) Incompatibiliy and the pollen-stigma interaction. Annu Rev Plant Physiol 26:403–425
Heslop-Harrison J (1975 b) Male gametophyte selection and the pollen-stigma interaction. In: Mulcahy DL (ed) Gamete competition in plants and animals. North-Holland/American Elsevier, New York, pp 177–180
Heslop-Harrison J (1978) Cellular recognition systems in plants. Arnold, London
Heslop-Harrison J (1979) An interpretation of the hydrodynamics of pollen. Am J Bot 66:737–743
Heslop-Harrison J (1982) Pollen-stigma interaction and cross-incompatibility in the grasses. Science (Wash DC) 215:1358–1364
Heslop-Harrison J, Heslop-Harrison Y (1975) Enzymic removal of the proteinaceous pellicle of the stigma papilla prevents pollen tube entry in the Caryophyllaceae. Ann Bot 39:163–165
Heslop-Harrison J, Heslop-Harrison Y (1980) The pollen-stigma interaction in the grasses. I. Finestructure and cytochemistry of the stigmas of *Hordeum* and *Secale*. Acta Bot Neerl 29:261–276
Heslop-Harrison J, Heslop-Harrison Y (1982) Pollen-stigma interaction in Leguminosae: Constituents of the stylar fluid and stigma secretion of *Trifolium pratense* L. Ann Bot 49:729–735
Heslop-Harrison J, Knox RB, Howlett B (1973) Pollen-wall proteins: "gametophytic" and "sporophytic" fractions in the pollen walls of the Malvaceae. Ann Bot 37:403–412
Heslop-Harrison J, Knox RB, Heslop-Harrison Y (1974) Pollen-wall proteins: Exine-held fractions associated with the incompatibility response in Cruciferae. Theor Appl Genet 44:133–137
Heslop-Harrison J, Heslop-Harrison Y, Barber J (1975 a) The stigma surface in incompatibility responses. Proc R Soc Lond B Biol Sci 188:287–297
Heslop-Harrison J, Knox RB, Heslop-Harrison Y, Mattisson O (1975 b) Pollen-wall proteins: Emission and role in incompatibility responses. In: Duckett JC, Racey PA (eds) The biology of the male gamete. Academic Press, London, pp 187–204
Heslop-Harrison Y (1977) The pollen-stigma interaction: Pollen-tube penetration in *Crocus*. Ann Bot 41:913–922
Heslop-Harrison Y (1981) Stigma characteristics and angiosperm taxonomy. Norw J Bot 1:401–420
Heslop-Harrison Y, Shivanna KR (1977) The receptive surface of the angiosperm stigma. Ann Bot 41:1233–1258
Hoefert LL (1969 a) Ultrastructure of *Beta* pollen I. Cytoplasmic constituents. Am J Bot 56:363–368
Hoefert LL (1969 b) Fine structure of sperm cells in pollen grains of *Beta*. Protoplasma 68:237–240
Hoefert LL (1971) Pollen grain and sperm cell ultrastructure in *Beta*. In: Heslop-Harrison J (ed) Pollen: Development and physiology. Butterworths, London, pp 68–69
Hoekstra FA (1979) Mitochondrial development and activity of binucleate and trinucleate pollen during germination in vitro. Planta 145:25–36
Hu SY, Zhu Ch (1979) The fusion of male and female nuclei in fertilization of higher plants. Acta Bot Sin 21:1–15
Jensen WA (1963) Cell development during plant embryogenesis. In: Meristems and differentiation. Brookhaven Symp Biol 16:179–202
Jensen WA (1964) Observations on the fusion of nuclei in plants. J Cell Biol 23:669–672
Jensen WA (1965) The ultrastructure and histochemistry of the synergids of cotton. Am J Bot 52:238–256
Jensen WA (1968 a) Cotton embryogenesis: The zygote. Planta 79:346–366
Jensen WA (1968 b) Cotton embryogenesis: Polysome formation in the zygote. J Cell Biol 36:403–406
Jensen WA (1972) The embryo sac and fertilization in angiosperms. Harold L Lyon Arbor Lect 3:1–32
Jensen WA, Fisher DB (1967) Cotton embryogenesis: Double fertilization. Phytomorphology 17:261–269
Jensen WA, Fisher DB (1968) Cotton embryogenesis: The entrance and discharge of the pollen tube in the embryo sac. Planta 78:158–183
Jensen WA, Fisher DB (1969) Cotton embryogenesis: The tissues of the stigma and style and their relation to the pollen tube. Planta 84:97–121
Jensen WA, Fisher DB (1970) Cotton embryogenesis: The pollen tube in the stigma and style. Protoplasma 69:215–235

Jensen WA, Schulz P, Ashton ME (1977) An ultrastructural study of early endosperm development and synergid changes in unfertilized cotton ovules. Planta 133:179–189

Kapil RN, Bhatnagar AK (1975) A fresh look at the process of double fertilization in angiosperms. Phytomorphology 25:334–368

Kapil RN, Vasil IK (1963) Ovule. In: Maheshwari P (ed) Recent advances in the embryology of angiosperms. Intl Soc Plant Morphologists, Univ Dehli, pp 41–67

Karas I, Cass DD (1976) Ultrastructural aspects of sperm cell formation in rye: Evidence for cell-plate involvement in generative cell division. Phytomorphology 26:36–45

Knox RB (1973) Pollen wall proteins: Pollen-stigma interactions in ragweed and *Cosmos* (Compositae). J Cell Sci 12:421–443

Knox RB, Heslop-Harrison J, Heslop-Harrison Y (1975) Pollen wall proteins: Localization and characterization of gametophytic and sporophytic fractions. In: Duckett JC, Racey PA (eds) The biology of the male gamete. Biol J Linn Soc, vol 7, Suppl I. Academic Press, London, pp 177–187

Konar RN, Linskens HF (1966a) The morphology and anatomy of the stigma of *Petunia hybrida*. Planta 71:356–371

Konar RN, Linskens HF (1966b) Physiology and biochemistry of the stigma fluid of *Petunia hybrida*. Planta 71:372–387

Korobova SN (1974) On the behaviour of sperms in the process of fertilization of higher plants. In: Linskens HF (ed) Fertilization in higher plants. North-Holland, Amsterdam, pp 261–270

Kroh M (1967) Bildung und Transport des Narbensekrets von *Petunia hybrida*. Planta 77:250–260

Kroh M (1973) Nature of the intercellular substance of stylar transmitting tissue. In: Loewus F (ed) Biogenesis of plant cell wall polysaccharides. Academic Press, New York, pp 195–205

Kroh M, Brakel CHJ van (1973) Incorporation of label into intercellular substance of stylar transmitting tissue from *Petunia* pistils labelled with tritiated myo-inositol. An electromicroscopic autoradiographic study. Acta Bot Neerl 22:106–111

Kroh M, Helsper JPFG (1974) Transmitting tissue and pollen tube growth. In: Linskens HF (ed) Fertilization in higher plants. North-Holland, Amsterdam, pp 167–178

Kroh M, Gorissen MH, Pfahler PL (1979) Ultrastructural studies on styles and pollen tubes of *Zea mays* L. General survey on pollen tube growth in vivo. Acta Bot Neerl 28:113–118

Labarca C, Kroh M, Loewus F (1970) The composition of stigmatic exudate from *Lilium longiflorum*. Labelling studies with myo-inositol, d-glucose and L-proline. Plant Physiol (Bethesda) 46:150–156

Larson D (1965) Fine structural changes in the cytoplasm of germinating pollen. Am J Bot 52:139–154

Linder R, Linskens HF (1972) Evolution des acides aminés dans le style d'*Oenothera missouriensis* vierge, autopollinisé et xénopollinisé. Theor Appl Genet 42:125–129

Linskens HF (1968) Egg-sperm interactions in higher plants. Acc Naz Lincei 104:47–60

Linskens HF (1973) Activation of the ovary. Caryologia 25:27–41

Linskens HF (1974a) Translocation phenomena in the *Petunia* flower after cross- and self-pollination. In: Linskens HF (ed) Fertilization in higher plants. North-Holland, Amsterdam, pp 285–292

Linskens HF (1974b) Some observations on the growth of the style. Incomp News Lett 4:4–15

Linskens HF (1974c) Study of growth of *Petunia* styles. Fyziologiya Rast 21:1059–1064

Linskens HF (1975a) Incompatibility in *Petunia*. Proc R Soc Lond B Biol Sci 188:299–311

Linskens HF (1975b) The physiological basis of incompatibility in angiosperms. In: Duckett JG, Racey PA (eds) The biology of the male gamete. Biol J Linn Soc, vol 7, Suppl I. Academic Press, London, pp 143–152

Linskens HF (1976a) Specific interactions in higher plants. In: Wood RKS, Graniti A (eds) Specificity in plant diseases. Plenum, New York, pp 311–325

Linskens HF (1976b) Stigmatic responses. In: Vardar Y, Sheikh KH, Oztürk MA (eds) Proc 3rd MPP Meeting Ege Univ Bornova, Izmir, pp 1–12

Linskens HF (1980a) Physiology of fertilization and fertilization barriers in higher plants. In: Subtelny S, Wessels NK (eds) The cell surface. Mediator of developmental process. Academic Press, London, pp 113–125

Linskens HF (1980b) Befruchtungs-Barrieren bei höheren Pflanzen. Naturwiss Rundsch 33:11–20

Linskens HF, Heinen W (1962) Cutinase Nachweis in Pollen. Z Bot 50:338–347

Linskens HF, Kroh M (1970) Regulation of pollen tube growth. In: Moscona AA, Monroy A (eds) Current topics in developmental biology, vol 5. Academic Press, New York, pp 89–113

Linskens HF, Linder R, Salden M, Havez R, Randoux A, Laniez D, Coustaut D (1970) Etude des glycoprotéines et glycannes-hydrolases au cours de la pollinisation chez *Petunia hybrida* auto-incompatible. Bull Soc Pharm 1:1–16

Loewus F, Labarca C (1973) Pistil secretion product and pollen tube wall formation. In: Loewus F (ed) Biogenesis of plant cell wall polysaccharides. Academic Press, New York, pp 175–193

Lord EM, Webster BD (1979) The stigmatic exudate of *Phaseolus vulgaris* L. Bot Gaz 140:266–271

Luxova M (1967) Fertilization of barley (*Hordeum distichum* L.). Biol Plant (Prague) 9:301–307

Maheshwari P (1950) An introduction to the embryology of angiosperms. McGraw-Hill, New York

Maheshwari P (ed) (1963) Recent advances in the embryology of angiosperms. Intl Soc Plant Morphologists, Univ Delhi

Mascarenhas JP (1973) Pollen tube chemotropism. In: Pérez-Miravette A (ed) Behaviour of micro-organisms. Plenum, London New York, pp 62–69

Mascarenhas JP (1975) The biochemistry of angiosperm pollen development. Bot Rev 41:260–314

Mascarenhas JP, Machlis L (1962a) The pollen tube chemotropic factor from *Antirrhinum majus:* Bioassay, extraction, and partial purification. Am J Bot 49:482–489

Mascarenhas JP, Machlis L (1962b) The hormonal control of directional growth of pollen tubes. Vitam Horm 20:347–371

Mascarenhas JP, Machlis L (1964) Chemotropic response of the pollen of *Antirrhinum majus* to calcium. Plant Physiol (Bethesda) 39:70–77

Miki-Hirosige H (1964) Tropism of pollen tubes to the pistils. In: Linskens HF (ed) Pollen physiology and fertilization. North-Holland, Amsterdam, pp 152–158

Miki-Hirosige H, Nakamura S (1982) Process of metabolism during pollen tube wall formation. J Electron Microsc 31:51–62

Maze J, Lin SC (1975) A study of the mature megagametophyte of *Stipa elmeri.* Can J Bot 53:2985–2977

Mogensen HL (1972) Fine structure and composition of the egg apparatus before and after fertilization in *Quercus gambelii:* The functional ovule. Am J Bot 59:931–941

Navashin MS (1969) On the nature of the movement of the generative cell in the pollen tube and the problem of the localization of cell elements. Rev Cytol Biol Vég 32:141–148

Nettancourt D de (1977) Incompatibility in angiosperms. Springer, Berlin Heidelberg New York

Nettancourt D de, Devreux H, Laneri N, Pacini E, Cresti M, Sarfatti G (1973a) Ultrastructural aspects of bilateral interspecific incompatibility between *Lycopersicum peruvianum* and *L. esculentum.* Caryologia 25:207–217

Nettancourt D de, Devreux M, Bozzini A, Cresti M, Pacini E, Sarfatti G (1973b) Ultrastructural aspects of the self-incompatibility mechanism in *Lycopersicum peruvianum* Mill. J Cell Sci 12:403–419

Nettancourt D de, Devreux M, Laneri U, Cresti M, Pacini E, Sarfatti G (1974) Genetical and ultrastructural aspects of self- and cross-incompatibility in interspecific hybrids between self-compatible *Lycopersicum esculentum* and self-incompatible *L. peruvianum.* Theor Appl Genet 44:278–288

Newcomb W (1973a) The development of the embryo sac of the sunflower *Helianthus annuus* before fertilization. Can J Bot 51:863–878

Newcomb W (1973b) The development of the embryo sac of the sunflower *Helianthus annuus* after fertilization. Can J Bot 51:879–890

Nishio T, Hinata K (1977) Positive PAS-reaction of S-specific proteins in stigma of *Brassica oleracea* L. Incomp News Lett 8:31–33

Noack R (1960) Die zwittrigen und eingeschlechtlichen Blüten von *Begonia cathayana.* 1. Die chemotropische Reaktionsfähigkeit der Pollenschläuche auf die Narbenstoffe der Blüten. Z Bot 48:463–487

Norstog K (1972) Early development of the barley embryo: Fine structure. Am J Bot 59:123–132

Ockendon DJ (1972) Pollen tube growth and the site of the incompatibility reaction in *Brassica oleracea.* New Phytol 71:519–522

Owens SJ, Kimmins FM (1981) Stigma morphology in Commelinaceae. Ann Bot 47:771–783

Pluym JE van der (1964) An electron microscopic investigation of the filiform apparatus in the embryo sac of *Torenia fournieri.* In: Linskens HF (ed) Pollen physiology and fertilization. North-Holland, Amsterdam, pp 6–16

Preil W, Keyser D (1975) Pollen germination on exposed ovules in *Begonia semper* florens. In: Mulcahy DL (ed) Gamete competition in plants and animals. North-Holland, Amsterdam, pp 219–226

Reiss HD, Herth W (1978) Visualization of the Ca^{2+}-gradient in growing pollen tubes of *Lilium longiflorum* with chlorotetracycline fluorescence. Protoplasma 97:373–377

Roberts IN, Stead DA, Dickinson HG (1979a) No fundamental changes in lipids of the pollen grain coating of *Brassica oleracea* following either self- or cross-pollinations. Incomp News Lett 11:77–83

Roberts IN, Stead DA, Ockendon DJ, Dickinson HG (1979b) A glycoprotein associated with the acquisition of the self-incompatibility system by maturing stigmas of *Brassica oleracea*. Planta 146:179–183

Roggen H (1974) Pollen washing influences (in)compatibility in *Brassica oleracea* varieties. In: Linskens HF (ed) Fertilization in higher plants. North-Holland, Amsterdam, pp 273–278

Rosen WG (1961) Studies on pollen-tube chemotropism. Am J Bot 48:889–895

Rosen WG (1964) Chemotropism and fine structure of pollen tubes. In: Linskens HF (ed) Pollen physiology and fertilization. North-Holland, Amsterdam, pp 159–166

Rosen WG (1971a) Pollen tube growth and fine structure. In: Heslop-Harrison J (ed) Pollen: Development and physiology. Butterworth, London, pp 177–189

Rosen WG (1971b) Pistil-pollen interactions in *Lilium*. In: Heslop-Harrison J (ed) Pollen: Development and physiology. Butterworth, London, pp 239–261

Rosen WG (1975) Pollen-pistil interactions. In: Duckett JG, Racey PA (eds) The biology of the male gamete. Biol J Linn Soc, vol 7, Suppl I. Academic Press, London, pp 153–164

Russell SD (1980) Participation of male cytoplasm during gamete fusion in an angiosperm, *Plumbago zeylanica*. Science (Wash DC) 210:200–201

Russell SD (1981) Structure and quantitative cytology of male gametes of *Plumbago zeylanica* L. Abstr No 042202, XIII Intl Bot Congr, Sydney, p 61

Russell SD (1982) Fertilization in *Plumbago zeylanica:* Entry and discharge of the pollen tube in the embryo sac. Can J Bot 60:2219–2230

Russell SD (1983) Fertilization in *Plumbago zeylanica:* Gametic fusion and fate of the male cytoplasm. Am J Bot 70:416–434

Russell SD, Cass DD (1981) Ultrastructure of fertilization in *Plumbago zeylanica*. Acta Soc Bot Pol 50:185–190

Sanger JW, Jackson WT (1971a) Fine structure study of pollen development in *Haemanthus katherinae* Baker. II. Microtubules and elongation of the generative cells. J Cell Sci 8:303–315

Sanger JW, Jackson WT (1971b) Fine structure study of the pollen development in *Haemanthus katherinae* Baker. III. Changes in organelles during development of the vegetative cell. J Cell Sci 8:317–329

Sassen MMA (1974) The stylar transmitting tissue. Acta Bot Neerl 23:99–108

Sastri DC, Shivanna KR (1980) Electrophoretic patterns of proteins and isozymes in developing pistils of *Petunia hybrida*. Incomp News Lett 12:24–29

Schildknecht H, Benoni H (1963) Über die Chemie der Anziehung von Pollenschläuchen durch die Samenanlagen von Oenotheren. Z Naturforsch 18:45–54

Schulz R, Jensen WA (1968a) *Capsella* embryogenesis: The synergids before and after fertilization. Am J Bot 55:541–552

Schulz R, Jensen WA (1968b) *Capsella* embryogenesis: The egg, zygote, and young embryo. Am J Bot 55:807–819

Schulz P, Jensen WA (1973) *Capsella* embryogenesis: The central cell. J Cell Sci 12:741–763

Schulz P, Jensen WA (1974) *Capsella* embryogenesis: The development of the free-nuclear endosperm. Protoplasma 80:183–205

Schulz P, Jensen WA (1977) Cotton embryogenesis: The early development of the free-nuclear endosperm. Am J Bot 64:384–394

Sedgley M (1976) Control by the embryo sac over pollen tube growth in the style of the avocado (*Persea americana* Mill). New Phytol 77:149–152

Sedgley M (1977) Reduced pollen tube growth and the presence of callose in the pistil of the male floral stage of the avocado. Sci Hortic (Canterbury) 7:27–36

Sedgley M, Buttrose MS (1978) Structure of the stigma and style of the avocado. Aust J Bot 26:663–682

Shivanna KR (1978) Elution of wall-bound proteins does not affect pollen germination and pollen tube growth in vitro in *Vicia faba*. Incomp News Lett 10:40–42

Shivanna KR (1980) Some correlations between the cytomorphology of the style and stigma and the details of pollen-pistil interaction. Incomp News Lett 12:16–23

Shivanna KR (1982) Pollen-pistil interaction and control of fertilization. In: Johri BM (ed) Experimental embryology of vascular plants. Springer, Heidelberg, pp 132–174

Shivanna KR, Heslop-Harrison J (1981) Membrane state and pollen viability. Ann Bot 47:759–770

Shivanna KR, Sastri DC (1981) Stigma-surface esterase activity and stigma receptivity in some taxa characterized by wet stigmas. Ann Bot 47:53–64

Shivanna KR, Heslop-Harrison Y, Heslop-Harrison J (1978 a) The pollen-stigma interaction: Bud pollination in the Cruciferae. Acta Bot Neerl 27:107–119

Shivanna KR, Heslop-Harrison Y, Heslop-Harrison J (1978 b) Inhibition of the pollen tube in the self-incompatibility response of grasses. Incomp News Lett 10:5–7

Shivanna KR, Johri BM, Sastri DC (1979) Development and physiology of angiosperm pollen. Today and Tomorrow's, New Delhi

Shivanna KR, Heslop-Harrison Y, Heslop-Harrison J (1982) The pollen-stigma interaction in the grasses. III. Features of the self-incompatibility response. Acta Bot Neerl 31:307–319

Singh AP, Mogensen HL (1975) Fine structure of the zygote and early embryo in *Quercus gambelii*. Am J Bot 62:105–115

Spanjers AW (1978) Voltage variation in *Lilium longiflorum* pistils induced by pollination. Experientia (Basel) 34:36–37

Stanley RG (1971) Pollen chemistry and tube growth. In: Heslop-Harrison J (ed) Pollen: Development and physiology. Butterworth, London, pp 131–155

Stanley RG, Linskens HF (1974) Pollen: Biology, biochemistry, management. Springer, Berlin Heidelberg New York

Stead AD, Roberts IN, Dickinson HG (1979) Pollen-pistil interaction in *Brassica oleracea*. Planta 146:211–216

Stead AD, Roberts IN, Dickinson HG (1980) Pollen-stigma interaction in *Brassica oleracea:* The role of stigmatic proteins in pollen grain adhesion. J Cell Sci 42:417–423

Steffen K (1951) Zur Kenntnis des Befruchtungsvorganges bei *Impatiens glanduligera* Lindl. Cytologische Studien am Embryosack der Balsamineen. Planta 39:175–244

Steffen K (1963) Fertilization. In: Maheshwari P (ed) Recent advances in the embryology of angiosperms. Intl Soc Plant Morphologists, Univ Dehli, pp 105–133

Thomas SM, Murray BG (1975) A new site for the self-incompatibility reaction in the Gramineae. Incomp News Lett 6:22–23

Townsend CE (1971) Advances in the study of incompatibility. In: Heslop-Harrison J (ed) Pollen: Development and physiology. Butterworth, London, pp 281–309

Tsao TH (1949) A study of chemotropism of pollen tubes in vitro. Plant Physiol (Bethesda) 20:494–504

Vasil IK (1973) The new biology of pollen. Naturwissenschaften 60:247–253

Vasil IK (1974) The histology and physiology of pollen germination and pollen tube growth on the stigma and in the style. In: Linskens HF (ed) Fertilization in higher plants. North-Holland, Amsterdam, pp 105–118

Vasil IK, Johri MM (1964) The style, stigma, and pollen tube. Phytomorphology 14:352–369

Vasilev AE (1970) Ultrastructure of stigmatoid cells in *Lilium*. Sov Plant Physiol 19:1035–1044

Vazart J (1969 a) Organisation et ultrastructure du sac embryonnaire du lin (*Linum usitatissimum* L.). Rev Cytol Biol Vég 32:227–240

Vazart J (1969 b) Degeneration of a synergid and pollen tube entrance into the embryo sac of *Linum usitatissimum* L. Ann Univ ARERS 9:89–97

Vazart J (1970) Aspects infrastructuraux de la reproduction sexuée chez le lin. Derniers stades de la différenciation du pollen. Structure inframicroscopique de la cellule génératrice et des gametes. Rev Cytol Biol Vég 33:289–310

Vijayaraghavan MR, Jensen WA, Ashton ME (1972) Synergids of *Aquilegia formosa*. Their histochemistry and ultrastructure. Phytomorphology 22:144–159

Vithanage HIMV, Heslop-Harrison J (1979) The pollen-stigma interaction: Fate of fluorescent-labelled pollen-wall proteins on the stigma surface in rye (*Secale cereale*). Ann Bot 43:113–114

Vries H de (1900) Het leven der bloem. Willink, Haarlem, p 142

Weisenseel MH, Nuccitelli R, Jaffé LF (1975) Large electrical currents traverse growing pollen tubes. J Cell Biol 66:556–567

Went JL van (1970 a) The ultrastructure of the synergids of *Petunia*. Acta Bot Neerl 19:121–132

Went JL van (1970 b) The ultrastructure of the fertilized embryo sac of *Petunia*. Acta Bot Neerl 19:468–480

Went JL van (1974) The ultrastructure of *Impatiens* pollen. In: Linskens HF (ed) Fertilization in higher plants. North-Holland, Amsterdam, pp 81–88

Went JL van, Linskens HF (1967) Die Entwicklung des sogenannten „Fadenapparatus" im Embryosack von *Petunia hybrida*. Gen Breed Res 37:51–56

Wilms HJ (1974) Branching of pollen tubes in spinach. In: Linskens HF (ed) Fertilization in higher plants. North-Holland, Amsterdam, pp 155–160

Wilms HJ (1980a) Ultrastructure of the developing embryo sac of spinach. Acta Bot Neerl 29:243–260

Wilms HJ (1980b) Ultrastructure of the stigma and style of spinach in relation to pollen germination and pollen tube growth. Acta Bot Neerl 29:33–47

Wilms HJ, Van Aelst AC (1978) Fertilization in spinach: The pathway of the pollen tubes in the style. Bull Soc Bot Fr 125:243–247

Woittiez RD, Willemse MTM (1979) Sticking of pollen on stigmas: The factors and a model. Phytomorphology 29:57–63

7 The Endosperm

M. R. Vijayaraghavan and K. Prabhakar

7.1 Introduction

Historically, the fate of the second male gamete involved in triple fusion was traced initially in *Lilium martagon* and *Fritillaria tenella* (Nawaschin 1898). Following this discovery, double fertilization was reported in a large number of angiosperms. The fusion product of one sperm with the secondary nucleus results in the formation of the primary endosperm nucleus. The nuclei or cells derived through mitoses of this nucleus and daughter nuclei constitute the endosperm. Whether the endosperm is haploid and a continuation of the gametophytic tissue, a diploid or triploid tissue, or a second embryo of abnormal size and shape, is still debatable. Strasburger (1900) even called the fusion of the two polar nuclei with the second male gamete as vegetative fertilization. The families which lack endosperm are: Orchidaceae, Podostemaceae, and Trapaceae.

The endosperm is essentially a triploid tissue, but can be diploid as in *Butomopsis* (Johri 1936) and even pentaploid as in *Fritillaria* (Maheshwari 1950). The division of the primary endosperm nucleus is followed by repeated divisions of the daughter nuclei. On the basis whether a wall is formed, or not, after the division of the primary endosperm nucleus, the endosperm is classified as of the Cellular or Nuclear type. There is also the intermediary Helobial type.

The endosperm tissue is responsible for the growth and germination of the embryo. The classical role of supplying nutrition to the developing embryo needs a reassessment. Recent studies, however, indicate that the endosperm does not have a significant role in nourishing the proembryo during early embryogenesis when the endosperm is in an actively-dividing stage. In seeds the persistent endosperm acts as the repository of reserve food materials, and in cereals it constitutes the edible portion.

7.2 Nuclear Endosperm

In Nuclear type the primary endosperm nucleus and the daughter nuclei undergo free-nuclear divisions (Fig. 7.1 A–E). These nuclei remain embedded in a cytoplasmic sheath around the central vacuole. The number of free-nuclear divisions, before wall formation, varies in different taxa. In the majority the endosperm, eventually, becomes cellular (Fig. 7.1 F). In *Limnanthes* (Mathur 1956) and *Oxyspora paniculata* (Subramanyam 1951), however, free-nuclear condition persists until the endosperm is almost completely consumed by the developing embryo.

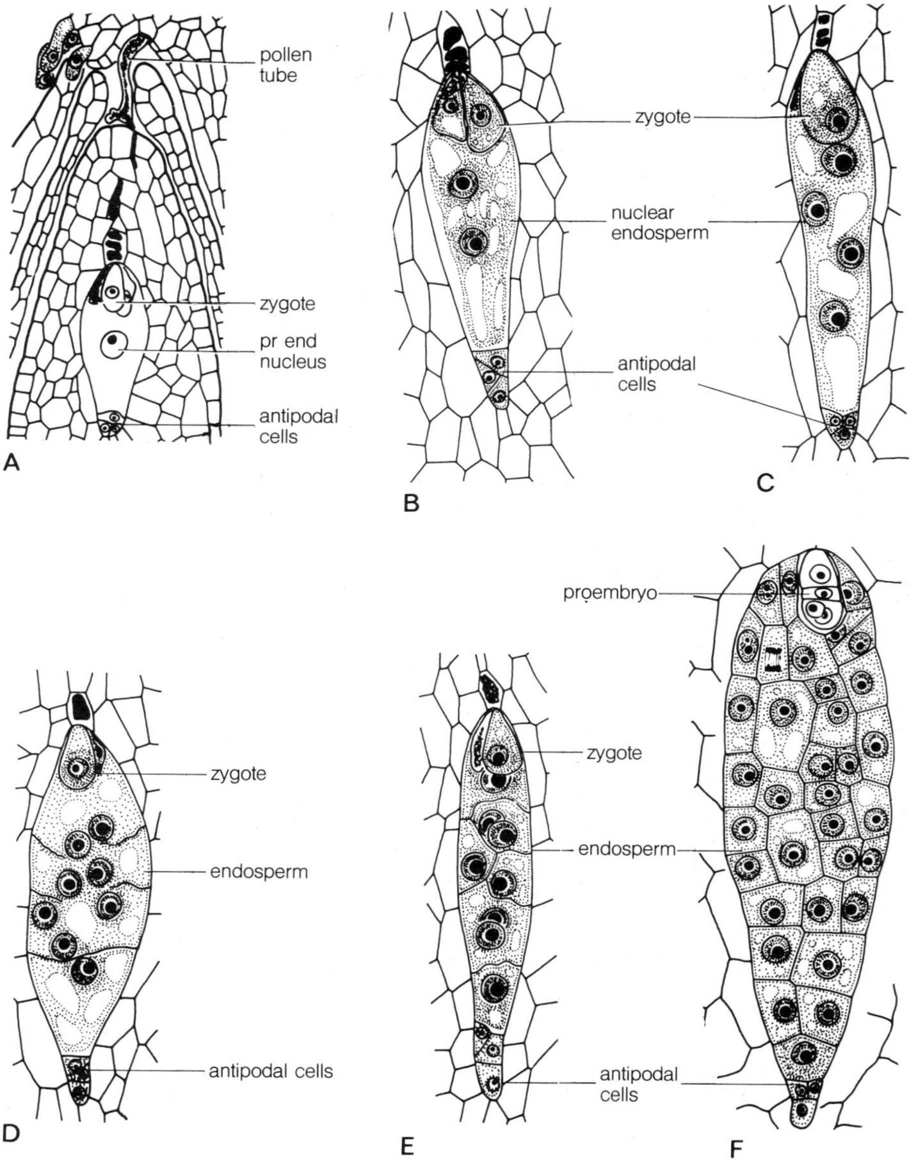

Fig. 7.1 A–F. *Diplacrum caricinum,* endosperm development. **A** Longisection of micropylar portion of ovule showing pollen tube, zygote, and primary endosperm nucleus. **B, C** Zygote, two- and four-nucleate endosperm. **D** Zygote and eight-nucleate endosperm; centripetal wall formation occurs at this stage. **E, F** Zygote (**E**) and proembryo (**F**), and cellular endosperm. Note persistent degenerated megaspores in **A–E,** and antipodal cells in **A–F.** (After Nijalingappa and Devaki 1978)

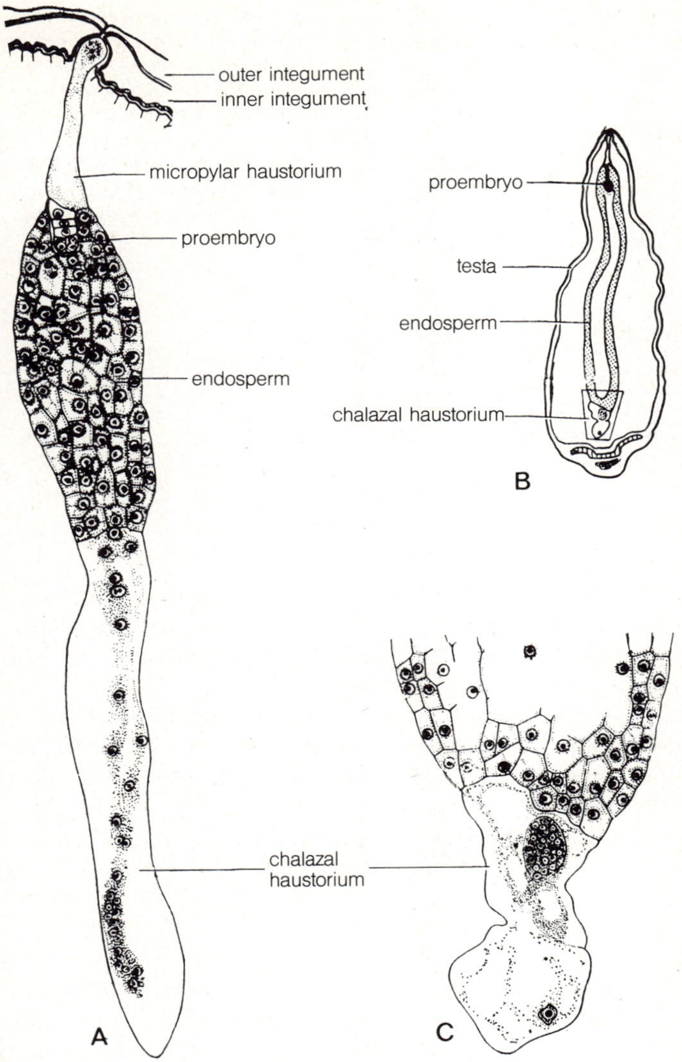

Fig. 7.2 A–C. *Scleria foliosa.* **A** Endosperm with micropylar and chalazal haustoria. **B** Longisection of seed showing a central cavity in the endosperm. **C** Portion of chalazal haustorium from **B.** (After Nijalingappa and Devaki 1979)

In *Scleria foliosa* (Nijalingappa and Devaki 1979), at the two-nucleate stage, a tubular extension is formed at the micropylar end of the embryo sac. It remains enucleate with scanty cytoplasm, and acts as a haustorium (Fig. 7.2 A). Centripetal wall formation is restricted only to the upper region of endosperm (Fig. 7.2 A). The chalazal portion remains coenocytic, and functions as a haustorium (Fig. 7.2 B, C).

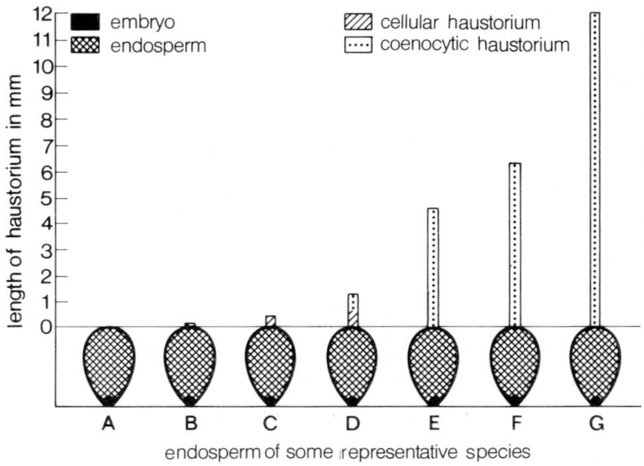

Fig. 7.3 A–G. Cucurbitaceae; endosperm at globular proembryo to show cellular portion and coenocytic haustorium. **A** *Blastania garcini,* **B** *Melothira heterophylla,* **C** *Citrullus fistulosus,* **D** *Benincasa cerifera,* **E** *Melothira maderaspatana,* **F** *Trichosanthes anguina,* **G** *Cucurbita ficifolia.* (After Chopra and Agarwal 1958)

7.2.1 Wall Formation

Wall formation is usually centripetal; it is initiated around the proembryo and gradually extends towards the chalazal region of the embryo sac. In most plants, wall formation once initiated continues until there are no more free nuclei. The cells divide until the maturity of the seed, and get filled with reserve food.

In a few families, Cruciferea, Cucurbitaceae, Leguminosae, and Proteaceae, wall formation is no doubt initiated but does not proceed beyond the central region. The chalazal region, thus, remains free-nuclear (Fig. 7.3). The free-nuclear zone not only shows much variation but also endopolyploid nuclei.

The development of endosperm in selected families is discussed.

7.2.2 Cucurbitaceae

Dissections and microtome sections have revealed many interesting types of chalazal haustoria in Cucurbitaceae (Fig. 7.3). The micropylar broader portion, which becomes cellular, is referred to as endosperm proper, thus demarcating it from the free-nuclear chalazal zone. Occasionally, the peripheral cells in the lower region of endosperm proper bulge out during later stages of the development of seed. In *Luffa aegyptica* and *Melothria maderaspatana* such bulges are usually referred to as "secondary" haustoria (Chopra and Agarwal 1958).

The lower coenocytic portion elongates gradually and forms a tubular process. This region contains dense cytoplasm with numerous nuclei of various sizes, shapes, and ploidy levels. The tip of the haustorium in *Cucurbita pepo* and *Cucumis sativus* becomes swollen. The development of the endosperm, with particular reference to the structure of the haustorium, has been studied in *Benincasa cerifera, Cucumis melo, Luffa aegyptica*, and *Melothria heterophylla*. In *Cucumis* the length of the haustorium varies from 1,170 to 3,006 μm; in *Benincasa* 270 to 1,170 μm; and in *Luffa* 360 to 990 μm. *Melothria* shows only a rudimentary haustorium which is represented by a short protuberance consisting of small cells. There is a correlation between the length of seed, length of haustorium, and haustorial activity (Fig. 7.3). The haustorium presumably transports metabolites from the circumjacent tissue to the developing embryo. Occasionally, in some taxa, the coenocytic haustorium cuts off multinucleate chambers which later segment into uninucleate cells. This aspect requires further study, both at the light microscopy and the ultrastructural level.

7.2.3 Leguminosae

The pioneering work of Rau (1951 a–c, 1953, 1955), and Johri and Garg (1959) have brought to light several interesting features. In many genera of the Phaseolae, after the free-nuclear phase, wall formation usually occurs in the vicinity of the embryo. The lower half of the endosperm remains free-nuclear, assumes vesicular contour, contains hypertrophied nuclei, and functions as a haustorium. The shape and size of the coenocytic haustorium varies in many taxa, and is clearly demarcated from the cellular region. In *Cassia* it is bulbous, in *Mimosa pudica* constricted, and in *Calliandra* coiled. *Cyamopsis* and *Desmodium* are exceptions, and cellularization occurs in the coenocytic portion.

7.2.4 Palmae

The correlation between the size of fruit and activity of endosperm is exemplified in *Cocos nucifera.* In 50 mm long fruits the embryo sac is still filled with liquid endosperm which contains many free nuclei and cytoplasm, usually referred to as "liquid syncytium". The endosperm nuclei, at this stage, are active and undergo mitotic divisions. When the fruit reaches 100 mm, many endosperm vesicles appear in the liquid syncytium. Such vesicles arise due to coalescence of cytoplasmic material that surrounds the free nuclei, are 10–30 μm in diameter, and contain as many as 40 nuclei. Occasionally, a single nucleus enlarges without undergoing any division, and fills the entire vesicle (Cutter et al. 1955). These vesicles often coexist with enucleate endosperm nodules. Several interesting features are reported in about 150 mm long fruits. The vesicles and nuclei migrate to the chalazal end and, finally, the vesicles disintegrate. The contents of vesicles mix with the existing free endosperm nuclei, and form a gelatinous syncytium. The nuclei in the syncytium elongate, and show amoeboid movement. Mitotic activity (in these nuclei) is rapid, and followed by wall formation. The cellular endosperm is referred to as the coconut meat which never fills the entire cavity, probably owing to the large size of the embryo sac.

The coconut milk, which is the liquid endosperm, contains growth hormones, protein granules, oil droplets, and many nuclei. Its use in tissue culture experiments dealing with growth and morphogenesis is well documented, and it also induces cell division in embryogenic pollen grains. Carrots and coconuts are made for each other, and their interdependence in experimental embryology is well known. The activity of the coconut milk at biochemical level, however, still remains an enigma.

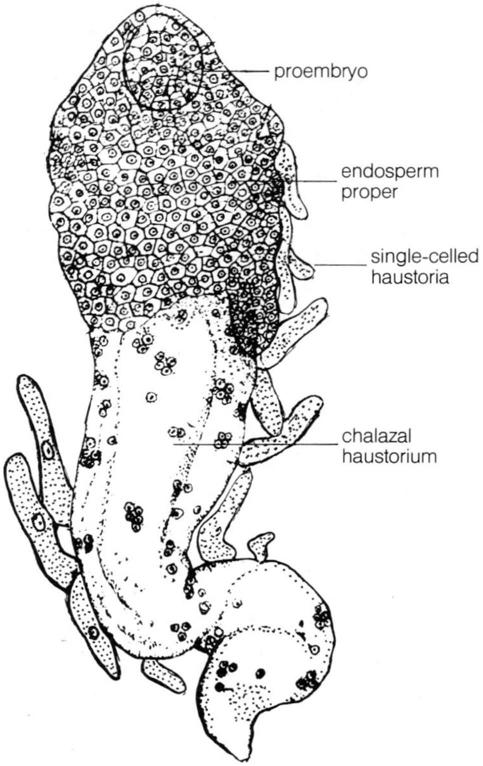

Fig. 7.4. *Lomatia polymorpha.* Wholemount showing proembryo, cellular endosperm proper, chalazal haustorium, and many finger-like nucleated outgrowths all over the surface of endosperm. (After Venkata Rao 1962)

7.2.5 Proteaceae

Dissections of endosperm in *Grevillea robusta* revealed that the chalazal region remains free-nuclear, shows streaming of cytoplasm, becomes coiled, and is aptly designated "vermiform appendage". This appendage invades the chalazal region of the ovule and functions as an aggressive haustorium (Kausik 1941). In *Lomatia polymorpha,* in addition to the main haustorium, numerous single-celled, finger-like outgrowths (Fig. 7.4) arise all over the surface of the endosperm and, thus, increase the absorbtive surface (Venkata Rao 1962).

7.2.6 Histochemistry and Ultrastructure

In the endosperm of *Linum usitatissimum* and *Ranunculus sceleratus* (Vijayaraghavan and Bhat 1982) the concentration of polysaccharides, total proteins, histones, DNA, and RNA increases during embryogeny and, at the dicotyledonous stage of the embryo, the endosperm cells accumulate protein bodies. The observations on *Capsella bursa-pastoris* (Schulz and Jensen 1974), and *Iberis amara*, and *Alyssum maritimum* (Prabhakar 1979) emphasize that the variations in the development of endosperm, at the micropylar and chalazal region, are induced by environmental differences. The development of the micropylar portion is influenced by the proximity of the embryo and suspensor, and the adjoining starch-filled integumentary cells. The development of the chalazal portion of endosperm is, however, affected by the adjacent chalazal "proliferating" tissue.

In *Capsella bursa-pastoris* the integumentary cells abutting the embryo sac wall projections gradually loose their starch reserves, and become filled with tannins (Schulz and Jensen 1974). Contrarily, in *Iberis amara* and *Alyssum maritimum* the integumentary cells around the micropyle and next to the embryo sac wall projections are the last to show accumulation of tannins (Prabhakar and Vijayaraghavan 1982). The presence of wall ingrowths indicates that the integumentary cells are involved in the short-distance transport of nutrients (see Gunning and Pate 1969). Schulz and Jensen (1974) postulated a strong possibility that the growing endosperm may absorb metabolites from the degenerating suspensor cells. Presence of smooth endoplasmic reticulum in close association with the wall projections supports their role in the transport of metabolites.

In *Iberis amara* and *Alyssum maritimum* (Prabhakar 1979) the micropylar portion of endosperm stains intensely for proteins, RNA, and DNA, but lacks starch or protein bodies during early embryogenesis.

In several members of Cruciferae the cytoplasm and endosperm nuclei aggregate at the two poles of the embryo sac, and both regions are rich in nucleolar and cytoplasmic RNA and proteins (Prasad 1974). In cotton the excised unfertilized ovules show two polar nuclei which go through a limited number of free-nuclear divisions after 2 or 3 days in culture. Each nucleus has two nucleoli, in contrast to the nuclei of fertilized triploid endosperm, which has three nucleoli (Jensen et al. 1977).

In *Capsella bursa-pastoris* chloroplasts have been observed, but starch is absent at the two-nucleate stage of endosperm (Schulz and Jensen 1974). The presence of plastids, mitochondria, and ribosomes in the endosperm of various taxa indicate a metabolically active state of this tissue (see Buttrose 1963, Marinos 1970, Newcomb 1973, Weinbaum and Simons 1974). In *Alyssum maritimum* (Prabhakar 1979) numerous dilated cisternae of RER join in such a way that they engulf a portion of cytoplasm. In *Quercus gambelii* (A.P. Singh and Mogensen 1976) too the endoplasmic reticulum undergoes modifications, and lobed portions of the ER cisternae contain cytoplasmic ribosomes along with a variety of other inclusions in the free-nuclear condition of the endosperm. Similar vesicles, or cytoplasmic islands, are reported in degenerated/persistent synergid (Mogensen 1972), and chalazal proliferating tissue (Schulz and Jensen 1971). The significance of these vesicles is not clear but, it appears, some kind of autolysis occurs in the endosperm around the embryo.

Crystals occur in the wide cisternae of rough ER in the micropylar portion of the endosperm in *C. bursa-pastoris* (Schulz and Jensen 1974). In *Alyssum maritimum*, however, paracrystalline bodies have been observed in the nucleoplasm of the nuclei located in the micropylar portion of the endosperm (Prabhakar et al. 1981).

In *Iberis amara* (Fig. 7.5) and *Alyssum maritimum* the chalazal region of endosperm differs from the micropylar and central portions. Before the initiation of wall formation, the endosperm is rich in many macromolecules like proteins,

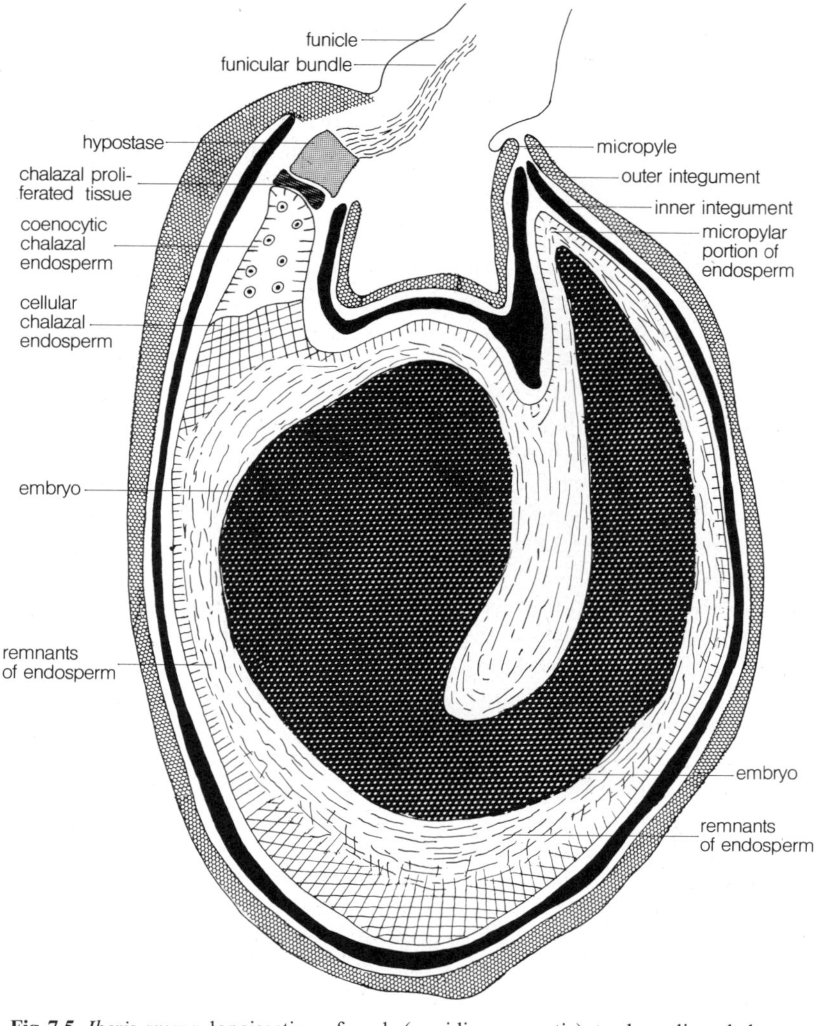

Fig. 7.5. *Iberis amara,* longisection of ovule (semidiagrammatic), to show dicotyledonous embryo, cellular endosperm, and chalazal endosperm caecum. (After Vijayaraghavan and Prabhakar, unpublished)

DNA, and RNA, but lacks storage polysaccharide reserves. Concomitant with cellularization, the micropylar and central endosperm tissue shows a decline in the level of proteins and RNA, but accumulates polysaccharides. In *A. maritimum* the chalazal portion continues to stain intensely for proteins (Fig. 7.6 A). The active metabolic state of the coenocytic chalazal portion of the endosperm is further emphasized by the asynchronous nuclear division (Fig. 7.6 B). However, with the initiation of wall formation, the level of proteins declines in this region. In *Iberis amara* the chalazal endosperm cells show polyploid nuclei that stain intensely for DNA (Fig. 7.6 C). The endosperm caecum, however, does not become compartmentalized even in mature seed and shows many degenerated nuclei (Fig. 7.6 D).

The absorptive capacity of the chalazal haustorium is supported by the presence of wall ingrowths in *Iberis amara* (Fig. 7.5, see also Fig. 7.24 C, D). The endosperm, at the micropylar region, is one-layered and shows polysaccharides only in the form of cell wall (Fig. 7.5, see also Fig. 7.24 B). The presence of modified embryo sac wall has been considered to play an important role in absorption by the chalazal proliferated tissue (Vijayaraghavan and Prabhakar 1981).

Schulz and Jensen (1973) reported that the developmental pattern of chalazal endosperm in *Capsella bursa-pastoris* changes dramatically when the chalazal proliferated cells begin to breakdown, and release intact organelles and degenerate cellular debris into the embryo sac. There is an increase in electron density and levels of proteins. The nuclei become deeply lobed and the nucleoli more "active", the number of dictyosomes and microbodies increases, RER becomes organized into long profiles, parallels nuclear contours, and shows increased connections with the nuclear envelope (Fig. 7.7 A). In *Alyssum maritimum* the peripheral region of endosperm does not differ significantly from the micropylar and chalazal portions of the endosperm. The single row of nuclei do not show much difference in ploidy level, or extrusion of nucleolar material. In this region there is negligible proliferation of ER which reflects a difference in metabolic status of the different portions, of the endosperm. The embryo sac wall also remains smooth in the central region, but numerous cisternae of some ER are associated with it. Thus, the entry of nutrients, presumed from either end of the embryo sac, may also be possible from the general surface as SER is associated with the function of transport.

The endosperm, during early embryogeny, is metabolically active and does not accumulate any food reserve (Newcomb 1973). It follows that, at this stage, the endosperm does not directly provide nutrition for the development of the embryo. In the cellular phase, however, accumulation of polysaccharide grains is reported in the endosperm (Evers 1970, Newcomb 1973). In wheat, as a result of difference in age, the cells of subaleurone layer and inner endosperm differ in the amount of starch they contain (Evers 1970). In *Iberis amara* and *Alyssum maritimum* the peripheral layer shows denser accumulation of polysaccharide grains than the inner layers. Gradually, the outermost layer of the endosperm is depleted of its polysaccharide grains (Fig. 7.5, see also Fig. 7.24 B) and shows accumulation of proteinaceous masses. This layer is comparable to the aleurone layer of the monocotyledons (Prabhakar 1979). In the mature seed, only the outermost layer of endosperm and haustorium persist (Fig. 7.5, see also Fig. 7.24 A, B), and several layers of endosperm rich in polysaccharides are used up by the growing embryo. Newcomb (1973) postulated that the endosperm represents a large pool of reserve materials which is certainly used during later stages of embryogenesis. Hu (1964) also speculated that the absence of starch grains in the region of endosperm around the embryo, during embryogenesis, may be connected with the digestion and absorption of such macromolecules by the maturing embryo. The classic role of storage and supply of nutrition to the embryo seems to be true only during later stages of embryogenesis. The role of endosperm during early/late embryogeny is discussed under endosperm–embryo relationship (see Sect. 10).

A peculiar feature in the Nuclear endosperm is the presence of endosperm nodules/cytoplasmic vesicles, as in *Capsella bursa-pastoris* (Maheshwari and

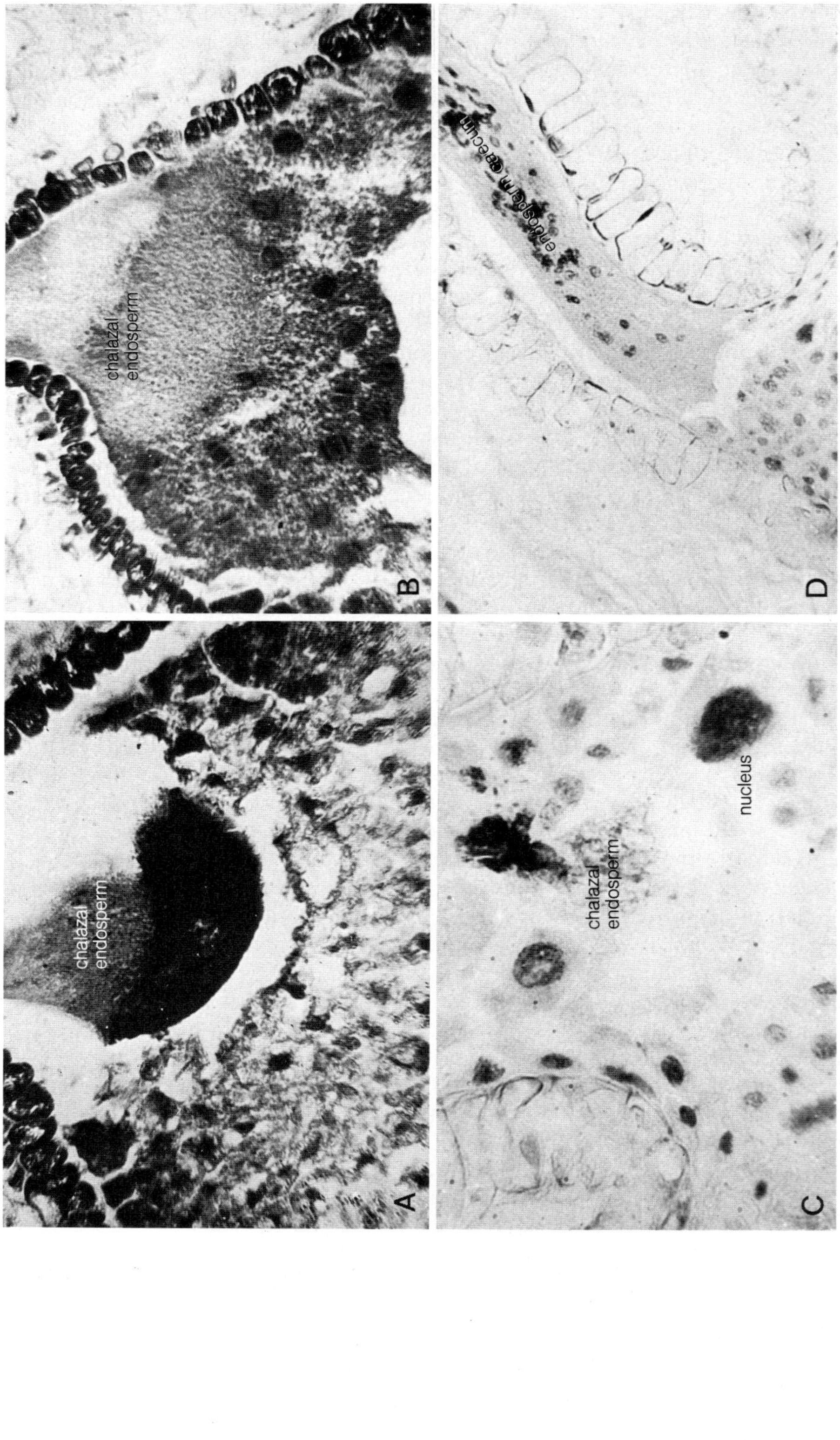
chalazal endosperm
A
chalazal endosperm
B
chalazal endosperm
nucleus
C
endosperm caecum
D

Fig. 7.6A–D

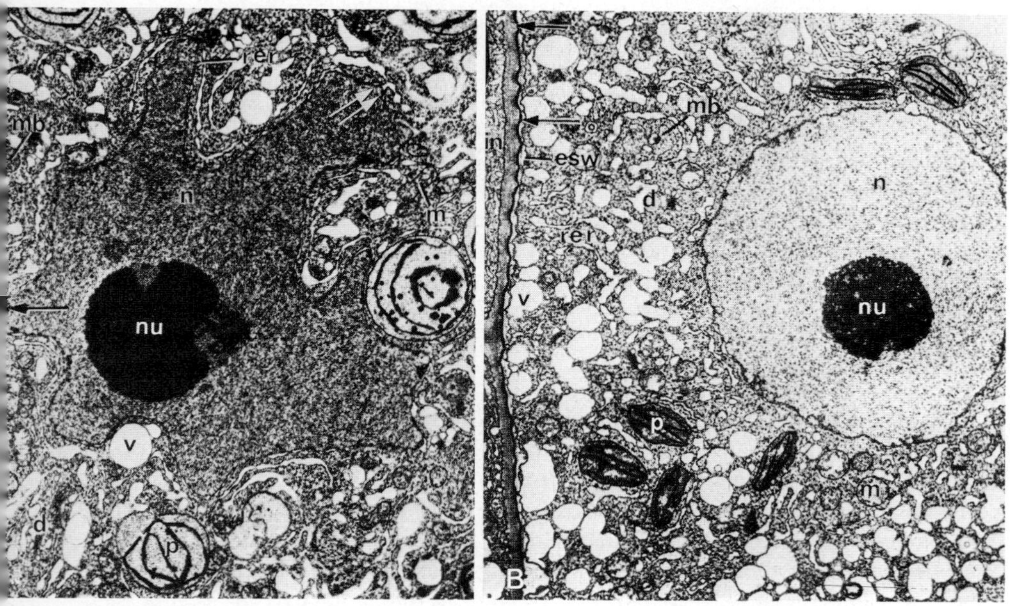

Fig. 7.7 A, B. *Capsella bursa-pastoris*. **A** Portion of chalazal endosperm at heart-shaped embryo, showing a deeply lobed nucleus (*n*) (*arrows*), nucleolus (*nu*), and clumps of nucleolar-like material (*arrows*) adpressed to nuclear envelope. Long profiles of rough endoplasmic reticulum (*rer*) parallel to the contours of nucleus are continuous with nuclear envelope. The cytoplasm shows active dictyosomes (*d*), mitochondria (*m*), microbodies (*mb*), many plastids (*p*), and vacuoles (*v*). **B** Portion of endosperm nodule at heart-shaped embryo. The cytoplasm contains dictyosomes (*d*), mitochondria (*m*), microbodies (*mb*), many plastids (*p*), and short vesiculate profiles of *rer*. The embryo sac wall projections (*esw*), nucleus (*n*), nucleolus (*nu*), and vacuoles (*v*) are prominent. (After Schulz and Jensen 1974)

Sachar 1954), *Coriandrum sativum* and *Foeniculum vulgare* (Gupta 1964), *Cuminum cyminum, Trachyspermum ammi* (Sehgal 1966), and *Thalictrum javanicum* (Vijayaraghavan and Bhandari 1970; see also D. Singh 1964). According to Schulz and Jensen (1974), in *Capsella bursa-pastoris,* the endosperm nodules grow as a result of increased mitotic activity of individual endosperm nuclei. The cytoplasm of these nodules is active (Fig. 7.7 B), but there are no wall projections. However, the ultrastructure of these nodules does not offer any clue about their specific function. The nodules disappear when the endosperm becomes cellular.

Fig. 7.6. A, B *Alyssum maritimum*. **A** Part of longisection of seed at early dicotyledonous embryo; endosperm cells filling the embryo sac cavity show low protein concentration. The coenocytic chalazal caecum of endosperm is, however, protein-rich. **B** Endosperm at a later stage to show asynchronous division in chalazal caecum accompanied by decline in the levels of cytoplasmic proteins. **C, D** *Iberis amara*. **C** Part of seed showing polyploid nuclei in a few cells derived from the coenocytic endosperm caecum at the chalazal end. **D** Chalazal portion of mature seed. Remnants of nuclei persist in the endosperm caecum (Vijayaraghavan and Prabhakar, unpublished)

7.3 Cellular Endosperm

In the Cellular type the initial and subsequent nuclear divisions are followed by cell-plate formation. During earlier development there is considerable variation in the method of cytokinesis in various families. Also, one or more cells at the micropylar, chalazal, or both ends become specialized to form a haustorium. The haustorial function is circumstantial, and presumed to be due to the hypertrophied nature of the nucleus and dense-staining of the cytoplasm. Only a few taxa that show special features are discussed below.

7.3.1 Acanthaceae

A row of three cells is formed following the division of the primary endosperm nucleus. The micropylar and chalazal chambers develop into micropylar and chalazal haustoria. The central cell (juxtaposed between the micropylar and chalazal chambers) produces the endosperm proper which may be ab initio cellular in some taxa, and free-nuclear in others. In *Dipteracanthus patulus* the wall formation is incomplete; a free-nuclear "basal apparatus" is formed below the cellular endosperm (Fig. 7.8 A). The wall formation takes place at the 32-nucleate stage in *Barleria cristata* (Fig. 7.8 B) and *Ruellia tuberosa*. In *Thunbergia alata* the micropylar chamber produces many branches which are coenocytic, and act as haustoria. The endosperm proper forms tube-like, secondary haustoria (Mohan Ram and Wadhi 1964).

7.3.2 Cyrillaceae

In *Cyrilla racemiflora* and *Cliftonia monophylla* the division of the primary endosperm nucleus is transverse followed by another transverse division in each daughter cell, resulting in four cells. Both micropylar and chalazal haustoria are formed (Fig. 7.8 C), and are one-celled and two-nucleate. Occasionally, the chalazal haustorium becomes four-nucleate. Embryological characters, especially the endosperm, support the assignment of Cyrillaceae to the Ericales (Vijayaraghavan and Dhar 1978).

7.3.3 Gesneriaceae

In this family the development of endosperm has been studied in *Klugia notoniana* (Arekal 1961) and *Platystemma violoides* (Adatia et al. 1971). Many embryologists regard this family as related to Orobanchaceae (see Adatia et al. 1971). Others consider it close to Scrophulariaceae. According to Tiagi (1963), the development of endosperm haustorium in *Boschniakia himalaica* (Orobanchaceae) and *Striga orobanchoides* (Scrophulariaceae) is essentially similar. He considers that the family Orobanchaceae can be derived from Scrophulariaceae through forms like *S. orobanchoides*.

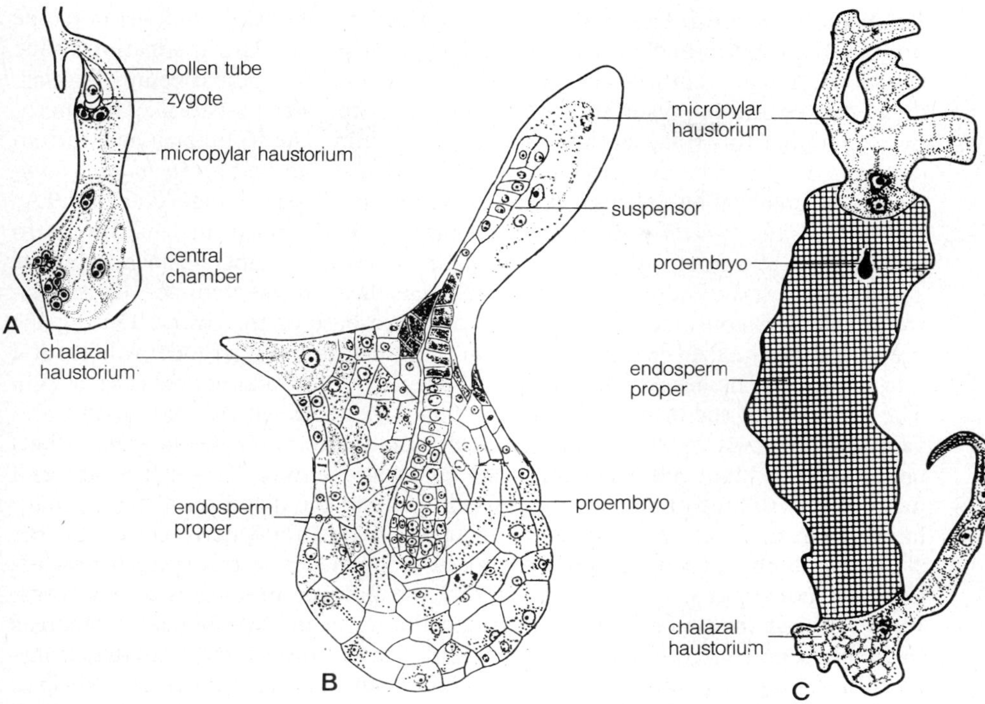

Fig. 7.8. A *Dipteracanthus patulus*, embryo sac shows zygote, and micropylar and chalazal haustoria. The central endosperm chamber has several free nuclei. (After Maheshwari and Negi 1955). **B** *Barleria cristata*, endosperm with micropylar and chalazal haustoria. The central chamber (initially free-nuclear) is completely cellular; note the globular proembryo with the suspensor extending into the micropylar haustorium. (After Mohan Ram 1962). **C** *Cyrilla racemiflora*, endosperm with one-celled, binucleate, branched, micropylar haustorium, central endosperm proper, and one-celled four-nucleate chalazal haustorium. The proembryo is globular. (After Vijayaraghavan and Dhar 1978)

7.3.4 Icacinaceae

In *Nothapodytes foetida* (Swamy and Ganapathy 1957) the division of the primary endosperm nucleus is transverse, forming two chambers. The micropylar chamber divides rapidly and produces the endosperm proper. The chalazal chamber has dense cytoplasm, hypertrophied nucleus, and functions as a haustorium. It gives rise to many, small, sac-like processes. The branches of the haustorium consume adjacent cell walls of the chalazal tissue and, thus, the protoplasts of the haustorium and affected cells intermingle. The nucleus of the haustorial cell enlarges followed by distortion of shape, and the chromatin material breaks into many lumps.

7.3.5 Santalaceae

According to Johri and Bhatnagar (1969), the parasitic angiosperms show several unusual features in the development of endosperm. Taxa like *Olax, Mida,* and

Santalum conform to Helobial type. In *Comandra umbellata* the division of the primary endosperm nucleus is followed by wall formation at the junction of the lateral caecum and embryo sac. A small cell is cut off from the caecum. This cell divides twice and the resultant four cells act as secondary haustoria. The micropylar chamber functions as the primary haustorium. The secondary haustorium elongates in the same direction as the primary one (Ram 1957). In *Iodina rhombifolia* the chalazal haustorium is aggressive and profusely branched (Fig. 7.9 A, B). In *Cansjera rheedii* (Opiliaceae) the primary endosperm nucleus divides to form a cyst-like micropylar and a tubular chalazal chamber. The micropylar chamber forms the endosperm proper, whereas the chalazal chamber has a haustorial role and continues to extend towards the base of the ovary. During the growth of the chalazal chamber, the tip of the haustorium undergoes proliferation into tenuous branches which ramify into deeper intercellular spaces (Fig. 7.9 C). The branches of the primary haustorium as well as the secondary haustoria invade and even destroy the tissue of the axial vasculature. In *Mida salicifolia* (Bhatnagar 1960) a small cell is cut off in the chalazal chamber. This cell produces a mass of cells. Three or four of the cells elongate in the direction of the primary haustorium and act as secondary haustoria. A few endosperm cells near the chalazal chamber give out finger-like processes that act as accessory absorbing organs. In *Exocarpus sparteus* the division of the endosperm nucleus is by a transverse wall. Of the two cells thus produced the lower divides vertically, whereas the upper segments transversely. The two cells of the lower tier do not divide further, but elongate considerably and produce many tubular processes that serve as efficient haustoria (Fig. 7.9 D).

7.3.6 Loasaceae

Following the division of the primary endosperm nucleus in *Blumenbachia hieronymi* (Garcia 1962 a, b), and *B. insignis* (Garcia 1963) two chambers – micropylar and chalazal – are formed. The micropylar chamber does not contribute to the formation of the endosperm proper, but develops into a haustorium of two, superposed, binucleate cells with hypertrophied nuclei (Fig. 7.10 A). The chalazal chamber undergoes two successive transverse divisions forming a row of four cells. The three upper cells contribute to the endosperm proper; the lowermost cell differentiates into the chalazal haustorium which branches profusely and digests even the integument (Fig. 7.10 A). In *Mentzelia laevicaulis* the micropylar haustorium is uninucleate, but the nucleus degenerates early. The chalazal haustorium, on the contrary, is multicellular.

7.3.7 Loranthaceae

In this family many embryo sacs develop concurrently, and their tips ascend to various heights in the style and stigma. The endosperm is formed simultaneously in several embryo sacs. Later, all the endosperms in an ovary fuse to form a composite structure – a feature so far restricted only to this family (see Chap. 1). In

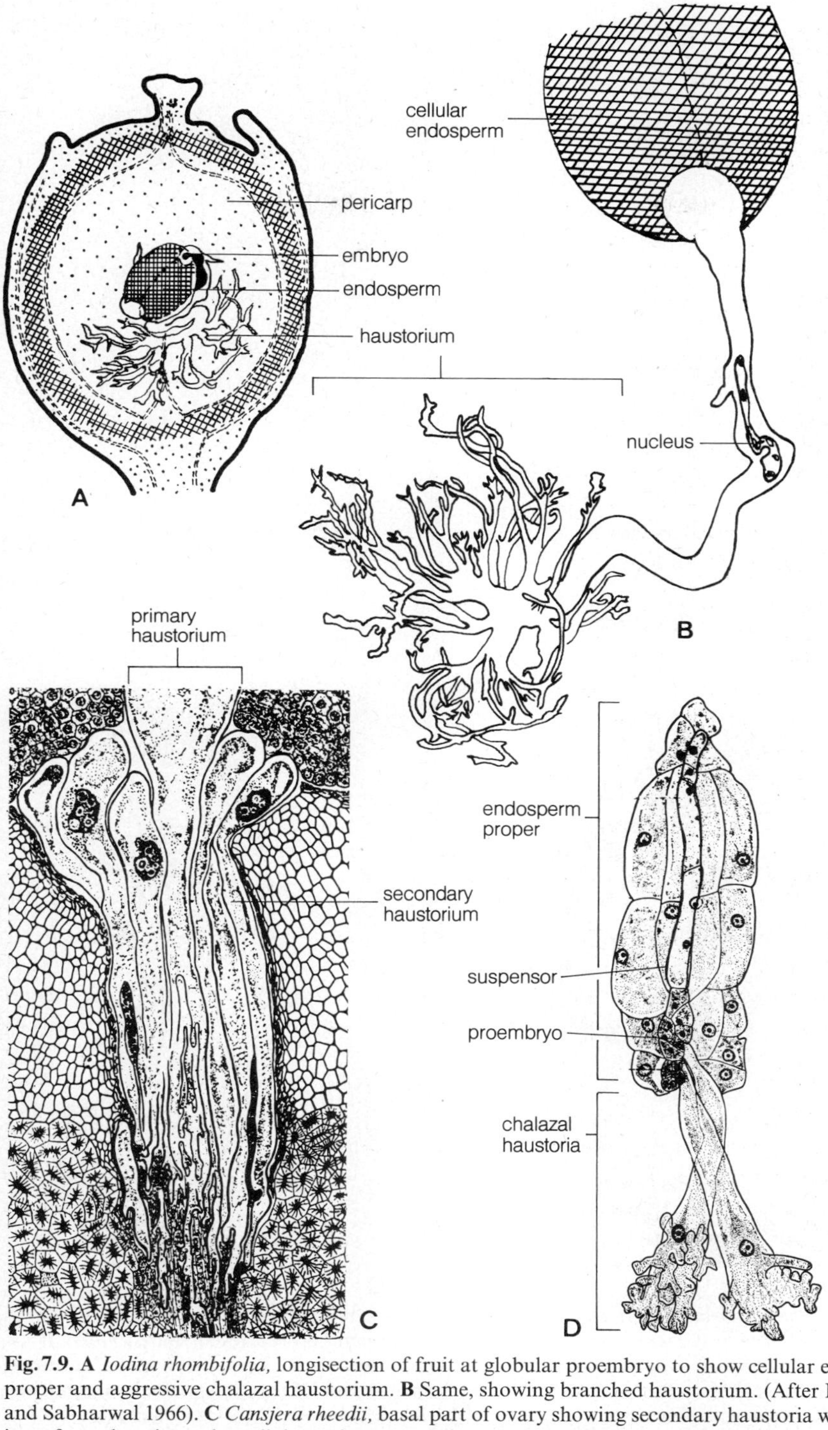

Fig. 7.9. A *Iodina rhombifolia,* longisection of fruit at globular proembryo to show cellular endosperm proper and aggressive chalazal haustorium. **B** Same, showing branched haustorium. (After Bhatnagar and Sabharwal 1966). **C** *Cansjera rheedii,* basal part of ovary showing secondary haustoria which originate from the micropylar cellular endosperm. (After Swamy 1960). **D** *Exocarpus sparteus,* proembryo with a long suspensor, endosperm proper, and two chalazal haustoria with finger-like outgrowths. (After Ram 1959)

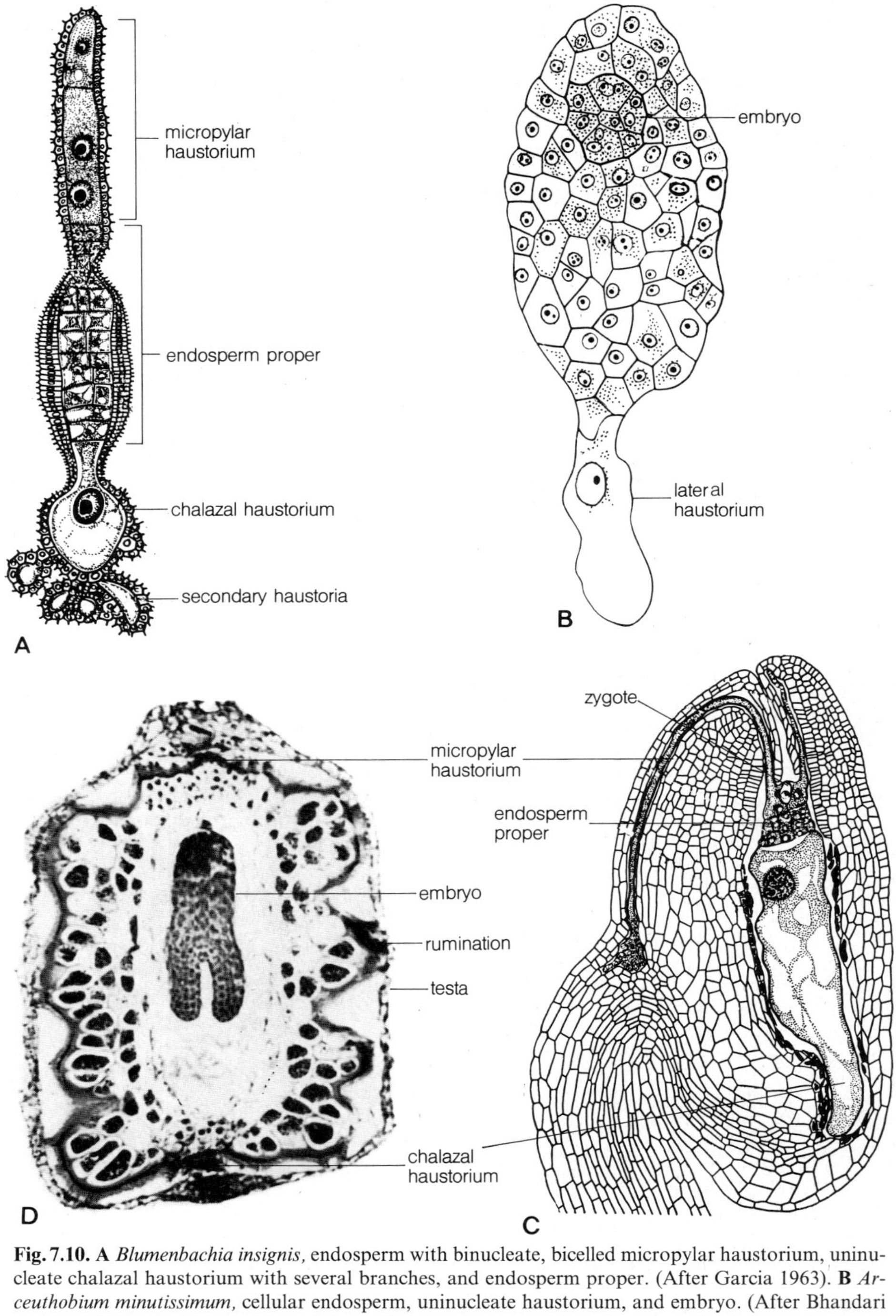

Fig. 7.10. A *Blumenbachia insignis,* endosperm with binucleate, bicelled micropylar haustorium, uninucleate chalazal haustorium with several branches, and endosperm proper. (After Garcia 1963). **B** *Arceuthobium minutissimum,* cellular endosperm, uninucleate haustorium, and embryo. (After Bhandari and Nanda 1968). **C** *Melampyrum lineare,* showing the penetration of one branch of micropylar haustorium as far as funiculus; endosperm proper, and binucleate, one-celled, chalazal haustorium. (After Arekal 1963a). **D** *Celsia coromandeliana,* longisection of seed to show dicotyledonous embryo and ruminate endosperm. The micropylar and chalazal haustoria have degenerated. (After Kapoor and Vijayaraghavan 1975)

Tapinostemma acaciae the basal cells of endosperm develop finger-like haustorial processes, and a lateral haustorium develops in *Arceuthobium minutissimum* (Viscaceae; Fig. 7.10 B). The fruits of *Psittacanthus* have a vase-shaped composite endosperm which persists in ripe fruits [Bhatnagar and Chandra 1968; see Johri and Bhatnagar (1969) for exhaustive review].

7.3.8 Scrophulariaceae

Arekal (1963 a) observed aggressive endosperm haustoria in three members of Scrophulariaceae. In *Melampyrum* the development of endosperm is Cellular, and a micropylar and a chalazal chamber are formed. The next division in the chalazal chamber is not accompanied by wall formation, whereas an incipient, vertical wall is formed in the micropylar chamber. The binucleate chalazal chamber acts as a haustorium. In *M. lineare* the endosperm differentiates into three regions: (1) a massive micropylar part composed of densely cytoplasmic cells with thick and pitted walls, (2) a narrow middle region of small cells, and (3) the chalazal part of thin-walled cells with large, intercellular spaces. During later stages one of the branches of the micropylar haustorium penetrates as far as the funicle (Fig. 7.10 C).

In *Celsia coromandeliana* remnants of both the micropylar and chalazal haustoria persist in mature seed (Fig. 7.10 D). The division of the primary endosperm nucleus is followed by a transverse wall, resulting in micropylar and chalazal chambers (Fig. 7.11 A–D). The first two divisions in the micropylar and chalazal chambers are vertical (Fig. 7.11 D). The four cells of the chalazal chamber directly form the chalazal haustorium (Fig. 7.11 E). The four eleongated micropylar cells divide transversely. The derivatives of the upper tier form the four-celled micropylar haustorium, whereas the lower four cells divide in both longitudinal and transverse planes to form the endosperm proper (Fig. 7.11 H). The four-celled micropylar haustorium with uninucleate cells is non-aggressive (Fig. 7.11 F, G). The four-celled chalazal haustorium is short and also non-aggressive, has one nucleus in each cell (Fig. 7.11 F, I), and is early to organize and degenerate.

Chelone glabra shows several interesting features. The primary endosperm nucleus divides by a transverse wall, forming two chambers. The chalazal chamber divides vertically but the wall remains incomplete. This incompletely divided, two-celled chalazal chamber functions directly as a haustorium (Arekal 1963 b). Such a haustorium is intermediate between the two-celled haustorium observed in *Scrophularia marylandica* (Schertz 1919), and the unicelled, binucleate haustorium in *Tetranema mexicanum*. In the micropylar chamber the division is vertical followed by another vertical division. The daughter cells undergo a transverse division and two tiers of four cells each are formed. The lower tier contributes to the endosperm proper, and the upper tier functions as micropylar haustorium. In some cells of the micropylar haustorium the nuclei undergo endopolyploidy and cut off micronuclei, resutling in a multinucleate condition.

In *Alectra thomsoni* (Vijayaraghavan and Ratnaparkhi 1972) the endosperm has well-developed micropylar, chalazal, and secondary haustoria. The micropylar haustorium is large, pear-shaped, bicelled, and binucleate. Secondary haus-

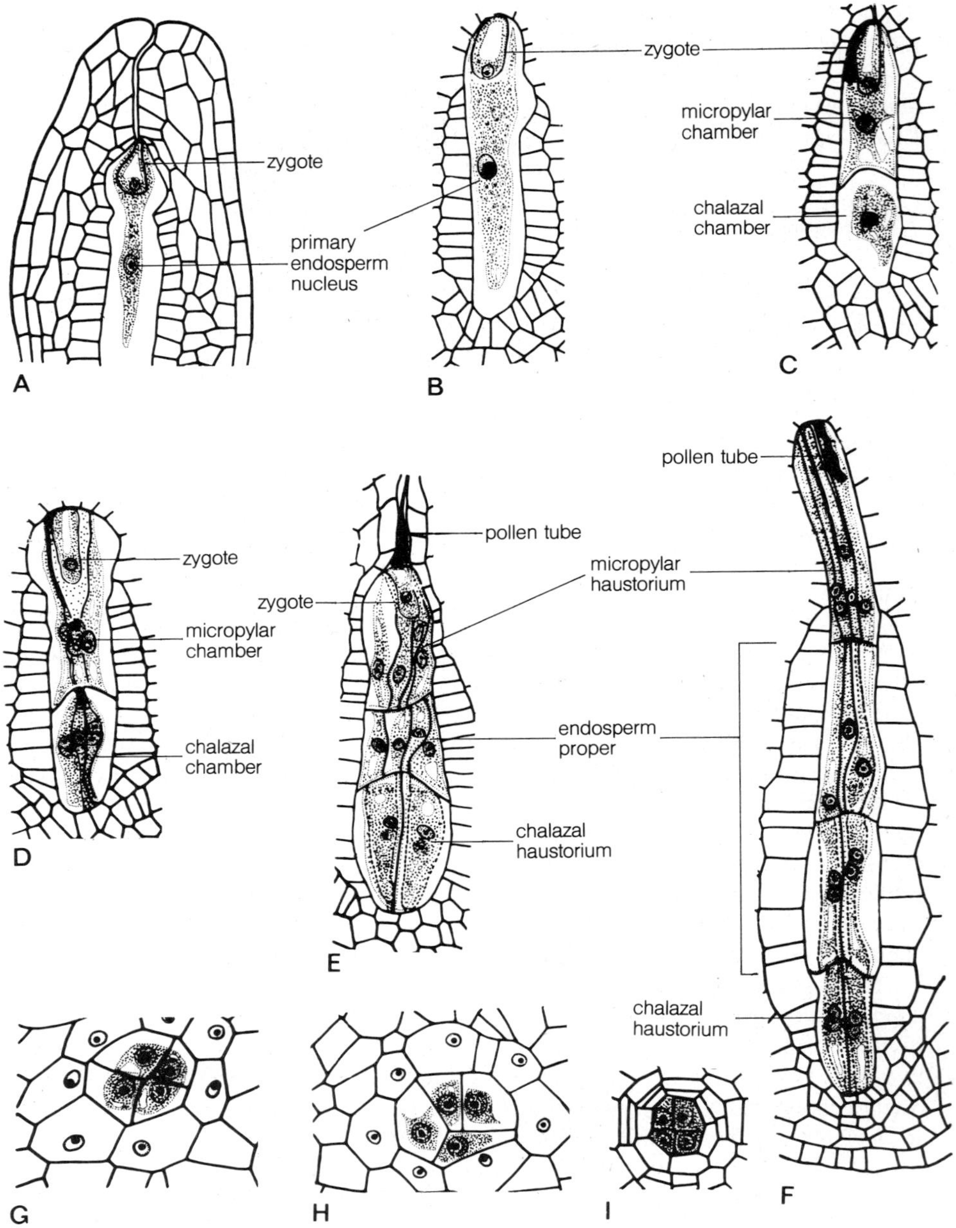

Fig. 7.11 A–I. *Celsia coromandeliana*. **A** Longisection of upper part of fertilized ovule. **B** Embryo sac surrounded by endothelium; **A, B** show zygote, and primary endosperm nucleus. Two- (**C**) and eight-celled (**D**) endosperm; fourcelled chalazal chamber in **D** forms a haustorium. **E** Micropylar chamber divides transversely and the upper tier forms micropylar haustorium. **F** By further transverse and longitudinal division the middle tier forms the endosperm proper. **G–I** Transections of endosperm at the levels of micropylar haustorium (**G**), middle region (**H**), and chalazal haustorium (**I**). (After Kapoor et al. 1975)

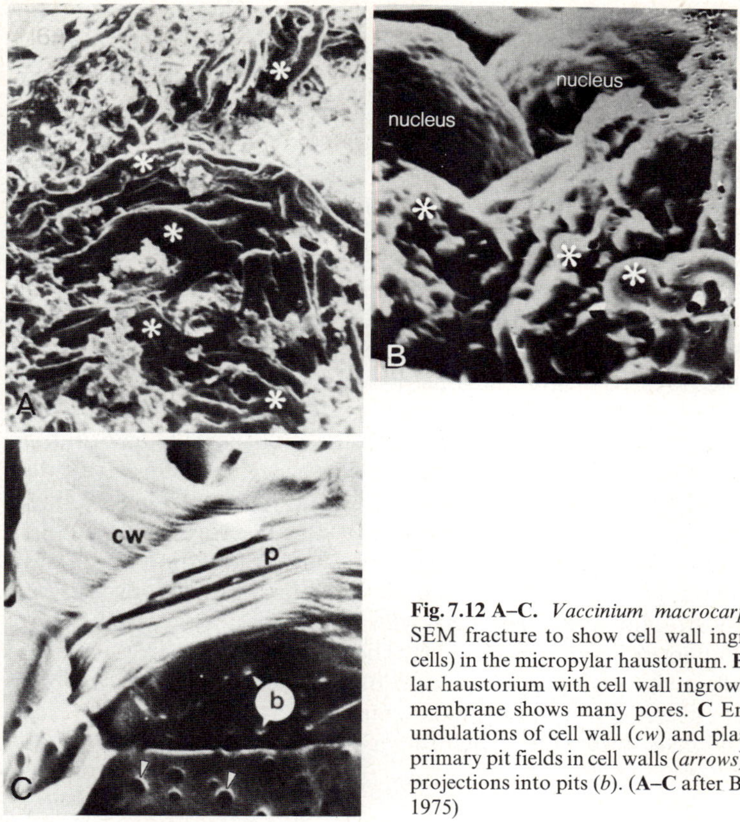

Fig. 7.12 A–C. *Vaccinium macrocarpum* endosperm. **A** SEM fracture to show cell wall ingrowths (*) (transfer cells) in the micropylar haustorium. **B** Adjacent micropylar haustorium with cell wall ingrowths (*). The nuclear membrane shows many pores. **C** Endosperm cells with undulations of cell wall (*cw*) and plasmalemma (*p*); note primary pit fields in cell walls (*arrows*), and plasmalemma projections into pits (*b*). (**A–C** after Brisson and Peterson 1975)

toria develop only from the micropylar region of the endosperm. These are small, uninucleate, and extend into tubular processes. The chalazal haustorium is conical, unicellular, and binucleate. Much variation is reported (in this family) in the structure of the chalazal haustorium. It is four-celled (Fig. 7.11 E, F, I) in *Celsia coromandeliana* (Kapoor et al. 1975) and *Isoplexis canariensis* (Krishna Iyengar 1939), two-celled in *Scrophularia marylandica* (Schertz 1919), and incompletely two-celled in *Chelone glabra* (Arekal 1963 b). In *Melampyrum lineare, Alectrolophus,* and *Orthocarpus luteus* the chalazal haustorium becomes binucleate (Arekal 1963 a).

Secondary chalazal endosperm haustoria are reported in *Veronica chamoedrys* (Schmidt 1906), and micropylar in *Centranthera hispida* (Krishna Iyengar 1942), and *Alectra thomsoni* (Vijayaraghavan and Ratnaparkhi 1972).

7.3.9 Histochemistry and Ultrastructure

Very few studies deal with the histochemistry and ultrastructure of Cellular endosperm. On the basis of cytochemical studies, a nutritional role has been assigned to the endosperm haustoria in *Linaria bipartita* (Bhatnagar and Kallarackal 1980) and *Scrophularia himalianensis* (Natesh 1983, pers. comm.). In *L. bipartita* both micropylar and chalazal haustoria show hypertrophied nuclei which undergo several cycles of DNA synthesis and increase of nucleolar and cytoplasmic RNA and proteins. A similar increase in the levels of RNA, total proteins, and histones in haustorial cells has been reported in *S. himalianensis*. In both *Linaria* and *Scrophularia* haustorial cells are devoid of starch grains. Other endosperm cells, however, store a large amount of starch and protein bodies. It is concluded that besides the absorptive role the endosperm haustoria are metabolically very active. The haustorial cells provide nutrients to the endosperm proper, and not to the embryo, at least during early embryogeny. Torosian (1971) reports that in *Lobelia dunnii* (Lobeliaceae) the walls of haustorial cells develop many ingrowths, increasing the surface area of membrane. She provides ample evidence in support of the absorptive role of haustoria. In *L. dunnii* even the basal cell of the embryo develops wall projections in areas of contact with the endosperm haustorium. In *Vaccinium macrocarpum* (Brisson and Peterson 1975) the wall of endosperm haustoria itself develops ingrowths (Fig. 7.12 A, B). These ingrowths act like "transfer cells", which are known to be associated with short-distance transport (Gunning and Pate 1969). A few endosperm cells (Fig. 7.12 C) show undulation of cell walls and of plasmalemma, primary pitfields, and plasmalemma projections into pits. In *Plantago lanceolata,* however, the micropylar haustorium is reported to have a smooth wall (Vannereau and Mestre 1975). The haustorial cells are considered to be metabolically active owing to a high ribosome content in *P. lanceolata,* and to the presence of mitochondria, spherosomes, and abundant rough ER in *L. dunnii.* The haustorial cells degenerate when the endosperm has developed a good deal.

7.4 Helobial Endosperm

The Helobial type of endosperm is characterized by the division of the primary endosperm nucleus, giving rise to two unequal chambers – the micropylar chamber usually larger than the chalazal. Free-nuclear divisions occur in the micropylar chamber before it undergoes cellularization. By contrast, the nucleus of the chalazal chamber either remains undivided, or divides once or twice. The chalazal chamber usually remains coenocytic but, in a few cases, it gets partitioned into cells. In *Halophila ovata* (Lakshmanan 1963) the endosperm is of the Helobial type. Although the basal apparatus has a single nucleus, the nucellar cells subjacent to the basal apparatus rupture and their nuclei and protoplast are incorporated into the haustorium.

In *Trillium undulatum* (Swamy 1948) the primary endosperm nucleus divides at the level of the antipodal cells so that a small chalazal and a large micropylar chamber are formed. Free-nuclear divisions occur in the micropylar chamber and, after about the 32-nucleate stage, the nucleus of the chalazal chamber also starts dividing. After about 13 nuclei in the chalazal chamber, and about 100 in the micropylar chamber, karyokinesis ceases. In the micropylar chamber the nuclei occupy the periphery of the embryo sac, and wall formation is initiated. In the chalazal chamber the nuclei are evenly distributed, wall formation is delayed, and the cells are small and densely cytoplasmic.

Zaman (1950) also reported wall formation in the chalazal chamber in *Juncus prismatocarpus*. According to Eunus (1952), in *Asphodelus tenuifolius*, the nucleus of the chalazal chamber divides only once or twice, whereas in the micropylar chamber many free-nuclear divisions occur. Centripetal wall formation in the micropylar chamber results in cellularization of the endosperm. The chalazal chamber shows hypertrophied nuclei and dense cytoplasm (see also Kaushik 1940). The basal portion even acts as a haustorium and persists in mature seed.

The Helobial endosperm in monocotyledons follows more or less a uniform pattern during early development, but several variations occur during later stages. Contrary to earlier reports of Nuclear endosperm, the Helobial type is confirmed, for the first time, in *Najas flexilis* and *N. marina* (Kapoor 1980).

7.4.1 Salient Features in Monocotyledons

The significant characteristics of Helobial endosperm in the monocots are:

1. The primary endosperm nucleus is situated in the chalazal region of the embryo sac adjacent to the antipodal cells.
2. The division of the primary endosperm nucleus is followed by a wall which separates a large micropylar and a small chalazal chamber. The cytoplasm of the chalazal chamber stains deeply.
3. Before undergoing division, the nucleus of the micropylar chamber always migrates to the upper part of the embryo sac.
4. The first two divisions in the micropylar chamber are free-nuclear; cell-wall formation, if any, occurs during later stages.
5. The chalazal endosperm chamber, also called the "basal apparatus", has a haustorial function. The division of the nucleus in this chamber, if it occurs, is always free-nuclear. If wall formation occurs, it is significantly belated as compared to the micropylar chamber.
6. The endosperm cells in the chalazal chamber lag behind in the rate of nuclear division as well as in number, as compared to the micropylar chamber.

Swamy and Parameswaran (1963) observed six developmental variations of Helobial endosperm in monocotyledons (Fig. 7.13).

Usually, the primary endosperm nucleus divides close to the antipodal region, followed by a wall separating a large micropylar and a small chalazal chamber. The next nuclear division occurs in the micropylar chamber, whereas the nucleus of the chalazal chamber remains undivided, becomes hypertrophied, and it functions as a haustorium. Wall formation occurs later in the micropylar chamber. The examples are: Aponogetonaceae, Hydrocharitaceae, Potomogetonaceae, and others with slight variations:

A Type. The variation is generally restricted to the chalazal chamber. At first synchronous free-nuclear divisions occur in micropylar and chalazal chambers, and

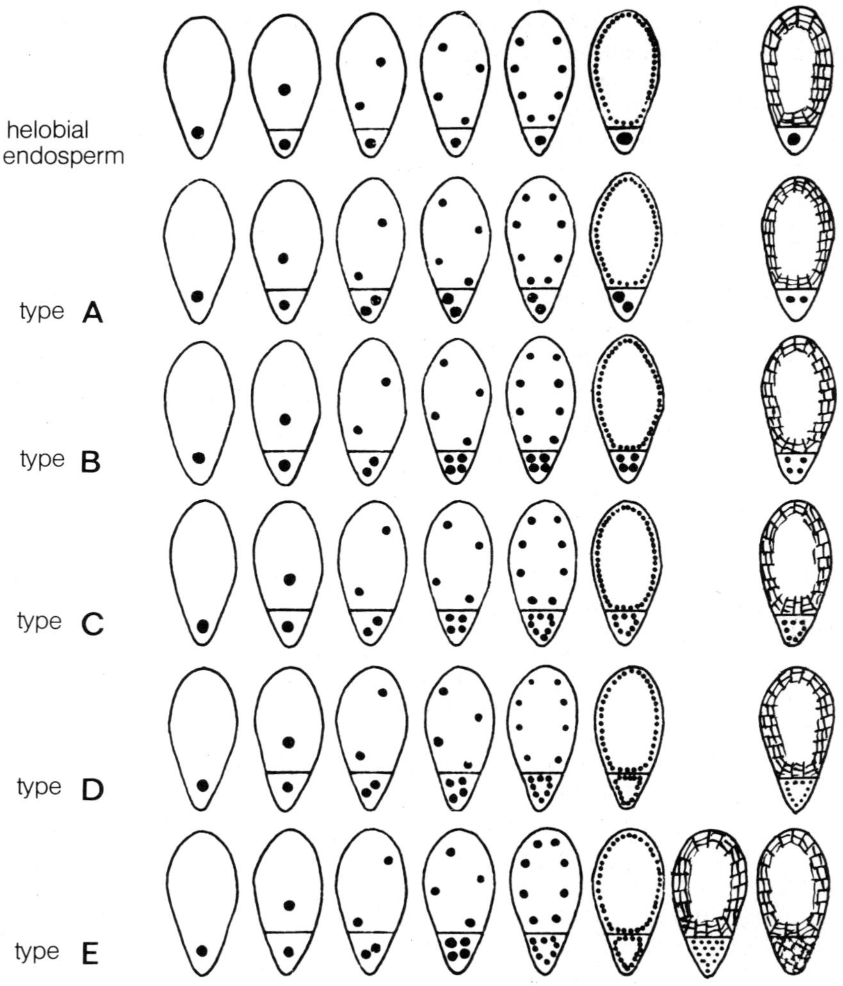

Fig. 7.13. Ontogenetic variability (Types *A–E*) in Helobial endosperm in Monocotyledons. (Modified after Swamy and Parameswaran 1963)

both have two nuclei each. While free-nuclear divisions continue in the micropylar chamber, in the chalazal chamber both nuclei become hypertrophied, and it acts as a haustorium.

B Type. In a few instances, the first three divisions in both the chambers are synchronous up to the four-nucleate stage. Subsequently, the chalazal chamber assumes haustorial function with cessation of nuclear divisions.

C Type. The nuclear divisions in both chambers are synchronous only up to the 8-nucleate stage. Thereafter, free-nuclear division continue in the micropylar chamber, while the eight nuclei in the chalazal chamber do not divide further. This chamber acts as a haustorium.

D Type. After the 16-nucleate stage, the nuclei of the chalazal chamber divide, increase in number at a retarded rate, and the coenocytic chalazal chamber acts as a haustorium.

E Type. In all the above Types, the chalazal chamber remains free-nuclear. Sometimes wall formation occurs in the chalazal chamber, though belatedly as compared to the micropylar chamber. The endosperm tissue produced in the chalazal chamber is scanty. In Types A–E the bulk of the endosperm is formed only by the micropylar chamber.

In contrast, the reports on Helobial endosperm in dicotyledons do not conform to any basic pattern as described for monocotyledons. Further, the developmental sequence cannot be entirely compared to the sequence, or its deviation, in monocotyledonous taxa (Swamy and Parameswaran 1963). The major differences are:

1. The position of the primary endosperm nucleus in the embryo sac, prior to division, is variable, and the two chambers are more or less equal, or the micropylar chamber is even smaller.
2. The nuclear divisions in the chalazal chamber are initially free-nuclear, but wall formation occurs before that of the micropylar chamber.
3. Wall formation is initiated in the chalazal chamber after the first nuclear division.

7.4.2 Histochemistry and Ultrastructure

Only a few plants with Helobial endosperm have been investigated histochemically and ultrastructurally. The two unequal chambers, formed after the first nuclear division, show different histochemical and histological composition. Kapoor (1980) observed that in *Potomogeton nodosus, Najas flexilis,* and *N. marina* polysaccharide granules are absent in both the chambers; the chalazal chamber stains more intensely for proteins, DNA, and RNA. The nucleus in the chalazal chamber becomes polyploid in all these taxa and shows extrusion of nuclear materials into the nucleoplasm and cytoplasm of the chalazal endosperm. She confirms that the chalazal chamber has a nutritive role, whereas the micropylar chamber is involved in the translocation of metabolites. In *Haemanthus katherinae* (Newcomb 1978) both the micropylar and the chalazal chamber of the endosperm show free-nuclear divisions (Fig. 7.14) and the presence of wall ingrowths over the entire surface of the embryo sac. Numerous mitochondria are reported in both the chambers, the nuclei are also uniformly spaced, but the pattern of vacuolation differs in the two chambers (Fig. 7.15 A). The micropylar portion of endosperm becomes cellular by "free" growth of cells (Fig. 7.15 B). The chalazal chamber shows a few large and many small vacuoles, whereas the micro-

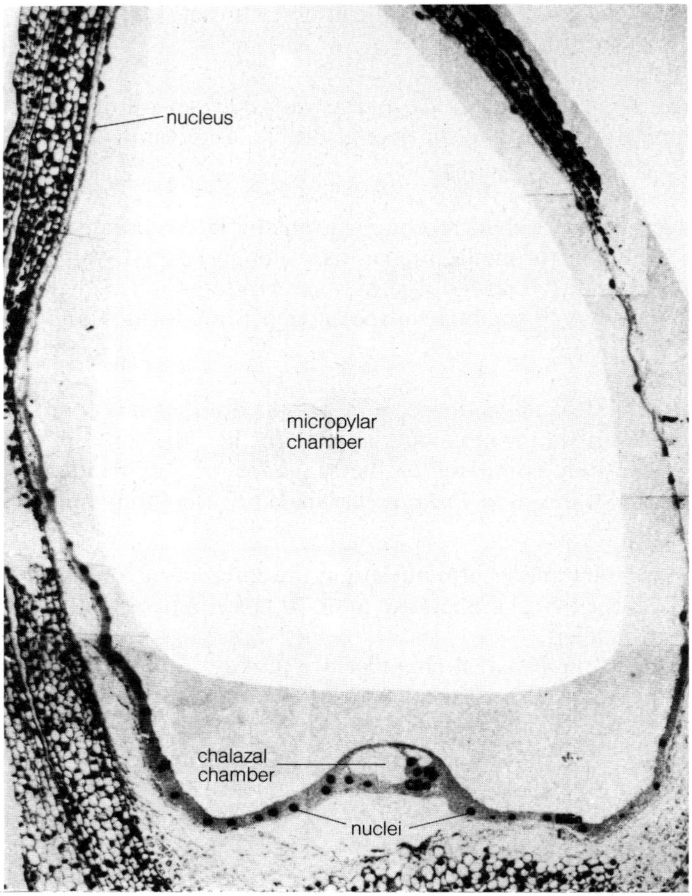

Fig. 7.14. *Haemanthus katherinae,* Helobial endosperm, both micropylar and chalazal chambers are coenocytic. (After Newcomb 1978)

pylar chamber contains a large central vacuole and numerous smaller ones. The endosperm cells in the micropylar region show a higher degree of vacuolation in the inner layers than in the outer first-formed layer. "Free" cells are also reported in the micropylar chamber, after ten or more layers of endosperm cells are formed. The free cells undergo karyokinesis forming binucleate cells; the final fate of the free cells is not known. The endosperm cells in the chalazal chamber have small vacuoles and, sometimes, the cells are binucleate.

7.5 Ruminate Endosperm

The ruminate endosperm in the mature seeds of angiosperms has attracted much attention. The word ruminate is derived from the Latin "ruminatus" meaning

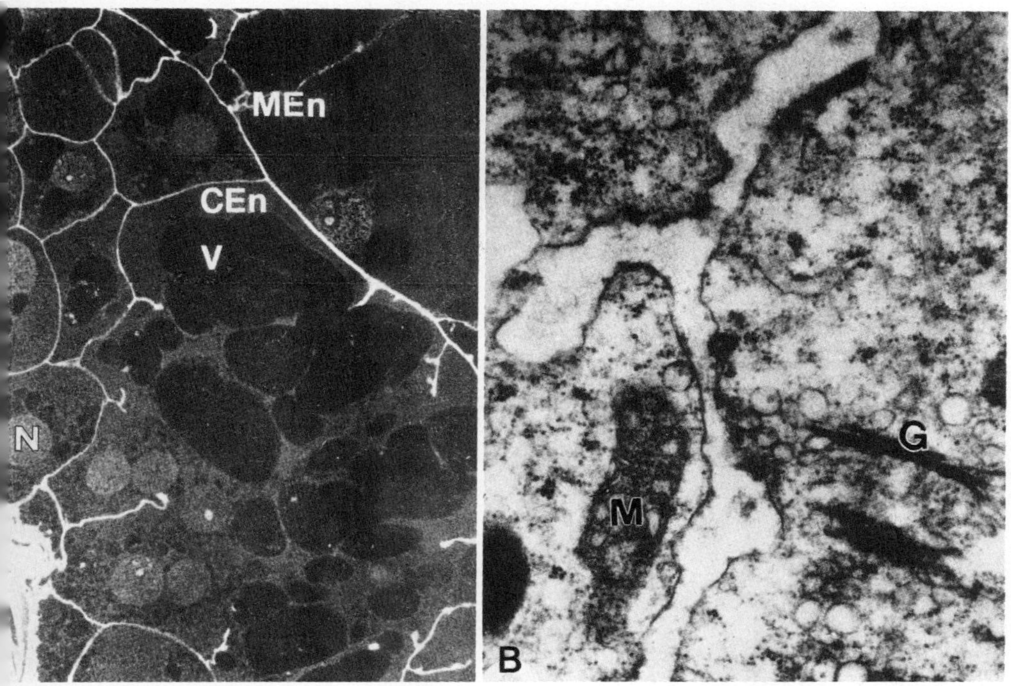

Fig. 7.15 A, B. *Haemanthus katherinae*. **A** Fluorescence light microscope montage of chalazal chamber of endosperm (*CEn*) which becomes cellular by means of freely growing walls. The nucleus (*N*) and vacuoles (*V*) are prominent. Cell walls also develop in the micropylar chamber of endosperm (*MEn*). **B** Same, electron micrograph of the end of a branched freely growing wall in micropylar chamber of endosperm; golgi bodies (*G*), with numerous vesicles, are located adjacent to cell wall. *M* mitochondria. (After Newcomb 1978)

chewed. The studies of Voigt (1888), however, indicate that there is no chewing up of the endosperm tissue, but the ruminate condition results owing to an irregular growth activity of the seed-coat as well as the endosperm. The endosperm surface, thus, becomes uneven. In *Celsia coramandeliana* (Kapoor and Vijayaraghavan 1975) the irregularity is brought about by unequal radial and tangential elongation of the cells of the endothelium (Fig. 7.10 D). This mode of rumination has been ascribed to the Scrophularia type (Hartl 1960) or Verbascum type (Periasamy 1962). According to Hartl (1960), in *Scrophularia,* most species of *Verbascum, Celsia,* and *Sutera,* the cell of the integumentary tapetum which enlarges and causes rumination is a "Bothroblast", and the cell which does not enlarge is a "Abothroblast".

In *Elytraria acaulis* (Johri and Singh 1959), *Andrographis serpyllifolia* (Mohan Ram 1960), and *A. echioides* (Mohan Ram and Masand 1962) rumination is caused by localized asymmetrical growth of the endosperm cells. In Annonaceae the inner or outer integument, or the middle integument, may produce ingrowths of the seed-coat. The cells invariably elongate in all directions without any regular pattern. In *Artabotrys odoratissimus,* however, the ingrowths of tegmen are plate-

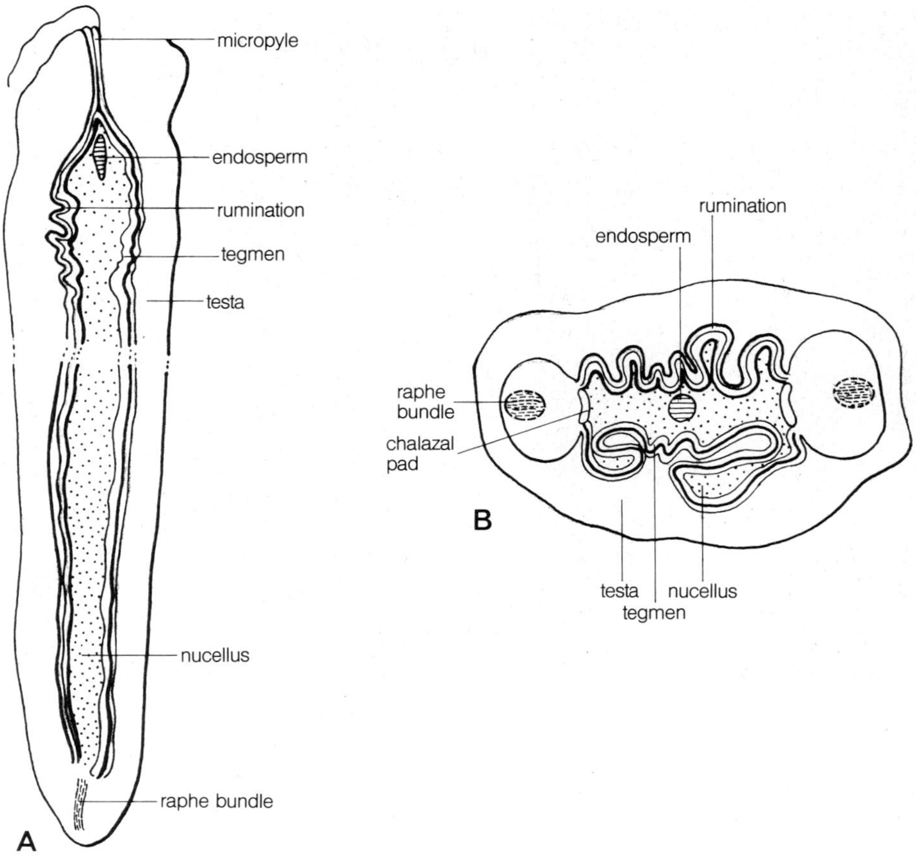

Fig. 7.16 A, B. *Artabotrys odoratissimus.* **A** Longisection and **B** transection of young seed to show rumination. (Adapted after Mohana Rao 1975)

like and lie one above the other (Fig. 7.16 A, B). The rumination is absent in the perichalazal region.

Seeds which reveal a network of lobes in sections in any plane are termed "labyrinth seeds" (van Heel 1970, 1971 a, b). This structure may be either due to rumination of the endosperm, or lobing and folding of the cotyledons. In seeds, thus, when lobing and folding of cotyledons occur together with rumination, the labyrinth structure is very prominent. In *Kingiodendron* sp. and *Erycibe griffithii* and *E. tomentosa* the seeds show excessive folding of cotyledons, followed by intrusion of the testa in between the folds and lobes, causing the labyrinth condition. In both taxa the mature seeds lack endosperm (van Heel 1970).

Ruminate endosperm could be an ancestral character still occurring in present-day seeds, belonging to both advanced and primitive taxa.

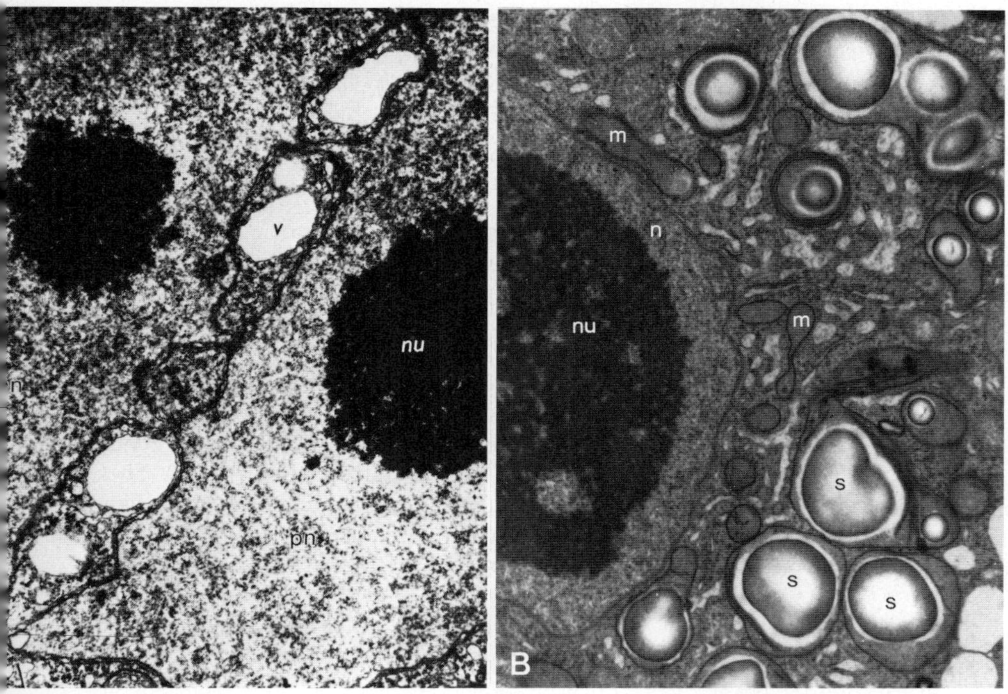

Fig. 7.17. A *Capsella bursa-pastoris,* polar nuclei (*pn*) separated by cytoplasmic pockets filled with vacuoles (*v*) between nuclear bridges. The nucleoli (*nu*) are cut through the periphery, and contain many dense granules. (After Schulz and Jensen 1973). **B** *Spinacea oleracea,* part of two-nucleate endosperm. Note the accumulation of starch (*s*) in plastids and abundant mitochondria (*m*); *n* nucleus; *nu* nucleolus. (After Wilms 1981)

7.6 Central Cell

The central cell is the largest cell in the embryo sac, generally with two haploid nuclei but variations in the number of nuclei are well known. Earlier studies were primarily focussed on the polar nuclei. Electron microscopic studies, however, reveal that this cell is a highly active constituent of the embryo sac (Jensen 1965, Mikulsa and Rodkiewicz 1967 a, b, Diboll 1968, Cocucci and Jensen 1969 a, b, Vazart 1969, van Went 1970 a, b, Plisko 1971).

Morphologically, both the polar nuclei are similar; occasionally, the upper (micropylar) may be larger than the lower (chalazal) (see Maheshwari 1950). The polar nuclei are usually completely fused at the time of fertilization although, in some cases, they are partially fused. Serial sectioning of these nuclei in *Petunia* revealed that they are connected by several narrow bridges and that, in a number of other places, only their outer membranes are continuous (van Went 1970 b). Both the nuclei evaginate and make contact at several points along their adjoining surfaces. This is followed by fusion of the outer nuclear membrane and, ultimately, the inner nuclear membrane. This is also true of cotton (Jensen 1964, Jensen and Fisher 1967) and *Capsella bursa-pastoris* (Fig. 7.17 A). The formation of fusion nucleus is accompanied by fusion of the two nucleoli.

7.6.1 Central Cell Cytoplasm

Plastids encircle the polar nuclei and the primary endosperm nucleus (Schulz and Jensen 1973). In *Petunia* plastids of the mature central cell (embryo sac) contain starch (van Went 1970 b); in the enlarging primary endosperm cell the plastids are small and lack starch (van Went 1970 b). Starch in the central cell is reported in many other taxa (see Jensen 1965, B. Vazart and J. Vazart 1965, Diboll and Larson 1966, Diboll 1968). In *Spinacea oleracea,* at the two-nucleate endosperm stage, the cytoplasm of endosperm contains plastids and abundant mitochondria (Fig. 7.17 B). Histochemically, insoluble polysaccharides encircle the primary endosperm nucleus in *Alyssum maritimum* (Vijayaraghavan et al. 1984). These granules are consumed during initial division of the primary endosperm nucleus. In cotton also the starch content declines during the division of the primary endosperm nucleus (Schulz and Jensen 1978). Yoffe (1952) observed chlorophyll in the central cell of certain cruciferous plants, and attributed to it physiological significance of intense metabolic activity. The presence of many large, well-developed, chloroplasts suggests that autotrophy may be an important factor in nutrition (Schulz and Jensen 1978).

A single, remarkably large, mitochondrion is sometimes present in the cytoplasm of the central cell, close to the polar nuclei, in both fertilized and unfertilized ovules of *Capsella bursa-pastoris* (Fig. 7.18 A). This mitochondrion has been correlated with megamitochondrion induced in animal cells by shock injury and specific nutritional deficiencies. There is, however, no evidence for the possible origin, function, or disposition of this unusual organelle. In *Petunia* the mitochondrion changes its shape following double fertilization (van Went 1970b).

There is a notable increase in the proliferation of endoplasmic reticulum in orchids (Cocucci and Jensen 1969 a, b), *Petunia* (van Went 1970 b) and *Hibiscus* (Ashley 1975), and an increase in dictyosome activity in *Zea mays* (Diboll 1968) and *Petunia* (van Went 1970 b). In cotton also dictyosomes and rough endoplasmic reticulum proliferate (Schulz and Jensen 1977). In *Petunia* masses of free ribosomes in the mature central cell, and the rapid formation of large polysomes after nuclear fusion, suggest that the synthesis of abundant protein is responsible for the rapid differentiation of the central cell. There is an increase in the formation of polysomes in orchids also (Cocucci and Jensen 1969 a, b). In *Capsella bursa-pastoris,* however, there is no notable change in the pattern and distribution of ribosomes from the time of formation of the fusion nucleus until the division of the primary endosperm nucleus (Schulz and Jensen 1973). In cotton two-nucleate endosperm shows plastids containing starch grains, and an aggregation of ribosomes forming polysomes (Fig. 7.19). These changes increase metabolic activity, and the organization of machinery for protein synthesis. In *Capsella* the mature central cell is already active and does not undergo any notable structural changes in organization following double fertilization (Schulz and Jensen 1973). In unfer-

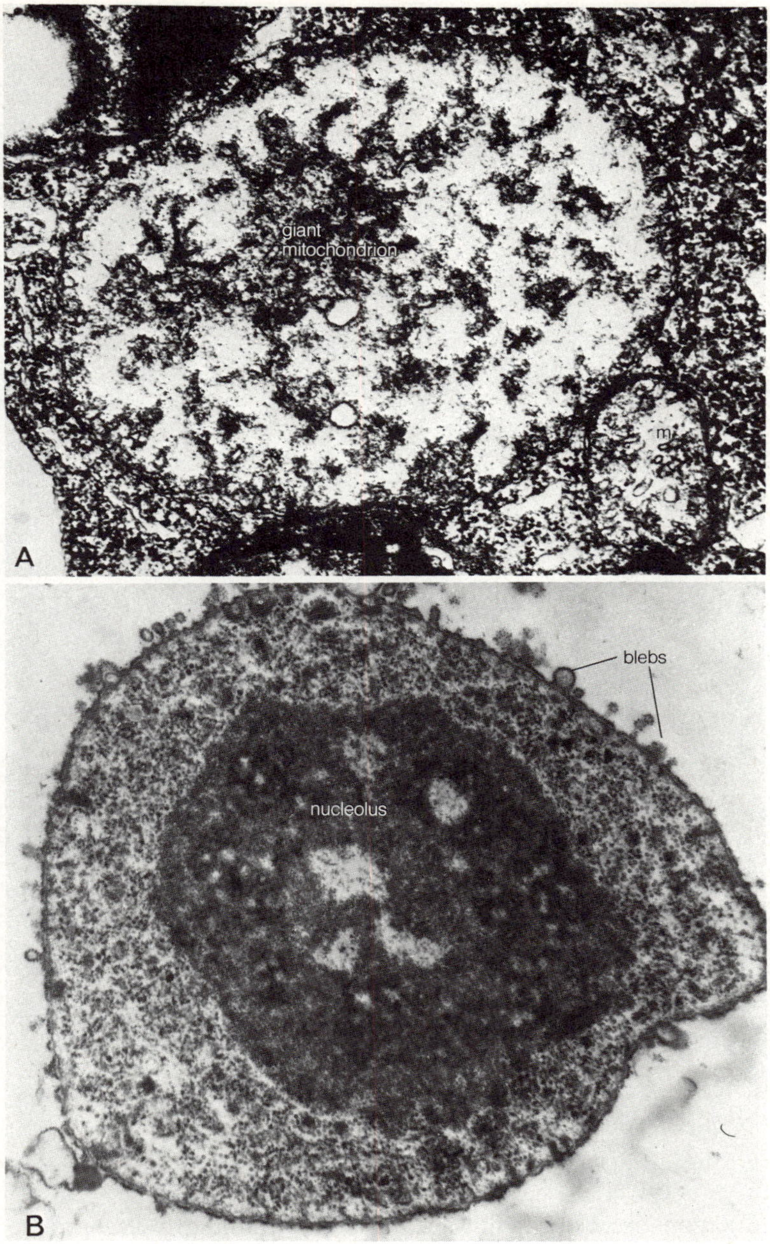

Fig. 7.18. A *Capsella bursa-pastoris,* giant mitochondrion in the central cell cytoplasm of fertilized embryo sac showing vesicular and tubular cristae, intramitochondrial granules, ribosomes, and fibrillar material; compare with neighbouring mitochondrion (*m*). (After Schulz and Jensen 1973). **B** *Alyssum maritimum,* portion of longisection of seed at heart-shaped stage of embryo showing nucleus from micropylar portion of endosperm. There is a dense and compact nucleolus interspersed with fibrillar zone; the blebbing of nuclear membrane is noteworthy. (After Prabhakar et al. 1981)

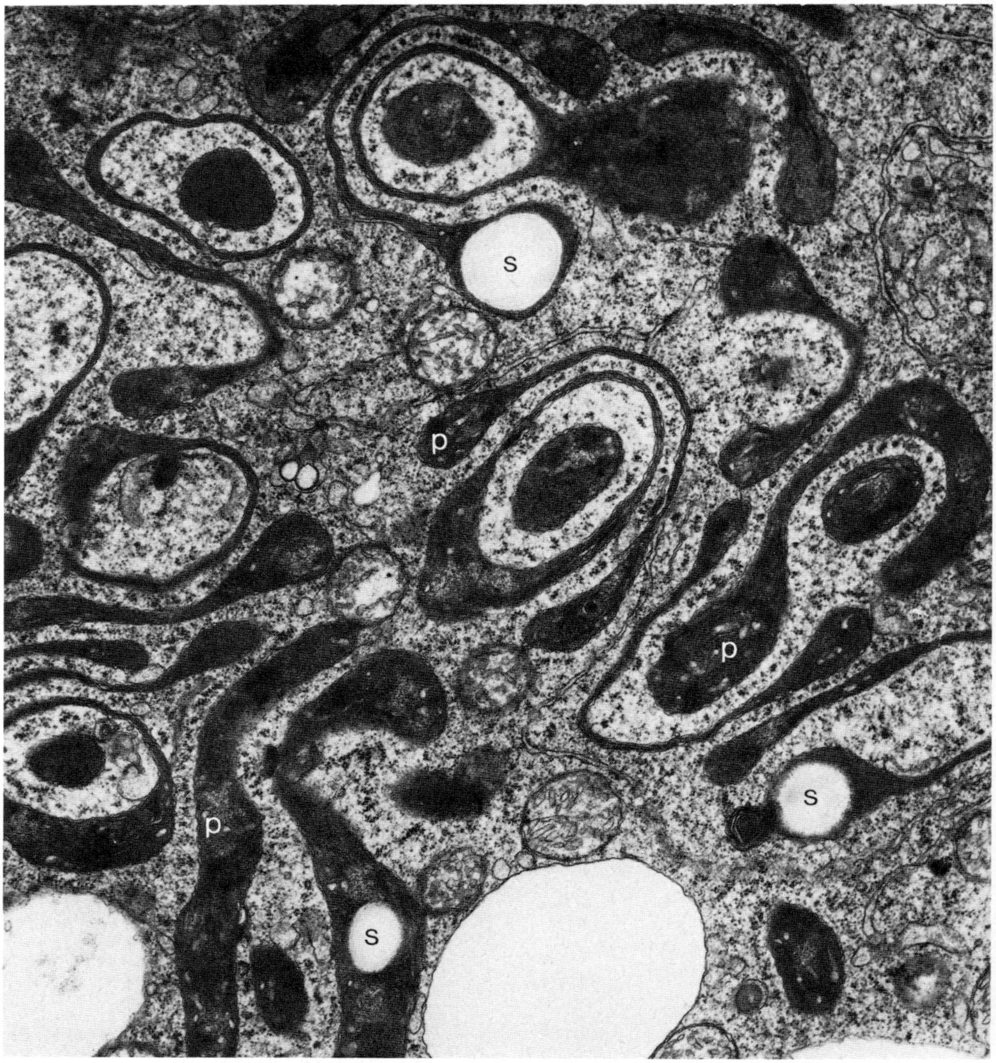

Fig. 7.19. *Gossypium*, portion of a two-nucleate endosperm of cotton with cup-shaped plastids (*p*) containing starch (*s*) and associated helical polysomes. (After Schulz and Jensen 1977)

tilized ovules of cotton, precocious cell walls are laid down but cytoplasm of the cellular endosperm resembles that of the central cell (Schulz and Jensen 1978).

In many taxa (Jensen 1965, Diboll and Larsen 1966, Diboll 1968) the central cell contains lipid reserves which, in *Capsella bursa-pastoris* (Schulz and Jensen 1974), are utilized during earlier development of the endosperm. The intimate association of microbodies with lipid droplets in the central cell of *Capsella* (Schulz and Jensen 1973) indicates the possible initiation of the conversion of stored fats to sugar to supply energy for subsequent development. In castor bean Vigil (1970) reported microbodies in close association with endosperm spherosomes which are utilized during germination.

7.6.2 Central Cell Nucleus

Incomplete fusion of the male nucleus with the central nucleus is much more frequent than with the egg (van Went 1970b). The primary endosperm nucleus has a large vacuolate, strongly pyroninophilic nucleolus in *Alyssum maritimum* (Vijayaraghavan et al. 1984). In *Capsella bursa-pastoris* the nucleolus is particulate and fibrillar, suspended in a dense matrix like the polar nuclei. In *Capsella* a unique feature of the primary endosperm nucleus is the presence of aggregate dense material in the perinuclear cytoplasm. These aggregates consist of granules embedded in a dense matrix, and strongly resemble the aggregates in the periphery of nucleolus and in the nucleoplasm. Perinuclear aggregates have also been observed in *Gossypium hirsutum* (Schulz and Jensen 1977). These grains disappear after the division of the primary endosperm nucleus, and a few divisions of daughter nuclei. In *Capsella* (Schulz and Jensen 1973) the dense aggregates probably represent masses of extruded nucleolar material which provides ribosomes and other kinds of RNA and proteins needed to support the initial rapid growth of the endosperm. Nucleolar extrusion is common in animal cells, especially in developing oocytes (Raven 1961, Scharrer and Wurzelmann 1969, Szollosi 1965). In cotton endosperm there is no evidence at present to suggest a relationship between these granular aggregates and helical polysomes (Fig. 7.19) (Schulz and Jensen 1977). In contrast to the appearance of perinuclear aggregates at the time of fertilization, nucleolar extrusion in the endosperm has been observed even at the heart-shaped stage of the embryo in *Alyssum maritimum* (Prabhakar et al. 1981). In this taxon the extrusion is exhibited mainly by the nuclei in the micropylar (Fig. 7.18B) and chalazal portions of the endosperm.

7.6.3 Covering of the Central Cell

The central cell has a plasma membrane of its own. The nature of the wall surrounding the central cell is complex and highly variable (Jensen 1974). In *Capsella bursa-pastoris* (Schulz and Jensen 1969) it shows projections from the wall similar to the filiform apparatus of synergids, whereas in *Gossypium hirsutum* (Jensen 1965) the wall is generally smooth. There are only a few studies on the structure of the embryo sac wall, at such an early stage, and it is difficult to generalize that wall ingrowths form a common feature of the central cell.

Modification of the embryo sac wall, forming ingrowths into the endosperm, are reported in *Linum usitatissimum* (B. Vazart and J. Vazart 1966), *Pisum sativum* (Marinos 1970), *Helianthus annuus* (Newcomb and Steeves 1971), *Lobelia dunnii* (Torosian 1971), *Hibiscus* spp. (Ashley 1975), *Gossypium hirsutum* (Schulz and Jensen 1977), *Haemanthus katherinae* (Newcomb 1978), *Phaseolus coccineus* (Yeung and Clutter 1978), *Iberis amara* and *Alyssum maritimum* (Fig. 7.20) (Prab-

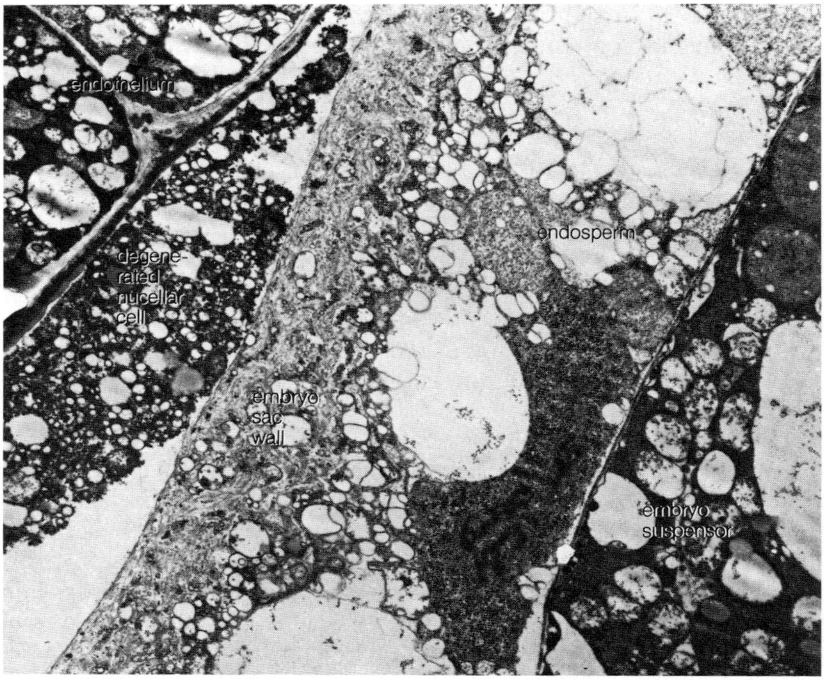

Fig. 7.20. *Alyssum maritimum*, portion of longisection of seed to show extensive labyrinth of wall projections from embryo sac wall. (After Prabhakar and Vijayaraghavan 1983)

hakar and Vijayaraghavan 1983). All the above instances provide enough evidence to support the view that the ingrowths facilitate the absorption of metabolites from the ovular tissue. In *Capsella bursa-pastoris, Iberis amara,* and *Alyssum maritimum* the wall ingrowths at the chalazal end probably facilitate the absorption of degraded chalazal nucellar tissue into the endosperm (Vijayaraghavan and Prabhakar 1981). In *Iberis amara*, during late embryogeny, the wall ingrowths at the chalazal end are thick, stain intensely with PAS reaction, and can be seen with a light microscope. These ingrowths divide the coenocytic chalazal portion of the endosperm into compartments and, thus, cellularization occurs without the involvement of spindle fibres (Prabhakar 1979).

7.7 Wall Formation in Endosperm

In the Nuclear type of endosperm there is no switching back and forth of karyokinesis and cytokinesis; the change-over to wall formation occurs once. In the Cellular type, karyokinesis is always coupled with cytokinesis. In the Helobial type, karyokinesis and cytokinesis are coupled in the first division. Subsequently, uncoupling results in the formation of coenocytic cells that later become cellular as a result of recoupling of karyokinesis and cytokinesis. Irrespective of the type of endosperm development, the process of wall formation is common in the coenocytic endosperm of either the Nuclear or the Helobial type. What factors regulate the control of cellularization in Nuclear and Helobial endosperm is not clear. The advantage of the endosperm becoming compartmentalized is interpreted as one way of storing carbohydrates in an insoluble form, which might be used later during digestion of the endosperm (Newcomb 1973).

The cellularization by freely growing cell walls during transition from coenocytic to cellular endosperm has been described in *Helianthus annuus* (Newcomb 1973), *Stellaria media* (Newcomb and Fowke 1973), *Quercus gambelii* (A. P. Singh and Mogensen 1976), *Triticum aestivum* (Mares et al. 1977), and *Haemanthus katherinae* (Fig. 7.15 A, B).

In *Helianthus annuus* wall formation in the endosperm starts at the micropylar end and extends towards the chalazal end of the embryo sac. The wall formation is not associated with the mitotic spindle apparatus, and the walls grow freely as the wall material is deposited by the coalescence of vesicles arising from golgi near the growing tip of the wall (Newcomb 1973). By contrast, the usual mode of wall formation is by cell plate, as reported in *Dianthus chinensis* (Buell 1952). The presence of a large number of dictyosomes and their smooth vesicles in close proximity to the free ends of wall supports the growth of walls. Bajer (1965) reported that in *Haemanthus katherinae* endosperm phargmoplast may form between sister and non-sister nuclei during any stage of karyokinesis, or between nuclei in interphase and membranes apparently formed between the nuclei before wall formation (see also Buttrose 1963, Mares et al. 1975). In fertilized cotton embryo sacs, wall formation in endosperm occurs approximately 10 days after pollination, whereas in cultured unfertilized embryo sacs wall formation in endosperm occurs on the third day of culture after dozens of free nuclei have been formed. The pattern of growth of walls in fertilized and unfertilized endosperm is comparable. The walls are initiated at the embryo sac wall, and grow towards the large central vacuole of the embryo sac, by apparent coalescence of dictyosome vesicles and, possibly, involvement of RER (Jensen et al. 1977).

The process of cellularization in wheat has been examined by Buttrose (1963), Evers (1970), Mares et al. (1975, 1977), and Morrison and O'Brien (1976). Mares et al. (1975) postulated that the cellularization of free-nucleate syncytium is accompanied by freely growing walls which, with the accompanying cytoplasm and organelles, extend centripetally from the periphery of the central cell. The central cell is gradually cleaved into smaller, highly vacuolate compartments. Following the formation of cellular endosperm, further increase in cell numbers results from the meristematic activity of the superficial layers of endosperm cells. Mares et al. (1977) further extended the studies to show that alveoli are not complete cells, as cell-wall material is absent in this thin layer of cytoplasm. After the division of the alveolar nucleolus, a cross-wall is laid down in the daughter nuclei by a process which is similar to normal cytokinesis, and a layer of endosperm cells is formed from the peripheral portion of alveoli (Fig. 7.21).

Morrison and O'Brien (1976) have proposed an alternative sequence of events. Wall ingrowths, at first recognizable as minute pegs, furrow inward from the edge of the embryo sac through the vacuolate cytoplasm, which is clearly de-

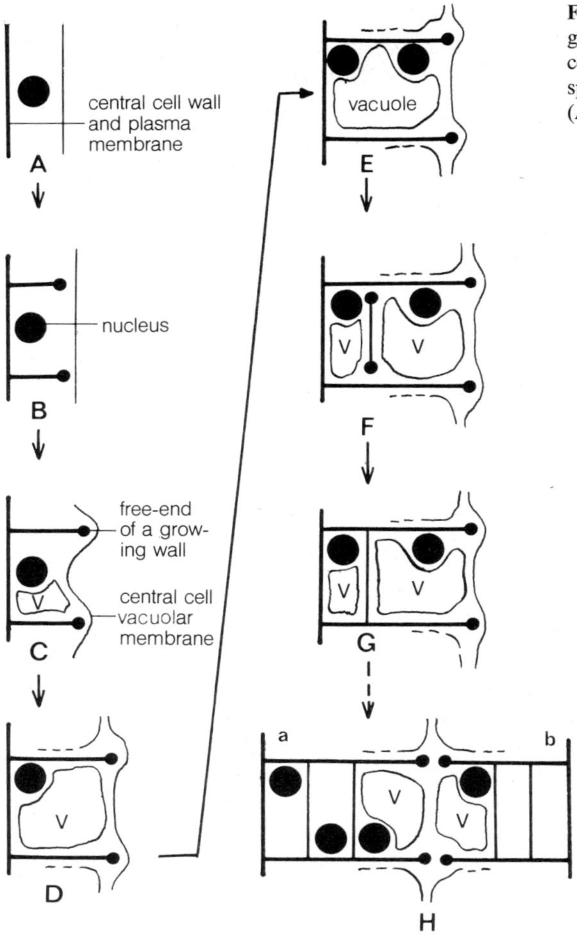

Fig. 7.21 A–H. *Triticum aestivum,* diagrammatic sketch of events leading to cell wall formation in free-nuclear endosperm; *a*, *b* endosperm cells; *v* vacuole. (Adapted after Mares et al. 1977)

limited on the inside by the central vacuole. Cellularization of the peripheral layer of endosperm cytoplasm is, thus, complete about 2 days after anthesis. Between 2 and 3 days after anthesis the peripheral layer of cells commences to divide both radially and tangentially, and by 4 days the entire embryo sac is cellular.

In wheat, microtubules have consistently been observed, aligned at right angles to the planar array of vesicles, and possibly with a role in orienting vesicles in the adjacent cytoplasm into the plane of the developing wall projections (Mares et al. 1977). Morrison and O'Brien (1976) did not observe any microtubules associated with the furrowing wall projections in developing wheat endosperm. However, Mares et al. (1977) report that the presence of microtubules is a regular feature of sections in which the ultrastructural appearance of other cellular organelles, particularly the vacuolar membranes, indicated that good fixation had been achieved. Some microtubules are perpendicularly oriented to the newly formed wall in the unfertilized cotton endosperm (Jensen et al. 1977). Many of the cross-walls, formed after four or more layers of endosperm cells develop, are derived from phragmoplast associated with cytokinesis (Newcomb 1978). A similar ontogeny of cross-walls has been observed in the endosperm of *Helianthus annuus* (Newcomb 1973) and *Stellaria media* (Newcomb and Fowke 1973). In *Haemanthus katherinae* (Hepler and Jackson 1968) the phargmoplast arises between the telophase nuclei, usually in association with a remnant strand of spindle element, and consists of an accumulation of microtubules oriented at right angles to the plane of the future cell-plate.

Fineran et al. (1982), contrary to the investigations of Mares et al. (1975, 1977), and Morrison and O'Brien (1976), observed cytokinesis when the first anticlinal walls are laid down during initial cellularization of wheat endosperm (Figs. 7.22 A–H, 7.23 A a–e, B a–d). A typical phragmoplast develops between the free endosperm nuclei that occupy the peripheral cytoplasm adjoining the central cell wall. The cell plates soon fuse with the wall of the central cell. The phragmoplast is discernible in optical preparations (Fig. 7.22 A), and in the electron microscope it shows many vesicles and microtubules aligned perpendicular to the wall of the central cell (Fig. 7.22 B). At this stage of development, the microtubules are associated with the entire phragmoplast (Fig. 7.22 C) followed by the penetration of tubular ER (Fig. 7.22 C). Cytochemical staining is positive for the localization of polysaccharide materials in the phragmoplast vesicles, and in the cell-plate. These plates extend and reach almost up to the wall of the central cell (Figs. 7.22 D, 7.23 A a, b). The inner margins of the walls, however, continue to extend towards the central vacuole within the advancing front of the cytoplasm. Anticlinal walls arise through cytokinesis, around the nuclei, approximately simultaneously at various mitotic phases, namely, prophase, metaphase, and telophase. Three-dimensionally, the nuclei come to lie within the open-ended compartments of the cytoplasm (Fig. 7.23 B a–d). Then the compartments undergo cell division and form periclinal walls through normal cytokinesis (Figs. 7.22 E, 7.23 A c). These walls, thus, delimit the first complete layer of endosperm cells centrifugally, and a new layer of open compartments centripetally (Fig. 7.23 A c). The inner compartments usually continue to extend centripetally until they meet those from the opposite side of the grain (Fig. 7.23 A d). The final closing phase is characterized by a further row of periclinal walls formed along the line where the compartments fuse (Figs. 7.22 F–H, 7.23 A d, e). Subdivisions in the first-formed large endosperm cells result in smaller cells.

Concerning cell-wall formation in wheat endosperm the controversy is that: (1) Mares et al. (1975, 1977) presumed that the anticlinal walls arose independently of the mitotic spindle, but were related to free growth of walls that followed previously delineated lines developed by ingrowths of the plasmalemma into the cytoplasm of the central cell. The cytoplasm, thus, is segmented into "open-ended cylinders" later designated as "alveoli" (Mares et al. 1977). (2) Morrison and O'Brien (1976) proposed essentially a similar view for the origin of the first anticlinal walls which, in the absence of a phragmoplast, they equated to the cleavage process common in lower plants. (3) Fineran et al. (1982) lend support to the explanation of Mares et al. (1977) but with a different interpretation: the first anticlinal wall that originates during cellularization of the endosperm nuclei is through phragmoplast and cell-plate formation.

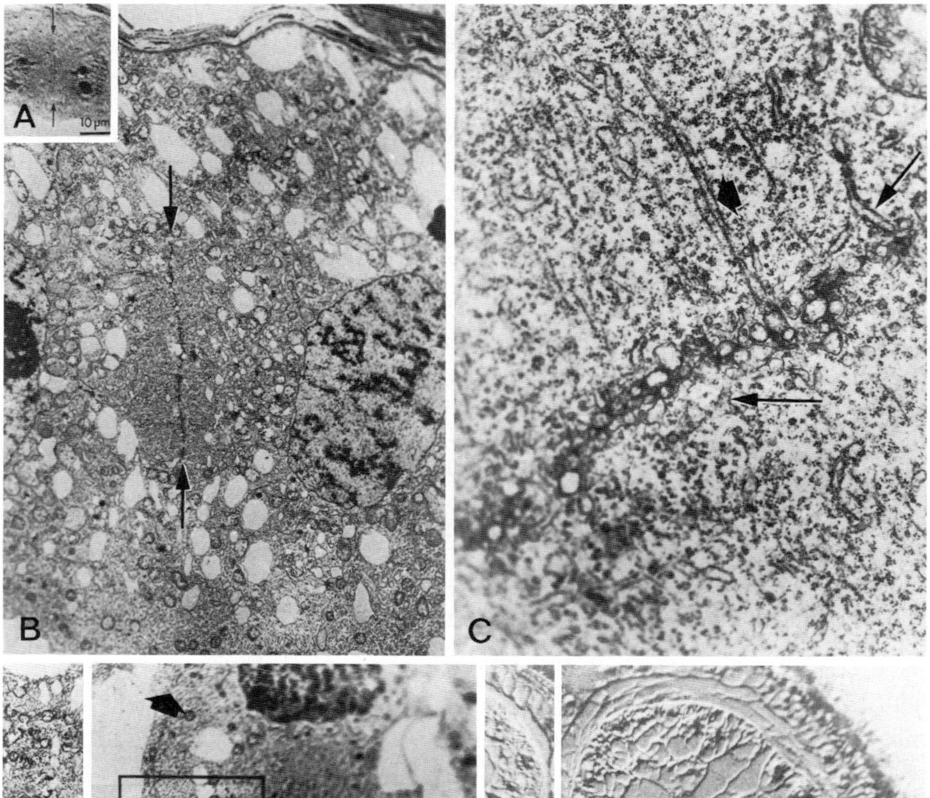

Fig. 7.22 A–H

In *Haemanthus katherinae* (Newcomb 1978) another interesting mode of compartmentalization of the coenocytic endosperm is by the formation of free cells. These cells arise from the innermost layer of endosperm cells, and lack a wall on the side facing the central vacuole. These free cells do not have a complete wall. The free cells, earlier reported in *Cocos nucifera*, are formed soon after fertilization, later migrate to the periphery, and then cellularization occurs in these portions (Cutter and Freeman 1955, Cutter et al. 1955).

Direct participation of wall ingrowths in wall formation has never been reported at the light microscopy level, and has been observed, for the first time, in *Iberis amara* (Prabhakar 1979). During the final stages of maturation of seed, the wall ingrowths show an increase in length, branch profusely, anastomose, and are incorporated to form cell walls. Initially, these wall ingrowths stain intensely with PAS reaction, but lose this property of staining as they become pro-parte cell wall of endosperm. All wall ingrowths do not form cell walls. When the rest of endosperm is cellular (Fig. 7.24 A, B), the chalazal caecum persists in mature seed and is covered with a web of wall ingrowths (Figs. 7.5, 7.24 C, D). These ingrowths continue to stain intensely with PAS reaction. The endosperm nuclei which fail to undergo cellularization lie in the centre of caecum and degenerate in situ (Fig. 7.24 C).

7.8 Cytology of the Endosperm

The endosperm tissue is basically triploid. As it matures, and with age, it becomes mixoploid due to endopolyploidy. Mitotic inhibition occurs either at prophase or metaphase. Tandon and Kapoor (1961 a, b, 1963) and Kapoor and Tandon (1963, 1964) published a series of papers on the cytology of the endosperm of *Nothoscordon fragrans*, *Zephyranthus ajax*, *Z. grandiflora*, and *Z. lancasteri*. They reported nuclei of various shapes – dumb-bell, spindle, star, and with whip-like extensions. Prabhakar et al. (1981), at ultrastructure level, observed lobing of nuclei in the endosperm of *Alyssum maritimum*. The degree of lobing varies and, occasionally, pseudopodia-like structures are formed. Such nuclei are known to be

Fig. 7.22 A–H. *Triticum aestivum*, endosperm. **A** Nomarski interference micrograph showing initiation of anticlinal wall between two endosperm nuclei in the parietal cytoplasm of central cell. **B** Electron micrograph from **A** showing the future anticlinal wall with phragmoplast (*arrows*) and fusing vesicles. The ER and microtubules are not visible at this magnification. The peripheral cytoplasm bulges towards the central cell vacuole where the wall develops. **C** Initial anticlinal wall (as in **B**) with groups of fusing vesicles, many tubular ER (*arrows*), and microtubules (*large arrow*). **D** Anticlinal wall (at a stage later than in **B**) suspended freely between a pair of endosperm nuclei (nuclei not shown). The cell-plate almost reaches up to the central cell wall (*) and the tonoplast bordering the large vacuole. **E** Electron micrograph showing proximal region of the compartment at telophase stage. The cell-plate is juxtaposed between the two nuclei, but has not yet reached the anticlinal wall. *Arrows* indicate the course of anticlinal walls from the central cell wall (*). **F** Nomarski interference micrograph of transection of the grain during completion of endosperm cells. The central cell vacuole is almost obliterated by the compartments that are about to meet. *Arrows* indicate where the fusion is complete and the line of periclinal walls which separate the original opposed compartments. **G** Still open region from **F**. **H** Endosperm shortly after completion of initial cellularization. (After Fineran et al. 1982)

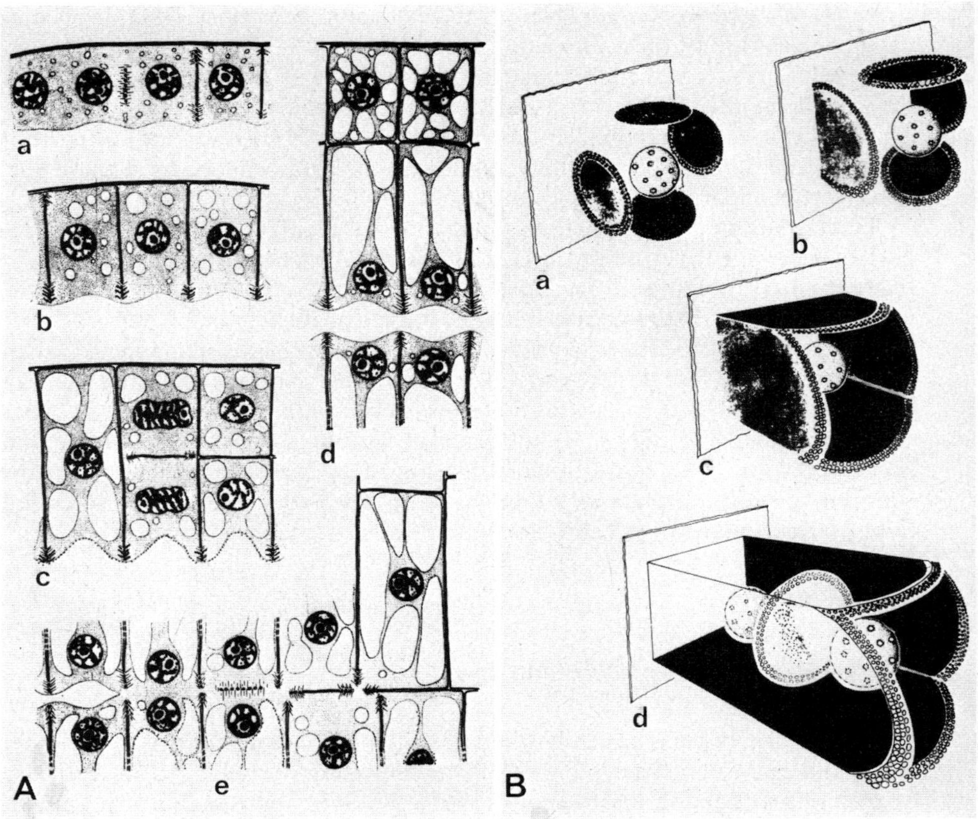

Fig. 7.23 A, B. *Triticum aestivum,* endosperm (**A** semidiagrammatic, **B** models). **A** Initiation of wall formation. *a* Phragmoplast is laid down between endosperm nuclei. *b* Fusion of cell plate to the wall of central cell. *c* Later stage; note a fully developed (*left*) open compartment, and periclinal division (*middle*) to complete endosperm cell centrifugally (*top*), and another compartment centripetally. *d* Complete endosperm cells (*top*), and inner layer of open compartments. The endosperm cells from the opposite end of the grain also approach each other. *e* Fusion of compartments from opposite side of the grain, during closing phase. The opposed compartments (*left*) are about to fuse, and in the middle region there is a fusion of "naked layers" of cytoplasm. In the fused zone phragmoplast arises (*middle*) followed by development of cell plate (*middle-right*). The cell plate grows to meet the approaching anticlinal wall and other developing periclinal walls between the opposed compartments (*right*). The anticlinal walls are aligned opposite to each other *on the left* and alternate *on the right*. This is due to cell plates meeting at various levels. **B** Three-dimensional representation of events occurring during initial wall formation. *a* Phragmoplasts and their first-formed cell plates, grouped around the nucleus, perpendicular to the wall of the central cell. *b* Later stage, cell plates are meeting and have fused with the wall of central cell. *c* Fusion to central cell is complete, and anticlinal walls continue to grow freely by their inner margin towards the central vacuole. *d* Fully-formed compartment ready for periclinal division. (After Fineran et al. 1982)

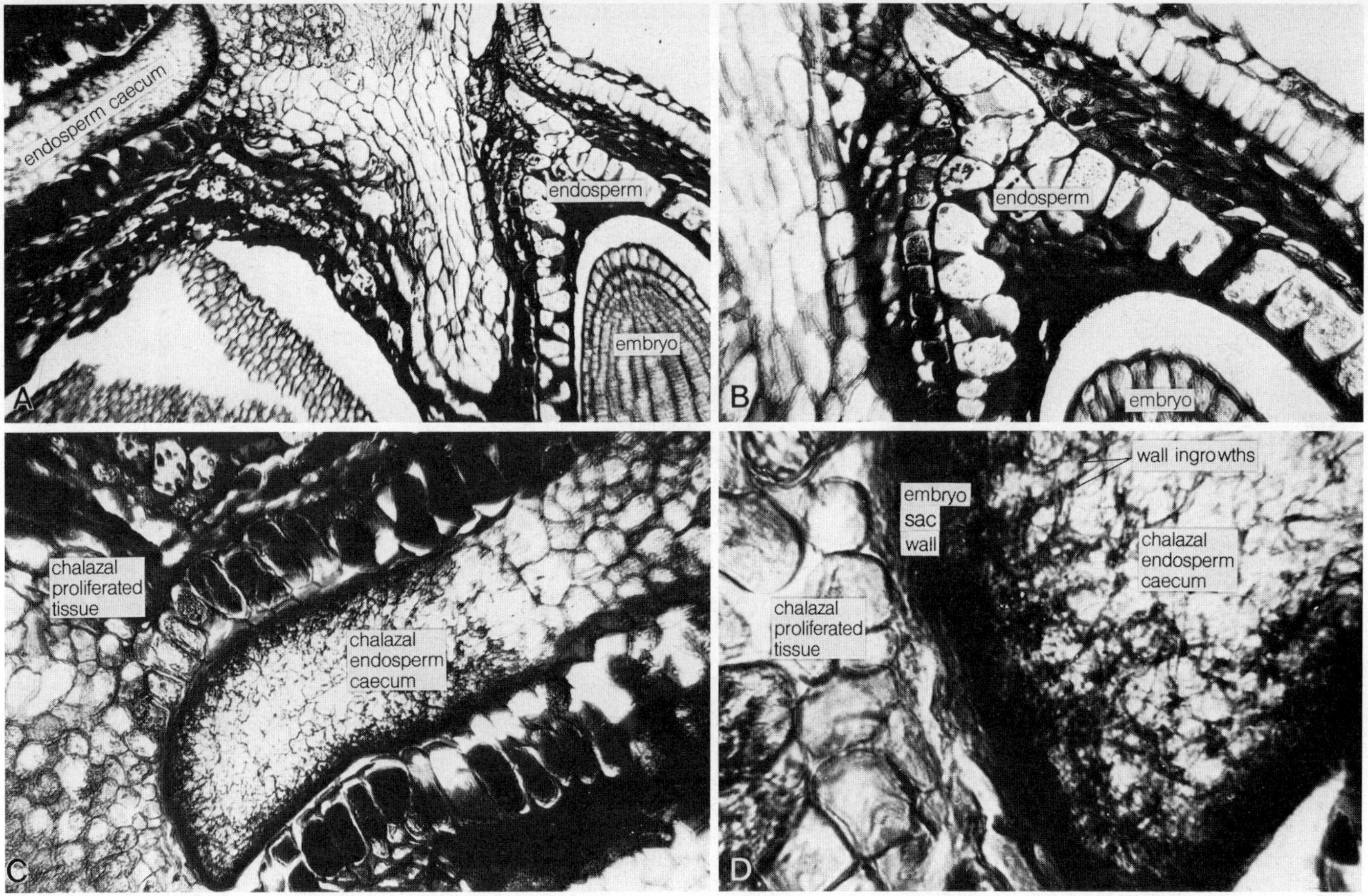

Fig. 7.24 A–D. *Iberis amara*. **A** Portion of longisection of seed to show coenocytic endosperm caecum at the chalazal end, and single-layered endosperm at the micropylar end. **B** Micropylar portion with a single layer of endosperm cells to show polysaccharides in the form of cell walls. **C** Chalazal endosperm caecum; note the web of PAS-positive wall ingrowths and thick deeply-stained embryo sac wall. (**A–C** after Vijayaraghavan and Prabhakar, unpublished). **D** Embryo sac wall separating endosperm from proliferated chalazal tissue. (After Vijayaraghavan and Prabhakar 1981)

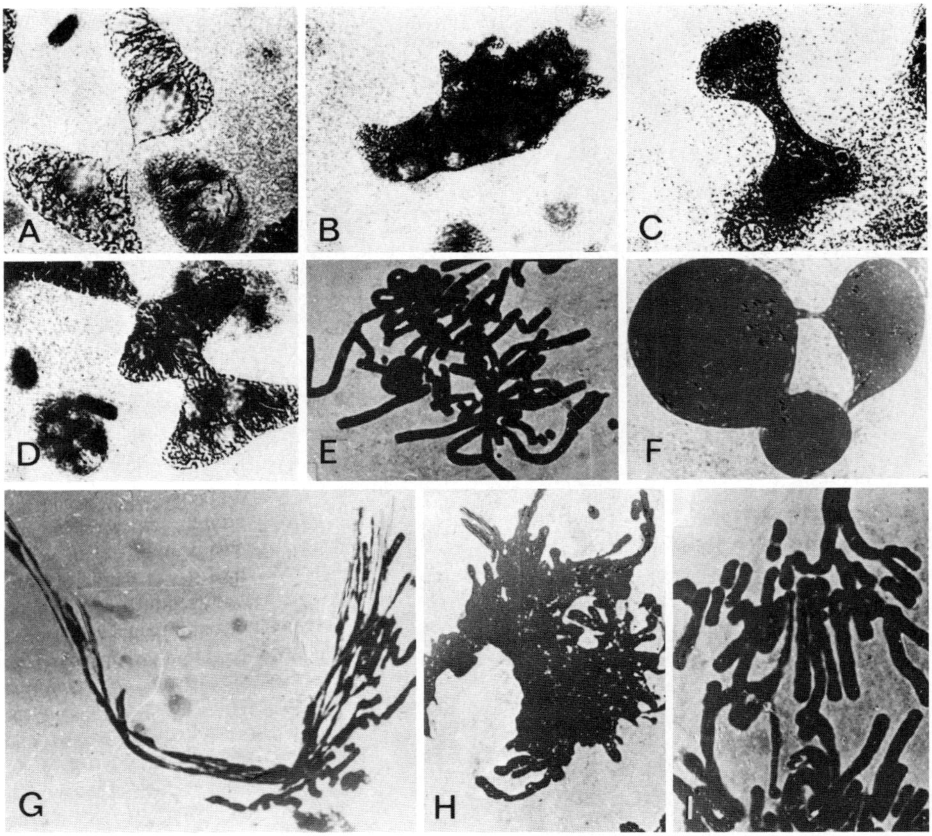

Fig. 7.25 A–I. Abnormal endosperm nuclei. **A–D** Triticale, mitotic abnormality. (After Kaltsikes 1973). **E–I** *Haemanthus katherinae*. **E** Endosperm cell with fragments and ring chromosomes at metaphase, along with sticky chromosomes. **F** Unusual mitotic behaviour often associated with degenerating endosperm. **G** Extreme unravelling of chromosomes into a very long hair-like structure. **H** Irregular giant nucleus. **I** Anaphase with sticky and swollen chromosomes. (After Stephen 1979)

associated with high metabolic activity. Kaltsikes (1973) stated that the nuclei at early stages of endosperm development may be concerned in some metabolic activity necessary for sustaining the production of metabolites essential for their own multiplication, and the growth of the parasitic embryo. He observed irregularly shaped or fused nuclei as an abnormality in various lines of triticale (Fig. 7.25 A–D).

The nucleoli in the endosperm nuclei of *Alyssum maritimum* show variation in shape and size, and are compact or vacuolate. The compact nucleolus may be fibrillar, and the vacuolated nucleolus shows distinct zones of par granulosa and par amorpha. The metabolically active state is further confirmed by blebbing of the nuclear membrane (Fig. 7.18 B). Extrusion of nucleolar material into the nucleoplasm, or the cytoplasm, is reported in the endosperm of *Phaseolus coccineus* (Cionini and Cremonini 1970), *Capsella bursa-pastoris* (Schulz and Jensen 1973),

Gossypium hirsutum (Schulz and Jensen 1977), *Alyssum maritimum* (Prabhakar et al. 1981) and *Potamogeton nodosus* (Vijayaraghavan and Kapoor 1984). Szollosi (1965) concluded that, in rat, the extrusion bodies are similar in composition to the nucleolus. Hay (1968) described that the dispersed nucleolar material passes through nuclear pores, and reaggregates in the perinuclear cytoplasm. Szollosi (1965), Hay (1968), and Schulz and Jensen (1973) believe that nucleolar extrusions probably represent the mass transport of ribosomal precursors or, possibly, other kinds of RNA into the cytoplasm. In *Zephyranthus lancasteri,* the nucleoli per nucleus vary from 1 to 18 (Kapoor and Tandon 1964). Kaltsikes (1973) mentions the formation of up to nine nucleoli per nucleus in different lines of triticale.

Kato (1957) reported a high degree of diversification in the size and shape of nuclei in the central region of the endosperm rather than in the peripheral area. According to Kapoor and Tandon (1964), the size and shape of nuclei do not vary in the central and peripheral region in *Zephyranthus lancasteri.* Prabhakar (1979) observed nuclei to show a higher degree of polyploidy in the portion of endosperm which accumulates at the chalazal end in *Iberis amara* (Fig. 7.6 C) and *Alyssum maritimum.* In *I. amara* the extreme chalazal region, which is coenocytic, becomes haustorial during later stages of embryogenesis. D'Amato (1952) suggests the demarcation of endosperm into two or more zones, on the basis of ploidy gradient. An increase in the size of endosperm nuclei can result from polyploidy or polyteny. Kaltsikes (1973), however, considers that cells with apparently the same amount of DNA are capable of 20-fold change in their volume. In maize polyteny is the cause for the endosperm nuclei to grow to 1,000 times the original volume in 24 days subsequent to pollination (Duncan and Ross 1950). In *Gagea* nuclear enlargement is accompanied with increase in the number of chromosomes (Geitler 1948).

Polyploidy is normally a more common cause for increase in the size of endosperm nuclei (see D'Amato 1952, Turala 1969). The chalazal haustorium generally has polyploid nuclei (Chopra and Sachar 1963). The micropylar haustorium also, occasionally, shows ploidy levels of 192–384 *n* in *Pedicularis palustris* (Steffen 1956). Amitosis (irregular mitotic behaviour seems to have been misinterpreted as amitosis) may not be a universal phenomenon, but is recorded in endosperm tissue (Hristov and Moskov 1957, Wang and Chien 1957, Tandon and Kapoor 1961 a).

Some of the common irregularities recorded in the endosperm are bridges, lagging chromosomes, and fragments (Fig. 7.25 A–D). The ovules of shrivelled fruits show either complete absence of endosperm, or collapsed endosperm with mitotic abnormalities. Kaltsikes (1973) correlated the rate of endosperm development with the amount of seed shrivelling. Irregular shape of fused nuclei is the most common abnormality in all lines of triticale (Fig. 7.25 A–D). Kaltsikes and Roupakias (1975) and Kaltsikes et al. (1975) established a correlation between mature shrivelled kernels and the number of aberrant endosperm nuclei in triticale.

Cytological analysis of endosperm disfunction has revealed several interesting features. The abortion of fruits in the hybrid clones of *Tradescantia* (Stephen 1977), *Borassus flabellifer* (Stephen 1974), and *Haemanthus katherinae* (Stephen 1979) is mainly due to endosperm senescence, aging and degeneration. It is well known that endosperm disfunction causes abortion of embryo, resulting in low fertility. The cytological analysis of this aspect is well understood (Fig. 7.25 E–I). The endosperm tissue is the storehouse of many macromolecules like carbohydrates, various types of proteins, lipids, and growth regulating substances. In most angiosperms, at first, the endosperm is predominantly triploid but, at maturity the cells show various levels of ploidy. For example, in *Borassus flabellifer* the mature endosperm (3n = 54) shows multiples of 6n, 12n, and up to 96n. The endopolyploidy is caused by the dissolution of the nuclear membrane, condensation of chromosomes, and occurrence of even metakinesis, and is comparable to C-mitosis (Stephen 1974).

In endosperm tissue endopolyploidy is brought about by various factors, and may be due to endoreduplication, dispersion endomitosis (Nagl 1970 a, b), C-mitosis (Stephen 1974), and nuclear division (Stephen 1980). Enzenberg (1961) described endopolyploidy in 13 angiosperm species, and calls it "restitution-kernbildung". It may be emphasized that nucleolar number is not an index of ploidy status (see Tandon and Kapoor 1961 b).

The endosperm in *Datura fusticosa* is basically triploid (3x = 36), but later it becomes mixoploid owing to progressive endopolyploidy and sporadic occurrence of a few aneuploid cells. The nuclei vary in size. Vigorous development of endosperm and growth of fruit are a positive adaptation (Stephen 1980). Endopolyploidy is a gene-controlled mechanism designed to enhance macromolecule synthesis, namely proteins and carbohydrates required not only for the growth of the endosperm but also for the growth of the embryo during development, differentiation, and germination. It is likely that there is a flow of nutrients to the maternal tissue, resulting in the overall growth of fruit.

The endosperm, in *Haemanthus katherinae* (3x = 27), with advancing maturity becomes mixoploid. There is mitotic inhibition either at prophase, or at metaphase, which causes endopolyploidy. The degenerating endosperm tissue exhibits spontaneous chromosomal aberrations and chromosomal inflamation followed by stickiness, extension of chromosomes into extremely long, hairlike structures, nuclear polymorphism, amitosis, and nucleolar fragmentation (Fig. 7.25 E–J). Such abnormalities cause fruit abortion in *Haemanthus*. Spontaneous abortion of horticultural fruits appears to be a common feature, and aberrant endosperm cytology may be the chief reason. Acute nuclear polymorphism is reported in hybrid endosperm of *Tradescantia* (Stephen 1977) and chromosomal aberrations in many interspecific hybrids (Brink and Cooper 1947, Brock 1954, 1955). Interestingly, abnormal and irregular nuclei with projecting chromosomal ends (as in endosperm tissue) can also be induced by dithiothreitol. This chemical perhaps prevents chromosome relaxation and telophasic reconstitution of the nuclear envelope (Wolfe and Rodriguez 1974). It is conjectured that similar inhibitors may be present in the degenerating endosperm, and such substances are perhaps responsible for the spindle abnormalities. These cytological abnormalities do reflect the molecular interaction between the endosperm and other tissues. Despite such abnormalities, the mature endosperm is quite normal in majority of the angiosperms.

7.9 Reserve Materials in Developing Endosperm

7.9.1 Protein Bodies

The storage proteins largely accumulate in the endosperm during seed formation. These proteins are synthesized on rough ER, and localized in specialized cell organelle – protein body. The protein body is bound by a single membrane, and contains a matrix within which only globoid, globoid and crystalloid inclusions, or none, may be present. The protein bodies are formed by various mechanisms which involve the participation of RER, vacuoles, and dictyosomes. The main constituents of protein bodies are storage proteins, and phytin. In addition, they also contain hydrolases, lectins, lipids, and carbohydrates. The storage proteins are characterized by (1) a large amount of amide nitrogen, (2) high molecular weight, and (3) heterogeneous nature.

The protein bodies in endosperm are degraded during germination. The type and amount of protein have a marked influence on the baking quality of wheat flour. In wheat endosperm the protein bodies are localized in vacuoles, and with progressive development these bodies occupy most of the expanding endosperm cells. The protein bodies do not have a true limiting membrane, but do have a definite external boundary. The rupture of this boundary releases the internal contents (Graham et al. 1962). In *Setaria lutescens* the endosperm protein bodies are membrane-bound. Each protein body shows internally a pattern of dark and light concentric layers; these stain with mercuric bromphenol blue (MBB). Such protein bodies have also been observed in rice and corn (Rost 1971). The internal concentric layers indicate differential deposition of proteins during formation, and more than one type of protein. In the aleurone grain of barley, the aleurone layers have two types of inclusions: (1) Globoids within globoid cavities without a membrane; these globoids stain red with Toluidine Blue (TB), indicating the presence of phytin, and they also stain for lipids. (2) Protein carbohydrate bodies which stain green with toluidine blue. The characteristics of globoid and protein carbohydrate bodies are interesting. The globoid bodies are anisotropic in polarized light, contain phytin and lipid, vary up to 3 μm in diameter, and in glutaraldehyde-osmium fixed tissues appear to be amorphous, electron-opaque, and shatter during sectioning. In potassium permanganate fixation they are mostly dissolved. The protein carbohydrate bodies are isotropic in polarized light, uniformly 1 or 2 μm in diameter; in glutaraldehyde-osmium tetraoxide material they are granular, electron-dense, and do not shatter during sectioning (Jacobsen et al. 1971).

Jones (1974) studied the endosperm of *Lactuca sativa,* and pointed out that two-layered persistent endosperm has a thick wall and dense cytoplasm. There are numerous peg-like projections from the endosperm cell wall, and these projections extend into the cytoplasm (Fig. 7.26). The dense cytoplasm contains protein bodies and spherosomes. During germination (12–15 h in light) the cell walls of the endosperm are extensively broken down, which increases with the time of imbibition. In *Alyssum maritimum* the seeds which fail to germinate show intact endosperm layer filled with proteinaceous masses, even up to 15 h after the soaking

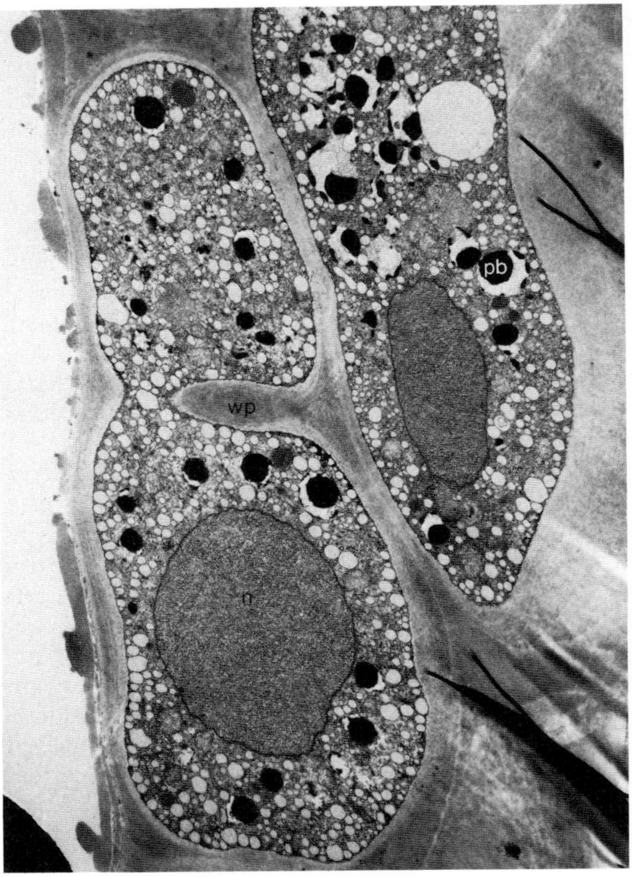

Fig. 7.26. *Lactuca sativa,* transection of endosperm cells to show prominent nucleus (*n*), protein bodies (*pb*), spherosomes, and wall protuberance (*wp*). (After Jones 1974)

of seeds (Prabhakar 1979). The inability to utilize storage proteins seems to be one of the factors causing failure of germination.

Bechtel and Pomeranz (1978 b) studied the ultrastructure of mature, ungerminated caryopsis of *Oryza sativa.* The endosperm, based on the number and type of protein bodies present, is divided into two regions: (1) the subaleurone, and (2) the central. The subaleurone region contains three different types of membrane-bound protein bodies: large spherical, small spherical, and crystalline. The crystalline protein bodies show crystal lattice fringes. The central region of endosperm lacks the small, spherical, and crystalline protein bodies. Protein content differs substantially between subaleurone and central endosperm of cereal grain. Variations in the amino acids between two regions in wheat endosperm have been interpreted to represent a large portion of storage protein in the subaleurone regions rather than in the central zone (Kent and Evers 1969). In *Alyssum maritimum* and *Iberis amara,* after cellularization, the endosperm cells are filled with polysaccharide grains. Gradually, the endosperm layer (adjacent to endothelium) shows depletion of insoluble polysaccharide grains (Fig. 7.24 B). All other endosperm cells are utilized as such. The single layer of endosperm cells, filled with proteinaceous masses, persists in mature seed giving an appearance similar to that of the aleurone layer (Prabhakar 1979).

Isolation techniques have to be perfected so that the three types of protein bodies can be separated for chemical analysis. Only then the specific function of each protein body can be ascertained. Jacobson et al. (1976) used gibberellin to breakdown the endosperm cells, in celery (*Apium graveolens*). With this treatment the aleurone grains swell, lose their contents, and the bulk of each cell is hydrolysed. However, a thin resistant layer of wall remains around each protoplast. The wall hydrolyses and causes endosperm to breakdown into individual cells which can be plasmolysed while still living. All the cells of the endosperm respond to gibberellin in a similar way, although the cells near the radicle elongate rapidly. The response is specific only to gibberellin, and does not occur in the presence of ethylene, kinetin, abscissic acid, and indole acetic acid. The results were similar both in light and darkness. The breakdown of endosperm during the germination of seed perhaps involves the release of hydrolases from the expanding embryo. The protein bodies in the endosperm of *Linum usitatissimum* originate in vacuoles, and have both globoid and crystalloid inclusions (Dhar and Vijayaraghavan 1979), whereas in *Ranunculus sceleratus* only globoid (Vijayaraghavan and Bhat 1982). Breakdown occurs by vacuolation and fragmentation, suggesting that the protein bodies are organelles, and contain hydrolytic enzymes.

Parker (1980) observed protein body inclusions in endosperm in two cultivars of wheat. Spherical, electron-dense inclusions were first embedded in the periphery of endosperm protein bodies (11 days after anthesis). These inclusions gave positive histochemical reaction for both proteins and lipids, persisted throughout endosperm development, and were present in the protein matrix of mature grain.

7.9.2 Starch

Corn flour, when finely ground, shows a wide variation in granulation, and size and content of endosperm tissue. A general outline (Fig. 7.27) of corn kernel shows hull and bran layers, embryonic axis (germ), scutellum, and various endosperm layers. A good understanding of the kernel structure, especially the endosperm tissue, is important so that there is minimum damage to the grain during harvest, drying, storage, and marketing (Bechtel and Pomeranz 1978 a, Pomeranz and Bechtel 1978). Scanning electron microscopic studies reveal that starch grains of wheat vary considerably in shape and surface structure (Fig. 7.28 A), and also change during sprouting (Fig. 7.28 B, C). Grains sprouted for 2 days show enzymatic degradation of starch granules near the aleurone layer. Most of the activity is confined to the larger (A-type) granules rather than to the smaller (B-type) ones, suggesting that the former granules differ in physical properties. Such studies on degradation of starch by enzymes are necessary for understanding the damage caused to the flour. Sprouted wheat-flour affect considerably bread-making quality (Dronzek et al. 1972).

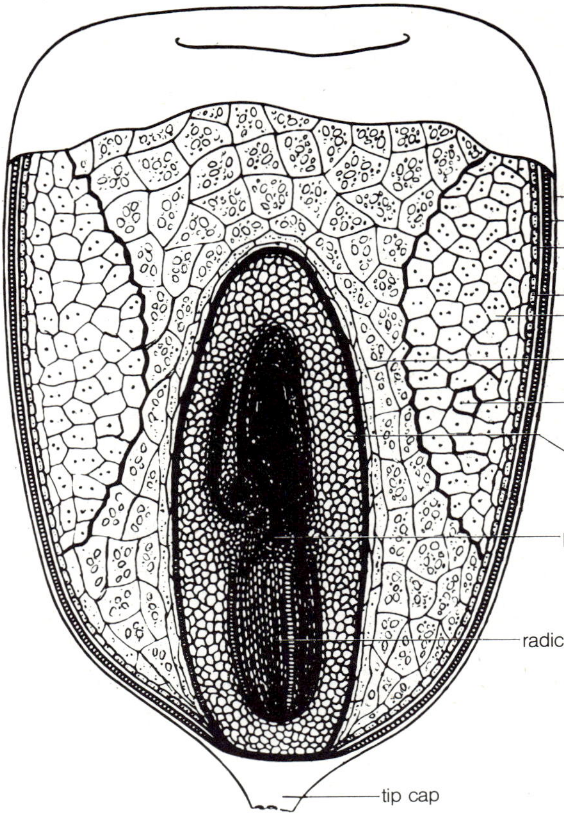

Fig. 7.27. *Zea mays*, longisection of grain to show the relationship between endosperm and embryo. Cornflour has wide variation in granulation, size, and content of endosperm. (After Pomeranz and Bechtel 1978)

7.10 Embryo-Endosperm Relationship

The endosperm has been classically assigned the function of nourishing the embryo (Brink and Cooper 1947, Maheshwari 1950). The intense metabolic activity of young endosperm may partially reflect the production of hormones and other substances which influence the early growth and morphogenesis of *Capsella* embryos in vitro (Raghavan 1966). With recent histochemical and ultrastructural studies the role of endosperm during early embryogenesis became questionable. In *Capsella bursa-pastoris* (Schulz and Jensen 1969) the endosperm develops up to the heart-shaped stage of the embryo, and probably does not contribute significantly to its nutrition. During early embryogeny in *Helianthus annuus* (Newcomb 1973) the endosperm seems to be of limited nutritional value. The persistent synergid (in this taxon) has cytoplasm similar to that present before fertilization, has complete cell wall, and shows increased lipid contents, and starch in plastids.

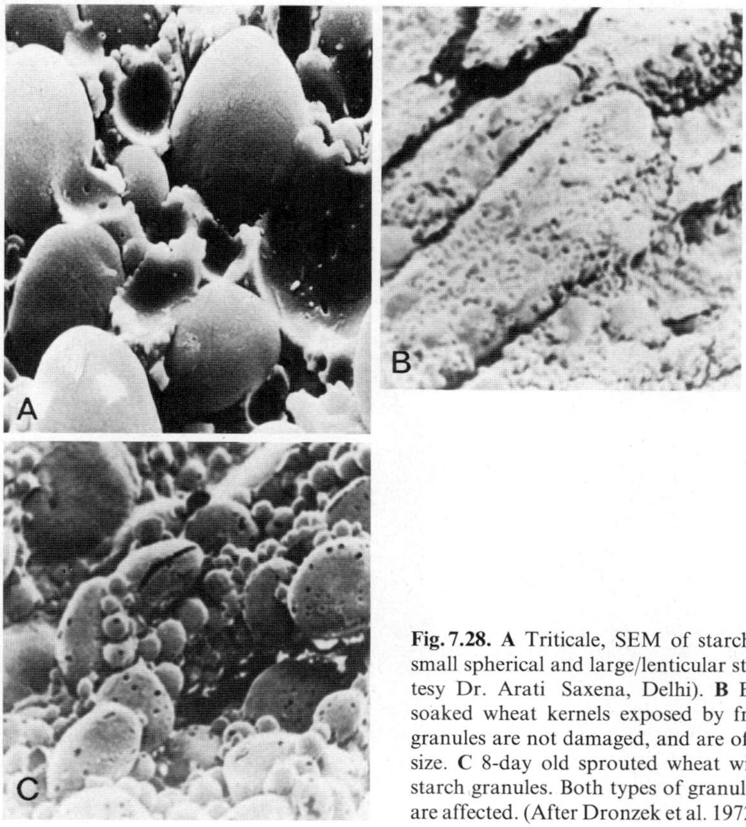

Fig. 7.28. A Triticale, SEM of starchy endosperm with small spherical and large/lenticular starch grains. (Courtesy Dr. Arati Saxena, Delhi). **B** Endosperm cells of soaked wheat kernels exposed by fracture. The starch granules are not damaged, and are of various shape and size. **C** 8-day old sprouted wheat with severely eroded starch granules. Both types of granules, large and small, are affected. (After Dronzek et al. 1972)

Thus, the persistent synergid is functional and contributes significantly to the growth and development of the embryo. The endosperm which is still developing, and has not reached a certain size, cannot provide nutrition to the embryo. This has been shown in *Phaseolus coccineus* (Yeung and Clutter 1978). In *Iberis amara* and *Alyssum maritimum*, up to the globular stage of proembryo, the endosperm is rich in proteins, RNA, and DNA, but there are no storage products (Prabhakar 1979). Thus, the young endosperm, during early stages of development of the embryo, needs adequate nutrients for its own growth. During late embryogeny, however, the endosperm does have a large pool of reserve materials which is utilized for later development of the embryo. In *Helianthus annuus* (Newcomb 1973) from early heart-shaped, and in *Iberis amara* and *Alyssum maritimum* from globular stage, a clear zone appears around the embryo. Ultrastructural studies in *H. annuus* (Newcomb 1973) and in *A. maritimum* (Fig. 7.29) show that this clear area represents particulate remnants of digested endosperm cells. It is reasonable to postulate that the embryo is not wholly dependent on the endosperm for its nutrition, but gradually consumes it. In *Quercus gambelii* the degenerating endosperm cells (Fig. 7.30) are probably involved in the transport of food material

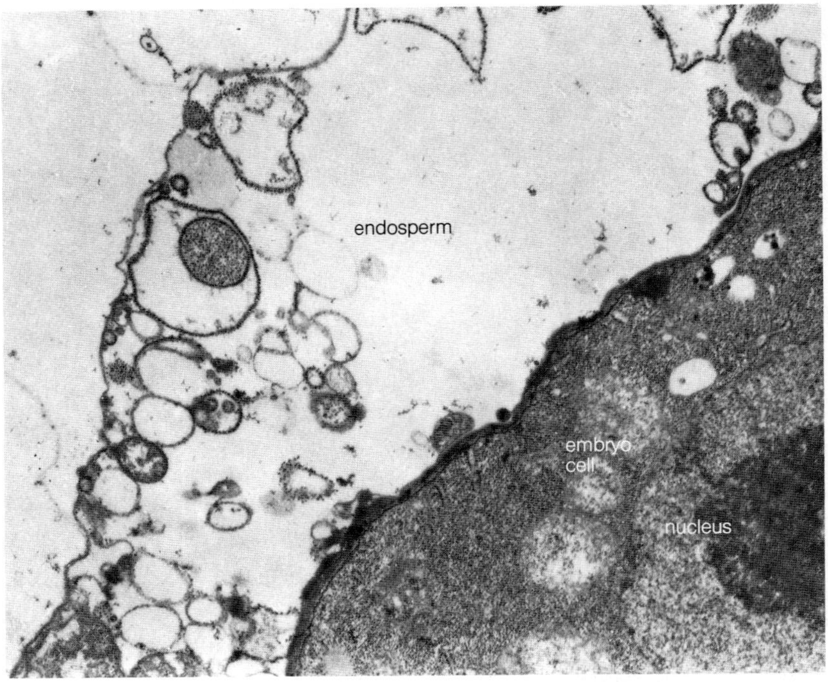

Fig. 7.29. *Alyssum maritimum*, portion of endosperm adjacent to the boundary of embryo; cytoplasmic organelles of endosperm lying close to embryonal cells have lysed, and are unrecognizable. (After Prabhakar 1979)

from other parts of the ovule, or from the vascular supply to the chalaza (A. P. Singh and Mogensen 1976). The degenerating cells are designated as "proximal cells", and show a very high electron density of the cytoplasm (Fig. 7.30). Initially, the identity of organelles is maintained but, gradually, the identity of cytoplasmic components is lost.

In *Phaseolus coccineus* the presence of wall ingrowths in the endosperm cells along the endothelial cells probably facilitate the transfer of nutrients from the maternal tissue to the developing embryo (Yeung and Clutter 1978). Yeung and Clutter emphasize that cellular endosperm–suspensor complex may function as the major site of entry for nutrients. The endosperm, besides serving as a nutrient source, may also have a morphogenetic role in the regulation of growth of the embryo (Cutter and Wilson 1954). In some leguminous plants different classes of hormones are reported in the nuclear endosperm (Eeuwens and Schwabe 1975, Gillard and Walton 1976). Since growth hormones are also reported in the suspensor (Alpi et al. 1975), the endosperm–suspensor complex may also control morphogenesis of the embryo.

Apart from other roles, the maintenance of osmotic gradient is also attributed to the endosperm. In *P. vulgaris* the osmotic pressure is 0.7 osmolar at the heart-shaped stage, and 0.5 osmolar at the late cotyledon stage (Smith 1973). The high osmotic environment may be essential for the normal development of the embryo,

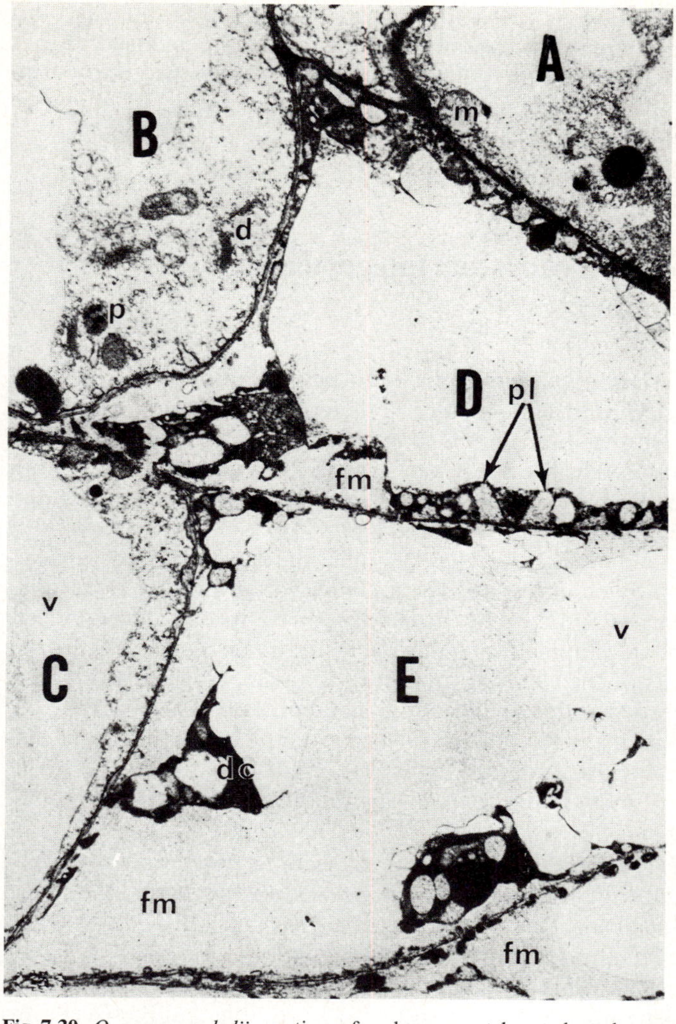

Fig. 7.30. *Quercus gambelii,* portion of endosperm at heart-shaped stage of embryo; the cells *A–C* (further away from the embryo) are intact, whereas the cells *D* and *E* (near the embryo) have collapsed; *d* dictyosome, *dc* dense cytoplasm, *fm* fibrillar material, *m* mitochondrion, *p* plastid, *pl* plasmalemma, *v* vacuole. (After A. P. Singh and Mogensen 1976)

and may also prevent it from precocious germination (Norstog 1967, Norstog and Klein 1972). The osmotic gradient in the endosperm tissue is, to some extent, determined by the concentration gradient of amino acids. Of the 23 free amino acids identified in the endosperm tissue, 12 or 13 are characterized by high concentration in the chalazal portion of the endosperm rather than in the micropylar portion, during different stages of development of the ovule (Ryczkowski 1971). Ryczkowski further states that the differentiation of embryo is influenced not only by the concentration gradient of low molecular compounds, but also by proper osmotic value of these compounds.

Jensen (1968) proposed a hypothesis to explain shrinkage of zygote based on a presumed osmotic gradient by the initial rapid growth of the endosperm that would cause water to move from the vacuole of the zygote to the endosperm. Thus, proper osmo-regulation by endosperm is one of the important functions during early embryogenesis.

7.11 Incorporation of Nucellus and Integument in Endosperm Formation

Brink and Cooper (1947) suggested that the endosperm continues to expand at the expense of the nucellus until a major portion thereof is digested and absorbed. Coe (1954) also reported that breakdown of nucellar cells provides nutrition to the developing embryo sac. The breakdown of nucellar cells and migration of nucellar nuclei into the embryo sac is reported in *Cocos nucifera* (Quisumbing and Juliano 1927) and *Tamarix erucoides* (Johri and Kak 1954).

Norstog (1974) pointed out that nucellar cells near the antipodes undergo lysis rather than destruction by compression. He concluded that nucellar lysate that surrounds the endosperm is absorbed by the endosperm for its own development, or is utilized during the development of aleurone wall. In *Iberis amara* and *Alyssum maritimum* the cells of the chalazal proliferating tissue, adjacent to the embryo sac wall, are degraded and depleted of the metabolites like proteins, DNA, and RNA (Vijayaraghavan and Prabhakar 1981). It is emphasized that as the embryo sac wall at this end shows wall ingrowths (Fig. 7.24 C, D), these cells are involved in the transfer of metabolites for the growth and development of the endosperm/embryo.

Schulz and Jensen (1971) also observed wall projections in the wall of the embryo sac, at the chalazal end, in *Capsella bursa-pastoris,* and suggested that these projections facilitate the absorption of degraded cytoplasm of crushed tissue and substances metabolized by the chalazal proliferating tissue.

There are no studies that directly show that nucellar tissue is utilized for the growth of endosperm or embryo. A few investigators presume that these cells are absorbed by the endosperm, but transferred to the embryo for its growth (Chopra 1955). According to others, the nucellar cells are utilized for the development of endosperm (Norstog 1974). Only autoradiographic studies can solve this problem; once the nutrients enter the embryo sac, whether these are utilized by the endosperm or embryo.

The incorporation of integumentary nuclei is an interesting aspect of embryo sac development. Johri and Tiagi (1952), on the basis of light microscopy in *Cuscuta reflexa,* observed that after fertilization the embryo sac continues to expand and consume the adjacent layers of integument. In *Eranthis hiemalis* the developing nuclear endosperm, before it becomes cellular, incorporates the adjoining integumentary nuclei. The nuclei become pycnotic while still in the integument, and enter into the endosperm once the entire cell wall or a greater part of it has disintegrated. The nuclei are surrounded with a good deal of ER stacks, and are disintegrated by this ER (Pacini et al. 1972). Whether such ER stacks originate in the cytoplasm, or are derived from the nuclear membrane, remains to be investigated.

7.12 Conclusions and Prospects

The study of endosperm, its morphology, histology, histochemistry, and ultrastructure still require much attention. Recently, an attempt has been made to investigate the histochemical and ultrastructural aspects, but it is not yet possible to generalize the significance of various phenomena occurring in the endosperm.

The so-called endosperm mother cell (Maheshwari 1950) is rather a complex structure. The fine structure of the central cell changes immediately after fertilization (Diboll 1968, Cocucci and Jensen 1969 a, b, van Went 1970 b). This change in ribosome, ER, and dictyosome formation has been interpreted to reflect the increased metabolic activity, and the organization of protein-synthesizing machinery for the rapid differentiation of the central cell (van Went 1970 b). Unlike these reports, in *Capsella bursa-pastoris* (Schulz and Jensen 1973) the central cell in the mature gametophyte is already quite active and, thus, does not undergo any notable structural reorganization as a result of fertilization. Much information is needed on this aspect to elucidate the cytochemical and fine structural changes associated with double fertilization.

The nucleolar extrusion exhibited by endosperm nuclei (Cionini and Cremonini 1970, Schulz and Jensen 1973, 1977) indicates that the endosperm is metabolically very active. Further studies are essential to determine the exact significance, composition, and final fate of nucleolar extrusion. There seems little doubt that the endosperm does not have a significant role in nutrition during early embryogeny.

There is no clear-cut demarcation between Nuclear, Cellular, and Helobial type of endosperm. What factor(s) control the endosperm to undergo karyokinesis followed by cytokinesis in the Cellular type, and the absence of cytokinesis in Nuclear and Helobial types? What factor(s) control cytokinesis, after a definite number of free-nuclear divisions, in the Nuclear type of endosperm? In *Iberis amara* and *Alyssum maritimum,* just before cellularization, the endosperm becomes highly vacuolate, shows decline in the concentration of cytoplasmic proteins and RNA, and the appearance and accumulation of polysaccharide grains. The nucleoli in the coenocytic condition are large and vacuolate, whereas those present after the laying down of walls are small and compact (Prabhakar 1979). There seems to be some reorganization in the machinery which replaces one process with another. The fine structural changes at a critical period of transition from the coenocytic to the cellular conditions may help in understanding this phenomenon. Such aspects have totally been ignored in the past.

If endosperm, in nuclear condition, has a nutritive role, why does it become cellular? Newcomb (1973) suggests that the advantage of endosperm becoming compartmentalized by the formation of cell walls could be one way of storing polysaccharides for use later by the embryo.

There are a number of light microscope studies dealing with endosperm tissue in numerous taxa, and most investigators attribute a general role of providing nutrition to the developing embryo (Brink and Cooper 1947, Maheshwari 1950). Studies on fine structure have dealt primarily with observations on specific organelles concerned with the production of storage products.

However, except for a few studies (Marinos 1970, Schulz and Jensen 1974, A. P. Singh and Mogenson 1976, Newcomb 1978), there is not much work on the general nature of organization of endosperm during early development. The endosperm is a highly heterogeneous tissue, and shows different features in different taxa. To be able to generalize any role/phenomenon, it is necessary to scrutinize a very large number of taxa, both histochemically and ultrastructurally.

Our information on Cellular and Helobial endosperm is rather meagre. Detailed ontogenetical, cytochemical, and ultrastructural studies are essential to understand the process of cellularization in the endosperm. A wide range of taxa, belonging both to advanced and primitive families, with Nuclear and Helobial endosperm development, should be carefully investigated. The ruminate endosperm has never been examined using histochemical techniques and electron microscopy. Also, the correlation between endosperm and embryo, at different stages of development, requires a comprehensive study. Autoradiographic studies would be helpful in understanding the transportation/translocation of nutrients to the embryo.

References

Adatia RD, Sharma YB, Vijayaraghavan MR (1971) Studies in the Gesneriaceae 1. Morphology and embryology of *Platystemma violoides* Wall. Bot Not 124:25–38

Alpi A, Tognoni F, D'Amato F (1975) Growth regulator levels in embryo and suspensor of *Phaseolus coccineus* at two stages of development. Planta 127:153–162

Arekal GD (1961) Embryology of *Klugia notoniana*. Bot Gaz 123:144–150

Arekal GD (1963a) Embryological studies in Canadian representatives of Rhinantheae, Scrophulariaceae. Can J Bot 41:267–302

Arekal GD (1963b) Contribution to the embryology of *Chelone glabra* L. Phytomorphology 13:376–388

Ashley T (1975) Fine structure of early endosperm development in *Hibiscus*. Caryologia 20:63–71

Bajer A (1965) Cine-micrographic analysis of cell plate formation in endosperm. Exp Cell Res 37:376–398

Bechtel DB, Pomeranz Y (1978a) Implications of the rice kernel structure in storage, marketing, and processing: A review. J Food Sci 43:1538–1542

Bechtel DB, Pomeranz Y (1978b) Ultrastructure of the mature ungerminated rice (*Oryza sativa*) caryopsis: The germ. Am J Bot 65:75–85

Bhandari NN, Nanda K (1968) Studies in Viscaceae. 1. Morphology and embryology of the Indian dwarf mistletoe: *Arceuthobium minutissimum*. Phytomorphology 18:435–450

Bhatnagar SP (1960) Morphological and embryological studies in the family Santalaceae. IV. *Mida salicifolia* A. Cunn. Phytomorphology 9:87–91

Bhatnagar SP, Chandra S (1968) Endosperm in *Psittacanthus*. Curr Sci (Bangalore) 37:704–706

Bhatnagar SP, Kallarackal J (1980) Cytochemical studies on the endosperm of *Linaria bipartita* (Vent.) Wild. with a note on the role of endosperm haustoria. Cytologia 45:247–256

Bhatnagar SP, Sabharwal G (1966) Female gametophyte and endosperm of *Iodina rhombifolia* Hook. and Arn. Phytomorphology 16:588–591

Brink RA, Cooper DC (1947) The endosperm in seed development. Bot Rev 13:423–541

Brisson JD, Peterson RL (1975) SEM of fractured plant material embedded in glycol methacrylate. Microscop Soc Can 2:64–65
Brock RD (1954) Spontaneous chromosome breakage in *Lilium* endosperm. Ann Bot 18:7–14
Brock RD (1955) Chromosome balance and endosperm failure in hyacinths. Heredity 9:199–222
Buell KM (1952) Developmental morphology in *Dianthus* 1. Structure of the pistil and seed development. Am J Bot 39:194–210
Buttrose MS (1963) Ultrastructure of the developing aleurone cells of wheat grain. Aust J Biol Sci 16:768–774
Chopra RN (1955) Some observations on endosperm development in the Cucurbitaceae. Phytomorphology 5:219–230
Chopra RN, Agarwal S (1958) Some further observations on the endosperm haustoria in the Cucurbitaceae. Phytomorphology 8:194–201
Chopra RN, Sachar RC (1963) Endosperm. In: Maheshwari P (ed) Recent advances in the embryology of angiosperms. Intl Soc Plant Morphologists, Univ Delhi, Delhi, pp 135–170
Coe GE (1954) Distribution of carbon-14 in ovules of *Zephyranthes drummondii*. Bot Gaz 115:342–346
Cionini PG, Cremonini R (1970) Alcuni aspetti citologici dell' endosperma di *Phaseolus coccineus*. Nuovo G Bot Ital (Nuova Ser) 104:483–491
Cocucci A, Jensen WA (1969a) Orchid embryology: Megagametophyte of *Epidendrum scutella* following fertilization. Am J Bot 56:629–640
Cocucci A, Jensen WA (1969b) Orchid embryology: The mature megagametophyte of *Epidendrum scutella*. Kurtziana 5:23–28
Cutter Jr VM, Freeman B (1955) Nuclear aberrations in the syncytial endosperm of *Cosos nucifera*. J Elisha Mitchell Sci Soc 71:49–58
Cutter Jr VM, Wilson KS (1954) Effect of coconut endosperm and other growth stimulants upon the development in vitro of embryos of *Cocos nucifera*. Bot Gaz 115:234–240
Cutter Jr VM, Wilson KS, Freeman B (1955) Nuclear behaviour and cell formation in the developing endosperm of *Cocos nucifera*. Am J Bot 42:109–115
D'Amato F (1952) Polyploidy in the differentiation and function of tissues and cells in plants: A critical examination of the literature. Caryologia 4:311–358
Dhar U, Vijayaraghavan MR (1979) Ontogeny, structure, and breakdown of protein bodies in *Linum usitatissimum* Linn. Ann Bot 43:107–111
Diboll AG (1968) Fine structural development of the megagametophyte of *Zea mays* following fertilization. Am J Bot 55:787–806
Diboll AG, Larson DA (1966) An electron microscopic study of the mature megagametophyte in *Zea mays*. Am J Bot 53:391–402
Dronzek BL, Hwang P, Bushuk W (1972) Scanning electron microscopy of starch from sprouted wheat. Cereal Chem 49:232–239
Duncan RE, Ross JG (1950) The nucleus in differentiation and development 3. Nuclei of maize endosperm. J Hered 41:259–268
Eeuwens CJ, Schwabe WW (1975) Seed and pod wall development in *Pisum sativum* L. in relation to extracted and applied hormones. J Exp Bot 26:1–14
Enzenberg U (1961) Beiträge zur Karyologie des Endosperms. Oesterr Bot Z 108:245–285
Eunus AM (1952) Contributions to the embryology of Liliaceae 3. Embryogeny and development of the seed of *Asphodelus tenuifolius* Cav. Lloydia (Cinci) 15:149–155
Evers AD (1970) Development of the endosperm of wheat. Am J Bot 34:547–555
Fineran BA, Wild DJC, Ingerfeld M (1982) Initial wall formation in the endosperm of wheat, *Triticum aestivum:* A re-evaluation. Can J Bot 60:1776–1795
Garcia V (1962a) Embryological studies in the Loasaceae with special reference to endosperm haustoria. In: Plant Embryology – A symposium. Council of Scientific and Industrial Research, New Delhi, pp 157–161
Garcia V (1962b) Embryological studies in the Loasaceae: Development of endosperm in *Blumenbachia hieronymi* Urb. Phytomorphology 12:307–312
Garcia V (1963) Embryological studies on the Loasaceae. Thesis Dr Biol Sci, Univ Buenos Aires
Geitler L (1948) Notizen zur endomitosischen Polyploidisierung in Trichozyten und Elaisosomen sowie über Kernstrukturen bei *Gagea lutea*. Chromosoma (Berl) 3:271–281
Gillard DF, Walton DC (1976) Abscisic acid metabolism by a cell free preparation from *Echinocystis lobata* liquid endosperm. Plant Physiol (Bethesda) 58:790–795

Graham JSD, Jennings AC, Morton RK, Palk BA, Raison JK (1962) Protein bodies and protein synthesis in developing wheat endosperm. Nature 196:967–969

Gunning BES, Pate JS (1969) Transfer cells: Plant cells with wall ingrowths, specialized in relation to short distance transport of solutes – Their occurrence, structure, and development. Protoplasma 68:107–133

Gupta SC (1964) The embryology of *Coriandrum sativum* L. and *Foeniculum vulgare* Mill. Phytomorphology 14:530–547

Hartl D (1960) Das alveolierte Endosperm bei Scrophulariaceen, seine Entstehung, Anatomie und taxonomische Bedeutung. Beitr Biol Pflanz 35:95–110

Hay ED (1968) Structure and function of the nucleolus in developing cells. In: Dalton AJ, Hagnenau F (eds) The nucleus. Academic Press, New York London, pp 1–79

Heel WA van (1970) Some unusual tropical labyrinth seeds. Proc K Ned Akad Wet 73C:298–301

Heel WA van (1971 a) The labyrinth seed of *Hernandia peltata* Meissn. Proc K Ned Akad Wet 70C:46–51

Heel WA van (1971 b) Note on some more tropical labyrinth seeds. Blumea 19:109–111

Hepler PK, Jackson WT (1968) Microtubules and early stages of cell plate formation in the endosperm of *Haemanthus katherine* Baker. J Cell Biol 38:437–446

Hristov M, Moskov IV (1957) The change of mitosis into amitosis during nuclear endosperm development in *Vicia faba* and *Lathyrus variegatus*. Nauch Trud Selsky Akad Georg, Dimitrov Agron Fak 4:27–36

Hu SY (1964) Accumulation and movement of starch during caryopsis development of the wheat. Acta Bot Sin 13:139–148

Jacobsen JV, Knox RB, Pyliotis NA (1971) The structure and composition of aleurone grains in the barley aleurone layer. Planta 101:189–209

Jacobsen JV, Pressman E, Pyliotis NA (1976) Gibberellin-induced separation of cells in isolated endosperm of celery seed. Planta 129:113–122

Jensen WA (1965) The ultrastructure and composition of the egg and central cell of cotton. Am J Bot 52:781–797

Jensen WA (1968) Cotton embryogenesis: The zygote. Planta 79:346–356

Jensen WA (1974) Reproduction in flowering plants. In: Robards AW (ed) Dynamic aspects of plant ultrastructure. McGraw-Hill, London, pp 481–503

Jensen WA, Fisher DB (1967) Cotton embryogenesis: Double fertilization. Phytomorphology 17:261–269

Jensen WA, Schulz P, Ashton ME (1977) An ultrastructural study of early endosperm development and synergid changes in unfertilized cotton ovules. Planta 133:179–189

Johri BM (1936) The life history of *Butomopsis lanceolata* Kunth. Proc Indian Acad Sci Sect B 4:139–162

Johri BM, Bhatnagar SP (1969) Endosperm in the Santalales. Rev Cytol Biol Vég 32:355–369

Johri BM, Garg S (1959) Development of endosperm haustoria in some Leguminoseae. Phytomorphology 9:34–46

Johri BM, Kak D (1954) The embryology of *Tamarix* Linn. Phytomorphology 4:230–247

Johri BM, Singh H (1959) The morphology, embryology, and systematic position of *Elytraria acaulis* (Linn. f.) Lindau. Bot Not 112:227–251

Johri BM, Tiagi B (1952) Floral morphology and seed formation in *Cuscuta reflexa* Roxb. Phytomorphology 2:162–180

Jones RL (1974) The structure of the lettuce endosperm. Planta 121:133–147

Kaltsikes PJ (1973) Early seed development in hexaploid triticale. Can J Bot 51:2291–2300

Kaltsikes PJ, Roupakias DG (1975) Endosperm abnormalities in *Triticum-Secale* combinations 2. Addition and substitution lines. Can J Bot 53:2068–2076

Kaltsikes PJ, Roupakias DG, Thomas JB (1975) Endosperm abnormalities in *Triticum-Secale* combinations 1. X. *Triticosecale* and its parental species. Can J Bot 53:2050–2067

Kapoor BM, Tandon SL (1963) Contribution to the cytology of endosperm in some angiosperms. 4. *Zephyranthes grandiflora* Lindl. Genetika 34:102–112

Kapoor BM, Tandon SL (1964) Contributions to the cytology of the endosperm in some angiosperms 5. *Zephyranthes lancasterii* Traub. Phyton (Argentina) 21:37–43

Kapoor T (1980) Embryological and histochemical studies in some aquatic monocotyledons – Potamogetonaceae and Najadaceae. Ph D Thesis, Univ Delhi

Kapoor T, Vijayaraghavan MR (1975) Ontogeny of ruminate endosperm in *Celsia coromandeliana*. Beitr Biol Pflanz 51:359–365
Kapoor T, Parulekar NK, Vijayaraghavan MR (1975) Contribution to the embryology of *Celsia coromandeliana* Vahl. with a discussion on its affinities with *Verbascum thaspus* L. Bot Not 128:438–449
Kato Y (1957) Cytological abnormalities in the embryo mother cell and endosperm tissue of *Lilium formolongo* Hort. and *Allium fistulosum* L. Cytologia 22:69–79
Kausik SB (1940) A contribution to the embryology of *Enalus acoroides* (L. f.) Steud. Proc Indian Acad Sci Sect B 11:83–99
Kausik SB (1941) Development of vermiform appendage in *Grevillea robusta* Cunn. Proc Indian Acad Sci Sect B 14:137–140
Kent NL, Evers AD (1969) Variation in protein composition within the endosperm of hard wheat. Cereal Chem 46:293–300
Krishna Iyengar CV (1939) Development of the embryo sac and endosperm haustoria in some members of Scrophulariaceae 2. *Isoplexis canariensis* Lindl. and *Celsia coromandeliana* Vahl. J Indian Bot Soc 19:5–17
Krishna Iyengar CV (1942) Development of seed and its nutritional mechanism in Scrophulariaceae. Part I. *Ramphicarpa longiflora* Benth., *Centranthera hispida* Br. and *Pedicularis zeylanica* Benth. Proc Nat Inst Sci India, Part B 8:249–261
Lakshmanan KK (1963) Embryological studies in the Hydrocharitaceae 2. *Halophila ovata* Gaudich. J Indian Bot Soc 42:15–18
Maheshwari P (1950) An introduction to the embryology of angiosperms. McGraw Hill, New York
Maheshwari P, Negi V (1955) The embryology of *Dipteracanthus patulus* (Jacq.) Nees. Phytomorphology 5:456–472
Maheshwari P, Sachar RC (1954) The endosperm nodules of *Capsella bursa-pastoris* Moench. Sci Cult 19:413–414
Mares DJ, Norstog K, Stone BA (1975) Early stages in the development of wheat endosperm 1. The change from free-nuclear to cellular endosperm. Aust J Bot 23:311–326
Mares DJ, Stone BA, Jeffery C, Norstog K (1977) Early stages in the development of wheat endosperm 2. Ultrastructural observations on cell wall formation. Aust J Bot 25:599–613
Marinos NG (1970) Embryogenesis of the pea (*Pisum sativum*) 1. The cytological environment of the developing embryo. Protoplasma 70:261–279
Mathur N (1956) The embryology of *Limnanthes*. Phytomorphology 6:41–51
Mikulska E, Rodkiewicz B (1967 a) Fine structure of four-nucleate stages and the central cell of *Lilium regale* embryo sac. Flora 157:365–372
Mikulska E, Rodkiewicz B (1967 b) Ultrastructure of the maturing embryo sac of *Lilium regale*. Acta Soc Bot Pol 36:555–566
Mogensen HL (1972) Fine structure and composition of the egg apparatus before and after fertilization in *Quercus gambelii*: The functional ovule. Am J Bot 59:931–941
Mohan Ram HY (1960) The development of seed in *Andrographis serpyllifolia*. Am J Bot 47:215–219
Mohan Ram HY (1962) Post-fertilization development of the ovule in *Barleria cristata* Linn. J Indian Bot Soc 41:288–296
Mohan Ram HY, Masand P (1962) Endosperm and seed development in *Andrographis echioides* Nees. Curr Sci 31:7–8
Mohan Ram HY, Wadhi M (1964) Endosperm in Acanthaceae. Phytomorphology 13:82–91
Mohana Rao PR (1975) Seed anatomy of *Artabotrys odoratissimus* with a discussion on chalaza, integumentary bundles and ruminate endosperm. Phytomorphology 25:215–228
Morrison IN, O'Brien TP (1976) Cytokinesis in the developing wheat grain division with and without a phargmoplast. Planta 130:57–67
Nagl W (1970 a) Temperature-dependent functional structures in the polytene chromosomes of *Phaseolus*, with special reference to the nucleolus organizers. J Cell Sci 6:87–107
Nagl W (1970 b) The mitotic and endomitotic nuclear cycle in *Allium carinatum* 2. Relations between DNA replication and chromatin structure. Caryologia 23:71–78
Nawaschin SG (1898) Resultate einer Revision der Befruchtungsvorgänge bei *Lilium martagon* und *Fritillaria tenella*. Bull Acad Imp Sci St. Petersburg 9:377–382
Newcomb W (1973) The development of the embryo sac of sunflower *Helianthus annuus* L. after fertilization. Can J Bot 51:879–890

Newcomb W (1978) The development of cells in the coenocytic endosperm of the African blood lily *Haemanthus katherinae*. Can J Bot 56:483–501
Newcomb W, Fowke LC (1973) The fine structure of the change from the free-nuclear to cellular condition in the endosperm of chickweed *Stellaria media*. Bot Gaz 134:236–241
Newcomb W, Steeves TA (1971) *Helianthus annuus* embryogenesis: Embryo sac wall projections before and after fertilization. Bot Gaz 133:367–371
Nijalingappa BHM, Devaki N (1978) Embryological studies in *Diplacrum caricinum* R. Br. Beitr Biol Pflanz 54:215–225
Nijalingappa BHM, Devaki N (1979) Endosperm of *Scleria foliosa* Hochstetter ex A. Richard. Curr Sci 48:451–452
Norstog K (1967) Studies on the survival of very small barley embryos in culture. Bull Torrey Bot Club 94:223–229
Norstog K (1974) Nucellus during early embryogeny in barley: Fine structure. Bot Gaz 135:97–103
Norstog K, Klein RM (1972) Development of cultured barley embryos 2. Precocious germination and dormancy. Can J Bot 50:1887–1894
Pacini E, Cresti M, Sarfatti G (1972) Incorporation of integumentary nuclei in *Eranthis hiemalis* endosperm and their disaggregation by the endoplasmic reticulum. J Submicrosc Cytol 4:19–31
Parker ML (1980) Protein body inclusions in developing wheat endosperm. Ann Bot 46:29–36
Periasamy K (1962) The ruminate endosperm: Development and types of rumination. In: Plant embryology – A symposium. Council of Scientific and Industrial Research, New Delhi, pp 62–74
Plisko NA (1971) An electron microscopic investigation of the characteristic features of megagametogenesis in *Calendula officinalis* L. Bot Zh (Leningr) 56:582–597
Pomeranz Y, Bechtel DB (1978) Structure of cereal grains as related to end-use properties. In: Hultin HO, Milner M (eds) Postharvest biology and biotechnology. Food and nutrition Press, Westport, Connecticut
Prabhakar K (1979) Histochemical and ultrastructural studies in some Cruciferae: Zygote to seedling. Ph D Thesis, Univ Delhi
Prabhakar K, Vijayaraghavan MR (1982) The endothelium in *Iberis amara* and *Alyssum maritimum* – Its histochemistry and ultrastructure. Phytomorphology 32:28–36
Prabhakar K, Vijayaraghavan MR (1983) Embryo sac wall in *Iberis amara* and *Alyssum maritimum:* Its histochemistry and ultrastructure. Phyton (Austria) 23:31–38
Prabhakar K, Vijayaraghavan MR, Puri BK (1981) The ultrastructure of endosperm nucleus in *Alyssum maritimum*. J Cytol Genet 16:105–110
Prasad K (1974) Studies in the Cruciferae gametophytes: Structure and development of seed in *Eruca sativa* Mill. J Indian Bot Soc 53:24–33
Quisumbing E, Juliano JB (1927) Development of ovule and embryo sac of *Cocos nucifera*. Bot Gaz 84:279–293
Raghavan V (1966) Nutrition, growth, and morphogenesis of plant embryos. Biol Rev 41:1–58
Ram M (1957) Morphological and embryological studies in the family Santalaceae 1. *Comandra umbellata* (L.) Nutt. Phytomorphology 7:24–35
Ram M (1959) Morphological and embryological studies in the family Santalaceae 2. *Exocarpus,* with a discussion on its systematic position. Phytomorphology 9:4–19
Rau MA (1951 a) The endosperm in *Crotalaria*. Curr Sci (Bangalore) 20:73–74
Rau MA (1951 b) The endosperm in *Rothia trifoliata* Pers. Ann Bot 15:175–177
Rau MA (1951 c) The endosperm in some species of *Cassia* L. Sven Bot Tidskr 45:516–522
Rau MA (1953) Some observations on the endosperm in Papilionaceae. Phytomorphology 3:209–222
Rau MA (1955) Embryological studies in the Leguminosae. J Mysore Univ Sect B 14:63–75
Raven CP (1961) Oogenesis: The storage of developmental information. Pergamon, New York
Rost RL (1971) Fine structure of endosperm protein bodies in *Setaria lutescens* (Gramineae). Protoplasma 73:475–479
Ryczkowski M (1971) Physico-chemical (chalaza-micropyle) gradient in the developing endosperm and embryo. Bull Acad Pol Sci Ser Sci Biol 19:801–806
Scharrer B, Wurzelmann S (1969) Ultrastructural study on nuclear-cytoplasmic relationships in Oocytes of the African lungfish *Protopterus aethiopicus* 1. Nucleolocytoplasmic pathways. Z Zellforsch Mikrosk Anat 96:325–343
Schertz FM (1919) Early development of floral organs and embryonic structures of *Scrophularia marylandica*. Bot Gaz 68:441–450

Schmidt E (1906) Beiträge zur Entwicklungsgeschichte der Scrophulariaceae. Beih Bot Zbl 20:175–299
Schulz P, Jensen WA (1969) *Capsella* embryogenesis: The suspensor and the basal cell. Protoplasma 67:139–163
Schulz P, Jensen WA (1971) *Capsella* embryogenesis: The chalazal proliferating tissue. J Cell Sci 8:201–227
Schulz P, Jensen WA (1973) *Capsella* embryogenesis: The central cell. J Cell Sci 12:741–763
Schulz P, Jensen WA (1974) *Capsella* embryogenesis: The development of free-nuclear endosperm. Protoplasma 80:183–205
Schulz P, Jensen WA (1977) Cotton embryogenesis: The early development of the free-nuclear endosperm. Am J Bot 64:384–394
Schulz P, Jensen WA (1978) Ultrastructural studies of early endosperm development in fertilized and unfertilized cotton ovules. Bull Soc Bot Fr 125:267–271
Sehgal CB (1966) The embryology of *Cuminum cyminum* L. and *Trachyspermum ammi* (L.) Sprague (= *Carum copticum* Clarke). Proc Natl Inst Sci India Part B Biol Sci 31:175–201
Singh AP, Mogensen HL (1976) Fine structure of early endosperm in *Quercus gambelii*. Cytologia 41:345–361
Singh D (1964) Cytoplasmic nodules in the endosperm of angiosperms. Bull Torrey Bot Club 91:86–94
Smith JG (1973) Embryo development in *Phaseolus vulgaris* 2. Analysis of selected inorganic ions, ammonia, organic acids, and sugar in the endosperm liquid. Plant Physiol (Bethesda) 51:454–458
Steffen K (1956) Endomitosen in Endosperm von *Pedicularis palustris* L. Planta 47:625–652
Stephen J (1974) Cytological investigations on the endosperm of *Borassus flabellifer* Linn. Cytologia 39:195–207
Stephen J (1977) Cytological analysis of endosperm failure in a hybrid clone of *Tradescantia*. Cytologia 42:137–145
Stephen J (1979) Cytological causes of spontaneous fruit abortion in *Haemanthus katherinae* Baker. Cytologia 44:805–812
Stephen J (1980) Mechanism of endopolyploidization in *Datura* endosperm. Cytologia 45:657–661
Strasburger E (1900) Einige Bemerkungen zur Frage nach der „doppelten Befruchtung“ bei Angiospermen. Bot Ztg 58:293–316
Subramanyam K (1951) Embryology of *Oxyspora paniculata* DC. Phytomorphology 1:205–212
Swamy BGL (1948) On the post-fertilization development of *Trillium undulatum*. Cellule 52:7–14
Swamy BGL (1960) Contributions to the embryology of *Cansjera rheedii*. Phytomorphology 10:397–409
Swamy BGL, Ganapathy PM (1957) A new type of endosperm haustorium in *Nothapodytes foetida*. Phytomorphology 7:331–336
Swamy BGL, Parameswaran N (1963) The Helobial endosperm. Biol Rev 38:1–50
Szollosi D (1965) Extrusion of nucleoli from pro-nuclei of the rat. J Cell Biol 25:545–562
Tandon SL, Kapoor BM (1961 a) Amitosis in the endosperm of *Zephyranthes ajax* Sprenger. Curr Sci (Bangalore) 30:72–73
Tandon SL, Kapoor BM (1961 b) Nucleolar number and polyploidy level. Curr Sci (Bangalore) 30:476–477
Tandon SL, Kapoor BM (1963) Contribution to the cytology of endosperm in some angiosperms 2. *Nothoscordum fragrans* Kunth. Caryologia 16:377–395
Tiagi B (1963) Studies in the family Orobanchaceae. 4. Embryology of *Boschniakia himalaica* Hook. and *B. tuberosa* (Hook.) Jepson, with remarks on the evolution of the family. Bot Not 116:81–93
Torosian CD (1971) Ultrastructural study of endosperm haustorial cells of *Lobelia dunnii* Greene (Campanulaceae, Lobelioideae). Am J Bot 58:456–457
Turala K (1969) Endopolyploidy and structure of the nuclei in the endosperm of the Cucurbitaceae. Oesterr Bot Z 116:95–111
Vannereau A, Mestre JCM (1975) Premières observations sur l'ultrastructure des cellules haustorials micropylaires du *Plantago lanceolata* L. CR Acad Sci Paris 281:699–702
Vazart B, Vazart J (1965) Infrastructure de l'ovule du lin, *Linum usitatissimum* L. les cellules du sac embryonnaire. CR Acad Sci Paris 261:3447–3450
Vazart B, Vazart J (1966) Infrastructure du sac embryonnaire du lin (*Linum usitatissimum*). Rev Cytol Biol Vég 29:251–266
Vazart J (1969) Organization et ultrastructure du sac embryonnaire du lin (*Linum usitatissimum*). Rev Cytol Biol Vég 32:227–240

Venkata Rao C (1962) Morphology and embryology of *Lomatia* Br. with a discussion on its probable origin. In: Plant embryology: A symposium. Council of Scientific and Industrial Research, New Delhi, pp 261–273
Vigil EL (1970) Cytochemical and developmental changes in microbodies (glyoxysomes) and related organelles of castor bean endosperm. J Cell Biol 41:435–454
Vijayaraghavan MR, Bhandari NN (1970) Studies in the family Ranunculaceae: Embryology of *Thalictrum javanicum* Blume. Flora 159:450–458
Vijayaraghavan MR, Bhat U (1982) Localization of macromolecules during achene development in *Ranunculus sceleratus* L. Beitr Biol Pflanz 57:105-118
Vijayaraghavan MR, Dhar U (1978) Embryology of *Cyrilla* and *Cliftonia* (Cyrillaceae). Bot Not 131:127–138
Vijayaraghavan MR, Kapoor T (1984) Nucleocytoplasmic interaction in the embryo basal cell and in the chalazal-endosperm chamber cell in *Potamogeton nodosus*. Cytologia (in press)
Vijayaraghavan MR, Prabhakar K (1981) Ontogenetical and histochemical studies on chalazal proliferating tissue in *Iberis amara* and *Alyssum maritimum*. Beitr Biol Pflanz 56:7–17
Vijayaraghavan MR, Prabhakar K, Puri BK (1984) Histochemical, structural and ultrastructural features of endosperm in *Alyssum maritimum* Lam. Acta Bot Neerl 33:111–122
Vijayaraghavan MR, Ratnaparkhi S (1972) Some aspects of embryology of *Alectra thomsonii*. Phytomorphology 22:1–8
Voigt A (1888) Untersuchungen über Bau und Entwicklung von Samen mit ruminiertem Endosperm aus den Familien der Palmen, Myristicaceen und Annonaceen. Ann Jard bot Buitenz 7:150–190
Wang FH, Chien NF (1957) On the development of the endosperm in cotton plants. Acta Bot Sin 6:29–38
Weinbaum SA, Simons RK (1974) An ultrastructural evaluation of the relationship of embryo/endosperm abortion to apple fruit abscission during the post-bloom period. J Am Soc Hortic Sci 99:311–314
Went JL van (1970a) The ultrastructure of the egg and central cell of *Petunia*. Acta Bot Neerl 19:313–322
Went JL van (1970b) The ultrastructure of the fertilized embryo sac of *Petunia*. Acta Bot Neerl 19:468–480
Wilms HJ (1981) Ultrastructure of the developing embryo sac of spinach. Acta Bot Neerl 30:75–99
Wolfe J, Rodriguez L (1974) Inhibition of nuclear reconstruction by dithiothrectol. J Cell Biol 60:497–501
Yeung EC, Clutter ME (1978) Embryogeny of *Phaseolus coccineus:* Growth and microanatomy. Protoplasma 94:19–40
Yoffe MD (1952) The presence of chlorophyll in the endosperm of cruciferous plants. Dokl Akad Nauk SSSR Ser Biol 82:473–476
Zaman B (1950) The embryology of *Juncus prismatocarpus* Br. and *J. effusus* Linn. Proc Indian Acad Sci Sect B 31:223–234

8 The Embryo

S. Natesh and M. A. Rau

8.1 Historical

The discovery of syngamy by Strasburger in 1884 is an important landmark in the history of angiosperm embryology. Investigations on the embryo, however, had been initiated much before this event, soon after the observation of the pollen tube by Amici in 1824. Amici contended that the embryo developed from the germinal vesicle present in the embryo sac, and the pollen tube merely provided some kind of stimulus for its development. This view was strongly challenged by Schleiden (1836), who argued that the tip of the pollen tube itself, once inside the embryo sac, got transformed into an embryo. The subject soon developed into a major controversy and several botanists joined the fray, arraying themselves into two groups, some endorsing Amici's contention, and others defending Schleiden's opinion. The controversy raged for nearly two decades and it was not until Hofmeister conclusively demonstrated, in a series of publications from 1847 to 1861, that the embryo originated from a cell present in the embryo sac prior to the arrival of the pollen tube, thus vindicating Amici's view, that the controversy came to an end [see Maheshwari (1950, 1963) for references].

A publication of considerable importance appeared in 1870. In this memoir Hanstein, for the first time, traced the sequence of cell segmentation during early embryogenesis in *Capsella bursa-pastoris* and *Alisma plantago*. Hanstein (1870) also introduced the concept of histogenesis which implies specific destiny for individual cells of the proembryo. Other pioneering works of this period include those of Fleischer (1874) who is considered to be the father of comparative embryology, and Strasburger (1879) who gave classic accounts of the embryogeny of several gymnosperms and angiosperms. In the meantime, L. Guignard (1882) published a valuable treatise on the Leguminosae in which the variation of suspensor of the embryo was vividly illustrated. Hitherto it had been assumed that all the dicotyledons followed the Capsella type of development of embryo, and all the monocotyledons the Alisma type. L. Guignard demonstrated that the course of embryogeny in legumes was quite different from that in *Capsella*. Treub (1879), likewise, had shown earlier that the orchid embryogeny was distinct from that of *Alisma*. The discovery of the unique phenomenon of double fertilization in angiosperms, independently by Nawaschin (1898) and L. Guignard (1899), in the closing years of the nineteenth century, gave a fresh impetus to embryological studies. The knowledge accumulated till then was summarized in 1903 by Coulter and Chamberlain in their book entitled *Morphology of Angiosperms*. In a subsequent edition of their book, published 9 years later (1912), the authors stated that the details recorded on embryogeny were very numerous but confusing, and that it was impossible at that time to formulate any laws for embryogenesis in angiosperms.

It was about this time that Souèges, the famous French embryologist, began his investigations which continued uninterrupted until his death in 1967. Using precise methods of study for the delineation of cell lineages in early embryogeny, Souèges suggested numerous patterns of development, and put forward an elaborate system of classification of embryogenic types. He also formulated the laws of embryonomy: (1) the law of origin, (2) the law of numbers, (3) the law of disposition, and (4) the law of destination [see Johansen (1950) for a detailed account]. Considerable embryological work was also carried out in other centres of Europe, particularly Sweden, at some centres in the United States, and in India during the first half of this century. Karl Schnarf at Vienna compiled a comparative account of the embryology of a number of families, and proposed a classification of the embryogenic types (Schnarf 1929, 1931). In a book entitled *Plant Embryology* Johansen (1950) confined the scope of his work to embryo alone, and summarized the available work, family by family, on embryogenesis in gymnosperms and angiosperms. He also suggested a classification of his own, and added a fifth law of embryonomy – the law of parsimony – to Souèges' four laws, as well as reviewed the system by Souèges. The latter system has been subjected to much modification removing the rigidity and ambiguities which marked it in its original form, and making it simpler.

The origin of the stem tip in monocotyledons had been a matter of considerable controversy, and a fresh look at the embryo in monocotyledons has helped in clearing up the doubts.

During the last decade or two, focus has shifted from a purely descriptive or comparative approach to embryo development to a more dynamic one, and this has resulted in regarding the embryo as a reaction system interacting in a special environment. The application of newer tools and techniques, such as the electron microscope, autoradiography, and cytochemistry have led to a deeper understanding of the process of embryogeny, as well as the role of suspensor. Some of these aspects are discussed hereafter.

8.2 Zygote

Although examples of other cells endowed with embryogenic potential are not unknown, the fertilized egg or zygote usually represents the forerunner of the embryo in almost all spermatophytes. The zygote is a unicellular system which, through a programmed sequence of events, gives rise to a multicellular embryo with well-differentiated organs and, as such, embodies within itself the multifarious properties of an adult plant. In angiosperms the zygote is located at the micropylar pole of the embryo sac with its basal portion attached to the embryo sac wall, while the apical portion projects into the central cell. Even under the light microscope, the cell presents a characteristic polarized appearance with the micropylar pole vacuolate, while the chalazal pole contains the nucleus and most of its cytoplasm. A good deal of new information has come to light about the structure of zygote with the application of electron microscopy. Before attempting to understand the process of embryogenesis in angiosperms, it is essential to look into the organization and environment of this unique cell.

8.2.1 Structure and Composition

Although the data on the changes undergone by the egg following fertilization are quite impressive, it is necessary to interpret these with some caution since detailed investigations are available on less than a dozen taxa. Nevertheless, we do have enough information to formulate a general picture of events, if only a somewhat tentative one. Much of the pioneering work has been carried out by Professor Jensen and his colleagues at the University of California, Berkeley, U.S.A, on *Gossypium, Capsella,* and *Epidendrum* (Jensen 1964, 1968 a–c, R. Schulz and Jensen 1968 b, Cocucci and Jensen 1969). The best-studied zygote todate is that of *Gossypium hirsutum* (Jensen 1964, 1968 c). In this species several dramatic changes have been observed soon after fertilization. The most striking of these involves the regrouping of the endoplasmic reticulum (ER), ribosomes, plastids, and mitochondria so that in a 24-h-old zygote they form a cluster round the apically located nucleus (Fig. 8.1). In the unfertilized egg of *Papaver nudicaule* the nucleus is sited towards the micropylar pole with a large vacuole towards the chalazal pole (cf. synergids). However, at fertilization there is a reversal of cytoplasmic polarity; the nucleus shifts to the chalazal pole and the micropylar pole is occupied by a vacuole (Olson and Cass 1981). A post-fertilization shift of nucleus and cytoplasmic contents towards the chalazal pole has also been observed in *Zea mays* (van Lammeren 1981). Such conspicuous organelle relocation does not characterize the fertilized eggs of other species such as *Capsella* (R. Schulz and Jensen 1968 b), *Epidendrum* (Cocucci and Jensen 1969), *Petunia* (van Went 1970), and *Linum* (D'Alascio-Deschamps 1973) where these organelles are so oriented in the unfertilized egg itself. However, the ER becomes more elaborate, and the ribosome density increases. Whereas, in most species, the ribosomes in the unfertilized egg occur singly, either in the cytoplasm or in association with ER, they now show a distinct tendency to aggregate into polysomes (cotton, *Capsella* and *Epidendrum*). In cotton, particularly, the bulk of the ribosomes forms helices of up to 20 subunits, and much of these represents a new population initiated after fertilization. In *Lycopersicum* (F. Kadej and A. Kadej 1981) the ribosomal system of the egg undergoes dissolution, and a new population is formed at a distance from the zygote nucleus; this region contains a considerable number of ER cisternae and resembles ergastoplasm. A large number of mitochondria and dictyosomes also accumulate in this area. During this period the cytoplasm shows a sharp increase in Azure B staining. These features could be interpreted as indication of mRNA synthesis in the zygote.

Other evidences suggestive of intense metabolic activity include internal differentiation of mitochondria and plastids, as well as increase in their number. While the increase in number is presumably due to a multiplication of the organelles present in the original egg cytoplasm, (at least) in some instances it is attributable to a contribution from the sperm cell (Hagemann 1976, 1981). In the

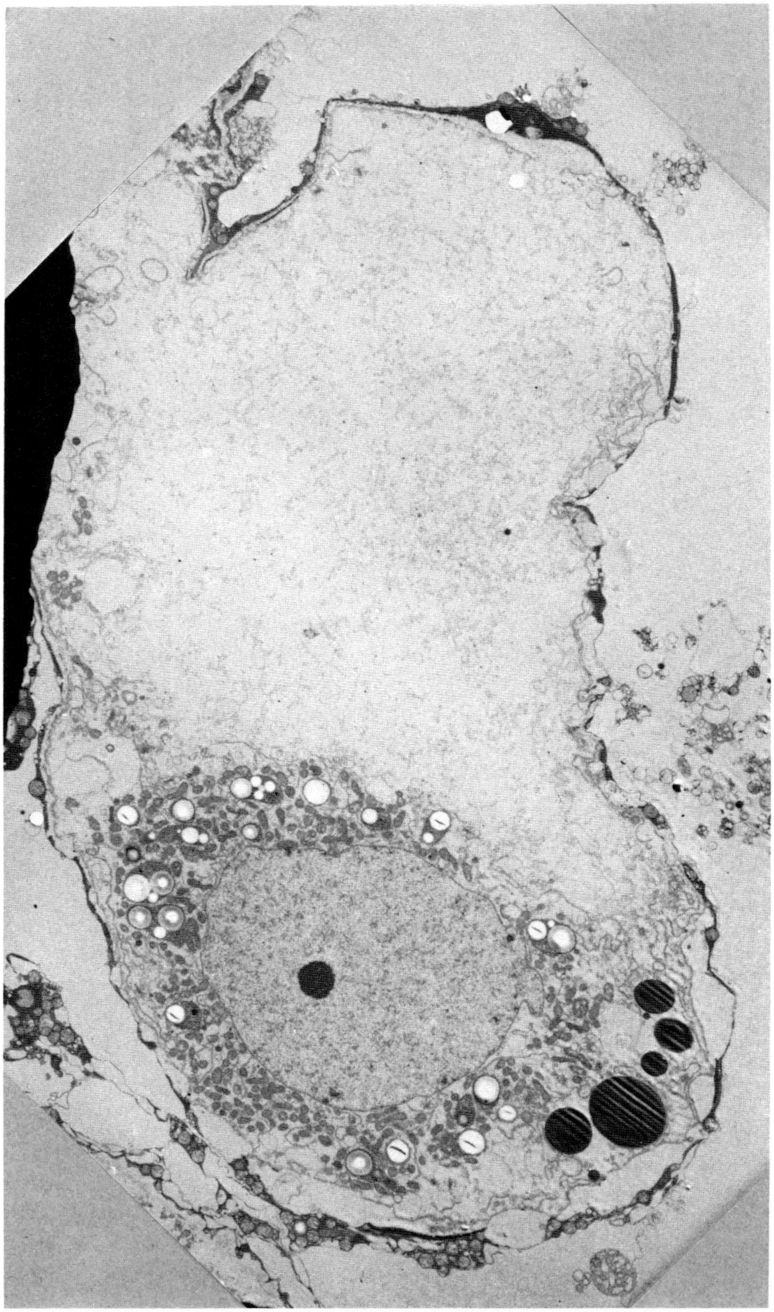

Fig. 8.1. *Gossypium hirsutum*, highly polarized zygote 4 h after fertilization; there is clustering of plastids and mitochondria around the nucleus at the chalazal pole. (After Jensen 1968c; photograph courtesy Professor W. A. Jensen, Berkeley, USA)

majority of cases, uniparental maternal inheritance is the rule for plastids as they are selectively excluded from the generative cell during pollen mitosis and, hence, the sperms are apoplastidic (Possingham 1980). However, evidence from genetic studies involving mutant plastids points to a biparental inheritance in some taxa. The generative cell in these, e.g. *Castilleja, Lobelia, Oenothera,* and *Pelargonium* has one or several plastids (see Hagemann 1976 for references). Plastids of sperm origin have been traced within the zygote of *Oenothera* and *Plumbago*. The unfertilized egg cell in *Oenothera* has between 14 and 39 plastids (Schötz 1954). In a young zygote of *O. erythrosepala* there are 33–42 plastids of which 8–13 are contributed by the sperm (Mayer and Stubbe 1974). An older zygote contains 68–82 plastids of which 14–18 are paternal, indicating that both types of plastids have undergone divisions (Mayer and Stubbe 1974). Recently, Russell (1980) observed plastids as well as mitochondria of male origin in both the egg and central cell of *Plumbago zeylanica,* soon after pollen tube discharge. The sperm organelles could be identified by their smaller size and different shape. It is important to realize that mere transmission of the male organelles into the egg during fertilization is not of much consequence, unless such organelles are capable of survival and multiplication in the zygotic environment, as well as during subsequent embryonic stages of development. That this is possible is indicated by cytological studies in *Oenothera* (Schötz 1954, Mayer and Stubbe 1974), and breeding experiments in *Pelargonium zonale* (Metzlaff et al. 1981, Tilney-Basset and Birky 1981). The fate of the male organelles in *Plumbago,* however, still awaits investigation.

An interesting transformation from the egg to the zygote involves the secretion of a Periodic acid-Schiff (PAS)-positive wall all over the cell boundary. The unfertilized eggs of most species have a cell wall confined to their basal (micropylar) portion, while the apical (chalazal) portion is bounded only by the plasmalemma. However, in *Capsella* (R. Schulz and Jensen 1968 b), *Epidendrum* (Cocucci and Jensen 1969), *Plumbago* (Cass and Karas 1974), *Agave* (Tilton and Mogensen 1979), and *Papaver* (Olson and Cass 1981), a discontinuous wall does exist in this region. In cotton (Jensen 1968 c) a complete wall is formed within 36 h after fertilization, whereas in *Nicotiana* the process requires 40–50 h (Mogensen and Suthar 1979). In *Quercus* (Mogensen 1972) the addition of new wall material begins at the micropylar end. At first, the new wall is discontinuous but soon surrounds the entire cell. Details of the mode of wall formation in other species are not available but, in all the species studied, a complete wall is formed following fertilization. In several taxa, including cotton, *Capsella, Quercus*, and *Zea,* dictyosome activity becomes prominent during the synthesis of wall material and, presumably, these organelles are involved in the process. However, in *Epidendrum* (Cocucci and Jensen 1969) and *Nicotiana* (Mogensen and Suthar 1979) at this stage the dictyosomes appear to be absent from the zygote cytoplasm. Significantly, in both these species, the ER undergoes a prominent increase after fertilization, and in *Nicotiana* lies in close proximity to the plasma membrane. Another interesting feature of this species is the presence of a large number of polysaccharide vesicles of pollen tube-origin in the penetrated synergid, and the possibility of their contribution to cell-wall synthesis in the zygote (Mogensen and Suthar 1979).

Several investigators have experienced difficulty in detecting DNA in the unfertilized egg by tinctorial methods, particularly the Feulgen (see Kapil and Bhatnagar 1981). However, soon after fertilization, the zygotic nucleus returns to a Feulgen-positive state. While some explanation has been offered to account for the aberrant response of the unfertilized egg, it may imply that the imposed restraints to staining, if any, are either removed or lost after fertilization. Other cytological changes apparent after fertilization include the accumulation of starch, lipids, and protein in the zygotic cytoplasm. A distinct oil droplet occurs in the basal part of the zygote in *Jasione montana,* and this droplet disappears after embryogeny has advanced (Erdelská 1969). In the egg of *Nicotiana rustica,* Sehgal and Gifford (1979) noticed two strongly RNA-positive bodies that increased in number after fertilization. The authors suggest that they play a role in the synthesis of key enzymes during early development of the proembryo. The precise mechanisms that set in motion all these changes are totally unknown.

8.2.2 Size Adjustments

The fertilized egg, in several species, undergoes changes in its size before embarking on division. This is particularly striking in *Gossypium* (Fig. 8.2) where the volume of the zygote decreases to half that of the egg in 8–10 h after fertilization (Pollock and Jensen 1964, Jensen 1968c). Similar shrinkage in volume is also characteristic of the zygote of *Hibiscus* (Ashley 1972). A less drastic reduction in volume has been observed in *Nicotiana tabacum,* the decrease being up to 27% in 40–50 h following fertilization (Mogensen and Suthar 1979). However, as the zygote embarks on division 7 days after fertilization, the possibility of further shrinkage in this species cannot be ruled out.

On the other side of the scale are taxa such as *Datura stramonium* (Satina and Rietsema 1959), *Cypripedium insigne* (Poddubnaya-Arnoldi 1967), and *Jasione montana* (Erdelská and Klasová 1978) in which the zygote enlarges during the pre-division phase. Elongation of the zygote before it divides is regularly seen in members of Lentibulariaceae, Orobanchaceae, Scrophulariaceae, and others. In *Capsella bursa-pastoris* (R. Schulz and Jensen 1968b) there is an immediate but temporary decrease, but soon the original volume is restored.

It is difficult to explain these intriguing changes in zygotic volume. One might conjecture that elongation of the zygote is a mechanism to place the future embryonal pole in a nutritionally more favourable environment. Similarly, it is possible to explain the shrinkage phenomenon on the basis of a simple change in osmotic balance between the zygote and its immediate milieu, i.e. the endosperm. In the zygote of cotton special ER containing tubular elements 300 Å in diameter and several micrometres in length become conspicuous during this period (Jensen 1968a). Whether this ER is related to the phenomenon is not clear. The zygotes of other species do not show such tube-containing ER. Whatever may be the mechanism, that such readjustments are crucial, at least in species where they occur, is borne out in certain interspecific crosses. The failure to produce a viable embryo appears to be associated with the absence of zygotic shrinkage (Ashley 1972). However, size changes involving either shrinkage or enlargement do not appear to be universal, and it is too early to generalize on this feature.

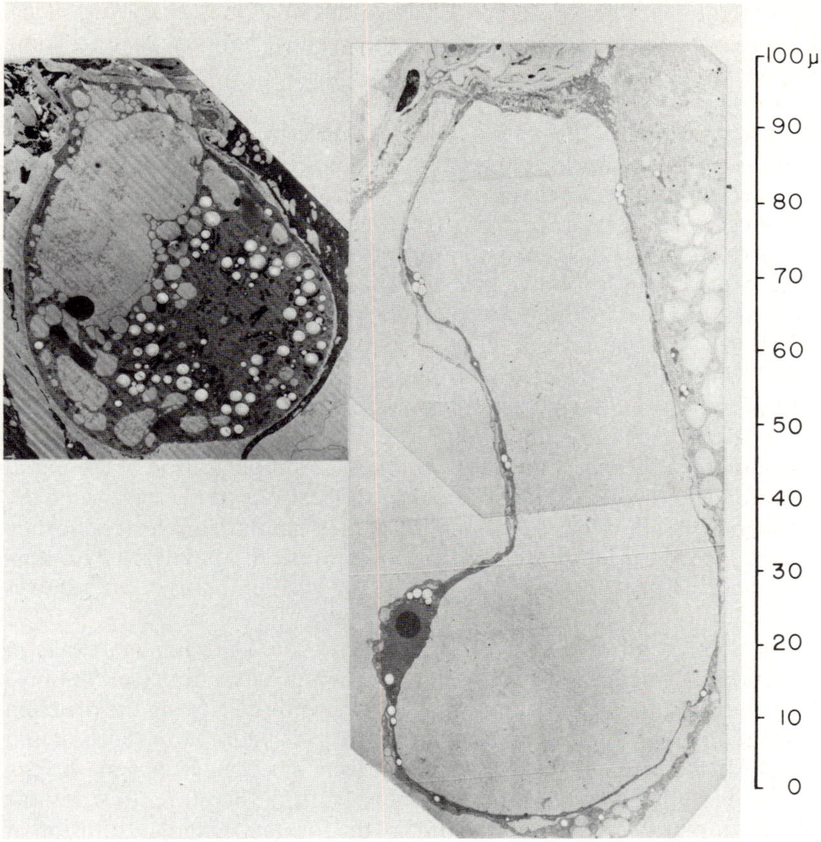

Fig. 8.2. *Gossypium hirsutum*, egg (*right*), and zygote (*left*) 72 h after fertilization. The shrinkage in egg after fertilization is about 50%. (After Jensen 1968c; photographs courtesy Professor W. A. Jensen)

8.2.3 Polarity

One of the striking features associated with the zygote is the polarized appearance of the cell, both under light and electron microscopes. The nucleus and most of the cytoplasm gather at the chalazal end, and a large vacuole occupies the micropylar end of the cell. At the ensuing mitosis the two daughter cells are, therefore, destined to inherit differentially distributed cytoplasmic elements and, therefore, the course of embryogenesis is already foreshadowed in the zygote itself: the chalazal daughter cell (apical cell) forms the embryo proper, while the basal cell forms the suspensor.

The timing, as well as the mechanism, of fixation of polar axis in the zygote are of great interest to a student of morphogenesis. Our knowledge is very inadequate not only in respect of the zygote of angiosperms but all the vascular plants as well. The zygote in Fucales – particularly *Fucus* – has been extensively used to probe the cytological, biochemical, and biophysical basis of determination of polar axis, and several reviews have appeared (see Jaffe 1968, Quatrano 1974, 1978). That the pattern of polar development during embryogenesis in *Fucus* bears some resemblance to that in angiosperms has been recognized for some time (Wardlaw 1968).

It is necessary to discuss the stages which characterize the development of a polar axis in the zygote of *Fucus;* concerning angiosperms Raghavan (1976) has already provided an excellent account. The spherical fucoid zygote has initially an apolar homogeneous appearance. The first sign of polarity is the localized accumulation of subcellular components in the cytoplasm, and the formation of a protuberance within 8–14 h following fertilization. An asymmetric division ensues resulting in a smaller rhizoidal cell, and a larger thallus cell. The events leading to the polarized reorganization of cytoplasmic elements has been explained by Jaffe (1968) on the basis of self-electrophoresis which includes: (1) decreased permeability of the plasma membrane on one side of the cell brought about by a variety of external agents, such as light, gradients of pH in the medium or in the concentration of substances, such as hormones; (2) amplification of the initial membrane difference through the accumulation of cations at the region of decreased permeability, followed by the establishment of a polarized electrophoretic gradient across the cell; (3) movement of electrophoretically polarized cytoplasmic elements (such as protein molecules) across the gradient, and often a movement of the nucleus.

As of now there is no evidence to suggest that such a mechanism exists in higher plants. Obviously, this is an area that requires much attention (see Stebbins 1974). Besides, one has to keep in mind other factors while discussing the problem of polarity in angiosperm zygote. Whereas in the fucoid zygote the structural and biochemical polarity occurs in a previously homogeneous egg cytoplasm, the egg is itself polarized in angiosperms. It may be more fruitful, therefore, to study the mechanism of cytoplasmic redistribution during the formation and maturation of the egg itself. Moreover, the fucoid eggs are released by the parent plant into the adjoining sea water, and fertilization is effected externally. By contrast, in angiosperms the egg is anchored to the embryo sac wall at the micropylar pole. In conclusion, the specific steps culminating in a highly polarized cytoplasm followed (in several taxa) by asymmetric mitotic division in the zygote are at present not well understood. Raghavan (1976) aptly remarks: "... the review of the basis of polarity ... might appear worth making if only because it may provide some perspective for dealing with this question."

8.3 Early Embryogenesis

8.3.1 Cell Patterns

In most investigated angiosperms, the zygote divides by a transverse wall resulting in two superposed cells, the apical and basal cells. These cells as well as their derivatives have been assigned specific abbreviations, originally by Souèges and the French School, and are now conventionally employed in embryological liter-

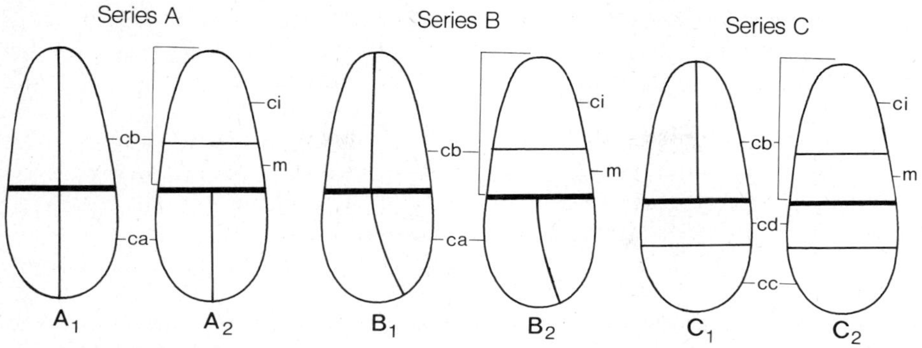

Fig. 8.3. Proembryonal tetrads; classification based primarily on the plane of division in apical cell: vertical in *Series A*, oblique in *Series B*, and transverse in *Series C*. Further subdivision takes into account the mode of division of basal cell which may be either vertical (A_1, B_1, C_1) or transverse (A_2, B_2, C_2)

ature. The apical cell is designated *ca* and the basal cell *cb* (Fig. 8.3, see also Fig. 8.6).

Rarely, as in Balanophoraceae, Dipsacaceae (*Scabiosa*), Loranthaceae, and Piperaceae, the division of the zygote is by a vertical or oblique wall which is so oriented as to be almost longitudinal.

The transverse division of the zygote and the respective contributions of the two daughter cells to the subsequent formation of the embryo and its suspensor are considered as the basis for the recognition of embryogenic types by Schnarf (1929) as well as Johansen (1950).

The division of the zygote in some taxa is such that the apical cell *ca* may be smaller than the basal cell *cb*. However, in a statistically significant number of angiosperms, the basal cell may be approximately equal to or appreciably smaller than the apical cell (Sivaramakrishna 1978). Any discussion on the nature of factors that control the specific size relationship between the apical and basal cell in a two-celled proembryo must, at the moment, remain largely conjectural for want of critical information concerning zygotic division in higher plants. It may involve shape and polarity of zygote itself, qualitative and/or quantitative biochemical differences in the microenvironment of the embryo sac adjoining the micropylar and chalazal end of zygote, the morphogenetic destination of the two cells, or some other factor.

8.3.2 Tetrad, Quadrant, and Octant Proembryos

The division of each of the two cells of the proembryo results in a four-celled tetrad arranged in specific patterns. The major patterns, or Series, are designated *A*, *B*, and *C* (Fig. 8.3). The partition wall in the apical cell is vertical in Series *A*, oblique in Series *B*, and transverse in Series *C*. In each Series two variations are possible, depending upon whether the basal cell initially undergoes a vertical or a transverse division and, accordingly, these tetrads are designated A_1 and A_2, B_1 and B_2, and C_1 and C_2 (Fig. 8.3). When the basal cell undergoes a transverse di-

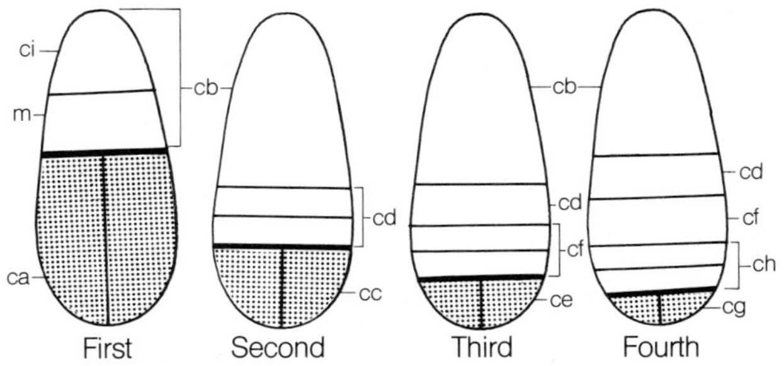

Fig. 8.4. Proembryonal tetrads of *Series C*, showing tetrads of first, second, third, and fourth order, respectively. The potential embryonal pole (*shaded*) shifts progressively to the extreme micropylar end

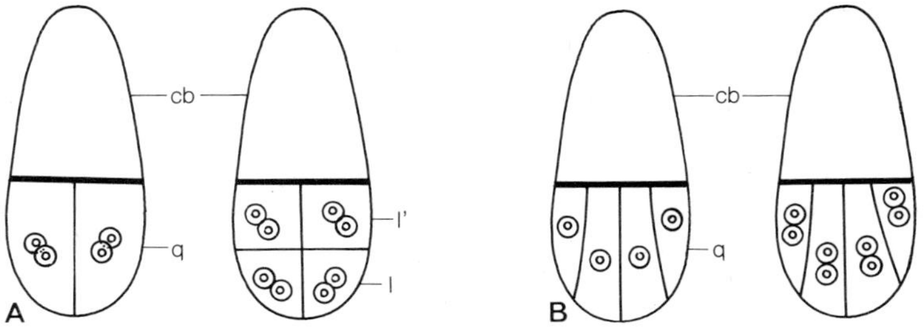

Fig. 8.5 A, B. Quadrants (four cells at distal end of proembryo) produce octants (eight cells) either in two tiers (**A**) or one tier (**B**). When in two tiers, the apical tier is designated *l* and the basal tier *l′*

vision (as in Series A_2, B_2, and C_2) the more basally placed daughter cell is designated *ci* and the intermediate daughter cell *m* (Fig. 8.3).

In a tetrad of Series *C*, when the apical cell divides transversely, two superposed cells *cc* and *cd* (Fig. 8.3) are formed. These cells may again divide and produce a tetrad of the second order (Fig. 8.4). Here the cell *cc* behaves like *ca*, and the cell *cd* as *cb*. The tetrad of the second order may conform to Series *A* (Fig. 8.4), *B* or *C*. Tetrads of the third and fourth order are also possible (Fig. 8.4) if the transverse divisions continue at the apical pole of the proembryo. The derivatives of *cc* are designated *ce* and *cf* (Fig. 8.4), and that of *ce* as *cg* and *ch* (Fig. 8.4). It should be emphasized that, in all these instances, the potential embryonal region is progressively shifted to the distal pole of the proembryo (Fig. 8.4); the basal part does not contribute to the embryo proper. Souèges (see Crété 1963) attached considerable importance to the constitution of the tetrads, and their contribution to the embryo proper.

The tetrad stage is followed by another vertical division in a plane usually at right angles to the first vertical division in the distal cell *ca, cc, ce,* or *cg*. These four juxtaposed cells constitute the quadrant *q* (Fig. 8.5) which through another division gives rise to an octant. The concept of quadrant and octant relates to the formation of four and eight cells at the distal end of the proembryo. The octant may be in two tiers *l* and *l′* (Fig. 8.5 A), or one tier (Fig. 8.5 B). The quadrant and octant are of much significance as the progenitors of the cotyledonary and epicotylary loci of the shoot pole and, in addition, partially or fully contribute to the hypocotyledonary portion of the embryo.

The derivatives of the basal cell, immediately next to embryonal mass, contribute to the hypophyseal region. Now the embryo is complete. The remaining derivatives of the basal cell [including the cells corresponding to the derivatives of the basal cell in tetrad of the second, third, and fourth order (see Fig. 8.4)] constitute the suspensor. It is noteworthy that the contribution of zygote to the embryo proper, volumewise, may range between 100% as in *Penaea* (where the suspensor does not differentiate and the apical and basal cell contribute equally to the embryo proper), to just about 6% as in some species of *Corydalis* (see Johansen 1950).

The foregoing account outlines the general course of development of the embryo in angiosperms. There is considerable variation in the sequence and chronology of the early divisions in various tiers of the proembryo, the orientation of the partition walls, and the contribution of the tiers to the further development of embryo and suspensor. A particular course of development may be specific to individual species, or genus, or even entire family, but variation may occur even within the same species. In *Utricularia coerulea* Kaushik and Raju (1955) observed tetrads of all the three Series. The variations during the formation of quadrant and octant in the same species have been highlighted in *Impatiens textorii* by Takao (1976); five patterns of octant formation are described.

8.3.3 Stages Leading to Mature Embryo

Further development from the octant stage following numerous cell divisions in various planes causes the proembryo to assume a globular configuration. The globular proembryo passes through a phase before the cotyledons *pco,* and epicotyl *pvt* become outwardly evident at specified loci, and there is a transition from radial to bilateral symmetry. The embryo is now heart-shaped, so characteristic of the dicotyledons.

In the monocotyledons, however, the early course of development is no doubt similar to that in dicotyledons, but there are fundamental departures at the time of differentiation in the globular proembryo. These are discussed later.

During early embryogenesis two characteristic cells, or groups of cells, become evident. These are generally referred to as hypophysis *h,* and epiphysis *e.* The intermediate cell *m,* between the embryonal mass and suspensor, cuts off a discoidal cell designated "hypophysis" (see Figs. 8.15 B–D and 8.17) by Hanstein (1870). Correspondingly, at the opposite end, i.e. at the apex of the embryonal region, a wedge-shaped cell may be cut off. This has been designated epiphysis by Souèges (1934). According to Souèges, these cells are so characteristic of the dicotyledonous embryos that they are considered to represent two polar morphogenetic units, the root pole and the shoot pole. The hypophysis may be a single cell, or a group of cells. The initials of the root cortex *iec* and the root cap *co* are derived from the hypophysis (Crété 1963). The hypophysis is pushed into the embryonal mass in the course of development. There is considerable variation in the origin and organization of hypophysis.

The epiphysis originates in the terminal tier of dicotyledonous embryos, and gives rise to the shoot tip. The epiphysis may be a single cell or a group of cells, and its organization also varies. The epiphysis may not be easy to recognize in several dicotyledons, and it remains to be seen whether it is differentiated during embryogenesis of all dicotyledons.

Irrespective of the variations in the sequence and disposition of cells in the tiers of the proembryo, ultimately there is hardly any difference in the organization of the embryo. Swamy and Padmanabhan (1962) made an overall comparative analysis of the assumed correspondence between the organization of the tiers and their destination. They have demonstrated that the corresponding tiers in diverse types of embryonal development do not follow comparable destinations. As stated earlier, even within the limits of a single species, such correlation may not exist.

8.4 Classification Based on Early Development of the Embryo

8.4.1 Systems Suggested by Schnarf and Johansen

The variations in the orientation of walls, and sequence of cell divisions in the proembryo, are the basis for the classification of embryogenic types. Schnarf (1929), for the first time, attempted to classify the embryogenic types. He recognized five basic types based on the respective contribution of the two-celled proembryo to the formation of the embryo proper and suspensor. These types were named after the taxa in which the embryogenic type was clearly demonstrated. The types are: Cruciferen, Astereen, Solanaceen, Chenopodiaceen, and Caryophyllaceen. Maheshwari (1950) recognized the same five types but gave the English connotation: Crucifer, Asterad, Solanad, Chenopodiad, and Caryophyllad. Johansen (1950), while adopting these five types, added a sixth type, the Piperad, to accommodate those cases where the zygote divides by a vertical wall, or an oblique wall with an essentially longitudinal disposition. He also changed the name of the Crucifer type to Onagrad type stating that the embryogeny of Onagraceae is highly simplified and the term Onagrad would be more appropriate. Apart from the Piperad type, the separation of one type from the other is based primarily on the destiny of the apical cell of the two-celled proembryo during the second and subsequent divisions, and whether the basal cell does, or does not, contribute to the construction of the embryo proper. Johansen also described a number of variations in each type, designating them by the names of the taxa characteristic of the particular variation. Figure 8.6 gives a schematic representation of the classificatory types based on the systems of Schnarf and Johansen.

Onagrad Type. A vertical wall in the apical cell (Fig. 8.6, see also Fig. 8.15 A) and a transverse wall in the basal cell (Fig. 8.6, see also Fig. 8.14 B) give rise to a T-shaped tetrad. The apical cell, in due course, contributes to the major portion of the embryo proper. The hypophyseal region (see Fig. 8.15 B–D), which engenders the initials of the root cortex and root cap and completes the embryo, is derived from the basal cell.

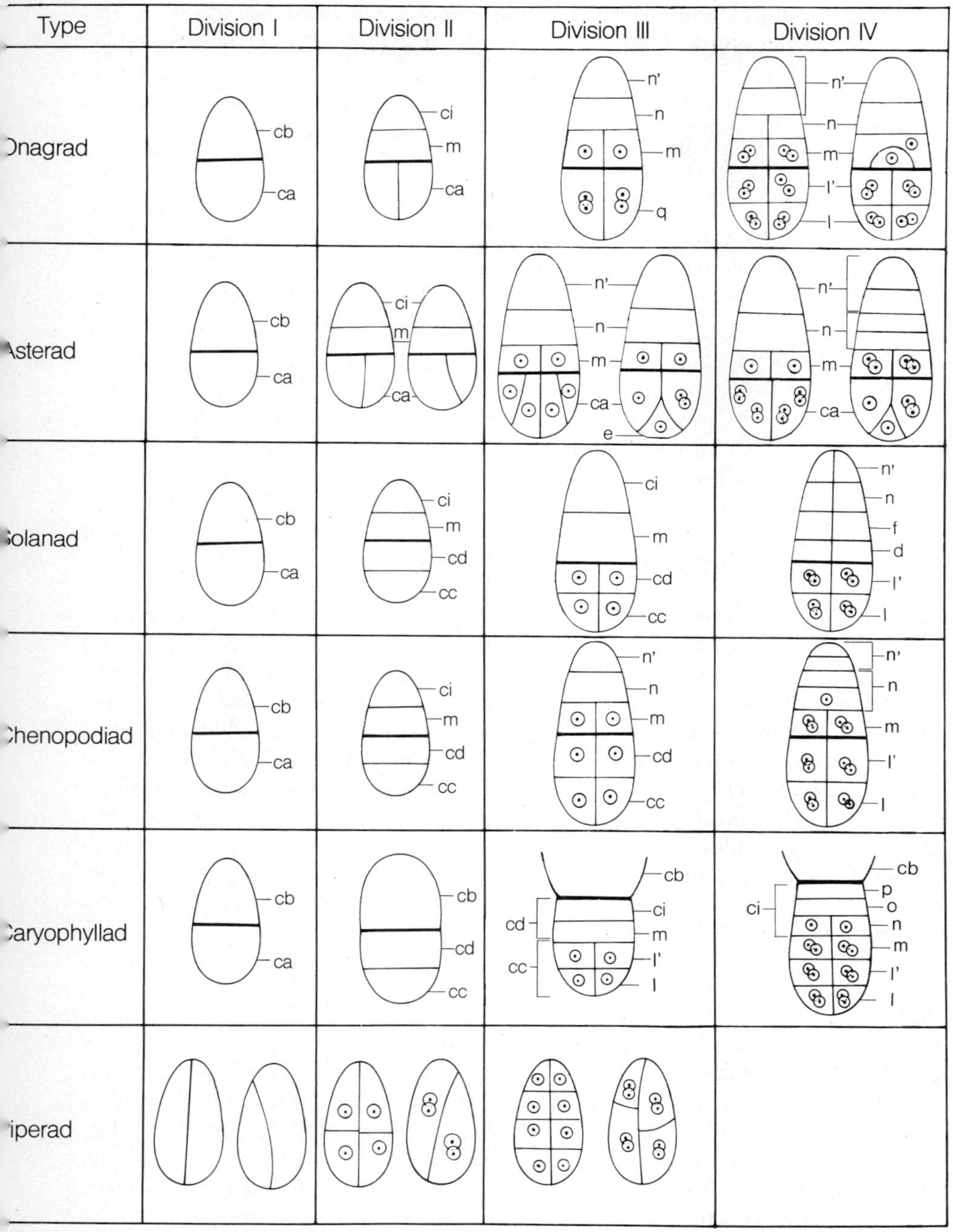

Fig. 8.6. Schematic representation of main types of embryogeny; based on Schnarf (1929) and Johansen's (1950) system of classification

In order to make comparisons between different embryogenic types easy, embryonomic formulae are employed. The formulae indicate, by means of standard abbreviations for different cells and embryonic regions during development, precise relationships between cells with respect to their origin, number, disposition, and histogenetic function. The general formula for the Onagrad Type is:

ca: *pco* + *pvt* + *phy* + *icc*
cb: *iec* + *co* + *s*.

The above formula indicates that the apical cell (*ca*) of the two-celled proembryo gives rise to the following: cotyledonary region (*pco*), stem tip (*pvt*), hypocotyledonary region (*phy, icc*). Similarly, the basal cell (*cb*) contributes to root cortex (*iec*), root cap (*co*), and suspensor (*s*). Therefore, the contribution of the basal cell to the embryo proper is not significant.

The Onagrad Type of embryogeny is characteristic of families such as Onagraceae and Brassicaceae, and of several genera in other families: Dicotyledons – Bignoniaceae (*Catalpa*), Lamiaceae (*Mentha*), Lythraceae (*Lythrum*), Papilionaceae (*Lotus*), Ranunculaceae (*Myosurus*), Rutaceae (*Ruta*), and Scrophulariaceae (*Veronica*); Monocotyledons – Juncaceae (*Juncus*) and Liliaceae (*Lilium*). The sequence of embryogeny in *Capsella bursa-pastoris,* which has been for long a favourite classroom material, is illustrated in Figs. 8.14–8.16.

Asterad Type. This is characteristic of the family Asteraceae, and is also known in several families of both dicotyledons and monocotyledons. Again, a three-tiered proembryo is formed with a vertical division in the apical cell and a transverse division in the basal cell. However, in some taxa conforming to this type, the apical cell may divide by an oblique wall and, at the time of formation of quadrant, a wedge-shaped epiphysis initial is cut off which later gives rise to the epicotylary region (Fig. 8.6). The remaining three cells of the quadrant produce the cotyledons. In this type the basal cell contributes substantially to the embryo proper; later the middle cell *m* of the proembryonic tetrad gives rise to the entire hypocotyledonary region. In extreme cases, as for example in *Penaea,* the suspensor does not differentiate, and the apical and basal cells contribute equally to the bulk of the embryo. The general formula is:

ca: *pco* + *pvt*
cb: *phy* + *icc* + *iec* + *co* + *s*.

Actual proof that the basal cell contributes considerably to the construction of the embryo proper has been provided through X-irradiation in *Crepis capillaris* by Husted (1941, see Johansen 1950, pp. 116–117). The study revealed that *ca* contributed 64% to 68% of the embryo while *cb* contributed 32% to 36%. The variations of the basic type are reported in several genera of Geraniaceae, Lamiaceae, Liliaceae, Oxalidaceae, Poaceae, Polygonaceae, Rosaceae, and Urticaceae. The embryogeny in *Penaea* needs reinvestigation.

Solanad Type. Solanad type is based on the formation of a linear tetrad in the proembryo. The derivatives of the basal cell do not take part in the organization of embryo proper, which is mostly derived from the apical cell of the two-celled proembryo (Fig. 8.6). The general formula is:

ca: *pco* + *pvt* + *phy* + *icc* + *iec*
cb: *co* + *s*.

The family Solanaceae, particularly, presents this type of embryogeny. It also occurs in Hydnoraceae, Linaceae, Papaveraceae, and some Rubiaceae. The proembryo of *Hydnora,* a parasitic member, is very peculiar and comprises a linear row of as many as 15 cells before the first vertical wall is laid. The mature embryo also does not show differentiation of organs.

Chenopodiad Type. In Chenopodiad type a linear tetrad of proembryo is formed, but the destiny of the cells *ca* and *cb* are as follows:

ca: *pco* + *pvt* + 1/2*phy*
cb: 1/2 *phy* + *icc* + *iec* + *co* + *s*.

Thus, the basal cell also contributes substantially to the hypocotyledonary region (Fig. 8.6). Among others, this type of embryogeny is common in the families Amaranthaceae, Chenopodiaceae, and Polemoniaceae. Usually, an epiphysis is not formed, but in the variants, e.g. *Polemonium* and *Myosotis,* an epiphyseal initial is cut off at the third cell generation in the proembryo.

Caryophyllad Type. The embryogeny of Caryophyllaceae is very different from the other types. Generally, the basal cell does not divide at all (Fig. 8.6) or, as in *Myriophyllum* and *Ruppia,* it may divide only once and both the cells undergo considerable enlargement. Many variations of the basic type occur in families such as Crassulaceae, Droseraceae, Fumariaceae, Haloragaceae, Portulacaceae, and Pyrolaceae among the dicotyledons, and Alismataceae, Araceae, Ruppiaceae, and Zannichelliaceae among the monocotyledons. Since the basal cell does not take part in the formation of the embryo proper, the apical cell shows the organization of a second tetrad of all the three Series, and further course of embryogenesis may follow any of the types described above. The apical cell *ca* may be considered equivalent to the zygote for purposes of describing the course of embryonal development in members exhibiting this type. In the system of classification of Souèges (see below), the Caryophyllad type corresponds to the Second Grand Division or the Second Period.

Piperad Type. Piperad type includes all those instances where the zygote undergoes a vertical or longitudinally oriented oblique division (Fig. 8.6). The second division is also in the longitudinal plane, but at right angles to the plane of the earlier division, and results in four cells. Transverse division may then take place in these cells. Till recently the Piperad type was known in only a few dicotyledonous families – Balanophoraceae, Dipsacaceae – *Scabiosa,* Loranthaceae and Piperaceae. Recently, however, Tohda (1974) has recorded a vertical or very oblique zygotic division in *Pogonia japonica,* an orchid. [In this connection it may be mentioned that Swamy (1943) had earlier observed a vertical division in the zygote of *Cymbidium bicolor,* also an orchid, but in this taxon the vertical wall, in some cases, also splits the zygote into two halves which develop independently into twin embryos.] In Loranthaceae the initial longitudinal division is followed by numerous transverse divisions and, with the elongation of cells, the biseriate suspensor of the proembryo assumes a long and tortuous course. *Scabiosa* of Dipsacaceae is unique in forming the first tetrad in a tetrahedral fashion with an oblique zygotic division followed by a transverse division in one daughter cell and an oblique division in the other. Except in Loranthaceae, a suspensor does not differentiate and the cotyledons may not be formed.

8.4.2 System Suggested by Souèges

The elaborate system of classification formulated by Souèges (see Crété 1963) essentially consists of the Grand Divisions or Periods, the embryogenic Groups, and the Megarchetypes. The Periods are based on the concept of the progressive shifting of the potentiality for embryo formation towards the extreme tip of the original apical cell *ca*. The Periods are designated First, Second, Third, Fourth, and so on, depending on the order of the proembryonal tetrad, but representatives are now known only for the first four Periods. In the First Period the apical cell of the two-celled proembryo contributes to the embryo proper, and the basal cell may also contribute to embryo proper to varying extent, or to suspensor alone. Souèges recognized six Megarchetypes:

Megarchetype I. *cb*: *pvt* + *phy* + *icc* + *iec* + *co* + *s*
II. *cb*: *phy* + *icc* + *iec* + *co* + *s*
III. *cb*: 1/2 *phy* + *icc* + *iec* + *co* + *s*
IV. *cb*: *iec* + *co* + *s*
V. *cb*: *co* + *s*
VI. *cb*: *s*

In each Period eight embryogenic groups based on the configuration of the tetrads of the three Series *A*, *B*, and *C* are recognized.

In the Second Period the apical cell of the two-celled proembryo undergoes a transverse division forming two superposed cells *cc* and *cd*. The destiny of *cc* and *cd* is similar to that of *ca* and *cb* in the First Period. Here again, the contribution of the cell *cd* to the embryo proper is classified under the same six Megarchetypes, and with the same destiny. As in the First Period, there are corresponding types of embryogenesis in the Second Period. For example, the embryogeny of *Levenhookia dubia* (Subramanyam 1950) placed in the Second Period is essentially similar to that of *Capsella bursa-pastoris* in the First Period, except that in *L. dubia* the cell *cb* contributes to suspensor alone. The apical cell *ca*, which forms a second order tetrad in Series *A*, functions exactly like the first tetrad derived from the zygote. Souèges considered these superpositions of embryogenic development in successive Periods as conforming to the "Law of Correspondence of Forms". Another example of this Law is clearly demonstrated in three members of Fabaceae: *Coronilla minima* in the First Period, *Medicago lupulina* in the Second Period, *Trigonella foenum-graecum* in the Third Period (Rau 1950). The embryogenesis is essentially of the same type, but the cells which contribute to the embryo proper are progressively shifted towards the distal end; the remaining cells together constitute the suspensor [see Rau (1954) for a detailed account of the framework of the classification of Souèges].

An important aspect of classification, as outlined above, is the application of "embryogenic laws" to the first four cell generations in embryogenesis: the two superposed cells *ca* and *cb* in the first generation, the tetrad in the second, the quadrant in the third, and octant in the fourth. Due to absence of synchrony in cell divisions, varying patterns of proembryo develop and, correspondingly, the destiny of cells varies in the successive generations. This is the basis for the classification of embryogenic types in this system.

8.4.3 Other Systems of Classification

In many investigated species the development conforms to the laws of embryogeny formulated by Souèges; not only are there exceptions but also too many variations and, hence, this system has met with much criticism. The numerous groups based on the nature of tetrad, and the destiny of cells, are not always specific for the species. The same species often figures in more than one group, on account of the variations in the configuration of cells as well as the destiny of its cells. Moreover, the prevalence of a large number of irregular types which do not conform to any group in this classificatory system necessitated a re-examination of this system. Mestre (1967), one of the pupils of Souèges, took a fresh look at this problem and its consequences as related to the laws of embryogeny. The irregularities in early embryogeny include cases where: (1) the sequence of cell divisions may not follow any definite pattern, (2) the partition walls are laid down in an oblique manner, (3) distinction into regular tiers may not be evident, and (4) the same species may show variation in the sequence of cell lineage and development. Mestre concluded that only the destiny of the two initial cells *ca* and *cb* had an absolute stability and, therefore, could serve as a criterion for the establishment of a satisfactory embryogenic law. Accordingly, he classified the embryogenic types into two categories, the regular types which conform to the embryogenic laws, and others which follow an irregular mode of embryogenesis. While retaining the main features of the classification of Souèges, Mestre proposed a simplified system which recognizes the four Periods for which representatives are known at present and, under each Period, only three Megarchetypes are recognized. These are designated *A, B,* and *C*. According to Mestre, Megarchetype *A* represents those cases where *ca* does not contribute to the embryo proper, and the formula is:

Megarchetype *A*: *ca*-no contribution
cb-stem + root.

Megarchetype *B* comprises cases where both *ca* and *cb* contribute to the formation of embryo:

Megarchetype *B*: *ca*-stem
cb-root.

Under the above Megarchetype, two subarchetypes are recognized and the formula is as follows:

Subarchetype B_1: *ca*-part or most of hypocotyl
Subarchetype B_2: *ca*-entire hypocotyl.

In Megarchetype *C, cb* does not contribute to the embryo at all, the formula being:

Megarchetype *C*: *ca*-stem + root
cb-no contribution.

Even in this revised and simplified system, Mestre retains the original view of Souèges that the stem tip is derived from the basal cell in Megarchetype *A* (monocotyledons). This view is erroneous as indicated by recent investigations (see Sect. 8.6). Therefore, J. L. Guignard (1975 a), another student of Souèges, further modified the classification. He suggested omission of the Megarchetype I in the original system of Souèges.

Yamazaki (1974) formulated a new system of clasification of embryogenic types based on the differentiation of organs and tissues during embryogeny, particularly the differentiation of the root cortex, and the stem apex. Four principal types – Polygonad, Solanad, Onagrad, and Asterad – are recognized. Instances in which the exact pattern of differentiation of root cortex is difficult to follow, are placed under the fifth type – Irregular type. The types are subdivided into several variations. Although the names of the types (except Polygonad) are those used by Johansen, their definitions are quite different. The Chenopodiad and Caryophyllad types of Johansen are merged into the Polygonad type as the pattern of initiation of root cortex is the same in both, and resembles that of *Polygonum persicaria*. The four principal types may be distinguished as follows:

A. The terminal cell *ca* itself, or its derivative, gives rise to the stem apex and the cotyledons:
   1. The cell *m* which produces the initials of root cortex does not contribute to the hypocotyl ... Onagrad type
   2. The cell *m, n* or *cd* which produces the initials of root cortex also contributes to the hypocotyl:
      a) Only a portion of the hypocotyl is derived ... Polygonad type
      b) The entire hypocotyl is derived ... Soland type

B. The quadrant formed directly from the terminal cell *ca* divides obliquely and gives rise to the stem apex and the cotyledons ... Asterad type

Based on this embryogenic typification, Yamazaki (1974) proposed a fresh classification of Gamopetalae, an assemblage of dicotyledonous plants known to be only partially natural. He groups the taxa under four Superorders, namely, Ericanae, Gentianae, Lamianae, and Asteranae.

1. Ericanae consists of seven orders (Ebenales, Diapensiales, Ericales, Primulales, Plumbaginales, Polemoniales, Oleales) and the type of embryo development in the Superorder conforms to the Polygonad, and rarely to the Solanad pattern.

2. Gentianae (Gentianales, Rubiales) conforms to the Solanad type, rarely the Irregular type.
3. Lamianae (Scrophulariales, Lamiales, Campanulales) conforms to the Onagrad type.
4. Asteranae (Menyanthales, Dipsacales, Asterales) conforms to the Asterad pattern.

According to Yamazaki (1974), the above classification of the Gamopetalae is more natural than the earlier ones by Melchior (1964), Cronquist (1968), and Takhtajan (1969).

Yamazaki (1982) has amplified his earlier classification (Yamazaki 1974) of embryogenic types to include the entire dicotyledons, and has suggested an additional type – the Geraniad type. In this type the cell *m* cuts off three or more tiers of several cells each. The centrally placed derivatives toward *ca* give rise to the initials of the root cortex, while those toward *ci* produce the root cap and also contribute to the suspensor. The Geraniad type of development has been observed in families such as Calycanthaceae, Geraniaceae, Leguminosae (partly), Magnoliaceae, Menispermaceae, and Ranunculaceae (partly). Yamazaki emphasizes that the pattern of differentiation of the root cortex initials is constant not only in the families but, in some instances, also within the Orders. Consequently, this is the "most reliable feature" in indicating phylogenetic relationship of the dicotyledons.

Periasamy (1977) presented a new approach to the classification of embryos. He points out that the existing classifications of Schnarf, Johansen, and Souèges are based solely on the early segmentation patterns of the proembryo. The criteria employed by them for recognizing the Types, Variations, Divisions, Periods, Megarchetypes, etc., appear to be mere arbitrary aspects without any morphological, physiological, or phylogenetic significance. During the early embryogeny in angiosperms there is a progressive reduction in surface-to-volume ratio of the individual cells produced. Whereas the outer surface of the zygote is wholly exposed to its external environment, the daughter cells derived from its division are relatively less exposed than the zygote itself. The succeeding generations of cells are exposed less and less to the external environment and more to the internal environment. At some stage cells are produced which are completely internal with no surface exposed to the exterior. Periasamy considers that the first formation of wholly internal cells is an important morphogenetic event at the cellular level in angiosperm embryogeny. According to him, an analysis of the proembryo with regard to the change in surface-to-volume ratio, and segmentation of internal cells, provide significant criteria for a more meaningful classification.

In his classification the total number of cell tiers that comprise the proembryo, when the first series of wholly internal cells are segmented, constitutes the basis for the recognition of Groups. A tier is defined as a layer of more than one cell arising by vertical divisions in a single cell. A distinction is made between the "truly embryonal part" and the suspensor. Counting from the terminal end towards the basal end, tiers that form the embryo proper are considered as ending with the first cell that has not undergone a vertical division, which means that a cell or a group of cells, longitudinally undivided, demarcate the embryonal part from the suspensor. Irrespective of the number of cells in the suspensor, even

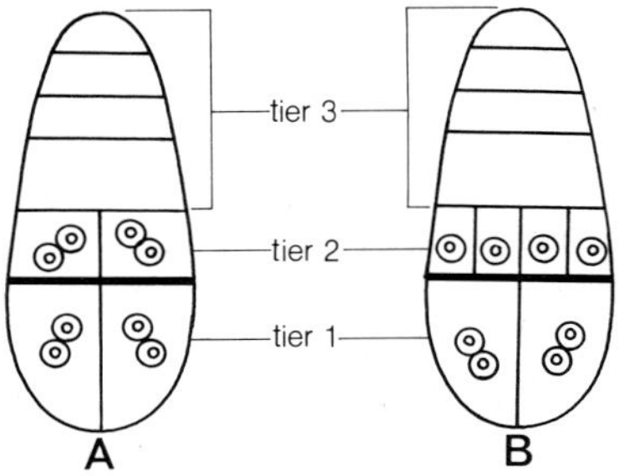

Fig. 8.7 A, B. *Papaver rhoeas*, formation of first "internal" cells during embryogeny. [Diagrammatic, after Souèges (1926), tiers in proembryo numbered after Periasamy (1977)]

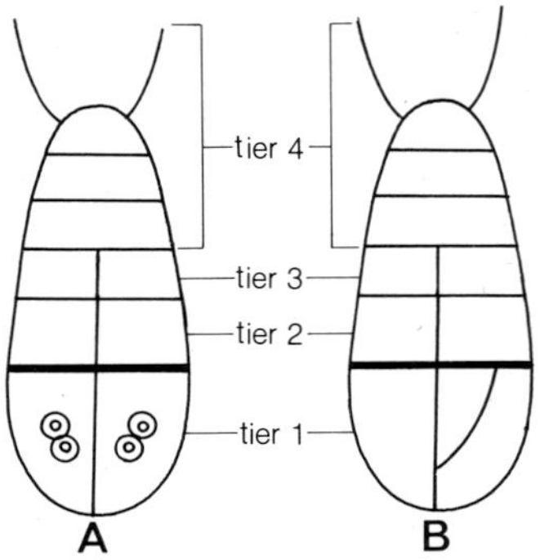

Fig. 8.8 A, B. *Halophila ovata*, formation of first "internal" cells during embryogeny. [Diagrammatic, after Swamy and Lakshmanan (1962); tiers in proembryo numbered after Periasamy (1977)]

where vertical divisions have taken place in its basal part, this region is collectively regarded as constituting a single tier. The minimum number of tiers before the wholly internal cells are segmented in the proembryo is two. This forms Group I. Since wholly internal cells are known to be segmented in the embryos only after three to six tiers, as defined above, these are classed as Groups II to V, based respectively on the number of tiers.

The number of cells in each tier of the proembryo immediately before the formation of the first internal cells is concerned with growth in girth. This is the criterion for the erection of the Series, under each Group.

The location of a tier in which the first internal cells are segmented forms the basis for recognizing the Variations under each Series. Two Variations are recognized on the basis whether the concerned tier is in the terminal, or the subterminal position.

The classification as proposed by Periasamy comprises five Groups: Group I with three Series, Group II with five Series, and Groups III, IV, and V with six Series each; under each Series are two Variations. Each category in the above three criteria is separately numbered, and an embryo can be characterized by its "Embryonic Number". This Embryonic Number is of three digits in which the first digit denotes the Group number (1 to 5), the second digit the Series number (1 to 6), and the third digit the Variation number (1 or 2). This may be illustrated with two specific examples, one each for the dicotyledons and monocotyledons.

The Embryonic Number for *Papaver rhoeas* (Fig. 8.7) is 222. The first digit refers to Group 2, as there are three tiers (Fig. 8.7 A), vertical division having taken place in the terminal and subterminal tiers; the second digit refers to Series 2 under Group 2 because the two terminal tiers have four cells each at the time the first wholly internal cells are formed (i.e. after the octant stage); and the third digit refers to Variation 2 because the first internal cells are formed in the subterminal tier (Fig. 8.7 B).

The Embryonic Number for *Halophila ovata* (Fig. 8.8) is 311. There are four tiers, three with vertical walls in the embryonal part when the first internal cell is formed (Fig. 8.8 A) and the first digit is, therefore, 3; there are four cells in the terminal tier when the first internal wall is laid down and the other tiers are less than four-celled (Fig. 8.8 A) and, thus, it is in Series 1 and, hence, the second digit is 1. The first internal cell is formed in the terminal tier (Fig. 8.8 B) and, therefore, it is classed as Variation 1 making the third digit also 1.

Periasamy's classification is applicable only when the embryogeny is regular. The Groups, Series, and Variations are also not easy of comprehension. It is equally arbitrary in the sense that it is not phylogenetic, though the morphogenetic implication attributed to it needs further study.

8.5 Differentiation in Embryo

The three-dimensional growth of the embryo is initiated when vertical walls are laid in the terminal cell of the linear proembryo. In due course, through divisions in various planes, the globular stage is attained. Differentiation sets in with the establishment of the cotyledonary and epicotylary loci in the shoot pole; the hypocotyledonary region and the root pole with the hypophysis can be distinguished. The organization of the loci, however, is not strictly correlated with the cell lineage, or the tier systems. Hanstein's (1870) concept of "histogenesis" supported by the French School of embryologists, assumed that the origin of the dermatogen, periblem, and plerome are traceable to specific cells in the proembryo and, as such, histogenesis and organogenesis were linked with cell lineage and destination. In fact, in this descriptive phase of embryogenic studies, much importance was given to the sequence of cell divisions and formation of the tiers and their destinies in early embryogenesis. The classificatory systems, as noted earlier, were also proposed on this basis. Swamy and Padmanabhan (1962), in a critical reference to this viewpoint, stated that "the entire framework of classification made by Souèges and by Johansen rest on principles that are not suited to explain the problems of developmental study".

In a new approach to the study of embryogenesis, which may be called the developmental approach, authors such as Foster (1949), Wardlaw (1955), Swamy and Padmanabhan (1962), and others have seriously questioned the validity of the embryonomic laws. According to them, there is a measure of plasticity underlying embryonal development, histogenesis, and organogenesis which become evident only at a comparatively late stage, i.e. the globular stage of proembryo and never at the earlier proembryonal level. Swamy and Padmanabhan (1962), citing examples of the inconsistencies in the specificity of destination of proembryonal cells, state that: "when angiosperm embryogenesis is subjected to an overall and comparative analysis, it appears as if no organ possesses any obligation to any particular proembryonic tier". In the globular proembryo the primary tissue systems, through differential vacuolation in the homogeneous mass of cells, become marked out. The cotyledonary, epicotylarly, and hypocotyledonary regions along with the hypophyseal region become identifiable through further cell divisions, growth, and development in the respective organs. The hypocotyledonary region, through cell increase and elongation, contributes to the main axial system of the later embryo. This differential growth phase of the embryo leads to the organization and activity of the respective meristems through the accentuation of the root-shoot polar differences. Swamy and Padmanabhan (1962), while discussing this phase of development of the embryo at the organizational level, point out that the existing systems of classification, based on the sequence of cell divisions in the proembryo, do not reflect the importance of underlying factors of growth, differentiation, and organization as applied to the dicotyledons as a whole. There are too many variants to give them any real status either from the morphogenetic or the phylogenetic point of view.

8.6 Dicot and Monocot Embryo

In the early phases of embryogenic studies, the concept prevailed that the dicotyledons and monocotyledons had their individualistic modes of embryo development. This is no more valid. Hanstein (1870), on the basis of his study of *Alisma plantago*, claimed that it represented the monocotyledonous type of development. When a large number of taxa had been investigated, particularly by Souèges and his collaborators, the essential similarity in the sequence of divisions in the young embryo in monocots and dicots was noted. Therefore, both mono- and dicotyledonous species figured in embryogenic classification. For example, Schnarf (1929) and Johansen (1950) included monocotyledonous species such as *Muscari Lilium, Juncus* and others along with dicotyledonous species in the Crucifer (Onagrad) type. Souèges (see Crété 1963), likewise, included several monocots in Period I though he designated a separate Megarchetype for them on the erroneous assumption that in all monocotyledons the stem tip was lateral, originating from the middle tier and, that, only the cotyledon was terminal. During early embryogenesis, quadrant and octant stages, and up to the globular proembryo, there is a similarity of lineage and configuration of the cells in monocots and dicots. At the time of differentiation in globular proembryo, fundamental differences arise. Swamy and Krishnamurthy (1975, 1977) point out that in the organization of embryonic shoot apex, the dicotyledons and monocotyledons follow entirely diverse morphogenetic patterns. In the dicotyledons the axial cells which constitute the epiphysis have a retarded growth in contrast to the faster rate of growth of circumaxial cells. In the monocots (Fig. 8.9 A–K), on the other hand, one-half of the terminal cell and its derivatives have the characteristic of the epiphysis with retarded growth, whereas the other half with a rapid rate of growth forms the cotyledonary locus. The apparent lateral position of the stem tip in later stages is due to the more rapid growth of the cotyledon. Recent studies on a number of monocotyledons by Haccius, Swamy, Lakshmanan and others (see Lakshmanan 1972) have clearly demonstrated that the epicotyl and cotyledon arise from one and the same terminal tier. These two loci are differentiated by the first vertical wall in the terminal tier. At the quadrant stage in the terminal region, the differences become quite apparent. Lakshmanan (1972) studied a series of transections of the terminal tier to explain the major difference between the dicotyledons and monocotyledons (Fig. 8.10 A–E). In the dicotyledons the two oposite cells of the

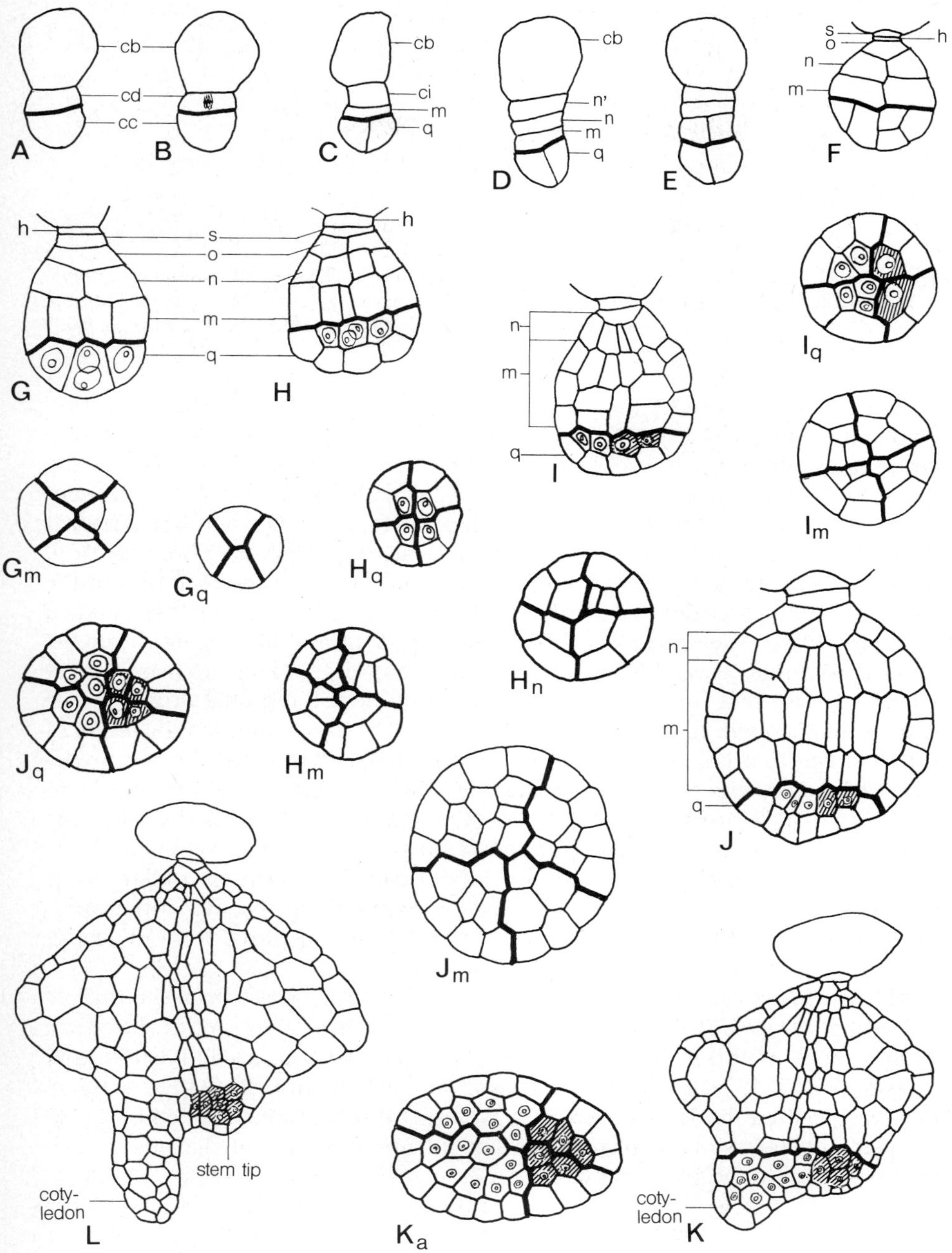

Fig. 8.9 A–K. *Halophila ovata,* stages in the development of embryo. In monocotyledons one-half of the apical cell exhibits retarded growth (*shaded regions* in **I–L**) and gives rise to the stem tip, whereas the other half grows rapidly by cell divisions and forms the cotyledon. Hence, the shoot tip is not lateral in origin. (After Swamy and Lakshmanan 1962)

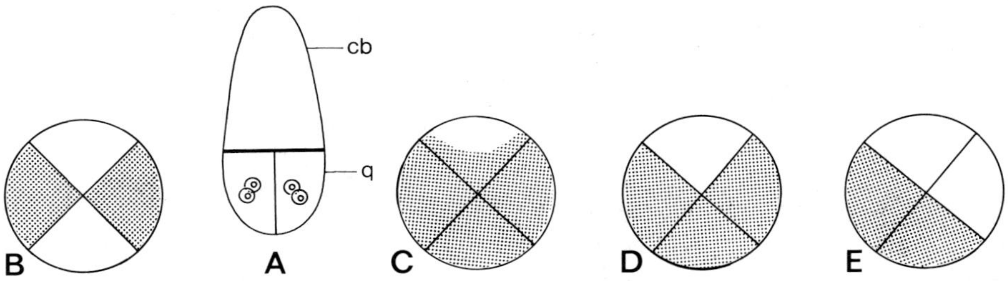

Fig. 8.10 A–E. Derivation of cotyledons in monocotyledons and dicotyledons. **A** Quadrant proembryo. **B–E** Transections of terminal tier of **A;** *stippled portion* represents progenitors of cotyledon. **B** Development in dicotyledons. **C–E** Development in various monocotyledonous taxa. (After Lakshmanan 1972)

terminal quartet engender the pair of cotyledons (Fig. 8.10 B); in the monocotyledons the number of cells of the quartet that gives rise to the cotyledon varies, with practically all the four cells (except a few derivatives of one of them) giving rise to the cotyledon as in Philydraceae (Fig. 8.10 C); three cells of the quadrant as in Pontederiaceae, Sparganiaceae, Iridaceae (Fig. 8.10 D); only the adjacent two cells as in Hydrocharitaceae, Potamogetonaceae, and Amaryllidaceae (Fig. 8.10 E). As far back as 1897, Campbell observed that in *Zannichellia* both the shoot apex and cotyledon arise from the terminal segment. Graves (1908) recalled this condition in *Ruppia maritima*. These findings remained unrecognized, and the Hanstein-Souèges view held the ground for a long time. The demonstration by Haccius (1952) that both stem tip and cotyledon have a terminal origin led the way for further investigations. J. L. Guignard (1975 b) also admitted the origin of the stem tip from the terminal tier and, accordingly, abandoned the Megarchetype I of the Souèges system of classification. In a recent investigation on the embryogeny of *Cheirostylis flabellata,* an orchid, Swamy (1979) has discussed the current status of the monocotyledonous embryo. He concludes that the basic features concerning the mode of origin of the cotyledon and the epicotylary locus are shared by all monocotyledons and that, essentially, irrespective of the variations noticed, both the cotyledon and the epicotyl are organized from the derivatives of apical cell.

There are very few studies on all aspects of embryogenesis from the time of formation of zygote to the organization of meristems in a single species. Two such studies merit attention: Ramji's (1975, 1976) on *Stellaria,* a dicotyledon, and Lakshmanan's (1977, 1978) on *Commelina*, a monocotyledon. Both these workers have emphasized the processes of change during different phases of embryogenesis leading finally to a completely coordinated, organized morphogenetic system, the forerunner of the future plant.

8.7 The Grass Embryo

Among the monocotyledons, the grasses present some remarkable features in their embryogeny as well as the structure of mature embryo and, therefore, deserve special attention. Within the caryopsis, the mature embryo is placed towards the base on the dorsal side and is relatively small in relation to the endosperm (Fig. 8.11 A, C). The scutellum is a long, shield-like vascularized structure laterally attached to the axis when seen in a sagittal longitudinal section (Fig. 8.11 B, D). The scutellar surface in contact with the endosperm bears a secretory epidermis. The point of attachment of the scutellum to the axis, known as the scutellar node, demarcates the axis into two regions. The upper region is the epicotyl which comprises a shoot apex and leaf primordia. The epicotyl is surrounded by a hollow sheath called the coleoptile (Fig. 8.11 B, C). During germination the coleoptile often becomes green and photosynthetic and the first foliage leaf breaks through it. In some grasses, for example *Triticum* and *Avena* (Fig. 8.11 B), there is an erect flap-like outgrowth devoid of vasculature on the antiscutellar side of the axis and is designated the epiblast; in others such as *Zea* (Fig. 8.11 D) the epiblast is not present. The area between the base of the coleoptile and the point of attachment of the scutellum has been referred to, by some authors, as the mesocotyl. Below the scutellar node the axis terminates in a radicle with a root cap, and is ensheathed at the root pole by a structure called coleorhiza (Fig. 8.11 B, D). The root emerges through the coleorhiza during germination.

The morphology of the different parts of the grass embryo has been the subject of controversy for over 150 years (see Brown 1960 for references). In brief, the scutellum is generally regarded as a cotyledon or part thereof (blade, ligule, bract, sucker), and the coleoptile as the foliage leaf. However, the latter has been variously interpreted as a scutellar sheath, an axillary bud, and the cotyledon itself (see Brown 1960). Opinion is also divided on the nature of mesocotyl which is supposed to represent a node, an internode, a fusion between hypocotyl and scutellum, or the middle part of cotyledon. After a careful analysis of various interpretations, Brown (1960, 1965) concluded that the scutellum, coleoptile, and mesocotyl are not homologous with any organs of the mature plant or other monocotyledons, but have evolved as useful and peculiar structures to the grass embryo alone. This view is also favoured by Swift and O'Brien (1970) in so far as the scutellum is concerned. J. L. Guignard and Mestre (1969, 1970) consider the scutellum to be a cotyledonary leaf transformed during the course of evolution into a suctorial organ with its base elongated along the hypocotyl.

The grass embryo is different from most other monocotyledonous embryos in that the origin of the first root primordium is endogenous. Hence, the coleorhiza has been interpreted as the suppressed primary root, the seminal root it encloses being adventitious (see Brown 1965, J. L. Guignard 1975 b, Philip and Haccius 1976, Tillich 1977). Interestingly, in cultured barley embryos, the scutellum and coleoptile developed leaf hairs showing green pigmentation, whereas the coleorhiza developed root hairs (Norstog 1969).

It has been suggested (see Brown 1960, 1965) that the epiblast denotes the remnant of a second cotyledon; no doubt because it is always positioned opposite the scutellum. Many regard it as an outgrowth of the coleorhiza.

As in most other monocotyledons, the shoot apex of the grass embryo also appears to be lateral. It was pointed out in the previous section that in the light of recent work on monocotyledons, both the stem tip and cotyledon arise from the terminal proembryonic tier, but due to overgrowth of the cotyledon, the stem

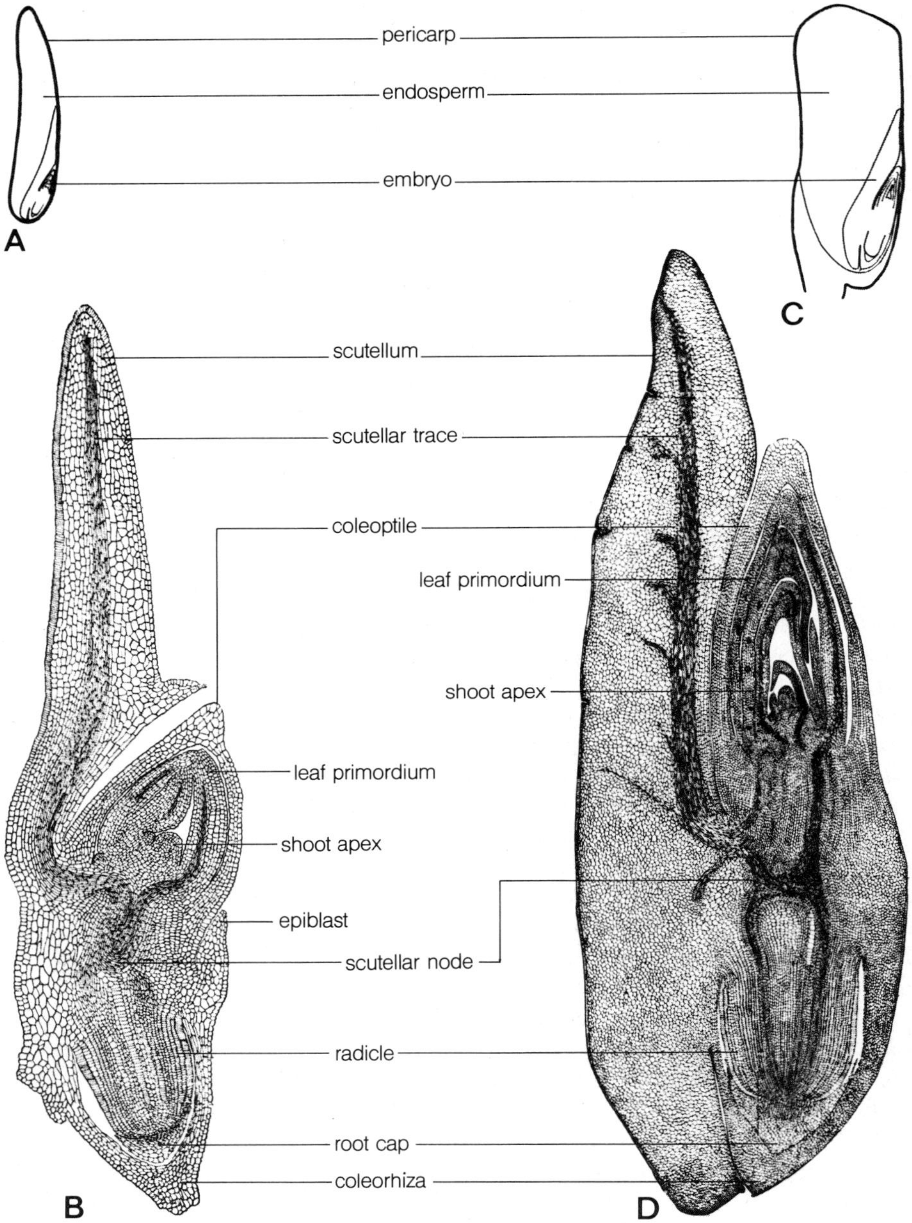

Fig. 8.11 A–D. The grass embryo. **A, B** *Avena sativa;* **C, D** *Zea mays.* **A** and **C** are longisections of caryopsis indicating the position of the embryo; **B** and **D** are longisections of the embryo. (After Avery 1930)

tip comes to occupy a lateral position. According to J. L. Guignard (1975 b), the asymmetry of the embryonic apex in grasses arises from the very beginning through partitioning of the proembryo by oblique walls. Batygina (1969) studied the development of embryo in four species of *Triticum* and she observed oblique walls during early segmentation of cells of the proembryo in all of them. This mode of segmentation continues even in later stages. The scutellum differentiates first, followed by the appearance of the primordia of the coleoptile and shoot apex. On the basis of embryo development, Batygina recognizes a new type of embryogeny – the Graminad Type.

The mature embryo of grasses with its special structures has proved to be of systematic value. Reeder (1957, 1962) investigated about 300 species of grasses representing over 150 genera from all the usually recognized tribes of the family. He recognized four characters of importance: (1) the course of the vascular system (whether the vascular trace to the scutellum and the leaf primordia diverge at the same point, or are separated by a more or less elongated internode), (2) the epiblast (whether present or absent), (3) the lower part of scutellum (whether free from the coleorhiza, or fused with it), and (4) the cross section of embryonic leaf (whether having many bundles and overlapping margins, or fewer bundles and margins which merely meet). The grass embryo conforms to two basic types, the "festucoid" and the "panicoid". Using embryo characters as the principal criteria, Reeder (1962) recognized six groups of grasses (festucoid, bambusoid – including oryzoid-olyroid, centothecoid, arundinoid-danthonioid, chloridoid-eragrastoid, and panicoid). The embryo of grasses has proved to be of promise at lower levels of taxonomic hierarchy also. Barkworth (1982) who investigated 39 species of Stipeae in order to distinguish *Stipa* from *Oryzopsis* found the nature of the epiblast and the length/width ratio of the embryo (as seen in ventral view) to be useful parameters. The epiblast is useful at the infrageneric level also. Tateoka (1964) classified the genus *Oryza* into three groups, based on the presence or absence of auricles, and the nature of the epiblast. The shape of the epiblast was employed by Baum (1968, 1978) to distinguish *Avena sativa* from *A. fatua,* and *Hordeum brachyantherum* from *H. secalinum.*

8.8 The Embryo in Palms

Judging from the literature, there is a good deal of disagreement on the embryogenesis in palms. Selvaratnam (1952) compared the palm embryo with the embryo of grasses, and suggested that it has originated from the latter. This, of course, is unlikely as palm embryos have neither a coleoptile nor a coleorhiza (von Guttenberg 1957).

The relative contribution of the apical and basal cell to the embryo proper is one of the features on which there is difference of opinion. On the one hand, in *Areca, Pritchardia, Actinophloeus,* and *Howea* (Rao 1955 a, b, 1956, 1959 a, b) the basal cell does not contribute to the formation of the embryo (Onagrad Type of Johansen). On the other hand, in *Chamaerops* (J. L. Guignard 1961), *Phoenix* (Biradar 1968, Biradar and Mahabale 1968, Mahabale and Biradar 1968),

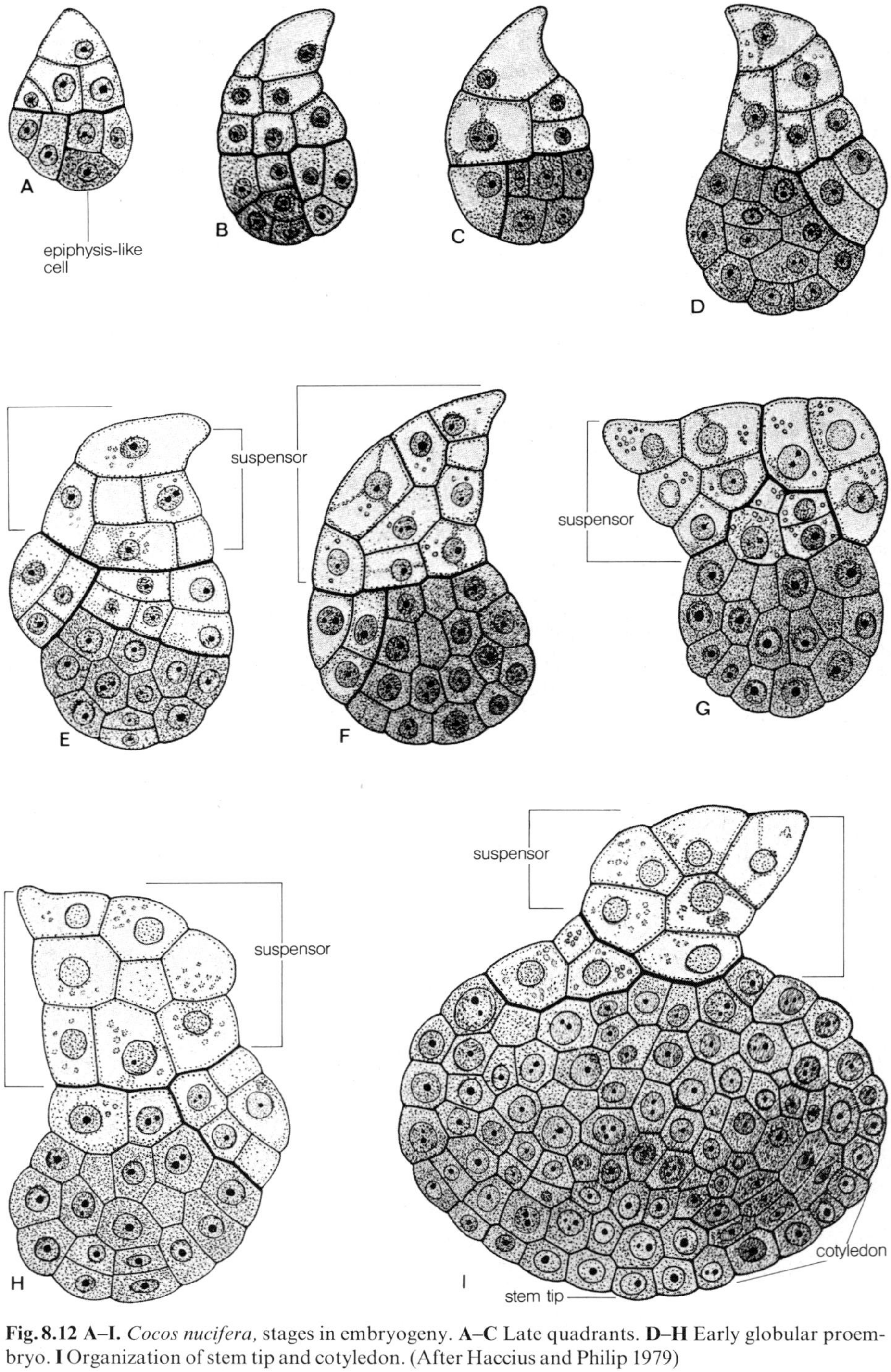

Fig. 8.12 A–I. *Cocos nucifera,* stages in embryogeny. **A–C** Late quadrants. **D–H** Early globular proembryo. **I** Organization of stem tip and cotyledon. (After Haccius and Philip 1979)

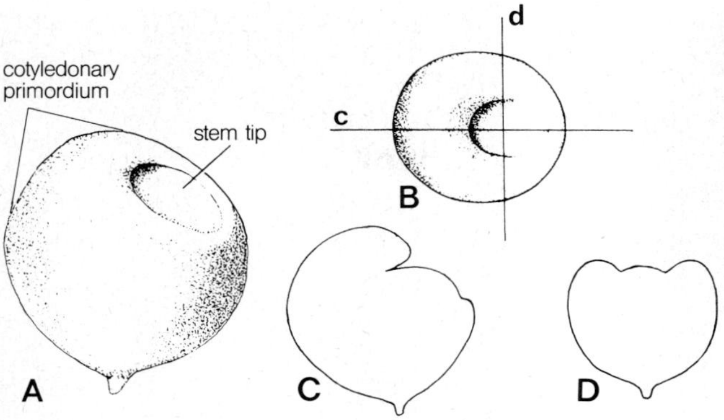

Fig. 8.13 A–D. *Cocos nucifera*. **A** Excised young embryo with cotyledonary primordium surrounding the stem tip. **B** Young embryo indicating planes *c* and *d* along which longisections can be cut to obtain configurations as in **C** and **D.** (After Haccius and Philip 1979)

Caryota (Shirke and Mahabale 1973), *Livistona* (Kulkarni and Mahabale 1974), and *Jubaeopsis* (Robertson 1976) the basal cell partially contributes to the construction of the embryo (Asterad Type of Johansen). However, in *Cocos* Haccius and Philip 1979) the entire embryo originates from one of the daughter cells of *ca* while its sister cells, along with the derivatives of *cb,* form the suspensor (Fig. 8.12 A–I). This cell is a wedge-shaped daughter cell resembling the "epiphysis initial" (Fig. 8.12 B) or, less often, a quadrant (Fig. 8.12 C, D).

The origin of stem tip is another point of contention. In *Chamaerops* (J. L. Guignard 1961) and *Jubaeopsis* (Robertson 1976) a lateral origin was indicated. However, in most of the investigated species the stem tip is considered to have a terminal origin from an epiphysis initial (Rao 1955 a, b, 1956, 1959 a, b, Biradar 1968, Biradar and Mahabale 1968, Mahabale and Biradar 1968, Kulkarni and Mahabale 1974). A terminal origin, but not from epiphysis, has been observed in *Cocos* (Haccius and Philip 1979).

Certain workers on embryogenesis of palms (Biradar 1968, Biradar and Mahabale 1968, Mahabale and Biradar 1968, Kulkarni and Mahabale 1974) have claimed that the embryo develops two cotyledonary primordia of which one overgrows and, consequently, overarches the other. Haccius and Philip (1979) made a thorough investigation of the origin of the cotyledon in *Cocos* and, according to them, the above conclusion is not well founded as it is based solely on the examination of serial longisections without supplementing these with observations of whole embryos under the microscope. Figure 8.13 demonstrates how the same embryo, when viewed in two different planes of sectioning (Fig. 8.13 B), can give the erroneous impression that there are two cotyledons of more or less different size (Fig. 8.13 C, D). On the basis of their own work, as well as a survey of published literature, Haccius and Philip feel that embryogenesis in palms ought to be more uniform than hitherto believed. A fresh look at some of the above taxa is desirable.

8.9 Ultrastructural and Cytochemical Aspects

Embryologists have been aware for long that cell division and cell destiny play a cardinal role in embryogenesis and, therefore, a large body of information has

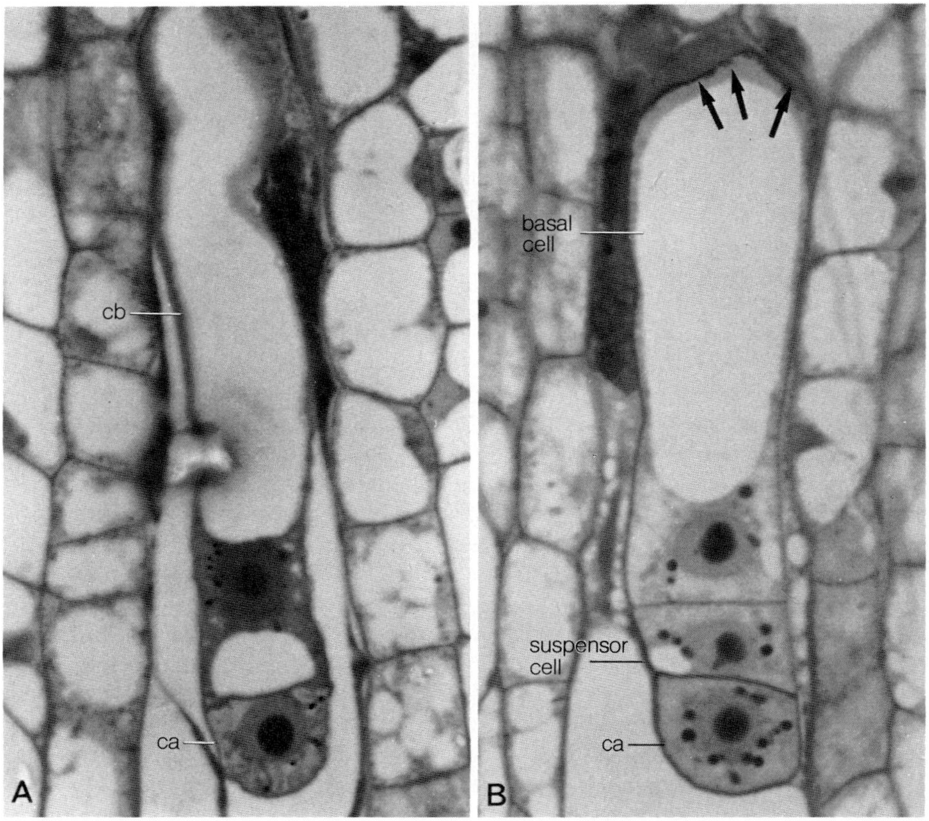

Fig. 8.14 A, B. *Capsella bursa-pastoris*. **A** Two-celled proembryo with a small, richly cytoplasmic apical cell (*ca*), and a large vacuolate basal cell (*cb*) which stains more intensely for protein. **B** Three-celled proembryo, apical cell stains more intensely than suspensor cells; note starch grains around the nuclei and wall ingrowths (*arrows*) at micropylar pole of basal cell. (After R. Schulz and Jensen 1968b; photographs courtesy Professor W. A. Jensen)

accumulated on the precise sequence of cell segmentation and organogenesis at the light microscope level. However, interest in the subcellular and molecular changes underlying these processes is relatively recent, and has resulted from the application of newer tools and techniques, such as electron microscopy, autoradiography, and cytochemistry among others. Admittedly, the number of taxa so studied is still very small, but there is enough to excite interest.

The apical cell and the basal cell resulting from the division of the zygote are cytologically as well as biochemically distinct from one another, and exhibit an amplification of the reorganization and polarity that characterized the egg following fertilization. The two cells exhibit a marked difference in their ultrastructure. The apical cell is generally more electron-dense than the basal cell, and probably contains more ribosomes per unit volume. Vacuoles are smaller and fewer in the apical cell but are larger in the basal cell in *Gossypium* (Jensen 1964), *Cap-*

sella (Fig. 8.14 A) (R. Schulz and Jensen 1968 b), *Linum* (D'Alascio-Deschamps 1973), *Quercus* (Singh and Mogensen 1975), and *Petunia* (Vallade 1980). In cotton, the plastids are perinuclear in the apical cell but their distribution is casual in the basal cell. Plastids are almost completely absent in the zygote of *Quercus gambelii* (Mogensen 1972), but they make a marked appearance and gradually increase in number through the 6- to 12-celled stage (Singh and Mogensen 1975). Although typically the zygote has plasmodesmatal connections with the synergids and the central cell, these connections are completely severed in the two-celled proembryo signalling, no doubt, a change in the nature of physiological interactions with the immediate milieu. However, plasmodesmata can be observed on the common wall separating the two cells of the proembryo. In cotton, the tube-containing ER present in the zygote (Jensen 1968 a, c) gets sequestered into the basal cell rather than the terminal cell (Jensen 1964).

As described earlier, the subsequent course of development depends upon the pattern of embryogeny specific to a particular species. At present, there is very little ultrastructural information on the sequence of developmental changes in a young proembryo. In *Capsella,* which follows the Onagrad pattern of development, the basal cell divides transversely (Fig. 8.14 B), and the daughter cells again undergo transverse divisions forming a long, filamentous suspensor (Fig. 8.15 A). The cell nearest to the micropyle becomes large and vesicular (P. Schulz and Jensen 1969), and is referred to as the basal cell of suspensor (Fig. 8.15 A; see also Fig. 8.20 B). This cell becomes highly vacuolate with a large, somewhat lobed nucleus. The apical cell, in the meanwhile, undergoes a vertical division (Fig. 8.15 A) and, after the quadrant and octant stages, produces a globular proembryo (Fig. 8.15 B–D).

During early stages of development the daughter cells become successively smaller in size because the embryo, as a whole, either increases slowly in size, may not increase at all, or actually decreases in size. In the embryo of cotton, Pollock and Jensen (1964) have shown that cell-size in the embryo shows a progressive decrease until, at the 100-cell stage, average cell size is about one-twentieth that of the zygote; thereafter cell size remains constant until the 1,000-cell stage when there is a further reduction by one-half. This feature appears to be characteristic of other species such as *Capsella* (R. Schulz and Jensen 1968 a) and *Quercus* (Singh and Mogensen 1975).

In a young proembryo the cells of the embryo proper contain few dictyosomes and scanty ER, but a large amount of polysomes. Mitochondria and plastids are common and the proembryo stains intensely for protein and nucleic acid. In *Capsella,* and in *Diplotaxis* (Simoncioli 1974), even when the octant cells form a protoderm, there is insignificant ultrastructural difference between the different cells of the embryo proper. However, a sharp contrast can be observed between cells of the embryo proper and those of the suspensor (Figs. 8.16 and 8.20). In *Hordeum* there is little or no indication of cytological distinction between the cells of the suspensor and the embryo proper, except for the progressive decrease in the number of vacuoles towards the suspensor in a young proembryo (Norstog 1972). The hypophysis bears ultrastructural resemblance to the suspensor rather than the embryonal cells. The cells of hypophyseal origin in *Capsella, Diplotaxis* (Fig. 8.16), and *Linum* have fewer ribosomes, more ER, and larger vacuoles than

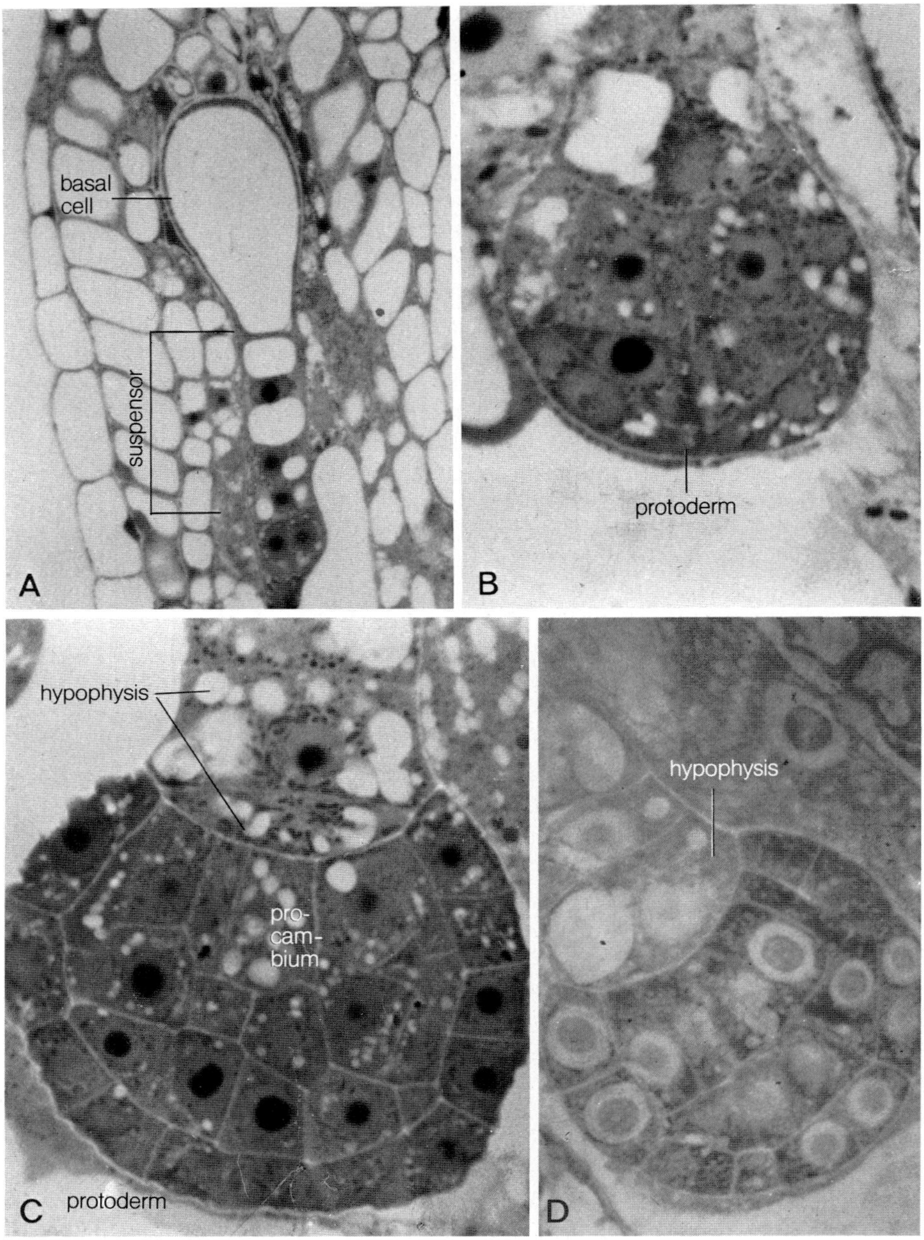

Fig. 8.15 A–D. *Capsella bursa-pastoris,* distribution of protein and nucleic acids during early embryogenesis. **A** Two juxtaposed derivatives of apical cell stain more intensely for protein than suspensor and its basal cell. **B** 16-celled proembryo in which the protoderm has been demarcated; all the cells exhibit similar staining reaction for protein. **C** Early globular embryo with protoderm, ground meristem, and future procambium showing similar distribution of protein. The hypophysis has produced a lenticular derivative. **D** Nucleic acids show a sharp contrast in staining, between embryonal region and rest of the proembryo. (After R. Schulz and Jensen 1968 b; photographs courtesy Professor W. A. Jensen)

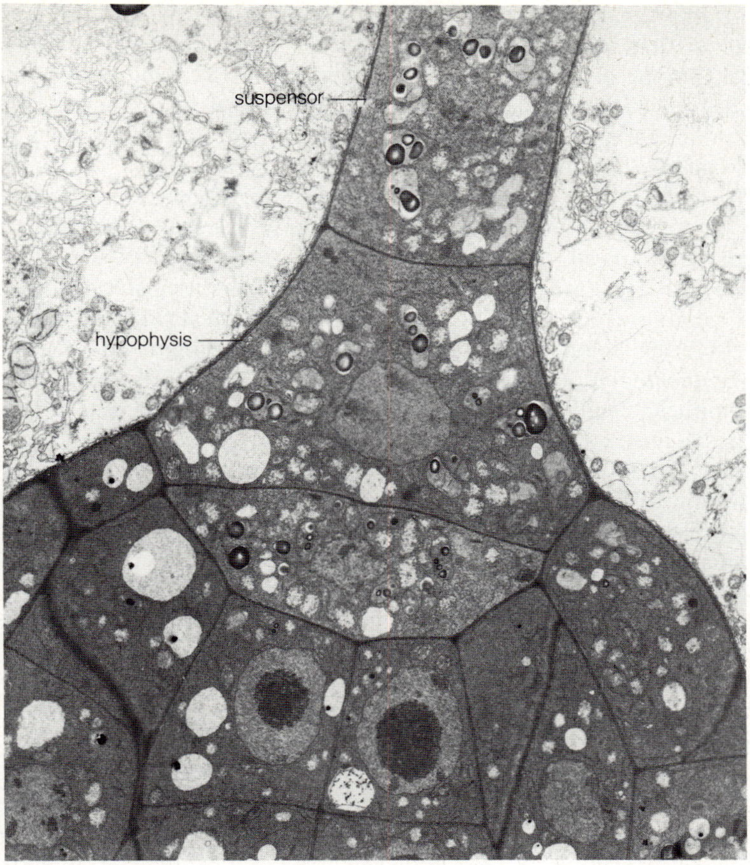

Fig. 8.16. *Diplotaxis erucoides,* ultrastructure of part of globular proembryo. Embryonal cells have a higher ribosome density than suspensor cells. The hypophysis has produced a lenticular derivative towards the embryonal cells. The hypophysis bears greater ultrastructural resemblance to the suspensor. (After Simoncioli 1974; electron micrograph courtesy Professor E. Pacini, Siena, Italy)

those of the embryo, and frequently contain starch and lipid. Plasmodesmatal connections are present in the common walls separating adjacent suspensor or embryonal cells. Often, the peripheral cells of the developing proembryo have thicker and more electron-dense walls on their outer tangential side as in *Linum* (Deschamps 1969) and *Helianthus* (Newcomb 1973).

Ultrastructural differences within the embryo become evident at the heart-shaped stage. The cells of the procambium and ground meristem are more vacuolate than those of protoderm, but the density of ribosomes is more or less comparable in all the regions. Plastids of protoderm and ground meristem show differentiation of lamellae, while those of procambial cells remain relatively undifferentiated.

During further growth the hypocotyl and the cotyledons elongate; the apical meristem of the epicotyl is organized. The root apical meristem is also formed at the radicular end. In several taxa the plastids in the embryonal cells differentiate into chloroplasts so that the embryo becomes green. Yakovlev and Zhukova (1980) surveyed 1,094 species belonging to 666 genera from 182 families of both dicotyledons and monocotyledons, and noted that 428 species belonging to 224 genera and 72 families have green embryos. The authors suggested that this character could be of systematic value.

The ultrastructural features observed during early embryogenesis are closely associated with cytochemical changes as well. Generally, the apical cell stains more darkly than the basal cell for ribonucleic acid and protein as in cotton (Jensen 1964), *Vanda* (Alvarez and Sagawa 1965), *Linum* (D'Alascio-Deschamps 1973), and *Coix* (Shah and Bhatt 1974). However, in *Stellaria* (Pritchard 1964), *Capsella* (Fig. 8.14 A) (R. Schulz and Jensen 1968 b), *Quercus* (Singh and Mogensen 1975), and *Limnophyton* (Shah and Pandey 1978) the basal cell appears to contain more ribonucleic acid and protein than the apical cell. In *Capsella* the staining reaction is reversed at the subsequent division of the basal cell (Fig. 8.14 B). It may be noted that in both *Stellaria* and *Limnophyton,* the basal cell never divides but enlarges into a suspensor cell and the stronger cytochemical reaction may be related to this fact. No dramatic differences with respect to RNA and protein distribution have been noted among the embryonal cells during the next few divisions (Fig. 8.15 A, B). In a globular proembryo, all the cells show equal staining, but as in the ultrastructure, there is a notable difference between the suspensor and proembryonal cells (Fig. 8.15 C, D). The hypophysis shows a cytochemical affinity with the cells of the suspensor (Fig. 8.15 C, D).

In the heart-shaped embryo an enhanced accumulation of nucleic acids and protein can be seen in the cotyledonary primordia. Derivatives of the hypophysis stain lightly and cells of suspensor weakly (Fig. 8.17 A, B). The radicular pole also records an increased staining intensity in *Capsella* (R. Schulz and Jensen 1968 a), *Lens, Myosurus, Alyssum* (Rondet 1958, 1961, 1962), and *Petunia* (Vallade 1970).

The difficulty experienced in localizing nuclear DNA within the egg by tinctorial methods is not encountered during embryogenesis. Pritchard (1964) observed nuclear DNA in early phase embryos of *Stellaria media.* He noted a decrease in nuclear staining in the 30- to 60-celled proembryo followed by the appearance of Feulgen-positive granules in the cytoplasm. In cotton nuclear DNA level remained more or less constant during embryogenesis with practically no cytoplasmic DNA (Fisher and Jensen 1972), although earlier Yoo and Jensen (1966) had demonstrated a decrease in total DNA as well as the presence of cytoplasmic DNA. Woodard (1956) also noticed similar differences in the content of DNA in the embryos of different ages in *Tradescantia.* Vallade et al. (1978) measured cytophotometrically the amount of DNA in the zygote, as well as in the first two cell generations of the proembryo in *Petunia hybrida.* There was an increase in DNA level, reaching nearly a 3C value, and the whole of this DNA was replicated during the real S-phase. However, during the first two cell generations, there was a fall in this level up to 5.92C in the G_1 zygote nucleus, and 4.94C in the G_2 nuclei of four-celled proembryo. Considered together with the fact that there is a decrease in nuclear and nucleolar volume during the first two cell generations, the authors suggested that there exists a metabolic DNA synthesized after fertilization, and that its quantity would decrease during the early stages of

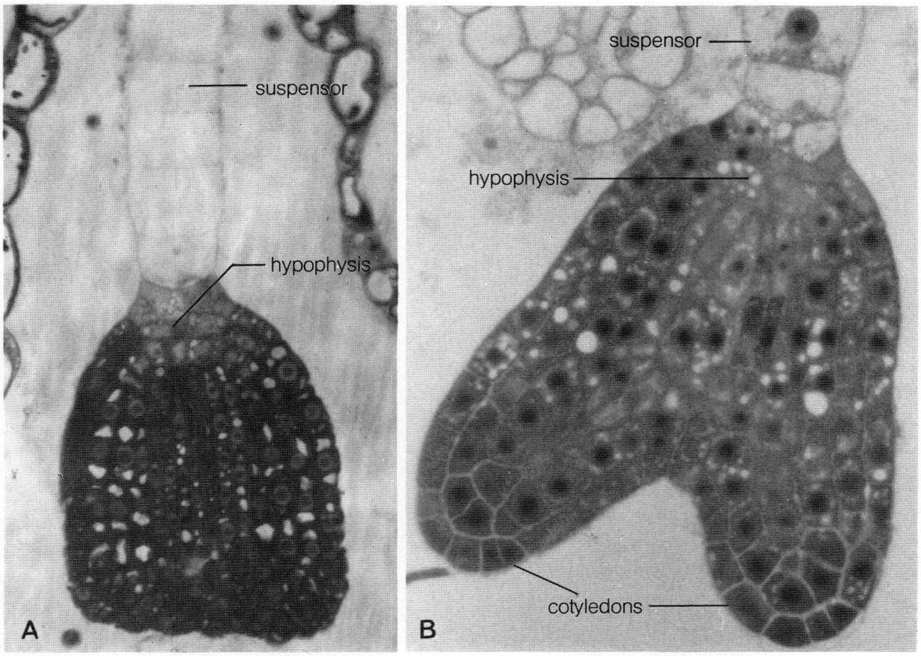

Fig. 8.17 A, B. *Capsella bursa-pastoris*, distribution of nucleic acid and protein in heart-shaped embryo. **A** Embryonal cells stain intensely for nucleic acids; derivatives of hypophysis less intensely and suspensor very faintly except for nucleoli which react strongly to RNA. **B** Protoderm, procambium, and ground meristem exhibit histochemical differences. Tips of cotyledons stain very intensely for protein, derivatives of hypophysis stain less intensely and suspensor very faintly. (After R. Schulz and Jensen 1968 a; photographs courtesy Professor W. A. Jensen)

embryogenesis. Vallade (1980) studied the duration of mitotic cycle, and the pattern of DNA synthesis in *Petunia hybrida*. In a four-celled proembryo the tier *l'* has the shortest G_1-phase and, hence, a faster mitotic cycle followed by rapid DNA synthesis compared with tiers *l*, *m*, and *ci*.

As the embryo matures, the cells of the embryo stop proliferating and are arrested in either the G_1 (presynthetic) phase or the G_2 (postsynthetic) phase of the cell cycle. Therefore, their nuclei may contain the presynthetic value (2C amount) or the postsynthetic value (4C amount) of DNA. In such a taxon as *Lactuca* (Brunori and D'Amato 1967) all the cells remain arrested in the G_1 and are of 2C type, whereas in others such as *Zea* (Stein and Quastler 1963), *Vicia* (Brunori 1967, Miah and Brunori 1970), *Triticum* (Avanzi et al. 1969), and *Pisum* (Jones 1977), the cells remain arrested both in G_1 and G_2 and are of 2C as well as 4C type. It would be interesting to know whether the discrepancy in the DNA amount is reflective of the specific developmental destinies of the different regions of the embryo.

Several investigators have directed their attention towards the localization of enzymes during embryogenesis. Enzymes are known to catalyze biochemical reactions in cell systems. Given the specificity of their reaction, the appearance

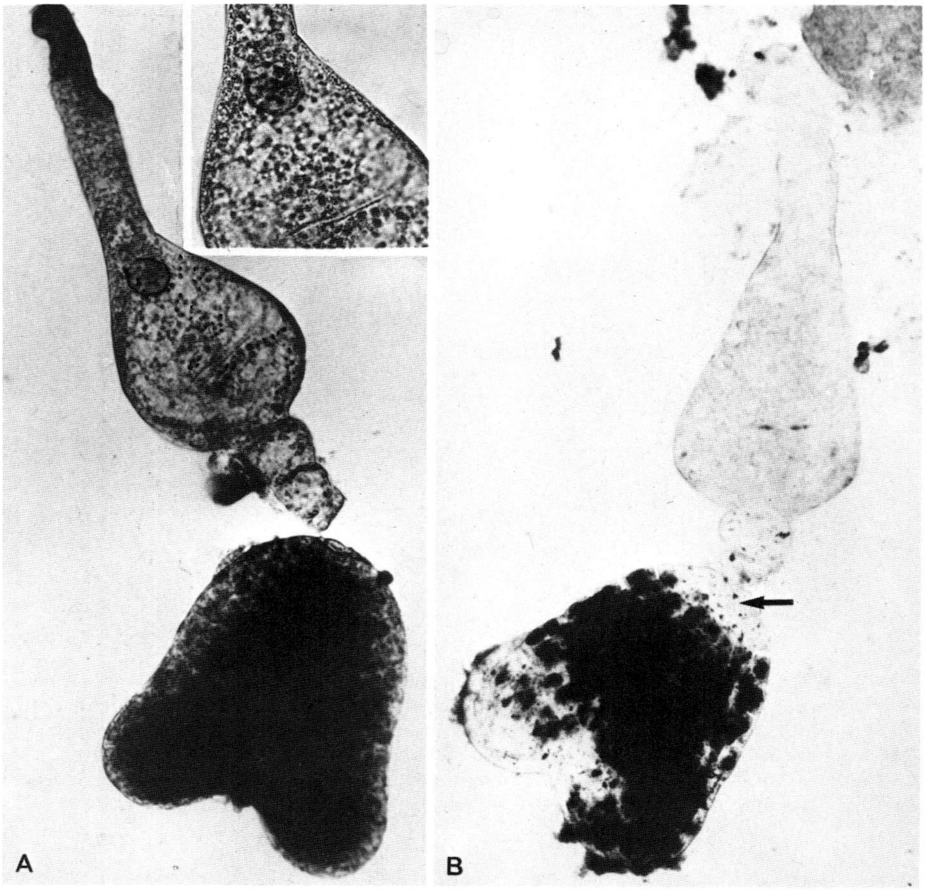

Fig. 8.18 A, B. *Stellaria media,* distribution of alkaline and acid phosphatase in heart-shaped embryo. **A** Alkaline phosphatase is present both in embryonal and suspensor portions, and is uniformly distributed in embryonal cells; in suspensor cells its activity is associated with proteinoplasts (*inset*). **B** Acid phosphatase is localized in protodermal cells of embryonal mass, but is absent in hypophysis (*arrow*) and suspensor. (After Pritchard and Bergstresser 1969; photographs courtesy Professor H. N. Pritchard, Bethlehem, USA)

and disappearance of enzymes during successive developmental stages at definite tissue sites could give us insights into the physiological mechanisms directing morphogenesis.

Hydrolytic enzymes, notably phosphatases and oxidative enzymes, chiefly oxido-reductases, have been investigated. Phosphatases are known to hydrolyze phosphate esters. Alkaline phosphatase has been localized in the embryos of *Stellaria media* (Pritchard and Bergstresser 1969). This enzyme occurs both in the embryo and suspensor (Fig. 8.18 A). Whereas in the embryonal cells the concentration of enzyme is uniform, in the suspensor region it is confined to the proteinoplasts (see inset in Fig. 8.18 A). On the contrary, acid phosphatase activity is mini-

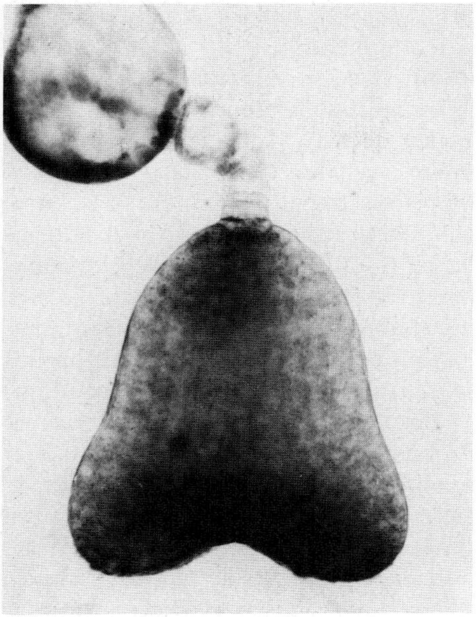

Fig. 8.19. *Stellaria media,* distribution of cytochrome oxidase in heart-shaped embryo. Although present in all cells of the embryo and suspensor, the enzyme is mainly concentrated at the tips of cotyledons, root meristem, and procambial strand. (After Pritchard and Bergstresser 1969; photographs courtesy Professor H. N. Pritchard)

mal in the suspensor, when compared with embryo proper (Fig. 8.18 B). Within the embryo most of the enzyme occurred in the cells of protoderm. However, the hypophysis did not show any enzyme activity. Acid phosphatase has also been localized in *Cocos* (Wilson and Cutter 1952, 1953). The earliest appearance of the enzyme is in the cells which form the future cotyledonary site; later the enzyme activity is confined to the epidermis and vascular trace of cotyledon. In the mature embryo the dermatogen, root, and shoot primordia, and the vascular trace showed enzyme activity. Acid phosphatase first appears in scutellum during embryogenesis in rice (Mizushima et al. 1955). The early appearance of acid phosphatase in the cotyledonary site in *Cocos* and the scutellum in rice points to its role in absorption as both these organs are considered to be haustorial in function. Raghavan (1976) studied the distribution of acid phosphatase in *Capsella bursa-pastoris.* The globular proembryo is the earliest stage showing evidence of activity, and the enzyme shows a uniform distribution. At the heart-shaped stage, the enzyme is preferentially localized in the cotyledons and root apex. In older embryos, the enzyme is mainly confined to the root and shoot apex, and cotyledons.

Of the oxidative enzymes, succinic dehydrogenase is a mitochondrial enzyme and, generally, indicates the sites of high respiratory activity. Forman and Jensen (1965) and Zinger and Poddubnaya-Arnoldi (1966) studied its distribution in the embryos of cotton and *Cypripedium.* Distribution of cytochrome oxidase, another key oxidative enzyme, has been investigated in *Stellaria* (Pritchard and Bergstresser 1969) and, predictably, its activity is higher in the meristematic regions (Fig. 8.19). *Capsella* (Raghavan 1976) shows a similar pattern. Bhalla et al. (1979) localized several respiratory enzymes in the globular, heart-shaped and mature embryos of *Tropaeolum majus*. They observed that the highest activity of the enzymes coincided with the globular stage of development.

Another enzyme involved in oxidation-reduction reactions is peroxidase which is supposed to regulate auxin levels in tissues (Galston and Davies 1969). Alvarez and King (1969) observed the highest activity of this enzyme in *Vanda* at younger stages of development, and progressively less during later stages. Using the technique of gel electrophoresis, they noticed that the number of isoenzyme bands also decreased during the development of the embryo. The results indicate that IAA levels are kept low during younger stages of development, whereas its concentration is allowed to increase during later stages.

What, in conclusion, have ultrastructural and histochemical studies on embryogenesis revealed? While these investigations confirm the differential morphogenetic potentialities of the apical and basal cell, already known from conventional light microscopic studies, we really have not learnt much regarding the causal factors that bring about differentiation in a homogeneous mass of proembryonal cells. Changes in total RNA or protein content as observed by qualitative or quantitative cytochemistry are, in all probability, likely to be portrayed at more subtle levels of expression (see also Raghavan 1976). It is also possible that the "position effect", or where exactly a cell is located within the developing embryo, may be crucial to its differentiation (R. Schulz and Jensen 1968a). Future investigations must be directed towards this problem.

8.10 Suspensor: Structure and Function

So far, attention has mainly been focussed on the development of the embryo proper, and not much has been said about the suspensor. For a long time the suspensor has been assumed to be an organ which merely pushes the embryo proper in a nutritionally more favourable medium of the endosperm (Maheshwari 1950), but a rethinking of its role has become necessary in the light of recent work. There is a great deal of variation in the organization of the suspensor. In some taxa as *Lycopsis, Penaea, Tilia,* and *Viola* there is no suspensor at all. A reduced suspensor characterizes *Bryonia, Euphorbia, Myosotis,* and *Ruta,* among others. Usually, in taxa in which the endosperm produces micropylar haustoria, the sus-

Fig. 8.20 A–K. Variation in organization of suspensor. **A** *Pedicularis silvatica,* the suspensor becomes nearly 20-cell long. (After Berg 1954). **B** *Dicraea stylosa,* the basal cell enlarges and sends out haustorial branches into neighbouring regions. (After Mukkada 1962). **C** *Orobus angustifolius,* the suspensor comprises large coenocytic cells in a biseriate filament. **D** *Cytisus laburnum,* the suspensor appears like a bunch of grapes. (After L. Guignard 1882). **E** *Dendrobium nobile,* seed enclosing a young proembryo with highly enlarged suspensor cell. (After Poddubnaya-Arnoldi 1976). **F–H** *Cymbidium bicolor,* the suspensor cells are tube-like. (After Swamy 1942). **I** *Haloragis micrantha,* the basal cell of the two-celled proembryo divides longitudinally to form a pair of juxtaposed cells which enlarge markedly and function as haustorial cells. They persist even in the mature embryo. (After Nijalingappa (1975). **J** *Machaerocarpus californicus,* the undivided basal cell enlarges into a vesicular structure containing hypertrophied nucleus. (After Maheshwari and Singh 1943). **K** *Cuscuta reflexa,* the suspensor has 2–4 vacuolate and multinucleate cells with up to 42 nuclei in each cell. (After Johri and Tiagi 1952)

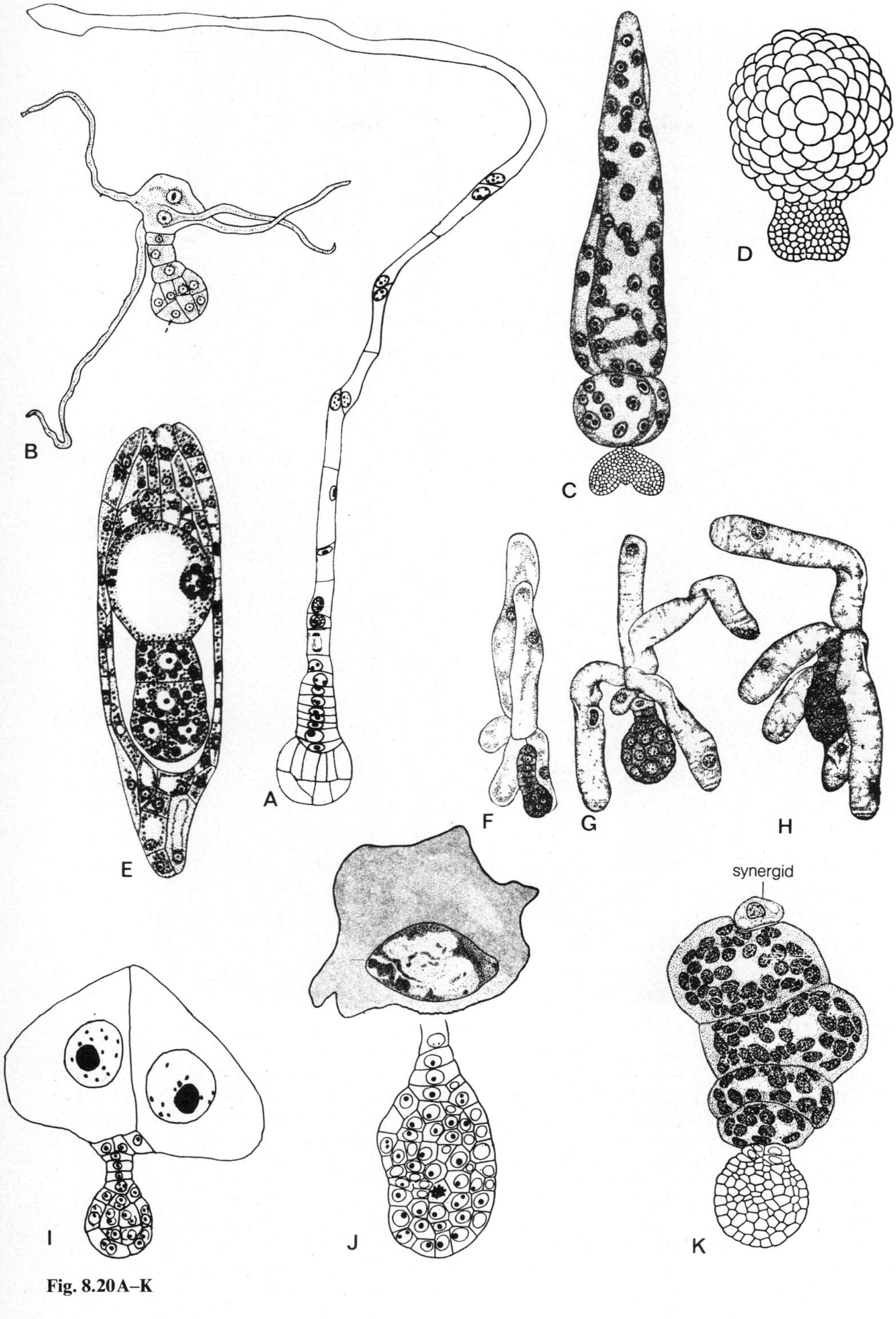

Fig. 8.20 A–K

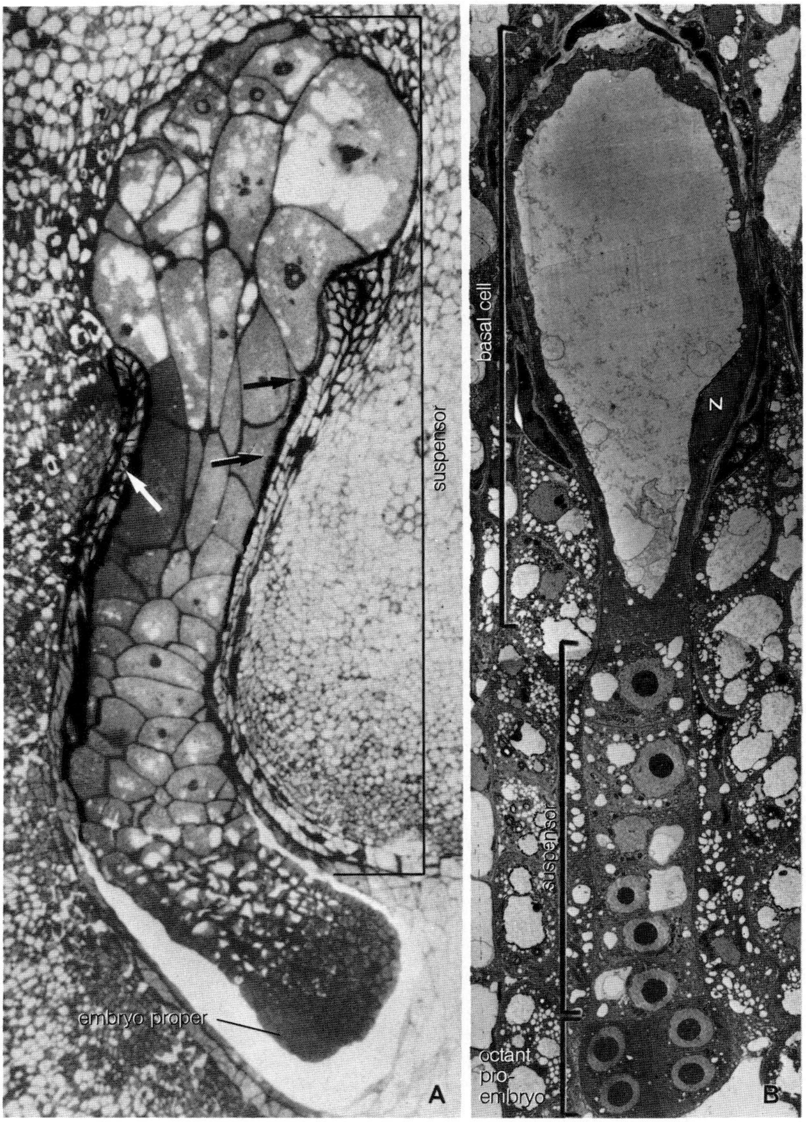

Fig. 8.21 A, B. Embryo and suspensor. **A** *Phaseolus coccineus,* early heart-shaped embryo with differential staining between embryo proper and massive suspensor; wall ingrowths (*arrows*) are conspicuous in suspensor cells; there is progressive increase in nuclear size towards the basal region of suspensor. (After Yeung and Clutter 1978; photograph courtesy Dr. E. C. Yeung, New Haven, USA. **B** *Capsella bursa-pastoris,* octant proembryo and six-celled suspensor with a large basal cell. The uppermost suspensor cell (basal cell) is very large and highly vacuolate; *N* nucleus. (After P. Schulz and Jensen 1969; photograph courtesy Professor W. A. Jensen)

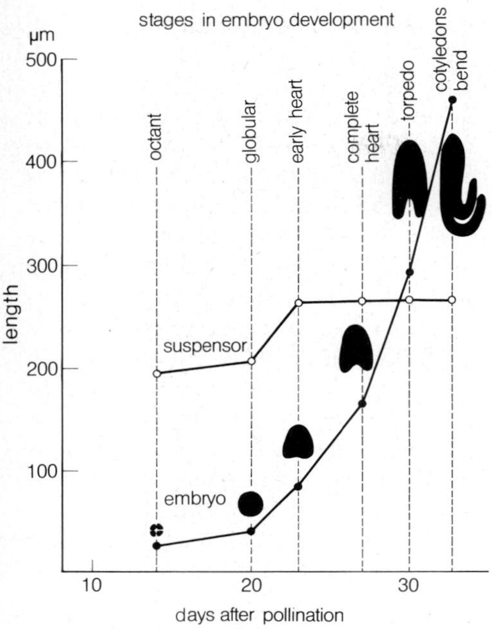

Fig. 8.22. *Diplotaxis erucoides,* comparative growth of suspensor and embryonal region during embryogenesis. (After Simoncioli 1974)

pensor is not very prominent and is usually short-lived. A long filamentous suspensor is characteristic of the crucifers; *Pedicularis* of Scrophulariaceae (Berg 1954) also has a long suspensor of about 20 cells (Fig. 8.20 A). In Crassulaceae, Fumariaceae, Podostemaceae (Fig. 8.20 B), Rubiaceae, Trapaceae, and Tropaeolaceae the suspensor becomes massive and haustorial. The family Fabaceae exhibits a wide variation in the organization of the suspensor (L. Guignard 1882, Rau 1955, Behl and Tiagi 1978) – from its virtual absence to very massive structures which may be filamentous (Fig. 8.20 C), or clustered like a bunch of grapes (Fig. 8.20 D). Orchidaceae are also characterized by a variable structure of the suspensor (Swamy 1949, Veyret 1974), and two types are illustrated in Fig. 8.20 E–H. Other interesting examples are shown in Fig. 8.20 I–K. The cells of the suspensor stain relatively sparsely compared to the embryo proper (Fig. 8.21 A, B).

In spite of its impressive diversity, the suspensor has not been given the attention it justly deserves. In recent years there have been a number of noteworthy studies: (1) ultrastructural (Clutter and Sussex 1968, P. Schulz and Jensen 1969, Schnepf and Nagl 1970, Ponzi and Pizzolongo 1972, 1973, Newcomb 1973, Newcomb and Fowke 1974, Simoncioli 1974, Nagl 1976a, b, Yeung and Clutter 1979), (2) cytochemical (Pritchard 1964, Pritchard and Bergstresser 1969, Malik et al. 1977, Shah and Pandey 1978), (3) biochemical (Walbot et al. 1972, Sussex et al. 1973, Clutter et al. 1974, Singh et al. 1980, Yeung 1980), (4) physiological (Alpi et al. 1975, Przybyllock and Nagl 1977, Lorenzi et al. 1978, Alpi et al. 1979), and (5) embryo culture (Corsi 1972, Cionini et al. 1976, Nagl 1974, Yeung and Sussex 1979). These studies have indicated that, in a number of angiosperms, the suspensor plays a more active and dynamic role than has been hitherto assumed. Some of these aspects have been reviewed by Raghavan (1976), Malik and Singh (1979), and Malik et al. (1980).

The suspensor comprises a numerically stabilized population of cells, whereas the embryo proper consists of a mitotically dividing population of cells which at later stages diversify by histodifferentiation (see Nagl 1974). The growth rate of the suspensor is usually faster during earlier stages of embryogeny. In *Diplotaxis* (Simoncioli 1974), for instance, the suspensor already has its maximal cell number between four and seven when the proembryo is still an octant. The growth is very rapid between the globular and the heart-shaped stages of development (Fig. 8.22). Thereafter, further increase in length ceases, and during later stages the suspensor begins to degenerate. The average length of the suspensor in this taxon is 200 μm, while that of the embryo is 40 μm. The faster rate of growth during early development may be accompanied by an increase in fresh weight as well. In *Phaseolus coccineus* (Yeung and Clutter 1978) the suspensor is heavier than the embryo for the first 6 days of development, but the rate of weight increase is slower in the suspensor. There is negligible increase in weight after the 11th day. After the 6th day the embryo weighs more than the suspensor. The suspensor in *P. coccineus* comprises nearly 200 cells which are all present by the 8th day. The maximum length attained by the suspensor is 1,500 μm while the mature embryo measures 2,200 μm.

Ultrastructurally, the supensor cells are usually filled with small vacuoles, are less electron-dense, and contain fewer ribosomes than the cells of the embryo proper. The cell walls separating individual suspensor cells are traversed by numerous plasmodesmata in *Capsella* (P. Schulz and Jensen 1969), *Helianthus* (Newcomb 1973), *Stellaria* (Newcomb and Fowke 1974), and *Diplotaxis* (Simoncioli 1974) (Fig. 8.23), but there is no protoplasmic continuity between the suspensor and the central cell. The suspensor cells, in several plants including *Phaseolus* (Schnepf and Nagl 1970, Yeung and Clutter 1979) and *Ipomoea purpurea* (Ponzi and Pizzolongo 1972), exhibit an abundance of stacked agranular endoplasmic reticulum and free ribosomes, but in *Helianthus* (Newcomb 1973) and *Diplotaxis* (Simoncioli 1974) the ER is rough. Typical undifferentiated leucoplasts occur in the suspensor cells of *Capsella, Diplotaxis,* and *Helianthus*. However, in *Pisum* (Marinos 1970), *Phaseolus* (Schnepf and Nagl 1970, Yeung and Clutter 1979), *Ipomoea* (Ponzi and Pizzolongo 1973), and *Tropaeolum* (Nagl and Kuhner 1976) the plastids are peculiar. These plastids, in most cases, are not present in the embryo proper. In *Phaseolus vulgaris* the plastids exhibit remarkable ultrastructural changes around the late heart-shaped stage of the embryo (Schnepf and Nagl 1970). They swell at the polar regions, and enclose portions of cytoplasm containing ER, polysomes, and lipid bodies. In the closely related *Phaseolus coccineus* (Yeung and Clutter 1979) the plastids have very few internal membranes. In *Ipomoea* (Ponzi and Pizzolongo 1973) the plastids exhibit a complex structure. They are often amoeboid with enclosed islands of cytoplasm, and their internal membranes comprise prolamellar bodies or tubules. Besides starch they contain phytopherritin, osmiophilic globules, and proteinaceous bodies. The plastids of *Pisum* (Marinos 1970) are similar. In *Tropaeolum* (Nagl and Kuhner 1976) there is a great diversity in the ultrastructure of plastids in various regions of the suspensor and embryo proper. The exact role of such unusual plastids is not yet de-

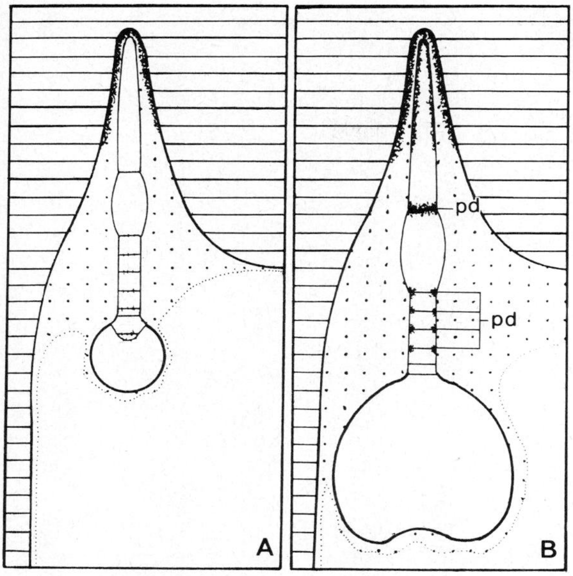

Fig. 8.23 A, B. *Diplotaxis erucoides*, diagrammatic representation of globular proembryo (**A**) and heartshaped embryo (**B**). Plasmodesmata (*pd*) occur on the transverse walls separating individual suspensor cells. (After Simoncioli 1974)

termined. Ultra-cytochemical studies indicate acid phosphatase activity in the plastids of *Phaseolus vulgaris* (Nagl 1977, Gartner and Nagl 1980). Such "plastolysomes" are converted into simple autophagic vacuoles during late stages, and it has been suggested that they are involved in the senescence of the suspensor. In *P. coccineus* (Yeung and Clutter 1979) the specialized plastids are particularly abundant at the late globular and heart stage, and disappear as the embryo matures. Presumably, they play some specific role in embryogenesis.

A karyological phenomenon of frequent occurrence in suspensor cells is endopolyploidy (Nagl 1962, 1974), mainly by endoreduplication. In *Phaseolus*, which has been studied a great deal (see Nagl 1974), there is a progressive increase in endopolyploidy of suspensor cells towards the micropylar end, the basal cells attaining very high levels. In *P. vulgaris* the ploidy level varies in different varieties between 8n and 2,048n. In *P. hysterinus* and *P. multiflorus*, the nuclei are 4,096n, and in *P. coccineus* the giant nuclei undergo as many as 12 endocycles, and cytophotometric measurements have established that the DNA content reaches 8,192C amount in the larger basal cells of the suspensor (Brady 1973). In the latter species the polytene nuclei are delimited by a highly lobed envelope associated with cytoplasmic invaginations, often in direct contact with the chromatin (Nagl 1976b). The nuclear envelope is provided with a large number of pores. The invaginations of the cytoplasm obviously increase the surface area of the nuclear membrane enhancing nucleo-cytoplasmic exchange. Polyteny in suspensor nuclei is associated with enhanced biosynthesis of nucleic acids and protein (see Chap. 11 for details).

A further feature of interest refers to the presence of cell-wall ingrowths (Figs. 8.14B, 8.21A, B, and 8.24) in the suspensor cells of several species (see Deschamps 1968, 1969, P. Schulz and Jensen 1969, Marions 1970, Schnepf and Nagl

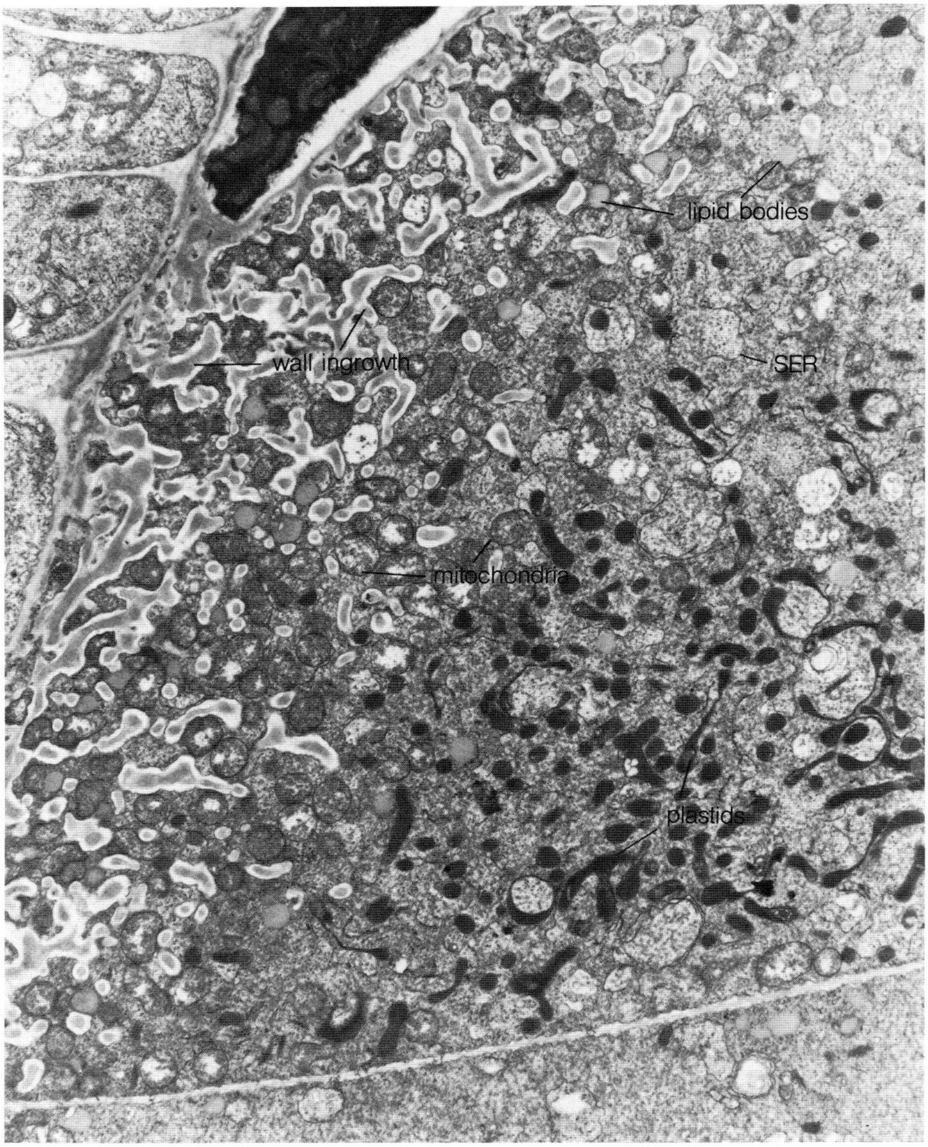

Fig. 8.24. *Phaseolus coccineus,* part of suspensor cell from late-globular proembryo to show wall ingrowths closely associated with mitochondria, smooth endoplasmic reticulum (*SER*), abundant plastids (plastids usually surround the nucleus), and some lipid bodies. (After Yeung and Clutter 1979; photograph courtesy Dr. E. C. Yeung)

1970, Gunning and Pate 1974, Newcomb and Fowke 1974, Simoncioli 1974, Nagl 1976a, Yeung and Clutter 1979). A number of organelles, particularly ER, mitochondria and dictyosomes have been observed adjacent to the wall embayments lined by plasma membrane (Fig. 8.24). The ultrastructure of this "wall-membrane apparatus" is characteristic of the so-called transfer cells (Gunning and Pate 1969, Pate and Gunning 1972) which are supposed to be involved in short-distance translocation of metabolites by enhancing the surface-to-volume ratio of cells. Taken together, these features have been interpreted as indicative of a role in the uptake or exchange of nutrients necessary for the growth of the embryo.

During late stages of embryo development, the suspensor undergoes degeneration and is apparently resorbed by the embryo. Previously it was assumed that the suspensor "becomes crushed and obliterated" (Maheshwari 1950), but it has recently been demonstrated in *Tropaeolum majus* (Nagl 1976a) that the death of suspensor cells is actuated by a basipetal wave of autolytic and autophagic process, and is characterized by the rupture of tonoplast and the disorganization of protoplast. Histochemical studies revealed a polar increase in the activity of hydrolases (Nagl 1976a). Reference has already been made to the plastolysomes in the suspensor of *Phaseolus* (Nagl 1977). In *Tropaeolum* a rapid decline in protein content, accompanied by an increase in the free amino acid pool, was observed in the senescing suspensor, but this was not correlated with corresponding increase in the specific activity of aminopeptidases (Singh et al. 1980). The implication is that protein breakdown is not controlled by proteolytic activity, but by some process which brings the cellular protein into contact with the hydrolases.

Currently, the suspensor is considered to function in two ways: (1) as an organ that absorbs nutrients from surrounding somatic tissues of the seed for onward transmission to the developing embryo (P. Schulz and Jensen 1969, Marinos 1970, Nagl 1974, Newcomb and Fowke 1974, Simoncioli 1974, Yeung and Clutter 1979, Yeung 1980), and (2) as a source of supply of important nutrients and growth regulators to the developing embryo (Satina and Rietsema 1959, Pritchard 1964, Corsi 1972, Ponzi and Pizzolongo 1972, Nagl 1974, Alpi et al. 1975, 1979, Cionini et al. 1976, Lorenzi et al. 1978, Yeung and Sussex 1979).

It has been pointed out earlier that in several families of angiosperms the suspensor produces haustorial extensions that are supposedly involved in absorbing nutritive material from the endosperm, nucellus, integument(s), placenta, and other tissues (for details see Subramanyam 1960, Masand and Kapil 1966). This idea appears quite attractive in the light of ultrastructural studies on the suspensor of several taxa such as *Capsella, Diplotaxis, Phaseolus, Pisum,* and *Stellaria.* The suspensor cells have highly convoluted labyrinthine wall projections characteristic of transfer cells involved in short-distance translocation. Further, plasmodesmata occur on the walls separating individual suspensor cells and, probably, betwen suspensor and the embryo proper. A more direct evidence in support has been provided by Yeung (1980). By administering [^{14}C]-sucrose to the developing embryos of *Phaseolus coccineus* via the base of the pods, he could demonstrate that the suspensor is the preferred major site of uptake for the embryo, particularly at the heart-shaped stage of development.

Table 8.1. The effect of suspensor on the formation of plantlets of *Phaseolus coccineus* embryos cultured in vitro. The results were scored after eight weeks in culture. (From Yeung and Sussex 1979)

Stage	Treatment	Number of plantlets/ total no. cultured	% precocious germination
Heart (4.5 mm seed)	Embryo proper only	37/89	42
	Embryo proper with suspensor attached	84/95	88
	Embryo proper with detached suspensor in direct contact	37/51	73
	Embryo proper with suspensor 1 cm away	10/30	33
	Embryo proper with heat-killed, detached suspensor in direct contact	16/43	37
Late heart (6.5 mm seed)	Embryo proper only	17/18	94
	Embryo proper with suspensor attached	17/18	94
Early cotyledon (7 mm seed)	Embryo proper only	18/18	100
	Embryo proper with suspensor attached	19/19	100
	Embryo proper with detached suspensor in direct contact	6/6	100
Cotyledon (1 cm seed)	Embryo proper only	22/22	100

Several investigators have implicated the suspensor in a secretory role. Satina and Rietsema (1959) observed that the basal cell of the proembryo in *Datura* had oil globules. Since oil was first detected in the basal cell and only later in other cells of the proembryo, the authors concluded that the suspensor acts as a source of oil to the developing embryo. Pritchard (1964) stated that the primary suspensor cell in *Stellaria* was responsible for the elaboration of proteinaceous material. Reference was earlier made (in Sect. 8.9) to the activity of several hydrolytic and oxidative enzymes in the suspensor cells, indicating that it is a metabolically highly active system.

There is evidence to indicate that the suspensor contains several growth regulators. Corsi (1972) cultured intact and suspensor-deprived embryos of *Eruca* in vitro at three stages of development, and noted that the presence of suspensor is greatly beneficial to the growth of the embryo proper. Similar observations were recorded by Yeung and Sussex (1979) an another variety of *P. coccineus* (Table 8.1). The qualitative differences shown in Table 8.1 are also reflected in fresh weight, with intact embryos registering a greater gain in fresh weight embryos deprived of suspensors. Cionini et al. (1976) demonstrated that in *Phaseolus coccineus* removal of the suspensor has no effect at late stages of embryogenesis, but at younger stages it results in reduced embryo development. The younger the embryo, the more severe is the effect. Gibberellic acid (GA) could partially replace the suspensor in vitro, indicating that this organ may provide GA to the young embryo. An analysis of the GA status of the suspensor and embryo proper in *Phaseolus* (Alpi et al. 1975) has shown that at the heart-shaped stage of development the suspensor has nearly 30 times greater GA-like activity than the embryo proper (Table 8.2). A dramatic decrease in the level of GA-like substances occurs

Table 8.2. Extractable GA-like substances from embryos and suspensors of *Phaseolus coccineus* at two developmental stages. (From Alpi et al. 1975)

Developmental stage	Mg. fresh weight per individual structure	μg GA_3 eq $\times 10^{-1}$/100 mg fresh weight			
		F-1[a]	F-2[b]	F-3[c]	Total
	Embryo				
Heart-shaped embryo (A)	0.60	0.035	0.001	0.001	0.037
Cotyledonary embryo (B)	5.00	0.171	0.141	0.130	0.422
	Suspensor				
Heart-shaped embryo (A)	0.40	0.007	0.117	1.002	1.126
Cotyledonary embryo (B)	0.25	0.032	0.240	0.100	0.372

[a] *F-1* = ethyl acetate fraction at pH 8.0. [b] *F-2* = free GAs. [c] *F-3* = bound GAs

in the suspensor of the cotyledonary stage embryo, but at this time the level of total GAs in the embryo proper is ten times that of heart-shaped embryo. Utilizing the technique of gas chromatography-mass spectrometry, this GA has been identified as gibberellin A_1 (Alpi et al. 1979). Przybyllock and Nagl (1977) studied the relative concentrations of auxin in the embryo proper and suspensor of *Tropaeolum majus* by mass fragmentation (single ion detection). The suspensor haustorium yielded significantly higher auxin amounts than the embryo proper. A study on the cytokinin status in *Phaseolus* (Lorenzi et al. 1978) has also demonstrated that in the heart-shaped embryo the suspensor has an abundance of biologically active cytokinins, while in the embryo proper cytokinins with low or scarce biological activity occur. On the other hand, at mid-cotyledonary stage the embryo proper seems to become autonomous for cytokinin supply as it has an abundance of biologically active cytokinins. Contrastingly, the suspensor contains, at this stage, cytokinins of low biological activity.

It is very clear from the above findings that there is overwhelming evidence to suggest that the suspensor acts as a stage-specific supplier of growth regulators to the embryo proper, during development. What is not certain at the moment, however, is whether it acts as a site of synthesis or a site of accumulation of these regulators. The ultrastructure of the suspensor has led several investigators (Marinos 1970, Schnepf and Nagl 1970, Ponzi and Pizzolongo 1972, Nagl 1974, Yeung and Clutter 1979) to believe that it is involved in the actual biosynthesis of phytohormones. The abundance of SER observed in suspensor cells is also characteristic of cells active in terpenoid biosynthesis (reviews by Schnepf 1974, Luttge and Schnepf 1976). Since GA is a class of substances synthesized from the terpenoid biosynthetic pathway, it may well be that the SER in suspensor cells is the site of GA synthesis. However, more direct evidence is necessary before a definite conclusion can be drawn.

In conclusion, the suspensor appears to play a very dynamic role not only in nourishing the embryo proper at specific developmental stages, but also in exercising control on its proper growth by supplying several important phytohormones. Nagl's (1973) comparison of the embryo suspensor to the mammalian trophoblast, therefore, seems to be very appropriate.

8.11 Deviations from Usual Development

In the foregoing discussion we have tried to present the usual course of events during embryogenesis. However, there are several deviations from the usual pattern in angiosperms. A few selected examples are reviewed below.

8.11.1 Embryogeny in *Paeonia*

It was stated earlier that zygotic division in angiosperms is accompanied, typically, by cytokinesis and results in a two-celled proembryo. Yakovlev and Yoffe (1957) reported a unique pattern of embryogeny in three species of *Paeonia* – *P. anomala, P. Moutan,* and *P. wittmanniana* – in which the division of the zygote is acytokinetic and a binucleate structure is produced. The next few divisions are also acytokinetic so that a free-nuclear phase ensues. The number of free nuclei is variable depending upon the species. According to Cave et al. (1961) there may be 64–128 nuclei as a concomitant of six or seven synchronous mitotic cycles. Carniel (1967) observed as many as 256 free nuclei in certain cases. Eventually, free-nuclear divisions discontinue and cell wall formation sets in, resulting in a globular cellular mass. Organ formation does not take place at this stage, instead numerous "embryonal primordia" begin to appear at the periphery of proembryo. However, only one of these is able to reach maturity and differentiates into a normal, dicotyledonous embryo. Yakovlev and Yoffe consider *Paeonia* as a unique example of proembryonal polyembryony.

The initial report of Yakovlev and Yoffe has been fully substantiated by a number of other investigations on different species of *Paeonia* (Cave et al. 1961, Matthiessen 1962, Moscov 1964, Wunderlich 1966, Carniel 1967). However, Murgai (1959, 1962) contradicted the views of Yakovlev and Yoffe on the basis of her work on *P. actiflora, P. albiflora, P. delavayi, P. suffruticosa,* and an undetermined species. According to her, the zygote divides by a transverse wall so that a two-celled proembryo comprising an apical and a basal cell is formed as in other angiosperms. Free-nuclear divisions then take place only in the basal cell to form a suspensor haustorium, whereas the apical cell divides by a vertical wall although its further development is delayed until after cell-wall formation in the basal cell. During later stages, it is difficult to distinguish the derivatives of *cb* from those of *ca,* but both contribute to the proembryonal mass. Further stages of development are identical to the observations of Yakovlev and Yoffe, and other workers.

The investigation of Cave et al. indicates the possibility that Murgai may have wrongly interpreted a persistent synergid adpressed to the chalazal end of zygote as the apical cell of the two-celled proembryo.

In any case, the unique features of *Paeonia* call for a critical and thorough investigation from a morphogenetic angle. It would be well worth examining the polarity, ultrastructure, and cytochemistry of the zygote, and early events during embryogenesis also Chap. 1; Sect. 9.3).

8.11.2 Embryos Devoid of Organs

In the majority of angiosperms the seed, at maturity, contains a bipolar embryo comprising an axis to which one or two cotyledons are attached at a node. The axis is surmounted by an apical meristem at each pole and is demarcated by the cotyledonary node into hypocotyl-radicle and plumule, or sometimes an epicotyl-plumule. However, in a number of taxa belonging to both dicotyledons and monocotyledons (see Table 8.3), the embryo remains small and reduced and lacks the cotyledon(s) and apical meristems; even the tissues may not be delineated when the seed is ready for dispersal. The simplest of these is represented by *Monotropa uniflora* in which the proembryo has just two cells (Fig. 8.25 A, B). The number of proembryonal cells in other taxa ranges from four, six (Fig. 8.26 A–F), ten to several (Fig. 8.27 A–I). Often, the proembryo assumes a globular configuration (Fig. 8.28 A–F, G), and in *Eriocaulon* (Fig. 8.27 G–I) it appears bell-shaped. A suspensor is mostly wanting, but may be present as in the orchids (Fig. 8.20 E, F–H). Besides the taxa listed in Table 8.3, intraseminal growth and after-ripening are reported in many taxa including *Actaea, Anthurium, Caltha, Clematis, Cocos, Corydalis, Crocus, Fraxinus, Hepatica, Orontium, Peltandra, Thalictrum,* and *Trillium* (see Eames 1961 for details). The duration of intraseminal growth is highly variable. Mention may be made here of a few examples: *Fumaria,* 8 days; *Clematis,* 17 days; *Actaea, Hepatica, Thalictrum,* 2 months; *Corydalis,* 10 months; *Trillium,* 12 months (Eames 1961). The factors that regulate and control this phenomenon are yet to be explained (see below).

A perusal of literature reveals that in almost all the taxa indicated in Table 8.3 the embryos have been regarded as "undifferentiated" or "unorganized". However, from a morphogenetic point of view this would be quite incorrect (see Rangaswamy 1967). Even if these embryos lack tissues and organs, it is obvious that both differentiation and organization are manifest in them at the cellular, subcellular, and even the molecular levels. In fact, it has been demonstrated in *Eriocaulon robusto-brownianum* (Ramaswamy et al. 1981) and *E. xeranthemum* (Ramaswamy and Arekal 1982) (Fig. 8.27) that the embryo does exhibit some degree of internal differentiation into radicular pole and shoot apex, the latter comprising cotyledonary and epicotylarly sectors. A histochemical study of embryo development in *Vanda,* an orchid, revealed that the apical cell of the two-celled proembryo is smaller and contains higher concentrations of RNA and total proteins (Alvarez and Sagawa 1965). Thus, as early as the two-celled stage, a biochemical and morphological dissimilarity is evident between the embryonal and suspensor portions. An ultrastructural investigation on *Cattleya aurantiaca,* another member of the Orchidaceae, by Harrison (1973, see Arditti 1979) has brought out the presence of different cell types within the dry embryo. The ovoid embryo of this species has cells of varying size depending on their position within the embryo. At the apical region are smaller meristematic cells, generally 8–10 μm in diameter, but the basally located cells are larger. All the cells are packed with oil globules, but protein bodies are restricted to cells toward the apical two-thirds of the embryo. Interestingly, plasmodesmata do not occur on the cell walls. The cytoplasmic organelles could not be identified because of poor internal differentiation, and poor state of hydration of the embryo. ER is also not discernible. Other

Table 8.3 List of taxa in which seeds at shedding stage contain embryos devoid of organs. (Based on Rangaswamy 1967)[a]

Family	Genus/species	Mode of nutrition	Structure of embryo	Investigators
Dicotyledons				
Balanophoraceae	*Balanophora* spp	Total root parasites	Embryo 4- to 12-celled; organs absent	Ekambaram and Panje (1935), Zweifel (1939), Arekal and Shivamurthy (1976)
Chloranthaceae	*Chloranthus japonicus* *C. serratus* *Hedyosmum nutans* *Sarcandra glabra* *S. irvingbaileyi*	Autotrophs	Embryo merely a mass of cells devoid of organs; in *Chloranthus* embryo lacks cotyledons at fruit shedding; in *H. nutans* organogenesis occurs during germination; in *S. glabra* embryo development is completed after shedding of seed; in *S. irvingbaileyi* the seed has a dicotyledonous embryo	Yoshida (1957, 1959a, b), Vijayaraghavan (1964)
Cynomoriaceae	*Cynomorium coccineum*	Total root parasite	Embryogeny described only up to 4-celled stage	Juel (1903), Steindl (1945)
Gentianaceae	*Cotylanthera* *Leiphaimos* *Voyria* *Voyriella*	Saprophytes	Embryo 5- to 24-celled; lacks organ formation at seed shedding	Oehler (1927)
Hydnoraceae	*Hydnora africana*	Leafless parasite	Oldest embryos without organs	Dastur (1922)
Lennoaceae	*Ammobroma* *Lennoa* *Pholisma*	Total parasites	Embryo spherical without organs	Willis (1973)
Lentibulariaceae	*Utricularia* spp	Aquatic, carnivorous	Embryo globular or ovoid; 2–13 "cotyledonoids" present; morphology of cotyledonoids debatable	Kaushik (1938), Kaushik and Raju (1955), Farooq (1965), Farooq and Bilquis (1966), Mohan Ram and Dore Swamy (1966), Kumazava (1967), Khan (1970)
Oleaceae	*Fraxinus excelsior*	Autotroph	Embryo undergoes 4 months of intra-seminal growth	Eames (1961), Villiers and Wareing (1960)

Orobanchaceae	*Aeginetia* *Boschniakia* *Christisonia* *Cistanche* *Conopholis* *Orobanche* *Phacellanthes*	Total root parasites	Embryo globular or ovoid; no histogens except dermatogen	Cassera (1935), Juliano (1935), Tiagi (1951, 1952, 1963, 1965), Tiagi and Sankhla (1963), Kadry (1955), Rangaswamy (1967), Rangan and Rangaswamy (1968), Tereokhin (1973), Nikiticheva and Tereokhin (1975), Kumar and Rangaswamy (1977), Olsen and Olsen (1980)
Papaveraceae	Several species (e.g. *Corydalis cava*)	Autotrophs	Seed contains only a proembryo and further growth is intraseminal	Eames (1961)
Piperaceae	*Heckeria* *Peperomia* *Piper*	Autotrophs	Seed with minute "undifferentiated" embryo; reports of such an embryo in *Piper nigrum* not substantiated	Kanta (1962), Nikiticheva (1981)
Pyrolaceae	*Allotropa* *Chimaphila* *Monotropa* *Monotropsis* *Pleuricospora* *Sarcodes*	Saprophytes	Cell number in embryo is 2 in *Monotropa uniflora*; 2 or 3 cells in *Allotropa* and *Pleuricospora*; 4 in *Monotropsis*; 9 in *Monotropa hypopithys*; several-celled with a globular structure in *Sarcodes*; in *Chimaphila* also embryo lacks organs	Copeland (1933, 1947), Francke (1934), Doyel and Goss (1941), Rangaswamy (1967), Pyykkö (1968), Olson (1980)
Rafflesiaceae	*Cytinus* *Mitrastemon* *Rafflesia*	Total root parasites	Embryo with 4 to 10 cells in 3 to 5 tiers	Ernst and Schmidt (1913), Watanabe (1936), Guzowska (1964)
Ranunculaceae	*Anemone* *Caltha* *Eranthis* *Ranunculus*	Autotrophs	Embryo lacks organs at shedding of seeds; further growth is intraseminal	Souèges (1913), Haccius (1953), Eames (1961)
Sauruaceae	*Houttuynia* *Saururus*	Autotrophs	Embryo minute and lacks organs; suspensor present in *Houttuynia*	Nikiticheva (1981)

[a] No attempt is made to present an exhaustive list. For additional examples consult Johansen (1950), Eames (1961) and Mayer and Poljakoff-Mayber (1963)

Table 8.3 (continued)

Family	Genus/species	Mode of nutrition	Structure of embryo	Investigators
Monocotyledons				
Burmanniaceae	*Apteria* *Burmannia* *Dictyostegia* *Gonianthes* *Gymnosiphon* *Thismia*	Saprophytes	Embryo 4- to 10-celled; in *Thismia clandestina* it is globular with 3-celled suspensor	Bernard and Ernst (1911), Ernst and Bernard (1912), Ramaswamy (1970), Arekal and Ramaswamy (1973)
Eriocaulaceae	*Eriocaulon* spp	Autotrophs	Embryo lacks organs but shows differentiation of dermatogen; bell-shaped	Hare (1950), Begum (1968), Arekal and Ramaswamy (1980), Ramaswamy and Arekal (1981, 1982), Ramaswamy et al. (1981)
Orchidaceae	Many taxa	Mostly autotrophs, some saprophytes	Seeds with a minute embryo devoid of organs; suspensor may be present or absent	Veyret (1974), Arditti (1979)
Pandanaceae	*Pandanus* spp	Autotrophs	Embryo lacks organs	Solms-Laubach (1878)
Xyridaceae	*Xyris* sp	Autotroph	Embryo lacks organs	Govindappa (1955)

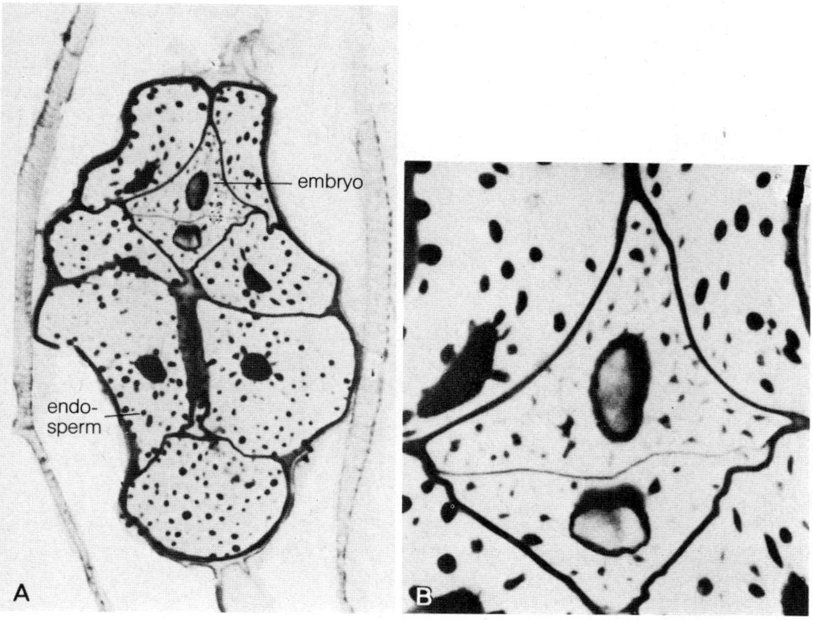

Fig. 8.25 A, B. *Monotropa uniflora.* **A** Longisection of seed at shedding stage; the two-celled proembryo is surrounded by 10–14 endosperm cells, and the seed-coat is very thin. **B** Proembryo. (After Olson 1980)

electron microscopic studies on orchid embryos (D. Schulz 1975, Zhukova and Savina 1978) have also revealed a distinct polarity coupled with differentiation of distinct cell types. Hence, it is advisable to discard the terms used by earlier investigators for taxa shown in Table 8.3, and refer to their embryos as those devoid of tissues and/or organs.

An important question is why, in these taxa, does the embryo remain reduced. What factor, or combination of factors, precludes their full development? The majority of taxa, mentioned in Table 8.3, are ecologically specialized as parasites or saprophytes. This is an important pointer to the mode of nutrition which may have (at least) an indirect role; however, the presence of several autotrophs in the list is certainly intriguing. A further point of interest, as mentioned above, is the absence of a suspensor in most of the species (except the orchids and a few other taxa). The role of the suspensor in controlling and regulating the growth of embryo, especially by absorbing and metabolizing nutritive substances as well as by supplying growth hormones, has already been discussed. The complex interrelationship between the embryo and its immediate milieu, the endosperm, is also worth probing. Hu et al. (1979) have demonstrated that in *Ilex* the endosperm is probably responsible for maintaining the embryo in a state of growth-arrest; the presence of endosperm inhibits the growth of embryo in vitro. Obviously, the growth-inhibiting factor(s) present in the endosperm must be removed, or suppressed, before the embryo is able to resume its normal growth. Intensive studies

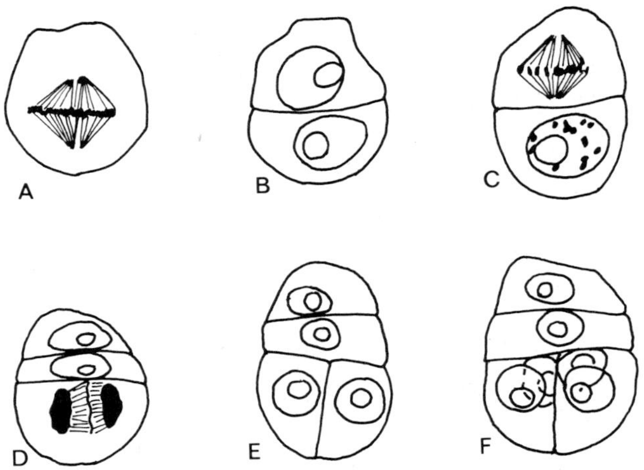

Fig. 8.26 A–F. *Burmannia pusilla,* embryogeny culminates in a quadrant proembryo at the time of seed shedding. (After Arekal and Ramaswamy 1973)

are urgenly needed to explore these and other probable factors before valid generalizations can be made.

8.11.3 Chimeral Embryos

While working on *Rudbeckia laciniata,* Battaglia (1945) observed a rather peculiar phenomenon: the sperm nucleus entered the egg cell, but the two nuclei did not fuse. Subsequently, both the nuclei divided independently giving rise to an embryo containing chimeral tissues of paternal and maternal origin. Battaglia named this abnormal process "semigamy". However, as this term is linguistically unsound (being made of both Latin and Greek roots), Solntzeva (1978) has suggested that the Greek form of the word, "hemigamy", is more appropriate (see also Battaglia 1981).

Hemigamy in its essential features is now known in several representatives of four families (see also Chap. 10). These are Asteraceae [*Rudbeckia* spp. (Battaglia 1963, Solntzeva 1978, Kościńska 1980)], Brassicaceae [*Arabidopsis thaliana* (Gerlach-Cruse 1970)], Malvaceae [*Gossypium barbadense* (Turcotte and Feaster 1967)], and Amaryllidaceae [*Cooperia drummondii, C. pedunculata* (Coe 1953) and *Zephyranthes* spp. (Battaglia 1963, Solntzeva 1978)]. However, in *Gossypium* the phenomenon is known only through genetic studies, and has not been investigated embryologically.

In all the above instances, the egg nucleus is diploid as a result of meiotic abnormalities, but the sperm is haploid. The sperm nucleus does not fuse with the egg nucleus but both of them remain in the egg cytoplasm and undergo division separately. The initiation of mitosis is synchronized in both the nuclei in *Rudbeckia laciniata* (Fig. 8.29 A, B) (Solntzeva 1978), but in *R. bicolor* (Kościńska 1980) the sperm nucleus occasionally undergoes several divisions before the egg nucleus divides. In *Zephyranthes macrosiphon* (Solntzeva 1978), on the other hand, the egg nucleus enters prophase earlier than the sperm nucleus. The deriv-

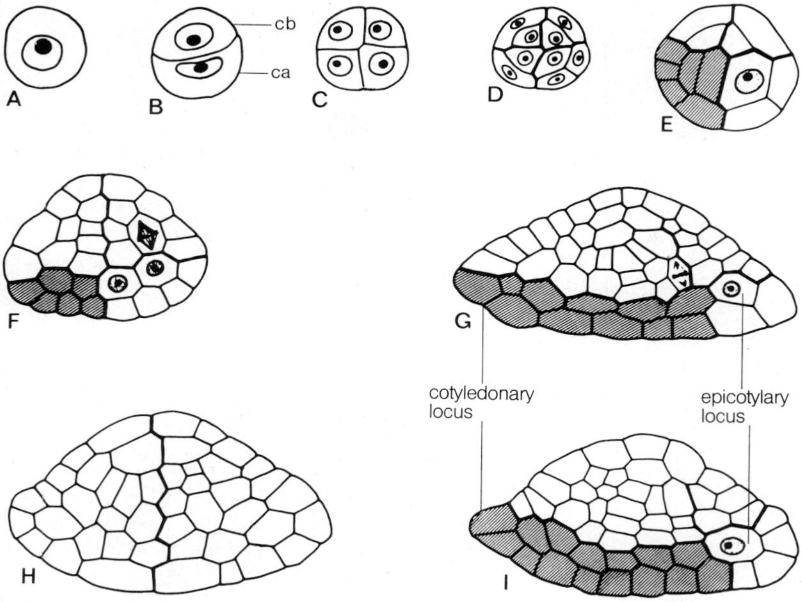

Fig. 8.27 A–I. *Eriocaulon xeranthemum*, embryogeny leading to the formation of a bell-shaped structure. There is some degree of internal differentiation into radicular pole and shoot apex; the latter can be distinguished into cotyledonary and epicotylary regions. (After Ramaswamy and Arekal 1982)

atives of the egg and sperm nuclei differ in size, those of egg-origin being larger by virtue of their diploid nature (Fig. 8.29 C). Cell wall formation occurs following nuclear division, and in the two-celled proembryo of *Rudbeckia laciniata* the apical cell usually contains only a female nucleus, whereas the basal cell has two male nuclei along with the female nucleus (Fig. 8.29 C). In *Cooperia drummondii* the male nucleus never divides but is cut off into the basal region of the egg, but in *C. pedunculata* it divides only once and the derivatives are sequestered laterally in the basal cell of the two-celled proembryo (Coe 1953). In *Rudbeckia sullivantii* four, at the most six, male nuclei are produced (Battaglia 1955), but more than this number may be present in other taxa. However, the number of male nuclei is always less than that of the egg derivatives. The earlier divisions of the male derivatives are normal; later ones are associated with mitotic abnormalities, such as restitution phenomena, and there is a tendency for them to become diploid (Gerlach-Cruse 1970, Vorsobina and Solntzeva 1973, Solntzeva 1978). During subsequent divisions the proembryo has various types of cells in which only male, only female, or a combination of both may be present. In *Rudbeckia laciniata* (Solntzeva 1978) and *R. bicolor* (Kościńska 1980) cells containing both male and female nuclei show spindle fusion during division, and a large triploid nucleus results. Such nuclei do not divide for a long time and, when they do, the divisions are irregular; often a hexaploid restitution nucleus may be formed.

Thus, the hemigamous embryo is chimeral, a mosaic of cells with nuclei of different ploidy and heredity; diploid male nuclei, diploid female nuclei, and hexa-

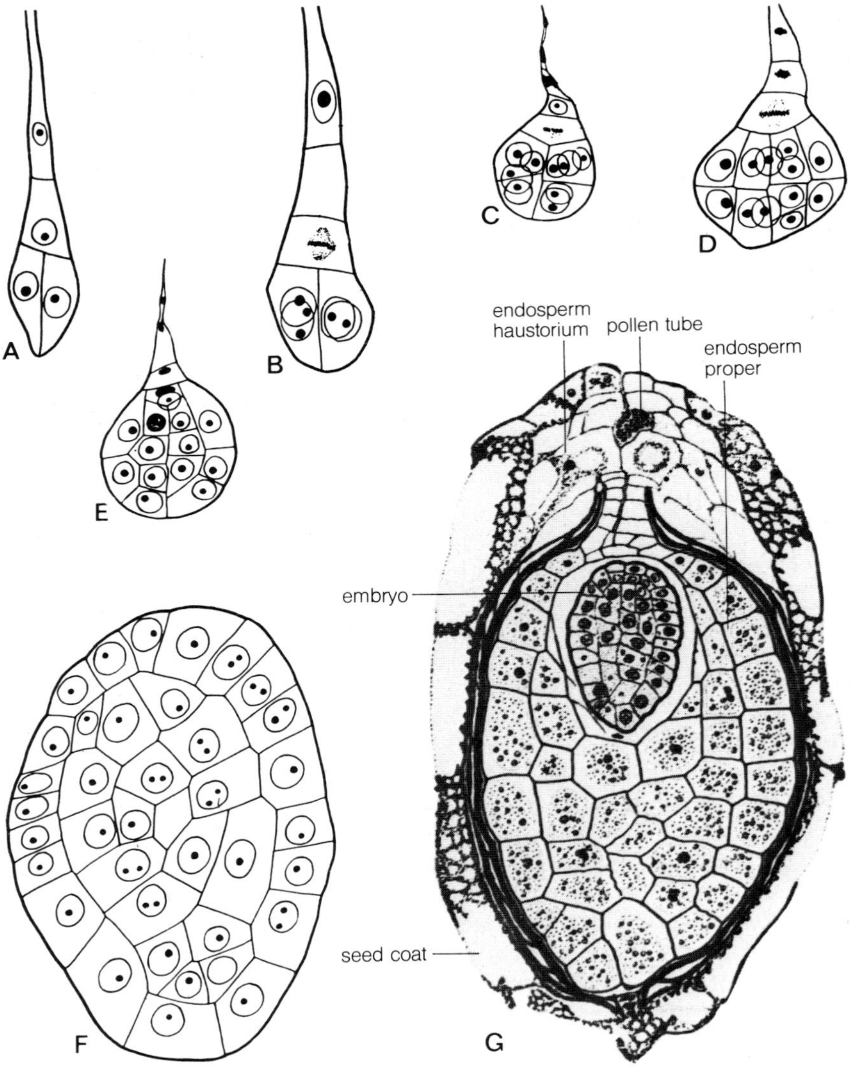

Fig. 8.28. A–F *Orobanche cernua,* embryogeny stops at the globular stage of development. **G** *Aeginetia indica,* longisecion of seed at shedding stage has a globular proembryo. (After Tiagi 1951, 1952)

ploid nuclei of combined heredity. The chimeral nature of the embryo appears at early stages of development, and may persist in the adult plant and affect the progeny.

That such embryos are capable of forming viable chimeral plants has been demonstrated by Turcotte and Feaster (1967, 1969, 1973) in *Gossypium barbadense* in which a doubled-haploid variety comprising dark-green leaves and oily glands was crossed with one having light-green leaves without glands. The result was chimeral plants in which some parts were dark-green and others light-green;

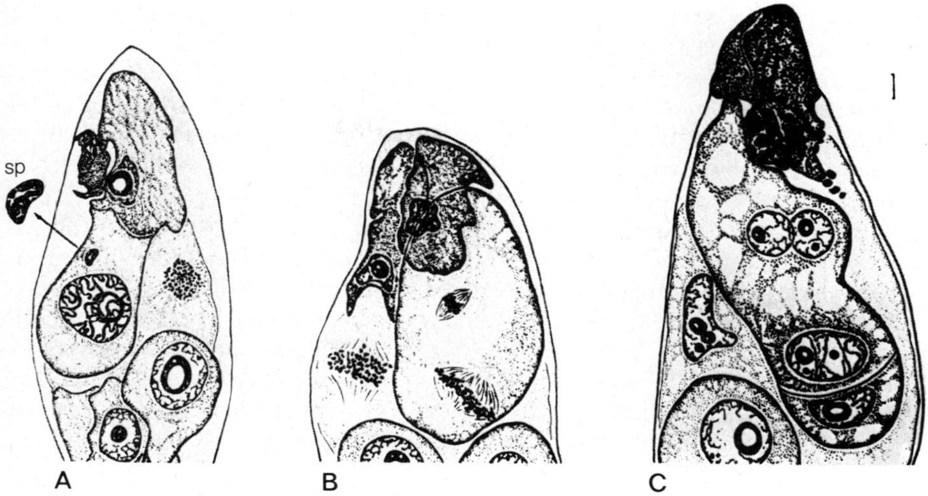

Fig. 8.29 A–C. *Rudbeckia laciniata,* hemigamy. **A** Egg cell with diploid nucleus and smaller sperm (*sp*) nucleus in prophase. **B** Both egg and sperm nuclei in metaphase. **C** Two-celled proembryo, apical cell is uninucleate; the basal cell has three nuclei of which the two smaller ones are derivatives of male gamete. (After Solntzeva 1978)

some with glands and others without. The tissues were of various levels of ploidy. According to Turcotte and Feaster (1975), hemigamy is a heritable trait whose expression is controlled by one dominant gene. This gene can be transmitted through both egg and pollen, but is functional only if the egg nucleus has the hemigametic factor. Consequently, hemigamy is a satisfactory procedure of obtaining haploids in cotton (Turcotte and Feaster 1975, Chaudhari 1978).

Hemigamy is not influenced by the mode of pollination, but by some inherent causes, probably incompatibility between gametic nuclei because of their different ploidy (Solntzeva 1978).

8.12 Concluding Remarks

Over the past century a great deal of information has been collected and collated on the morphological aspects of embryo development in angiosperms, and it has been possible to arrive at several generalizations. However, it is only recently – particularly in the last two decades – that the process of embryo development has been approached from fresh angles. The application of the electron microscope has proved to be highly rewarding. However, the number of taxa so investigated is still very small, and makes generalizations rather risky. Additional information is necessary concerning the events leading to the formation of the zygote – particularly the subcellular reorganization of the zygotic protoplasm and the process and pattern of wall synthesis in this cell. Details of embryogenesis beginning with the division of the zygote have also to be documented at the ultrastructural level for many more plants, if our concepts on this score are to be placed on a more confident footing. The observed ultrastructural facts should be increasingly correlated with physiological investigations emanating from in vitro culture experiments. Cytochemical studies in the recent past have yielded a wealth of useful information, but are still fragmentary; these must be further augmented, but there is an urgent need to work at more refined levels, such as different species of RNA and specific proteins, etc. In this connection the recent finding of specific "embryonic proteins" in somatic embryos of cultured carrot cells by Sung and Okimoto (1981) is noteworthy. The orderly progression of events during embryogenesis is intimately and inseparably associated with their immediate environment – notably the endosperm. A fresh look must be taken at this old problem (cf. Sheridan and Neuffer 1982). Wherever possible, living material must be employed as has been very successfully done by Erdelská (1969, 1983). After all, as Professor Jensen (1974) has remarked, the modern study of angiosperm embryogenesis has just begun, and one would expect the coming decades to be very exciting indeed.

References

Alpi A, Tognoni F, D'Amato F (1975) Growth regulator levels in embryo and suspensor of *Phaseolus coccineus* at two stages of development. Planta 127:153–162

Alpi A, Lorenzi R, Cionini PG, Bennici A, D'Amato F (1979) Identification of gibberellin A_1 in the embryo suspensor of *Phaseolus coccineus*. Planta 147:225–228

Alvarez MR, King DO (1969) Peroxidase localization, activity and isozyme patterns in the developing seedling of *Vanda* (Orchidaceae). Am J Bot 56:180–186

Alvarez MR, Sagawa Y (1965) A histochemical study of embryo development in *Vanda*. Caryologia 18:251–261

Amici GB (1824) Observations microscopiques sur diverses espèces de plantes. Ann Sci Nat Bot 2:41–70; 211–248

Arditti J (1979) Aspects of the physiology of orchids. In: Woolhouse HW (ed) Adv Bot Res 7:422–655

Arekal GD, Ramaswamy SN (1973) Embryology of *Burmannia pusilla* (Wall. ex Miers) Thw. and its taxonomic status. Beitr Biol Pflanz 49:35–46

Arekal GD, Ramaswamy SN (1980) Embryology of *Eriocaulon hookerianum* Stapf. and the systematic position of Eriocaulaceae. Bot Not 133:295–309

Arekal GD, Shivamurthy GR (1976) "Seed" germination in *Balanophora abbreviata*. Phytomorphology 26:135–138

Ashley T (1972) Zygote shrinkage and subsequent development in some *Hibiscus* hybrids. Planta 108:303–317

Avanzi S, Brunori A, D'Amato F (1969) Sequential development of meristems in the embryo of *Triticum durum*. A DNA autoradiographic and cytophotometric analysis. Dev Biol 20:368–377

Avery GS Jr (1930) Comparative anatomy and morphology of embryos and seedlings of maize, oats, and wheat. Bot Gaz 89:1–39

Barkworth ME (1982) Embryological characters and the taxonomy of the Stipeae (Gramineae). Taxon 31:233–243

Battaglia E (1945) Fenomeni citologici nuovi nella embriogenesi („semigamia") e nella microsporogenesi (Doppionucleo de restituzione) di *Rudbeckia laciniata* L. (nota preventiva). Nuovo G Bot Ital (Nuova Ser) 52:34–38

Battaglia E (1955) Unusual cytological features in the apomictic *Rudbeckia sullivantii* Boynton et Beadle. Caryologia 8:1–32

Battaglia E (1963) Apomixis. In: Maheshwari P (ed) Recent advances in the embryology of angiosperms. Intl Soc Plant Morphologists, Univ Delhi, Delhi, pp 221–264

Battaglia E (1981) Embryological questions: 3. Semigamy, hemigamy, and gynandroembryony. Ann Bot (Roma) 39:173–175
Batygina TB (1969) On the possibility of a new type of embryogenesis in angiosperms. Rev Cytol Biol Vég 32:335–341
Baum BR (1968) On some relationships between *Avena sativa* and *A. fatua* (Gramineae) as studied from Canadian material. Can J Bot 46:1013–1024
Baum BR (1978) The status of *Hordeum brachyantherum* in Eastern Canada, with discussions. Can J Bot 56:107–109
Begum M (1968) Embryological studies on *Eriocaulon quinquangulare* Linn. Proc Indian Acad Sci Sect B 67:148–156
Behl HM, Tiagi B (1978) The haustorial embryonal suspensor and endosperm in Fabaceae (Phaseoleae and Vicieae). In: Malik CP, Srivastava AK, Bhattacharya NC, Singh R (eds) Physiology of sexual reproduction in flowering plants. Kalyani, Ludhiana, pp 527–531
Berg RY (1954) Development and dispersal of the seed of *Pedicularis silvatica*. Nytt Mag Bot (Oslo) 2:1–60
Bernard C, Ernst A (1911) Beiträge zur Kenntnis der Saprophyten Javas. 6. Embryologie von *Thismia clandestina* und *T. verstegii*. Ann Jard Bot Buitenz 24:70–77
Bhalla PL, Singh MB, Malik CP (1979) Physiology of sexual reproduction VI. Embryogenesis in *Tropaeolum majus* L. Enzyme changes. Acta Bot Indica 7:72–86
Biradar NV (1968) Studies on palms: Embryology of *Phoenix pusilla* Gaertn., *P. acaulis* Buch. and *P. reclinata* Jacq. Proc Indian Acad Sci Sect B 67:165–173
Biradar NV, Mahabale TS (1968) Studies on palms: Embryology of *Phoenix robusta* Hook. Proc Indian Acad Sci 68B:1–8
Brady T (1973) Feulgen cytophotometric determination of the DNA content of the embryo proper and suspensor cells of *Phaseolus coccineus*. Cell Differ 2:65–75
Brown WV (1960) The morphology of the grass embryo. Phytomorphology 10:215–223
Brown WV (1965) The grass embryo: A rebuttal. Phytomorphology 15:274–284
Brunori A (1967) Relationship between DNA synthesis and water content during ripening of *Vicia faba* seed. Caryologia 20:333–338
Brunori A, D'Amato F (1967) The DNA content of nuclei in the embryo of dry seeds of *Pinus pinea* and *Lactuca sativa*. Caryologia 20:153–161
Campbell DH (1897) A morphological study of *Najas* and *Zannichellia*. Proc Calif Acad Sci 1:1–67
Carniel K (1967) Über die Embryobildung an der Gattung *Paeonia*. Oesterr Bot Z 114:4–19
Cass DD, Karas I (1974) Ultrastructural organization of the egg of *Plumbago zeylanica*. Protoplasma 81:49–62
Cassera JD (1935) Origin and development of the female gametophyte, endosperm, and embryo in *Orobanche uniflora*. Bull Torrey Bot Club 62:455–466
Cave MS, Anolt HJ, Stanton AC (1961) Embryogeny of the Californian peonies, with reference to their systematic position. Am J Bot 20:358–364
Chaudhari HK (1978) Use of semigamy in the production of cotton hybrids. Bull Torrey Bot Club 105:98–103
Cionini PG, Bennici A, Alpi A, D'Amato F (1976) Suspensor, gibberellin, and in vitro development of *Phaseolus coccineus* embryos. Planta 131:115–117
Clutter M, Sussex IM (1968) Macromolecular synthesis during plant embryogeny. Cellular rates of RNA synthesis in diploid and polytene cells in bean embryo (Abstr). J Cell Biol 39:26 a
Clutter M, Brady T, Walbot V, Sussex IM (1974) Macromolecular synthesis in diploid and polytene cells in bean embryo. J Cell Biol 63:1097–1102
Cocucci A, Jensen WA (1969) Orchid embryology: Megagametophyte of *Epidendrum scutella* following fertilization. Am J Bot 56:629–640
Coe GE (1953) Cytology of reproduction in *Cooperia penduculata*. Am J Bot 40:335–343
Copeland HF (1933) The development of seeds in certain Ericales. Am J Bot 20:513–517
Copeland HF (1947) Observations on the structure and classification of the Pyroleae. Madroño 9:65–102
Corsi G (1972) The suspensor of *Eruca sativa* Miller (Cruciferae) during embryogenesis in vitro. Giorn Bot Ital 106:41–54
Coulter JM, Chamberlain CJ (1912) Morphology of angiosperms. 2nd edn. Chicago Univ Press, Chicago

Crété P (1963) Embryo. In: Maheshwari P (ed) Recent advances in the embryology of angiosperms. Intl Soc Plant Morphologists, Univ Delhi, Delhi, pp 171–220
Cronquist A (1968) The evolution and clasification of flowering plants. Mifflin, Boston
D'Alascio-Deschamps R (1973) Modification ultrastructurales liées à la fécundation chez le Lin. Memoires Coll Morphologie. Bull Soc Bot Fr, pp 137–146
Dastur RH (1922) Notes on the development of the ovule, embryo sac, and embryo of *Hydnora africana* Thunb. Trans RS South Africa 10:27–31
Davis GL (1966) Systematic embryology of the angiosperms. Wiley, New York
Deschamps R (1968) L' embryon du Lin, *Linum usitatissimum* L., étude au microscope électronique. Rev Cytol Biol Vég 31:357–360
Deschamps R (1969) Premiers stades du développement de l'embryon et de l'albumen du Lin: Etude au microscope électronique. Rev Cytol Biol Vég 32:379–390
Doyel BE, Goss LM (1941) Some details of the structure of *Sarcodes*. Madroño 6:1–7
Eames AJ (1961) Morphology of the angiosperms. McGraw-Hill, New York
Ekambaram T, Panje RR (1935) Contributions to our knowledge of *Balanophora*. 2. Life history of *B. dioica*. Proc Indian Acad Sci Sect B 1:522–543
Erdelská O (1969) Untersuchung der ersten Entwicklungsphasen von Embryo und Endosperm der Art *Jasione montana* L. an lebendem Material. Planta 84:43–47
Erdelská O (1983) Microcinematographical investigation of the female gametophyte, fertilization, and early embryo and endosperm development. In: Symposium on "Fertilization and embryogenesis in ovulated plants". High Tatra Račkova dalina, Czechoslovakia, pp 49–54
Erdelská O, Klasová A (1978) La région micropylaire du sac embryonnaire de *Jasione montana* L. avant et après la fécondation. Actual Bot Bull Soc Bot Fr 125:249–252
Ernst A, Bernard C (1912) Beiträge zur Kenntnis der Saprophyten Javas. 12. Entwicklungsgeschichte des Embryosackes, des Embryos und des Endosperms von *Burmannia coelestis*. Ann Jard Bot Buitenz (2) 11:234–257
Ernst A, Schmid E (1913) Über Blüte und Frucht von *Rafflesia*. Morphologische-biologische Beobachtungen und entwicklungsgeschichtliche-zytologische Untersuchungen. Ann Jard Bot Buitenz (2) 12:1–58
Farooq M (1965) Studies in the Lentibulariaceae. 3. The embryology of *Utricularia uliginosa*. Phytomorphology 15:123–131
Farooq M, Bilquis S (1966) Studies in Lentibulariaceae 7. The embryology in *Utricularia scandens* Benj. Beitr Biol Pflanz 42:127–131
Fisher DB, Jensen WA (1972) Nuclear and cytoplasmic DNA synthesis in cotton embryos: A correlated light and electron microscope autoradiographic study. Histochemie 32:1–22
Fleischer E (1874) Beiträge zur Embryologie der Monokotylen und Dikotylen. Flora 57:369–375, 384–394, 401–411, 417–448
Forman M, Jensen WA (1965) Respiration and embryogenesis in cotton. Plant Physiol (Bethesda) 40:765–769
Foster AS (1949) Practical plant anatomy. Van Nostrand, New York
Francke HL (1934) Beiträge zur Kenntnis der Mykorrhiza von *Monotropa hypopitys* L. Flora 129:1–52
Galston AW, Davies PJ (1969) Hormonal regulation in higher plants. Science 163:1288–1297
Gartner PJ, Nagl W (1980) Acid phosphatase activity in plastids (plastolysomes) of senescing embryo suspensor cells. Planta 149:341–349
Gerlach-Cruse D (1970) Experimentelle Auslösung von Semigamie bei *Arabidopsis thaliana* (L.) Heynh. Biol Zentralbl 89:435–456
Govindappa DA (1955) Embryological studies on *Xyris pauciflora*. Proc Indian Acad Sci Sect B 42:47–57
Graves AH (1908) The morphology of *Ruppia maritima*. Trans Conn Acad Arts Sci 10:59–170
Guignard JL (1961) Embryogénie des Palmiers. Développement de l'embryon chez le *Chamaerops humilis* L. CR Acad Sci Paris 253:1834–1836
Guignard JL (1975 a) Embryogénie et classification embryogenique. Bull Soc Bot Fr 122:281–294
Guignard JL (1975 b) Du cotylédon des monocotylédons. Phytomorphology 25:193–200
Guignard JL, Mestre JC (1969) L'origine du cotylédon et du cône végétatif chez les monocotylédons. Bull Soc Bot Fr 116:207–214
Guignard JL, Mestre JC (1970) L'embryon des Graminées. Phytomorphology 20:190–197
Guignard L (1882) Recherches anatomiques et physiologiques sur des Legumineuses. Diss Univ Paris

Guignard L (1899) Sur les anthérozoides et la double copulation sexuelle chez les végétaux angiosperms. Rev Gen Bot 12:5–166

Gunning BES, Pate JS (1969) "Transfer cells". Plant cells with wall-ingrowths, specialized in relation to short distance transport of solutes: Their occurrence, structure, and development. Protoplasma 68:107–133

Gunning BES, Pate JS (1974) Transfer cells. In: Robards AW (ed) Dynamic aspects of plant ultrastructure. McGraw-Hill, London, pp 441–480

Guttenberg H von (1957) Embryologische und histogenetische Untersuchungen an Monokotyledonen. Bot Stud H 7. Fischer, Jena

Guzowska I (1964) Reinvestigation of embryo sac development, fertilization, and early embryogeny in *Cytinus hypocistis*. Acta Soc Bot Polon 33:157–166

Haccius B (1952) Die Embryoentwicklung bei *Ottelia alismoides* und das Problem des terminalen Monokotylenkeimblatts. Planta 40:443–460

Haccius B (1953) Histogenetische Untersuchungen an Wurzelhaube und Kotyledonarscheide geophiler Keimpflanzen (*Podophyllum* und *Eranthis*). Planta 41:439–458

Haccius B, Philip VJ (1979) Embryo development in *Cocos nucifera* L.: A critical contribution to a general understanding of palm embryogenesis. Plant Syst Evol 132:91–106

Hagemann R (1976) Plastid distribution and plastid competition in higher plants and the induction of plastom mutations by nitroso-urea-compounds. In: Bucher Th, Neuper W, Siebald W, Werner S (eds) Genetics and biogenesis of chloroplasts and mitochondria. North Holland, Amsterdam, pp 331–338

Hagemann R (1981) Unequal plastid distribution during the development of male gametophyte of angiosperms. Acta Soc Bot Pol 50:321–327

Hanstein J (1870) Die Entwicklung des Keimes der Monocotylen und der Dikotylen. Bot Abhandl (Bonn) 1:1–112

Hare CL (1950) The structure and development of *Eriocaulon septangulare* With. J Linn Soc Bot 53:422–448

Hu CY, Rogalski F, Ward C (1979) Factors maintaining *Ilex* rudimentary embryos in the quiescent state and the ultrastructural changes during activation. Bot Gaz 140:272–279

Jaffe LF (1968) Localization in the developing *Fucus* egg and the general role of localizing currents. In: Abercrombie M, Brachet J, King TJ (eds) Advances in morphogenesis. Academic Press, New York, pp 295–328

Jensen WA (1964) Cell development during plant embryogenesis. In "Meristems and differentiation". Brookhaven Symp Biol 16:179–202

Jensen WA (1968 a) Cotton embryogenesis: The tube-containing endoplasmic reticulum. J Ultrastruct Res 22:296–302

Jensen WA (1968 b) Cotton embryogenesis: Polysome formation in the zygote. J Cell Biol 36:403–406

Jensen WA (1968 c) Cotton embryogenesis: The zygote. Planta 79:346–366

Jensen WA (1974) Reproduction in flowering plants. In: Robards AW (ed) Dynamic aspects of plant ultrastructure. McGraw-Hill, London, pp 481–503

Johansen DA (1950) Plant Embryology. Waltham, Massachusettes

Johri BM, Tiagi B (1952) Floral morphology and seed formation in *Cuscuta reflexa* Roxb. Phytomorphology 2:162–180

Jones PA (1977) Nuclear DNA content of embryonic radicles and cultured stationary phase root tips of pea (*Pisum sativum* L.). Am J Bot 64:455–460

Juel HO (1903) Zur Entwicklungsgeschichte des Samens von *Cynomorium*. Beih Bot Zbl 13:194–202

Juliano JB (1935) Anatomy and morphology of the Bunga, *Aeginetia indica*. Philipp J Sci 56:405–451

Kadej F, Kadej A (1981) Ultrastructural organization of the tomato embryo sac at the time of fertilization. Acta Soc Bot Pol 50:139–142

Kadry AER (1955) The development of endosperm and embryo in *Cistanche tinctoria* (Forssk.) G. Beck. Bot Not 108:231–243

Kanta K (1962) Morphology and embryology of *Piper nigrum*. Phytomorphology 12:207–221

Kapil RN, Bhatnagar AK (1981) Ultrastructure and biology of the female gametophyte in flowering plants. Intl Rev Cytol 70:291–341

Kaushik SB (1938) Pollen development and seed formation in *Utricularia coerulea*. Beih Bot Zbl 58A:365–378

Kaushik SB, Raju MVS (1955) Variations in the development of proembryo in *Utricularia coerulea* L. Curr Sci (Bangalore) 25:296–297

Khan R (1970) Lentibulariaceae. In: Comparative embryology of angiosperms. Bull Indian Natl Sci Acad 41:290–297

Kościńska M (1980) Embryo and endosperm in *Rudbeckia bicolor* Nutt. Acta Biol Cracov Ser Bot 22:71–80

Kulkarni KM, Mahabale TS (1974) Studies on palms: Embryology of *Livistona chinensis* R Br. Proc Indian Acad Sci Sect B 80:1–17

Kumar U, Rangaswamy NS (1977) Regulation of seed germination and polarity in seedling development in *Orobanche aegyptiaca* by growth substances. Biol Plant (Prague) 19:353–359

Kumazava M (1967) An experimental study on the seedling of *Utricularia pilosa* Makino. Phytomorphology 17:494–498

Lakshmanan KK (1972) The monocot embryo. In: Varghese TM, Grover RK (eds) Vistas in plant sciences, vol 2. Intl Bio Sci, Hissar, pp 61–110

Lakshmanan KK (1977) Studies on the development of *Commelina benghalensis* L. 1. Zygote to globular proembryo. Proc Indian Acad Sci Sect B 86:167–174

Lakshmanan KK (1978) Studies on the development of *Commelina benghalensis* 3. Cotyledon. Phytomorphology 28:253–261

Lammeren AAM van (1981) Early events during embryogenesis in *Zea mays* L. Acta Soc Bot Pol 50:289–290

Lorenzi R, Bennici A, Cionini PG, Alpi A, D'Amato F (1978) Embryo suspensor relations in *Phaseolus coccineus:* Cytokinins during seed development. Planta 143:59–62

Luttge U, Schnepf E (1976) Elimination processes by glands. Organic substances. In: Lüttge U, Pitman MG (eds) Encyclopedia of plant physiology. New Ser, vol II B. Springer, Berlin Heidelberg New York, pp 224–277

Mahabale TS, Biradar NV (1968) Studies on palms: Embryology of *Phoenix sylvestris* Roxb. Proc Indian Acad Sci Sect B 67:1–17

Maheshwari P (1950) An introduction to the embryology of angiosperms. McGraw-Hill, New York

Maheshwari P (ed) (1963) Recent advances in the embryology of angiosperms. Intl Soc Plant Morphologists, Univ Delhi, Delhi

Maheshwari P, Singh B (1943) Studies in the family Alismaceae 5. The embryology of *Machaerocarpus californicus* (Torr.) Small. Proc Natl Inst Sci India Part B 9:311–322

Malik CP, Singh MB (1979) Embryo suspensor in angiosperms: Structural and functional aspects. In: Malik CP (ed) Current advances in plant reproductive biology. Kalyani, Ludhiana, pp 257–292

Malik CP, Singh MB, Thapar N (1977) Physiology of sexual reproduction 4. Histochemical characteristic of suspensor of *Brassica campestris*. Phytomorphology 27:384–389

Malik CP, Singh MB, Bhalla PL (1980) Embryo suspensor no more a botanical enigma. In: Nagaraj M, Malik CP (eds) Current trends in botanical research. Kalyani, New Delhi, pp 1–17

Marinos NG (1970) Embryogenesis of the pea (*Pisum sativum*) 2. An unusual type of plastid in the suspensor cells. Protoplasma 71:227–233

Masand P, Kapil RN (1966) Nutrition of the embryo sac and embryo – A morphological approach. Phytomorphology 16:158–175

Matthiessen A (1962) A contribution to the embryology of *Paeonia*. Acta Horti Bergiani 20:57–61

Mayer AM, Poljakoff-Mayber A (1963) The germination of seeds. Pergamon, Oxford

Mayer B, Stubbe W (1974) Das Zahlenverhältnis von mütterlichen und väterlichen Plastiden in den Zygoten von *Oenothera erythrosepala* Borbas (syn. *O. lamarckiana*). Ber Dtsch Bot Ges 87:29–38

Melchior H (1964) A Engler's Syllabus der Pflanzenfamilien. Bd 2, Borntraeger, Berlin

Mestre JC (1967) La signification phylogénétique de l' embryogénie. Rev Gen Bot 74:273–324

Metzlaff M, Borner T, Hagemann R (1981) Variations of chloroplast DNAs in the genus *Pelargonium* and their biparental inheritance. Theor Appl Genet 60:37–41

Miah AJ, Brunori A (1970) Chromosomal aberrations induced in meristematic regions of the embryos in X-irradiated dry seeds of *Vicia faba*. Radiat Bot 10:87–94

Mizushima U, Murakami K, Hozyo Y (1955) Influence of anomalous environmental conditions upon embryogenesis in the rice plant *Oryza sativa* L. Histochemical observation of normal embryogenesis. Tôhuku J Agric Res 6:1–19

Mogensen HL (1972) Fine structure and composition of the egg apparatus before and after fertilization in *Quercus gambelii:* The functional ovule. Am J Bot 59:931–941

Mogensen HL, Suthar HK (1979) Ultrastructure of the egg apparatus of *Nicotiana tabacum* (Solanaceae) before and after fertilization. Bot Gaz 140:168–179

Mohan Ram HY, Dore Swamy R (1966) Growth and flowering of *Utricularia inflexa* Forsk. var. *inflexa* Taylor in axenic cultures. Naturwissenschaften 53:387–388
Moscov IV (1964) On the development of the embryo in some species of *Paeonia* (in Russian). Bot Zh 49:887–894
Mukkada AJ (1962) Some observations on the embryology of *Dicraea stylosa* (Wight). In: Plant embryology – A Symposium. Council of Scientific and Industrial Research, New Delhi, pp 139–145
Murgai P (1959) The development of the embryo in *Paeonia*. Phytomorphology 9:275–277
Murgai P (1962) Embryology of *Paeonia* with a discussion on its systematic position. In: Plant embryology – A symposium. Council of Scientific and Industrial Research, New Delhi, pp 215–223
Nagl W (1962) Über Endopolyploidie, Restitutionskernbildung und Kernstrukturen in Suspensor von Angiospermen und einer Gymnosperme. Oesterr Bot Z 109:431–494
Nagl W (1973) The angiosperm suspensor and the mammalian trophoblast: Organs with similar cell structure and function. Memoires Coll Morphologie. Bull Soc Bot Fr, pp 289–301
Nagl W (1974) The *Phaseolus* suspensor and its polytene chromosomes. Z Pflanzenphysiol 73:1–44
Nagl W (1976a) Early embryogenesis in *Tropaeolum majus* L.: Ultrastructure of the embryo-suspensor. Biochem Physiol Pflanz (BPP) 170:253–260
Nagl W (1976b) Ultrastructure of the polytene nuclei in the embryo-suspensor of the bean, *Phaseolus coccineus:* Toothing of cytoplasm and nucleoplasm. Nature and Life (Kyungpook) J Biol Sci 6:11–17
Nagl W (1977) Plastolysomes – Plastids involved in the autolysis of the embryo-suspensor in *Phaseolus*. Z Pflanzenphysiol 85:45–51
Nagl W, Kuhner S (1976) Early embryogenesis in *Tropaeolum majus* L.: Diversification of plastids. Planta 133:15–19
Nawaschin SG (1898) Resultate einer Revision der Befruchtungsvorgänge bei *Lilium mortagon* und *Fritillaria tenella*. Bull Acad Imp Sci St. Petersburg 9:377–382
Newcomb W (1973) The development of the embryo sac of sunflower *Helianthus annuus* after fertilization. Can J Bot 51:879–890
Newcomb W, Fowke LC (1974) *Stellaria media* embryogenesis: The development and ultrastructure of the suspensor. Can J Bot 52:607–614
Nijalingappa BHM (1975) Embryological studies in *Haloragis micrantha*. Phytomorphology 25:146–155
Nikiticheva ZI (1981) Embryological features of some Piperales. Acta Soc Bot Pol 50:329–332
Nikiticheva ZI, Tereokhin ES (1975) The structure of the embryo and the postseminal development of some parasite angiosperms. (Abstr). XII Intl Bot Congr, Leningrad, p 228
Norstog K (1969) Morphology of the coleoptile and scutellum in relation to tissue culture responses. Phytomorphology 19:235–241
Norstog K (1972) Early development of the barley embryo: Fine structure. Am J Bot 59:123–132
Oehler E (1927) Entwicklungsgeschichtlich-zytologische Untersuchungen an einigen saprophytischen Gentianaceen. Planta 3:641–733
Olsen S, Olsen ID (1980) The seed of *Boschniakia hookeri* (Orobanchaceae). Bot Tidskr 75:159–172
Olson AR (1980) Seed morphology of *Monotropa uniflora* L. (Ericaceae). Am J Bot 67:968–974
Olson AR, Cass DD (1981) Changes in megagametophyte structure in *Papaver nudicaule* following in vitro placental pollination. Am J Bot 68:1338–1341
Pate JS, Gunning BES (1972) Transfer cells. Annu Rev Plant Physiol 23:173–196
Periasamy K (1977) A new approach to the classification of angiosperm embryos. Proc Indian Acad Sci Sect B 86:1–13
Philip VJ, Haccius B (1976) Embryogenesis in *Bambusa arundinacea* Willd. and structure of the mature embryo. Beitr Biol Pflanz 52:83–100
Poddubnaya-Arnoldi VA (1967) Comparative embryology of the Orchidaceae. Phytomorphology 17:312–320
Poddubnaya-Arnoldi VA (ed) (1976) Cytoembryology of angiosperms, Moscow (in Russian)
Pollock EG, Jensen WA (1964) Cell development during early embryogenesis in *Capsella* and *Gossypium*. Am J Bot 51:915–921
Ponzi R, Pizzolongo P (1972) The ultrastructure of suspensor cells of *Ipomoea purpurea* Roth. J Submicrosc Cytol 4:199–204
Ponzi R, Pizzolongo P (1973) Ultrastructure of plastids in the suspensor cells of *Ipomoea purpurea* Roth. J Submicrosc Cytol 5:257–263

Possingham JV (1980) Plastid replication and development in the life cycle of higher plants. Annu Rev Plant Physiol 31:113–129

Pritchard HN (1964) A cytochemical study of embryo development in *Stellaria media*. Am J Bot 51:472–479

Pritchard HN, Bergstresser KA (1969) The cytochemistry of some enzymatic activities in *Stellaria media* embryos. Experientia (Basel) 25:1116–1117

Przybyllock T, Nagl W (1977) Auxin concentration in the embryo and suspensors of *Tropaeolum majus*, as determined by mass fragmentation (single ion detection). Z Pflanzenphysiol 84:463–465

Pyykkö M (1968) Embryological and anatomical studies on Finnish species of the Pyrolaceae. Ann Bot Fenn 5:153–165

Quatrano RS (1974) Developmental biology: Development in marine organisms. In: Mariocal R (ed) Experimental marine biology. Academic Press, New York, pp 303–346

Quatrano RS (1978) Development of cell polarity. Annu Rev Plant Physiol 29:487–510

Raghavan V (1976) Experimental embryogenesis in vascular plants. Academic Press, New York

Ramaswamy SN (1970) Burmanniaceae. In: Comparative embryology of angiosperms. Bull Indian Natl Sci Acad 41:375–379

Ramaswamy SN, Arekal GD (1981) Embryology of *Eriocaulon setaceum*. Plant Syst Evol 138:175–188

Ramaswamy SN, Arekal GD (1982) Embryology of *Eriocaulon xeranthemum* Mart. (Eriocaulaceae). Acta Bot Neerl 31:41–54

Ramaswamy SN, Swamy BGL, Govindappa DA (1981) From zygote to seedling in *Eriocaulon robusto-brownianum* Ruhl. (Eriocaulaceae). Beitr Biol Pflanz 55:179–188

Ramji MV (1975) Histology of growth with regard to embryos and apical meristems in some angiosperms 1. Embryogeny of *Stellaria media*. Phytomorphology 25:131–145

Ramji MV (1976) The histology of growth with regard to apical meristems in some angiosperms 2. Differentiation of apical meristems in *Stellaria media*. Phytomorphology 26:124–135

Rangan TS, Rangaswamy NS (1968) Morphogenetic investigations on parasitic angiosperms 1. *Cistanche tubulosa* Wight (Family Orobanchaceae). Can J Bot 46:263–266

Rangaswamy NS (1967) Morphogenesis of seed germination in angiosperms. Phytomorphology 17:477–487

Rao CV (1955 a) Embryo development in areca nut. Nature 175:432–433

Rao CV (1955 b) Embryological studies in Palmae I–III (Abstr). Proc 42 nd Indian Sci Congr Pt 3:230–232

Rao CV (1956) Embryological studies in Palmae IV–VI (Abstr). Proc 43 rd Indian Sci Congr Pt 3:225–227

Rao CV (1959 a) Contributions to the embryology of Palmae I. Sabaleae. Proc Natl Inst India Part B 25:143–169

Rao CV (1959 b) Contributions to the embryology of Palmae II. Ceroxylineae. J Indian Bot Soc 38:46–75

Rau MA (1950) Development of the embryo in *Trigonella foenum-graecum* L. J Indian Bot Soc 29:210–213

Rau MA (1954) The development of the embryo in *Cyamopsis, Desmodium,* and *Lespedeza* with a discussion of the position of the Papilionaceae in the system of embryogenic classification. Phytomorphology 14:418–430

Rau MA (1955) Embryological studies in the Leguminosae. J Mysore Univ Sect B 14:63–75

Reeder JR (1957) The embryo in grass systematics. Am J Bot 44:756–768

Reeder JR (1962) The bambusoid embryo: A reappraisal. Am J Bot 49:639–641

Robertson BL (1976) Embryology of *Jubaeopsis caffra* Becc. III. Endosperm and embryogeny. JS Afr Bot 42:285–299

Rondet P (1958) Répartition et signification des acides ribonucléiques au cours de l'embryogenése chez *Lens culinaris* L. CR Acad Sci Paris 246:2396–2399

Rondet P (1961) Répartition et signification des acides ribonucléiques au cours l' embryogenése chez *Myosurus minimus* L. CR Acad Sci Paris 253:1725–1727

Rondet P (1962) L'organogenése au cours de l'embryogenése chez l'*Alyssum maritimum* Lamk. CR Acad Sci Paris 255:2278–2280

Russel SD (1980) Participation of male cytoplasm during gamete fusion in an angiosperm, *Plumbago zeylanica*. Science 2010:200–201

Satina S, Rietsema J (1959) Seed development. In: Avery AG, Satina S, Rietsema J (eds) Blakeslee: The genus *Datura*. Ronald, New York, pp 181–195

Schleiden MJ (1836) Einige Blicke auf die Entwicklungsgeschichte des vegetalischen Organismus bei den Phanerogamen. Arch Biol Naturgeschichte III 1:289–320
Schnarf K (1929) Embryologie der Angiospermen. Borntraeger, Berlin
Schnarf K (1931) Vergleichende Embryology der Angiospermen. Borntraeger, Berlin
Schnepf E (1974) Gland cells. In: Robards AW (ed) Dynamic aspects of plant ultrastructure. McGraw-Hill, London, pp 331–357
Schnepf E, Nagl W (1970) Über einige Strukturbesonderheiten der Suspensorzellen von *Phaseolus vulgaris*. Protoplasma 69:133–143
Schötz F (1954) Über Plastidenkonkurrenz bei *Oenothera*. Planta 43:182–240
Schulz D (1975) The orchid embryo: A revision on the base of several electron microscopical preparation methods (Abstr). XII Intl Bot Congr, Leningrad, p 233
Schulz P, Jensen WA (1969) *Capsella* embryogenesis: The suspensor and the basal cell. Protoplasma 67:139–163
Schulz R, Jensen WA (1968 a) *Capsella* embryogenesis: The early embryo. J Ultrastruct Res 22:376–392
Schulz R, Jensen WA (1968 b) *Capsella* embryogenesis: The egg, zygote, and young embryo. Am J Bot 55:541–552
Sehgal CB, Gifford EM Jr (1979) Developmental and histochemical studies of the ovules of *Nicotiana rustica* L. Bot Gaz 140:180–188
Selvaratnam EM (1952) Embryo of the coconut. Nature 169:714–715
Shah CK, Bhat PN (1974) Cytophotometric analysis of ascorbic acid (AA), ribonucleic acid, and sulfhydryl proteins during embryogenesis in *Coix lacryma-jobi* L. Curr Sci (Bangalore) 43:573–575
Shah CK, Pandey SN (1978) Histochemical studies during embryogenesis in *Limnophyton obtusifolium* (L.) Miquel. Phytomorphology 28:31–42
Sheridan WF, Neuffer MG (1982) Maize developmental mutants: Embryos unable to form leaf primordia. J Hered 73:318–329
Shirke NS, Mahabale TS (1973) Studies on palms: Embryology of *Caryota urens* L. In: Murthy YS, Johri BM, Mohan Ram HY, Varghese TM (eds) Advances in plant morphology. Sarita Prakashan Meerut, pp 218–232
Simoncioli C (1974) Ultrastructural characteristics of *Diplotaxis erucoides* (L.) DC suspensor. Giorn Bot Ital 108:175–189
Singh AP, Mogensen HL (1975) Fine structure of the zygote and early embryo in *Quercus gambelii*. Am J Bot 62:105–115
Singh MB, Bhalla PL, Malik CP (1980) Activity of some hydrolytic enzymes in autolysis of the embryo suspensor in *Tropaeolum majus* L. Ann Bot 45:523–527
Sivaramakrishna D (1978) Size relationships of apical cell and basal cell in 2-celled embryos in angiosperms. Can J Bot 56:1434–1439
Solms-Laubach HG (1878) Über monocotyle Embryonen mit scheitelbürtigem Vegetationspunkt. Bot Zeit 36:65–74, 81–93
Solntzeva MP (1978) Apomixis and hemigamy as one of its forms. Proc Indian Nat Sci Acad Part B 44:78–90
Souèges R (1913) Recherches sur l'embryogénie des Renonculacées-Renonculées. Genre *Ranunculus*. Bull Soc Bot Fr 60:506–514, 542–549, 615–621
Souèges R (1926) Embryogénie des Papavéracées. Développement du proembryon chez le *Papaver rhoeas*. CR Acad Sci Paris 183:902–904
Souèges R (1934) L' hypophyse et l' épiphyse; les problèmes d' histogenèse qui leur sont liés. Bull Soc Bot Fr 81:737–748, 769–778
Stebbins GL (1974) Evolution of morphogenetic patterns. In: Basic mechanisms in plant morphogenesis. Brookhaven Symp Biol 25:227–243
Stein OL, Quastler H (1963) The use of tritiated thymidine in the study of tissue activation during germination in *Zea mays*. Am J Bot 50:1006–1011
Steindl F (1945) Beiträge zur Pollen- und Embryobildung bei *Cynomorium coccineum*. Arch Julius Klaus Stift 20:342–355
Strasburger E (1879) Die Angiospermen und die Gymnospermen. Fischer, Jena
Strasburger E (1884) Neue Untersuchungen über den Befruchtungsvorgang bei den Phanerogamen. Fischer, Jena
Subramanyam K (1950) An embryological study of *Levenhookia dubia* Sond. Proc Natl Inst Sci India Part B 15:245–253
Subramanyam K (1960) Nutritional mechanism of the seed 2. Nutritional mechanism of the embryo. J Madras Univ Sect B 30:45–56

Sung ZR, Okimoto R (1981) Embryonic proteins in somatic embryos of carrot. Proc Natl Acad Sci USA 78:3683–3687
Sussex I, Clutter M, Walbot V, Brady T (1973) Biosynthetic activity of the suspensor of *Phaseolus coccineus*. Caryologia (Suppl) 25:261–272
Swamy BGL (1942) Female gametophyte and embryogeny in *Cymbidium bicolor* Lindl. Proc Indian Acad Sci Sect B 15:194–201
Swamy BGL (1943) Embryology of Orchidaceae. Curr Sci (Bangalore) 12:13–17
Swamy BGL (1949) Embryological studies in the Orchidaceae II. Embryogeny. Am Midl Nat 41:202–232
Swamy BGL (1979) Embryogenesis in *Cheirostylis flabellata*. Phytomorphology 29:199–203
Swamy BGL, Krishnamurthy KV (1975) Certain conceptual aspects of meristems. I. On hypophysis and quiescent centre. Phytomorphology 25:60–75
Swamy BGL, Krishnamurthy KV (1977) Certain conceptual aspects of meristems. II. Epiphysis and shoot apex. Phytomorphology 27:1–8
Swamy BGL, Lakshamanan KK (1962) The origin of epicotylary meristem and cotyledon in *Halophila ovata* Gaudich. Ann Bot 26:243–249
Swamy BGL, Padmanabhan D (1962) A reconnaissance of angiosperm embryogenesis. J Indian Bot Soc 41:422–437
Swift JG, O'Brien TP (1970) Vascularization of the scutellum of wheat. Austr J Bot 18:45–53
Takao S (1976) Embryogenic study of *Impatiens textorii*. Phytomorphology 26:200–210
Takhtajan A (1969) Flowering plants, origin, and dispersal. Oliver and Boyd, Edinburgh
Tateoka T (1964) Notes on some grasses. 16. Embryo structure of the genus *Oryza* in relation to the systematics. Am J Bot 51:539–543
Tereokhin ES (1973) The embryology of Orobanchaceae. 2. On the embryology and ecology of *Phacellanthus tubiflorus* Sieb et Zucc. Bot Zh 58:1775–1788
Tiagi B (1951) Studies in the family Orobanchaceae 3. A contribution to the embryology of *Orobanche cernua* Loeffl. and *O. aegyptiaca* Pers. Phytomorphology 1:158–169
Tiagi B (1952) Studies in the family Orobanchaceae 2. A contribution to the embryology of *Aeginetia indica* Linn. Bull Torrey Bot Club 79:63–78
Tiagi B (1963) Studies in the family Orobanchaceae. 4. Embryology of *Boschniackia himalaica* Hook. and *B. tuberosa* (Hook.) Jepson with remarks on the evolution of the family. Bot Not 116:81–93
Tiagi B (1965) Studies in the family Orobanchaceae 6. Development of the seed in *Conopholis americana* (L. Fil) Wallr. Acta Bot Hung 11:253–261
Tiagi B, Sankhla NS (1963) Studies in the family Orobanchaceae 5. Contribution to the embryology of *Orobanche leucorum*. Bot Mag Tokyo 76:81–88
Tillich HJ (1977) Vergleichend-morphologische Untersuchungen zur Identität der Gramineen-Primarwurzel. Flora 166:415–421
Tilney-Basset RAE, Birky CW Jr (1981) The mechanism of the mixed inheritance of chloroplast genes in *Pelargonium*. Evidence from gene frequency distributions among the progeny of crosses. Theor Appl Gen 60:43–53
Tilton VS, Mogensen HL (1979) Ultrastructural aspects of the ovule of *Agave parryi* before fertilizaton. Phytomorphology 29:338–350
Tohda H (1974) Development of the embryo of *Pogonia* (Orchidaceae). Sci Rep Tôhoku Univ Fourth Ser (Biol) 37:89–93
Treub M (1879) Notes sur l' embryogénie de quelques Orchidées. Nat Verh Koninkl Akad Amsterdam 19:1–50
Turcotte EL, Feaster CV (1967) Semigamy in *Pima* cotton. J Hered 58:55–57
Turcotte EL, Feaster CV (1969) Semigametic production of haploids in *Pima* cotton. Crop Sci 9:653–655
Turcotte EL, Feaster CV (1973) The origin of 2*n* and *n* sectors in chimeral *Pima* cotton plants. Crop Sci 13:111–113
Turcotte EL, Feaster CV (1975) Inheritance of semigamy in American Pima cotton (*Gossypium barbadense* L.). Agron Abstr, p 65
Vallade J (1970) Développement embryonnaire chez un *Petunia hybrida* hort. CR Acad Sci Paris 270:1893–1896
Vallade J (1980) Données cytologiques sur la proembryogenèse du *Petunia;* intérêt pour une interpretation morphogénétique du développement embryonnaire. Actual Bot Bull Soc Bot Fr 127:19–37

Vallade J, Cornu A, Essad S, Alabouvette J (1978) Niveaux de DNA dans les noyaux zygotiques chez le *Petunia hybrida* hort. Actual Bot Bull Soc Bot Fr 125:253–258

Van Went JL (1970) The ultrastructure of the fertilized embryo sac of *Petunia*. Acta Bot Neerl 19:468–480

Veyret Y (1974) Development of the embryo and the young seedling stages of orchids. In: Withner CL (ed) The orchids. Wiley, New York, pp 223–265

Vijayaraghavan MR (1964) Morphology and embryology of a vesselless dicotyledon – *Sarcandra irvingbaileyi*, and systematic position of the Chloranthaceae. Phytomorphology 14:429–441

Villiers TA, Wareing PF (1960) Interaction of growth inhibitor and a natural germination stimulator in the dormancy of *Fraxinus excelsior* L. Nature 185:112–114

Vorsobina LI, Solntzeva MP (1973) Fertilization and early embryogenesis in *Zephyranthes candida* Herb. (in Russian). Proc Symp Polovoj Process i Chmriogenz, pp 41–42

Walbot V, Brady T, Clutter M, Sussex I (1972) Macromolecular synthesis in plant embryos: Rates of RNA synthesis in *Phaseolus coccineus* embryos and suspensors. Dev Biol 29:104–111

Wardlaw CW (1955) Embryogenesis in plants. Methuen, London

Wardlaw CW (1968) Morphogenesis in plants: A contemporary study. Methuen, London

Watanabe K (1936) Morphologisch-biologische Studien über die Gattung *Mitrastemon*. 4. J Jap Bot 12:848–858

Willis JC (1973) A dictionary of the flowering plants and ferns. 8 th edn Revised by Airy Shaw HK, Cambridge Univ Press, Cambridge

Wilson KS, Clutter VM Jr (1952) The distribution of acid phosphatases during development of the fruit of *Cocos nucifera*. Am J Bot 39:57–58

Wilson KS, Clutter VM Jr (1953) Acid phosphatase distribution patterns in the developing embryo of *Cocos nucifera*. J Elisha Mitchell Sci Soc 69:170–174

Woodard JW (1956) DNA in gametogenesis and embryogeny in *Tradescantia*. J Biophys Biochem Cytol 2:765–777

Wunderlich R (1966) Zur Deutung der eigenartigen Embryoentwicklung von *Paeonia*. Oesterr Bot Z 113:395–407

Yakovlev MS, Yoffe MD (1957) On some peculiar features in the embryogeny of *Paeonia* L. Phytomorphology 7:74–87

Yakovlev MS, Zhukova GYA (1980) Chlorophyll in embryos of angiosperm seeds: A review. Bot Not 133:323–336

Yamazaki T (1974) A system of Gamopetalae based on embryology. J Fac Sci Univ Tokyo Sect III Bot 11:263–281

Yamazaki T (1982) Recognized types in early development of the embryo and phylogeny in the dicotyledons (in Japanese). Acta Phytotax Geobot 33:240–249

Yeung EC (1980) Embryogeny of *Phaseolus:* The role of the suspensor. Z Pflanzenphysiol 96:17–28

Yeung EC, Clutter ME (1978) Embryogeny of *Phaseolus coccineus:* Growth and microanatomy. Protoplasma 94:19–40

Yeung EC, Clutter ME (1979) Embryogeny of *Phaseolus coccineus:* The ultrastructure and development of the suspensor. Can J Bot 57:120–136

Yeung EC, Sussex IM (1979) Embryogeny of *Phaseolus coccineus:* The suspensor and the growth of the embryo proper in vitro. Z Pflanzenphysiol 91:423–433

Yoo B, Jensen WA (1966) Changes in nucleic acid content and distribution during cotton embryogenesis. Exp Cell Res 42:447–459

Yoshida O (1957) Embryologische Studien über die Ordnung Piperales. 1. Embryologie von *Chloranthus japonicus*. J. Coll Arts Sci Chiba Univ (Nat Sci) 2:172–178

Yoshida O (1959 a) Embryologische Studien über die Ordnung Piperales. 2. Embryologie von *Chloranthus serratus*. J Coll Arts Sci Chiba Univ (Nat Sci) 2:295–303

Yoshida O (1959 b) Embryologische Studien über die Ordnung Piperales. 3. Embryologie von *Sarcandra glabra*. J Coll Arts Sci Chiba Univ (Nat Sci) 3:55–60

Zhukova GYA, Savina GI (1978) Electron microscope study of early embryo in *Epipactis atrorubens* (Hoffm.) Schult. (Orchidaceae). Bot Zh 63:1241–1246

Zinger NV, Poddubnaya-Arnoldi VA (1966) Application of histochemical techniques to the study of embryonic processes in certain orchids. Phytomorphology 16:111–124

Zweifel R (1939) Zytologisch-embryologische Untersuchungen *Balanophora abbreviata* und *B. indica*. Vjschr Naturf Ges Zürich 84:245–306

9 Polyembryony

K. K. Lakshmanan and K. B. Ambegaokar

9.1 Introduction

Angiosperms show the unique phenomenon of double fertilization. The egg, after fertilization, develops into an embryo. It is nourished in the younger stages by the suspensor, and in later stages by the endosperm – a post-fertilization tissue resulting from the fertilization of polar nuclei. Besides the egg, other constituent cells of the embryo sac, with or without fertilization, produce embryos. Moreover, the activity of one or more cells of the ovule, outside the embryo sac, also leads to the development of accessory embryos. Such embryos are asexual. Thus, in addition to the zygotic embryo, sexual and/or asexual non-zygotic embryo(s) may also develop in the same seed. Such a condition is referred to as "polyembryony".

Although the occurrence of twin embryos in orange was first reported by Leeuwenhoek in 1719, it was Strasburger (1878) who first demonstrated the nature of polyembryony in *Citrus, Coelebogyne, Funkia,* and *Nothoscordum*. Since then, polyembryony has been reported in many taxa. The types of polyembryony are diverse, and the probable causes, advocated from time to time, are far from satisfactory. Moreover, the genetical background and the ontogeny of accessory embryos have not been thoroughly investigated. During the last 25 years, our knowledge of polyembryony has not progressed as much as in other areas of plant sciences.

The term polyembryony has been defined as:

1. "die Bildung mehrerer Embryonen in einem Samen" (Schnarf 1929);
2. "... production of two or more embryos within an ovule" (Webber 1940);
3. "... occurrence of more than one embryo in a seed ..." (Maheshwari 1950);
4. "... all cases in which there are clear indications of the potential or actual occurrence of two or several proembryos or embryos in a single seed, irrespective of the question whether they may ultimately give rise to viable seedlings ..." (Bouman and Boesewinkel 1969).

Johansen (1950) questions the validity of including, under polyembryony, (1) embryos produced by various types of apomixis, (2) embryos produced by cells other than the egg cell, and (3) plural embryos resulting from multiple embryo sacs in an ovule. Thus, polyembryony should include only the accessory embryos arising from the zygotic proembryo. However, it is customary to refer to a seed as polyembryonate if two or more embryos occur therein, regardless of their origin. Which seed is truly polyembryonate? Maheshwari and Sachar (1963), while admitting the views of Johansen (1950), state that a truly polyembryonate seed is one in which the embryos are produced only from the zygotic proembryo, or from a mass of embryogenic cells. They also include embryos which develop from the synergids, antipodal cells, and sporophytic nucellar or integumentary cells.

The reports on the development of one or more embryos from antipodal cells as in *Paspalum scrobiculatum* (Narayanaswami 1954), *Ulmus campestris* (Guignard and Mestre 1966), and *Solidago canadensis* (Pullaiah 1978), or endosperm as in *Isomeris arborea* (Billings 1937) and *Brachiaria setigera* (Muniyamma 1977) are incorrect, and are discussed in Chapter 1.

9.2 Classification

Braun (1859) mentioned 60 cases of polyembryony under four categories: (1) fusion of two or more ovules, (2) several embryo sacs in the nucellus of the same ovule, (3) several cells of the same embryo sac developing into embryos, and (4) divisions of the proembryo. The findings of Strasburger (1878) on the cytology of the embryo sac and, subsequently, the process of fertilization, led to an additional category, namely, nucellar polyembryony. Ernst (1901) proposed a hierarchic system of classification which has generally been accepted, with some modifications. Alternative schemes have been suggested by Archibald (1939), Robijns and Louis (1942), Leroy (1947), Maheshwari (1950), Lebègue (1952), Johri (1965), Yakovlev (1967), and Bouman and Boesewinkel (1969). Archibald and Leroy stressed the role of true fertilization, and the development of endosperm. Robijns and Louis considered the development of an embryo within the embryo sac as "intrasaccal", or outside the embryo sac but within the same ovule as "extrasaccal". The views of Yakovlev and Johri are genetically based.

Taking advantage of the earlier classification, we propose a revised scheme, taking into account fertilization, presence or absence of embryo sac in the developmental pathway of a cell into an embryo, and morphological and cytological characteristics of the cell destined to give rise to the additional embryo.

Polyembryony can be broadly classified into "simple" and "multiple" depending upon the presence of one or more embryo sacs in the same ovule. The proposed classification is schematically represented below (induced polyembryony by chemical treatment as in *Eranthis*, and origin of multiple embryos in vitro are not included):

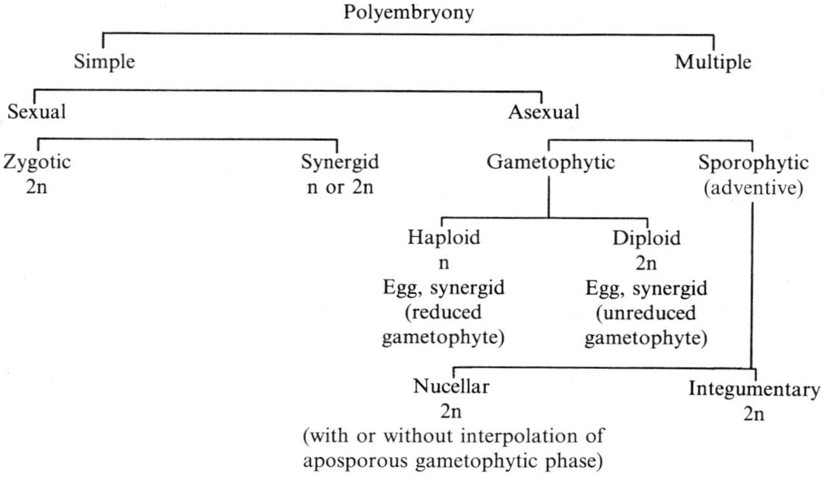

9.2.1 Simple Polyembryony

By its nature, that is whether it develops with or without fertilization, simple polyembryony can be sexual or asexual. The examples of sexual polyembryony are the embryos originating from the egg cell and synergid; supernumerary embryos also arise from the proembryo and suspensor. Asexual embryos develop within the embryo sac without fertilization. If the embryos develop from the nucellar and integumentary cells without the interpolation of the gametophytic phase, the polyembryony is said to be "adventive" or "sporophytic".

The failure of meiosis in megaspore mother cell, or the development of embryo sac in sporophytic tissue (nucellar or integumentary), result in asexual diploid gametophyte; then the egg (and synergid) can produce diploid embryos. Haploid embryos develop parthenogenetically from the synergid, or egg, of a reduced female gametophyte.

9.2.2 Multiple Polyembryony

In multiple polyembryony accessory embryos are produced from two or more embryo sacs in the same ovule. The development of embryo in each embryo sac conforms to the method described for simple polyembryony. Multiple gametophytes in an ovule may develop if, during ontogeny, two or more ovules (one embryo sac in each ovule), due to spatial demand, before or after the differentiation of integuments, fuse and function as a single ovule; fusion or division of two or more nucelli with common integuments, and one embryo sac in each nucellus; multiple archesporium leading to the formation of several embryo sacs. In sea-island cotton (*Gossypium barbadense*), Harland (1936) described the development of a haploid-diploid twin which resulted from multiple embryo sacs in an ovule. The fertilization of egg cell in one embryo sac induced the parthogenetic development of the egg cell in the other embryo sac.

In multiple embryo sacs within an ovule, whereas the egg cell (diploid) of the unreduced embryo sac may develop parthenogenetically, the egg cell (haploid) of the reduced embryo sac may develop after fertilization forming a diploid-diploid twin. If the unreduced egg is fertilized and the reduced egg develops parthenogenetically, the twin would be triploid-haploid.

Asker (1979) refers to four possible modes of reproduction in facultative apomicts: (1) the unreduced egg cells, by parthenogenesis, produce uniform maternal offsprings; (2) the unreduced egg cells, by fertilization, give rise to autotriploids (U hybrids); (3) the reduced egg cells, by parthenogenesis, give rise to (poly)haploids, and (4) the reduced egg cells, through fertilization, produce variable sexual offspring (R hybrids) (see also Chap. 10).

9.3 Nucellar Polyembryony

Nucellar (adventive) polyembryony is reported in 16 out of 172 crassinucellate families of angiosperms (Rangaswamy 1981). The embryos develop from the nucellar cells with (*Cortaderia jubata* – Philipson 1978) and without the interpolation of the gametophytic phase.

The nucellar cells destined to develop into embryos contain dense cytoplasm and starch. They divide actively, develop into small embryonic masses, and push their way into the embryo sac. Some of the embryonic masses take the lead, grow further, and differentiate into embryos. Those towards the micropylar region are better organized than those situated farther down laterally. These embryos exhibit various stages of development, as well as irregular organization. They usually lack a suspensor. Nevertheless, a well-defined suspensor has been observed in *Citrus microcarpa* (Webber and Batchelor 1948, Rangaswamy 1959). The cotyledons are mostly unequal. Some embryos may show tricotyly, and fasciation at the radicular end, so that the seedlings have multiple shoots but a single root.

The origin of the embryo may be autonomous, sporadic, induced, or stimulated. When the embryos develop autonomously, the entry of the pollen tube, or stimulation due to degenerated aposporous megagametophytes, accelerates the induction and development of nucellar embryos as in *Ochna serrulata* (Chiarugi and Francini 1930). Plural embryos can be induced in the nucellar cells by distant hybridization, application of physical or chemical stimuli, or pollination of flowers with abortive or foreign pollen. Stimulation by pollination in *Epilobium hirsutum* with the pollen of *E. dodonei, E. montanum,* and *E. angustifolium* (Michaelis 1925) induces nucellar embryos. This requires reinvestigation.

The successful development of nucellar embryos depends upon (1) fertilization of egg, and formation of endosperm in the sexual embryo sac, (2) development of normal endosperm but without the fertilization of egg, and (3) rarely, even in the absence of fertilization of the polars or the egg. There is considerable information on nucellar embryos in fruit trees like *Citrus, Eugenia, Mangifera,* and some others.

9.3.1 Rutaceae

Nucellar polyembryony is a common feature. In *Aegle marmelos* (Johri and Ahuja 1957) 75% to 80% of the mature embryo sacs degenerate completely. In the remaining embryo sacs the egg apparatus and the antipodals degenerate, but triple fusion occurs and the endosperm is well developed. One of the nucellar cells (at the micropylar end) enlarges, becomes richly cytoplasmic, and forms an embryonal mass which gives rise to two or three embryos. These develop concurrently in the embryo sac containing endosperm, exhibit varying degrees of fusion and, ultimately, only one reaches maturity. The cotyledons are unequal.

In *Toddalia asiatica* (Narmatha Bai and Lakshmanan 1982) the nucellar embryos develop in the middle region of the embryo sac, when the zygotic embryo is 32- to 64-celled. The growth of the zygotic embryo is arrested, whereas the nucellar embryo enlarges considerably and shows well-differentiated procambial strands. Further details are not available.

Polyembryony in *Citrus* has attracted much attention. In *Citrus microcarpa,* during post-fertilization development, Rangaswamy (1959) observed that several nucellar cells in the micropylar region become conspicuous by denser cytoplasm and large nuclei. These cells divide and redivide to produce embryonal structures which project into the embryo sac (Fig. 9.1 A–D). In an embryo sac, at the cellular stage of endosperm, there can be 9 to 21 proembryos at different stages of development. These structures pass through the globular, heart-shaped (Fig. 9.1 B, C), and torpedo stages and, finally, attain complete differentiation. The nucellar embryos do have a suspensor (Fig. 9.1 D), and are difficult to distinguish from the zygotic embryo. The cotyledons of nucellar embryos are unequal in length, or thickness, or both. The embryos are well supplied with vascular elements, and oil glands are distributed all along the margin of cotyledons.

The clones, by repeated vegetative multiplication, tend to deteriorate. The vigour can be restored by growing plants from nucellar embryos. The nucellar tissue of *Citrus* is often employed in experimental studies to obtain uniform seedlings. These seedlings are virus-free and are preferred by horticulturists because of their well-developed tap root system, which serves as a better orchard stock. Natural hormones, perhaps, bring about vegetative invigoration, or "neophysis", exercised by the "magic bath" of the embryo sac (Swingle 1932, Hodgson and Cameron 1938).

The number of embryos in each seed is related to (1) the age of plant – mature nucelli are more embryogenic than younger nucelli (Furusato et al. 1957), (2) nutritional status of fruit (Treub 1936), (3) pollen parent, and (4) environmental factors (Frost and Soost 1968). The number of seedlings per seed varies from species to species: *Citrus microcarpa* (Maheshwari and Rangaswamy 1958) produces 21 seedlings, and *C. unshiu* (Furusato et al. 1957) as many as 40. The nucellar cells of polyembryonate species can continuously produce genetically uniform embryos (Maheshwari and Rangaswamy 1958) which is a boon to horticulturists.

The observations of Furusato (1951) and Furusato et al. (1957) indicate that in *Citrus* mono- and polyembryonate seeds may occur side by side. The number of embryos per seed (1) is not affected by the characters of pollen parent, (2) depends upon the nutritional and physiological status of the plant, and (3) upon the position of the branches: northern branches yielding more embryos. The number of embryos produced varies if the fruits are situated somewhat apart. The production of embryos is influenced by the age of the tree, and tends to be higher during the year of prolific yield.

It is difficult to identify the zygotic embryos from nucellar embryos. This is a major hurdle in *Citrus* breeding as the breeder has to wait for a long period before he can isolate the hybrid from nucellar seedlings. Also, for clonal propagation the seedlings derived from the zygotic embryos need to be identified and removed. It would be a great advantage for citriculturists if a dependable method is standardized for an early detection of zygotic and nucellar embryos. Genetic markers expressing clear morphological traits in the seedlings are rather rare in *Citrus*. The dominant trifoliate leaf character in *Poncirus trifoliata* has been used as a marker in determining zygotic seedlings. This is not fully reliable; trifoliate character is determined by a single gene and, when the parent is heterozygous, half the zygotic embryos would not show this character.

Other methods for early identification of zygotic embryos have been explored. Pieringer and Edwards (1967) attempted to identify zygotic and nucellar embryos

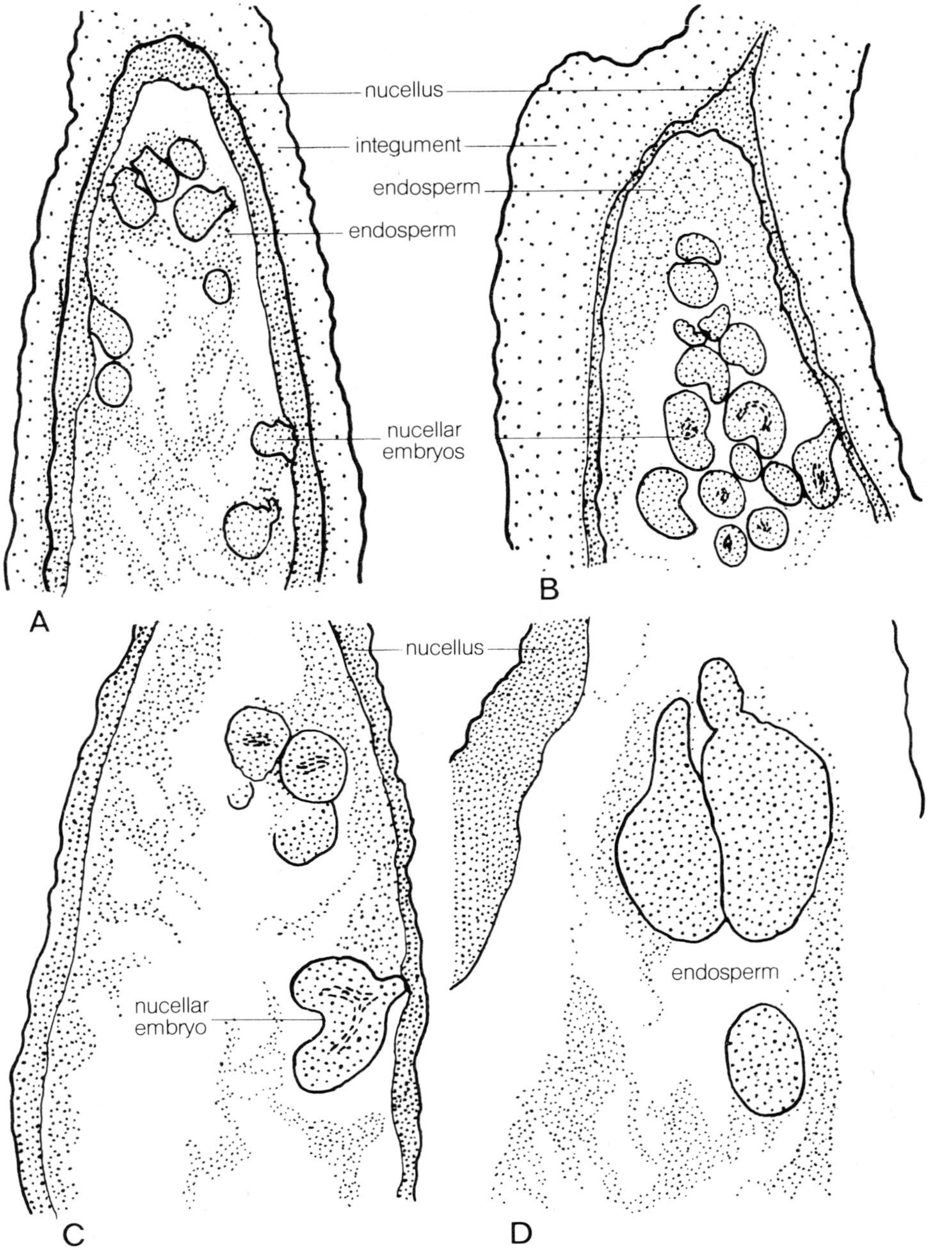

Fig. 9.1 A–D. Nucellar polyembryony in *Citrus microcarpa* (semi-diagrammatic). **A, B** Micropylar portion of ovule to show embryos of nucellar origin. **C** Of the several embryos one is early dicotyledonous and has a distinct suspensor. **D** Two embryos, with suspensor, are dicotyledonous; only one cotyledon is seen in each due to the plane of sectioning. (Adapted from Rangaswamy 1959)

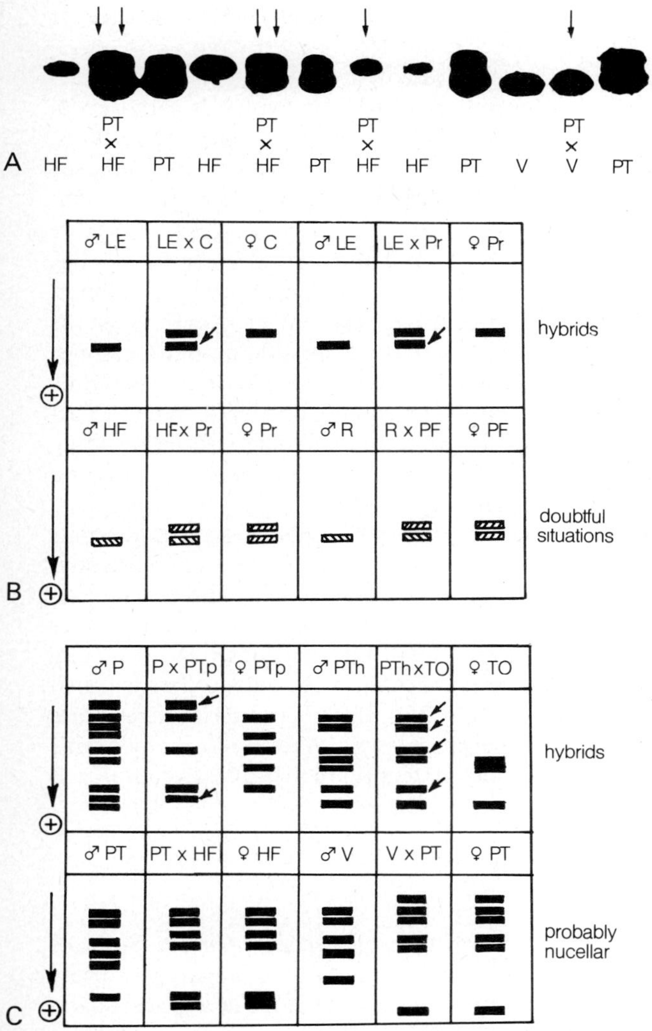

Fig. 9.2 A–C. Isoenzyme identification of zygotic and nucellar seedlings in *Citrus*. **A** Electrophoretic profiles of peroxidase isozyme from 20-day-old seedlings of parents and crosses. *Poncirus trifoliata* (*PT*) seedlings show two bands, while *Citrus limon* (Hibrido florida, *HF*) and *C. sinensis* (Valencia Orange, *V*) show only one band. Two of the seedlings from crosses *PT* × *HF* (*double arrows*) are of zygotic origin (as they show two bands), and two other seedlings from crosses *PT* × *HF* and *PT* × *V* (*single arrow*) are of nucellar origin as they show only one band corresponding to the female parent. **B** Peroxidase isozyme profiles from parents and interspecific and intervarietal crosses. In hybrid seedlings peroxidase band from each parent shows different mobility and, thus, zygotic seedlings can be identified by the presence of both the bands. *Arrows* indicate the bands contributed by the male parent. The origin of the crosses (doubtful situations) are not clear, as the parent seedlings do not show distinct differences. *C* Mandarin Cleopatra (*C. reticulata*); *LE* Lemon Eureka (*C. limon*); *Pr* Lemon Perrine (*C. limon*); *HF* Hibrido Florida (*C. limon*); *R* Mandarin Reina; *PF* Pummelo Foster (*C. paradisi*). **C** Esterase isozyme profiles of parents and crosses. The hybrids are of zygotic origin as the seedlings show bands from both the parents. *Arrows* indicate the band contributed by male parent. The profiles in *lower column* are probably of nucellar origin as they show the bands of only the female parent. *P* Orange Pina (*C. sinensis*); *PTp* Pummelo Triumph; *PTh* Pummelo Thompson (*C. paradisi*); *TO* Tangelo Orlando (*C. paradisi* var. Duncan × *C. reticulata* var. Dancy); *PT Poncirus trifoliata; HF* Hibrido Florida (*C. limon*); *V* Valencia Orange (*C. sinensis*); *PF* Pummelo Foster. (Adapted from Iglesias et al. 1974)

from 18-month-old seedlings on the basis of essential oils. Rangan et al. (1969) were able to identify and excise zygotic embryos from the ovules, before the development of nucellar embryos, and grow them into seedlings in aseptic cultures. Under certain environmental conditions, however, zygotic and nucellar embryos develop simultaneously and it is not feasible to identify and excise zygotic embryos for culture (see Iglesias et al. 1974).

Analysis of isozyme patterns following polyacrylamide gel electrophoresis has shown great promise for early identification of zygotic and nucellar seedlings. Iglesias et al. (1974) could identify zygotic seedlings in several crosses through the analysis of peroxidase and esterase isozyme patterns, from 20-day-old seedlings. The seedlings of *Poncirus trifoliata* showed two bands of peroxidase isozyme, whereas those of *Citrus limon* and *C. sinensis* showed only one band. When *P. trifoliata* was used as the male parent and *Citrus* spp. as female, all the resulting seedlings that showed two bands of peroxidases could be identified as zygotic, and those with a single band as of nucellar origin (Fig. 9.2 A).

Peroxidase band in many species and varieties of *Citrus* shows variation in its mobility. When two such species/varieties are crossed, the zygotic seedlings would show two bands corresponding to the bands present in the male and female parents (Fig. 9.2 B). However, in many species and varieties such clear-cut differences in peroxidase pattern are absent and, hence, the results are not categorical (Fig. 9.2 B). Similarly, zygotic seedlings, in some combinations, could also be identified on the basis of esterase isozyme profiles (Fig. 9.2 C). Both peroxidase and esterase analysis are useful only in a limited number of crosses. Analysis of other isozyme profiles over the whole range of crosses may be helpful in identification of nucellar and zygotic seedlings.

9.3.2 Anacardiaceae

Sachar and Chopra (1957) investigated 19 varieties of *Mangifera indica*. Some of these like Cambodia, Higgins, Olour, Paho, and an unknown Singapore variety are polyembryonate, and contain almost 50 embryos in a seed. The zygote usually degenerates except in Paho. The nucellar embryos arise from the micropylar region (Fig. 9.3 A–F), and grow up to the middle of the embryo sac. In Carabao even the epidermal cells of the nucellus proliferate and produce supernumerary embryos. When the zygotic embryo is also present, as in *M. odorata,* it is difficult to distinguish it from the nucellar embryos. Occasionally, the nucellar embryo may show three or more cotyledons. Some of the nucellar embryonal masses fuse with each other to varying degrees. The fusion in the radicular region is probably responsible for the appearance of seedlings with multiple shoots, but only one tap-root. Such seedlings have been recorded in Bambai, Langra, and Fazli varieties (Sen and Mallik 1940).

9.3.3 Myrtaceae

Van der Pijl (1934) reported the formation of nucellar embryos in *Eugenia jambos,* prior to pollination. However, they attained full size only after pollination, fertil-

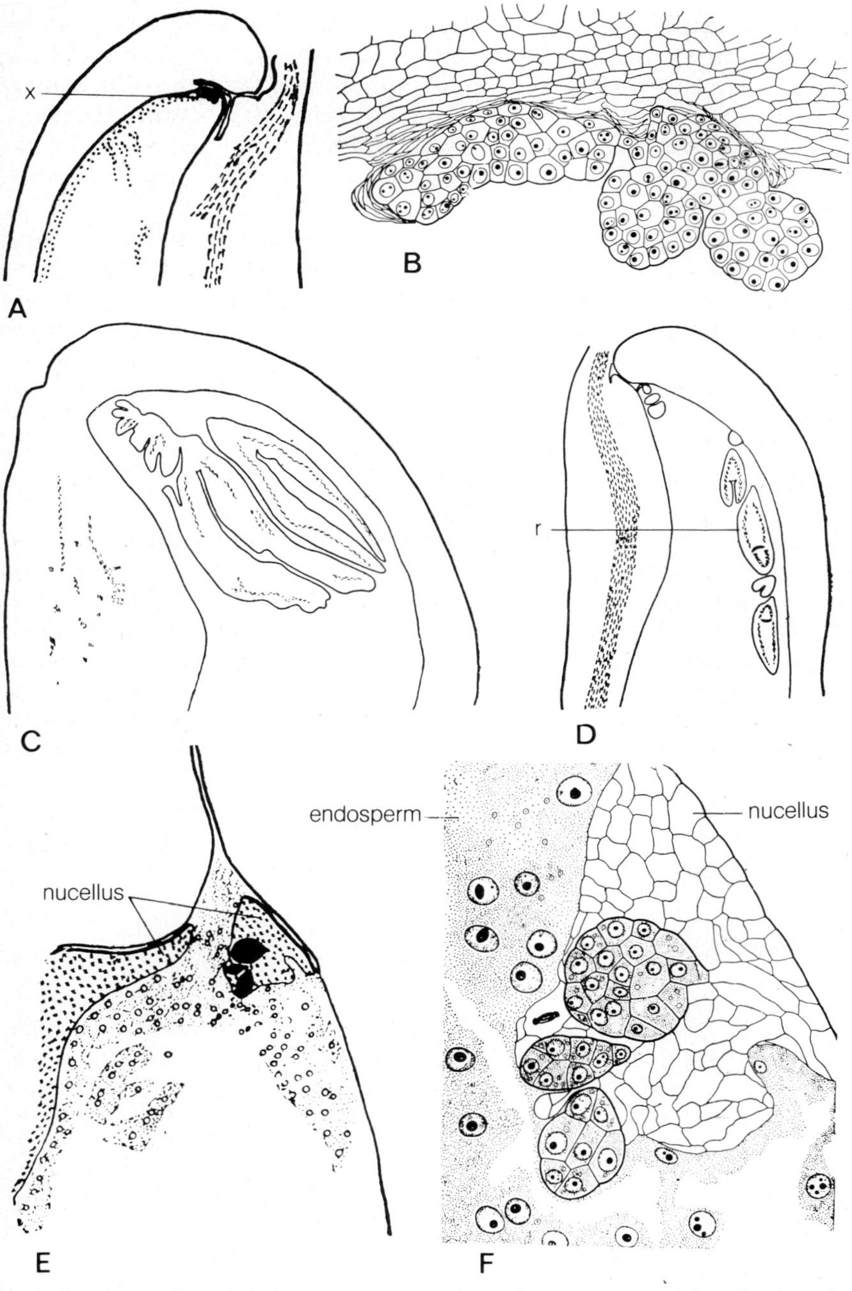

Fig. 9.3 A–F. Nucellar polyembryony. **A, B** *Magnifera indica* var. *Higgins*. **A** Longisection of micropylar part of ovule showing adventive proembryos at *x*. **B** Adventive proembryos at *x*. **C** *M. indica* var. Olour, micropylar portion of ovule with multiple adventive embryos; endosperm is not shown. **D** *M. indica* var. Paho, micropylar part of ovule with eight adventive embryos at different stages of development; the one marked *r* shows reverse polarity. **E, F** *M. odorata*. **E** Micropylar part of ovule with adventive proembryos (*black*). **F** Adventive proembryos from **E.** (After Sachar and Chopra 1957)

ization, and endosperm formation. In *Eugenia jambos* (Roy 1953), *Syzygium cumini* (Narayanaswami and Roy 1960), and *S. caryophyllifolium* (Roy and Sahai 1962) the nucellar embryos develop only after fertilization of the egg cell, and development of endosperm; the zygote degenerates. The initials of the nucellar embryos differentiate in the micropylar region, and undergo several divisions to produce embryonal masses which project into the embryo sac. Then they develop further in the presence of endosperm.

In *S. caryophyllifolium* 45%–50% of seeds are polyembryonate, and two to three embryos develop in each seed; the immature seed of *S. cumini* usually contains nine embryos. There is a competition for survival and, finally, the mature seed contains only three to five embryos. An embryonal mass of nucellar origin is reported in the chalazal region in *S. cumini*. Swamy (1948 a) also observed nucellar embryos in the chalazal region in *Trillium undulatum* (Liliaceae). This feature is rather rare.

9.3.4 Cactaceae

Maheshwari and Chopra (1955) pointed out that in *Opuntia dillenii* the pollen tube is guided towards the ovule by the papillate hairs on the inner surface of the ovule. Pollen tubes have been seen in some ovules, but double fertilization was not observed. The authors are not sure whether it does occur. The development of endosperm is autonomous, perhaps due to the stimulation resulting from the entry of pollen tube. The endosperm is consumed in free-nuclear state. The egg almost always collapses. The synergids also degenerate. The nucellar embryos with suspensor develop in the upper-half of the embryo sac. Only one or two of these reach maturity. Due to the fusion of embryonal masses, polycotyly is of common occurrence. In *Mammillaria tenuis* (Tiagi 1957, 1970) while several nucellar embryos develop after fertilization, only one reaches maturity. The zygotic as well as nucellar embryos show a well-defined suspensor.

9.3.5 Orchidaceae

Nucellar polyembryony may occur before, or after, pollination and fertilization of the egg cell. In *Zeuxine sulcata* (Swamy 1946 b) there is a high degree of pollen sterility. The ovule shows a row of only three to four nucellar cells. The megaspore mother cell, due to irregular meiotic activity, remains arrested. Two or three adventive embryos ($2n = 44$, 48) originate from the nucellar epidermis in the micropylar region. *Spiranthes cernua* (Swamy 1948 b) comprises sexual, asexual, and intermediate races. Nucellar embryos develop in sterile ovules of intermediate individuals, but their development is delayed until after the fertilization of fertile ovules. The mature seed contains two to three embryos.

9.4 Integumentary Polyembryony

The development of embryos from the integumentary cells is not dependent on the fertilization of the egg cell, and there is no interpolation of the gametophytic

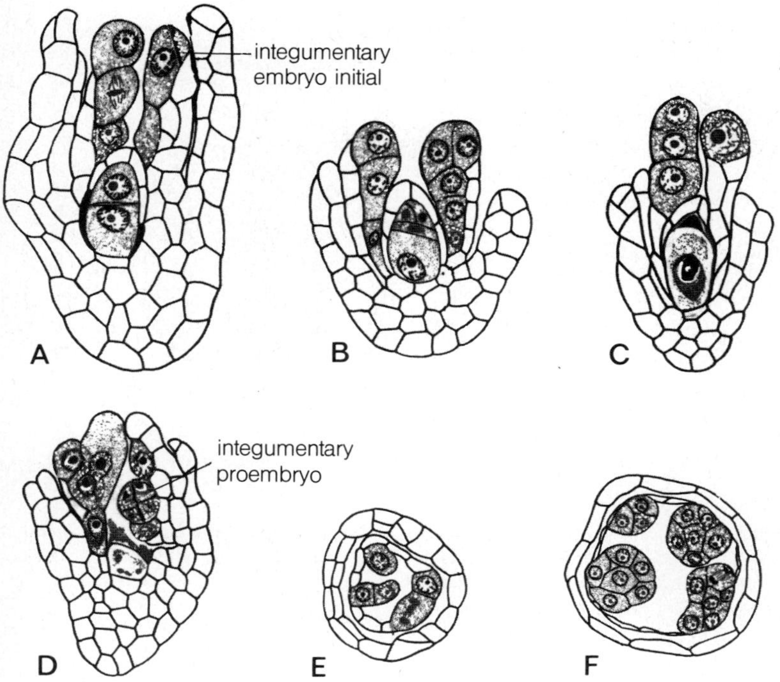

Fig. 9.4 A–F. *Spiranthes cernua.* **A–D** Various stages of adventive integumentary proembryos in micropylar region of a bitegmic ovule; longisections. **A** Dyad. **B** T-shaped tetrad. **C** Functional megaspore. **D** Degenerated nucellus, and degenerated four-nucleate embryo sac. **E, F** Transections of ovules, in micropylar region, to show adventive proembryos. (After Swamy 1948 b)

phase. In the asexual individuals of *Spiranthes cernua* (Swamy 1948 b) the inner integument exhibits an unusual phenomenon. The terminal cell, as well as the lower adjoining cells, enlarge and become richly cytoplasmic (Fig. 9.4 A–C). By the time the sexual embryo sac shows four degenerated nuclei (Fig. 9.4 D), the richly cytoplasmic integumentary cells divide and produce two to six proembryos in the micropylar region of the ovule (Fig. 9.4 E, F). Such embryos remain arrested and do not grow very much, nor do they differentiate.

9.4.1 Endothelial Polyembryony

When the innermost layer of the integument becomes distinct by its morphological and cytological features, it is referred to as "endothelium", and is a common feature in unitegmic, tenuinucellate ovules. According to Maheswari Devi and Pullaiah (1976 a), in *Melampodium divaricatum* in a number of ovules the cells of the innermost layer of the integument become richly cytoplasmic, and undergo repeated divisions (Fig. 9.5 A, B). Thus, one finds several developmental stages of the proembryo; some may simulate the earlier stages of a zygotic embryo. None of the proembryos develop far enough to differentiate. In *Carthamus tinctorius*

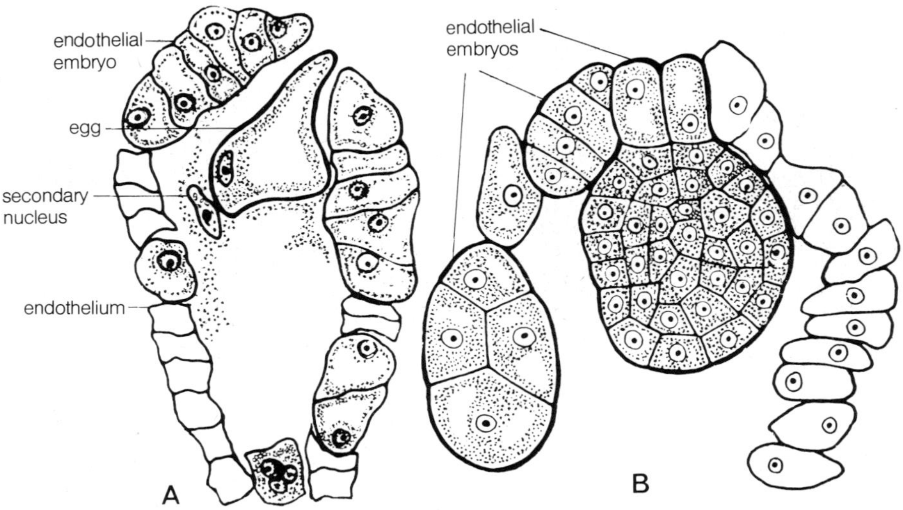

Fig. 9.5 A, B. Endothelial polyembryony in *Melampodium divaricatum*. **A** Embryo sac with degenerated egg cell, and secondary nucleus; endothelium shows adventive proembryos at different stages of development. **B** Portion of endothelium with one initial, two young, and one globular adventive proembryo. (After Maheswari Devi and Pullaiah 1976a)

(Maheswari Devi and Pullaiah 1977) supernumerary embryos develop from the endothelial cells during post-fertilization stages. They also do not show any differentiation, and do not reach maturity.

In *Vincetoxicum nigrum* Guignard (1921) had reported the development of additional embryos due to the cleavage of the group of embryonic cells formed by the zygote. Kordyum (1961) points out that in *V. officinale* the supernumerary embryos develop from the inner cells of the integument. This may also be true of *V. nigrum*, and Guignard may not have carefully followed the ontogeny of the adventive embryo. A reinvestigation may prove rewarding.

It is questionable whether such multicelled and globular structures should at all be considered as proembryos? So far, not a single instance of a polyembryonic seed due to differentiated integumentary or endothelial embryos is known. It would be interesting to induce, in vitro, embryos from integumentary cells.

9.5 Zygotic and Suspensor Polyembryony

9.5.1 Zygotic Polyembryony

In *Erythronium americanum* (Jeffrey 1895) the development of supernumerary embryos results from the cleavage of the apical cells of the embryogenic mass of cells produced by the zygote. A similar condition occurs in *Tulipa gesneriana* (Ernst 1901), *E. dens-canis* (Guérin 1930), *Cocos nucifera* (Whitehead and Chap-

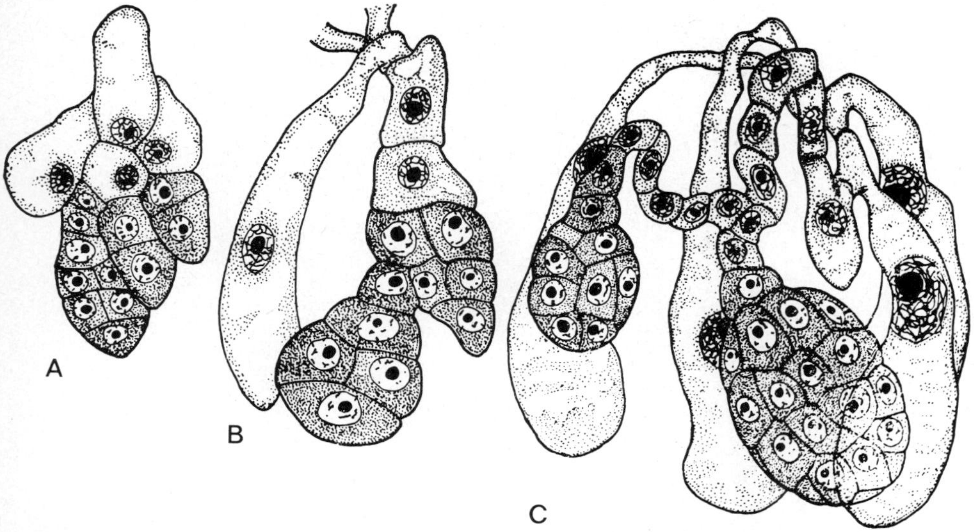

Fig. 9.6 A–C. Zygotic and suspensor polyembryony in *Eulophia epidendraea* (whole mounts). **A** The zygote has produced a group of cells, and three proembryos have developed therefrom. **B** Budding of an embryo from globular proembryo. **C** Twin proembryos presumed to have arisen by splitting of a single proembryo. Note the haustorial suspensor cells in **A–C.** (After Swamy 1943)

man 1962), *Primula auricula* (Veillet-Bartoszewska 1957), and *Corydalis cava* (Lakshmanan, unpublished).

In *Cymbidium bicolor* (Swamy 1942) the zygote divides vertically or obliquely, the daughter cells get somewhat separated, and divide further to form two independent proembryos. As is common in the Orchidaceae, the proembryos develop only up to the globular stage in mature seed. The frequency of polyembryony is 2%, and the monozygotic twin is unattached.

Like *Erythronium,* in *Eulophia epidendraea* (Swamy 1943), *Habenaria platyphylla* (Swamy 1946a), and *Geodorum densiflorum* (Swamy 1949) also the zygote produces an irregular mass of cells, and some cells at the apical end develop simultaneously into multiple proembryos (Fig. 9.6A). The additional embryo may also arise due to budding of the proembryo (Fig. 9.6B). Mathew (1980) observed such a behaviour in *Hamamelis virginiana.*

Swamy (1943) states that in *Eulophia* "the filamentous proembryo becomes branched and an embryo may be formed at the tip of each branch" (Fig. 9.6C). In one case as many as three proembryos developed.

Due to incomplete cleavage of the zygotic proembryo, in *Stachyurus chinensis* (Mathew and Chaphekar 1977), two proembryos develop side by side (Fig. 9.7 A, B).

Sedum spectabile (Fig. 9.7C) (Vignon-Fétré 1966) is an unusual instance, where the additional embryo is produced by cleavage of the apical cells of the three-celled proembryo. This should be reinvestigated.

Of the 49,903 maize seeds examined, Sarkar and Coe (1966) observed 32 pairs of twin embryos. The authors state that this condition was either due to cleavage

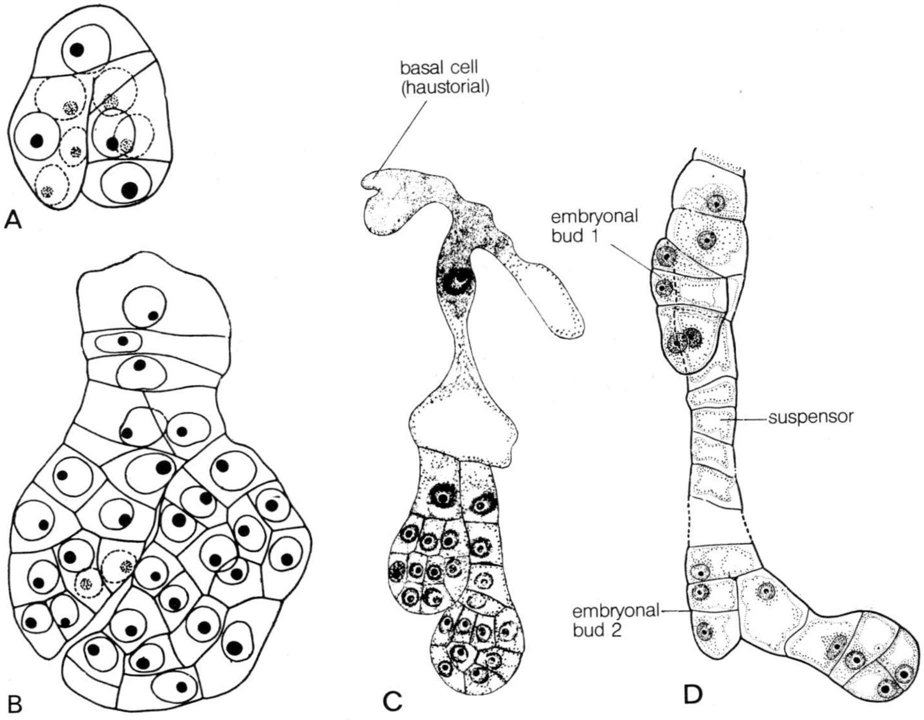

Fig. 9.7 A–D. Zygotic (**A–C**) and suspensor (**D**) polyembryony due to cleavage. **A, B** *Stachyurus chinensis,* **C** *Sedum spectabile.* Note the haustorial suspensor cell in **C**. **D** *Zygophyllum fabago,* proembryo with embryonal buds arising from the uniseriate suspensor. (**A, B** after Mathew and Chaphekar 1977, **C** after Vignon-Fétré 1966, **D** after Masand 1963)

of the proembryo, or a result of the fertilization of two reduced female cells in the embryo sac by two sperms. Besides these diploid-diploid twins, 15 were haploid-haploid, and 2 haploid-diploid (see also Sect. 9.6.1.1).

9.5.2 Suspensor Polyembryony

The cells of the uniseriate suspensor (Fig. 9.7 D) develop buds in *Zygophyllum fabago* (Masand 1963). Suspensor polyembryony is of common occurrence in Acanthaceae: *Dipteracanthus patulus* (Maheshwari and Negi 1955), *Acanthus ilicifolius,* and *Sericocalyx scaber* (Ambegaokar, unpublished).

The plural embryos, which develop from the proembryonal mass, or a suspensor cell, have the diploid number of chromosomes.

9.6 Synergid Polyembryony

Tretjakow (1895) and Hegelmaier (1897) were the first to describe the development of additional embryos from the synergids. These embryos develop frequent-

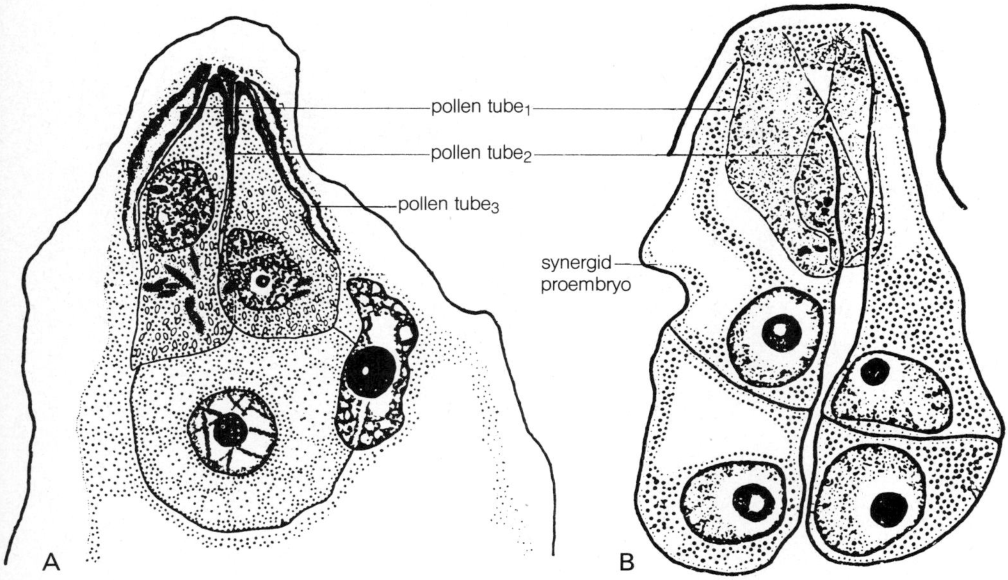

Fig. 9.8 A, B. Synergid polyembryony. **A** *Cuscuta reflexa,* micropylar part of fertilized embryo sac with zygote and primary endosperm nucleus; note remnants of three pollen tubes, and several male gametes in each synergid. (After Johri and Tiagi 1952). **B** *Tamarix ericoides,* upper portion of embryo sac with two-celled zygotic (*right*) and synergid (*left*) proembryos (presumably both are diploid); note remnants of two pollen tubes. (After Johri and Kak 1954)

ly, after fertilization. When the synergid and egg cell are fertilized by two sperm cells of the same pollen tube, there is no possibility for the secondary nucleus to be fertilized. The embryos collapse in the absence of endosperm. When more than one pollen tube is involved in fertilization, the development of the embryos from the egg and synergid is accompanied by well-organized endosperm. The synergid and egg cell are both fertilized in *Najas major* (Guignard 1901), the embryos develop equally, and resemble each other. Such diploid gametophytic polyembryony has been observed in several other members. Entry of additional pollen tubes has been observed in *Cuscuta reflexa* (Fig. 9.8 A) (Johri and Tiagi 1952), *Tamarix ericoides* (Fig. 9.8 B) (Johri and Kak 1954), and *Pennisetum squamulatum* (Sindhe et al. 1980). In the latter two taxa the diploid synergid embryo develops along with the zygotic embryo.

Sachar (1955) reported synergid polyembryony in *Argemone mexicana.* He observed that the synergids are ephemeral but, sometimes, one synergid persists after fertilization. He has illustrated a series of developmental stages of zygotic and synergid proembryo to establish the correct identity of the proembryo originating from the synergid (Fig. 9.9 A–C). This is smaller than the zygotic proembryo; both have independent suspensors, and the proembryos lie juxtaposed.

Souèges (1949) had pointed out earlier that, in *A. mexicana,* the additional proembryo is formed due to the segmentation of the massive suspensor. Both the proembryos have a common suspensor. Sachar (1955) remarks that Souèges

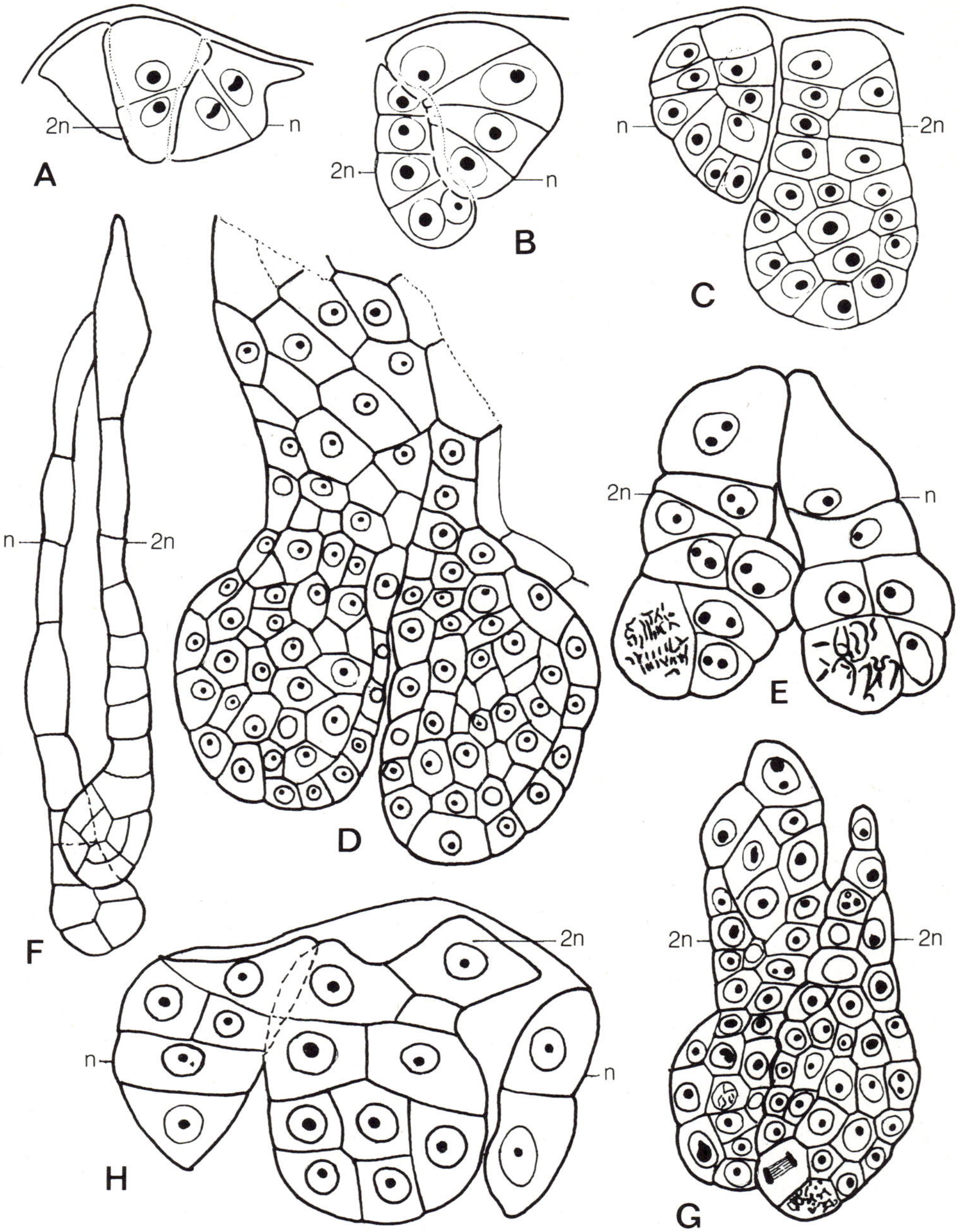

Fig. 9.9 A–H. Synergid polyembryony. **A–C** *Argemone mexicana,* synergid n and zygotic 2n proembryos. **D** *Dioscorea composita,* zygote (*left*) and synergid (*right*) proembryos (presumably both diploid). **E** *Lilium martagon*. **F** *Arabis lyallii,* synergid n and zygotic 2n proembryos. **G** *Pennisetum squamulatum,* synergid (*right*) and zygotic (*left*) proembryo, both diploid. **H** *Azadirechta indica,* two lateral synergid n and central zygotic 2n proembryos. (**A–C** after Sachar 1955, **D** after Thirumaran and Lakshmanan 1982, **E** after Cooper 1943, **F** after Lebègue 1948, **G** after Sindhe et al. 1980, **H** after Nair and Kanta 1961)

(1949) misinterpreted the synergid proembryo as a cleavage product of the zygotic proembryo. Crété (1956) also failed to observe the occasionally persistent synergid in *A. mexicana,* and confirmed the observations of Souèges. Sachar (1955) reexamined the description and figures by Souèges and Crété, and confirmed his earlier finding: the twin comprises a diploid zygotic and a haploid synergid proembryo. The zygotic embryo alone develops further so that the seed is monoembryonate.

In *Dioscorea composita* (Fig. 9.9 D) (Thirumaran and Lakshmanan 1982) the zygotic and synergid embryos show comparable development until a late stage. They follow a uniform pattern of development up to organogenesis, in *Nechamandra alternifolia* (Lakshmanan 1963).

The development of haploid embryo from the synergid, along with the diploid zygotic embryo, is reported in several taxa: *Lilium martagon* (Fig. 9.9 E) (Cooper 1943), *Arabis lyallii* (Fig. 9.9 F) (Lebègue 1948), *Erythraea centaurium* (Crété 1949), and *Aristolochia bracteata* (Johri and Bhatnagar 1955). The suspensor of the zygotic as well as of the synergid embryo of *Arabis lyallii* (Fig. 9.9 F) elongates into similar filamentous structures with a small embryonal mass at the apical end. Both the embryos have a reduced suspensor in *Erythraea centaurium,* and develop up to the heart-shaped stage. The zygotic and the synergid embryos are simultaneously nourished by the endosperm.

In *Pennisetum squamulatum* (Sindhe et al. 1980) the diploid synergid embryo develops along with the zygotic embryo. The suspensor of the zygotic embryo is typically multicellular as in the Gramineae, and that of the synergid is partly uniseriate and partly biseriate (Fig. 9.9 G).

At times, one or both the synergids (Fig. 9.9 H) develop into haploid embryo(s) as in *Azadirachta indica* (Nair and Kanta 1961). In *Carthamus tinctorius* (Maheswari Devi and Pullaiah 1976 b) triplets develop due to fertilization of the egg and both the synergids.

It is difficult to explain the mechanism of stimulus that induces the unfertilized synergid to develop into an embryo. Such embryo complexes occur frequently either in distant crosses, or as a result of a special pretreatment with pollen (e.g. X-ray treatment), or exposing the ovaries, just after pollination, to varying temperature. To begin with the zygotic and synergid proembryos are morphologically similar. However, the haploid proembryo remains arrested since the diploid proembryo (perhaps being more compatible) utilizes most of the endosperm, starving the weaker one.

9.6.1 Twins and Triplets

The multiple embryos inside a seed can be of varied origin and with different ploidy. But the frequency of polyembryony is extremely low and, therefore, it becomes difficult to assess the origin of additional embryos. In *Capsicum frutescens* (Morgan and Rappleye 1954) the frequency of polyembryony is 0.06% to 0.65% in the different varieties tested.

The following combinations are reported.

9.6.1.1 Twins

1. Haploid-Haploid: In nature the occurrence of haploid-haploid twins is rare. The twin may develop from an unfertilized synergid and an unfertilized egg cell (of the same embryo sac) as in *Orchis maculata* (Hagerup 1944). Haploid twins are also reported in *Gossypium barbadense* (Beasley 1940) and *Sesbania aculeata* (Haque 1946); it was not ascertained whether these twins originated from the parthenogenetic development of the egg and synergid, or both synergids. A reinvestigation is desirable.

An examination of 49,903 dry seeds of maize, by Sarkar and Coe (1966), showed 49 twins. Of these 15 pairs were haploid-haploid. Due to complete similarity, the authors concluded that each twin originated from a single embryo sac. Whether the twin developed by cleavage of the haploid embryo, or from two cells of the embryo sac was not determined. However, as each constituent of the twin embryo in seeds, and many twin seedlings, were developmentally identical, and mutually oriented, the authors presumed that cleavage of the haploid embryos resulted in twins.

2. Haploid-Diploid: The haploid-diploid twins are of common occurrence. The haploid embryo develops from an unfertilized synergid, while the diploid originates from the fertilized egg: *Plantago lanceolata* (Souèges 1926), *Linum usitatissimum* (Kappart 1933), *Solanum tuberosum* and *Phleum pratense* (Müntzing 1937), *Orchis maculatus, Listera ovata,* and *Platanthera chlorantha* (Hagerup 1944, 1947).

Of the seven spp. of *Lilium,* Cooper (1943) noted a high frequency of haploid-diploid twins in *Lilium martagon* and *L. henryi.* Nearly 1% of the ovules examined showed stimulated synergids and their development into haploid embryos. The nuclear division in synergid may precede, accompany, or follow that of the zygote. The haploid embryo grows slowly and degenerates early, while the diploid zygotic embryo shows vigorous growth and consumes most of the endosperm.

Yamamoto (1936) reports that in *Triticum vulgare,* of the twin which develops in the same embryo sac, the haploid member arises from the unfertilized egg, and the diploid from the fertilized synergid. A reinvestigation is necessary to confirm this feature. However, it may be mentioned that the investigations of Batygina (1969) on *Triticum aestivum, T. dicoccum, T. monococcum,* and *T. zhukovsky,* and Bhatnagar and Chandra (1975) on *T. aestivum,* do not make any mention of polyembryony.

3. Haploid-Triploid: In *Phleum* (Nissen 1937) the unfertilized synergid gives rise to the haploid embryo, while the triploid embryo arises due to the fertilization of a diploid (unreduced) egg of aposporous embryo sac, or fertilization of the haploid egg cell by two haploid male gametes, or one diploid (unreduced) male gamete.

4. Diploid-Diploid: Conjoined and unattached monozygotic twin seedlings, developing from the cleavage of the proembryo, are reported in *Asparagus officinalis* (Randall and Rick 1945).

The diploid-diploid twin can also arise by fertilization of a synergid and egg cell as in *Geum rivale* (Juguet 1966). If the two male gametes from the same pollen tube fertilize the haploid egg and haploid synergid, there is a possibility of the formation of identical twins as in some orchids (Hagerup 1947). Diploid twins are also reported in *Nicotiana tabacum* (Cameron 1949) and *Medicago sativa* (Greenshields 1951).

Two more pathways are likely for the development of diploid-diploid twins. There is a possibility of the formation of more than two male gametes in the same pollen tube, so that the supernumerary male gametes can fertilize the egg, synergid(s), as well as effect triple fusion, producing a diploid twin or triplet and normal endosperm. Johri and Tiagi (1952) observed 10 (or 11) male gametes lying in the synergids, and three pollen tubes in an embryo sac, in *Cuscuta reflexa* (Fig. 9.8 A). The authors mention that polyspermy may occur in the pollen grain or the pollen tube, and one of the synergids may be fertilized.

Secondly, a diploid-diploid twin can also result due to the endoduplication of chromosomes of a haploid synergid, and fertilization of the haploid egg.

5. *Diploid-Triploid:* Such twins are reported in *Triticum* (Yamamoto 1936). According to Yamamoto the diploid embryo develops from the zygote, while the triploid embryo is of endospermic origin. As yet there is no authentic report of the development of embryo from endosperm (see Chap. 1, Sect. 8.2). However, the following possibilities are worthy of consideration: (1) In an unreduced embryo sac the diploid synergid may develop parthenogenetically, while the diploid egg is fertilized by one haploid sperm. (2) In a reduced embryo sac, the haploid synergid is fertilized by a haploid male gamete, and the haploid egg by two haploid male gametes.

9.6.1.2 Triplets

In *Sagittaria graminea* (Johri 1936) double fertilization is a normal feature, and one of the synergids collapses during the process. In a few cases the synergids simulate the egg cell with the vacuole in the upper part and nucleus in the basal part, and two pollen tubes enter the embryo sac. One embryo sac showed three two-celled proembryos, and remnants of two pollen tubes at the micropylar end. The author concludes that the two additional proembryos developed from the two synergids, both fertilized by the male gametes from the second pollen tube. Thus, there is a diploid-diploid-diploid triplet.

The haploid-diploid-haploid triplet in *Azadirachta indica* (Nair and Kanta 1961) and diploid-diploid-diploid triplet in *Carthamus tinctorius* (Maheswari Devi and Pullaiah 1976 b) are discussed in Section 9.6.

A diploid-triploid-diploid triplet can arise by the fertilization of an unreduced diploid egg by a haploid male gamete, and parthenogenetic development of the diploid synergids.

9.7 Polyembryony in Hybrids

The occurrence of polyembryony in hybrid seeds has not attracted much attention.

Table 9.1. Origin of multiple seedlings of *Capsicum frutescens* L. arising from individual polyembryonic seeds, following reciprocal crosses of two varieties. (Adapted after Morgan and Rappleye 1954)

Parents	Twins			Triplet	
	Unattached			Conjoint	Unattached
	Sexual-sexual	Maternal-sexual	Maternal-maternal	Sexual-sexual	Sexual-sexual-sexual
	2n 2n	n 2n	n n	2n 2n	2n 2n 2n
Non-pungent, non-elongated, dark-green immature fruit × pungent, elongated, yellow immature fruit	72	11	1	7	1
Pungent, elongated, yellow immature fruit × non-pungent, non-elongated, dark-green immature fruit	10	4	0	3	0
Total 109	82	15	1	10	1

In two instances, in material collected 144 h after pollination, Cooper (1943) observed the development of haploid-triploid twin in a cross *Nicotiana glutinosa* (♀ n = 12) × *N. tabacum* (♂ n = 24). Both the proembryos were four-celled, the larger one was triploid and had originated by fertilization of the egg cell. The smaller haploid embryo developed parthenogenetically from the synergid. The ratio of average volume of nuclei of the larger embryo to the smaller embryo was 3.18:1. The author considers this difference to indicate that the twin is triploid-haploid.

In reciprocal crosses of two varieties of *Capsicum frutescens,* Goliath and Hungarian yellow-wax, Morgan and Rappleye (1954) noticed that the seedlings were haploid-haploid, haploid-diploid, diploid-diploid twins, and a diploid-diploid-diploid triplet (Table 9.1).

The haploid-haploid twin resulted from parthenogenetic development of both synergids, haploid-diploid from the development of one synergid and fertilized egg, and diploid-diploid from cleavage of the zygotic proembryo. The last-named identical twins developed into conjoined or unattached seedlings. The conjoined seedlings apparently originated from incomplete cleavage, and the unattached seedlings due to complete cleavage during early phases of embryonic development. The "unattached triplet", stated as sexual-sexual-sexual (Table 9.1), developed from the fertilized synergids and zygote of the same embryo sac, or multiple embryo sacs in the same ovule, presuming that more than one pollen tube was involved. The exact origin was not determined.

Following cross-pollination in *Melilotus,* polyembryony has been reported (Jaranowski 1961). In *Melilotus alba* (Hubam) × *M. messanensis* (Fig. 9.10 A) the diploid zygotic embryo, 6 days after pollination, showed budding resulting in an additional embryo from the massive globular proembryo. When a cross with changed partners was made, that is, *M. messanensis* × *M. alba* (Hubam) (Fig. 9.10 B), sometimes haploid-diploid twins developed. It is assumed that the diploid zygotic embryo and the haploid synergid embryo would have started developing at the same time. But, at this stage, 140 h after pollination, the zygotic embryo takes

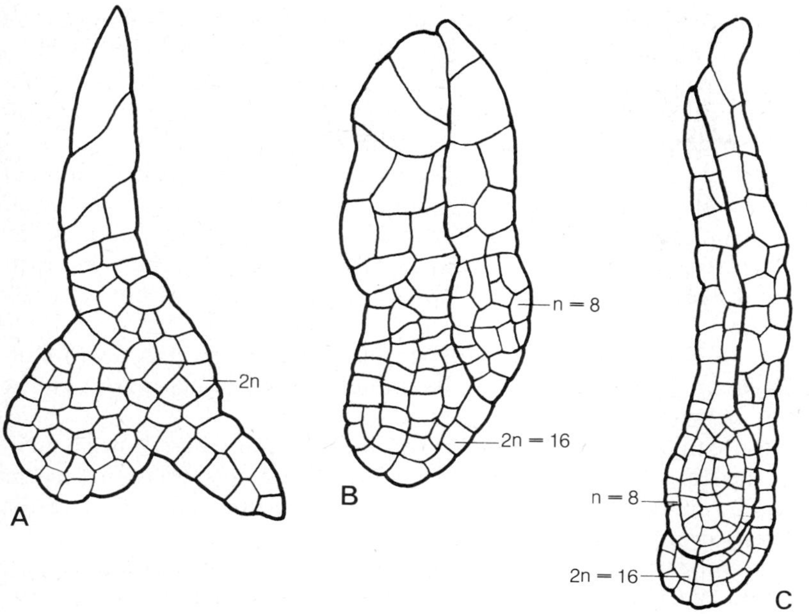

Fig. 9.10 A–C. Polyembryony in hybrids. **A** *Melilotus alba* (Hubam) × (*M. messanensis*), 144 h after pollination, a budded proembryo from diploid proembryo. **B** *M. messanensis* × *M. alba* (Hubam), 140 h after pollination, haploid-diploid twin. **C** *M. messanensis* × (*M. polonica* × *M. alba*), 120 h after pollination, haploid-diploid twin. (After Jaranowski 1961)

the lead and grows faster than the synergid embryo. The two dissimilar embryos show the somatic pattern of 8 and 16 chromosomes. Similar haploid-diploid twins have been observed in a cross between *M. messanensis* and (*M. polonica* × *M. alba*) (Fig. 9.10 C). Keeping the same male parent (*M. polonica* × *M. alba*) and using *M. alba* (Hubam) as female, two haploid and one diploid proembryo developed.

The embryos produced with *M. polonica* × *M. alba* as male parent had an elongated suspensor, and a higher potential for the development of additional embryos. Figure 9.10 B, C are haploid-diploid twins but with a different structure of the suspensor.

Lee (1970) reported a haploid-diploid hybrid twin from the cross *Gossypium barbadense* (♀) × *G. davidsonii* (♂). The zygotic embryo was diploid, while the haploid embryo developed parthenogenetically from the synergid.

Thus, the twins in which one embryo is haploid and the other haploid, diploid, or triploid, and triplets with haploid-diploid-haploid, diploid-diploid-diploid and other combinations, usually develop in intervarietal, interspecific, and intergeneric crosses. All such cases should be critically verified cytologically, and the sequence of development should also be carefully followed.

9.8 Causes of Polyembryony

Polyembryony, for a long time, was regarded as a freak, an abnormality, an exception, and explained as a primitive feature, gradual replacement of sexual reproduction caused by weakening of sexuality, early stages of development of "something" new, a phenomenon to increase the capacity of the species for dispersal, or mere somatic buds due to stimulation developed into embryos. Subsequently, with ontogenetic, morphological, physiological, cytological, and genetical studies, several causes of polyembryony have been suggested.

Maheshwari and Rangaswamy (1958), while discussing the in vitro culture of nucellar embryos of *Citrus* and *Mangifera,* raise the following questions: (1) What factors determine the inception of nucellar embryos? (2) Can the stimulus of fertilization be replaced, or overcome, by any hormone, or suitable chemical treatment? (3) Can the adventive embryos be induced in plants which are not naturally polyembryonate?

Nucellar embryony does not occur in nature in *Vitis vinifera*. However, Mullins and Srinivasan (1976) succeeded in obtaining plantlets of nucellar origin by in vitro culture of ovules (see Chap. 16).

The results obtained during the last 20 years offer some answers. Nucellar embryos show genetical identity. However, phenotypic variability between nucellar seedlings is quite marked. Perhaps, pollination and fertilization influence the genomic expression in adventive embryos.

The association of meiotic abnormalities, endoduplication of chromosomes in haploid synergid/egg, parthenogenetic development of unreduced/reduced synergid/egg, cleavage/budding of zygotic proembryo, budding of suspensor, occurrence of multiple embryos in hybrids, and nucellar and integumentary polyembryony have not provided any clues to the causes of this interesting phenomenon. Polyembryony is considered to be a recessive hereditary character, controlled by a series of multiple genes (Kappert 1933, Maheshwari and Rangaswamy 1958), or by one or more recessive genes (Leroy 1947) which come together in suitable recombinations following hybridization. When the supernumerary embryos follow the normal pathway, they can attain maturity. Rhoedes (1961) offered a cytological interpretation for the successful development of the seed with several mature embryos. The chromosomal balance between the embryo, endosperm, and maternal tissue is very important. A ratio of 2:3:2 is the successful formula comprising diploid embryos, triploid endosperm, and diploid sporophytic tissue. In an unreduced embryo sac, the egg may develop parthenogenetically into an embryo (2n), the endosperm formed after fertilization (4n+n) is 5n, and the maternal tissue remains diploid (2n). Such an unbalanced ratio results in the degeneration of embryo.

9.9 Induction of Polyembryony

In a seed, the embryo develops heralding the new sporophytic generation of the plant. It is usually the sexual embryo which undergoes dramatic changes, differentiates various organs, and attains maturity. The constituent cells of the embryo sac apparently have an identical physical and physiological environment. This is really not so. In a live and metabolically active embryo sac the physical and chemical environment keeps changing all the time. It is, therefore, only natural that, after fertilization, the synergids, egg, antipodals, and polars should behave differently. Each cell has to perform a distinct function, and the ultrastructural studies reveal enough evidence for their specialized role. Nevertheless, one or both of the synergids sometimes produce accessory embryos. Is it that their physiological and genetic make up, and position in the embryo sac, are responsible for this behaviour?

The nucellar cells also produce multiple embryos, and their physiological and biochemical status must also be quite different from the adjoining cells, as well as from the embryo sac.

Besides the physiological causes that help induction of polyembryony, the physical factors are also important. Treub (1936) pointed out that environmental conditions influence the production, development, and survival of the number of seedlings per seed. Juliano (1937) reported that a mango tree that is monoembryonate in one locality may behave as polyembryonate in another locality, thus stressing more the habitat than genetical factors. However, according to Leroy (1947), the monoembryonate forms carry dominant genes, whereas the polyembryonate condition prevails in areas where the forms with recessive genes grow.

The application of various hormones has failed to achieve artificial induction of adventive embryony. Injection of naphthalene-acetic acid (NAA) and indolebutyric acid (IBA) in the ovaries of *Datura fastuosa* (Chopra and Sachar 1957) resulted in the formation of parthenocarpic fruits. The embryo sac cavities, in these fruits, were filled with irregular masses of densely cytoplasmic cells which did not show any differentiation into an embryo. This tissue originated from the endothelium. Van Overbeek et al. (1941) also reported similar outgrowths which filled the embryo sac, in *D. stramonium*.

Eranthis hiemalis is an unusual plant. At the shedding stage the mature seed has a 0.1 mm long, pear-shaped, undifferentiated embryo with a conspicuous suspensor (Haccius 1953, 1963). The cotyledonary primordia and the radicle develop in the soil. The process of intraseminal growth and differentiation continues for about 8 months, and the mature embryo attains a length of 2 mm. It comprises 90% cotyledons, and 10% hypocotyl and radicle. In contrast to other angiosperms, the cotyledons are closely adpressed to each other in the upper-half region (called "cotyledonary sheath" by the author); the lower half of the cotyledons is free.

When young undifferentiated pear-shaped embryos with a long suspensor, excised from freshly harvested seeds of *Eranthis hiemalis* (Haccius 1957), are treated with 2,4-dichlorophenoxyacetic acid (500–1,000 ppm), differentiation processes are temporarily inhibited without affecting the high mitotic activity. After a lapse of time, when organ differentiation is resumed, additional primordia are initiated, or the entire embryo cleavages into twins (Haccius 1960). Haccius successfully induced plural embryos by the application of other chemicals as well. Young undifferentiated embryos of *Eranthis hiemalis* (Haccius 1963), when treated with Mc Ilwaine's citric acid-phosphate buffer mixture in acidic solution, selectively killed the fully meristematic cells of the embryonal mass while the suspensor and the endosperm survived. The physiologically isolated but vital suspensor cells are thus released from the inhibiting influence of the original embryo. They regain mitotic potency and proliferate into groups of cells developing into normal or irregular

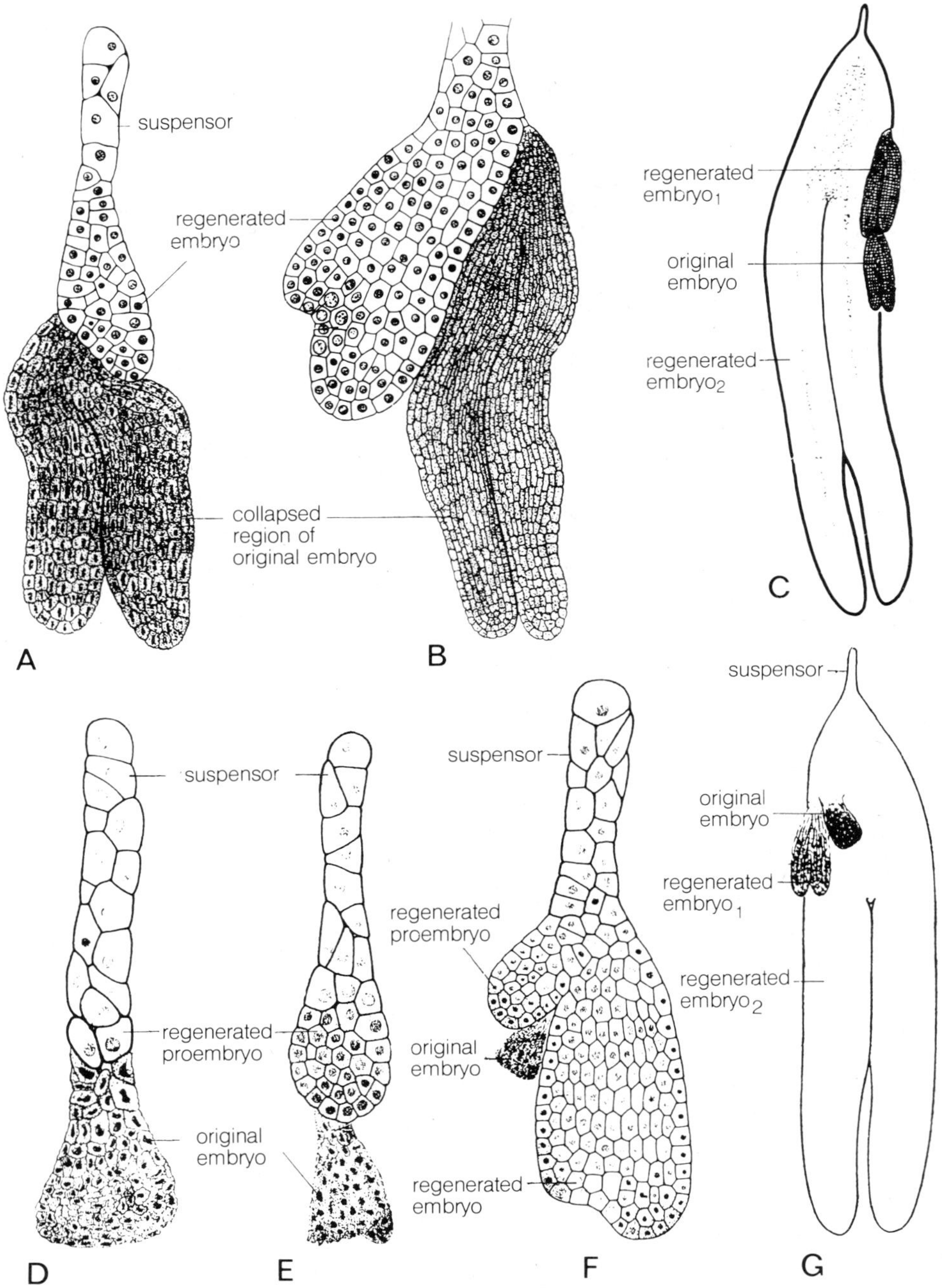

Fig. 9.11 A–G. Induced suspensor polyembryony in *Eranthis hiemalis*. **A–C** Chemical treatment resulting in the collapse of the original embryo (*dark*), and regeneration of embryos from suspensor cells. **D–G** Differentiation of suspensor cells into embryos. The original embryo is selectively killed by acidic buffer solutions. **D–F** Results obtained after 8 (**D**), 14 (**E**), and 15 (**F**) days. **G** Another treatment after 8 weeks produces secondary adventive embryos. (**A–C** after Haccius 1965, **D–G** after Haccius 1978)

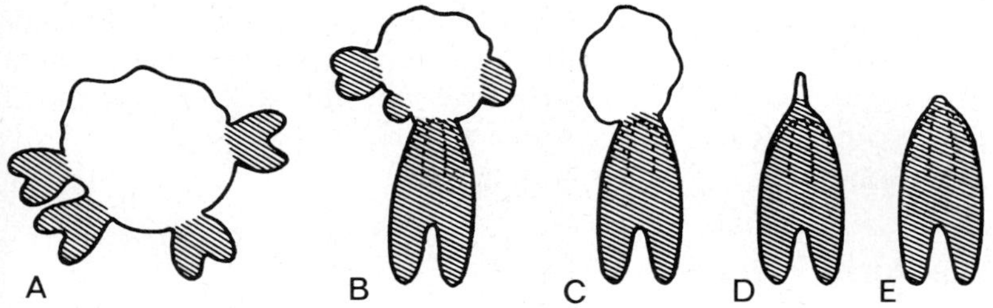

Fig. 9.12 A–E. Induced suspensor polyembryony; **A–C** in vitro, **D, E** in vivo. All figures diagrammatic. **A** Four dicotyledonous embryos developed from the proembryonal cell complex (indeterminate cleavage polyembryony). **B** Four embryos at different stages of development from the proembryonal cell complex; the terminal embryo is dicotyledonous, while the growth of others has been arrested due to inhibitory effect of the terminal embryo. **C** Dicotyledonous embryo arising from proembryonal complex, or massive suspensor. **D** Embryo with filiform suspensor. **E** Embryo with inconspicuous suspensor. (After Haccius 1978)

embryos. Additional suspensor cells develop, and give rise to supernumerary embryos.

The suspensor cells of fully differentiated embryos are also potentially capable of proliferation after treatment with acidic buffer (pH 4) solutions (Haccius 1965). Repetition of the treatment, after a suitable time interval, results in the formation of secondary adventive embryos (Fig. 9.11 A–C).

When the seeds at different stages of embryo development are treated with buffer solutions, pH 3.5 and 4.5, the embryo proper is killed (Fig. 9.11 D), whereas the suspensor and endosperm survive. The inhibitory effect of the original embryo is lost after a few days, the suspensor cells regain their mitotic activity, and regenerate one or two additional embryos (Fig. 9.11 E, F). When the treatment is repeated, another embryo is regenerated (Fig. 9.11 G).

The suspensor cells, by virtue of their altered physiological function (compared to a normal embryo), do not lose their morphogenetic potentiality to develop into additional embryo. A relatively small suspensor, as in *Eranthis,* can be restored to embryogenic potentiality by such artificial methods (Haccius 1978).

Haccius (1978) formulated an hypothetical evolutionary series of the growth of suspensor in vitro and in vivo. She defines suspensor as a "proembryonal cell-complex". A portion of suspensor grown in vitro develops four, equal-sized dicotyledonous embryos (Fig. 9.12 A) indicating "indeterminate cleavage polyembryony". In "determinate cleavage polyembryony" the embryo which is terminal in position in the seed grows further, whereas the others remain retarded. A similar condition is represented in Fig. 9.12 B; a portion of suspensor was grown in vitro. Figure 9.12 C is a hypothetical figure to show the terminal embryo with the irregular outline of the proembryonal cell complex. Such a feature occurs mostly in primitive angiosperms like *Leitneria, Moringa* and some others (Haccius and Bhandari 1975). In vivo the angiospermous dicotyledonous embryos may have a filiform suspensor (Fig. 9.12 D), or an inconspicuous suspensor (Fig. 9.12 E).

Polyembryony is common in gymnosperms. In angiosperms when a portion of proembryonal cell complex (Fig. 9.12 A, B) is grown in vitro, cleavage polyembryony (suspensor polyembryony?) occurs. The author suggests that this indicates "regressive evolution". In our opinion the in vitro behaviour of tissues does not reflect any evolutionary significance. However, the ontogeny of regenerated tissues and organs can be further studied with great care.

9.10 Utilization of Plural Embryos

Polyembryony has been utilized to increase the developmental process of intraspecific forms. An enhanced capacity to produce nucellar seedlings in individual varieties helps to select parental genotypes for crossing, with the object of increasing the number of additional embryos and variation in forms in the progeny.

The multiple embryos within a seed can be haploid, diploid, or triploid in varied combinations. Haploid seedlings are used in the exploitation of totally homozygous diploid forms by doubling the chromosome complements. Such haploids and double haploids can be successfully used as tools in plant breeding. Production of genetically uniform plants, haploids and polyploids, are useful for agricultural and horticultural practices. Most of the agricultural plants like cotton, tobacco, wheat, and potato are undoubtedly polyploids which originated through polyembryony. This has opened up new avenues for experimental exploitation in nature. The tissue culture techniques have been widely exploited for induction of polyembryony from somatic and reproductive tissues (see Chap. 16).

The cause(s) of polyembryony in nature are still unknown. The applications of modern techniques of microscopy, histochemistry, genetics, physiology, and biochemistry may result in new information, in the near future.

References

Archibald EEA (1939) The development of the ovule and seed of jointed cactus, *Opuntia aurantiaca* Lindley. S Afr J Sci 36:195–211

Asker S (1979) Progress in apomixis research. Hereditas 91:231–240

Batygina TB (1969) On the possibility of separation of a new type of embryogenesis in angiospermae. Rev Cytol Biol Vég 32:335–341

Beasley JO (1940) The production of polyploids in *Gossypium*. J Hered 31:39–48

Bhatnagar SP, Chandra S (1975) Reproductive biology of *Triticum* 3. Unfertilized ovule and embryo sac; fertilization; postfertilization changes in embryo sac and transformation of pistil into caryopsis in relation to time. Phytomorphology 25:471–477

Billings FH (1937) Some new features in the reproductive cytology of angiosperms, illustrated by *Isomeris arborea*. New Phytol 36:301–326

Bouman F, Boesewinkel FD (1969) On a case of polyembryony in *Pterocarya fraxinifolia* (Juglandaceae) and on polyembryony in general. Acta Bot Neerl 18:50–57

Braun (1859) Über Polyembryonie und Keimung von *Coelebogyne*. Ein Nachtrag zu der Abhandlung über Parthenogenesis bei Pflanzen. Abh Kon Akad Wiss Berlin, S. 109–263

Cameron DR (1949) Inheritance in *Nicotiana tabacum*. XXII. Investigations on multiple seedlings. Am J Bot 36:526–529

Chiarugi A, Francini E (1930) Apomissia in *Ochna serrulata* Walp. Nuovo G Bot Ital (Nuova Ser) 37:1–250

Chopra RN, Sachar RC (1957) Effect of some growth substances on fruit development. Phytomorphology 7:387–397
Cooper DC (1943) Haploid-diploid twin embryos in *Lilium* and *Nicotiana*. Am J Bot 30:408–413
Crété P (1949) Un cas de polyembryonie chez une Gentianacées, l' *Erythraea centaurium* Pers. Bull Soc Bot Fr 96:113–115
Crété P (1956) A propos de l'embryologie de l'*Argemone mexicana* L. Phytomorphology 6:144–148
Ernst A (1901) Beiträge zur Kenntnis der Entwicklung des Embryosackes und des Embryos (Polyembryonie) von *Tulipa gesneriana* L. Flora 88:37–77
Frost HB, Soost RK (1968) Seed reproduction: Development of gametes and embryos. In: Reuther W, Batchelor LD, Webber HJ (eds) The citrus industry. Univ Calif, Riverside 2:290–324
Furusato K (1951) Studies on polyembryony in *Citrus*. (in Japanese) Jpn J Genet 26:223–224
Furusato K, Ohta Y, Ishibashi K (1957) Studies on polyembryony in *Citrus*. Seiken Ziho 8:40–48
Greenshields JER (1951) Polyembryony in alfalfa. Sci Agric 31:212–222
Guérin P (1930) Le développement de l'oeuf et la polyembryonie chez *Erythronium dens-canis* L. CR Acad Sci Paris 191:1369–1372
Guignard L (1901) La double fécondation dans le *Najas major*. J Bot Paris 15:205–213
Guignard L (1921) La fécondation et la polyembryonie chez les *Vincetoxicum*. Mem Acad Sci Paris II 57:1–25
Guignard JL, Mestre JC (1966) Sur le développement d' embryons à partir des antipodes chez l'*Ulmus campestris* L. Bull Soc Bot Fr 113:227–228
Haccius B (1953) Histogenetische Untersuchungen an Wurzelhaube und Kotyledonarscheide geophiler Keimpflanzen (*Podophyllum* und *Eranthis*). Planta 41:439–458
Haccius B (1957) Regenerationserscheinungen an pflanzlichen Embryonen nach Behandlung mit antimitotisch wirksamen Substanzen. Beitr Biol Pflanz 30:3–18
Haccius B (1960) Experimentell induzierte Einkeimblättrigkeit bei *Eranthis hiemalis*. II. Monokotylie durch Phenylborsäure. Planta 54:482–497
Haccius B (1963) Restitution in acidity-damaged plant embryos: Regeneration or regulation? Phytomorphology 13:107–115
Haccius B (1965) Weitere Untersuchungen über Somatogenese aus den Suspensorenzellen von *Eranthis hiemalis:* Embryonen. Planta 64:219–224
Haccius B (1978) Question of unicellular origin of non-zygotic embryos in callus cultures. Phytomorphology 28:74–81
Haccius B, Bhandari NN (1975) Delayed histogen differentiation as a common primitive character in all types of non-zygotic embryos. Phytomorphology 25:91–94
Hagerup O (1944) On fertilization, polyploidy, and haploidy in *Orchis maculatus* L. Dan Bot Ark 11:1–26
Hagerup O (1947) The spontaneous formation of haploid, polyploid, and aneuploid embryos in some orchids. K Dan Vidensk Selsk Biol Skr 20:1–22
Haque A (1946) Haploid-haploid polyembryony in *Sesbania aculeata* Pers. Curr Sci (Bangalore) 15:287
Harland SC (1936) Haploids in polyembryonic seeds of Sea Island cotton. J Hered 27:229–231
Hegelmaier F (1897) Zur Kenntnis der Polyembryonie von *Allium odorum* L. Bot Z 55:133–140
Hodgson RW, Cameron SH (1938) Effects of reproduction by nucellar embryony on clonal characteristics in *Citrus*. J Hered 29:417–419
Iglesias L, Lima H, Simon JP (1974) Isoenzyme identification of zygotic and nucellar seedlings in *Citrus*. J Hered 65:81–84
Jaranowski J (1961) Haploid-diploid twin embryos in *Melilotus*. Genet Pol 2:129–137
Jeffrey EC (1895) Polyembryony in *Erythronium americanum*. Ann Bot 9:537–541
Johansen DA (1950) Plant embryology. Waltham Mass
Johri BM (1936) Studies in the family Alismaceae. IV. *Alisma plantago* L., *Alisma plantago-aquatica* L. and *Sagittaria graminea* Mich. Proc Indian Acad Sci Sect B 4:128–138
Johri BM (1965) Chemical induction of polyembryony. In: Ramakrishnan CV (ed) Proc "Tissue culture" Seminar Baroda. W. Junk, The Hague, pp 330–337
Johri BM, Ahuja MR (1957) A contribution to the floral morphology and embryology of *Aegle marmelos* Correa. Phytomorphology 7:10–24
Johri BM, Bhatnagar SP (1955) A contribution to the morphology and life history of *Aristolochia*. Phytomorphology 5:123–137
Johri BM, Kak D (1954) The embryology of *Tamarix* Linn. Phytomorphology 4:230–247

Johri BM, Tiagi B (1952) Floral morphology and seed formation in *Cuscuta reflexa* Roxb. Phytomorphology 2:162–180

Juguet M (1966) Polyembryonie chez le *Geum rivale* L. Bull Soc Bot Fr 113:133–137

Juliano JB (1937) Embryos of carabao mango, *Mangifera indica* L. Philipp J Agric 25:749–760

Kappert H von (1933) Erbliche Polyembryonie bei *Linum usitatissimum*. Biol Zentralbl 53:276–307

Kordyum EL (1961) On polyembryony in *Vincetoxicum officinale* Moench. Ukr Bot J 18:54–65

Lakshmanan KK (1963) Embryological studies in the Hydrocharitaceae. III. *Nechamandra alternifolia*. Phyton (Argentina) 20:49–58

Lebègue A (1948) Embryologie des Crucifères: Polyembryonie chez l'*Arabis lyallii* S. Wats. Bull Soc Bot Fr 95:250–252

Lebègue A (1952) La polyembryonie chez les Angiospermes. Bull Soc Bot Fr 99:329–367

Lee JA (1970) On origin of haploid/diploid twinning in cotton. Crop Sci 10:453

Leeuwenhoek A von (1719) Epistolae physiologicae super compluribus naturae arcanis. Langerak

Leroy JF (1947) La polyembryonie chez les *Citrus*, Son intérêt dans la culture et l'amélioration. Rev Intl Bot Appl Agric Trop 27:483–495

Maheswari Devi H, Pullaiah T (1976a) Embryological investigations in the Melampodinae I. *Melampodium divaricatum*. Phytomorphology 26:77–86

Maheswari Devi H, Pullaiah T (1976b) Embryology of safflower (*Carthamus tinctorius*). Botanique (Nagpur) 7:63–70

Maheswari Devi H, Pullaiah T (1977) Embryological abnormalities in *Carthamus tinctorius* Linn. Acta Bot Indica 5:8–15

Maheshwari P (1950) An introduction to the embryology of angiosperms. McGraw-Hill, New York

Maheshwari P, Chopra RN (1955) The structure and development of the ovule and seed of *Opuntia dillenii* Haw. Phytomorphology 5:112–122

Maheshwari P, Negi V (1955) The embryology of *Dipteracanthus patulus* (Jacq.) Nees. Phytomorphology 5:456–472

Maheshwari P, Rangaswamy NS (1958) Polyembryony and in vitro culture of embryos of *Citrus* and *Mangifera*. Indian J Hortic 15:275–282

Maheshwari P, Sachar RC (1963) Polyembryony. In: Maheshwari P (ed) Recent advances in the embryology of angiosperms. Intl Soc Plant Morphologists, Univ Delhi, Delhi, pp 265–296

Masand P (1963) Embryology of *Zygophyllum fabago* Linn. Phytomorphology 13:293–302

Mathew CJ (1980) Embryological studies in Hamamelidaceae: Development of female gametophyte and embryogeny in *Hamamelis virginiana*. Phytomorphology 30:172–180

Mathew CJ, Chaphekar M (1977) Development of female gametophyte and embryogeny in *Stachyurus chinensis*. Phytomorphology 27:68–78

Michaelis P (1925) Zur Zytologie und Embryoentwicklung von *Epilobium*. Ber Dtsch Bot Ges 43:61–67

Morgan DT Jr, Rappleye RD (1954) A cytogenetic study of the origin of multiple seedlings of *Capsicum frutescens*. Am J Bot 41:576–585

Mullins MG, Srinivasan C (1976) Somatic embryos and plantlets from an ancient clone of the grapevine (cv Cabernet-Sauvignon) by apomixis in vitro. J Exp Bot 27:1022–1030

Muniyamma M (1977) Triploid embryo from endosperm in vivo. Ann Bot 41:1077–1079

Müntzing A (1937) Polyploidy from twin seedlings. Cytologia, (Fujii Jubl Vol), pp 211–227

Nair NC, Kanta K (1961) Studies in Meliaceae 4. Floral morphology and embryology of *Azadirachta indica* A. Juss.: A reinvestigation. J Indian Bot Soc 40:382–396

Narayanaswami S (1954) The structure and development of the caryopsis in some Indian millets III. *Paspalum scrobiculatum* L. Bull Torrey Bot Club 81:288–299

Narayanaswami S, Roy SK (1960) Embryo sac development and polyembryony in *Syzygium cumini* (Linn.) Skeels. Bot Not 1960:273–284

Narmatha Bai V, Lakshmanan KK (1982) Occurrence of polyembryony in *Toddalia asiatica* Lamb. Curr Sci (Bangalore) 51:374

Nissen O (1937) Spålteapringenes störrelse has twillingsplanter med Ulike kromosomtall. Bot Not 1937:28–34

Overbeek J van, Conklin ME, Blakeslee AF (1941) Chemical stimulation of ovule development and its possible relation to parthenogenesis. Am J Bot 28:647–656

Philipson MN (1978) Apomixis in *Cortaderia jubata* (Gramineae). N Z J Bot 16:45–59

Pieringer AP, Edwards GJ (1967) Identification of nucellar and zygotic *Citrus* seedlings by infra-red spectroscopy. Proc Am Soc Hortic Sci 86:226–234

Pijl L van der (1934) Über die Polyembryonie bei *Eugenia*. Rec L Trav Bot Neerl 31:113–187
Pullaiah T (1978) Studies in the embryology of Compositae 3. The tribe Astereae. Bot Mag Tokyo 91:197–205
Randall TE, Rick CM (1945) A cytogenetic study of polyembryony in *Asparagus officinalis* L. Am J Bot 32:560–569
Rangan TS, Murashige T, Bitters WP (1969) In vitro studies of zygotic and nucellar embryogenesis in *Citrus*. Proc First Intl Citrus Symp 1:225–229
Rangaswamy NS (1959) In vitro studies on the ovules of *Citrus microcarpa* Bunge. Ph D Thesis, Univ Delhi
Rangaswamy NS (1981) Nucellus as an experimental system in basic and applied tissue culture research. In: Rao AN (ed) Proc Costed Symp on "Tissue culture of economically important plants". Singapore, pp 269–286
Rhoedes MM (1961) Unusual events in seed development. USDA Year Book, pp 75–79
Robijns W, Louis A (1942) Beschouwingen over Polyembryonie en Polyspermie bij de Bedektzadigen. Verh K Vlaam Acad Wet Lett Schone Kunsten Belg Kl Wet 4:5–112
Roy SK (1953) Embryology of *Eugenia jambos* L. Curr Sci (Bangalore) 22:249–250
Roy SK, Sahai R (1962) The embryo sac and embryo of *Syzygium caryophyllifolium* DC. J Indian Bot Soc 41:45–51
Sachar RC (1955) The embryology of *Argemone mexicana* L. – A reinvestigation. Phytomorphology 5:200–218
Sachar RC, Chopra RN (1957) A study of the endosperm and embryo in *Mangifera* L. Indian J Agric Sci 27:219–228
Sarkar KR, Coe EH Jr (1966) A genetic analysis of the origin of maternal haploids in maize. Genetics 54:453–464
Schnarf K (1929) Embryologie der Angiospermen. Borntraeger, Berlin
Sen PK, Mallik PC (1940) The embryo of the Indian mangoes (*Mangifera india* L.). Indian J Agric Sci 10:750–760
Sindhe ANR, Swamy BGL, Govindappa DA (1980) Synergid embryo in *Pennisetum squamulatum*. Curr Sci (Bangalore) 49:914–915
Souèges R (1926) Polyembryonie chez le *Plantago lanceolata* L. Bull Soc Bot Fr 73:955–957
Souèges R (1949) Embryogénie des Papavèracées. Développement de l'embryon chez l'*Argemone mexicana*. CR Acad Sci Paris 229:573–576
Strasburger E (1878) Über Polyembryonie. Jena Z Naturwiss 12:647–670
Swamy BGL (1942) Female gametophyte and embryogeny in *Cymbidium bicolor* Lindl. Proc Indian Acad Sci Sect B 15:194–201
Swamy BGL (1943) Gametogenesis and embryogeny of *Eulophia epidendraea* Fischer. Proc Natl Inst Sci India Part B 9:59–65
Swamy BGL (1946a) Embryology of *Habenaria*. Proc Natl Inst Sci India Part B 12:413–426
Swamy BGL (1946b) The embryology of *Zeuxine sulcata* Lindl. New Phytol 45:132–136
Swamy BGL (1948a) On the post-fertilization development of *Trillium undulatum*. Cellule 52:7–14
Swamy BGL (1948b) Agamospermy in *Spiranthes cernua*. Lloydia (Cinci) 11:149–162
Swamy BGL (1949) Embryological studies in the Orchidaceae 2. Embryogeny. Am Midl Nat 41:202–232
Swingle WT (1932) Recapitulation of seedling characters by nucellar buds developing in the embryo sacs of *Citrus*. Proc 6th Intl Congr Genet 2:196–197
Thirumaran K, Lakshmanan KK (1982) Synergid embryo in *Dioscorea composita* Hemsl. Curr Sci (Bangalore) 57:428–429
Tiagi YD (1957) Studies in floral morphology III. A contribution to the floral morphology of *Mammillaria tenuis* DC. J Univ Saugar Sect B 6:7–31
Tiagi YD (1970) Cactaceae. In: Symposium on "Comparative embryology of angiosperms". Bull Indian Natl Sci Acad 41:29–35
Tretjakow S (1895) Die Beteiligung der Antipoden in Fällen der Polyembryonie bei *Allium odorum* L. Ber Dtsch Bot Ges 13:13–17
Treub HP (1936) Artificial control of nucellar embryony in *Citrus*. Science (Wash D.C.) 83:165–166
Veillet-Bartoszewska M (1957) La polyembryonie chez le *Primula auricula* L. Bull Soc Bot Fr 104:473–475
Vignon-Fétré J (1966) Polyembryonie chez le *Sedum spectabile Boreau*. Bull Soc Bot Fr 113:137–140

Webber JM (1940) Polyembryony. Bot Rev 6:575–598
Webber HJ, Batchelor LD (1948) The citrus industry 1. History, botany and breeding. Univ Calif Publ Bot
Whitehead RA, Chapman GP (1962) Twinning and haploidy in *Cocos nucifera* Linn. Nature 195:1228–1229
Yakovlev MS (1967) Polyembryony in higher plants and principles of its classification. Phytomorphology 17:278–282
Yamamoto Y (1936) Über das Vorkommen von triploiden Pflanzen bei Mehrlings-Keimlingen von *Triticum vulgare* Vill. Cytologia 7:431–436

10 Gameptophytic Apomixis*

A third factor in apomixis influences the formation of endosperm: the relation of polar nuclei to fertilization. There are two possibilities:

1. The polar nuclei must be fertilized so that only the egg cell is capable of parthenogenetic development; the endosperm formation is pseudogamous (unreduced "pseudogamy").
2. Both egg cell and polar nuclei are independent of fertilization; both embryo and endosperm formation are parthenogenetic, i.e., autonomous ("autonomous apomixis" = "diploid, unreduced parthenogenesis").

Maternal offspring are, to summarize, the result of a cooperation of different, independent processes which can be conceived as "basic components of gametophytic apomixis". The two most important of these components are diplosporous/aposporous formation of embryo sac, and parthenogenetic formation of embryo. If 100% of the offspring were maternal, then "obligate apomixis" would be the case. However, most apomicts retain traces of sexual reproduction in that one or both of these basic components are not always completely effective so that, rarely, aberrant offspring occur; in extreme cases so rarely that they are hardly detectable. These "facultative apomicts" can produce, besides maternal offspring, aberrants in three different ways:

1. Fertilization of unreduced egg cells results in "B_{III} hybrids" (Rutishauser 1948) = U-hybrids (Asker 1977), usually formed as "2n+n hybrids" and, rarely, as "2n+2n hybrids" (Harlan and deWet 1975).
2. Fertilization of reduced egg cells, i.e., normal, sexual reproduction, results in "B_{II} hybrids" (Rutishauser 1948) = R-hybrids (Asker 1977), formed as "n+n hybrids" or, exceptionally, as "n+2n hybrids".
3. Parthenogenetic development of reduced egg cells ("haploid parthenogenesis" or "haploid pseudogamy") results in "polyhaploids" (n+0), e.g., in dihaploids from tetraploid mother plants, or in (mono)haploids from diploid mother plants. "Haploid parthenogenesis" is the most important case of "nonrecurrent apomixis", counterpart of "recurrent apomixis" = "stable apomixis" which includes "diploid parthenogenesis" and results in maternal offspring (2n+0). Another instance is "androgenesis" (see Sect. 10.2.2.3). For a general review on haploid angiosperms see Kimber and Riley (1963).

10.1.2 Synonyms

The terminology used here is that simplified by Gustafsson (1946) and Rutishauser (1967) and now in general use. Those of the following synonyms which are not in *italics* are dispensable – some of them are even misleading:

Apomixis sensu stricto (s.s.): *agamospermy*, seed apomixis.
Adventitious embryony: nucellar embryony.
Gametophytic apomixis: apogamogony.
Diplospory: generative apospory sensu lato (s.l.).
Antennaria type: *mitotic diplospory*, somatic, direct diplospory, gonial apospory, generative apospory s.s.
Taraxacum and Ixeris types: (restitutional) aneuspory, indirect diplospory, diplospory s.s., apomeiosis s.s. (to be avoided in this restricted sense, see below and Sect. 10.2.1.5).

Apospory: somatic apospory. Certain authors use the term "apospory" erroneously to designate adventitious embryony. Khokhlov (1976) further adds to the confusion by using apospory instead of diplospory. For apospory he proposes the superfluous synonym "apoarchespory".
Apomeiosis (= diplospory + apospory, original definition by Renner 1916): no synonym.
Parthenogenesis: apozygoty.
Haploid parthenogenesis: reduced, generative parthenogenesis.
Diploid parthenogenesis: unreduced, somatic parthenogenesis.
Pseudogamous apomixis: (unreduced) *pseudogamy*. In adventitious embryony this phenomenon is called "induced adventitious embryony".
Autonomous apomixis: diploid, unreduced parthenogenesis, diploparthenogenesis.

Apogamy, in its original meaning (de Bary 1878), is synonymous with apomixis (in a wider sense, i.e., including vegetative propagation), but it was soon restricted to designate the development from a vegetative cell of the female gametophyte (not from an egg cell). Consequently, Winkler (1908) replaced apogamy by apomixis. Apogamy is an important phenomenon in ferns (see review by Döpp 1967), but not in angiosperms where the concept apogamy as well as its synonym "apogamety" are dispensable (see Sect. 10.2.1.6).

10.1.3 Remarks on Terminology

For the sake of a clear terminology, the misleading expression "triploid hybrid" for a higher polyploid B_{III} hybrid must be avoided, since almost all apomicts are polyploid. The same applies to the illogical terms "reduced hybrid" and "unreduced hybrid". While designations such as "apomictic hybrid", "sexual offspring", etc. naturally describe the mode of reproduction of the plant to which these terms are applied, they do not say anything about its origin; the latter should be referred to by circumscriptions such as "apomictically produced, formed ..." etc.

In connection with apomixis, embryo sacs and gametes should always be defined as meiotic (i.e., of meiotic origin), or as diplosporous/aposporous (apomeiotic), or even better as (cytologically) *reduced* (n) or *unreduced* (2n), but never as "haploid" or "diploid" unless they really have x or 2x chromosomes, respectively. On no account may embryo sacs be called "sexual" or "apomictic", as sexuality or apomixis have nothing to do with the origin of gametophytes.

The symbol "n" should be used for the meiotically reduced, the symbol "2n" for the unreduced, somatic chromosome number (and for unreduced gametophytes and gametes) of any plant, be it diploid or polyploid, euploid or aneuploid. Reduced gametes "n" do not necessarily have exactly half-the-reduced number of "2n" chromosomes. Since not only gametes with identical values of "n" are compatible, it follows that the zygotic values of "2n" do not necessarily comprise the exact two-fold values of "n". Higher multiples than "2n" are used, e.g., for endopolyploid cells or tissues (see Chap. 11) but never to symbolize the level of polyploidy of a whole plant. For this purpose "x" (basic number) is the correct symbol. Only the precise distinction between "n" and "x", as used in specialized literature, allows the distinction of hybrids such as "n+n", "2n+n", etc. (see Harlan and deWet 1975), and avoids confusion – especially when working with anorthoploids (polyploids with uneven genome numbers such as triploids, $2n=3x$, with gametes $n=x...2x$; enneaploids, $2n=9x$), or with aneuploids (e.g., disomic triploids, $2n=3x-1$).

10.2 Embryology of Gametophytic Apomicts

10.2.1 Development of Unreduced Embryo Sacs

The different types of unreduced embryo sacs are defined on the basis of their origin: diplosporous embryo sacs originate from generative cells (female arche-

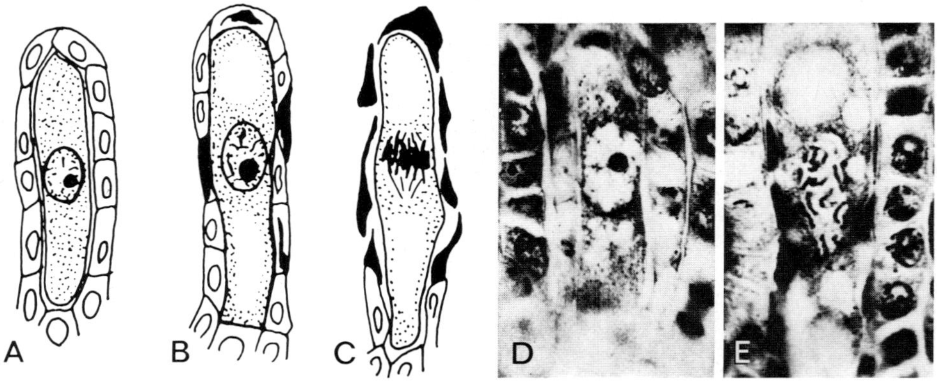

Fig. 10.1 A–E. Diplospory (Antennaria type) in *Antennaria porsildii* (**A–C**) and *A. alpina* (**D, E**). The megaspore mother cell omits meiosis and develops directly into an unreduced embryo sac. **A** Exceptionally long interphase. **B** Prophase, vacuolation. **C** First mitosis with unreduced chromosome number; further development as in Polygonum type. **D, E** Earliest photomicrographic documentation of diplospory. (**A–C** From Urbańska-Worytkiewicz 1974; **D, E** from Juel 1900)

sporium, megaspore mother cells), aposporous ones from somatic cells. If the origin of unreduced embryo sacs cannot be established, apomeiosis in its original definition (Renner 1916) should be used as a covering term.

10.2.1.1 Diplospory (Earlier Synonym: "Generative Apospory")

The simplest and most frequent means of development of unreduced embryo sacs from generative cells was first clarified by Juel (1900) in *Antennaria alpina* and is, therefore, called "Antennaria type".

Antennaria Type (= Mitotic Diplospory). The megaspore mother cell doeas not enter into meiosis but, after a long interphase (Fig. 10.1 A, D), growth, and pronounced vacuolation (Fig. 10.1 B), characteristic of functional megaspores, proceeds directly to the first mitosis (Fig. 10.1 C, E). The female meiosis is omitted, and the megaspore mother cell functions as an unreduced ("diploid") megaspore. Two further mitoses lead to the mature, eight-nucleate embryo sac. The Antennaria type is widely distributed, and some of the important genera comprising apomictic species are: *Antennaria* (not all the apomictic species – Juel 1900, Stebbins 1932, Urbańska-Worytkiewicz 1974), *Burmannia coelestis* (Ernst and Bernard 1912), *Erigeron* and *Eupatorium* (Holmgren 1919, Sparvoli 1960), *Hieracium* (*Euhieracium* – Bergman 1941), *Poa alpina* (Håkansson 1943) and *P. nervosa* (Grun 1955), *Parthenium* (Esau 1946), *Calamagrostis* (Nygren 1946), *Zephyranthes* (Brown 1951), *Cooperia* (Coe 1953), *Nardus stricta* (Rychlewski 1961). In *Elatostema eurhynchum* (Fagerlind 1944) a parietal cell is formed as in sexual relatives. For apomictic Rosaceae with multicellular female archesporium see Section 10.2.1.5).

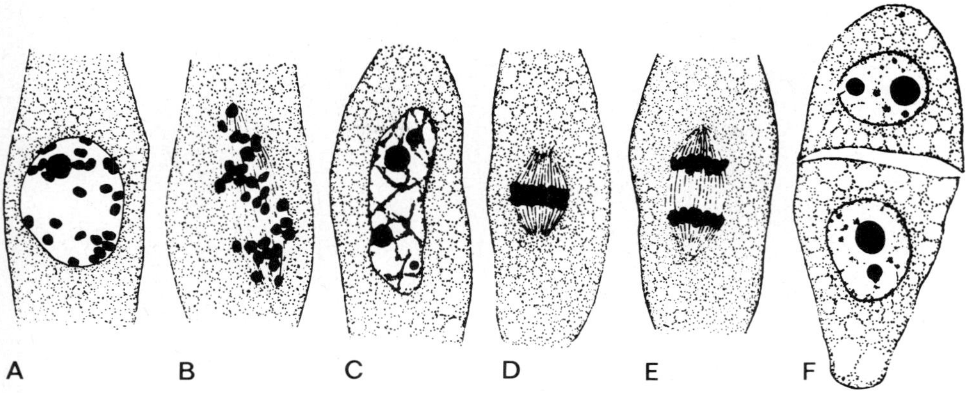

Fig. 10.2 A–F. Diplospory (Taraxacum type) in *Taraxacum megalorrhizon*. **A, B** The megaspore mother cell enters meiosis, but because of asyndesis the univalents (**A**) remain scattered over the whole spindle of metaphase I (**B**). **C** Instead of chromosome distribution, all the univalents are enclosed in one common restitution nucleus with unreduced chromosome number. **D–F** The next division (**D, E**), analogous to meiosis II, followed by formation of a cell wall, leads to two unreduced cells (**F**); usually the chalazal dyad cell gives rise to an unreduced, eight-nucleate embryo sac. (From Battaglia 1948)

Taraxacum Type. In this type, discovered in *Taraxacum* (Juel 1906, Osawa 1913), the megaspore mother cell enters into the meiotic prophase but, because of asynapsis, there is mostly no pairing (Fig. 10.2 A); the univalents remain scattered over the whole spindle of metaphase I (Fig. 10.2 B) and, therefore, are not shared between the poles. The first meiotic "division" results in a meiotic restitution nucleus (Fig. 10.2 C) enclosing the complete somatic chromosome complement[1]. Therefore, the second meiotic division (Fig. 10.2 D, E) continues with the unreduced chromosome number and results, after cytokinesis (cell wall formation), in an unreduced dyad (Fig. 10.2 F) instead of a reduced tetrad. Usually, the chalazal dyad cell undergoes three mitoses leading to an eight-nucleate embryo sac. The restitution nucleus formation in the Taraxacum type applies, primarily, only to the female meiosis. The important representatives of the Taraxacum type are among the Compositae: *Taraxacum* (Osawa 1913, Fagerlind 1947 a, Battaglia 1948, Małecka 1971), *Erigeron* (Holmgren 1919, Fagerlind 1947 b, Battaglia 1950), *Chondrilla* (Bergman 1950), certain strains of *Antennaria carpatica* (Bergman 1951). Examples from other families are scarce: *Arabis Holboellii* (Böcher 1951), *Agropyron scabrum* (Hair 1956), *Paspalum* (only certain species – Chao 1974, 1980; most are aposporous).

Other Types of Diplospory. Both the Antennaria type and the Taraxacum type are met with in such diplosporous apomicts the sexual relatives of which show mono-

1 Restitution nuclei also develop when univalents, instead of being scattered over the spindle of metaphase I, are arranged in the metaphase plate, as in metaphase I no distribution of chromatids takes place. This finding (Fagerlind 1947 a, b, Battaglia 1948, Bergman 1950) makes the existence of the disputed "pseudohomeotypic division" ("semiapospory") appear very improbable. (In this hypothetical type of diplospory the chromatids of each univalent are distributed in anaphase I, forming an unreduced dyad. The second meiotic division is omitted.) For other mechanisms leading to meiotic restitution nuclei compare Ramanna (1979) and Hermsen (1980)

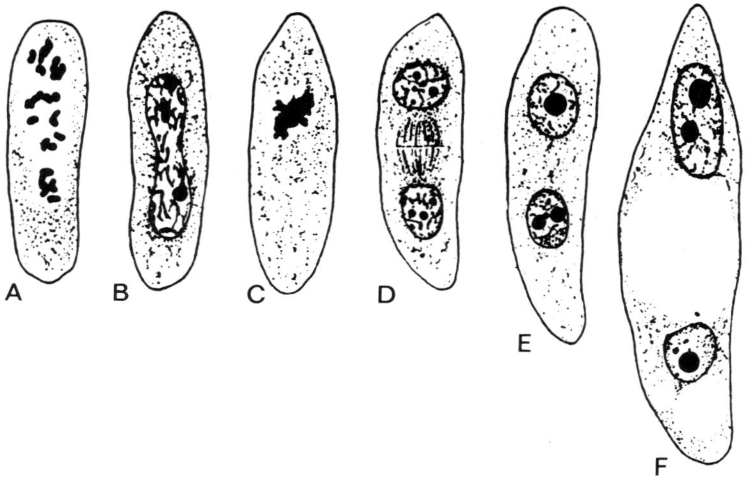

Fig. 10.3 A–F. Diplospory (Ixeris type) in *Ixeris dentata*. **A, B** Meiotic restitution nucleus (as in Taraxacum type). **C** Metaphase with unreduced chromosome number. **D, E** A cell wall is not formed; both daughter nuclei take part in the formation of unreduced embryo sac. **F** Two-nucleate embryo sac; two further mitoses lead to the formation of an eight-nucleate, unreduced embryo sac. (From Okabe 1932)

sporic embryo sac development (Polygonum type). Only a few diplosporous apomicts belong to groups in which sexual plants show tetrasporic development (Drusa type, Fritillaria type). As in the Taraxacum type, an asyndetic meiotic prophase leads to a restitution nucleus (Fig. 10.3 A, B) which undergoes a division corresponding to the second meiotic division (Fig. 10.3 C, D), but this nuclear division is not followed by a cell division. Therefore, the resulting coenomegaspore (Fig. 10.3 E, F) contains two unreduced instead of four reduced nuclei. Two further mitoses lead to an eight-nucleate embryo sac. This so-called Ixeris type has been reported not only in *Ixeris dentata* (Okabe 1932), but also (although sometimes with slight deviations) in a few other tetrasporic Compositae such as *Rudbeckia* (Battaglia 1946 a, Fagerlind 1946), and *Erigeron* (Fagerlind 1947 b, Battaglia 1950), and in Plumbaginaceae in *Statice oleaefolia* (D'Amato 1949).

Only two diplosporous species are known in groups with disporic embryo sac development (Allium type): *Allium nutans* (Håkansson 1951) and *A. odorum* (Håkansson and Levan 1957). In these species the chromosome number is doubled by a premeiotic endomitosis and, therefore, a normal female meiosis (with pairing of identical chromosomes forming "autobivalents") leads to a tetrad of unreduced nuclei. As in the Allium type, there is no cell wall formation in the chalazal dyad so that only two further mitoses are necessary to achieve maturity. This "Allium nutans type" as well as the Ixeris type show strikingly how apomicts conserve the embryological mechanisms of their sexual relatives as best as possible (Rutishauser 1967). Premeiotic chromosome doubling plays an important role in apomictic ferns (refer to Döpp 1967). Postmeiotic chromosome doubling by endomitoses and reduplications are rarely reported, e.g., in *Potentilla collina* (Müntzing 1958 b), *Saccharum* (Bremer 1963), and *Ranunculus auricomus* (Izmaiłow 1976).

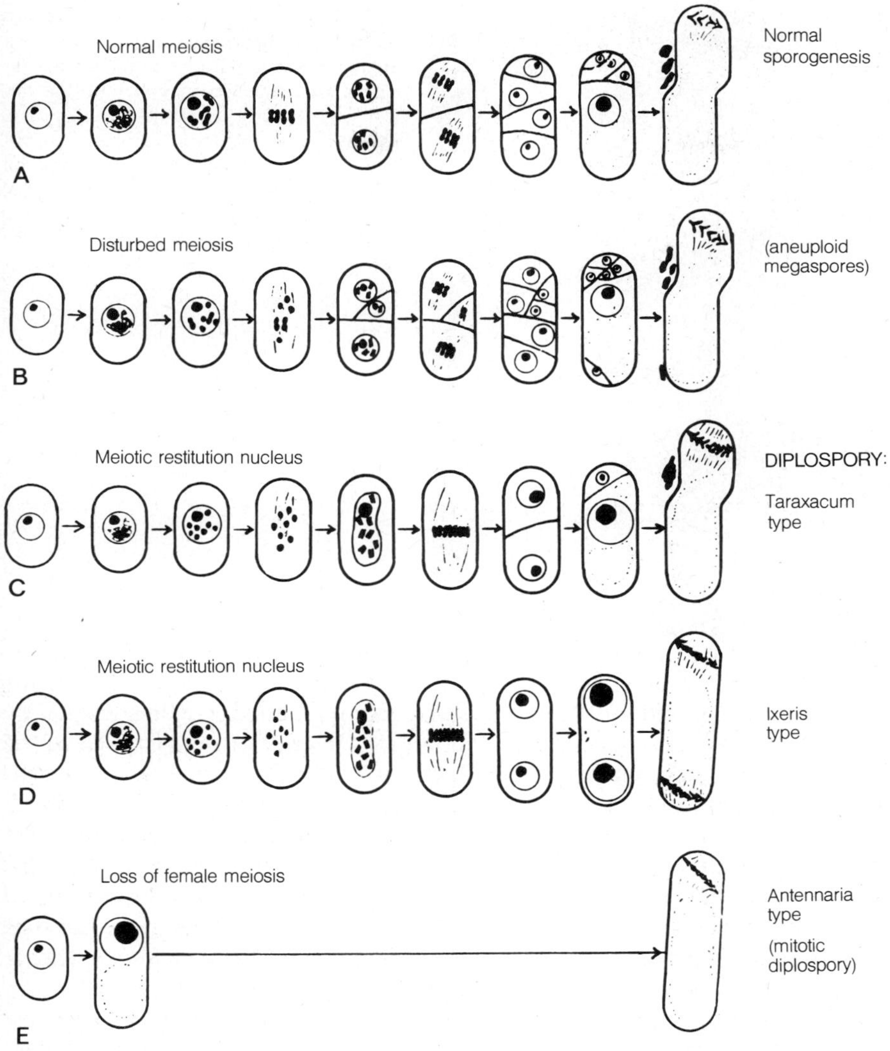

Fig. 10.4 A–E. Female meiosis in diplosporous apomicts. (Modified from Fagerlind 1944)

10.2.1.2 Female Meiosis in Diplosporous Apomicts

Figure 10.4 A–E gives a survey of the various possibilities of development of megaspore mother cells. In Fig. 10.4 A a normal meiosis is presented for comparison. Asyndesis can result either in a disturbed meiosis with irregular distribution of chromosomes which leads to aneuploidy (Fig. 10.4 B), or in a meiotic restitution nucleus which leads to an unreduced dyad (Fig. 10.4 C – Taraxacum type; Fig. 10.4 D – Ixeris type). Finally, meiosis can be entirely lacking (Fig. 10.4 E – Antennaria type, mitotic diplospory).

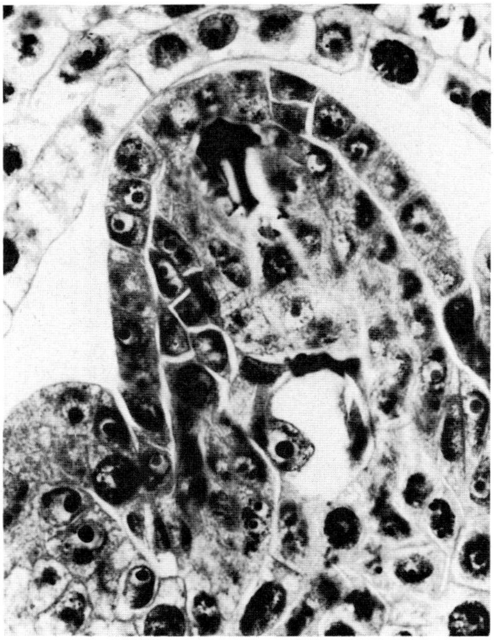

Fig. 10.5. Apospory in *Ranunculus auricomus*. The initial of the unreduced embryo sac is a somatic cell, usually from the center of the nucellus. After intense vacuolation (one-nucleate stage), three mitoses lead to a mature, eight-nucleate, unreduced embryo sac (Hieracium type). The functional megaspore, still visible in the photograph, degenerates sooner or later and is crushed by the growth of the aposporous embryo sac. (Photograph by the author)

As was first reported in 1912 by Ernst and Bernard, and since then frequently confirmed, meiosis can, occasionally, be induced instead of mitosis even in the Antennaria type. Then, extreme or total asyndesis usually occurs, leading to restitution nuclei as typical for the Taraxacum type. Exceptionally, even a completed, more or less regular, meiosis is still possible in both Antennaria and Taraxacum types, at best resulting in a reduced megaspore tetrad. For the importance of this potential see Section 10.3.1. In an important publication which has not received due attention, Sparvoli (1960) showed that in *Eupatorium riparium* there is a shift from the typical *"tipo mitotico"* (Antennaria type) to the *"tipo meiotico"* (either Taraxacum type or a complete meiosis). This shift can be influenced by the conditions of cultivation, confirming Gustafsson's suppositions (1947b). Sparvoli's results, as well as investigations on *Antennaria,* a genus comprising species of either Antennaria or Taraxacum types (see Bergman 1951, Urbańska-Worytkiewicz 1974), further suggest that there are no fundamental differences between these two kinds of diplospory.

10.2.1.3 Apospory (Earlier Designation: "Somatic Apospory")

The unreduced embryo sacs develop from somatic (vegetative) cells lying, as a rule, in the center of the nucellus, often adjoining the chalazal pole of the megaspore mother cell, dyad, or megaspore (Fig. 10.5). Or, rarely, deeper in the chalaza, or even in the inner integument. Often there is only a single aposporous initial per ovule. Even if there are more, usually only one of them matures into an aposporous embryo sac.

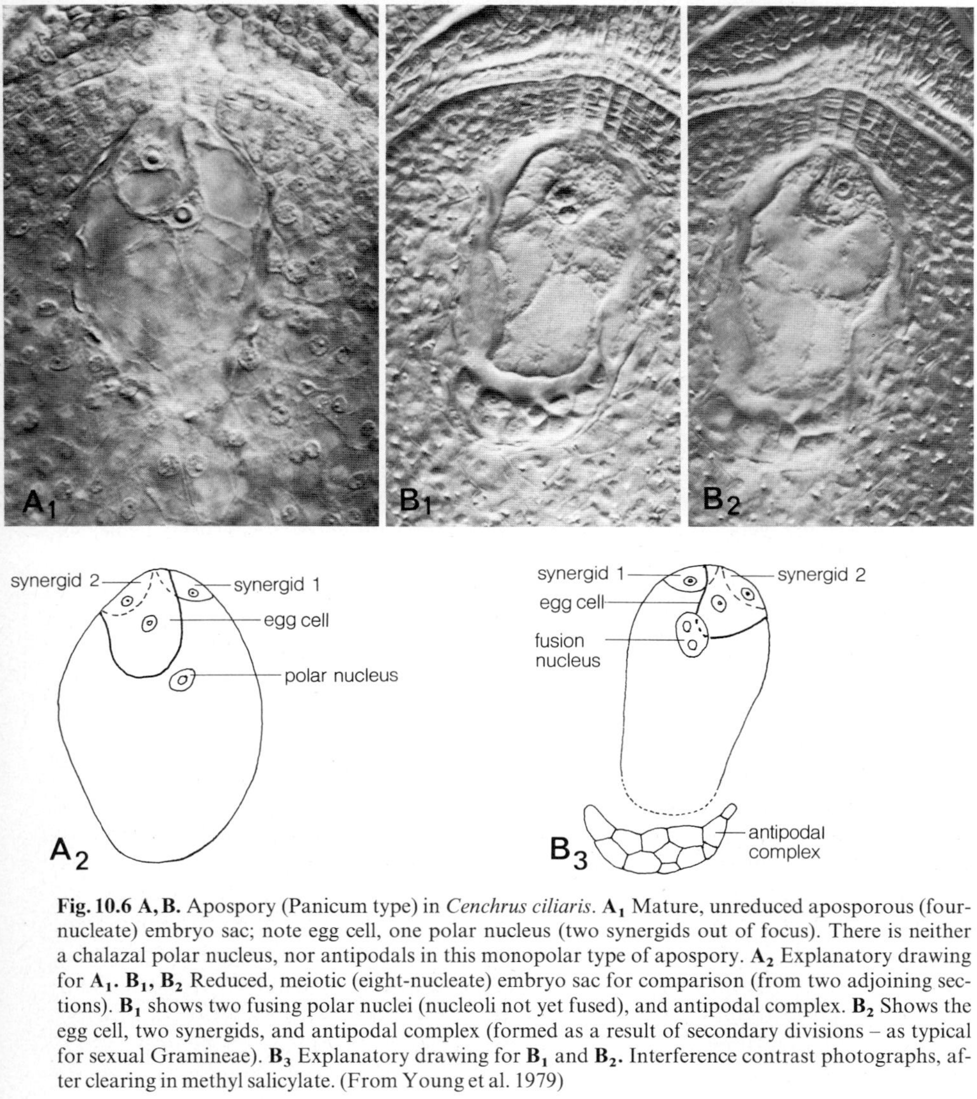

Fig. 10.6 A, B. Apospory (Panicum type) in *Cenchrus ciliaris*. **A_1** Mature, unreduced aposporous (four-nucleate) embryo sac; note egg cell, one polar nucleus (two synergids out of focus). There is neither a chalazal polar nucleus, nor antipodals in this monopolar type of apospory. **A_2** Explanatory drawing for **A_1**. **B_1, B_2** Reduced, meiotic (eight-nucleate) embryo sac for comparison (from two adjoining sections). **B_1** shows two fusing polar nuclei (nucleoli not yet fused), and antipodal complex. **B_2** Shows the egg cell, two synergids, and antipodal complex (formed as a result of secondary divisions – as typical for sexual Gramineae). **B_3** Explanatory drawing for **B_1** and **B_2**. Interference contrast photographs, after clearing in methyl salicylate. (From Young et al. 1979)

Aposporous initials are first recognizable by their growth and the enlargement of their nucleus, especially the nucleolus, and by vacuolation causing compression and resorption of adjacent generative as well as vegetative cells. The darker staining of certain nucellar cells before vacuolation, especially with iron-hematoxylin, has turned out to be a very unreliable criterion for potential aposporous initials (Nogler 1978). Polarization and orientation of the spindle of the first mitosis are usually parallel to the nucellus, as in reduced embryo sacs.

This bipolar aposporous type is called the "Hieracium type" by Rosenberg (1908), after the first aposporous plant analyzed embryologically. It is widely distributed, but is rather rare in Compositae: *Hieracium* subg. Pilosellae (Rosenberg 1908, 1930, Euhieracium follows the Antennaria type), *Crepis* (some American species – Stebbins and Jenkins 1939), and some Australian Compositae (Davis 1967). Important examples from other families are *Hypericum perforatum* (Noack 1939), *Poa pratensis* (Tinney 1940), *P. granitica* (Skalińska 1959), *Elatostema acuminatum* (Fagerlind 1944), *Ranunculus cassubicus* (Izmaiłow 1965), *R. auricomus* (Nogler 1971), *Hierochloë* (Weimarck 1967, 1970), and *Beta* sect. Corollinae (Jassem 1976), etc. For apospory in multicellular archesporium see Section 10.2.1.5.

A peculiar form of apospory occurs in Panicoideae (Gramineae). There is no initial polarization in the embryo sac, and the vacuole develops at the chalazal end. The spindle of the first mitosis lies crosswise at the micropylar end. A second mitosis leads to the mature, four-nucleate embryo sac. Thus, this monopolar "Panicum type" of apospory (Fig. 10.6 A, B) contains the three-celled egg apparatus and single polar nucleus, while antipodals are absent. Such four-nucleate, monopolar embryo sacs (not to be confused with the Oenothera type) are always unreduced; if reduced embryo sacs also occur in the same plant, they are always eight-nucleate and bipolar.

The Panicum type, discovered in *Panicum maximum* (see Warmke 1954), is known only in Panicoideae and Andropogoneae, but is very widespread (see reviews by Brown and Emery 1958, Savidan 1978, Connor 1979). Important genera comprising apomictic species are: *Bothriochloa-Dichanthium-Capillipedium, Cenchrus, Chloris, Digitaria, Eriochloa, Heteropogon, Hyparrhenia, Panicum, Paspalum, Pennisetum, Sorghum, Themeda, Urochloa,* etc. Occasionally, four-nucleate embryo sacs have two polar nuclei and one synergid (e.g., *Setaria* – Emery 1957), or may arise by diplospory (*Eragrostis* – Voigt and Bashaw 1972, Brix 1974).

In Panicoideae the unreduced embryo sacs are usually four-nucleate, and the reduced ones eight-nucleate. However, exceptions have been reported. In certain species unreduced, aposporous embryo sacs can also be bipolar and, therefore, eight-nucleate (Reddy and d'Cruz 1969, Moskova and Yakovlev 1974). Vice versa, Gildenhuys and Brix (1959) and Brix (1974) noted that, in certain cases, reduced embryo sacs can also give the impression of being four-nucleate due to fusion of the two polar nuclei and early degeneration of the antipodal nuclei, which is atypical for grasses. The genus *Paspalum* is especially instructive. Besides usual apomictic species (Burson 1975) there are intermediate species with unreduced embryo sacs sometimes four-nucleate, sometimes eight-nucleate (Reusch 1961), or with a variable number of nuclei (Quarín et al. 1982). The other extreme is demonstrated by a few species of *Paspalum* in which also the unreduced embryo sacs are always bipolar and eight-nucleate (Snyder 1957, and others). There are even reports of diplosporous species in *Paspalum* (Chao 1974, 1980).

Such exceptional conditions must be reckoned with when Panicoideae are investigated by routine embryological methods based on squash preparations, and clearing techniques for phase-contrast or interference-contrast microscopy (Fig. 10.6), as used in plant breeding or in cytogenetics of apomixis.

In most aposporous apomicts (Hieracium type and Panicum type) asynchronous development of unreduced embryo sacs in the individual ovules of the same bud is typical, contrary to the conditions in most sexual plants.

10.2.1.4 The Female Meiosis in Aposporous Apomicts

The meiosis is independent of the aposporous embryo sac development, a fact largely verified for various apomicts (see Davis 1967, Nogler 1971, and others). In principle both these processes can occur simultaneously in the same ovule but, usually, in wild apomicts the megaspore mother cell, or the products of its meiotic divisions, are sooner or later eliminated so that only unreduced, aposporous embryo sacs persist. This elimination can take place in various ways. The simplest and most effective, but surprisingly rare, is the interruption in development and the degeneration of the megaspore mother cell during meiotic prophase (e.g., Stebbins and Jenkins 1939, Brown and Emery 1957). More often, in aposporous apomicts, the female meiosis can, in principle, be completed, but mostly the megaspores degenerate. However, further development is still possible, at least in some ovules. The meiotic development has to compete with aposporous embryo sacs which are usually more or less advanced in development, not having been delayed by meiosis (Fig. 10.5).

The timing of the induction of apospory is decisive. The sooner an aposporous initial is induced, the fewer are the chances of development of the megaspore, as already shown in 1943 in the classical investigation by Christoff and Papasova in aposporous *Potentilla*. *P. adscharica* with very early induction of apospory is totally aposporous; *P. recta* with delayed induction produces meiotic embryo sacs in about 3% of the ovules. Similar observations have been reported for various other apomicts, e.g., *Ranunculus auricomus* where apomictic wild forms show early induction of apospory, in certain ovules even before meiotic prophase. Crossing with sexual *R. cassubicifolius* and each backcrossing increasingly delays the induction of apospory until, in certain hybrids, apospory can only be determined in older buds in which the female meiosis has been completed (Nogler 1971, 1973). According to Krylova (1976), in potentially apomictic *Malus domestica,* aposporous initials start their development later than the meiotic embryo sacs so as to replace these if they fail.

Effects similar to those caused by such variations in timing can arise from variations in the number of developing aposporous initials in each ovule, owing to an enlargement of the induced nucellar region (see e.g., Harlan et al. 1964). However, even in these cases of "multicellular apospory" (Stebbins 1941) usually only one aposporous embryo sac completes development. The combination of these two effects (in time and space) on apospory has been described by Stebbins and Jenkins (1939), Liljefors (1953), and Snyder et al. (1955).

The question is still open whether the cessation of development and the successive degeneration of megaspores are purely passive, i.e., caused only by the growth of aposporous embryo sacs, or not. In aposporous *Panicum maximum* (Warmke 1954), *Themeda triandra* (Brown and Emery 1957) and *Ranunculus auricomus* (Nogler 1971) some of the ovules do not show induction of apospory but, in spite of this, most megaspores degenerate even in these ovules. This points to the probability that the majority of the megaspores are not capable of embryo sac development.

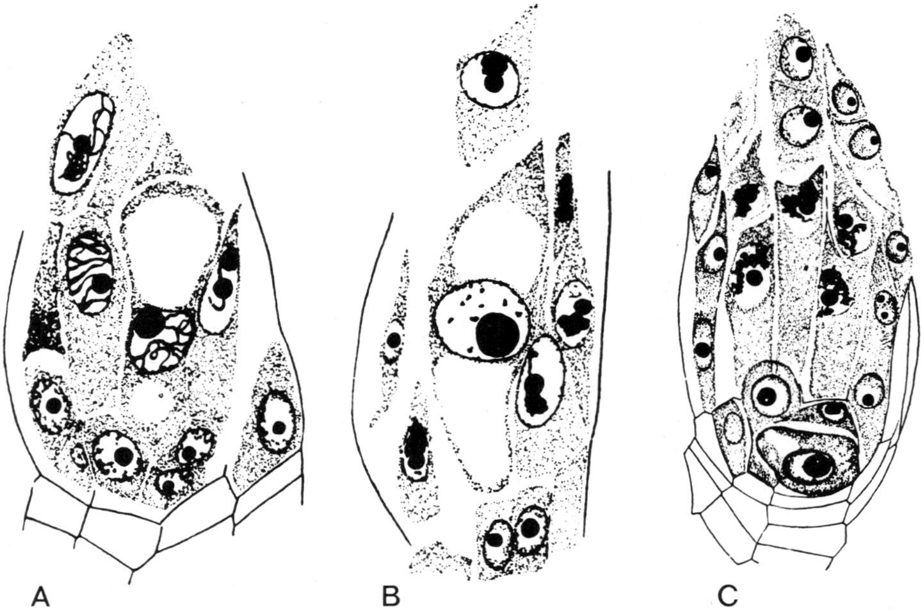

Fig. 10.7 A–C. Multicellular archesporium in *Potentilla*. **A, B** Diplospory; the apomeiotic megaspore mother cells are in interphase, or early mitotic prophase, and show vacuolation. **C** Apospory; beginning of vacuolation in aposporous initial. (**A, B** After Rutishauser 1948, **C** after Hunziker 1954)

10.2.1.5 Apomicts with Multicellular Female Archesporium

These occur especially in Rosaceae. Some are purely aposporous, such as certain species of *Potentilla* (Christoff and Papasova 1943, Rutishauser 1943), *Sorbus* (Liljefors 1953), *Malus* (Hjelmqvist 1957, Schmidt 1964, Krylova 1976), *Aphanes* (Hjelmqvist 1959), and, probably, *Alchemilla* (cf. Rutishauser 1967, p. 61). *Potentilla verna* is an example for diplospory (see Rutishauser 1943). In other *Potentilla*, and especially in *Rubus*, diplospory occurs besides apospory[2]. This often makes the interpretation of embryological preparations difficult because, exactly as in sexual plants, it is not always easy to delimit the generative region since there are often transitions between ordinary megaspore mother cells, potential ones, and adjoining somatic nucellar cells. Many cases are indeed clear (see Smith 1963 a, b, Schwendener 1969; and those quoted above; compare also Fig. 10.7 A–C). In others, e.g., *Potentilla* (Rutishauser 1948, Smith 1963 a) and *Rubus* (Christen 1950, Berger 1953) it is sometimes hardly possible to distinguish between diplospory and apospory. Therefore, a term which covers both of these pathways of producing unreduced embryo sacs is indispensable. Such a term is apomeiosis as originally defined by Renner (1916), viz., embryo sac formation without reduction of the number of chromosomes. This definition has been adopted by Rosen-

2 For reports of "automixis" in certain meiotic *Rubus* strains see Section 10.4.1

berg (1930), Stebbins (1941), Rutishauser (1967, 1969), Doll (1971), and is also quoted in relevant technical dictionaries. (Unfortunately, the term apomeiosis is sometimes restricted to designate only displospory, or certain types thereof.)

10.2.1.6 Mature Unreduced Embryo Sacs

Diplosporous as well as aposporous (apart from the four-nucleate Panicum type) unreduced mature embryo sacs are in principle of the same pattern as the reduced embryo sacs typical for the related sexual species (Rutishauser 1967), even in such details as secondary division of antipodals (Davis 1967, Chao 1974). In apomictic *Crepis* these divisions are fewer than in sexual relatives (Stebbins and Jenkins 1939). However, structure and organization of unreduced, especially aposporous embryo sacs, show more or less characteristic variability, chiefly in the number of mitoses of certain embryo sac nuclei (e.g., Philipson 1978). In particular, unreduced egg cells and polar nuclei can deviate from the expected number. Murbeck (1902) reported that in *Alchemilla* the embryo sacs sometimes showed supernumerary egg cells, or more than two polar nuclei. Since then, this has been confirmed for many species, in particular detail by Fagerlind (1944), Rutishauser (1954a), Hair (1956), and Izmaiłow (1967). Variations can also occur in the four-nucleate Panicum type (Emery 1957, Voigt and Bashaw 1972, and others). In the autonomous apomict *Burmannia coelestis,* where the synergids have no function, Ernst and Bernard (1912) found egg apparatuses with one or even two additional egg cells instead of synergids.

Supernumerary egg cells easily explain the occurrence of twin embryos from one and the same embryo sac. According to their position in the embryo sac such additional embryos are sometimes interpreted as "synergid embryos", i.e., as cases of apogamy. However, this is hardly justified as in angiosperms the extremely reduced megagametophytes have no vegetative prothallial cells but only highly specialized synergid and antipodal cells; their dedifferentiation is difficult to imagine.

10.2.2 Formation of Endosperm and Embryo

Maternal offspring originate only when the unreduced egg cells, containing the complete maternal genetic information, are capable of parthenogenetic development, as a zygote would lead to B_{III} offspring with increased polyploidy. The differentiation of a normal embryo presupposes a normal differentiation of the endosperm. For this, pollination and fertilization of the polar nuclei are necessary only in "pseudogamous apomicts", while in "autonomous apomicts" the endosperm develops autonomously, i.e., both the egg cell and polar nuclei develop parthenogenetically.

10.2.2.1 Autonomous Endosperm Development

Plants showing autonomous apomixis (synonym: diploid parthenogenesis) do not depend on pollination for maternal reproduction because neither the unreduced egg cell nor the polar nuclei require fertilization; nevertheless, they are only rarely completely male-sterile. A very instructive example for the independence from pollination is the dioecious *Antennaria alpina* s.l. in which hundreds of taxa are known only as female plants. Most apomictic Compositae show autonomous endosperm development; in other families autonomous apomicts occur only sporadically, e.g., *Burmannia coelestis* (Ernst and Bernard 1912), *Elatostema* (Fagerlind 1944; apomictic species produce almost only female inflorescences), *Calamagrostis* (Nygren 1946), *Cortaderia jubata* (Philipson 1978, Costas-Lippmann 1979), *Alchemilla* (Izmaiłow, unpublished results).

Cytological analyses of endosperm in autonomous apomicts, although not very numerous, show that the degrees of polyploidy are rather variable, due mainly to the variability in number and extent of fusion of polar nuclei. Moreover, there are, as in sexual plants, endomitotic duplications. Stebbins and Jenkins (1939) observed, in *Crepis,* 2n as well as 4n endosperm, arising from free or fused polar nuclei, respectively. This means that fusion of polar nuclei is not a prerequisite for mitotic activity in endosperm. Similar reports have been published by Holmgren (1919) and Okabe (1932). In apomictic *Hieracium* both the synergids remain intact, whereas in sexual species one synergid is always destroyed (Skalińska and Kubień 1972). In *Taraxacum scanicum* Małecka (1973) noted that, exceptionally, even in autonomous apomicts occasional fertilization of polar nuclei is still possible. Most embryo sacs show, as expected, intact synergids; only rarely one of them is destroyed (obviously, in such cases fertilization of polar nuclei has taken place). It is remarkable that synergids, as well as pollen grains, remain latently functional, at least in certain autonomous apomicts, although they are not needed to ensure fertility.

10.2.2.2 Pseudogamous Endosperm Development (Pseudogamy)

The initiation of endosperm development requires fertilization of polar nuclei. Possibly, an exception could be *Arabis Holboellii* (Böcher 1951) where the stimulus of pollination is said to be sufficient for endosperm development; this should, however, be reinvestigated.

The most reliable evidence that fertilization of the polar nuclei is necessary is provided by cytological analyses of metaphases in Nuclear endosperm, or in early stages of Cellular endosperm (Noack 1939, Håkansson 1943, Rutishauser 1954a, and others). Moreover, fertilization has been confirmed by direct observation:

1. of the triple fusion (Pace 1913, Coe 1953),
2. of the penetration of one male gamete into fused or free polar nuclei, whereas the other male gamete "is arrested at the boundary of the egg cell without penetrating into the egg cell cytoplasm" (Moskova and Yakovlev 1974),
3. of the synergids, one of which is destroyed (Håkansson 1943, Izmaiłow 1967, 1973), and
4. of the growth of pollen tube into the embryo sac and its penetration and discharge into the degenerating synergid (Chao 1980).

An indirect method based on determining the lapse of time between pollination and fertilization is described by Warmke (1954) in *Panicum maximum* and

Snyder (1957) in *Paspalum secans*. Electrophoresis of proteins shows that the endosperm in different ovules of pseudogamous apomicts is genetically not identical and, therefore, must have resulted after fertilization, as only the male gametes are reduced and, consequently, show recombination (Souciet 1978).

Pseudogamy is typical for most apomictic Rosaceae (exception: *Alchemilla/Aphanes* – Hjelmqvist 1959, Izmaiłow, pers. comm. 1982), and apomictic Gramineae (exceptions: *Calamagrostis* – Nygren 1946, *Cortaderia jubata* – Philipson 1978, Costas-Lippmann 1979; *Lamprothyrsus* – Connor 1979). In apomictic Compositae, on the other hand, pseudogamy occurs only exceptionally as in *Parthenium* (Esau 1946). Important examples from other families are *Zephyranthes texana* (Pace 1913), *Hypericum perforatum* (Noack 1939), *Arabis Holboellii* (Böcher 1951), *Cooperia pedunculata* (Coe 1953), and *Ranunculus auricomus* (Rutishauser 1954a).

Generally, all apomictic species within the same genus show either autonomous or pseudogamous endosperm development. Genera comprising both alternatives are so rare that there is no reason to suppose that pseudogamy is a precursor of autonomous apomixis (in the sense of a complete breakdown of double fertilization). Such exceptions are known in *Poa* (all apomicts except *P. nervosa* are pseudogamous – Grun 1955), and *Malus* (Schmidt 1964). This is important in the cultivation of *Malus,* since only autonomous species can set fruit without pollination.

Facultative pseudogamy has been noted only by Rychlewski (1961) in a male-sterile strain of *Nardus stricta:* when pollination occurs, polar nuclei are fertilized; but when pollen is not available endosperm can develop autonomously. Unfortunately, Rychlewski does not give any data on the effect of nonpollination on seed-set; therefore, this case should be reinvestigated.

Male-sterile strains of pseudogamous apomicts depend on the pollen supply from neighboring plants. Smith (1963a) quotes the examples of *Poa arctica* subsp. *caespitosa* (after Nygren 1950), and a male-sterile strain of *Potentilla tabernaemontani,* and writes: "It is ironical that both plants are dependent on cross-pollination for the production of maternal progeny".

The degree of polyploidy of the endosperm of pseudogamous apomicts is even more variable than that of autonomous apomicts. Certain levels of polyploidy usually predominate, but not always those expected from the embryological data. For instance, Rutishauser (1954a) found, after having selfed 4x *Ranunculus auricomus,* that only 22% of the 170 endosperms analyzed were 10x, as expected ([4x+4x] + 2x), whereas 69% were 12x (double fertilization: ([4x+4x] + [2x+2x]), and the remaining 9% varied between 4x and 16x. In the endosperm of hybrids between apomictic and sexual plants (Nogler 1972), the variability can be even greater. Some other apomicts show a relative uniformity, as for instance *Hypericum perforatum* (Noack 1939), the first apomict analyzed from this point of view.

The causes of this variability of the levels of polyploidy are, on the one hand, the number of unreduced polar nuclei and their extent of fusion; on the other, the number of fertilizing male gametes (Rutishauser 1954a, Hair 1956, Izmaiłow 1967, Nogler 1972). It is possible that polar nuclei (fused or free) can be fertilized by *two* male gametes. Development without fertilization seems to be exceptional, and it is questionable whether this can result in fertile seeds. Variability can be further increased by endomitosis in certain regions of endosperm. One must always bear in mind that the endosperm can only be analyzed in free-nuclear, or early cellular stages, when there is still mitotic activity. A breakdown at a later stage in the development of the endosperm in certain ovules may always be expected, so that the level of polyploidy of mature endosperm may be more uniform also in pseudogamous apomicts.

The seed-setting in pseudogamous apomicts is often poor and can furthermore depend on the male parent, chiefly on its level of ploidy. Pseudogamous apomicts can also show a kind of seed incompatibility (endosperm incompatibility) after crossing between plants of different levels of ploidy, although the sterility (viability) barriers are not as effective as in sexual plants. Similarly, a misproportion in the genetic composition of the endosperm can lead to its degeneration and, as a consequence, also of the embryo. As yet, the results are far too fragmentary to give a clear picture (cf. Rutishauser 1954b, Nogler 1978). Seed incompatibility, at least in pseudogamous *Ranunculus auricomus*, is an effective mechanism to isolate apomicts from their sexual relatives.

In connection with pseudogamous endosperm formation, I should like to note a point which hitherto has not received due attention. In normal, sexual plants showing Polygonum type of embryo sac, crossings $4x \times 2x$ are, as a rule, incompatible because the endosperm receives from the mother plant $2x + 2x$ (reduced polar nuclei), plus only 1x from the male parent. This causes premature cell wall formation, and degeneration of endosperm and embryo follows. In pseudogamous apomicts the same ratio in endosperm results after $4x \times 4x$ crossings (and selfing), as the endosperm contains $(4x + 4x) + 2x$ chromosomes (only the polar nuclei are unreduced). Pseudogamous $4x \times 4x$ crossings are, of course, compatible. Therefore, correlated with the change to apomictic reproduction, an adaptation in the optimal endosperm ratio must have taken place, in order to avoid seed incompatibility. Pseudogamous plants have different mechanisms to ensure a genomic endosperm ratio similar to that typical for sexual plants:

1. Fertilization of one unreduced polar nucleus by one reduced male gamete, occurring in the four-nucleate, monopolar Panicum type (apospory – Warmke 1954, Brown and Emery 1958, Savidan 1975, and others; diplospory – Voigt and Bashaw 1972).
2. Fertilization of two free unreduced polar nuclei, each by one reduced male gamete (Reddy and d'Cruz 1969).
3. Fertilization of two fused unreduced polar nuclei by one unreduced male gamete (diplospory – Böcher 1951, Hair 1956, Chao 1974, 1980; apospory – Snyder 1957).
4. Double fertilization of fused unreduced polar nuclei by *two* reduced male gametes (Rutishauser 1954a, Izmaiłow 1967). This feature is genetically fixed and not connected with apospory (Nogler 1972).

To what extent such mechanisms are a prerequisite for normal development of endosperm and, therefore, influence seed-setting and fertility, should be investigated more closely.

"Pseudogamous heterosis", i.e., the peculiar phenomenon that vigor of maternal offspring can depend on the male parent, has been observed in certain apomictic strains of *Rubus* (Haskell 1960) and *Malus* (Schmidt 1964). In pseudogamous *Panicum maximum* (Souciet 1978) seed dormancy and the rates of germination can depend on the male parent. Such influences can only originate from the endosperm, as it is the only genetically variable tissue in the seeds of pseudogamous plants.

As already stressed by Rutishauser (1967, 1969) and Asker (1971), scant attention has been paid to endosperm development in pseudogamous apomicts. Clarification of mechanisms of seed incompatibility would be very desirable, above all for plant breeders. Unfortunately, the endosperm development is still the least known aspect of embryology of gametophytic apomicts.

10.2.2.3 Embryo Development

In many pseudogamous apomicts the embryo starts to develop, in most ovules, only after the initiation of the endosperm, as in sexual plants: e.g., *Rubus* (in which Focke discovered pseudogamy in 1881 – see Gustafsson 1943, Berger 1953), *Hypericum perforatum* (Noack 1939), *Ranunculus auricomus* (Rutishauser 1954b), *R. cassubicus* (Izmaiłow 1967), *Hierochloë* (Weimarck 1967), and many aposporous Panicoideae such as *Panicum maximum* (Warmke 1954), *Pennisetum ciliare* (Snyder et al. 1955), *P. dubium* (Gildenhuys and Brix 1959).

Other pseudogamous apomicts (possibly the majority), while still in the bud stage, before anthesis, show precocious embryo development in most ovules ("precocious embryony"). The embryo precedes the endosperm which develops only after fertilization of polar nuclei. Examples for this derived feature are *Poa* (Tinney 1940, Skalińska 1959), *Parthenium* (Esau 1946), *Tripsacum dactyloides* (Farquharson 1955), and others.

Transitions between these two features are known. Certain genera contain species with both features, e.g., *Potentilla* (Rutishauser 1948, Smith 1963b) and *Paspalum* (Snyder 1957, Reusch 1961). There are also species in which only single strains deviate, e.g., *Bothriochloa ischaemum* (Brown and Emery 1957) and *Potentilla tabernaemontani* (Smith 1963a).

In autonomous apomicts both features are known, but precocious embryony prevails (examples: *Hieracium* – Skalińska 1973; *Taraxacum* – Małecka 1971, Richards 1973). Of course, here also the endosperm can be initiated before anthesis. In others, such as *Crepis* (Stebbins and Jenkins 1939) and *Cortaderia jubata* (Costas-Lippmann 1979), the division of the egg cell takes place only after endosperm formation has begun.

The importance of precocious embryony is obvious. The division of egg cell before anthesis makes fertilization impossible, and prevents formation of B_{III} hybrids. They, however, can still be produced from occasional ovules with delayed development containing persistent undivided egg cells even after anthesis. Richards (1970a) even claims that "precocious embryony may well be the only mechanism preventing sexuality in polyploid Taraxaca", and similar statements

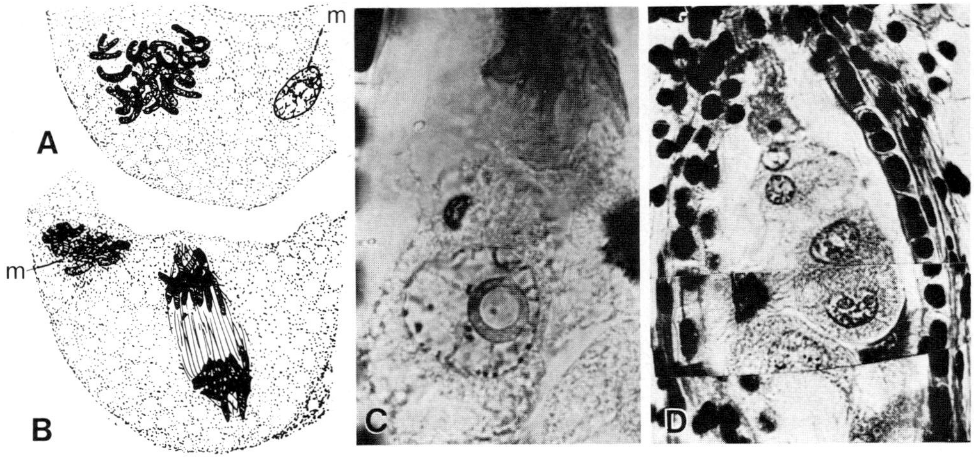

Fig. 10.8 A–D. Hemigamy in *Zephyranthes (Atamosco) texana* (**A, B**) and *Rudbeckia laciniata* (**C, D**). The nucleus of male gamete is in the cytoplasm of the egg cell but fails to fuse with its nucleus. **A, B** Asynchronous division; the female nucleus precedes the division of the male (*m*) which usually degenerates. **C, D** Simultaneous division of the female (large) and male (small) nuclei. **C** Both nuclei in prophase. **D** Participation of the nucleus of the male gamete in the formation of the proembryo; the two larger nuclei result from the division of the egg nucleus, the two smaller ones from the male gamete. These are situated at the micropylar pole and form the suspensor. *At the bottom,* note one endosperm nucleus and a large vacuole. Combined photograph from two adjacent sections. (**A, B** From Pace 1913, **C, D** courtesy M. P. Solntzeva, Leningrad)

have repeatedly been made by other investigators. But precocious embryony cannot be in causal connection with parthenogenesis – a view demonstrated by the fact that those pseudogamous apomicts showing division of the egg cell only after fertilization of polar nuclei do not produce more B_{III} hybrids than those showing precocious embryony. The actual cause of parthenogenesis of unreduced egg cells, and what prevents their fertilization, i.e., why "the other sperm is arrested at the boundary of the egg cell" (Moskova and Yakovlev 1974), is still unknown. As it is certain that the synergids remain functional even in certain autonomous apomicts (see Sect. 2.2.1), one can assume that the barrier preventing fertilization may be the wall of the egg cell. However, this is uncertain due to the lack of electron microscopic and other investigations and, above all, on account of the phenomenon of hemigamy which has repeatedly been observed in pseudogamous apomicts.

In "hemigamy" (a linguistically wrong synonym: "semigamy") the nucleus of the second male gamete enters the cytoplasm of the egg cell but does not fuse with its nucleus, and degenerates sooner or later (Fig. 10.8 A, B). Hemigamy was first observed by Navashin in 1898 (see Solntzeva 1974) in *Rudbeckia speciosa,* but unaware of the pseudogamy of this species he interpreted it as a prestage of double fertilization. Pace (1913) showed that in pseudogamous *Atamosco (Zephyrantes) texana* the nucleus of the second male gamete degenerates in the cytoplasm of the egg cell without ever fusing with its nucleus. Later, this phenomenon was described by Battaglia (1946b) who coined the term semigamia, Fagerlind (1946), Emery (1957), and Solntzeva (1974). Solntzeva (1978; see Fig. 10.8C, D), Coe (1953), and Kośińska (1980) reported cases in which the nucleus of the male gamete in the two-nucleate "zygote" also divided once or several times before degenerating, or even participated in the formation of the embryo (especially the suspensor) which can become a chimera (Rao and Narayana 1980). Hemigamy is not a prerequisite for embryo differentiation but only an "additional feature in certain pseudogamous apomicts" (Coe 1953) as well as in mutants of sexual plants. Hemigamy is of theoretical interest because it shows that the fertilization of the egg cell is not prevented by a premature completion of the cell wall between synergid and egg cell, but by some mechanism within the egg cell – possibly by the membrane of the egg cell nucleus itself.

In "androgenesis" (male parthenogenesis) the embryo originates from the nucleus of the male gamete, whereas the egg cell nucleus in the two-nucleate (one ♂ and one ♀) "zygote" degenerates. Androgenetic offspring contain, therefore, only the paternal chromosome complement. In nature, androgenesis is just another additional feature occurring in certain apomicts (e.g., Rao 1974), as well as in certain sexual plants. The importance of androgenesis lies in its application in plant breeding.

10.2.3 Male Meiosis and the Male Gametophyte

Diplospory as well as apospory are phenomena concerning only the female reproductive structures. The male meiosis is not primarily affected but is often more or less normally concluded, similar to the female meiosis in apospory. Irregularities are secondary. It is noteworthy that even in the Antennaria type the male meiosis can be more or less normal (e.g., see Rutishauser 1948).

Unreduced pollen from restitution nuclei is relatively rare, even in diplosporous apomicts of the Taraxacum type. The formation of female restitution nuclei is caused by a special factor which only affects the female, but not necessarily also the male sporogenesis (Battaglia 1950, Bergman 1951, Richards 1973, Iwanaga and Peloquin 1979). Giant pollen grains are occasionally reported. In aposporous *Potentilla tabernaemontani* they occur especially in anorthoploid (5x, 7x) biotypes (Asker, pers. comm. 1981).

Autonomous apomicts, which do not need viable pollen for endosperm initiation, show the whole range from completely normal male meiosis (with only bivalent formation, e.g., Mendel's *Hieracium aurantiacum* – see Ostenfeld 1910, Christoff 1942) to complete male sterility. A male-sterile dioecious apomict is *Cortaderia jubata* (Costas-Lippmann 1979). Certain aposporous species of *Cortaderia* and *Lamprothyrsus* are known from female plants only (Connor 1979). Apparently, all possible disturbances, up to complete degeneration of the male meiosis, can accumulate (e.g., in *Taraxacum* – Richards 1970a). There are reports to show that even male gametogenesis can be disturbed in that the second pollen mitosis fails (Davis 1967).

Pseudogamous apomicts, which depend on pollen for the fertilization of polar nuclei, have, on an average, pollen of a much better quality; but even here completely normal male meioses are surprisingly rare (e.g., see Combes 1975). Obviously, pollen with a more or less reduced capability to germinate is sufficient to fertilize polar nuclei, whereby a considerable proportion of aneuploid pollen is tolerated (Nogler 1972). Completely male-sterile strains are an exception in pseudogamous apomicts; they are, of course, dependent on pollen from neighboring plants (see, among others, Smith 1963a).

Embryo sac-like structures in anthers can be formed "aposporously" from anther wall or from tapetal cells (Berger 1953, Davis 1967, Weimarck 1967), or "diplosporously" from male archesporial cells (Weimarck 1970).

For further details on male meiosis in apomicts and for a comprehensive survey of literature see Rutishauser (1967, pp. 62–121).

10.3 Apomixis and Sexuality

10.3.1 Sexual Potentials in Gametophytic Apomicts

As stated earlier, neither can diplospory/apospory totally exclude meiosis, nor can the capacity of parthenogenesis totally exclude fertilization. In apomictic wild types both these components of sexuality must be suppressed by auxiliary mechanisms such as failure of stages of megasporogenesis or gametogenesis, aposporous competition (Sect. 10.2.1.4), and precocious embryony (Sect. 10.2.2.3). In spite of this, exceptionally, maturation of reduced embryo sacs, or fertilization of reduced or unreduced egg cells can occur. Later, during embryogenesis, and especially during endosperm development, decisive hindrances have to be overcome before an aberrant offspring can develop. Unfortunately, however, these stages are difficult to approach methodically and, therefore, have hardly been investigated. Even during germination, and afterward, aberrant embryos often show signs of weakness compared to the vigorous maternal ones and, eventually, some of them die. This applies not only to polyhaploid but also to B_{II} and B_{III} hybrid embryos and seedlings.

In principle, therefore, gametophytic apomicts still preserve the possibility of producing aberrant offspring, although this is exceptional. Thus, it becomes clear that apomixis and sexuality are not alternatives, as was formerly assumed, but independent modes of reproduction which can occur side by side. Apomixis and sexuality are in a state of balance which, during the course of development between megasporogenesis and seed germination, or even later, shifts more and more in favor of apomixis. The quantitative expressions of this state of balance between apomixis and sexuality are:

1. the "degree of diplospory/apospory" (common designation: "degree of apomeiosis"), i.e., the percentage of ovules with unreduced embryo sac initials or stages, and
2. the "degree of apomixis", i.e., the percentage of maternal offspring (offspring of apomictic origin). From a theoretical point of view, the degree of apomixis (as well as the degree of pseudogamy) should also include polyhaploid offspring.

The degree of diplospory/apospory (apomeiosis) reaches values which are, as a rule, lower when determined by embryological analyses, higher when calculated from cytological embryo or endosperm analyses, and still higher when calculated from cytogenetical offspring analyses.

A rise to degrees of apomixis approximating 100% is the quantitative indication of the suppression of sexuality in the course of development. Such an increase in the degree of apomixis was first described in 1939 by Noack and, since then, has been confirmed by Warmke (1954), Harlan et al. (1964), Nogler (1972, 1973), Combes (1975), Savidan (1975), and others. In populations with an extremely high degree of apomixis, traces of sexuality are often difficult to prove. Aberrants can, if necessary, be detected by means of chemical methods (e.g., Spoor and Hay 1978, Matzel 1982).

Recently, it has become questionable whether, theoretically, "total" apomixis exists at all, as traces of sexuality have been detected even in plants hitherto regarded as obligate apomicts: *Ranunculus auricomus, R. cassubicus* (Marklund and Rousi 1961, Izmaiłow 1970, 1976), *Taraxacum* sect. Palustria (Fürnkranz 1961, Małecka 1973), and *Hieracium aurantiacum* (Skalińska 1973, 1976). Thus, even autonomous apomicts retain traces of sexuality, a decisive finding in connection with the evolutionary possibilities within apomictic species complexes. From a purely theoretical point of view, the distinction between obligate apomixis and facultative apomixis as two fundamentally different types in hardly justifiable.

10.3.2 Possibilities of Influencing the Degree of Apomixis

The degree of diplospory/apospory, the degree of apomixis, and the extent of suppression of sexuality are genetically determined but can often, to a certain extent, be influenced quantitatively by factors beyond the mother plant.

10.3.2.1 Male Parent

The male parent may influence whether a (reduced or unreduced) egg cell develops parthenogenetically or after fertilization. An early and especially well-documented cytological investigation by Noack (1939) showed that selfing of tetraploid *Hypericum perforatum* resulted in 87 maternal and 33 (=27%) B_{III} hybrid embryos, whereas pollination with diploid *H. quadrangulum* resulted in 54 maternal and 113 (68%) B_{III} embryos. Later, similar results were obtained by Müntzing (1958a), Dowrick (1961), and others. Such influences by the male parent may often take place via the fertilized endosperm, but it is not known how. However, the level of ploidy of the male parent is not always decisive, as shown, for example, by recent pollination experiments in *Ranunculus auricomus* (Nogler 1982). Obviously, certain plants produce pollen with a greater tendency to fertilize egg cells inclined by nature to parthenogenetic development.

10.3.2.2 Environmental Factors

Environmental factors influence the degree of apomixis in many cases. Nygren (1946) reported that *Calamagrostis purpurea,* instead of mitotic embryo sacs

(Antennaria type, typical for this species), under certain circumstances has meiotic embryo sacs in some of the inflorescences (which, because of genetic self-incompatibility, remain sterile after selfing). Similar, but more convincing, results were obtained by Sparvoli (1960) with *Eupatorium riparium* under varying conditions of cultivation (see Sect. 10.2.1.2). A classical investigation was published in 1963 by Knox and Heslop-Harrison on *Dichanthium annulatum,* a partially aposporous short-day grass in which the photoperiod is the decisive factor. Continuous short-day conditions lead to a significantly higher degree of apospory than long-day conditions, in that not only more unreduced but also fewer reduced embryo sacs mature – obviously the result of greater and earlier aposporous competition. According to Gupta et al. (1969) both short photoperiods and low temperatures facilitate apospory in *Dichanthium annulatum.* Investigations by Evans and Knox (1969) on *Themeda triandra* show that short-day conditions favor apospory in both short-day and long-day races. In the facultatively diplosporous subtropical grass *Eragrostis curvula,* short-day conditions decrease the degree of diplospory in that many megaspore mother cells degenerate instead of differentiating directly (i.e., mitotically) into unreduced (four-nucleate) embryo sacs typical for the species. This gives the impression of increased frequencies of reduced embryo sacs but, in reality, only the fertility is reduced (Brix, pers. comm. 1980). Hjelmqvist and Grazi (1964), Saran and deWet (1970), and others, also showed environmental influences on the degree of apomixis. Saran and deWet (1976) found, in a partially sexual hybrid of *Dichanthium intermedium,* that the manifestation of apospory can even be completely suppressed under certain photoperiods. In contrast, the investigations of McWilliam et al. (1970), and Burton (1982) show that apospory is not always influenced by environmental factors.

Not only the degree of diplospory/apospory but also the timing of the initiation of the embryo is reported to be, to some extent, prone to environmental influences. *Taraxacum* sect. Palustria belongs to those apomicts that show precocious embryo development initiated even before anthesis. Normally, i.e., in spring florets, ovules with delayed development and thus having still undivided egg cells at anthesis occur only sporadically. According to Małecka (1973), higher percentages of such delayed ovules are found in autumn florets, which increases their chances of fertilization.

10.3.2.3 Other Factors

X-ray and other ionizing radiation experiments (some including mutagenic agents such as ethyl methane sulfonate) did not affect the state of balance between apomixis and sexuality in *Taraxacum vulgare* (von Hofsten 1954, cited from Asker 1966), *Paspalum dilatatum* (Bashaw and Hoff 1962, Burton and Jackson 1962), *Potentilla* (Asker 1966), and *Pennisetum pedicellatum* (Saran and Narain 1982). Only in two predominantly apomictic varieties of *Poa pratensis* ("Fylking" – Julén 1961; "Merion" – Hanson and Juska 1962) X rays caused a slight increase in the number of sexually formed aberrant offspring which, however, reverted again rather rapidly to apomixis. Instances of mutagenic breakdown of apomixis, or reversal to sexuality, are not known. A case of mutagenic induction of a tendency toward apospory leading to facultative apomixis in diploid *Pennisetum typhoides* was reported by Hanna and Powell (1973). Further research on mutagenesis concerning genes involved in reproduction would be highly desirable.

Chromosome doubling by means of colchicine has been carried out on many apomicts, but only in a few cases was any influence on apomixis detected. Müntzing (1958a) and Asker (1967, 1970) investigated a diploid, partially apomictic strain of *Potentilla argentea,* where either reduced or unreduced egg cells developed, depending on the male parent. Autotetraploids produced after colchicine treatment became phenotypically sexual in that probably only some reduced egg cells remained functional. The reason why apomeiosis is no longer manifest in autotetraploids is unknown, due to lack of embryological investigations. Perhaps, a parallel case is a desynaptic tetraploid strain of diplosporous *Paspalum longifolium* (Chao 1974), where an octoploid cytotype produced by colchicine treatment becomes synaptic on the female side, cancelling diplospory. Instead of formation of restitution nuclei a complete meiosis leads to reduced embryo sacs and to B_{II} hybrids. This alteration seems to be rather subtle since the simultaneously investigated hexaploid *P. commersonii* did not become synaptic after colchicine treatment: the dodecaploid cytotype remained diplosporous. M. and M.A. Christoff (1948) and Müntzing (1958b) reported an occasional restoration of meiosis after spontaneous or induced chromosome doubling leading to partial haploid parthenogenesis. M. and M.A. Christoff's investigation should, however, be checked considering its peculiar interpretation. In contrast to these isolated cases, many investigations show that normally chromosome doubling has no influence on the mode of reproduction: apomicts remain apomictic, sexual plants sexual. For a recently reported case of facultative apospory appearing after chromosome doubling (Quarín and Hanna 1980) see Section 10.4.3.2.

In predominantly sexual *Rubus saxatilis* the two ovules of each carpel are, according to Czapik (1981), different in their apomeiotic tendencies. As a rule, in the lower, larger ovule a reduced embryo sac differentiates from the multicellular archesporium, whereas diplosporous or, rarely, aposporous initials are more frequent in the upper, smaller ovule which usually degenerates.

The effects of plant growth regulators on gametophytic apomixis are inadequately known. Attempts to modify the degree of apomixis with phytohormones and other substances (Sherwood and Young 1979, Matzel 1982) have not as yet brought results clear enough to gain an insight into the physiology of gametophytic apomixis.

10.4 Causes and Consequences of Apomixis

10.4.1 Heterozygosity and Variability

Maternal reproduction means genetic fixation of any heterozygous gene combination and leads, in spite of hybridity, to uniform offspring, to populations of genetically exact copies of the mother plant. An extreme heterozygosity is almost characteristic of all apomicts investigated genetically as, after crossing with sexual mother plants, manifold segregations occur [compare the notheworthy lithography in Ostenfeld (1910) illustrating one of Mendel's *Hieracium* crossings]. Crossings are possible as nearly all apomicts produce a certain amount of viable pollen. On the other hand, even in "obligate" apomicts a B_{II} hybrid or a B_{III} hybrid formation, or haploid parthenogenesis, is not totally excluded (see Sect. 10.3.1). Therefore, in spite of apomixis there is, in principle, still the possibility of a limited gene recombination and variability and, consequently, a restricted adaptability. Gametophytic apomixis is rather a suppression than a loss of sexuality, as was formerly assumed.

There are also some *asexual* sources of variability, besides traces of sexuality, which both increase heterozygosity. Mutation is one of them; however, its importance should not be overestimated, above all that of recessive mutations, as apomicts are in general polyploid. Autosegregation, i.e., changes or rearrangements of the genotype during egg cell formation in plants with parthenogenesis, is probably more important in diplosporous apomicts which form restitution nuclei, such as the Taraxacum type. There are two possibilities: (1) In "subsexual" reproduction (Darlington 1937) a partially asyndetic meiotic prophase with a few bivalents leads to a restitution nucleus with very limited crossing-over. The best investigated cases are *Parthenium* hybrids (Rollins 1945), *Rubus nitidioides* (Haskell 1953), and *Agropyron scabrum* (Hair 1956). (2) Individual chromosomes can be lost during the formation of restitution nuclei. This possibility enabled Sørensen (1958) to establish in triploid apomictic *Taraxacum* a complete series of eight different disomic aberrants with $2n = 3x - 1 = 23$ chromosomes. Other anomalies such as occasional gains or losses of individual chromosomes through nondisjunction occur in apomicts as well as in sexual plants, and will not be discussed here. Doubling of the whole chromosome complement is sometimes observed, as mentioned in Sect. 10.2.1.1. An unusual instance of autosegregation is reported by Dowrick (1961) and Gerlach (1965) in certain partially sexual, tetraploid *Rubus* strains with normal syndesis which – unlike apomictic, pseudogamous blackberries (see Sect. 10.2.1.5) – are free from apospory. The synergids do not always discharge the sperm nuclei, thus preventing, sometimes, double fertilization. If the reduced egg cell develops parthenogenetically, the spindle may lie crosswise to the axis of the embryo sac so that the daughter nuclei "are not placed in the gradient of polarity in the egg cell plasm" and may fuse ("automixis"). Nevertheless, total homozygosity does not result, as the original, tetraploid strains are highly heterozygous.

The effects of heterosis in sexual crop plants show strikingly significant advantages of heterozygosity. The fixation of a pronounced heterozygosity is, however, not compatible with sexual reproduction – apart from complex heterozygotes. No doubt, the well-known superiority of many apomictic strains is a further example of the positive effects of heterozygosity. Another parallel to heterosis is that hybrid offspring of apomicts, if any, are almost always inferior.

10.4.2 Causes of Gametophytic Apomixis; Components of Apomixis in Sexual Plants

In his book entitled *Bastardierung als Ursache der Apogamie [= Apomixis] im Pflanzenreich,* Ernst (1918) stipulated hybridization as a cause of apomixis. At that time hybridization and heterozygosity were already known to be typical in apomicts. But shortly afterward, Holmgren (1919) disputed Ernst's hypothesis, stating that hybridity is only one prerequisite for apomixis. Ernst's hypothesis has never been supported experimentally, and has been abandoned, notwithstanding a few reports on aposporous tendencies in the hybrids *Raphanobrassica* (Ellerström and Zagorcheva 1977) and *Festuca pratensis* × *F. arundinacea* (Gröber, pers. comm. 1980). Hybridity and heterozygosity are now better understood as a consequence, as a typical concomitant phenomenon, rather than as a cause of apomixis, which is sought in genetic factors (see Sect. 5.3). At least the basic components of apomixis could originate somehow or other by mutation. A definitive statement about the causes of gametophytic apomixis cannot yet be made, as many more details must be collected. However, some pieces of the puzzle are already within our grasp. Among the mutants of genes concerning meiosis, those are of especial interest that cause asyndesis and restitution nuclei only in the female meiosis (see Bergman 1951, Richards 1973, Iwanaga and Peloquin 1979). Aposporous initials or embryo sac-like structures in certain sexual individuals are usually functionless, especially in diploids, but may, in some cases, become functional in polyploids of hybrid origin, as reported by Bergman (1935), Hjelmqvist (1959), Davis (1967), Weimarck (1967), Quarín et al. (1982), and others. However, when interpreting irregularities in the nucellus as signs of apospory, or even of apomixis, one cannot be careful enough. Latent tendencies to parthenogenesis, which can become manifest under certain circumstances, are known to occur in many sexual plants (see review by Kimber and Riley 1963). Recently, G. and A. Hagberg (1981) investigated a mutant *hap* in sexual barley which, when homozygous, produces haploid offspring parthenogenetically in a rather high frequency.

The frequent occurrence of latent tendencies to parthenogenesis in normal, sexual plants agrees well with the new concept (compare Asker 1980, Asker and Hagberg 1983) according to which "the different constituents, or elements (Petrov 1976) of apomixis, lie within the reproductive potentialities of sexual plants". In my opinion, however, it is questionable whether this also applies to the other basic component of apomixis, i.e., the capacity to form unreduced embryo sacs. It is indeed well known that many sexual plants have a latent capacity to form unreduced gametophytes occasionally, a capacity which can become functional under certain circumstances. This is the normal way of polyploidization in sexual plants (see reviews by Franke 1975, Harlan and deWet 1975). But the conditions are quite different. In sexual plants meiosis and gametophyte formation are bound together so that unreduced gametophytes result, at most, from meiotic restitution nuclei. In apomictic plants unreduced embryo sacs are formed from meiotic restitution nuclei only in the Taraxacum and Ixeris types, which are remarkably rare, except in Compositae. In the majority of apomicts, on the contrary, the bonds between female meiosis and embryo sac formation are broken up. This applies to both mitotic diplospory (Antennaria type) and apospory, but it hardly ever occurs in sexual plants. Even in aposporous *Panicum maximum* the *sexual* hybrids formed unreduced embryo sacs, if at all, not by apospory but through meiotic restitution nuclei (Savidan 1982).

All these facts suggest the hypothesis that the bonds one has to imagine linking female meiosis and embryo sac formation in normal sexual plants must be opened somehow in most gametophytic apomicts. I consider that this opening is a decisive step in the formation of most apomictic species.

10.4.3 Apomixis and Polyploidy

10.4.3.1 Polyploidy

Allopolyploidy is just as typical for gametophytic apomicts as hybridity and heterozygosity. Triploidy is relatively rare and, in most cases, restricted to the Taraxacum and Ixeris types of diplospory. Examples: *Erigeron* (Holmgren 1919), *Chondrilla* (Bergman 1950), *Taraxacum* (Richards 1970a) in Compositae; *Statice oleaefolia* (D'Amato 1949) in Plumbaginaceae; *Arabis Holboellii* (Böcher 1951) in Cruciferae. An aposporous triploid is *Pennisetum setaceum* (Simpson and Bashaw 1969). More common are higher levels of polyploidy up to dodecaploidy, and, sometimes, even higher. However, tetraploids are the most frequent. Very often different levels of polyploidy are realized within the same species. Some apomicts, especially with high levels of polyploidy, show a tendency to revert by haploid parthenogenesis (e.g., Müntzing 1958 b).

10.4.3.2 Diploid Apomicts

Diploid apomicts are very rare in nature. The most important are single strains of *Potentilla argentea* (Müntzig 1945, 1958 a), *Arabis Holboellii* (Böcher 1951), and, possibly, *Hierochloë australis* (Weimarck 1967). Earlier statements to the effect that diploid *Ranunculus cassubicifolius* is apomictic are erroneous, as has repeatedly been confirmed (Rutishauser 1954 b, Nogler 1978). In *R. auricomus,* diploid apomictic plants are obtained only experimentally (see below). *Nardus stricta* ($2n=26$) is, according to Rychlewski (1967), actually a tetraploid in which two chromosomes have been lost. On the contrary, sexual relatives are usually diploid, a fact which does not mean that purely sexual species cannot be polyploid as well – for instance apomictic *Paspalum commersonii* which includes dodecaploid sexual strains (Chao 1974). Sexuality is also maintained after polyploidization by colchicine treatment. This proves that apomixis and polyploidy are not causally connected. However, there are a few possible exceptions. In two phenotypically sexual strains of *Dichanthium aristatum* and *D. caricosum* (both $2x=10$) apospory becomes apparent in tetraploid hybrids resulting from the summation of the unreduced gametes ($2n+2n$) of both strains (d'Cruz and Reddy 1971). In another phenotypically sexual grass, the diploid *Paspalum hexastachyum* ($2x=12$), mere chromosome doubling by colchicine treatment allows facultative apospory (Quarín and Hanna 1980). It is unknown why in these exceptional cases apospory does not become manifest at the diploid level.

Recent results on the cytogenetics of apomixis in *Ranunculus auricomus* (Nogler 1982) can possibly provide an answer to the question why hardly any diploid apomicts can be found in nature. It could be proved that apospory cannot be transmitted by haploid (monoploid) gametes but only by diploid (reduced or unreduced) or polyploid ones (see Sect. 10.5.3). Diploid apomicts can, therefore, be formed only as dihaploids but not as hybrids – at least in *R. auricomus*. On the other hand, apomictic dihaploids can give rise at most to sexual B_{II} hybrids, and to B_{III} hybrids, but are generally too weak to survive by maternal reproduction.

The results obtained by d'Cruz and Reddy (1971) with a phenotypically sexual diploid strain of *Dichanthium aristatum* would compare very well with this finding. Diplosporous *Taraxacum* species seem to be another comparable case, as crossing experiments by Richards (1970b) and Müller (1972) suggest. Moreover, the peculiar results of a disappearance of apomixis (diplospory) after crossing diploid *Parthenium argentatum* reported by Gerstel and Mishanec (1950) and Gerstel et al. (1953) become more intelligible on this basis, rather than by the authors' interpretation (cf. Nogler 1982).

10.4.3.3 Dihaploids

Dihaploids, formed by parthenogenetic development of reduced (i.e., diploid) egg cells of partially apomictic tetraploids, are known from many investigations. Most dihaploids are sexual, very often rather weak and more or less sterile, but not invariably, as shown, among others, by Savidan (1980) who obtained very vigorous and fertile sexual dihaploids in apomictic Panicoideae. DeWet (1965) presumes that sterility in certain dihaploids may be due to apospory. Whether sterile dihaploids are aposporous can be detected only by embryological investigations (see Combes 1975, Savidan and Pernès 1982). Weak but fairly fertile aposporous dihaploids have recently been discovered in *Ranunculus argoviensis* (Nogler, unpublished). To my knowledge, aposporous dihaploids are formed in large numbers only from certain tetraploid *R. auricomus* hybrids (Nogler 1973). They are, most surprisingly, very vigorous and fertile. Most of them are aposporous to a very high degree and, as expected, produce almost exclusively maternal and – depending on the male parent – B_{III} hybrid offspring. Therefore, it is certain that apomictic dihaploids are possible, and are not necessarily weak and sterile. Some of these apomictic dihaploid *R. auricomus* plants are no doubt vigorous enough to survive in natural competition. May be one or another of the known diploid apomictic wild strains mentioned above has resulted from tetraploid apomicts by haploid parthenogenesis.

10.4.4 Agamic Species Complexes and Their Evolution

Remnants of mechanisms of sexual reproduction enable the formation of aberrants, however rare this may be, even in "totally" apomictic strains. Aberrants are usually more or less inferior, and hardly ever capable of competing. Exceptionally, they can, especially after alien pollination, indirectly give rise to new, successfully adapted strains which can again be apomictic, since the single components of gametophytic apomixis are genetically fixed. Thus, actual "agamic species complexes" (Babcock and Stebbins 1938), even "agamic genus complexes", can originate through gradual expansion of an apomictic species, and through hybridization, i.e., incorporation of genes of one species into the gene pool of another (see Stebbins 1950). In this connection Turesson (1929) coined the term "agamospecies" for apomictic populations of common origin. Polymorphism is typical for agamic species complexes, which often develop an enormous abundance of forms. The most striking example is *Hieracium*. Agamic complexes include distinct strains connected by all possible intermediary, true-breeding hybrids. Besides agamospecies, which are generally polyploid, such complexes include sexual species. These are generally diploid and, usually, more or less isolated geographically and by genetic sterility barriers. For example, in pseudogamous *Ranunculus auricomus* seed incompatibility "between tetraploid apomicts and their diploid sexual relatives prevent the introgression of genes responsible for pseudogamous reproduction. In this way apomictic reproduction may be restricted to sharply defined groups in an otherwise sexual genus" (Rutishauser 1961).

Some especially instructive agamic complexes are: Compositae – *Crepis* (Babcock and Stebbins 1938), *Taraxacum* (Richards 1973, Doll 1976, including *Chondrilla,* den Nijs and Sterk 1980), *Antennaria* (Urbańska-Worytkiewicz 1974); Gramineae – *Poa* (Müntzing 1940, 1966, Nygren 1950), *Setaria macrostachya* (Emery 1957), *Calamagrostis* (Nygren 1962, Tateoka et al. 1977), *Dichanthium-Bothriochloa-Capillipedium* (deWet and Harlan 1970), *Panicum maximum* (Combes 1975, Pernès et al. 1975a); other families – *Potentilla* (Müntzing 1958b, Smith 1963a, b, Schwendener 1969, Asker 1977), *Rubus* (Gustafsson 1943, Haskell 1966), *Ranunculus auricomus* (Rousi 1956, Marklund and Rousi 1961, Rutishauser 1961), *Aphanes/Alchemilla* (Hjelmqvist 1959).

Taxonomic implications of apomixis as well as aspects of distribution and origin cannot be dealt with here. For detailed information see Gustafsson (1947b), Löve (1960), Valentine (1960), Walters (1966), Pernès et al. (1975b); for statistical models, methods, etc. see Marshall and Brown (1974), Williamson and Killick (1978), and Marshall and Weir (1979).

The evolutionary significance of apomixis is discussed by Stebbins (1950), deWet and Stalker (1974), Băra and Ghiorghiţă (1980), Marshall and Brown (1981), and Grant (1981). The rudimentary adaptability of apomicts makes it at least possible that in general "agamospecies remain evolutionarily active despite apomixis" (deWet and Stalker 1974) – contrary to plants with purely vegetative reproduction. Thus, apomixis provides for many highly heterozygous hybrids an "escape from sterility" (Darlington 1937), but not an "escape into a blind alley of evolution". According to Stebbins (1950), "there is no evidence that apomicts have ever been able to evolve a new genus or even subgenus" and "... agamic complexes are destined to produce only new variations on an old theme".

A new model combining different possibilities of evolution within agamic complexes applies to "diploid-tetraploid-di(ha)ploid cycles" ("reversible tetraploidy"). Such a cycle begins with a diploid sexual species which can become tetraploid via unreduced (2n) gametes as may occasionally be produced by any sexual plant on both the female and the male side, through meiotic disturbances (Harlan and deWet 1975). Tetraploidization can take place directly as a B_{III} hybrid formation by addition of two unreduced gametes ($2n + 2n = 4x$). Such tetraploids are, of course, again sexual and can (as a sterility barrier no longer exists) be fertilized by pollen of tetraploid apomicts, and so may possibly lead directly or indirectly to new tetraploid apomictic strains. "... even partially sterile but fully sexual autotetraploids may provide a temporary genetic bridge between apomictic biotypes" (deWet 1968). Or, as an alternative, such tetraploids can revert by haploid parthenogenesis to dihaploids, whereby the cycle is completed. If these dihaploids produce unreduced egg cells, they can again be the basis for new tetraploid strains arising as B_{III} hybrids, i.e., either as "2n + 2n hybrids" when fertilized by unreduced pollen from diploids, or as "2n + n hybrids" when fertilized by reduced pollen from tetraploids. Furthermore, dihaploids can also arise indirectly, from 2x × 4x crosses between diploid sexual and tetraploid apomictic strains, provided that genetic sterility barriers do not prevent the formation of triploid B_{II} hybrids in the F_1. After backcrossing of such triploid hybrids with the diploid parent, among other offspring, some tetraploid B_{III} hybrids may be formed, and can then give rise to new dihaploids if they have a sufficient tendency toward haploid parthenogenesis (Nogler 1973 and unpublished results, Cleij et al. 1976, and others).

Such diploid-tetraploid-dihaploid cycles enable the testing of many new tetraploid apomictic forms and "make possible extensive recombination within the apomictic complex", as discussed by deWet (1968), deWet and Harlan (1970), and deWet (1971) for *Dichanthium-Bothriochloa,* by Asker (1977) for *Potentilla argentea,* by Nogler (1978) for *Ranunculus auricomus,* by Savidan and Pernès (1982) for *Panicum maximum*. DeWet (1965) and Savidan (1982) suppose that at least some of the small sexual populations within tetraploid agamic complexes may be derived from vigorous sexual dihaploids.

10.5 Genetics of Gametophytic Apomixis

Gametophytic apomixis is genetically fixed, as is proved by many crossing experiments. Phenotypic influences are, if at all present, only slight (see Sect. 10.3.2.2) As discussed in Section 10.3.1, apomixis and sexuality are not alternatives but are in a state of balance, in wild types mostly with an extreme shift toward apomixis. This decisive finding has to be considered in every genetic investigation of apomixis versus sexuality, even if "totally" apomictic plants are available for crossing with sexual mother plants.

10.5.1 Dominant or Recessive Inheritance?

Early genetic analyses of crossings between sexual mother plants and apomicts, especially with aposporous *Hieracium,* which were already attempted by Mendel (cf. Correns 1905), indicated a tendency toward a dominant inheritance of apomixis versus sexuality: *Hieracium* (Ostenfeld 1910, Christoff 1942), aposporous *Potentilla* (Christoff and Papasova 1943), aposporous *Malus* (Sax 1959). Later investigations showed rather an intermediate (*Rubus* – Christen 1952, Berger 1953) to a recessive inheritance with quantitative dosage effects in polyploids. An increasing number of genomes carrying genes for apomixis become more and more dominant over one or more sexual genomes (*Poa* – Müntzing 1940, *Rubus* – Gustafsson 1943, *Potentilla* – A. and G. Müntzing 1945, Rutishauser 1948, Hunziker 1954). The results on *Poa* published by Åkerberg and Bingefors (1953) are somewhat contradictory, and Almgård (1966, p. 54) stated: "Evidently apomixis as a functioning system can here [in *Poa*] be described as completely recessive, while its basic prerequisite, the presence in this case of an aposporous ES, rather showed a dominant pattern of inheritance in F_1".

It is known from many apomicts that partially or totally sexual hybrids often result from crossings between two apomictic plants. To explain this surprising fact Müntzing (1940) suggested the hypothesis that apomixis "is due to a rather delicate genetic balance" of a few recessive genes with dosage effects which may easily be disturbed after crossing, but is often difficult to reestablish. However, he did not specify the number and the function of these recessive apomixis genes. After it was realized that gametophytic apomixis is a complex process, it became necessary to split it into its components, especially concerning the two deviations from the normal, sexual life cycle (alternation of generations), viz. the failure of meiotic reduction, and the failure of fertilization. In 1945, Powers (on diplosporous *Parthenium*) published a much quoted but purely hypothetical model, postulating three recessive genes, one for failure of reduction, and two for failure of fertilization and replacement by parthenogenesis. Application of this model to crosses in the same *Parthenium* material not being possible, Gerstel et al. (1953) assumed the number of recessive apomixis genes to be at least four, without establishing their function. (These results can be interpreted more easily on quite a different basis – see Sect. 10.4.3.2).

The chief difficulty met with by these models, based on recessive apomixis genes, is the necessity to assume homozygosity, at least for some of the apomixis genes. However, in the entire literature there are no convincing arguments for the assumption of homozygosity in apomicts; on the contrary, in all cases examined extreme heterozygosity has been proved. Moreover, it is not easy to understand why, in nature, apomixis, once established, should always gain the upper hand after hybridization as a recessive characteristic, even in mixed populations. The reason can hardly be the mere superiority of certain apomictic hybrids (see Sect. 10.4.1).

10.5.2 Some Critical Remarks

Before giving a short account of some – partly rather contradictory – results, I should like to stress some points, the frequent disregard of which contributes greatly to the confused picture of the genetics of gametophytic apomixis.

In crossing experiments involving apomicts, it is desirable to establish their degree of apomixis and heterozygosity. Above all, one must ascertain by embryological and progeny analyses that the sexual plant is free from tendencies to apomixis. If the mother plant is not purely sexual, particularly when polyploid, the degrees of heterozygosity or of apomixis must be exactly determined.

Another important point is the necessity to analyze separately the different components involved in gametophytic apomixis, or at least of diplospory/apospory (apomeiosis) as its basic prerequisite. If this is done, in most cases recessivity cannot be confirmed (for examples and references see Sect. 10.5.3). Although not carried out by all investigators, separate analysis should be performed as a matter of course, since it has been established in many apomicts that these components are genetically independent. In *Taraxacum,* for instance, they are localized on two different chromosomes (Sørensen 1958).

Embryological investigations are essential to determine the true degree of apomeiosis (diplospory/apospory) as reduced embryo sacs have often – mainly due to aposporous competition – little or no chance of developing into aberrant offspring. Thus, even in "totally" apomictic wild strains the degree of apomeiosis can be far from 100%. In contrast to this, Saran and deWet (1970) described a phenotypically "sexual" hybrid with considerable frequencies of functionless aposporous embryo sacs. A serious gap in certain analyses is that female-sterile plants are often not examined embryologically. However, the reasons for the female sterility are often not connected with megasporogenesis or gametogenesis and, therefore, the segregation of apomeiotic versus meiotic embryo sac formation can, under certain circumstances, be decisively misleading if the female-sterile offspring are disregarded. Not only meiotic but also apomeiotic plants can sometimes be highly or totally female-sterile: dihaploids (e.g., in *Panicum maximum* – Combes 1975), as well as polyploids (e.g., in *Antennaria* – Bergman 1951, Urbańska-Worytkiewicz 1974; in *Pennisetum/Cenchrus* – Read and Bashaw 1974). Two highly sterile *Potentilla* hybrids, which reproduce mainly vegetatively by runners, are especially instructive. The few offspring from their seeds are mostly (Czapik 1975), or exclusively (Schwendener 1969, and pers. comm. 1980) B_{II} hybrids, although embryological investigations clearly show that apomeiotic embryo sacs do occur. In *Ranunculus auricomus* two weak and poorly fertile aneuploid dihaploids (2n = 17) were found to produce – although highly aposporous – nevertheless mainly B_{II} hybrids, obviously because maternal embryos have even fewer chances of survival than hybrid ones (Nogler 1982). Most informative is the progeny of the crossing between sexual *Pennisetum americanum* and apomictic *P. orientale* (Dujardin and Hanna 1983a). All backcross hybrids obtained – be they B_{II} or B_{III} hybrids – frequently have ovules with maturing aposporous embryo sacs, sometimes even without a meiotic one developing simultaneously. Nevertheless, only B_{II} hybrids develop.

Cytological investigations of *all* the offspring (or at least of a reliable random selection) is imperative as aberrants may originate either apomeiotically (B_{III} hybrids), or meiotically (B_{II} hybrids and polyhaploids which, in some cases, are rather frequent and vigorous). Therefore, mere counting of maternal versus aberrant offspring does not give a realistic picture. Such summary methods give, at most, a rough estimate of the degree of apomixis and may be useful for practical breeding purposes. They are on no account sufficient for genetic research as they give neither an idea of apomeiosis (diplospory/apospory) nor of parthenogenesis.

The necessity of both embryological and cytological investigations is best illustrated by the occurrence of apomeiotic hybrids with unreduced egg cells incapable of parthenogenetic development which, therefore, produce chiefly or exclusively aberrants, i.e., B_{III} hybrids (Powers 1945, Müntzing 1958a, Jassem 1980, Nogler 1984). Such individuals are often highly apomeiotic and may not be regarded as normally sexual, as pointed out by Rutishauser (1967).

Finally, one should not forget the possibilities of quantitative influences of environmental factors and the male parent on the degrees of apomeiosis and apomixis.

10.5.3 Some Results, Conclusions, and Prospects

Recent genetic analyses of apospory suggest a monogenic inheritance for this form of apomeiosis. The results of cytogenetic investigations on "totally" apomictic *Ranunculus auricomus* (Rutishauser 1965, Nogler 1971, 1975, 1984) agree very well with the simplest possible assumption that the induction of aposporous embryo sacs from somatic nucellar cells is made possible by an apospory factor A^-. Its wild allele A^+ does not contribute to apospory but may function in the normal, sexual life cycle, most likely in connection with the initiation of the meiotic embryo sac development. The apospory factor A^-, which could recently be localized on a particular chromosome (Nogler, unpublished), is not recessive to the wild allele A^+, as has been shown by independent methods (Nogler 1973, 1975, 1978). Sexual plants – and of course the diploid, sexual parent – are homozygous for A^+; the apomictic parent and other aposporous hybrids are heterozygous (A^+/A^-). Plants homozygous for A^- are unknown. The frequently reported formation of sexual hybrids after crossing two apomictic plants, established for *R. auricomus* and for many other apomicts, is convincingly explained by recombination of genomes carrying the wild allele A^+ from the two heterozygous apomictic parents. As already pointed out (see Sect. 10.4.3.2), the apospory factor A^- cannot be transmitted by haploid (monoploid) gametes in *R. auricomus,* and by diploid gametes only when in a heterozygous condition (A^+/A^-). Probably diploid gametes homozygous for A^- are also functionless, but this has not yet been proved. If this were certain, the apospory factor A^- would be a recessive lethal factor. Considering these results, those cases where diploid (2x) progenies arising from apomeiotic diploids or triploids lack apomeiosis do not necessarily point to recessivity and homozygosity of the genes involved – apospory (diploid *Leontodon hispidus* – Bergman 1935; diploid *Dichanthium aristatum* – d'Cruz and Reddy 1971) – diplospory (diploid *Parthenium argentatum* – Gerstel et al. 1953; triploid *Taraxacum* species – Richards 1970b, Müller 1972).

The genetics of apospory in "totally" apomictic *Panicum maximum* were chiefly clarified by Savidan (1975, 1980, 1981) in extensive crossing and backcrossing experiments. In this important subtropical forage grass apospory is a dominant factor (whereas in *R. auricomus* certain dosage effects can be found). The sexual plants are homozygous for the wild allele and the aposporous ones all heterozygous. In *Panicum maximum* apospory and parthenogenesis are strictly linked. Savidan (1982) postulates that apomixis is inherited monogenically, as one dominant supergene, whereas in *R. auricomus,* and no doubt in most apomicts, the linkage is not as close. Finer quantitative differences in the degree of apospory, which in some cases turned out to be somewhat susceptible to environmental influences, may not depend directly on the dosage of the apospory factor. In *Panicum maximum* they concern rather the number of aposporous initials and embryo sacs, in *R. auricomus* rather the timing of induction of apospory.

Similar cases are, as far as can be judged: *Hypericum perforatum* (Noack 1939), aposporous species of *Hieracium* (Christoff 1942), some aposporous species of *Potentilla* (Christoff and Papasova 1943), *Bothriochloa-Dichanthium* (Harlan et al. 1964, d'Cruz and Reddy 1971, Saran, pers. comm. 1977), *Beta lomatogona* (Cleij et al. 1976), and possibly *Poa pratensis* (Åkerberg and Bingefors 1953,

Almgård 1966). In diplosporous *Eragrostis curvula* recent results (Voigt and Burson 1981) "suggest that apomixis may be dominant to sexuality and indicate a possible dosage effect". In obligate apomictic *Cenchrus ciliaris* (=*Pennisetum ciliare*) Taliaferro and Bashaw (1966) postulated a dominant gene conditioning apospory with simultaneous suppression of the sexual mechanism, and epistasis of a second gene the dominant allele of which restores sexuality. In *Pennisetum orientale* and *P. squamulatum* apospory is dominant, as becomes evident from crossings and backcrossings with sexual *P. americanum* (Dujardin and Hanna 1983a, b). Different patterns of inheritance have been reported mainly in *Leontodon hispidus* (Bergman 1935, compare, however, Nogler 1982), *Paspalum notatum* (Burton and Forbes 1960), and possibly the above-mentioned *Parthenium, Poa, Potentilla, Rubus,* and *Malus*. Certain results, however, can be explained in different ways.

It is an open question whether the different forms of diplospory are, from a genetic point of view, fundamentally different from apospory, or whether there is a common basis. The following hypothetical concept is based on normal conditions in sexual plants, in which meiosis and gametophyte formation are closely bound together, and therefore lead to reduced gametophytes, unless a disturbed meiosis results in a restitution nucleus. Those nucellar cells which are not determined for meiosis do not form embryo sacs; they are obviously blocked by some kind of barrier. In aposporous plants this barrier is open, and the bonds between meiosis and embryo sac development are broken up (see Sect. 4.2), thus making the way free for embryo sac development also from somatic cells adjoining the female archesporium. In other words, the physiological conditions for the activation of genes involved in the differentiation of the embryo sac, which normally prevail only in meiospores, prevail also in certain unreduced cells. In aposporous *Panicum maximum* and *Ranunculus auricomus* this is due to the apospory factor A^- (Nogler 1971, p. 168). In addition to this qualitative factor, apospory can, quantitatively, be promoted by the timing of the induction of the embryo sacs which, in aposporous wild strains, usually takes place so early that a developing unreduced embryo sac can compete successfully with an eventually developing meiotic embryo sac. If this timing is so extreme that the induction of embryo sac development takes place before the induction of meiosis, the premeiotic megaspore mother cell itself is directly determined to develop mitotically into an unreduced embryo sac (Antennaria type) instead of undergoing meiosis, so that meiosis occurs at most in occasional ovules with delayed development, and in anthers. However, these purely quantitative variations in timing, which are often subject to environmental influences, are only supporting elements. The decisive factor is to be sought in the qualitative factor (A^-) giving the capability to form unreduced embryo sacs regularly, and independently of meiosis. May be the common basis for both apospory and mitotic diplospory (Antennaria type) lies here. This hypothesis agrees well with the observations of more or less pronounced tendencies to apospory in diplosporous plants such as *Parthenium* (Esau 1946), *Rubus* (Christen 1950), *Antennaria* (Bergman 1951), *Potentilla* (Smith 1963a).

There is scant information about the genetics of parthenogenesis, which is hardly based on a simple Mendelian factor as was assumed by Powers (1945). Apomeiotic hybrids lacking the capability to parthenogenesis occur very rarely.

Such individuals are known in *Parthenium argentatum* (Powers 1945), *Potentilla argentea* (Müntzing 1958a), and *Ranunculus auricomus* (Nogler 1984, and unpublished results) where one single aposporous hybrid (hyperdiploid, 2n = 17) produces, after backcrossing with the sexual parent (2n = 16), only B_{III} hybrids (2n = 25). In their offspring from selfing, however, less than half are again (higher polyploid) B_{III} hybrids; the others are maternal. This unexpected reappearance of parthenogenesis, not present in either of the diploid parents, cannot be caused by Mendelian genes. According to Asker (pers. comm. 1981), parthenogenesis is probably, to a large extent, a quantitative trait under polygenic control.

The genetic results obtained so far have much in common but, nevertheless, show important discrepancies. In spite of this I still consider it possible that one will be able to formulate a concept for the genetics of gametophytic apomixis in its different forms. Before this can be achieved, however, many new facts from as many apomicts as possible must be collected. The main difficulty is that the major genes responsible for the single components of gametophytic apomixis may not show only dosage effects (due to polyploidy), but may also be influenced, to a greatly varying degree, by their genetic background, and be subject to purely modifying factors. Furthermore, plasmatic factors may play a role in certain cases (Nitzsche, pers.comm. 1982). Between the different genes involved there is undoubtedly much greater interaction in certain apomicts than in others and, therefore, crossing disturbs or reestablishes their balance more easily in some cases than in others.

10.6 Apomixis and Breeding. Conclusions

Gametophytic apomixis is the result of a delicately balanced interaction of different, genetically independent components. Although many details are known, there remain nonetheless important gaps in the theoretical knowledge of apomixis. The most serious gaps concern the physiological background of its components, particularly with respect to the formation of megagametophyte (diplospory/apospory), sporophyte (parthenogenesis), and endosperm.

At present, the chief interest lies in its practical applications in plant breeding. One first thinks of cultivated apomictic plants such as: Compositae – *Parthenium argentatum* and *Taraxacum kok-saghyz* (for rubber), Rosaceae – *Malus, Fragaria × ananassa,* and *Rubus caesius,* Gramineae – *Poa, Panicum, Paspalum, Pennisetum,* and other subtropical and tropical forage grasses. (*Citrus,* mango and many other taxa which show adventitious embryony are dealt with in Chap. 9 on "Polyembryony".)

Any breeding program stands or falls with the availability of hybrids. In highly apomictic plants one can try different possibilities of influencing the degree of apomixis (see Sect. 10.3.2) to increase the rate of hybridization. Different breeding research programs in course for *Poa* will serve as an illustration (see Williamson and Watson 1980, Williamson 1978, 1981, Nitzsche 1981, van Dijk and Winkelhorst 1982). If the suppression of the mechanisms of sexual reproduction (see Sect. 10.3.1) is so effective that "obligate apomicts" result, their pollen can be used to pollinate sexual relatives which, however, are usually diploid (see Sect. 10.4.3). If no sexual relatives are available, facultative apomicts of the same level of polyploidy may do as female parent. Every successful hybridization involving apomicts causes abundant variability in offspring, due to the heterozygosity typical of apomicts – not least for reproductive mechanisms (see Sect. 10.4.1), but highly apomictic new types can often be selected only in succeeding generations following hybridization. In certain especially favorable cases such as *Pennisetum ciliare* (Taliaferro and Bashaw 1966) or *Panicum maximum* (Savidan 1981) obligate apomicts occur. It is not always easy to identify aberrants in apomictic populations and to define the stability of each line without genetic markers. Subjective visual assessments made in the field can be supported by chemical (Wehner et al. 1976, Spoor and Hay 1978) or statistical methods (Williamson and Killick 1978). Special problems on the breeding of apomictic plants, on the search for related sexual plants and complications due to polyploidy, on breeding schedules, on selection and improvement of new varieties, etc., cannot be discussed here. Reference may be made to Khokhlov (1976), Savidan (1978), Voigt and Burson (1981), and Petrov (1984). Bashaw (1980) gives an excellent survey with a detailed account of methods to detect and confirm apomixis for breeding and other purposes.

The most topical aspect at the moment is the attempt to introduce apomixis into cultivated sexual plants, not only in those wherein apomixis may play a certain role (*Coffea, Musa, Saccharum* – see Bremer 1963, Rutishauser 1967, p. 162), but quite generally in sexual plants which are not suitable for vegetative multiplication on a large scale, particularly cereals (Matzk 1982) and other crop plants. In these cases one can try to find "elements" of apomixis within natural sexual populations – such as mutants with disturbed female meiosis or increased tendencies to form unreduced embryo sacs, or to develop parthenogenetically – and to collect and bring them together by suitable breeding programs. (In this context I should like to refer to the objection I have made in Section 10.4.2 concerning the concept of "elements" of gametophytic apomixis latently present in sexual plants.)

Tendencies towards pseudogamy are reported in certain hybrids from *Triticum* × *Avena* (Kruse 1969) and *Hordeum* × *Triticum* (Mujeeb-Kazi 1981). In analogy to the pseudogamous development in apomicts, remote hybridizations in sexual plants can lead, in some cases, to the fertilization only of the polar nuclei but not of the egg cell. Such an "induced pseudogamy" can be largely dependent on the male parent (see Sect. 10.3.2.1). However, in most cases the porportion of parthenogenetically developing (reduced or unreduced) egg cells can, at mist, only be slightly increased by remote hybridization or by late pollination, so that one cannot speak of "facultative apomixis" or of hereditary pseudogamy. Remote hybridization is a useful method for detecting individuals with an increased frequency of unreduced embryo sacs. Remote hybridizations and their results, especially concerning wheat and rye, are discussed by Kandelaki (1976). Extensive crossing experiments involving sexual *Bromus, Festuca, Lolium,* and apomictic *Poa pratensis,* intended to induce apomixis by remote hybridization, are being carried out by Gröber et al. (1974, 1976, 1978).

More promising than the search for individual "elements" of apomixis might be the search for apomictic relatives which may be crossed with the cultivated sexual species. This is being tried with apples, raspberries, strawberries (Krylova 1976, Sukhareva 1976, 1978, Schmidt 1977), sugar beet (Cleij et al. 1976, Jassem 1976, 1980), wheat (apomictic relative: *Agropyron scabrum* – Hair 1956), sorghum (tendency to aposporous apomixis, e.g., in *Sorghum bicolor* – Hanna et al. 1970, Tang et al. 1980), pearl millet (*Pennisetum typhoides* = *P. americanum* – Hanna and Powell 1973, Dujardin and Hanna 1983a, b), and others. Especial efforts have been concentrated on maize, in spite of great difficulties. In Maydeae apomixis has been found in *Tripsacum* (Farquharson 1955) and *Coix* (Venkateswarlu and Rao 1975, Rao, pers. comm. 1980) from which the transfer of genes for apomixis to the diploid, sexual maize is being attempted. An idea of the difficulties met with in these experiments can be gathered from the contributions in the book of Khokhlov (1976), from Petrov et al. (1978) and Harlan et al. (1978). In potatoes – a crop plant in which no apomictic relatives are known – possibilities and prospects of provoking apomixis for reproduction by seeds are discussed by Hermsen (1980).

The reasons for the practical interest in apomixis, and for the great efforts in making its manipulation possible, are obvious. Apomictic reproduction would allow one to fix any desired heterozygous gene combination and to maintain new, heterozygous "varieties" with valuable agronomical properties, genes for resistancy to diseases, etc. Most alluring is the prospect of being able to perpetuate heterosis effects by apomixis and to produce seeds of "hybrid varieties" (F_1 hybrids) without constant crossings of the parent lines. Moreover, the manipulation of gametophytic apomixis would profit by the fact that different virus infections can be prevented by seed reproduction, thanks to the double barrier of meristem and megagametophyte – in contrast to vegetative propagation. This might enable one to invigorate and rejuvenate old varieties (Petrov 1976).

The hitherto rather modest successes in introducing apomixis into sexual plants, in spite of the great efforts undertaken, are at least partly due to the fact that there is as yet no other practicable way than that of proceeding empirically. A real advance can only be expected from a thorough theoretical knowledge concerning the causes and the genetics of gametophytic apomixis and the exact functioning of the genes involved and, above all, the physiological background of the components of gametophytic apomixis. Extensive mutation research would seem to me to be particularly promising. It is to be hoped that an increasing practical need will give new impetus to fundamental and systematic research, as this most important and widespread anomaly of reproduction in angiosperms is still not really understood.

References

Åkerberg E, Bingefors S (1953) Progeny studies in the hybrid *Poa pratensis* × *Poa alpina*. Hereditas 39:125–136

Almgård G (1966) Experiments with *Poa* III. Ann Agric Coll Sweden 32:3–64

Asker S (1966) Effects of mutagen treatment on some apomictic *Potentilla* species. Hereditas 55:249–265

Asker S (1967) Induced sexuality after chromosome doubling in an apomictic *Potentilla argentea*-biotype. Hereditas 57:339–342

Asker S (1970) Apomixis and sexuality in the *Potentilla argentea* complex II. Crosses within the species. Hereditas 66:189–204
Asker S (1971) Apomixis and sexuality in the *Potentilla argentea* complex III. Eutriploid and aneuploid derivatives (including trisomics) of some apomictic biotypes. Hereditas 67:111–142
Asker S (1977) Pseudogamy, hybridization, and evolution in *Potentilla*. Hereditas 87:179–184
Asker S (1980) Gametophytic apomixis: Elements and genetic regulation. Hereditas 93:277–293
Asker S, Hagberg A, Hagberg G (1983) Apomixis in barley? Sver Utsädesfören Tidskr 93:75–76
Babcock EB, Stebbins GL (1938) The American species of *Crepis:* Their relationships and distribution as affected by polyploidy and apomixis. Carnegie Inst Wash Publ 504:1–200
Băra II, Ghiorghită GI (1980) Din enigmele evolutiei. Apomixia şi rolul ei in evoluţie. Editura ştiinţifică şi enciclopedică. Bucureşti, pp 73–149
Bary A de (1878) Über apogame Farne und die Erscheinung der Apogamie im Allgemeinen. Bot Z 36:449–487
Bashaw EC (1980) Apomixis and its application in crop improvement. In: Fehr WR, Hadley HH (eds) Hybridization of crop plants. Am Soc Agron Press, Madison, Wis, pp 45–63
Bashaw EC, Hoff BJ (1962) Effects of irradiation on apomictic common dallisgrass (*Paspalum dilatatum*). Crop Sci 2:501–504
Battaglia E (1946a) Ricerche cariologiche ed embriologiche sul genere *Rudbeckia* (Asteraceae). VI. Apomissia in *Rudbeckia speciosa*. Nuovo G Bot Ital (NS) 53:27–69
Battaglia E (1946b) Ricerche cariologiche ed embriologiche sul genere *Rudbeckia*. VIII. Semigamia in *Rudbeckia laciniata*. Nuovo G Bot Ital (NS) 53:483–511
Battaglia E (1948) Ricerche sulla parameiosi restitutionale nel genere *Taraxacum*. Caryologia 1:1–47
Battaglia E (1950) L'alterazione della meiosi nella reproduzione apomittica di *Erigeron karwinskianus* DC. var. *mucronatus* DC. Caryologia 2:165–204
Berger X (1953) Untersuchungen über die Embryologie partiell apomiktischer *Rubus* bastarde. Ber Schweiz Bot Ges 63:224–265
Bergman B (1935) Zytologische Studien über die Fortpflanzung bei den Gattungen *Leontodon* und *Picris*. Svensk Bot Tidskr 29:155–301
Bergman B (1941) Studies on the embryo sac mother cell and its development in *Hieracium*, subg. *Archieracium*. Svensk Bot Tidskr 35:1–42
Bergman B (1950) Meiosis in two different clones of the apomictic *Chondrilla juncea*. Hereditas 36:297–320
Bergman B (1951) On the formation of reduced and unreduced gametophytes in the females of *Antennaria carpatica*. Hereditas 37:501–518
Böcher TW (1951) Cytological and embryological studies in the amphiapomictic *Arabis Holboellii* complex. K Dan Vidensk Selsk Biol Skr VI 7:1–59
Bremer G (1963) Problems in breeding and cytology of sugar cane. VI. Additional contemplations on fertilization and parthenogenesis in *Saccharum*. Euphytica 12:178–188
Brix K (1974) Sexual reproduction in *Eragrostis curvula* (Schrad.) Nees. Z Pflanzenzücht 71:25–32
Brown WV (1951) Apomixis in *Zephyranthes texana* Herb. Am J Bot 38:697–701
Brown WV, Emery WHP (1957) Apomixis in the Gramineae, tribe Andropogoneae: *Themeda triandra* and *Bothriochloa ischaemum*. Bot Gaz 118:246–253
Brown WV, Emery WHP (1958) Apomixis in the Gramineae: Panicoideae. Am J Bot 45:253–263
Burson BL (1975) Cytology of some apomictic *Paspalum* species. Crop Sci 15:229–232
Burton GW (1982) Effect of environment on apomixis in bahiagrass. Crop Sci 22:109–111
Burton GW, Forbes I (1960) The genetics and manipulation of obligate apomixis in common Bahia grass (*Paspalum notatum* Flugge). Proc VIII Intl Grassland Congr Reading (England) 1960:66–71
Burton GW, Jackson JE (1962) Radiation breeding of apomictic prostrate dallisgrass, *Paspalum dilatatum* var. *pauciciliatum*. Crop Sci 2:495–497
Chao C-Y (1974) Megasporogenesis and megagametogenesis in *Paspalum commersonii* and *P. longifolium* at two polyploid levels. Bot Not 127:267–275
Chao C-Y (1980) Autonomous development of embryo in *Paspalum conjugatum* Berg. Bot Not 133:215–222
Christen HR (1950) Untersuchungen über die Embryologie pseudogamer und sexueller *Rubus*arten. Ber Schweiz Bot Ges 60:153–198
Christen HR (1952) Die Embryologie von *Rubus idaeus* und von Bastarden zwischen *Rubus caesius* und *Rubus idaeus*. Z Indukt Abstamm Vererbungslehre 84:454–461

Christoff M (1942) Die genetische Grundlage der apomiktischen Fortpflanzung bei *Hieracium aurantiacum* L. Z Indukt Abstamm Vererbungslehre 80:103–125
Christoff M, Christoff MA (1948) Meiosis in the somatic tissue responsible for the reduction of chromosome number in the progeny of *Hieracium hoppeanum* Schult. Genetics 33:36–42
Christoff M, Papasova G (1943) Die genetischen Grundlagen der apomiktischen Fortpflanzung in der Gattung *Potentilla*. Z Indukt Abstamm Vererbungslehre 81:1–27
Cleij G, Bock TSM de, Lekkerkerker B (1976) Crosses between *Beta vulgaris* L. and *Beta lomatogona* F. et M. Euphytica 25:539–547
Coe GE (1953) Cytology of reproduction in *Cooperia pedunculata*. Am J Bot 40:335–343
Combes D (1975) Polymorphisme et modes de reproduction dans la section des Maximae du genre *Panicum* (Gramineae) en Afrique. Mém ORSTOM 77:1–99
Connor HE (1979) Breeding systems in grasses: A survey. N Z J Bot 17:547–574
Correns C (1905) Gregor Mendels Briefe an Carl Nägeli 1866–1873. Abh Math-Phys Kl Kgl Sächs Ges Wiss 29:189–265
Costas-Lippmann M (1979) Embryogeny of *Cortaderia selloana* and *C. jubata* (Gramineae). Bot Gaz 140:393–397
Czapik R (1975) Apomixis in a sterile hybrid species of *Potentilla*. In: Walters SM (ed) European floristic and taxonomic studies. Classey, Cambridge, pp 38–47
Czapik R (1981) Embryology of *Rubus saxatilis* L. Acta Biol Cracov Ser Bot 23:7–13
D'Amato F (1949) Triploidia e apomissia in *Statice oleaefolia* Scop. var. *confusa* Goodr. Caryologia 2:71–84
Darlington CD (1937) Recent advances in cytology. Churchill, London
Davis GL (1967) Apomixis in the Compositae. Phytomorphology 17:270–277
d'Cruz R, Reddy PS (1971) Inheritance of apomixis in *Dichanthium*. Indian J Genet Plant Breed 31:451–460
DeWet JMJ (1965) Diploid races of tetraploid *Dichanthium* species. Am Nat 99:167–171
DeWet JMJ (1968) Diploid-tetraploid-haploid cycles and the origin of variability in *Dichanthium* agamospecies. Evolution 22:394–397
DeWet JMJ (1971) Reversible tetraploidy as an evolutionary mechanism. Evolution 25:545–548
DeWet JMJ, Harlan JR (1970) Apomixis, polyploidy, and speciation in *Dichanthium*. Evolution 24:270–277
DeWet JMJ, Stalker HT (1974) Gametophytic apomixis and evolution in plants. Taxon 23:689–697
Dijk GE van, Winkelhorst GD (1982) Interspecific crosses as a tool in breeding *Poa pratensis* L. 1. *Poa longifolia* Trin. × *P. pratensis* L. Euphytica 31:215–223
Doll R (1971) Probleme der Apomixis bei höheren Pflanzen. Biol Rundsch 9:215–239
Doll R (1976) Apomixis und Speziation in den Gattungen *Taraxacum* and *Chondrilla*. Feddes Repert 87:483–492
Döpp W (1967) Apomixis bei Archegoniaten. In: Ruhland W (ed) Handbuch der Pflanzenphysiologie Bd. XVIII. Springer, Berlin Heidelberg New York, pp. 531–550
Dowrick GJ (1961) Biology of reproduction in *Rubus*. Nature 191:680–682
Dujardin M, Hanna WW (1983a) Meiotic and reproductive behavior of facultative apomictic BC_1 offspring derived from *Pennisetum americanum-P. orientale* interspecific hybrids. Crop Sci 23:156–160
Dujardin M, Hanna WW (1983b) Apomictic and sexual pearl millet × *Pennisetum squamulatum* hybrids. J Hered 74:277–279
Ellerström S, Zagorcheva L (1977) Sterility and apomictic embryo-sac formation in *Raphanobrassica*. Hereditas 87:107–120
Emery WHP (1957) A study of reproduction in *Setaria macrostachya* and its relatives in the south-western United States and northern Mexico. Bull Torrey Bot Club 84:106–121
Ernst A (1918) Bastardierung als Ursache der Apogamie im Pflanzenreich. Fischer, Jena
Ernst A, Bernard Ch (1912) Entwicklungsgeschichte des embryosackes, des Embryos und des Endosperms von *Burmannia coelestis* Don. Ann Jard Bot Buitenzorg II, 11:234–257
Esau K (1946) Morphology of reproduction in guayule and certain other species of *Parthenium*. Hilgardia 17:61–101
Evans LT, Knox RB (1969) Environmental control of reproduction in *Themeda australis*. Austral J Bot 17:375–389

Fagerlind F (1944) Die Samenbildung und die Zytologie bei agamospermischen und sexuellen Arten von *Elatostema* und einigen nahestehenden Gattungen nebst Beleuchtung einiger damit zusammenhängender Probleme. K Sven Vetenskapsakad Handl 21(4):1–130

Fagerlind F (1946) Sporogenesis, Embryosackentwicklung und pseudogame Samenbildung bei *Rudbeckia laciniata* L. Acta Hoti Bergiani 14:39–90

Fagerlind F (1947a) Makrosporogenese und Embryosackbildung bei agamospermischen *Taraxacum*-Biotypen. Svensk Bot Tidskr 41:365–390

Fagerlind F (1947b) Macrogametophyte formation in two agamospermous *Erigeron* species. Acta Horti Bergiani 14:221–247

Farquharson LI (1955) Apomixis and polyembryony in *Tripsacum dactyloides*. Am J Bot 42:737–743

Franke R (1975) Über das Auftreten von unreduzierten Gameten bei Angiospermen. Arch Züchtungsforsch Berlin 5:201–208

Fürnkranz D (1961) Cytogenetische Untersuchungen an *Taraxacum* im Raume von Wien II. Hybriden zwischen *T. officinale* und *T. palustre*. Österr Bot Z 108:408–415

Gerlach D (1965) Befruchtung und Autogamie bei *Rubus caesius*. Biol Zentralbl 84:611–633

Gerstel DU, Mishanec WM (1950) On the inheritance of apomixis in *Parthenium argentatum*. Bot Gaz 112:96–106

Gerstel DU, Hammond BL, Kidd C (1953) An additional note on the inheritance of apomixis in guayule. Bot Gaz 115:89–93

Gildenhuys PJ, Brix K (1959) Apomixis in *Pennisetum dubium*. S Afr J Agric Sci 2:231–245

Grant V (1981) Plant speciation. Columbia Univ Press, New York

Gröber K, Matzk F, Zacharias M (1974, 1976, 1978) Untersuchungen zur Entwicklung der apomiktischen Fortpflanzung bei Futtergräsern I–III. Kulturpflanze 22:159–180, 24:349–364, 26:303–327

Grun P (1955) Cytogenetic studies in *Poa* III. Variation within *Poa nervosa*, an obligate apomict. Am J Bot 42:778–784

Gupta PK, Roy RP, Singh AP (1969) Aposporous apomicts: Seasonal variation in tetraploid *Dichanthium annulatum*. Port Acta Biol Ser A 11:253–260

Gustafsson Å (1943) The genesis of the European blackberry flora. Lunds Univ Årsskr N F II 39(6):1–200

Gustafsson Å (1946) Apomixis in angiosperms. I. Lunds Univ Årsskr N F II 42(3):1–67

Gustafsson Å (1947a) Apomixis in angiosperms. II. Lunds Univ Årsskr N F II 43(2):71–179

Gustafsson Å (1947b) Apomixis in angiosperms. III. Lunds Univ Årsskr N F II 43(12):183–370

Hagberg G, Hagberg A (1981) Haploidy initiator gene in barley. Proc X Intl Barley Gen Sympos Edinburgh 1981:686–689

Hair JB (1956) Subsexual reproduction in *Agropyron*. Heredity 10:129–160

Håkansson A (1943) Die Entwicklung des Embryosacks und die Befruchtung bei *Poa alpina*. Hereditas 29:25–61

Håkansson A (1951) Parthenogenesis in *Allium*. Bot Not 1951:143–179

Håkansson A, Levan A (1957) Endo-duplicational meiosis in *Allium odorum*. Hereditas 43:179–200

Hanna WW, Powell JB (1973) Stubby head, an induced facultative apomict in pearl millet. Crop Sci 13:726–728

Hanna WW, Schertz KF, Bashaw EC (1970) Apospory in *Sorghum bicolor* (L.) Moench. Science 170:338–339

Hanson AA, Juska FV (1962) Induced mutations in Kentucky bluegrass. Crop Sci 2:369–371

Harlan JR, deWet JMJ (1975) On Ö. Winge and a prayer: The origins of polyploidy. Bot Rev 41:361–390

Harlan JR, Brooks MH, Borgaonkar DS, deWet JMJ (1964) Nature and inheritance of apomixis in *Bothriochloa* and *Dichanthium*. Bot Gaz 125:41–46

Harlan JR, deWet JMJ, Newell CA (1978) Apomixis and pseudo-apomixis in *Tripsacum*. Abstr Proc XIV Intl Congr Genet, Moscow, vol II, p 166

Haskell G (1953) Quantitative variation in subsexual *Rubus*. Heredity 7:409–418

Haskell G (1960) Role of the male parent in apomictic *Rubus* species. Heredity 14:101–113

Haskell G (1966) The history, taxonomy and breeding system of apomictic British Rubi. In: Hawkes JG (ed) Reproductive biology and taxonomy of vascular plants. Pergamon Press, Oxford, pp 141–151

Hermsen JGTh (1980) Breeding for apomixis in potato: Pursuing a utopian scheme? Euphytica 29:595–607

Hjelmqvist H (1957) The apomictic development in *Malus Sieboldii*. Bot Not 110:455–467
Hjelmqvist H (1959) Studien über Embryologie und Variabilität bei einigen *Aphanes*-Arten. Bot Not 112:17–64
Hjelmqvist H, Grazi F (1964) Studies on variation in embryo sac development. Bot Not 117:141–166
Holmgren I (1919) Zytologische Studien über die Fortpflanzung bei den Gattungen *Erigeron* und *Eupatorium*. K Sven Vetenskapsakad Handl 59(7):1–118
Hunziker HR (1954) Beitrag zur Aposporie und ihrer Genetik bei *Potentilla*. Arch Julius Klaus-Stift Vererbungsforsch 29:135–222
Iwanaga M, Peloquin SJ (1979) Synaptic mutant affecting only megasporogenesis in potatoes. J Hered 70:385–389
Izmaiłow R (1965) Megasporogenesis in the apomictic species *Ranunculus cassubicus*. Acta Biol Cracov Ser Bot 8:183–195
Izmaiłow R (1967) Observations in embryo and endosperm development in various chromosomic types of the apomictic species *Ranunculus cassubicus* L. Acta Biol Cracov Ser Bot 10:99–111
Izmaiłow R (1970) Cytogenetic studies in the apomictic species *Ranunculus cassubicus* L. Acta Biol Cracov Ser Bot 13:37–50
Izmaiłow R (1973) Cyto-embryological studies in experimental hybrids of the apomictic species *Ranunculus cassubicus* L. Acta Biol Cracov Ser Bot 16:99–120
Izmaiłow R (1976) Problem of apomixis in the *Ranunculus auricomus* group. Acta Biol Cracov Ser Bot 19:15–28
Jassem B (1976) Embryology and genetics of apomixis in the section *Corollinae* of the genus *Beta*. Acta Biol Cracov Ser Bot 19:149–172
Jassem B (1980) Origin and reproduction of higher polyploids within the *Corollinae* section of the genus *Beta*. Genet Pol 21:17–27
Juel O (1900) Vergleichende Untersuchungen über typische und parthenogenetische Fortpflanzung bei der Gattung *Antennaria*. K Sven Vetenskapsakad Handl 33(5):1–59
Juel O (1906) Die Tetradenteilung bei *Taraxacum* und anderen Cichorieen. K Sven Vetenskapsakad Handl 39(4):1–21
Julén G (1961) The effects of X-rays on the apomixis in *Poa pratensis*. IAEA, Vienna 1961:527–532
Kandelaki GV (1976) Remote hybridization and the phenomenon of pseudogamy. In: Khokhlov SS (ed) Apomixis and breeding. Amerind, New Delhi, pp 179–189
Khokhlov SS (1976) Apomixis and breeding. Amerind, New Delhi
Kimber G, Riley R (1963) Haploid angiosperms. Bot Rev 29:480–531
Knox RB, Heslop-Harrison J (1963) Experimental control of aposporous apomixis in a grass of the Andropogoneae. Bot Not 116:127–141
Kościńska M (1980) Embryo and endosperm in *Rudbeckia bicolor*. Acta Biol Cracov Ser Bot 22:71–79
Kruse A (1969) Intergeneric hybrids between *Triticum aestivum* L. (v. Koga II, 2n = 42) and *Avena sativa* L. (v. Stal, 2n = 42) with pseudogamous seed formation. Royal Vet Agric Univ Yearbook, Copenhagen, pp 188–200
Krylova VV (1976) Apospory and polyembryony in apple. In: Khokhlov SS (ed) Apomixis and breeding. Amerind, New Delhi, pp 124–129
Liljefors Á (1953) Studies on propagation, embryology, and pollination in *Sorbus*. Acta Horti Bergiani 16:277–329
Löve A (1960) Biosystematics and classification of apomicts. Feddes Repert 63:136–149
Małecka J (1971) Cyto-taxonomical and embryological investigations on a natural hybrid between *Taraxacum kok-saghyz* Rodin and *T. officinale* Web. and their putative parent species. Acta Biol Cracov Ser Bot 14:179–197
Małecka J (1973) Problems of the mode of reproduction in microspecies of *Taraxacum* section Palustria Dahlstedt. Acta Biol Cracov Ser Bot 16:37–84
Marklund G, Rousi A (1961) Outlines of evolution in the pseudogamous *Ranunculus auricomus* group in Finland. Evolution 15:510–522
Marshall DR, Brown AHD (1974) Estimation of the level of apomixis in plant populations. Heredity 32:321–333
Marshall DR, Brown AHD (1981) The evolution of apomixis. Heredity 47:1–15
Marshall DR, Weir BS (1979) Maintenance of genetic variation in apomictic plant populations. Heredity 42:159–172
Matzel J (1982) Untersuchungen zur Enwicklung der apomiktischen Fortpflanzungsweise bei Futtergräsern. IV. Versuche zur modifikativen Beeinflussung der Apomixis. Kulturpflanze 30:167–180

Matzk F (1982) Vorstellungen über potentielle Wege zur Apomixis bei Getreide. Arch Züchtungsforsch 12:183–195

McWilliam JR, Shanker K, Knox RB (1970) Effects of temperature and photoperiod on growth and reproductive development in *Hyparrhenia hirta*. Austral J Agric Res 21:557–569

Moskova RD, Yakovlev MS (1974) Apomixis of *Bothriochloa ischaemum* L. Flowering, pollination, fertilization. In: Linskens HF (ed) Fertilization in higher plants. North-Holland Publ, Amsterdam, pp 341–349

Mujeeb-Kazi A (1981) Apomictic progeny derived from intergeneric *Hordeum-Triticum* hybrids. J Hered 72:284–285

Müller U (1972) Zytologisch-embryologische Beobachtungen an *Taraxacum*-Arten aus der Sektion Vulgaria Dahlst. in der Schweiz. Ber Geobot Inst Eidg Tech Hochsch Stift Rübel Zürich 41:48–55

Müntzing A (1940) Further studies on apomixis and sexuality in *Poa*. Hereditas 26:115–190

Müntzing A (1958a) The balance between sexual and apomictic reproduction in some hybrids of *Potentilla*. Hereditas 44:145–160

Müntzing A (1958b) Heteroploidy and polymorphism in some apomictic species of *Potentilla*. Hereditas 44:280–329

Müntzing A (1966) Apomixis and sexuality in new material of *Poa alpina* from middle Sweden. Hereditas 54:314–337

Müntzing A, Müntzing G (1945) The mode of reproduction of hybrids betwen sexual and apomictic *Potentilla argentea*. Bot Not 1945:237–278

Murbeck SV (1902) Über Anomalien im Baue des Nucellus und des Embryosackes bei parthenogenetischen Arten der Gattung *Alchemilla*. Lunds Univ Årsskr 38(2):1–11

Nijs JCM den, Sterk AA (1980) Cytogeographical studies of *Taraxacum* sect. *Taraxacum* (= sect. *Vulgaria*) in Central Europe. Bot Jahrb Syst 101:527–554

Nitzsche W (1981) Interspecific hybrids between apomictic forms of *Poa palustris* L. × *Poa pratensis* L. Proc XIV Intl Grassland Congr Lexington Ky 1981:155–157

Noack KL (1939) Über Hypericum-Kreuzungen VI. Fortpflanzungsverhältnisse und Bastarde von *Hypericum perforatum* L. Z Indukt Abstamm Vererbungslehre 76:569–601

Nogler GA (1971) Genetik der Aposporie bei *Ranunculus auricomus* s.l.W.Koch. I. Embryologie. Ber Schweiz Bot Ges 81:139–179

Nogler GA (1972) Genetik der Aposporie bei *Ranunculus auricomus*. II. Endospermzytologie. Ber Schweiz Bot Ges 82:54–63

Nogler GA (1973) Genetik der Aposporie bei *Ranunculus auricomus*. III. F_2-Rückkreuzungsbastarde. Ber Schweiz Bot Ges 83:295–305

Nogler GA (1975) Genetics of apospory in *Ranunculus auricomus*. IV. Embryology of F_3 and F_4 backcross offspring. Phytomorphology 25:485–490

Nogler GA (1978) Zur Zytogenetik der Apomixis bei *Ranunculus auricomus*. Unpubl. Habilitationsschrift ETH Zürich (mimeographed)

Nogler GA (1982) How to obtain diploid apomictic *Ranunculus auricomus* plants not found in the wild state. Bot Helv 92:13–22

Nogler GA (1984) Genetics of apospory in *Ranunculus auricomus*. V. Conclusions. Bot Helv 94 (2) (in press)

Nygren A (1946) The genesis of some Scandinavian species of *Calamagrostis*. Hereditas 32:131–262

Nygren A (1950) Cytological and embryological studies in arctic *Poae*. Symb Bot Ups 10(4):1–64

Nygren A (1962) Artificial and natural hybridization in European *Calamagrostis*. Symb Bot Ups 17(3):1–105

Okabe S (1932) Parthenogenesis bei *Ixeris dentata*. Bot Mag (Tokyo) 46:518–523

Osawa J (1913) Studies on the cytology of some species of *Taraxacum*. Arch Zellforsch 10:450–469

Ostenfeld CH (1910) Further studies on the apogamy and hybridization of the Hieracia. Z Indukt Abstamm Vererbungslehre 3:241–285

Pace L (1913) Apogamy in *Atamosco*. Bot Gaz 56:376–394

Pernès J, Combes D, Réné-Chaume R, Savidan Y (1975a) Biologie des populations naturelles du *Panicum maximum* Jacq. Cah ORSTOM Sér Biol 10:77–89

Pernès J, Savidan Y, Réné-Chaume R (1975b) *Panicum:* structures génétiques du complexe des „*Maximae*" et organisation de ses populations naturelles en relation avec la spéciation. Boissiera 24:383–402

Petrov DF (1976) Genetically regulated apomixis as a method of fixing heterosis and its significance in breeding. In: Khokhlov SS (ed) Apomixis and breeding. Amerind, New Delhi, pp 18–28

Petrov DF (1984) Apomixis and its role in evolution and breeding. D.K. Agencies, New Delhi
Petrov DF, Belousova NI, Fokina ES (1978) The inheritance pattern of apomictic elements in interspecific hybrids between maize and *Tripsacum*. Abstr Proc XIV Int Congr Genet, Moscow, vol II, p 184
Philipson MN (1978) Apomixis in *Cortaderia jubata* (Gramineae). N Z J Bot 16:45–59
Powers L (1945) Fertilization without reduction in guayule (*Parthenium argentatum* Gray) and a hypothesis as to the evolution of apomixis and polyploidy. Genetics 30:323–346
Quarín CL, Hanna WW (1980) Effect of three ploidy levels on meiosis and mode of reproduction in *Paspalum hexastachyum*. Crop Sci 20:69–75
Quarín CL, Hanna WW, Fernández A (1982) Genetic studies in diploid and tetraploid *Paspalum* species. J Hered 73:254–256
Ramanna MS (1979) A re-examination of the mechanisms of 2n gamete formation in potato and its implications for breeding. Euphytica 28:537–561
Rao PN (1974) Male parthenogenesis in tetraploid Job's tears. Heredity 32:412–414
Rao PN, Narayana DS (1980) Occurrence and identification of semigamy in *Coix aquatica*. J Hered 71:117–120
Read JC, Bashaw EC (1974) Intergeneric hybrid between pearl millet and buffelgrass. Crop Sci 14:401–403
Reddy PS, d'Cruz R (1969) Mechanism of apomixis in *Dichanthium annulatum* (Forssk) Stapf. Bot Gaz 130:71–79
Renner O (1916) Zur Terminologie des pflanzlichen Generationswechsels. Biol Zentralbl 36:337–374
Reusch JDH (1961) The relationship between reproductive factors and seed set in *Paspalum dilatatum*. S Afr J Agric Sci 4:513–530
Richards AJ (1970 a) Eutriploid facultative agamospermy in *Taraxacum*. New Phytol 69:761–774
Richards AJ (1970 b) Hybridization in *Taraxacum*. New Phytol 69:1103–1121
Richards AJ (1973) The origin of *Taraxacum* agamospecies. Bot J Linn Soc 66:189–211
Rollins RC (1945) Evidence for genetic variation among apomictically produced plants of several F_1 progenies of guayule (*Parthenium argentatum*) and mariola (*P. incanum*). Am J Bot 32:554–560
Rosenberg O (1908) Cytological studies on the apogamy in *Hieracium*. Bot Tidsskr 28:143–170
Rosenberg O (1930) Apogamie und Parthenogenesis bei Pflanzen. In: Baur E, Hartmann M (eds) Handbuch Vererbungswiss II (12). Borntraeger, Berlin
Rousi A (1956) Cytotaxonomy and reproduction in the apomictic *Ranunculus auricomus* complex. Ann Bot Soc Vanamo 29(2):1–64
Rutishauser A (1943) Untersuchungen über die Fortpflanzung und Bastardbildung apomiktischer Potentillen. Ber Schweiz Bot Ges 53:1–83
Rutishauser A (1948) Pseudogamie und Polymorphie in der Gattung *Potentilla*. Arch Julius Klaus-Stift Vererbungsforsch 23:267–424
Rutishauser A (1954 a) Die Entwicklungserregung des Endosperms bei pseudogamen *Ranunculus*arten. Mitt Naturforsch Ges Schaffhausen 25:1–45
Rutishauser A (1954 b) Entwicklungserregung der Eizelle bei pseudogamen Arten der Gattung *Ranunculus*. Bull Schweiz Akad Med Wiss 10:491–512
Rutishauser A (1961) Pseudogamous reproduction and evolution. Recent advances in botany. Univ Toronto Press, pp 699–702
Rutishauser A (1965) Genetik der Pseudogamie bei *Ranunculus auricomus* s.l.W.Koch. Ber Schweiz Bot Ges 75:157–182
Rutishauser A (1967) Fortpflanzungsmodus und Meiose apomiktischer Blütenpflanzen. Protoplasmatologia VI/F/3, Springer, Wien
Rutishauser A (1969) Embryologie und Fortpflanzungsbiologie der Angiospermen. Springer, Wien
Rychlewski J (1961) Cyto-embryological studies in the apomictic species *Nardus stricta* L. Acta Biol Cracov Ser Bot 4:1–23
Rychlewski J (1967) Karyological studies on *Nardus stricta* L. Acta Biol Cracov Ser Bot 10:55–72
Saran S, deWet JMJ (1970) The mode of reproduction in *Dichanthium intermedium* (Gramineae). Bull Torrey Bot Club 97:6–13
Saran S, deWet JMJ (1976) Environmental control of reproduction in *Dichanthium intermedium*. J Cytol Genet 11:22–28
Saran S, Narain PD (1982) X-irradiation studies in *Pennisetum pedicellatum*. J Cytol Genet 18: (in press)

Savidan Y (1975) Hérédité de l'apomixie. Contribution à l'étude de l'hérédité de l'apomixie sur *Panicum maximum* Jacq. (analyse des sacs embryonnaires). Cah ORSTOM Sér Biol 10:91–95

Savidan Y (1978) L'apomixie gamétophytique chez les Graminées et son utilisation en Amélioration des Plantes. Ann Amélior Plant (Paris) 28:1–9

Savidan Y (1980) Chromosomal and embryological analyses in sexual × apomictic hybrids of *Panicum maximum* Jacq. Theor Appl Genet 57:153–156

Savidan Y (1981) Genetics and utilization of apomixis for the improvement of Guinea Grass (*Panicum maximum* Jacq.). Proc XIV Intl Grassland Congr Lexington Ky 1981:182–184

Savidan Y (1982) Nature et hérédité de l'apomixie chez *Panicum maximum* Jacq. Trav Doc ORSTOM 153:1–159

Savidan Y, Pernès J (1982) Diploid-tetraploid-dihaploid cycles and the evolution of *Panicum maximum* Jacq. Evolution 36:596–600

Sax K (1959) The cytogenetics of facultative apomixis in *Malus* species. J Arnold Arbor (Harv Univ) 40:289–297

Schmidt H (1964) Beiträge zur Züchtung apomiktischer Apfelunterlagen I. Zytogenetische und embryologische Untersuchungen. Z Pflanzenzücht 52:27–102

Schmidt H (1977) Contributions on the breeding of apomictic apple stocks IV. On the inheritance of apomixis. Z Pflanzenzücht 78:3–12

Schwendener J (1969) Experimente zur Evolution von *Potentilla procumbens* Sibth. Ber Schweiz Bot Ges 79:49–92

Sherwood RT, Young BA (1979) Effect of plant growth regulators on embryo sac development in buffelgrass. Agron Abstr 1979:C-1

Simpson CE, Bashaw EC (1969) Cytology and reproductive characteristics in *Pennisetum setaceum*. Am J Bot 56:31–36

Skalińska M (1959) Embryological studies in *Poa granitica* Br.Bl., an apomictic species of the Carpathian range. Acta Biol Cracov Ser Bot 2:91–112

Skalińska M (1973) Further studies in facultative apomixis of *Hieracium aurantiacum* L. Acta Biol Cracov Ser Bot 16:121–133

Skalińska M (1976) Cytological diversity in the progeny of octoploid facultative apomicts of *Hieracium aurantiacum*. Acta Biol Cracov Ser Bot 19:39–46

Skalińska M, Kubień E (1972) Cytological and embryological studies in *Hieracium pratense* Tausch. Acta Biol Cracov Ser Bot 15:39–50

Smith GL (1963 a) Studies in *Potentilla* L. I. Embryological investigations into the mechanism of agamospermy in British *P. tabernaemontani* Aschers. New Phytol 62:264–282

Smith GL (1963 b) Studies in *Potentilla* L. II. Cytological aspects of apomixis in *P. crantzii* (Cr.) Beck ex Fritsch. New Phytol 62:283–300

Snyder LA (1957) Apomixis in *Paspalum secans*. Am J Bot 44:318–324

Snyder LA, Hernandez AR, Warmke HE (1955) The mechanism of apomixis in *Pennisetum ciliare*. Bot Gaz 116:209–221

Solntzeva MP (1974) Disturbances in the process of fertilization in angiosperms under hemigamy. In: Linskens HF (ed) Fertilization in higher plants. North-Holland Publ, Amsterdam, pp 311–324

Solntzeva MP (1978) Apomixis and hemigamy as one of its forms. Proc Indian Natl Sci Acad 44B:78–90

Sørensen Th (1958) Sexual chromosome-aberrants in triploid apomictic *Taraxaca*. Bot Tidsskr 54:1–22

Souciet JL (1978) Contrôle de l'inhibition de la germination chez une espèce apomictique, le *Panicum maximum*. Thèse 2385, Univ Paris-Sud

Sparvoli E (1960) Osservazioni cito-embriologiche in *Eupatorium riparium* Reg. II. Megasporogenesi e sviluppo del gametofito femminile. Ann di Bot (Rome) 26:481–504

Spoor W, Hay JM (1978) Identification of *Poa pratensis* lines using electrophoresis of seed extracts. Seed Sci Technol 7:467–474

Stebbins GL (1932) Cytology of *Antennaria* II. Parthenogenetic species. Bot Gaz 94:322–344

Stebbins GL (1941) Apomixis in the angiosperms. Bot Rev 7:507–542

Stebbins GL (1950) Variation and evolution in plants. Columbia Univ Press, New York, pp 380–419

Stebbins GL, Jenkins JA (1939) Aposporic development in the North American species of *Crepis*. Genetica 21:191–224

Sukhareva NB (1976) Elements of apomixis in strawberry. In: Khokhlov SS (ed) Apomixis and breeding. Amerind New Delhi, pp 120–123

Sukhareva NB (1978) Apomixis in remote hybridization of *Fragaria*. Abstr Proc XIV Intl Congr Genet Moscow, vol II, p 195

Taliaferro CM, Bashaw EC (1966) Inheritance and control of obligate apomixis in breeding buffelgrass, *Pennisetum ciliare*. Crop Sci 6:473–476

Tang CY, Schertz KF, Bashaw EC (1980) Apomixis in *Sorghum* lines and their F_1 progenies. Bot Gaz 141:294–299

Tateoka T, Hiraoka A, Tateoka TN (1977) Natural hybridization in Japanese *Calamagrostis*. II. Bot Mag (Tokyo) 90:193–209

Tinney FW (1940) Cytology of parthenogenesis in *Poa pratensis*. J Agric Res (Washington) 60:351–360

Turesson G (1929) Zur Natur und Begrenzung der Arteinheiten. Hereditas 12:323–333

Urbańska-Worytkiewicz K (1974) L'agamospermie, système de reproduction important dans la spéciation des Angiospermes. Bull Soc Bot Fr 121:329–346

Valentine DH (1960) The treatment of apomictic groups in Flora Europaea. Feddes Repert 63:119–127

Venkateswarlu J, Rao PN (1975) Apomixis in *Coix aquatica* Roxb. Ann Bot 39:1131–1136

Voigt PW, Bashaw EC (1972) Apomixis and sexuality in *Eragrostis curvula*. Crop Sci 12:843–847

Voigt PW, Burson BL (1981) Breeding of apomictic *Eragrostis curvula*. Proc XIV Intl Grassland Congr Lexington Ky 1981:160–163

Walters SM (1966) The taxonomic implications of apomixis. In: Hawkes JG (ed) Reproductive biology and taxonomy of vascular plants. Pergamon Press, Oxford, pp 162–168

Warmke HE (1954) Apomixis in *Panicum maximum*. Am J Bot 41:5–11

Wehner DJ, Duich JM, Watschke TL (1976) Separation of Kentucky bluegrass cultivars using peroxidase isoenzyme banding patterns. Crop Sci 16:475–480

Weimarck G (1967) Apomixis and sexuality in *Hierochloë australis* and in Swedish *H. odorata* on different polyploid levels. Bot Not 120:209–235

Weimarck G (1970) Apomixis and sexuality in *Hierochloë alpina* (Gramineae) from Finland and Greenland and in *H. monticola* from Greenland. Bot Not 123:495–504

Williamson CJ (1978) Problems in the utilization of interspecific hybrids of *Poa* in a plant breeding programme. Proc 8th EUCARPIA Congr Madrid, pp 289–298

Williamson CJ (1981) The influence of light regimes during floral development on apomictic seed production and on variability in resulting seedling progenies of *Poa ampla* and *P. pratensis*. New Phytol 87:769–797

Williamson CJ, Killick RJ (1978) Multivariate methods as an aid in identifying *Poa ampla* × *P. pratensis* hybrids from maternal-type offspring. Heredity 41:215–225

Williamson CJ, Watson PJ (1980) Production and description of interspecific hybrids between *Poa pratensis* and *P. longifolia*. Euphytica 29:715–725

Winkler H (1908) Über Parthenogenesis und Apogamie im Pflanzenreiche. Progr Rei Bot 2:293–454

Young BA, Sherwood RT, Bashaw EC (1979) Cleared-pistil and thick-sectioning techniques for detecting aposporous apomixis in grasses. Can J Bot 57:1668–1672

11 Role of Polyploidy in Reproductive Organs and Tissues

F. D'AMATO

11.1 Introduction

In the older literature on angiosperm morphology many examples of very large cells with giant or "hypertrophied" nuclei within differentiated tissues, including reproductive tissues, have been reported, and the connection between nuclear size and trophic activity of the cell has been stressed (references in Schnarf 1929, Tischler 1944, Maheshwari 1950). Goldstein (1928) even attempted to establish a correlation between nuclear form and functional activities of normal and pathological cells. In the absence of adequate knowledge of the mechanisms responsible for the multiplication of the genome, the large size and the variable form (irregular, crenate, lobate, constricted, furrowed, etc.) of nuclei were generally assumed to result from fusion of nuclei and/or amitosis. The significance of restitutional mitosis as a mechanism of doubling the chromosome number was not realized, and was generally regarded as a pathological process. The situation began to change in the late 1930's–early 1940's following the discovery of endomitosis in Homoptera, e.g., the pondskaters of the genus *Gerris* (Geitler 1939), and the proposition that the tetraploid mitoses "with paired chromosomes" (now called "diplochromosomes") in polysomatic root tips of *Spinacia oleracea* are due to a double chromosome reproduction at interphase (Berger 1941). After more than 30 years of research on the nuclear cytology of differentiated tissues, it is now clear that the "supernumerary chromonemal reproduction" at interphase (Lorz 1947), better called "chromosome endoreduplication" (Levan and Hauschka 1953), is the commonest and most widespread process of cell polyploidization in both plants and animals (Brodsky and Uryvaeva 1977, D'Amato 1977a, Nagl 1978).

11.2 Methods of Polyploidization and Ploidy Determination

In agreement with Brodsky and Uryvaeva (1977) the term "cell polyploidy" is used here as a general term to define a multiple (2^n) doubling of nuclear DNA independently of the mechanisms responsible for the doubling. The methods of cell polyploidization in both plants and animals may be summarized as follows (for more details, see Brodsky and Uryvaeva 1977, D'Amato 1977a).

11.2.1 Acytokinetic Mitosis (AM)

The absence of cytokinesis at the end of mitosis results in a binucleate cell. If the two nuclei undergo one or more additional acytokinetic mitoses a multinucleate cell is formed. In sufficiently large cells, which allow a spatial distribution of the mitoses, a doubling of the nuclear number (2, 4, 8, 16, etc.) takes place in the cell at each mitosis; in smaller cells, the limited space available favors mitotic disturbance, especially spindle fusion (see Sect. 11.2.2). In any case, it seems justified to regard binucleate and multinucleate cells as polyploid cells (D'Amato 1952, 1977a); thus, e.g., a binucleate diploid cell (2n+2n) can be regarded as a tetraploid (4n) cell. The acytokinetic mitosis is a constant feature in the development of the multinucleate anther tapetum and of the nuclear type of endosperm; in the latter case, very large coenocytes may be formed.

11.2.2 Spindle Fusion (SF)

Since the two nuclei of binucleate cells generally divide synchronously, there is opportunity, especially in smaller cells, for spindle fusion (SF), viz., the fusion of metaphase and/or anaphase plates. In the case of a 2n+2n cell, SF at metaphase or anaphase will produce a 4n+4n and a 2n+4n+2n cell, respectively. The process of spindle fusion may be repeated during the next synchronous nuclear division(s): the result will be a cell with a number of nuclei always lower, or much lower, than that expected from the number of mitoses which took place. Spindle fusion is one of the modes of polyploidization in antipodal cells, and in multinucleate tapetal cells.

11.2.3 Restitutional Mitosis (RM)

Restitutional mitoses result from spindle disturbances at anaphase or, rarely, at metaphase; they lead to the production of polyploid restitution nuclei. Spindle disturbances may frequently be associated with the defective movement of individual chromosomes (laggards) and/or chromosome stickiness which favors sticky bridges at anaphase or chromosome clumping at metaphase. Mitotic restitution at anaphase or metaphase is reflected in the form of the restitution nuclei: dumb bell-shaped in the former case and round in the latter. Restitutional mitoses are not rare in such short-lived tissues as the anther tapetum and the endosperm. Whether some, at least, of the metaphase-blocked mitoses which have been reported in the tapetum of some species are endometaphases of an endomitosis (see Sect. 11.2.4) is not yet clear.

11.2.4 Endomitosis (Em)

In the heteropteran *Gerris lateralis,* in which the ploidy level of the interphase nuclei can be easily determined by counting the number of the heterochromatic X chromosomes, Geitler (1939) showed that cell polyploidization at different times occurred via an endonuclear mitosis which he called "endomitosis". Endomitosis starts with a normal prophase (endoprophase), the chromosomes contract further (endometaphase), their chromatids separate parallel to each other (endoanaphase) and, finally, decondense to take the interphase nuclear structure. Thus, endomitosis – a process which produces a doubling of the chromosome number (endopolyploidy) at each endomitotic cycle (Fig. 11.1) – essentially differs from mitosis for the absence of a mitotic spindle; so, it has been likened to a colchicine-arrested mitosis (Geitler 1953). Endomitosis, sometimes with variations especially in the degree of chromosome contraction, occurs in several animal groups (Geitler 1953); by contrast, it is very rare in angiosperms. According to Witkus (1945), Brown (1949), Avanzi (1950), D'Amato (1952, 1954), Scarascia (1952), Chiarugi (1954), and Oksala and Therman (1977), endomitoses comparable to those of *Gerris* occur in the anther tapetum of some species (see Table 11.2). This view is opposed by the Viennese school (cf. Geitler 1953, Tschermak-Woess 1971) who claim that the stages resembling endomitosis are, in reality, phases of "inhibited mitoses" („gehemmte Mitosen"), i.e., mitoses arrested at different stages of the mitotic process, thus resulting in restitution nuclei. In the present chapter, I will use the term endomitosis in its narrow sense (cf. D'Amato 1954 for further discussion) and maintain that endomitosis occurs in tapetal cells of some species (Em in Table 11.2).

11.2.5 Chromosome Endoreduplication (Er)

Chromosome endoreduplication (Er) (Levan and Hauschka 1953) is an endonuclear chromosome duplication which occurs in the absence of any obvious condensation and decondensation stages, and leads to the production of chromosomes with 2^n (2, 4, 8, 16, 32, etc.) chromatids while leaving the chromosome number unchanged (polyteny) (Fig. 11.1). It is generally assumed that endoreduplication is the commonest mode of endonuclear chromosome duplication in both animals and plants including angiosperms (cf. D'Amato 1954, 1977a, Barlow 1976, 1978, Brodsky and Uryvaeva 1977), but the Viennese school (Geitler 1948, 1953, Nagl 1970a, Tschermak-Woess 1971) regard the endoreduplication in angiosperms as a particular type of endomitosis which they call "angiosperm type". The justification for using the term endomitosis is based on the assumption that endonuclear chromosome duplication is frequently associated with a phase during which the chromocenters disperse („Zerstäubungsstadium", viz., "dispersion phase" or Z-phase). The Z-phase was described by Heitz (1928, 1929) in his studies on the mitotic cycle in plants and interpreted by him as an early prophase which preceded the beginning of chromosome spiralization. Accepting Heitz's interpretation of the Z-phase, the Viennese school consider the "angiosperm endomitosis" as a mitosis arrested at the Z-stage, viz., early or very early prophase. Recent investigations have, however, shown that in several plants the Z-phase occurs during the DNA replication phase (S-phase), and in barley it may even occur several hours before the onset of the S-phase (cf. Barlow 1976, 1978). Since the Z-phase may be independent of the S-phase (Barlow 1978), the possibility cannot

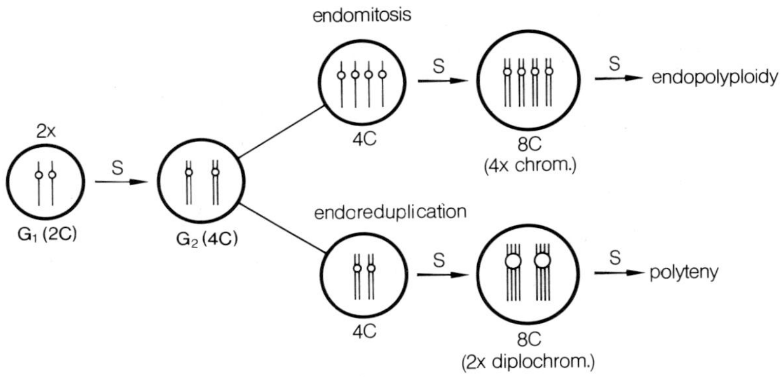

Fig. 11.1. Comparison of endomitosis and chromosome endoreduplication in an initial diploid 2 × nucleus (only one pair of chromosomes is shown); nuclear DNA contents are given in C values. (After D'Amato 1977a)

be ruled out that the Z-phase might also represent a condition of overall transcription in the nucleus. So, it might be concluded with Nagl's (1978, p. 103) words: "grave doubts must be expressed as to whether the Z-phase can be generally interpreted as a type of endomitosis". For the above reasons, the term "endoreduplication" as used in this chapter will also include the "masked endomitosis" („maskierte Endomitose") of Resch (1952) and Steffen (1955), the "Angiosperm Endomitose" of the Viennese school (Geitler 1953, Nagl 1970a, Tschermak-Woess 1971), the "Kryptoendomitose" of Tschermak-Woess (1971), and the term "endomitosis" of several authors who do not make distinction between endomitosis and endoreduplication.

11.2.6 Ploidy Determination

There are two direct methods for the determination of ploidy levels: (1) chromosome counts in cells undergoing nuclear division (e.g., anther tapetum, endosperm and antipodal cells), and (2) cytophotometric determination of DNA content in nuclei of nondividing cells. Nuclear DNA contents are expressed in C values: 1C is the DNA content of haploid (n) nuclei in the period (G_1) preceding the DNA synthesis phase (S), whilst 2C represents the DNA content of a haploid nucleus in the period (G_2) which follows DNA duplication or the DNA content of a diploid (2n) nucleus in the G_1-period. Thus, 16C is the DNA content of a 16n nucleus in G_1 or an 8n nucleus in G_2 (for more details on DNA values, see D'Amato 1977a, Nagl 1978). Despite its usefulness, DNA cytophotometry has so far found a relatively limited application in the study of reproductive organs and tissues in angiosperms.

DNA cytophotometry has proved the validity of a commonly used indirect method of ploidy determination: the volumetric analysis of nondividing nuclei. This is explained by the fact that each doubling of the genome – be it due to doubling of the chromosome number or to doubling of the number of chromatids within each chromosome (polyteny) – is reflected in a proportionate increase in the nuclear volume. Because of the rather good correspondence between the results obtained with the measurement of nuclear volumes and those obtained with DNA cytophotometry (cf. Nagl 1978), the maximum ploidy levels reported in Tables 11.1–11.4, most of which were determined by volumetric analysis, have a rather good degree of reliability.

In the case of true polyploidy (increase in chromosome number), somewhat reliable information on ploidy levels may be obtained from counting the nucleoli (Kubien 1968) or the chromocenters.

11.3 Anther

The anther is a two-lobed, four-loculed organ in which each loculus contains a microsporangium (microsporogenous tissue). At maturity, the two microsporangia of each side become confluent as a consequence of the breaking down of the partition between them. The microsporogenous tissue is surrounded by ground tissue which forms the anther wall and the connective, the tissue between the lobes, where the vascular strand of the stamen filament ends blindly (for further details, see Chap. 2).

11.3.1 Anther Wall and Connective

The anther wall and connective are made up, at full development, of parenchymatous cells which may attain large size, indicating polyploidization. In the connective of the anther in *Camelina sativa, Iberis amara* (Jaretzky 1928) and a few other species (cf. Tischler 1944) giant cells were observed. The presence, in them, of cells with a diakinesis-like appearance suggests a situation of chromosome endoreduplication. In contrast with the unsatisfactory knowledge of the anther wall, a large body of observations are available on the nuclear cytology of the innermost cell layer of the anther wall, the tapetum.

11.3.2 Tapetum

The anther tapetum is a usually uniseriate, rarely multiseriate, cell layer which envelops the microsporogenous tissue. The tapetum, which is generally of parietal origin (it forms through a series of mitotic divisions from the innermost cell layer of the anther wall) and only rarely originates from sporogenous cells, comprises uninucleate diploid cells at the time when meiosis begins in the pollen mother cells (Carniel 1952, 1963a, Wunderlich 1954). During further development, the tapetum may remain uninucleate or, much more commonly, it becomes multinucleate; so two categories of cellular ("secretory") tapetum are distinguished: uninucleate and multi-nucleate (Carniel 1952, 1963a). The distribution of these two types of cellular tapetum among angiosperms has been dealt with by Wunderlich

Table 11.1. Maximum ploidy levels in nuclei of the uninucleate cellular tapetum. Polyploidization occurs via chromosome endoreduplication in all species but *Oxalis* (polyploidization via metaphase arrested mitoses)

Species (family)	Maximum ploidy level	References
Bryonia dioica (Cucurbitaceae)	8n[a]	Turala (1963)
Carex hirta (Cyperaceae)	8n	Carniel 1962)
Cucumis sativus (Cucurbitaceae)	8n	Turala (1963)
Cucurbita pepo (Cucurbitaceae)	16n	Turala (1958)
Cyperus alternifolius (Cyperaceae)	8n	Carniel (1962)
Echinocystis lobata (Cucurbitaceae)	8n	Turala (1963)
Eleocharis mammillata (Cyperaceae)	8n	Carniel (1962)
Helleborus foetidus (Ranunculaceae)	4n	Carniel (1952)
Kyllingia triceps (Cyperaceae)	8n	Carniel (1962)
Lupinus regalis (Papilionaceae)	8n	Carniel (1952)
Manettia bicolor (Rubiaceae)	8n	Carniel (1952)
Oxalis pubescens and *O. rosea* (Oxalidaceae)	8n	Carniel (1969a)
Schoenoplectus tabernaemontanum (Cyperaceae)	8n	Carniel (1962)
Sicyos angulata (Cucurbitaceae)	8n	Turala (1963)
Vicia faba (Papilionaceae)	8n	Carniel (1952)

[a] 4n+4n in 20% tapetal cells

(1954). In both the uninucleate and multinucleate cellular tapetum, the tapetal cells remain in situ until they degenerate, generally at the time of separation of the microspores from each other. In representatives of a few families (Wunderlich 1954, Carniel 1963 a), however, the wall of the tapetal cells breaks down precociously, viz., at zygotene-pachytene in the pollen mother cells, and the protoplasts invade the cavity of the anther loculus, where they may fuse to form a syncyte called the tapetal periplasmodium or the "amoeboid" tapetum (Maheshwari 1950, Carniel 1963 a).

In dealing with the nuclear cytology of the tapetum, we will consider the three types separately.

11.3.2.1 Uninucleate Tapetum

Chromosome endoreduplication is the commonest, though not exclusive, method of cell polyploidization in the uninucleate tapetum. Among the species listed in Table 11.1, only *Oxalis pubescens* and *O. rosea* seem to polyploidize their tapetal cells via metaphase arrested mitoses which, because of the omission of anaphase movement, result in rounded restitution nuclei (Carniel 1969 a). In *Bryonia dioica* (Turala 1963), which represents an exception within the Cucurbitaceae, the tapetal cells generally (80%) undergo two endoreduplications (8n), the other possibility (20%) being an acytokinetic mitosis (2n+2n) which is followed by only one endoreduplication (4n+4n). The equivalence between the two cellular conditions, 8n uninucleate and 4n binucleate, is rightly stressed by Turala (1963).

As is the rule in every tissue, endoreduplication in tapetal cells is evidenced by an increase in nuclear volume at each endoreduplication cycle, to which a proportionate increase in cell volume coresponds. Within one and the same tapetum, the level of endoreduplication varies from cell to cell and some cells do not endoreduplicate, so that diploid and polyploid cells coexist in the tapetum. Thus, Turala (1958) found the four classes of nuclear volumes corresponding to 2n, 4n, 8n, and 16n in one and the same tapetum in *Cucurbita pepo*. In the tapetum of *Vicia faba*, Carniel (1952) reported the occurrence of typical diplochromosome mitoses, but was unable to ascertain whether they were spontaneously occurring mitoses, or mitoses induced by a concomitant plant infestation by coccids.

In addition to the species reported in Table 11.1, many other species undergo up to two endoreduplications in the nuclei of their uninucleate tapetum; they are listed by Carniel (1952).

11.3.2.2 Multinucleate Tapetum

The multinucleate tapetum is the most widespread type of tapetum among angiosperms (Wunderlich 1954). The multinucleate condition results from acytokinetic mitoses, whose number varies between two and four in different species. Within one and the same tapetum, the number of mitoses is not constant (Wunderlich 1954, Carniel 1963a), and in the case of four mitoses, the fourth mitosis is restricted to a few cells only. The nuclear divisions start at the time of very early, or early, prophase of meiosis in the pollen mother cells and continue up to diakinesis or, in isolated cells, up to the pollen tetrad stage. Because of spindle disturbances due to chromosome stickiness and spindle fusion (fusion of metaphase and/or anaphase plates), restitution nuclei are frequently produced in the multinucleate type of tapetum. In some species, restitutional mitoses may variously combine with endomitosis and/or diplochromosome mitoses (endoreduplication) to produce a rather complex tapetal cell development. In the case of restitutional mitoses, a correlation has been reported between increasing degree of chromosome stickiness and "age" of tapetal cell, and between frequency of spindle fusions and size and shape of tapetal cells. Thus, in *Crepis zacyntha* chromosome stickiness is absent during the first mitosis and becomes more and more intense and frequent during the second and third mitoses (Battaglia 1949). In the same species, in *Knautia drymeia*, *K. arvensis*, *Scabiosa kladnikiana* (Carniel 1963b), *Valeriana officinalis* (Skalinska 1958), *Sambucus nigra* and *S. racemosa* (Carniel 1952) in which the tapetum develops through three mitoses, an eight-nucleate condition ($2n \times 8$) is common only in the larger tapetal cells, whereas anaphase restitutions and/or spindle fusions frequently occur in smaller cells.

In members of the Solanaceae and Scrophulariaceae, the tapetum frequently differentiates into an internal tapetum, facing the anther connective where the vascular bundles terminate, and an external tapetum; the former is generally better developed and its cells have a higher level of ploidy than those of the latter (Carniel 1963a, Greilhuber 1974). For species characterized by an internal and an external tapetum, the maximum ploidy levels reported in Table 11.2 refer to the internal tapetum. For a better evaluation of the data reported in the table, the nuclear cytology of the multinucleate tapetum will be discussed in relation to the number of mitoses needed for its formation.

Table 11.2. Maximum ploidy level and methods of polyploidization of nuclei in the multinucleate cellular tapetum

Species (family)	Maximum ploidy level	Methods of polyploidization	References
Aconitum variegatum (Ranunculaceae)	8n, 4n+4n	AM, SF, RM	Trela (1958)
Aloë ciliaris and *A. obscura* (Liliaceae)	16n, 8n+8n	AM, SF, RM	Carniel (1952)
Antirrhinum majus (Scrophulariaceae)	16n, 8n+8n	AM, Em, RM, DM	Berger et al. (1951), Mechelke (1952)
Bartsia alpina (Scrophulariaceae)	8n+8n	1AM+2Er	Greilhuber (1974)
Bellardia trixago (Scrophulariaceae)	16n+16n	1AM+1RM+2Er	Greilhuber (1974)
Cassia laevigata and *C. australis* (Caesalpiniaceae)	8n+8n	AM, RM, Er	Carniel (1952)
Cestrum elegans (Solanaceae)	16n, 8n+8n	AM, SF, Em, RM, DM?	Carniel (1952), Pavari (1957)
Crepis zacyntha (Compositae)	2n×8, 2n×4+8n, etc.[a]	AM, SF, RM	Battaglia (1949)
Eremurus sp. (Liliaceae)	8n+8n, 4n×4	AM, Em, RM, DM?	Oksala and Therman (1977)
Euphorbia pulcherrima (Euphorbiaceae)	4n+4n, 2n×2+4n, 2n+6n	AM, SF, RM	Carniel (1952)
Euphrasia picta and *E. rostkoviana* (Scrophulariaceae)	4n+4n	1AM+1Er	Greilhuber (1974)
Knautia drymeia and *K. arvensis* (Dipsacaceae)	2n×8, 2n×4+4n×2	AM, RM	Carniel (1963b)
Lathraea squamaria (Scrophulariaceae)	8n+8n	1AM+2Er	Greilhuber (1974)
Lycopersicum esculentum (Solanaceae)	32n, 16+16n	AM, RM, Em, DM	Brown (1949)
Nicotiana tabacum (Solanaceae)	16n, 8n+8n	AM, RM, Em, DM	Scarascia (1952)
Odontites rubra (Scrophulariaceae)	8n+8n	AM+RM+Er	Greilhuber (1974)
Oxalis acetosella (Oxalidaceae)	4n+4n, 2n×4, 2n+2n+4n	AM, RM	Carniel (1969b)
Paeonia tenuifolia (Ranunculaceae)	2n×8	AM	Marquardt et al. (1968)
Pedicularitis sylvatica (Scrophulariaceae)	4n, 2n+2n	AM, RM	Greilhuber (1974)
Rhinanthus alectorolophus and *R. minor* (Schrophulariaceae)	8n+8n	1AM+2Er	Tschermak-Woes (1967)
Sambucus ebulus, S. nigra and *S. racemosa* (Caprifoliaceae)	2n×4, 6n+2n, 4n+2n×2	AM, SF	Carniel (1952)
Scabiosa kladnikiana (Dipsacaeae)	2n×8, 2n×4+4n×2	AM, RM	Carniel (1963b)
Silene nemoralis (Caryophyllaceae)	8n+8n	AM, SF, RM, DM	Favarger (1944), Carniel (1952)
Solanum dulcamara (Solanaceae)	16n+16n	AM, Em, RM	Turala and Urbanska-Worytkiewicz (1964)
S. jasminioides (Solanaceae)	8n+8n	AM, Em, RM, DM?	Carniel (1952)
S. nigrum (Solanaceae)	8n+8n	AM, Em, RM	Turala and Urbanska-Worytkiewicz (1964)
S. tuberosum (Solanaceae)	16n+16n, 8n×4	AM, SF, Em, RM, DM	Avanzi (1950)
Spinacia oleracea (Chenopodiaceae)	4n+4n, 8n	AM, Em, RM	Witkus (1945), Carniel (1952)
Tozzia alpina (Scrophulariaceae)	4n+4n	1AM+1Er	Greilhuber (1974)
Valeriana officinalis (Valerianaceae)	2n×8, 8n+8n, 16n	AM, SF, RM	Skalinska (1958)
Zea mays (Gramineae)	2n+2n, 4n, 8C	AM, RM, Er?	Bianchi (1959), Carniel (1961), Moss and Heslop-Harrison (1967)

[a] For further data, see text

Abbreviations: *AM* acytokinetic mitosis, *DM* diplochromosome mitosis, *Em* endomitosis, *Er* endoreduplication, *RM* restitutional mitosis, *SF* spinde fusion

In *Zea mays* (Bianchi 1959, Carniel 1961) and some *Pedicularis* species including *P. sylvatica* (Greilhuber 1974), only one mitosis occurs in the tapetal cells: it produces binucleate (2n + 2n) or uninucleate (4n) cells depending on whether the mitotic process is normal or restitutional. In *Pedicularis* an identical level of ploidy has been observed in the internal and external tapetum; this condition contrasts with that occurring in other species of Scrophulariaceae in which the ploidy level in the internal tapetum is higher than in the external tapetum (Greilhuber 1974). In some species of Scrophulariaceae, *Tozzia alpina, Euphrasia picta, E. rostkoviana, Bartsia alpina, Lathraea squamaria* (Greilhuber 1974), *Rhinanthus alectorolophus* and *R. minor* (Tschermak-Woess 1967), the two diploid nuclei resulting from the single mitosis undergo one or two endoreduplications to produce 4n + 4n and 8n + 8n cells, respectively. In the case of tapetal development through two mitoses, the second mitosis frequently ends with nuclear restitution at anaphase in *Aconitum variegatum* (Trela 1958) and *Oxalis acetosella* (Carniel 1969 b), or is characterized by fusion of metaphase and/or anaphase plates (SF in the table) in *Euphorbia pulcherrima, Sambucus nigra,* and *S. racemosa* (Carniel 1952). In both cases the tapetal cells, with a total ploidy of 8n, are rarely (larger cells) four-nucleate (2n × 4), more frequently three-nucleate or two-nucleate. In *Odontites rubra* and *Bellardia trixago* the first acytokinetic mitosis in the tapetal cells is followed by a restitutional mitosis (4n + 4n), and the two nuclei undergo endoreduplication to attain the 16n + 16n and 8n + 8n conditions, respectively (Greilhuber 1974). Another tapetum whose development is based on two nuclear divisions is that of *Spinacia oleracea.* Here both mitoses may be endomitotic or the first mitosis may be either normal or restitutional, the final condition being 4n + 4n or 8n (Witkus 1945, Carniel 1952).

We now come to the most complex types of multinucleate tapetum, those differentiating after three or, more rarely, four mitoses. As seen from Table 11.2, the 2n × 8 (total ploidy: 16n) condition which is expected as a result of three mitoses, is a constant feature in the tapetum of *Paeonia tenuifolia.* In other species the final number of nuclei in tapetal cells is lower or much lower than eight, due to restitutional mitoses (RM) (restitutional nucleus formation at anaphase or, rarely, metaphase) concomitant, or not, with the fusion of metaphase and/or anaphase plates (SF). Thus, e.g., in *Crepis zacyntha* (Battaglia 1949), the final 16n condition of the tapetal cells can also be represented by six-nucleate (2n × 5 + 6n; 2n × 4 + 4n × 2), five-nucleate (4n × 3 + 2n × 2; 2n × 4 + 8n), four-nucleate (2n × 2 + 4n + 8n), and three-nucleate (2n × 2 + 12n) cells. In more extreme cases, the 16n condition may be represented by binucleate and/or uninucleate cells (see Table 11.2).

In some types of tapetum, whose development results from three or four mitoses, restitutional mitoses and/or spindle fusions are combined with endomitosis (Em) and/or diplochromosome mitosis (DM). Since diplochromosomes (four-chromatid chromosomes) result from a double chromosome reproduction at interphase, inclusion of one DM among the three, or four, nuclear divisions leads to one-step increase in the final ploidy of the tapetal cells. This situation occurs in the internal tapetum of *Solanum tuberosum* (Fig. 11.2) whose cells attain the final ploidy of 32n (16n + 16n; 8n × 4) within the range of three nuclear divisions (Avanzi 1950). Diplochromosome mitoses also occur in the development of the tapetal cells in *Antirrhinum majus.* They were called mitoses

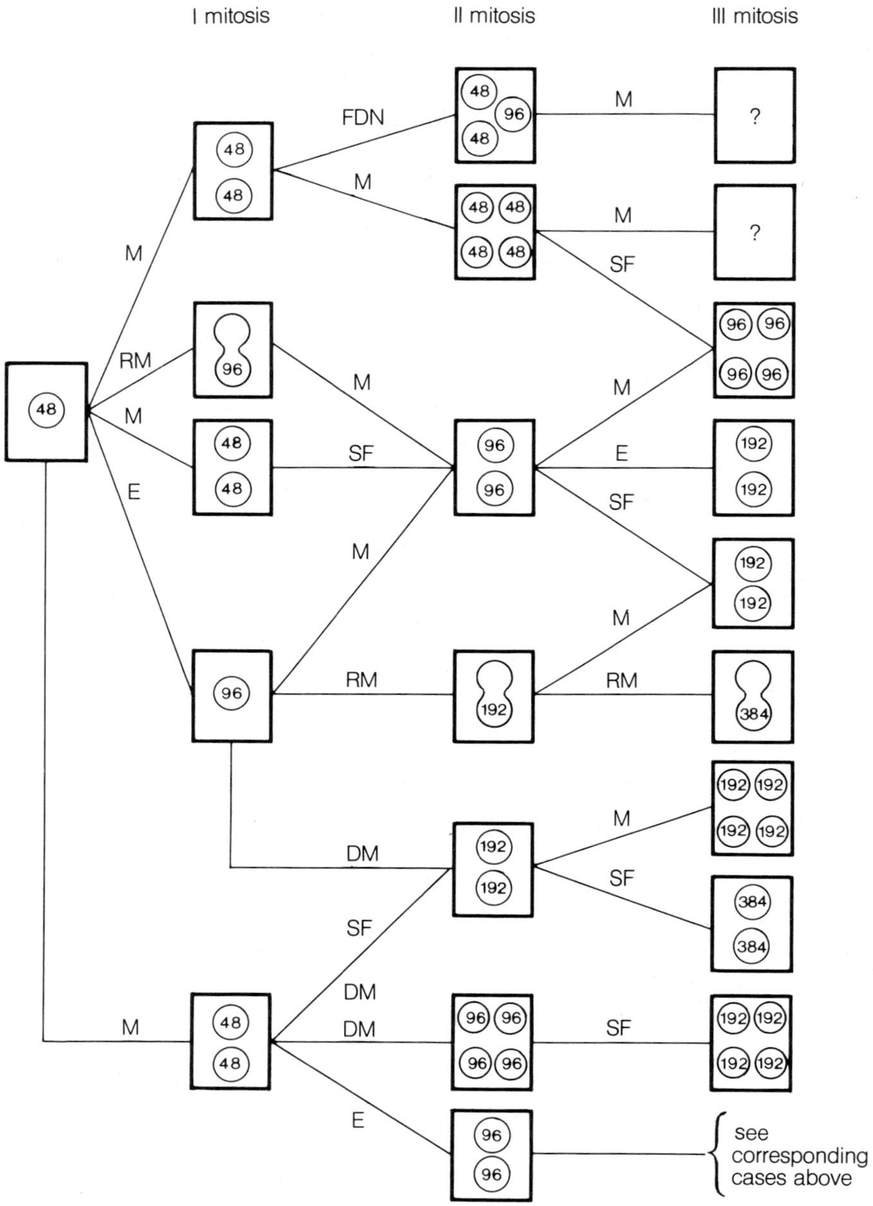

Fig. 11.2. Schematic representation of the methods of polyploidization in the cells of the internal anther tapetum in *Solanum tuberosum* (2n = 48). *DM* diplochromosome mitosis, *E* endomitosis, *FDN* fusion of daughter nuclei, *M* mitosis, *RM* restitutional mitosis, *SF* spindle fusion, *SF/DM* spindle fusion in diplochromosome mitosis. (Condensed from Avanzi 1950, after D'Amato 1977a)

"with paired chromosomes" by Berger et al. (1951) or "polymer" mitoses by Mechelke (1952). The polymer chromosomes were assumed by Mechelke (1952) to result from "an internal chromosome division" („innere Chromosomenteilung") of the chromosomes not accompanied by an increase in the number of chromatids; however, the appearance of the chromosomes in the original microphotographs denotes DMs (D'Amato 1954). Diplochromosome mitoses occur, though rarely, in the tapetal cells of *Nicotiana tabacum* (Scarascia 1952) and of other species, but they were not interpreted as such by the respective authors. Cases of DM which the present writer regards as still dubious are reported as DM? in the Table.

11.3.2.3 Periplasmodial Tapetum

This type of tapetum appears as of marginal interest from the viewpoint of nuclear cytology. Carniel (1952) reported that in some Commelinaceae (*Spironema fragrans, Rhoeo discolor, Tradescantia fluminensis,* and *Palisota harteri*), during meiotic prophase in the pollen mother cells, there occurs in the periplasmodium a wave of synchronous mitoses, during which spindle fusions and/or nuclear restitution occur. The result is the contemporary presence in the periplasmodium of 2n and 4n nuclei. Other families with a periplasmodial tapetum are the Araceae (Wunderlich 1954) and the Scheuchzeriaceae (Carniel 1963a). According to Carniel (1961, 1963a), the tapetum of *Zea mays* makes a bridge between the nuclear (secretory) and periplasmodial tapetum: although the tapetum develops according to the scheme of a nuclear tapetum, its nuclei divide synchronously.

11.3.3 Pollen Grains

Pollen grains (male gametophytes) are usually haploid as are the microspores from which they form through mitotic division(s). There are, however, several types of formation of nonhaploid (diploid and tetraploid) pollen grains, which have been discussed by Battaglia (1951), where the pertinent literature is reported. The most significant types are the following:

Hieracium boreale type. Due to asynapsis in the pollen mother cells (PMC) anaphase I ends with a 2n restitution nucleus which undergoes the second meiotic division to produce a dyad of unreduced (2n) microspores and pollen grains.

Hieracium laevigatum type. This type differs from the preceding one in that the restitution nucleus is formed at metaphase I.

Rudbeckia laciniata type. The asynaptic meiosis I, which leads to the production of a 2n restitution nucleus, is followed by a second division which is also restitutional. The result is the production from each PMC of a single 4n microspore.

Allium amplectens type. Univalents are formed at meiosis I as a result of desynapsis. The univalents congregate on the metaphase plate of a normal looking spindle but do not divide and produce a 2n restitution nucleus. The mitotic division of these diploid nuclei produces dyads of 2n microspores.

Allium schoenoprasum type. The first division of meiosis is normal (2 haploid nuclei are formed). It is followed by a long interphase during which the chromosomes (= half bivalents) undergo reduplication: thus, dyads of microspores, each containing diplochromosomes (four-chromatid chromosomes) in the haploid number are formed. At the first pollen mitosis, the n diplochromosomes divide to produce vegetative and generative nuclei which are both 2n.

11.4 Ovule

The megasporangium or ovule consists of the nucellus and one or two integuments (unitegmic and bitegmic ovules). Depending on the number of cell layers surrounding the embryo sac mother cell, ovules are classified as tenuinucellate and crassinucellate (see Chap. 3).

11.4.1 Nucellus

Information on the nuclear cytology of nucellar cells is not extensive. In *Campanula pyramidalis* (Enzenberg 1961) the nucellus remains diploid and it is probable that the same condition occurs in other angiosperms, but direct evidence is lacking. More frequently, nucellar cells may undergo chromosome endoreduplication, which starts after cessation of the initial phase of growth by cell division. The following maximum ploidy levels were determined by measurements of nuclear volumes: 8n (two endoreduplications) in the three to four nucellar cell layers facing the internal tapetum (endothelium) in *Pedicularis palustris* (Steffen 1955), 16n (three endoreduplications) in the nucellus of *Ecballium elaterium, Echinocystis lobata* (Turala 1966a), and *Cucumis sativus* (Enzenberg 1961), 32n (four endoreduplications) in the nucellus of *Daphne mezereum* (Enzenberg 1961) and 256n (seven endoreduplications) in the nucellus of *Trapa natans* (Nagl 1962).

11.4.2 Integument(s) and Endothelium

Chromosome endoreduplication may also occur during development of the integument(s), sometimes in connection with the differentiation of an integumentary tapetum or endothelium (Kapil and Tiwari 1978). The integumentary tapetum differentiates from the inner epidermis of the inner integument in bitegmic ovules, or from that of the single integument in unitegmic ovules. It is a specialized layer of radially stretched cells with dense cytoplasm and, sometimes, large nuclei. The endoreduplicated nuclei of the endothelium have been estimated to attain the 32n level in *Pedicularis palustris* (Steffen 1955), and the 512n level (eight endoreduplications) in the tapetal cells which closely surround the embryo suspensor in *Ge-*

ranium phaeum (Nagl 1962). In some species endothelial cells polyploidize via binucleation or multinucleation (Kapil and Tiwari 1978).

In the genus *Allium* most of the inner integument differentiates into a "water tissue" (water-storing parenchyma) whose nuclei grow by chromosome endoreduplication and attain the 8n condition as in *A. angulosum, A. ammophilum,* and *A. ursinum*. Interestingly enough, the water tissue in the ovules of *A. pulchellum* differentiates in the diploid state (Hasitschka-Jenschke 1958).

Within an integument, as is the case in all plant tissues, the level of polyploidization is not uniform and, sometimes, striking ploidy differences occur between cells. Thus, e.g., in *Tropaeolum majus* cells of the inner integument adjoining the embryo sac do not surpass the 8n level, those of the integumentary epidermis attain 32-ploidy, whereas the cells of the micropylar half of the integument – to be regarded as a physiological continuation of the suspensor – may have nuclei up to 1,024-ploid (Nagl 1962).

11.5 Embryo Sac

Typically, the embryo sac consists of an egg apparatus (an egg and two synergids), a proendospermic (Chiarugi 1927), or endosperm mother cell (Maheshwari 1950), or central cell which contains one or more polar nuclei, and three antipodal cells.

11.5.1 Egg Cell

In sexually reproducing angiosperms the egg is haploid (n); in apomictic species, with unreduced (2n) embryo sacs, the egg is diploid (2n). According to Battaglia (1963), the following mechanisms are responsible for the production of unreduced female gametophytes (see also Chap. 10):

1. After a normal meiotic division I a long interphase follows during which an endoreduplication occurs leading to the formation of n diplochromosomes (four-chromatid chromosomes); these segregate normally during the next mitosis (2n chromatids to each pole). Since this division is acytokinetic, two further mitoses suffice to produce the eight-nucleate unreduced embryo sac – *Datura type*.
2. Asynapsis is responsible for the formation of a 2n restitution nucleus at the end of division I. Division II may or may not be followed by cytokinesis, so that three or two further divisions are needed for the production of the unreduced eight-nucleate embryo sac – *Taraxacum* and *Ixeris types,* respectively.
3. Chromosome doubling via endoreduplication (diplochromosome formation) produces 4n embryo sac mother cells which then undergo chromosome reduction and produce a dyad of 2n cells. The lower cell of the dyad produces (three mitoses) the eight-nucleate unreduced embryo sac – *Allium nutans type*.
4. Meiosis is omitted in the embryo sac mother cell; the cell undergoes three mitoses to produce the eight-nucleate unreduced embryo sac – *Antennaria type*.

5. The eight-nucleate or the four-nucleate unreduced embryo sac is produced by a somatic (nucellar) cell which undergoes either three or two mitoses – *Hieracium and Panicum types*, respectively.

The fertilized eggs (zygotes) of *Hordeum distichum* cv Hannchen and *Petunia hybrida* have been observed to have a nuclear DNA content clearly higher than the maximum expected DNA value of 4C, corresponding to the DNA postsynthesis phase (G_2) of 2n nuclei; namely, 8C and occasionally 16C in *Hordeum* (Mericle and Mericle 1970) and 4.94C in *Petunia* (Vallade et al. 1978). In both species the nuclear DNA amount is progressively reduced during proembryo development, and stabilizes at the diploid values, 2C to 4C. The most plausible explanation for the origin of the DNA amount in excess of 4C is a process of DNA amplification: but no evidence for this is presently available in either species. It must, however, be pointed out that recent cytophotometric measurements on the zygotes of *Hordeum vulgare* cvs "Vada" and "Sultan" have given nuclear DNA values within, or close to, the diploid range, 2C–4C (Bennett and Smith 1976).

11.5.2 Synergids

Usually the synergids are ephemeral cells which degenerate soon after fertilization, or even before it. When synergids persist up to the time of fertilization, usually only one synergid is destroyed by the impact of the pollen tube, but in some species both synergids or neither of them seem to be affected by the fertilization process (Maheshwari 1950; see also Chaps. 4 and 6). Persistence of one or both synergids for some time after fertilization, and during early embryo development, as occurs in several species, is always manifested with an enormous increase in cellular and nuclear size and show very clear signs of high activity (Håkansson 1951, 1957, Hasitschka 1956, Hasitschka-Jenschke 1959b, Syamasundar and Panchkasharappa 1975, see Maheshwari 1950 for older literature).

It is not known at present whether ephemeral synergids, which seem to serve several functions before and during fertilization (Linsken 1969), remain haploid, at least generally. In the female gametophyte of *Allium ursinum,* with one ephemeral and one persistent synergid, the ephemeral cell has been shown to undergo two cycles of chromosome endoreduplication, thus becoming "tetraploid" (n diplochromosomes). By contrast, the other synergid, which enlarges enormously and persists till the beginning of cellularization in the endosperm, has undergone four cycles of endoreduplication (16n) already before fertilization, and grows further to attain the 32n level (five endoreduplications). The endoreduplicated nuclei invariably show the polytene chromosome structure (Hasitschka-Jenschke 1957).

Polytene ("giant") chromosomes in persistent synergids were first described by Håkansson (1951) in a pentaploid ($2n = 5\times = 40$) parthenogenetic biotype of *Allium nutans*. When its flowers were emasculated, the endosperm did not develop, one or both synergids were stimulated into active growth and their nuclei acquired the polytene chromosome structure. Nuclei with polytene chromosomes were also described in the persistent synergid of *Allium odorum* by Håkansson (1957) and in the persistent synergid of *A. cepa* by Syamasundar and Panchaksharappa (1975); but neither of these authors, nor Håkansson, attempted to evaluate the degree of polytenation in their materials. More precise information is provided by the observations of Hasitschka-Jenschke (1958) on three more species of *Allium*. In *A. pulchellum* and *A. angulosum* both synergids enter an endoreduplication phase, but only one of them, the persistent synergid (which in *pulchellum* degenerates at the 20-celled embryo stage) grows further to attain the 8n condition (three endoreduplications) in both species. In *A. ammophilum* one synergid remains haploid while the other persists and its nucleus endoreduplicates to attain at least the 16n condition (four endoreduplications). As to the structure of endoreduplicated nuclei, typical "giant chromosome" structure was common in *A. ammophilum;* other structures included loose chromosome bundles, endoreduplicated chromocenters (endochromocenters), and more or less contracted chromatin threads (Hasitschka-Jenschke 1958). This variability in structure of endoreduplicated nuclei is a general phenomenon at all levels of polytenization; it has been discussed comprehensively by Nagl (1978).

Information on the physiological role of persistent synergids is quite meager. Their extensive growth before endosperm development, their close association with the zygote and the developing embryo, or embryos in the case of polyembryony (D'Amato 1949), and their high metabolic activity, as indicated by the cytochemical analyses of Syamasundar and Panchaksharappa (1975) in *Allium cepa,* support the contention of their involvement in the nutrition of the embryo up to the time when the endosperm takes over this nutritive function.

11.5.3 Proendospermic Cell

The proendospermic cell of the embryo sac is destined to produce the endosperm. In embryo sacs of the *Oenothera* type the proendospermic cell is haploid (one polar nucleus); in the remaining types of embryo sac it contains two or more polar nuclei which fuse to form the secondary nucleus of the embryo sac (see also Chaps. 4 and 7). The ploidy of the secondary nucleus varies with the number and the genomic constitution of the polar nuclei (Battaglia 1951): it is diploid (n+n) in the embryo sac types Polygonum, Scilla, Pyrethrum parthenifolium, Adoxa; triploid (n+2n) in Pyrethrum cinerariaefolium; tetraploid in the types Fritillaria and Plumbagella (n+3n) and the types Penaea and Plumbago in which four polar nuclei fuse with each other. The highest ploidy levels are known in the secondary nucleus of *Peperomia pellucida* (8n) and *P. hispidula* (14n) due to fusion of, respectively, eight and 14 polar nuclei (Maheshwari 1950). In apomictic unreduced (2n) embryo sacs secondary nuclei are generally 4n (2n+2n). In rare cases of unreduced embryo sacs the secondary nucleus is either 2n (one polar nucleus) or 8n (4n+4n) (Battaglia 1963).

The fusion of polar nuclei to produce a diploid or polyploid secondary nucleus is regarded by Chiarugi (1950) as one of the most outstanding features in the construction of the angiosperm embryo sac. In his view, the central polyploidization in the embryo sac is an adaptation to a dehydration gradient starting from the micropylar and chalazal poles, which may be regarded as the zones of nutrient supply to the embryo sac. By this method, the haploid gametophytic generation is presumed to correct its physiological inferiority toward the diploid sporophytic tissues in which it is immersed.

11.5.4 Antipodal Cells

Although usually short-lived, antipodal cells often persist for some time after fertilization; e.g., until the formation of a six-celled proembryo in *Papaver rhoeas* (Hasitschka 1956), or the formation of nuclear endosperm in *Ammophila arenaria* (Kubien 1968). Ephemeral and persistent antipodal cells may occur in one and the same genus, or even in one and the same species. Thus, *Allium angulosum* and *A. ammophilum* produce embryo sacs with three antipodals which remain haploid and degenerate during or soon after fertilization, whereas one and two antipodals, respectively, in *Allium ursinum* and *A. pulchellum,* show a considerable increase in size. In *A. ursinum* the persistent antipodal cell degenerates only when cellularization of the endosperm begins (Hasitschka-Jenschke 1957, 1958).

Conspicuous growth processes in uninucleate antipodals are known since the classical papers of Osterwalder (1898) and Huss (1906). As pointed out by D'Amato (1952), the appearance of heterochromatin in the giant nuclei in the figures of antipodals of *Aconitum napellus, Papaver heldreichii,* and *Hypecoum procumbens* are suggestive of repeated chromosome endoreduplications (polytenization). Indeed, this view has been substantiated by recent investigations (see below). To my knowledge, polytene ("giant") chromosomes in antipodal cells were first observed in *Hordeum jubatum* (Brink and Cooper 1944). A few years later, Beaudry (1951) observed that, in the cross *Elymus virginicus* × *Agropyron repens,* at about 18 h after syngamy, mitoses started in the antipodal cells; they showed either polychromosomes in the haploid number ($n=14$) or more than 100 monochromosomes. Beaudry (1951) rightly concluded that the hybridity merely stimulated mitosis in *Elymus* antipodals thereby revealing their pre-existent nuclear conditions.

Polytenization of antipodal cells has been described for many species, e.g., *Cimicifuga japonica, Asarum europaeum, Muscari racenosum, Ornithogalum boucheanum, Bellevalia romana, Scilla biflora, S. sibirica, Ranunculus sceleratus, R. ficaria, Chionodoxa luciliae, Narcissus pseudonarcissus, N. rotundifolia, Podophyllum peltatum, Mahonia aquifolium, Crocus aureus* (Hasitschka-Jenschke 1962), *Nothoscordum fragrans* (Turala 1966b), and *Triticum durum* (Ivanovskaja 1973) but no information is available on the possible level of antipodal cell polyteny in these species.

In Table 11.3 the maximum ploidy levels of antipodal cells as determined either directly (DNA cytophotometry, chromosome counts in exceptional mitoses) or indirectly (measurements of nuclear volumes) have been reported. The highest ploidy level so far determined in antipodal nuclei is 1024C DNA content corresponding to ten endoreduplication cycles (*Scilla bifolia*). On the basis of the labelling pattern obtained after [^{3}H]-thymidine incorporation, it seems possible that some of the polytene nuclei undergo extrareplication (amplification) of rDNA, the DNA sequences coding for ribosomal RNA (Nagl 1976a). As seen from

Table 11.3. Maximum ploidy level and methods of polyploidization of nuclei in antipodal cells

Species (family)	Antipodals involved in polyploidization	Maximum ploidy level	Methods of polyploidization	References
Achillea millefolium (Compositae)	3	16n, 8n+8n	AM, SF, RM	Titz (1965)
Aconitum neomontanum, A. ranunculifolium, A. variegatum, A. vulparia (Ranunculaceae)	3	128n	Er	Tschermak-Woess (1956)
Allium pulchellum (Liliaceae)	2	8n	Er	Hasitschka-Jenschke (1958)
Allium ursinum (Liliaceae)	1 (central)	16n	Er	Hasitschka-Jenschke (1957)
Ammophila arenaria (Gramineae)	Most (peripheral)	256n	AM, SF, RM	Kubien (1968)
Anemone hepatica (Ranunculaceae)	3	32n	AM, SF, RM	Hasitschka-Jenschke (1959b)
Arum maculatum (Araceae)	3	32n	Er	Erbrich (1965)
Caltha palustris (Ranunculaceae)	3	8n+8n	1AM+2SF	Grafl (1941)
Chrysanthemum alpinum (Compositae)	All	32n, 16n+16n	AM, SF, RM	Titz (1965)
Clivia miniata (Amaryllidaceae)	3	32n	Er	Tschermak-Woess (1957a)
Corydalis cava and *C. nobilis* (Fumariaceae)	Not stated	64n	Er	Hasitschka-Jenschke (1959b)
Crocus suaveolens (Iridaceae)	Not stated	64n	Er	Turala (1966b)
Dicentra spectabilis (Fumariaceae)	1	16n	Er	Hasitschka-Jenschke (1959b)
Eranthis hiemalis (Ranunculaceae)	3	64n+64n	1AM+Er	Hasitschka-Jenschke (1959b)
Eupatorium glabrum (Compositae)	2	1n×6 (six haploid nuclei)	AM	Hasitschka-Jenschke (1959b)
Helleborus niger (Ranunculaceae)	3	8n	Er	Hasitschka-Jenschke (1959b)
Hordeum distichum (Gramineae)	Not stated	512n	Er	Erdelská (1966)
Hyacinthus orientalis (Liliaceae)	3	32n	Er	Turala (1966b)
Kleinia ficoides (Compositae)	All (8)	64n	Er	Hasitschka-Jenschke (1959b)
Othonna crassifolia (Compositae)	3	16n(1), 4n(2)	Er	Hasitschka-Jenschke (1959b)
Papaver rhoeas (Papaveraceae)	3	128n	Er	Hasitschka (1956)
Scilla bifolia (Liliaceae)	3	1024C	Er	Nagl (1976a)
Triticum aestivum (Gramineae)	Several	196C	Er	Bennett et al. (1973)

Abbreviations: *AM* acytokinetic mitosis, *Er* endoreduplication, *RM* restitutional mitosis, *SF* spindle fusion

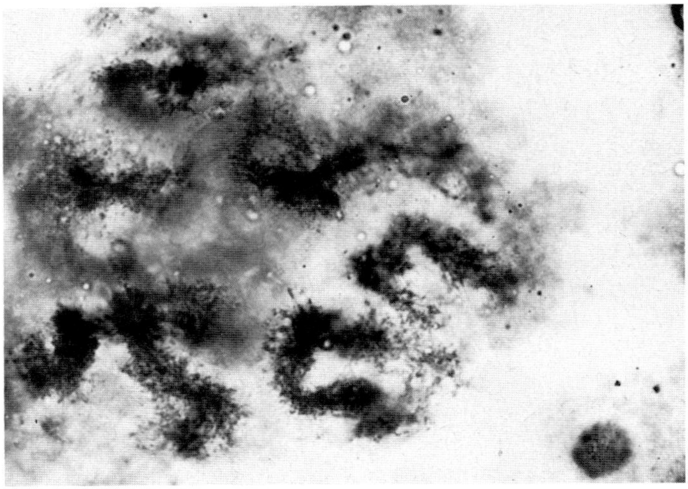

Fig. 11.3. Antipodal cell nucleus of *Papaver rhoeas* with polytene chromosomes in the haploid number (n = 7); compare with a diploid nucleus from the integument (*bottom right*). (After Hasitschka 1956, courtesy E. Tschermak-Woess, Vienna)

Table 11.3, the number of endoreduplication cycles in the antipodal cells of other species ranges from 3(8n) to 10(1024C). Due to variation in structure of polytene nuclei, referred to in the section on persistent synergids, polytene chromosomes are not always visible. They have always been observed in haploid number in favourable materials, e.g., in four species of *Aconitum* (Tschermak-Woess 1956), *Corydalis cava, C. nobilis* (Hasitschka-Jenschke 1959 b), and *Papaver rhoeas* (Fig. 11.3) (Hasitschka 1956) (see Table 11.3).

Another mechanism of polyploidization of antipodal cells is bound to the occurrence of supernumerary mitoses in the antipodal apparatus. When some or all of the supernumerary mitoses are followed by cell wall formation, an antipodal complex consisting of from six to eight to several dozen cells may be formed (Maheshwari 1950, Johri 1963). In *Ammophila arenaria,* after formation of an "antipodal tissue" consisting of about 30 cells, the centrally located antipodals remain haploid while the peripheral ones become highly polyploid through acytokinetic mitoses combined with spindle fusion (formation of a common metaphase plate), fusion of anaphasic or telophasic chromosome groups, and restitutional mitoses at anaphase (Table 11.3). As seen from Table 11.3, similar phenomena occur in the antipodal cells of other species. Of particular interest, because of its regularity, is the polyploidization process of *Caltha palustris* antipodals (Table 11.3). After an acytokinetic mitosis which produces a cell with two haploid nuclei (n + n), there occur three successive bimitoses (two synchronous mitoses) which are constantly characterized by the fusion of the two spindles at metaphase, so that the antipodals finally contain two octoploid nuclei each (8n + 8n).

When acytokinetic mitoses are not complicated by spindle fusions (the mitoses are not synchronous), or mitotic disturbances, polyploid antipodal cells with two, four, or more haploid nuclei are formed. An example is *Eupatorium gla-*

brum whose antipodals may contain a maximum of six haploid nuclei (Table 11.3); other examples can be found in literature (Maheshwari 1950, Johri 1963).

Finally, mention must be made of several plants in which the antipodal nuclei do not show any change in structure and, most probably, remain haploid (no endoreduplication). They are: *Allium ammophilum, A. angulosum* (Hasitschka-Jenschke 1958), *Eupatorium glabrum* (Hasitschka-Jenschke 1959b), *Convallaria majalis, Smilacina stellata, Kniphofia meleagris, Uvularia grandiflora, Paris quadrifolia, Trillium grandiflorum, Fritillaria meleagris* (Hasitschka-Jenschke 1962).

11.6 Endosperm

11.6.1 Origin and Types of Endosperm

As discussed in another section of this chapter, the ploidy-level of the proendospermic cell is different in different types of embryo sac development. The ploidy level is further increased at fertilization as a consequence of the fusion of one of the male gametes with the central (secondary) nucleus of the embryo sac. The primary endosperm nucleus is 2n in the embryo sac type Oenothera; 3n in the types Polygonum, Scilla, Pyrethrum parthenifolium, Adoxa; 4n in the type Pyrethrum cinerariaefolium; 5n in the types Fritillaria, Plumbagella, Penaea, Plumbago; 9n in Peperomia pellucida, and 15n in P. hispidula. Confirmation of the expected chromosome number in the endosperm has been obtained in some, but not all the above-mentioned embryo sac types; chromosomes counts have been facilitated by the application of the squash technique to the endosperm (Geitler 1948, Rutishauser and Hunziker 1950). Cases are known in which the secondary nucleus of the embryo sac develops autonomously, viz., independently of the fusion of the secondary nucleus with one of the male gametes. This condition occurs in some apomicts (see Chap. 10) or interspecific hybrids. For a discussion on the subject, reference is made to Brink and Cooper (1947), Malecka (1967), and Jankun (1973).

There are three general types of endosperm formation, namely: Nuclear, Cellular, and Helobial. For a detailed description of these endosperm types, see Chapter 7.

11.6.2 Polyploidy in the Endosperm

Polyploidy is of common occurrence in endosperm. In Table 11.4 the maximum ploidy levels of endosperm nuclei have been reported. Although chromosome endoreduplication (Er) is a common process of polyploidization in the endosperm, there are examples of species where the endosperm differentiates in the absence of endoreduplication. Thus, the endosperm of *Nigella damascena, N. sativa* (Kapoor 1966), *Phytolacca acinosa, Rivina humilis,* and *Solidago virgaurea* (Enzenberg 1961) remain 3n, and those of *Gagea lutea* (Geitler 1948) and *Lilium* sp. (Swift 1953) remain 5n.

Table 11.4. Maximum ploidy level and methods of polyploidization of nuclei in the endosperm

Species (family)	Structures involved in polyploidization	Maximum ploidy level	Methods of polyploidization	References
Alisma lanceolatum (Alismataceae)	E	6n, 12n	Er	Hasitschka-Jenschke (1959a), Frey (1966)
Allium ursinum (Liliaceae)	E	24n	Er	Geitler (1955), Turala (1966b)
Anemone nemorosa (Ranunculaceae)	E	32n[a]	Er	Trela (1963)
Arum maculatum (Araceae)	EP	3n		Erbrich (1965)
Arum maculatum (Araceae)	CH, 1-nucleate	24, 576n	Er	Erbrich (1965)
Bartsia alpina (Scrophulariaceae)	CH, 2-nucleate	96n	Er	Erbrich (1965)
Bartsia alpina (Scrophulariaceae)	MH, 4-nucleate	768n	Er	Erbrich (1965)
Borassus flabellifer (Palmaceae)	E	96n	Er	Stephen (1974)
Bryonia dioica (Cucurbitaceae)	EP	48n	Er	Turala-Szybowska (1974)
Bryonia dioica (Cucurbitaceae)	CH, coenocytic	6n	Er	Turala-Szybowska (1974)
Campanula pyramidalis (Campanulaceae)	CH, 1-nucleate	96n	Er	Enzenberg (1961)
Campanula pyramidalis (Campanulaceae)	MH, 2-nucleate	96n	Er	Enzenberg (1961)
Citrullus colocynthis (Cucurbitaceae)	EP	96n	Er	Turala (1969)
Citrullus colocynthis (Cucurbitaceae)	CH, coenocytic	12n	Er	Turala (1969)
Codonopsis clematidea (Campanulaceae)	CH and MH both 2-nucleate	192n	Er	Erbrich (1965)
Cucumis sativus (Curcurbitaceae)	EP	24n	Er	Enzenberg (1961)
Cucumis sativus (Curcurbitaceae)	CH, coenocytic	48n	Er	Enzenberg (1961)
Cucurbita pepo (Cucurbitaceae)	EP	384n	Er	Turala-Szybowska (1974)
Cucurbita pepo (Cucurbitaceae)	CH, coenocytic	24n		Turala-Szybowska (1974)
Cyclanthera pedata (Cucurbitaceae)	EP	192n	Er	Turala (1969)
Cyclanthera pedata (Cucurbitaceae)	CH, coenocytic	24n	Er	Turala (1969)
Daphne mezereum (Thymelaeaceae)	E	12n	RM	Enzenberg (1961)
Delphinium kotulae (Ranunculaceae)	E	192n	RM and SF+Er	Jankun (1970)
Ecballium elaterium (Cucurbitaceae)	EP	192n	Er	Turala (1971)
Ecballium elaterium (Cucurbitaceae)	CH, multicellular	3n		Turala (1971)
Echinocystis lobata (Cucurbitaceae)	EP, central	3072n	Er	Turala (1966a)
Echinocystis lobata (Cucurbitaceae)	EP, chalazal	48n	Er	Turala (1966a)
Echinocystis lobata (Cucurbitaceae)	CH, coenocytic	24n	Er	Turala (1966a)
Echium vulgare (Boraginaceae)	EP	12n	Er	Malecka (1975)
Echium vulgare (Boraginaceae)	UEC	96n	Er	Malecka (1975)

Echium vulgare (Boraginaceae)	LEC	24n	Er	Malecka (1975)
Euphrasia minima and *E. rostkoviana* (Scrophulariaceae)	EP	12n	Er	Erbrich (1965)
Euphrasia minima and *E. rostkoviana* (Scrophulariaceae)	CH, 2-nucleate	96n	Er	Erbrich (1965)
Euphrasia minima and *E. rostkoviana* (Scrophulariaceae)	MH, 4-nucleate	192n	Er	Erbrich (1965)
Geranium phaeum (Geraniaceae)	E	384n	Er	Nagl (1962)
Hypericum olympicum (Hypericaceae)	E	12n	RM	Enzenberg (1961)
Iris bucharica, I. orchioides and *I. reticulata* (Iridaceae)	E	192n	RM	Enzenberg (1961)
Melampyrum cristatum (Scrophulariaceae)	MH, 8-nucleate	64n[b]	Er	Speta (1975)
Melampyrum nemorosum, M. pratense and *M. silvaticum* (Scrophulariaceae)	EP	48n	Er	Erbrich (1965)
Melampyrum nemorosum, M. pratense and *M. silvaticum* (Scrophulariaceae)	CH, 2-nucleate	1536n	Er	Erbrich (1965)
Melampyrum nemorosum, M. pratense and *M. silvaticum* (Scrophulariaceae)	MH, 16-nucleate	12n	Er	Erbrich (1965)
Nothoscordum fragrans (Liliaceae)	E	6n	Er, RM	Tandon and Kapoor (1963)
Pedicularis palustris (Scrophulariaceae)	EP	24n	Er	Steffen (1965)
Pedicularis palustris (Scrophulariaceae)	CH, 2-nucleate	96n, 192n	Er	Steffen (1956), Erbrich (1965)
Pedicularis palustris (Scrophulariaceae)	MH, 4-nucleate	384n	Er	Steffen (1956), Erbrich (1965)
Phaseolus coccineus (Papilionaceae)	E	192n	Er	Nagl (1970b)
Phaseolus vulgaris (Papilionaceae)	E	96n	Er	Nagl (1970b)
Phlomis viscosa (Labiatae)	EP	12n	Er	Enzenberg (1961)
Phlomis viscosa (Labiatae)	MH, 2-nucleate	384n	Er	Enzenberg (1961)
Plantago atrata (Plantaginaceae)	EP	12n	Er	Czapyka-Dziekanowska (1965)
Plantago atrata (Plantaginaceae)	CH, 2-nucleate	1536n	Er	Czapyka-Dziekanowska (1965)
Plantago atrata (Plantaginaceae)	2MH, both 1-nucleate	384n	Er	Czapyka-Dziekanowska (1965)

[a] Endosperm development without fertilization (autonomous)
[b] Primary endosperm nucleus: 2n (micropylar polar nucleus fertilized by male gamete)
Abbreviations: *CH* chalazal haustorium, *E* endosperm, *EP* endosperm proper, *Er* endoreduplication, *LEC* lower endosperm chamber, *MH* micropylar haustorium, *RM* restitutional mitosis, *SF* spindle fusion, *UEC* upper endosperm chamber

Table 11.4 (continued)

Species (family)	Structures involved in polyploidization	Maximum ploidy level	Methods of polyploidization	References
Potamogeton densus (Potamogetonaceae)	E	192n	Er	Hasitschka-Jenschke (1959a)
Rhinanthus alectorolophus (Scrophulariaceae)	CH, 2-nucleate and MH, 4-nucleate	384n	Er	Tschermak-Woess (1957b)
Salvia coccinea, S. pratensis and *S. splendens* (Labiatae)	E	48n	Er	Erbrich (1965)
Sicyos angulata (Cucurbitaceae)	EP	384n	Er	Turala (1969)
Sicyos angulata (Cucurbitaceae)	CH, coenocytic	24n	Er	Turala (1969)
Thesium alpinum and *T. linophyllum* (Santalaceae)	EP	3n		Erbrich (1965)
Thesium alpinum and *T. linophyllum* (Santalaceae)	CH, 1-nucleate	384n	Er	Erbrich (1965)
Tozzia alpina (Scrophulariaceae)	CH, 2-nucleate	192n	Er	Erbrich (1965)
Tozzia alpina (Scrophulariaceae)	MH, 4-nucleate	384n	Er	Erbrich (1965)
Tradescantia sp. (Commelinaceae)	E	24C	Er	Swift (1953)
Zea mays (Graminaceae)	E	384n	Er	Tschermak-Woess and Enzenberg-Kunz (1965)
Zephyranthes lancasteri (Amaryllidaceae)	E	12n	RM	Kapoor and Tandon (1964)

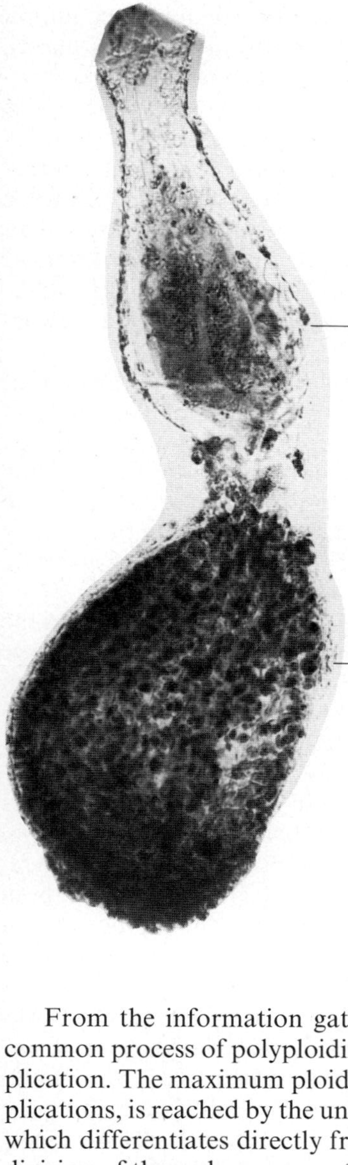

Fig. 11.4. *Phlomis viscosa,* excised embryo sac (the embryo was removed) with endosperm proper (cellular), and micropylar haustorium with two giant nuclei. (After Enzenberg 1961, courtesy E. Tschermak-Woess, Vienna)

From the information gathered in Table 11.4, it is apparent that the most common process of polyploidization in the endosperm is chromosome endoreduplication. The maximum ploidy level, viz., 24,576 corresponding to 13 endoreduplications, is reached by the uninucleate chalazal haustorium of *Arum maculatum,* which differentiates directly from the primary chalazal cell formed after the first division of the endosperm mother cell. These giant nuclei are filled with chromosomes in a prophase-like condition; no giant chromosomes have been seen (Erbrich 1965). Giant (polytene) chromosomes or chromosome bundles are, on the other hand, seen in the endosperm nuclei of *Allium ursinum* (Turala 1966b), *Loasa papaverifolia* (Hasitschka-Jenschke 1962), *Phlomis viscosa* (Enzenberg 1961) (Fig. 11.4), *Rhinanthus alectorolophus* (Tschermak-Woess 1957b), *Salvia coccinea, S. pratensis, S. splendens, Thesium alpinum* (Erbrich 1965), and *Zea mays* (Duncan and Ross 1950, Tschermak-Woess and Enzenberg-Kunz 1965). When countable, the polytene chromosomes in the endosperm cells are triploid.

It must be recalled that even in these species the polytene chromosome condition is not the exclusive nuclear structure but, as is generally the case in plants, polytene nuclei may show a variety of nuclear structures (Nagl 1978).

An analysis of the maximum ploidy levels in complex endosperms, viz., those consisting of the endosperm proper and one or two haustoria, offers some hints of their reciprocal relations. In the Cucurbitaceae, for which considerable information is available at present (Table 11.4), the most common trend seems to be a higher, or much higher, level of polytenization in the endosperm proper than in the haustoria. The opposite condition is, however, reported in *Cucumis sativus* (Enzenberg 1961) and *Tozzia alpina* (Erbrich 1965). How frequently this condition of low ploidy in the endosperm proper and high ploidy in haustoria may occur remains to be investigated.

In the family Scrophulariaceae, which is characterized by the occurrence of both chalazal and micropylar haustoria (Maheshwari 1950), the chalazal and micropylar haustoria are generally two-nucleate and four-nucleate, respectively. In the species so far investigated (Table 11.4), the ploidy level of the nuclei of the endosperm proper is clearly lower than that of the haustorial nuclei. An example of the inverse relation between number of nuclei in a coenocyte and their ploidy level is offered by the micropylar haustorium of *Melampyrum,* a 16-nucleate coenocyte with a total ploidy of 192n ($= 12n \times 16$) (Table 11.4).

As to the other species with endosperm haustoria, which are reported in the Table, they present isolated examples within the families. These cases clearly serve the purpose of enlarging the picture of the role of polytenization in endosperm differentiation and function. An indirect aspect of the possible physiological significance of cell endoreduplication appears to be the gradient in ploidy levels, which has been observed in some endosperms. In 1939 Wanscher had already observed in the endosperm of the apple "Woldike Pigeon" (*Pyrus malus*) two zones of increasing polyploidy (3n, 6n, 12n) which were symmetrically distributed from the micropylar and chalazal ends to the central zone of the embryo sac. More recently, other examples of gradients of polyploidization in endosperm have been reported. In *Zea mays,* after a phase of free-nuclear division throughout the endosperm, mitotic activity is retained by only the peripheral cells after 16 to 18 days after pollination. At the end of mitotic divisions, the endosperm nuclei undergo chromosome endoreduplication attaining high ploidy levels (Table 11.4). It was observed by Duncan and Ross (1950) that nuclei in the adgerminal layer of surface cells were generally smaller than those on the abgerminal layer. In the endosperm proper of *Bryonia dioica,* different ploidies are distributed as follows: the peripheral cells are 3n, the chalazal cells are 12n, and the central cells are 24n or 48n (Turala-Szybowska 1974). Finally, in the endosperm of *Ecballium elaterium,* the central cells undergo endoreduplication up to the 192n level, while the peripheral cells, which polyploidize through restitutional mitosis or spindle fusion, attain only the 12n level (Turala 1971).

From Table 11.4 it might appear that restitutional mitosis and spindle fusion would represent a polyploidization process rarely operating in the endosperm. It must, however, be emphasized that such phenomena have been observed in the endosperm of a number of species and interspecific hybrids (Brink and Cooper 1947, Maheshwari 1950, Jankun 1973), but no attempt was generally made to evaluate ploidy levels. Since restitutional mitoses and fusions of metaphase and/ or anaphase plates tend to occur at the last mitoses preceding differentiation of endosperm cells, the polyploidy resulting from these disturbed mitoses might be further increased by endoreduplication cycles. Thus, in the endosperm of *Delphinium kotulae* disturbed mitoses produce 24n nuclei which undergo further polyploidization up to the 192n level via chromosome endoreduplication (Jankun 1970).

11.7 Embryo

After syngamy the zygote undergoes a rest period of different duration in different species, probably depending for the most part on the DNA replicative state of the nuclei of male and female gametes, and of the primary endosperm nucleus. Among the Gramineae, characterized by a rather short resting period of the zygote, it has been reported that the trinucleate pollen grains of *Hordeum vulgare* (D'Amato et al. 1965) and of seven species of *Lolium* (Manthriratna and Hayward 1973) bear both the vegetative and the two sperm nuclei in the DNA-doubled condition (2C) corresponding to the DNA postsynthesis (G_2) phase of the haploid cell cycle. Since the interval between fertilization and division of the endosperm nucleus and the zygote is short, it seems probable that both the egg nucleus and the secondary nucleus of the embryo sac have undergone DNA duplication before being fertilized. However, this assumption needs confirmation by direct DNA measurements. Whether fusion of reproductive cells in the DNA-doubled state is a more general phenomenon than hitherto assumed, both in plants and animals, awaits further investigations (D'Amato 1977a).

The division of the zygote, which is generally unequal and transversal, forms the two daughter cells; the chalazal (apical) cell is destined to produce the embryo proper while the micropylar (basal) cell, apart from some exceptions, produces the suspensor.

11.7.1 Embryo Proper

At the time of seed shedding, the embryo has generally attained its characteristic differentiation into three main parts, viz., root tip, stem tip, and cotyledons (or cotyledon). Highly differentiated embryos occur in the Gramineae, in which the embryo also possesses a coleoptile, a coleorrhiza, the first internode (mesocotyl), and the primordia of some adventitious roots (Esau 1965). By contrast, the embryo of the Orchidaceae is quite rudimentary at the time of seed shedding. For this reason, the term embryo is retained to describe the developmental stages between germination and the green seedling. In *Vanda sanderiana* the embryo at germination appears as an ovoid structure which consists of three histologically distinct regions: meristematic, parenchymatous, and suspensor. The rudimentary suspensor consists of 8 or 16 cells and degenerates soon after germination. The parenchymatous portion apparently differentiates from the meristematic region

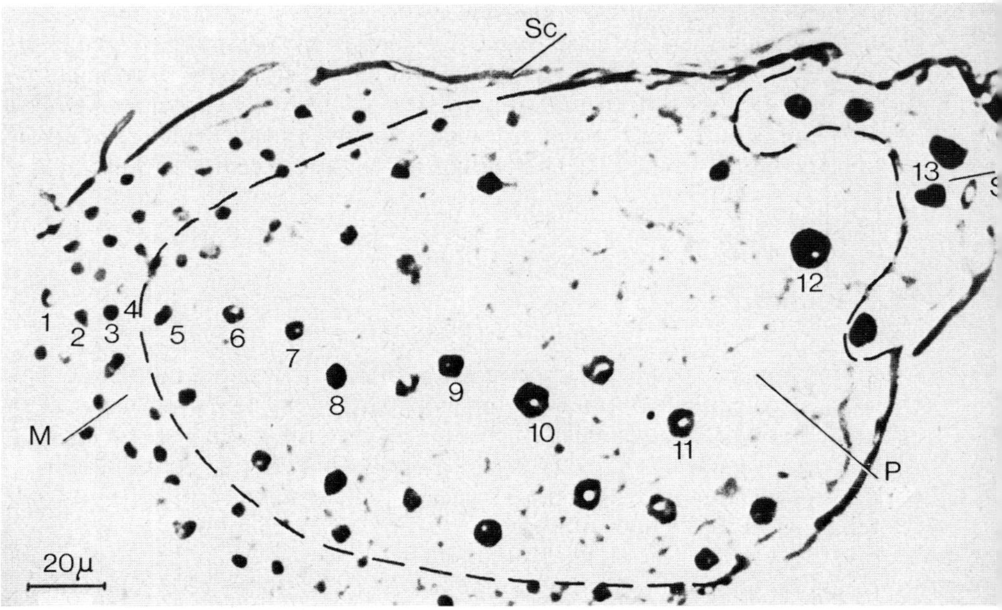

Fig. 11.5. *Vanda sanderiana,* longitudinal section of embryo at germination in which meristematic (*M*), parenchymatous (*P*), and suspensor (*Su*) regions, and adhering seed-coat (*Sc*) fragments are seen. *Numbered* nuclei were measured for DNA content. Note increase in nuclear volume with distance from meristem. (After Alvarez 1968, courtesy M. R. Alvarez, Tampa)

and shows a gradual increase in cell and nuclear volume with increasing distance from the meristem (Fig. 11.5). DNA cytophotometric measurements have shown a close correlation between nuclear volume and DNA content: the parenchymatous nuclei can be classified as 2C, 4C, and 8C and higher. Since these last DNA contents, which are the least numerous, occur in unequal fractional multiples of the C value (Alvarez 1968), it is probable that the parenchymatous cells undergo DNA extrareplication (amplification) after completing two rounds of chromosome endoreduplication (=8C).

In species of angiosperms with well-differentiated mature embryos, chromosome endoreduplication also occurs in those regions of the embryo proper which comprise differentiating and/or adult tissues, e.g., cotyledon, scutellum, hypocotyl, mesocotyl. Tetraploid mitoses with diplochromosomes or "chromosome pairs" were first observed during the development of the seedling in the mesophyll of the cotyledon in *Allium cepa* (Berger and Witkus 1946), and at the base of the cotyledons in *Brassica oleracea* (Netroufal 1927). More recently, data on the polytenization of cotyledonary cells, especially in the storage parenchyma, have been presented for a few species.

By using DNA cytophotometry, Smith (1971) showed that, in seeds of *Pisum arvense* pregerminated for 1 day, the epidermal cells of the cotyledons were diploid (2C nuclear DNA content), the hypodermal nuclei varied from 2C to 4C, with most at about 4C; the outer storage-cell nuclei wre mainly in the range 4C to 8C, with the majority at the higher level, and the inner storage-cell nuclei ranged from 8C to 16C, with the majority apparently between 8C and 16C. That cotyledonary epidermal cells remain predominantly diploid is also known for the cotyledons of *Pisum sativum* (Scharpe and van Parijs 1973) and *Gossypium hirsutum* (Walbot and Dure 1976). In *Pisum sativum* mitotic growth of the cotyledons ended at 12 days from flowering, when growth by chromosome endoreduplication began. At full development, the maximum ploidy level, viz., 64C DNA content corresponding to five endoreduplication cycles, was present in the most centrally located cells of the cotyledon parenchyma.

In *Gossypium hirsutum* cotyledonary growth through mitosis ended at 27 days after anthesis; in full-grown cotyledons the average DNA-content per nucleus (as measured by chemical analysis) corresponded to ca. 12C. Since the diploid epidermal cells formed about one-third of all cotyledonary cells, the average DNA-content per parenchymatous cell was calculated as 16C. Since analyses of DNA-DNA reassociation and DNA-RNA hybridization demonstrated that in polyploid cotyledonary nuclei the different DNA sequences were represented in equal proportion as in diploid cells, Walbot and Dure (1976) concluded that polyploidization of cotton cotyledonary cells was due to "endoreduplication of the entire genome", thus excluding selective amplification of certain sequences. Biochemical data demonstrating endoreduplication to the 16C level and excluding DNA amplification had previously been presented by Millerd and Whitefeld (1973) for the cotyledons of *Vicia faba*. DNA values up to 32C and higher have been measured in the immature cotyledons of the variety Aguadulce of *V. faba* (Cionini et al. 1978). A maximum DNA value of 32C is also known for the scutellum of *Zea mays* (Swift 1953). Finally, 8C, 16C, and 32C DNA values have been measured in the parenchymatous cells of *Phaseolus vulgaris* cotyledons (Smith 1974).

Very recently, Marks and Davies (1979) have shown that by appropriate culture techniques the highly polyploid nuclei (up to 64C DNA content) of *Pisum sativum* cotyledons can be induced into division and reach prophase and metaphase. Polytene chromosomes in the diploid number ($2n = 14$) are then observed. As pointed out by Marks and Davies (1979), the appearance of polytene chromosomes might prove useful for molecular biological analyses on cotyledons.

In contrast with the species discussed above, all nuclei in the cotyledons of mature seeds of *Lactuca sativa* rest in the 2C condition corresponding to the DNA presynthesis phase (G_1) of the diploid cell cycle (Brunori and D'Amato 1967). It is probable that in other species of angiosperms, which differentiate their tissues in a purely diploid state (D'Amato 1977 a), cotyledons at maturity remain diploid. Indeed, in *Pinus pinea,* a species whose tissues differentiate in the diploid state, mature cotyledons wholly consist of 2C cells (Brunori and D'Amato 1967).

Isolated cases of polyploidy due to chromosome endoreduplication have been reported in such differentiated embryonic regions as the epicotyl, hypocotyl, and mesocotyl. The following examples may be cited: the epicotyl of *Cucumis melo* (Ervin 1941), the hypocotyl of *Allium cepa* (Berger and Witkus 1946), and the mesocotyl of *Triticum durum* (D'Amato et al. 1980).

The treatment of polyploidy in embryos would be incomplete if the phenomenon of polysomaty (contemporary occurrence of diploid and polyploid cells) in the shoot and/or root meristems in the mature embryo of some angiosperms were

Table 11.5. Maximum ploidy level of nuclei in embryo suspensor cells; polyploidization is due to chromosome endoreduplication

Species (family)	Maximum ploidy level	References
Alisma lanceolata (Alismataceae)	128n	Hasitschka-Jenschke (1959a)
A. plantago-aquatica (Alismataceae)	1024n	Bohdanowicz (1973)
Cucubalus baccifer (Caryophyllaceae)	128n	Nagl (1962)
Dianthus chinensis (Caryophyllaceae)	128n	Nagl (1962)
Echinodorus tenellus (Alismataceae)	128n	Nagl (1962)
Eruca sativa (Cruciferae)	75C	Corsi et al. (1973)
Geranium phaeum (Geraniaceae)	32n	Nagl (1962)
Lupinus sp. (Papilionaceae)	64n	Geitler (1941)
Melandrium album, M. rubrum (Caryophyllaceae)	128n	Nagl (1962)
Phaseolus acutifolius (Papilionaceae)	1024n	Nagl (1974)
P. coccineus (Papilionaceae)	8192C	Brady (1973)
P. hysterinus (Papilionaceae)	4096n	Nagl (1974)
P. lunatus (Papilionaceae)	256n	Nagl (1974)
P. mungo (Papilionaceae)	512n	Nagl (1974)
P. tuberosus (Papilionaceae)	256n	Nagl (1974)
P. vulgaris (Papilionaceae)	4096C	Nagl (1974)
Potamogeton densus (Potamogetonaceae)	128n	Hasitschka-Jenschke (1959a)
Sophora flavescens (Papilionaceae)	16n	Nagl (1962)
Trapa natans (Onagraceae)	256n	Nagl (1962)
Tropaeolum majus (Tropaeolaceae)	2048C	Nagl (1976b)
Tunica saxifraga (Caryophyllaceae)	32n	Nagl (1962)
Vigna capensis (Papilionaceae)	256n	Nagl (1974)

not considered. In these cases, polyploidy originates via endoreduplication cycles: one, two or three to produce 4n, 8n, and 16n cells, respectively (review in D'Amato 1952).

11.7.2 Suspensor

The development of a suspensor, a short-lived differentiated structure, in association with the embryo proper, is a very common, though not universal, feature of embryogenesis in angiosperms. An enormous variation in suspensor structure occurs both between and within species (Maheshwari 1950, Nagl 1962, 1974): from rudimentary suspensors, without any apparent role during embryogenesis, to suspensors whose cells remain diploid, to more or less complex and massive suspensors characterized by conspicuous cell growth through chromosome endoreduplication or, much less frequently, multinucleation. Already in 1919 Tison, and later Schnarf (1929), called attention to the scarcity of endosperm development in species characterized by massive suspensors, and suggested that in these species the suspensor might take over the functions of the endosperm. As we will see in the following pages, recent work on suspensors strongly supports this old suggestion. Although the greatest attention has been devoted to the most favorable material, the suspensor of *Phaseolus coccineus,* the other investigations on the nuclear cytology of suspensors of other species have provided useful informa-

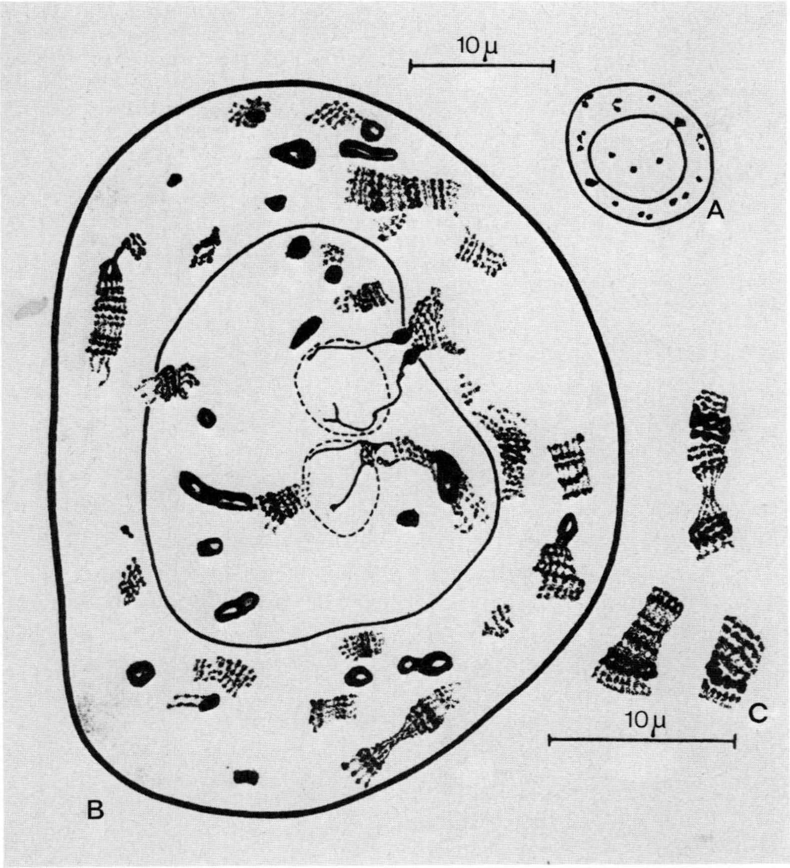

Fig. 11.6 A–C. *Potamogeton densus.* **A** Diploid nucleus from the embryo. **B** Presumably 64n nucleus from the basal cell of the suspensor with banded polytene chromosomes. **C** Segments of polytene chromosomes with banded structure. (After Hasitschka-Jenschke 1959a)

tion which helps to better realize the variety of structures and the possible developmental significance of the suspensor. In Table 11.5, the maximum level of ploidy attained by suspensor cells has been reported. In all the species listed in the table, polyploidization occurs via chromosome endoreduplication, which remains the most common process of cell polyploidization in suspensors. Indeed, many other suspensors which have been described in literature (cf. Maheshwari 1950, Viegi et al. 1976), provide indications for chromosome endoreduplication in their nuclei.

In *Alisma plantago-aquatica* (Bohdanowicz 1973), *A. lanceolata, Potamogeton densus* (Hasitschka-Jenschke 1959a), and *Echinodorus tenellus* (Nagl 1962), as in the other Helobiae, the basal larger cell, which originates by unequal transverse division of the zygote, grows directly into a large unicellular suspensor. Its nucleus undergoes up to six endoreduplications (128n) in *A. lanceolata* and *P. den-*

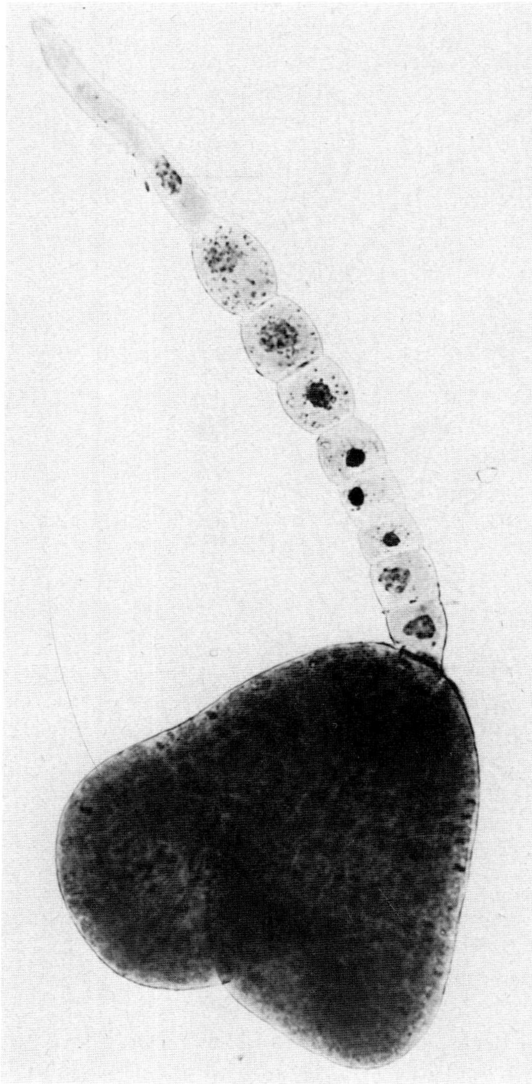

Fig. 11.7. *Eruca sativa,* heart-shaped embryo with suspensor. Note the size differences of nuclei in the terminal cell and in the two immediately preceding cells. (After Corsi et al. 1973, courtesy G. Corsi, Pisa)

sus, and nine endoreduplications (1,204n) in *A. plantago-aquatica*. Here, the endoreduplication process occurs very rapidly before endosperm development, suggesting that the suspensor replaces functionally the endosperm in the first stages of embryogenesis (Bohdanowicz 1973). The highly endoreduplicated nuclei of the *Potamogeton* suspensor frequently show polytene chromosomes with a clearly banded structure (Fig. 11.6) while the highly endoreduplicated nuclei of *Alisma* are most frequently characterized by more or less lax chromosome bundling, endochromocenters, or an overall granular structure of their chromatin.

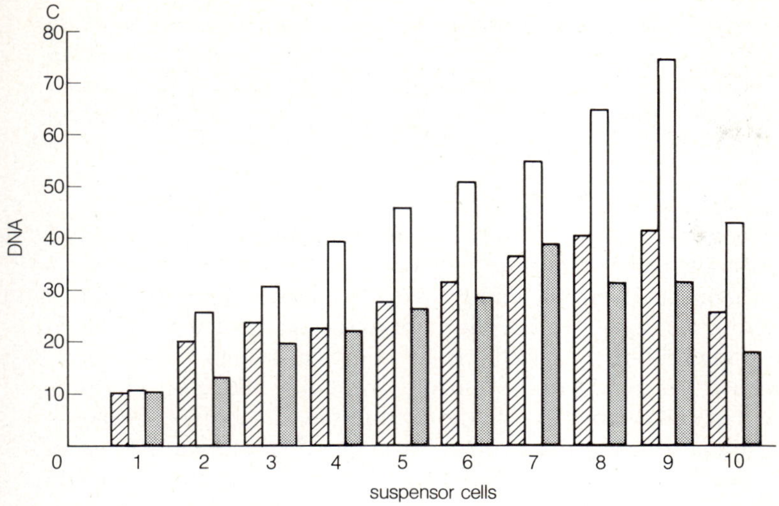

Fig. 11.8. *Eruca sativa,* DNA values (arbitrary units) of the different suspensor cells belonging to the globular (*dashed bars*), early heart-shaped (*open bars*) and heart-shaped (*dotted bars*) embryo stage. (After Corsi et al. 1973)

Direct differentiation of the basal cell into a large suspensor cell with haustorial function also occurs in *Melandrium album* and *M. rubrum,* but here the giant cell is superposed on a filament of seven cells (suspensor stalk), which are of proembryonic origin as in all Caryophyllaceae. Six of these cells remain diploid, the other cell (adjoining the suspensor basal cell) becomes 4n. The basal cell reaches the maximum level of endoreduplication (128n) when the embryo consists of about 100 cells (Nagl 1962).

In *Eruca sativa,* as in other species of Cruciferae, the suspensor is produced by the divisions of the basal cell which forms after the division of the zygote. At full differentiation, the suspensor appears as a filament of cells; the number is usually 10, but may vary from 9 to 13 (Fig. 11.7). DNA cytophotometry in 10-celled suspensors has shown that the nuclear DNA content of embryogenic stages rises from the first, embryo-adjoining cell (ca. 10C) to cell 9 of the suspensor corresponding to the globular proembryo stage (41.6C) and the early heart-shaped stage (75C), and in cell 7 of the suspensor corresponding to the heart-shaped stage (39.2C). At all three developmental stages, the DNA amount of cell 10 is less than in the immediately preceding cells (Fig. 11.8). Measurements of DNA amounts in the very few suspensors with 11, 12, and 13 cells showed the same trend in endoreduplication. The level of endoreduplication is always highest in the last-but-one or last-but-two cell, and is invariably relatively less in the most distal cell. As pointed out by Corsi et al. (1973), the nongeometric increase of nuclear DNA amounts in suspensor cells may indicate processes of differential DNA replication. Moreover, the high DNA levels in suspensor cells of the globular and especially the heart-shaped embryo are indications of high metabolic activity, which might well be correlated with some fostering action of the suspensor on the developing early embryo (Corsi et al. 1973).

Among species with massive suspensors, a brief mention may first be made of the suspensors of *Sophora flavescens* and *Geranium phaeum*. Their cells attain the maximum ploidy of 16n and 32n, respectively (Nagl 1962). In *Geranium phaeum,* 16n and 32n cells make the central portion of the suspensor, flanked on both sides by a zone of 4n cells which are followed, toward both the micropylar and embryo end, by two zones of 2n cells.

Massive suspensors also develop in *Trapa natans* and *Tropaeolum majus*. In *Trapa natans* (Tison 1919, Nagl 1962) the suspensor consists of several cell filaments which push the embryo into the embryo sac cavity. Of the two suspensor appendages, which start from the embryo–suspensor boundary, the shorter appendage runs along the dorsal wall of the embryo sac. The much more developed appendage, after crossing the nucellar tissue complex, Tison's "talon nucellaire", reaches the vascular bundle in the funiculus, probably taking up a haustorial function. The maximum ploidy level estimated in the *T. natans* suspensor is 256n, the nuclei being of the endochromocenter structure (Nagl 1962). In *Tropaeolum majus,* with a much-reduced endosperm development, the suspensor is made up of three threads which radiate from a cell group lying outside the micropyle: one suspending the embryo into the endosperm cavity (suspending thread), one running along the ovule and penetrating the carpel wall a little opposite the chalaza (dorsal thread or carpel haustorium), and the third growing through the integument and placenta to the point of entry of the vascular bundle of the raphe (Walker 1947, Nagl 1962, 1976b). The highest level of endoreduplication ascertained in the polytene nuclei of the suspensor ($2n=28$ polytene chromosomes) is 2,048C. Several elongated suspensor cells have nuclei deficient in heterochromatin, with rather low DNA values, which do not fall under the geometrical order of chromosome polytenization. This condition results from heterochromatin underreplication (Nagl 1976b, Nagl et al. 1976).

We now come to the most extensively investigated angiosperm suspensor, the *Phaseolus* suspensors. As seen from Table 11.5, suspensors with highly polytene cells occur in *P. coccineus* (Fig. 11.9), the species which has received most attention, and other species of *Phaseolus*. Comparatively lower levels of chromosome endoreduplication occur in some species of *Phaseolus* (Table 11.5) and in *P. vulgaris* cultivars with small suspensors (8n and 128n nuclei), and cultivars with medium-large suspensors (512n and 1,024n nuclei) have also been observed; even a *Phaseolus* species in which a suspensor is not formed has been described (Nagl 1974).

The embryo suspensor of *P. coccineus* was first described by Guignard (1882) in his thesis, and then reinvestigated by Nagl (1962). The literature, nearly 40 papers, on the *P. coccineus* suspensor, which has accumulated between 1962 and 1974, has been reviewed by Nagl (1974). In the last 6 years, knowledge of the *Phaseolus* suspensor has progressed further. The suspensor of *P. coccineus* is produced by repeated division of the basal cell resulting from the division of the zygote (the apical cell will form the embryo proper). When completely formed, the suspensor consists of 200 ± 25 cells which undergo chromosome endoreduplication, the level progressively increasing from the embryonal to the basal pole. At maximal growth, the suspensor is a club-shaped structure (Fig. 11.10). The "handle" portion, made of 4C to 128C cells (Bennici et al. 1976), grades into the

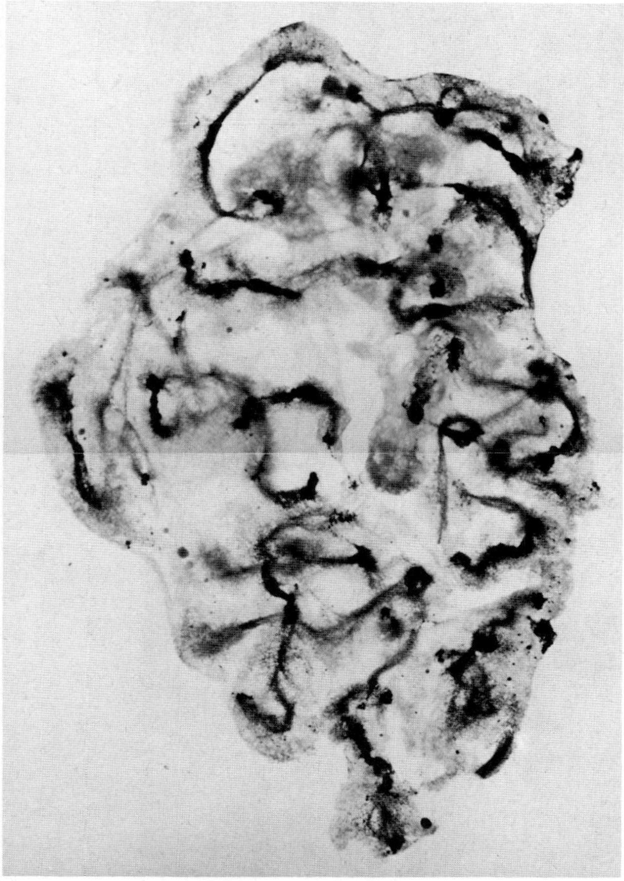

Fig. 11.9. *Phaseolus coccineus*, highly polytene chromosome nucleus of a suspensor. (After Schweizer and Ambros 1979, courtesy D. Schweizer, Vienna)

"knob" portion whose terminal cells attain the highest level of polytenization, viz., up to 12 endoreduplications, or 8,192C DNA amount (Brady 1973). The nuclei of the giant suspensor cells are in a permanent polytene condition. The polytene chromosomes are in the diploid number (2n = 22, no somatic pairing) and show heterochromatic and euchromatic segments which, together with centromere position and chromosome arm length, have helped in drawing the idiogram of the eleven giant chromosomes (Nagl 1967, Forino et al. 1979, Schweizer and Ambros 1979).

The highly polytene structure has favored the application, to the *Phaseolus* suspensor, of such techniques as specific staining for nucleic acids and proteins, autoradiography, quantitative DNA determination at the nuclear and chromosomal level, RNA-DNA hybridization in situ, etc. These have been used very successfully in analyzing the structure and function of Dipteran polytene cells. In situ rRNA-DNA hybridization has not only allowed the localization of the polytene chromosome regions bearing the genes for the four classes of ribosomal RNA (rRNA), viz., 25S, 18S, 5.8S, and 5S (Avanzi et al. 1971, 1972, Brady and Clutter

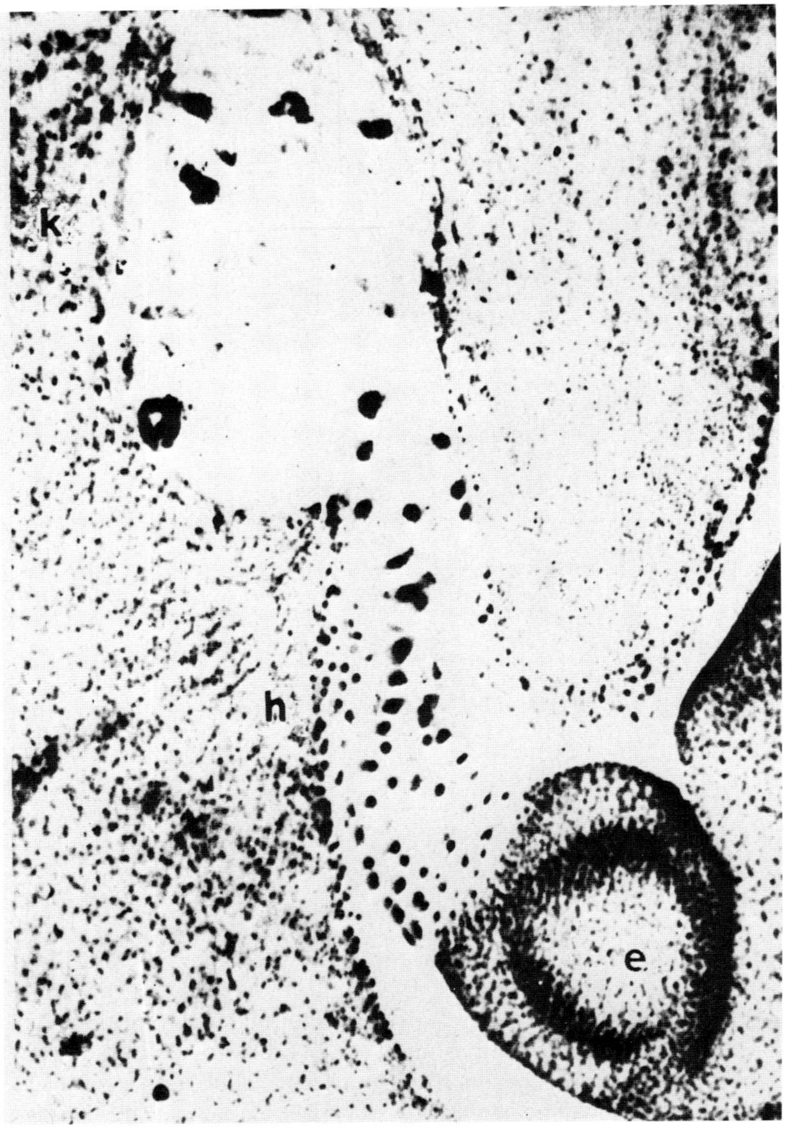

Fig. 11.10. *Phaseolus coccineus*, longitudinal section of a developing seed showing the embryo (*e*) and the club-shaped suspensor with its "handle" (*h*) and "knob" (*k*) portion, where highly polytene nuclei occur

1972, Durante et al. 1977), but also shown that rDNA (ribosomal genes)-containing supernumerary micronucleoli are produced at the nucleolus-organizing regions of polytene chromosome pair I(S_1) and released into the nucleoplasm and the cytoplasm (Avanzi et al. 1972). Moreover, the in situ rRNA-DNA hybridization has demonstrated that ribosomal genes are present not only in the nucleolus-organizing regions of the chromosome pairs I(S_1) and V(S_2), but also in the ter-

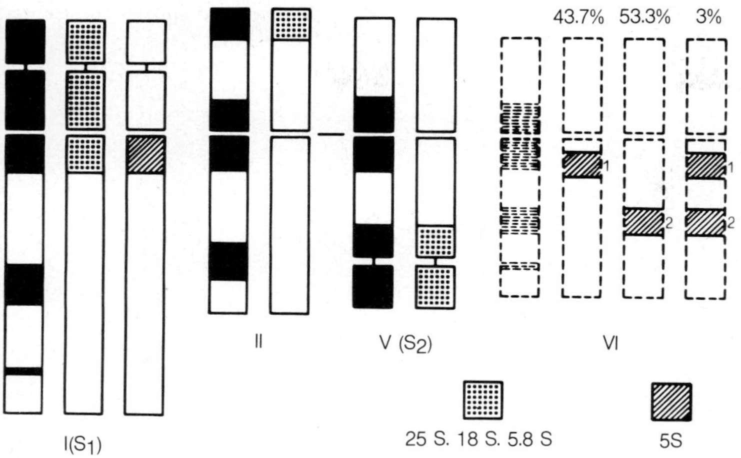

Fig. 11.11. *Phaseolus coccineus*, ideograms of polytene chromosomes *I* (S_1), *II*, *V*(S_2), and *VI* showing the localization of the genes for the four classes of ribosomal RNA (rRNA – *25S*, *18S*, *5.8S*, and *5S*). The frequency with which sites *1* and *2* of chromosome *VI* were labelled in the *5S* rRNA-DNA hybridization in situ is indicated. The ideograms *on the left* present the distribution of euchromatin (*blank*) and heterochromatin (*black*, *dotted* in chromosome *VI*) as reported by Nagl (1967). (After Durante et al. 1977)

minal heterochromatic segment of chromosome pair II (Fig. 11.11). The presence, in *P. coccineus*, of three nucleolar chromosome pairs (Avanzi et al. 1972, Durante et al. 1977) is now confirmed by recent analyses by Schweizer and Ambros (1979).

Cytochemical and autoradiographic tests first indicated the occurrence of differential RNA synthesis (transcription) on polytene chromosome regions (Avanzi et al. 1970, Nagl 1970c), and the localized accumulation of extra DNA (amplification) resulting in "DNA puffs" (Avanzi et al. 1970). Extra DNA replication in localized chromosomal regions has been recently confirmed by an autoradiographic study (Cremonini and Cionini 1977). On the other hand, evidence for DNA amplification had been previously provided by a thorough analysis of the DNA extracted from suspensors, shoots, and roots of *P. coccineus*, which showed that suspensor cells contain a satellite DNA (density: 1.692 g/ml) which is not detectable in either shoot or root (Fig. 11.12). The investigation also revealed that the rRNA genes (rDNA) are underrepresented (underreplication) in the suspensor DNA as compared to shoot, root, and integument DNA's (Lima-de-Faria et al. 1975). The underreplication of rDNA shows that the rDNA-containing segments of *P. coccineus* are left behind to other chromosome regions during the polytenization process: a similar condition also applies to the polytene chromosomes of some Diptera (see D'Amato 1977a).

The course of the amplification-underreplication events has been studied at the chromosomal level in relation to embryogenic stages, both on the whole chromosome complement (Forino et al. 1976), and at the level of the two nucleolus-organizing polytene chromosome pairs I(S_1) and V(S_2) (Figs. 11.13–11.15). The analyses of Forino et al. (1979) have shown that: (1) differential DNA replication in bands bearing rRNA genes occurs in the two chromosome paris and

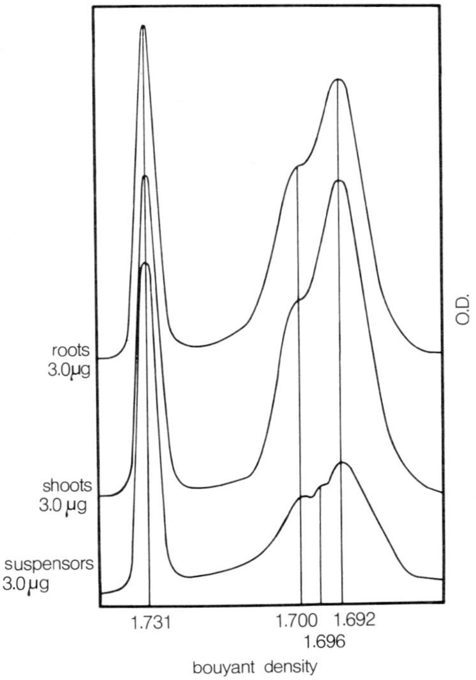

Fig. 11.12. *Phaseolus coccineus,* ultracentrifugation profile of the DNA isolated from suspensors, shoots, and roots. The suspensor DNA contains a satellite DNA (density: 1.696 g/ml) which is not detected in either shoots or roots. (After Lima-de-Faria et al. 1975)

within the same chromosome; (2) within chromosome S_1, there is a positive correlation between extrareplication (amplification) of band p 13 and underreplication of band p 12; (3) within chromosome S_2bands q 21 and q 22 often manifest a simultaneous amplification; (4) there is a negative correlation between amplification of band p 13 of chromosome S_1 and amplification of bands q 21 and q 22 of chromosome S_2; (5) within chromosome S_2 there is no correlation between amplification of bands p 11 and q 11, and amplification of bands q 21 and q 22. How far the observations of Forino et al. (1979) correlate with the functional activity of individual suspensor cells, or of suspensors taken as a whole, remains to be elu-

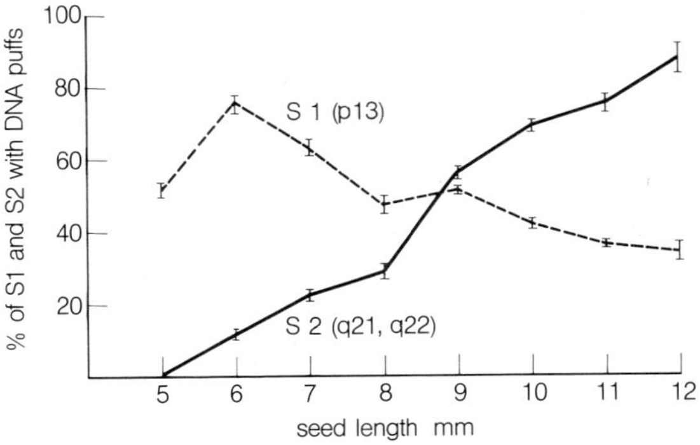

Fig. 11.13. *Phaseolus coccineus,* percentage of polytene chromosomes $I(S_1)$ (*dotted line*) and of polytene chromosomes $V(S_2)$ (*solid line*) showing DNA puffs in band p 13 and in bands q 21 and q 22, respectively, in seeds of different length. (After Forino et al. 1979)

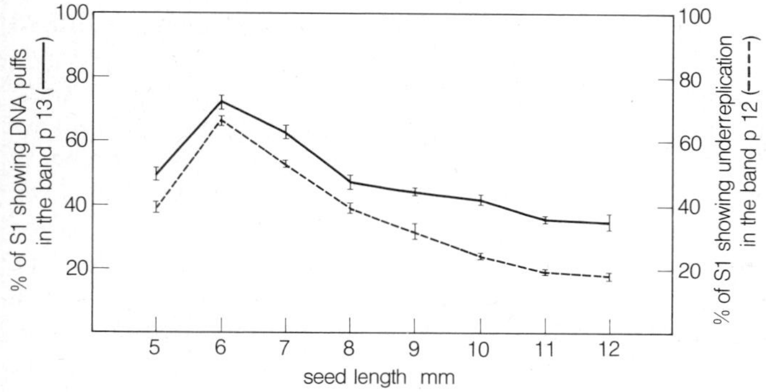

Fig. 11.14. *Phaseolus coccineus*, percentage of polytene chromosomes I(S_1) showing DNA amplification in band p 13 (*solid line*) and DNA underreplication in band p 12 (*dotted line*) in seeds of different length. (After Forino et al. 1979)

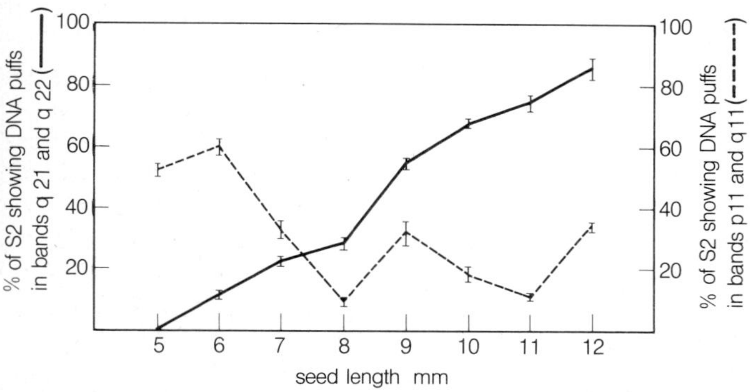

Fig. 11.15. *Phaseolus coccineus*, percentage of polytene chromosomes V(S_2) showing DNA puffs in bands q 21 and q 22 (*solid line*) and in bands p 11 and q 11 (*dotted line*) in seeds of different length. (After Forino et al. 1979)

cidated. In this connection, it is worth considering that (1) in the polytene suspensor cells of *P. coccineus* the rate of RNA synthesis (transcription) is hundreds of times that of the embryonic cells, and (2) the transcriptional activity per unit amount of DNA in the suspensor is twice that in the embryo (Clutter et al. 1974).

The above observations, together with the occurrence of wall ingrowths and several other structural and ultrastructural features (Nagl 1974, Yeung and Clutter 1978), suggest a major role of the *Phaseolus* suspensor in embryo development. Indications for a role of the suspensor in embryogenesis were first reported by Corsi (1972) who showed, with in vitro experiments, that the growth of *Eruca* embryos is stimulated by the presence of the suspensor. More recently, it has been observed that the suspensor of *P. coccineus* heart-shaped embryos is much richer in auxins (Alpi et al. 1975) and gibberellins than the embryo proper; the major free gibberellin of the suspensor is GA_1 (Alpi et al. 1975, Cionini et al. 1976, Alpi et al. 1979). Evidence has also been obtained that suspensors may provide gibberellins to the young embryo: suspensor-deprived heart stage embryos cultured on a medium supplemented with GA_3 concentrations of 10^{-6} to 10^{-8}M show a development comparable to that of intact embryos (with suspensor) of the same stage cultured on hormone-free medium (Cionini et al. 1976). High cytokinin activity has also been noted in the seeds and embryos of *P. coccineus* (Lorenzi et al. 1978) and *P. vulgaris* (Nesling and Morris 1979). In *P. coccineus* it has been noted that: (1) at the heart stage, the suspensor shows cytokinin activity at the level of zeatin (Z), 2-isopentenyladenosine (2iPA) and zeatin riboside (Zr), whereas the more polar cytokinins Zr and zeatin glucoside (Z_9G) are present in the embryo; (2) at the middle cotyledonary stage, when the embryo seems to become autonomous for cytokinin supply, there is a relative abundance of active cytokinins (Z, 2iPA) in the embryo to which Z_9G activity in the suspensor corresponds (Lorenzi et al. 1978). Since selfed *P. vulgaris* ovules have been reported to contain much higher levels of cytokinins than the *P. vulgaris* × *P. acutifolius* F_1 values (Nesling and Morris 1979), a reduced suspensor development in the hybrid embryo cannot be ruled out. Indeed, in crosses between *P. coccineus* and *P. vulgaris* embryo abortion is associated with functional abnormalities in the basal zone of the suspensor, including reduced RNA synthesis (Haq et al. 1973). The last point of interest concerning the hormonal embryo–suspensor relations is the recent demonstration that the heart-shaped embryos of *P. coccineus* react with an increased growth to zeatin and zeatin riboside concentrations which are not effective, or are only slightly effective, on middle cotyledonary embryos (Bennici and Cionini 1979).

Concerning suspensors made of multinucleate (coenocytic) cells, they were already described by Guignard (1882) in some Leguminosae, including *Pisum sativum*. In this species the suspensor consists of two superposed pairs of multinucleate cells which result from acytokinetic mitoses of the four initially uninucleate cells. The free-nuclear divisions, which are synchronous, end with the production of as many as 64 nuclei in the much elongated basal cells, and as many as 32 nuclei in the spherical subbasal (apical) cells (those adjoining the embryo) (Cooper 1938).

A four-celled suspensor made of two multinucleate tubelike basal cells and two multinucleate spherical apical cells (Fig. 11.16) also differentiate in four species of *Lathyrus* (Nagl 1962). In *L. vernus* the free-nuclear divisions frequently produce 32 nuclei (five mitotic cycles) in each of the two apical cells, and 64 nuclei (six mitotic cycles) in each of the two basal cells. The number of the mitotic cycles may, however, be much greater and some mitoses, especially at late stages of development, are restitutional ones. In both *L. vernus* and the other *Lathyrus* species studied by Nagl (1962), *L. latifolius, L. silvestris,* and *L. luteus,* differences occur in the number of mitotic cycles as well as in the number of mitoses ending in nuclear restitution and in their time of occurrence. This results in suspensor cells with irregular numbers of nuclei and different nuclear sizes, which become most clear in full-grown suspensors. Thus, the full-grown apical cells of the suspensor of *L. vernus* usually have 18–22 nuclei each, which are estimated to be about 64n and 128n, whereas in each of the basal cells, 20–35 probably 64n nuclei are common. This means that the total ploidy of each *Lathyrus* suspensor cell must be in the order of thousands: e.g., 2,240n (64n × 35) or 2,816n (128n × 22).

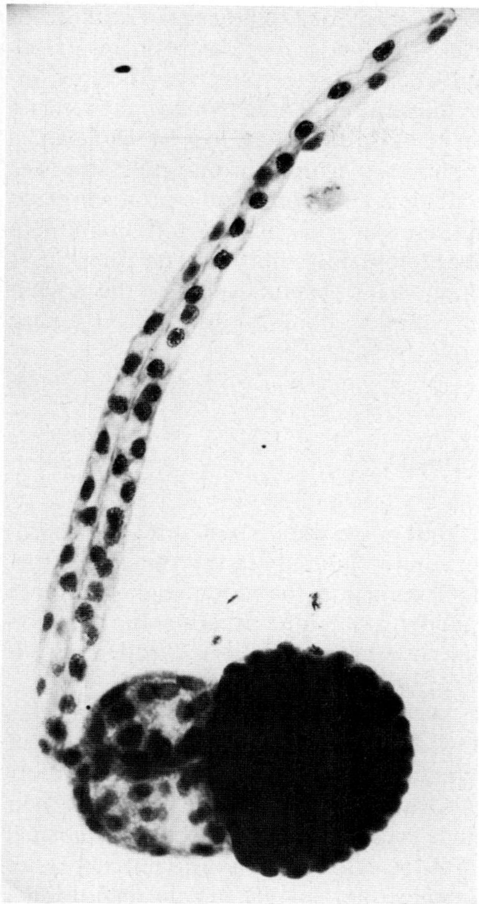

Fig. 11.16. *Lathyrus latifolius,* globular proembryo with suspensor consisting of four coenocytic cells, two spherical (apical) and two tubelike (basal). (After Nagl 1962, courtesy W. Nagl, Kaiserslautern)

As to suspensors made of diploid cells, information is meager. Nagl (1962) reports diploidy in the bunch-like suspensor of *Laburnum* sp., in the filamentous suspensor of *Verbena* sp., in the tubelike cells of the greatly ramified suspensor of *Galium* sp., and in the suspensor cells of *Ruellia* sp.

11.7.3 Nonzygotic Embryos

Nonzygotic embryos are dealt with in Chapter 9. Here, we will consider only the aspect of the increased ploidy which is frequently manifested by nonzygotic (adventitious) embryos which may be regenerated in vitro from single cells, either somatic or gonial (microspores and/or pollen grains): see the reviews of Murashige (1974) and D'Amato (1977 b, 1978 a). Limiting the discussion to microspore or pollen-derived embryos, species are now known in which not only haploid, but also diploid and polyploid embryos (and plants) (2n, 3n, 4n, 5n, 6n) have been obtained (cf. D'Amato 1977b; Table 2). These different ploidies may result from fusion of vegetative and generative nuclei in pollen grains, development of unreduced microspores produced by meiotic abortion in the pollen mother cell, fusion of spindles, and chromosome endoreduplication. These two last phenomena certianly combine in the production of 3n embryos in *Petunia hybrida* and *Datura innoxia*. The endoreduplicated generative nucleus (n diplochromosomes) and the vegetative nucleus (n chromosomes) divide on a common spindle producing two triploid nuclei. Failure of the spindle mechanism or chromosome doubling in a 3n nucleus can explain the formation of hexaploid (6n) embryos or 6n–3n mosaics which have been observed rarely. An important genetic aspect of the pollen-derived diploids and polyploids is their complete homozygosity (cf. D'Amato 1977b).

11.8 Physiological Role of Polyploidy

From the preceding pages it is apparent that polyploidy is widespread in reproductive organs and tissues. The occurrence of polyploidy in reproductive structures is just one aspect of a more general phenomenon, the widespread occurrence and possible physiological role of polyploidy in the differentiation and function of cells and tissues in the plant and animal: for reviews, reference is made to Brodsky and Uryvaeva (1977), D'Amato (1977a, 1978a), and Nagl (1978).

If we first consider the commonest method of cell polyploidization, viz., chromosome endoreduplication, we know that only a minority (about 20%) of the angiosperms investigated so far differentiate their cells and tissues in a purely diploid state, generally in the $2C(G_1)$ nuclear condition. This means that their diploid genome is able to perform the many cellular functions needed for normal development, growth, and reproduction of the organism. In these species, other methods of cell polyploidization (multinucleation, spindle fusion, restitutional mitosis) may sometimes operate in such ephemeral tissues as the anther tapetum and the endosperm where they produce relatively low ploidy levels. Whether the increased ploidy is really needed to enhance cellular activity in these tissues is not known, however. It might, therefore, appear that polyploidy is not essential in tissue or organ differentiation of higher plants (Evans and Van't Hof 1975).

Although polyploidy is not essential for cell differentiation and, in several instances, it appears as the consequence rather than the cause of cell differentiation (cf. D'Amato 1964, 1977a, Brodsky and Uryvaeva 1977), the multiplication of the genome brought about by polyploidization must be favorably reflected in the growth of the protoplast and its functional activity. Indeed, the widespread occurrence of polyploidy in the soma of plants and animals suggests that somatic polyploidy has been used in evolution as an adaptation for increasing the availability of templates to particular cells. Since the transcriptional and translational activity increases proportionately at each doubling of the genome (cf. Brodsky and Uryvaeva 1977, Nagl 1978), a highly polyploid cell is functionally equivalent to hundreds or thousands of diploid cells. Consequently, restriction of a given function to single or relatively few highly polyploid cells might well be economical for the organism, because it spares the whole machinery of the several to many mitotic divisions needed to produce the number of diploid cells functionally equivalent to one or a few highly polyploid cells. This is more so because of the spatial and temporal limitations which may be imposed, sometimes stringently, upon the growth by cell division of some tissues or organs.

Limiting our discussion to reproductive organs and tissues, we see that high polyploidy due to chromosome endoreduplication occurs in single cells, or in a relatively low number of cells, within the reproductive structures (antipodal cells, synergids, endosperm haustoria, suspensors). The high functional activity of those cells becomes obvious at a simple microscopic examination. Apart from the rather extensively studied suspensor of *Phaseolus coccineus,* direct evidences of cellular activities, based on appropriate analytical methods, are meager or lacking. For structures other than suspensors, we may cite: the correlation between the high degree of polyploidy in the parenchymatous cells of developing cotyledons of *Pisum sativum* and the high rate of RNA and protein synthesis (Scharpe and van Parijs 1973) (for the cotyledons of *Vicia faba,* see Millerd and Whitfeld 1973); the results of cytochemical tests applied to the synergids of *Allium cepa* (Syamasundar and Panchaksharappa 1975); the ultrastructural observations of Vannerau and Mestre (1975) on the micropylar haustorium of *Plantago lanceolata.*

Indirect evidences for the physiological role of polyploidy in reproductive organs and tissues are numerous, and some of them are rather striking. Let us consider first the gradient of cell polyploidization, which probably means a gradient of functional activities, in tissues of such specialized structures as suspensors, e.g., the suspensor of *Phaseolus coccineus* (Nagl 1962, 1974, Brady 1973) and *Eruca sativa* (Corsi et al. 1973), and endosperm, e.g., the endosperm of *Pyrus malus* (Wanscher 1939) and that of *Zea mays* (Duncan and Ross 1950, Tschermak-Woess and Enzenberg-Kunz 1965). Although the establishment of a polyploidy gradient might indicate some involvement of phytohormones (Nagl 1978), the direction of the gradient, its placement within a tissue or an organ, its connection with particular structures, etc., are more suggestive of an ordered distribution of one or more specific functions in response to particular physiological demands. Also, the pattern of distribution of ploidy levels in different regions of many endosperms (Table 11.4) indicates a spatial distribution of degrees of nuclear activity. For several endosperm haustoria and suspensors, their close association with other structures within the ovule or ovary [funiculus, integument(s), placenta] are obvious.

Another aspect of the physiological role of polyploidy is certainly the vicariousness of structures in the ovular range (antipodal cells, synergids, endosperm, suspensor). Thus, species with massive, sometimes highly polyploid, suspensors have reduced endosperm and, conversely, species with reduced suspensors, or none at all, develop highly polyploid endosperm nuclei (Enzenberg 1961, Nagl 1962, 1978). This situation, together with the results of investigations on *Eruca* and *Phaseolus* dealt with previously, suggest some interchangeability of the two structures which the angiosperms have apparently developed for the control of early differentiation and growth of the embryo, viz., the endosperm and the suspensor (Nitsch 1971, D'Amato 1978 b).

Other observations suggest physiological correlations among endosperm, antipodal cells, and synergids. Thus, in parthenogenetic *Allium nutans* flower emasculation, while impeding endosperm formation, favors further persistence of one or both synergids which increase their polytenization level (Håkansson 1951). Reduced or defective endosperm development in interspecific crosses is frequently, if not generally, accompained by an increased polyploidization in the antipodals (cf. Brink and Cooper 1947, Kubien 1968).

11.9 Future Research

The application of such diverse methods as cytochemistry, autoradiography, electron microscopy, extraction and characterization of DNA's, molecular RNA-DNA hybridization in vitro and in situ, hormone bioassay and identification, to embryo suspensors – especially the *Phaseolus* suspensor – has provided sound bases for an elucidation of the physiological role of the suspensor in embryogenesis. It is expected that future work on suspensors will also use other molecular biological techniques with high resolution, e.g., extraction and characterization of messenger RNA's and in vitro protein synthesis. In view of the recent observations on coordinated amplification-underreplication processes in segments of single polytene chromosomes at different embryonic stages of *Phaseolus coccineus* (Forino et al. 1979), there is hope that specific functions may be visualized at the cellular level.

Because of the similarities in function between suspensors and endosperm, which have emerged in the last few years, investigations on the endosperm using the techniques which have been applied successfully to suspensors would be of particular interest. Since isolation of endosperm from seeds at different developmental stages is rather easy in some species, a few analytical methods (e.g., isolation and characterization of DNA, isolation of messenger RNA's) would certainly involve much less labor for collecting fresh material than the time-consuming excision of thousands of suspensors: remember, e.g., that in the experiments of Lima-de-Faria et al. (1975) on the DNA of *P. coccineus* over 10,000 excised suspensors were used.

Another tissue which deserves a renewed research interest is the anther tapetum whose cells should be intensively studied at two crucial points of their functional activity, viz., during meiosis I in the pollen mother cells, and during wall deposition in pollen grains. Because the tapetum cannot be isolated from the other tissues in the anther, the best analytical methods for tapetal cells remain cytochemistry, autoradiography, and in situ molecular hybridization, both at the light and the electron microscope levels. Further work is also needed for the cytological analysis of multinucleate tapetal cells in which both endomitosis and diplochromosome mitosis occur (see examples in Table 11.2). Since some stages in endomitosis and diplochromosome mitosis are not easily distinguished (e.g., an endotelophase and a diplochromosome prophase), combined cytological and DNA cytophotometric analyses of tapetal cells are to be recommended.

Finally, there is need for a better knowledge of the functional activity of the component cells of the embryo sac, especially the egg cell, the synergids, and the antipodal cells. Being deeply seated within the ovule, these cells are amenable to analysis with the same methods as those needed for the anther tapetum. One interesting aspect to be approached is whether egg cells and/or zygotes may undergo extra DNA replication (amplification); a condition which, as we have seen, is still controversial.

In conclusion, much is to be expected from a continued use of molecular biological techniques both at the cellular and at the histological level.

References

Alpi A, Tognini F, D'Amato F (1975) Growth regulator levels in embryo and suspensor of *Phaseolus coccineus* at two stages of development. Planta (Berl) 127:153–162

Alpi A, Lorenzi R, Bennici A, Cionini PG, D'Amato F (1979) Identification of gibberellin A_1 in the embryo suspensor of *Phaseolus coccineus*. Planta (Berl) 147:225–228

Alvarez MR (1968) Quantitative changes in nuclear DNA accompanying post-germination embryonic development in *Vanda* (Orchidaceae). Am J Bot 55:1036–1041

Avanzi MG (1950) Endomitosi e mitosi a diplocromosomi nello sviluppo delle cellule del tappeto di *Solanum tuberosum* L. Caryologia 2:205–222

Avanzi S, Cionini PG, D'Amato F (1970) Cytochemical and autoradiographic analyses on the embryo suspensor cells of *Phaseolus coccineus*. Caryologia 23:605–638

Avanzi S, Buongiorno-Nardelli M, Cionini PG, D'Amato F (1971) Cytological localization of molecular hybrids between rRNA and DNA in the embryo suspensor cells of *Phaseolus coccineus*. Atti Accad Naz Lincei Rend Cl Sci Fis Mat Nat Ser VIII 50:357–361

Avanzi S, Durante M, Cionini PG, D'Amato F (1972) Cytological localization of ribosomal cistrons in polytene chromosomes of *Phaseolus coccineus*. Chromosoma 39:191–203

Barlow PW (1976) The relationship of the dispersion-phase of chromocentric nuclei in the mitotic cycle to DNA synthesis. Protoplasma 90:381–392

Barlow PW (1978) The interrelationship of the cycles of chromosome condensation and reduplication in cell growth processes. Nucleus (Calcutta) 21:1–11

Battaglia E (1949) Agglutinazione cromosomica („stickiness") quale causa di eccezionali condizioni nucleari nelle cellule del tappeto di *Crepis zacyntha* (L.) Babc. Caryologia 1:248–268

Battaglia E (1951) The male and female gametophytes of angiosperms: An interpretation. Phytomorphology 1:1–30

Battaglia E (1963) Apomixis. In: Maheshwari P (ed) Recent advances in the embryology of angiosperms. Intl Soc Plant Morphologists, Univ Delhi, pp 221–264

Beaudry JR (1951) Seed development following the mating *Elymus virginicus* L. × *Agropyron repens* (L.) Beauv. Genetics 36:109–133

Bennett MD, Smith JB (1976) The nuclear DNA content of the egg, the zygote and the young proembryo cells in *Hordeum*. Caryologia 29:435–446

Bennett MD, Rao MK, Smith JB, Bayliss MW (1973) Cell development in the anther, the ovule, and the young seed of *Triticum aestivum* L. var Chinese Spring. Phil Trans R Soc Lond Ser B 266:39–81

Bennici A, Cionini PG (1979) Cytokinins and in vitro development of *Phaseolus coccineus* embryos. Planta (Berl) 147:27–29

Bennici A, Cionini PG, D'Amato F (1976) Callus formation from the suspensor of *Phaseolus coccineus* in hormone-free medium: A cytological and DNA cytophotometric study. Protoplasma 89:251–261

Berger CA (1941) Reinvestigation of polysomaty in *Spinacia*. Bot Gaz 102:759–769

Berger CA, Witkus ER (1946) Polyploid mitosis as a normally occurring factor in the development of *Allium cepa*. Am J Bot 33:785–787

Berger CA, Witkus ER, Joseph TC (1951) Tapetal cell and meiotic divisions in *Antirrhinum majus* L. and *Linaria vulgaris* Hill. Caryologia 4:110–114

Bianchi A (1959) Mitosi nel tappeto di antere di Mais. Caryologia 12:338–340

Bohdanowicz J (1973) Karyological anatomy of the suspensor of *Alisma* L. I. *Alisma plantago-aquatica*. Acta Biol Cracow Ser Bot 16:235–246
Brady T (1973) Feulgen cytophotometric determination of the DNA content of the embryo proper and suspensor cells of *Phaseolus coccineus*. Cell Differ 2:65–75
Brady T, Clutter ME (1972) Cytological location of ribosomal cistrons in plant polytene chromosomes. J Cell Biol 53:827–832
Brink RA, Cooper DC (1944) The antipodals in relation to abnormal endosperm behaviour in *Hordeum jubatum* × *Secale cereale* hybrid seeds. Genetics 29:391–406
Brink RA, Cooper DC (1947) The endosperm in seed development. Bot Rev 13:423–541
Brodsky WY, Uryvaeva IV (1977) Cell polyploidy: Its relation to tissue growth and function. Intl Rev Cytol 50:275–332
Brown SW (1949) Endomitosis in the tapetum of tomato. Am J Bot 36:703–716
Brunori A, D'Amato F (1967) The DNA content of nuclei in the embryo of dry seeds of *Pinus pinea* and *Lactuca sativa*. Caryologia 20:153–161
Carniel K (1952) Das Verhalten der Kerne im Tapetum der Angiospermen mit besonderer Berücksichtigung von Endomitosen und sogenannten Endomitosen. Oesterr Bot Z 99:318–362
Carniel K (1961) Das Antherentapetum von *Zea mays*. Oesterr Bot Z 108:89–96
Carniel K (1962) Die Entwicklung des Antherentapetums der Cyperaceen und Endomitosen an Kernen mit diffus-centromerischen Chromosomen. Oesterr Bot Z 109:168–173
Carniel K (1963a) Das Antherentapetum. Oesterr Bot Z 110:145–176
Carniel K (1963b) Beiträge zur Entwicklungsgeschichte des sporogenen Gewebes und des Tapetums bei einigen Dipsacaceen. Oesterr Bot Z 110:547–555
Carniel K (1969a) Beiträge zur Entwicklungsgeschichte des Antherentapetums in der Gattung *Oxalis*. I. *Oxalis rosea* und *O. pubescens*. Oesterr Bot Z 116:423–429
Carniel K (1969b) Beiträge zur Entwicklungsgeschichte des Antherentapetums in der Gattung *Oxalis*. II. *Oxalis acetosella*. Oesterr Bot Z 117:201–204
Chiarugi A (1927) Il gametofito femminile delle Angiosperme nei suoi vari tipi di costruzione e di sviluppo. Nuovo G Bot Ital (Nuova Ser) 34:1–133
Chiarugi A (1950) La poliploidia della generazione aploide femminile delle Fanerogame. Caryologia 3:149–155
Chiarugi A (1954) La poliploidia somatica nelle piante. Caryologia Suppl 6:488–520
Cionini PG, Bennici A, Alpi A, D'Amato F (1976) Suspensor, gibberellin, and in vitro development of *Phaseolus coccineus* embryos. Planta (Berl) 131:115–117
Cionini PG, Bennici A, D'Amato F (1978) Nuclear cytology of callus induction and development in vitro. I. Callus from *Vicia faba* cotyledons. Protoplasma 96:101–112
Clutter M, Brady T, Walbot V, Sussex I (1974) Macromolecular synthesis during plant embryogenesis: Cellular rates of RNA synthesis in diploid and polyploid cells in bean embryo. J Cell Biol 63:1097–1102
Cooper DC (1938) Embryology of *Pisum sativum*. Bot Gaz 100:123–132
Corsi G (1972) The suspensor of *Eruca sativa* Miller (Cruciferae) during embryogenesis in vitro. Giorn Bot Ital 106:41–54
Corsi G, Cela-Renzoni G, Viegi L (1973) A DNA cytophotometric investigation on the suspensor of *Eruca sativa* Miller. Caryologia 26:531–540
Cremonini R, Cionini PG (1977) Extra DNA synthesis in embryo suspensor cells of *Phaseolus coccineus*. Protoplasma 91:303–313
Czapyka-Dziekanowska D (1965) Studies in the mode of reproduction and the differentiation of the endosperm of *Plantago atrata* var. *carpatica*. Acta Biol Cracow Ser Bot 8:101–112
D'Amato F (1949) Risultati di una analisi cario-embriologica in una popolazione di *Nothoscordum fragrans* Kunth. Caryologia 1:194–200
D'Amato F (1952) Polyploidy in the differentiation and function of tissues and cells in plants. A critical examination of the literature. Caryologia 4:311–358
D'Amato F (1954) A brief discussion on "endomitosis". Caryologia 6:341–345
D'Amato F (1964) Nuclear changes and their relationships to histological differentiation. Caryologia 17:317–325
D'Amato F (1977a) Nuclear cytology in relation to development. Cambridge Univ Press, Cambridge
D'Amato F (1977b) Cytogenetics of differentiation in tissue and cell cultures. In: Reinert J, Bajaj YPS (eds) Applied and fundamental aspects of plant cell, tissue, and organ culture. Springer, Berlin Heidelberg New York, pp 343–357

D'Amato F (1978 a) Chromosome number variation in cultured cells and regenerated plants. In: Thorpe TA (ed) Frontiers of plant tissue culture 1978. Univ Calgary, pp 287–295
D'Amato F (1978 b) Sviluppo del seme nelle Angiosperme con particolare riguardo al genere *Phaseolus*. Giorn Bot Ital 112:407–417
D'Amato F, Devreux M, Scarascia-Mugnozza GT (1965) The DNA content of the nuclei of the pollen grains of tobacco and barely. Caryologia 18:377–382
D'Amato F, Bennici A, Cionini PG, Baroncelli S, Lupi MC (1980) Nuclear fragmentation followed by mitosis as mechanism for wide chromosome number variation in tissue cultures: Its implications for plant regeneration. In: Sala F, Parisi B, Cella R, Ciferri O (eds) Plant cell cultures: Results and perspectives. Elsevier/North-Holland Biomedical Press, Amsterdam, pp 67–72
Duncan RE, Ross J (1950) The nucleus in differentiation and development. III. Nuclei of maize endosperm. J Hered 41:259–268
Durante M, Cionini PG, Avanzi S, Cremonini R, D'Amato F (1977) Cytological localization of the genes for the four classes of ribosomal RNA (25S, 18S, 5.8S, and 5S) in polytene chromosomes of *Phaseolus coccineus*. Chromosoma 60:269–282
Enzenberg U (1961) Beiträge zur Karyologie des Endosperms. Oesterr Bot Z 108:245–285
Erbrich P (1965) Über Endopolyploidie und Kernstrukturen in Endospermhaustorien. Oesterr Bot Z 112:197–262
Erdelská O (1966) Einfluß der Befruchtung auf die Entwicklung der Antipodenkerne der Gerste. Biologia Plant 21:857–864
Ervin E (1941) A study of polysomaty in *Cucumis melo*. Am J Bot 28:113–124
Esau K (1965) Plant anatomy. Wiley, New York
Evans LS, Van't Hof J (1975) Is polyploidy necessary for tissue differentiation in higher plants? Am J Bot 62:1060–1064
Favarger C (1944) Sur quelques phénomènes de pseudo-appariement des chromosomes dans les tissus somatiques. Ber Schweiz Bot Ges 54:579–590
Forino LMC, Cremonini R, Tagliasacchi AM (1976) Frequenza di cellule con „DNA puffs" nel sospensore di *Phaseolus coccineus* in momenti diversi dello sviluppo embrionale. Giorn Bot Ital 110:337–346
Forino LMC, Tagliasacchi AM, Avanzi S (1979) Different structure of polytene chromosomes of *Phaseolus coccineus* suspensors during early embryogenesis. I. Nucleolus organizing chromosome pairs S_1 and S_2. Protoplasma 101:231–246
Frey L (1966) Embryological studies on *Alisma lanceolata* With. Acta Biol Cracow Ser Bot 9:125–135
Geitler L (1939) Die Entstehung der polyploiden Somakerne der Heteropteren durch Chromosomenteilung ohne Kernteilung. Chromosoma 1:1–22
Geitler L (1941) Das Wachstum des Zellkerns in tierischen und pflanzlichen Geweben. Ergeb Biol 18:1–54
Geitler L (1948) Notizen zur endomitotischen Polyploidisierung in Trichocyten und Elaiosomen sowie über Kernstrukturen bei *Gagea lutea*. Chromosoma 3:271–281
Geitler L (1953) Endomitose und endomitotische Polyploidisierung. Protoplasmatologia VI/C. Springer, Wien
Geitler L (1955) Riesenkerne im Endosperm von *Allium ursinum*. Oesterr Bot Z 102:460–475
Goldstein B (1928) Nuclear form as related to functional activities of normal and pathological cells. Bot Gaz 86:365–383
Grafl I (1941) Über das Wachstum der Antipodenkerne von *Caltha palustris*. Chromosoma 2:1–11
Greilhuber J (1974) Uncommon karyological features in the tapetum of some Pedicularieae (Scrophulariaceae). Caryologia 27:169–182
Guignard L (1882) Recherches anatomiques et physiologiques sur les légumineuses. Diss Paris
Håkansson A (1951) Parthenogenesis in *Allium*. Bot Not H2:143–179
Håkansson A (1957) Notes on the giant chromosomes of *Allium nutans*. Bot Not 110:196–200
Haq ANMN, Smarth J, Lane GR (1973) Causes of interspecific hybrid abortion in *Phaseolus*. Ann Rep Bean Improv Coop 16:31–32
Hasitschka G (1956) Bildung von Chromosomenbündeln nach Art der Speicheldrüsenchromosomen, spiralisierte Ruhekernchromosomen und andere Struktureigentümlichkeiten in den endopolyploiden Riesenkernen der Antipoden von *Papaver rhoeas*. Chromosoma 8:87–113
Hasitschka-Jenschke G (1957) Die Entwicklung der Samenanlage von *Allium ursinum* mit besonderer Berücksichtigung der endopolyploiden Kerne in Synergiden und Antipoden. Oesterr Bot Z 104:1–24

Hasitschka-Jenschke G (1958) Zur Karyologie der Samenanlage dreier *Allium*-Arten. Oesterr Bot Z 105:71–82
Hasitschka-Jenschke G (1959a) Bemerkenswerte Kernstrukturen in Endosperm und in Suspensor zweier Helobiae. Oesterr Bot Z 106:301–314
Hasitschka-Jenschke G (1959b) Vergleichende karyologische Untersuchungen an Antipoden. Chromosoma 10:229–267
Hasitschka-Jenschke G (1962) Notizen über endopolyploide Kerne im Bereich der Samenanlage von Angiospermen. Oesterr Bot Z 109:125–137
Heitz E (1928) Das Heterochromatin der Moose. I. Jahrb Wiss Bot 69:762–818
Heitz E (1929) Heterochromatin, Chromozentren, Chromomeren. Ber Dtsch Bot Ges 47:276–284
Huss HA (1906) Beiträge zur Morphologie und Physiologie der Antipoden. Beih Bot Zbl 20:77–174
Ivanovskaja HV (1973) Functional morphology of polytene chromosomes of wheat antipodal cells. Tsitologya 15:1445–1452
Jankun A (1970) Studies in endosperm development of *Delphinium kotulae* Pawl. Acta Biol Cracow Ser Bot 13:51–64
Jankun A (1973) Further studies in the development of hybrid endosperm in representatives of the genus *Delphinium* L. Acta Biol Cracow Ser Bot 16:215–225
Jaretzky R (1928) Histologische und karyologische Studien an Poligonaceen. Jahr Wiss Bot 96:357–490
Johri BM (1963) Female gametophyte. In: Maheshwari P (ed) Recent advances in the embryology of angiosperms. Intl Soc Plant Morphologists, Univ Delhi, pp 69–103
Kapil RN, Tiwari C (1978) The integumentary tapetum. Bot Rev 44:457–528
Kapoor BM (1966) Contributions to the cytology of endosperm in some angiosperms. X. *Nigella damascena* and *N. sativa*. Caryologia 19:73–83
Kapoor BM, Tandon SL (1964) Contributions on the cytology of the endosperm in some angiosperms. V. *Zephyranthes lancasteri* Traub. Phyton (Argentina) 21:37–43
Kubien E (1968) Cytological processes during the development of antipodals in *Ammophila arenaria* Link. Acta Biol Cracow Ser Bot 11:21–29
Levan A, Hauschka TS (1953) Endomitotic reduplication mechanisms in ascites tumors of the mouse. J Natl Cancer Inst 14:1–43
Lima-de-Faria A, Pero R, Avanzi S, Durante M, Ståhle U, D'Amato F, Granström H (1975) Relation between ribosomal RNA genes and the DNA satellites of *Phaseolus coccineus*. Hereditas 79:5–20
Linskens HF (1969) Fertilization mechanisms in higher plants. In: Metz CB, Monroy A (eds) Fertilization. Academic Press, New York, pp 189–253
Lorenzi R, Bennici A, Cionini PG, Alpi A, D'Amato F (1978) Embryo-suspensor relations in *Phaseolus coccineus:* Cytokinins during seed development. Planta (Berl) 143:59–62
Lorz A (1947) Supernumerary chromosomal reproduction, polytene chromosomes, endomitosis, multiple chromosome complexes, polysomaty. Bot Rev 13:597–624
Maheshwari P (1950) An introduction to the embryology of angiosperms. McGraw-Hill, New York Toronto London
Malecka J (1967) Cyto-embryological studies in *Taraxacum scanicum* Dt. Acta Biol Cracow Ser Bot 10:195–206
Malecka J (1975) The course of differentiation of the endosperm of *Echium vulgare*. Acta Biol Cracow Ser Bot 18:13–26
Manthriratna MAPP, Hayward MD (1973) Pollen development and variation in the genus *Lolium*. 3. Pollen histochemistry. Z Pflanzenzuecht 70:11–21
Marks GE, Davies DR (1979) The cytology of cotyledon cells and the induction of giant polytene chromosomes in *Pisum sativum*. Protoplasma 101:73–80
Marquardt H, Barth OM, Rahden U von (1968) Zytophotometrische und elektronenmikroskopische Beobachtungen über die Tapetumzellen in den Antheren von *Paeonia tenuifolia*. Protoplasma 65:407–421
Mechelke F (1952) Die Entwicklung der Polyploiden Zellkerne des Antherentapetums bei *Antirrhinum majus* L. Chromosoma 5:246–295
Mericle LW, Mericle RP (1970) Nuclear DNA complement in young proembryos of barley. Mutat Res 10:515–518
Millerd A, Whitfeld PR (1973) DNA and RNA synthesis during the expansion phase of cotyledon development in *Vicia faba* L. J Plant Physiol 51:1005–1010
Moss GI, Heslop-Harrison J (1967) A cytochemical study of DNA, RNA, and protein in the developing maize anther. II. Observations. Ann Bot NS 31:555–572

Murashige T (1974) Plant propagation through tissue cultures. Annu Rev Plant Physiol 25:135–166
Nagl W (1962) Über Endopolyploidie, Restitutionskernbildung und Kernstrukturen im Suspensor von Angiospermen und einer Gymnosperme. Oesterr Bot Z 109:431–494
Nagl W (1967) Riesenchromosomen von *Phaseolus coccineus* L.: Bau, Strukturmodifikationen, zusätzliche Nukleolen und Vergleich mit den mitotischen Chromosomen. Oesterr Bot Z 114:171–182
Nagl W (1970a) Inhibition of polytene chromosome formation in *Phaseolus* by polyploid mitoses. Cytologia 35:252–258
Nagl W (1970b) Karyologische Anatomie des Endosperms von *Phaseolus coccineus*. Oesterr Bot Z 118:566–571
Nagl W (1970c) Temperature-dependent functional structures in the polytene chromosomes of *Phaseolus* with special reference to the nucleolus organizers. J Cell Sci 6:87–107
Nagl W (1974) The *Phaseolus* suspensor and its polytene chromosomes. Z Pflanzenphysiol 73:1–44
Nagl W (1976a) The polytenic antipodal cells of *Scilla bifolia:* DNA replication pattern and possibility of nuclear DNA amplification. Cytobiologie 14:165–170
Nagl W (1976b) Early embryogenesis in *Tropaeolum majus* L.: Evolution of DNA content and polyteny in the suspensor. Plant Sci Lett 7:1–6
Nagl W (1978) Endopolyploidy and polyteny in differentiation and evolution. North-Holland Co, Amsterdam New York Oxford
Nagl W, Peschke C, Gyseghem R van (1976) Heterochromatin underreplication in *Tropaeolum* embryogenesis. Naturwissenschaften 63:198
Nesling FA, Morris DA (1979) Cytokinin levels and embryo abortion in interspecific *Phaseolus* crosses. Z Pflanzenphysiol 91:345–358
Netroufal F (1927) Zytologische Studien über die Kulturrasse von *Brassica oleracea*. Oesterr Bot Z 76:101–115
Nitsch JP (1971) Perennation through seeds and other structures: Fruit development. In: Steward FC (ed) Plant physiology: A treatise, vol VI/A. Academic Press, New York, pp 413–501
Oksala T, Therman E (1977) Endomitosis in tapetal cells of *Eremurus* (Liliaceae). Am J Bot 64:866–872
Osterwalder A (1898) Beiträge zur Embryologie von *Aconitum napellus* L. Flora 85:254–292
Pavari F (1957) Ricerche embriologiche e cariologiche su *Cestrum elegans* L. (Solanaceae). Caryologia 9:436–452
Resch A (1952) Untersuchungen über Kerndifferenzierung in peripheren Zellschichten der Sproßachse einiger Blütenpflanzen. Chromosoma 3:296–316
Rutishauser A, Hunziker HR (1950) Untersuchungen über die Zytologie des Endosperms. Arch Julius Klaus-Stift 25:477–483
Scarascia GT (1952) Sviluppo del tappeto dell'antera in *Nicotiana tabacum* L. var. Bright. Caryologia 5:25–42
Scharpe A, Parijs R van (1973) The formation of polyploid cells in ripening cotyledons of *Pisum sativum* in relation to ribosome and protein synthesis. J Exp Bot 24:216–222
Schnarf K (1929) Embryologie der Angiospermen. Handbuch der Pflanzenanatomie 10/2. Borntraeger, Berlin
Schweizer D, Ambros P (1979) Analysis of nucleolus organizer regions (NOR) in mitotic and polytene chromosomes of *Phaseolus coccineus* by silver staining and Giemsa C-banding. Plant Syst Evol 132:27–51
Skalinska M (1958) Studies in the karyological differentiation of the tapetum in *Valeriana officinalis* L. Acta Biol Cracow Ser Bot 1:45–54
Smith DL (1971) Nuclear changes in the cotyledons of *Pisum arvense* L. during germination. Ann Bot 35:511–521
Smith DL (1974) A histological and histochemical study of the cotyledons of *Phaseolus vulgaris* L. during germination. Protoplasma 79:41–57
Speta F (1975) Die Entwicklung des Endosperms von *Melampyrum cristatum* und *M. bihariense*. Linzer Biol Beitr 7:393–402
Steffen K (1955) Kern- und Nukleolenwachstum bei endomitotischer Polyploidisierung. (Ein Beitrag zur karyologischen Anatomie von *Pedicularis palustris* L.). Planta (Berl) 45:379–394
Steffen K (1956) Endomitosen im Endosperm von *Pedicularis palustris* L. Planta (Berl) 47:625–652
Stephen J (1974) Cytological investigations on the endosperm of *Borassus flabellifer* Linn. Cytologia 39:195–207
Swift H (1953) Quantitative aspects of nuclear nucleoproteins. Intl Rev Cytol 2:1–76

Syamasundar J, Panchaksharappa MG (1975) A cytochemical study of the hypertrophied synergids of *Allium cepa*. Cytologia 40:371–376
Tandon SL, Kapoor BM (1963) Contributions to the cytology of endosperm in some angiosperms. II. *Nothoscordum fragrans* Kunth. Caryologia 16:377–395
Tischler G (1944) Allgemeine Pflanzenkaryologie (Kernteilung und Kernverschmelzung) 2. Aufl: Bd II, 2.H, Handbuch der Pflanzenanatomie. Borntraeger, Berlin
Tison MA (1919) Sur le suspenseur du *Trapa natans*. Rev Gen Bot 1919:219–228
Titz W (1965) Vergleichende Untersuchungen über den Grad der somatischen Polyploidie an nahe verwandten diploiden und polyploiden Sippen einschließlich der Cytologie von Antipoden. Oesterr Bot Z 112:101–172
Trela Z (1958) Cyto-histological processes during the differentiation of the tapetal cells of *Aconitum variegatum* L. Acta Biol Cracow Ser Bot 1:35–43
Trela Z (1963) Cytological studies in the differentiation of the endosperm in *Anemone nemorosa* L. Acta Biol Cracow Ser Bot 6:177–183
Tschermak-Woess E (1956) Notizen über die Riesenkerne und Riesenchromosomen in dem Antipoden von *Aconitum*. Chromosoma 8:114–134
Tschermak-Woess E (1957a) Über Kernstrukturen in den endopolyploiden Antipoden von *Clivia miniata*. Chromosoma 8:637–649
Tschermak-Woess E (1957b) Über das regelmäßige Auftreten von „Riesenchromosomen" im Chalazalhaustorium von *Rhinanthus*. Chromosoma 8:523–544
Tschermak-Woess E (1967) Über das zweikernige endopolyploide Antherentapetum von *Rhinanthus*. Phytomorphology 17:188–195
Tschermak-Woess E (1971) Endomitose. In: Altman HW (Hrsg) Der Zellkern. Handbuch der allgemeinen Pathologie, Bd II/2. Springer, Berlin Heidelberg, S 569–625
Tschermak-Woess E, Enzenberg-Kunz U (1965) Die Struktur der hoch endopolyploiden Kerne im Endosperm von *Zea mays*, das auffallende Verhalten ihrer Nukleolen und ihr Endopolyploidiegrad. Planta 64:149–169
Turala K (1958) Endomitosis in the tapetal cells of *Cucurbita pepo*. Acta Biol Cracow Ser Bot 1:25–34
Turala K (1963) Studies in endomitotical processes during the differentiation of the tapetal layer of the Cucurbitaceae. Acta Biol Cracow Ser Bot 6:87–102
Turala K (1966a) Endopolyploidie im Endosperm von *Echinocystis lobata*. Oesterr Bot Z 113:235–244
Turala K (1966b) Strukturen endopolyploider Kerne im Bereich der Samenanlage einiger Monokotylen. Oesterr Bot Z 113:529–541
Turala K (1969) Endopolyploidy and structures of the nuclei in the endosperm of the Cucurbitaceae. Oesterr Bot Z 116:95–111
Turala K (1971) Disturbed mitotic divisions and endomitoses during the differentiation of the endosperm in *Ecballium elaterium* (Cucurbitaceae). Acta Biol Cracow Ser Bot 14:27–35
Turala K, Urbanska-Worytkiewicz K (1964) Cytological processes during the differentiation of the tapetal layer in *Solanum dulcamara* L. and *S. nigrum* L. Acta Biol Cracow Ser Bot 7:171–183
Turala-Szybowska K (1974) Karyological anatomy of the endosperm of *Bryonia dioica* and *Cucurbita pepo*. Acta Biol Cracow Ser Bot 17:75–90
Vallade J, Cornu A, Essad S, Alabouvette J (1978) Niveaux de DNA dans les noyaux zygotique chez le *Petunia hybrida* hort. Soc Bot Fr Actualités Bot 1–2:253–258
Vannerau A, Mestre JC (1975) First ultrastructural observation on the haustorial micropylar cells of *Plantago lanceolata*. C R Acad Sci Paris Ser D 281:699–702
Viegi L, Pagni AM, Corsi G, Cela-Renzoni G (1976) Il sospensore embrionale delle Cruciferae. I. Morfologia e struttura. Giorn Bot Ital 110:347–357
Walbot V, Dure LS (1976) Developmental biochemistry of cotton seed embryogenesis and germination. J Mol Biol 101:503–536
Walker RI (1967) Megasporogenesis and embryo development in *Tropaeolum majus*. Bull Torrey Bot Club 74:279–282
Wanscher JA (1939) Contributions to the cytology and life history of apple and pear. Roy Veter Agric Coll Yearb 1939:21–70
Witkus ER (1945) Endomitotic tapetal cell division in *Spinacia*. Am J Bot 32:326–330
Wunderlich R (1954) Das Antherentapetum mit besonderer Berücksichtigung seiner Kernzahl. Oesterr Bot Z 101:1–63
Yeung EC, Clutter ME (1978) Embryogeny of *Phaseolus coccineus:* Growth and microanatomy. Protoplasma 94:19–40

12 The Seed: Structure

F. D. BOESEWINKEL and F. BOUMAN

12.1 Introduction

12.1.1 Importance of Seeds

The seed functions as the reproductive unit of the Spermatophyta (seed plants), and links the successive generations. Other important seed functions concern dispersal and survival under cold, dry or other inclement conditions. There is an immense diversity in the internal and external structure of seeds. These differences are, to an appreciable extent, related to dispersal and germination strategies, and may involve the size and position of endosperm and embryo, structure, texture, and color of seed-coat, and the shape and dimensions of the seed as a whole. The minute seeds of orchids may weigh as little as 0,000002 g each. The seeds of the legume *Mora oleifera* are possibly the heaviest, and the weight of each exceeds 1,000 g, whereas a one-seeded fruit of the palm *Lodoicea maldivica* weighs more than 20,000 g. The seed consists of three components: embryo, endosperm (sometimes perisperm), and seed-coat. Both endosperm and embryo are the products of double fertilization, whereas the seed-coat develops from the maternal, ovular tissues. The seed habit is a significant advancement in the evolution of higher plants. Seed plants show several evolutionary advantages over spore-producing plants. Fertilization takes place within the protective tissues of the mother plant, and during its development the embryo is nourished by the mother plant. Besides, the embryo is covered by the protective seed-coat and often provided with storage material. These factors made the seed plants so successful that they became the dominant component in the terrestrial environments of the earth.

The evolutionary development of the seed habit in spermatophytes, and the evolution of the viviparous mammals from egg-producing vertebrates, which took place more or less in the same geological era, show several parallels. Apart from the importance of seeds for the plants themselves, they are most important in the food chains of both man and animals. Probably, seeds are more valuable to man than any other plant part. The large cultivated grasses, cereals, produce wheat, oat, barley, rice, maize, etc. Seeds of legumes such as peanuts, beans, and soybeans are rich either in proteins or in carbohydrates. Seeds are also used as spices, condiments, and beverages (coffee, chocolate, and strong drinks), seed extracts as medicines, and edible and industrial oils are expressed from soybean, peanut, coconut, sunflower, and linseed, etc. (Vaughan 1970). Other industrial uses of seeds are mentioned by Gassner (1931). The storage capacity of the food reserve in seeds is a special advantage.

Identification of seeds by means of morphological and anatomical characters is very important for testing seed samples in agri- and horticulture, and in paleobotany and archeology.

12.1.2 Historical Survey

The earlier inquiries into seed-coat structure took place simultaneously with the beginning of microscopic studies of plants. Grew (1671) discovered in the outer coat of leguminous seeds a very small "foramen" (micropyle) situated opposite to the radicle of the embryo. He thought that the seed-coats function as a kind of filter for the nutrients of the embryo, and compared the foramen with the bunghole of wine or beer barrels, which serve as a passage for gases during the process of fermentation. According to Grew, the foramen promotes also the emergence of the radicle. Malpighi (1675) described the seed-coat development in several plants. He was so fascinated by the apparent agreement between the seed-coat layers and the membranes of the animal embryo that he applied the terms "secundine", "chorion", and "amnion" to various layers in the seed. Grew described details of the cells in the leguminous seed-coat and, for this reason, Targioni-Tozzetti (1855) named the characteristic palisade cells in the epidermis as "Malpighian cells". In most cases Grew's outer, middle, and innermost membrane, and Malpighi's secundine, chorion, and amnion are comparable to the integuments, nucellus, and endosperm, respectively.

In the eighteenth century hardly any substantial publication on the anatomy of seed-coats appeared. Gaertner's *De Fructibus et Seminibus Plantarum* (1788–1798) deals with the morphology of mature seeds and fruits, and does not contain original observations on development and anatomy of seeds.

In the nineteenth century there was a renewed interest in the anatomy and physiology of seeds. Before 1826 the structure of the ovule was not correctly known and, therefore, the layers of the seed-coat cannot be related with that of the ovule. After 1826, the year of discovery of the ovule, the study of the seed-coat remains nevertheless mostly stratigraphical. Most contributions are purely descriptive, and restricted to the mature seed-coat. Mainly, after the second half of the nineteenth century, the seed-coat became an object of study by investigators with a different background:

1. The agriculturists, with Nobbe (1876) as their chief representative, proposed a classification of the various testa layers based on anatomical and physiological criteria. He introduced the terms Hartschicht (hard or bony layer), Quellschicht (swelling layer), Stickstoff- or Proteinschicht (protein layer) and Pigmentschicht (pigment layer). This classification was not universally adopted because an absolute association between anatomy and function appeared to be untenable. According to Lohde (1874) and Harz (1885), these subdivisions went too far because these layers are only represented in a part of the seed-coats. Other agriculturists like Sempolowski (1874) and Schroeder (1871) may also be mentioned.
2. The pharmacists, of whom the important exponents were Tschirch and Oesterle (1893–1900) in Bern – others Berg (1865), Koch (1908), and van Wisselingh (1918–1924) – introduced the term Nährschicht (feeding layer) in contrast to Hartschicht, next to other elements of the seed-coat.
3. The botanists, mainly situated in Paris (Brandza 1891, Guignard 1893, Guérin 1899) and scattered over Germany (Lohde 1874, Bachmann 1882) recognized

the necessity of ontogenetic studies for a complete description of the seed-coat. It was especially the botanists who started inquiries into the developmental aspects of ovules and seed-coats.

Netolitzky (1926) presented an excellent review of the available data till the end of 1923, relevant to development and structure of seed-coats. After the appearance of this book, the interest in seed anatomy declined. In recent years seed-coat anatomy is mainly being investigated by Indian embryologists (see Bhatnagar and Johri 1972). The work of B. Singh (1964) and some of his pupils (R.P. Singh 1970, D. Singh 1967) is more exclusively dedicated to seed anatomy.

Corner (1976) published an extensive treatise containing many original new data on dicotyledonous seeds, dedicated to the memory of Netolitzky. According to him, Netolitzky came to the brink of the discovery that seed structure should be the basis of the natural classification of flowering plants. He wonders why Netolitzky's work did not attract wide attention. Corner thinks that the great exponents of classification failed to perceive the importance of seed structure, and their authority masked Netolitzky's findings.

12.2 Structure of Seeds

12.2.1 General Morphology

The ovule is the starting point for the development of the seed. Sometimes, seeds can be the result of a prolonged and complicated process by which the volume of the ovule increases many-fold. However, in other cases, the development is of rather short duration and simple. The cells of the integuments do not undergo further divisions, but enlarge and differentiate, thus giving rise to seeds which are merely differentiated ovules (microspermous seeds of Orchidaceae, Begoniaceae).

The term seed is often used in a loose sense, as for dry one-seeded fruits in, e.g., Gramineae, Compositae.

From the outside to the inside of the seed, the following parts can be recognized (Fig. 12.1):

1. The seed-coat of bitegmic seeds consists of the testa (outer integument) and tegmen (inner integument). Each integument has its own opening, that of the outer integument is called exostome, and of the inner endostome. Exostome and endostome together form the micropyle. Unitegmic seeds have only one opening, also called micropyle, and the seed-coat is called testa. The raphe is the prolongation of the funicle running along the seed and ends at the chalaza, where the raphal bundle often terminates in a funnel-shaped structure below the nucellar base. The raphe and chalaza contain the vascular supply for the seed but, functionally, they partake in the formation of the seed-coat. The scar resulting from the disconnection of the seed from the funicle is called hilum. Within the seed-coat sometimes remnants of nucellar tissue persist, often separated from the seed-coat by a more or less conspicuous cuticle. When this part of the nucellus is rather large and contains food reserves, it functions as a storage tissue called perisperm.

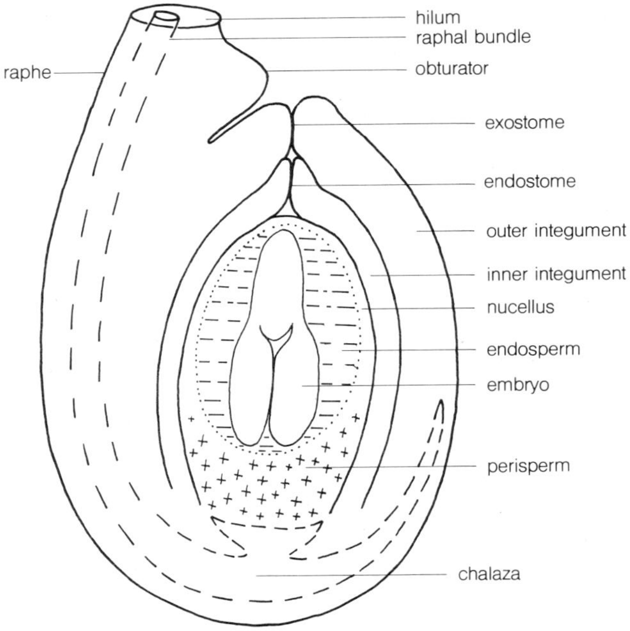

Fig. 12.1. General seed morphology

2. In most seeds there is a storage tissue called endosperm. In some cases the endosperm may be (almost) completely used up by the developing embryo which is then rather large and fills the entire interior of the seed. In this case the seeds are called exalbuminous or exendospermous.
3. The embryo is situated opposite the micropyle, thus promoting the penetration of the seed-coat. The portion of the stem is called hypocotyl, and the two first leaves (in dicots) cotyledons. Monocots have only one cotyledon. The terminal meristem of the embryo is called plumule. When the endosperm, or perisperm, is scanty or absent, only the embryo contains storage materials.

During the development the volume ratios of ovule, endosperm, and embryo do not remain constant, and differ in endospermous and exendospermous seeds. Rijven (1952) described the volume changes in the seed of *Capsella bursa-pastoris,* and Marinos (1970) in the seed of *Pisum sativum.* After fertilization the ovule starts to grow, soon followed by the growth of the endosperm. The enlargement of the volume of the endosperm is related to that of the embryo sac. The volume of the embryo is the last to start increasing, its rapid growth is at the expense of the endosperm. After half the seed is more or less formed, the embryo already contains the ultimate number of cells which is apparently the case in most dicotyledons. The later growth in bulk of the pea embryo is the result of cell enlargement alone. When the seed is mature most of the endosperm is consumed by the embryo, which then completely fills the interior of the seed.

In endospermous seeds the growth of the embryo starts much later than that of the endosperm; the embryo remains relatively small, however, and the endosperm is not so exhaustively consumed as in exendospermous seeds. The nutrients of the endosperm, in the latter case, are not used up before the germination has started. According to Wardlaw (1965), the endosperm is very important for the growth of the embryo. The absorption of nutrients from nucellus and embryo sac may somehow be unsuitable, or inadequate, and the endosperm acts as a necessary mediator between the sporophytic tissue of the mother plant and the embryo. The participation of the male gamete at the inception of the endosperm seems to have the consequence that the embryo and endosperm are more compatible than the embryo and sporophytic tissue of the mother plant.

It seems that, in most plants, the embryo draws its metabolites mainly from the endosperm which obtains nutrients from the embryo sac and adjacent tissues, this leads to the crushing and resorption of the nucellus and, eventually, of the inner layers of the seed-coat. To what extent the embryo is also supplied more directly by nutrients from the surrounding ovular tissues is uncertain. The embryo sac, endosperm, and embryo may form haustorial structures to promote the absorption of nutrients for their growth. The transport of nutrients for the new sporophyte is facilitated by the occurrence of transfer cells (Pate and Gunning 1972). These cells show a highly specialized adaptation in which ingrowth of cell wall material increase the surface area of the plasma membrane, facilitating transmembrane flux of solutes. The transfer cells are situated between donor and receptor compartments, e.g., in integumentary endothelium, outer and inner faces of endosperm and outer face of embryo (see Chap. 1).

12.2.2 Vascular Supply of the Seed

Although in ovules the vascular tissues are either represented by procambial strands, or by more differentiated bundles, the vascular system develops mainly during seed development (Fig. 12.2). The nutrients to developing seeds are transported through a vascular system which usually consists of the placental, funicular, and raphal bundles. Seeds of more primitive families are, on the whole, relatively large and have a more extensively developed vascular system. More advanced taxa (e.g., Sympetalae) often have simple, small seeds with poorly differentiated vascular bundles or none at all. Especially in large seeds, the funicular and raphal bundles are inadequate to transport enough nutrients, and a more elaborate system of vasculature has developed. Different patterns of vascular bundles occur in such taxa, mostly in the outer integument. These patterns consist of connected or unconnected bundles running parallel, dichotomously branched, reticular or otherwise (Corner 1976). The elaborate systems of vasculature may consist of:

1. A massive or compound raphal bundle with well-differentiated xylem and phloem elements, ending fanwise in the chalazal region (Fig. 12.2 A).
2. Vascular bundles running from the chalazal supply to the outer integument (Fig. 12.2. B, C) (so-called postchalazal bundles). These bundles may even extend as far as the micropyle (Meliaceae, Connaraceae, Dichapetalaceae). In certain families the seed is surrounded by a single postchalazal bundle in the

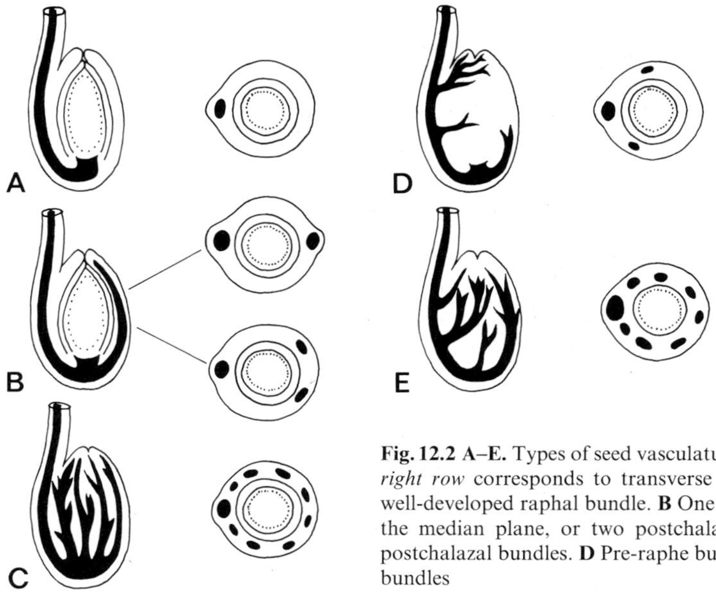

Fig. 12.2 A–E. Types of seed vasculature. *Left row* side views, *right row* corresponds to transverse sections. **A** Seed with well-developed raphal bundle. **B** One postchalazal bundle in the median plane, or two postchalazal bundles. **C** Many postchalazal bundles. **D** Pre-raphe bundles. **E** Pachychalazal bundles

median plane (Fig. 12.2B) (Annonaceae, Cucurbitaceae, Vitaceae, Polygalaceae).

3. A network of vascular bundles originates from the chalaza which, by intercalary growth, replaces the seed-coat partly or wholly. Such a condition is termed pachychalaza (Fig. 12.2E) (Meliaceae, Tropaeolaceae, *Glycosmis*).

Next to the raphal, chalazal, and testal bundles, pre-raphe bundles run from the beginning of the raphe in the direction of the micropyle (Fig. 12.2D) (Meliaceae: *Dysoxylon, Trichilia grandifolia,* and Connaraceae). Also, sometimes, recurrent bundles run from other parts of the raphe to the testa. Branches to the tegmen are rare (Euphorbiaceae and Meliaceae) as nucellar tracheids (certain Amentiferae, Capparidaceae). The question whether, during germination, the vascular system transports water to the embryo is debatable.

12.2.3 Seed Types

For understanding the structure of a seed, an adequate knowledge of the development of the ovule is essential. During development there may be considerable changes in the relative position of the various ovular structures. The shape of the seed, however, usually depends on the shape of the ovule. According to Davis (1966), about 80% of the families of angiosperms have anatropous ovules. This type of ovule gives rise to the so-called normal seed in which the hilum is situated close to the micropyle, and the embryo is straight (see Fig. 12.3E). Of this general structure several variations are possible, owing to local growth, which lead to de-

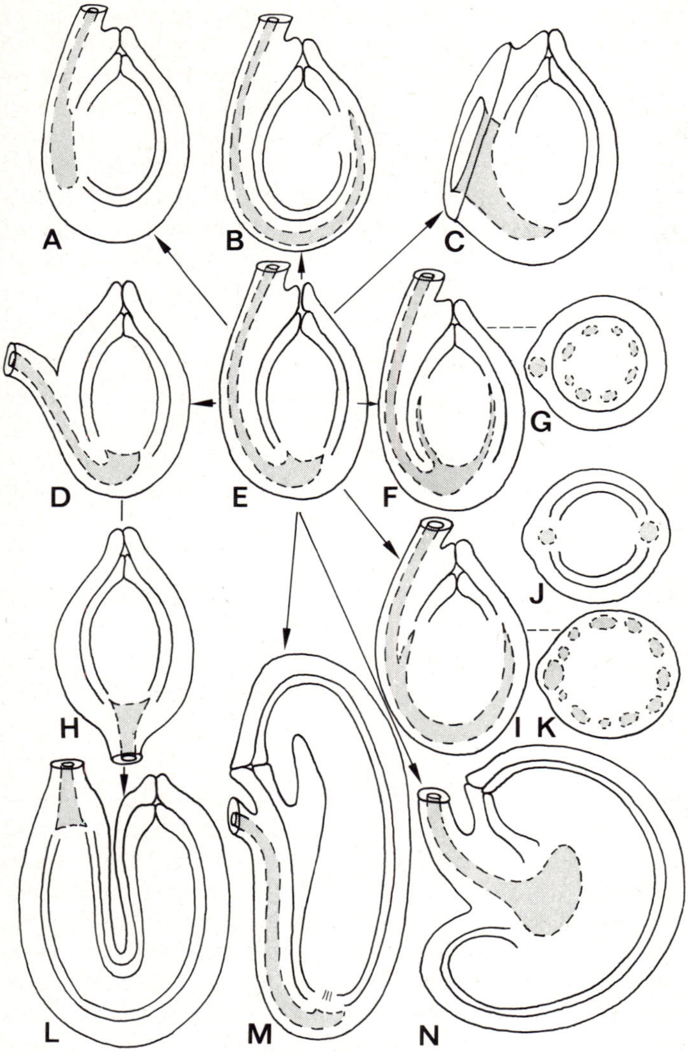

Fig. 12.3 A–N. Possible derivation of the seed types from the normal anatropous seed; mostly longisections, **G, J,** and **K** in transection. **A** Chalaza shifted in raphal direction. **B** Obcampylotropous seed, chalaza shifted in antiraphal direction. **C** Hilar seed. **D** Hemianatropous seed. **E** Normal anatropous seed. **F** Pachychalazal seed with short inner integument. **G** Transverse section of **F.** **H** Orthotropous seed. **I** Pachychalazal seed with short inner and outer integuments. **J** Perichalazal seed in transverse section. **K** Transverse section of **I.** **L–N** Different types of campylotropous seeds

rived seed types. These variations can be interpreted as extensions of the external, or internal, parts of the ovule.

12.2.3.1 Hilar Seeds. In the hilar seed the hilum (place of attachment of seed to placenta) is enlarged (Fig. 12.3 C). A greater, or smaller, part of the surface of the seed-coat consists of the extended hilum (some Meliaceae, *Erythrina, Aesculus*).

12.2.3.2 Pachychalazal Seeds. In pachychalazal seeds the place of attachment of the integument(s) is extended (Fig. 12.3 F, I, K). The intercalary growth of the chalaza occurs below the place of attachment of both integuments, so that a new container (pachychalaza) for endosperm and embryo arises. Pachychalazy can occur in different gradations. In the extreme case the integumentary part of the ovule hardly undergoes any development, and results in a massive chalaza; a scanty, or hardly any, portion of the integuments persists in the mature seed. The seed-coat is mainly of chalazal origin, and, consequently, mostly exotestal. This seed type indicates a derived condition, and has developed independently in many families (Meliaceae, Sapindaceae, Lauraceae, Tropaeolaceae). Due to lack of ontogenetic details, the pachychalaza is often overlooked.

12.2.3.3 Perichalazal Seeds. In perichalazal seeds the extension of the chalaza takes place only in the median plane of the seed (Fig. 12.3 J), and one of the vascular bundles extends around the periphery of the seed from funicle to micropyle. Perichalazal seeds are rather rare, and known mainly in Annonaceae and Meliaceae.

12.2.3.4 Endopachychalazal Seeds. In some Euphorbiaceae, particularly *Ricinus,* and also in *Myristica* (van Heel 1982), the zone below the nucellus and above the attachment of the outer integument proliferates and results in a short inner and a long outer integument (Fig. 12.3 F, G). In these seed types the seed-coat may remain either endotestal or exotegmic.

12.2.3.5 Campylotropous Seeds. Another case of local growth of a seed component occurs in campylotropous seeds. In this type the antiraphal side is strongly extended (Fig. 12.3 L–N). Further, a one-sided pressure is exerted on the embryo sac, mostly from the chalazal or raphal direction, or from the inner integument at the raphal side. Therefore, the embryo sac (and later the embryo) becomes U- or V-shaped. Campylotropous seeds can be the result of diverse ontogenetic processes, and occur in Papaveraceae, Capparidaceae, Cactaceae, Leguminosae, Geraniaceae, Malvaceae and other taxa. Because of the U- or V-shaped embryo sac, the embryo can be twice as large as the length of the seed, which is advantageous for the seedling.

12.2.3.6 Obcampylotropous Seeds. The obcampylotropous seed is the converse of the campylotropous seed. The raphal side extends, and the antiraphe remains small (Fig. 12.3 B) (Leguminosae, Vitaceae).

12.2.3.7 Orthotropous Seeds. The orthotropous seed develops from the orthotropous ovule. This ovule does not bend, there is no raphe, and the hilum is situated opposite the micropyle (Fig. 12.3 H). The orthotropous seed is mostly found in one-seeded fruits, and occurs in Piperaceae, Polygonaceae, Urticaceae, and Juglandaceae.

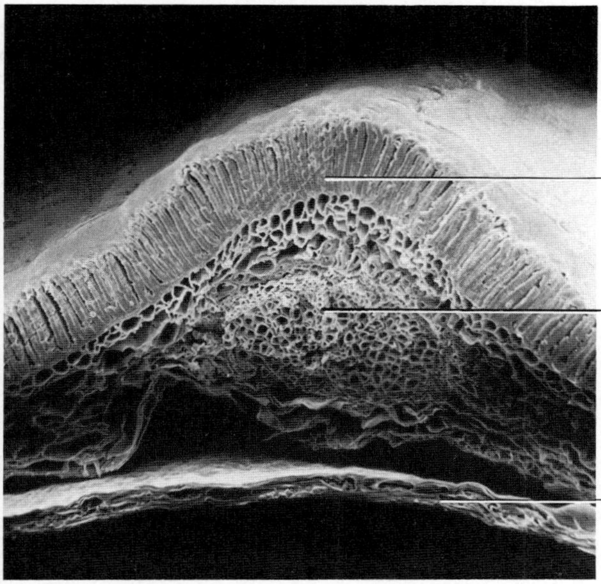

Fig. 12.4. SEM photograph of transverse section of seed of *Poncirus trifoliata* showing raphe and overlying exotestal seed-coat

12.2.4 Development and Structure of Seed-coats

The mode of development of the seed-coat provides an important set of characters of the species, but its knowledge is especially important for a correct interpretation of the mature seed-coat. Seemingly identical structures may arise along different ontogenetic pathways. This is, for instance, the case with the seed-coat in which the only mechanical layer is the outer layer of the outer integument (exotestal seed-coat) (Fig. 12.8 A). This type has developed in different unrelated taxa: Fagaceae, Anacardiaceae, Aceraceae, Umbelliflorae, Begoniaceae. A seed with the mechanical layer in the outer layer of the inner integument (exotegmic seed-coat) is (Figs. 12.5 C,D and 12.6 D) also characteristic of many unrelated families: Celastraceae, Geraniaceae, Meliaceae, Sapindaceae. Another example is the vascularized seed-coat which can arise from a dermally or subdermally initiated outer integument – dermal outer integument: Dichapetalaceae, Meliaceae; subdermal outer integument: Juglandaceae, Magnoliaceae; or pachychalaza: Tropaeolaceae. The seed-coat development is triggered when the ovule is fertilized. The seed-coat is formed by the integuments, the chalazal and raphal tissue, and often also by the included nucellar tissue. All these parts consist of diploid tissues of the mother plant. The raphal, and to a lesser degree the chalazal, tissue resembles the testa in its ultimate structure. For instance, when the outer layer of the outer integument differentiates as the mechanical layer, the so-called exotesta, the outer layer of the raphe and chalaza also undergo more or less a similar differentiation (Fig. 12.4). When the inner layer of the outer integument differ-

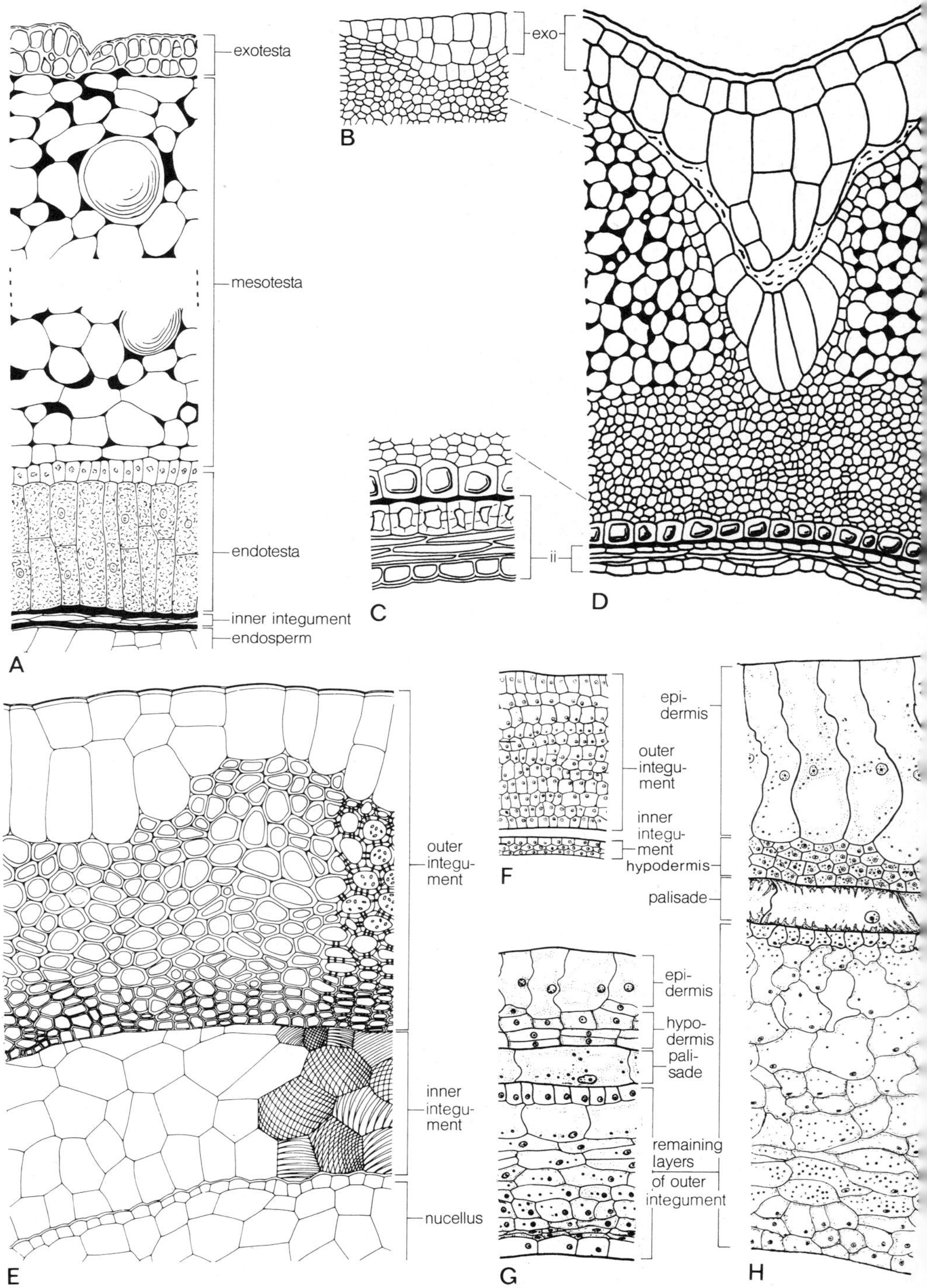
exotesta
mesotesta
endotesta
inner integument
endosperm
A
exo
B
ii
C
D
outer
integu-
ment
inner
integu-
ment
nucellus
E
epi-
dermis
outer
integu-
ment
inner
integu-
ment
hypodermis
palisade
F
epi-
dermis
hypo-
dermis
pali-
sade
remaining
layers
of outer
integument
G
H

Fig. 12.5 A–H

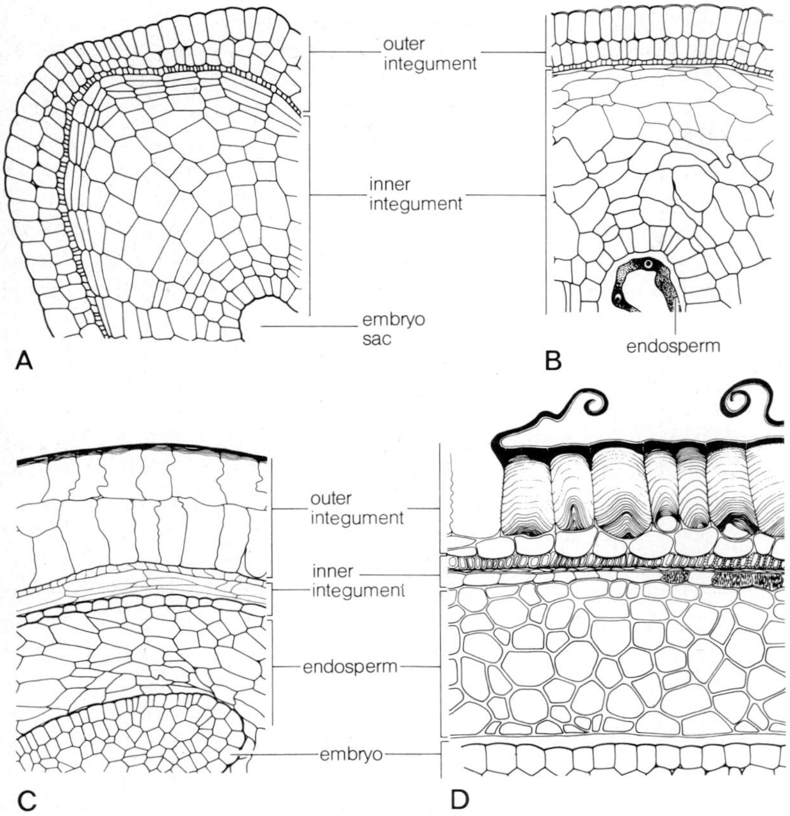

Fig. 12.6 A–D. *Linum usitatissimum*, development of seed-coat with multiplicative inner integument shown in transverse section. (From Boesewinkel 1980). **A–C** Developing seed-coat. **D** Mature seed-coat

entiates as the mechanical layer (endotesta), this layer also develops on the inner side of the raphe. In seeds derived from unitegmic ovules the seed-coat is also termed testa, which is not quite correct; only in a portion of such ovules is the single integument fully homologous with the outer integument (see Chap. 3). In the earlier literature one comes across the rather general use of the term testa for the entire seed-coat. In the seed-coat classification by Corner (1976), this term is restricted to the derivatives of the outer, or the single, integument. In spite of the above-mentioned inconsistencies, Corner's terminology is adopted here for practical reasons.

The ontogenetic potential of the bitegmic ovule is generally much larger than that of the unitegmic ovule, which implies that most unitegmic seeds have a rather

Fig. 12.5 A–H. Transverse sections of seed-coats. **A** *Magnolia soulangeana*, multilayered exotesta and endotesta. **B–D** *Carica papaya*, outer part of developing seed-coat (**B**) with multilayered exotesta (*exo*), inner layer of outer integument and inner integument (*ii*) with exotegmen (**C**), and mature seed-coat (**D**). (From Corner 1976). **E** *Zanthoxylum simulans*, multilayered exotesta and tracheidal inner integument. (From Boesewinkel 1978). **F–H** *Cucurbita moschata*, developing inner and outer integument (**F**), and developing outer integument (**G, H**). (From D. Singh and Dathan 1972) simple seed-coat. As a result thereof, the extent of change in ovules during seed development can differ widely. In small seeds none or hardly any cell divisions take place after fertilization, and the seed-coat is formed largely through the elongation and differentiation of integumentary cells. These seeds are, in fact, not much more than differentiated ovules. Large ovules develop into large seeds, often with complicated seed-coats. For example, as extremes, the ovules and mature seed of *Trichilia grandifolia,* which measure about 1 mm and 10 cm, respectively (a hundred-fold increase), can be compared with those of *Begonia squamulosa,* 0.5 mm and 1 mm, respectively (only a two-fold increase). Some of the larger seeds have lost the possibility of tissue differentiation, and have rather simple seed-coats.

Developing integuments show both anticlinal and periclinal divisions. The periclinal divisions are mainly responsible for the increase in the number of layers of the concerned integument, whereas anticlinal divisions mainly cause the growth in length. Complicated seed-coats are mostly the result of both anti- and periclinal divisions in integuments. Simple seed-coats often show only anticlinal divisions. When periclinal divisions predominate, Corner speaks of a multiplicative integument. These integuments often develop in multilayered complicated seed-coats. Periclinal divisions can be diffuse in the integument, or more localized and restricted to certain layers. Periclinal divisions in the outer layer of the outer integument are not very common. In *Zanthoxylum simulans* and Caricaceae the outer layer shows isolated periclinal divisions (Fig. 12.5 B, D, E). In species of Magnoliaceae the epidermis of the seed becomes two to six cells thick, and contributes to the formation of the sarcotesta (Fig. 12.5 A). This behavior of the outer integument is very characteristic. In Cucurbitaceae (Fig. 12.5 F–H) the outer epidermis, by two successive periclinal divisions, forms three characteristic layers (D. Singh and Dathan 1972). An outer layer – called seed epidermis – which is one- to multilayered, the second layer – the seed hypodermis – also one- to multilayered, and an inner layer – the seed palisade, or main sclerotic layer – which is one- or two-layered. Also, the hypodermal palisade layer in Paeoniaceae is a derivative of the epidermal layer of the ovule (see Corner 1976).

Multiplicative middle layers in the outer integument are very common, and often produce aerenchyma. They occur in Burseraceae, Bombacaceae, Connaraceae, Dichapetalaceae, Meliaceae, Simaroubaceae, and Leguminosae. The inner layer of the outer integument mostly divides only anticlinally. A multilayered endotesta occurs in the zoochorous seeds of Magnoliaceae, Vitaceae, and Leeaceae. Taxa showing periclinal divisions in the outer, or inner, layer of the inner integument are unknown. Multiplicative inner integuments are mainly formed by periclinal divisions of the middle layers. Also, in the inner integument the multiplicative middle layers lead to aerenchymatous tissue, and they often get crushed at maturity (Connaraceae, Cruciferae, Euphorbiaceae, Malvales, and Linaceae) (Fig. 12.6). In many taxa the inner integument remains very thin, unspecialized and ultimately gets crushed or resorbed (Magnoliaceae, Annonaceae, Rosaceae, Leguminosae, Polygalaceae, and most monocotyledons). Integuments with mainly anticlinal divisions are called nonmultiplicative. In these integuments mainly the number of cells per layer increases (hardly the number of layers). The nonmultiplicative integuments generally have a simple structure, and may be two- or three- or paucilayered, and persist in this condition.

After cell divisions cell enlargement occurs, followed by cell differentiation. Cell enlargement may take place in all directions (isodiametric cells) or be unequal, giving rise to stretched, fibrous, tangentially flattened or stellate cells. Although the different components of the seed-coat are mostly packed closely, their borderlines can be distinguished due to the presence of cuticles. A cuticle can be especially observed on the outer surface of seeds, also between the integuments, integument and nucellus, and seed-coat and endosperm. Cuticles can easily fuse, for instance, between the integuments, and between integument and nucellus. As a result of the hydrophobic character of these cuticles, they function as barriers in the diffusion of water and dissolved products. The function of the outer cuticle is more complex, and is discussed later.

In the last phases of seed maturation there is a drastic reduction of metabolic activity accompanied by the drying out and, eventually, compression of the seed-coat.

12.2.5 Differentiation

Because the integuments are usually built up of cell layers, the differentiated elements in the seed-coat are mostly also arranged in layers (Fig. 12.6 D). However, it is possible that they occur in clusters, or as idioblastic elements. Elements which, at a later stage, become sclerized often show dense cytoplasm. The following cell types may contribute toward the differentiation and buildup of the seed-coat: parenchyma cells, tannin cells, crystal cells, mucilage cells, cork cells, and sclerenchyma or collenchyma cells.

12.2.5.1 Parenchyma Elements. These cells function during their development, at least to some extent, as storage cells, and the reserve materials are used up at maturity. Parenchyma cells may also transform into chlorenchyma, or aerenchyma which is concerned in gas exchange. Ultimately, the parenchyma elements often get crushed or resorbed.

Secretory tissues in seed-coats are rare. Latex tubes and gum channels occur in Clusiaceae; oil-containing idioblasts in Winteraceae, Monimiaceae, Canellaceae, Myristicaceae, Magnoliaceae, Lauraceae, and Bixaceae.

12.2.5.2 Tannin Cells. Tannin is formed in tannin vacuoles, and consists of polyphenols, secondary compounds which are difficult to metabolize. Tannin cells occur especially in the outer layers of seeds, but sometimes in internal layers too, e.g., inner integument (*Linum*) and around vascular bundles. About the function of tannin not much concrete information is available. The most probable functions are: protection against predators (herbivores, insects), microorganisms (bacteria, fungi) (Swain 1979), and light, increase of the hardness of seed-coat, and imparting color to the seeds.

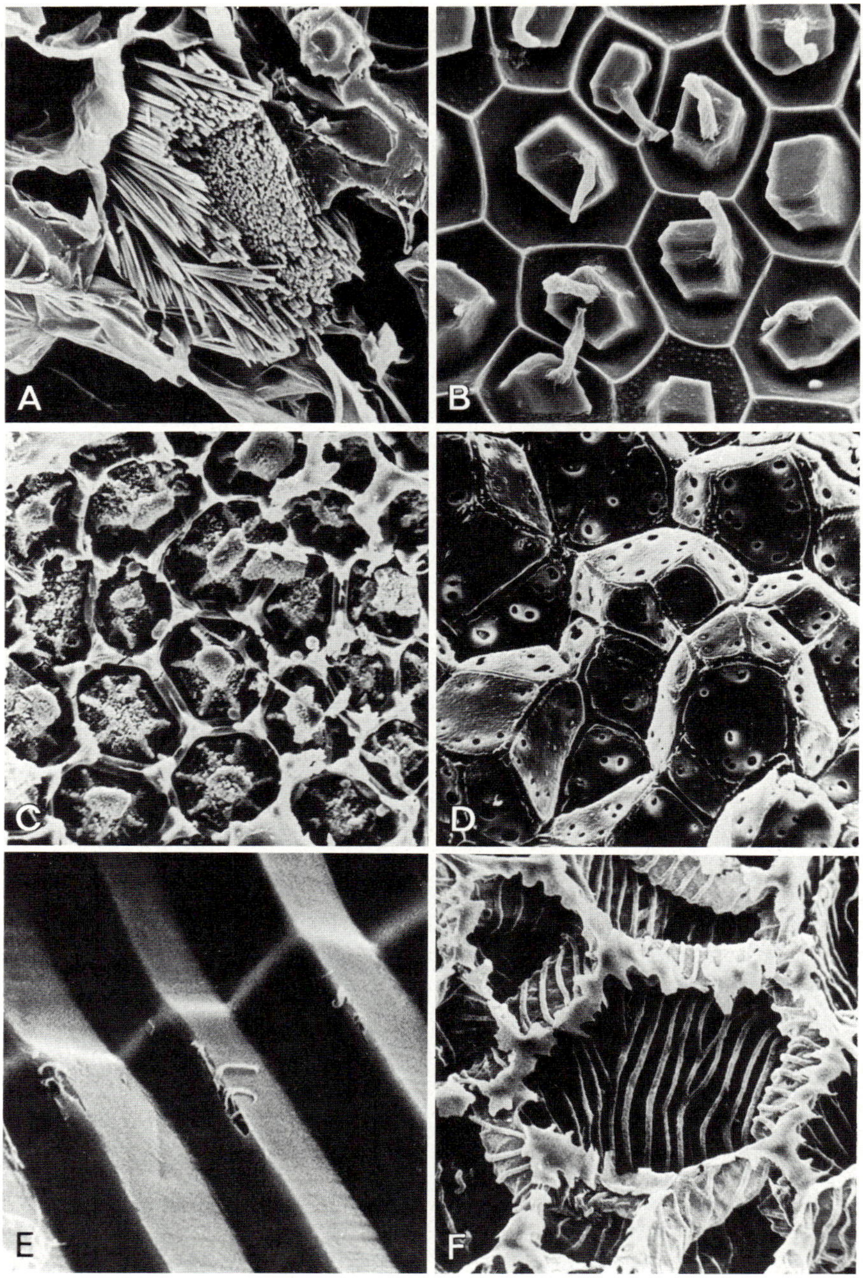

Fig. 12.7 A–F. SEM photographs of internal seed structures. **A** *Impatiens parviflora,* raphide bundles in seed-coat. **B** *Oxalis europaea,* polygonal crystals in the endotesta after seed dispersal. **C** *Tripogandra pflanzii,* silica bodies in endotesta. (From Grootjen and Bouman 1981). **D** *Toddalia asiatica,* mechanical cells of the mesotesta with pits. **E** *Canna jaegeriana,* exotestal Malpighian cells with light-line. **F** *T. asiatica,* tracheidal cells of the inner integument

12.2.5.3 Crystal Cells (Fig. 12.7 A–C). The form of the crystals can be variable, but is mostly polygonal (prismatic), or stellate (Dichapetalaceae, Sapindaceae, Araliaceae, Pedaliaceae, and Caprifoliaceae). They consist mostly of calcium oxalate, sometimes of calcium carbonate (Euphorbiaceae) or of silicate (Monocotyledons-Zingiberales, Commelinaceae). There can be one large crystal, many small crystals (crystal sand), or raphide bundles (many monocots – Typhaceae, Pandanaceae, Araceae, Liliaceae; dicots – Balsaminaceae, Vitaceae, Dilleniaceae, Onagraceae). Styloids occur in Melianthaceae. Crystals can be dispersed in the mesophyll or restricted to certain layers. Several taxa are characterized by crystal cells in the inner layer of outer integument: Parietales, Rutales, Geraniales, Magnoliaceae, Myristicaceae, Papaveraceae, Resedaceae, and Centrospermae. In several cases crystal cells show cell wall thickenings and form a sclerotic layer.

Generally, the occurrence of crystals seems to have arisen polyphyletically and does not necessarily have taxonomic implications. Much is not known about the function of crystals in seed-coats. They may serve as storage organs of waste products, defend against animal feeders or assist in mechanical protection, especially in the case of silicate.

12.2.5.4 Mucilage Cells. Mucilage cells are mostly epidermal. The mucilage-forming substance consists of secondary walls swelling up after hydration. The most important function of mucilage cells is the fixing of the seeds to the substrate [see special structures and adaptations (Sect. 12.3)]. Besides, mucilage sacs occur in the testa and tegmen of Sterculiaceae, and in the tegmen of Ebenaceae.

12.2.5.5 Cork Cells. Suberine may impregnate into the cell walls, or be present as distinct lamellae within cell walls. Cork cells often occur as a barrier tissue in the chalaza, blocking the interruption in the protective tegmic or testal layers, or as a distinct layer in the seed-coat. Integumentary cork is present in the floating tissue of hydrochoric seeds, preventing penetration of water. Cork lamellae may develop against the inner walls of epidermal slime cells (slime-cork cells sensu Netolitzky). The cork (lamellae) separates the slime epidermis from the other tissues of the seed.

12.2.5.6 Mechanical Cells (Sclerenchyma Elements). Mechanical cells lending hardness and rigidity to seeds are very common in seed-coats (Fig. 12.7 D, E). These mechanical cells may contain lignins, which provide additional protection against pathogenic attack and consumption by herbivores (Swain 1979). Sclerenchyma may be present in the form of sclereids or fibers. Macrosclereids in their most typical form are the Malpighian cells in the exotesta of Leguminosae. Malpighian cells are mostly radially stretched, often lignified palisade cells polygonal in cross section with a light-line (linea lucida). Much research has been conducted on Malpighian cells in hard legume seeds, because it is supposed that a high degree of impermeability is a property associated with their structure (Reeve 1946, Spurny 1973). During earlier stages Malpighian cells are radially stretched, and the thickenings on the wall in pea (*Pisum sativum*) seed-coat are uneven, leaving a star-shaped lumen in mature seeds. The apical regions of the cells show specific suberized, star-shaped encrustations (a suberine matrix) which reinforce the walls and also influence water permeability. The light-line does not seem to play a major role in water permeability, and shows a high density of cellulose microfibrils, without interfibrillar spaces. Which layer of the seed-coat is concerned with water exclusion is controversial (Ballard 1973). It is striking that the reports of extreme longevity in seeds concern mainly taxa with Malpighian cells (Leguminoseae and Malvaceae). Malpighian-like cells with a linea lucida also occur in the outer layer of the seed-coat of Rhamnaceae, Elaeagnaceae, and Cannaceae (Fig. 12.7 E), in the inner layer of the outer integument of Geraniaceae (Fig. 12.8 D), in the outer layer of the inner integument of Cistaceae, Malvaceae, Tiliaceae, and Euphorbiaceae. Other macrosclereids do not show a light-line, and may have a more cubic, isodiametric, or stellately lobed appearance. Sclerenchyma fibers are common in the outer layer of the inner integument, but also occur in the outer layer of the outer integument in various genera of different families, and in the mesophyll of Annonaceae. Also, transitions between sclereids and fibers seem to exist. Corner (1976) is of the opinion that they are separate categories. Collenchyma-like tissue is rare in seed-coats, and restricted to the middle layers of testa and tegmen. It occurs in some species of Rutaceae (Gallet 1913).

12.2.5.7 Tracheidal Cells and Fibers. Corner categorizes tracheidal cells in the seed-coat with localized, spiral, annular or reticulate wall thickenings (Figs. 12.5 E and 12.7 F). Tracheidal tegmic cells occur in Rutaceae, Dilleniaceae, Vitaceae, Erythroxylaceae, and Combretaceae. At present, the phylogenetic significance of tracheidal elements is rather obscure. It is uncertain whether they should be segregated as a separate category, or interpreted as reduced sclerenchyma elements with localized wall thickenings. Perhaps, in some cases, tracheidal layers may represent water-storing or conducting tissue.

These mechanical layers are the most characteristic of seed-coats. Corner drew up a classification of seed-coat types, based on the origin of mechanical layers, and the distinguishing features of the constituent elements. He distinguished exo-, meso-, and endotestal, and exo-, meso-, and endotegmic seeds. These seed-coat types can be further divided according to the form of mechanical elements.

12.2.5.8 Exotestal Seed-coats. In exotestal seeds the mechanical layer originates from the outer epidermis of the outer integument (Fig. 12.8 A). Exotestal seeds are of wide occurrence, and are found in derived families, especially unitegmic taxa of the Sympetalae. Exotestal seeds are known in Begoniaceae, Berberidaceae, Centrospermae, Leguminosae, Melianthaceae, Rhamnaceae, and Sapindaceae. An exotestal tendency occurs in many seed-coats with other main mechanical layers. There is a general trend of transference of function of seed-coat layers to the periphery of the seed-coat, or to the fruit wall.

12.2.5.9 Mesotestal Seed-coats. In mesotestal seeds the middle layers of the outer integument form the mechanical elements which are mostly cuboid or tangentially elongated (Fig. 12.8 B). The mesotesta is a characteristic feature of Hamamelidaceae, Lecythidaceae, Myrtaceae, Rosaceae, Theaceae, Cucurbita-

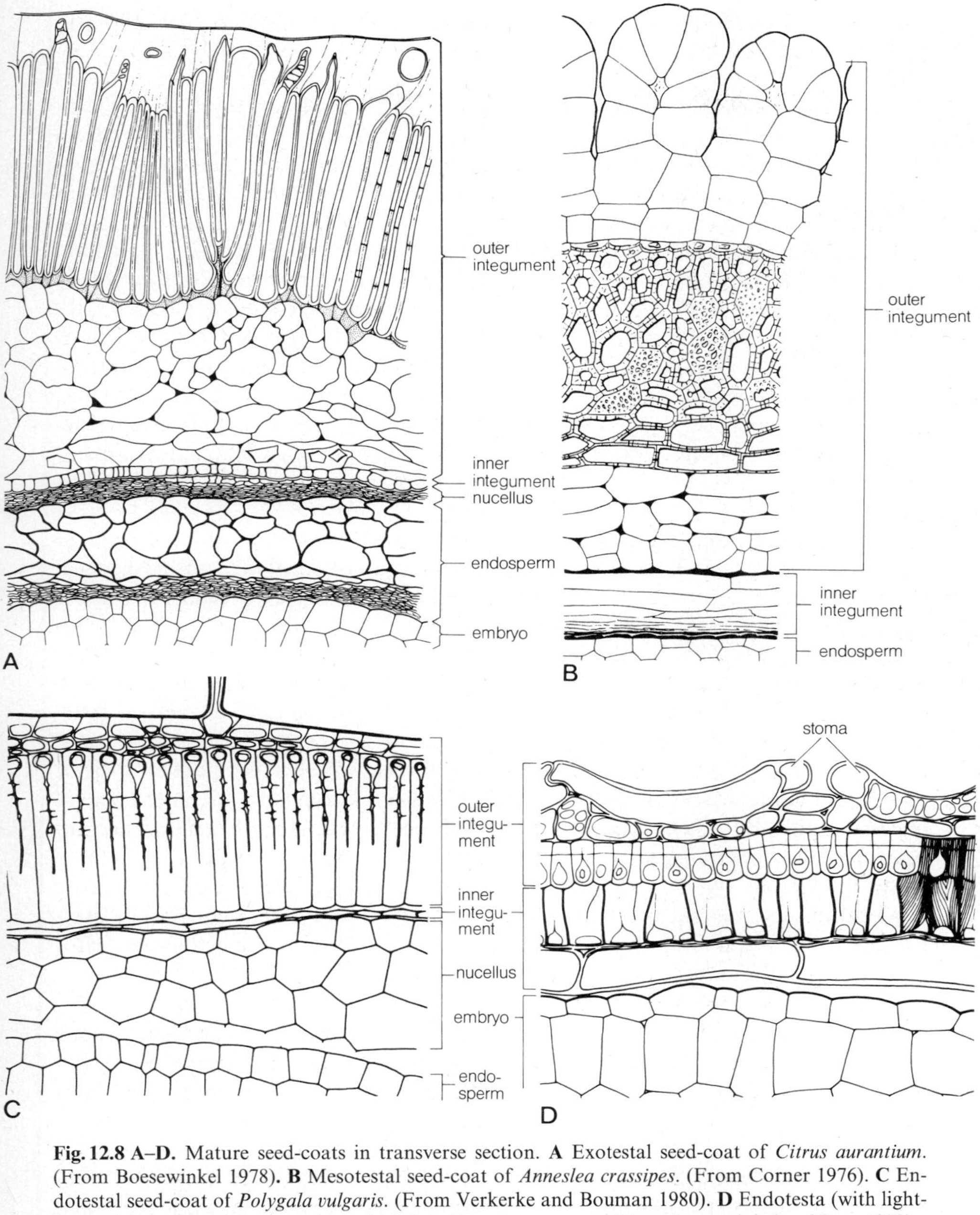

Fig. 12.8 A–D. Mature seed-coats in transverse section. **A** Exotestal seed-coat of *Citrus aurantium*. (From Boesewinkel 1978). **B** Mesotestal seed-coat of *Anneslea crassipes*. (From Corner 1976). **C** Endotestal seed-coat of *Polygala vulgaris*. (From Verkerke and Bouman 1980). **D** Endotesta (with light-line and crystals) and exotegmen combined, in *Geranium pratense*. (From Boesewinkel and Been 1979)

ceae, and Paeoniaceae. Paeoniaceae have an outer hypodermal palisade, Annonaceae a fibrous mesotesta.

12.2.5.10 Endotestal Seed-coats. In endotestal seeds the inner layer of the outer integument forms the mechanical layer (Figs. 12.5 A and 12.8 C). The endotesta is mostly single-layered or, sometimes, multilayered (Magnoliaceae, Degeneriaceae, Vitaceae). The endotestal cells show much variation in the form of their cell wall thickenings and cell contents.

The endotesta consists of a palisade layer in Myristicaceae and Polygalaceae, of cuboid cells in Grossulariaceae, of stellate ones in Onagraceae or longitudinally elongated in Lauraceae. The cell wall thickenings are orbicular, U-shaped with the unthickened wall facing the periphery of the seed (Cruciferae), or show annular, spiral or reticular thickenings. Endotestal cells often contain crystals, or other specific contents: separate crystals of calcium oxalate in Geraniaceae (Fig. 12.8 D), Oxalidaceae (Fig. 12.7 B), Malvaceae, Violaceae, and Rutaceae-Aurantioideae, silicate bodies in Commelinaceae (Fig. 12.7 C), Zingiberales, or amorphous calcium carbonate in Euphorbiaceae.

12.2.5.11 Exotegmic Seed-coats. In this type the outer layer of the inner integument comprises mechanical cells. The exotegmen can be combined with other mechanical layers (Geraniaceae, Meliaceae, etc.) (Fig. 12.8 D). There are two types of exotegmic cells – the fibrous type in which the fibers are arranged in a longitudinal direction of the seed (Fig. 12.6 D), and the sclereid type, often radially stretched in the form of a palisade (Fig. 12.8 D). Corner assumes that these two types of exotegmic seeds (with a palisade, and with fibers) have developed independently from their origin in the Magnoliales, and that their occurrence does not point to a phylogenetic relationship. The exotegmic seed is an important type of seed-coat characteristic of many families. The sclereid type of exotegmen occurs in Malvaceae, Bombacaceae, Clusiaceae, Hypericaceae, Geraniaceae, and Euphorbiaceae (Crotonoideae).

In the fibrous type of exotegmen the fibers are rather variable and, in a cross section, may be mistaken for a palisade. This type of seed-coat occurs in Celastraceae, Linaceae, Oxalidaceae, Meliaceae, Violaceae, Capparidaceae, Malpighiaceae, and Euphorbiaceae (Phyllanthoideae).

12.2.5.12 Mesotegmic Seed-coats. This type, with the mechanical cells restricted to the middle layers of the tegmen, is rare (Capparidaceae, Chloranthaceae). Exclusively mesotegmic seeds in which neither the outer nor the inner epidermis takes part in the mechanical construction do not seem to exist.

12.2.5.13 Endotegmic Seed-coats. There are only a few instances in which the inner layer of the inner integument is the principal mechanical layer of the seed-coat (Nandinaceae, some species of Zygophyllaceae and Clusiaceae). The endotegmen does not seem to be a primary structure. It is known only in derived taxa with indehiscent fruits, and a more or less degraded seed-coat. An endotegmic layer is common in combination with an exotegmen in Piperaceae, Saururaceae, Rafflesiaceae, Hydnoraceae, and Podostemonaceae; or with both an exotegmen and endotesta or an exotesta only as in Capparidaceae, Resedaceae, Violaceae, Caricaceae, Papaveraceae.

12.2.5.14 Undifferentiated Seed-coats. The seed-coats lack a mechanical layer or show a slight exotestal tendency only. They often occur in advanced families with indehiscent or drupaceous fruits (Aceraceae, Anacardiaceae, Cornaceae, Menispermaceae, and Umbelliferae). Families which show occasionally taxa with undifferentiated seed-coats are Celastraceae, Meliaceae, Proteaceae, Rosaceae, Rutaceae, and Thymelaeaceae. However, other families with indehiscent fruits still show a differentiated seed-coat: Aquifoliaceae, Burseraceae, Combretaceae, Elaeocarpaceae, and Tiliaceae.

12.2.5.15 Seed-coatless Seeds. In extreme cases, the seed-coat is completely lacking in the mature seed. This occurs in taxa with reduced ovules, Santalaceae and Loranthaceae, and in *Crinum* (Amaryllidaceae). Such a condition is also known in families with well-developed ovules (in some species of Apocynaceae, Menispermaceae, and Rubiaceae).

Although Corner's classification of seeds on the basis of the principal mechanical layer of the seed-coat is very useful, it is often very difficult to classify seeds in one or the other category. Frequently, there is more than one sclerenchymatous layer in the seed-coat, and several combinations are possible. There are no absolute criteria to distinguish the various layers. A slightly thickened outer layer of the testa of a relatively undifferentiated seed-coat is described as exotesta, whereas a similar layer in a seed-coat with well-developed endotesta (or other layer) is disregarded. Moreover, the function of layers with thickened cell walls is not restricted to mechanical protection, but may also be the exclusion of water, oxygen, microorganisms; and water storage or slime production.

12.2.6 The Seed-coat Surface

In the last decade the SEM technique has opened an entire new field of micromorphological research on seed-coats. Brisson and Petersen (1976) reviewed the use of SEM in the study of seed-coats, and remark that the SEM has a greater flexibility in observing qualitative relationships of the seed-coat than the light microscope. The SEM till now is mostly used in solving taxonomic problems. The current SEM research has proved that the seed-coat surface provides rather stable characters, and can be of great taxonomic significance in many families, especially those with relatively small seeds which in a more traditional method of investigation show only a few, or hardly any useful characters. The SEM also shows the way the seed-coat morphology is influenced by contact with the endocarp (Fig. 12.9 A, B). Conceivably, SEM studies of seeds in future will become a routine procedure in comparable taxonomic studies.

SEM studies are mostly restricted to the surface of the seed. Seeds are often mounted and viewed intact, but the internal structure can also be studied at the SEM level by cutting seeds, or by peeling off seed-coat layers (Fig. 12.7). Also,

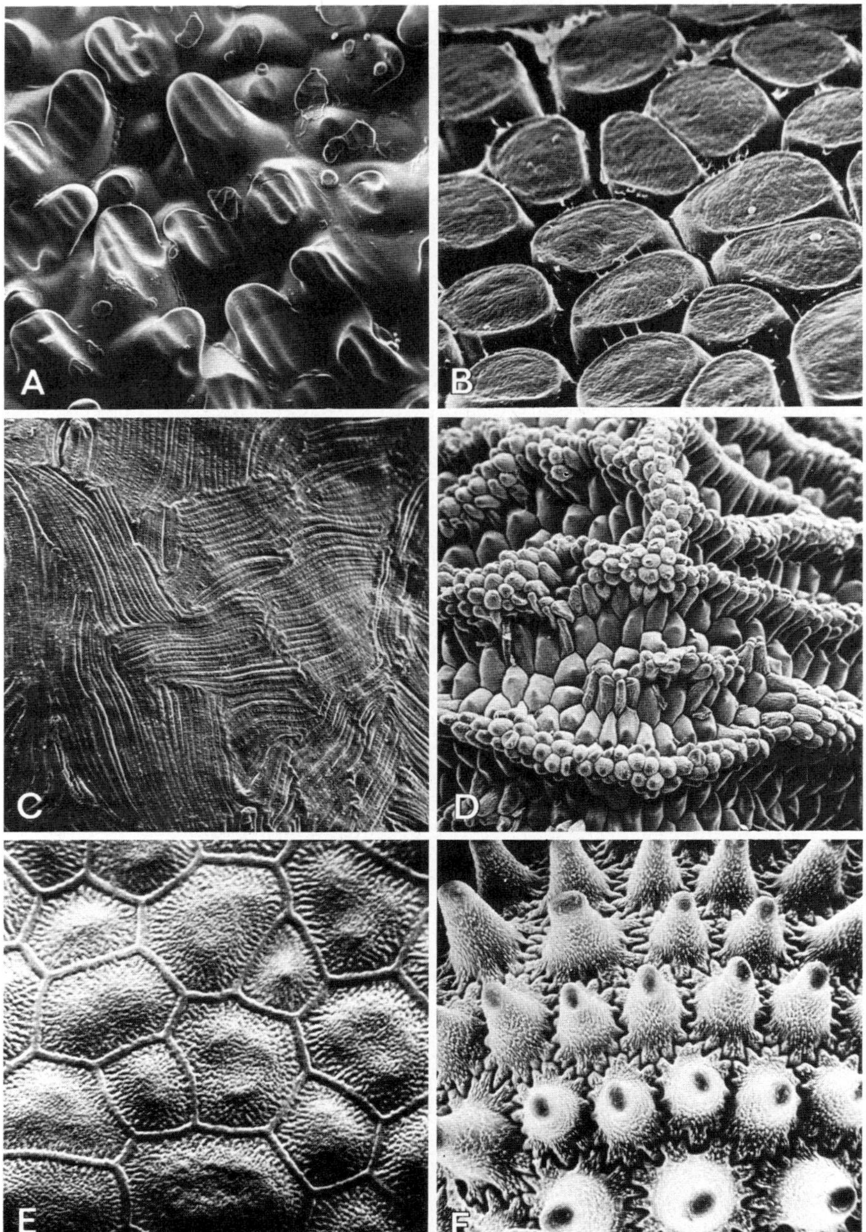

Fig. 12.9 A–F. SEM photographs of seed-coat surfaces. **A, B** *Ptelea trifoliata* and *Vepris elliotti*, papils of seed-coat flattened by contact with endocarp. **C, D** *Atalantia racemosa* and *Nigella damascena*, different types of arrangement of seed-coat cells. **E** *Melia azedarach*, isodiametric cells with raised anticlinal cell boundaries. **F** *Melandrium rubrum*, papillate cells and undulated anticlinal walls

treatments with chemicals of waxy substances, cuticles, cellulose, or pectins are possible.

In the description of seed-coats a consistent and generally accepted terminology is indispensable. Barthlott and Ehler (1977), and Barthlott (1981) studied the epidermis of seeds, and other plant parts with the SEM. They introduced a new terminology and grouped the surface characters into four categories:

12.2.6.1 Cellular Arrangement (Fig. 12.9 C, D). Although the cellular arrangement can be useful in defining the appearance of the seed, it is usually of minor taxonomic importance. In some families specific patterns or idioblastic elements, such as trichomes and stomata, can be of high systematic significance.

12.2.6.2 Shape of Cells (Primary Sculpture). The most important cause of surface characters is usually the cell shape and the curvature of the outer periclinal wall. Important aspects of this primary sculpture are:

1. The outline of cells which can be isodiametric or elongated, both cell types are mostly tetra- or hexagonal.
2. The anticlinal walls (cell boundaries) which may be straight, or irregularly curved to more or less regularly lobed. Anticlinal undulations are often of high taxonomic significance, and mostly characterize taxa at the genus or species level (Fig. 12.9 F).
3. The relief of the boundary which may be obscure, channeled, raised (Fig. 12.9 E), or show special structures such as a pearl-like appearance, or a splitting along the middle lamellae as in some species of *Begonia*.
4. The curvature of the outer periclinal wall, which can be conical (Fig. 12.9 F), domed, papillate (Fig. 12.9 A, B, D), or transitional toward unicellular trichomes. Since most epidermal cells of seed-coat are dead and do not exert osmotic pressure, the outer periclinal walls collapse unless supported by some thickenings on the wall. In concave cells with thin outer walls, this wall can even come in contact with and follow the sculpture of the inner periclinal wall (Fig. 12.10 A).

12.2.6.3 Fine Relief of the Cell Wall (Secondary Sculpture). The surface of the outer wall can be smooth, striate, reticulate, or micropapillate (verrucose or warty). These structures have a different origin, which must be exactly determined if utilized in comparative systematics. The following possibilities exist:

1. Cuticular sculptures which show a great micromorphological diversity (Fig. 12.10 B, C).
2. Secondary wall thickenings in reticulate or helical patterns on the inner side of outer, or inner, periclinal walls. These are not surface features, but become visible only with the shrinkage of the testa cells.
3. Secondary sculpturing as micropapillae are often caused by subcuticular or cuticular inclusions.

12.2.6.4 Epicuticular Secretions (Tertiary Sculpture). Epicuticular secretions are not very common on seed surfaces. They are known, e.g., in Aizoaceae, Dros-

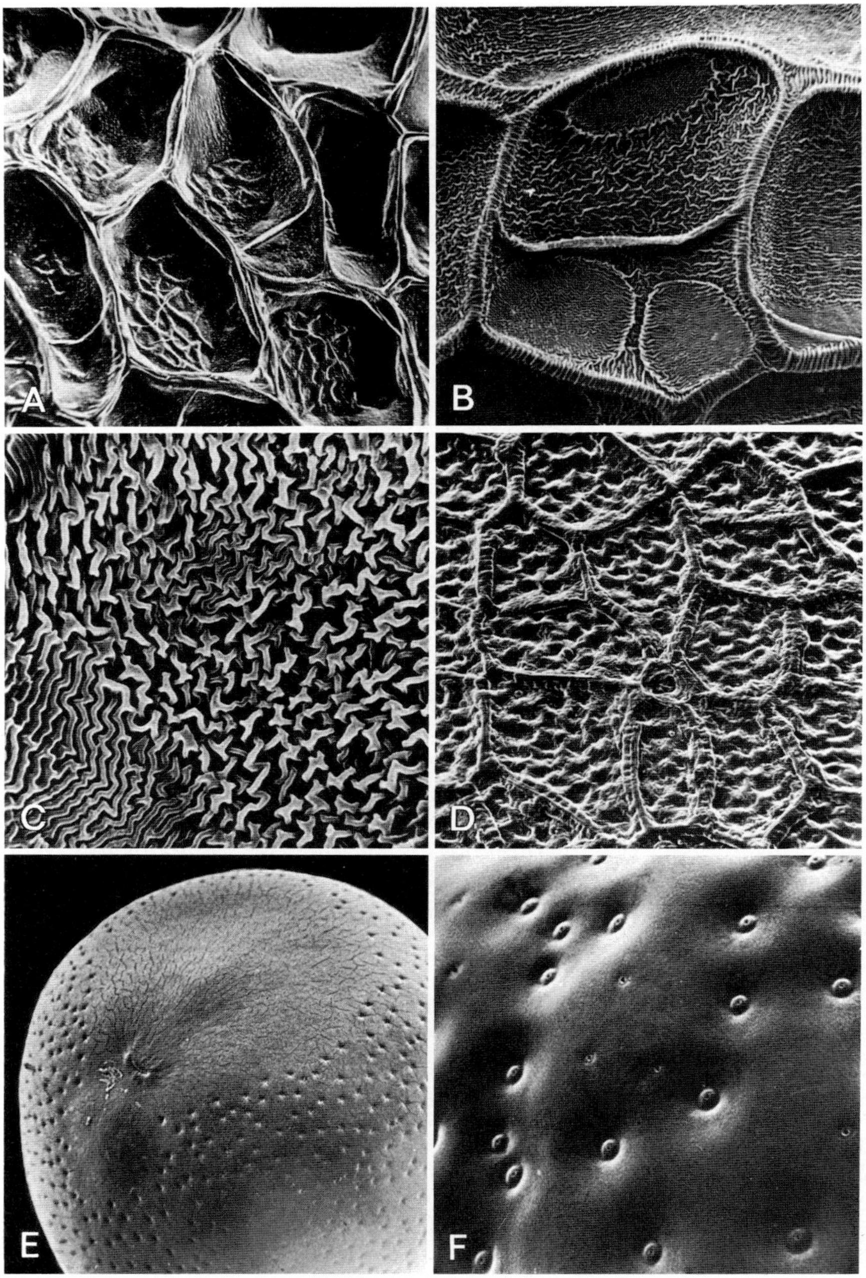

Fig. 12.10 A–F. SEM photographs of seed-coat surface. **A** *Gentiana asclepiadea,* collapsed outer periclinal walls showing reticulate structure of the inner wall. **B, C** *Begonia epipsila* and *B. quadrialata,* cuticular pattern. **D** *Geranium sylvaticum,* stoma in seed-coat with radially arranged accessory cells. **E, F** *Canna paniculata,* stomata on pachychalazal part of the seed

eraceae, Lentibulariaceae, and Orchidaceae, and represent chemically very different epicuticular substances, such as waxes and related products secreted by epidermal cells. They often occur in thin layers, and are hardly detectable by SEM techniques. They may appear in the form of local wax projections, such as irregular particles, crystalloid rodlets, filaments, flakes, or plates, etc. (Barthlott and Wollenweber 1981).

12.2.7 Stomata in Seed-coats

Schleiden's (1839) report on *Canna maculata* was the first observation on occurrence of stomata in seed-coats. At present ±30 families are known to have stomata on the seed-coat, e.g., Cochlospermaceae, several families of Malvales, Geraniaceae, Magnoliaceae, Papaveraceae, Polygalaceae, and Amaryllidaceae (Figs. 12.8 D and 12.10 D–F). The stomata are mostly situated in the outer epidermis of the outer integument; *Purskia* (Rosaceae) is the only taxon known in which stomata occur on the inner layer of the outer integument. Sometimes, the stomata are not evenly distributed over the seed-coat. The stomata are often poorly developed as compared with those on vegetative parts. The modifications include: lack of subsidiary cells, distortion of guard cells, or occlusion of the stomatal chamber.

It is uncertain if the majority of stomata are capable of opening and closing their aperture. Also, much is not known about the function of these stomata. Stomata on seed-coats may allow gas exchange for photosynthesis; some seed-coats do contain chlorophyll as in *Hymenocallis occidentalis* (Flint and Moreland 1943). The integuments, in this species, develop into an elaborate chlorophyllous tissue with vascular strands and stomata, capable of considerable photosynthetic activity. After shedding from the fruit, both development of the embryo and germination are dependent upon the photosynthetic activity of the seed-coat. Also, *Eschscholzia californica* has stomata in the chlorophyllous seed-coat (Jernstedt and Clark 1979). It is assumed that the stomata function in photosynthetic gas exchange. The stomata in the outer fruit wall can open and close, and control transpiration and gas exchange. The stomata in the inner fruit wall and seed-coat always remain open, even under plasmolyzing conditions.The seeds are illuminated by light, passing some regions of the fruit.

When the seed-coat is not green and rather thin, there is a possibility that gas exchange through stomata on the seed-coat to the green embryo takes place. Stomata also occur in seeds with a nonchlorophyllous seed-coat and/or embryo. On the other hand, seeds may be photosynthetic without the presence of stomata in the seed-coat. Leguminosae are known to have stomata in the seed-coat. Rugenstein and Lersten (1981) studied stomata in the seeds of eight species of *Bauhinia* (Leguminosae-Caesalpinioideae). The stomata are often modified or abnormal, and these authors could not ascertain if the stomata are functional; the inner fruit wall does not contain stomata. They suggest photosynthetic function after the shedding of the seeds. Netolitzky (1926) suggested the possibility of water release in developing seeds, and of absorption of water during germination (see Boesewinkel and Been 1979). The occurrence of stomata seems to be correlated with well-developed seed-coats, and tegumentary bundles (e.g., Magnoliaceae, Bombacaceae, Juglandaceae, Myristicaceae, and Euphorbiaceae).

Generally, the stomata in seed-coats occur in families which are not closely related, and they seem to be relicts or may have originated de novo. In Caryocaraceae and Euphorbiaceae, intercellular spaces may be present between the epidermal cells of the outer integument.

12.3 Special Structures, Adaptations, and Functions

After shedding of the seed from the mother plant, a successful future for the offspring as a separate entity has to be ensured. During this period the seed must rely on its own recources, must survive unfavorable environmental conditions, and be dispersed (in some way) to a site suitable for subsequent germination. Special structures have developed on the seeds, mostly related to some type of dispersal.

12.3.1 Sarcotestal Seeds

The term sarcotesta is applied to the pulpy and edible part of the seed-coat. Sarcotestal seeds are often attractively colored, the sarcotesta being a clear adaptation to zoochory. A sarcotesta is always followed by hard, sclerenchymatous layers as a protection against damage through their gut when eaten by animals.

There is not much agreement on the question whether the sarcotesta is, or is not, primitive. Attention is invited to the discussion between Corner (1976) and van der Pijl (1982) on the primitiveness of the arillate and sarcotestal seed. According to Corner, the sarcotesta and arillus are different organs, and the arillus is more primitive. Van der Pijl visualizes a transition between sarcotesta and arillus. The sarcotesta being the most primitive from which different arillate forms can be derived. The sarcotesta can have the following origins (Corner 1976):

1. The mesotesta forms the sarcotesta (*Magnolia,* Vitaceae); the sclerenchymatic layer is endotestal. This type is generally interpreted as the most primitive.
2. The entire testa forms the sarcotesta (*Aporosa,* Euphorbiaceae); the exotegmen forms the protective layer.
3. The sarcotesta is a derivative of only the outer layer of testa. In Ribesiaceae and Punicaceae the epidermal cells undergo extreme enlargement. In some Cucurbitaceae the epidermis becomes multilayered.
4. The sarcotesta is of chalazal origin (Meliaceae, Connaraceae, and Sapindaceae).

From this heterogenous origin one can conclude that the sarcotesta is not a homologous structure, and has arisen independently in different taxa.

12.3.2 Arillate Seeds

Arils are fleshy seed appendages, often with vivid colors to attract animals. The seeds are eaten and pass through the digestive tract (endozoochory), or they are

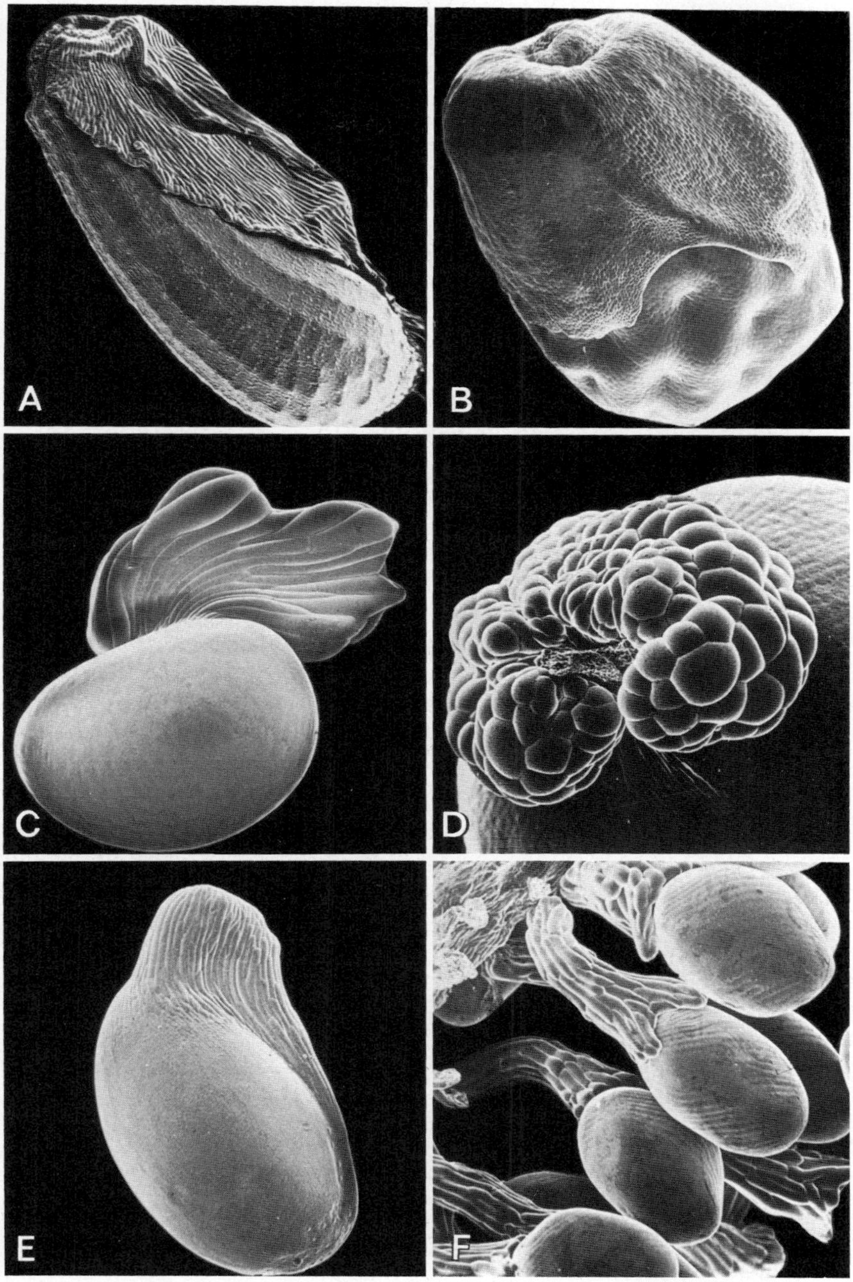

Fig. 12.11 A–F. Types of seed arils, SEM photographs. **A** *Turnera ulmifolia,* raphal aril. **B** *Passiflora suberosa,* funicular outgrowth. (From Kloos and Bouman 1980). **C** *Chelidonium majus,* strophiole. **D** *Chionodoxa luciliae,* caruncle. **E** *Viola odorata,* exostomal elaiosome. **F** *Episcia melittifolia,* funicular aril

collected to eat the aril (synzoochory). Also, arillate seeds are often protected by sclerotic layers.

Until recently, the use of morphological terms concerning seed appendages has been very inconsistent. Much attention has not been paid to the origin and histogeny of seed appendages and, for that reason, the earlier terms are only of historical interest. Gaertner (1788) defined the aril, for the first time, as an accessary integument attached to the hilum and enveloping the seed completely or incompletely. He also distinguished a number of small outgrowths. Planchon (1845), Pfeiffer (1891), and van der Pijl (1982) distinguish the arils according to their mode of attachment, or development, for instance:

1. Funicular or hilar outgrowth (true arils) (Fig. 12.11 B, F).
2. Outgrowths or swellings of the exostome (arillodes, caruncles) (Fig. 12.11 D, E).
3. Small swellings of the raphe (strophioles) (Fig. 12.11 A, C).

Baillon (1876) and Corner (1976) rejected the previous morphological subdivisions and terminology, and prefer to use the term aril in a more general sense. Corner designates the term aril as a general name for pulpy structures which grow from some part of the ovule, or funicle, after fertilization, and invest a part or the entire seed. Many other authors also doubt the usefulness of a morphological distinction. The question whether each seed appendage denotes a specific category, or has a polymorphic nature, remains as yet unanswered. Kapil et al. (1980) argue that the distinction between raphe, chalaza, and outer integument serves only a topographical purpose. A distinction between the arils protruding from these zones is of interest only for description; there are numerous intermediate structures. The diversity in size, shape, structure, and location of arils may, thus, be interpreted as an expression of the enormous plasticity of the testa.

Sernander (1906) introduced the ecological term elaiosome for all fleshy and edible parts of seeds dispersed by ants. The elaiosome attracts ants and gets easily detached from the remainder of the seed which is hard and inedible due to a thick-walled often crystal-containing tissue. The seeds are not eaten but only collected for the elaiosome. The color of the elaiosome is white or yellow and differs from that of the rest of the seed which is generally darker. Bresinsky (1963) described the development and structure of elaiosomes. According to him, the ants are attracted by the unsaturated fatty acid ricinolid acid. The elaiosome also contains nutrients such as proteins, lipids, starch, and vitamins.

Morphologically, the elaiosome can be an exostomal or raphal aril (*Polygala, Turnera, Calathea*), a caruncle (*Euphorbia, Scilla, Viola*), or a chalazal outgrowth (*Galanthus, Luzula*).

Rarely, the aril may have an additional function, and contribute to the opening of fruit valves as in Maranthaceae.

The widespread occurrence among both primitive and advanced families, and the great diversity in structure of the arils render it plausible that the aril is the result of polyphyletic evolution, as is true of many other structures concerned in seed dispersal.

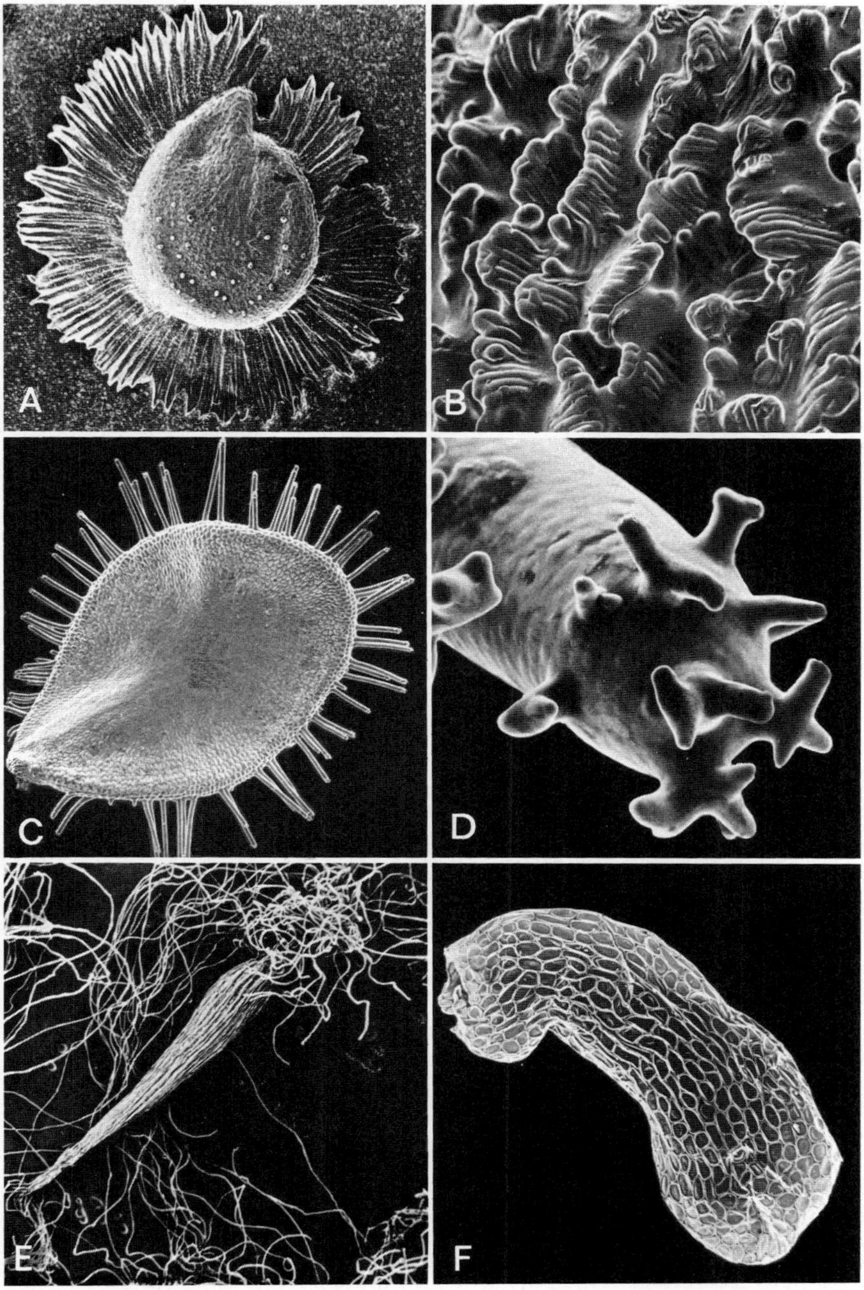

Fig. 12.12 A–F. SEM photographs. **A** *Spergularia salina,* winged seed. (From Kapil et al. 1980). **B** *Murraya paniculata,* developing multicellular hairs. (From Boesewinkel 1978). **C, D** *Nymphoides peltata,* seed and detail of hairtip. (From Kapil et al. 1980). **E** *Tillandsia recurvata,* haired seed. **F** *Parnassia palustris,* small balloon seed

12.3.3 Jaculator

The jaculator, which occurs only in Acanthaceae, is another remarkable structure about which little is known. It is the dorsal, lignified, hook-like outgrowth of the funicle which remains attached to the fruit after seed expulsion. It puts the seed in a favorable position for ejection. Therefore, the term jaculator is functionally incorrect. It has no active part in the expulsion of the seed which is caused by the fruit wall. Sell (1969) rejects the name jaculator, and prefers the term "hook". Corner interprets this structure as a modified funicular aril, although it is not a genuine seed appendage.

12.3.4 Winged Seeds

Winged seeds are generally anemochorous (Fig. 12.12 A). The number of winged seed types is very large, but the anatomical structure of the wing shows much resemblance. The wing of the seed is a local outgrowth of the seed-coat. A wing can be formed as an outgrowth of epidermal cells and it is then mostly one- or two-layered; or the outer integument and/or inner integument are involved in its construction. It can be one-sided, completely peripheral, or consist of several separate parts, and is mostly medianly situated. Seed wings provide optimum strength combined with a minimum of material. The epidermal cells of the wings often have a reticulate pattern and the thickenings are restricted to the radial walls but, sometimes, also to the inner tangential walls. Seed wings are rarely provided with vascular bundles, the raphe bundle excepted. They are less controversial than the arils, but equally diverse in origin, structure, and distribution among the angiosperms. Guttenberg (1971) distinguishes four types of seed flyers:

1. Disc flyers have thin, flat seeds with a circular or oval outline. The function becomes clear when the seed-coat forms a discrete, small or large, wing as in the seeds of *Tulipa, Lilium, Campanula, Gentiana.*
2. Sail flyers have large wings as in *Bignonia,* Rubiaceae, *Flindersia, Macrozanonia macrocarpa,* Meliaceae, etc.
3. Propeller flyers with only one wing and the embryo forming the excentric center of gravity at one of the small ends. This type is of wide occurrence (Proteaceae, *Vochysia,* Sterculiaceae, Meliaceae, Dioscoreaceae).
4. Spinning seeds, with three equidistant wings, occur in *Moringa oleifera* and *Crowfurdia.*

12.3.5 Haired Seeds

Seed hairs are not of general occurrence, and do not show the diversity of hairs as on vegetative parts of the plant. Usually, seed hairs are simple and unicellular. Multicellular hairs occur in the rutaceous genera *Aegle* and *Murraya* (Fig. 12.12 B) (Boesewinkel 1978) and Theaceae; peltate scales occur in *Scaphium;* stellate hairs in *Elaeocarpus;* glandular trichomes in *Impatiens* and some Labiatae. According to their distribution on the seed surface, Guttenberg (1971) distinguishes three types of hairs:

1. Dispersed seed hairs in woolly seeds. Their economic importance is best illustrated by *Gossypium* (Beasly 1975). Other instances are Cochlospermaceae, Convolvulaceae, Loganiaceae, Rutaceae, and Polygalaceae.
2. One or two groups of hairs present in the form of tufts or parachutes (coma). Hairs of placental-funicular origin occur in Salicaceae (*Salix, Populus*), of chalazal origin in Tamaricaceae and Onagraceae, and of micropylar origin in Apocynaceae and Asclepiadaceae. The hair tufts of Tillandsioideae have a very remarkable origin, and are formed by the splitting of the testa. The splitting starts at the chalazal end, but the hairs remain attached at the micropylar end of the seed (Fig. 12.12 E).
3. Hairs are mostly studied in relation to the anemochorous way of dispersal (enlargement of surface), but may also function in hydrochory as in the air-filled hairs of *Nymphoides* and *Villarsia,* or in epizoochory (anchoring of seeds by hooked hairs to coats of animals as in some Menyanthaceae) (Fig. 12 C, D).

12.3.6 Dust and Balloon Seeds (Figs. 12.12 F and 12.13 A, B)

Dust seeds are mainly anemochoric, and may fly passively over large distances. Thus representing a very effective adaptation to wind dispersal (Rauh et al. 1975). By the reduction of embryo and endosperm they minimize their weight. The embryo usually remains underdeveloped, lacks cotyledons, and is globular at the shedding stage of seeds. The endosperm is completely absent in Orchidaceae, or reduced to a single layer. Even the chemical constitution of the storage material may be restricted to low-molecular weight components (Youle and Huang 1981). The weight of many of the minute seeds is about 0.000003–0.000005 g; in Orobanchaceae each seed weighs 0.000001 g only. The seed-coat is thin, mostly with dead, air-filled exotestal cells showing surface sculpturing due to collapsed outer periclinal walls, sometimes also due to micro-ornamentation of surface areas (Angriffsfläche) useful for wind dispersal and increasing the resistance to air. In *Boschniaka hookeri* (Orobanchaceae) the cells of radially elongate testa often lose their outer walls, which results in a honeycombed appearance (S. Olsen and J. D. Olsen 1980). The low weight, relatively large surface area, and surface sculpturing result in a decrease in sink velocity and increase in air buoyancy.

Minute seeds have arisen independently in different monocot and dicot families with remarkable convergences. It is striking that such seeds are common in saprophytic-mycotropic, or otherwise nutritionally specialized families like Burmanniaceae, Droseraceae, Lentibulariaceae, Monotropaceae, Pyrolaceae, and Orchidaceae.

The spindle-shaped minute seeds are very common, and are termed "scobiform". The character complex of minute seeds may be augmented by the formation of wings, hairs, or air bladders. Seeds with distinct air spaces are called "balloons". The air bladders may be present in the form of very large, empty cells, or of intercellular spaces in the chalazal region, around the micropylar canal, or as a space between the seed-coat and endosperm/embryo. Balloon seeds are known in Ericaceae, Orchidaceae, Parnassiaceae, and Saxifragaceae (Fig. 12.12 F).

12.3.7 Hydrochoric Seeds

As compared to animal and wind dispersal, hydrochory is of less importance as a factor controlling evolutionary trends in the dispersal ecology of angiosperms. Adaptations of seeds to hydrochory mainly occur in aquatic and marshy plants. Very spectacular examples are the tropical drift seeds which can be dispersed by sea currents even to other continents. Adaptations to displacement by rain are also important. The rain can splash the seeds out of their containers and/or take them away in the water stream. To increase their buoyancy, hydrochoric seeds have adaptations which diminish the specific gravity, and enlarge the surface area. The hydrophobic character is emphasized by cuticles, or epicuticular excretions. The adaptations can be present in the following forms:

1. An often corky tissue with air enclosed in the intercellular spaces, or in the lumen of cells of the testa diminishes specific gravity as in *Pancratium maritimum, Menyanthes, Pistia, Calla palustris, Iris pseudacorus* (Fig. 12.13 C, D).
2. One or more air-filled cavities within the seed, e.g., the swim bladder formed by the aril of *Nymphaea, Victoria,* and *Euryale,* and cavities between seed-coat, endosperm, and embryo, or cavities between the cotyledons.
3. The presence of light-weight cotyledonary tissue.
4. The enlargement of the surface in flattened seeds, and by hair lashes as in *Nymphoides* (Fig. 12.12 C, D).
5. The increase of water repellency of the seed-coat surface, taking advantage of the surface tension of water, as is common in small seeds.
6. A combination of the above-mentioned factors.

12.3.8 Mucilage Seeds

Until the publications of Grubert (1974, 1981) on the distribution of myxospermy among the fruits and seeds of Angiospermae, scanty and scattered information was available on the external production of mucilage by wetted diaspores.

Myxospermy is a rather widespread phenomenon. Although there are many divergent ideas about the ecological importance of myxospermy, the most important functions of secreted mucilage seem to be the adherence of seeds to animals, and fixing of seeds to soil. Attachment of the seed to the soil in a moist place gives a greater chance of germination in arid regions. Other functions attributed to mucilage are highly controversial. These include water retention, regulation of germination, oxygen barrier, storage substance, and as a converging lens. The mucilage-forming substance is mostly present as secondary wall thickenings which swell after absorption of water. The pressure built up by the swelling slime layers

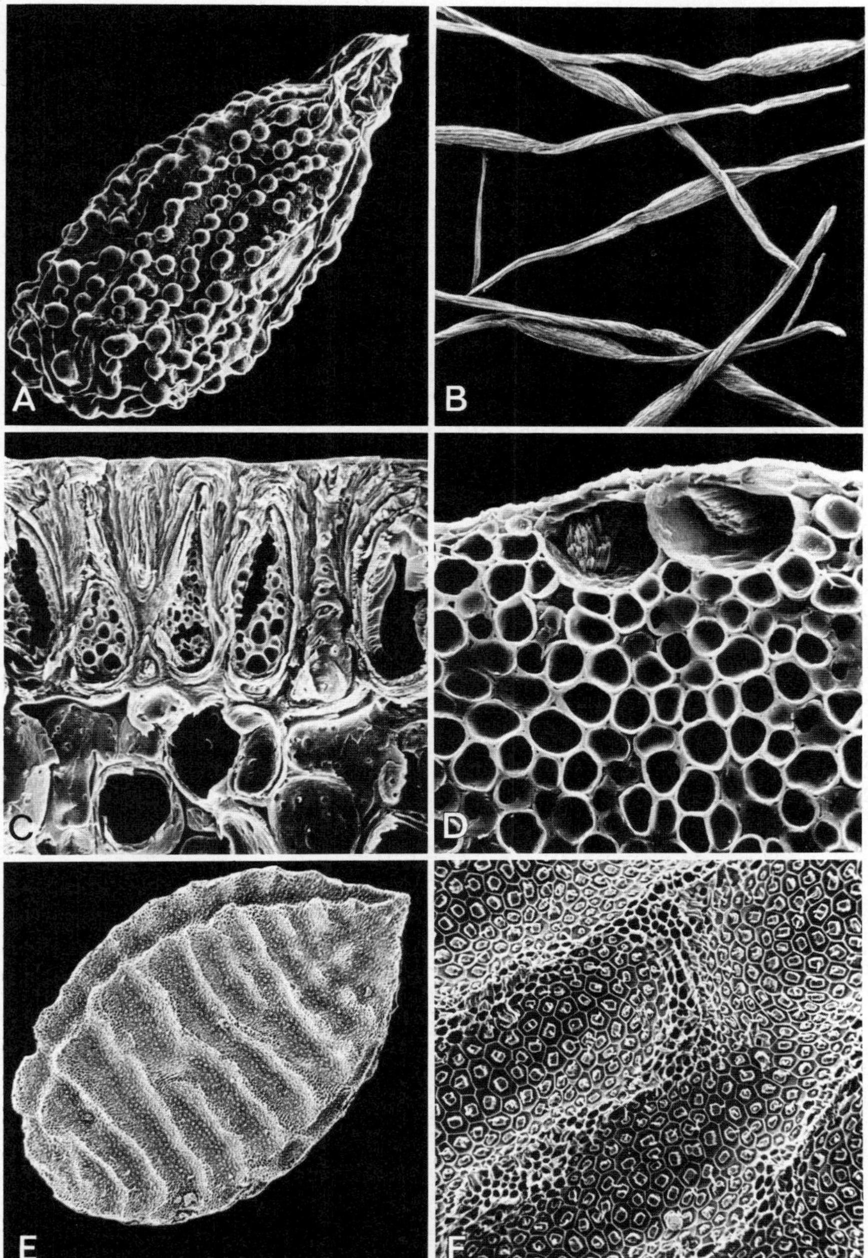

Fig. 12.13 A–F. SEM photographs. **A, B** *Voyria spruciana* and *Narthecium ossifragum,* "dust" seeds. **C, D** *Menyanthes trifoliata* and *Calla palustris,* transverse section of outer part of seed-coat with floating tissue. **E, F** *Oxalis europaea,* dispersed seed and detail of its surface

leads to rupture of the cuticle and exudation of slime. Mucilage cells often contain distinct internal structures as threads or spirals of cellulose embedded in mucilage. The spirals uncoil after the absorption of water.

Three types of mucilage production can be distinguished:

1. Epidermal mucilage production occurs in seeds of *Citrus* (Fig. 12.8 A); some Leguminosae have a subcuticular mucilage stratum which functions in endozoochory. A typical mucilage epidermis exists in seeds of Podostemaceae, Plantaginaceae, Cruciferae, and Linaceae (Fig. 12.6 D). After wetting, mucilage completely covers the seed as a result of a radial elongation of mucilage cells. Many Lythraceae have peripheral introversible mucilage hairs which turn inside out after some hours of soaking. Seeds of *Hydrocharis morsus-ranae* and *Stratiotes aloides* are covered by papillose mucilage cells. In many Acanthaceae the surface of the seed consists of unicellular mucilage hairs. The seed-coat of *Blepharis* is provided with hair tufts which excrete mucilaginous substances from their tips.
2. Epidermal and subepidermal mucilage production occurs in the seed-coats of *Helianthemum*.
3. Subepidermal mucilage production is reported in seeds of *Magonia glabrata*. The subepidermal mucilage-producing layers swell, destroy the epidermis of seed-coat, and excrete copious mucilage.

12.3.9 Active Dispersal of Seeds (Autochoric Seeds)

Autochory mechanisms are nearly always situated in fruit walls, and the only examples in which the seed-coat is responsible are *Oxalis* and *Biophytum* (Oxalidaceae). In Oxalidaceae the thick elastic cuticular membrane is excessively stretched in girth by the growth of the seed. Moreover, the parenchymatous cells of the outer integument become strongly turgescent by enzymatic degradation of starch immediately before the maturation of seed. Finally, the counteracting forces in the cuticular membrane and those exerted by the parenchymatous cells become so excessive that the seed-coat bursts at the antiraphal side. The outer part of the outer integument is very rapidly reversed, thus ejecting forcefully the seed without the outer part of the outer integument. The rupture takes place in the crystal-bearing inner layer of the outer integument, so that the ejected central part of the seed consists of the embryo, the inner integument, adjacent tangential walls, bases of the radial walls, and crystals of the crystal-bearing layer (Fig. 12.13 E, F) (Bouman 1974).

12.3.10 Color of Seeds

Brown, grey, and dim colors are by far the most common in seeds. Brown and black make up over half of the seed colors. Bright, often contrasting, colors, such as red, green, and yellow are infrequent. If present, they are mostly concerned in zoochorous dispersal, e.g., the brilliantly colored sarcotesta of *Magnolia*, red-orange arils in Meliaceae and Celastraceae.

Seeds imitating an aril are called mimetic seeds. Mimetic seeds with pseudoarils in legumes often have red and black colors, due to the presence of tannin, carotenoids, flavonoids, and anthocyanids, and chlorophyll and its derivatives. Flavonoids and anthocyanides are water-soluble pigments, and more frequent than colors in plastids. The black color of the seeds of many representatives of the Asparagales is brought about by a special phytomellin-containing layer. The color of seeds can also be the result of interference phenomena, especially when the outer wall is very thin as in Solanaceae and *Papaver*.

12.4 Adaptations to Germination

12.4.1 Opercula

There is scanty information on the structural aspects of seed germination. That is because the seed anatomists look at mature seeds, not at their germination, whereas vegetative morphologists and anatomists start with the seedling. Thus, the moment at which the embryonal root bursts the seed-coat is neglected. According to Klebs's summary (1884), there are three types of emergence of the germ root:

1. Commonly, the root penetrates the micropyle which bursts with slits or teeth, or in two or three valves.
2. The embryo bursts from its seed-coat, tearing the seed-coat to pieces.
3. The embryo finds its freedom by lifting a kind of seed lid (Fig. 12.15 A–C). Seed lids are mostly called operculum, sometimes embryostega, Samendeckel (German), stopper, or plug. Opercula have been known for a long time. At the end of the 18 th century, Gaertner (1788–1798) figured opercula in several families. Opercula show differences in their ontogeny, structure, as well as in mechanism of opening. Opercula are formed by the micropylar and hilar region, exo- and endostome, endostome alone, or by the endosperm haustorium (Campanulaceae). Opercula are more common in monocotyledons, and are known in about 25 families. In the monocots we also find a rather complicated type, in which the operculum is combined with the so-called micropylar collar (Commelinaceae, Costaceae, Maranthaceae, Musaceae, and Zingiberaceae). The micropylar collar is formed at the apex of the outer integument, and protrudes as an orbicular wedge into the nucellus. At the site of the micropylar collar the sclerotic layer of the seed-coat is discontinuous and, in some cases, the collar shows a special rupture layer (Fig. 12.14).

The other extreme case is that of *Asparagus setacea* (Liliaceae) with a false operculum. The embryonal root does not correspond with the micropyle, and a part of the seed-coat is pushed aside by the emerging root. Other monocot families with opercula are Araceae, Bromeliaceae, Palmae, Philydraceae, Pontederiaceae, and Typhaceae. Opercula are not so well-known in dicotyledons, and occur in about 20 families, half the number being families with parietal placentation (Begoniaceae, Datiscaceae, Elatinaceae, Hypericaceae, Turn-

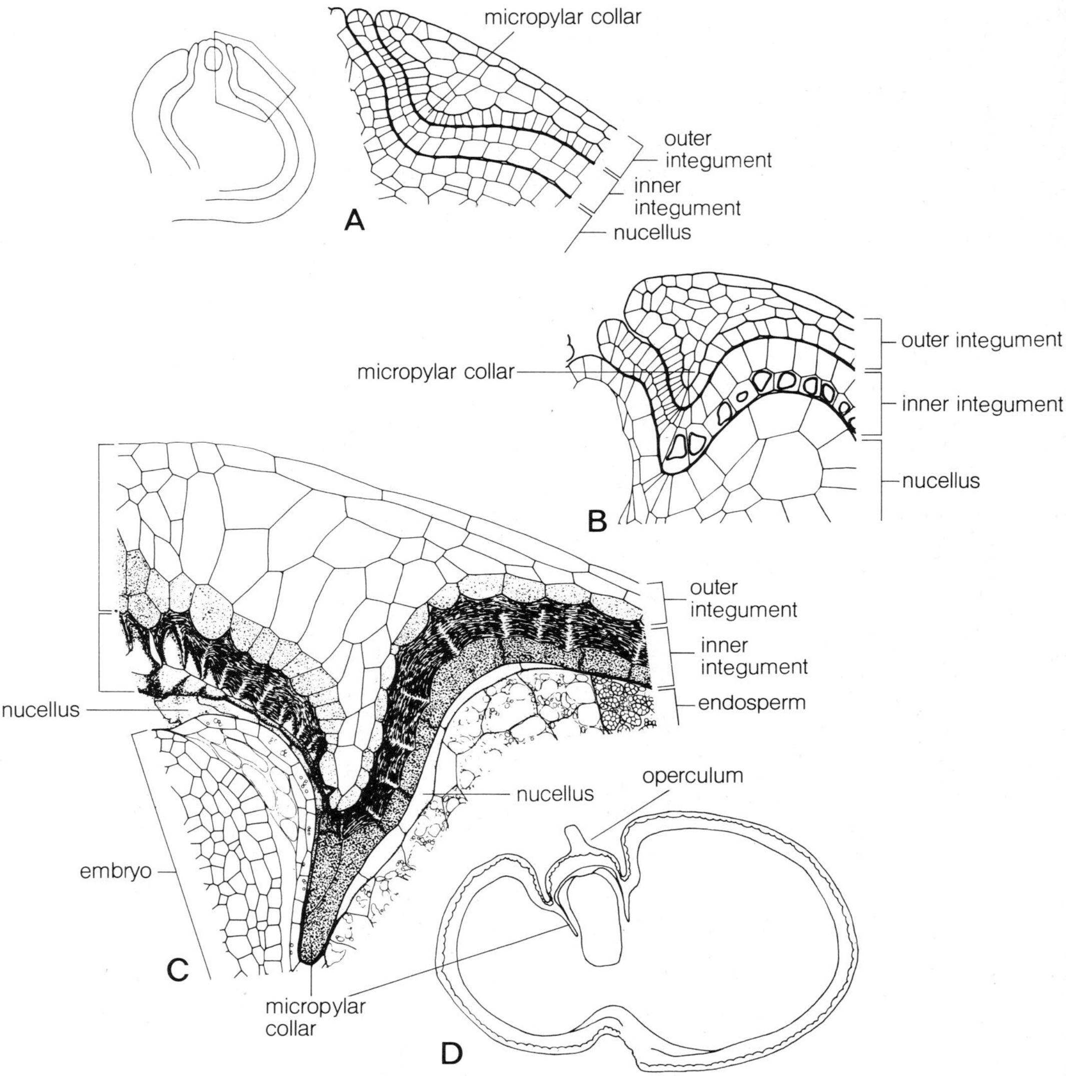

Fig. 12.14 A–D. *Stanfieldiella imperforata,* development of micropylar collar and operculum. **A** Mature ovule with initiation of micropylar collar. *Outline diagram* shows site of **A. B** Developing seed. **C** Almost mature seed showing interruption of the sclerotic exotegmen. **D** Longisection of mature seed after splitting off of the outer and middle layers of the testa. (Compare Fig. 12.15 B). (From Grootjen and Bouman 1981)

eraceae); Cactaceae, Lythraceae, Melastomataceae, Myrtaceae, and Nymphaeaceae. The opercula are more common than is generally accepted.

The opercula in angiosperms are possibly polyphyletic in origin, and of limited use for indicating phylogenetic relationships between the families. Also, the function of the opercula may be diverse. Sometimes they facilitate germination, i.e., they enable the development of strong, mechanically protective layers

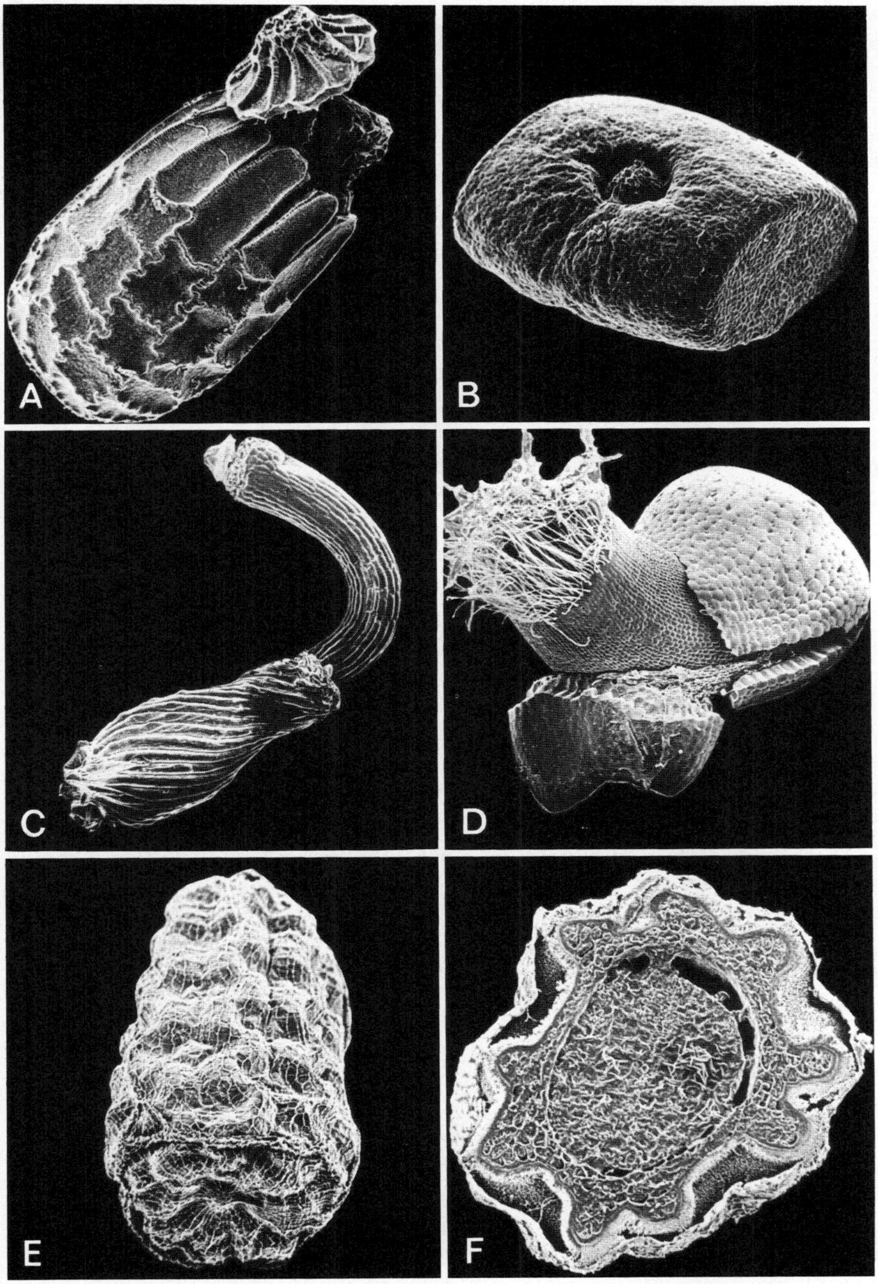

Fig. 12.15 A–F. SEM photographs. **A** *Begonia leptotricha,* **B** *Stanfieldiella imperforata.* (From Grootjen and Bouman 1981). **C** *Philydrum lanuginosum,* **D** *Cereus aethiops:* seeds showing opercula. **A, C, D** Germinated seeds. **E, F** *Verbascum densiflorum,* entire seed and transverse section showing rumination

in the seed-coat without any necessity to develop extreme pressure during germination. In other cases, opercula provide only an extra protection of the micropylar region of seed which is generally a weak zone in the seed-coat.

12.4.2 Imbibition Lids

The impermeability of seed-coats to water allows seeds of several families to remain dormant in the soil for long periods (see Sect. 2.3.6). Seeds can show special structural adaptations to regulate water loss as well as water absorption. In some Papilionaceae the hilum contains a cleft which functions as a one-way hygroscopic valve. It opens when the relative humidity of the atmosphere is rather low, thus permitting water loss and the seed to dry out, and closes in the reverse situation. A low water content of the seed tissue is one of the requirements for dormancy. However, under conditions of gradually increasing relative humidity, the fissure remains open and the seed imbibes water. Lersten (1982) speculates that the tracheid bar, a strip of tracheid-like cells occurring in the hilum of Papilionaceae, also enhances the efficiency of gas exchange. In imbibed seeds of Cannaceae, Malvaceae, and some Leguminosae, a part of the raphal region – the imbibition lid – sometimes containing the chalazal plug can be lifted, resulting in an interruption of the impermeable palisade layer and allowing water intake. In *Albizia lophanta* the imbibition lid is ejected after heating (Dell 1980). By this mechanism the seeds are enabled to imbibe and germinate after a forest fire, and to function as one of the pioneers in the destructed area. In Malvaceae the imbibition lid is formed by the tissue of raphe and chalaza. It may be present in both preripened and afterripened seeds. The subsequent release of the palisade at the place of the imbibition lid initiates imbibition of water (Egley and Paul 1981). The lid is lifted after special treatments, such as soaking in sulphuric acid or in hot water, high temperature, and pressure. In *Canna* the imbibition lid is of raphal origin (Grootjen and Bouman, personal observation). Mucilage hairs in *Blepharis ciliaris* (Gutterman and Witztum 1977) direct the micropylar end of the seed to the soil so that the seed root can easily penetrate the soil at germination (see also Sect. 12.3.8).

12.4.3 Epistase and Hypostase

Both terms were suggested by van Tieghem (1901). The hypostase is a specialized tissue situated in the chalazal region at the level of the origin of the two integuments and embryo sac. Its cytological characters are diverse, and may change during development. In young stages the thin-walled cells may be protoplasmic, whereas, in later stages, the cells are thick-walled and often sclerotic or lignified with tanniferous contents. Several functions have been ascribed to the hypostase. It is interpreted as a sort of glandular tissue secreting hormones eventually required for the growth of the embryo sac, as a tissue connecting the chalazal bundle with the embryo sac and thus facilitating transport of food materials, as a tissue regulating the water economy of the dormant seed, or as a barrier tissue preventing the extension of the embryo sac. In any case, in mature seeds, the hypostase seems to function as a blocking tissue of the chalazal opening. The hypostase seems to be a polymorphic and multifunctional structure. Detailed research is needed to elucidate its true nature.

The epistase consists of similar thick- or thin-walled cells in the micropylar part of the seed and, usually, develops from the apical cells of the nucellar epidermis. The nucellar cells may show radial elongation, occasionally by periclinal divisions. It forms a caplike structure of often thick-walled, and sometimes cutinized or sclerified cells covering the apex of the embryo sac and blocking the micropyle. In *Costalia* the epidermal cells show very pronounced sclerification, and in *Costus* the inner tangential walls of these cells become thickened.

12.5 Rumination and Labyrinth Seeds

Seeds in which the mature endosperm shows any degree of irregularity in its surface contour are called ruminate (Periasamy 1962). The term rumination is derived from the latin "ruminatus", which means chewed. The original concept was that the endosperm is eaten away by ingrowths of the periphery of the seed which is not true, but the term is in general use and has been retained.

Rumination is one of the peculiar features in the seeds of angiosperms. It occurs in about 30 mainly dicotyledonous families. Rumination is due to activity of the seed-coat, or the endosperm itself, during later stages of seed development. Periasamy distinguishes seven types of rumination, on morphological grounds:

1. Passiflora Type (Passifloraceae, Oxalidaceae) (Figs. 12.13 E, F, 12.16 B). In bitegmic ovules, the mature seed shows unequal radial elongation by any one of the layers of the seed-coat. In Passifloraceae the inner layer of the outer integument, or the outer layer of the inner integument, or a combination of both, is responsible.

2. Verbascum Type (Scrophulariaceae) (Fig. 12.15 E, F). This type resembles the Passiflora type, but occurs only in unitegmic ovules. In Scrophulariaceae the endothelial cells show unequal radial elongation.

3. Myristica Type (Myristicaceae, Palmae) (Fig. 12.16 C, D). The plurilayered seed-coat shows, by localized meristematic activity, definite ingrowths or infoldings which are vascularized.

4. Annona Type (Annonaceae, Aristolochiaceae, Degeneriaceae, Dipterocarpaceae, Ebenaceae, Menispermaceae, Vitaceae, and Palmae) (Fig. 12.16 A). This type resembles the Myristica type, but the ingrowths, in contrast to the integument, are not vascularized.

5. Coccoloba Type (Polygonaceae) (Fig. 12.16 E). This type occurs in one-layered seed-coats of bitegmic ovules.

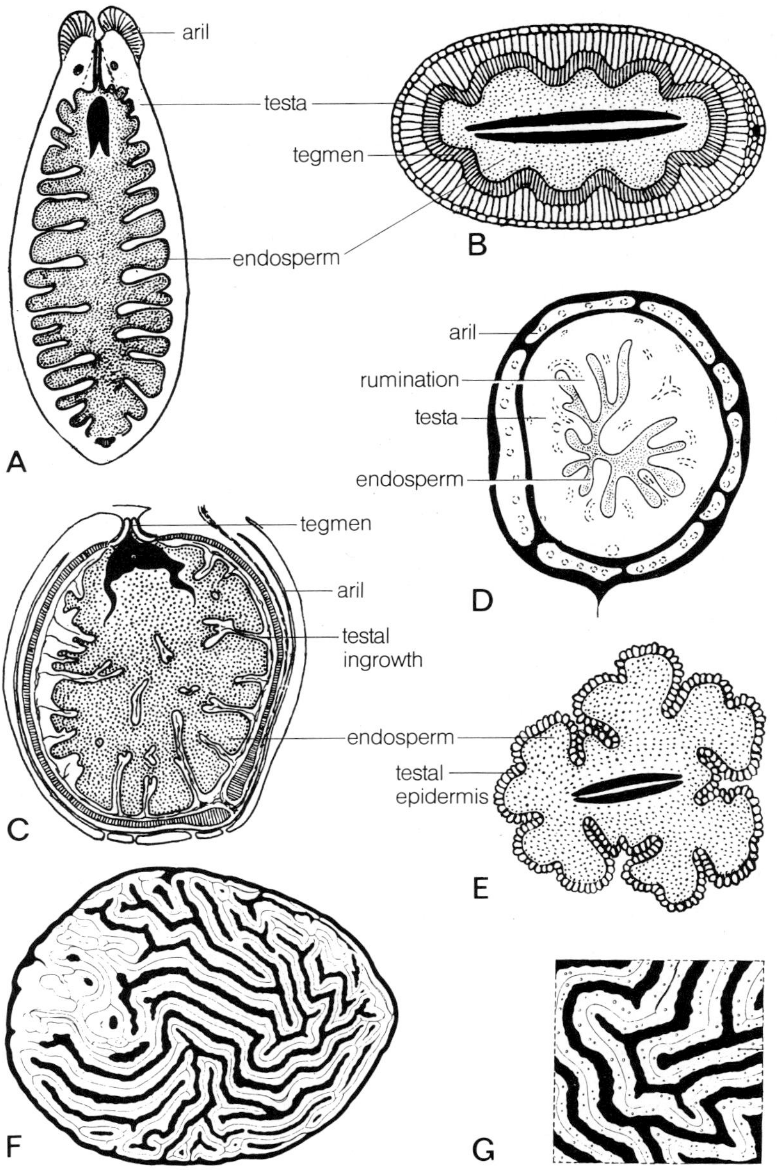

Fig. 12.16 A–G. Ruminated (**A–E**) and labyrinth (**F, G**) seeds. **A** *Annona squamosa* (l.s.). **B** *Passiflora calcarata (t.s.)*. **C, D** *Myristica fragans* (**C** l.s., **D** t.s.). **E** *Coccoloba uvifera* (t.s.). (**A, B, E** from Periasamy 1962, **C** from Voigt 1888, **D** adapted from Mohana Rao 1974). **F** *Kingiodendron pinnatum,* longisection of labyrinth seed. **G** Detail of testa intruding the cotyledons. (From van Heel 1970)

6. Spigelia Type (Apocynaceae, Loganiaceae, Araliaceae, Caprifoliaceae, and Rubiaceae). The rumination is the result of meristematic activity in the tissues of the seed-coat of unitegmic ovules.

7. Elytraria Type (Acanthaceae). The rumination is the result of unequal peripheral activity of the endosperm, during later stages of development.

Not much is known about the functions of rumination. Possibly, the surface enlargements of the endosperm assist in the intake of water and/or nutrients from the integuments.

Rumination has polyphyletically arisen in primitive and derived families, and occurs in both small and large seeds.

12.5.1 Labyrinth Seeds

The extreme form of rumination in which the lobing affects the embryo also is defined by van Heel (1970, 1971) as labyrinth seeds. A labyrinth seed is one which shows a network of folds or lobes in sections in every plane. Such a structure originates probably along different ontogenetic pathways (polyphyletically) but the end stages show much resemblance.

The testa first encroaches on the endosperm, and later the cotyledons, and in this way the labyrinth structure becomes pronounced. When the testa encroaches on the cotyledons, it may be located between portions of folded and lobed, mostly flat cotyledons, sometimes surrounded by a small amount of endosperm. Some Convolvulaceae and Caesalpiniaceae have storage cotyledons (Fig. 12.16 F, G), or the testa finds its way in many crevices of the massive cotyledons: *Hernandia peltata* (Hernandiaceae), *Mangifera* sp. (Anacardiaceae). Hardly anything is known about the possible ecological significance of labyrinth seeds. In Convolvulaceae it is of an advantage and, after germination, the photosynthetic surface is very large due to the unfolding of the cotyledons. Probably, in most labyrinth seeds, the cotyledons do not expand at germination, and by the large surface area between the cotyledon and endosperm, the short transport of nutrients to the cotyledons is effected.

12.6 Storage Tissue

The economic importance of seeds depends largely on the presence of reserve nutritive materials such as lipids, carbohydrates, and proteins. One of the paramount functions of seeds is energy storage. Germinating seeds have two critical requirements, namely, a source of carbon skeleton precursors, and a source of energy to assemble these precursors. These requirements are most effectively met in lipids which are, in this respect, superior to carbohydrates (Levin 1974). Lipids occur in the form of oil bodies in endosperm and embryo. The oil content varies from about 30% in sunflower to 50% in castor bean, maize, and peanuts, and even more in copra (*Cocos nucifera*) and oilpalm (*Elaeis guayanensis*).

Carbohydrates are stored as starch or as thickened walls mostly composed of hemicelluloses. Thickened walls are common in endosperm (many families of the Liliiflorae, Palmae, *Coffea*, Ebenaceae, Loganiaceae), or in cotyledons (*Impatiens, Lupinus*, and *Tropaeolum*). Cereals contain 70%–80% starch, beans ±50% starch.

Nearly all seeds contain protein reserves for the supply of nitrogenous compounds needed by the young seedlings before they are able to absorb nitrogen by their roots. Storage proteins occur as discrete protein bodies often called aleurone grains. These grains are the main location of protein and mineral storage in seeds. In addition to structurally homogeneous proteinaceous matrix, the aleurone grains may contain protein crystalloids, electron-dense globoid crystals, electron-transparent soft globoids, and crystals of calcium oxalate. Phosphorus and various cations (K, Mg, Ca, Fe, Ba, Mn) are located in globoid crystals, and can be studied with an energy dispersive X-ray analysis system (Lott 1981).

Protein bodies occur in the entire embryo or endosperm, or they are restricted to a specialized layer as in Gramineae and Leguminosae. This layer is active only during seed germination. The protein reserves are needed for the synthesis of enzymes used in the digestion of starch. Peanuts contain about 25% and soybean and walnut about 40% proteins.

Nutrient storage tissues in seeds can be endosperm, perisperm, or chalazosperm. Endosperm is the most common nutritive tissue for the embryo during development and germination. The endosperm is mostly triploid, and originates from the fusion product of the polar nuclei and the second male gamete. The development and structure of endosperm are dealt with in Chap. 7. The second type of storage tissue is perisperm. This is the general name for storage tissue of nucellar origin. It occurs in the monocots, mainly in Zingiberalean families, and in the dicots especially in the families of Centrospermae, Piperaceae, and Nymphaeaceae.

A nutritive tissue of chalazal origin has been termed chalazosperm. It is known in *Cyanastrum* (Cyanastraceae) where chalazal cells divide repeatedly and develop into a starch-containing tissue. Also, the storage tissue in pachychalazal seeds of *Canna* should be designated as chalazosperm. The mature seed-coat has no important storage function. Storage materials present in the seed-coat, during development, are mostly used up in the construction of the seed-coat itself (secondary wall thickenings, or slime-forming layers). While storage tissue is important for the survival of the embryo, on the other hand, a large quantity of storage materials has consequences for the weight and size of seeds and may conflict with dispersal strategies. The presence or absence of nutrient storage tissues is related to the choice between a few larger well-equipped seeds, or many small seeds with a low chance of survival. According to Stebbins (1974), the number and diversity of the compromises between these conflicting demands of the amount of stored food and dispersal strategies account, to a great extent, for the diversity in seed characters of angiosperms. Especially, in anemochorous seeds the weight greatly influences the potential distance of dispersal.

Seeds containing endosperm or perisperm are called albuminous, and those lacking these storage tissues exalbuminous. The terms endospermous and exendospermous concern only the endosperm. In endospermous seeds the endosperm grows ahead of the embryo. If, for any reason, the endosperm fails in its function of supplying nutrition to the embryo, abortive or defective seeds are likely to result. No absolute distinction exists between these two seed types. Also, in endospermous seeds a well or poorly developed transitional endosperm occurs during development. Often, only one or a few layers of endosperm can be detected in a mature seed. Combinations of storage in the embryo, endosperm, and perisperm are frequent. The endosperm is absent in Orchidaceae, Podostemonaceae, and Trapaceae.

Depending on the type of storage material, endosperm can be characterized as mealy (starch), or horny or stony (wall thickenings).

12.7 Concluding Remarks

Our knowledge of the structure of seeds has gradually increased during the last three decades. The external and internal structures of seeds are very diverse. Nevertheless, there are still many gaps to be filled. There are several angiospermous families, in particular some smaller ones with a restricted distribution, of which not even a single member has been investigated. Also, at the tribal level the seed characters of many of the more common families are not sufficiently known.

Some of the main trends in seed evolution are gradually becoming clear. There is a general trend from more complicated to more simplified seeds. Opposite trends leading to more specialized seeds also exist, however. For example, seeds can become more complicated by adaptations to special modes of dispersal (hairs, wings, or edible appendages). These adaptations may be the result of divergent courses of natural selection (anemochory vs. zoochory, etc.). Seeds can also become more complicated by adaptations to special germination processes (opercula, imbibition lids, etc.). The original angiospermous seed must have been rather complicated, and must have combined functions related to protection and dispersal, and also to germination. Their seeds must have had essentially testal seed-coats which did not show the variation existing among several of the more advanced families.

On the other hand, in many derived families the function of the seed-coat is transferred to the fruit wall, or to other peripheral floral parts by which the seed-coat became of lesser importance. The research is hampered by many pitfalls due to parallel and convergent evolution. Seemingly identical structures may have originated independently along different ontogenetic pathways.

The study of seed structure can, moreover, also be very helpful at lower taxonomic levels. Comparative anatomical and micromorphological studies of seed structures, especially in cooperation with systematics, palynology, and other related disciplines may elucidate many evolutionary relationships.

More attention should be paid to the different functions of the seed. Many functions are still obscure. An integration of the anatomical inquiry with the work of seed physiologists and ecologists may enhance our knowledge concerning the mechanisms involved in dispersal, dormancy, imbibition, and germination.

TEM records concerning the development and structure of seeds are still very meagre. In combination with histochemical research, it may solve problems regarding the flow of nutrients within seeds.

References

Bachmann E (1882) Darstellung der Entwicklungsgeschichte und des Baues der Samenschalen der Scrophulariaceen. Verh Kais Leop Carol Akad Naturf 43:1–179

Ballard LAT (1973) Physical barriers to germination. Seed Sci Technol 1:285–303

Baillon H (1876) Sur l'origine du macis de la muscade et des arilles en général. Adansonia 11:329–340

Barthlott W (1981) Epidermal and seed surface characters of plants: systematic applicability and some evolutionary aspects. Nord J Bot 1:345–355

Barthlott W, Ehler N (1977) Raster-Elektronenmikroskopie der Epidermis-Oberflächen von Spermatophyten. Trop Subtrop Pflanzenwelt 19:1–110

Barthlott W, Wollenweber E (1981) Zur Feinstruktur, Chemie und taxonomischen Signifikanz epicuticularer Wachse und ähnlicher Sekrete. Trop Subtrop Pflanzenwelt 32:1–67

Beasly CA (1975) Developmental morphology of cotton flowers and seed as seen with the scanning electron microscope. Am J Bot 62:584–592

Berg O (1865) Anatomischer Atlas zur pharmaceutischer Warenkunde. Gaertner, Berlin

Bhatnagar SP, Johri BM (1972) Development of angiosperm seeds. In: Kozlowski TT (ed) Seed biology, vol 1. Academic Press, New York London, pp 77–150

Boesewinkel FD (1978) Development of ovule and testa in Rutaceae III. Some representatives of the Aurantioideae. Acta Bot Neerl 27:341–354

Boesewinkel FD, Geenen J (1980) Development of ovule and seed-coat of *Erythroxylum coca* Lamk. Acta Bot Neerl 29:231–241

Boesewinkel FD, Been W (1979) Development of ovule and testa of *Geranium pratense* L. and some other representatives of the Geraniaceae. Acta Bot Neerl 28:335–348

Bouman F (1974) Developmental studies of the ovule, integuments and seed in some angiosperms. Thesis Univ Amsterdam

Brandza M (1891) Développement des téguments de la graine. Rev Gén Bot 3:227–240

Bresinsky A (1963) Bau, Entwicklungsgeschichte und Inhaltsstoffe der Elaiosomen. Bibl Bot 126:1–54

Brisson JD, Peterson RL (1976) A critical review of the use of scanning electron microscopy in the study of the seed-coat. Scanning Electron Microsc 2:477–495

Corner EJH (1976) The seeds of dicotyledons, 1 st edn, 2 vols. Cambridge University Press, Cambridge London New York Melbourne

Davis GL (1966) Systematic embryology of the angiosperms. Wiley, New York London Sydney

Dell B (1980) Structure und function of the strophiolar plug in seeds of *Albizia lophantha.* Am J Bot 67:556–563

Egley GH, Paul RN (1981) Morphological observations on the early imbibition of water by *Sida spinosa* (Malvaceae) seed. Am J Bot 68:1056–1065

Flint LH, Moreland CG (1943) Notes on the photosynthetic activity in seeds of the spider lily. Am J Bot 30:315–317

Gaertner J (1788–1798) The fruits and seeds of plants (in Latin), 3 vols. Stuttgart Tübingen. Carpologia, III vol. Lipsiae 1805–1807

Gallet F (1913) Développement et structure anatomique du tégument séminal des Rutacées. Thèse Univ Paris

Gassner G (1931) Mikroskopische Untersuchung pflanzlicher Nahrungs- und Genußmittel. 1. Aufl. Fischer, Jena

Grew N (1671) The anatomy of plants begun. London. Also in: The anatomy pf plants. 2nd edn, 1682. Royal Society, London

Grootjen CJ, Bouman F (1981) Development of ovule and seed in *Stanfieldiella imperforata* (Commelinaceae). Acta Bot Neerl 30:265–275

Grubert M (1974) Studies on the distribution of myxospermy among seeds and fruits of Angiospermae and its ecological importance. Acta Biol Venez 8:315–551

Grubert M (1981) Mucilage or gum in seeds and fruits of angiosperms: A review. In: Minerva Publikation München. Minerva Fachserie Naturwissenschaften
Guérin P (1899) Recherches sur le développement du tégument séminal et du péricarpe des Graminées. Ann Sci Nat Bot (8) 9:1–59
Guignard L (1893) Note sur l'origine et la structure du tégument séminal chez les Capparidées, Résédacées, Hypericacées, Balsaminées et Linacées. Bull Soc Bot Fr 40:56–59
Guttenberg H von (1971) Bewegungsgewebe und Perzeptionsorgane. In: Zimmermann W, Carlquist S, Ozenda P, Wulff HD (Hrsg) Handbuch der Pflanzenanatomie, 2. Neubearb. Aufl. Band V/5. Borntraeger, Berlin Stuttgart
Gutterman Y, Witztum A (1977) The movement of integumentary hairs in *Blepharis ciliaris* (L.). Burtt. Bot Gaz 138:29–34
Harz CD (1885) Landwirthschaftliche Samenkunde. Bd II. Parey, Berlin
Heel WA van (1970) Some unusual tropical labyrinth seeds. Proc K Ned Akad Wet C73:298–301
Heel WA van (1971 a) The labyrinth seed of *Hernandia peltata* Meissn. in DC. Proc K Ned Akad Wet C74:46–51
Heel WA van (1971 b) Notes on some more tropical labyrinth seeds. Blumea 19:109–111
Heel WA van (1982) Note on the structure of developing seeds of *Knema* and *Horsfieldia* (Myristicaceae). Blumea 28:53–60
Jernstedt JA, Clark C (1979) Stomata on the fruits and seeds of *Eschscholzia* (Papaveraceae). Am J Bot 66:586–590
Kapil RN, Bor J, Bouman F (1980) Seed appendages in angiosperms. Bot Jahrb Syst 101:555–573
Klebs G (1884) Beiträge zur Morphologie und Biologie der Keimung. Unters Bot Inst Tübingen 1:536–635
Kloos A, Bouman F (1980) Case studies in aril development: *Passiflora suberosa* L. and *Turnera ulmifolia* L. Beitr Biol Pflanz 55:49–66
Koch L (1908) Die mikroskopische Analyse der Drogenpulver. Bd IV. Die Samen und Früchte. Borntraeger, Leipzig
Lersten NR (1982) Tracheid bar and vestured pits in legume seeds (Leguminosae: Papilionoideae). Am J Bot 69:89–107
Levin DA (1974) The oil content of seeds: An ecological perspective. Am Nat 108:193–206
Lohde G (1874) Über die Entwicklungsgeschichte und den Bau einiger Samenschalen. Thesis Univ Leipzig. Also in: Schenk A, Luerssen C (1875) (Hrsg) Mittheilungen aus dem Gesammtgebiete der Botanik. Bd II:43–80
Lott JNA (1981) Protein bodies in seeds. Nord J Bot 1:421–432
Malpighi M (1675) Anatomy of plants (in Latin). Royal Society London. Also in: Collected works (in Latin), 1687. Venice
Marinos NG (1970) Embryogenesis of the Pea (*Pisum sativum*). I. The cytological environment of the developing embryo. Protoplasma 70:261–279
Mohana Rao PR (1974) Nutmeg seed: Its morphology and developmental anatomy. Phytomorphology 24:262–273
Netolitzky F (1926) Anatomie der Angiospermen-Samen. In: Linsbauer K (Hrsg) Handbuch der Pflanzenanatomie. Bd X,4. Borntraeger, Berlin
Nobbe F (1876) Handbuch der Samenkunde. Wiegand Hempel Parey, Berlin
Olsen S, Olsen JD (1980) The seed of *Boschniaka hookeri* (Orobanchaceae). Bot Tidsskr 75:159–172
Pate JS, Gunning BES (1972) Transfer cells. Annu Rev Plant Physiol 23:173–196
Periasamy K (1962) The ruminate endosperm: Development and types of rumination. In: Plant embryology: A symposium. CSIR, New Delhi, pp 62–74
Pfeiffer A (1891) Die Arillargebilde der Pflanzensamen. Bot Jahrb Syst 13:492–540
Pijl L van der (1982) Principles of dispersal in higher plants. 2 nd edn. Springer, Heidelberg New York
Planchon JE (1845) Développements et caractères des vrais et des faux arilles. Ann Sci Nat Bot (3) 3:275–312
Rauh W, Barthlott W, Ehler N (1975) Morphologie und Funktion der Testa staubförmiger Flugsamen. Bot Jahrb Syst 96:353–374
Reeve RM (1946) Structural composition of the sclereids in the integument of *Pisum sativum* L. Am J Bot 33:191–204
Rijven AHGC (1952) In vitro studies on the embryo of *Capsella bursa-pastoris*. Acta Bot Neerl 1:157–200

Rugenstein SR, Lersten NR (1981) Stomata on seeds and fruits of *Bauhinia* (Leguminosae: Caesalpinioideae). Am J Bot 68:873–876

Schleiden M (1839) Über die Bildung des Eichens und Entstehung des Embryos bei den Phanerogamen. Verh Kais Leop Carol Akad Naturf 11:27–58. Also in: Sur la formation de l'ovule et de l'origine de l'embryon dans les Phanérogames. Ann Sci Nat Bot (2) 11:129–141

Schroeder J (1871) Untersuchung der Samen der *Brassica*-Arten und Varietäten. Landwirthschaftl. Versuchs-Stationen 14:179–194

Sell Y (1969) La dissémination des Acanthacées. Variations sur le type xérochasique fondamental. Rev Gén Bot 76:417–493

Sempolowski A (1874) Über den Bau der Schalen landwirthschaftlich wichtiger Samen. Landwirthschaft Jahrb 3:823–866. Also in: Thesis Leipzig

Sernander R (1906) Entwurf einer Monographie der Europäischen Myrmekochoren. K Sven Vetenskapsakad Handl 41(7):1–410

Singh B (1964) Development and structure of angiosperm seed. I. Bull Natl Bot Gard (Lucknow) 89:1–115

Singh D (1967) Structure and development of seed-coat of the Cucurbitaceae. I. Seeds of *Biswarea* Cogn., *Edgaria* Clarke and *Herpetospermium* Hook. f. Proc Indian Acad Sci Sect B 65:267–274

Singh D, Dathan ASR (1972) Structure and development of seed-coat in Cucurbitaceae. VI. Seeds of *Cucurbita*. Phytomorphology 22:29–45

Singh RP (1970) Structure and development of seeds in Euphorbiaceae: *Jatropha* species. Beitr Biol Pflanz 47:79–90

Spurny M (1973) The imbibition process. In: Heydecker W (ed) Seed ecology. Butterworths, London, pp 367–389

Stebbins GL (1974) Flowering plants. 1 st edn. Harvard Univ Press, Cambridge Mass

Swain TS (1979) Tannins and lignins. In: Rosenthal GA, Janzen DH (eds) Herbivores, their interaction with secondary plant metabolites. Academic Press, New York, pp 657–675

Targioni-Tozzetti A (1855) Saggio di studi intorno al guscio dei semi. Memoria della Reale Academia delle Sciense di Torino, Serie seconda 15:359–444

Tieghem P van (1901) L'oeuf des plantes considéré comme la base de leur classification. Ann Sc Nat Bot (8) 14:213–390

Tschirch A, Oesterle O (1893–1900) Anatomischer Atlas der Pharmakognosie und Nahrungsmittelkunde. Tauchnitz, Leipzig

Vaughan JG (1970) The structure and utilization of oil seeds. 1 st edn. Chapman and Hall, London

Verkerke W, Bouman F (1980) Ovule ontogeny and its relation to seed-coat structure in some species of *Polygala* (Polygalaceae). Bot Gaz 141:277–282

Voigt A (1888) Untersuchungen über Bau und Entwicklung von Samen mit ruminiertem Endosperm aus den Familien der Palmen, Myristaceen und Anonaceen. Ann Jard Bot Buitenz 7:150–190

Wardlaw CW (1965) Physiology of embryonic development in Cormophytes. In: Ruhland E (ed) Encyclopedia of Plant Physiology, vol XV,1. Springer, Berlin Heidelberg New York, pp 844–965

Wisselingh C van (1918–1924) Contributions to the knowledge of the seed-coat (in Dutch). Pharm Weekbl

Youle RJ, Huang HC (1981) Occurrence of low molecular weight and high cysteine-containing albumin storage proteins in oilseeds of diverse species. Am J Bot 68:44–48

13 The Seed: Germination

J. V. Jacobsen

13.1 Introduction

Knowledge of the mature seed and its germination is of importance in the study of seed formation. Indeed, embryogeny and germination are extensions of each other separated by a period of relative metabolic inactivity called quiescence, and are essentially different phases of the continuing process of embryo growth and development. Obviously, the nature of the pre-quiescence embryo bears very heavily on early germination behaviour and its control, and answers to questions relating to germination behaviour are increasingly being sought during the period of seed development.

Germination has been studied from a number of viewpoints. For many years, seed dormancy and its causes were of prime concern to many seed physiologists and much has been written about it (for reviews see Koller et al. 1962, Vegis 1964, Barton 1965a, c, Evenari 1965, Lang 1965, Stokes 1965, Wareing 1965, Amen 1968, Roberts 1972b, Thomas 1972, Villiers 1972, Khan 1977, Nikolaeva 1977, Bewley and Black 1978). In the last 15 years, a large part of seed research was devoted to understanding events of the normal germination process and seedling establishment. As a result of this, our understanding of the structure and composition of the mature seed, the structural changes occurring during germination, and the molecular events and controls associated with germination has increased greatly. Storage organs (cotyledons and endosperm), in particular, have received considerable attention in relation to the breakdown of storage reserves and its control.

Although germination is usually defined as protrusion of the radicle from the seed-coat, it would be artificial to limit this discussion to that event. Radicle emergence is not necessarily an indication of the extent of structural and physiological change within the seed. In seeds with very large embryos like pea, germination can occur after very little change, but it occurs only after considerable growth, morphogenesis, and cellular reorganization in seeds with small undeveloped embryos like *Fraxinus*. In the former, reserve breakdown and reorganization of embryonic cells is primarily a post-germination phenomenon but, in the latter, it can occur before germination. This discussion is concerned with changes occurring within seeds associated with the resumption of embryo growth and development following hydration regardless of whether they are pre- or post-germination changes.

In line with recent developments in the study of seed germination, the emphasis of this chapter is on cell structure and function during normal germination, particularly in relation to breakdown of stored food within the embryo (axis and cotyledons) and in the extra-embryonic storage tissues (endosperm and perisperm), and on interrelations between the embryo and extraembryonic storage tissues.

13.2 General Considerations

13.2.1 Internal Morphology

In any study of germination it is important to identify the seed components and their relative locations within the seed. The seed is not an amorphous structure but it has component parts with a high degree of structural and physiological differentiation. An understanding of the physiology of the whole seed can only come from an understanding of the functions and divisions of labour of its component parts. Information on internal and external morphology of seeds is available from papers by Netolitzky (1926) and Martin (1946), and from a variety of reviews and textbooks (Haywood 1938, Corner 1951, 1976, Fahn 1974, Mayer and Poljakoff-Mayber 1975, Esau 1977, Bewley and Black 1978). Many of the studies, which will be discussed later, provide more recent contributions to the subject of seed structure, but much of the recent work involves ultrastructural studies and rarely provides information on additional species.

The analysis of the internal morphology of seeds by Martin (1946) is particularly useful and interesting (see also Chap. 12 by Boesewinkel and Bouman). It is primarily a study of the relative size, shape, and position in the seed of the embryo and extra-embryonic storage tissues, and covers representatives from 1887 genera. It not only catalogues internal morphology but provides a system of classification as well as a phylogenetic analysis. From this and other studies mentioned above, a good overall picture of seed structure can be formed. However, perhaps this applies more to the embryo than to other seed tissues.

The structure and composition of the endosperm and probably the perisperm, and their location relative to the embryo, are of great importance in germination. Except for a few species, detailed information on endosperm is not available. However, interpreting some of the descriptive terminology, the following generalizations emerge from the studies mentioned above but particularly those of Martin. Endosperm occurs, to some extent, in the seeds of almost all families varying in amount from tissue which occupies most of the seed volume to a tissue which is composed of only a few cell layers. Those with copious endosperm (and probably perisperm as well) are called endospermic or albuminous seeds, and those with little (or none) are referred to as non-endospermic or exalbuminous. In most plant families the endosperm appears to consist of living cells, although in some cases it consists of both living and dead cells, as in Gramineae and perhaps Leguminosae. The endosperm not only provides nutrition during the early resumption of growth and differentiation of the embryo, but it may physically restrict embryo expansion and particularly radicle emergence. Where endosperm is copious, the embryo is often embedded in and completely surrounded by it, and even when the endosperm is reduced almost to obscurity, as in many Leguminosae (Corner 1951), it is located around the radicle and may play a role in radicle emergence. Thus, the structure and composition of the endosperm as well as the control of its function are likely to be fundamental to the study of germination.

13.2.2 Composition of Seeds and Reserve Foods

The role of seeds in the plant life cycle requires that they store materials which can be made available to support the early growth and development of the embryo, and to aid establishment of the seedling. The reserves are stored commonly as polymers including protein, lipid, and carbohydrate (starch or various cell wall polymers), and in addition compounds like phytin and calcium oxalate. All of these reserves occur in discrete cellular organelles.

Studies of the composition of seed reserves show that there is considerable variation in oil, protein, and starch content (Crocker and Barton 1957, Earle and Jones 1962, Mayer and Poljakoff-Mayber 1975, Bewley and Black 1978). Figure 13.1 A, B shows that reserve levels can vary widely not only between families, but within families and even within genera (Earle and Jones 1962) but, in general, a gradation can be found among families from high to low contents of oil and protein. It would be of interest to know if such gradations exist within genera.

Reserves are commonly stored throughout the seed but not uniformly. Some reserves tend to be localized in different tissues. Living endosperm rarely contains starch but it commonly stores lipid (as oleosomes), protein (in protein bodies), phytin (in protein bodies), calcium oxalate (as crystals in protein bodies), and carbohydrate is usually stored in the form of thickened cell walls. In dicotyledons the endosperm walls are often thickened with mannans, as in Leguminosae (Bailey 1971), *Lactuca sativa* (lettuce) (Halmer et al. 1975), *Carum carvi* (caraway) (Hopf and Kandler 1977), and coffee (Wolfrom et al. 1961). In monocotyledons the walls of living endosperm are thickened either with mannan, as in *Iris* (Andrews et al. 1953), *Scilla nonscripta* (Thompson and Jones 1964), and *Phytelephas macrocarpa* (ivory nut), and *Phoenix dactylifera* (date) (Meier 1958), or with β-glucan and arabinoxylan as in barley and wheat aleurone (Bacic and Stone 1979). The dead endosperm of the cereals stores protein (in protein bodies) and carbohydrate mainly as starch grains but also as cell wall β-glucans and heteroxylans (Mares and Stone 1973, Fincher 1975, Ballance and Manners 1978, Bacic and Stone 1979). Cotyledons store protein (in protein bodies), lipid (as oleosomes), and starch grains although the proportions vary in different species. For example, in broad bean (*Vicia faba*) the ratio of protein : lipid : starch is 23 : 1 : 56, but in *Arachis hypogaea* (peanut) it is 31 : 48 : 12 (Bewley and Black 1978). Perisperm, at least in *Yucca*, stores protein and lipid but not starch (Horner and Arnott 1966). Thus, the composition of the reserves can provide information about the internal structure of the seed. Oily seeds can have almost any internal structure but starchy seeds, if not cereals, usually have large embryos and little extra-embryonic reserve.

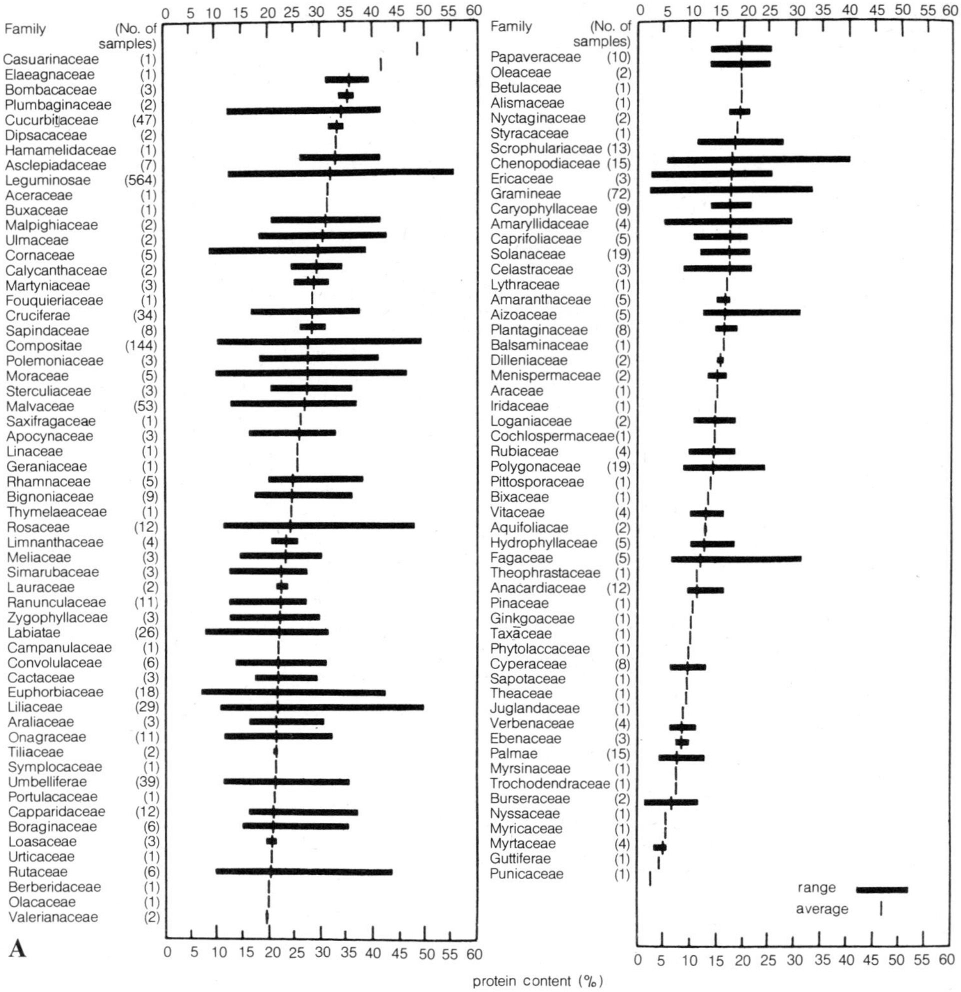

Fig. 13.1 A, B. Protein (**A**) and oil (**B**) content of seeds. Range and mean are shown for the samples analyzed for each family. (From Earle and Jones 1962)

13.3 Germination: Structural, Compositional, and Physiological Changes

It is not the intention of this section to cover gross morphology of seeds during germination and subsequent seedling growth. The general nature of epigeal and hypogeal germination has been discussed and illustrated in various places (Berlyn 1972, Foster and Gifford 1974, Mayer and Poljakoff-Mayber 1975, Bewley and Black 1978). Anatomy of growth and differentiation of the young seedling has also been discussed in various reviews and textbooks (see for example Berlyn 1972,

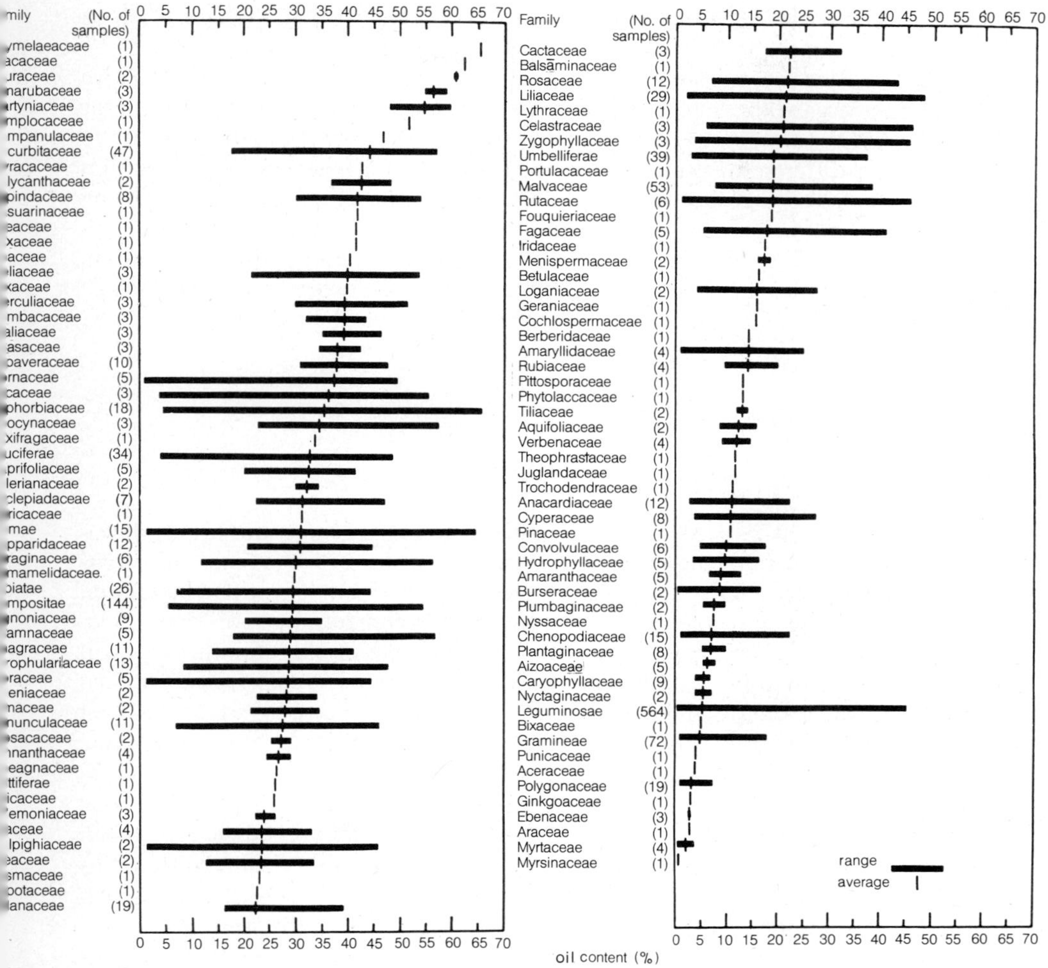

Fig. 13.1 B

Foster and Gifford 1974, Esau 1965 a, b, 1977). The purpose of this section is to review those aspects of seed germination and seedling growth that have not received any extensive treatment previously.

Germination involves resumption of growth and development of the embryo and the transition of cells from a desiccated state having very low metabolic activity to a hydrated, metabolically active one. Water is taken up by many seeds in a triphasic manner (for discussion see Bewley and Black 1978). The first phase is rapid involving hydration of the tissues, the second phase is a relatively long period of full hydration during which major metabolic and structural changes begin, and the third phase, a period of rapid and sustained water uptake, is usually associated with germination and continued embryo growth. This section is concerned primarily with the events of phase 2 and perhaps the early part of phase 3 during which major structural, compositional, and functional changes occur in almost all cells of the seed, in some cases before germination occurs.

Many cells of seeds, both embryonic and extra-embryonic, contain reserve food. Other organelles characteristic of metabolically active cells are often either not present (or cannot be visualized), or are present in small numbers, or are in some poorly differentiated and ill-defined form. The early stages of germination are devoted to the degradation of food reserves, the development of pre-existing organelles, and the synthesis of new organelles to build up a complement of cell structures normally associated with metabolism like endoplasmic reticulum, mitochondria, microbodies, vacuoles, dictyosomes, ribosomes, and plastids of various types, all of which contribute to the maintenance of seedling growth.

13.3.1 Dicotyledons

13.3.1.1 Growth of Embryo Within the Seed

Embryo size in mature seeds ranges from those which are large and highly developed occupying most of the seed volume to those which are very small and commonly occupy only a small part of the seed (Martin 1946). The latter have been referred to as poorly or incompletely developed, rudimentary, undifferentiated, or immature. Rudimentary embryos occur in a number of plant families (Martin 1946) and some have been studied in detail, for example, *Heracleum sphondylium* (Stokes 1952a, b, 1953a) and *Apium graveolens* (Jacobsen and Pressman 1979) in Umbelliferae, *Anemone coronaria* (Bullowa et al. 1975) in Ranunculaceae, *Fraxinus excelsior* (Villiers and Wareing 1964) in Oleaceae, *Ilex opaca* (Ives 1923) in Aquifoliaceae, *Viburnum* (Giersbach 1937) in Caprifoliaceae, and various members of the Orchidaceae (Wirth and Withner 1959). Before radicle protrusion occurs in these species, the embryos usually grow and develop within the seed. Immaturity of the embryo has been listed as a type of dormancy (Crocker 1916, Villiers 1972), but immaturity of the embryo and dormancy of the seed can be separated. In *Heracleum* and *Apium graveolens,* low temperature and light, respectively, are necessary for the protrusion of the radicle and these seeds show true dormancy. However, embryo enlargement within the seed can occur following hydration. In *Fraxinus* the embryo enlarges within the seed at normal temperatures, but the radicle protrudes only after a period of low temperature (Villiers and Wareing 1964). Embryos of *Anemone coronaria* (Bullowa et al. 1975) and *Ilex opaca* (Ives 1923), on the other hand, are quiescent, and embryo enlargement and radicle emergence, albeit delayed, occur following hydration. Therefore, embryo enlargement can occur within seeds after hydration alone and, thus, morphological immaturity is not always associated with dormancy.

Intraseminal growth (growth within the seed) is a phenomenon which has only been described in a few dicotyledonous species. The rudimentary embryos are surrounded by fleshy (living) endosperm, and the embryos enlarge at the expense of the surrounding tissue. In *Heracleum sphondylium,* at low temperature, embryo growth is accompanied by degradation of the endosperm cells adjacent to it (Stokes 1952a, b, 1953a, b). The cell walls and endosperm protein breakdown and

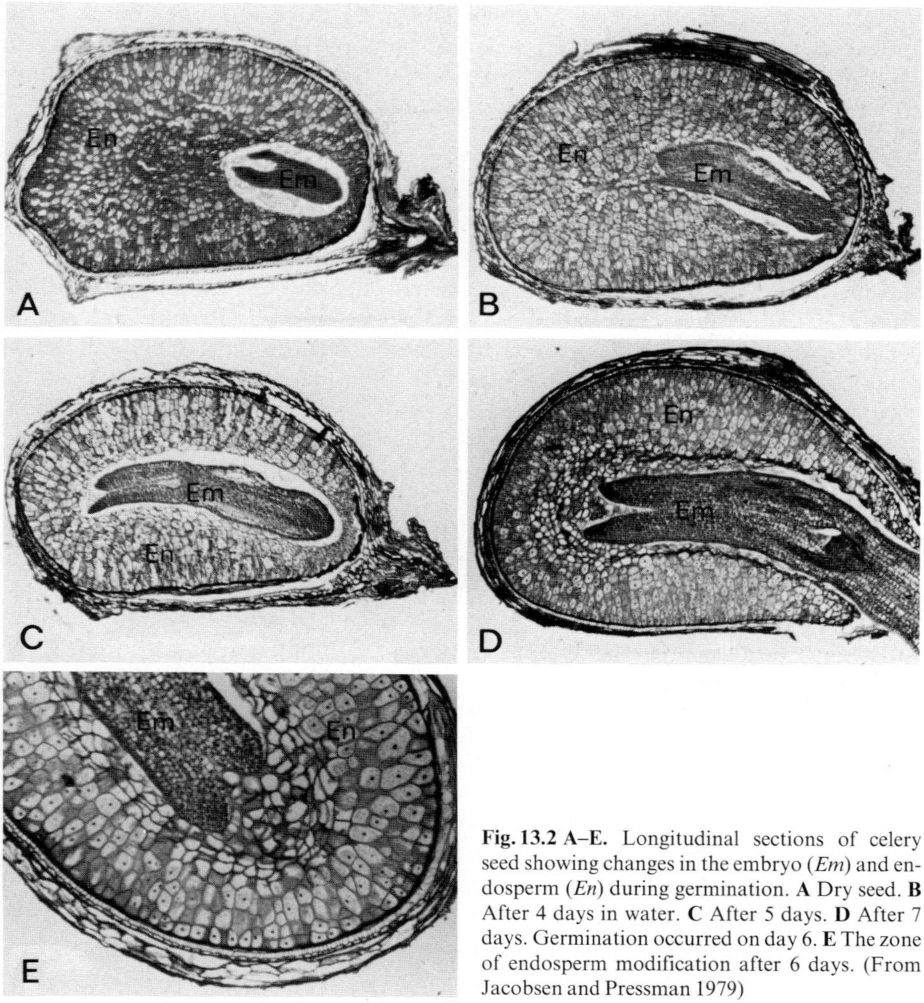

Fig. 13.2 A–E. Longitudinal sections of celery seed showing changes in the embryo (*Em*) and endosperm (*En*) during germination. **A** Dry seed. **B** After 4 days in water. **C** After 5 days. **D** After 7 days. Germination occurred on day 6. **E** The zone of endosperm modification after 6 days. (From Jacobsen and Pressman 1979)

the products flow to the expanding embryo where, among other things, starch accumulates. Embryo growth is apparently dependent on nutrition provided by endosperm breakdown which, in turn, is dependent on some factor (probably not enzymatic) released by the embryo during cold treatment. Intraseminal embryo growth in *Apium graveolens* appears to follow a very similar course (Jacobsen and Pressman 1979). The embryo is originally about one-third of the length of seed and, over a period of 6 days, elongates to about three-fourths the length of the seed before the radicle emerges (Fig. 13.2 A–E). The zone of depleted endosperm cells adjacent to the embryo is clearly visible. Endosperm breakdown may be under the control of gibberellin emanating from the embryo, since isolated endosperm can be caused to degrade by addition of the hormone (Jacobsen et al. 1976).

Intraseminal embryo growth occurs uniformly throughout the embryo. In both *Apium graveolens* (Jacobsen and Pressman 1979) and *Anemone coronaria* (Bullowa et al. 1975) embryo elongation begins 1–2 days following hydration, and growth occurs both in the axis and cotyledons.

13.3.1.2 The Sequence of Cell Activation

From the meager information available it appears that during hydration the cells become activated in an ordered sequence throughout the embryo. In lettuce (Srivastava and Paulson 1968) and *Sinapis* (Werker and Vaughan 1974) changes begin in the hypocotyl and proceed basipetally towards the radicle and acropetally through the cotyledons. At all levels in the embryo, gradients of activation occur across the organs. The sequence of cell activation in cotyledons, studied mostly in legumes, is very complex and varies from species to species (see Bewley and Black 1978). In one group of species, protein body degradation begins in the epidermis and in the cells close to the vascular bundles. In another group it begins in cells farthest from the vascular bundles, and in a third group, the pattern of activation is not related to the vascular bundles and can begin almost anywhere. Questions relating to the mechanisms underlying the formation of such patterns are unanswered. Do the patterns represent physiological programming within the embryo, or do they reflect artifactual patterns resulting from hydration or fixation and embedding of tissue for microscopy (see also Yoo 1970)? If the former, why do such patterns occur and what are the mechanisms involved, and do the differences in cotyledons reflect fundamental differences in control of reserve breakdown? Much more work on a wide variety of species is required before these and other questions can be answered.

13.3.1.3 Changes in Cotyledons

At the beginning of hydration, cotyledonary cells are characteristically packed with starch grains, protein bodies, and oleosomes. The overall changes in cotyledonary composition reflect breakdown of reserves and transport of the products to the embryonic axis. The major changes in cotyledons of *Pisum sativum* are shown in Fig. 13.3. The fresh weight increases during the first few days of germination and then gradually declines. The concentration of soluble sugars rises rapidly and this is associated with starch hydrolysis and a reduction in starch content, which begins a day or so following hydration. Protein hydrolysis begins at the same time as starch hydrolysis, and protein nitrogen falls in parallel with starch content. After 5 or 6 days, transport of materials out of the cotyledons begins and this is reflected in a rapid fall in soluble sugars, total nitrogen and total dry weight. Since radicle emergence in *Pisum sativum* occurs on day 2, starch and protein breakdown are only in their very early stages at that time. The rate of cotyledonary respiration shows an initial rapid rise and then maintains a high rate. The overall loss of sugar from the cotyledons, therefore, is probably accounted for primarily by transport to the axis and by providing energy for maintenance of coty-

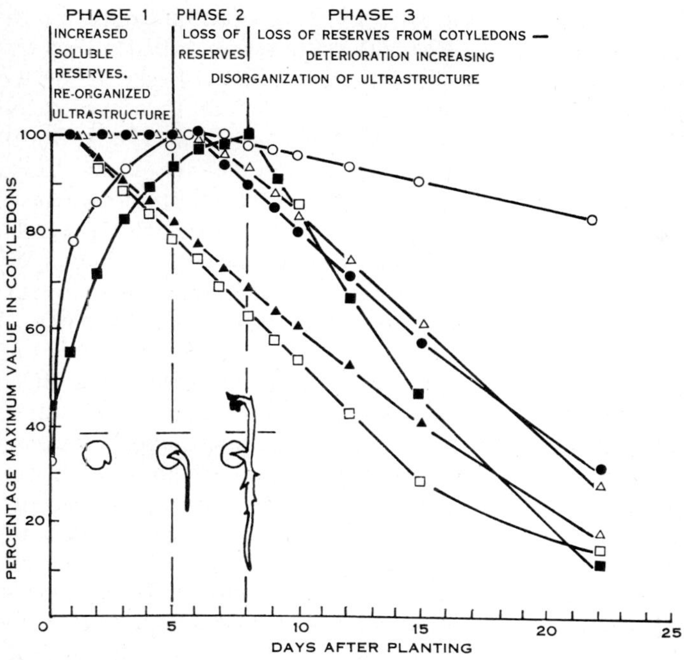

Fig. 13.3. Changes in fresh weight (○), dry weight (●), total nitrogen (△), protein nitrogen (▲), starch (□), and sugar (■) in the cotyledons during germination of pea seed, and establishment of seedling. Growth of seedling has been divided into three morphological phases. Data are expressed as a percentage of their maximum value in the cotyledons during 22 days after planting. (From Bain and Mercer 1966)

ledonary metabolism. Changes in cotyledons of other species show basically the same trends (Bewley and Black 1978).

Associated with these compositional changes there are striking structural changes in protein bodies. In *Vicia faba* protein bodies initially swell to about twice their size during hydration. Further increase in size is associated with the commencement of rapid protein hydrolysis and is due to coalescence of protein bodies. Similar size changes have been reported for protein bodies of *Vigna unguiculata* (Harris et al. 1975). During the initial swelling, which occurs usually over 1–1½ days, protein body contents become more diffuse and with protein hydrolysis the contents disappear, usually after 3–5 days. The bodies become vacuoles which coalesce to form larger vacuoles and, ultimately, often one large central vacuole. The original unit membrane of the protein body becomes the tonoplast of the cell although presumably its nature could change in the process. The same general picture of protein body change has been reported in a number of cases which have been reviewed at length by Ashton (1976) and Lott (1980).

There are some aspects of protein body change that vary from species to species. Protein breakdown can occur internally, leading to cavitation and ultimately fragmentation, or it can occur peripherally. Also, fusion of protein bodies can occur at any time during protein hydrolysis. In *Pisum arvense* (Smith and Flinn 1967), *Yucca* (Horner and Arnott 1965), *Vicia faba* (Briarty et al. 1970). *Arachis hypogaea* (Bagley et al. 1963) and wheat scutellum (Swift and O'Brien 1972), fusion occurs when the protein bodies are still full of protein and then hydrolysis occurs in the enlarged body. On the other hand, in *Phaseolus vulgaris* (Opik 1966), *Lactuca sativa* (Srivastava and Paulson 1968) and *Lupinus albus* (Lott and Vollmer 1973) protein bodies lose their contents before fusion occurs. In some species, like *Lupinus luteus* (Mlodzianowski 1978) it occurs both ways. In all cases the end-result in the same – coalescence of protein bodies to form the central vacuole of the cell. A further case involves the myrosin cells of *Sinapis* where the myrosin grains coalesce producing one large central grain, but the grains retain visible contents (Werker and Vaughan 1974).

The disappearance of oleosomes as a result of triglyceride hydrolysis is often described only in general terms probably because there is no notable structural change, or there are difficulties in preserving structure for microscopy. In many cases, oleosomes appear to have no internal structure but although they are not bounded by a unit membrane, evidence from studies on oleosomes from *Ricinus* endosperm (Ory et al. 1968, Muto and Beevers 1974), pea and bean cotyledons (Mollenhauer and Totten 1971a, b), and from *Arachis hypogaea* (Jacks et al. 1967, Yatsu and Jacks 1972) indicates that oleosomes are surrounded by a proteinaceous coat or modified membrane. Visualization of the coat appears to depend on preparative technique. Oleosomes are commonly arranged around the periphery of the cell and around protein bodies, and during germination they are often described as simply becoming more randomly distributed and fewer in number.

In general, oleosomes persist much longer than protein bodies, as in lettuce (Srivastava and Paulson 1968), *Pisum sativum* (Bain and Mercer 1966), *Bidens cernua* (Simola 1973), and *Vicia faba* (Briarty et al. 1970), but they have been described as disappearing simultaneously with other reserves in *Phaseolus vulgaris* (Opik 1966) and in *Vigna* (Harris et al. 1975).

Recently, oleosome structure during germination has been examined in some detail in *Sinapis* cotyledons (Bergfeld et al. 1978). The most rapid lipid breakdown occurs after most of the protein has been depleted, and the central vacuole has formed (Werker and Vaughan 1974). The lipid is digested centrifugally, apparently by the organelle zone rich in microbodies and mitochondria located at the periphery of the central vacuole. The oleosome coats collapse as lipid digestion occurs. Although other studies of oleosome breakdown in *Sinapis* (Werker and Vaughan 1974) and *Yucca* (Horner and Arnott 1965) have not been so detailed, it is clear that structural aspects of oleosome breakdown are at variance with those described above. It is difficult to assess oleosome studies because of the apparent problems in preservation of structure.

It appears that the cell biology of oleosomes has not been treated in any great detail. Whether the coat always exists and is just not visualized, and what is the nature of the coat where present, are questions deserving further attention. Variation in appearance of oleosomes during degradation might indicate that lipid hydrolysis occurs in different ways, although the main problem seems to be variability of preservation of structure.

Mitochondria are generally not well defined in dry cotyledons. Soon after hydration they are variously described as not discernible, sparse, or ill-defined, or with poorly developed cristae. Commonly, over 24–48 h they become more clearly defined. In some cases cristae have been seen to increase, as in lettuce (Srivastava and Paulson 1968) and *Pisum sativum* (Bain and Mercer 1966). A detailed study of mitochondria of *Phaseolus vulgaris* cotyledons by Opik (1965a) showed that although there seemed to be a well-developed mitochondrial system in the quiescent embryo, the pattern of mitochondrial development depended on the tissue.

The endoplasmic reticulum (ER) undergoes interesting but poorly understood changes. By using thick sections, Harris (1979) observed that most of the ER in hydrated quiescent embryos of *Vicia faba* was in a tubular rather than a cisternal or lamellate configuration. Tubular ER has also been described in hydrated cotyledons of *Phaseolus vulgaris* (Opik 1966). ER in the cisternal form usually becomes prominent in 24 h or more, as in *Sinapis alba* (Werker and Vaughan 1974), in *Phaseolus vulgaris* (Opik 1966) where it coexists with the cisternal form, in *Pisum sativum* (Bain and Mercer 1966), and in *Vicia faba* (Briarty et al. 1970).

From a comparison of thick and thin sectioning techniques for electron microscopy (Harris 1979) it was concluded that it is very difficult to identify tubular ER by conventional thin sectioning. It appears mostly as vesicles. It would seem now that application of thick sectioning techniques to germinating tissues would be very profitable in order to determine how extensive tubular ER is in quiescent seeds, whether it is a storage form of ER, or whether it plays a role in early germination (see Harris 1979), and whether it is related structurally and/or functionally to cisternal ER, which appears to arise some time after hydration.

Of all cotyledonary organelles plastids probably undergo the most extensive changes. In lettuce embryos plastids are initially poorly represented and ill defined, but on hydration they become well defined (Srivastava and Paulson 1968). The same appears to occur in *Sinapis* (Werker and Vaughan 1974). Subsequently, starch grains may appear within plastids, as in *Sinapis* (Werker and Vaughan 1974), or they may develop lamellae and grana stacks and become normal chloroplasts with or without starch grains, as in *Bidens cernua* (Simola 1973). In *Bidens* cotyledons, lamellae appear after 1 day from hydration, thylakoids after 2 days, and grana after 5 days (see Fig. 13.4A–D).

Dictyosomes are never obvious in dry or newly hydrated cotyledons. They are either absent or poorly developed. They appear commonly a day or two after hydration but, even then, while they are prominent in root tip provascular cells and epidermal cells, they are rare (although developed) in vacuolated cotyledonary storage cells (Bain and Mercer 1966, Opik 1966, Srivastava and Paulson 1968, Briarty et al. 1970, Werker and Vaughan 1974).

Microbodies (glyoxysomes and peroxisomes) also develop in seed tissues during germination, in association with reserve breakdown (see Sect. 13.3.1.6). Commonly, the organelles and their enzyme complements occur at low levels in dry seed, and both increase during germination. This subject has been reviewed recently (Beevers 1979), and will not be discussed further.

Ribosomes appear to be numerous at all stages of germination, and do not show any detectable changes in amount. They may occur singly or in clusters (per-

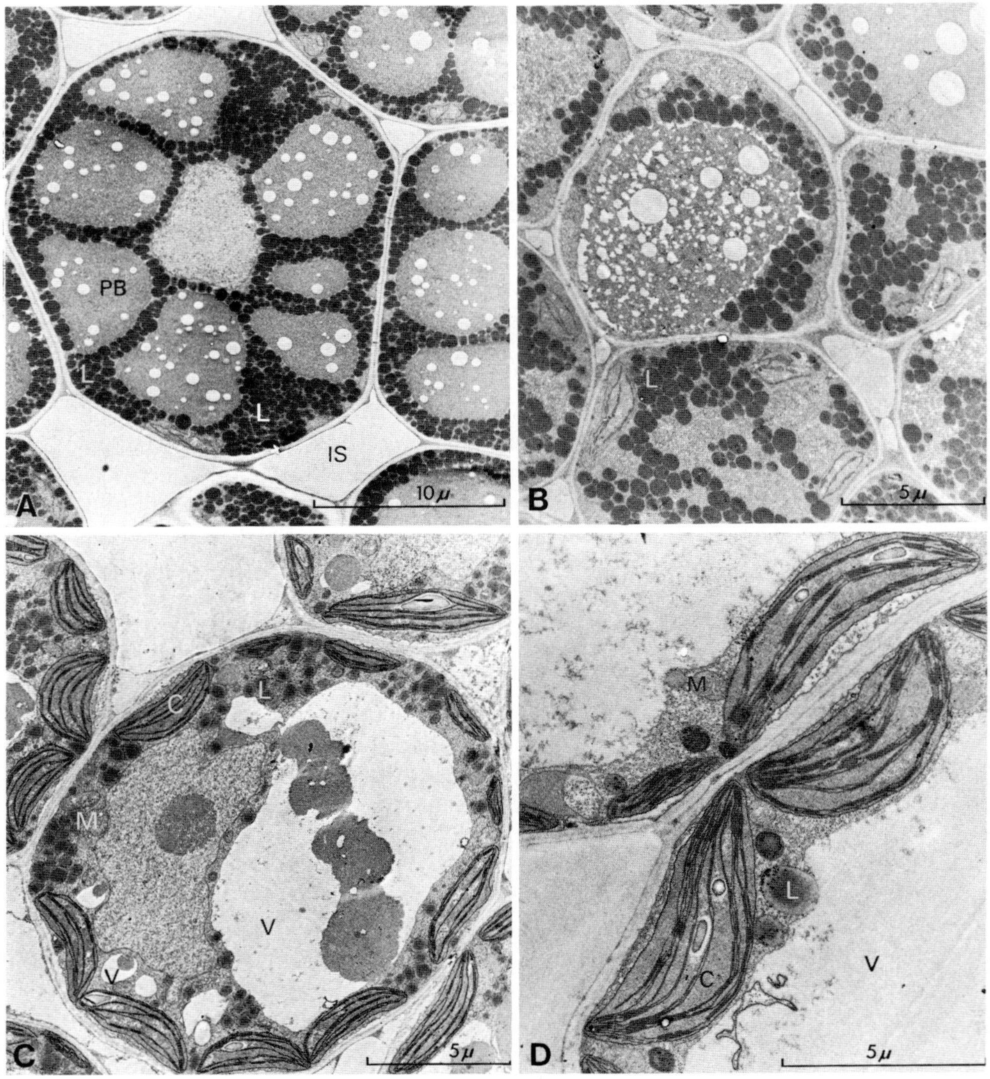

Fig. 13.4 A–D. Electron micrographs of cotyledonary cells of *Bidens cernua* during germination. **A** After day's hydration, showing protein bodies (*PD*), lipid droplets [oleosomes (*L*)], and intercellular space (*IS*). **B** After 2 days, showing disruption of protein bodies. **C** After 5 days, showing large and small vacuoles (*V*), remaining lipid droplets (*L*), mitochondria (*M*), and chloroplasts (*C*). **D** Higher magnification micrograph of parental cytoplasm after 5 days. (From Simola 1973)

haps polysomes) and, in some cases, as in *Phaseolus vulgaris* (Opik 1966), large numbers of the ribosomes become attached to the ER during reserve digestion in storage cells.

The occurrence of starch grains in cotyledons depends on the species. They occur rarely in the oil seeds but are common in some cotyledons, notably the legumes. During germination, diverse patterns of starch grain digestion have been observed. In *Pisum sativum* starch grains became smaller and fewer after a lag of 4 days (Bain and Mercer 1966). No sequence of digestion was noticed. However, in *Pisum arvense* degradation began at the outside of the cotyledons and proceeded inwards in parallel with protein body changes (Smith and Flinn 1967). In some species such as *Pisum arvense* (Smith and Flinn 1967) and *Arachis hypogaea* (Bagley et al. 1963) there is a transient increase in small starch grains before degradation begins. The increase in starch could result from gluconeogenesis (sugar synthesis) from lipids. In many cases starch is the last reserve to be mobilized.

Other than changes in shape, very few changes in nuclear structure have been noted. The shape of the nucleus appears to depend on the tissue in which it occurs. In the epidermis, hypodermis, and procambium, the nuclei are round or oval, small, and compact (13 μm in diameter in *Phaseolus vulgaris*) and stay that way during germination, as in *Phaseolus vulgaris* (Opik 1965 b, 1966) and *Pisum arvense* (Smith and Flinn 1967). In the storage cells the nuclei are usually large, diffuse and lobed, as in *Phaseolus vulgaris* (Opik 1965 b, 1966), *Pisum sativum* (Bain and Mercer 1966), *Pisum arvense* (Smith and Flinn 1967), and *Vicia faba* (Briarty et al. 1970), although no such lobing was observed in lettuce (Srivastava and Paulson 1968). One day from the beginning of hydration of *Pisum sativum,* the nuclei contained ribosome-like particles, and over the next 4 days they became greatly enlarged and highly lobed with aggregated particles in the nucleoplasm. As reserves were hydrolysed and the cells became vacuolate, the nuclei became spherical. Essentially, the same changes occurred in nuclei of *Pisum arvense* cotyledons (Smith and Flinn 1967). In addition, Feulgen staining (for DNA) and Fast-Green positive material (histone ?) did not change with increase in the size of nuclei, but staining became less intense as storage compounds disappeared. Following Azure B staining of *Pisum arvense* nuclei, nucleoli of the epi- and hypodermis were most prominent, presumably reflecting high RNA content. After 2 days, the staining of nucleoli in other cotyledonary cells, progressed inwards in a wave which may reflect the trend of ribosome synthesis.

During the period of storage protein deposition in legumes, there is a parallel increase in the level of DNA which is not associated with cell division (for review see Spencer and Higgins 1979). The increase in cotyledonary cell DNA is accompanied by enlargement of the nuclei (and nucleoli). The function of the additional DNA is unknown, but it arises by replication of the whole genome (endoreduplication) rather than selected parts of it (Millerd and Whitfeld 1973). It is probably the non-transcribed DNA, and may serve as a "reserve". During germination cotyledonary DNA has been noted either to decrease as in *Pisum arvense* (Smith 1971), or to increase as in garden pea (Beevers and Guernsey 1966), or to increase and then decrease as in *Pisum sativum* (Lehman and Garz 1962) and *Arachis hypogaea* (Cherry 1963). Changes in Feulgen staining of nuclei (Smith and Flinn 1967) have not been correlated well with changes in DNA levels. Because of this and because DNA content during germination does not seem to show a consistent pattern from species to species, future studies of nuclear size, staining characteristics, and DNA content should perhaps take into account the observation that DNA content changes to different extents in different tissues of the same organism (Bewley and Black 1978, p. 172).

Vascular differentiation also occurs in cotyledons of germinating seeds. Many cotyledons have a complex reticulate vascular system which is relatively undifferentiated in the quiescent embryo. It undergoes differentiation during the earlier part of germination 2–4 days from hydration in *Vicia faba* (Briarty et al. 1970), and 12–48 h in *Pisum arvense* (Smith and Flinn 1967). Mature sieve elements and secondarily thickened xylem appear. In *Vicia faba* transfer cells were observed bordering both xylem and phloem cells. Transfer cells are probably involved in short-distance transfer of metabolites (Gunning and Pate 1969), and probably facilitate transport by increasing the plasmalemma-wall interface area. It is notable that the vascular bundle cells remain full of cytoplasm much longer than other cotyledonary cells. In *Vicia faba* (Briarty et al. 1970) even 21 days after hydration, when all storage parenchyma cells were competely vacuolate, the vascular bundle parenchyma cells appeared active and contained numerous ribosomes, plastids, ER, and mitochondria with numerous cristae. This ensures that all the products of reserve hydrolysis can be transported into the developing axis.

13.3.1.4 The Role of Hydrolytic Enzymes in Reserve Breakdown

Mobilization of seed storage proteins has often been associated with an increase in extractable endopeptidase activity (for reviews see Ashton 1976, and Minamikawa 1979). However, the difficulty has been to establish that such enzymes are responsible for breakdown of protein body contents. One approach to the problem has been to establish coincidence between protease activity and protein breakdown in the cotyledon using gelatin films and tissue slices. Films have been used to localize protease activity in germinating seeds of *Phaseolus vulgaris* (Yomo and Taylor 1973), *P. mungo* (Minamikawa and Koshiba 1979), and in all cases the protease detected during the major rise in activity has shown a distribution matching the breakdown of protein.

Another approach to detecting protease active in protein breakdown has been to isolate protein bodies and assay their contents. Although a number of hydrolases have been detected in protein bodies isolated from dry seeds (Ashton 1976), protein bodies from germinating seeds are fragile and attempts to isolate them have been generally unsuccessful. It is of interest to note that although the protein bodies of dry seed of *P. aureus* contained various protease activities, these enzymes were unable to hydrolyse the storage protein (Harris and Chrispeels 1975).

Recent studies by Chrispeels and colleagues have shown that the enzyme required for storage protein breakdown in *Phaseolus* is an endopeptidase which is synthesized de novo during germination (Chrispeels and Boulter 1975, Chrispeels et al. 1976, Baumgartner and Chrispeels 1977). The endopeptidase has been purified, antibodies made to it and used for localization of the enzyme in germinating tissue by immunofluorescence (Baumgartner et al. 1978). The enzyme occurred initially in the cytoplasm and, subsequently, it appeared in the protein bodies. Apparently, the enzyme is synthesized in the cytoplasm and transported into the protein bodies perhaps by vesicles which form at the ends of rough ER, and fuse with protein bodies at about the right time (Chrispeels et al. 1976).

The distribution of α-amylase has also been determined using a starch film method in *Phaseolus mungo* cotyledons (Minamikawa and Koshiba 1979). Amylase activity showed the same distribution pattern as protease, and was coincident with degradation of starch. However, nothing is known of the site of production and the distribution of α-amylase within the cell.

13.3.1.5 Changes in the Embryonic Axis

A unique aspect of the axis is its elongation leading to radicle protrusion which requires cell elongation, cell division, or both. From reviews on the subject (Berlyn 1972, Bewley and Black 1978) it is clear that cell elongation usually precedes cell division. However, by varying the conditions of germination (Haber and Luippold 1960) mitosis can be made to follow or precede cell elongation, and so it appears that initiation of cell expansion and cell division are under different controls during germination, and that each can occur independently of the other. Furthermore, it has been shown that germination can occur in seeds in which cell division has been prevented, demonstrating that radicle protrusion can be the result of cell elongation only (Haber and Luippold 1960).

Subcellular structural changes occurring in the axes of germinating seeds have been studied in relatively few cases including lima bean (Klein and Ben-Shaul 1966), lettuce (Srivastava and Paulson 1968, Nabors et al. 1974), and radicles of *Pisum sativum* (Yoo 1970), *Zea mays* (Berjak and Villiers 1970), and *Bidens cernua* (Simola 1973). In general, the early events involve activation of pre-existing organelles, and the development of vacuoles, dictyosomes, and perhaps microbodies and the hydrolysis of reserves. Many axial cells contain oleosomes, protein bodies, and starch grains. The structural and compositional changes occurring in such cells are essentially the same as in cotyledons although the changes occur at different rates and to different extents in different tissues. For example, in lettuce, lipid and protein-body protein is lost first from the root tip cells, and later from the hypocotyl cells, but while a large central vacuole appears in the epidermal cells, numerous small vacuoles persist in root tip cells. Plastid development is also tissue-dependent. In roots of lettuce, pea, and *Bidens* plastids remain as proplastids with poorly developed lamellae, but in the hypocotyl (e.g., lettuce) plastids develop grana. The proliferation of ER is also of interest. In *Fraxinus* proliferation occurs firstly in the apical cells and spreads from there, but whereas the proliferating form was sheet-like, the form existing in freshly imbibed cells was tubular (Villiers 1971). Tubular ER has also been described in lima bean axis (Klein and Ben-Shaul 1966) which was the first such description.

Finally, it should be pointed out that although all the axial and cotyledonary changes described above are events occurring in germinating seed, they may also occur in dormant seed. The embryo of *Fraxinus* grows extensively within the seed at high temperatures, but germination does not occur. The seeds are dormant. However, protein bodies lose their contents, oleosomes decrease, ER, microbodies, and dictyosomes appear, microfibrils appear around the cell periphery, and while some plastids develop internal vesicular and lamellar systems, others become amyloplasts (Villiers 1971). Therefore, dormancy does not involve a blockage of these cellular rearrangements.

13.3.1.6 Changes in Endosperm and Perisperm

Mature endosperm in dicotyledons is commonly a fleshy tissue consisting of living cells varying from one to many cells in thickness. The cells commonly contain large numbers of protein bodies and oleosomes. The protein bodies contain globoids which are usually spherical and contain phytate, crystalloids which are often angular and are thought to consist primarily of protein in crystalline form, and calcium oxalate crystals. The cells commonly have walls thickened with carbohydrates of different kinds which provide rigidity to the endosperm, may be water-holding capacity, and a carbohydrate reserve. During germination these reserves are broken down, the organellar machinery to accomplish this develops and, ultimately, the tissue is totally degraded as the embryo expands. At the same time, the radicle often must penetrate the endosperm to emerge. Structural and physiological studies have been performed primarily to understand these cytolytic events and their control, since they are fundamental to embryo nutrition and radicle emergence.

Cell Walls. The breakdown of endosperm cell walls during germination was described long ago by Netolitzky (1926), and has since been examined in some detail in lettuce, *Ricinus communis* (castor bean), some legumes of the Fabaceae, and *Apium graveolens* (celery).

The endosperm of lettuce consists of two layers of cells increasing to four or five at the radicle tip (micropylar region) (Borthwick and Robbins 1928, Foard and Haber 1966, Jones 1974), and there is evidence that the endosperm (but not the epidermis) acts as a mechanical restraint to radicle emergence (Evenari and Neumann 1952, Koller et al. 1962, Ikuma and Thimann 1963, Pavlista and Haber 1970, Globerson et al. 1974, Speer 1974). Therefore, it is of interest to examine the changes occurring in endosperm structures during germination in order to understand how the mechanical strength of the endosperm might be weakened. Using light and electron microscopy, Jones (1974) has shown that initially the endosperm cell walls are thick and show numerous outgrowths extending into the cytoplasm as in transfer cells. On the basis of periodic acid-Schiff (PAS) staining and electron density, the wall appeared to be a layered structure. The layer of wall 1.0–3.0 μm wide next to the plasmalemma was heavily stained in PAS, and was less electron-dense than the bulk of the wall. After 12–15 h of germination the walls had lost some PAS stainability and were presumably extensively broken down. The loss of stainability became more extensive with time, but PAS staining of the wall adjacent to the integument did not change and, presumably, this part of the wall remained intact. These studies imply a relationship between germination and hydrolytic events in the cell walls. This is supported by the fact that endosperm from freshly hydrated seed is very strong but endosperm from seed hydrated for 24 h can be torn very easily. However, to assist radicle emergence, it would be necessary to degrade cells in the micropylar region of the endosperm, and observations of these cells during germination do not yet appear to have been made.

In a more recent study using scanning electron microscopy, Pavlista and Valdovinos (1978) examined the mycropylar region of lettuce before germination occurred at 12–16 h. The surface of the endosperm remained intact until 9–10 h when cracks and pits appeared extensively throughout the micropylar region. At this time there was disruption of endosperm at the cotyledonary end, or the mid-section of the seed. Although there was no real evidence that the changes were caused by enzymatic digestion rather than by physical stress, the latter appeared to be unlikely since the seeds were fully hydrated after 3 h, some 4–5 h before changes in the endosperm appeared.

The possibility that cytolytic events in the endosperm of lettuce might be involved in germination was first indicated by the early experiments of Ikuma and Thimann (1963) who showed that hydrolases injected into the root tip region facilitated germination. The studies described above strengthen that belief. In the studies of Jones (1974), wall modification commenced near the plasmalemma and proceeded outwards, indicating that the wall degrading enzymes might originate from the endosperm cell cytoplasm rather than from the embryo. On the other hand, it has been shown that hydrated lettuce embryos can develop sufficient expansive force to break through the micropylar region (Nabors and Lang 1971 a, b), in which case endosperm hydrolysis would be unnecessary. In fact, the two mechanisms need not be mutally exclusive, and it would seem likely that further studies will reveal that radicle emergence is the result of both phenomena (see also Pavlista and Haber 1970).

Recently, the composition of the lettuce endosperm cell walls has been determined and enzymes probably capable of digesting it have been detected. The wall is composed primarily of mannose (58%), galactose (9.5%), and glucose (10%), and the principal carbohydrate is therefore probably a galactomannan (Halmer et al. 1975). The walls are broken down rapidly following radicle emergence (Halmer et al. 1978), and this is associated with the appearance of endo-β-mannanase activity in the endosperm (Halmer et al. 1978, Halmer and Bewley 1979). However, these studies have been concerned with wall degradation and enzyme production in the whole endosperm and although they establish that there is massive digestion of cell-wall mannans during germination which probably provides carbohydrate for the embryo early in germination (Leung et al. 1979), it remains to be determined whether wall breakdown and endo-β-mannanase appearance play any role in germination. It is possible that endo-β-mannase could arise and wall hydrolysis begin earlier in the micropylar end of the endosperm, and a cytochemical study of the area would seem worthwhile.

The cell walls of celery endosperm also appear to be degraded both during germination (Jacobsen and Pressman 1979) and during incubation of de-embryonated seed (isolated endosperm) with gibberellin (Jacobsen et al. 1976). In the latter case, the walls become thin and irregular in outline, and the cells separate from each other but remain intact, bounded by a thin wall layer which is apparently resistant to degradation. These studies imply that (hormone controlled) cytolytic events occur, but studies of the cell wall composition, ultrastructure and the occurrence of cell wall hydrolysing enzymes are needed.

In legumes the occurrence of endosperm varies from almost complete absence, as in *Pisum* and *Phaseolus*, to copious, as in *Cyamopsis* and *Ceratonia*. In the latter, the endosperm commonly contains large quantities of storage polysaccharide, usually galactomannan, in the form of heavily thickened cell walls (Bailey 1971).

Galactomannan occurs to the extent of more than 1% of the seed in 2 out of 5 tribes of Caesalpiniaceae, 4 of the 6 tribes of Mimosaceae, and about half of the 50 tribes of Fabaceae. The structural aspects of endosperm wall breakdown have been examined in *Trigonella foenum-graecum* (fenugreek) and *Ceratonia siliqua* (carob). In both cases wall breakdown occurs after germination.

In fenugreek endosperm the only living cells at maturity are those in the outer layer – the so-called aleurone layer. The lumina of all internal cells are occluded at maturity by the extensive deposition of galactomannan. Cell wall breakdown begins adjacent to the living cells and proceeds inwards, suggesting that enzymes for galactomannan breakdown may originate in the outer living cells. In carob many of the endosperm cells retain living cytoplasm at maturity, and wall breakdown appears to occur more uniformly throughout the endosperm (Seiler 1977). This constitutes circumstantial evidence that enzymes for galactomannan degradation arise in the living cells, and not in the embryo. Degradation begins as pockets of digestion adjacent to the fibrillar layer surrounding each cell. Only at the end does the fibrillar layer breakdown but the primary wall persists. Once again the cells are surrounded by a digestion-resistant coat.

Although it has not yet been demonstrated unequivocally, the studies of legume endosperm mobilization strongly indicate that enzymes for galactomannan breakdown originate in the living cells of the tissue. In *Trigonella* the enzymes required to hydrolyse galactomannan completely, namely, α-galactosidase, β-mannosidase, and β-mannanase, have been detected (Courtois and Percheron 1961, Clermont-Beaugiraud and Percheron 1968, Reid and Meier 1973, Reid and Davies 1977), and are known to be produced in the endosperm. β-Mannanase has also been isolated from *Trifolium repens* (Villaroya et al. 1978). The accumulation of these enzymes is associated with galactomannan hydrolysis. A similar correlation between galactomannan breakdown and endo-β-mannanase activity occurs during germination of lucerne and carob seed (McCleary and Matheson 1975), but increases in β-mannosidase have not been detected in guar, lucerne, carob, and honey locust (Sømme 1970, 1971, McCleary and Matheson 1975).

Although the number of studies is limited, it appears that in dicotyledonous endosperm the production of cell wall hydrolysing enzymes in the cytoplasm and their secretion into the walls might be a common process. Also, commonly, the cell walls of dicotyledonous endosperm are heavily thickened with mannans, often galactomannans. Galactomannans occur in abundance in the endosperm of a number of Palmae, Leguminosae, Rubiaceae, and Convolvulaceae, presumably in the cell walls. It appears that this system constitutes a well-developed and widespread method of carbohydrate storage and mobilization in dicotyledonous endosperm. The thickened walls are rigid, affording the enclosed embryo a considerable degree of physical protection, and its hydrolysis not only provides nutrition for the embryo but it is also necessary for embryo emergence in many cases. Finally, the cell walls seem to be layered structures which are differentially subject to hydrolysis. The nature of the layering deserves further attention.

Cytoplasm. As already noted, endosperm cytoplasm is commonly packed with aleurone grains with inclusions (globoids, crystalloids, calcium oxalate) and oleosomes, and the major changes concern the hydrolysis of these reserves and metabolism of the products. Indeed, much of our knowledge of lipid breakdown and gluconeogenesis in plant tissues has been obtained from the studies of *Ricinus communis* endosperm by Beevers and colleagues.

The process of lipolysis and gluconeogenesis has been reviewed extensively elsewhere (Marré 1967, Ching 1972, Beevers 1975, 1979). Triglycerides which form the matrix of the oleosomes are hydrolysed by lipases in the oleosomes, and the fatty acids are processed sequentially through glyoxysomes, mitochondria, and the cytosol to sucrose which is then stored in the vacuole. Generally, changes in cellular structure have been well correlated with these events (see Vigil 1970, and above reviews). The cells of *Ricinus* endosperm are originally packed with aleurone grains and oleosomes. Glyoxysomes and mitochondria are rare and undeveloped. As gluconeogenesis progresses, the oleosomes become small and fewer in number, well-developed mitochondria and glyoxysomes become more numerous, and plastids, ribosomes, rough ER (which seems to give rise to glyoxysomes), dictyosomes and, occasionally, starch grains appear. Aleurone grains swell and vacuolate. In the advanced state cells contain few oleosomes, but numerous glyoxysomes and mitochondria. The aleurone grains coalesce to form large vacuoles. Ultimately, the cell contents and the cell walls disappear. Both turnover and decay of organelles appear to occur inside the (autophagic) vacuoles. These changes do not occur simultaneously throughout the endosperm; they occur adjacent to the embryo first and spread outwards as the embryo enlarges. Why this happens is not understood. The decay of cell contents is very abrupt so that the line between cells with and without contents is very sharp. A very similar series of structural events has been shown to occur in *Trigonella foenum-graecum* endosperm cells (Reid and Meier 1972).

13.3.2 Monocotyledons

13.3.2.1 Changes in Cereal and Other Embryos

Although it contributes very little to the total amount of reserve food in the cereal grain, the embryo is nevertheless well endowed with stored compounds, containing about 10% of lipid and 25%–35% protein. Such high levels reflect the large numbers of protein bodies and oleosomes occurring in many of the embryonic cells (scutellum, coleoptile, coleorhiza). The structure and ultrastructure of the cereal embryo has been studied quite extensively, and not only in relation to reserve mobilization. It has been studied in the general framework of embryonic growth and morphogenesis, and perhaps particularly because the haustorial function of the scutellum has long fascinated anatomists and physiologists. In addition, the embryo gives rise to gibberellin which controls aleurone layer function and endosperm degradation. Although major changes in the endosperm of cereal grains occur only on germination 1 or 2 days after hydration, many changes in embryonic cells, particularly root cells, occur rapidly in a few hours.

Studies of the germinating cereal embryo axis have focused primarily on events occurring in the coleorhiza, root cap, and root meristem area. Events in the coleorhiza-root zone of *Setaria lutescens* (Rost 1972) and *Secale cereale* (Hallam et al. 1972) have much in common. The cells become active after only a few hours of hydration and, as they elongate, protein (body) and lipid reserves are digested and organelles such as glyoxysomes, mitochondria, rough ER, and others characteristic of metabolically active cells become prominent. In *Setaria* the coleorhiza cells die and slough off following radicle emergence. In *Secale* cell divisions were observed only after 12–15 h, so that considerable cell elongation occurred before cell division began.

Similar changes occur in *Zea mays* roots during germination (Deltour and Bronchart 1971, Deltour and Jacqmard 1974) although studies have focussed on changes occurring within the nuclei (Deltour and Bronchart 1971, de Barsy et al. 1974, Deltour et al. 1979). Initially, the chromatin is condensed and the nuclei compact and fibrillar. Following hydration, chromatin becomes dispersed, then condensed again and nucleoli become vacuolate. Dispersion of the nucleolar organizer region is associated with the appearance of DNA in the nucleolar matrix and vacuoles. These changes are complex and poorly understood.

The scutellum of germinating grain has received considerable attention because it is probably involved in metabolite transfer from the endosperm to the embryonic axis, and in enzyme release to the endosperm. Further, the peculiar elongation of epithelial cells accompanying germination has fascinated many workers and has been described a number of times (Torrey 1902, Reed 1904, Sargant and Robertson 1905, Guillermond 1908, Toole 1924, O'Brien 1942).

For wheat, the fine structural details of scutellum over the first day or so of germination have been described, using light and electron microscopy, by Swift and O'Brien (1972). Initially, the epithelial cells are closely packed and joined laterally. By 24 h they have elongated and some cell walls have separated from each other. In barley the cells elongate to about twice their original length, become cylindrical in shape and completely separate from each other (Nieuwdorp and Buys 1964). The separation occurs by dissolution of the middle lamella region. In wheat starch grains appear in the epithelium and the parenchyma cells and become very abundant over 24 h. After a few days, the starch grains disappear from the epithelium while in the parenchyma they remain large and abundant. Most of the protein bodies of the epithelium have lost their contents by 24 h, becoming vacuoles which merge to form large vacuoles and these occupy most of the cell lumen after a few days.

Protein-body breakdown in the scutellum parenchyma occurs over essentially the same time span as in the epithelium, but it occurs most rapidly in large parenchyma cells in the central region of the scutellum, and slowly next to the epithelium. Oleosomes are originally abundant and concentrate around protein bodies and the cell periphery, but after 24 h they begin to decline slowly in size and number. As for starch grains, other organelles go through cyclic changes. Dictyosomes become prominent and active, and then seem to become less abundant. Mitochondrial cristae increase after 3 days in barley and 1 day in wheat, and at least in barley they return to their original condition after some time. Ribosomes apparently free and bound are always abundant, and microbodies are present.

In the quiescent cereal embryo the vascular tissue is in the provascular state. During germination it differentiates into xylem and phloem in the scutellum, the coleoptile, and the primary leaf. In wheat, changes first begin in the scutellum 3 h

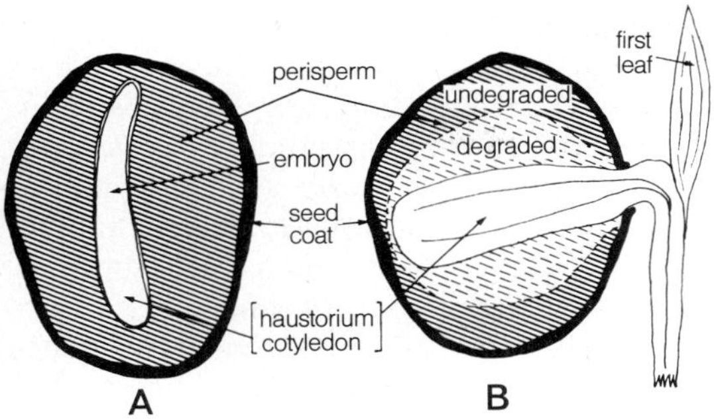

Fig. 13.5. A Longitudinal section of dry *Yucca* seed. **B** Seed which has germinated after 8 days in water. (From Horner and Arnott 1966)

after soaking when sieve elements appear (Swift and O'Brien 1971). Then they appear in the first leaf after 6 h, and in the coleoptile after 18 h. Xylem differentiation occurs in the same sequence but lags behind by 12–18 h, the first mature tracheary elements appear in the scutellum by 18 h. The large central vascular bundle extending from the scutellar node into the apex of the scutellum divides into many smaller vascular bundles at its apex, and these curve over and downwards into the wings towards the epithelial face of the scutellum (Swift and O'Brien 1970). Barley scutellum is vascularized in essentially the same way (MacLeod and Palmer 1966). Therefore, it would seem that efficient translocation of nutrients out of the endosperm and around the embryo cannot occur for 18 h, during which time the phloem develops. Fully developed xylem is not present until about 30 h after hydration. Thus, vascular differentiation in the scutellum is very advanced by 24–30 h, and this precedes extensive endosperm breakdown and growth and differentiation of the axis. Perhaps scutellar vascular differentiation is a prerequisite for these events to occur. It may also be necessary for hormone release to the endosperm (see Sect. 13.3.2.2).

Other than for the cereal embryos, descriptions of changes in monocot embryos are rare. Only two other species appear to have been examined, *Yucca* (Horner and Arnott 1965, 1966) and *Elaeis guineensis* (Boatman and Crombie 1958). In both cases the cotyledon remains behind in the seed to function as an haustorium (Fig. 13.5 A, B). The haustorium enlarges and the storage tissue, endosperm in *Elaeis* and perisperm in *Yucca,* is degraded in a centrifugal wave. Ultimately, all the storage tissue is degraded and the haustorium breaks down. Initially, *Yucca* embryo cells contain abundant protein bodies and lipid droplets but there are no starch grains. Other organelles are difficult to visualize. The protein bodies breakdown in an ordered sequence over 4 days. Comparison of the changes occurring in the embryo and perisperm (see Table 1 of Horner and Arnott 1966) shows that changes in the embryo are well under way before anything happens in the perisperm.

It is worth noting that although there have been very few structural studies of germinating gymnospermous seeds, an excellent series of articles on *Welwitschia mirabilis* by Bornman and colleagues has appeared recently (see references by Bornman et al. 1979a–d, Butler et al. 1979a–e). This species has a "feeder" which remains in the nutritive tissue (female gametophyte) of the seed, and plays a similar role to the haustorium of *Yucca* and other angiosperms.

13.3.2.2 Changes in Endosperm and Perisperm

Cereal Endosperm. The cereal endosperm is unusual in consisting of two very different tissues at maturity. The outer aleurone layer consists of living cells and is usually one cell layer in wheat, rye, maize, sorghum, triticale, and oats, and two to four cells in barley. The inner starchy endosperm is dead at maturity. On rehydration it does not respire or synthesize protein and does not have the metabolic machinery of a living cell. It does, however, contain some enzymes.

The endosperm usually occupies the bulk (87%) of the grain, and about 10% of the endosperm dry weight is aleurone. The aleurone layer stores lipid (as oleosomes), which contributes about 90% of the total endosperm lipid (which is about two-thirds of the total grain lipid), and it also contains about 20% of protein. The aleurone grains which are the storage organelles contain protein (some enzymatic), phytic acid, and a crystalloid-like organelle. Carbohydrate is stored in the aleurone as thickened cell walls (see further, this section) as well as some intracellular sugars like sucrose. The starchy endosperm stores starch, which constitutes about 85% of the tissue, and protein which makes up about 10%. The cell walls contain β-glucans and pentosans (see further, this section), but little or no cellulose.

During germination, the major events in the endosperm involve the production of hydrolytic enzymes primarily in the aleurone but also in scutellum cells, and their release into the starchy endosperm where the reserves are hydrolysed. As in cotyledons, endosperm breakdown is largely a post-germination event (Chen and Varner 1970, Chen and Chang 1972) concerned not with radicle protrusion but with nutrition of the enlarging embryo. The literature concerning these events is voluminous (for reviews see MacLeod 1969, Yomo and Varner 1971, Briggs 1973, Palmer and Bathgate 1976, Raghavan 1976, Varner and Ho 1976, Jacobsen 1977, Varner 1977, Bewley and Black 1978, Briggs 1978, Higgins and Jacobsen 1978, Jacobsen and Higgins 1978a, b), and only sufficient background for the structural studies will be presented here.

Following hydration of the barley grain, gibberellin (GA_3 and GA_1) is released from the scutellum and diffuses into the endosperm. The target tissue for the hormone is the aleurone which responds by breaking down its own reserves (including the cell walls), by undergoing extensive cytoplasmic changes, and by secreting enzymes (mostly hydrolytic) into the starchy endosperm (for list see Palmer and Bathgate 1976). Some of these enzymes are newly synthesized (e.g., α-amylase), and some (e.g., β-glucanase) pre-exist. Probably all cereals, except sorghum, have aleurone that responds similarly. A group of hydrolases, similar to those of the aleurone layer, are also released by the scutellum soon after hydration, but this probably represents a small fraction of the hydrolytic power degrading the endosperm. Starchy endosperm degradation begins adjacent to the scutellum, and proceeds towards the apex of the grain, hydrolysis proceeding more rapidly immediately beneath the aleurone and most rapidly in the dorsal part of endosperm (Briggs 1978). This probably provides controlled release of endosperm nutrients to the embryo.

The appearance of some of the hydrolytic enzymes, notably α-amylase, has been examined using cytochemical techniques. Using a starch film method, Oka-

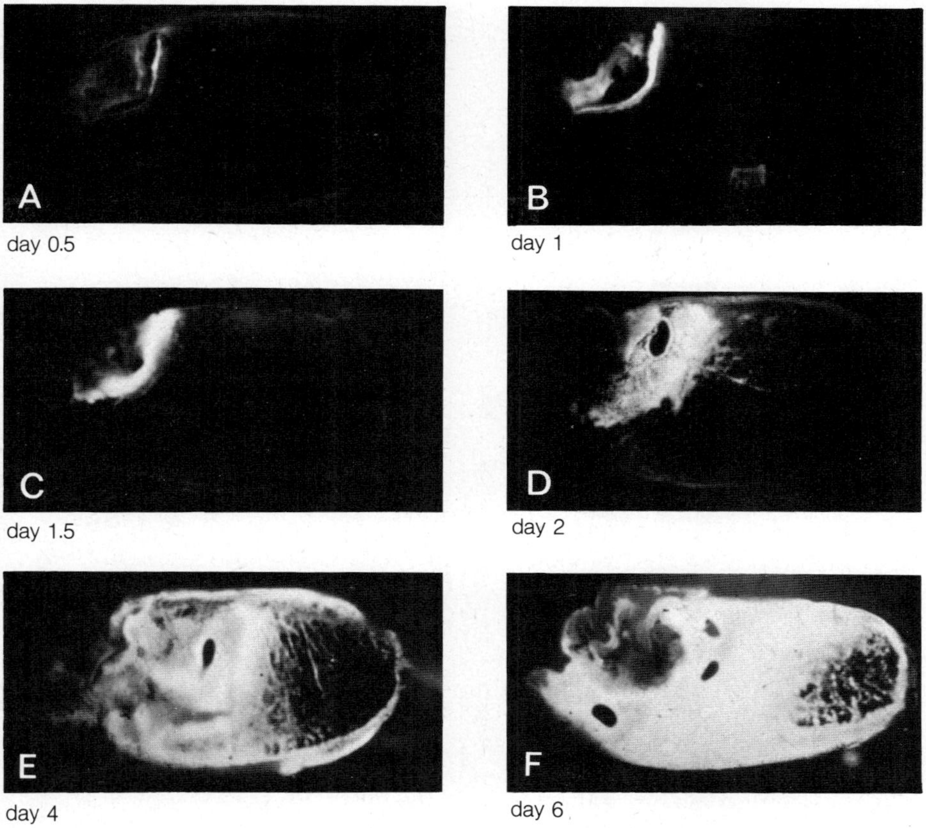

Fig. 13.6 A–F. Amylase localization in germinated rice seed, using the starch film method. At various stages of germination seeds were freeze-sectioned and the median longitudinal sections were applied to acrylamide gel film containing soluble starch. After incubation tissue sections were removed from the film which was stained in I-KI solution. Amylase activity is shown by unstained (*light*) areas. The numbers stand for days of germination. (From Okamoto and Akazawa 1979)

moto and Akazawa (1979) working with rice showed that α-amylase appeared firstly in the scutellar epithelium, then in the starchy endosperm adjacent to the scutellum, and then in the aleurone becoming intense and spreading away from the embryo (Fig. 13.6 A–F). A similar pattern has been reported for barley (Gibbons 1980). These data are compatible with the pattern of endosperm hydrolysis described above.

Very limited information is available from attempts to localize α-amylase on a cellular level. Use of starch film and immuno-fluorescence techniques on barley aleurone (Jacobsen and Knox 1973) indicated that α-amylase was initially associated with aleurone grains; its significance is not clear. In GA_3-treated tissue, α-amylase increased dramatically and was located in the cytoplasm, apparently concentrated in the perinuclear region (Jones and Chen 1976) and in the inner layers of cell wall.

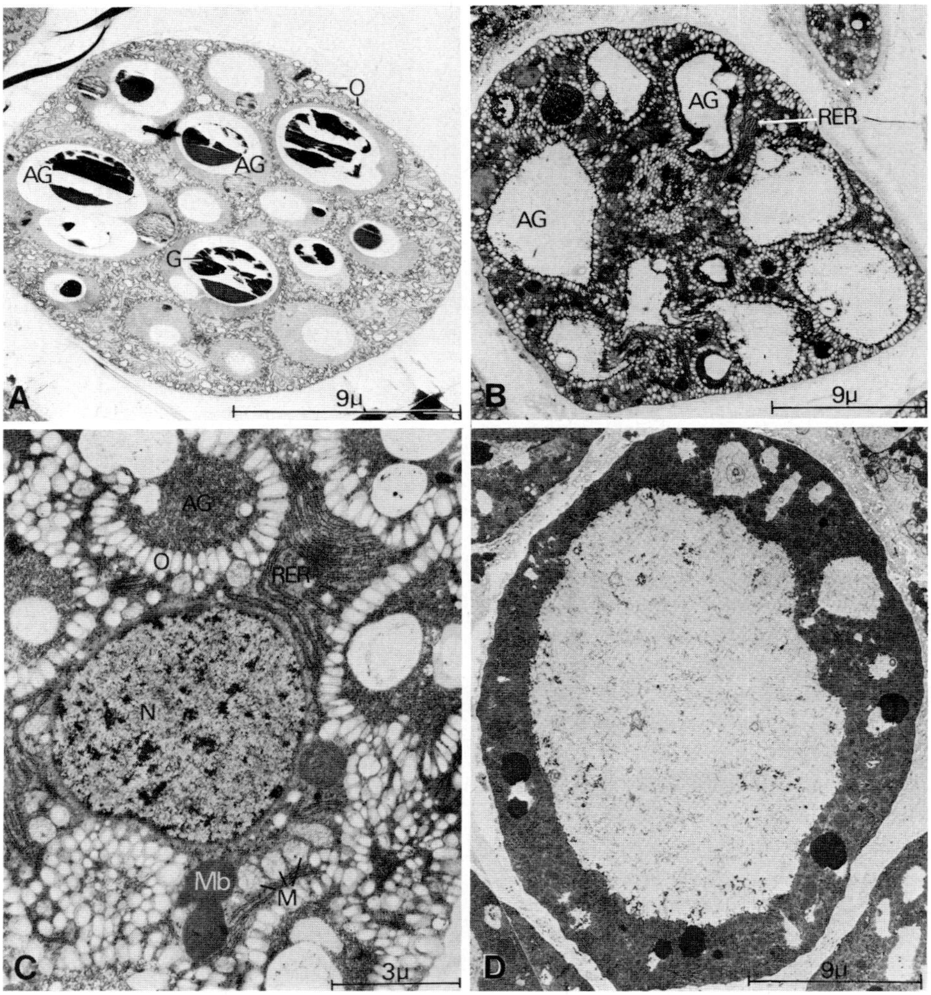

Fig. 13.7 A–D. Electron micrographs of barley aleurone cells, showing the changes induced by gibberellic acid. **A** Fully hydrated aleurone cell showing oleosomes (*O*) and aleurone grains (*AG*) containing globoids (*G*). **B** Exposed to GA_3 for 10 h. Aleurone grains have swollen and lost contents, and rough endoplasmic reticulum (*RER*) has appeared. **C** Higher magnification micrograph of tissue treated as in **B**, showing *RER*, mitochondria (*M*), and microbodies (*Mb*) around the nucleus (*N*). **D** Exposed to GA_3 for 36 h. The cytoplasm surrounds the large central vacuole. (**A–C** from Jones 1969a; **D** from Jones and Price 1970)

The changes occurring in aleurone cells of barley during germination, examined by transmission electron microscopy (Fig. 13.7 A–D) (Jones 1969a, Jones and Price 1970, Vigil and Ruddat 1973, Jones and Chen 1976, Obata 1979), are comparable to those which occur in endosperm cells of dicotyledonous seeds. The aleurone grains swell and their contents fragment. Ultimately, large vacuoles are formed by coalescence of smaller ones. Vacuolating protein bodies commonly appear crenate or angular in transmission microscopy preparations, although freeze-etching of vacuolating cells (Buttrose 1971) shows that they are probably round. Oleosomes become randomly distributed and decrease in number, and stacked ER (concentrated around the nucleus), microbodies, mitochondria, and dictyosome vesicles become prominent. Ultimately, the cell contents disintegrate. Essentially, the same kind of events occur in wheat aleurone tissue during germination (Laidman et al. 1974, Colbourne et al. 1976) although control of the events in wheat is different. ER appears to proliferate with hydration only, and cytokinin, auxin, and glutamine as well as gibberellin appear to play a role.

Studies of enzyme secretion have employed a considerable amount of electron microscopy and even more guesswork, and the pathway of enzyme release is still unknown. Indeed, newly synthesized and pre-existing enzymes may be released in different ways. Secretory vesicles may (Vigil and Ruddat 1973), or may not (Jones 1969a, b, Jones 1972, Obata 1979) be involved in the enzyme release process. Clearly, a technique which permits the specific cytochemical detection of α-amylase by electron microscopy is required. Such techniques have been applied to acid phosphatase (Pyliotis et al. 1979). Although not much is known about this enzyme in aleurone, it is known to increase and to be released from aleurone tissue following GA_3 treatment (Ashford and Jacobsen 1974b, Bailey et al. 1976). The study showed that acid phosphatase arising in the cell after GA_3 treatment was confined to the lumen of the ER. There were no vesicles detected at any stage. These results may indicate that there is some direct transfer of newly formed enzyme from the lumen of ER through the plasmalemma. A similar conclusion was reached in a recent structural study of barley aleurone (Obata 1979) in which secretion (but not synthesis) of α-amylase was associated with the formation of stacked rough ER, some of which extended to the plasmalemma with its end opening to the outside of the membrane.

The ultrastructure of cell wall breakdown has been described a number of times (Jones 1969b, Taiz and Jones 1970, Jones 1972, Taiz and Jones 1973, Vigil and Ruddat 1973). Corrosion of the walls begins adjacent to the cytoplasm in patches (after about 16 h of GA_3 treatment), and is often associated with plasmodesmata. Although the bulk of the wall is digested, a thin layer of wall around each protoplast is resistant (Taiz and Jones 1970, Ashford and Jacobsen 1974a, b). Digestion of the wall appears to give rise to channels which act as preferential routes through which enzyme is conduced out of the aleurone layer (Ashford and Jacobsen 1974b). In the early stages acid phosphatase can be seen to accumulate in channels associated with plasmodesmata (Figs. 7 and 21 of Ashford and Jacobsen 1974b), but the relation between the two is not clear.

Early analyses of composition of aleurone cell wall (McNeil et al. 1975, Morrall and Briggs 1978) indicated that the walls were very high in pentosans, particularly arabinoxylans. However, in both cases, the walls were treated in such ways that polymeric material was probably lost (Bacic and Stone 1979). Taking this into account, more recent analyses of the aleurone walls of wheat and barley show that although the walls do contain large amounts of arabinose and xylose (about 20% and 40%, respectively), they also contain about 30% of glucose (Bacic and Stone 1979). The pentose is present as arabinoxylans, and the glucose as a mixed link 1,3–1,4 β-glucan. Enzymes capable of degrading both arabinoxylans (Mac-

Leod and Millar 1962, Taiz and Honigman 1976, Dashek and Chrispeels 1977) and β-glucans (MacLeod and Millar 1962, Taiz and Jones 1970, Ballance et al. 1976) have been reported to increase, and to be secreted by GA_3-treated barley aleurone. However, the composition of the inner wall and the reason for its resistance are unknown.

The starchy endosperm and its degradation have been studied extensively since they are fundamental to the malting and brewing industries, and to bread-making. The starchy endosperm has large, thin-walled cells packed with starch grains and reserve protein, sometimes in definable protein bodies (without inclusions), and sometimes not (see reviews cited above especially Bewley and Black 1978, p. 19, Barlow et al. 1973a). The starch grains usually fall into two size categories, large and small. In barley the groups have mean diameters of 5 μm and 25 μm. They consist of amylose and amylopectin. The grains are coated with water-soluble protein in wheat and barley (Barlow et al. 1973a, Barlow et al. 1973b, Jacobsen and Knox 1974).

The walls of barley endosperm seem to have a proteinaceous middle lamella (Palmer and Bathgate 1976), an adjacent layer of β-glucan and arabinoxylan, and an inner layer (next to the cytoplasm) of mostly β-glucan. There is evidence that β-glucan may form massive thickenings in the walls of the subaleurone cells of barley (Fulcher et al. 1977). The overall composition of the matrix of barley walls is about 25% of arabinoxylan and 75% of β-glucan (Fincher 1975, Ballance and Manners 1978), while the major proportion of wheat endosperm wall is arabinoxylan (Mares and Stone 1973, Bacic and Stone 1979). Endosperm walls appear to contain little or no cellulose and pectin polysaccharides.

During germination and early seedling growth, all components of most of the starchy endosperm cells are broken down beginning near the scutellum as previously described. The biochemistry of the process is described extensively in the above reviews (see especially Briggs 1973, 1978, Palmer and Bathgate 1976) and others (Manners 1974). The early events probably involve hydrolysis of middle lamella protein by protease, and of the wall matrix (β-glucans and pentosans) by β-glucanase and pentosanases originating in the aleurone layer. Not much is known of the breakdown of proteins in starchy endosperm except that it is probably an endopeptidase originating in the aleurone which initiates large-scale hydrolysis. The hydrolysis of starch grains has been the subject of considerable debate, and seems to be still incompletely understood. It appears to be hydrolysed primarily by β-amylase, which pre-exists in the endosperm, and by α-amylase which arises de novo in the aleurone.

Light microscopy (Dickson and Shands 1941), and more recently scanning electron microscopy (Morrall and Briggs 1978), have been used to study endosperm digestion. Generally, the cell walls breakdown first. This is followed by disappearance of small starch grains and, progressively, more extensive pitting of large ones until complete hydrolysis occurs. The morphology of starch grain degradation is complex and variable. It can differ between wheat and barley, and between grains in different parts of the same endosperm (Palmer 1972).

Other Monocotyledons. Changes occurring in living storage cells of other monocotyledonous seeds are very poorly documented. Only two species, *Elaeis* *guineensis*, which has endosperm (Boatman and Crombie 1958), and *Yucca*, which has perisperm (Horner and Arnott 1965, 1966), appear to have been studied. The storage cells of both species contain abundant oleosomes and, in addition, those of *Yucca* contain aleurone grains with inclusions as in cereal aleurone. In both cases the endosperm is digested centrifugally by an expanding haustorial cotyledon originally encased in the storage tissue (Fig. 13.5). In *Yucca* the cell walls were degraded initially by localized pocketing, and in the advanced stages only a framework described as "cellulosic" remained. Protein bodies enlarged and lost their contents, but it is noteworthy that in *Yucca* they broke down in different ways depending on the species. Starch grains commonly appeared as lipid was slowly broken down, but their appearance was transient.

What controls the centrifugal degradation of perisperm is unknown. It could be diffusion of enzymes, or a hormonal trigger from the embryo. It could also be an intrinsic property of the tissue as may be the case in celery endosperm (Jacobsen and Pressman 1979).

13.4 Storage and Aging of Seeds: Structure and Physiology

During storage of seeds it is often desirable to extend the life of seeds for as long as possible. Seeds must not only maintain viability but be able to produce normal, vigorous seedlings. Storage usually involves manipulation of temperature and seed-water content. The principles and practices involved have been reviewed previously and will not be discussed here (see Barton 1961, Barton 1965b, Harrington 1972, Roberts 1972c, 1975, Bewley and Black 1982). In most cases the best conditions involve low relative humidity, low temperature, seed-moisture content of about 2%–5%, low oxygen, high carbon dioxide, and darkness. However, seeds of tropical and some temperate species (termed recalcitrant) cannot be dried (below 12%–31%) without losing viability, and cannot be stored for long periods (for review see Roberts 1975). Aging and deterioration of seeds is expressed in a number of ways. Commonly, there are changes in seed colour, delayed germination, reduced germinability, decreased tolerance to suboptimal germination conditions, retarded seedling growth, and the presence of abnormal seedlings.

That such functional aberrations are associated with structural changes was first demonstrated in aged rye (Hallam 1973, Hallam et al. 1973). In root primordia cells of dry seed, the only differences between non-viable and viable embryos (Hallam et al. 1972) involved mitochondrial structure and withdrawal of the plasmalemma from the cell wall but, on hydration, the cytoplasm of aged seed was rapidly degraded until only coalesced lipid droplets remained. Similar changes have been described, although in less detail, for artificially aged root meristem cells of lettuce (Villiers 1975).

Depending on their extent, aging changes can be reversed. Studies on root cap cells of *Zea mays* have shown that if seed was aged to a point where germination was very slow but embryos were still viable, structural abnormalities occurring in the first 24 h following seed hydration were rectified in the ensuing 24 h, and the reversal corresponded with the ability of the seed to germinate and produce normal seedlings (Berjak and Villiers 1970, 1972). Thus, the lag time for germination was a recovery time.

Nuclear changes have also been associated with aging. They have been detected cytochemically as differences in chromatin staining (Villiers 1975), and as changes in chromosome morphology during mitosis (for reviews see D'Amato and Hoffman-Osterhoff 1956, Barton 1961, Roberts et al. 1967, Abdul-Baki and Anderson 1972, Roberts 1972a, 1975). In root tips of artificially aged seeds of pea, bean, and barley, Roberts and colleagues have detected a variety of chromosomal aberrations whose incidence increased with age regardless of the factor causing loss of viability. Visible chromosomal damage is normally associated with mutation so that loss of viability may be an index of mutation.

There have been numerous studies of physiological and compositional changes associated with slow germination and loss of vigour in aging seeds, and these have been the subject of a number of reviews (Abdul-Baki and Anderson 1972, Roberts 1972a, Abdul-Baki and Baker 1973, Hallam et al. 1973). These studies have involved quantitative and qualitative changes in enzymes, leakage of solutes, degradation of storage compounds, effects of fungi during storage, accumulation of growth inhibitors and mutagens, changes in respiratory metabolism, and changes in synthesis of RNA, DNA, and protein. Total cellular protein, RNA, and DNA do not usually change measurably with aging, but ribosomal RNA and DNA are cleaved and polyA-RNA has been found to decrease in rye (Osborne et al. 1977). As for the structural studies, there seems to be no apparent single biochemical lesion associated with loss of vigour and viability. Roberts and Osborne (1973) concluded that "the whole molecular structure of the cells passes from one of an organized living system to one of molecular disorganization of the non-living state". In their case it applied to integrity of DNA and the function of protein synthesis.

It is possible that DNA breakage and activity of DNA repair enzymes may be at the core of aging damage and recovery. In parallel with chromosome aberrations, DNA has been reported to contain increasing numbers of breaks with aging (Roberts and Osborne 1973, Osborne et al. 1974, Cheah and Osborne 1978), and it has been proposed that faulty transcription and disruption of protein synthesis may result. Although DNA repair occurs in plants, it has been proposed that in dry seed breaks in DNA may accumulate to very high levels during aging and, if repair enzymes were also labile in dry seeds, breaks might reach a level which could not be repaired on hydration resulting in death of the cell (Cheah and Osborne 1978). The possibility that hydration is necessary for repair to occur could account for the observation that germination of aged seed and seedling establishment can be improved by seed hydration (Heydecker 1972). It also forms the basis of an hypothesis developed by Villiers (1975) to explain delayed germination of aged seed, and why some hydrated seeds can exist in the soil for very long periods of time with no apparent damage.

In general, these studies show that associated with loss of vigour and viability there is increased damage to nuclear and cytoplasmic organelles, and thus far there is no evidence for a single causal lesion. Most of the cytoplasmic structural changes are consistent with loss of integrity and permeability of the cellular membranes. Such changes, which would undoubtedly be associated with a general breakdown in physiology, would lead to disruption of cellular compartmentation and the mixing of cellular enzymes and substrates. If hydrolases like protease, ribonuclease, deoxyribonuclease, lipase and others are contained in vacuoles, their release could lead to the macromolecular degeneration which occurs with aging and, in extreme cases, complete degeneration of the cell.

Acknowledgements. The author wishes to thank Dr. Rod King and Dr. Russell Jones for their suggestions for improving the manuscript, and Miss Rosemary Metcalf for her assistance during all stages of manuscript preparation.

References

Abdul-Baki AA, Anderson JD (1972) Physiological and biochemical deterioration of seeds. In: Kozlowski TT (ed) Seed biology, vol II. Germination control, metabolism, and pathology. Academic Press, New York, pp 283–315

Abdul-Baki AA, Baker JE (1973) Are changes in cellular organelles or membranes related to vigor loss in seeds? Seed Sci Tech 1:89–125

Amen RD (1968) A model of seed dormancy. Bot Rev 34:1–31

Andrews P, Haugh L, Jones JKN (1953) Mannose-containing polysaccharides. III. The polysaccharides in the seeds of *Iris ochroleuca* and *I. sibirica*. J Chem Soc 1186–1192

Ashford AE, Jacobsen JV (1974a) Cytochemical localisation of acid phosphatase in isolated aleurone layers treated with gibberellic acid and in germinating barley grains. In: Bieleski RL, Ferguson AR, Creswell MM (eds) Mechanisms of regulation of plant growth. Royal Society New Zealand Bull 12 (Wellington), pp 591–599

Ashford AE, Jacobsen JV (1974b) Cytochemical localisation of phosphatase in barley aleurone cells: The pathway of gibberellic acid-induced enzyme release. Planta (Berl) 120:81–105

Ashton FM (1976) Mobilization of storage proteins of seeds. Annu Rev Plant Physiol 27:95–117

Bacic A, Stone BA (1979) Chemical composition of aleurone cell walls from wheat and barley. Proc 29th Annu Conf Royal Australian Chem Inst, Cereal Chem Div, Adelaide, pp 36–38

Bagley BW, Cherry JH, Rollins ML, Altschul AM (1963) A study of protein bodies during germination of peanut (*Arachis hypogaea*) seed. Am J Bot 50:523–532

Bailey KM, Phillips IDJ, Pitt D (1976) Effects of gibberellic acid on the activation, synthesis, and release of acid phosphatase in barley seed. J Exp Bot 27:324–336

Bailey RW (1971) Polysaccharides in the Leguminosae. In: Harborne JB, Boulter D, Turner BL (eds) Chemotaxonomy of the Leguminosae. Academic Press, New York London, pp 503–541

Bain JM, Mercer FV (1966) Subcellular organization of the cotyledons in germinating seeds and seedlings of *Pisum sativum* L. Aust J Biol Sci 19:69–84

Ballance GM, Manners DJ (1978) Structural analysis and enzymic solubilization of barley endosperm cell walls. Carbohydr Res 61:107–118

Ballance GM, Meredith WOS, Laberge DE (1976) Distribution and development of endo-β-glucanase activities in barley tissues during germination. Can J Plant Sci 56:459–466

Barlow KK, Buttrose MS, Simmonds DH, Vesk M (1973a) The nature of the starch protein interface in wheat endosperm. Cereal Chem 50:443–454

Barlow KK, Simmonds DH, Kendrick KG (1973b) The localization of water soluble proteins in the wheat endosperm as revealed by fluorescent antibody techniques. Experientia 29:229–231

Barton LV (1961) Seed preservation and longevity. Interscience, New York

Barton LV (1965a) Dormancy in seeds imposed by the seed-coat. In: Ruhland W (ed) Encyclopedia of Plant Physiology, vol XV/2. Springer, Berlin, pp 727–745

Barton LV (1965b) Longevity in seeds and in the propagules of fungi. In: Ruhland W (ed) Encyclopedia of Plant Physiology, vol XV/2. Springer, Berlin, pp 1058–1085

Barton LV (1965c) Seed dormancy: General survey of dormancy types in seeds and dormancy imposed by external agents. In: Ruhland W (ed) Encyclopedia of Plant Physiology, vol XV/2. Springer, Berlin, pp 699–720

Baumgartner B, Chrispeels MJ (1977) Purification and characterization of vicilin peptidohydrolase, the major endopeptidase in the cotyledons of mung-bean seedlings. Eur J Biochem 77:223–233
Baumgartner B, Tokuyasu KT, Chrispeels MJ (1978) Localization of vicilin peptidohydrolase in the cotyledons of mung-bean seedlings by immunofluorescence microscopy. J Cell Biol 79:10–19
Beevers H (1975) Organelles from castor bean seedlings: Biochemical roles in gluconeogenesis and phospholipid biosynthesis. In: Galliard T, Mercer EI (eds) Recent advances in the chemistry and biochemistry of plant lipids. Proc Phytochem Soc 12:287–299
Beevers H (1979) Microbodies in higher plants. Annu Rev Plant Physiol 30:159–193
Beevers L, Guernsey FS (1966) Changes in some nitrogenous components during the germination of pea seeds. Plant Physiol 41:1455–1458
Bergfeld R, Hong Y-N, Kühnl T, Schopfer P (1978) Formation of oleosomes (storage lipid bodies) during embryogenesis and their breakdown during seedling development in cotyledons of *Sinapis alba*. Planta (Berl) 143:297–307
Berjak P, Villiers TA (1970) Ageing in plant embryos 1. The establishment of the sequence of development and senescence in the root cap during germination. New Phytol 69:929–938
Berjak P, Villiers TA (1972) Ageing in plant embryos 11. Age-induced damage and its repair during early germination. New Phytol 71:135–144
Berlyn GP (1972) Seed germination and morphogenesis. In: Kozlowski TT (ed) Seed biology, vol 1. Academic Press, New York, pp 223–312
Bewley JD, Black M (1978) Physiology and biochemistry of seeds, vol 1. Springer, Berlin Heidelberg New York
Bewley JD, Black M (1982) Physiology and biochemistry of seeds, vol 2. Springer, Berlin Heidelberg New York
Boatman SG, Crombie WM (1958) Fat metabolism in the West African oil palm. J Exp Bot 9:52–74
Bornman CH, Butler V, Jensen WA (1979a) *Welwitschia mirabilis:* Fine structure of the germinating seed I. Orientation. Z Pflanzenphysiol 91:189–196
Bornman CH, Butler V, Jensen WA (1979b) *Welwitschia mirabilis:* Fine structure of the germinating seed V. The quiescent and imbibed gametophyte interface. Z Pflanzenphysiol 92:115–132
Bornman CH, Butler V, Jensen WA (1979c) *Welwitschia mirabilis:* Fine structure of the germinating seed VI. The quiescent embryo. Z Pflanzenphysiol 92:355–365
Bornman CH, Butler V, Jensen WA (1979d) *Welwitschia mirabilis:* Fine structure of the germinating seed VIII. Interface of the developing feeder. Z Pflanzenphysiol 92:417–430
Borthwick HA, Robbins WW (1928) Lettuce seed and germination. Hilgardia 3:275–305
Briarty LG, Coult DA, Boulter D (1970) Protein bodies of germinating seeds of *Vicia faba*. J Exp Bot 21:513–524
Briggs DE (1973) Hormones and carbohydrate metabolism in germinating cereal grains. In: Milborrow BV (ed) Biosynthesis and its control in plants. Academic Press, London New York, pp 219–277
Briggs DE (1978) Barley. Chapman and Hall, London
Bullowa S, Negbi M, Ozeri Y (1975) Role of temperature, light, and growth regulators in germination in *Anemone coronaria* L. Aust J Plant Physiol 2:91–100
Butler V, Bornman CH, Jensen WA (1979a) *Welwitschia mirabilis:* Fine structure of the germinating seed II. Cytochemistry of gametophyte and feeder. Z Pflanzenphysiol 91:197–210
Butler V, Bornman CH, Jensen WA (1979b) *Welwitschia mirabilis:* Fine structure of the germinating seed III. The quiescent gametophyte. Z Pflanzenphysiol 91:399–413
Butler V, Bornman CH, Jensen WA (1979c) *Welwitschia mirabilis:* Fine structure of the germinating seed IV. The imbibed gametophyte. Z Pflanzenphysiol 92:95–113
Butler V, Bornman CH, Jensen WA (1979d) *Welwitschia mirabilis:* Fine structure of the germinating seed VII. The imbibed embryo. Z Pflanzenphysiol 92:407–415
Butler V, Bornman CH, Jensen WA (1979e) *Welwitchia mirabilis:* Fine structure of the germinating seed IX. Inner cells of the developing feeder. Z Pflanzenphysiol 93:45–61
Buttrose MS (1971) Ultrastructure of barley aleurone cells as shown by freeze-etching. Planta (Berl) 96:13–26
Cheah KSE, Osborne DJ (1978) DNA lesions occur with loss of viability in embryos of ageing rye seed. Nature 272:593–599
Chen SSC, Chang JLL (1972) Does gibberellic acid stimulate seed germination via amylase synthesis? Plant Physiol 49:441–442
Chen SSC, Varner JE (1970) Respiration and protein synthesis in dormant and after-ripened seeds of *Avena fatua*. Plant Physiol 46:108–112

Cherry JH (1963) Nucleic acid, mitochondria, and enzyme changes in cotyledons of peanut seeds during germination. Plant Physiol 38:440–446
Ching TM (1972) Metabolism of germinating seeds. In: Kozlowski TT (ed) Seed biology, vol II. Germination control, metabolism, and pathology. Academic Press, New York, pp 103–218
Chrispeels MJ, Boulter D (1975) Control of storage protein metabolism in the cotyledons of germinating mung beans: Role of endopeptidase. Plant Physiol 55:1031–1037
Chrispeels MJ, Baumgartner B, Harris N (1976) Regulation of reserve protein metabolism in the cotyledons of mung bean seedlings. Proc Natl Acad Sci USA 73:3168–3172
Clermont-Beaugiraud S, Percheron F (1968) Etude du mode d'action de la β-mannanase de semences germées du fenugrec. Bull Soc Chim Biol 50:633–640
Colborne AJ, Morris G, Laidman DL (1976) The formation of endoplasmic reticulum in the aleurone cells of germinating wheat: An ultrastructural study. J Exp Bot 27:759–767
Corner EJH (1951) The leguminous seed. Phytomorphology 1:117–150
Corner EJH (1976) The seeds of dicotyledons, vol 1. Cambridge Univ Press, Cambridge London New York Melbourne
Courtois JE, Percheron F (1961) L'α-galactosidase de semences germées de fenugrec. Bull Soc Chim Biol 43:167–175
Crocker W (1916) Mechanisms of dormancy in seeds. Am J Bot 3:99–120
Crocker W, Barton LV (1957) Physiology of seeds. Chron Bot
D'Amato F, Hoffman-Ostenhoff O (1956) Metabolism and spontaneous mutations in plants. Adv Genet 8:1–28
De Barsy Th, Deltour R, Bronchart R (1974) Study of nucleolar vacuolation and RNA synthesis in embryonic root cells of *Zea mays*. J Cell Sci 16:95–112
Dashek WV, Chrispeels MJ (1977) Gibberellic acid-induced synthesis and release of cell-wall-degrading endoxylanases by isolated aleurone layers of barley. Planta (Berl) 134:251–256
Deltour R, Bronchart R (1971) Changements de l'ultrastructure de cellules radiculaires de *Zea mays* au début de la germination. Planta (Berl) 97:197–207
Deltour R, Jacqmard A (1974) Relation between water stress and DNA synthesis during germination of *Zea mays* L. Ann Bot 38:529–534
Deltour R, Gautier A, Fakan J (1979) Ultrastructural cytochemistry of the nucleus in *Zea mays* embryos during germination. J Cell Sci 40:43–62
Dickson JG, Shands HL (1941) Cellular modification of the barley kernel during malting. Proc Am Soc Brew Chem 1–10
Earle FR, Jones Q (1962) Analyses of seed samples from 113 plant families. Econ Bot 16:221–250
Esau K (1965a) Plant anatomy. Wiley, London
Esau K (1965b) Vascular differentiation in plants. Rinehard and Winston, New York
Esau K (1977) Anatomy of seed plants, 2nd edn. Wiley, New York
Evenari M (1965) Light and seed dormancy. In: Ruhland W (ed) Encyclopedia of Plant Physiology, vol XV/2. Springer, Berlin, pp 804–847
Evenari M, Neumann G (1952) The germination of lettuce seed 11. The influence of fruit coat, seed coat, and endosperm upon germination. Bull Res Counc Isr Sect D Bot II:75–78
Fahn A (1974) Plant anatomy. Pergamon, Oxford New York Toronto Sydney Braunschweig
Fincher GB (1975) Morphology and chemical composition of barley endosperm cell walls. J Inst Brew 81:116–122
Foard DE, Haber AH (1966) Mitosis in thermodormant lettuce seeds with reference to histological location, localized expansion, and seed storage. Planta (Berl) 71:160–170
Foster AS, Gifford EM (1974) Comparative morphology of vascular plants. Freeman, San Francisco
Fulcher RG, Setterfield G, McCully ME, Wood PJ (1977) Observations on the aleurone layer. 11. Fluorescence microscopy of the aleurone-sub-aleurone junction with emphasis on possible β-1,3-glucan deposits in barley. Aust J Plant Physiol 4:917–928
Gibbons GC (1980) Immunohistochemical determination of the transport pathways of α-amylase in germinating barley seeds. Cereal Res Commun 8:87–96
Giersbach J (1937) Germination and seedling production of species of *Viburnum*. Contrib Boyce Thompson Inst 9:79–90
Globerson D, Kadman-Zahavi A, Ginzburg C (1974) A genetic method for studying the role of the seed coats in the germination of lettuce. Ann Bot 38:201–203
Guillermond A (1908) Recherches cytologiques sur la germination des graines de quelque Graminées et contribution à l'étude des graines d'aleurone. Arch Anat Microsc Morphol Exp 10:141–226

Gunning B, Pate JS (1969) Transfer cells – Plant cells with wall ingrowths, specialized in relation to short distance transport of solutes – Their occurrence, structure, and development. Protoplasma 68:107–133

Haber AH, Luippold H (1960) Separation of mechanisms initiating cell division and cell expansion in lettuce seed germination. Plant Physiol 35:168–172

Hallam ND (1973) Fine structure of viable and non-viable rye and other embryos. In: Heydecker W (ed) Seed Ecology. Butterworths, London (Nottingham University Easter School in Agricultural Science No 19), p 115

Hallam ND, Roberts BE, Osborne DJ (1972) Embryogenesis and germination in rye (*Secale cereale* L.) II. Biochemical and fine structural changes during germination. Planta (Berl) 105:293–309

Hallam ND, Roberts BE, Osborne DJ (1973) Embryogenesis and germination in rye (*Secale cereale* L.) III. Fine structure and biochemistry of the non-viable embryo. Planta (Berl) 110:279–290

Halmer P, Bewley JD (1979) Mannanase production by the lettuce endosperm: Control by the embryo. Planta (Berl) 144:333–340

Halmer P, Bewley JD, Thorpe TA (1975) An enzyme to degrade the lettuce endosperm cell wall during giberellin- and light-induced germination. Nature 258:716–718

Halmer P, Bewley JD, Thorpe TA (1978) Degradation of the endosperm cell walls of *Lactuca sativa* L. cv Grand Rapids: Timing of mobilization of soluble sugars, lipid and phytate. Planta (Berl) 139:1–8

Harrington JF (1972) Seed storage and longevity. In: Kozlowski TT (ed) Seed biology, vol III. Insects and seed collection, storage, testing, and certification. Academic Press, New York, pp 145–245

Harris N (1979) Endoplasmic reticulum in developing seeds in *Vicia faba*. Planta (Berl) 146:63–69

Harris N, Chrispeels MJ (1975) Histochemical and biochemical observations on storage protein metabolism and protein body autolysis in cotyledons of germinating mung beans. Plant Physiol 56:292–299

Harris N, Chrispeels MJ, Boulter D (1975) Biochemical and histochemical studies on protease activity and reserve protein metabolism in cotyledons of germinating cowpeas (*Vigna unguiculata*). J Exp Bot 26:544–554

Haywood HE (1938) The structure of economic plants. Macmillan, New York

Heydecker W (1972) Vigour. In: Roberts EH (ed) Viability of seeds. Chapman and Hall, London, pp 209–252

Higgins TJV, Jacobsen JV (1978) Phytohormones and subcellular structural modification. In: Letham DS, Goodwin PB, Higgins TJV (eds) Phytohormones and related compounds: A comprehensive treatise, vol 1. The biochemistry of phytohormones and related compounds. Elsevier/North Holland, Amsterdam Oxford New York, pp 419–465

Hopf H, Kandler O (1977) Characterization of the "reserve cellulose" of the endosperm of *Carum carvi* as a β (1-4)-manan. Phytochemistry 16:1715–1717

Horner HT, Arnott HJ (1965) A histochemical and ultrastructural study of *Yucca* seed proteins. Am J Bot 52:1027–1038

Horner HT, Arnott HJ (1966) Histochemical and ultrastructural study of pre- and post-germinated *Yucca* seeds. Bot Gaz 127:48–64

Ikuma H, Thimann KV (1963) The role of the seed coats in germination of photosynthetic lettuce seeds. Plant Cell Physiol 4:169–185

Ives SA (1923) Maturation and germination of seeds of *Ilex opaca*. Bot Gaz 76:60–77

Jacks TJ, Yatsu LY, Altschul AM (1967) Isolation and characterization of peanut spherosomes. Plant Physiol 42:585–597

Jacobsen JV (1977) Regulation of ribonucleic acid metabolism by plant hormones. Annu Rev Plant Physiol 28:537–564

Jacobsen JV, Higgins TJV (1978a) Post-transcription, translational, and post-translational effects of plant hormones. In: Letham DS, Goodwin PB, Higgins TJV (eds) Phytohormones and related compounds: A comprehensive treatise, vol 1. The biochemistry of phytohormones and related compounds. Elsevier/North Holland, Amsterdam Oxford New York, pp 583–621

Jacobsen JV, Higgins TJV (1978b) The influence of phytohormones on replication and transcription. In: Letham DS, Goodwin PB, Higgins TJV (eds) Phytohormones and related compounds: A comprehensive treatise, vol 1. The biochemistry of phytohormones and related compounds. Elsevier/North Holland, Amsterdam Oxford New York, pp 515–582

Jacobsen JV, Knox B (1973) Cytochemical localization and antigenicity of α-amylase in barley aleurone tissue. Planta (Berl) 112:213–224

Jacobsen JV, Knox B (1974) The proteins released by isolated barley aleurone layers before and after gibberellic acid treatment. Planta (Berl) 115:193–206
Jacobsen JV, Pressman E (1979) A structural study of germination in celery (*Apium graveolens* L.) seed with emphasis on endosperm breakdown. Planta (Berl) 144:241–248
Jacobsen JV, Pressman E, Pyliotis NA (1976) Gibberellin-induced separation of cells in isolated endosperm of celery seed. Planta (Berl) 129:113–122
Jones RL (1969a) Gibberellic acid and the fine structure of barley aleurone cells. I. Changes during the lag-phase of α-amylase synthesis. Planta (Berl) 87:119–133
Jones RL (1969b) Gibberellic acid and the fine structure of barley aleurone cells II. Changes during the synthesis and secretion of α-amylase. Planta (Berl) 88:73–86
Jones RL (1972) Fractionation of the enzymes of the barley aleurone layer: Evidence for a soluble mode of enzyme release. Planta (Berl) 103:95–109
Jones RL (1974) The structure of the lettuce endosperm. Planta (Berl) 121:133–146
Jones RL, Chen R-F (1976) Immunohistochemical localization of α-amylase in barley aleurone cells. J Cell Sci 20:183–198
Jones RL, Price JM (1970) Gibberellic acid and the fine structure of barley aleurone cells III. Vacuolation of the aleurone cell during the phase of ribonuclease release. Planta (Berl) 94:191–202
Khan AA (1977) Seed dormancy: Changing concepts and theories. In: Khan AA (ed) The physiology and biochemistry of seed dormancy and germination. North-Holland, Amsterdam, pp 29–50
Klein S, Ben-Shaul Y (1966) Changes in cell fine structure of lima bean axes during early germination. Can J Bot 44:331–340
Koller D, Mayer A, Poljakoff-Mayber A, Klein S (1962) Seed germination. Annu Rev Plant Physiol 13:437–457
Laidman DL, Colborne AJ, Doig RI, Varty K (1974) The multiplicity of induction systems in wheat aleurone tissue. In: Bieleski RL, Ferguson AR, Creswell MM (eds) Mechanisms of regulation of plant growth. Royal Society New Zealand Bull 12, Wellington, pp 581–590
Lang A (1965) Effects of some internal and external conditions on seed germination. In: Ruhland W (ed) Encyclopedia of Plant Physiology, vol XV/2. Springer, Berlin, pp 848–893
Lehman K, Garz J (1962) Untersuchungen über den Umsatz der phosphathaltigen Substanzen und des Eiweißes in den Kotyledonen keimender Erbsensamen. Flora (Jena) 152:516–522
Leung DWM, Reid JSG, Bewley JD (1979) Degradation of the endosperm cell wall of *Lactuca sativa* L. cv Grand Rapids in relation to the mobilization of proteins and the production of hydrolytic enzymes in the axis, cotyledons, and endosperms. Planta (Berl) 146:335–341
Lott JNA (1980) Protein bodies. In: The biochemistry of plants. A comprehensive treatise, vol 1. The plant cell. Academic Press, New York
Lott JNA, Vollmer CM (1973) Changes in the cotyledons of *Cucurbita maxima* during germination. IV. Protein bodies. Protoplasma 78:255–271
MacLeod AM (1969) The utilization of cereal seed reserves. Sci Prog 57:99–112
MacLeod AM, Millar AS (1962) Effects of gibberellic acid on barley endosperm. J Inst Brew 68:322–332
MacLeod AM, Palmer GH (1966) The embryo of barley in relation to modification of the endosperm. J Inst Brew 72:580–589
Manners DJ (1974) The structure and metabolism of starch. In: Campbell PN, Dickens F (eds) Essays in Biochemistry, vol 10. Academic Press, London
Mares DJ, Stone BA (1973) Studies of wheat endosperm. I. Chemical composition and ultrastructure of the cell walls. Aust J Biol Sci 26:793–812
Marré E (1967) Ribosome and enzyme changes during maturation and germination of the castor bean seed. Curr Top Dev Biol 11:75–105
Martin AC (1946) The comparative internal morphology of seeds. Am Midl Nat 36:513–660
Mayer AM, Poljakoff-Mayber A (1975) The germination of seeds. Pergamon, Oxford
McCleary BV, Matheson NK (1975) Galactomannan structure and β-mannanase and β-mannosidase activity in germinating legume seeds. Phytochemistry 14:1187–1194
McNeil M, Albersheim P, Taiz L, Jones RL (1975) The structure of plant cell walls. VII. Barley aleurone cells. Plant Physiol 55:64–68
Meier H (1958) On the structure of cell walls and cell wall mannans from ivory nuts and from dates. Biochim Biophys Acta 28:229–240
Millerd A, Whitfeld PR (1973) Deoxyribonucleic acid and ribonucleic acid synthesis during the cell expansion phase of cotyledon development in *Vicia faba* L. Plant Physiol 51:1005–1010

Minamikawa T (1979) Hydrolytic enzyme activities and degradation of storage components in cotyledons of germinating *Phaseolus mungo* seeds. Bot Mag Tokyo 92:1–12
Minamikawa T, Koshiba T (1979) Histochemical studies on mobilization of storage components in cotyledons of germinating *Phaseolus mungo* seeds. Bot Mag Tokyo 92:325–332
Mlodzianowski F (1978) The fine structure of protein bodies of lupin cotyledons during the course of seed germination. Z Pflanzenphysiol 86:1–13
Mollenhauer HH, Totten C (1971 a) Studies on seeds 1. Fixation of seeds. J Cell Biol 48:387–394
Mollenhauer HH, Totten C (1971 b) Studies on seeds 11. Origin and degradation of lipid vesicles in pea and bean cotyledons. J Cell Biol 48:395–405
Morrall P, Briggs DE (1978) Changes in cell wall polysaccharides of germinating barley grains. Phytochemistry 17:1495–1502
Muto S, Beevers H (1974) Lipase activities in castor bean endosperm during germination. Plant Physiol 54:23–28
Nabors MW, Lang A (1971 a) The growth physics and water relations of red light-induced germination in lettuce seeds. I. Embryos germinating in osmoticum. Planta (Berl) 101:1–25
Nabors MW, Lang A (1971 b) The growth physics and water relations of red light-induced germination in lettuce seeds. II. Embryos germinating in water. Planta (Berl) 101:26–42
Nabors MW, Kugrens P, Ross C (1974) Photodormant lettuce seeds: Phytochrome-induced protein and lipid degradation. Planta (Berl) 117:361–365
Netolitzky F (1926) Anatomie der Angiospermen-Samen. In: Linsbauer K (ed) Handbuch der Pflanzenanatomie, Band X. Borntraeger, Berlin
Nieuwdorp PJ, Buys MC (1964) Electron microscope structure of the epithelial cells of the scutellum of barley. II. Cytology of the cells during germination. Acta Bot Neerl 13:559–565
Nikolaeva MG (1977) Factors controlling the seed dormancy pattern. In: Khan AA (ed) The physiology and biochemistry of seed dormancy and germination. North-Holland, Amsterdam New York Oxford, pp 51–76
Obata T (1979) Fine structural changes in barley aleurone cells during gibberellic acid-induced enzyme secretion. Ann Bot 44:333–337
O'Brien JA (1942) Cytoplasmic inclusions in the glandular epithelia of the scutellum of *Triticum sativum* and *Secale cereale*. Am J Bot 29:479–491
Okamoto K, Akazawa T (1979) Enzymic mechanisms of starch breakdown in germinating rice seeds. Plant Physiol 63:336–340
Opik H (1965 a) Respiration rate, mitochondrial activity, and mitochondrial structure in the cotyledons of *Phaseolus vulgaris* L. during germination. J Exp Bot 16:667–682
Opik H (1965 b) The form of nuclei in the storage cells of the cotyledons of germinating seeds of *Phaseolus vulgaris* L. Exp Cell Res 38:517–522
Opik H (1966) Changes in cell fine structure in the cotyledons of *Phaseolus vulgaris* L. during germination. J Exp Bot 17:427–439
Ory RL, Yatsu LY, Kircher HW (1968) Association of lipase activity with the spherosomes of *Ricinus communis*. Arch Biochem Biophys 264:255–264
Osborne DJ, Roberts BE, Payne PI, Sen S (1974) Protein synthesis and viability in rye embryos. In: Bieleski RL, Ferguson AR, Creswell MM (eds) Mechanisms of regulation of plant growth. Royal Society of New Zealand Bull 12 Wellington, pp 805–812
Osborne DJ, Dobrzanska M, Sen S (1977) Factors determining nucleic acid and protein synthesis in the early hours of germination. In: Jenning DH (ed) Symp Soc Exp Biol No 31. Cambridge Univ Press, Cambridge, pp 177–194
Palmer GH (1972) Morphology of starch granules in cereal grains and malts. J Inst Brew 78:326–332
Palmer GH, Bathgate GN (1976) Malting and brewing. In: Pomeranz Y (ed) Advances in cereal science and technology, vol 1. American Association of Cereal Chemists, St Paul Minnesota, pp 237–324
Pavlista AD, Haber AH (1970) Embryo expansion without protrusion in lettuce seeds. Plant Physiol 46:636–637
Pavlista AD, Valdovinos JG (1978) Changes in the surface appearance of the endosperm during lettuce achene germination. Bot Gaz 139:171–179
Pyliotis NA, Ashford AE, Whitecross MI, Jacobsen JV (1979) Localization of gibberellic acid-induced acid phosphatase activity in the endoplasmic reticulum of barley aleurone cells with the electron microscope. Planta (Berl) 147:134–140
Raghavan V (1976) Experimental embryogenesis in vascular plants. Academic Press, London New York

Reed HS (1904) A study of the enzyme-secreting cells in the seedling of *Zea mays*. Ann Bot 19:115–122
Reid JS, Davies C (1977) Endo-β-mannanase, the leguminous aleurone layer and the storage galactomannan in germinating seeds of *Trigonella foenum-graecum* L. Planta (Berl) 133:219–22
Reid JS, Meier H (1972) The function of the aleurone layer during galactomannan mobilization in germinating seeds of fenugreek (*Trigonella foenum-graecum* L.), crimson clover (*Trifolium incarnatum* L.), and lucerne (*Medicago sativa* L.): A correlative biochemical and ultrastructure study. Planta (Berl) 106:44–60
Reid JS, Meier H (1973) Enzymic activities and galactomannan mobilization in germinating seeds of fenugreek (*Trigonella foenum-graecum* L., Leguminosae): Secretion of α-galactosidase and β-mannosidase by the aleurone layer. Planta (Berl) 112:301–308
Roberts BE, Osborne DJ (1973) Protein synthesis and viability in rye grains. In: Heydecker W (ed) Seed ecology. (Nottingham University Easter School in Agricultural Science No 19) Butterworths, London, pp 99–114
Roberts EH (1972a) Cytological, genetical, and metabolic changes associated with loss of viability. In: Roberts EH (ed) Viability of seeds. Chapman and Hall, London, pp 253–306
Roberts EH (1972b) Dormancy: A factor affecting seed survival in the soil. In: Roberts EH (ed) Viability of seeds. Chapman and Hall, London, pp 321–359
Roberts EH (1972c) Storage environment and the control of viability. In: Roberts EH (ed) Viability of seeds. Chapman and Hall, London, pp 14–58
Roberts EH (1975) Problems of long-term storage of seeds and pollen for genetic resources conservation. In: Frankel OH, Hawkes JG (eds) Crop genetic resources for today and tomorrow. Cambridge University Press, Cambridge London New York Melbourne, pp 269–296
Roberts EH, Abdulla FH, Owen RJ (1967) Nuclear damage and the ageing of seeds with a model for seed survival curves. Symp Soc Exp Biol 21:65–100
Rost TL (1972) The ultrastructure and physiology of protein bodies and lipids from hydrated dormant and non-dormant embryos of *Setaria lutescens* (Gramineae). Am J Bot 59:607–616
Sargant E, Robertson A (1905) The anatomy of the scutellum in *Zea mays*. Ann Bot 19:115–122
Seiler A (1977) Galactomannanabbau in keimenden Johannisbrotsamen (*Ceratonia siliqua* L.). Planta (Berl) 134:209–221
Simola LK (1973) The origin and development of organelles in germinating embryos of *Bidens cernua*: Ultrastructural effects of cycloheximide, actinomycin D and chloramphenicol. Ann Bot Fenn 10:71–88
Smith DL (1971) Nuclear changes in cotyledons of *Pisum arvense* L. during germination. Ann Bot 35:511–521
Smith DL, Flinn AM (1967) Histology and histochemistry of the cotyledons of *Pisum arvense* L. during germination. Planta (Berl) 74:72–85
Sømme R (1970) Glycosidases and monosaccharides in germinated leguminous seeds. Acta Chem Scand 24:72–76
Sømme R (1971) The correlation between the mono-, oligo-, and polysaccharides and the glycosidases present in clover seeds. Acta Chem Scand 25:759–761
Speer HL (1974) Some aspects of the function of the endosperm during the germination of lettuce seeds. Can J Bot 52:1117–1121
Spencer D, Higgins TJV (1979) Molecular aspects of seed protein biosynthesis. Curr Adv Plant Sci 34:1–15
Srivastava LM, Paulson RE (1968) The fine structure of the embryo of *Lactuca sativa* II. Changes during germination. Can J Bot 46:1447–1453
Stokes P (1952a) A physiological study of embryo development in *Heracleum sphondylium* L. I. The effect of temperature on embryo development. Ann Bot 16:441–447
Stokes P (1952b) A physiological study of embryo development in *Heracleum sphondylium*. II. The effect of temperature on after-ripening. Ann Bot 64:572–576
Stokes P (1953a) A physiological study of embryo development in *Heracleum sphondylium* L. III. The effect of temperature on metabolism. Ann Bot 65:157–169
Stokes P (1953b) The stimulation of growth by low temperature in embryos of *Heracleum sphondylium* L. J Exp Bot 4:222–234
Stokes P (1965) Temperature and seed dormancy. In: Ruhland W (ed) Encyclopedia of Plant Physiology, vol XV/2. Springer, Berlin, pp 746–803
Swift JG, O'Brien TP (1970) Vascularization of the scutellum of wheat. Aust J Bot 18:45–53
Swift JG, O'Brien TP (1971) Vascular differentiation in the wheat embryo. Aust J Bot 19:63–71

Swift JG, O'Brien TP (1972) The fine structure of the wheat scutellum during germination. Aust J Biol Sci 25:469–486

Taiz L, Honigman WA (1976) Production of cell wall hydrolyzing enzymes by barley aleurone layers in response to gibberellic acid. Plant Physiol 58:380–386

Taiz L, Jones RL (1970) Gibberellic acid, β-1,3-glucanase and the cell walls of barley aleurone layers. Planta (Berl) 92:73–84

Taiz L, Jones RL (1973) Plasmodesmata and an associated cell wall component in barley aleurone tissue. Am J Bot 60:67–75

Thomas H (1972) Control mechanisms of the resting seed. In: Roberts EH (ed) Viability in seeds. Chapman and Hall, London, pp 360–396

Thompson JL, Jones JKN (1964) The glucomannan of bluebell seed (*Scilla nonscripta* L.). Can J Chem 42:1088–1091

Toole EH (1924) The transformation and course of development of the germinating maize. Am J Bot 11:325–350

Torrey JC (1902) Cytological changes accompanying the secretion of diastase. Bull Torrey Bot Club 29:421–435

Varner JE (1977) Hormonal control of protein synthesis. In: Bogorad L, Weil J (eds) Nucleic acids and protein synthesis in plants. Plenum, New York London, pp 293–307

Varner JE, Ho DT-H (1976) The role of hormones in the integration of seedling growth. In: Papaconstantinou J (ed) The molecular biology of hormone action. Academic Press, New York San Francisco London, pp 173–194

Vegis A (1964) Dormancy in higher plants. Annu Rev Plant Physiol 15:185–224

Vigil EL (1970) Cytochemical and developmental changes in microbodies (glyoxysomes) and related organelles of castor bean endosperm. J Cell Biol 46:435–454

Vigil EL, Ruddat M (1973) Effect of gibberellic acid and actinomycin D on the formation and distribution of rough endoplasmic reticulum in barley aleurone cells. Plant Physiol 51:549–558

Villaroya H, Williams J, Dey P, Villaroya S, Petek F (1978) Purification and properties of β-mannanases I and II from the germinated seeds of *Trifolium repens*. Biochem J 175:1079–1087

Villiers TA (1971) Cytological studies in dormancy I. Embryo maturation during dormancy in *Fraxinus excelsior*. New Phytol 70:751–760

Villiers TA (1972) Seed dormancy. In: Kozlowski TT (ed) Seed biology, vol II. Academic Press, New York, pp 219–281

Villiers TA (1975) Genetic maintenance of seeds in imbibed storage. In: Frankel OH, Hawkes JG (eds) Crop genetic resources for today and tomorrow. Cambridge University Press, Cambridge London New York Melbourne, pp 297–316

Villiers TA, Wareing PF (1964) Dormancy in fruits of *Fraxinus excelsior* L. J Exp Bot 15:359–367

Wareing PF (1965) Endogenous inhibitors in seed germination and dormancy. In: Ruhland W (ed) Encyclopedia of Plant Physiology, vol XV/2. Springer, Berlin, pp 909–924

Werker E, Vaughan JG (1974) Anatomical and ultrastructural changes in aleurone and myrosin cells of *Sinapsis alba* during germination. Planta (Berl) 116:243–255

Wirth M, Withner CL (1959) Embryology and development in the Orchidaceae. In: Withner CL (ed) The orchids: A scientific survey. Ronald Press, New York, pp 155–188

Wolfrom ML, Laver ML, Patin DL (1961) Carbohydrates of the coffee bean. II. Isolation and characterization of a mannan. J Org Chem 26:4533–4535

Yatsu LY, Jacks TJ (1972) Spherosome membranes. Half unit-membranes. Plant Physiol 49:937–943

Yomo H, Taylor MP (1973) Histochemical studies on protease formation in the cotyledons of germinating bean seeds. Planta (Berl) 112:35–43

Yomo H, Varner JE (1971) Hormonal control of a secretory tissue. In: Moscona AA, Monray A (eds) Current topics in developmental biology. Academic Press, New York, pp 111–144

Yoo BY (1970) Ultrastructural changes in cells of pea embryo radicle during germination. J Cell Biol 45:158–171

14 Embryology and Taxonomy

J. M. Herr, Jr.

14.1 Introduction

The quest for a natural system of classification of flowering plants has involved consideration of many botanical disciplines to include embryology. Indeed, comparative embryology has been an important factor in revealing the relationships of taxa at all levels, and many contributions from this field have been summarized in the writings of Maheshwari (1950, 1963), Davis (1966), Johri (1967, 1970), Palser (1975), Poddubnaya-Arnoldi (1976), Kapil and Bhatnagar (1980) and Bhojwani and Bhatnagar (1983). To discover true phylogenetic relationships remains an obsession among botanists in spite of the recent exciting advancements in other aspects of plant sciences, and new contributions from embryology toward this end are replete.

The contributions of embryology to systematics may be regarded as significant as those offered by other disciplines. However, the pathway to these contributions has been a difficult one. As, according to Bhojwani and Bhatnagar (1983), the late Professor P. Maheshwari stated: "It ... takes many years of laborious and painstaking research to study (the embryology of) even a few representatives of a family". The analysis of embryological characters in comparative studies, while sufficiently broad, often lacks in depth and in sufficient attention to detail. Thus, the potential value of embryology in the solution of disputed taxonomic problems has not yet been fully realized, and for an insight into that potential, an examination of embryology in the ancestors of angiosperms would be of great value. Unfortunately, the ancestors have eluded identity. Nevertheless, a wealth of embryological data accumulated from present-day flowering plants permits conjecture on features in a hypothetical proangiosperm. That is, an assemblage of primitive characters would serve to identify an embryological archetype for modern angiosperms.

A diagnosis of the embryological archetype would include the following features: microsporangia separate on the abaxial (*Degeneria*) or adaxial (*Magnolia*) surface of the stamen; thick-walled with secretory tapetum of several layers; microspores numerous in tetrads of variable form; ovules with three enveloping layers, two integuments, and an arillus; orthotropous; massively crassinucellate with nucellar growth from periclinal divisions of epidermis and proliferation of parietal tissue; arrangement in megaspore tetrads variable; all megaspores in a tetrad with equal potential to function; development of female gametophyte a generalized expression of Polygonum type; cellular endosperm copious; massive dicotyledonous embryo with a high frequency of polyembryony.

From the embryological archetype the great variety of present structure among angiosperms can be derived, as can be illustrated by development of the

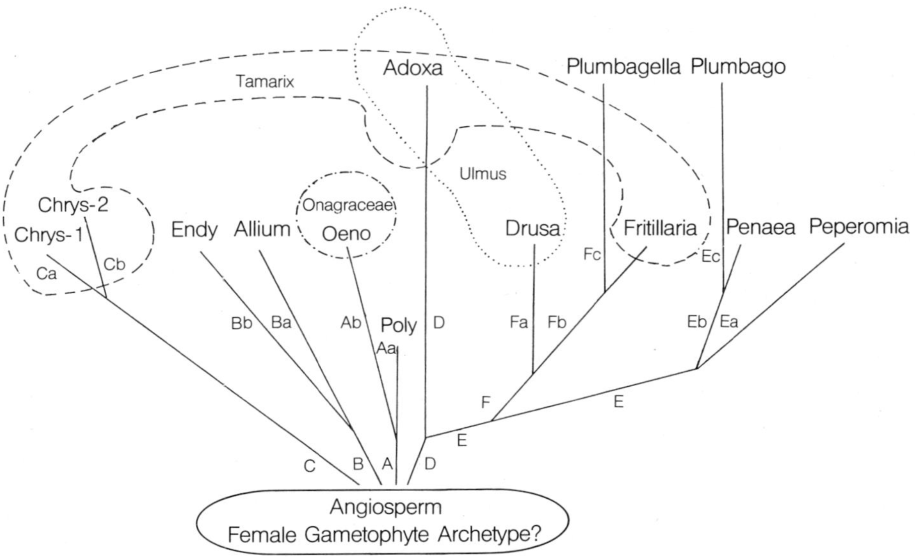

Fig. 14.1. Proposed phylogenetic relationships of female gametophyte types in angiosperms (*Chrys* Chrysanthemum cinerariaefolium type, variations 1 and 2; *Endy* Endymion type; *Oeno* Oenothera type; *Poly* Polygonum type). *A* Cytokinesis I and II; one megaspore functions – *Aa* chalazal megaspore, triphasic, bipolar; *Ab* micropylar megaspore, biphasic, unipolar. *B* Cytokinesis I only; one dyad cell functions – *Ba* chalazal dyad cell, megaspore nuclei biphasic, bipolar; *Bb* micropylar dyad cell, megaspore nuclei biphasic, bipolar. *C* Coenomegaspore; 1+2+1; micropylar and chalazal megaspore nuclei biphasic – *Ca* fusion of central megaspore nuclei, fusion nucleus monophasic, bipolar; *Cb* central megaspore nuclei become polar nuclei, bipolar. *D* Coenomegaspore; 2+2; megaspore nuclei monophasic, bipolar. *E* Coenomegaspore; 1+1+1+1; megaspore nuclei tetrapolar (partial migration) – *Ea* megaspore nuclei biphasic, tetrapolar during development, bipolar at maturity; *Eb* megaspore nuclei biphasic, tetrapolar through mature stage; *Ec* megaspore nuclei monophasic; tetrapolar through mature stage. *F* Coenomegaspore; 1+3; megaspore nuclei bipolar (migration) – *Fa* megaspore nuclei free, biphasic, bipolar; *Fb* chalazal nuclei fuse, megaspore nucleus and fusion nucleus biphasic, bipolar; *Fc* megaspore nucleus and fusion nucleus monophasic, bipolar

female gametophyte (Fig. 14.1). The female gametophyte is an especially apt example, for, as is the case with most embryological features, the phylogeny of the various types does not correlate with the overall phylogenetic schemes for orders, families, or genera. That is, primitive taxa are not always close to the form of the archetype, nor are advanced taxa always removed from it. In plants following Polygonum type, for example, the tendency for any megaspore in the tetrad to become functional may be expressed not only in certain Magnoliaceae but also in some members of Compositae. It should also be noted that major evolutionary steps illustrated in Fig. 14.1 may take place within a genus (e.g., *Ulmus; Tamarix*). Accordingly, the value of the female gametophyte in revealing taxonomic relationship will be realized differently for different taxa. Relationship within *Ulmus* may be revealed by the frequency of occurrence of Adoxa and Drusa types in each elm species. On the other hand, some relationships within Onagraceae may be clarified on the basis of variation within the Oenothera type of development. In either of these cases, however, the value of the female gametophyte as a taxonomic tool is dependent upon a thorough investigation and attention to detail.

An understanding of present applications of embryology to taxonomic problems and a realistic view of those of the future require one to search further into the nature of embryological evidence. For inquiry in detail the reader is referred to the viewpoints of Takhtajan (1969, 1976), Gerassimova-Navashina (1958, 1971), Cave (1953), and Herr (1967). A summary of the principles expressed in these writings is presented here.

The factors for maturation emerge early in floral ontogeny for both generations, sporophyte, and gametophyte. Turning to the female gametophyte once more as an example, the simplicity of the mature structure is the direct consequence of imposed maturation during the 8- or at most 16-nucleate stage. In gymnosperms, on the other hand, the factors promoting maturation do not appear until as many as several thousand free nuclei are produced. The imposing of maturation early in ontogeny (neoteny) produced apparent uniformity in structure as indicated by the occurrence of the Polygonum type of female gametophyte throughout angiosperms. The obvious conclusion is not that all families showing Polygonum type are closely related, but rather that Polygonum type has been derived many times. The same conclusion would follow for other embryological features as well. The relationship between two taxa should be assessed on the basis of the total embryological characters of each; the features each one has derived ultimately from the archetype. Two taxa would be considered closely related if they share a large number of embryological features expressed in a similar manner especially where these features correlate with other characteristics of taxonomic importance.

For two taxa to express Polygonum type development is of itself not an indicator of close relationship, since Polygonum type has been derived along several lines. However, for two taxa to produce closely similar expressions of Polygonum type by parallel evolution is not likely; to do so within ovules of similar structure is less likely, and with the addition of other similar features the likelihood of parallel evolution diminishes, thus increasing the probability of close relationship.

The feature of neoteny, and the consequence of reduced structural complexity, makes necessary a thorough approach to an investigation to include a close monitoring of developmental stages. The proper approach in comparative studies will lead to the differences between or among taxa, differences that otherwise might be hidden in the simplicity of form.

14.2 Embryological Features of Taxonomic Importance

An examination of embryological characters of taxonomic importance will open avenues to the solution of many disputed taxonomic problems. The features presented here are partly derived from earlier commentaries on the application of embryology to taxonomy (Maheshwari 1950, 1963, Cave 1962, Johri 1963, Palser 1975, Poddubnaya-Arnoldi 1976, Bhojwani and Bhatnagar 1983). These include the characters that have been traditionally and most frequently applied to taxonomic problems. Those additional features which have not been widely applied are italicized so as to set them apart.

14.2.1 Anther. Number of microsporangia; number of vertical rows of cells in each archesporium; type of wall development (basic, monocotyledonous, dicotyledonous, reduced); number of layers and level of differentiation in the endothecium; persistence of the epidermis; glandular or amoeboid tapetum; number and ploidy of nuclei in tapetal cells; delimitation of microspores by simultaneous or successive cytokinesis and by furrowing or cell-plate formation; form of tetrad (tetrahedral, decussate, isobilateral, linear, T-shaped, rhomboidal); microspores separate, or remain in tetrads, or larger clusters.

14.2.2 Male Gametophyte (Pollen Grain). Number and position of apertures; sculpturing pattern of exine; location of generative cell; number and shape of cells in pollen grains at the shedding stage.

14.2.3 Ovule. Characteristic pattern of development and maturation (orthotropous, anatropous, hemianatropous, campylotropous, amphitropous, circinotropous); number of integuments; micropyle formed by one (single inner or outer) or two integuments with micropylar passage straight or zigzag; nucellus crassinucellate, pseudocrassinucellate, or tenuinucellate; *growth of the nucellus throughout ovule ontogeny in terms of extent and location (below, above, or lateral to the female gametophyte)*; presence or absence of endothelium, hypostase, epistase, or associated ovular structures (aril, caruncle, obturator, etc.); extent of vascular supply, whether confined to the funiculus or extending into the integuments.

14.2.4 Archesporium. Number of archesporial cells separately differentiated; position in nucellus; extent of parietal tissue, if present; origin and number of sporogenous cells (by direct enlargement or division of archesporial cells); enlargement of the functional sporogenous cells to form the megasporocytes; persistence or degree of degeneration of defunct sporogenous cells.

14.2.5 Megasporogenesis. Changes in the megasporocyte prior to meiosis *(nuclear movement, degree of vacuolation, alteration of size and/or shape)*; location and orientation of spindle during meiosis I; cytokinesis in relation to the nuclear division (concomitant with telophase I, variously delayed, or absent); *size and shape of dyad cells or of the coenocyte containing dyad nuclei prior to meiosis II;* meiosis II synchronous, nonsynchronous, or resticted to one dyad nucleus; presence or absence of cytokinesis to establish a megaspore tetrad (monospory), coenocytic cells containing two megaspore nuclei (bispory) or a coenocyte containing four megaspore nuclei (tetraspory); in monospory, the arrangement and relative initial size of megaspores; position of functional megaspore; persistence and functional tendency of defunct megaspores.

14.2.6 Early Female Gametophyte. Developmental pattern as indicated by the various monosporic, bisporic, and tetrasporic types (Polygonum type, Allium type, Adoxa type, etc.); *initial and final size and shape of each stage;* degree of synchrony in the nuclear divisions; division planes of nuclei (arrangement of nuclei in relation to one another *and to the overall symmetry of the ovule*); *size (range) of nuclei (volumes) throughout entire course of development.*

14.2.7 Mature Female Gametophyte. *Initial and final size and shape;* differentiation of egg apparatus and antipodals (synchronous or nonsynchronous); structure of antipodals, synergids, and egg (position of nuclei and vacuoles); form and behavior of synergids and antipodals (ephemeral, persistent in original form or persistent in variable degrees of disintegration); position of polar nuclei; fusion of polar nuclei (before or after entry of sperm); *orientation of egg apparatus in relation to symmetry of the ovule.*

14.2.8 Fertilization. Entrance of the pollen tube (porogamous, chalazogamous, mesogamous); interval between pollination and fertilization.

14.2.9 Endosperm. Pattern of development (Nuclear type, Cellular, or Helobial); position of nuclei (Nuclear and Helobial types) during early stages and orientation of cell walls (Cellular type); origin of haustoria, where present; type and amount of food reserve; persistence in mature seed (albuminous or exalbuminous).

14.2.10 Embryogeny. Developmental type (Asterad, Onagrad, Chenopodiad, Solanad, Caryophyllad, Piperad); form of mature embryo; presence or absence of suspensor haustoria.

14.2.11 Seed-coat. Number of cell layers and their derivation (from one or two integuments); pattern of cell differentiation.

14.2.12 Special Features. Apomixis; polyembryony.

An assessment of the above embryological features has been of considerable value in determining the systematic position of families and subfamilies. The most significant contributions are those studies where many embryological characters have been correlated with data from other disciplines of accepted value to taxonomy. A variety of examples of the application of embryology to several kinds of taxonomic problems are discussed here.

14.3 Applications of Embryology to Problems of Family Relationships

14.3.1 Podostemaceae

An exclusively tropical family, Podostemaceae display many unique morphological, anatomical, and ecological features, and stand clearly apart from all other angiospermous families. The family, usually treated as the sole member of Podostemales, is regarded as well defined by its peculiar features and yet as one of uncertain affinities. Close relationship has been proposed with Hydrostachyaceae (Hutchinson 1973), Saxifragaceae (Engler 1930, Rendle 1952), and Crassulaceae (Mauritzon 1933, 1939). The embryology has been rather extensively investigated and reinvestigated, and at first view apparently contributes little to our knowledge of the family's relationships. Development of the ovule and female gametophyte, first studied by Went (1908) and Magnus (1913), has especially been the subject of much controversy, largely resolved through comprehensive investigations (Battaglia 1971, Nagendran 1974, Arekal and Nagendran 1975, Nagendran et al. 1977, Nagendran et al. 1980 a). These studies clearly define the variations in an ontogenetic theme which is itself sufficiently unique to justify ordinal status for this family. At the same time, embryological features clearly indicate a close relationship of Podostemaceae to Crassulaceae and their direct derivation from these.

Throughout the family the ovules are anatropous, bitegmic, and tenuinucellate. A single hypodermal archesporial cell differentiates in the nucellus and enlarges into the megasporocyte. The nucellus below the megasporocyte elongates markedly. The apical portion extends beyond the bounds of the inner integument (Fig. 14.3 A). During the megasporocyte stage (*Polypleurum, Zeylanidium*) or after fertilization (*Indotristicha, Terniola*), the cell walls of the extended chalazal nucellus breakdown to produce a multinucleate cytoplasmic mass termed the nucellar plasmodium (Fig. 14.3 B, C).

The development of the female gametophyte follows several well-defined and related patterns. The megasporocyte, after meiosis I, produces unequal dyad cells. Meiosis II takes place in both dyad cells without cytokinesis, but the smaller micropylar member disintegrates. It persists as a dark cap over the developing female gametophyte. In one type first characterized in *Oenone* (= *Apinagia*) by Went (1908), the smaller chalazal megaspore nucleus becomes pycnotic while the micropylar nucleus undergoes two divisions to produce four nuclei. Subsequent cleaveage of the cytoplasm establishes two synergids, a larger, central egg cell, and a polar cell which contains the polar nucleus and disintegrated chalazal megaspore nucleus. Accordingly, the female gametophyte is often described as five-nucleate, four-celled. This pattern has been designated as the Apinagia type, form A (Fig. 14.2 A–D). Three minor variations of this pattern have been recognized as forms B, C, and D. In form B, the defunct chalazal megaspore nucleus disintegrates rapidly so that the female gametophyte never attains a five-nucleate stage (Fig. 14.2 E). In form C, the partially disintegrated chalazal megaspore nucleus becomes separated from the polar cell by a wall so as to establish a single antipodal (Fig. 14.2 F). Form D is the most variable in Apinagia type development. The chalazal megaspore nucleus divides. Both daughter nuclei may occupy the polar cell, or one or both of them may become isolated in separate antipodals (Fig. 14.2 G–I). The Apinagia type forms A and B are clearly monosporic, since only one megaspore nucleus participates in the formation of the female gametophyte. Nagendran (1974) regards the participation of the megaspore nucleus in forms C and D at most very limited, since the establishment of antipodals and disintegration of the nuclei are concomitant. To him, the C and D forms need confirmation

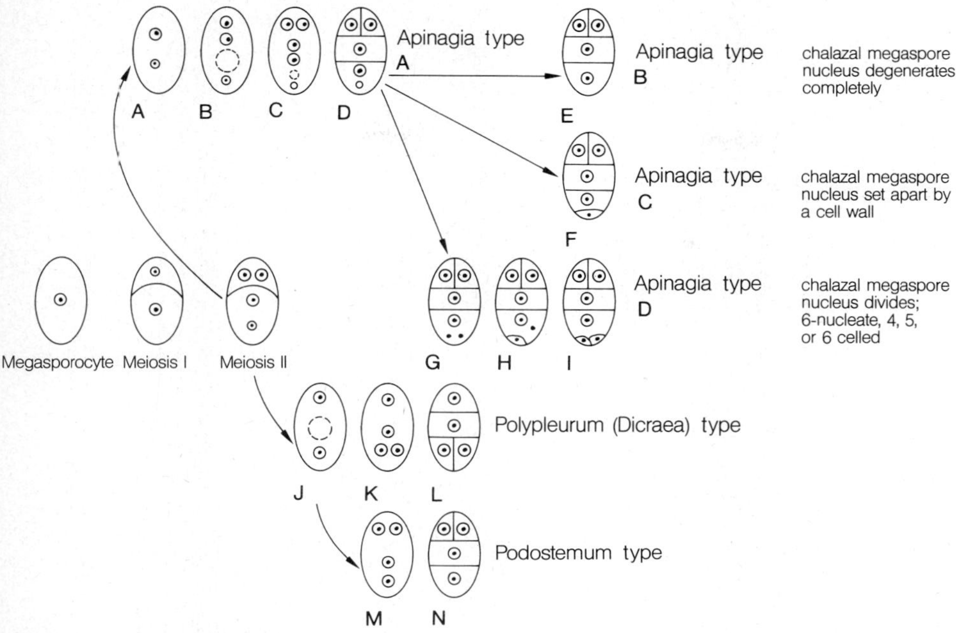

Fig. 14.2 A–N. Development of female gametophyte in Podostemaceae. (Adapted after Battaglia 1971)

and certainly do not represent a true bisporic production of the organized female gametophyte. Two typical bisporic patterns do occur whereby both megaspore nuclei in the chalazal dyad cell contribute equally to the establishment of the female gametophyte. In one pattern the micropylar megaspore nucleus divides vertically while the chalazal one divides horizontally. Cytokinesis establishes two chalazal synergids, a central egg cell, and a micropylar polar cell. The polarity of the female gametophyte in this case has been reversed. Initially described in *Dacraea elongata* by Magnus (1913), this pattern is designated as Dacraea type. However, since *Dacraea* has been merged with *Polypleurum* (Hall 1971), Nagendran et al. (1977) have renamed the pattern Polypleurum type (Fig. 14.2 J–L). In the Podostemon type, the micropylar megaspore nucleus divides horizontally and the chalazal one vertically. Cytokinesis sets apart two micropylar synergids, a central egg cell, and a chalazal polar cell (Fig. 14.2 J, M, N). This bisporic pattern of development has been controversial largely because, unlike other types, it does not exist independently in any of the species (of the family) investigated.

A close examination of available embryological data leaves little doubt of the close relationship between Podostemaceae and Crassulaceae. Of several forms of nucellar structure in Crassulaceae recognized by Mauritzon (1933), the Sedum type and Crassula type are both characterized by marked growth of the hyponucellus, i.e., the nucellus below the developing female gametophyte. A similar nucellar growth pattern occurs through Podostemaceae (Fig. 14.3 B cf. D). It is the hyponucellus that uniformly produces the nucellar plasmodium in all members

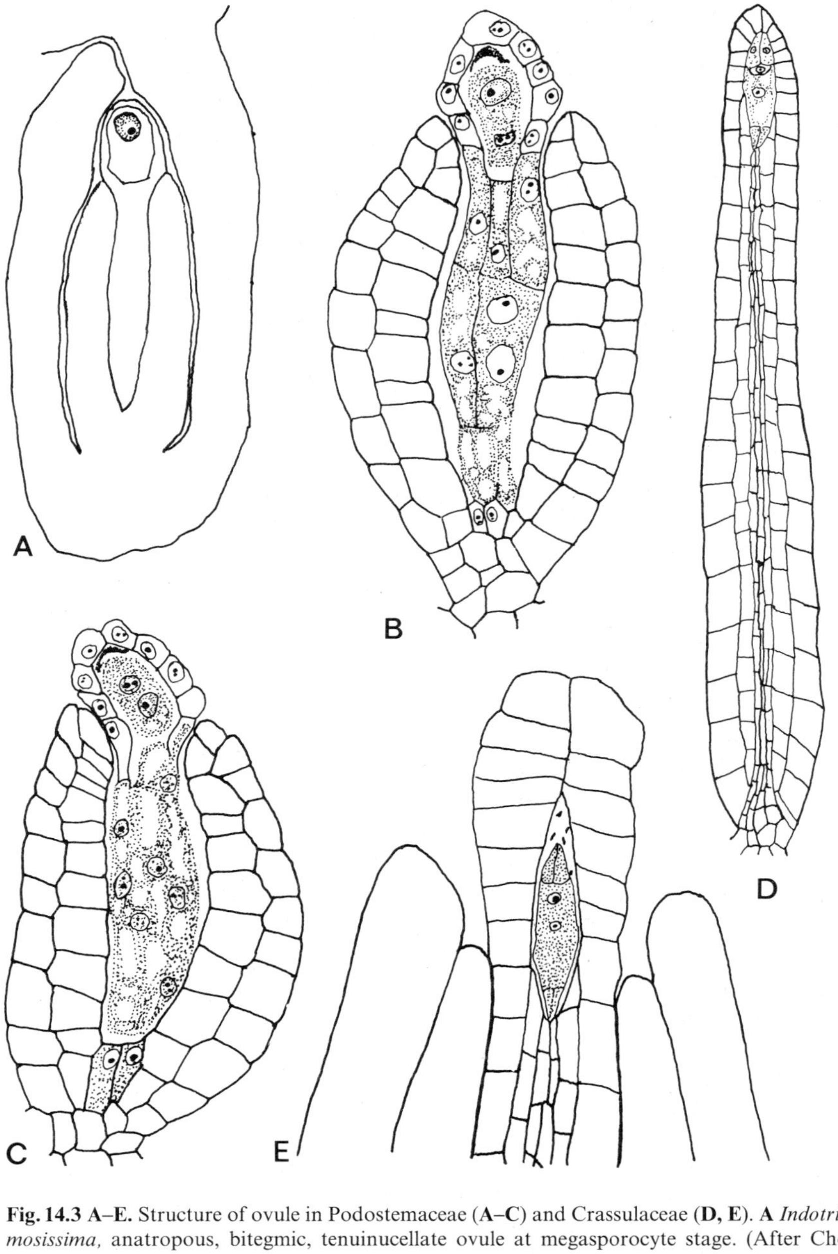

Fig. 14.3 A–E. Structure of ovule in Podostemaceae (**A–C**) and Crassulaceae (**D, E**). **A** *Indotristicha ramosissima,* anatropous, bitegmic, tenuinucellate ovule at megasporocyte stage. (After Chopra and Mukkada 1966). **B, C** *Farmeria indica,* l.s. ovules (outer integument omitted); female gametophyte located in the apical portion of nucellus. (After Arekal and Nagendran 1975). **B** Disorganization of cell walls and enlargement of nuclei in the lower portion of nucellus. **C** Fusion of protoplasts and formation of nucellar plasmodium (pseudo-embryo sac). Note two persistent nucellar cells in **B** and **C**. **D** *Sempervivum alpinum;* Sedum type of nucellus with female gametophyte located in the apical portion. (After Mauritzon 1933). **E** *Sedum acre,* the apex of nucellus (containing female gametophyte) extends beyond the integuments. (After Mauritzon 1933)

of the family so far investigated (Fig. 14.3 B cf. C). Consideration of the nucellar plasmodium as derived from an intact hyponucellus of a form similar to that described for some Crassulaceae appears reasonable. Throughout Podostemaceae the apex of the nucellus, the portion that encloses the developing female gametophyte, extends beyond the inner integument. Although not of universal occurrence, similar development does occur in Crassulaceae (Mauritzon 1933). In *Sedum arce,* for example, the nucellar apex regularly extends beyond the inner integument and in some cases beyond the outer integument (Fig. 14.3 A cf. E).

Mauritzon (1933) reports Scilla type (=Endymion type) of female gametophyte in *Sedum populifolium* and *S. fabaria,* although his figures indicate that the development is of the Allium type. In any case, bisporic development is recorded for Crassulaceae and provides another link with Podostemaceae. In addition, *Crassula aquatica* displays reduced endosperm. The chalazal chamber of the two-celled stage does not divide and becomes haustorial. The micropylar chamber produces only a few cells. This reduced endosperm may be a transitional stage leading to the elimination of endosperm in Podostemaceae. Finally, the highly developed suspensor haustorium observed throughout both families suggests their close relationship.

In each case where structural and developmental parallels can be drawn between these two families, Crassulaceae appear generalized while Podostemaceae occupy the derived position. Comparative embryology thus seems to document the statement by Maheshwari (1945) that, "the Podostemaceae are much reduced apetalous derivatives of the Crassulaceae".

14.3.2 Basellaceae

Several divergent views have been expressed about the systematic position of Basellaceae, a small family of herbaceous twining vines that occur chiefly in tropical America and Asia. Eichler (1878), Volkens (1893), and Lawrence (1951) include the family in Centrospermales close to Portulacaceae. A similar assignment is offered by Bessey (1915), Gunderson (1950), and Takhtajan (1980), although these systematists have adopted the alternative ordinal name, Caryophyllales. Hutchinson (1973) includes the family among a group of advanced caryophyllaceous families which he elevates to a separate order, Chenopodiales. Hooker (1880) includes it in Chenopodiaceae as the subfamily Baselleae, and Franz (1908) lowers it to tribal rank, placing it in subfamily Montioideae of Portulacaceae.

A thorough investigation of the life history of *Basella rubra* by Maheswari Devi and Pulliah (1975) includes a detailed analysis and summarization of the embryological features of the family. They also offer a comparison of these features, together with other morphological characteristics, with those of other families included in Centrospermales (Caryophyllales). The embryological features are compared in Table 14.1.

Basellaceae and Chenopodiaceae differ markedly embryologically, and the inclusion of Basellaceae as a subfamily of Chenopodiaceae is untenable. This conclusion is also supported by anatomical features summarized by Metcalfe and Chalk (1950). For example, stems in Basellaceae contain a single ring of vascular

Table 14.1. Embryological features in selected families of Caryophyllales. (Adapted from Maheswari Devi and Pullaiah 1975)

	Portulacaceae	Basellaceae	Chenopodiaceae	Caryophyllaceae
Anther	Tetrasporangiate	Tetrasporangiate	Tetrasporangiate	Tetrasporangiate
Anther wall	Monocotyledonous	Monocotyledonous	Monocotyledonous	Monocotyledonous or Dicotyledonous
Anther tapetum	Glandular with multinucleate cells. Ubisch granules present	Glandular with multinucleate cells	Glandular or periplasmodial	Glandular with 2-nucleate cells
Endothecium	Fibrous	Fibrous	Fibrous	Fibrous
Epidermis	Persistent	Persistent	Degenerates	–
Pollen tetrads	Tetrahedral or isobilateral	Tetrahedral or isobilateral	Tetrahedral, isobilateral or decussate	Tetrahedral or decussate
Pollen grains	2- or 3-celled; contain starch grains	3-celled	3-celled	3-celled
Ovules	Ana- to amphitropous, bitegmic and crassinucellate; nucellular beak absent	Amphitropous, bitegmic and crassinucellate; nucellar beak present	Campylotropous, bitegmic and crassinucellate; nucellar beak present	Hemianatropous or campylotropous, bitegmic and crassinucellate; nucellar beak present
Micropyle	Formed by inner integument	Formed by inner integument	Formed by inner integument	Formed by inner integument
Air space	Present	Absent	Present	–
Archesporium	1–3-celled; cuts off parietal cell	2- or 3-celled; cuts off parietal cell	1-celled; cuts off parietal cell	Multicellular; only one develops further; cuts off parietal cell
Megaspore tetrad	Linear	Linear	Linear	Linear
Female gametophyte	Polygonum type; starch grains present	Polygonum type; starch grains present	Polygonum or Allium type; develops chalazal caecum	Polygonum, Allium or Adoxa type; develops chalazal caecum
Synergids	Hooked	Not hooked	Hooked	Hooked
Polar nuclei	Fuse before or at the time of fertilization	Fuse before fertilization	Fuse before fertilization	Fuse before fertilization
Antipodals	2 ephemeral	3 ephemeral	3 ephemeral, sometimes more than 3 due to further division	3 ephemeral
Endosperm	Nuclear, later becomes cellular; perisperm present	Nuclear, later becomes cellular; perisperm present	Nuclear, later becomes cellular; perisperm present	Nuclear, later becomes cellular; perisperm present
Embryogeny	Solanad or Caryophyllad type	Chenopodiad type	Chenopodiad type	Caryophyllad type

strands with intraxylary phloem in the larger bundles. Chenopodiaceae, on the other hand, are characterized by the development of polycyclic, anomalous secondary vascular bundles.

Basellaceae and Portulacaceae share many embryological features (Table 14.1) supporting a close relationship, as also sugested by Takhtajan (1980). However, to include Basellaceae as a tribe in Portulacaceae does not appear justified in view of certain differences.

The embryological features which predominate in Caryophyllales are: tetrasporangiate anthers, monocotyledonous type of wall development, glandular tapetum with multinucleate cells, simultaneous cytokinesis of microsporocytes, occurrence of amphitropous, bitegmic, crassinucellate ovules, formation of a micropyle by the inner integument, occurrence of an air space at the base of the two integuments, formation of parietal tissue, Polygonum type of female gametophyte, ephemeral antipodals, Nuclear endosperm, and Caryophyllad or Chenopodiad embryogeny. Deviation of Basellaceae from these features is minor, and their retention as one of the independent families in Caryophyllales seems to be fully justified on embryological grounds.

14.3.3 Stachyuraceae

The genus *Stachyurus* was first assigned to a separate family, Stachyuraceae, by Gilg (1893). Modern taxonomists, such as Hutchinson (1973), Melchior (1964), Cronquist (1981), and Takhtajan (1980), recognize the monotypic Stachyuraceae, but there has been disagreement about its systematic position. Partly, the controversy is based on anatomical studies. Tippo (1938) stated that the anatomy of Stachyuraceae is very similar to that of Hamamelidaceae and Flacourtiaceae. However, in modern taxonomic systems Hamamelidaceae and Flacourtiaceae are not treated as closely related families. Both Takhtajan (1980) and Cronquist (1981) associate Stachyuraceae with Flacourtiaceae in Violales. According to them, therefore, the affinities of Stachyuraceae lie within the subclass Dilleniidae far removed from Hamamelidaceae in the subclass Hamamelidae. Hutchinson (1973), on the other hand, includes Stachyuraceae in the order Hamamelidales.

With a view toward elucidating the relationships of Stachyuraceae with Theaceae, Flacourtiaceae, and Hamamelidaceae, Mathew and Chaphekar (1977) presented a detailed embryological study of two species of *Stachyurus*. They pointed out that the embryology of Hamamelidaceae (see Flint 1957a, b, 1959) is markedly dissimilar to that of Stachyuraceae. In addition, the families are dissimilar in certain vegetative and floral features. Hamamelidaceae have exstipulate leaves and tetralocular ovaries with axile placentation in contrast with Stachyuraceae which possess distinct, ephemeral stipules and unilocular ovaries with parietal placentation. The association of Stachyuraceae with Hamamelidaceae is not justifiable.

Stachyuraceae and Theaceae share a number of important embryological features: (1) absence of integumentary tapetum, (2) bisporic female gametophyte, (3) fusion of polar nuclei before fertilization, (4) Solanad type of embryogeny, (5) similar structure of suspensor, and (6) polyembryony. Mathew and Chaphekar point out that since Stachyuraceae show both monosporic (Polygonum) and bisporic (Endymion and Allium) types of female gametophyte, the family shows an intermediate condition between Flacourtiaceae (monosporic) and Theaceae (bisporic). They conclude that Stachyuraceae should be assigned to Theales close to Theaceae in agreement with the classification of Cronquist (1981).

14.3.4 Frankeniaceae

Of the five genera comprising Frankeniaceae, *Frankenia* is the most widely distributed through the north and south temperate zones. The family, unsettled in its systematic position, has been associated with both polypetalous and sympetalous families. Bentham (1862a) first noted resemblances between Frankeniaceae and Caryophyllaceae. Gundersen (1927) regarded Frankeniaceae as a link between Caryophyllaceae and Parietales. Because of the similarity in structure of the calyx in *Frankenia* and *Plumbago,* and, according to Metcalfe and Chalk (1950), the presence of characteristic epidermal glands, relationship of Frankeniaceae to the gamopetalous Plumbaginaceae seems a possibility. However, Rendle (1952) followed the reasoning of Wettstein (1935) and placed Frankeniaceae between Elatinaceae and Tamaricaceae in Parietales. More recently, Cronquist (1981) placed Elatinaceae in Theales and included Frankeniaceae in Violales in close association with Tamaricaceae. Takhtajan (1980) treated these families similarly, except that he removed Tamaricaceae and Frankeniaceae from Violales to a separate order Tamaricales. It should be noted that Theales and Violales of Cronquist (1981) and Theales, Violales, and Tamaricales of Takhtajan (1980) are mostly included in Parietales (see Engler and Diels 1936).

In an attempt to elucidate the systematic position of the family, Walia and Kapil (1965) studied the embryology of *Frankenia hirsuta* and *F. pulverulenta*. Their attention to the principal genus of the family has indeed contributed to a better understanding of the true affinities of Frankeniaceae, especially in relation to Elatinaceae and Tamaricaceae. As clearly indicated by Walia and Kapil, the embryology of *Frankenia* departs markedly from that of Tamaricaceae and, thus, there is no relationship of Frankeniaceae with that family (Takhtajan 1980, Cronquist 1981).

In *Frankenia* the mature pollen grains are tricolpate and three-celled (Fig. 14.4A–E). The archesporial cell functions directly as the megasporocyte, and development of the female gametophyte follows the Polygonum type (Fig. 14.4F–

Fig. 14.4. **A–Q** *Frankenia hirsuta;* **R–V** *F. pulverulenta.* **A–D** Formation of three-celled pollen grain; note an extranuclear body in generative cell in **B,** and starch grains in **C** and **D. E** Palynogram; pollen grains tricolpate. **F–H** Stages in the development of anatropous, bitegmic ovule. **I–M** L.s. nucelli. **I** Hypodermal sporogenous cell. **J** Megasporocyte, parietal cell absent. **K** Dyad cells. **L, M** Tetrads of megaspores. Note lack of uniformity in the order of degeneration of three micropylar megaspores. **N** Four-nucleate gametophyte. **O** Eight-nucleate gametophyte. **P, Q** Mature gametophytes, the antipodal cells have degenerated in **Q. R–V** Stages in embryogeny. **R** Zygote. **S–U** Two-, four- and eight-celled proembryo. **V** Proembryo at quadrant stage, the suspensor is long and uniseriate. Conventional labelling of embryonal tiers in Figs. 14.4, 14.6, 14.7, and 14.10. (After Walia and Kapil 1965)

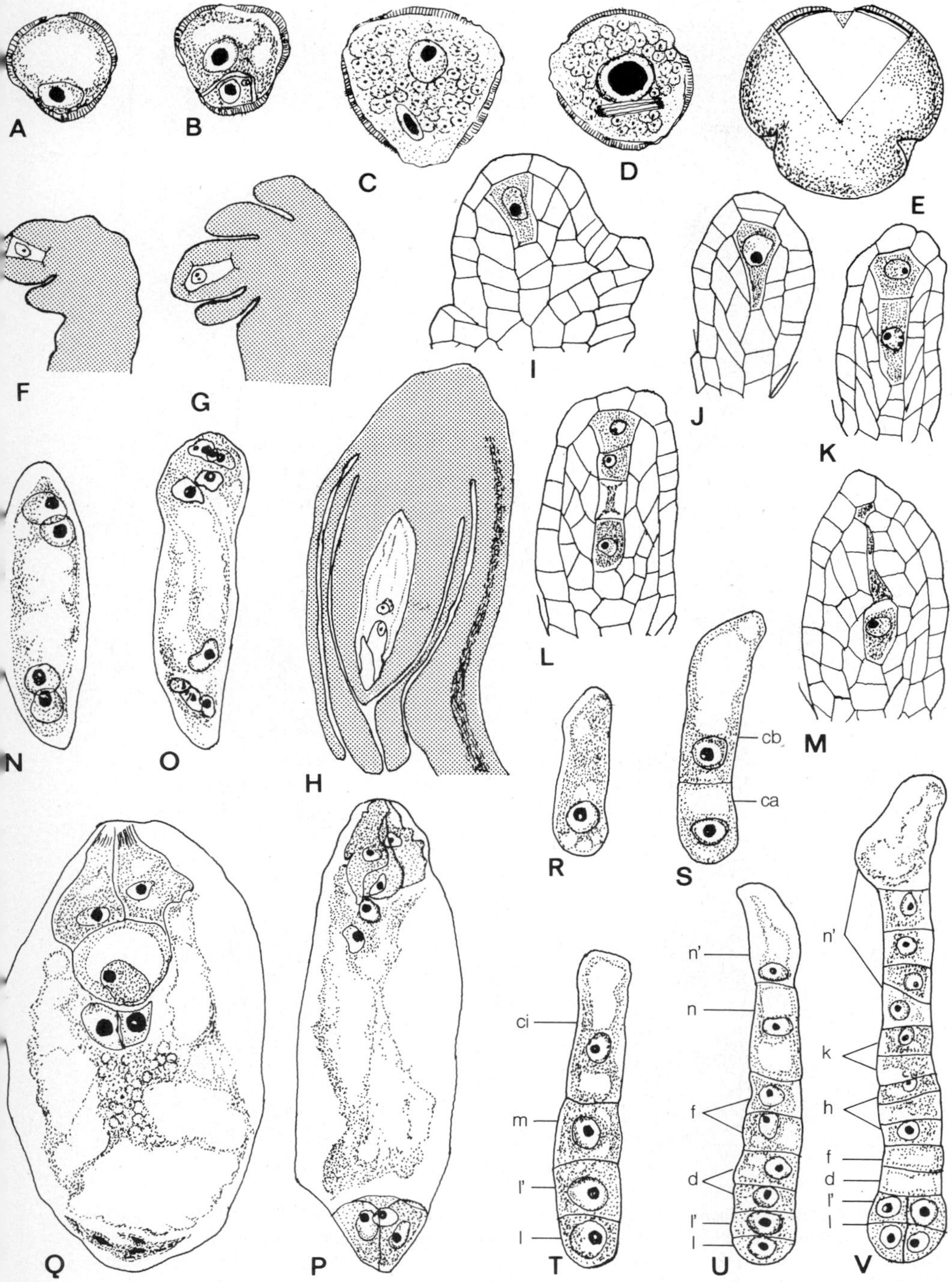

Fig. 14.4 A – V

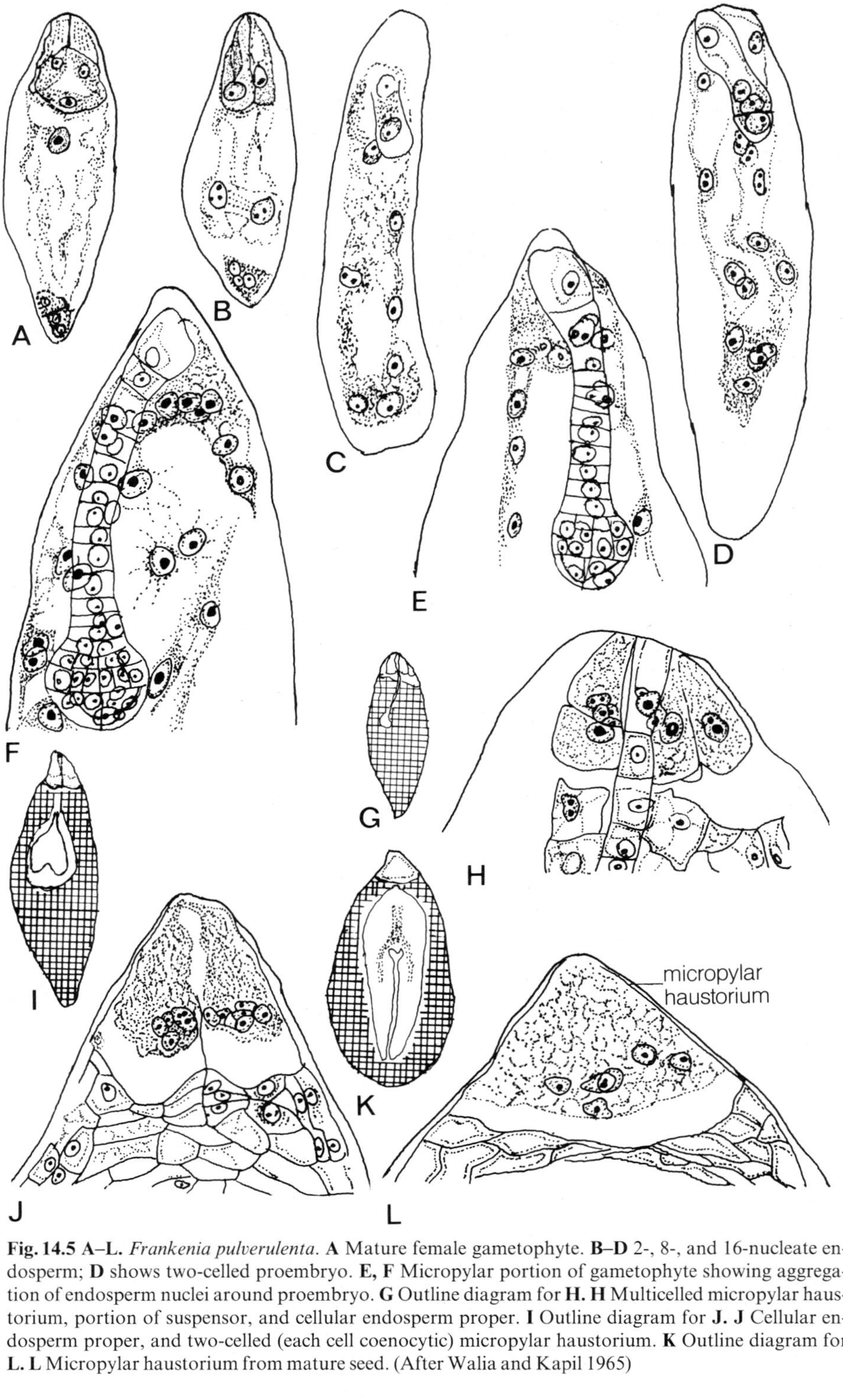

Fig. 14.5 A–L. *Frankenia pulverulenta.* **A** Mature female gametophyte. **B–D** 2-, 8-, and 16-nucleate endosperm; **D** shows two-celled proembryo. **E, F** Micropylar portion of gametophyte showing aggregation of endosperm nuclei around proembryo. **G** Outline diagram for **H. H** Multicelled micropylar haustorium, portion of suspensor, and cellular endosperm proper. **I** Outline diagram for **J. J** Cellular endosperm proper, and two-celled (each cell coenocytic) micropylar haustorium. **K** Outline diagram for **L. L** Micropylar haustorium from mature seed. (After Walia and Kapil 1965)

Q). The development of endosperm is of the Nuclear type (Fig. 14.5 A–F) and is characterized by a prominent micropylar haustorium (Fig. 14.5 G–L). Embryogeny follows the Solanad type (Fig. 14.4 R–V); polyembryony is absent. The mature embryo is straight without any remnants of suspensor, and the seed is albuminous. The seed-coat is derived from both integuments.

Erdtman (1952) stated that tricolpate pollen grains with baculate exine, reticulate in surface view, in *Tamarix* are structurally very similar to the pollen grains of Frankeniaceae. With the exception of the tricolpate pollen grains (Fig. 14.4 E), Tamaricaceae differs in each of the characters illustrated here for *Frankenia*. Tamaricaceae have two-celled pollen grains, ovule with parietal tissue, tetrasporic female gametophyte with several variations, Nuclear type of endosperm without haustoria, Solanad type of embryogeny with frequent polyembryony, straight mature embryo with a massive suspensor, exalbuminous seeds, and the seed-coat is derived only from the outer integument.

Caryophyllaceae are characterized by free central placentation, three- to six- or twelve-porate pollen grains, multicellular archesporium, parietal tissue in the nucellus, Caryophyllad type of embryogeny, curved mature embryo, exalbuminous seeds, perisperm present, endosperm haustoria absent, and seed-coat formed only by the outer integument. None of these features occur in Frankeniaceae, and assignment of the family to Caryophyllales as indicated by Bentham (1862 a) and Gundersen (1927) is untenable.

The possibility of a close relationship between Frankeniaceae and Plumbaginaceae is also negated by their great dissimilarity in embryological features. Plumbaginaceae are characterized by amoeboid tapetum, circinotropous ovules, parietal cells in the nucellus, tetrasporic female gametophyte, and chalazal endosperm haustoria. These features do not occur in Frankeniaceae.

On the other hand, Elatinaceae resemble Frankeniaceae in several embryological characters: three-celled pollen grains, bitegmic ovules, absence of parietal tissue in the nucellus, monosporic development of female gametophyte, nuclear endosperm, and wall formation proceeding from the chalazal toward the micropylar portion of endosperm. Walia and Kapil (1965) conclude that the appropriate place for Frankeniaceae is in the order Parietales (Theales) near Elatinaceae as also suggested by Wettstein (1935) and Rendle (1952).

14.3.5 Tropaeolaceae

The family Tropaeolaceae, now accepted by most systematists, although once included in Geraniaceae is, according to Cronquist (1981), no more disputed. This family of two genera, *Tropaeolum* and *Magallana*, is included in the order Geraniales along with Geraniaceae, Oxalidaceae, and Balsaminaceae (see Lawrence 1951, Takhtajan 1980, Cronquist 1981). However, as Takhtajan (1980) points out, Tropaeolaceae differ from Geraniaceae in tenuinucellate ovules, and from Balsaminaceae in Nuclear endosperm. They differ from both families in a prominent suspensor haustorium which penetrates the seed-coat and pericarp.

Tiwari et al. (1977), in their study of ovule ontogeny in *Tropaeolum majus*, have called attention to other doubts expressed about the inclusion of Tropaeolaceae within Geraniales. In addition, they point out that Tropaeolaceae differ from the families of Geraniales in the following ovular features: outer integument is subdermal, whereas it is of dermal origin in the other families; the seeds are pachychalazal, i.e., the massive chalaza becomes meristematic and forms the seed-coat; the outer layer of inner integument does not develop into a sclerotic layer; and crystalliferous and sclerotic cells are absent in the inner epidermis of outer integument. Tiwari et al. conclude that on the basis of ovular characteristics the inclusion of Tropaeolaceae in Geraniales is not justified.

Hegnauer (1973) also points out that there is no significant similarity between Tropaeolaceae and the families of Geraniales on the basis of phytochemical data.

Chadefaud (1974), on the basis of his investigation of floral morphology, placed Tropaeolaceae in Capparales, as did Dahlgren (1975). Dahlgren, however, presented an especially strong case based on phytochemistry, embryology, anatomy, cytology, palynology, and gross morphology.

There now seems to be little doubt that the proper place for Tropaeolaceae is with Capparales, and a number of embryological features, those pointed out by Takhtajan (1980), Tiwari et al. (1977), and Dahlgren (1975), support this assignment.

14.3.6 Salvadoraceae

A small family of shrubs and trees distributed chiefly in tropical Asia and Africa, Salvadoraceae include three genera, *Azima, Dobera,* and *Salvadora,* comrising 12 species. Its systematic position has been much disputed. Bentham (1876b) and Bessey (1915) assigned the family to Gentianales, while Rendle (1952) associated it with Oleaceae in Oleales. Rendle's treatment is not very different from that of Bentham and Hooker, since their Gentianales includes Oleaceae. Both treatments, however, ignored or discounted the importance of polypetalous corollas and hypogynous stamens in *Azima* and *Dobera.* Engler and Diels (1936) placed the family in Polypetalae in Celastrineae (Sapindales), a position similar to that recommended by Hutchinson (1973) and Gundersen (1950) who included it as one of the families in Celastrales. Placement among the polypetalous orders Sapindales and Celastrales is based on vegetative and floral characters. Takhtajan (1980) and Cronquist (1981) also include Salvadoraceae in Celastrales, and Takhtajan accepts its close relationship with Celastraceae.

The embryology of Salvadoraceae, based on a study of *Azima* and *Salvadora,* is presented by Maheswari Devi (1972). She also studied the embryology of Gentianalean families (Maheswari Devi 1962, 1964, 1971). Salvadoraceae are charac-

Fig. 14.6. A–C, F–J, L–N, P–V *Salvadora persica;* **D, E, K, O** *Azima tetracantha.* **A** Microspore. **B** Two-celled pollen grain. **C** Unilocular ovary with a single ovule. **D** Bilocular ovary, one ovule in each locule. **E** L.s. nucellus showing multicellular archesporium. **F, G** Same, with two and three megasporocytes. **H** Linear tetrad. **I** Functional and degenerated megaspores. **J, K** Mature, eight-nucleate gametophytes. **L** Twin gametophytes. **M** Two-celled proembryo and nuclear endosperm. **N** Proembryo and initiation of wall formation in endosperm; note nuclear fusions in chalazal region. **O** Embryo and cellular endosperm. **P–V** Various stages in Asterad type of embryogeny. *s* suspensor. (After Maheswari Devi 1972)

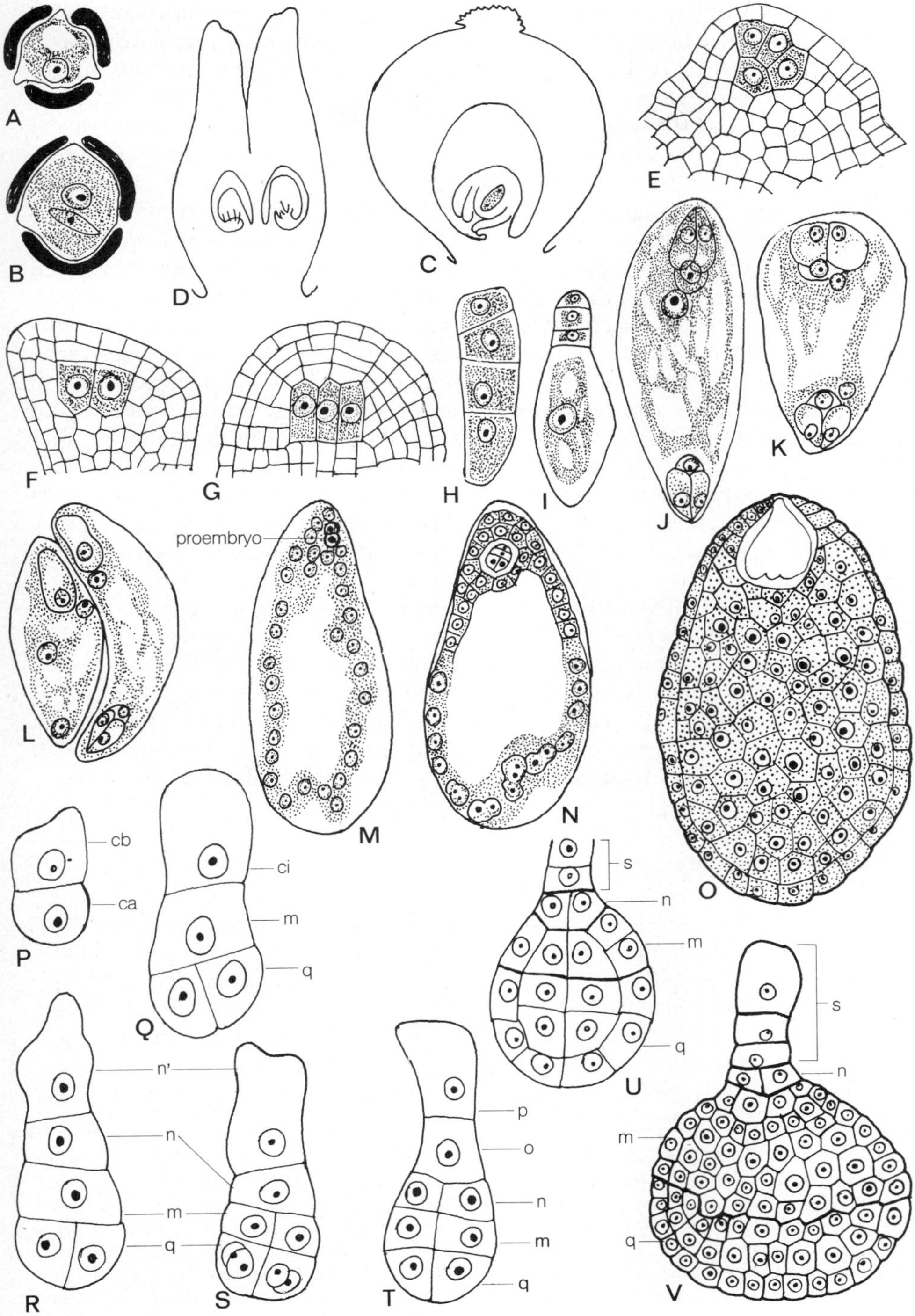
A
B
D
C
E
F
G
H
I
J
K
L
proembryo
M
N
O
P
cb
ca
Q
ci
m
q
U
s
n
m
q
R
n'
n
m
q
S
T
p
o
n
m
q
V
s
n
m
q

Fig. 14.6 A–V

terized by two-celled pollen grains (Fig. 14.6 A, B), and uni- or bilocular ovaries (Fig. 14.6 C, D). The ovules are bitegmic, crassinucellate with a massive parietal tissue (Fig. 14.6 E–G), and the micropyle is formed by the inner or both integuments. The female gametophyte is of the Polygonum type (Fig. 14.6 H–L); the endosperm of Nuclear type (Fig. 14.6 M–O), and nuclear fusions occur in the chalazal region (Fig. 14.6 N). Embryogeny follows the Asterad type (Fig. 14.6 P–V). The seed-coat is formed by both integuments.

In contrast, the families of Gentianales (to include Oleaceae) are embryologically characterized by two- or three-celled pollen grains, unitegmic and tenuinucellate ovules, absence of parietal tissue, direct development of archesporial cell into megasporocyte, and development of seed-coat from the single integument. Although the female gametophyte is usually of the Polygonum type, it is sometimes bisporic in Oleaceae. Whereas embryogeny in Gentianales is highly variable, Solanad type predominates among the representatives investigated.

Collectively, the embryological features indicate that Salvadoraceae stand apart from Gentianales as well as Oleales, and the inclusion of the family in either order is not justified. On the other hand, the inclusion of Salvadoraceae in Celastrales is supported on the basis of embryology, since both *Azima* and *Salvadora* share with representatives of this order two-celled pollen grains, anatropous, bitegmic, crassinucellate ovules, formation of parietal tissue, and Nuclear type of endosperm.

The placement of Salvadoraceae in Celastrales also gains support on the basis of floral anatomy as reported by Kshetrapal (1970). A feature which seems to contradict this position, gamopetaly exhibited by *Salvadora,* should be regarded as an exceptional feature for the family in the same way as the polypetalous condition in *Fraxinus* is exceptional for the otherwise gamopetalous Oleaceae. Such a view (apparently) is expressed by Melchior (1964) who, like Takhtajan (1980) and Cronquist (1981), treats Salvadoraceae as one of the families of Celastrales.

14.3.7 Centrolepidaceae

Predominantly Australasian, Centrolepidaceae represent a small family of annuals distributed among five genera which have been assigned to various taxonomic associations. In most classificatory systems, the family is associated with Restionaceae and Flagellariaceae; Benson (1957) lists all three families among those of uncertain taxonomic position. Takhtajan (1980) states that the most notable features shared by these families are some typical graminoid characters. Perhaps these features led Bessey (1915) and other systematists to include these

Fig. 14.7 A–Y. *Centrolepis fascicularis.* **A** L.s. young anther showing archesporium; some cells are in division or have divided. **B** Portion of anther at microsporocyte stage; note differentiation of wall layers. **C** Anther at dehiscence. **D–F** Pollen grains at uni-, bi-, and three-celled stages. **G** L.s. carpel with single orthotropous ovule. **H, I** Nucelli show archesporial cell and megasporocyte. **J** Functional and degenerated megaspores. **K, L** Two- and four-nucleate gametophytes. **M** Mature gametophyte; note radial elongation of nucellar epidermal cells in **J–M**. **N** Mature gametophyte; note starch grains in central cell. **O–V** Derivation of proembryonal tetrad from zygote. **W, X** Early and late globular proembryos. **Y** L.s. embryo at shedding of seed. (After Prakash 1969)

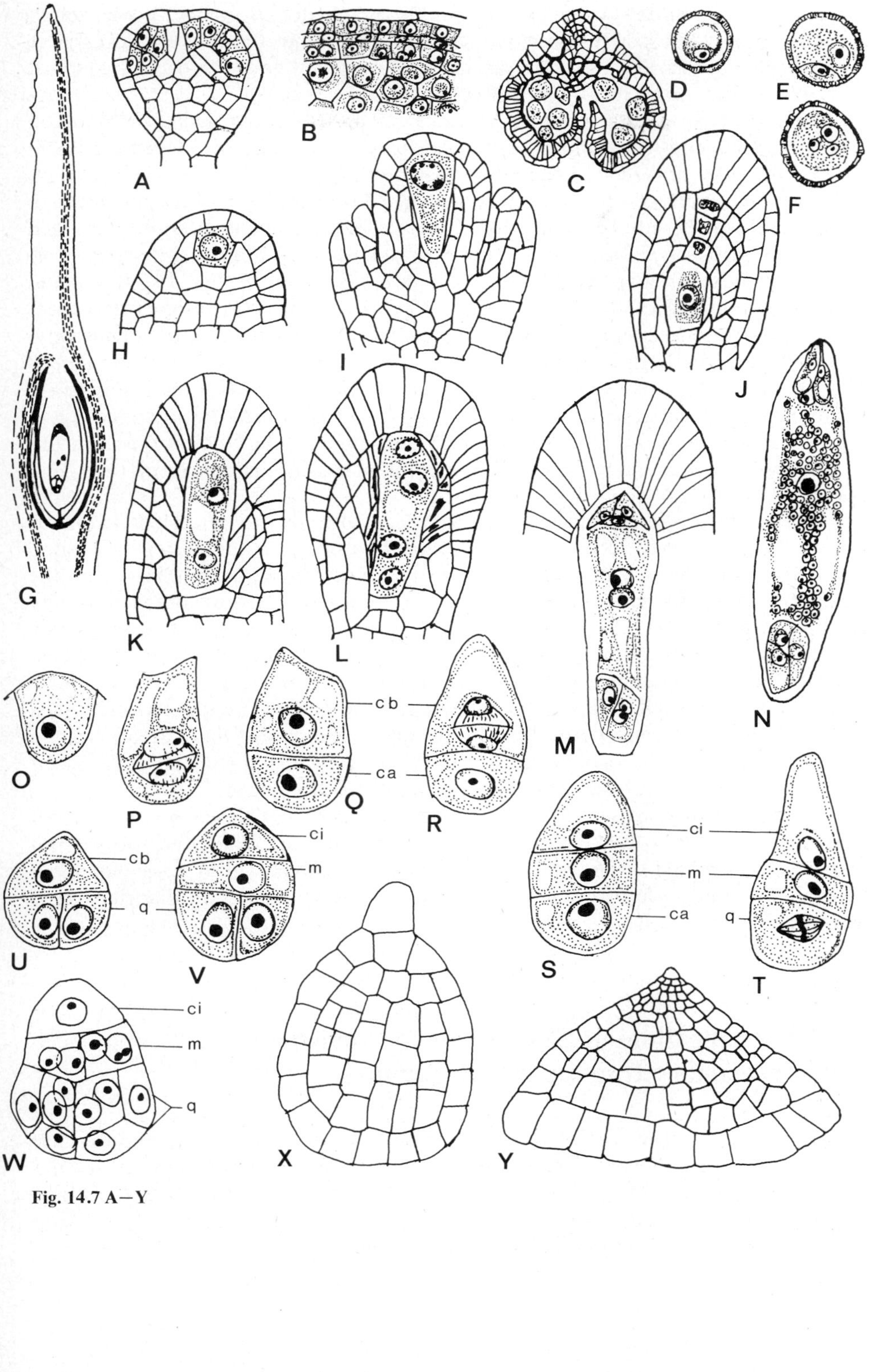

Fig. 14.7 A–Y

families in the order Graminales (Glumiflorae). Hutchinson (1973), who segregates grasses and sedges into separate orders (Graminales and Cyperales), places Centrolepidaceae and Restionaceae in Juncales. Takhtajan (1980) and Cronquist (1981) include all three families in Restionales.

On the basis of extensive embryological studies of Centrolepidaceae, Hamann (1962, 1975) confirmed the close relationship of this family with Gramineae and Restionaceae. Prakash (1969), however, in his embryological study of *Centrolepis fascicularis,* pointed out an especially close relationship of Centrolepidaceae to Restionaceae. A number of common features between the two families emerged in his study. For example, the anthers are usually bisporangiate and have a persistent epidermis (Fig. 14.7 A–C). The tapetum is secretory (Fig. 14.7 B, C), and the pollen grains are three-celled (Fig. 14.7 D–F). Each of the three carpels bears an orthotropous, bitegmic, tenuinucellate ovule (Fig. 14.7 G–J), and the micropyle is formed by both integuments (Fig. 14.7 G). The development of the female gametophyte is of the Polygonum type (Fig. 14.7 J–N), and starch grains are abundant in the mature gametophyte (Fig. 14.7 N). Fusion of the polar nuclei precedes fertilization. The development of endosperm is of the Nuclear type. A striking similarity is found in the radial elongation of the nucellar epidermis (Fig. 14.7 J–M).

Gramineae share a few features with Centrolepidaceae: monocotyledonous type of development of anther wall, secretory anther tapetum, bitegmic ovules, Polygonum type of female gametophyte, and Nuclear endosperm. However, the differences exhibited by Gramineae are much more marked in comparison with the similarity shown by Restionaceae to Centrolepidaceae: (1) tetrasporangiate anthers; (2) anatropous to campylotropous ovules; (3) micropyle usually formed by the inner integument; (4) periclinal divisions in the nucellar epidermis to form a prominent nucellar cap; and (5) meristematic activity in the peripheral cell layers of mature endosperm. Finally, the embryogeny is of the Asterad type; in Centrolepidaceae it corresponds to the Onagrad type (Fig. 14.7 O–Y). Polyembryony is rare in Centrolepidaceae but common in Gramineae.

Consequently, embryological evidence indicates that the closest relative of Centrolepidaceae is Restionaceae rather than Gramineae. This conclusion is supported by Chanda (1966) who observed a distinct palynological series indicating probable relationship from Centrolepidaceae to Restionaceae to Flagellariaceae to Gramineae.

14.4 Applications of Embryology to Problems of Taxonomic Structure Within Families

The preceding examples of the application of embryology to taxonomy have focused attention on problems of relationship between or among families. Another important type of contribution of this discipline relates to its use in settling questions of taxonomic structure within families, i.e., organization at the subfamily level. Contributions of this kind are briefly examined here.

14.4.1 Aizoaceae

The status of the subfamilies of Aizoaceae has fluctuated in various systematic treatments. Hooker (1867) divides the family into three tribes – Mesembryeae, Aizoideae, and Mollugineae. Pax (1889) assigns the members of this family to two subfamilies, Molluginoidea and Ficoidea; Hutchinson (1959) elevates them to family status – Molluginaceae and Ficoidaceae. Friedrich (1955) assigns *Tetragonia* and *Tribulocarpus*, from Ficoidaceae of Hutchinson, to a new family, Tetragoniaceae. According to Jacobsen (1960), Schwantes (1947), and Herre and Volk (1950) separate the mesembryanthamums from Ficoidaceae to form the family Mesembryanthamaceae. Takhtajan (1980) and Cronquist (1981) recognize three subfamilies in Aizoaceae, viz., Aizoideae, Mesembryanthemoideae, and Tetragonioideae, and elevate Molluginoideae to Molluginaceae.

Embryological investigation of Aizoaceae to include Molluginoideae has been extensive, and has been critically examined by Prakash (1967) in an attempt to settle the taxonomic structure within the family. He points out that among the investigators who support delimiting Aizoaceae from Molluginaceae, Joshi and Rao (1936) have done so on the basis of embryological characters. According to them a two-layered outer integument, periclinal divisions in the nucellar epidermis, absence of starch in the female gametophyte, and absence of an aril characterize Molluginaceae. These features are in contrast as Aizoaceae have a three-layered outer integument and aposing features for each of the other four characters. More recent studies, however, indicate that these features are not as distinctly categorical as suggested by Joshi and Rao. *Gisekia* and *Orygia* of Molluginaceae (Molluginoideae) have a three-layered outer integument and an aril (Narayana 1962). *Aptenia, Delosperma, Hereroa,* and *Tetragonia* (Prakash 1967) of Aizoaceae (Ficoideae) have a two-layered outer integument and lack an aril. Periclinal divisions in the cells of the nucellar epidermis occur in both Ficoideae and Molluginoideae (Prakash 1967).

Although there are no significant embryological differences between the mesembryanthemums and other Aizoaceae, they are clearly distinct in other morphological features. Therefore, they should be accorded the status of a subfamily – Mesembryanthemoideae. Furthermore, *Tetragonia,* in addition to other morphological distinctions, differs from other Aizoaceae in some embryological characters: (1) a single pendulous ovule, (2) druses in the nucellus, (3) inner layer of the inner integument well developed, and (4) indehiscent fruit. Establishment of a separate subfamily Tetragonioideae is, therefore, warranted.

Prakash (1967) has, thus, clearly justified the conclusion that (1) elevation of Molluginoideae to the status of family is not in order, and (2) differences from other members of Aizoaceae displayed by mesembryanthemums and *Tetragonia* warrant the recognition of two subfamilies, Mesembryanthemoideae and Tetragonioideae, but not the elevation of either one to the rank of family. Therefore, Aizoaceae are now grouped into four subfamilies, viz., Aizoideae (Ficoideae), Mesembryanthemoideae, Tetragonioideae, and Molluginoideae.

14.4.2 Vacciniaceae

Hooker (1876a) places *Vaccinium* and allied genera in Vacciniaceae under Ericales, and this family is recognized by Hutchinson (1973). Engler and Diels

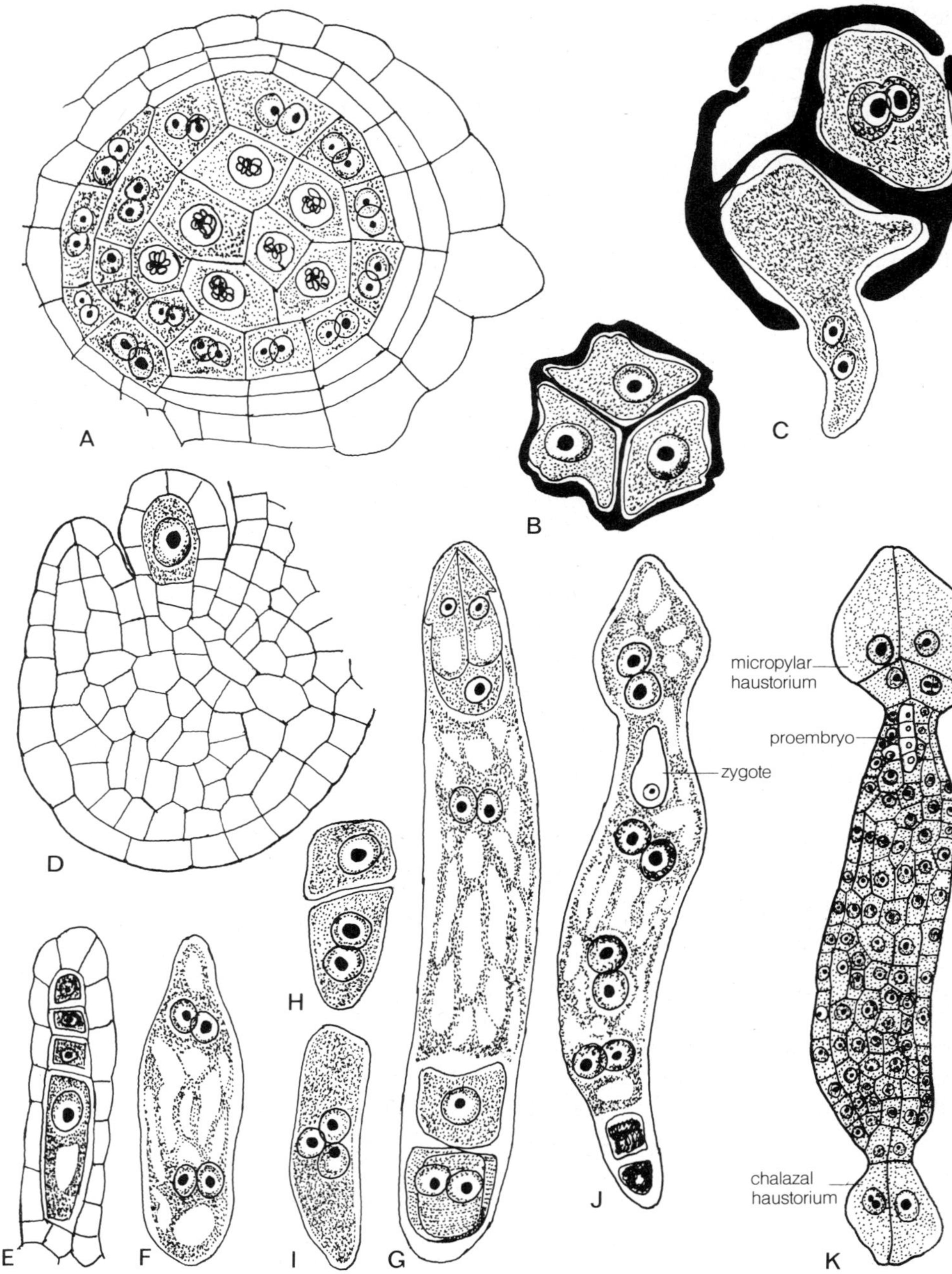

Fig. 14.8. **A–C** *Vaccinium serratum;* **D–G** *V. retusum;* **H** *V. nummularia;* **I–K** *V. serratum.* **A** T.s. anther lobe. Note wall layers, binucleate tapetal cells, and microsporocytes in meiotic prophase I. **B** Microspore tetrad. **C** Same, with one microspore two-celled, and another germinated. **D** L.s. ovule showing megasporocyte. **E** Tetrad of megaspores. **F, G** Four-nucleate and mature gametophytes. **H, I** Antipodal cells. **J** Gametophyte with zygote, nuclear endosperm, and degenerated antipodal cells. **K** Proembryo, cellular endosperm, and four-celled micropylar and two-celled chalazal haustoria. (After Venkateswarlu and Maheswari Devi 1973)

(1936), Rendle (1952), and Gundersen (1950) assign *Vaccinium* to the subfamily Vaccinioideae of Ericaceae along with Rhododendroideae, Arbutoideae, and Ericoideae. Takhtajan (1980) recognizes Vaccinioideae as a subfamily of Ericaceae and merges Arbutoideae with it.

Embryological studies in Ericaceae have not been extensive except for *Vaccinium* in which species of different ploidy levels, including some cultivars of hybrid origin, have been investigated by Bell and Burchill (1955), Bell (1957), Rousi (1967), and Stushnoff and Palser (1969). This latter study was directed to three species, *Vaccinium atrococcum, V. ashei,* and *V. constablaei,* collected in North America and Europe, and it inspired an investigation by Venkateswarlu and Maheswari Devi (1973) of three species growing in India, viz., *V. nummularia, V. retusum,* and *V. serratum.*

These investigators examined and compared the embryological data previously collected for *Vaccinium* and other ericaceous plants in an attempt to settle the taxonomic status of Vacciniaceae. The salient features of the three species of *Vaccinium* investigated by Venkateswarlu and Maheswari Devi (1973) typify also those species studied by Stushnoff and Palser (1969) and serve generally to characterize the genus. The microsporangium has two middle layers and a secretory tapetum of binucleate cells (Fig. 14.8 A). Although an endothecium does not differentiate, the walls of the epidermal cells develop secondary thickenings and project outwardly. The pollen grains are two-celled but remain united in tetrads (Fig. 14.8 B, C) which are liberated through apical pores in the anther. The anatropous, unitegmic, tenuinucellate ovule (Fig. 14.8 D) develops a prominent integumentary tapetum. The development of the female gametophyte is of the Polygonum type (Fig. 14.8 E–G), and the antipodals are one- to three-nucleate depending on their number from one to three in each female gametophyte (Fig. 14.8 H, I). The endosperm is of the Nuclear type (Fig. 14.8 J) and has distinctive micropylar and chalazal haustoria (Fig. 14.8 K). Embryogeny follows the Solanad type.

These embryological features and those of other Ericaceae are summarized in Table 14.2. Variation in the embryological characters is clearly minimal especially in consideration of the total diversity of the subfamilies compared. These data support retention of Vaccinioideae as a subfamily rather than its elevation to the rank of a distinct family, Vacciniaceae. Anatomical features reported by Watson (1965) also support this position. Vaccinioideae are, then, one of the six subfamilies of Ericaceae recognized by Stevens (1971) and incorporated in the classificatory system of Takhtajan (1980).

14.4.3 *Zannichellia*

Zannichellia is a genus of submerged aquatic herbs which occur in either fresh or brackish water and includes two species, one of broad distribution (*Z. palustris*) and one confined to South Africa (*Z. aschersoniana*). It has been placed variously in Potamogetonaceae, Najadaceae, and Zannichelliaceae. Gleason (1952) includes the genus in Najadaceae, whereas Magnus (1889) recognizes only one genus, *Najas,* for that family. Hutchinson (1973) segregates *Zannichellia* along with *Althenia* and *Cymadocea* to Zannichelliaceae, and concludes a derivation for the

Table 14.2. Comparison of morphological and embryological features in four subfamilies of Ericaceae. (Adapted after Venkateswarlu and Maheswari Devi 1973)

	Rhododendroideae	Arbutoideae	Vaccinioideae	Ericoideae
Anther	Appendages absent	Appendaged	Awned anthers; appendages absent	Appendaged
Fibrous endothecium	Present in some members and absent in others	Present in some members and absent in others	Absent	Absent
Tapetum	Secretory with binucleate cells	Secretory	Secretory with binucleate cells	Secretory
Pollen at liberation	Both tetrads and free pollen grains	Both tetrads and free pollen grains	Tetrads	Both tetrads and free pollen grains
Pollen grains	2-celled	2-celled	2-celled	2-celled
Ovary	Superior	Superior	Inferior	Superior
Ovule	Anatropous, unitegmic and tenuinucellate	Anatropous, unitegmic and tenuinucellate	Anatropous, unitegmic and tenuinucellate	Anatropous, unitegmic and tenuinucellate
Embryo sac	Polygonum type	Both Polygonum and bisporic types	Polygonum type	Polygonum type
Antipodals	1–3-celled	1–3-celled	1–3-called	1–3-celled
Endosperm	Nuclear	Cellular	Both nuclear and cellular	–
Embryogeny	Solanad type	Solanad type	Solanad type	Solanad type
Fruit	Septicidal capsule	Loculicidal capsule or berry	Berry or drupe	Loculicidal capsule
Seed	Winged	Not winged	Not winged	Not winged

family perhaps from *Potamogeton* or *Zostera*. Zannichelliaceae was also recognized by Willis and Airy Shaw (1973) and Cronquist (1981). Acherson and Graebner (1907), however, include the genus in Potamogetonaceae as does Lawrence (1951).

Paucity of embryological data led to the investigation of *Zannichellia palustris* by Vijayaraghaven and Kumari (1974). These authors reinterpreted some of the earlier fragmentary information and also compared their findings with the embryological features reported for *Cymadocea,* Potamogetonaceae, and Najadaceae. Data for *Althenia* are not available.

The comparative data indicate that *Zannichellia* differs from Potamogetonaceae in: bisporic Allium type rather than monosporic Polygonum type of female gametophyte, persistent instead of ephemeral antipodals, anatropous as apposed to orthotropous ovules, absence of fibrous thickenings in the endothecium, and absence of exine in the pollen grains.

Najadaceae differ from *Zannichellia* in: uninuclear as apposed to multinuclear tapetal cells, Polygonum type female gametophyte, persistent antipodals, Nuclear or Cellular instead of Helobial type of endosperm, and a protective layer around the embryo formed by the prominent inner integument instead of the specially thickened hypodermis of the pericarp.

The genus *Cymadocea* can be distinguished from *Zannichellia* on the basis of uninucleate tapetal cells, filiform, coiled pollen grains as apposed to round pollen grains, orthotropous ovules, ephemeral antipodals, and Nuclear type of endosperm. The suggestion by Takhtajan (1980) that *Cymadocea* be removed to a separate family Cymadoceaceae is justified on the basis of embryological data.

Embryological features support the assignment of *Zannichellia* to a separate family Zannichelliaceae which should be regarded as closely related to Potamogetonaceae. Both share some common characters: amoeboid anther tapetum, Helobial type of endosperm, and Caryophyllad type of embryogeny with a prominent vesicular suspensor. Bisporic Endymion type of female gametophyte in *Potamogeton foliosus* (Wiegand 1898) may indicate this species as a link between Potamogetonaceae and Zannichelliaceae. A close relationship with Potamogetonaceae is also strongly supported on the basis of floral morphology as disclosed by Singh (1965), and Posluszny and Sattler (1976).

14.5 Applications of Embryology to Problems of Generic Affinity and Assignment

Of the several categories of applications of embryological data to taxonomic issues, none has been more extensive or effective than those related to the genus.

14.5.1 *Ploiarium*

A genus of three species found in Southeast Asia extending to New Guinea, *Ploiarium* has been assigned variously to Bonnetiaceae and Theaceae. Willis and

Airy Shaw (1973), and Hutchinson (1973), include the genus in Bonnetiaceae which actually forms the tribe Bonnetieae in Engler's Theaceae (Melchior 1964). Whereas Cronquist (1981) too includes Bonnetiaceae in Theaceae, Takhtajan (1980) retains it as a separate family intermediate between Theaceae and Clusiaceae. Metcalfe and Chalk (1950) recognize Bonnetiaceae as separate from Theaceae but assign *Ploiarium* to Theaceae.

Prakash and Lau (1976) in their study of *Ploiarium alternifolium* included a very thorough embryological investigation. They pointed out several morphological and anatomical features common to *Archytaea, Bonnetia,* and *Ploiarium* (Bonnetiaceae). They also compared the embryology of *Ploiarium* with that of *Camellia,* the only genus in Theaceae accorded a detailed embryological investigation (Kapil and Sethi 1963, Sethi 1965).

Important differences between *Ploiarium* and *Camellia* include Polygonum type of female gametophyte in *Ploiarium* as apposed to Allium type in *Camellia.* The shape of the ovule and structure of the mature embryo markedly differ in the two taxa. A hypostase, present in *Camellia,* does not occur in *Ploiarium.* On the other hand, there is similarity in the structure and development of the anther, in the anatropous, bitegmic, tenuinucellate ovules, and Nuclear type of endosperm.

Prakash and Lau conclude that in consideration of both anatomical and embryological features, *Ploiarium* should be separated from Theaceae and assigned to Bonnetiaceae. The relationship of Theaceae and Bonnetiaceae suggested by Takhtajan (1980) finds support in the embryological similarity between *Ploiarium* and *Camellia.*

14.5.2 *Filicium*

Filicium decipiens was assigned to Burseraceae by Hooker (1862) and to Anacardiaceae by Wight (see Hooker 1876b). *Filicium* resembles Anacardiaceae in a solitary pendulous ovule in each locule, and Burseraceae in the presence of a dorsal raphe on the ovule and an embryo with foliaceous cotyledons. Radlkofer (1933) assigned this genus to Sapindaceae in an advanced tribe, Doratoxyleae, subfamily Dyssapindaceae (= Dodonaeoideae) as did Muller and Leenhouts (1976) who, however, regarded the tribe as primitive.

Gulati and Mathur (1977) undertook an investigation with the intent of using embryological data to ascertain its taxonomic position. *Filicium* is characterized by a bitegmic, anatropous, crassinucellate ovule which becomes campylotropous after fertilization. The micropyle is formed by both the integuments, the nucellus is distinctly beaked, and a placental obturator is present. Development of the female gametophyte is of the Polygonum type, and the endosperm is of Nuclear type. The embryogeny conforms to the Asterad type, and the foliaceous cotyledons are unequal. In all these features, *Filicium* is similar to Sapindaceae from which it differs in but a few exomorphic features. These are, for *Filicium,* the presence of a superior micropyle, foliaceous cotyledons, and ovules attached at the apex of the locule.

Although Burseraceae also have a superior micropyle and foliaceous cotyledons, the family is unlike *Filicium* in the occurrence of an intrastaminal disc, a micropyle formed by only the inner integument, and in the absence of obturator and nucellar beak.

Among other morphological distinctions, Anacardiaceae differ from *Filicium* in the absence of a nucellar beak, in having hooked synergids, showing early wall formation in the endosperm, and in the frequent occurrence of adventive polyembryony.

Accordingly, it is clear that the association of embryological and morphological features indicates that *Filicium* should be retained in the family Sapindaceae. Its assignment as a primitive genus by Muller and Leenhouts (1976) on the basis of their studies of pollen types finds support in the presence of widely accepted primitive embryological features, such as the obturator, two integuments, a nucellar beak, massive nucellus, prominent hypostase, and Nuclear endosperm.

14.5.3 *Nyctanthes-Jasminum*

Most taxonomists, including Bentham (1876a) and Knoblauch (1897), place *Nyctanthes* in Oleaceae. Rendle (1952) recognizes two subfamilies, Jasminoideae and Oleoideae, and, as Lawrence (1951) states, of the 22 genera in the family only 3 (*Jasminum, Nyctanthes,* and *Menodora*) are assigned to Jasminoideae. Airy Shaw (1952), largely on the basis of the quandrangular nature of the stem, removed *Nyctanthes* from Oleaceae to a new subfamily Nyctanthoideae under Verbenaceae. On the basis of anatomical evidence, Stant (1952) supported this position, and in the revised edition of Engler's Syllabus der Pflanzenfamilien *Nyctanthes* is placed in Verbenaceae (Melchior 1964). Hutchinson (1973) also assigns *Nyctanthes* to Verbenaceae.

In their investigation of *Nyctanthes arbor-tristis,* Kapil and Vani (1966) compared the embryological and other morphological features of *Nyctanthes* with those of Oleaceae and Verbenaceae. *Nyctanthes* is characterized by an anther wall composed of an epidermis, fibrous endothecium, one middle layer, and a glandular tapetum. Cytokinesis following meiosis is simultaneous, and the tetrads are decussate or tetrahedral. Pollen is shed at the two-celled stage. The anatropous, unitegmic ovule develops an integumentary tapetum (Fig. 14.9 H–K), and usually there is a single hypodermal archesporial cell which develops directly as the megasporocyte (14.9 A). The development of the female gametophyte is of the monosporic Polygonum type (Fig. 14.9 A–D, H–K), or bisporic Endymion type (Fig. 14.9 E, F), or Allium type (Fig. 14.9 G). The antipodals are small and ephemeral, and twin female gametophytes are present (Fig. 14.9 I). The endosperm is of Cellular type (Fig. 14.9 L, M), and embryogeny conforms to the Megarchetype III, Period II (Souèges system) or the Solanad type (Johansen system).

In most of the features compared by Kapil and Vani, *Nyctanthes* resembles Oleaceae rather than Verbenaceae (see Table 14.3). Kapil and Vani also point out that the only anatomical character, among those studied by Stant (1952), in which *Nyctanthes* stands apart from Oleaceae is the presence of inversely oriented vascular bundles at the angles of the quadrangular stem. Additional support is furnished by Das and Rao (1966) who have shown that *Nyctanthes,* like other members of Oleaceae, lacks phloretic, O-pyrocatechuic, and syringic acids which are

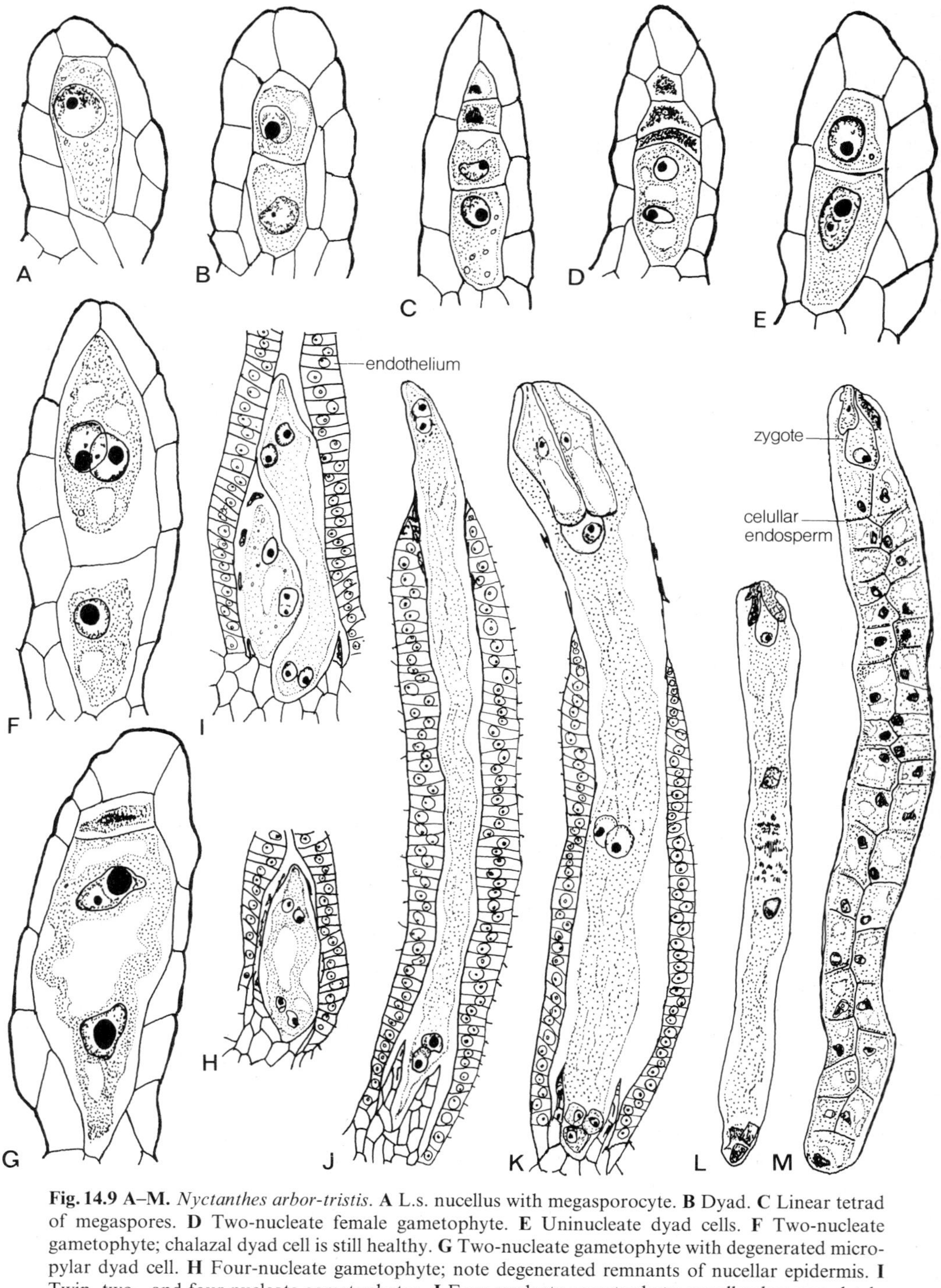

Fig. 14.9 A–M. *Nyctanthes arbor-tristis.* **A** L.s. nucellus with megasporocyte. **B** Dyad. **C** Linear tetrad of megaspores. **D** Two-nucleate female gametophyte. **E** Uninucleate dyad cells. **F** Two-nucleate gametophyte; chalazal dyad cell is still healthy. **G** Two-nucleate gametophyte with degenerated micropylar dyad cell. **H** Four-nucleate gametophyte; note degenerated remnants of nucellar epidermis. **I** Twin, two-, and four-nucleate gametophytes. **J** Four-nucleate gametophyte; nucellus has completely degenerated except for a few cells in the chalazal region. **K** Organized gametophyte, remnants of nucellus form a chalazal cup-shaped structure; note the conspicuous endothelium in **H–K. L** Zygote and two-celled endosperm. **M** Zygote and biseriate endosperm. (After Kapil and Vani 1966)

Table 14.3. Floral and embryological features in *Nyctanthes*, Oleaceae, and Verbenaceae. (Adapted after Kapil and Vani 1966)

	Nyctanthes	Oleaceae	Verbeneaceae
Corolla	Contorted	Contorted	Quincuncial
Stamens	2, filaments short; carpelloid anthers present	2, filaments short; carpelloid anthers present	4, didynamous, filaments long; carpelloid anthers not reported
Pollen	Spheroidal, tricolpate, tenuinexinous, sexine reticulate with piloid excrescences	Spheroidal, tricolpate, tenuinexinous, sexine reticulate with piloid excrescences	Euripalynous peroblate-prolate, 2–5-colpate or tricolpate, sexine and nexine of equal thickness, sexine reticulate without piloid excrescences
Nucellar archesporium	Single-celled, rarely 2-celled	Single-celled, rarely 2- or 3-celled	Single-celled
Embryo sac	Mono- or bisporic, antipodal cells small and ephemeral, twin embryo sacs present	Mono- or bisporic, antipodal cells small and ephemeral, twin embryo sacs occasionally present	Monosporic, antipodal cells large and persistent, twin embryo sacs not known
Endosperm	Cellular, first two divisions transverse, chalazal haustorium absent	Cellular, first two divisions transverse, chalazal haustorium absent	Cellular, first division transverse and second longitudinal or transverse, chalazal haustorium common
Embryogeny	Solanad Type	Solanad Type	Onagrad Type
Seed-coat	Vascularized	Vascularized	Nonvascularized
Fruit	Capsule	Capsule, drupe, berry or samara	Mostly drupaceous

consistently present in Verbenaceae. Ferulic acid occurs in Oleaceae as well as in *Nyctanthes* but is negligible in Verbenaceae.

In addition to anatomical and cytochemical features, *Nyctanthes* resembles Oleaceae in some floral characters, pollen structure, and in the structure and development of the female gametophyte, endosperm, and embryo. Its assignment to Oleaceae rather than Verbenaceae is, accordingly, assured; a position reflected in the recent classification presented by Takhtajan (1980).

In recognition of the paucity of information on the embryology of *Jasminum*, Maheswari Devi (1975) undertook an investigation of *Jasminum angustifolium, J. calophyllum*, and *J. sambac*. She primarily compared the embryological features but also other morphological characters of *Jasminum* with those of *Nyctanthes;* these taxa belong to the same subfamily. *Menodora*, the remaining genus in Jasminoideae, was not included in the comparison because of insufficient embryological data for that genus.

The comparison yielded a number of significant differences: (1) six to eight rows of archesporial cells in the microsporangia of *Jasminum* as apposed to two rows in *Nyctanthes:* (2) in *Jasminum* binucleate cells without nuclear fusions in the secretory anther tapetum, while in *Nyctanthes* both nuclear divisions and fusions occur; (3) a nucellus consisting of a single layer of epidermal cells which disintegrate during the two-nucleate stage of female gametophyte in *Jasminum*,

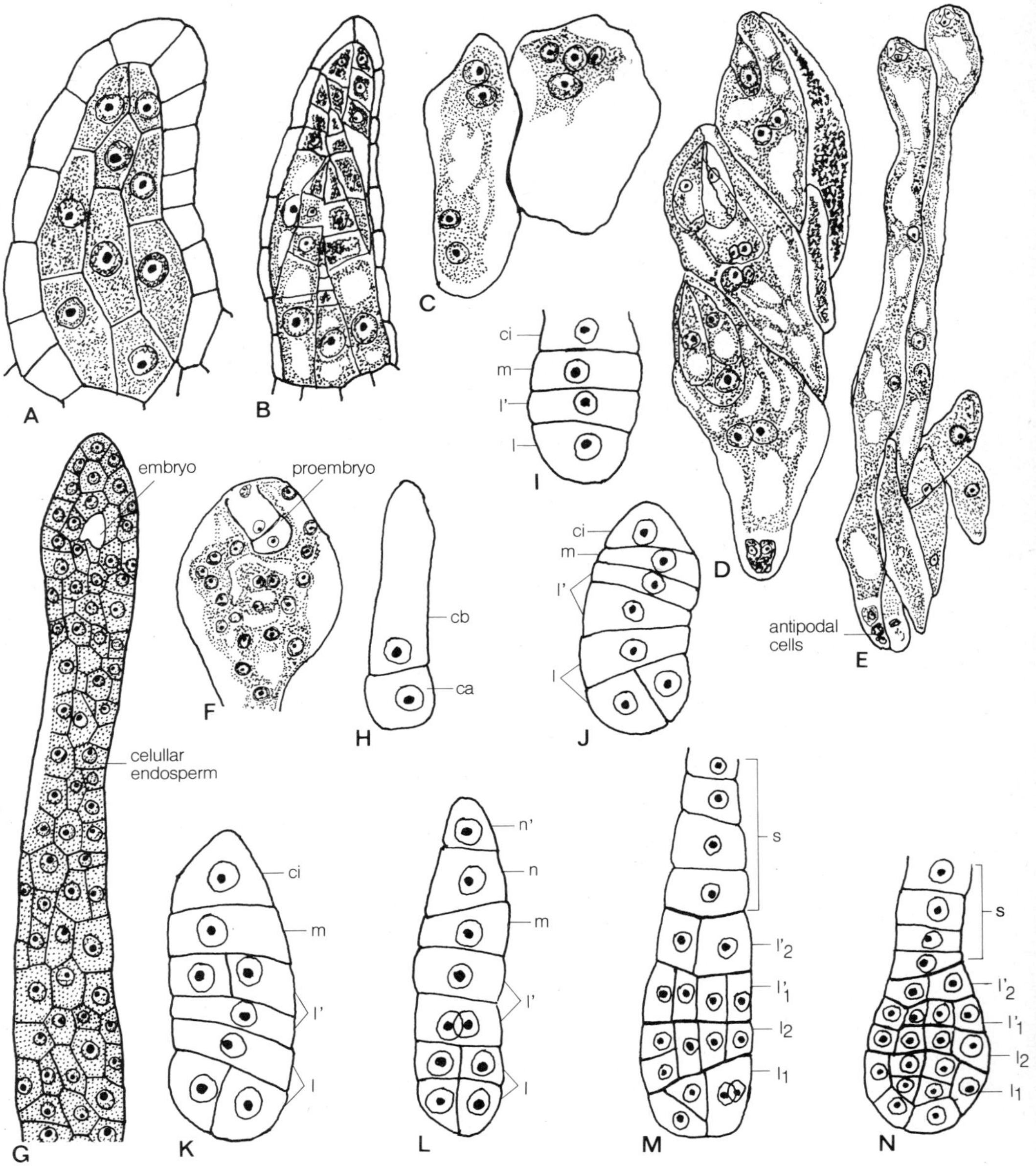

Fig. 14.10. A, E *Jasminum calophyllum;* **B, F** *J. sambac;* **C, D** and **G–N** *J. angustifolium.* **A** L.s. nucellus with multicelled archesporium. **B** Multiple megaspore tetrads. **C** Twin gametophytes. **D, E** Multiple gametophytes. **F** Upper part of gametophyte sac with two-celled proembryo and nuclear endosperm. **G** Cellular endosperm with early dicotyledonous embryo, lower part of endosperm omitted. **H–N** Development of proembryo. *s* suspensor. (After Maheswari Devi 1975)

while in *Nyctanthes* the lateral and basal nucellus persists as a cup-shaped structure below the mature female gametophyte; (4) a multicellular archesporium in the ovules of *Jasminum* (Fig. 14.10 A, B), but only one or two archesporial cells in *Nyctanthes;* (5) Polygonum type of female gametophyte with multiple gametophytes common in *Jasminum* (Fig. 14.10 C–E) as apposed to Polygonum, Allium, or Endymion type of female gametophyte in *Nyctanthes* with usually one female gametophyte in each ovule; (6) Nuclear type of endosperm in *Jasminum* (Fig. 14.10 F, G) and Cellular type in *Nyctanthes;* (7) embryogeny of Solanad type, Nicotiana variation in *Jasminum* (Fig. 14.10 H–N), and Solanad type, Physalis variation in *Nyctanthes*.

A comparative study shows that *Nyctanthes* differs significantly from *Jasminum* in the nature of the archesporium in the ovule, development of female gametophyte, endosperm, and embryogeny. Other differences occur as well in habit, form of inflorescence, structure of fruit, and some features of anatomy. Maheswari Devi suggests that it would be reasonable to treat *Nyctanthes* as a distinct taxon separate from Jasminoideae while still retaining it in Oleaceae. Accordingly, she recommends that a third subfamily, Nyctanthoideae, be established to accomodate the solitary genus *Nyctanthes*.

14.5.4 *Ophiopogon*

Bentham (1883) assigned *Ophiopogon* to Haemodoraceae in the tribe Ophiopogonae on the basis of semi-inferior ovary. Kraüse (1930), on the other hand, placed the genus in Liliaceae in recognition of its short rhizome, thin pericarp, and fleshy seed-coat. Hutchinson (1973) retained it in Liliaceae because of its racemose inflorescence, and the splitting of the ovary wall to expose developing seeds. Clearly, the genus *Ophiopogon* must be included among the group of genera with disputed affinity.

In their embryological study of *Ophipogon intermedius,* Mohana Rao and Kaur (1979) have presented a comparison of morphological and embryological characters of *Ophiopogon* with those of Haemodoraceae and Liliaceae (see Table 14.4).

Ophiopogon resembles Liliaceae in several features: successive delimitation of microspores, glandular tapetum, monocolpate pollen grains, anatropous, bitegmic, crassinucellate ovules, thick outer integument, presence of an obturator, presence or absence of the primary parietal cell, T-shaped tetrad of megaspores, Polygonum type of female gametophyte, Helobial type of endosperm, and copious endosperm in the mature seed.

Although the genus resembles Haemodoraceae in some respects, *Ophiopogon,* unlike genera in Haemodoraceae, is characterized by: glandular tapetum, anatropous ovules, thick outer integument, presence of obturator, presence or absence of primary parietal cell, and a T-shaped tetrad of megaspores. *Ophiopogon* does not share many features in common with Haemodoraceae and agrees in most embryological characters with the genera of Liliaceae. Embryology, thus, supports the opinions of Kraüse (1930), Pax (1930), and Hutchinson (1973) in favor of assigning *Ophiopogon* to Liliaceae rather than to Haemodoraceae.

Table 14.4. Morphological and embryological features in *Ophiopogon*, Haemodoraceae, and Liliaceae. (Adapted after Mohana Rao and Kaur 1979)

	Ophiopogon	Haemodoraceae	Liliaceae
Inflorescence	Terminal raceme	Panicle, subumbellate cyme, raceme	Solitary, axillary or terminal raceme, cyme
Trichomes	Absent	Present on perianth, ovary, glandular	Rare
Ovary	Semi-inferior	Superior to inferior	Superior, ralely semi-inferior or inferior
Seed	Albuminous	Albuminous	Albuminous
Fruit	Ovary splits before maturity to expose developing seeds	Loculicidal capsule	Loculicidal or septicidal capsule, felshy berry, or rupturing to expose seeds
Anther tapetum	Glandular	Amoeboid	Glandular
Pollen	Monocolpate, 2-celled	Monocolpate, 2-celled	One-, two- or more colpate or porate, 2- or 3-celled
Cytokinesis	Successive	Successive	Successive or simultaneous
Microspore tetrad	Isobilateral	Tetrahedral, isobilateral	Isobilateral, decussate, tetrahedral
Ovule	Anatropous, bitegmic and crassinucellate	Hemianatropous, orthotropous, bitegmic and crassinucellate	Anatropous, orthotropous, hemianatropous, bitegmic and crassinucellate
Outer integument	3- to 5-layered	2-layered	3- to 5-layered
Obturator	Present	Absent	Present
Parietal cell	Present or absent	Always present	Present or absent
Megaspore tetrad	T-shaped	Linear	Linear of T-shaped
Embryo sac development	Polygonum type	Polygonum type	Polygonum, Allium, Adoxa, Drusa, Fritillaria or Endymion type
Endosperm	Helobial type	Helobial type	Helobial or Nuclear type
Embryo	Onagrad type	Asterad type	Onagrad, Asterad, Caryophyllad or Chenopodiad type

14.5.5 *Kyllinga*

The taxonomic position of *Kyllinga* in the tribe Eucypereae of Cyperaceae should be regarded as controversal. Some investigators such as Cooke (1908), Sedgwick (1918), Gamble (1931), Blatter and McCann (1935), and Hutchinson (1973) considered it a separate genus from *Cyperus* while others, Santapau (1960) and Koyama (1961) merged *Cyperus* with *Kyllinga* and four other genera.

Padhye (1971) in his study of *Kyllinga brevifolia* took into account *K. triceps* (Padhye 1960) and *K. melanospora* (Khanna 1965) in an attempt to evaluate the assignment of the genus on the basis of embryology. The majority of embryological features reported by Khanna and Padhye are shared by *Cyperus* and *Kyllinga*. One should not, however, conclude that the uniformity in embryology indicates that these taxa should be combined. The same features are shared with other genera from both subfamilies of Cyperaceae (Cyperoideae and Caricoideae) and, indeed, with related Juncaceae as well. An excellent example is found where meiosis in the microsporangia occurs without cytokinesis. One microspore nucleus segregates from the other three and is later separated by a wall to form the single functional microspore in the tetrad. The remaining nonfunctional microspore nuclei usually disintegrate. This pattern is followed in *Kyllinga* and in all the genera of Cyperaceae so far studied and in some members of Juncaceae. In *Kyllinga brevifolia* and *K. melanospora* the nonfunctional microspore nuclei sometimes divide, as also in *Bulbostylis, Carex, Cyperus,* and *Scirpus.* Yet, this phenomenon is apparently absent in *Kyllinga triceps.*

Throughout Cyperaceae and in adjacent Juncaceae, the distribution of each embryological character follows the same random and often disjunct pattern of the example cited. On the basis of embryology, therefore, it would be as reasonable to combine *Kyllinga* with any one of several genera as it would be to combine it with *Cyperus.*

Very early in embryogeny in *Kyllinga* the cells in the quadrant that form the embryo proper, instead of producing the octant stage, divide periclinally to form the initials of dermatogen. Early differentiation of the dermatogen forms the basis for the Juncus variation of Onagrad type of embryogeny. It is a character of wide distribution and occurs in *Bulbostylis, Cyperus, Fimbristylis, Fuirena, Scirpus,* and certain taxa of Juncaceae. In *Kyllinga,* however, there is not only the early differentiation of dermatogen, but also of the plerome and periblem. This feature, although reported in *Bulbostylis barbata* (Shah 1965) and in *Juncus bufonius* (Shah 1963), sets *Kyllinga* apart from *Cyperus* where the plerome and periblem are differentiated later in embryogeny. Additionally, *Kyllinga* differs markedly from *Cyperus* in pollen structure. In most species of *Cyperus,* the pollen grains are monocolpate, whereas in *Kyllinga* they show a zonal distribution of four to six apertures. Erdtman (1952) refers to this condition of zonal distribution as uncommon not only in Cyperaceae but in Monocotyledoneae.

The controversy regarding the status of *Kyllinga* has arisen because the morphological characters used for generic delimitation are not sharply defined. However, as Padhye (1971) has shown in his synthesis of important recent findings the palynological and embryological characters strongly support the retention of *Kyllinga* as a separate genus in Cyperaceae.

14.5.6 *Burmannia pusilla*

A number of examples have been presented here where the findings of a broad assessment of embryology have been applied to specific taxonomic problems at the ordinal, family, and general levels. One final example relates a case where a question of species definition and limit has been settled on the basis of embryology.

The taxonomic status of *Burmannia pusilla* (Wall ex Miers) Thw, has been repeatedly questioned by taxonomists. Hooker (1888) considered it as a part of *B. coelestis* as did Gamble (1928). Cooke (1908) gave it varietal rank under *B. coelestis.*

The embryology of *B. coelestis* was studied by Ernst (1909) and Ernst and Bernard (1912). Arekal and Ramaswamy (1973) worked out the embryology of *B. pusilla* and compared their findings with data from *B. coelestis.*

The sporophyte of *Burmannia coelestis* is more robust than that of *B. pusilla*. Its inflorescence is composed of sessile flowers, while in *B. pusilla* the lateral flowers have conspicuous pedicels. The shape of the perianth is orbicular and broader at the upper part in *B. pusilla* as compared to that in *B. coelestis*.

In addition, *B. coelestis* is an apomict with displospory. It is characterized by: diploid, nonfunctional pollen, unreduced female gametophyte, absence of fertilization, diploid parthenogenesis, and polyembryony. *B. pusilla,* on the other hand, is a sexual taxon with haploid, functional pollen and a haploid Polygonum type of female gametophyte. Double fertilization occurs, and apogamy and polyembryony do not occur. Special note should be taken of the endosperm which is of the Helobial type not only in *B. coelestis* and *B. pusilla* but also in *B. candida, B. championii,* and *B. tuberosa*. However, in *B. pusilla* the chalazal chamber becomes four- to six-nucleate as apposed to uninucleate in *B. championii* and binucleate in the other species including *B. coelestis*.

The totality of differences in morphological and embryological features justifies the status of *B. pusilla* as a distinct species, separate from *B. coelestis*.

14.6 Contributions of Embryology to Taxonomic Problems Still Unsolved

As stated by Johri (1963), embryology without providing the final solution may yet make a definite contribution toward solving taxonomic problems and confirming relationships of genera and families. A number of studies have been reported where embryological features provide some indications of taxonomic affinity, but for variable reasons more intensive study is recommended. Several examples of this kind are presented here.

14.6.1 Erythroxylaceae-Linaceae

A close relationship between Erythroxylaceae and Linaceae is generally accepted. However, differences between their anatomical (Heimsch 1942, Metcalfe and Chalk 1950) and embryological features (Narayana and Rao 1978) indicate that these taxa should be treated as separate families. Bentham (1862b) and Hallier (1923) are among the few taxonomists who classify Erythroxylaceae as a tribe of Linaceae.

The development of the ovule and seed-coat in *Erythroxylum coca* (Boesewinkel and Geenen 1980) and *Linum usitatissimum* (Boesewinkel 1980) is strikingly similar. The common salient features include: bitegmic crassinucellate ovule, slender nucellus which is resorbed, dermal integuments, well differentiated raphe bundle, differentiation of the inner layer of inner integument into pigmented endothelium, fibrous exotegmen in seed-coat, Nuclear type of endosperm, and a conspicuous cuticle between seed-coat and endosperm.

Although *Linum usitatissimum* is a derived species (not representative of its family), and on the basis of development of the ovule more advanced than *Ery-*

throxylum, the similarity in embryology and anatomy of seed-coat between these genera casts doubts on the propriety of placing them in separate families. In addition, there is disagreement as to the taxonomic assignment of Linaceae and Erythroxylaceae. Most systematists place these taxa in Geraniales (Schulz 1931, Lawrence 1951, Takhtajan 1980). However, Hutchinson (1973) placed them in Malpighiales and Cronquist (1981) segregated them to Linales.

Embryological data provide indications of both the relationship of Erythroxylaceae to Linaceae and ordinal placement of these taxa. However, embryological information, especially regarding the development of the ovule and seed-coat is limited to the generic relatives of *Erythroxylum* and *Linum* and, generally, for Malpighiales and Geraniales. Accordingly, considerable investigation will have to be undertaken before the settlement of the taxonomic status and ordinal placement of *Erythroxylum* and *Linum* can be regarded as definitive.

14.6.2 Theligonaceae

The status and relationships of the monotypic Theligonaceae have long been controversial. *Theligonum* has been included in Urticaceae (Bentham 1880), and Phytolaccaceae (Warming 1895). According to Takhtajan (1959), Dumortier in 1829 elevated the genus to the rank of a family, a view supported by Poulsen (1894). As a family, it has been placed in Centrospermales (Caryophyllales) (Eichler 1878, Poulsen 1894, Takhtajan 1959), Myrtales in alliance with Haloragaceae (Hallier 1912, Ulbrich 1933, Melchior 1964), and in Chenopodiales with Amaranthaceae (Hutchinson 1973).

Kapil and Mohana Rao (1966a) examined all the proposed family assignments and ordinal relationships of *Theligonum*. They conclude that the proper place for Theligonaceae is in Centrospermales (Caryophyllales) near Phytolaccaceae and Amaranthaceae on the basis of features common to these families: monocarpellate ovary with solitary ovule, secretory anther tapetum, three-celled pollen grains, Polygonum type of female gametophyte, Nuclear type of endosperm, and albuminous seeds with curved embryos.

Cronquist (1968) with reluctance favored the alliance with Haloragaceae and pointed out that Theligonaceae differ from Caryophyllales in basal style, unitegmic, tenuinucellate ovules, distinctive sculpturing of pollen, absence of perisperm, and apparent absence of both betalains and enthocyanins.

Wunderlich (1971) states that the embryological and anatomical features of *Theligonum* indicate its relationship with Rubiaceae. She recommends that the genus be placed in a separate tribe Theligoneae in the subfamily Rubioideae, Thorne (1976) also regards *Theligonum* as undeserving of family status and assigned it to Rubioideae. Takhtajan (1980) retained Theligonaceae as a family in Gentianales, but states that *Theligonum* is very closely related to Rubiaceae and probably should not retain a family status. Cronquist (1981) abandoning his earlier view, agrees with Takhtajan as to the close relationship of *Theligonum* to Rubiaceae. He finds it useful, however, to treat Theligonaceae as a distinct family. On the other hand, Praglowski (1973) pointed out that on the basis of pollen morphology the affinity of Theligonaceae is with Caryophyllales rather than Rubiales.

In spite of various intensive investigations including the embryological studies of Kapil and Mohana Rao (1966a), and Wunderlich (1971), it is apparent that the taxonomic status and relationships of *Theligonum* are still unsettled.

14.6.3 *Blandfordia*

Kraüse (1930) placed *Blandfordia* in the tribe Hemerocallideae of Liliaceae along with *Hemerocallis, Hesperocallis, Hosta, Leucocrinum,* and *Phormium.* Hutchinson (1973), however, associated *Blandfordia* with *Kniphofia* in the tribe Kniphofieae.

Di Fulvio and Cave (1964) compared the embryological features of *Blandfordia nobilis* with those of other genera in Hemerocallideae and Kniphofieae. *Blandfordia* differs markedly from *Kniphofia* in anatropous ovules, a nucellar cap of several layers, and absence of an aril. In *Kniphofia* the arillate, hemianatropous ovule has a single layer of nucellus covering the mature female gametophyte. Helobial type of endosperm typifies both genera, but in *Blandfordia* the chalazal chamber is multinucleate as compared to binucleate in *Kniphofia.* In all these features *Blandfordia* resembles *Phormium,* although in other characters the genera are dissimilar. In addition, *Blandfordia* is unlike the other genera of Hemerocallideae in its embryological characteristics. Di Fulvio and Cave assert that the embryological evidence clearly indicates that *Blandfordia* should not be taxonomically associated with *Kniphofia.* However, much more study both of embryology and other disciplines must be accomplished before the taxonomic relationships within the whole liliaceous complex can be settled.

This conclusion is echoed by Takhtajan (1980) who assigned *Blandfordia* with *Phormium* to Phormiaceae, but in recognition of the dissimilarities in these genera segregates them to different tribes (Blandfordieae and Phormieae). He, as also Di Fulvio and Cave, recognizes *Blandfordia* as especially isolated and in need of further study before its taxonomic affinities can be firmly established.

14.6.4 Philydraceae

Regarded as an isolated family, Philydraceae have been associated with various orders. Pilger (1930) placed the family in the order Farinosae, Wettstein (1935) in Liliflorae, Deyl (1955) in Orchidales, and Hutchinson (1973) in Haemodorales.

Kapil and Walia (1965) and Hamann (1963, 1966) compared the embryological features of *Philydrum* with those of taxa proposed to be related. With each comparison, characters of both similarity and dissimilarity emerged. They pointed out, for example, that while Philydraceae resemble Pontederiaceae in structure of the anther wall, pollen grains, ovule, female gametophyte, and development of endosperm, the two families are dissimilar in floral structure, structure of seed, and embryogeny. They cannot, therefore, be considered closely related. Provisionally, these workers conclude that the family should be dissociated from Haemodorales and raised to the rank of Philydrales. On the other hand, Hamann in an evaluation of similar data considers Philydraceae closely related to

Pontederiaceae and secondarily to Haemodoraceae. An affinity with Pontederiaceae is proposed also by Hallier (1912), Huber (1969), and Cronquist (1981). Takhtajan (1980) includes Philydraceae, Pontederiaceae, and Haemodoraceae in Liliales but places the families in separate suborders. In this assessment, morphological and anatomical features in addition to the embryological data were taken into account. Takhtajan too continues to hold Pontederiaceae as the family closest in relationship to Philydraceae, yet recognizes prominent differences, especially in embryological characters, as indicative of the need for further study.

14.6.5 Eriocaulaceae

Taxonomists have not found a general consensus as to the systematic position of Eriocaulaceae. Hieronymus (1887) placed the family with Xyridaceae and Mayacaceae in the same suborder of the order Farinosae. Hutchinson (1973) separated the families to closely related orders and considered Eriocaulales as considerably advanced over Xyridales. Benson (1957) accepted the treatment of Bessey (1915) who placed all three families in Liliales along with Commelinaceae. Takhtajan (1980) and Cronquist (1981) separated Eriocaulaceae to the order Eriocaulales regarded as derived from Commelinales which includes Commelinaceae, Xyridaceae, and Mayacaceae.

In their embryological study of *Eriocaulon hookerianum,* Arekal and Ramaswamy (1980) called attention to the individuality and distinctiveness of the family especially as regards the following features: three-celled siraperaturate pollen, tenuinucellate ovules, micropyle organized by the inner integument, organization of a unique antipodal cyst in the female gametophyte, and Nuclear type of endosperm associated with the chalazal location of the primary endosperm nucleus. For the uniqueness of this combination of features, the systematic assignment of the family as highly derived, expressed by Takhtajan (1980) and Cronquist (1981), is in the view of Arekal and Ramaswamy most appropriate. However, as they have stated further, embryological investigations have been directed to only a few genera of Commelinaceae, one genus in Xyridaceae, and three genera of Eriocaulaceae. Of the Englerian Farinosae which comprise over 5,000 species in 180 genera, embryological data are available for only 25 genera and often for only one species in each genus (Davis 1966). Therefore, Arekal and Ramaswamy conclude that a better understanding of the phylogeny and systematic position of the family will require a more detailed study of many more taxa in Eriocaulaceae and in the families proposed to be of close relationship.

14.7 Confirmation or Alteration of Earlier Conclusions

There are numerous examples where embryological data have led to firm conclusions concerning taxonomic relationships. Much of the investigation that has taken place since the recording of these examples tends either to confirm or to refute those conclusions, and a few of the well-documented cases are reviewed here.

14.7.1 Cactaceae

Maheshwari (1950) summarized embryological data from a number of investigations which (to him) assured a place for the family Cactaceae in Centrospermales. More specifically, he concluded, "Present evidence ... seems to be in favor of regarding the Cactaceae as a sort of bridge between the Aizoaceae and Portulacaceae". In their study of *Cereus jamacaru* and *Ferocactus wislizeni,* Kapil and Prakash (1969) reviewed previous proposals for the systematic position of Cactaceae and compared the salient embryological features of the family with those of families proposed to be related, viz., Nymphaeaceae, Cucurbitaceae, Passifloraceae, and Aizoaceae. They conclude that Cactaceae should be placed in Centrospermales (Caryophyllales) close to Aizoaceae. The extensive scanning electron microscopic survey of seed morphology in Cactaceae by Barthlott and Voit (1979) fully supports this conclusion, thus to demonstrate the firm basis for the conclusion put forward by Maheshwari (1950).

14.7.2 Garryaceae

Largely in consideration of the study by Hallock (1930) on the embryology of *Garrya elliptica,* Maheshwari (1950) concurs with earlier views that Garryaceae represent an advanced family closely related to Cornaceae. A contention of Hallock that the synergids and antipodals serve a haustorial function in *G. elliptica* has been convincingly refuted by Kapil and Mohana Rao (1966b) who observed that these cells degenerate soon after fertilization in *G. veatchii* and *G. flavescens.* They agree with Maheshwari in the systematic position of Garryaceae as closely related to Cornaceae in the Englerian order Umbelliflorae. The anatomical studies of Moseley and Beeks (1955) support the same view. Takhtajan (1980) also acknowledges the close relationship of Garryaceae to Cornaceae and points out that in pollen morphology (Eramian 1971) and phytochemistry (Bate-Smith et al. 1975) *Garrya* resembles *Aucuba,* which is usually included in Cornaceae. Thus, the affinity of Garryaceae with Cornaceae designated by Maheshwari (1950) on rather fragmentary evidence has survived the close scrutiny provided by a more detailed examination of embryological characters and by other types of study as well.

14.7.3 Orobanchaceae

Although a close relationship of Orobanchaceae to Scrophulariaceae has been long recognized (Boeshore 1920, Glišic 1929), Maheshwari (1963) cites this affinity as one of the noteworthy examples for the value of embryology to taxonomy. Tiagi (1956) investigated *Striga euphrasioides* and *S. orobanchoides* and observed that the mode of development of the endosperm in these Scrophulariaceae is essentially like that in Orobanchaceae. In addition, *Striga* resembles Orobanchaceae in: extent of curvature in the female gametophyte, form of the micropylar and chalazal endosperm haustoria, Onagrad type of embryogeny, and structure of seed. Tiagi concludes that Orobanchaceae are derived from the subfamily Rhinanthoideae of Scrophulariaceae, and Maheshwari (1963) accordingly regarded the matter as settled. This position is supported by the work of Nagendran et al. (1980 b) on *S. densiflora*. They showed that there is consistency and uniformity in development of the female gametophyte and endosperm among the species of *Striga*. Also Teriokhin (1980), who finds similarity in the major embryological features within Orobanchaceae, supports the conclusions of Tiagi and Maheshwari.

14.7.4 *Pentaphragma*

Johri (1963) includes the genus *Pentaphragma* among those taxa whose relationships are confirmed on the basis of embryology. He pointed out that the preponderance of evidence favors assigning the genus to a separate subfamily, Pentaphragmoideae, under Campanulaceae. Kapil and Vijayaraghavan (1965) stated that although *Pentaphragma* shows a few distinctive features (e.g., extramicropylar female gametophyte, uninucleate micropylar endosperm haustorium, and short suspensor), this aberrant genus correctly belongs to Campanulaceae.

A contrary view has been expressed by Airy Shaw (1941) who assigned *Pentaphragma* to a separate family, Pentaphragmataceae. He justified this move largely on the basis of floral morphology and stated: "Almost every author who has dealt with *Pentaphragma* has admitted its highly anomalous position in the Campanulaceae, and it is therefore surprising that it has not been removed earlier." Dunbar (1978), in a study of pollen morphology in *Pentaphragma*, finds that the trilobate shape and pleurotreme condition of pollen grains of this genus are not evident in Campanulaceae. The exine surface in Campanulaceae is covered with spinules or is variously reticulate whereas the surface of the grain in *Pentaphragma* has no supratectal sculpture. Dunbar concludes that palynologically *Pentaphragma* belongs to a separate family in accordance with Airy Shaw (1941). Cronquist (1981) also recognizes the distinctiveness of *Pentaphragma* and places it in a separate family. The variation in the interpretations registered so far would seem to indicate that the taxonomic position of *Pentaphragma* is still unsettled and awaits further investigation.

14.8 Embryology in Relation to Taxonomy – the Prospect

Two taxa which share a greater number of features than either one shares with other similar taxa are likely to be more closely related to one another than to any other taxon of the same category. This statement presents a basic postulate that is operative in most considerations of taxonomic relationships. It is certainly a fundamental principle in each example of the application of embryology presented here. To conjecture about the direction future applications will take is not difficult, for the kinds of approach of proven worth in the past assuredly will continue in the future. That embryology as a taxonomic tool, however, has not reached its potential has been emphasized at the beginning of this chapter. Certainly a major cause for this deficiency, as pointed out by Kapil and Bhatnagar (1980), is the availability of data from only a small fraction of the total number of flowering plants. Accordingly, data of the kinds applied in the past must be collected from a much broader range of flowering plants. One may ask at this point, what is the prospect of innovation in the application of embryology to taxonomy? A measure of innovation will surely accrue as investigations incorporate improved and innovative techniques and thus become more thorough. The improved approaches will permit a closer monitoring of developmental stages and in turn a critical, analytical treatment of the data collected.

Among the various kinds of investigation, those directed to the female gametophyte perhaps have the greatest potential value in taxonomic considerations. So far, a relating of the female gametophyte to taxonomy has been minimal, usually not progressing beyond the association of taxa through a particular type of development. This limited application, if continued, cannot be expected to become more fruitful, since 70% of angiosperms follow the Polygonum type. In addition, many taxa are not defined by a specific type of development (e.g., *Tamarix* and *Ulmus*, Fig. 14.1). However, a closer monitoring of developmental stages and generally a greater attention to detail will permit better understanding of variation within each type and concomitantly a broader application of gametophytic features to the solution of disputed taxonomic problems. Several approaches to a more thorough understanding of the female gametophyte are already in progress and hold great promise for the future. Recent studies have been enhanced by an improved microscopy afforded by phase contrast, differential interference, and electron microscopes.

One of the new approaches expected to have taxonomic significance is the statistical analysis of growth of the developing female gametophyte. Studies reported so far collectively indicate that growth pattern indeed presents a variable feature among plants that follow the Polygonum type. Size is assessed through measurements of the length (a–a′) and width (b–b′) of recognized stages of development (e.g., functional megaspore, two-, four-, and eight-nucleate stages). Length and width intercepts (a′–0, b′–0) are also recorded for their potential use in deriving a quantitative expression of the symmetry characteristic for each stage. The two axes are oriented in a standard manner (Fig. 14.11). Generally, between 10 and 30 examples of each stage are measured for accurate assessment of the size range and calculation of the mean and confidence interval (i.e., the size interval which would include 95% of all representatives of a particular stage). The results of several studies are summarized here to convey the potential value as well as the limitations of this approach for taxonomic considerations.

14.8.1 *Cornus*

Smith (1973) described the development of the ovule and female gametophyte in five species of *Cornus*, viz., *C. florida* of the subgenus *Benthamidia*, and *C. alternifolia*, *C. amomum*, *C. asperifolia*, and *C. stricta* of the subgenus *Thelycrania*. Development of the female gametophyte follows Polygonum type in all five species.

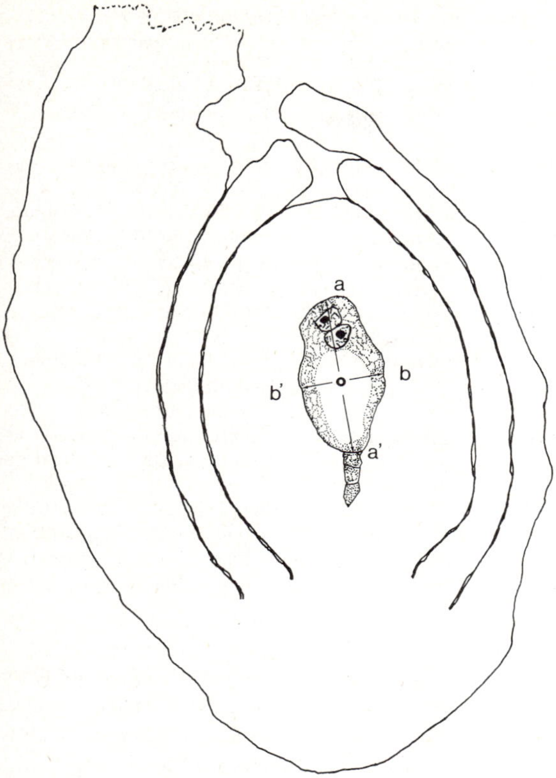

Fig. 14.11. *Ludwigia uruguayensis,* l.s. crassinucellate ovule with two-nucleate female gametophyte deep in the nucellus. For the female gametophyte length (a–a′), width (b–b′), and length and width intercepts (a′–0; b′–0) are indicated

However, *C. florida* stands apart from the others in several features (Table 14.5). Whether these differences represent distinction between the two subgenera awaits investigation of other species of *Benthamidia*. It is not surprising that this descriptive study indicates a high degree of uniformity among the four species of *Thelycrania*. Studies of this traditional kind often do not reveal differences within the Polygonum type among closely related species. However, in a quantitative analysis of the female gametophyte in these five species of *Cornus,* Smith (1975) found prominent differences among the four species of *Thelycrania* and even more marked distinction between these four species and *C. florida*. The size of the functional megaspore, two-, four-, and eight-nucleate female gametophytes for each species was assessed through measurements of length (a–a′) and width (b–b′). Length and width intercepts (a′–0 and b′–0) were also recorded for potential use in the characterization of symmetry for each stage (Fig. 14.11). As many examples of each stage (5–26) as could be found in good condition were measured in order to assess the size range. From these measurements, means and confidence intervals were calculated. The study reveals several features which are beyond the scope of traditional descriptive studies. Differences in the means for successive stages, for example, indicate the amount of growth that occurs between each stage. For the subgenus *Thelycrania,* the time of maximal growth is variable and

Table 14.5. Embryological fatures in five species of *Cornus*. (Adapted after Smith 1973)

	C. florida	*C. alternifolia*	*C. amomum*	*C. asperifolia*	*C. stricta*
Number of archesporial cells	Many	Two or more	One	Two or more	One
Number of parietal cells	None	One	One	Two	One
Nucellar growth	Tenuinucellate	Crassinucellate	Crassinucellate	Crassinucellate	Crassinucellate
Origin of megasporocyte	Archesporial cells (many)	Primary sporogenous cell (one)	Primary sporogenous cell (one)	Primary sporogenous cell (two)	Primary sporogeneous cell (one)
Type of tetrad	Linear (many)	Linear and T-shaped	Linear and T-shaped	Linear (two)	Linear
Degeneration of defunct megaspores	Variable	Variable	Variable	Second and third then uppermost	Variable
Duration of synergids	Persist until polar nuclei fuse	Degenerate rapidly	Degenerate	Persist until polar nuclei fuse	Persist until polar nuclei fuse
Duration of antipodals	Persist until polar nuclei fuse	Persist until polar nuclei fuse	Degenerate	Persist until polar nuclei fuse	Persist until polar nuclei fuse

may be recorded as follows: *C. stricta,* between the four- and eight-nucleate stages; *C. alternifolia,* between the functional megaspore and two-nucleate stage; *C. amomum,* between the two- and four-nucleate stages; and *C. asperifolia,* uniformly between each set of stages from the functional megaspore to the eight-nucleate stage. In *C. florida* the confidence intervals for length of the four- and eight-nucleate stages were too broad to permit accurate conclusions about differences in the means for successive stages. However, the pattern of growth in the developing female gametophyte is clearly dissimilar from that of any species in *Thelycrania*. Maximal increase in length occurs during the four-nucleate stage. Accordingly, the statistical analysis of growth of the female gametophyte confirms the separation of the subgenera *Thelycrania* and *Benthamidia*. The degree of separation, however, awaits the study of additional species in both subgenera especially in *Benthamidia*.

14.8.2 *Glycine* and *Phaseolus*

In their comparative study of ovule and female gametophyte development in field-grown and greenhouse-grown plants of *Glycine max* and *Phaseolus aureus,* George et al. (1979) found several common features: (1) a multiple archesporium; (2) origin of the megasporocyte by enlargement of the primary sporogenous cell; (3) unequal division of the megasporocyte during meiosis I; (4) a functional chalazal megaspore; (5) Polygonum type of development; and (6) a hypostase. The several marked differences in development for *Glycine* and *Phaseolus* are summarized in Table 14.6. However, the chief objective of this investigation was

Table 14.6. Embryological differences between *Glycine max* and *Phaseolus aureus*. (Data from George et al. 1979)

Glycine max	*Phaseolus aureus*
Two or three sporogenous cells	One sporogenous cell
Meiosis II nonsynchronous	Meiosis synchronous
Linear and T-shaped megaspore tetrads	Linear megaspore tetrads
Micropylar and chalazal nuclei perpendicular to long axis of early 4-nucleate female gametophyte	Micropylar and chalazal nuclei parallel to long axis of early 4-nucleate female gametophyte
Micropylar and chalazal nuclei parallel to long axis of late 4-nucleate stage	Micropylar and chalazal nuclei perpendicular to long axis of late 4-nucleate stage
Three antipodals, one broad, two narrow	Three antipodals of equal size
Starch present in mature female gametophyte	Starch absent in mature female gametophyte

to determine for each genus whether development of the female gametophyte manifests differences in response to the field and greenhouse habitats. It has been often stated in the past that development of the female gametophyte in flowering plants follows a uniform pattern, unaffected by the variable environment of the sporophyte. Protection offered by the enclosing layers of ovary and ovule is cited as the cause. The findings of George et al. in this traditional, descriptive study confirm those of the past, viz., that development of the female gametophyte in plants grown under field conditions is markedly similar to development in plants grown in the more uniform conditions of the greenhouse.

The investigation of George et al. (1979) represents a summation of portions of earlier studies by R. George (1976) and G. George (1976) on *Glycine max* and *Phaseolus aureus,* respectively. In each of these original investigations a quantitative analysis of growth in the developing female gametophyte was undertaken similar in procedure and scope to the study reported by Smith (1975) for *Cornus*. Accordingly, the a–a′, b–b′, a′–0, and b′–0 axes were measued (Fig. 14.11) for numerous examples of each stage (megasporocyte, functional megaspore, and two-, four-, eight-nucleate and mature female gametophytes). T-tests were used to determine whether the means for each stage produced in the greenhouse differed significantly from the means of comparable stages produced in the field. For *Phaseolus aureus,* the T-test shows a significant difference between stages of greenhouse- and field-origin for: (1) length (a–a′) of the megasporocyte; (2) length and length intercept (a–a′, a′–0) of both the two- and four-nucleate stages; and (3) length (a–a′) of the eight-nucleate stage. The T-test indicates that differences between stages of greenhouse- and field-origin for *Glycine max* are confined to length (a–a′) and length and width intercepts (a′–0, b′–0) of the megasporocyte, and length (a–a′) of the mature female gametophyte.

The data, thus, indicate that development in greenhouse and field conditions is for the most part uniform. Yet, the few differences of statistical significance suggest that caution should be brought to bear on choosing material for comparative studies. In the collection of embryological data, especially those of a quantitative nature, to be applied in taxonomic assessments, material for all taxa studied should be collected from natural habitats.

14.8.3 *Eustachys petraea* and *E. glauca*

As is often the case with grass species, *Eustachys petraea* and *E. glauca* have very similar sporophytes. *Eustachys* was once included in the genus *Chloris* and is still regarded as closely related to it. In her comparative study of ovule and female gametophyte development, Aulbach-Smith (1979) found marked similarity in the two species until the maturation process during the eight-nucleate stage. In *E. petraea,* as in most grasses investigated, the antipodals differentiate first followed by establishment of the egg apparatus. Next, the polar nuclei migrate to the center, the antipodals divide, and the synergids degenerate. Fusion of the polar nuclei then occurs to form a secondary nucleus. For *E. glauca,* migration of the polar nuclei to the center is the first event followed by antipodal differentiation and formation of the egg apparatus. Next, the antipodals divide, the synergids degenerate, and the polar nuclei fuse. *E. petraea* produces 12 antipodals, and the central polar nuclei are well separated from the egg apparatus and the antipodals. *E. glauca,* on the other hand, almost always produces 6 (rarely 9) antipodals, and the polar nuclei are actually in contact with the egg and antipodals.

Aulbach-Smith also carried out a statistical study of growth of the female gametophyte in these two species by the procedure (Fig. 14.11) introduced by Smith (1975). This portion of her study revealed a number of differences previously not disclosed. In consideration of the length parameter (a–a′), she confirmed through T-tests that the functional megaspore and two-nucleate female gametophyte of *E. petraea* are significantly smaller than those stages in *E. glauca.* However, the eight-nucleate and mature stages of *E. petraea* are significantly larger than those stages in *E. glauca.* Only the four-nucleate stage shows no significant size difference in these two species.

This study of *Eustachys* is the first of its kind to include a comparison of nuclear volumes in the developing female gametophyte. Nuclear volumes were measured for three substages of the enlarging functional megaspore (D_1, D_2, and D_3, based on degree of degeneration of the defunct megaspores), and the two- and four-nucleate female gametophyte. In *E. petraea,* the volume of the megaspore nucleus increased markedly between the D_2 and D_3 substages. In *E. glauca,* this increase occurs evenly during all three substages. T-tests reveal that the nucleus in the enlarged megaspore and both nuclei of the two-nucleate stage in *E. petraea* are significantly smaller in volume than those nuclei in *E. glauca.* The nuclei of the four-nucleate stage, however, are essentially of the same size in both species.

The value of size differences in the developmental stages of the female gametophyte emerges clearly from this study. As mentioned earlier, the sporophytes of *E. petraea* and *E. glauca,* structurally complex as they are, are difficult to distinguish. The spikelets are nearly identical in size, shape, and coloration. The species differ, however, in plant size and in number of spikes per culm. On the other hand, the female gametophytes are very distinct especially if their differences are considered in relation to their simple and phylogenetically reduced structure. Presently, little is known of ovule ontogeny and development of the female gametophyte in other members of the tribe Chlorideae. However, the information reported by Aulbach-Smith (1979) for *Eustachys* warrants the prediction that with the expansion of study in other Chlorideae, development of the female gametophyte, especially the quantitative aspects of its growth, will prove useful in the determination of taxonomic relationships within the tribe.

That embryology cannot alone provide the ultimate solution to taxonomic problems is certainly clear. At the same time, its important contributions to this end are replete and well documented at every level of taxonomic consideration. The continuing improvement of embryological techniques and their application across the full range of flowering plants brightens the prospect for a closer and more significant relationship between embryology and taxonomy.

References

Acherson PFA, Graebner P (1907) Potamogetonaceae. In: Engler A (ed) Das Pflanzenreich, vol IV, pt 31. Engelmann, Leipzig, S. 171–184

Airy Shaw HK (1941) Additions to the flora of Borneo and other Malay Islands: The Pentaphragmataceae of the Oxford University Expedition of Sarawak, 1932. Bull Misc Inf Kew 1941:233–236

Airy Shaw HK (1952) Note on the taxonomic position of *Nyctanthes* L. and *Dimetra* Kerr. Kew Bull 1952:271–272

Arekal GD, Nagendran CR (1975) Is there a Podostemum type of embryo sac in the genus *Farmeria?* Caryologia 28:229–235

Arekal GD, Ramaswamy SN (1973) Embryology of *Burmannia pusilla* (Wall. ex Miers) Thw and its taxonomic status. Beitr Biol Pflanz 49:35–46

Arekal GD, Ramaswamy SN (1980) Embryology of *Eriocaulon hookerianum* Stapf. and the systematic position of Eriocaulaceae. Bot Not 133:295–309

Aulbach-Smith CA (1979) A comparative study of ovule development, megasporogenesis and megagametogenesis in *Eustachys petraea* (Swartz) Dsevaux and *E. glauca* Chapman (Gramineae). Thesis Univ South Carolina, Columbia

Barthlott W, Voit G (1979) Seed coat morphology and taxonomy of Cactaceae. A scanning electron microscopic survey. Plant Syst Evol 132:205–230

Bate-Smith EC, Ferguson IK, Hutson K, Jensen SR, Nielson BJ, Swain T (1975) Phytochemical interrelationships in the Cornaceae. Biochem Syst Ecol 3:79–89

Battaglia E (1971) The embryo sac of Podostemaceae: An interpretation. Caryologia 24:403–420

Bell HP (1957) The development of the blueberry seed. Can J Bot 35:139–153

Bell HP, Burchill J (1955) Flower development in the low bush blueberry. Can J Bot 33:251–258

Benson L (1957) Plant classification. DC Heath, Boston

Bentham G (1862a) Frankeniaceae. In: Bentham G, Hooker JD (eds) Genera Plantarum, vol I, pt 1. Lovell Reeve, London, pp 140–141

Bentham G (1862b) Linaceae. In: Bentham G, Hooker JD (eds) Genera Plantarum, vol I, pt 1. Lovell Reeve, London, pp 241–246

Bentham G (1876a) Oleaceae. In: Bentham G, Hooker JD (eds) Genera Plantarum, vol II, pt 2. Lovell Reeve, London, pp 672–680

Bentham G (1876b) Salvadoraceae. In: Bentham G, Hooker JD (eds) Genera Plantarum, vol II, pt 2. Lovell Reeve, London, pp 680–681

Bentham G (1880) Urticaceae.In: Bentham G, Hooker JD (eds) Genera Plantarum, vol III, pt 1. Lovell Reeve, London, pp 341–395

Bentham G (1883) Haemodoraceae. In: Bentham G, Hooker JD (eds) Genera Plantarum, vol III, pt 2. Lovell Reeve, London, pp 671–681

Bessey CE (1915) The phylogenetic taxonomy of flowering plants. Ann Mo Bot Gard 2:109–164

Bhojwani SS, Bhatnagar SP (1983) Embryology of angiosperms, 4th edn. Vikas Publishing House, Delhi

Blatter E, McCann C (1935) Cyperaceae in revision of the flora of the Bombay presidency. J Bombay Nat Hist Soc 37:15–35

Boesewinkel FD (1980) Development of ovule and testa of *Linum usitatissimum* L. Acta Bot Neerl 29:17–32

Boesewinkel FD, Geenen J (1980) Development of ovule and seed-coat of *Erythroxylum coca* Lamk. Acta Bot Neerl 29:231–241
Boeshore I (1920) The morphological continuity of the Scrophulariaceae and Orobanchaceae. Contr Bot Lab Univ Pa 5:139–177
Cave MS (1953) Cytology and embryology in the delimitation of genera. Chron Bot 14:140–153
Cave MS (1962) Embryological characters of taxonomic significance. Lilloa 31:171–181
Chadefaud M (1974) Sur la formule florale de la Capucine (*Tropaeolum majus* L.). Bull Soc Bot Fr 121:347–357
Chanda S (1966) On the pollen morphology of the Centrolepidaceae, Restionaceae, and Flagellariaceae, with special reference to taxonomy. Grana Palynol 6:355–415
Chopra RN, Mukkada AJ (1966) Gametogenesis and pseudo-embryo sac in *Indotristicha ramosissima* (Wight) van Royen. Phytomorphology 16:182–188
Cooke T (1908) The flora of the presidency of Bombay, vol 2. Taylor and Francis, London
Cronquist A (1968) The evolution and classification of flowering plants. Houghton Mifflin, Boston
Cronquist A (1981) An integrated system of classification of flowering plants. Columbia Univ Press, New York
Dahlgren R (1975) A system of classification of the angiosperms to be used to demonstrate the distribution of characters. Bot Not 128:119–147
Das VSR, Rao KN (1966) Chemotaxonomical investigation of *Nyctanthes*. Naturwissenschaften 53:439
Davis GL (1966) Systematic embryology of the angiosperms. John Wiley Sons, New York
Deyl M (1955) The evolution of the plants and the taxonomy of the monocotyledons. Acta Mus Natl Pragae Ser B Hist Nat 11:1–143
Di Fulvio TE, Cave MS (1964) Embryology of *Blandfordia nobilis* Smith (Liliaceae), with special reference to its taxonomic position. Phytomorphology 14:487–499
Dunbar A (1978) Pollen morphology and taxonomic position of the genus *Pentaphragma* Wall. (Pentaphragmaceae). Grana 17:141–147
Eichler AW (1878) Blütendiagramme, vol II. Engelmann, Leipzig
Engler A (1930) Podostemaceae. In: Engler A, Prantl K (eds) Die natürlichen Pflanzenfamilien, vol 18a. Engelmann, Leipzig, S. 483–484
Engler A, Diels L (1936) Syllabus der Pflanzenfamilien. Borntraeger, Berlin, S. 1–419
Eramian EN (1971) Palynological data on systematics and phylogeny of the Cornaceae Dumort. and related families. In: Kuprianova LA, Yakovlev MS (eds) Morphology of the pollen of Cucurbitaceae, Thymelaeaceae, Cornaceae. Izd "Nauka" Leningradskoe Otdel, Leningrad, pp 235–273
Erdtman G (1952) Pollen morphology and plant taxonomy. Chronica Bot, Waltham, Mass
Ernst A (1909) Apogamie bei *Burmannia coelestis*. Ber Dtsch Bot Ges 27:157–168
Ernst A, Bernard Ch (1912) Entwicklungsgeschichte des Embryosackes, des Embryos und des Endosperms von *Burmannia coelestis*. Ann Jard Bot Buitenz Ser 2 11:231–257
Flint FF (1957a) Megasporogenesis and megagametogenesis in *Forthergilla gardeni* Murr. and *F. major* Lodd. Trans Am Microsc Soc 76:307–311
Flint FF (1957b) Megasporogenesis and megagametogenesis in *Hamamelis virginiana* L. Va J Sci 8:185–189
Flint FF (1959) Development of the megagametophyte in *Liquidambar styraciflua* L. Madrono 15:25–29
Franz E (1908) Beiträge zur Kenntnis der Portulacaceen und Basellaceen. Bot Jahrb Syst 42:1–48
Friedrich HC (1955) Beiträge zur Kenntnis einiger Familien der Centrospermae. Mitt Bot Staatssamml Muench 2:56–66
Gamble JS (1928) Flora of the presidency of Madras, vol III. London
Gamble JS (1931) Flora of the presidency of Madras. Adlard and West Newman, London
George GP (1976) A comparative study of early ovule development, megasporogenesis, megagametogenesis, and an analysis of size variations in the megasporocyte and megagametophyte stages of field-grown and greenhouse-grown ovules of *Phaseolus aureus* Roxb. Thesis Univ South Carolina, Columbia
George GP, George RA, Herr JM Jr (1979) A comparative study of ovule and megagametophyte development in field-grown and greenhouse-grown plants of *Glycine max* and *Phaseolus aureus* (Papilionaceae). Am J Bot 66:1033–1043

George RA (1976) A comparative study of early ovule development, megasporogenesis, megagametogenesis and an analysis of size variations in the megasporocyte and megagametophyte stages of field-grown and greenhouse-grown ovules of *Glycine max* (L.) Merrill. Thesis Univ South Carolina, Columbia

Gerassimova-Navashina EH (1958) On the gametophyte and on the salient features of development and functioning of reproducing elements in angiospermous plants. Probl Bot (Leningrad) 3:125–167

Gerassimova-Navashina EH (1971) Double fertilization in angiosperms and some of its theoretical aspects. In: Zosimovich VP (ed) Problems in embryology. Naukova dumka, Kiev, pp 113–152

Gilg E (1893) Stachyuraceae. In: Engler A, Prantl K (eds) Die natürlichen Pflanzenfamilien, vol III, pt 6. Engelmann, Leipzig, S. 192–193

Gleason HA (1952) The new Britton and Brown illustrated flora of the northeastern United States and adjacent Canada. Hafner, New York

Glišic LM (1929) Über die Endosperm- und Haustorienbildung bei *Orobanche hederae* Duby and *O. gracilis* Sm. Bull Inst Jard Bot Univ Belgrade 2:106–141

Gulati N, Mathur S (1977) Embryology and taxonomy of *Filicium decipiens*. Phytomorphology 27:261–266

Gundersen A (1927) The Frankeniaceae as a link in the classification of dicotyledons. Torreya 27:65–71

Gundersen A (1950) Families of dicotyledons. Chronica Bot Co, Waltham Mass USA

Hall JB (1971) New Podostemaceae from Ghana with notes on related species. Kew Bull 26:125–136

Hallier H (1912) L'origine et le système phylétique des Angiosperms exposés à l'aide de leur arbre gégàalogique. Archs Neerl Sci 1:146–234

Hallier H (1923) Beiträge zur Kenntnis der Linaceen. Beih Bot Zbl 39:1–178

Hallock FA (1930) The relationship of *Garrya*. The development of the flowers and seeds of *Garrya* and its bearing on the phylogenetic position of the genus. Ann Bot 44:771–812

Hamann U (1962) Beitrag zur Embryologie der Centrolepidaceae mit Bemerkungen über den Bau der Blüten, Blütenstände und die systematische Stellung der Familie. Ber Dtsch Bot Ges 75:153–171

Hamann U (1963) Die Embryologie von *Philydrum lanuginosum* (Monocotyledoneae-Philydraceae). Ber Dtsch Bot Ges 76:203–208

Hamann U (1966) Embryologische, morphologisch-anatomische und systematische Untersuchungen an Philydraceen. Willdenowia Beih 4:1–178

Hamann U (1975) Neue Untersuchungen zur Embryologie und Systematik der Centrolepidaceae. Bot Jahrb Syst 96:154–191

Hegnauer R (1973) Chemotaxonomie der Pflanzen. Birkhaeuser, Basel

Heimsch CJ (1942) Comparative anatomy of the secondary xylem in the Gruinales and Terebinthales of Wettstein with reference to taxonomic groupings. Lilloa 8:83–198

Herr JM Jr (1967) On the nature of variation. Phytomorphology 17:200–207

Hieronymus G (1887) Eriocaulaceae. In: Engler A, Prantl K (eds) Die natürlichen Pflanzenfamilien, vol II, pt 4. Engelmann, Leipzig, pp 21–31

Hooker JD (1862) Burseraceae. In: Bentham G, Hooker JD (eds) Genera Plantarum, vol I, pt 1. Lovell Reeve, London, pp 321–327

Hooker JD (1867) Aizoaceae. In: Bentham G, Hooker JD (eds) Genera Plantarum, vol I, pt 3. Lovell Reeve, London, pp 854–856

Hooker JD (1876a) Vacciniaceae. In: Bentham G, Hooker JD (eds) Genera Plantarum, vol II, pt 2. Lovell Reeve, London, pp 564–577

Hooker JD (1876b) Anacardiaceae. In: Flora of British India, vol 2, pt 4. Lovell Reeve, London, pp 7–44

Hooker JD (1880) Chenopodiaceae. In: Bentham G, Hooker JD (eds) Genera Plantarum, vol III, pt 1. Lovell Reeve, London, pp 76–78

Hooker JD (1888) Burmanniaceae. In: Flora of British India, vol 5, pt 4. Lovell Reeve, London, pp 664–667

Huber H (1969) Die Samenmerkmale und Verwandtschaftsverhältnisse der Liliifloren. Mitt Bot Staatssamml Muench 8:219–538

Hutchinson J (1959) The families of flowering plants, vol 1, Dicotyledons. Clarendon, Oxford

Hutchinson J (1973) The families of flowering plants, 3rd edn. Clarendon, Oxford

Jacobsen H (ed) A handbook of succulent plants. Blandford, London

Johri BM (1963) Embryology and taxonomy. In: Maheshwari P (ed) Recent advances in the embryology of angiosperms. Intl Soc Plant Morphologists, Univ Delhi, Delhi, pp 395–444
Johri BM (1967) Angiosperm embryology and taxonomy. In: Symposium on newer trends in taxonomy, Bull No 34, Natl Inst Sci India, pp 263–268
Johri BM (Convener) (1970) Symposium on comparative embryology of angiosperms. Bull No 41, Indian Natl Sci Acad, pp 1–385
Joshi AC, Rao VR (1936) The embryology of *Gisekia pharnaceoides* Linn. Proc Indian Acad Sci Sect B3:71–92
Kapil RN, Bhatnagar AK (1980) Embryology in relation to angiosperm taxonomy. In: Nair PKK (ed) Glimpses in plant research, vol V. Vikas, New Delhi, pp 272–295
Kapil RN, Mohana Rao PR (1966a) Embryology and systematic position of *Theligonum* Linn. Proc Natl Inst Sci India Part B 32:218–232
Kapil RN, Mohana Rao PR (1966b) Studies of the Garryaceae 2. Embryology and systematic position of *Garrya* Douglas ex Lindley. Phytomorphology 16:564–578
Kapil RN, Prakash N (1969) Embryology of *Cereus jamacaru* and *Ferocactus wislizeni* and comments on the systematic position of the Cactaceae. Bot Not 122:409–426
Kapil RN, Sethi SB (1963) Development of male and female gametophytes in *Camellia sinensis* (L.) O. Kuntze. Proc Natl Inst Sci India Part B 29:567–574
Kapil RN, Vani RS (1966) *Nyctanthes arbor-tristis* Linn.: Embryology and relationships. Phytomorphology 16:553–563
Kapil RN, Vijayaraghavan MR (1965) Embryology of *Pentaphragma horsfieldii* (Miq.) Airy Shaw with a discussion on the systematic position of the genus. Phytomorphology 15:93–102
Kapil RN, Walia K (1965) The embryology of *Philydrum lanuginosum* and the systematic position of the Philydraceae. Beitr Biol Pflanz 41:381–404
Khanna P (1965) A contribution to the embryology of *Cyperus rotundus* L., *Scirpus mucrinatus* L., and *Kyllinga melanospora* Nees. Can J Bot 43:1539–1547
Knoblauch E (1897) Oleaceae. In: Engler A, Prantl K (Hrsg) Die natürlichen Pflanzenfamilien, vol IV, pt 2. Engelmann, Leipzig, S. 1–16
Koyama T (1961) Classification of the family Cyperaceae I. J Fac Sci Univ Tokyo 8:37–148
Kraüse K (1930) Liliaceae. In: Engler A, Prantl K (Hrsg) Die natürlichen Pflanzenfamilien, vol 15a. Engelmann, Leipzig, S. 227–386
Kshetrapal S (1970) A contribution of the vascular anatomy of the flower of certain species of the Salvadoraceae. J Indian Bot Soc 49:92–98
Lawrence GH (1951) Taxonomy of vascular plants. Macmillan, New York
Magnus P (1889) Najadaceae. In: Engler A, Prantl K (Hrsg) Die natürlichen Pflanzenfamilien, vol II, pt 1. Engelmann, Leipzig, S. 214–218
Magnus W (1913) Die atypische Embryonalentwicklung der Podostemaceen. Flora 195:275–336
Maheshwari P (1945) The place of angiosperm embryology in research and teaching. J Indian Bot Soc 24:25–41
Maheshwari P (1950) An introduction to the embryology of angiosperms. McGraw-Hill Book, New York
Maheshwari P (1963) Embryology in relation to taxonomy. In: Turrill WB (ed) Vistas in Botany. IV. Recent research in plant taxonomy. Pergamon, Oxford, pp 55–97
Maheswari Devi H (1962) Embryological studies in Gentianaceae (Gentianoideae and Menyanthoideae). Proc Indian Acad Sci Sect B 56:195–216
Maheswari Devi H (1964) Embryological studies in Asclepiadaceae. Proc Indian Acad Sci Sect B60:52–65
Maheswari Devi H (1971) Embryology of Apocynaceae I. Plumiereae. J Indian Bot Soc 50:74–85
Maheswari Devi H (1972) Salvadoraceae: A study of its embryology and systematics. J Indian Bot Soc 51:56–62
Maheswari Devi H (1975) Embryology of jasminums and its bearing on the composition of Oleaceae. Acta Bot Indica 3:52–61
Maheswari Devi H, Pulliah T (1975) Life history of *Basella rubra* Linn. and taxonomic status of the family Basellaceae. J Indian Bot Soc 54:154–166
Mathew CJ, Chaphekar M (1977) Development of female gametophyte and embryogeny in *Stachyurus chinensis*. Phytomorphylogy 27:68–79
Mauritzon J (1933) Über die systematische Stellung der Familien Hydrostachyaceae und Podostemaceae. Bot Not 1933:172–180

Mauritzon J (1939) Contribution to the embryology of the orders Rosales and Myrtales. Acta Univ Lund 35:1–121
Melchior H (1964) A Engler's Syllabus der Pflanzenfamilien, vol II. Borntraeger, Berlin
Metcalfe CR, Chalk L (1950) Anatomy of dicotyledons. Clarendon, Cambridge
Mohana Rao PR, Kaur A (1979) Embryology and systematic position of *Ophiopogon intermedius*. Proc Indian Natl Sci Acad Part B 45:175–187
Moseley MF Jr, Beeks RM (1955) Studies of the Garryaceae I. The comparative morphology and phylogeny. Phytomorphology 5:314–346
Muller J, Leenhouts PW (1976) A general survey of pollen types in Sapindaceae in relation to taxonomy. Linn Soc Symp Ser No 1:407–455
Nagendran CR (1974) Is the embryo sac of Podostemaceae bisporic? Curr Sci (Bangalore) 43:259–260
Nagendran CR, Arekal GD, Subramanyam K (1977) Embryo sac studies in three species of *Polypleurum* (Podostemaceae). Plant Syst Evol 128:215–226
Nagendran CR, Anand VV, Arekal GD (1980a) The embryo sac of *Podostemum subulatus* (Podostemaceae). Plant Syst Evol 134:121–125
Nagendran CR, Ramakrishna TM, Narayana BM (1980b) Development of embryo sac and endosperm haustoria in *Striga densiflora* Benth. (Scrophulariaceae). Indian J Bot 3:51–54
Narayana HS (1962) Seed structure in the Aizoaceae. In: Maheshwari P, Johri BM, Vasil IK (eds) Proc Summer School Bot Darjeeling. Min Sci Res Cultural Affairs, New Delhi, pp 220–230
Narayana LL, Rao D (1978) Systematic position of Humiriaceae, Linaceae, and Erythroxylaceae in the light of their comparative floral morphology and embryology: A discussion. J Indian Bot Soc 57:258–266
Padhye MD (1960) A contribution to the life history of *Kyllinga triceps* Rottb. Bot Soc Coll Sci Nagpur Bull 1:1–15
Padhye MD (1971) Studies in Cyperaceae 3. Life history of *Kyllinga brevifolia* Rottb. with a brief discussion on systematic position of *Kyllinga*. Bot Gaz 132:172–179
Palser BF (1975) The bases of angiosperm phylogeny: Embryology. Ann Mo Bot Gard 62:621–646
Pax F (1889) Aizoaceae. In: Engler A, Prantl K (Hrsg) Die natürlichen Pflanzenfamilien, vol III, pt 1b. Engelmann, Leipzig, S. 33–51
Pax F (1930) Haemodoraceae. In: Engler A, Prantl K (Hrsg) Die natürlichen Pflanzenfamilien, vol 15a. Engelmann, Leipzig, S. 386–391
Pilger R (1930) Philydraceae. In: Engler A, Prantl K (Hrsg) Die natürlichen Pflanzenfamilien, vol 15a. Engelmann, Leipzig, S. 190–191
Poddubnaya-Arnoldi VA (1976) Cytoembryology of angiosperms: Principles and perspectives (in Russian). Acad Sci SSR Nauka, Moscow
Posluszny U, Sattler R (1976) Floral development of *Zannichellia palustris*. Can J Bot 54:651–662
Poulsen VA (1894) Theligonaceae. In: Engler A, Prantl K (Hrsg) Die natürlichen Pflanzenfamilien, vol III, pt 1a. Engelmann, Leipzig, S. 121–124
Praglowski J (1973) The pollen morphology of the Theligonaceae with reference to taxonomy. Pollen Spores 15:385–396
Prakash N (1967) Aizoaceae: A study of its embryology and systematics. Bot Not 120:305–323
Prakash N (1969) The floral development and embryology of *Centrolepis fascicularis*. Phytomorphology 19:285–291
Prakash N, Lau YY (1976) Morphology of *Ploiarium alternifolium* and the taxonomic position of *Ploiarium*. Bot Not 129:279–285
Radlkofer L (1933) Sapindaceae. In: Engler A (Hrsg) Das Pflanzenreich, vol IV, pt 98d–f. Engelmann, Leipzig, S. 801–1274
Rendle AB (1952) The classification of flowering plants, vol II. Univ Press, Cambridge
Rousi A (1967) Cytological observations on some species and hybrids of *Vaccinium*. Zuchter Genet Breed Res 36:352–359
Santapau H (1960) The flora of Khandala, edn 2. Bot Survey India, Calcutta
Schulz OE (1931) Erythroxylaceae. In: Engler A, Prantl K (Hrsg) Die natürlichen Pflanzenfamilien, vol 19a. Engelmann, Leipzig, S. 130–143
Sedgwick LJ (1918) The Cyperaceae in Bombay Presidency. J Bombay Nat Hist Soc 25:192–209
Sethi SB (1965) Structure and development of seed in *Camellia sinensis* (L.) O. Kuntze. Proc Natl Inst Sci India Part B 31:24–33
Shah CK (1963) The life history of *Juncus bufonius* Linn. J Indian Bot Soc 42:238–251

Shah CK (1965) Embryogeny in some Cyperaceae. Phytomorphology 15:1–9
Singh V (1965) Morphological and anatomical studies in Helobiae 2. Vascular anatomy of the flower of Potamogetonaceae. Bot Gaz 126:137–144
Smith BB (1973) The use of a new clearing technique for the study of early ovule development, megasporogenesis and megagametogenesis in five species of *Cornus* L. Am J Bot 60:322–338
Smith BB (1975) A quantitative analysis of the megagametophyte of five species of *Cornus* L. Am J Bot 62:387–394
Stant MY (1952) Anatomical evidence for including *Nyctanthes* and *Dimetra* in the Verbenaceae. Kew Bull 1952:273–276
Stevens PF (1971) A classification of Eriaceae: Subfamilies and tribes. J Linn Soc Lond Bot 64:1–53
Stushnoff C, Palser BF (1969) Embryology of five *Vaccinium* taxa including diploid, tetraploid, and hexaploid species or cultivars. Phytomorphology 19:312–331
Takhtajan AL (1959) Die Evolution der Angiospermen. Fischer, Jena
Takhtajan AL (1969) Flowering plants: Origin and dispersal. Oliver and Boyd, Edinburgh
Takhtajan AL (1976) Neoteny and the origin of flowering plants. In: Beck CB (ed) Origin and early evolution of angiosperms. Columbia Univ Press, New York, pp 207–219
Takhtajan AL (1980) Outline of the classification of flowering plants (Magnoliophyta). Bot Rev 46:225–359
Teriokhin ES (1980) Embryology of Orobanchaceae 3. *Mannagettaea hummelii*. Bot Zh (Leningr) 64:1390–1400
Thorne RF (1976) A phylogenetic classification of the Angiospermae. Evol Biol 9:35–106
Tiagi B (1956) A contribution to the embryology of *Striga orobanchoides* and *S. euphrasioides*. Bull Torrey Bot Club 83:154–170
Tippo O (1938) Comparative anatomy of the Moraceae and their presumed allies. Bot Gaz 100:1–99
Tiwari SC, Bouman F, Kapil RN (1977) Ovule ontogeny in *Tropaeolum majus*. Phytomorphology 27:350–358
Ulbrich E (1933) Gattung *Theligonum*. Notizbl Bot Gart Mus Dahlem 11:889–902
Venkateswarlu J, Maheswari Devi H (1973) An embryological approach to the taxonomic status of Vacciniaceae. Proc Indian Acad Sci 78B:282–291
Vijayaraghavan MR, Kumari AV (1974) Embryology and systematic position of *Zannichellia palustris* L. J Indian Bot Soc 53:292–302
Volkens G (1893) Basellaceae. In: Engler A, Prantl K (Hrsg) Die natürlichen Pflanzenfamilien, vol III, pt 1a. Engelmann, Leipzig, S. 124–128
Walia K, Kapil RN (1965) Embryology of *Frankenia* Linn. with some comments on the systematic position of the Frankeniaceae. Bot Not 118:412–429
Warming E (1895) A handbook of systematic botany. Sonnenschein, London
Watson L (1965) The taxonomic significance of certain anatomical variations among Ericaceae. J Linn Soc Lond Bot 59:111–125
Went FAFC (1908) The development of the ovule, embryo sac and egg in Podostemaceae. Recl Trav Bot Neerl 5:1–16
Wettstein R (1935) Handbuch der systematischen Botanik, edn 4. Deuticke, Leipzig Wien
Wiegand KM (1898) Notes on the embryology of *Potamogeton*. Bot Gaz 25:116–117
Willis JC, revised by Airy Shaw HK (1973) A dictionary of the flowering plants and ferns, edn 8. Cambridge Univ Press, Cambridge
Wunderlich R (1971) Die systematische Stellung von *Theligonum*. Oesterr Bot Z 119:329–394

15 Homologies and Phylogeny

M. Favre-Duchartre

15.1 Introduction

The structures and functions described in earlier chapters give a good idea about our present knowledge of sexual reproduction in angiosperms.

Where did angiosperms come from? Are they monophyletic as considered by Takhtajan (1969), or polyphyletic as proposed by Emberger (1949)? What are the equivalents of these sexual and seed elements in taxa which preceded the angiosperms? In these taxa, are there phenomena comparable to double fertilization and endosperm formation in angiosperms?

Within the plant kingdom, sexual reproduction appears to have a long history, and the living plants afford testimonies of different evolutionary levels from the most primitive up to the most advanced taxa.

The cytological, anatomical, and morphological structure of the cells, organs, and organisms concerned in sexual reproduction, as well as the biological processes of their individual development, mutual competition, and reciprocal influences are adequate bases to trace the phylogeny of plants.

The first approach to that reconstruction consists in establishing evolutionary levels (cf. Emberger 1971). Schizophytes, thallophytes, and cormophytes (Fig. 15.1) are, respectively, characterized by: (1) asexuality, (2) sporocytes and gametocytes being naked cells, (3) sporangia and gametangia (antheridia, archegonia) in which, respectively, sporocyte(s) and gametocyte(s) differentiate, associated with at least one sterile cell.

Archegonia have been estimated to be sexual organs so typical of the cormophytes that the latter are designated "archegoniates"; but they might just as well have been called "antheridiates" because the antheridium in cormophytes is characteristic in the same way as are their archegonia.

Schizophytes, which are protokaryotes, are quite apart, much below eukaryotes.

Through the eukaryotes, the only speciality yet acquired is that amongst the Chlorophyceae the cormophytes did find their stock in Palaeozoic epoch: the "green algae" and the "plants with a stem" together constitute the enormous phylum chlorophytes (from green algae to angiosperms). The specificity of chlorophytes is supported by three characteristics: (1) unmasked chlorophylls, (2) intraplastidial starch as glucide reserve, and (3) when zooids are present, their flagella are of equal size. Graham and McBride (1979) write: "The structure of the spermatozoid of *Coleochaete* supports the hypothesis that this alga may be relatively close to the phylogenetic line which led directly to archegoniates". According to Robbins and Carothers (1978), the spermatozoid of *Lycopodium* most closely resembles that of *Coleochaete*.

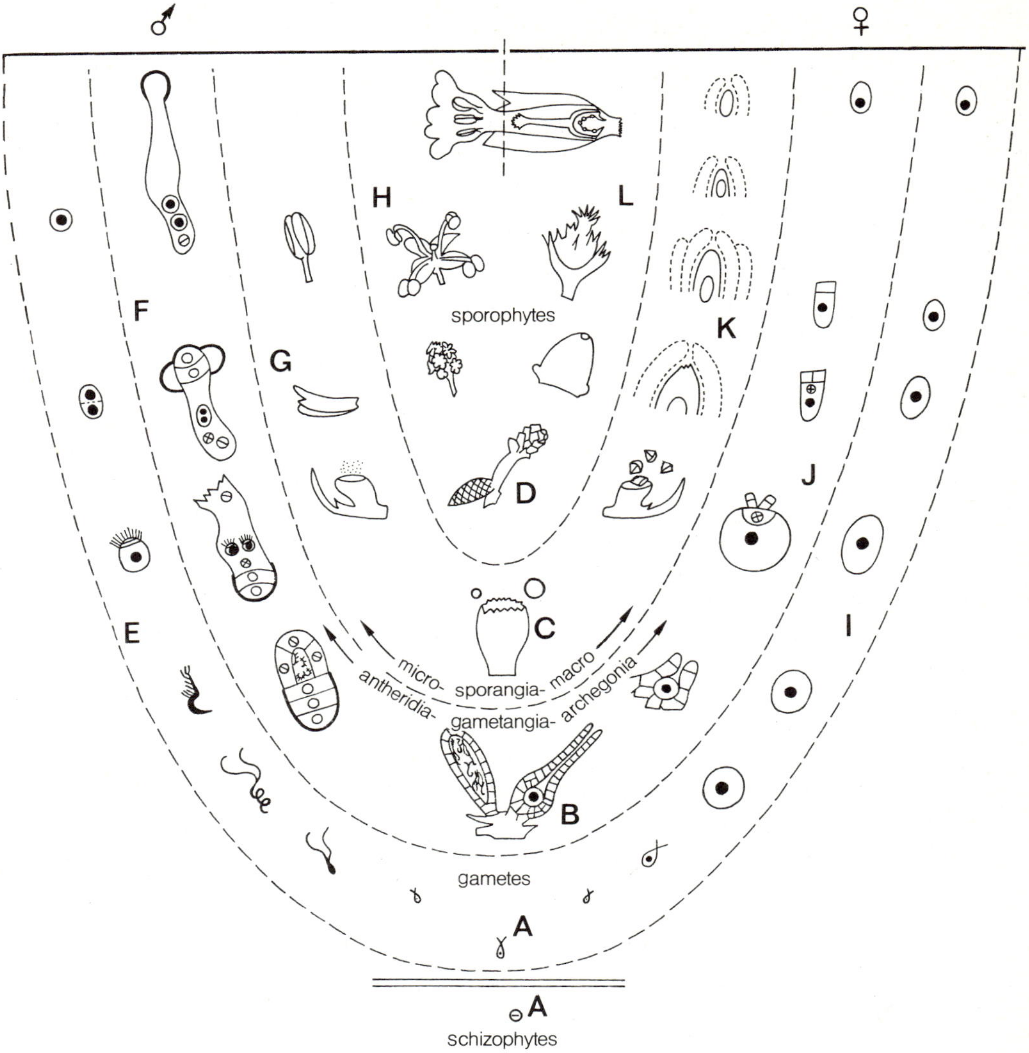

Fig. 15.1 A–L. Plant sexuality and proposed phylogeny. Relative enlargements have no correlation. **A** Asexuality in Schizophytes (aplanospore) and some Chlorophyceae (zoospore). **B** Organization of gametophytes; bisexuality is represented by antheridia and archegonia on one and the same gametophyte (*Atrichum*). **C** Organization of sporangia; one sporangium produces simultaneously micro- and megaspores (*Macromitrium*). **D** Organization of sporophytes; one sporophyte bears male and female strobili (*Pinus*); hermaphrodite flower (*Primula*). Evolution of male structures. **E** Male sexual cells (spermatozoids and sperm cells). **F** Male haploid organisms (male gametophytes, pollen tubes). **G** Microsporangia (*Selaginella, Pinus,* angiosperms). **H** Unisexual male sporophytes of gymnosperms (*Taxus*), angiosperms (*Urtica*), and androecium of self-incompatible angiosperms (*Primula*). Evolution of female structures. **I** Female sexual cells (anisogamous flagellate receptive female gamete and egg cells). **J** Archegonial structures (Polypodiaceae, *Ginkgo, Cephalotaxus, Taxus, Gnetum* and angiosperms). **K** Megasporangia (*Selaginella*) and nucelli in gymnosperms and angiosperms. **L** Unisexual female sporophytes (*Taxus, Urtica*) and gynoecium of self-incompatible angiosperms (*Primula*). After a marked specialization through Chlorophyceae, bryophytes, pteridophytes, and prespermaphytes, simplification occurs through gymnosperms and angiosperms due to neoteny as documented in Figs. 15.4, 15.8, and 15.9

We should look for evolutionary scales and, if possible, recognize the more precise trends *before* the evolution of angiosperms, and *within* the angiosperms (220,000 species). The search for homologous structures and biological organizations in chlorophytes, more ancient than angiosperms [Chlorophyceae, Bryophyta, Pteridophyta, Prespermaphyta (*Ginkgo,* cycads) and Gymnospermae], is considered "extrinsic". And within angiosperms themselves, evolution has been and is still at work: the comparison between more or less advanced structures and developments is considered "intrinsic" (Fig. 15.1).

The sexuality in angiosperms vis-a-vis the plant kingdom is represented in Fig. 15.1, and will be commented upon in the following discussion.

15.1.1 Sexuality and Overall Phylogeny

Some Cyanophyceae produce aplanospores, but schizophytes seem to lack true fertilization.

In Chlorophyceae the evolutionary sequence appears to be (1) isogamy, (2) anisogamy, and (3) oogamy. Oogamy has been acquired by all cormophytes in which spermatozoids are protected in antheridia, and egg cells in archegonia.

Once these gametic specializations are phylogenetically established, the next step is the unisexual condition, i.e., dioecy. In mosses *Atrichum* gametophytes bear both antheridia and archegonia; *Polytrichum* gametophytes are dioecious but they still develop from isospores, i.e., meiospores morphologically similar whether they are male or female; *Macromitrium salakanum* is heterosporous (Augier 1966), but microspores and megaspores are produced in one and the same sporangium (unique for a sporophyte) (Fig. 15.1).

In pteridophytes heterospory is more common and, except in *Equisetum arvense,* the sporangia attain unisexuality.

In ovulate plants specialization into microsporangia and megasporangia is quite common. In plants with naked ovules, megasporangia are often aggregated into female strobili and microsporangia into male strobili; in teratologic cases the same strobilus is male at the base and female upward (*Araucaria*). (This disposition is opposite to the condition in the strobili of ligulate lepidophytes.) With sporophytes the female strobili are located higher up on the tree, male ones lower down (*Pinus*). During the entire lifetime of the sporophyte, the latter remains sterile at first, for several years; then it produces male strobili only, and after 1 or 2 years, also female strobili (*Pinus*). The maximum sexual separation occurs in dioecious sporophytes (*Ginkgo, Taxus* ...) (Fig. 15.1). In plants with naked ovules, about 30% of the species are dioecious (Favre-Duchartre 1968).

In angiosperms the gradient, male at the base, female upward, encountered in gymnospermous trees as well as in teratologic strobili, occurs in hermaphrodite flowers.

In angiosperms 13% of the species are dioecious (*Urtica*), but self-incompatibility (*Primula*) favors cross-pollination. Self-incompatibility does not exist in plants with naked ovules, probably owing to the absence of stigma and style. This acquisition in angiospermous species has the advantage of a mixing of genomes without the loss of seed production, which is implied in dioecy by the existence of strictly male sporophytes (Fig. 15.1). At most levels of the phylum Chlorophyta, some taxa have lost sexual reproductive ability, and propagate vegetatively.

Thus, it appears that (1) gametes, (2) gametophytes, (3) mega- and microspores, (4) sporangia, and (5) sporophytes are successively becoming unisexual. However, several self-fertilization barriers in angiosperms lead to outcrossing problems which represent a general evolutionary trend. Finally, the separation of sexes is no longer established between individual sporophytes, but inside one and the same sporophyte (*Primula*) (Fig. 15.1).

We may now focus our attention on angiosperm micro- and megasporangia which produce male and female gametophytes. These produce gametes, fertilization occurs, and zygote and primary endosperm nucleus are formed leading to the development of embryo and endosperm in the ovule, which becomes the seed protected by ovary wall (fruit).

15.2 Microsporangia

In angiosperms microsporangia are grouped into anthers. A comparison between ontogeny of the microsporangium wall and the spore content reflects their phylogeny.

15.2.1 Microsporangium Wall

Figure 15.2 is based on the descriptions of Davis (1966), and Foster and Gifford (1974).

The angiosperm microsporangia are of the eusporangiate type; they originate from the periclinal divisions of archesporial cells. These archesporial cells are situated in the superficial layer of microsporangia in *Ophioglossum* (Fig. 15.2 A) and *Marattia* (eusporangiate pteridophytes), as well as in cycads and some primitive gymnosperms such as *Cedrus* and *Pseudotsuga;* while in *Cupressus, Taxus* (Fig. 15.2 B) and all angiosperms the archesporial cells are hypodermal.

The mechanical layer (exothecium) differentiates superficially in naked-ovulate plants, and hypodermally (and for that reason it is called endothecium) in angiosperms (Fig. 15.2 D′–G′). Therefore, in angiosperms the epidermis of anther is poorly differentiated.

As for tapetum, Audran (1977, 1978) has shown that it differentiates from peripheral layers of the sporogenous tissue in cycads, while Maheshwari (1950) asserts that in angiosperms the tapetum differentiates from the internal wall layers. Gupta and Nanda (1978) point out that, in angiosperms, the tapetum shows structural and functional homogeneity, although that layer has a dual origin: "inner parietal layer of microsporangium on the outer side (toward epidermis), while on the inner side it is contributed by connective cells". This is the usual feature in angiosperms. Eames (1961) remarks that the tapetum is typically uniseriate, rarely biseriate, and often irregular. I presume that the tapetum essentially orig-

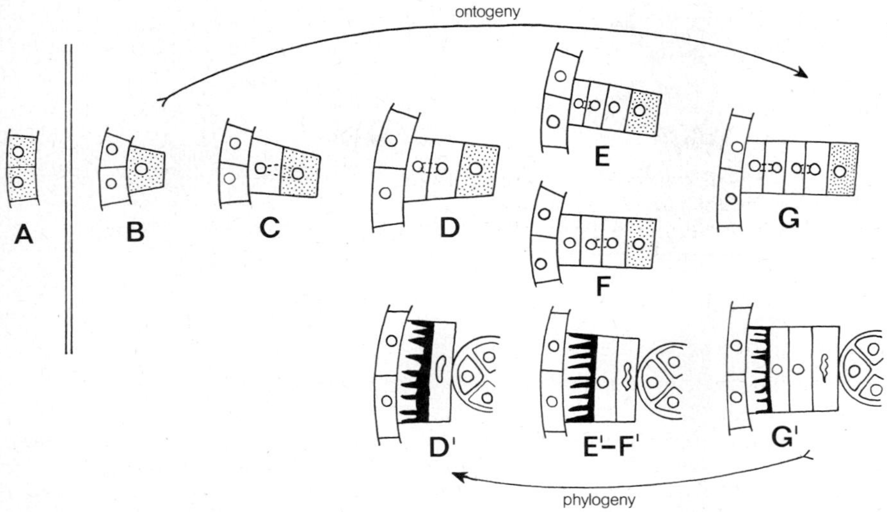

Fig. 15.2 A–G. Ontogeny of microsporangium wall in different taxa (*upper arrow*). **A** *Ophioglossum* and *Cedrus,* archesporial cells (*dotted*) are marginal. **B** Subepidermal in *Taxus* and angiosperms. **C** Ephemeral periclinal division of archesporial cell. **D–G** Four types of ontogeneses giving rise to four types **G′–D′** of microsporangia: **G′** Basic type, **E′** Dicot, **F′** Monocot (**E′** and **F′** are the same convergent result of both **E** and **F** ontogenic processes), and **D′** Reduced type. The *lower arrow,* labelled "Phylogeny", stresses the fact that the more advanced is the microsporangium type the earlier it is achieved

inates owing to a reaction of degeneration of the cells adjoining the functional microsporocytes.

Projections of sterile plates into the sporogenous tissues are commonly met with in Lamiaceae, and some other gamopetalous families. Such plates divide, more or less completely, the male sporogenous tissue as is characteristic of the microsporangia of *Isoetes*.

However, Eames (1961; p. 127) emphasizes that in angiosperms the microsporangia are "morphologically naked", being sunken into the mesophyll so that some of the sterile tissue of the laminar sporophyll would constitute the protective layers. Periasamy and Swamy (1964) state that "the chief attribute of a morphologically true sporangial wall is that it should arise from the same initial as the spore mother cell". Figure 15.2 confirms that this is the case in the microsporangia of angiosperms.

15.2.2 Microsporangium Content

The numerous pollen mother cells give rise (theoretically) to four microspores each which usually dissociate, and each divides unequally to produce a generative and a tube cell (Fig. 15.3).

Let us point out some deviations: (1) In monads (Fig. 15.3 A) (Cyperaceae) there is one functional microspore without any postmeiotic wall formation. (2) In a bicelled pollen grain (Fig. 15.3 B) (*Cuscuta*) two haploid nuclei are functional.

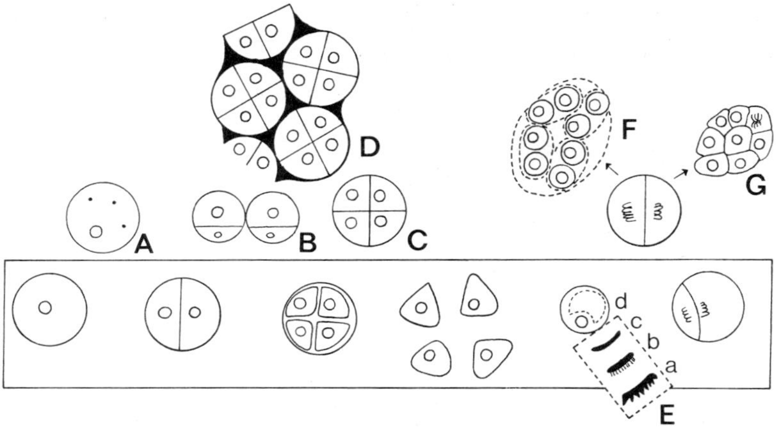

Fig. 15.3 A–G. Development of microsporocyte to microspores and pollen grains (*rectangular frame*); some deviations may occur. **A** Monad (Cyperaceae). **B** Bicelled pollen grains (*Cuscuta*). **C** Coherent tetrad (Ericaceae). **D** Pollinia (orchids). **E** The thinner the exine around the uninucleate microspores, the more advanced is the taxon; *a* Pteridophytes, *b* Prespermaphytes, *c* Gymnosperms, *d* Angiosperms. **F** Octads (*Hyphaene*). **G** Embryoids are initiated by an equipotential partition of microspore (Solanaceae); in vitro behaviour

(3) When the microspores remain attached (Fig. 15.3 C) in tetrads (Ericaceae), they represent a neotenic condition as compared to the separated four microspores. (4) The neotenic process is further accentuated in pollinia (Fig. 15.3 D) where the tetrads remain attached permanently. This disposition represents an advanced character as compared to individual pollen grain. Comparing the number of pollen grains (P) to the number of ovules (O) in the same flower, the ideal (never attained) P/O ratio would be equal to one; it has been estimated to be 57,310 for *Secale* with powdery pollen, less than 11 in *Asclepias* producing pollinia (Cruden 1977). So the nonfunctional pollen grains are much fewer with agglomerated (tetrad, pollinia) pollen grains, and this represents an obvious progress. (5) Comparing the sporoderm of dissociated microspores of lower taxa to those of angiosperms, Audran and Masure (1977) pointed out that microspores are liberated from tetrads (a) ready for germination in pteridophytes, (b) covered with a thick exine and giving rise to pollen grains inside the microsporangia in prespermaphytes, (c) with a thinner exine in gymnosperms, and (d) only with primexine and acquiring, lately, their true exine after vacuolation in microspores in angiosperms (Fig. 15.3 E). (6) Instead of dividing unequally to form generative and tube cells, each microspore may divide equally, giving rise to four pairs of microspores (Fig. 15.3 F) grouped into an octad (*Hyphaene*). (7) In in vitro cultures of anthers in Solanaceae, some microspores divide equally, undergo coherent proliferation, and develop into embryoids (Fig. 15.3 G) which grow into plants. This is reported in many other taxa.

15.2.3 Microsporangium Number and Position

As for the number and disposition of microsporangium in archegoniate microsporophylls, there is (1) one adaxial microsporangium in ligulate lepidophytes, (2) two or more abaxial microsporangia in prespermaphytes and gymnosperms, and (3) generally four microsporangia per microsporophyll in angiosperms. Eames (1961) has pointed out that in the primitive Ranalian taxa the microsporangia are in adaxial, marginal, or abaxial position.

The structure, number, position, or dehiscence of microsporangia of angiosperms do not give any indication of their unknown ancestors, except of the eusporangiate character.

15.3 Megasporangia

At the angiosperm phylogenetic level, megasporangia are called nucelli.

As with microsporangia, let us consider the (1) general structure, (2) ontogenesis, (3) tapetal layer (endothelium), and (4) spore content of the female eusporangia.

15.3.1 General Structure

Figure 15.4 represents the evolutionary tendencies of the megasporangium (nucellus), (1) extrinsic to angiosperms, at lower archegoniate levels, (2) intrinsic to angiosperms. These structural comparisons do not imply any relationship between the considered taxa. A naked megasporangium in *Selaginella* (Fig. 15.4 A), with apical dehiscence, may be proposed as a primitive structure.

At the prespermaphyte level, the carboniferous *Lagenostoma* (Fig. 15.4 B) presents an integumented megasporangium, i.e., an ovule which is no longer able to liberate its unique megaspore. However, the nucellar apex in *Lagenostoma* is differentiated into a vestigial apparatus for dehiscence. The latter consists of a central parenchymatous core free in the axis of a cylindric pit, the pollen chamber, surrounded by the persistent, apically sclerified marginal nucellar layer. The so-called lagenostome "appears to represent the earliest type of pollen-receiving mechanism that functioned in passive manner" (Taylor and Millay 1979, p. 333). From the latter primitive nucellar apex, two trends may be considered (cf. Favre-Duchartre 1963).

15.3.1.1 Nucellar Apex: Central Parenchymatous Core

Lagenostoma, with its parenchymatous core, may represent the starting point of this series. The parenchymatous core extrudes through the micropyle of the ovule of *Araucaria* (Fig. 15.4 C); some tanniferous epidermal cells are present as a nucellar epidermis in the region covered by the integument (cf. Favre-Duchartre 1960). *Saxegothaea* (Fig. 15.4 D) and *Cedrus* (Fig. 15.4 E) also have extruded nucellar apices which receive pollen grains or pollen tubes. In some angiosperms too, such as *Malva* (Fig. 15.4 F), *Hibiscus* and teratologic cases in *Rosa,* extruded nu-

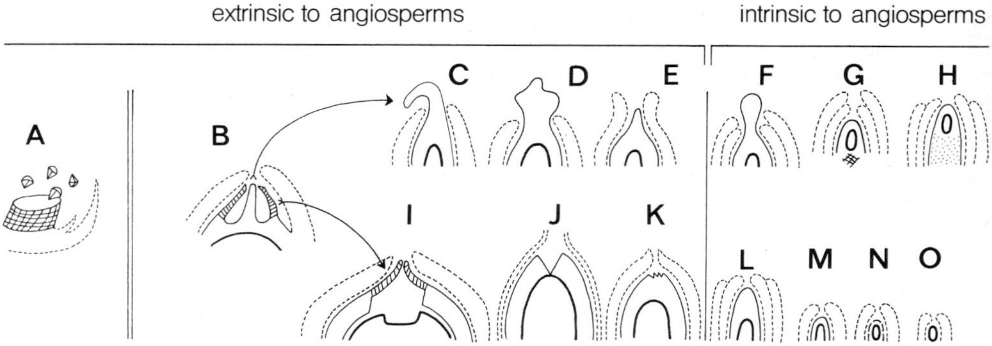

Fig. 15.4 A–O. Phylogenetic trends in megasporangia. **A** *Selaginella,* naked megasporangium liberates four megaspores. **B** *Lagenostoma,* integumented megasporangium; only one megaspore produces female gametophyte in situ. **C** *Araucaria,* **D** *Saxegothaea,* **E** *Cedrus,* and **F** *Malva* indicate the evolution of nucellus. **G, H** Nucellus with a hypostase, or a perisperm, demonstrate persistence of parenchymatous tissue. **I** *Cycas,* **J** *Ephedra,* **K** *Taxus* demonstrate evolutionary reduction in empty pollen chamber. **L–O** Illustrate evolutionary reduction from crassinucellate to tenuinucellate, ephemeral, and anucellate ovules

celli are met with (cf. Favre-Duchartre 1963). Extension of nucellus beyond the micropyle is a regular feature in some taxa as *Acalypha* (Johri and Kapil 1953).

There are also basal nucellar differentiations: (1) A cup-shaped hypostase has been described in *Ephedra* by Moussel (1971), and is present in a number of angiosperms (Fig. 15.4G). Loranthaceae (Johri and Bhatnagar 1972) is a good example. (2) Podostemaceae do not develop any endosperm at all (as we shall discuss later; cf. Fig. 15.15G_1) but "The nucellar cells below the megaspore mother cell enlarge longitudinally, and their walls become thin and fragile" (Bhojwani and Bhatnagar 1983, p. 181), soon constituting a nutritive structure (nucellar plasmodium) wrongly called pseudo-embryo sac. (3) In Piperaceae and some other families the nucellar tissue situated below the embryo sac enlarges and persists up to the ripe-seed stage, constituting an actual nutritive tissue called perisperm (Fig. 15.4H).

15.3.1.2 Marginal Cells Delimit an Empty Pollen Chamber in Nucellar Apex

In living prespermaphytes, i.e., cycads and *Ginkgo,* there is only a superficial lignified layer at the nucellar apex, while most of the subjacent parenchymatous tissue is destroyed, leaving an empty pollen chamber (Fig. 15.4I).

Gymnosperms are siphonogamous and, curiously, a pollen chamber is still well represented in *Ephedra* (Fig. 15.4J), one of the advanced genera. However, this apical pit is no longer wide at its base but infundibuliform, its narrow base touching the edge of the female prothallium, and there is no trace of the superficial lignified nucellus. In some other gymnosperms the nucellar apices are poorly eroded; the small excavation left by the collapsed cells represents the vestiges of a pollen chamber (Fig. 15.4K).

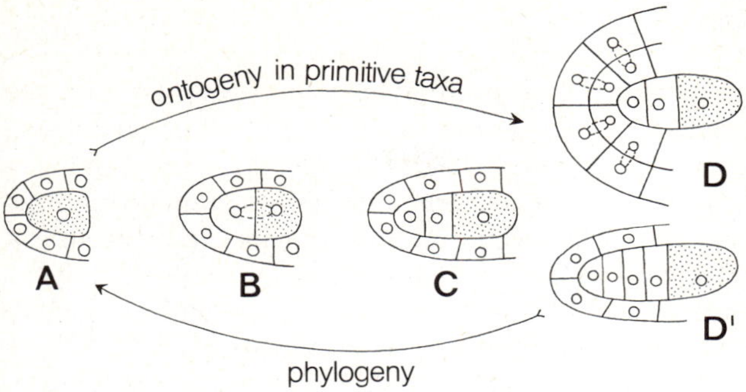

Fig. 15.5 A–D′. In angiosperms subepidermal female archesporium (*dotted*) gives rise to megaspore mother cell; *upper arrow* indicates more and more delayed differentiation of megasporocyte. The earlier the differentiation, the more advanced the taxon; *lower arrow* indicates ontogenic reduction

In some angiosperms the nucellus represents what I interpret as an archaic character. In *Ornithogalum caudatum* (Tilton and Lersten 1981) the radial and tangential walls of nucellar epidermis are thickened at their proximal end and polarizing optics indicates that they are probably lignified ... reminding of a mechanical layer of pteridophyte megasporangium. However, in angiosperms, there is a progressive elimination of nucellus up to complete disappearance, which is in direct relation to taxonomic advancement: (1) in primitive taxa the ovules are crassinucellar because their domelike apices are thick; (2) in most gamopetalous dicotyledons the ovules are tenuinucellar and a single nucellar epidermal layer covers the embryo sac; (3) in Asteraceae (*Cichorium* – cf. Favre-Duchartre et al. 1979) this thin nucellar layer collapses by the time the megaspore mother cell has undergone meiosis; (4) Guignard (1917) has shown in Apocynaceae, and Fagerlind (1937) in Rubiaceae, how through some genera a series may be proposed, suggesting the phylogenetic reduction of nucellus up to its complete disappearance (see Fig. 15.4 L–O).

15.4 Ontogeny of Nucellus

Figure 15.5 shows how, during phylogeny, ontogeny becomes more and more reduced. The periclinal divisions of epidermal layer may create what Davis (1966) calls a “pseudo-crassinucellar” ovule (*Anemone*) which may be associated with the division of primary parietal cell (*Nigella*). In *Glottidium* (Papilionaceae) (Prakash and Herr 1979) about eight layers of epidermal origin and five of subepidermal origin distinctly separate the mature embryo sac from the micropyle.

When the phylogeny of female sporogenesis is compared to the male (Fig. 15.2), it appears that there are two steps toward precocious development of tenuinucellar ovules since the hypodermal archesporial cell functions directly as megasporocyte. Whereas, in microsporangia, the archesporial cell always produces the endothecium, middle layer/s, tapetum, and sporogenous cells. The anucellate condition (Fig. 15.4 O) shows the maximum reductional advance of the ovule.

15.4.1 Tapetum/Endothelium

Prespermaphytes and most gymnosperms show a "spongy tissue" around the megaspore and female gametophyte. In *Araucaria* several nucellar layers surround the female prothallium; their contents are consumed and, finally, they become empty but their walls persist. In angiosperms the endothelium around the embryo sac commonly disappears, whereas, in microsporangia the tapetum is a conspicuous feature. However, Davis (1966, p. 16) points out that "persisting lateral nucellar cells elongate radially and form a jacket around the embryo sac" in all Begoniaceae and Droseraceae. In tenuinucellar ovules, such as *Cichorium* (Favre-Duchartre et al. 1979), the inner epidermis of integument precociously comes into contact with the embryo sac and differentiates into a jacket-like "endothelium".

15.4.2 Megasporangium Content

The megasporangium of *Selaginella* liberates 16, 8, or 4 megaspores (Fig. 15.4). In ovulate plants the megaspores are no longer free. Figure 15.6 represents a proposed series, from left to right, of primitive to more advanced megaspore formation. This series first takes into account a reduction in the number of megasporocytes: in *Casuarina* (Fig. 15.6 A) several megasporocytes are functional; in *Rubus* (Fig. 15.6 B) any one among the four megaspores produces the embryo sac. Then specialization occurs. The micropylar position of functional megaspore(s) is considered more advanced than the chalazal because (1) it is associated with ontogenetic reduction of embryo sac (*Oenothera* is tetracellular); (2) there appears to be a fundamental tendency for the differentiation of progressively superficial megaspores (cf. Fig. 15.6 C–G). In bisporic (Fig. 15.6 E–G), still more in tetrasporic (Fig. 15.6 H), the absence of individual megaspore is considered a neotenic advancement since the megaspores become functional progressively earlier in their own development. The progress of tetrasporic development is confirmed by the fact that it is met with in *Gnetum* and *Welwitschia,* the most advanced genera of gymnosperms.

The history of megasporangia may reflect on the problem of angiosperm ancestry. Their contribution is similar to that of microsporangia: the universal occurrence of unisexual integumented nucelli (eusporangia) confirms the highly advanced level of angiosperms.

15.4.3 A Proposed Macroevolution of Meiosporic Structures

Figure 15.7 represents an overall evolution of some meiosporic levels of organization (cf. Favre-Duchartre 1982 a). Naked unisexual sporocyte in dioecious Charophyceae (Fig. 15.7 A), bisexual sporangium of heterosporous Muscinae (*Macromitrium;* Fig. 15.7 B), bisexual sorus of Marsileaceae with apical microsporangium and subjacent macrosporangia (Fig. 15.7 C), and unisexual sori of *Azolla* (Fig. 15.7 D) form a series. In ovulate plants megasporangia (called nucelli)

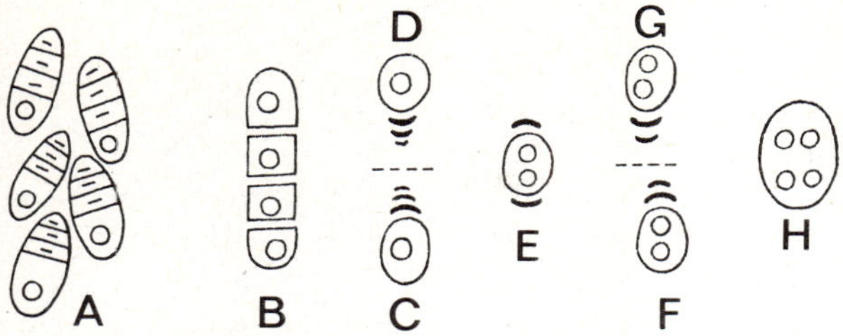

Fig. 15.6 A–H. Specialization and neoteny during megasporogenesis in angiosperms. **A** *Casuarina*, several megaspore mother cells differentiate and produce tetrads of monosporic type. **B** *Rubus*, megaspore mother cell produces four potential megaspores. **C** *Polygonum*, chalazal megaspore functions. **D** *Oenothera*, micropylar megaspore functions. **E** *Tridax*, two median megaspore nuclei function. **F** *Allium*, two lower megaspore nuclei function. **G** *Endymion*, two upper megaspore nuclei function. **H** *Adoxa*, four megaspore nuclei function (coenomegaspore)

are always separated from microsporangia (Fig. 15.7 E–K). The nucelli are considered the more advanced the more neotenically reduced they are, until the nucellus does not differentiate at all or disorganizes at a very early stage (cf. Fig. 15.7 E–K). In Hydropteridae (*Marsilea* C′, *Azolla* D′) microspororangia are protected in sporocarps, and they have only a two-layered wall, i.e., epidermis and periplasmodium. However, in ovulate plants (Fig. 15.7 F′–H′), the microsporangial wall is rarely less than four layers, i.e., epidermis, endothecium, middle layer, and tapetum; it appears three-layered if the middle layer is lacking. The leptosporangiate character of Filicinae, the fact that their microsporangial wall is thinner than those of the highly advanced angiosperms, *forbids* looking for the origin of angiosperms among the leptosporangiate ferns.

15.5 Male Gametophyte

In angiosperms male gametophytes develop as "pollen grains" inside the microsporangia, and they produce "pollen tubes" inside the receptive carpels which enclose the ovules. Before discussing their structure, let us try to understand the main steps of evolution of the male gametophytes.

15.5.1 Extrinsic to Angiosperms

In the living plants let us examine (Fig. 15.8) the evolution of male gametophyte in earlier groups. (1) Naked condition of spermatocytes in Chlorophyceae (*Coleochaete*, Fig. 15.8 A). (2) Structure of antheridium in isosporous Cormophytes (*Lycopodium*; Fig. 15.8 B). (3) The entire male gametophyte is protected by the microspore exine and is viable for a long time (*Selaginella*; Fig. 15.8 C). (4) Prothallial cells persist under the microspore wall, while the antheridium protrudes (*Azolla*; Fig. 15.8 D). (5) In prespermaphytes the male gametophytes have be-

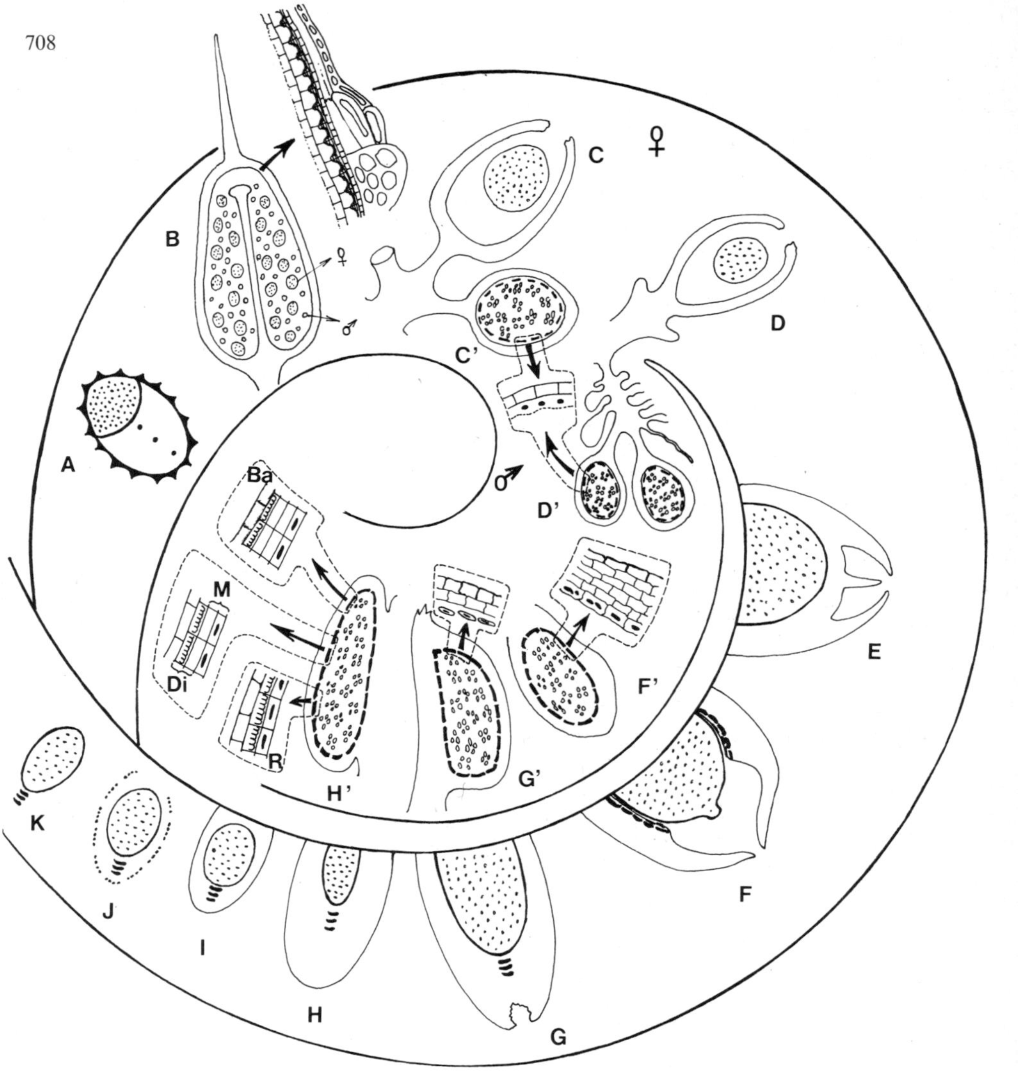

Fig. 15.7 A–K. Organization of meiosporic structures in some plants. **A** Naked potential mega- or microspore of a higher thallophyte. **B** Bisexual sporangium of a heterosporous moss. **C, C′** Bisexual, and **D, D′** unisexual sori of heterosporous Filicinae. **E–J** Megasporangia of ovulate plants, designated nucelli. Apical vestigial dehiscence in **E–G**. There are several layers of wall in **H,** one layer in **I**, and an ephemeral layer in **J. K** Naked female tetrad. In ovulate plants the microsporangia **F′–H′** have at least a three-layered wall.

A *Nitella,* tetrad of meiospores, the functional one (*dotted*) gives rise either to female or male gametophyte. **B** *Macromitrium,* unique apical sporangium with large megaspores and small microspores which will be liberated by apical dehiscence (predehiscent stage indicated by *arrow*). **C, C′** *Marsilea,* portion of a bisexual sorus with apical megasporangium and subjacent microsporangium. *Arrow* points to detail of wall. **D** *Azolla,* apical megasporangium and subjacent aborted sporangia. **D′** *Azolla,* portion of a sorus in microsporocarp. Apical sporangium aborted and basipetally maturing microsporangia. Wall as in microsporangia of *Marsilea.* **E** *Lagenostoma,* nucellus with female gametophyte (*dotted*), and apical circular pollen chamber. **F** *Ginkgo,* megasporangium-nucellus. **F** *Gingko,* microsporangium with details of several-layered wall. **G** *Thuja,* nucellus with eroded apex. **G′** *Thuja,* microsporangium with detail of three-layered wall including exothecium. **H′** Angiosperms, pollen sacs pertaining to four types: basic (*Ba*), dicotyledon (*Di*), monocotyledon (*M*), and reduced (*R*) come simplified and produce a unicellular antheridial envelope, the "tube cell", so-called because it elongates a great deal and anchors the entire gametophyte to the megasporangium (nucellus). Although the tube cell functions as a haustorium, it represents the true homologue of the cover cell of a gametangium. Since the latter is a sexual organ, to designate its cover cell as "vegetative cell" appears unsatisfactory. In the pollen tube of prespermaphytes two multiciliate sperm cells are produced, as well as a "stalk cell" considered as sterile spermatogenous cell (Fig. 15.8E). (6) At the evolutionary scale in gymnosperms (*Pinus;* Fig. 15.8F) the tube cell demonstrates its sexual character because it grows toward the female gametophyte, carrying the two unciliate sperm cells: this is siphonogamy common to all spermaphytes. (7) Prothallial vegetative cells are no longer formed; male gametes are independent cells (*Cupressus;* Fig. 15.8G). (8) Sperm cells remain partially separated (*Cephalotaxus;* Fig. 15.8H) (Gianordoli 1978). (9) Sperm nuclei are unequal (*Taxus;* Fig. 15.8I).

The structure of exine, Walker and Skvarla (1975) have pointed out, is homogenous in *Lycopodium* and ancient gymnosperms. Whereas the exine is alveolate in *Selaginella* (Robert 1971) and advanced gymnosperms, it is tectate in Cycadales (Audran 1977, 1978, Audran and Masure 1977).

15.5.2 Intrinsic to Angiosperms

In angiosperms the male gametophytes have undergone a drastic neotenic reduction, initiated among naked ovulate plants, leading to a three-cell condition, i.e., a tube cell and two sperm cells (Fig. 15.8J). I do not think that there is a true suppression of the stalk cell, but the latter has assumed sexual function at the same time as its sister spermatogenous cell (or body cell); both cells then neotenically function as gametes comparable to one-cell generation later in naked ovulate plants (cf. Fig. 15.8 I′, J′). However, the morphological similarity of both sperm cells may hide a fundamental difference: one male gamete in angiosperms is the progenitor of two sexual cells, the other one (equivalent to stalk cell) is without any descendant (except in *Microcycas*).

Theoretically, male gametophytes might show further neotenic simplification to a single sperm cell, and the tube cell (Fig. 15.8K): then they would retain the status of gametangium. In case the microspore itself functions as sperm cell (Fig. 15.8L), as is actually the case in animals, the male "gametophyte" would have disappeared and become a "gametocyte" comparable to the ones produced in thallophytes (Fig. 15.8A). The general production of two sperm cells in angiosperms is linked to the process of double fertilization.

The precocious division of the generative cell inside the pollen grain is characteristic of numerous families, including some advanced ones: Asteraceae, Apiaceae. The three-cell condition is a variable character, occasionally, within the same species, in 51 families reviewed by Davis (1966). It is met with in primitive species such as *Anemone nemorosa*. In a letter (dated 11 February 1980), Prof. R. B. Knox (Univ. Melbourne), makes the following remark: "The time between arrival on the stigma and fertilization is exceedingly short (being approximately 10 min in the grasses and 40 min in sunflower and *Cosmos*) so perhaps the precocious division has made this possible".

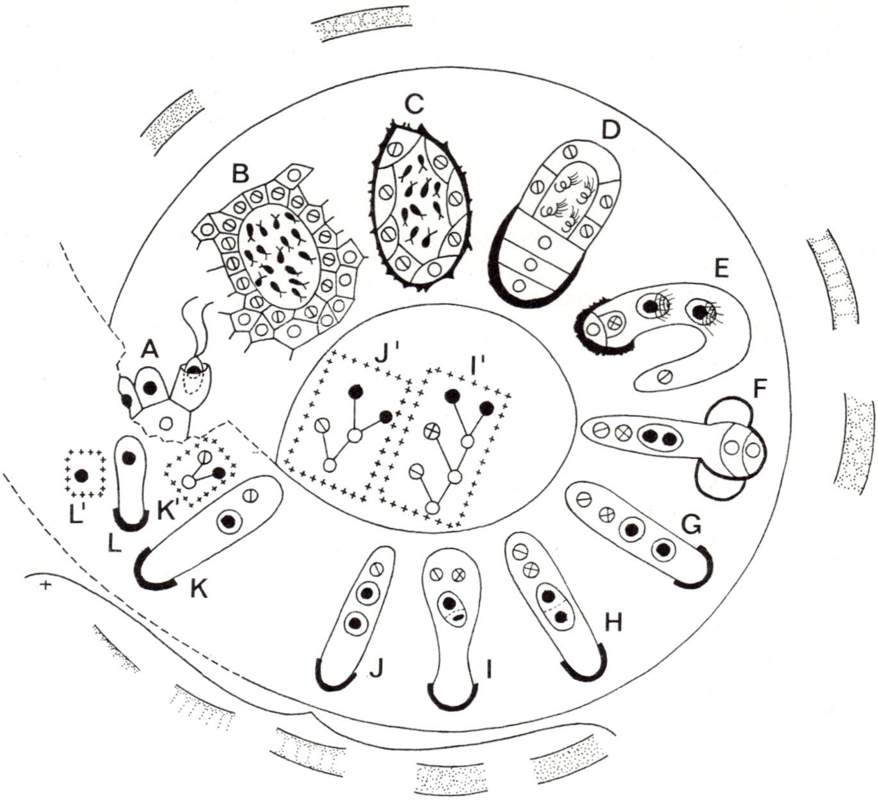

Fig. 15.8 A–L′. Proposed evolutionary trend from naked spermatocytes in Chlorophyceae through more and more reduced antheridia up to a naked male gametocyte (sexual nuclei *black*, stalk nuclei *crossed*, antheridial envelope or tube cell nuclei *barred*, prothallial and intermediary nuclei *white*). **A** *Coleochaete*, **B** *Lycopodium*, **C** *Selaginella*, **D** *Azolla*, **E** *Cycas*, **F** *Pinus*, **G** *Cupressus*, **H** *Cephalotaxus*, **I** *Taxus*, **J** Angiosperms, **K, L** future hypothetical male structures. Inside *rectangular frames limited by crosses*, the comparison of *Taxus* (**I′**) and present-day angiosperms (**J′**) demonstrates that male nuclei of angiosperms correspond to stalk and body nuclei of *Taxus*. From the microspore nucleus (*white*) at the *bottom* up to both sexual nuclei (*black*), there are three generations in **I′**, and two in **J′**. As for the hypothetical still more neotenized male structures **K** and **L**, they would be attained after only one (**K′**) or no (**L′**) microspore nucleus mitosis. *Outside the helicoidal enclosure* are figured some exine structures concerning taxa under consideration. Angiosperms show primitive as well as more and more advanced structures of exine. *Right to left* (*bracketed*): Homogeneous thick exine in *Degeneria* (Degeneriaceae), granular in *Magnolia* (Magnoliaceae), columellate-tectate in *Desmopsis* (Annonaceae), columellate-intectate in *Horsfieldia* (Myristicaceae), rudimentary in *Zostera* (Zosteraceae), and absent in *Amphibolis* (Helobiae)

As for the structure of exine in angiosperms, Walker and Skvarla (1975) point out that columellaless exine is again met with in the most primitive Magnoliaceae, as is the case in *Lycopodium* and ancient gymnosperms. Walker (1974) suggests this phylogenetic trend in the structure of exine in angiosperms (cf. Fig. 15.8 J): tectate-imperforate, tectate-perforate, semitectate, intectate, and, finally, in aquatics, parasites, saprophytes, advanced taxa, there is a secondary loss of columellate structure. According to Ducker et al. (1978), the pollen of *Amphibolis antarctica* (Helobiae) appears to be unique since it lacks exine.

Since in primitive angiosperms the exine is thick and homogenous, it seems possible to eliminate the most advanced lepidophytes (cf. *Selaginella*), prespermaphytes (cf. Cycadales), gymnosperms (cf. *Ephedra*) with lighter exine structures as possible ancestors of the angiosperms.

The universal three-cell condition of angiosperm male gametophytes, associated to siphonogamy, and neotenic structural simplification of exine and male gametes, appear as characters of the highest evolutionary level attained in cormophytes. These characteristics do not give any indication about the ancestry(ies) of angiosperms.

15.6 Female Gametophyte

In angiosperms the female gametophyte is called embryo sac. As previously mentioned (see Sect. 15.1), angiosperms may be designated "archegoniates", but shall we be able to find archegonia (Favre-Duchartre 1966) in such a reduced organism as the embryo sac?

To help us assign a position to haploid female sexual structures of angiosperms, let us arrange the other chlorophytes into a series (Fig. 15.9).

15.6.1 Female Sexual Structures in Chlorophyta Extrinsic to Angiosperms

Figure 15.9 symbolizes, in a spiral path, the sequence starting from a naked oocyte containing one egg in a green alga (Fig. 15.9 A), such as *Coleochaete,* to an oocyte of *Nitella* (Fig. 15.9 B) with three pericentral cells (cf. Favre-Duchartre 1971, Goëtz 1899).

In trying to fill the gap between naked oocytes of algae, and obtain the status of archegonium (1) Church (1919, p. 12) suggests "that the archegonium is a derivative only of a truly parenchymatous soma", the egg cell being imbedded in "a small lateral ramulus". (2) According to the theory of Davis, an archegonium may be considered as an aggregation of eggs, the marginal ones being sterile. None of the two views find the least confirmation in ontogeny of archegonium which may actually be followed in Fig. 15.9, in the reverse sequence: K (archegonium initial); J (female cell covered by neck initial); I, H, G (egg cell, ventral canal cell, two-layered neck). To my mind, following Chadefaud (1960, p. 418, Fig. 311), the female structure of *Nitella* would be a missed archegonium: the pericentral cells are set aside by the growing egg and fail to enclose wholly and permanently the female gamete. (The protecting spiral cells of *Nitella* are not taken into consideration.) A true archegonium may be obtained, without any transition, from a marginal female cell cutting off, in stable positions, surrounding sterile cells.

In some teratologic archegonia of moss (Fig. 15.9 C), several eggs differentiate in the axis but, normally, only one situated at the base of the axis is functional;

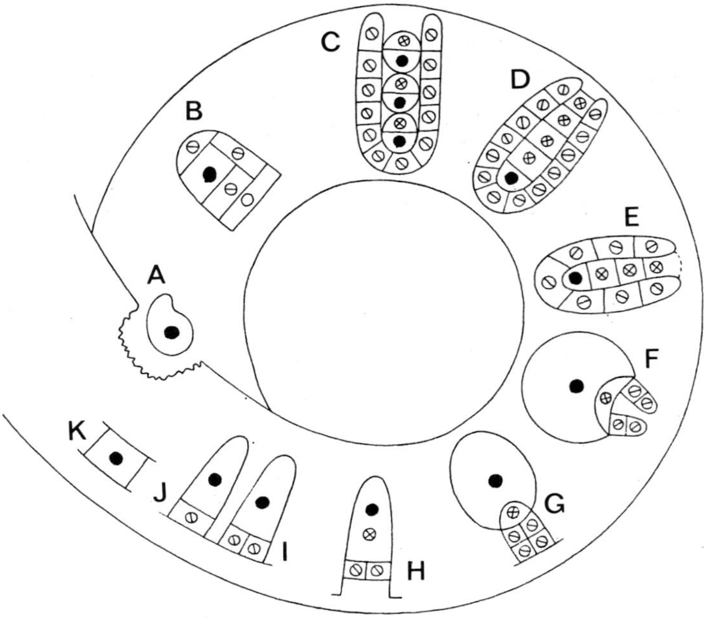

Fig. 15.9 A–K. Female sexual structures met with at different evolutionary levels in green plants (sexual nuclei *black*, *crossed* in ventral canal or canal cells, *barred* in neck cells). **A** *Coleochaete*, agal oocyte with one naked egg. **B** *Nitella*, young egg cell and three pericentral cells. **C** Moss, teratologic archegonium. **D** Moss, normal archegonium. **E** Filicean archegonium. **F** *Ginkgo*, archegonium. **G** *Pinus*, archegonium. **H** *Cephalotaxus*, archegonium. **I** *Taxus*, mature archegonium. **J** *Taxus*, immature archegonium with unicelled neck. **K** *Gnetum*, marginal egg cell. According to the present author, the female haploid structures of angiosperms have developed from organizational levels at **J** and **K**

the others function as neck canal nuclei (Fig. 15.9 C, D). Later, in pteridophytes (Fig. 15.9 E), and prespermaphytes, the archegonia have become simplified but their neck still protrudes (Fig. 15.9 F). A ventral canal cell persists in Pinaceae (Fig. 15.9 G); only a ventral canal nucleus in *Cephalotaxus* (Fig. 15.9 H), *Ephedra*, Taxodiaceae and Cupressaceae. In *Taxus* (Fig. 15.9 I, J) as well as *Torreya taxifolia* and *Widdringtonia cupressoides* a ventral canal nucleus is not produced, and the central cell functions as egg; in *Gnetum* (Fig. 15.9 K) and *Welwitschia* the egg is a marginal cell, in the position of archegonium initial, and is fertilized. Thus, phylogenetically oogenesis has undergone (Fig. 15.9 J, K) the last two steps in reduction which are considered (Fig. 15.8 K, L) as future trends in spermatogenesis. I wish to emphasize the lack of synchronization in *Taxus* during the maturation of archegonia: in Fig. 15.9 I the archegonium has a bicellular neck; in Fig. 15.9 J unicellular; in one and the same female prothallium one may also find archegonium initials (Fig. 15.9 K).

Extrinsic to angiosperms, the trends of synchrony and asynchrony are obvious in Fig. 15.10 A–C. Synchrony leading to simultaneous maturation of all the archegonia in a gametophyte is common in *Pinus* (Fig. 15.10 A). The lack of synchronization during maturation is evident in *Cephalotaxus* (Fig. 15.10 B) and

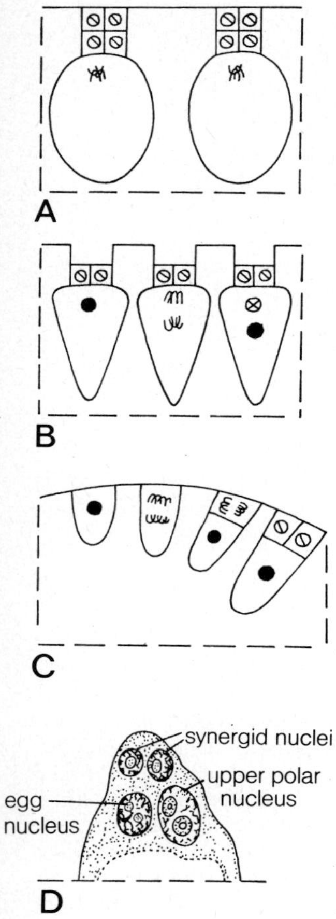

Fig. 15.10 A–D. Time relations in completion of oogenesis. **A** *Pinus*, synchronous oogenesis is primitive. **B** *Cephalotaxus*, **C** *Taxus*, nonsynchronous oogenesis. **D** Angiosperms, aspects of oogenesis. [**D** *Nicotiana*, after Bannikova (1971), who did demonstrate that, though the eight haploid female nuclei are formed quite simultaneously, during the short period they are free they differentiate along with their surrounding cytoplasm]

Taxus (Fig. 15.10 C); even an archegonial initial is seen in Fig. 15.10 C. Nonsynchrony is considered an evolutionary progress.

Intrinsic to angiosperms, usually there is a synchronous division at the four-nucleate stage of embryo sac. However, the short duration preceding wall formation represents the crucial moment of angiosperm female gametogenesis: during this stage and depending on the different situations they acquire in the sac, the eight free nuclei (and their surrounding cytoplasm) differentiate from one another (Fig. 15.10 D).

15.6.2 Embryo Sac

Figure 15.11 represents how different authors consider their models of the evolution of embryo sac through the series Algae (A, A′), Filicineae (B, B′), Pinaceae (C, C′; D; E, E′), *Taxus* (F, F′), *Gnetum* (G, G′) and even angiosperms (H) when the embryo sac is considered a sui generis structure.

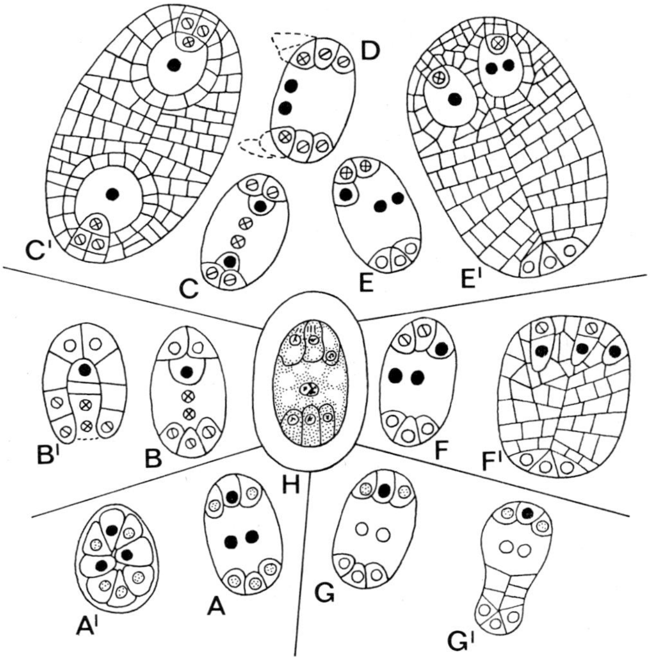

Fig. 15.11 A–H. Eight interpretations for the origin of Polygonum type of embryo sac (*central figure*). Cells of six models (*at the periphery of figure*) which are not supposed to persist in embryo sac lack nuclei; all figures oriented with micropylar pole upward. Although embryo sacs are ovoid, for the sake of clarity they are drawn here (as well as in Figs. 15.12, 15.13, and 15.15) stretched in a plane like a planisphere (sexual nuclei *black*, potentially sexual *dotted*, canal and ventral canal nuclei *crossed*, neck cells *barred*, vegetative nuclei *white*). According to Eames (1961) and Swamy (1974), the embryo sac is considered to be an original structure (*central figure*). **A** According to Yakovlev (1974), the embryo sac is supposed to be derived from algal oocyte **A'**. **B** According to Tourte (1969), from a Filicean archegonium **B'**. **C** According to Porsch (1907) and Schnarf (1942), from two opposite tetracelled (Pinaceae) archegonia **C'**, in which the egg and ventral canal nuclei have interchanged their positions. **D** According to Chadefaud (1941), from two opposite tetra-celled archegonia **C'**, in which ventral canal and egg nuclei have retained their positions. Two neck cells are absent in each archegonium. **E** According to Schürhoff (1919), the embryo sac is supposed to have a chalazal vegetative prothallial portion (antipodals) and a micropylar generative portion composed of two archegonia **E'**, one made up of a synergid (as ventral canal cell) and egg cell (its own egg cell). The second archegonium comprises the other synergid (as a ventral canal cell) and binucleate central cel (as an egg cell). **F** According to Favre-Duchartre (1964), the embryo sac is polarized in the same way as in **E,** with a chalazal vegetative antipodal portion and a micropylar archegonium complex of *Taxus* model **F'**. The marginal egg cell in embryo sac is interpreted as fertilizable archegonium initial, and two bicelled archegonia comprise one synergid each (as a unicellular neck) and central (polar) nucleus as an egg. These central egg nuclei pertaining to both archegonia are not separated by a wall. **G** According to Fagerlind (1941) and Battaglia (1951, 1980), the embryo sac is compared to the female gametophyte of *Gnetum* **G'**. Every marginal cell is actual or potential egg cell, whereas the two central (polar) nuclei are supposed to be vegetative. **H** Polygonum type of embryo sac

Favre-Duchartre (1964, 1978) compared the embryo sac to a female prothallium of *Taxus* in which fertilization takes place at an early stage of development (Fig. 15.11 F). When antipodals occur in embryo sac, they represent the vegetative part of prothallium. At the micropylar pole of most embryo sacs, there is an archegonial complex composed of three elements: the marginal egg cell (Favre-Duchartre 1975) equivalent to fertile archegonium initial; the other two elements represent bicellular archegonia in which one synergid is equivalent to an unicellular neck, and one central nucleus comparable to a fertile central cell (Fig. 15.11 F).

As proposed by Schürhoff (1919) (Fig. 15.11 E), the general polarity of a female gametophyte in gymnosperms also occurs in the embryo sac with a chalazal vegetative pole and a micropylar generative one. The archegonia with a unicellular neck and a central cell are due to simplification (Fig. 15.9 J) (Favre-Duchartre 1970). As for the theories of Porsch (1907) and Chadefaud (1941) (Fig. 15.11 C, D), the synergids are equivalent to neck cells. It is not so because of their secretory character and receptive function of male gametes, but because of their position. The lack of separation between both central nuclei (so-called upper and lower polar nuclei) is due to the fact that centripetal wall formation does not occur in female gamtophyte. The originality in the archegonial complex of embryo sac is that the archegonium initial never divides to give rise to the bicelled stage because the embryo sac becomes arrested during early development.

Arguments for the support of Taxalean theory may appear in the structural diversity and ontogenesis of the embryo sac, as compared to that of *Taxus*.

15.6.3 Diversity in Embryo Sacs

Figure 15.12 shows how *Taxus* may be regarded as the theoretical stock with four trends:

1. The disappearance of prothallial vegetative cells up to their total elimination, as is the case in the Oenothera type (Fig. 15.12 A g). It is worth pointing out that the diploid embryo sac of Poaceae develops aposporously from a nucellar cell, and conforms to the tetracellular Oenothera type. In angiosperms the pollen grains also lack any vegetative prothallial cell (Fig. 15.8 J). Exceptions are known. A prothallial cell occurs in *Cuscuta epithymum* (Fedortschuk 1931) and probably also in *Atriplex hymenelytra* (Billings 1934).
2. If we admit the existence of three-celled archegonia in embryo sacs, there are four archegonia in Penaea, and one in Oenothera and Apinagia types (Fig. 15.12 B).
3. The two-celled archegonia would be seven, three, two, one, respectively in (a) *Peperomia,* (b) *Plumbago* (with a supernumerary central nucleus), (c) *Polygonum* (along with *Allium, Endymion, Adoxa* representing convergence of embryo sacs), (d) *Acalypha* and *Dicraea* (Fig. 15.12 C).
4. More than one marginal archegonium initial appears to be teratological. One is the common number, and none differentiates in *Dicraea* and *Apinagia* (Fig. 15.12 D).

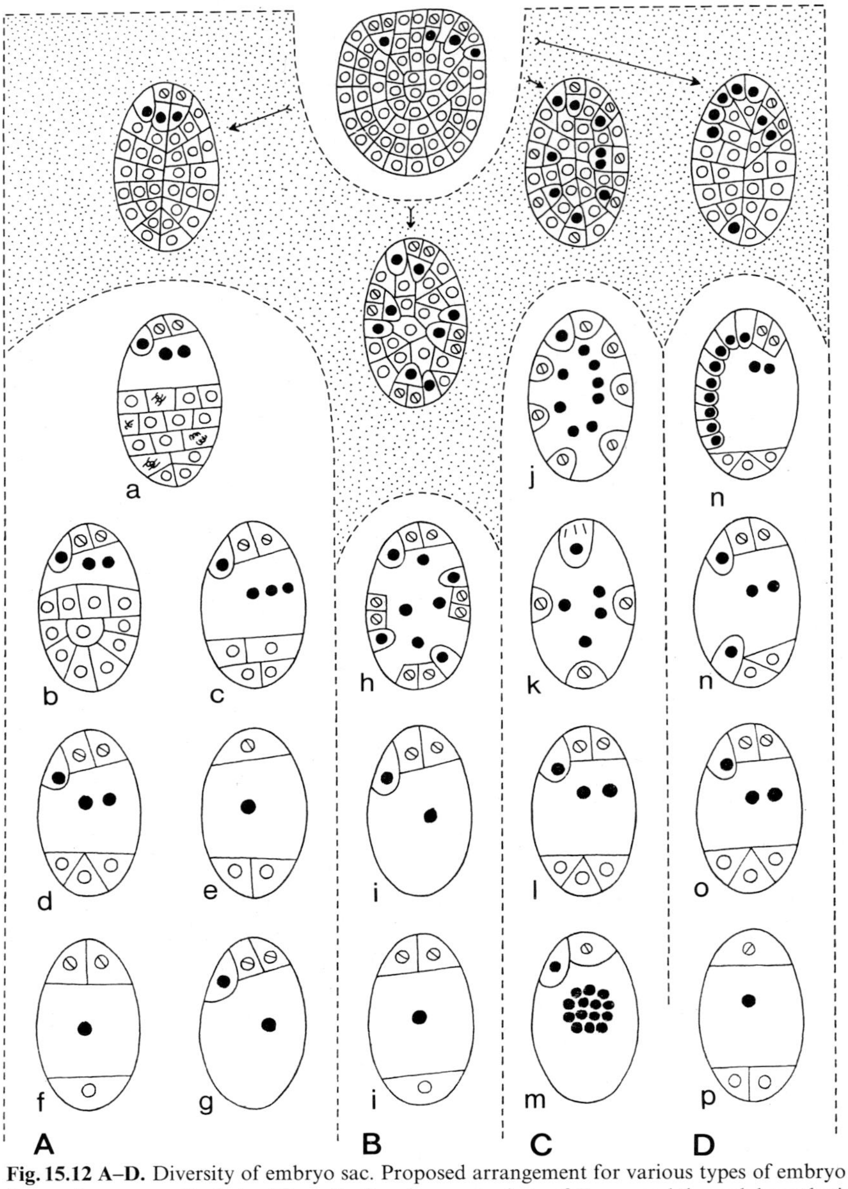

Fig. 15.12 A–D. Diversity of embryo sac. Proposed arrangement for various types of embryo sac into four series supposedly derived from the female gametophyte of *Taxus*, and through hypothetical intermediaries suggested as facultative missing links (sexual nuclei *black*, neck cells and synergids *barred*, vegetative nuclei *white*). **A** Progressive decrease of antipodal (prothallial) cells: numerous (*a*) in Poaceae, 11 (*b*) in *Drusa*, four (*c*) in *Chrysanthemum* I, three (*d*) in *Polygonum*, two (*e*) in *Dicraea*, one (*f*) in *Apinagia*, none (*g*) in *Oenothera*. **B** *Penaea*, four (*h*) tricellular archegonia; *Oenothera* and *Apinagia* only one (*i*) tricellular archegonium. **C** Bicellular archegonia, seven (*j*) in *Peperomia*, three (*k*) in *Plumbago*, two (*l*) in *Polygonum*, one (*m*) with a 14-ploid sexual nucleus in *Acalypha*, and one (*p*) with haploid nucleus in *Dicraea*. **D** More than one (*n*) archegonial initials differentiate in teratologic cases; only one (*o*) in *Polygonum*, and none (*p*) in *Dicraea* since the central egg cell and its covering synergid represent a bicellular archegonium

The structural diversity in embryo sacs is still more accentuated than it is represented in Fig. 15.12, when we consider the free-nuclear fusions (*Fritillaria, Plumbagella, Chrysanthemum* II) which appear as deviations.

15.6.4 Ontogeny of Embryo Sac as Compared to Female Gametophyte of *Taxus*

Since ontogeny recapitulates phylogeny, the latter, in its turn, may throw some light on incomplete cell formation in the embryo sac. Figure 15.13 indicates that the development of a Polygonum type of embryo sac takes less than 10 days, while the female gametophyte of *Taxus* requires about 2 months (cf. Fig. 15.17 B). The chalazal megaspore and the two- and four-nucleate stages (Fig. 15.13 A–C) are similar in both taxa. In the young embryo sac the two central nuclei lie in common cytoplasm. Their movement in the embryo sac is due to the increase of peripheral vacuoles. These are rhythmically produced between the free nuclei at the coenocytic stages of development of female gametophyte in gymnosperms and, finally, they migrate to the central vacuole (cf. Moussel 1971). The latter no longer exists in a young eight-nucleate embryo sac which is rather of small size. Therefore, the cytoplasm around peripheral vacuoles fuses in the center of the embryo sac. In my opinion, since the central nuclei are not confined to the periphery (as egg nuclei are in *Gnetum*), their position below the synergids, where they are fertilized, makes them comparable to egg cells covered by neck cells (Fig. 15.13 D–H).

The marginal egg cell (characterized by its position, poor cytological differentiation, persistence of a peripheral vacuole, absence of a cell wall toward the center of the embryo sac) has maintained the main features of an alveolus, but these are now linked with female sexuality.

Antipodals are usually the first cells to be recognized in an embryo sac, and they are separate from each other. In Poaceae they even proliferate (cf. Fig. 15.12 A a) and can be considered to be equivalent to vegetative prothallial cells. Reversely, a more precocious reduction in embryo sac differentiation in *Acalypha lanceolata* (Euphorbiaceae) (Thathachar 1952) leads to a three-celled female gametophyte made up of one marginal egg cell flanked by only one synergid and a 14n central (polar) nucleus (Fig. 15.12 C m).

The embryo sac appears to be the result of a drastic hastening of the development of female gametophyte. This neotenic evolution has the common consequence to produce two kinds of female gametes, because of the precocious interruption of their disparate developments. Such embryo sacs may be attained convergently from different lower taxa without any intermediaries.

15.7 Pollen Reception by Female Organ

In naked ovulate plants the pollen is commonly received by the ovular micropyle with the help of a mucilaginous drop exuded by the nucellus. However, in *Araucaria*, the reception process may be considered preangiospermic; the pollen grains germinate on the adaxial surface of the scale-bract complexes of female strobilus,

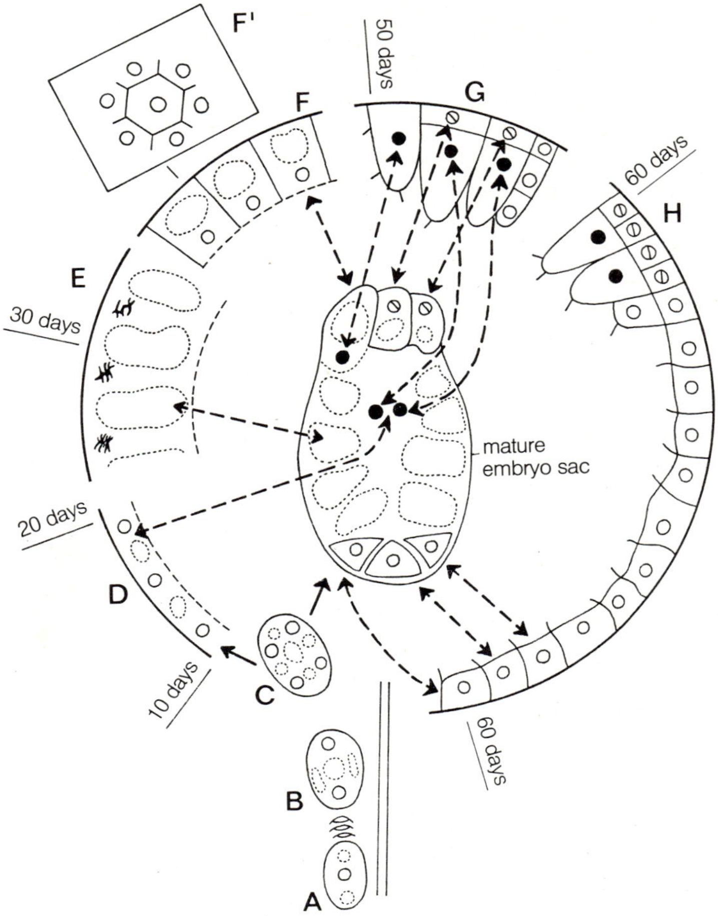

Fig. 15.13 A–H. Comparison between oogenesis of *Taxus* (*outside*) during two months, and about 10 days in *Polygonum* (*from bottom to center of figure*). **A** Functional megaspore. **B** Two-nucleate embryo sac. **C** Four-nucleate stage (these three first stages are common to monosporic female gametogenesis in gymnosperms and angiosperms). Early mature embryo sac *in center*. **D** Free-nuclear stage in *Taxus* with peripheral vacuoles between nuclei. **E** The peripheral vacuoles enlarge considerably when free nuclei undergo mitosis. **F** Alveoli, the peripheral vacuoles, are now against each nucleus. **F′** Transection of an alveolus with six adjoining alveoli. **G, H** Near-mature stage of female gametophyte of *Taxus*. The micropylar part has an archegonial complex with one marginal archegonium initial, and two bicellular archegonia. **H** Mature stage attained approximately 60 days after megaspore formation. Archegonia are tricellular and many prothallial cells represent vegetative tissue. *Dashed arrows* join the structures suggested to be homologous (sexual nuclei *black*, neck cells and synergids *barred*, vegetative nuclei *white*)

at a distance from the nucellar apex extruding through the micropyle (cf. Fig. 15.4 C). Of course, gymnosperms do not develop stigma (stigmata) and van Tieghem (1891) called them "Astigmatae"; nevertheless *Araucaria* ovules receive pollen tubes and not pollen grains.

Angiosperms differentiate stigma (stigmata) and were called "Stigmatae" by van Tieghem (1891). However, in some genera (*Reseda, Viola, Butomopsis* – cf.

Johri 1936) carpels are not closed at the tip and pollen grains have been observed at the ovule surface (see Chapt. 1, Sect. 1.9.2).

In cases of cleistogamy (in *Viola*) the stigma is present but is not pollinated since the pollen grains germinate inside the pollen sacs, and the pollen tubes reach the ovules through the multicellular envelopes of both pollen sacs and carpels.

Commonly, the angiosperm pollen grains are received at the stigmatic surface; but incompatibility barriers may prevent either their germination (cf. *Primula;* see Sect. 15.1.1; Fig. 15.1), or pollen tube growth within the conducting tissue at different levels of the style. The prevention of self-fertilization represents an important advantage afforded by angiospermy.

This phyletic tendency to the protection of female gametophytes by (1) nucellus, (2) ovular integument(s), (3) carpels, and (4) inferior ovary (produced either by caulinar proliferation or, more often, by concrescence of staminal and perianth bases with the surface of ovary) has been compared by Mangenot (1973, p. 53) to Russian dolls (my translation): "the female organ of angiosperms calls to mind the Russian dolls (those series of wooden dolls, the larger ones containing smaller and smaller ones): inside the gynoecium are found ovules; inside each ovule one discovers a megasporangium-nucellus, and within the latter a female prothallium".

Mangenot suggests that this process may be further promoted if the carpels themselves are completely protected. That "angiocarpy", supposed to exist in *Tambourissa,* a Madagascarian plant of Monimiaceae pertaining to the primitive Magnoliideae stock, is vital in evolutionary potentialities (cf. Mangenot 1973). Endress (1979) observed that "noncarpellary pollination" actually occurs in *Tambourissa religiosa,* the "hyperstigma" represented by a "mucilage barrier" appearing as "an advantageous protection against predation damage of the carpels by pollinating insects". Endress (1980) has also revealed the presence of a hyperstigma in *Hennecartia, Wilkiea,* and *Kibara* (see Chap. 1, Fig. 1.18 A–F).

15.7.1 Fertilization

Angiosperms are characterized by double fertilization. Eames (1961) and Takhtajan (1969) consider this feature so unique that it did not happen again; hence the conclusion that angiosperms are monophyletic. Let us discuss the eventuality of *intra*-archegonial double fertilization and analyze the occurrences of *inter*-archegonial double fertilizations in cormophytes in general, and in angiosperms in particular, drawing attention to cases of single fertilization in angiosperms, if any.

15.7.2 Double Ferilization, Extrinsic to Angiosperms

15.7.2.1 Intra-archegonial Double Fertilization

Let us first consider the eventual division of the mature gametes, female on the one hand, male on the other, within the cavity of the archegonium of a gymnosperm.

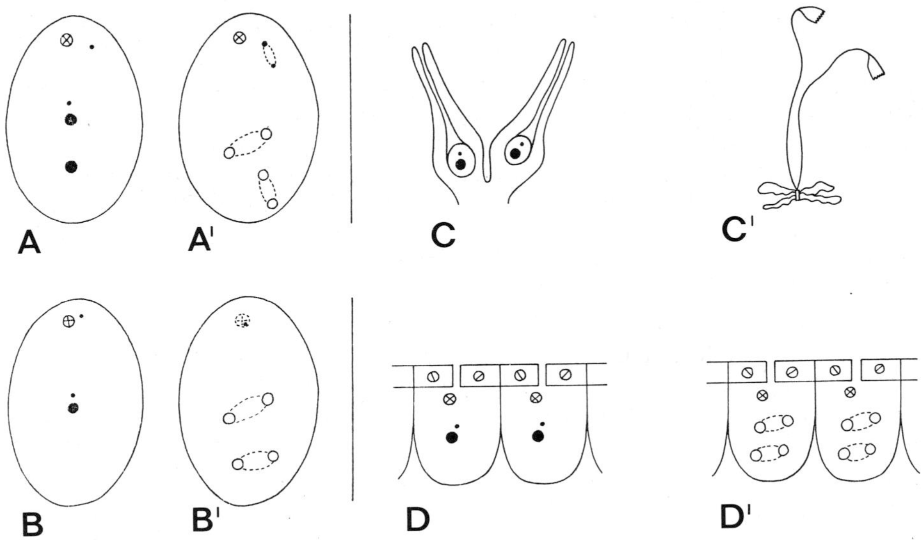

Fig. 15.14 A–D′. Single (**A**) and double fertilization – intra- (**B**) and inter-archegonial (**C, D**) and some of its results (**A′–D′**), in gymnosperms and a moss (sexual nuclei *black*, ventral canal nuclei *crossed*, neck cell nuclei *barred*,proembryo nuclei *white*). **A** *Cephalotaxus*, in the archegonial cavity the egg nucleus may divide before fertilization. **A′** Supernumerary male nucleus commonly proliferates during early embryogenesis. **B** *Ephedra*, in exceptional cases the ventral canal nucleus may be fertilized. **B′** Supernumerary zygote most probably degenerates, **C, D** *Mnium undulatum* (**C**) and *Callitris* (**D**), two adjacent archegonia are commonly fertilized. They give rise to two similar sporophytes (**C′**) or proembryos (**D′**)

In some gymnosperms (*Cephalotaxus*) two similar egg nuclei have been observed in the center of a mature archegonium (Fig. 15.14 A). Since the ventral canal nucleus is clearly demonstrated by Feulgen reaction as a collapsed mass, the upper one of both the female nuclei cannot be interpreted as an abnormal, deeply situated ventral nucleus. The only conclusion is that the egg nucleus has divided. A drawing by Coker (1907) demonstrates that this nucleus can be fertilized.

Also, one supernumerary male nucleus may accompany fertilization inside the archegonium cavity. I have observed such a behavior in *Cephalotaxus*, and this also occurs in *Cedrus*. In no case has the supernumerary male nucleus been observed to fuse with the ventral canal nucleus which appears to degenerate below the neck of archegonium.

Moussel (1979) has analyzed an exceptional case of a deeply seated ventral canal nucleus of *Ephedra* which shows all the characters of a fertilized nucleus; but the upper (second zygote) nucleus did not appear to have any chance of development perhaps because of the surrounding highly vacuolate cytoplasm (Fig. 15.14 B, B′). In *Cephalotaxus* (Fig. 15.14 A′) I have observed the mitotic division of the haploid male nucleus, and it is suggested that the few nuclei below the archegonial neck of *Cedrus* and *Taxus* are of the same origin. In *Taxus*, which lacks a ventral canal nucleus (Fig. 15.9 I), the small nuclei below the neck are obviously the male haploid nuclei.

15.7.2.2 Inter-archegonial Double Fertilization

Two or even more than two archegonia in the same female gametophyte may be simultaneously fertilized and produce two sporophytes (Fig. 15.14 C, C′) in mosses such as *Mnium undulatum* or *Dicranum majus,* in prespermaphytes such as *Ginkgo* (Fig. 15.16) (Favre-Duchartre 1956), in gymnosperms such as *Callitris* (Fig. 15.14 D, D′) (Baird 1953) and *Gnetum* (Fig. 15.16) (Vasil 1959).

15.7.3 Intrinsic to Angiosperms

15.7.3.1 Intra-archegonial

According to Favre-Duchartre (1964), in angiosperms the intra-archegonial double fertilization is not conceivable since there is but one female sexual nucleus in each archegonium.

15.7.3.2 Inter-archegonial

Again, in angiosperms, (1) both similar marginal egg cells are most probably the result of a double marginal fertilization (Favre-Duchartre 1974); (2) rarely, both the central (polar) nuclei are individually fertilized in *Euphorbia dulcis* (Fig. 15.15 B, B_1).

However, in most angiosperms, double fertilization implies a central nucleus which is haploid (Oenotheraceae) (Fig. 15.15 C, C_1), mostly diploid (Fig. 15.15 D), or polyploid (*Peperomia, Fritillaria*), and a marginal one which is always haploid (Fig. 15.12).

In rare cases in *Anemone nemorosa,* when the embryo sac is rather long (220 µm), the second sperm nucleus may not fertilize the polar fusion nucleus (Brouland and Jeanny 1970). The latter, however, divides and develops parthenogenetically into an apparently normal endosperm (Fig. 15.15 E, E_1). Where the embryo sac is rather short (64 µm) (Fig. 15.15 D, D_1), the diploid polar fusion nucleus is fertilized normally (Yoffe 1969).

In Monandrae (orchids) where the polar nuclei are not fertilized, the endosperm is absent; but the seed germinates with the aid of a fungus (mycorrhiza), another nutritive mechanism (Fig. 15.15 F, F_1).

Nevertheless, in Podostemaceae there is a unique egg cell situated centrally and covered by one synergid as in *Dicraea* (Fig. 15.15 G), or two synergids as in *Apinagia* (Fig. 15.12 A f).

Thus, in Podostemaceae and Monandrae there is single fertilization (see Sect. 15.7.1).

15.7.4 Cytology of Zygote Formation

In ovulate plants the zygote is produced as a result of the fusion of sperm and egg cell. We may consider the cytology of the relative participants to zygote for-

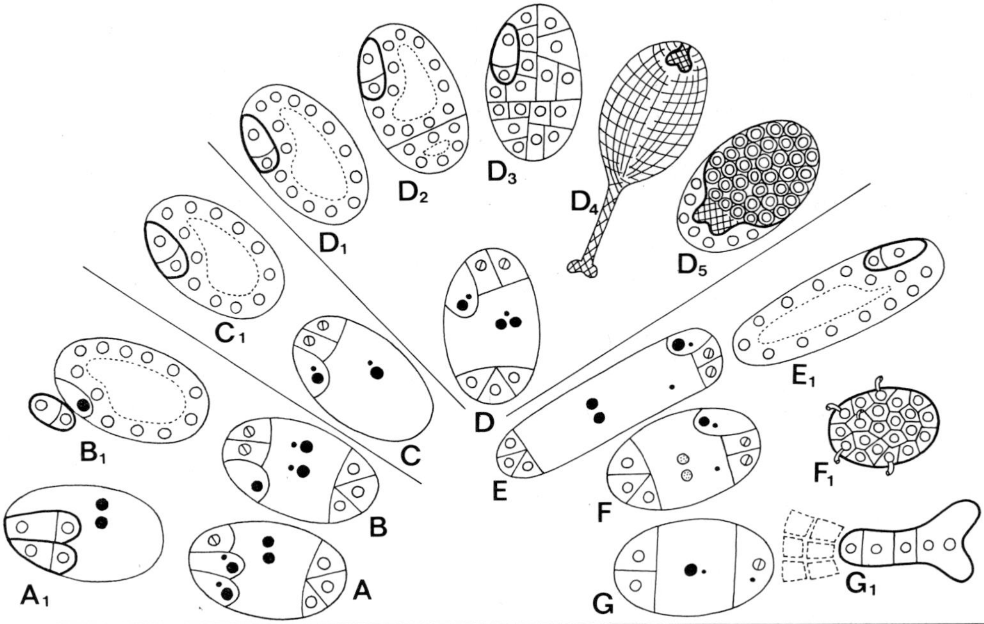

Fig. 15.15 A–G. Double fertilization accomplished with similar female gametes (both marginal in **A**, central in **B**), with dissimilar female gametes (one marginal, the other central in **C** and **D**), and single fertilization (in **E–G**) in angiosperms, and respective embryogenesis (**A_1–G_1**). Embryos are bicellular except in **D_4, D_5, F_1** and **G_1**. **A, A_1** *Malva,* two lateral fertilized egg cells (the supernumerary one being sometimes called "egg-like synergid") give rise to two similar proembryos. **B, B_1** *Euphorbia dulcis,* central (polar) fertilized "egg" nuclei produce endosperm, and a nucellar adventitious embryo also develops. **C, C_1** *Oenothera,* a good example where the egg and central (polar) nucleus are sister nuclei. Thus the endosperm and embryo have the same genotype. **D** *Polygonum,* double fertilization accomplished with dissimilar female gametes (one marginal = egg cell, two central = polar nuclei) is common. In different taxa, the endosperm may be of the Nuclear type (**D_1**), Helobial type (**D_2**), or Cellular type (**D_3**), showing, in each case, an ontogenic advance as compared to the development of a two-celled embryo (cf. Fig. 15.16). **D_4** *Prunus,* chalazal endosperm haustorium. **D_5** *Cytisus,* massive suspensor. **E** *Anemone nemorosa,* long embryo sac (common in Reims regions). The polar nuclei remain unfertilized, they fuse and produce a parthenogenetic 2n ♀ endosperm. **E_1** *A. nemorosa,* with short embryo sac (common in Leningrad region), is like **D** and **D_1**. **F** Monandrae (orchids). Polar nuclei are not fertilized. **F_1** A fungus (mycorrhiza) promotes germination of seed in the absence of endosperm. **G** *Dicraea,* the embryo sac comprises one egg cell in central position. **G_1**, the embryo grows in a nucellar plasmodium replacing the nutritive function of endosperm. (**D_4** After Surfatti 1961)

mation in different taxa. Let us analyze the contribution of cytoplasm s.l., and association of gametic nuclei which are called female and male pronuclei.

15.7.4.1 Cytoplasm s.l.

Camefort (1967) in *Larix*, Chesnoy (1969) in *Biota,* and Gianordoli (1974a, b) in *Cephalotaxus, Taxus,* and *Sciadopitys* established that the entire stock of female plastids and mitochondria degenerate. How do the organelles then appear in the zygote, from the cytoplasm of the fertilizing sperm cell (or by both associated sperms as in *Cephalotaxus, Taxus,* and *Sciadopitys*)? In *Ephedra,* a much more advanced genus than the previously quoted taxa, Moussel (1979) has shown the reverse situation; all plastids and mitochondria of the zygote are of exclusive female origin.

Jensen and Fischer (1968) demonstrated that in some angiosperms, as in *Gossypium,* the male cytoplasm is not involved in fertilization. The male gametes release their cytoplasm into the receptive degenerated synergid, and neither the primary endosperm nucleus (consequently the endosperm), nor the marginal zygote (consequently the embryo), may inherit any male cytoplasm.

Russell and Cass (1981) observed that, at least in the mature pollen grains of *Plumbago zeylanica,* the organelles of the generative cell are polarized so that the first sperm cell (the leading one), which fertilizes the marginal egg cell, inherits the plastids, and the second sperm cell, ready to fertilize the polar nuclei, inherits the mitochondria.

15.7.4.2 Gametic Nuclei

In all naked ovulate plants, except *Gnetum* and *Welwitschia,* the egg cell is covered by an archegonial neck. Caryogamy (fusion of nuclei) occurs when the sperm and egg nuclei are in preprophasic, or even prophasic, phase. In *Pinus* the female and male gametes fuse during haploid mitoses, and the actual integration of chromosomes occurs at the telophase of diploid nucleus. This behavior reflects a tendency for the gametic nuclei to undergo mitoses within the archegonial cavity (cf. Fig. 15.14A, A′).

The novelty of fertilization in angiosperms is found not so much in the fact that it is most often double (this phenomenon exists in some other archegoniates also), but that it concerns the differently organized egg cells. That female gametic disparity goes along with an immediate development of the primary endosperm nucleus, while the zygote does not show any apparent activity.

Let us examine the consequences of the development of the zygote and primary endosperm nucleus.

15.8 Embryogenesis

Since double fertilization commonly occurs in one and the same embryo sac, two embryogeneses are to be expected and, actually, this does happen.

In the first instance, one "embryogenesis" gives rise to endosperm. The angiosperm endosperm should preferably be called "albumen", as proposed by Swamy (1974), to avoid confusion with the female haploid tissue in gymnosperms. It stores abundant reserve food. The seed appears white (=albus) because of the presence of starch and/or oil droplets, and/or the aleurone grains which it contains when mature.

The following ontogenetic processes will help in understanding the morphological nature of endosperm.

15.8.1 Albuminogenesis (or Endosperm Formation) in Angiosperms

15.8.1.1 Genetic Point of View

In *Oenothera* embryo sac (Figs. 15.12 and 15.15 C, C_1) (Favre-Duchartre 1972) both egg and polar nucleus are haploid sister nuclei with strictly the same genome, and they are fertilized by the sister male gametes. Thus, the endosperm and embryo, in the same seed, are genetically identical and, yet, the endosperm will remain undifferentiated while the embryo initiates a radicle, cotyledons, and plumule. Another example is *Butomopsis lanceolata* (Johri 1936) where also the zygote and primary endosperm nucleus (2n) are genetically identical, but the monocot embryo and Helobial endosperm develop as usual.

In some cases diploid endosperm may originate when it develops parthenogenetically, e.g., *Anemone nemorosa* (Fig. 15.15 E, E_1).

Most frequently, the primary endosperm nucleus is triploid (in Polygonum – Fig. 15.15 D, D_1, Allium, Endymion, Adoxa, and Drusa type), or tetraploid (Chrysanthemum II type), or pentaploid (Plumbago, Plumbagella, Penaea, Fritillaria type) or nine-ploid (Peperomia type). Moreover, let us remember that triploid and other kinds of plant sporophytes are known to be somewhat morphologically similar to diploids.

One might ask the question whether the endosperm remains without organogenesis due to the genetic constitution of primary endosperm nucleus? Even when undifferentiated, the endosperm may express its paternal heredity. As early as 1899, de Vries pointed out that the pollen from a starchy variety of *Zea mays* transmits the dominant character to the endosperm developing on a sugary variety. This is a rare instance.

In naked ovulate plants, the endosperm is a maternal haploid tissue which is the source of nourishment to the embryo. However, this parental relationship provides to the nutritive tissue only half the genome of the embryo. When angiospermous endosperm (albumen) is considered, the paternal genome integrates with the maternal one for the related embryo and endosperm of the same seed.

15.8.1.2 Dynamic Growth of Endosperm

We have considered (see Sect. 15.7.4.2) that triple fusion gives rise to primary endosperm nucleus, and is immediately followed by mitosis. This is followed by a second mitosis, and so on without any interruption.

Whether wall formation (1) does not take place (Nuclear type), or (2) occurs transversely followed by free-nuclear divisions in both chambers (Helobial), or (3) is a regular feature (Cellular) (Fig. 15.15 D_1, D_2, D_3), does not appear to make much difference, as also pointed out by Wunderlich (1959) and Swamy (1974). There is no important relationship between the type of endosperm and taxonomic assignment.

15.8.1.3 Interpretation of Morphology of Endosperm

In my opinion, the endosperm is to be considered a "proembryo" s.s.

The concept that the endosperm of angiosperms is a new structure (though incorporated in the ovule) is not novel at all. (1) This was suggested as early as 1868 by Le Maout and Decaisne who pointed out (p. 88) that the endosperm "ne tient à rien" (is attached to nothing). (2) In 1887, i.e., even before the discovery of double fertilization, Le Monnier considered the endosperm to be "an accessory plant, independent from the mother plant and associated to the embryo". The arguments put forward by Le Monnier (1887) for such a "proper development" of the endosperm are: (A) In the sporophyte of *Papaver* laticifers are present in every parenchymatous tissue, but are absent in the endosperm. (B) In Apiaceae secretory canals interupt their course in the pericarp, and are never present in the endosperm. (C) While sucrose accumulates in the tuber of *Beta,* the endosperm is starchy. (3) In 1900, Sargant considered the endosperm as a second "maimed" embryo transformed into a kind of "monster".

However, when dealing with the endosperm, most authors refer to it as a "nutritive tissue".

15.8.1.4 Nutritive Seed Structures: History

(1) In prespermaphytes the entire development of the endosperm, including accumulation of reserve food, occurs previous to fertilization (Favre-Duchartre 1943). (2) In primitive gymnosperms such as *Pinus* and *Araucaria,* the endosperm develops before fertilization, but the storage of reserve food is initiated only when embryogenesis begins. (3) In more advanced gymnosperms such as *Taxus, Torreya,* and *Gnetum,* the female gametophyte is very small at the time of fertilization, and the endosperm increases during embryogenesis (cf. Gianordoli 1978). This appears as a forerunner (since naked ovulate plants preceded angiosperms) of the behavior of the angiospermous endosperm. (4) Since the latter did not exist before fertilization, the endosperm of angiosperms accumulates reserve food only after fertilization. However, because of its development, more rapid than that of the embryo, the angiospermous endosperm maintains the character of an "organism" providing its first nutrition. (5) Successful as it appears to be in the phylogeny of ovulate plants, the angiospermous endosperm gives signs of evolutionary evanescence up to total absence: (A) In Leguminosae, the suspensor of the embryo is well developed and provides nutrition to the growing embryo; the latter consumes most of the endosperm too (Fig. 15.15 D_5). (B) When a perisperm, i.e., nucellar tissue with reserve food, is present, the endosperm is poorly developed. (C) In Podostemaceae only the upper polar nucleus differentiates. In such a condition the single fertilization gives rise to the true embryo (with a suspensor haustorium) which is able to develop without the endosperm, because the underlying nucellus degenerates into a nutritive syncytial pseudo-embryo sac in which the embryo develops (Fig. 15.15 G, G_1).

The endosperm appears to be a true "organism" without organs, and is present in the majority of angiosperms. Some angiospermous taxa, such as Asterales, are rather well defined. It appears that mature seeds are albuminous in the small primitive families (Goodeniaceae, Lobeliaceae, Campanulaceae, Dipsaceae, Calyceraceae), whereas they are exalbuminous in the large Asteraceae and the still more advanced Ambrosiaceae. Ranales, which are albuminous, gave rise to the exalbuminous Rosales. However, a common feature of development of the endosperm is that it lacks a resting period, except when mature (Fig. 15.16).

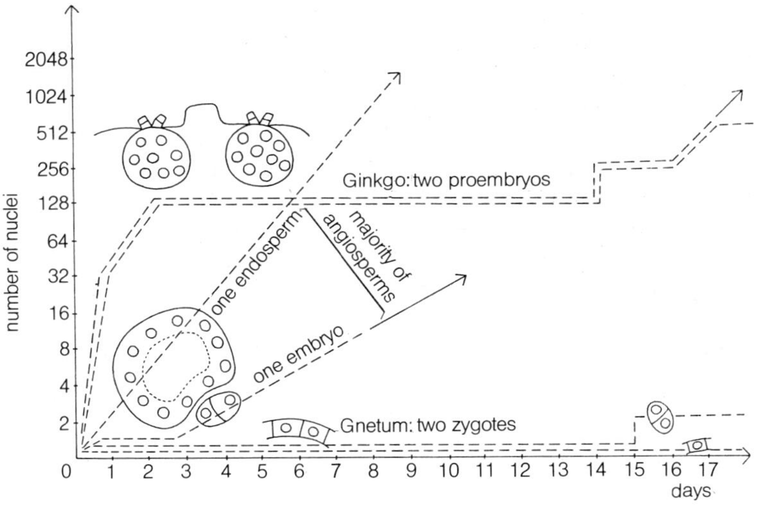

Fig. 15.16. Time relations in the activity of zygote and primary endosperm nucleus, and development of proembryos, endosperm, and embryos. (1) The marginal zygotes of *Gnetum* and angiosperms do not show any mitotic activity for two or more days after fertilization. Then a new embryonic sporophyte differentiates. (2) The zygotes covered by archegonial neck (*Ginkgo*) produce proembryos, and primary endosperm nucleus covered by synergids produces endosperm. In *Ginkgo* the proembryos remain dormant for 12 days before resuming their development leading to progressive differentiation of organs

15.8.2 Embryo

In the embryo sac the central egg cell is covered by synergids (= neck cells according to Taxalean theory). The primary endosperm nucleus undergoes mitoses more or less immediately after formation. This appears to be quite common as reviewed by Favre-Duchartre (1980).

By contrast, in the marginally situated egg cell of *Gnetum*, Vasil (1959) revealed a rest period of the zygote nucleus for 15 days (Fig. 15.16).

In angiosperms the marginal zygote, which gives rise to the embryo, usually remains in a resting phase for some days. Jensen (1974) has pointed out a renewal of polyribosomes during the resting period of marginal zygote. May be the unfertilized egg cell polyribosomes degenerate, and new polyribosomes in the zygote allow initiation of a new sporophytic organism.

Jalouzot (1982) has demonstrated that the marginal zygote of *Colchicum autumnale,* which remains in a resting stage during 5 months in the natural rather cold condition, may divide after about 30 days when the temperature of the last 12 days is 20 °C.

The early rapid growth of the proembryo occurs in prespermaphytes and most gymnosperms, except in the marginal proembryo of *Gnetum* (see Sect. 15.8.2.1) (Fig. 15.16). During the development of the proembryo of *Ginkgo,* I observed (Favre-Duchartre 1956) a resting period of 12 days at the 128-nucleate stage. If such a resting period in the growth of the proembryo is common to all naked ovulate plants, one may presume that (1) during the first phase of uninterrupted mitoses, in the proembryos of naked ovulate plants and in the endosperm of angiosperms, there is no translation of DNA in any intra- and extranuclear form of RNA. During this period, there is no formation of any meristem which later gives rise to organs. Brachet (1977, p. 87) writes: "on admet généralement qu'il existe une opposition entre la multiplication des cellules et leur différenciation" (my translation: there exists an opposition between multiplication of cells and their differentiation). (2) The endosperm of angiosperms, which at first develops below the synergids, affects rapid development of a proembryo.

Typically, in angiosperm seeds there is only one functional embryo. It has a sexual origin, a marginal position since the egg cell is attached to the margin of the embryo sac, an endoscopic development, and differentiates a radicle, two or one cotyledon, and a plumule.

Before discussing these characters of the embryos, let us consider, for comparison, some characters common to other cormophytes.

15.8.2.1 Some Embryogenic Characters of Bryophytes, Pteridophytes, and Naked Ovulate Plants

(1) The double fertilization (Fig. 15.14) is followed by development of two embryos but, being identical at first, they grow concurrently; only one commonly survives. (2) Apospory and apogamy, sometimes, occur in Filicinae. (3) The zygote is commonly covered by an archegonial neck, and the earlier development of the embryo is very rapid; whereas the marginal zygote of *Gnetum* remains in a resting phase for 15 days before it gives rise to an embryo. (4) Endoscopy and eusporangy are linked characters in cormophytes. However, in bryophytes, the endoscopic penetration of the young sporophyte into the gametophytic tissue is not as significant as exoscopic growth. (5) Embryos of bryophytes and psilophytes do not differentiate any radicle or cotyledons, but a basal haustorium and an aerial axis (seta of bryophytes).

15.8.2.2 Some Exceptional Characters of Angiosperm Embryos

(1) Two similar sexual embryos appear as a result of double fertilization, when more than one marginal egg cell differentiates in one and the same embryo sac (Figs. 15.12 D n, and 15.15 A_1). Developing concurrently, both embryos abort due to lack of endosperm. (2) Embryos with asexual origin develop parthenogenetically from a haploid or diploid marginal egg cell of the embryo sac, or diploid nucellar cells (Fig. 15.15 B_1). (3) Centrally situated zygote in Podostemaceae develops into embryo; this unique egg cell is associated with one (*Dicraea*) or two (*Apinagia*) synergids (Figs. 15.12 A e and 15.15 G) considered as neck cells by me. (4) Endoscopy is common in angiosperms. However, in Piperales, Loranthaceae and some other taxa, the division of zygote is vertical; but further embryonic development is endoscopic. (5) It is well known that in angiosperms the embryo with one or two cotyledons is a valuable taxonomic character linked to many others:

thus the classical taxonomic distinction between Monocotyledons and Dicotyledons. Doyle (1973) has confirmed, through fossil evidence, the most ancient independence of these two subclasses. But neoteny may also affect these young sporophytes and interrupt their early intraseminal development: in orchids the embryo is only a multicellular ovoid body lacking organs as an endosperm does (Fig. 15.15 F_1). This is an argument to interpret the endosperm as an embryo.

15.9 Seed

A true seed commonly contains an embryo of sexual origin with a nutritive haploid female gametophyte in naked ovulate plants; triploid endosperm, or diploid perisperm, in angiosperms. This seed remains viable for months or years, in a resting stage, protected by the seed-coat.

There is much diversity in the time relations of the development of different elements of the seed, as well as in the final structure of seeds.

15.9.1 Time Relations in the Development of Seed

Emberger (1942), Mangenot (1945), and Favre-Duchartre (1943, 1980) have pointed out that *Ginkgo,* cycads, as well as carboniferous Cordaites and pteridosperms have large differentiated ovules with (1) an integument distinguishable into three regions simulating a drupe, and (2) a female gametophyte with abundant reserve starch, lipids, and proteids. Both these characters are independent of fertilization and embryogenesis (although the ovule needs pollination). It is not possible to recognize externally fertile and sterile ovules. Moreover, the embryo, when present, and its adjoining female gametophyte (which contains abundant water, 66% in *Ginkgo*) are unable to survive more than a few weeks, in dry conditions. Emberger (1944) created the taxon "prephanerogams" ("prespermaphytes" appears more suitable according to the rules of nomenclature) for those plants which have not yet acquired the true status of seed. Favre-Duchartre (1958) has compared the ripe ovule of *Ginkgo* to a bird's egg which may possess a shell and reserve food even in the absence of an embryo.

Figure 15.17 represents time relations for the maturation of seed in three taxa: a prespermaphyte (*Ginkgo*, A, A′), an advanced gymnosperm (*Taxus,* B, B′), and an advanced angiosperm (*Cichorium,* C, C′). The shorter and shorter duration for the maturation of seed is obvious. It appears (Favre-Duchartre 1982 b) that, when double fertilization is accomplished, (1) the division of marginal zygote occurs some hours (*Cichorium, Zea*), days (*Aleurites, Ornithogalum*), or months (*Thea, Colchicum*) later; (2) at the time of initiation of embryogenesis, the development of endosperm is commonly much more advanced in taxa with albuminous seeds (Magnoliales, Umbellales) than in exalbuminous ones (Asteraceae, Orchidaceae); (3) interesting exceptions to the last rough rule are to be found in parasitic taxa (Loranthaceae, Cuscutaceae). In different genera of Euphorbiaceae, where the seeds are albuminous, there is a great variety of endosperm development at the time the marginal zygote divides.

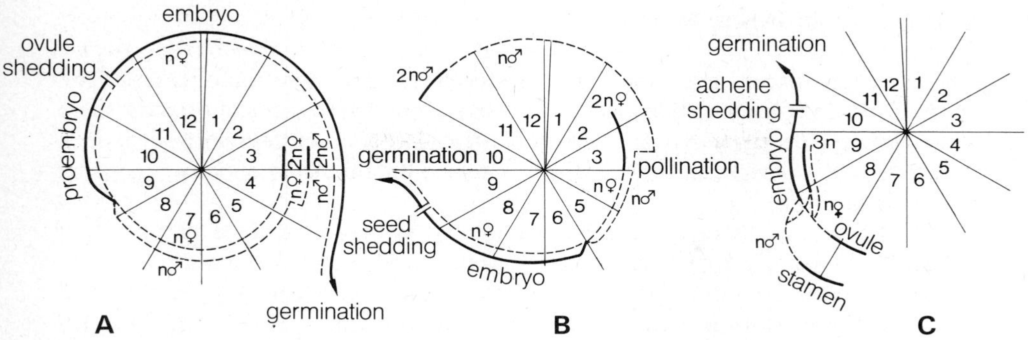

Sexual phenomena	A′ *Ginkgo* in Paris	B′ *Taxus*	C′ *Cichorium*
microsporangia formation	March	November	July
ovule formation	March	February	July
male meiosis	April	November	August
female meiosis	April	March	August
pollination	April	March	August
intraovular development of male and female gametophytes	about 145 days	about 60 days	some hours
fertilization	September	May	August
embryogenesis (duration)	about 240 days	about 100 days	about 45 days
shedding from the mother sporophyte	November	September	October

Fig. 15.17. A *Ginkgo*, **B** *Taxus*, **C** *Cichorium*. Comparison of time relations between male and female meiosis, pollination, fertilization, formation of seed, its dispersal, and germination. The 12 months of the year are represented as a *dial*. Haploid phases are shown by *interrupted lines;* pollination is symbolized by the *centripetal proximity* of the haploid male gametophyte; female gametophyte persists during embryogenesis in *Ginkgo* and *Taxus*. In angiosperms (*Cichorium*) the female gametophyte does not persist after fertilization, the endosperm (symbolized by *one full line* and *one interrupted line* since it is triploid) is completely consumed by the growing embryo before the shedding of achenes. The separation of the new sporophyte (intraseminal embryo) from the maternal parent is symbolized by *two perpendicular lines*. Germination is marked by *centrifugal arrow*. A more classical, but less expressive, calendar **A′, B′, C′** is added to help understanding the three *dials* **A, B, C**

15.9.2 Diversity in Seed Structure

Let us briefly discuss some exceptions to the proposed definition of a seed. The sterile seeds lack the embryo; there is no reserve food in the seeds of orchids; there is no resting period in cases of vivipary (*Rhizophora; Poa pratensis* var. *vivipara*); the seed-coat is absent in the seeds of *Crinum*. These deficiencies have occurred during evolution.

Whatever the diversity in the morphology, structure, and development of seeds, they typically appear as apparatuses made of one or two maternal sporophytic integument(s) protecting one nutritive organism which is to be absorbed by the embryo during seed maturation or germination. These manifold apparatuses have obviously originated along independent phyla.

15.10 Concluding Remarks

What about the "abiding mystery of the evolutionary relationship of flowering plants" (Eames 1961, p. 635)?

As far as formation of seeds is concerned, I propose that embryologically angiosperms have attained a high level of evolution probably along several lines, and through neotenic processes.

F. Jacob wrote (in "Le Monde" dated 11 February 1979; my translation): "In scientific studies, theory always takes precedence. Relevant experimental data can only be obtained with reference to that theory".

The above consideration is quite pertinent as far as the "embryology of angiosperms" is concerned. A better theoretical understanding would provide a better knowledge of the sexual structures of angiosperms, and processes concerning practical improvements to the nutritive value of seeds, seed conservation, control of seed germination, and elimination of weeds (if the reproductive processes are altered specifically).

For further studies, let me suggest two main fields for investigation: (1) a more complete and accurate observation of natural structures, and associated developments through the taxa; (2) experiments using physical, chemical, and biological agents to modify the natural conditions.

15.10.1 Questions Raised by Natural Diversities

(1) Why do the megasporocytes differentiate singly, or in multiples? Why are they deep-seated or subepidermal? (2) The single functional megaspore is chalazal, median, or micropylar in position; but two or four female haploid nuclei may be involved in bi- and tetrasporic gametogenesis. To which other characters are these linked? (3) In mature embryo sac the antipodals may be absent (*Oenothera*); the receptive function of synergids may be assumed by the egg cell itself (*Plumbago*); either marginal (egg cell) or central (polar) female cell may be absent or non functional (Figs. 12, 15), but one or both have to fulfil their sexual function otherwise

the embryo sac would no longer deserve the designation "female gametophyte". What influences are responsible for differentiation or suppression of these cells? (4) Pollen grains, or tubes, become tricellular more or less early, but are both sperm cells fundamentally equipotent? Is the unequal distribution of plastids and mitochondria, observed in sperm cells of *Plumbago,* a general character (see Russell and Cass 1981). Is this associated to a difference in chromatin values in both sperm nuclei? (5) Fertilization in angiosperms presents most specialized characters; the fundamental problem of gametic attraction and fusion remains unsolved. (6) The extent of comparison of endosperm and embryo development in different taxa may afford some light as to their relative differentiation.

15.10.2 Questions Raised by Suggested Experiments

Human intervention to try to modify the normal course of sexual reproduction in angiosperms appears limited to destroying particular cells, or modifying their environment. (1) It appears difficult to interfere with megasporogenesis and oogenesis unless through specific action of parasites (fungus, animals), or pharmacodynamic products. (2) Pollen grains and in vitro cultured pollen tubes would be more accessible to physical agents (laser, X rays ...) to kill the tube nucleus and/or one sperm cell as future evolutionary steps (Fig. 15.8 K, L). Would fertilization be then possible? Would it be the marginal egg cell the one to be fertilized by that single sperm cell? (3) Would suppression of either marginal or central zygote allow the development of the preserved one?

References

Audran JC (1977) Recherches cytologiques et cytochimiques sur la genèse des grains de pollen et des exines chez les Cycadales (Préspermaphytes). Thèse, Univ Reims

Audran JC (1978) Cytobiologie de la microsporogenèse de la palynogenèse et du tapis chez *Ceratozamia mexicana* (Cycadacées). Ann Univ ARERS 15:1–26

Audran JC, Masure E (1977) Contribution à la connaissance de la composition des sporodermes chez les Cycadales (Préspermaphytes). Etude en microscopie électronique à transmission (M.E.T.) et à balayage (M.E.B.). Palaeontographica Beitr zur Naturgeschichte der Vorzeit 162:115–158

Augier J (1966) Flore des Bryophytes. Morphologie, Biologie, Ecologie, distribution geographique. P Lechevalier, Paris

Baird A (1953) The life history of *Callitris*. Phytomorphology 3:258–284

Bannikova VP (1971) Development features of female gametophyte of *Nicotiana*. Ann Univ ARERS 9:73–77

Battaglia E (1951) The male and female gametophytes of angiosperms: An Interpretation. Phytomorphology 1:87–116

Battaglia E (1980) Embryological questions: 2. Is the endosperm of angiosperms sporophytic or gametophytic? Ann Bot (Rome) 39:9–30

Bhojwani SS, Bhatnagar SP (1983) The embryology of angiosperms, 4th edn. Vikas, New Delhi

Billings FH (1934) Male gametophyte of *Atriplex hymenelytra*. Bot Gaz 95:477–484

Brachet J (1977) Aspects cellulaires (croissance et différenciation). In: Colloque de morphogenèse: La multiplication végétative chez les Végétaux. Bull Soc Bot Fr 124:97–107

Brouland M, Jeanny JC (1970) Nouvelles recherches cytologiques et Cytophotométriques sur le rôle des gamètes mâles et femelles dans la formation de la plantule et de l'albumen chez l'*Anemone nemorosa* L. Ann Univ ARERS 9:228–237

Camefort H (1967) Fécondation et formation d'un néocytoplasme chez le *Larix decidua* Mill. (*Larix europea* D.C.). CR Acad Sci Paris 265:1784–1787

Chadefaud M (1941) Un problème de botanique classique mais non résolu: celui des anthéridies et des archégones des Phanérogames et plus particulièrement des Angiospermes. Rev Sci 79:479–482

Chadefaud M (1960) Traité de Botanique systématique. Paris

Chesnoy L (1969) Sur l'origine du cytoplasme des embryons chez le *Biota orientalis* (Endl.) (Cupressacées). CR Acad Sci Paris 268:1921–1924

Church AH (1919) Thallasiophyta and subaerial transmigration. Bot Memoirs 3. Oxford UK

Coker WC (1907) Fertilization and embryogeny in *Cephalotaxus furtunei*. Bot Gaz 43:1–10

Cruden RW (1977) Pollen ovule ratio: A conservative indicator of breeding system in flowering plants. Evolution 31:32–46

Davis GL (1966) Systematic embryology of the angiosperms. John Wiley, New York

De Vries H (1899) Sur la fécondation hybride de l'albumen. CR Acad Sci Paris 129:973–975

Doyle JA (1973) The monocotyledons: Their evolution and comparative biology V. Fossil evidence on early evolutions of the monocotyledons. Q Rev Biol 48:399–413

Ducker SC, Pettitt JM, Knox RB (1978) Biology of Australian seagrasses: Pollen development and submarine pollination in *Amphibolis antarctica* and *Thalassodendron ciliatum* (Cymodoceaceae). Aust J Bot 26:265–285

Eames AJ (1961) Morphology of the angiosperms. McGraw-Hill, New York

Emberger L (1942) Sur les Ptéridospermées et les Cordaitales. Bull Soc Bot Fr 89:202–213

Emberger L (1944) Les plantes fossiles dans leurs rapports avec les régétaux vivants. Masson, Paris

Emberger L (1949) Les Préphanérogames. Ann Sci Nat Bot, 11ème Ser 10:131–144

Emberger L (1971) Travaux de Paléobotanique et d'Ecologie. Paris

Endress PK (1979) Noncarpellary pollination and "hyperstigma" in an angiosperm (*Tambourissa religiosa,* Monimiaceae). Experientia 35:45

Endress PK (1980) Ontogeny, function, and evolution of extreme floral construction in Monimiaceae. Plant Syst Evol 134:79–120

Fagerlind F (1937) Embryologische, zytologische und bestäubungsexperimentelle Studien in der Familie Rubiaceae. Acta Hortic Bergiani 11:195–470

Fagerlind F (1941) Bau und Entwicklung der *Gnetum* Gametophyten. K Sven Ventenskaps Akad Handl 19:1–55

Favre-Duchartre M (1943) Sur le comportement des ovules de *Ginkgo biloba*. Bull Soc Bot Fr 90:7–9

Favre-Duchartre M (1956) Contribution à l'étude de la reproduction chez le *Ginkgo biloba*. Rev Cyt Biol Vég 17:1–218

Favre-Duchartre M (1958) *Ginkgo,* an oviparous plant. Phytomorphology 9:377–390

Favre-Duchartre M (1960) Contribution à l'étude de la reproduction sexuée chez *Araucaria araucana*. CR Acad Sci Paris 250:4435–4437

Favre-Duchartre M (1963) A propos de nucelles des plantes à ovules. Ann Univ ARERS 1:35–42

Favre-Duchartre M (1964) Nouvelles précisions sur l'organisation des gamétophytes ♀ de *Torreya myristica* (= *T. californica*) envisagées comme éléments d'interprétation des sacs embryonnaires des Angiospermes. CR Acad Sci Paris 258:661–664

Favre-Duchartre M (1966) Les angiospermes sont-elles des Archégoniates? Bull Soc Bot Fr 113:368–375

Favre-Duchartre M (1968) Point de vue sur l'apparition et la répartition des sexes au cours de l'évolution des plantes. Bull Soc Bot Fr 115:515–522

Favre-Duchartre M (1970) Des ovules aux graines. Aspects cytologiques de la reproduction sexuée chez les plantes supérieures. Masson, Paris

Favre-Duchartre M (1971) A propos de la notion d'archégone. Bull Soc Bot Fr 118:525–536

Favre-Duchartre M (1972) Interpretation de la reproduction sexuée des Angiospermes à la lumiére d'organisations gamétophytiques et embryonnaires d'autres Archégoniates. Phytomorphology 21:353–361

Favre-Duchartre M (1974) Phylogenetic aspect of the spermaphytes' double fertilization. In: Linskens HF (ed) Fertilization in higher plants. North-Holland, Amsterdam, pp 243–252

Favre-Duchartre M (1975) About the marginal situation of angiospermous egg cell. Phytomorphology 25:66–68

Favre-Duchartre M (1978) Eight interpretation of the embryo sac. Phytomorphology 28:407–418

Favre-Duchartre M (1980) Time relations and sexual reproduction in *Cichorium* and other angiosperms as compared with archegoniates. Phytomorphology 29:166–178

Favre-Duchartre M (1982a) Macroévolution des structures méiosporangiales. Bull Soc Bot Fr Lettres Bot 129:217–222

Favre-Duchartre M (1982b) Débuts des développements des albumens et des embryons selon les taxons. Ann Univ ARERS 18:59–60

Favre-Duchartre M, Louant BP, Tilquin JP (1979) Durée de vie et organisation du sac embryonnaire au sein de l'ovule de *Cichorium intybus* L. (Compositae) et la théorie taxalienne. CR Acad Sci Paris 288:331–334

Fedortschuk W (1931) Embryologische Untersuchung von *Cuscuta monogyna* Vahl. und *Cuscuta epithymum* L. Planta 14:94–111

Foster AS, Gifford EM (1974) Comparative morphology of vascular plants. WH Freeman, San Francisco

Gianordoli M (1974a) A cytological investigation on gametes and fecundation among *Cephalotaxus drupacea*. In: Linskens HF (ed) Fertilization in higher plants. North-Holland, Amsterdam, pp 221–232

Gianordoli M (1974b) Ultrastructure des spermatozoides de trois gymnospermes: *Cephalotaxus drupacea, Sciadopitys verticillata, Taxus baccata*. CR Acad Sci Paris 278:2637–2640

Gianordoli M (1978) Gamétogenèses mâles et femelles, mises en contact des gamètes et formation du cytoplasme proembryonnaire chez trois Gymnospermes: *Sciadopitys verticillata, Cephalotaxus drupacea, Taxus baccata*. Etude ultrastructurale et cytochimique. Thèse, Univ Reims

Goetz G (1899) Über die Entwicklung der Eiknospe bei den Characeen. Bot Zeit 57:1–13

Graham L, McBride G (1979) The occurrence and phylogenetic significance of a multilayered structure in *Coleochaete* spermatozoids. Am J Bot 66:887–894

Guignard L (1917) L'ovule chez les Apocynacées et Asclépiadacées. Mém Acad Sci Inst Fr 55:1–34

Gupta SC, Nanda K (1978) Ontogeny and histochemistry of dimorphic tapetum in *Tecoma stans* anthers. Soc Bot Fr Actualités Bot No 1–2, pp 129–134

Jalouzot MF (1982) Cytochronologie expérimentale des développements de l'albumen et de l'embryon de *Colchicum autumnale*. Ann Univ ARERS Reims 18:21–24

Jensen WA (1974) Reproduction in flowering plants. In: Robards AW (ed) Dynamic aspects of plant ultrastructure. McGraw-Hill, London, pp 481–503

Jensen WA, Fischer DB (1968) Cotton embryogenesis: Double fertilization. Phytomorphology (Maheshwari Memorial Vol) 18:261–269

Johri BM (1936) The life history of *Butomopsis lanceolata* Kunth. Proc Indian Acad Sci Sect B4:139–162

Johri BM, Bhatnagar SP (1969) Endosperm in the Santalales. Rev Cytol Biol Vég 32:353–369

Johri BM, Bhatnagar SP (1972) Loranthaceae. Bot Monograph No 8. CSIR, New Delhi

Johri BM, Kapil RN (1953) Contribution to the morphology and life history of *Acalypha indica* L. Phytomorphology 3:137–151

Le Maout E, Decaisne J (1868) Traité général de Botanique descriptive et analytique. Paris

Le Monnier G (1887) Sur la valeur morphologique de l'albumen chez les Angiospermes. J Bot 1:140–142

Maheshwari P (1950) An introduction to the embryology of angiosperms. MacGraw-Hill, New York

Mangenot G (1945) A propos de la notion de graine. Rev Sci 83:117–119

Mangenot G (1973) Données élémentaires sur l'angiospermie. Ann Univ Abidjan 6:1–245

Moussel B (1971) Sur l'existence et l'évolution rhythmique d'un système vacuolaire périphérique dans le prothalle femelle cénocytique de l'*Ephedra distachya*. CR Acad Sci Paris 273:1929–1932

Moussel B (1978) Double fertilization in the genus *Ephedra*. Phytomorphology 28:336–345

Moussel B (1979) Etude cytobiologique de la reproduction sexuée de l'*Ephedra distachya* L. Ultrastructure et cytochimie. Thèse, Univ Reims

Periasamy K, Swamy BGL (1964) Is the microsporogangium of angiosperms wall-less? Curr Sci (Bangalore) 33:735–738

Porsch D (1907) Versuch einer phylogenetischen Erklärung des Embryosackes und der doppelten Befruchtung der Angiospermen. Verh Zool Bot Ges Oesterr 57:120–134

Prakash N, Herr JM Jr (1979) Embryological study of *Glottidium viscarium* through the use of clearing technique. Phytomorphology 29:71–77

Robert D (1971) Etude en microscopie électronique des modalités d'édification des parois microsporales chez le *Selaginella selaginoides* (L.). Mise en place du feuillet externe. CR Acad Sci Paris 273:332–335

Robbins RR, Carothers ZB (1978) Spermatogenesis in *Lycopodium:* The mature spermatozoid. Am J Bot 65:533–540

Russell SD, Cass DD (1981) Ultrastructure of the sperms of *Plumbago zeylanica*. I. Cytology and association with the vegetative nucleus. Protoplasma 107:85–107

Sarfatti G (1958) Ricerche sull'endosperma a l'austorio endospermatico di *Prunus communis*. Nuovo G Bot Ital 65:371–374

Sarfatti G (1961) Accrescimento del pericarpo, seme, endosperma ed embrione in *Prunus amygdalus* Stockes. Nuovo G Bot Ital (Nuova Ser) 68:118–135

Sargant E (1900) Recent work on the results of fertilization in angiosperms. Ann Bot 14:689–712

Schnarf K (1942) Archegonium und Archegontheorie: kritischer Versuch einer Geschichte des Pflanzenarchegons. Biologia Gen 16:198–224

Schürhoff P (1919) Zur Phylogenie des Angiospermenembryosackes. Ber Dtsch Bot Ges 37:160–168

Swamy BGL (1974) On the presumed ancestry of angiosperm embryo sac. Phytomorphology 24:102–106

Takhtajan A (1969) Flowering plants. Oliver and Boyd, Edinburgh

Taylor TN, Millay MA (1979) Pollination biology and reproduction in early seed plants. Rev Palaeobot Palynol 27:329–355

Thathachar T (1952) Morphological studies in *Acalypha lanceolata* (Euphorbiaceae). Phytomorphology 2:197–201

Tieghem P van (1891) Traité de Botanique. F Savy, Paris

Tilton VR, Lersten NR (1981) Ovule development in *Ornithogalum caudatum* (Liliaceae) with a review of selected papers on angiosperm reproduction. III. Nucellus and megagametophyte. New Phytol 88:477–504

Tourte Y (1969) Interprétation des cellules du sac embryonnaire des angiospermes à la lumière des études de l'oogenèse chez les Ptéridophytes. Rev Cytol Biol Vég 32:241–251

Vasil V (1959) Morphology and embryology of *Gnetum ula* Brongh. Phytomorphology 9:167–215

Walker JW (1974) Evolution of exine structure in the pollen of primitive angiosperms. Am J Bot 61:891–902

Walker JW, Skvarla JJ (1975) Primitively columellaless pollen: A new concept in the evolutionary morphology of angiosperms. Science 18:445–447

Wunderlich R (1959) Zur Frage der Phylogenie der Endospermtypen bei den Angiospermen. Oesterr Bot Z 106:203–293

Yakovlev M (1974) Gametogenesis, embryo sac and pollen grains (contribution to the origin in Angiospermae). In: Linskens HF (ed) Fertilization in higher plants. North-Holland, Amsterdam, pp 253–260

Yoffe M (1969) Fertilization and formation of the primary endosperm nucleus in *Anemone nemorosa* L. Rev Cytol Biol Vég 32:319–323

16 Experimental Embryology

B. M. Johri and P. S. Rao

16.1 Introduction

The beginnings of experimental embryology date back to the observations of Massart (1902) who obtained swelling of the ovary by treating the ovaries of certain plants with spores of *Lycopodium,* dead pollinia, and aqueous extracts of pollen. It was suspected that the pollen tube, apart from carrying the sperm nuclei, also contributes to the growth of ovary by supplying auxins. Rechinger (1893) placed root pieces of beet and dandelion, and stem segments of poplar, on moist filter paper. He attempted to induce growth in a minimal number of cells, although slices thinner than 1.5 mm gave no callus. Haberlandt (1902) attempted to cultivate cells from the leaves of several angiosperms. He experimented with isolated palisade cells of *Lamium purpureum,* and hair cells of *Tradescantia virginica* and *Pulmonaria mollissima.* His assumption that chlorophyllous cells supply themselves with organic substances for nutrition proved erroneous. However, Haberlandt's unsuccessful efforts paved the way to finding suitable nutrient composition and optimal conditions to grow organs, tissues, and cells. Knudson (1919) cultured the root cap cells which remained viable for 70 days but did not divide. Molliard (1921) adopted a different approach. He used segments of roots and hypocotyl of young radish shoots which had embryonic activity and, hence, grew in culture but did not divide to form new tissues.

Botanists who worked with excised tissues and cells were impressed by the success of contemporary zoologists (Carrel and Burrows 1911) who cultivated mammalian tissue on complex nutrient media consisting of blood plasma or embryonic fluids.

Prompted by these observations, Prat (1927) and others cultured excised plant tissues on plant extracts. Their experiments failed due to the improper selection of material. A suitable choice of the material enabled Robbins (1922) and Kotte (1922) to grow simultaneously, and independently, apical meristems in synthetic nutrient medium.

The pioneers in the development of tissue culture techniques were Philip R. White (USA) and R. Gautheret (France). White (1934) succeeded, for the first time, in repeatedly subculturing excised tomato roots without diminution of growth rate. He was also able to grow lateral root tips that developed in the cultures, and establish a clone of tomato roots in continuous culture.

Following White's success, Gautheret (1938) and Nobécourt (1938) established cultures from the cambial tissues of *Salix capraea* and carrots, respectively. Gautheret (1939) then cultured the cambial and phloem explants of carrot roots on a medium containing inorganic salts, glucose, thiamine, cystine hydrochloride, and indoleacetic acid (IAA). Simultaneously, White (1939) also reported the continuous cultivation of undifferentiated callus mass obtained from the stem of a hybrid of *Nicotiana* (*N. glauca* × *N. langsdorfii*) on a medium devised for tomato roots but jelled with 0.5% agar. Thus, the basic technique of tissue culture described by White and Gautheret laid the foundation for the successful establishment of cultures of many species (see also Gautheret 1982).

16.2 The Problems: Role of Experimental Embryology

Since the pioneering work of White and Gautheret there have been significant advances in tissue culture research, and we now understand much better the development of anther, pollen, ovule, ovary, endosperm, embryo, and seed. Control of pollination and fertilization is yet another aspect. Thus, from an era of observations and inferences, experimental embryology has introduced an era of experiments and deductions to unravel the physiological, biochemical, and genetical processes involved in growth, development, and differentiation of reproductive organs.

The in vitro culture of ovaries has led to the successful culture of ovules into mature seeds. This not only offers immense opportunities for overcoming barriers to incompatibility, but also cuts short the time between pollination and maturation of seeds and, thus, hastens the production of seedlings. In members of Rutaceae, as in *Citrus,* the nucellus offers an excellent experimental material because of its capacity to produce embryos. This is of great advantage in horticultural practice, and can be utilized for clonally propagating desirable varieties.

The culture of floral organs is another area of study which has attracted much attention. Haploid plants derived from anther/pollen have great significance in crop improvement. Their importance in studies on mutations, and for the production of homozygous plants is well recognized. Conventional breeding methods to produce haploids is a tedious and time-consuming task. The first report of Guha and Maheshwari (1964), on the induction of haploids from pollen in *Datura,* introduced a new method for producing haploids in large numbers. Since then, during the last decade, haploid plants have been obtained in over a hundred species. Some of the factors for successful culture of anthers appear to be the developmental stage of microspores, the culture media, physiological stage of the plant at the time of culture, and seasonal variations. The successful production of haploid plants is achieved either through the regeneration of embryoids, or through the development of callus and its subsequent differentiation (see C. Nitsch 1981).

To date, anther culture has been extensively used in many parts of the world, especially in China, in programmes related to genetics and plant breeding. The isolation of nullihaploid plants from cultured anthers of monosomic tobacco plants, and the regeneration of these nullisomics afford a good system for determining linkage groups and the genetic map of the plant. The feasibility for producing breeding lines of haploid tobacco plants differing in alkaloid content is yet another milestone in the field of anther culture.

Although success with the technique of in vitro fertilization has been reported for a few plants (Rangaswamy 1977), its usefulness has not been widely recognized. The method of placental pollination should prove promising in such systems in which selfing or hybridization has been unsuccessful. The technique of test-tube fertilization which has been so elegantly demonstrated in *Papaver somniferum* and *Argemone mexicana* (Kanta et al. 1962) has important implications for the plant breeder. Thus, rearing of hybrids and homozygous plants can be accomplished much more easily than has been possible until now.

The artificial culture of excised embryos is of considerable usefulness to the plant breeder and the horticulturist: it enables them (1) to rear hybrid embryos which may otherwise fail to survive, due to the baneful influence of maternal tissues, (2) to overcome the period of dormancy in certain plants, (3) to make viability tests of the seeds and, most important, (4) it will serve as a tool for understanding the nutritional requirements of the growing embryo (see Raghavan 1980).

The endosperm, which is produced by a unique event of double fertilization, contains the triploid number of chromosomes, and is the main nutritive tissue of the embryo. During the maturation of seed, the endosperm may be completely consumed, or may persist as a massive tissue containing reserve food in the form of starch, fat, or proteins. The morphological nature of endosperm has been a subject of much discussion (P. Maheshwari 1950). Whether it is a "second embryo" or a "maimed embryo" is among the problems for theoretical considerations. Endosperm is the only ovular tissue which does not form an embryo in situ. Investigations have provided ample evidence that triploid plants can be raised from endosperm culture. This would be extremely useful to plant breeders who can apply the technique for raising economically important plants where conventional breeding methods prove futile.

Embryogenesis in plants begins with the fertilization of the egg cell by a male gamete. Initially, the zygote does not show any evident differentiation but, soon after, polarity sets in and it develops into an organized embryo through a sequence of orderly events characteristic for a particular species. However, during the past three decades, emphasis has shifted from essentially morphological studies to experimental induction of embryos in order to understand the processes involved in embryogenesis. Embryogenesis in somatic and reproductive tissues has been reported in many taxa. Asexual embryogenesis has implications in large-scale clonal propagation of some species.

Some of the factors which regulate embryogenesis are: the nitrogen and potassium levels in the medium, inorganic ions such as calcium, osmotic relation, carbohydrates, vitamins, amino acids and amides, and growth-regulating substances.

One of the most significant developments in experimental embryology has been the isolation, culture, and fusion of plant protoplasts which have far-reaching implications in studies on plant improvement by cell modification and somatic hybridization. Protoplasts in culture can be regenerated into an entire plant. They can be induced to undergo interspecific, intraspecific, and intergeneric fusions resulting in somatic hybrids. Successful fusion of protoplasts resulting in somatic hybrids has been achieved in tobacco, *Petunia,* and carrot. More recently, intergeneric fusion of protoplasts from potato and tomato has been demonstrated (Melchers 1982). Considerable work is in progress on developing selection systems which is an important parameter for the identification of somatic hybrids.

16.3 Techniques of Cell, Tissue, and Organ Culture

The earlier techniques were fairly simple, and employed semi-solid media with agar for establishing and maintaining callus cultures. However, much of the critical work on nutrition, metabolism, and growth has been conducted with liquid media; many tissues are grown either on stationary or agitated media. The advantages of growing tissues in agitated liquid media are that it facilitates gaseous exchange, reduces nutrient gradients, and is amenable to manipulation in experiments where radioisotopes are employed. Depending upon the need of the experiment, tissues are grown either under continuous agitation, or alternately immersed in the liquid medium and exposed to air. The latter method ensures adequate mixing as well as provides efficient gaseous exchange. Cultures are generally incubated in darkness, or in low-intensity light.

Suspension cultures are usually initiated by transferring a vigorously growing tissue to an agitated liquid medium in Erlenmeyer flasks. Dispersion of cells from the tissue placed in suspension can be enhanced by using high auxin concentrations (Torrey and Reinert 1961), or an appropriate balance of auxin and kinetin (Earle and Torrey 1963), or between auxin and coconut milk. Sometimes, deficiencies of certain vitamins in the nutrient medium are useful for cell separation (Torrey and Reinert 1961).

Suspension cultures are subcultured by transfer at regular intervals, of either untreated or fractionated aliquots of suspension, to fresh medium. It is desirable that a high cell density is used for subculture, since there is a correlation between population density and growth of cells into colonies.

Callus cultures require periodic transfer to a fresh nutrient medium; this process is usally repeated every 4–6 weeks.

Cell divisions occur more readily and actively in cell aggregates than in free cells. Against this background, various experimental approaches have been suggested to culture single, isolated cells. Muir et al. (1954) obtained single cells from friable crown gall tissue, and placed them singly on the upper surface of filter-paper squares which were kept on small pieces of actively growing callus tissue. The callus served as "nurse tissue". Although they observed growth of the single cells to few-celled colonies, this method has certain disadvantages. Firstly, only a limited number of cells can be cultured at a time and, secondly, the manipulation of a single cell is extremely difficult. Also, observations of the growth of single cells on the "nurse tissue" is rather tedious.

Another alternative technique, adopted by De Ropp (1955), is to grow single cells in microchambers containing liquid medium. This could be observed continuously under a phase-contrast microscope. Torrey (1957) used microchamber, i.e. Maximow's double cover-slip method. He placed the cells on agar around a small "nurse" callus tissue. Although some cells divided, the largest aggregate derived contained only seven cells. The most successful experiment with the microchamber was conducted by Jones et al. (1960). They sealed the microchamber with inert mineral oil, and obtained the growth of single cells isolated from a callus of hybrid tobacco (*Nicotiana tabacum* × *N. glutinosa*) using "conditioned" medium which had previously supported the growth of cell population.

Bergmann (1960) developed the technique of plating cell suspensions on agar. It involves spreading of a free-cell suspension on the surface of agar in Petri dishes, sealing the petri dishes, and incubating them in diffuse light. Alternatively, the suspension can be mixed with sterile medium containing 0.6% agar, and cooled to 35 °C before mixing.

Raghavan (1976) has given the constitution of 18 nutrient media that are commonly used for the culture of reproductive organs. Thorpe (1981) refers to media and characteristics of cell and tissue culture, and organogenesis and embryogenesis in cultures (see also Dodds and Roberts 1982). Reinert and Yeoman (1982) give procedural details of how to raise cultures, maintain them, and study growth behaviour in protoplast, cell, tissue, and organ cultures, bioassay systems for cytokinins, and production of secondary metabolites. Wetter and Constabel (1982), in their *Plant Tissue Culture Methods,* devote attention to composition of media, use of vitamins, hormones, other supplements, callus and cell culture, organogenesis and embryogenesis, protoplast culture and fusion of protoplasts, cryopreservation, and several other related topics. During the last several years the refinements in techniques and synthesis of more exacting nutrient media have significantly advanced our knowledge of growth, development, and differentiation in tissue cultures (see Bhojwani and Razdan 1983).

The recent publications provide a good deal of information on "experimental embryology", and attention is invited to *Perspectives in Plant Cell and Tissue Culture* (I. K. Vasil 1980), *Experimental Embryology of Vascular Plants* (B. M. Johri 1982), *Plant Tissue Culture 1982* (Fujiwara 1982), and *Tissue Culture of Economically Important Plants* (A. N. Rao 1982).

16.4 Floral Organs and Buds

16.4.1 In Vivo Studies

Gynoecium. Walker (1978) tested the role of physical confinement on epidermal differentiation – dedifferentiation of gynoecium of *Catharanthus roseus.* Plants were grown at 32 °C and 12 h light (ca. 20,000 lx), and 21 °C and 12 h dark period. In young floral buds, remaining attached to parent plant, a gold-foil (150 μm × 200 μm × 15 μm) barrier (ca. 250 μm high) was inserted between the unfused carpel primordia. The gynoecium was allowed to grow up to 14 days. Epidermal cells adjacent to the barrier failed to dedifferentiate as they would do in normal carpel fusion. Dedifferentiation occurs only when the cells are in direct contact; the stimulus of dedifferentiation is not transferred to adjacent cells that lack direct contact. In this system the "diffusible messenger molecule or some kind of cell surface interaction" can be effective only when there is a cell-to-cell contact. A physical impermeable barrier prevents the movement of the morphogenetic regulator and, thus, dedifferentiation does not occur.

Stamen. Sawhney and Greyson (1979) studied stamen development in a tomato mutant-stamenless 2(Sl_2/Sl_2). Clones of normal (+1+) and mutant (Sl_2/Sl_2) plants were maintained through cuttings. For hormone application experiments, plants were grown in growth chambers at 23 °C ± 2 °C during day and 18 °C ± 2 °C during night. They were illuminated (16 h/day; 1,000 ft-c). Gibberellic acid (GA_3 10 µl of 10^{-3} M) was applied in the axil of the youngest leaf, or at the base of a floral bud. GA_3 solution contained 0.05% Tween-20. Control plants were treated with 0.05% Tween-20 only. GA_3 was applied to Sl_2/Sl_2 plants with floral buds at different stages of development.

These experiments were designed to obtain information as to when, during ontogeny, the staminal primordia of Sl_2/Sl_2 flowers (1) were sensitive to GA_3 treatment, i.e. whether the treatment could induce all or some of the features of normal stamens, or (2) the primordia of SL_2/Sl_2 did not respond to GA_3 treatment.

1. Bud with sepal 0.3 mm long, stamen of 0.08 mm length – normal stamens develop on mutant plants but differentiation progressively decreases if GA_3 is applied to buds with sepals longer than 0.3 mm.
2. Sepals 0.5 mm long, stamens of 0.15 mm length – stamens yellow, pubescent, length significantly more than in control, fuse to form a cone, ovules also differentiate at the junction of filament and anther.
3. Sepal 0.8 mm, stamen of 0.3 mm length – GA_3 failed to induce the formation of staminal cone, the number of yellow and pubescent stamens per flower was reduced.
4. Sepals 1.5–2.0 mm long, mean stamen length 0.3–0.6 mm – marked increase in filament length as compared to stamens of normal or Sl_2/Sl_2 control flower, or GA_3-treated younger bud; staminal cone does not develop; a few ovules develop at the junction of anther and filament.
5. Sepals 3.0 mm long, stamen of 0.77 mm length – buds with stamens longer than 0.77 mm did not show any significant change.

Thus, the ability of Sl_2/Sl_2 flower to produce normal characteristics, when treated, is restricted to early stages of bud. There is a positive correlation between the sepal length and stamen length of Sl_2/Sl_2 flowers.

When GA_3 was applied to plants with enlarged, flattened, bolting apices, or to floral buds with sepal primordia of 0.1–0.3 mm length, stamens were yellow and pubescent as in normal flowers. Ovules failed to develop at the junction of filament and anther. The results are interpreted on the basis of "determination" of the tissue which is also committed to a certain pattern of development.

16.4.2 In Vitro Studies

Petals. McHughen (1977) cultured the petals of *Nicotiana tabacum* L. cv Wisconsin 38 on modified Shenck and Hildebrandt liquid medium. The cultures were maintained at 25 °C, on a 12 h light: 12 h dark cycle. Petals were cultured between the time they become visible and when they fuse with adjoining petals (each petal 1 mm long). In vivo each mature petal is ca. 53 mm long, in vitro (after 3–4 weeks) only 17.4 mm. In cultures the petal merely expands, and neither forms callus nor undergoes organogenesis. It appears to be mature with trichomes on adaxial surface as in petals in vivo. The author concludes that even at its initiation the petal is "determined", and develops normal morphology in isolation, without adding any exogenous hormones to the nutrient medium.

Stamens. Staminal primordia of *Nicotiana tabacum* var. Wisconsin 38 (Hicks 1979) were excised at three stages of floral buds: (1) primordial sepals, petals, and stamens (PS stamens); (2) buds with emergent carpel primordia, and staminal primordia had begun to differentiate anthers (OC stamens); (3) buds with almost fused carpel primordia (CC stamens). The plants were reared and brought into flower at 20 °C and a 18 h light: 6 h dark cycle. The explants were cultured on Linsmaier and Skoog (LS) liquid medium + 1.0 ppm kinetin + 4% sucrose. Light-grown cultures were terminated at weekly intervals up to a total of 4 weeks.

On LS medium, PS, OC, and CC staminal primordia did not grow at all, or only to a limited extent, in light or dark.

On LS + kinetin medium, of the 250 PS staminal primordia, after 7 days, 45 explants formed outgrowths on the adaxial surface, distal tip of explant, or at both sites. In other experiments, after 14 days in light, the tips of outgrowths became stigmatoid (stigma-like) and bright-green. In dark the tips became yellow.

In PS, OC, and CC staminal primordia, 3- to 4-week-old cultures showed lowest (%) survival rate in PS explants. In CC explants the survival rate was more than double. Only the PS staminal primordia showed a high frequency of feminization (stigma-like outgrowths), whether in the light (78%) or in the dark (67%). The OC and CC staminal primordia developed anthers only.

The staminal primordia "show a progressive loss with age in their ability to express alternate development pattern". "... some tissues of the young stamen are very labile and the mitotic progeny of these cells can reorganize rapidly into organs of the 'other' sex". It is obvious that the young staminal primordia are not "determined" and, therefore, show change in sex. The older primordia become "determined" and cannot change their commitment.

Floral Buds/Flowers. The culture of floral buds has made it possible to determine the mutual relationship of individual parts. The technique was first initiated by LaRue (1942) who cultured pollinated flowers of tomato, *Kalanchoe* and *Caltha* which resulted in normal fruits. This work was extended to gherkin, bean, strawberry, and tobacco (J. P. Nitsch 1951). Aseptically cultured fruits described by J. P. Nitsch were relatively small in size, whereas fruits of *Iberis amara* (N. Maheshwari and Lal 1958), *Althaea rosea* (Chopra 1962), *Ranunculus sceleratus* (Sachar and Guha 1962), and *Allium cepa* (Johri and Guha 1963) were as large as, or larger than, those in nature. Auxins (IAA or phenoxy or naphthoxyacetic acids), gibberellic acid and kinetin, particularly in combination, frequently stimulated fruit enlargement (Street and Henshaw 1966).

In *Nigella damascena* (Raman and Greyson 1974) the pedicel callus sloughed off into single cells and produced embryoids, while the inflorescence axis and floral buds of tuberose (*Polyanthes tuberosa*) produced imperfect embryos in culture (Narayanaswamy and Prabhudesai 1979). In *Begonia franconis* (Berghoef and Bruinsma 1979c) culture of inflorescence pedicels yielded adventitious flower buds.

For flower-bud culture of *Begonia franconis* liquid medium proved better than solid, for complete development. Growth required both nitrates and ammonium (Berghoef and Bruinsma 1979a); cytokinins and auxins were required at specific concentrations. After callus induction auxin generally decreased the number of flower buds. Ringe and Nitsch (1968) suggested that the function of auxin is to maintain the vitality of the explants until meristems are organized. Gibberellins promoted organ initiation in the buds (Berghoef and Bruinsma 1979b). Auxins and cytokinins did not affect the ratio of male and female flowers, although the nutrient as well as hormonal requirements for induction of male and female flowers exhibited minor differences (Berghoef and Bruinsma 1979c). A high sucrose level was required for the differentiation of female buds. Besides, optimum cytokinin concentration for the female bud was 10- to 30-times higher than that for male buds. While male buds were produced from callus, pistillate buds never differentiated directly on the callus; they invariably originated from the axillary buds in the bracts of the pedicels of staminate buds. This indicates that either a generative meristem must be developing before female organs can differentiate, or that the male buds produced a specific factor required for female differentiation (Berghoef and Bruinsma 1979c).

Shoot production in in vitro-cultured flower buds of *Allium porrum* has been reported by Novak and Havel (1981), and their technique could be of value as a micropropagation method, especially for selected strains of *Allium*. Plantlet regeneration has been reported in callus tissues obtained from flower petal parts of *Hemerocallis* cv Chipper Cherry (Heuser and Apps 1976), immature inflorescences of *Hosta plantaginea* (Papachatzi et al. 1980), and catkin explants of European birch (*Betula pendula*) (Srivastava and Steinhauer 1981). Embryogenesis has also been reported in callus cultures derived from cultured immature inflorescences and nodes of *Lolium multiflorum* (Dale et al. 1981).

Salix shows a dormant phase (at Chandigarh, Punjab) between autumn (August–September) and next spring (February–March). During this period the dormant buds "get determined" into reproductive buds. In *S. tetrasperma* (Nanda et al. 1981) the pre- and post-dormant female buds sprout when cultured on modified Murashige and Skoog (MS) medium; female catkins attain full development only when kinetin is added to the medium. Angrish and Nanda (1982a) state that when the buds were cultured towards the end of the dormant period (January–February), they could trace sequential stages of development of female fertile catkins with well-developed ovaries. The post-dormant male buds of *S. babylonica* (Nanda et al. 1981) sprouted and formed male catkins in cultures; the pre-dormant buds developed into catkins only when MS medium contained kinetin (KN). According to Angrish and Nanda (1982b), only towards November, anthesis was observed in the cultured buds.

Lord and Mayers (1982) investigated whether GA_3 can switch over a floral primordium "fated" to be cleistogamous (CL) to chasmogamous (CH) form. They worked on *Lamium amplexicaule* which develops CL flowers at node 4 (N_4)

and CH at N_7 and N_8. The two types of flowers are indistinguishable from initiation to anther differentiation. In vivo, 0.2 mm primordium at N_4 and N_7 were treated with 0.1 mM GA_3 by a single application of a 20 μl drop to the node. For in vitro studies, 0.2 mm primordia at N_4 and $N_{7,8}$ attached to a segment of leaf tissue were grown on modified MS medium supplemented with filter-sterilized GA_3 (1 ppm) alone, or with KN (10 ppm). The cultures were maintained at 27 °C, 16 h light (3,000 lx) and 16 h dark period. Occurrence of anthesis was monitored in all samples.

In vivo, GA_3 caused cell expansion in corolla converting a N_4(CL) corolla base into $N_{7,8}$(CH) corolla base. In the upper region of corolla also there was cell expansion, but no cell divisions to convert a N_4(CL) into a $N_{7,8}$ (CH) corolla. The cultured explants showed similar results.

GA_3 reduced significantly pollen counts in $N_{7,8}$ anthers, in vivo, apparently inducing a N_4 (CL) type reduced anther. In cultures there was no effect on pollen contents in N_4 as compared to in vivo controls. Pollen counts in $N_{7,8}$ flowers in vitro were more reduced as in the in vivo GA_3-treated flowers; normal growth of CH flowers was not attained in cultures.

There was reduced seed-set in GA_3-treated flowers in vivo, perhaps due to pollen sterility. In cultures of CL flowers also there was pollen sterility and no seed-set.

The application of GA_3 to a potentially CL floral primordium results in production of a "mimic" CH floral form in vivo as well as in vitro. Therefore, GA_3 is not the only factor responsible for the switch over from CL to CH flower in an inflorescence. This investigation needs further study to determine whether it is, or is not, possible for the switch over of CL to CH flower. Perhaps, the primordia are already "determined" and the commitment cannot be changed.

16.5 Anther and Pollen Culture

Anther culture is a means of obtaining haploids from pollen grains. Since haploids possess only a single set of chromosomes, they are of special interest to geneticists and plant breeders. Natural haploids of angiosperms are of rare occurrence, and are limited to a few species (Kimber and Riley 1963).

Since the discovery of anther culture by Guha and Maheshwari in 1964, this technique of haploid induction (Fig. 16.1 A–F) has been successfully applied to over a hundred taxa including several economically important species. The potential value of haploids to plant breeding, genetics and other disciplines of plant sciences has been extensively reviewed (Vasil 1980, S. C. Maheshwari et al. 1982, Narayanaswamy and George 1982).

16.5.1 Cultural Procedure and Nutritional Requirements

The procedure for culturing the anthers is relatively simple. Young flower buds at the right stage of microspore development are surface-sterilized, and the an-

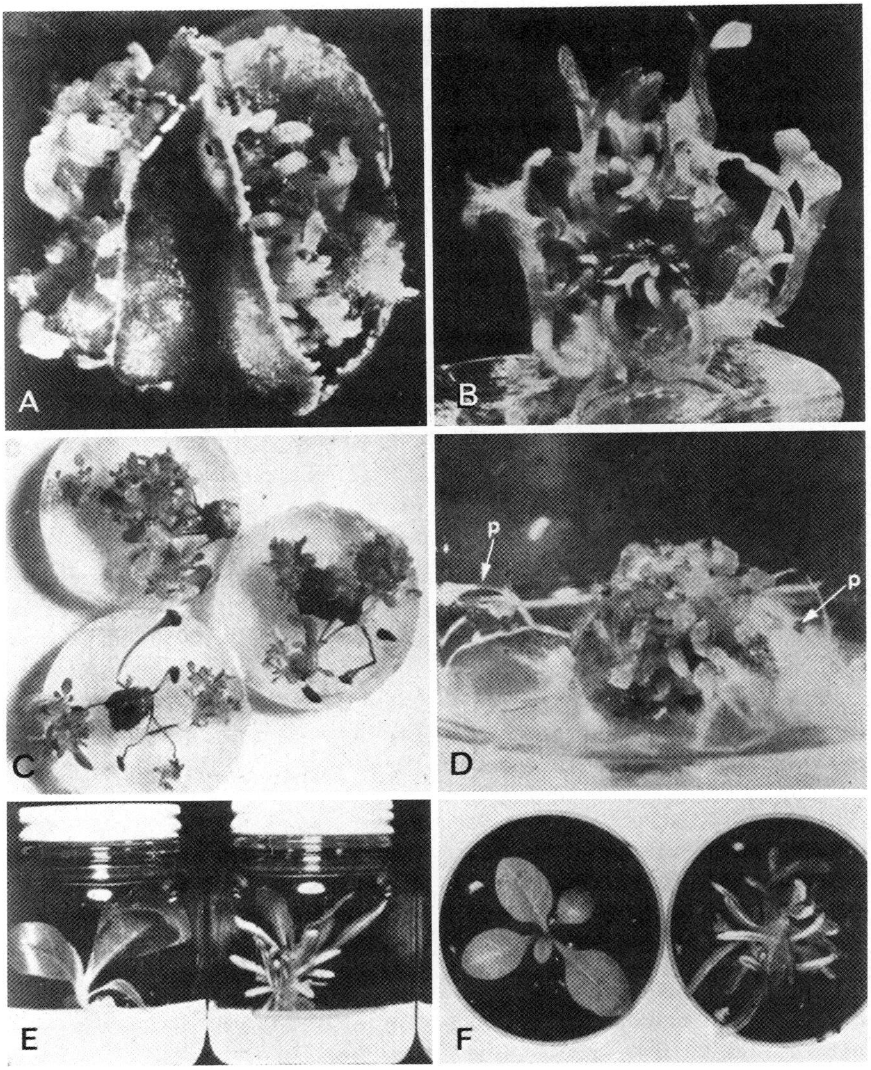

Fig. 16.1 A–F. *Nicotiana tabacum* cv White Burley, anther culture. **A** 28-day-old culture of anther on Bourgin and Nitsch medium showing the development of embryos from pollen grains. **B** Small anther as in **A** showing the development of plantlets. **C** Stamens cultured intact with ovary and style showing the development of plantlets. **D** Single anther cultured in Nitsch medium + NAA (5 mg/l), anther tissues have formed callus from which numerous roots have developed, *p* haploid plantlet. **E, F** Individual plantlets; in **F** plantlets are transferred to a pot with a mixture of peat and sand. (After Sunderland and Dunwell 1977)

thers are carefully removed and cultured on a nutrient medium. For most species semi-solid agar medium has been used (Guha and Maheshwari 1964, J. P. Nitsch 1969). However, in certain species, a better yield of pollen-plantlets is obtained by using liquid medium (Wernicke and Kohlenbach 1977, see also Dodds and Roberts 1982). When the dissection of buds poses problems, intact whorl of sta-

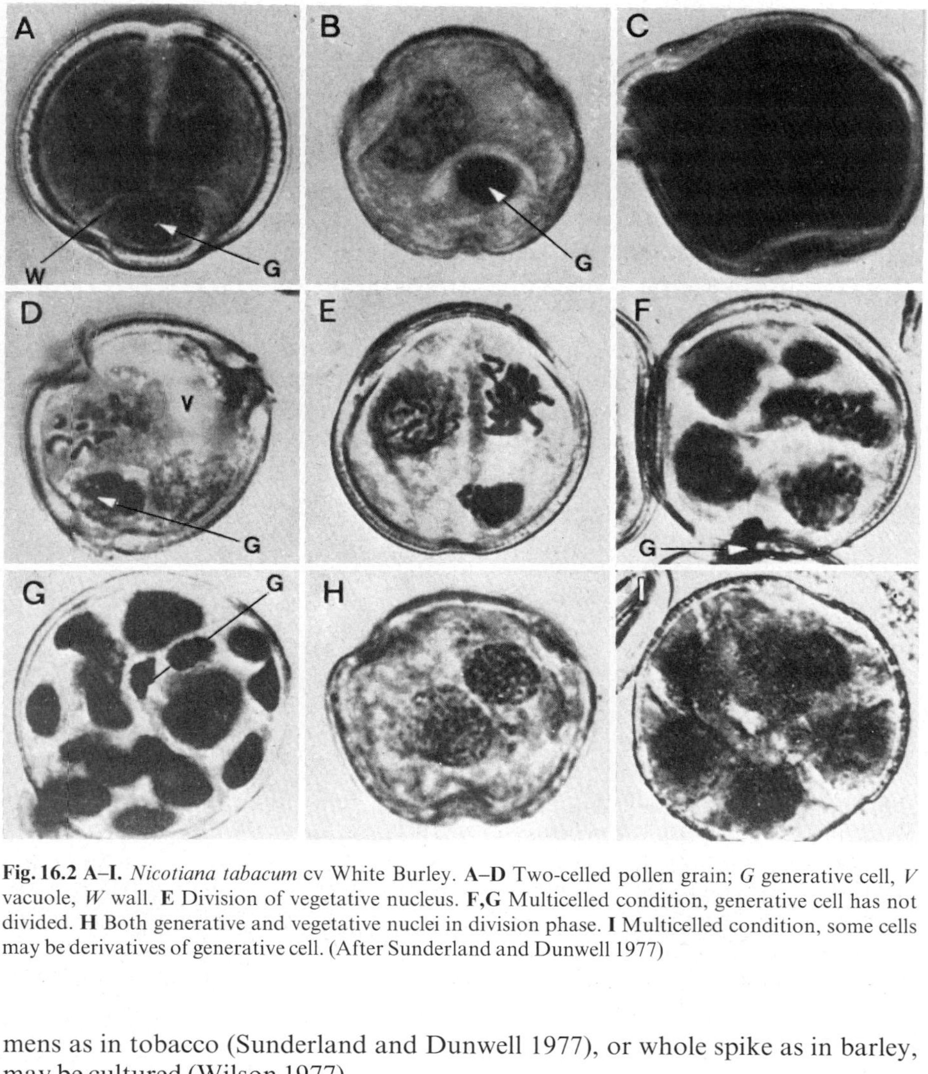

Fig. 16.2 A–I. *Nicotiana tabacum* cv White Burley. **A–D** Two-celled pollen grain; *G* generative cell, *V* vacuole, *W* wall. **E** Division of vegetative nucleus. **F,G** Multicelled condition, generative cell has not divided. **H** Both generative and vegetative nuclei in division phase. **I** Multicelled condition, some cells may be derivatives of generative cell. (After Sunderland and Dunwell 1977)

mens as in tobacco (Sunderland and Dunwell 1977), or whole spike as in barley, may be cultured (Wilson 1977).

Culture medium plays an important role in inducing divisions in microspores. Basal media of Murashige and Skoog (1962) and Gamborg et al. (1968) are especially advantageous for many species (Keller and Armstrong 1978, Sunderland 1978). Although usually 2% sucrose is added to the medium, anthers of barley (Clapham 1971) and *Brassica* gave better results with higher levels of sucrose (Keller and Armstrong 1978, Wenzel 1978). Iron in the medium is reported to be indispensable for the production of plantlets from microspores (J. P. Nitsch 1969). The hormonal requirements of cultured anthers vary from species to species. Most species need an additional supply of auxin, cytokinins, or natural products in the basal medium. Cytokinins are very effective in *Datura innoxia* (Sopory and Maheshwari 1976) and *Solanum tuberosum* (Sopory et al. 1978), while in tobacco (J. P. Nitsch 1969), *Lycopersicon esculentum* (Debergh and Nitsch 1973), and *Hyoscyamus niger* (Raghavan 1978) auxins seem to be relatively more effective. For cereals, the presence of auxins and cytokinins are necessary (Clapham 1977). Recently, the potato extract medium has been used for anther culture, particularly of cereals (Chuang et al. 1978). Activated charcoal is used for augmenting haploid production in anther culture of *Nicotiana* species (Anagnostakis 1974), *Secale cereale* (Wenzel et al. 1977), *Solanum tuberosum* (Sopory et al. 1978), and *Datura innoxia* (Tyagi et al. 1980). Charcoal possibly adsorbs the inhibitory substances present in the anther wall.

The developmental stage of microspores at the time of culture is also a critical factor influencing embryoid formation (J. P. Nitsch 1969). Stages between late meiosis and late binucleate stage have been responsive in one species or another (Engvild et al. 1972, Gresshoff and Doy 1972a, b, Sunderland 1978). Figure 16.2 A–I shows stages in embryo formation in pollen grains of tobacco.

Factors like growth conditions of the donor plants also influence the response of cultured anthers (Foroughi-Wehr and Mix 1979). In *Nicotiana* the embryogenic potential of anthers could be increased by keeping detached flowering stems in nutrient solution (Sunderland and Dunwell 1977), and by subjecting the plants to nitrogen starvation (Sunderland 1978). By subjecting the anthers to varying degrees of chilling, an enhancement in plantlet yield could be achieved in pollen cultures of *Nicotiana, Atropa* (C. Nitsch 1974a, Reinert et al. 1975, Sunderland 1978), and *Datura innoxia* (C. Nitsch and Norreel 1973). Cold treatment maintains pollen in a viable condition, and increases the frequency of embryoid formation by maintaining symmetrical division in a pollen grain (C. Nitsch 1974b). High temperature treatments increased the response of cultured anthers in *Brassica napus* (Keller and Armstrong 1978), whereas in *B. juncea* (George and Rao 1982) a combination of cold pre-treatment and high-temperature shock was essential for inducing embryogenesis (Fig. 16.3 A–E).

Genotype of the plant is another parameter which has a role in the induction of androgenetic haploids. Polyploid species are more responsive in tobacco (J. P. Nitsch 1969), in rice only certain cultivars responded (Guha-Mukherjee 1973), and in *Arabidopsis thaliana, Lycopersicon esculentum,* and *Vitis vinifera* only certain lines showed the potential for pollen embryogenesis (Gresshoff and Doy 1972a, b, 1974). This is also true of rye (Wenzel et al. 1977) and *Petunia* (Mitchell et al. 1980).

16.5.2 Anther Culture

The techniques for anther culture are being refined very rapidly. Johansson and Eriksson (1982) suggest a double-layer method with the addition of activated charcoal (AC). The Petri dish contains semi-solid medium with AC and, above this, liquid medium without AC. Anthers of *Anemone canadensis, A. dichotoma, A. vitifolia, A. hupehensis, Papaver setigerum, P. radicatum,* and *Clematis viticella,* at uninucleate pollen grain stage, were cultured on H-medium. After inoculation the cultures were treated at 7 °C for 7 days. Then the cultures were kept at 25 °C in darkness in normal air or air with 2% CO_2. Sometimes, the cultured anthers

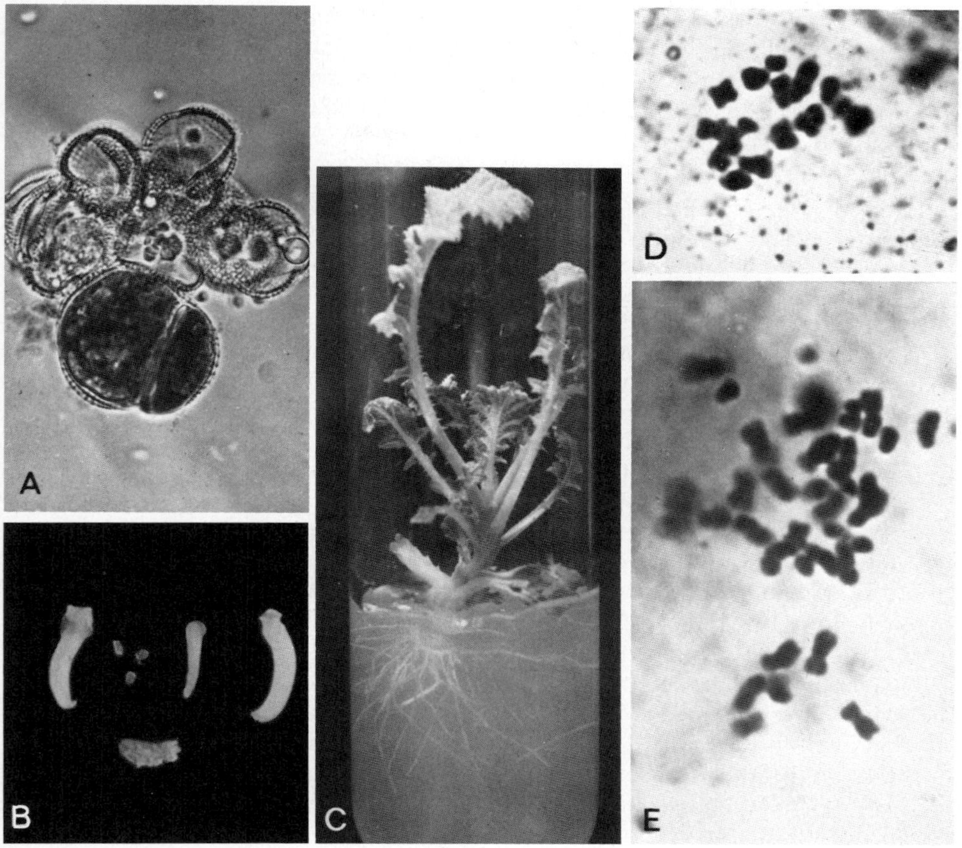

Fig. 16.3 A–E. *Brassica juncea* var. TM-4, pollen embryogeny. **A** First microspore division. **B** Embryos removed from an anther. **C** Shoot-bud growing on rooting medium. **D** Root-tip cell of a haploid plant, n = 18. **E** Root-tip cell of a diploid plant, 2n = 36. (After George and Rao 1982)

were transferred from the double-layer medium to liquid medium without AC. Compared to the controls, anthers incubated on agar medium with AC, with the double-layer method, produced three-times more embryos. Anthers cultured on agar medium treated with AC, but AC removed before inoculation, produced more embryos than the controls (incubated on untreated medium). Cultures floating on a liquid medium should be preferred to incubation on solid medium. In the double-layer method AC promotes embryogenesis. AC probably adsorbs abscisic acid (ABA) produced by the anthers as well as the phenolic compounds produced by the developing embryos.

Zhenghua et al. (1982) kept the inflorescences (at the dyad-tetrad stage) of *Saccharum officinarum* at 1 °C–5 °C for 5 to 10 days. Murashige and Skoog (MS) medium supplemented with 0.5–2.0 mg/l 2,4-D (2,4-dichlorophenoxy-acetic acid) and 0–2.0 mg/l kinetin was adequate to induce callus in cultured anthers, while 20% sucrose promoted development of dyads and tetrads into uninucleate microspores (in 3 to 7 days), and multicelled masses. In 7- to 15-day-old cultures the microspore (pollen grain) nucleus divided into two equal or unequal nuclei. MS medium with 5% sucrose, 2.0 mg/l 6-benzyl-aminopurine (BAP) and 2.0 mg/l NAA (naphthaleneacetic acid) was used as "differentiation" medium. White's medium with 1.0 mg/l of IAA or 0.5–1.0 mg/l IBA (indolebutyric acid) was used as "plant-forming" medium, and it supported the formation of root and leaves of plantlets. Casein hydrolyzate (CH) promoted the number of green plantlets. The cells of root tip of plantlets showed 50–90 chromosomes (in donor sugarcane plant $2n = 120$). The androgenic characters of pollen plants segregated intensively. A plant with higher content of sugar has been obtained and propagated.

Zhenghua et al. (1982, see also Rahaman et al. 1982) cultured the anthers of *Hevea braziliensis* at the uninucleate stage of pollen grains. After 5 days the nucleus divided and produced two daughter nuclei. After 25 days about 20% pollen grains formed multicellular masses; somatic tissue also proliferated. After 50 days the calli and small embryoids of pollen-origin showed vigorous divisions. After transfer of callus-forming anthers to "differentiation" medium, about 15 days later, the embryos could be distinguished. Cotyledonary primordia developed rapidly 30 to 60 days later, and primordia of lateral roots differentiated. At this time the embryoids were transferred to the "plant-forming" medium, and main and lateral roots and green cotyledons differentiated. Sixty to 90 days later the embryoids showed marked increase, and shoot apex and terminal bud primordia were formed. Finally, intact plantlets developed, but the frequency was only 3%–6% (plantlets/100 anthers). Cytological observations of embryoids, root tips, and leaves of transplanted pollen-plants confirmed their haploid nature.

Anand et al. (1980a, b, 1981a, b) cultured the anthers of *Nicotiana tabacum* cv flue-cured Virginia Special at two-celled stage of pollen, on MS medium + 1 mg/l BAP. The vegetative cell only, or the generative cell only, or both, divided repeatedly followed by embryogenesis, and development of three distinct types of haploid embryoids: (1) Derivatives of vegetative cell; cells vacuolate and lightly stained. (2) Derivatives of generative cell; cells deeply stained. (3) Derivatives of both vegetative and generative cells, resulting into a chimera.

The embryoids produced plantlets which were transferred to pots, and maintained until they flowered. On the basis of a comparative study of the morphology and anatomy of the leaf, and morphology of the flower, the three types of plants could be clearly distinguished.

There are reports of pollen haploids from the derivatives of the vegetative cell, of the generative cell, and of derivatives of both the cells. The segregation of plants has been reported for the first time.

In *Hyoscyamus niger* (Raghavan 1978, 1982) buds (5–7 mm long) were cultured when the anthers contained uninucleate, nonvacuolate pollen, on Bourgin and Nitsch semi-solid medium with 2% sucrose. The 3-day-old cultures kept in 12 h photoperiod (ca. 500 ft-c) showed multicellular pollen, and seedlings in 3-week-old cultures. The embryogenic induction is equally good in liquid medium; and even in complete darkness. Within 24 h after inoculation, the pollen showed generative and vegetative cells. Further divisions occurred in the generative cell; the vegetative cell appeared colourless and undivided at the proximal end of embryoid, forming a suspensor-like structure. Finally, the embryoids broke through the anther wall, grew on the agar medium, and produced plantlets. Pollen embryogenesis by the exclusive division of the generative cell, in anther cultures, is rare.

Dodds and Reynolds (1980) used SEM to study pollen embryogenesis in *H. niger*. They cultured anthers at uninucleate pollen-grain stage on semi-solid Bourgin and Nitsch medium. The cultures were incubated at 25 °C with 12 h photoperiod (500 ft-c). Many grains doubled their volume in 24 h and, after 48 h, the volume was nearly eight-times that of the original volume. Embryogenic pollen was detected in 4- or 5-day-old cultures. Some pollen grains break open at the raphe revealing a papillate structure (probably the generative cell). Pollen embryos of globular (7-day-old cultures), heart (9 days), and torpedo (10 days) stage were observed. These simulated the stages in zygotic embryogeny. In 14-day-old cultures the embryoids burst through the anther wall. The pollen surface undergoes some breakdown concurrent with the development of the embryo; this does not happen in the pollen grains, which collapse.

The embryoids develop mostly from the derivatives of the generative cell; derivatives of the vegetative cell may also be involved, sometimes.

16.5.3 Pollen Culture

Anther culture has two disadvantages. Firstly, the plants obtained may originate not only from the pollen grain but also from various parts of the anther, resulting in mixed ploidy. Secondly, the anther wall may produce certain inhibitory substances which are not conducive to pollen embryogenesis. Isolated pollen culture is advantageous, except in certain lines of rye (Wenzel et al. 1976) which produce unreduced gametes. Pollen culture ab initio has been successful only in a few species, and the response is rather low. C. Nitsch and Norreel (1973) were the first to demonstrate the induction of embryoids from isolated pollen grains of *Datura innoxia*. Subsequently, similar success was reported in microspore cultures of *Petunia* (Sangwan and Norreel 1975), *Secale cereale* (Wenzel et al. 1975), and *Solanum tuberosum* (Sopory 1977). For pollen culture of *Nicotiana* and *Atropa*, a liquid medium is more favourable (Wernicke and Kohlenbach 1977, Sunderland 1978).

Pre-culture of anthers (taken from cold-treated buds) on liquid or agar medium, and using such anthers for culture of pollen grains, has a marked effect in enhancing embryoid formation in tobacco, *Hyoscyamus*, and other taxa (C. Nitsch 1974a, Reinert et al. 1975, Wernicke and Kohlenbach 1977).

Sunderland and Roberts (1977) improvised another technique for tobacco pollen. This requires pre-culture of anthers in a liquid medium, where they dehisce and shed pollen into the medium, and serial, periodic transfer of anthers to a fresh medium. Thus, a series of cultures of free pollen is obtained which keep developing into embryoids with much greater frequency, and in all fractions of shed pollen. This technique has been used with success in *Datura innoxia* (Tyagi et al. 1979) and rice (Chen and Lin 1976).

Wenzel et al. (1975) have further refined the pollen culture technique in the case of rye (*Secale cereale*), and they were able to separate the enriched fraction of viable pollen grains by placing a filtered suspension of macerated anthers over a 30% sucrose solution and centrifuging it for 5 min at 1,200 g. Active, embryogenic pollen grains formed a band over the sucrose solution, while the dead pollen grains sedimented. Wernicke et al. (1978) also employed a similar technique with tobacco pollen by using density gradient centrifugation and combining "percoll" with sucrose. The separated fraction showed a 40-fold enrichment in plantlet-forming microspores over the controls.

Rashid and Reinert (1980) selected buds of *Nicotiana tabacum* var. Badischer Burley at uninucleate pollen, from plants grown in 8 h photocycles at 18 °C and supplied with mineral salts. The buds were given a cold treatment, 10 °C for eight, or more, days, and embryogenic pollen selected by density gradient centrifugation using 55% percoll and 4% sucrose as suspension medium. The embryogenic pollen is small, starch-free, with clear cytoplasm; non-embryogenic pollen is large and starch-filled. It is, thus, possible to have androgenic embryos in appreciable frequencies in cultures of cold-treated anthers.

Embryogenic pollen were selectively isolated from buds after cold treatment at 10 °C for 10 days (Rashid and Reinert 1981 a). In buds from short-day and low-temperature (SD 8 h light; LT 18 °C) plants the embryogenic pollen could be detected in 7 days after cold treatment, and the frequency of pollen embryos was higher (up to 4% of cultured pollen) than those (up to 1% only) from long-day and high-temperature plants (LD 16 h light; HT 24 °C). The ratio of embryogenic pollen was about one-fifth of the total population, and they readily developed into plantlets. The isolated embryogenic pollen was cultured in a filter-sterilized liquid nutrient medium containing mineral salts of Nitsch medium, Fe-EDTA 10^{-4}M, and 2% sucrose, pH 6.8.

Cold treatment initiates differentiation of pollen dimorphism on the plant, and differentiation of embryogenic pollen is completed in culture (Rashid and Reinert 1981 b). In plants induced to flower at 18 °C, 8 h light, pollen from buds (petal length 3.4 ± 0.1 cm) fails to form embryos, but embryos are formed if the cultures are kept at 10 °C for 10 days. Low temperature treatment at two stages – to the plant and to the culture – is essential, and complementary for embryogenesis in pollen cultures. Unlike gametophytic pollen, the embryogenic pollen is not fully differentiated. Cold treatment represses gametophytic differentiation, resulting in embryogenic potential. The isolated embryogenic pollen was cultured in a filter-sterilized liquid medium which comprised mineral salts, 2% sucrose; pH 6.8. The cultures were kept at 4 °C and 10 °C for different duration, and then in the culture room at 24 ° or 25 °C.

Rashid et al. (1981) studied ultrastructure (TEM) of embryogenic pollen of *Nicotiana tabacum* var. Badischer Burley. In cultures, the frequency of pollen embryos from isolated pollen grains from near-mature buds [from plants induced to flower in short days (8 h light) and low temperature (18 °C)] was very low. At the subcellular level they showed different degrees of regression of cytoplasm and mitochondria. Pollen from young buds from these plants, further treated at 10 °C for 10 days, showed high frequency of pollen embryos, up to 25% of cultured pollen. They did not show regression of cytoplasm in organelles, had condensed mitochondria, and attenuated cytoplasm which was not rich in ribosomes. The gametophytic pollens were very distinctive with a thick exine and dense cytoplasm rich in ribosomes. The embryogenic pollens indicate that they are repressed for gametophytic function. The close similarity of embryogenic grains to thin exine and sparse cytoplasm, as in microspores, suggests an indeterminate state. The thin exine does not seem to be a common feature in embryogenic pollen. Dodds and Reynolds (1980) report that in *Hyoscyamus niger* the pollen surface (exine) undergoes some breakdown concurrent with the development of embryo. This aspect requires further attention.

16.5.4 The Pollen Plants

The division in the microspore is usually unequal, forming a small generative cell and a large vegetative cell, as in *Nicotiana* (Sunderland and Wicks 1969). Embryoid formation may result from the derivatives of the vegetative cell only (Sunderland and Wicks 1969), generative cell only (Raghavan 1978), or derivatives of both cells may contribute to the embryoid (Sunderland et al. 1974, Anand et al. 1980a). When the first division is equal, either one or both cells divide and develop into embryoid (Sunderland et al. 1974).

Once the globular stage of development is reached, further growth depends upon the species and the cultural environment. In plants like *Nicotiana* the microspores develop into embryoids and plantlets directly (J. P. Nitsch 1969), whereas in most other species, such as rice, there is the intervention of a callus phase before plantlets are formed (Niizeki and Oono 1968).

In several taxa, like *Nicotiana* and *Atropa,* most of the plantlets obtained through anther culture are normal (J. P. Nitsch 1969, Zenkteler 1971). However, in most monocots, especially cereals, albinos are produced in a very high proportion (Clapham 1977).

Although one would expect all haploids through anther culture, it is not so and the yield is very low (George and Narayanaswami 1973, Wenzel 1978). The occurrence of non-haploids among pollen plantlets is a regular phenomenon (Narayanaswamy and Chandy 1971, Engvild et al. 1972, George and Rao 1979). Various factors, like endoduplication of chromosomes, nuclear fusion, or the functioning of unreduced microspores, may all contribute to polyploidy (Sunderland et al. 1974).

Once haploids are produced, homozygous diploids may be obtained by spontaneous diploidization (Narayanaswamy and Chandy 1971), artificial doubling by colchicine (Burk et al. 1972), or by callus cultures (Kasperbauer and Collins 1972).

The Bulbosum method of haploid induction is used in cereal breeding (Jensen 1977). It is based on interspecific hybridization and somatic chromosome elimination, and is very efficient for cereals like barley (Kasha and Kao 1970), and wheat (Barclay 1975).

16.5.5 Use of Haploids

Haploids are an interesting experimental material for studies in genetics and plant breeding. Homozygosity can be achieved quickly through haploids, and doubled haploids have been utilized for crop improvement. In many countries, more so in Japan and China, extensive trials on the use of haploids have been undertaken. In Japan a new variety of tobacco, *Nicotiana tabacum* "F 211", has been produced by anther culture (Nakamura et al. 1974). This is resistant to bacterial wilt, and has mild-smoking quality. In China high-yielding and superior varieties of tobacco: Tanyu 1, Tanyu 2, and Tanyu 3, new varieties of rice: Huayu 1, Huayu 2, and Tanfong 1, and of wheat: Haupei 1 and Lunghu 1, have been produced through anther culture technique (Chen and Li 1978, Hu et al. 1978). In Canada anther-derived double-haploid lines of *Brassica napus* have been released (Keller and Stringham 1978). Haploid breeding of rape, rye, and potato has been undertaken in West Germany. Wenzel et al. (1979) and Wenzel (1980) have combined anther culture technique, somatic cell hybridization, and conventional methods of plant breeding in potato (*Solanum tuberosum*). Autotetraploid potato could be reduced to monohaploid through anther culture and, by combinations, a heterozygous tetraploid potato with desirable traits could be produced.

In cereal breeding the full potential of anther culture could not be exploited mainly because of the occurrence of albinos in large numbers, and due to the presence of non-haploids. The Bulbosum method seems to be more promising for cereals because only green haploids are obtained by this technique. Haploid *Hordeum* plants regenerated from callus cultures are all green (Orton 1980). Chilling of the panicles prior to culture resulted in the production of all green plants, mainly haploids, in tall fescue (*Festuca arundinacea*) (Kasperbauer et al. 1980). These techniques, if adaptable to other grasses and cereals, might solve the problem of albinos.

Free-cell cultures and protoplasts of haploids provide useful material for various kinds of genetic manipulations. Cell lines resistant to streptomycin (Maliga et al. 1973), 5-bromodeoxyuridine (Binding et al. 1970), and aminopterin (Mastrangelo and Smith 1977) have been isolated from haploid cell cultures. Haploids raised through anther culture have been used for protoplast culture, and somatic hybridization (Melchers and Labib 1974, Schieder 1978 a), and for the induction of mutants (Schieder 1978 b).

16.6 Ovary, Ovule, and Nucellus Culture

When embryo and ovule culture do not succeed, ovary culture may give the desired result (see Rangan 1982). The triploid hybrid of *Brassica* (*B. chinensis* × *B. pekinensis*) could be made feasible only through ovary culture since embryo culture was unsuccessful (Inomata 1968). The placenta and ovary wall have a promotive effect; the retention of calyx also promotes the growth of ovary. Studies on cereals, especially certain species of wheat, have revealed that retention of lemma and palea (the "hull" factor) played an important role in the successful culture of ovaries (G. Rédei and G. Rédei 1955 a). J. P. Nitsch (1963) suggested that accessory floral lobes like hull, perianth, and calyx render inorganic nitrates into ammonia, a readily utilizable form.

The time of excision of ovaries influences nutritional requirements in culture. Ovaries excised several days after pollination show satisfactory growth on a simple medium of sucrose and mineral salts. The unpollinated ovaries usually fail to develop in cultures; sometimes parthenocarpic fruits are produced, as in *Althea rosea* (Chopra 1962). The pollinated ovaries show two types of responses. At an early stage with the ovules at zygote/young proembryo and free-nuclear endosperm, many abnormalities occur. In older ovaries with undifferentiated proembryo and well-developed endosperm, the ovules grow into normal mature seeds comparable to that in nature, and usually do not skip any developmental stage, as in *Zephyranthes* (Sachar and Kapoor 1959).

The probable factors attributed to the success of ovary culture are the promotive effect of GA and cytokinins but, in some instances, growth adjuvants prove totally ineffective, or even inhibitory. Various abnormalities in the growth of embryos occurred when the ovaries of certain taxa of Gramineae were grown in nutrient medium containing IAA, KN, and AD (adenine) individually, or in combination (Narayanaswami 1963). A combination of IAA and KN enhances the production of viable seeds in cultured ovaries of *Reseda odorata* (N. Sankhla and D. Sankhla 1967).

16.6.1 Ovary Culture

Besides the seeded fruits, parthenocarpic fruits also develop in vivo, in *Althea rosea* (Chopra 1962). The unpollinated ovaries cultured on modified NM+IBA (20 mg/l) and NM+KN (0.5 mg/l)+IAA (5 mg/l) produced parthenocarpic fruits of the same size as in vivo.

Chopra (1962) studied the effects of IAA, IBA, KN, GA, and colchicine on the growth of ovaries, 3 days after pollination, at the globular proembryo and free-nuclear endosperm stage. Maximal growth of the fruit occurred on NM+IBA (20 ppm) in darkness; 50% of fruits matured in 25–30 days.

On NM+colchicine (50 ppm) the ovules showed initiation of cotyledons in embryos, and cellular endosperm only in the micropylar region. At higher concentrations of colchicine, the growth of embryo was inhibited, and the endosperm remained free-nuclear. The other supplements also did not promote the development of endosperm or embryo. The endosperm remained free-nuclear, and the embryo remained arrested at the heart-shaped stage.

The growth of fruit, in vitro, is significantly affected by the retention of calyx. When calyx was removed, the fruit remained much smaller (12 mm diam; 19 mm with calyx). The development of endosperm and embryo was also normal when calyx was retained.

Guha and Johri (1966) cultured flowers of *Allium cepa* excised 2 days after pollination, on NM. The fruits (diam 3 mm) attained a diameter of 5.4 mm in 5 weeks, but were smaller than natural size. Approximately 5% of ovules matured into viable seeds and germinated in situ, 6 to 7 weeks after inoculation. The seeds remained attached to the ovary during germination and, as the cotyledonary sheath elongated, the fruit was carried upward. The primary root developed in 2 weeks after germination and, simultaneously, the first foliage leaf emerged.

NM+IAA (0.5 or 1.0 ppm) promoted growth of fruits and seeds. IAA, NAA, and 2,4-D caused a change in the orientation of seedlings, and the root-end curved upward. IAA also stimulated bulb formation.

NM + GA (4.0 ppm) promoted development of fruits, and there was maximal seed-set. Nearly 20% of ovules developed into seeds which germinated in situ. NM + GA (4.0 ppm) + IAA (0.5 ppm) supported optimal development of fruit (diam 5 mm), but the seed-set was only 10%. Nitsch medium + tryptophan (2.0 or 5.0 ppm) resulted in 30% of seed-set.

When ovaries denuded of perianth were planted, their growth was markedly inhibited.

The ovaries of *Anethum graveolens, Foeniculum vulgare,* and *Trachyspermum ammi* (Johri and Sehgal 1966) were cultured on modified White's medium (WM), 3, 7, and 6 days after pollination, respectively. The ovules of *Anethum* contained zygote and free-nuclear or early cellular endosperm, of *Foeniculum* globular proembryo and cellular endosperm, and of *Trachyspermum* two- to four-celled proembryo and early cellular endosperm. The growth of fruits, comparable to that in nature, occurred on WM + coconut milk (CM 20%–40%). GA, IAA, or 2,4-D did not stimulate the growth of fruits. In vivo the embryo of *Anethum* matures in 25 days after pollination, but in cultures in 31 days after inoculation. On WM + IAA (1.0 ppm) the embryos were larger than those formed in vivo. *Anethum* and *Trachyspermum* showed normal development of endosperm and embryo on WM + colchicine (1.0 ppm), and *Anethum* and *Foeniculum* on WM + CM (4.0%, 10%), WM + CH (100, 500, 1,000 ppm), WM + yeast extract (YE 100, 500, 1,000 ppm) – with or without IAA (1.0 ppm). In *Trachyspermum* 5% –15% of ovules developed into fertile seeds, on WM + CH (100, 500, 1,000 ppm). In 13- to 14-week-old cultures, the in vitro-produced seeds of *Anethum, Foeniculum,* and *Trachyspermum* germinated in situ. In *Anethum* some of the seedlings also produced rudimentary floral buds.

Bo-ji and Kuo-Chang (1982) cultured (in H medium) unpollinated ovaries of one cv of *Nicotiana rustica* and four cv of *N. tabacum,* at the one- or two-celled stage of "pollen cell". The ovary callused and embryoids differentiated, and then produced plantlets. The medium supplemented with 6-benzylaminopurine (BAP) and IAA resulted in direct induction of higher frequency of embryoids. The frequency of embryoid formation on the callus, or directly from the ovary, increased from 55% to 100% when the concentration of thiamine, pyridoxine, ascorbic acid, nicotinic acid, inositol, and folic acid was raised. The authors claim that haploid plantlets developed from the embryoids derived from the megaspore, egg cell, synergids, and antipodals cells. The diploid plants originated directly from the cells of the young ovule, nucellus, integument, and even placenta. To say the least, the observations are interesting, but one requires more facts, and sequential stages of development of embryoids and plantlets from the various cells and tissues mentioned above. Thus far, there is no such report in this or any other plant. In Solanaceae the nucellus is scanty and is not known to produce adventive embryos in vivo or in vitro. A detailed reinvestigation would be very rewarding.

16.6.2 Ovule Culture

The physiological state of the embryo contained in the ovule constitutes a critical factor in the culture of isolated ovules. Haynes (1954) observed that in *Solanum tuberosum* the relatively small embryos, if pre-grown for a couple of days by culturing ovules, responded more readily than freshly excised embryos.

Demonstration of excised ovules containing zygote, or two-celled proembryo, developing into mature seeds in cultures, in *Papaver somniferum,* was first reported by N. Maheshwari (1958). KN, CH, or YE enhanced the growth of proembryos in ovules, whereas IAA and GA inhibited growth.

Sachar and Kapoor (1959) report that, in *Zephyranthes,* 6 days after pollination, 50%–55% of ovules abort but remain attached to the placentae even after dehiscence of the fruit. Within 16 days after pollination the seeds mature. The seeds germinate readily on moist filter paper.

The pollinated ovaries, cultured on modified NM, developed mature seeds in 2 weeks. The size of seeds was larger than obtained in ovule culture, but slightly smaller than in vivo. The seeds germinated in situ, and the shoots emerged through the fruit wall. The roots grew vigorously and, sometimes, through the pedicel.

On NM, with various supplements, the unpollinated ovules developed into seeds, much smaller than in nature, but did not contain endosperm or embryo.

Pollinated ovules, two days after pollination at the zygote and primary endosperm nucleus stage, were cultured on NM fortified with niacin (1.25 mg/l), thiamine HCl (0.25 mg/l), pyridoxine HCl (0.25 mg/l), calcium pantothenate (0.25 mg/l), glycine (7.5 mg/l), and kinetin (0.5 ppm). Some ovules matured but the seeds were smaller than in vivo. Ovules, 2 to 4 days after pollination, cultured on NM+KN (0.5 ppm)+IBA (2 ppm) matured into seeds of normal size in 4 weeks. Ovules excised 4 or 5 days after pollination, at early globular and free-nuclear endosperm, cultured on NM + vitamins + glycine + KN (0.5 ppm) + IAA (4 ppm) + GA (5 ppm), matured normally and the seeds germinated in situ. The development of seed-coat in unpollinated and pollinated ovules was comparable to that in vivo.

In the ovules cultured at the zygote and primary endosperm nucleus, the embryo did not develop beyond the late globular stage. Kapoor (1959), therefore, cultured such ovules on NM with seitz-filtered CM (25% v/v). The ovules developed a globular embryo and cellular endosperm in 4-day-old cultures; more or less this is the case in vivo. Within 20 days the seeds germinated in situ. On NM+CH (200, 400, 600, 800, 1,000 ppm) the growth of ovules was slower, the seeds were also smaller, and germinated in 24 days.

Usually, the development of endosperm and embryo was normal comparable to that in nature. Frequently, the endosperm was absent; the embryo had a hypertrophied suspensor with uninucleate cells, and a lobed embryonal mass caused by localized meristematic activity of the peripheral cells. With CM, the embryos showed a strong tendency for cleavage, whether the endosperm was or was not present. Occasionally, two embryos at different stages of development were observed. With CH (800, 1,000 ppm) also the embryos appeared to undergo cleavage.

With CH (200 ppm) the development of endosperm was normal, but higher concentration resulted in supernumerary divisions and nuclear fusions. The cellular endosperm became quite massive due to meristematic cambium-like activity in peripheral cells. With CM such an activity was less pronounced. However, the cells increased in size and thickenings (tracheid-like) appeared on the walls.

To further analyse the growth requirements of ovules, various amino acids (present in 200 ppm CH) were added individually to NM. Only arginine (7.6 mg/l), histidine (4.6 mg/l), and leucine (9.2 mg/l) gave a positive response.

It is significant to note that in appropriate nutrient medium the excised ovules, at the zygote and primary endosperm nucleus, develop into normal seeds.

Aerva tomentosa is a dioecious species, male plants are rare, and the seeds develop apomictically. The inflorescence is a compound spike, and the flower has five to eight hairy tepals enclosing a monocarpellary gynoecium.

Puri (1963) cultured excised ovaries and open flowers at the mature embryo sac stage, and pieces of inflorescence, on modified NM. In ovary cultures on NM+CM (20% v/v) and NM+YE (500 ppm), 5%–7% of ovules showed dicotyledonous embryos; the endosperm did not develop at all. Due to proliferation of nucellar cells into the embryo sac, in some ovules the growth of embryo was arrested. In cultures of flowers on NM, 15% of ovules produced mature seeds, 20% on NM+YE (1,000 ppm), and 25% on NM+CH (400 ppm) and NM+CM (20%). The embryo showed maximal growth on NM+CM (30%) even though it was smaller than that in vivo. The endosperm also developed as in vivo. When pieces of spikes (flowers at anthesis) were cultured, there was 20% seed-set on NM+IAA (0.5 ppm). On NM+YE (100 ppm) and NM+CH (400 ppm) the seed-set was comparable to that in vivo. With CH the embryo surpassed the natural size, and the uninucleate cells of perisperm were healthy as compared to dead cells in natural seeds. The seed-set was higher than in vivo on NM+CM (20% v/v); however, the seed as well as the embryo remained small. There was enhanced seed-set on NM+CM (1% v/v)+KN (1 ppm); the cotyledons in embryos were unequal. With KN (0.5 ppm) the embryos were larger than in nature. Endosperm development was normal on all media and, 2 weeks after culture, cellular endosperm surrounded the dicotyledonous embryo. The retention of perianth was essential for satisfactory growth of the ovary.

Due to the scarcity of male plants in *Aerva tomentosa* pollination does not occur, and the development of endosperm is autonomous, i.e. the polar nuclei fuse and assume the role of primary endosperm nucleus. In cultures also the growth of endosperm is comparable to that in vivo.

The ovules of *Allium cepa* (Guha and Johri 1966), 1 day before anthesis (containing mature embryo sacs), and 2 days after pollination (containing zygote and two to four endosperm nuclei), cultured on NM + IAA + GA + 2% sucrose, failed to grow and collapsed within 7 days. However, the immature seeds 13 days after pollination (containing 2- or 3-mm-long well-differentiated embryo and cellular endosperm), or later, cultured on NM, developed normally and germinated in situ after 20 days.

Joshi and Johri (1972) attempted to grow 6-day-old cotton ovules which contained 12-celled proembryo. In the presence of KN in the medium, the embryos developed folded cotyledons. Eid et al. (1973) succeeded in growing 5-day-old ovules of *Gossypium* (10-celled proembryo) on a medium enriched with nitrogen. Presumably, the cotton ovules require high amounts of inorganic nitrogen. Proembryo-bearing ovules of *Abelmoschus esculentus* developed into seeds and even germinated in situ, on a medium containing IAA, CH, or CM (Bajaj 1964).

Gadwal et al. (1968) used the ovule culture technique to obtain interspecific hybrids of *Abelmoschus*.

The role of floral organs (such as calyx) in the growth of fruit, and that of endosperm in the development of embryo, is an established fact. In the growth of ovules also, placental tissue appears to play a significant role. The stimulatory effect of placental tissue enhanced the growth of embryos in the ovules of *Gynandropsis gynandra,* cultured with the placenta intact (Chopra and Sabharwal 1963). In contrast, Sachar and Iyer (1959) did not observe any stimulation of embryonic growth in ovules of *Opuntia dillenii* when grown together with the placenta. This aspect needs further investigation.

Beasley (1971, 1973) developed a phytohormone-supplemented medium which stimulated the growth of ovules of *Gossypium,* and fibre elongation was maximal in ovules collected 2 days post-anthesis from actively growing plants. ABA inhibited the growth of fibre (Dhindsa et al. 1976). Beasley also cultured the unfertilized ovules of cotton on a kinetin-enriched medium. The immature ovules enlarged, and produced fibres if the medium contained IAA or GA, or both. The exogenously supplied growth regulators probably substitute for pollination, fertilization, and subsequent development of embryo in cotton. This aspect needs further study.

Wakizuka and Nakajima (1974) investigated the in vitro development of ovules of *Petunia hybrida.* The ovules with early globular proembryo developed into mature seeds, on a medium supplemented with cucumber-fruit juice, adjusted to favourable osmotic value. They inferred that osmotic value of the medium influenced young ovules which developed to maturity.

Differentiation of somatic embryos culminating in plantlets has been reported in callus tissues derived from fertilized ovules of red currant, *Ribes rubrum* (Zatyko et al. 1975), and *Paulowania tomentosa* (Radojevic 1979). Litz and Conover (1981) reported somatic embryogenesis in the cultured ovules of papaya. Optimum conditions for somatic embryogenesis were obtained when papaya ovules derived from *Carica papaya* × *C. cauliflora* crosses were plated on WM with 20% CM, 20–40 days after pollination. Callus from ovules from interspecific crosses was more embryogenic, and remained so for a longer period, than the callus from the ovules of the parents.

16.6.3 Nucellus Culture

Nucellar embryony is of common occurrence in *Citrus* and *Mangifera* (see Chap. 9). Nucellar embryos provide genetically uniform seedlings which reproduce the maternal genotype. Another attractive feature of nucellar embryos is that whereas the cuttings are often infected by various pathogens, the nucellar embryos are free from them. Rangaswamy (1961) exploited the potential of nucellar tissue by in vitro methods. He established tissue cultures of the nucellus of *Citrus microcarpa* which differentiated into several "embryos". Eventually, plantlets were recovered from the embryos. Sabharwal (1963) reported similar results in two species of *Citrus*. In three monoembryonic species of *Citrus* (*C. grandis, C. limon,* and *C. reticulata* × *C. sinensis*), polyembryony can be induced in the cul-

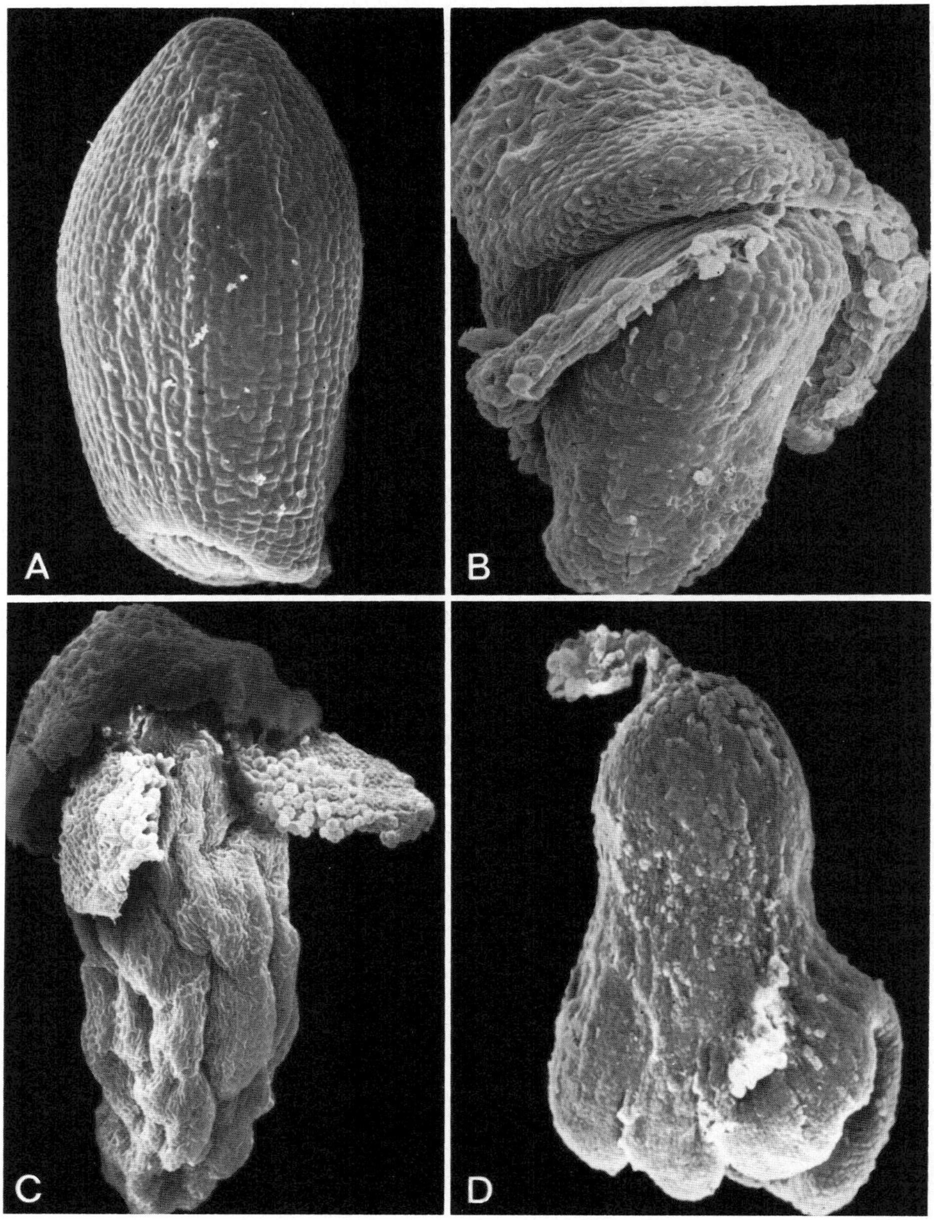

Fig. 16.4 A–H. Grapevine (*Vitis vinifera* cv Cabernet-Sauvignon), somatic embryos and plantlets from nucellar culture. **A** SEM of ovule. Note the prominent micropyle. **B** An ovule 2 weeks after culture with BA (5 μm). Note the protrusion of nucellus from the integumental enclosures. **C** Ovule showing the growth and development of nucellus, on medium containing BA (5 μM) and NOA (5 μM). **D** Pluricotyledonous embryo. Note enlarged hypocotyl and the radicle. **E** Pluricotyledonous embryos with primary and adventitious roots. **F, G** Plantlets produced from somatic embryos. **H** Potted plant. (After Mullins and Srinivasan 1976; **B–F, H** unpublished; courtesy Prof. M. G. Mullins, Sydney)

Fig. 16.4 E–H

tures of nucellar tissue (Rangan et al. 1969). In this investigation the nucellus was free from zygotic embryo, and cultured on a medium supplemented with malt extract. The plantlets obtained through nucellar embryos were successfully transferred to soil. Nucellar tissue isolated from unfertilized ovules of *Citrus sinensis* and *C. aurantifolia* showed embryoid differentiation (Mitra and Chaturvedi 1972).

Kochba and Spiegel-Roy (1977) made extensive investigations on the factors governing embryogenesis in ovular callus of *C. sinensis*. While auxins such as NAA and IAA strongly inhibited embryogenesis, a marked stimulation of embryogenesis was observed after treatment with inhibitors of auxin synthesis, 5-hydroxynitrobenzyl bromide (HNB), and 7-aza indole (AZI). Kinetin and benzyladenine, and 2-isopentyladenine (2, i-P) also reduced embryogenesis, whereas azaguanine (AG) stimulated the embryogenic process. Kochba et al. (1978) observed that abscisic acid, ethephon (2-chloroethyl-phosphonic acid), CCC (2-chloroethyl trimethylammonium chloride), and succinic acid 2,2-methyl hydrazide (Alar) stimulated embryogenesis, while gibberellic acid suppressed embryo formation. Bitters et al. (1972) and Button and Bornman (1971) induced embryos in nucellus obtained from abortive ovules of seedless fruits. It is, thus, possible to propagate *Citrus* clonally, as well as to obtain virus-free plants.

Vitis vinifera L. cv Cabernet-Sauvignon is a non-apomictic plant. Apomixis, if it occurs in Nature, is extremely rare, and it is a monoembryonate species. Mullins and Srinivasan (1976) cultured unfertilized ovules (Fig. 16.4A), 5 to 15 days before anthesis, in 100 ml flasks, each with 25 ml liquid medium. The flasks were kept in a gyro-rotary incubator (80–90 osc. min^{-1}) and maintained at 27 °C. Illumination (16 h) was provided by Gro-Lux fluorescent tubes (2.5 W m^{-2}).

On NM containing casein hydrolyzate (CH 0.1%) and benzyladenine (BA 5–10 μM), the ovules turned green and nucellar callus protruded through the micropyle (Fig. 16.4B). NM with BA (5–10 μM) and β-naphthoxyacetic acid (NOA 5–25 μM) promoted vigorous growth of integuments and formation of green compact nodular outgrowths, but nucellar callus showed scanty growth. While BA is essential for the optimal growth of nucellar callus, a combination of BA and NOA is necessary for "organized development". Therefore, the excised ovules were first cultured on NM+BA (5 μM), and then transferred after 3 weeks to NM+BA (5 μM)+NOA (5 μM). After 2 weeks the ovules were transferred to a similar medium but with a reduced concentration of BA (2.5 μM). During 3 to 4 weeks protuberances developed on the nucellar callus projecting from the micropyle (Fig. 16.4C). Ultimately, they differentiated into proembryonic masses which gave rise to embryoids.

Nucellar embryos of grapevine were pluricotyledonous (Fig. 16.4D, E), and slightly larger than the zygotic embryo. The hypocotyl was short and thick with well-developed cotyledons. The embryoids showed growth of primary and adventitious roots in the same medium but, for the formation of shoots, the embryoids were transferred first to semi-solid media containing 2-isopentenyladenine (2, i-P 5 μM) and GA (1 μM), and finally to WM (Fig. 16.4F–G). The plantlets were transferred to pots. The shoots showed spiral phyllotaxy, but absence of tendrils which are present in normal grapevine seedlings.

The nucellus alone, in grapevine, possesses regenerative competence. The embryoids were formed when subjected to a sequence of auxin (NOA) and cytokinin (BA) treatments. The development of plantlets from the embryoids (somatic) was difficult, and required the treatment with both GA and cytokinin. This is the first report of somatic embryo formation in a cultivar of a temperate fruit.

In *Citrus* and mango nucellar seedlings produced in vivo are usually virus-symptomless (due to exclusion of virus particles), as there is no vascular contact between the nucellus and its contiguous tissues. Similarly, it appears, that the grapevines regenerated from nucellus are likely to be virus-symptomless.

Singh (1963) cultured the nucelli of six polyembryonic species of *Citrus, Poncirus trifoliata,* and *Aegle marmelos.* The nucelli of *C. aurantifolia* and *C. reticulata* callused, and callus differentiated into embryonal masses, but plantlets were not formed. The nucelli of two species of *Citrus* and *P. trifoliata* produced embryos which developed into well-rooted seedlings.

Rangaswamy (1982) refers to nucellar cultures of several cucurbits (*Coccinea, Cucumis, Lagenaria, Luffa, Trichosanthes*) on 62 media involving 12 supplements in 20 combinations. The nucelli cultured whole, or as transverse segments and, with and without the young embryo and endosperm, produced profuse callus which could be subcultured, but embryogenesis was completely absent. The nucelli of four cv of *Gossypium arboreum* and one cv each of *G. barbadense* and *G. hirsutum* were cultured on 105 media fortified with 16 supplements in 49 combinations. The nucelli invariably succumbed to all the treatments. Rangaswamy concludes that lack of embryonic capacity in nucellar cultures in Cucurbitaceae and cotton are due to a chemical block, cucurbitacins in the former, and gossypol in the latter.

16.7 Control of Fertilization

The success of fertilization depends on the compatibility of male and female plants, while survival of the resultant embryo is concerned with the relationship between the endosperm and embryo. The barriers to effective fertilization can be (1) short life-span of pollen, (2) inability of pollen to germinate, or arrested growth of pollen tube so that it does not reach the embryo sac, (3) failure of the male gamete to fuse with the egg, and (4) genetic incompatibility between the parents, which can be operative at stigmatic, stylar, or ovular level.

16.7.1 Intra-ovarian Pollination and Fertilization

The direct introduction of pollen into the ovary is a convenient method since the pollen tube does not have to pass through the stigma and style (see Shivanna 1982). P. Maheshwari and Kanta (1964) experimented with *Argemone mexicana, A. ochroleuca, Eschscholtzia californica,* and *Papaver rhoeas* and *P. somniferum.* A pollen suspension was prepared in 2 ml of sterile, double-distilled water with 0.01% boric acid. The flowers were emasculated, and the surface of ovary wiped with cotton soaked in ethanol. Two punctures were made on the ovary wall, one to allow the air to escape, and the other to inject the suspension of pollen. After the operation the punctures were sealed with petroleum jelly. In the injected ovaries, germination of pollen, entry of pollen tubes into embryo sacs, and fertilization occurred almost as in nature. The ovaries developed normally, and the capsules contained viable seeds. Intra-ovarian pollination has limitations and can be used preferably with unilocular ovaries as in Caryophyllaceae and Papaveraceae. This technique has not attracted much attention.

16.7.2 In Vitro Pollination of Ovules and Fertilization

A still more promising method is to pollinate, in vitro, the stigma, tip of ovary (after removing the stigma and style), or directly the excised ovules (see Yeung et al. 1981). The rearing of both ovules and pollen in an artificial medium, eliminates the gynoecial tissue altogether. Fertilization occurs as usual and normal viable seeds are produced. Success was obtained in *Argemone mexicana, Dicranostigma franchetianum, Eschscholtzia californica, Nicotiana tabacum, Papaver somniferum* and some other plants (P. Maheshwari and Kanta 1964, Rangaswamy and Shivanna 1969). The entire pistil was cultured and aseptic pollen deposited on the stigma. Such stigmatic pollinations in vitro led to successful fertilization, and seed-set in cross-pollination in *Nicotiana rustica* (P. S. Rao and Rangaswamy 1972), *Petunia violacea* (Shivanna 1965), and self-pollination in *Antirrhinum majus* (Usha 1965) and *Nicotiana tabacum* (Dulieu 1966). Zenkteler (1980) and coworkers raised several interspecific and intergeneric hybrids otherwise unknown in nature (Figs. 16.5 A–D and 16.6 A, B).

However, self-incompatibility in some species could not be overcome by adopting stigmatic pollination, or by culture of excised ovules and pollen (Shivanna 1965). The method of placental pollination devised by Rangaswamy and Shivanna (1967) proved a remarkable success in overcoming self-incompatibility in *Petunia axillaris* and *P. hybrida* (see also Niimi 1970). The procedure of placental pollination has implications in plant breeding research. Wagner and Hess (1973) utilized the technique of two-site pollination for *Petunia,* and concluded that pollen put on stigma had a much better chance of effecting fertilization, despite the longer way the pollen tubes had to traverse, than pollen deposited on the ovules. In 1974 Hess and Wagner reported production of haploids with the ovule parent of *Mimulus lutens* followed by in vitro pollination by pollen of *Torenia fournieri.* The pollen grains with their store of growth-promoting factors are brought into much closer contact with the embryo sacs than in other pollen methods, thereby stimulating the egg cell to undergo parthenogenesis.

Besides its practical applications in plant breeding and genetics, the technique of in vitro pollination and fertilization has been employed to study the problems of pollen physiology and fertilization. The index for the success of fertilization is the degree of viable seed-set. And this is dependent on the age of plants, particularly the ovules, adequate pollen germination, proper growth of pollen tubes and microgametogenesis, pollen tube entry into ovules, high incidence of fertilization, and development of viable seeds (see Shivanna 1982).

We have meagre information concerning the effects of physical factors such as light, temperature, and humidity on seed-set consequent to in vitro pollination and fertilization. It is reported that keeping cultures in light or dark does not seem to affect seed-set. The success also depends on the composition of nutrient medium (P. Maheshwari and Kanta 1964, Balatkova and Tupý 1972). The growth of

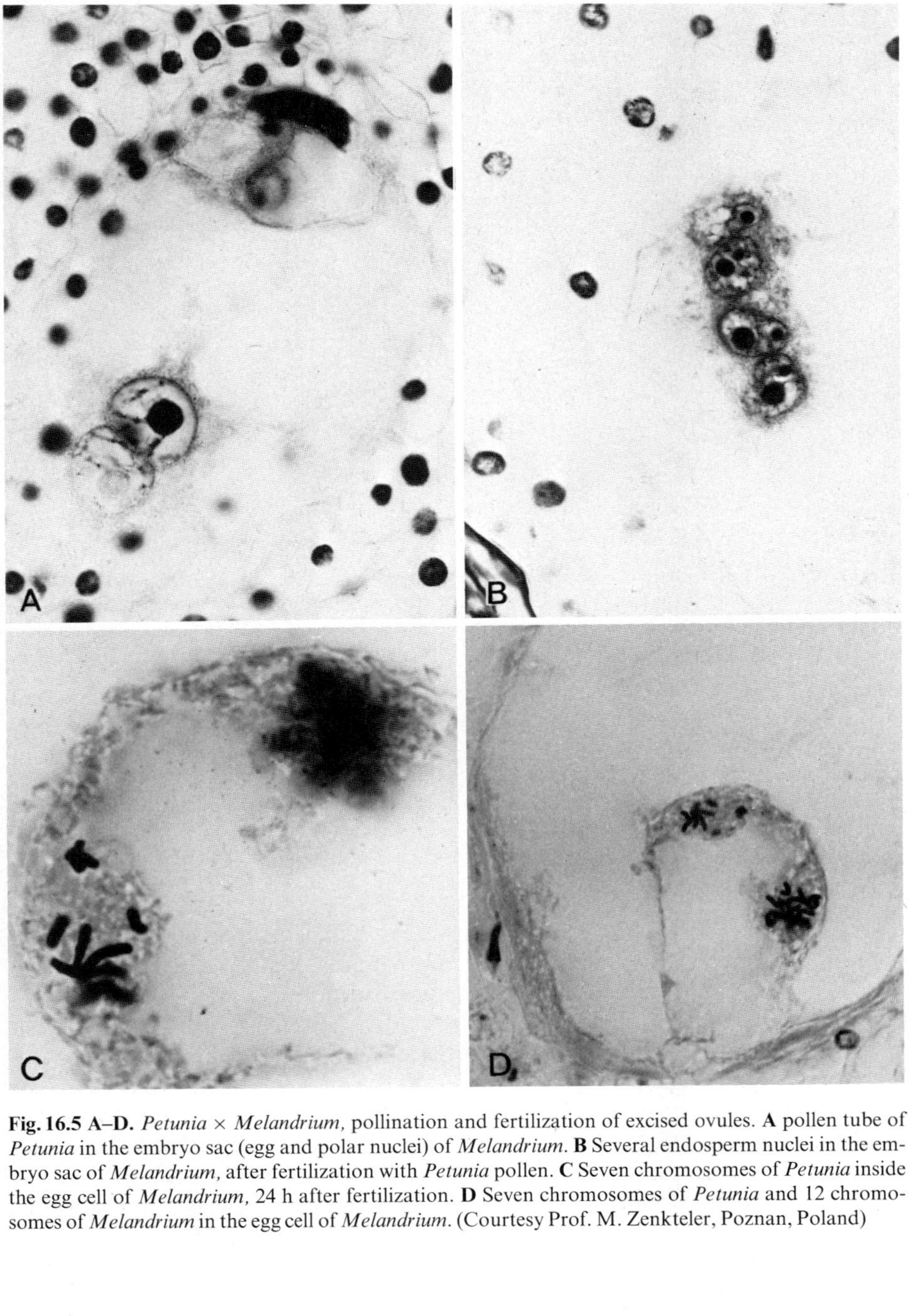

Fig. 16.5 A–D. *Petunia* × *Melandrium*, pollination and fertilization of excised ovules. **A** pollen tube of *Petunia* in the embryo sac (egg and polar nuclei) of *Melandrium*. **B** Several endosperm nuclei in the embryo sac of *Melandrium*, after fertilization with *Petunia* pollen. **C** Seven chromosomes of *Petunia* inside the egg cell of *Melandrium*, 24 h after fertilization. **D** Seven chromosomes of *Petunia* and 12 chromosomes of *Melandrium* in the egg cell of *Melandrium*. (Courtesy Prof. M. Zenkteler, Poznan, Poland)

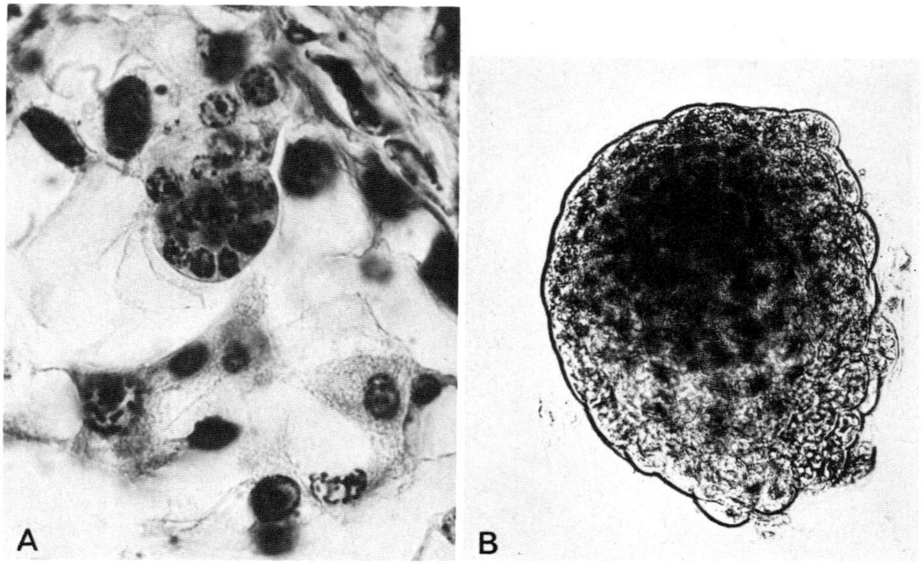

Fig. 16.6 A, B. Hybrid embryos. **A** Hybrid embryo, 7 days after fertilization, in *Nicotiana tabacum* × *Physochlaina fracalta*, surrounded by endosperm. **B** Hybrid embryo of *Nicotiana tabacum* × *Hyoscyamus niger*, 15 days after pollination of ovules in vitro. (Courtesy Prof. M. Zenkteler)

embryo and its development are affected by critical shifts in hormonal balance at various stages (Reinert 1973). Therefore, periodical modification of culture medium at the right stage of growth is essential. This may overcome the problem of embryo degeneration in test-tube hybridization (Zenkteler 1980).

16.7.3 Inter- and Intrageneric Hybrids

In in vivo systems, the post-fertilization events – like development of embryo, relationship of the embryo, endosperm, and maternal tissue – also play decisive roles in obtaining a progeny. Endosperm is the nutritive tissue of the growing embryo. In many instances the embryo thrives better on foreign endosperm than on its own: *Triticum* embryo on *Secale* endosperm (Stingel 1907), and *T. vulgare* embryo on the endosperm of *T. durum* and *T. turgidum*. Camara (1943), for the first time, observed changes in the plants derived from embryos grown on foreign endosperm. He made interspecific embryo-endosperm grafts in *Aegilops, Avena, Hordeum, Secale,* and *Triticum* at the milk-stage of endosperm, and noted differences in the texture of seeds of the first generation. Sanders and Ziebur (1963) observed that embryos of plants belonging to diverse families can grow in identical nutrient media. Apart from the nursing role, endosperm has a role in chemical regulation of morphogenesis. The endosperm extracts of coconut, *Aesculus hippocastanum, Camellia japonica, Ginkgo biloba, Sechium edule, Thea sinensis,* and *Zea mays* (van Overbeek 1942, Steward and Shantz 1959) induce division and enlargement of cells.

The technique of in vitro pollination and fertilization has great potential in hybridization programmes, induction of parthenogenesis, and in mutation research. It is a hopeful substitute for delicate operations such as style grafting, tandem grafting, and stump pollination. These are difficult to accomplish, and are often unsuccessful in field conditions. The success of haploid induction through in vitro pollination would be advantageous when anther cultures are unsuccessful.

16.8 Endosperm Culture

Endosperm tissue, formed during the development of an angiospermous seed, provides the essential food materials required for the growth of embryo and, in many cases, seedlings. In non-endospermic seeds the endosperm is consumed by the developing embryo whereas, in endospermic seeds, copious amount of endosperm is present in mature seeds. The histological structure is highly variable, and the cells are often densely packed with starch grains, protein granules, or oils. Endosperm formation is initiated by mitotic division of the primary endosperm nucleus, usually prior to the division of zygote. Comparative studies reveal that there is considerable variation in the mode of development of the endosperm (see Chap. 7).

Except endosperm, in nature non-zygotic somatic embryos originate from the ovular tissue, e.g. nucellus and integument (see Chap. 9).

The endosperm of angiosperms has been cultured on nutrient media of defined composition. One of the goals is to understand the nutritional requirements of the immature and mature endosperm, to undergo morphogenesis leading to the formation of callus followed by organogenesis, and differentiation of roots, shoots, and plantlets. To a plant breeder and horticulturist, regeneration of triploid plants from endosperm, different from that of the sporophyte, is of much interest (see Srivastava 1982).

Attempts to grow the endosperm tissue in vitro began in the 1930's. After trials extending over 10 years, Johri and Bhojwani (1965) succeeded in inducing morphogenesis and shoot-bud differentiation in *Exocarpus cupressiformis*. Until now, among the autotrophs, endosperm of nine taxa proliferates and produces callus; two taxa embryoids; three taxa callus, roots, and shoots; one taxon each only callus and roots, callus, embryoids, and plantlets, and callus, roots, and plantlets. Of the semi-parasites, callus develops from the endosperm of two taxa; callus and shoot buds in one taxon; shoot-buds in two taxa; shoot-bud and haustoria in three taxa; callus and plantlets in one taxon, and callus, embryoids and plantlets in one taxon (see Johri et al. 1980). There is an additional report on the formation of callus, roots and shoot-bud-like structures in *Juglans regia* (Cheema and Mehra 1982).

LaRue (1949), for the first time, succeeded in establishing a continuously growing tissue culture from immature maize endosperm. He also reported that "in *Zea mays* a few specimens less than one in a thousand developed roots and one formed a shoot-root axis and miniature leaves". Several workers have attempted to culture the endosperm of cereals (wheat, maize, barley), but have failed to induce organogenesis (see Johri et al. 1980, Srivastava 1982).

The endosperm tissue of several species has shown more or less similar nutrient requirements; differences have been noticed in the cultures of immature and mature endosperm. Of the waxy, starchy, and sugary varieties of maize, the endosperm of only the sugary variety can be cultured (see Sehgal 1969).

LaRue (1947) cultured maize endosperm on media supplemented with tomato juice, extracts of grape, green-corn kernels, coconut milk, and yeast extract. Twenty per cent tomato juice induced maximal growth of the tissue. Subsequently, Straus (1960) reported that the active component in tomato juice was similar to asparagine. He substantiated it by demonstrating better callus growth with asparagine than with tomato juice. A combination of aspartic acid and asparagine, and aspartic and glutamic acid also induced sustained growth. For proper growth of immature endosperm of *Triticum* (Sehgal 1974) and *Vitis* (Mu et al. 1977), the addition of coconut milk in the medium was essential.

16.8.1 Immature Endosperm

Tamaoki and Ullstrup (1958) obtained satisfactory growth of maize endosperm on NM supplemented with YE, whereas that of *Cucumis* (Nakajima 1962) was grown on WM supplemented with 1,3-diphenylurea, YE and/or IAA. Yeast extract appears to be essential for the growth of mature endosperm of many flowering plants.

The immature endosperm also requires complex substances like YE, CH, or CM, and cell division factors.

Another important finding emerged from early studies. The success of cultured endosperm depended upon the variety of the grain and its age. For example, in maize, endosperm excised from half-grown and large grains did not proliferate, nor did the young endosperm (LaRue 1947). The endosperm of *Lolium* (Norstog 1956) and *Cucumis* (Nakajima 1962) grows in cultures only when excised 7 to 10 days after pollination. There are no reports on the proliferation of free-nuclear endosperm; Nakano et al. (1975) pointed out that in rice only the cellular endosperm proliferated.

16.8.2 Mature Endosperm

Satisfactory growth of callus from mature endosperm of *Santalum, Jatropha,* and *Ricinus* occurred on a medium fortified with 2,4-D + KN + YE. In *Dendrophthoe, Taxillus,* and *Leptomeria* callus growth was optimal on a medium supplemented with IBA, KN, and CH (see Srivastava 1982).

16.8.3 Embryo-Endosperm Relationship

The role of embryo in the induction of morphogenesis, callus, shoot-bud, and plantlet formation in endosperm cultures has been much debated. According to one view, there is absolute necessity of the so-called embryo factor for the induction of proliferation and/or organogenesis. However, there are others who suggest that the presence of embryo is not absolutely necessary. GA_3, however, replaces the embryo factor and stimulates the endosperm cells to divide in *Croton, Putranjiva,* and *Ricinus* (see Srivastava 1982). Cheema and Mehra (1982) have once again questioned the role of embryo (embryo factor) or GA_3 in evoking morphogenetic response in endosperm. They emphasize that in *Juglans regia* the endosperm does not require the association of embryo or GA_3 for the induction of callus and/or differentiation. Whether the endosperm requires the embryo factor or GA_3, or not, entirely depends on the endogenous level of the promotory factor; if it is deficient, the exogenous supply is essential.

16.8.4 Organogenesis in Endosperm Cultures

In *Scurrula pulverulenta* regeneration of buds occurred directly on the cultured endosperm on a kinetin-supplemented medium. In *Dendrophthoe falcata, Leptomeria acida, Taxillus cuneatus*, and *T. vestitus* differentiation of shoot-buds as well as callus formation occurred on an auxin and a cytokinin-supplemented medium. Among the autotrophic plants, regeneration of shoot-buds has been reported mainly from the mature endosperm of Euphorbiaceae. Casein hydrolysate was effective in inducing bud differentiation in *Jatropha panduraefolia* and *Croton bonplandianum*. Organogenesis in endosperm callus has been demonstrated in *Croton, Nigella damascena, Petroselinum hortense, Citrus,* apple, and rice (see Johri et al. 1980).

Cheema and Mehra (1982) cultured the endosperm of *Juglans regia,* with and without the embryo, on modified MS medium supplemented with various growth promoters. The cultures were maintained at 25 °C in dark, and 16 h light (3,500 lx): 8 h dark. The immature endosperm, without the embryo, callused and regenerated roots, or the roots regenerated directly from the endosperm. Shoot-bud-like structures were also observed. The mature endosperm, again without the embryo, cultured on MS+CH (500 mg/l)+2,4-D (concentration not mentioned) +KN (1 mg/l), also produced callus. When transferred to MS, roots and shoot-bud-like structures differentiated from the callus. The growth response of immature endosperm was much better than that of the mature endosperm. The endosperm, with the embryo intact, showed a more or less similar response but shoot-bud-like structures did not differentiate. The presence of the embryo delayed/inhibited morphogenesis and organogenesis. The endosperm requires growth promoters to evoke optimal growth response. The cells of the callus and root tips showed $3n=48$ which confirms their triploid nature.

Differentiation of triploid plantlets from the callus of endosperm is reported in *Putranjiva roxburghii* (see Johri et al. 1980, Srivastava 1982).

The endosperm of *Santalum album* (Lakshmi Sita et al. 1980), cultured without the embryo, on MS+BAP+NAA, produced callus. Embryoids differentiated on MS+GA (1.0–2.0 mg/l), MS+BAP (0.3–2.0 mg/l) +IAA (1.0 mg/l) and MS+KN (0.3–2.0 mg/l) +GA (1.0 mg/l); profuse embryogenesis on MS+ BAP (0.3 mg/l), +IAA (1.0 mg/l) and MS+GA (1.0 mg/l) +KN (0.3 mg/l). The

Fig. 16.7 A–D. Sandalwood (*Santalum album*), embryogenesis and plantlet development in endosperm tissue. **A** Mature endosperm at culture. **B** Endosperm showing proliferation on MS+2,4-D (5 mg/l). **C** Differentiation of "embryos" in endosperm callus on MS+IAA (1 mg/l)+BAP (1 mg/l). **D** Triploid plantlets with well-developed roots. (After P. S. Rao and Raghava Ram 1983)

embryoids developed further upon transfer to WM, and triploid plants developed. These were transferred to pots, nursery and, finally, to the field. This is the first report of a successful transfer of triploid plants, from endosperm, to field. In due course it would be possible to compare these plants with normal field-grown plants.

P.S. Rao and Raghava Ram (1983) also obtained the differentiation of triploid plantlets in the endosperm culture of sandalwood (*Santalum album*). Mature endosperm (Fig. 16.7A) (devoid of embryo) cultured on MS+2,4-D (1 mg/l) +KN (0.2 mg/l) proliferated into a callus tissue (Fig. 16.7B). Embryogenesis (Fig. 16.7C) occurred when the callus was transferred to MS+IAA (1 mg/l) +BAP (1 mg/l) with 5% sucrose. The "embryos" developed into plantlets (Fig. 16.7D) on MS+IAA (1 mg/l) +IBA (0.5 mg/l) +GA (0.5 mg/l).

Several chemical and physical factors affect the growth and differentiation in endosperm callus. For example, in *Taxillus* the position of the half-split endosperm on the medium influenced bud formation. pH of the medium also plays an important role. In *Asimina* pH 4.0 was optimal for 100% increase in growth, whereas in *Ricinus* prolific growth occurred at pH 5.0. The endosperm showed good growth at pH 5.6 in *Jatropha* and *Putranjiva,* at pH 6.0 in apple, and pH 7.0 in maize (see Johri et al. 1980). Other physical factors like light and temperature also affect the growth of endosperm callus. In maize (Straus and LaRue 1954), *Jatropha* and *Ricinus* (Johri and Srivastava 1973) the suitable temperature for the growth of endosperm was 25 °C. Maize endosperm callus grew well in dark, in contrast to castor where continuous light was required, whereas light had no effect in *Lolium* (Norstog 1956).

Investigations on the effect of sugar on endosperm callus reveal that increase in concentration decreases growth. Sucrose, glucose, and fructose promote growth in maize (Straus and LaRue 1954), *Lolium* (Smith and Stone 1973), and *Ricinus* (Srivastava 1971). Amino acids like glutamic and aspartic acid, and glutamine, support appreciable growth in maize (Straus and LaRue 1954).

Histological investigations in *Exocarpus, Taxillus,* and *Dendrophthoe* indicate that the shoot-buds originate from epi- and hypodermal zone of the endosperm. In *Putranjiva* growth centres in peripheral cells develop within 30 days of culture. Differentiation of tracheidal cells was observed in localized peripheral cells in *Osyris, Jatropha, Putranjiva,* and *Ricinus.* Nakajima (1962) in *Cucumis,* and Rangaswamy and Rao (1963) in *Santalum,* did not observe any tracheidal cells in callus.

Occurrence of different ploidy levels in cultured cells of endosperm is a common feature in maize (Straus and LaRue 1954), *Lolium* (Norstog 1956), *Croton* and *Jatropha,* whereas endosperm callus of *Dendrophthoe, Taxillus,* and *Putranjiva* retained triploidy and did not show any variation in chromosome number even after several serial subcultures (see Johri et al. 1980).

Triploid plants of endosperm origin have implications in plant breeding programmes. The conventional breeding methods for raising triploids is through a cross between tetraploid and diploid. This, however, has two disadvantages: (1) the cross may not be successful due to early abortion of the embryo, and (2) the process is cumbersome. In contrast, triploid plants from endosperm tissue, particularly of *Citrus* and rice, is a very significant advancement. It is advisable to intensify research on the culture of endosperm, especially of economically important plants (Srivastava 1982).

16.9 Embryo Culture

The cultures of embryos of younger stages are not always successful, because their nutritional requirements are very exacting. In such cases the pre-culture of ovules is very helpful.

It should be emphasized that it is not always the size of the embryo that matters, but the number of cells and the status of endogenous metabolites. Since the latter cannot be investigated, at least at present, the size as well as the approximate number of cells should be the criteria to decide the youngest stage (in any taxon) at which the embryo can be cultured. It is also important that the in vitro development should be comparable to that in vivo. It should be possible to control callusing and differentiation of embryoids and plantlets but, as yet, our knowledge about the control of embryogenesis is very inadequate.

The most significant aspect of embryo culture is raising interspecific and intergeneric hybrids, especially when the hybrid seed is sterile due to abortion of the embryo.

There are well-documented reviews on embryo culture (see Raghavan 1976, 1980, Raghavan and Srivastava 1982), and the readers may refer to them for citations left out in the text.

The mature embryo (6–7 mm length) of *Allium cepa* (Guha and Johri 1966), when cultured on NM with 5% sucrose, germinated in situ after five days. When the embryo was cultured with the endosperm intact, there was no difference in growth response. NM + IAA (1 ppm) caused proliferation of the radicle, and the roots were abnormal; 0.5 ppm IAA caused inverted orientation of the seedling and bulb formation, and development of leaves and fibrous roots was inhibited. However, healthy seedlings were obtained on NM + tryptophan (5 ppm) or tomato juice (20% v/v). NM alone and NM + tryptophan promoted the growth of undifferentiated embryo (0.1 mm length) into normal seedlings. The undifferentiated embryo produced only callus on NM + IAA, and the roots differentiated in 6 weeks. The globular proembryo does not grow on NM, but, with IAA (1 ppm), it forms a massive callus. It shows differentiation of bud-like structures which fail to grow further.

The embryo of mature coconut (*Cocos nucifera*) grows well on MS medium (Iyer 1982). The addition of activated charcoal is essential to prevent rapid browning of tissue which otherwise arrests further growth. The cotyledon (= haustorium), in vivo, enlarges into a spongy sweet tissue drawing nourishment from the endosperm for the growth of the embryo. In cultures the cotyledon turns dark-brown, and in transplanted seedlings it gets infected bringing about a high mortality. Root growth in vitro was stimulated by decapitating the first large root which emerged from the sprouted embryo, along with many laterals and sublaterals. Liquid cultures, with filter paper platforms for supporting the embryo, showed better rooting response.

The seeds of *Costus speciosus* (Pal and Sharma 1982), removed from young fruits 15–20 days after anthesis, failed to germinate. The embryos were excised at different stages of development: (1) seeds white, soft, 1 mm diam, (2) seeds greyish-black, endosperm viscous, not separable from embryo, (3) seeds blackish at apex, greyish below, embryo separable from endosperm, embryo dumb-bell-shaped, 1.5 mm long, (4) seeds mature, brownish-black, seed-coat soft, embryo large and dumb-bell-shaped, endosperm single-layered at micropylar end, and several cells thick below the micropylar collar, (5) seeds fully mature, black, seed-coat hard. The embryo lacks a suspensor; the embryos of stages 3 to 5 are morphologically identical. The cultures were maintained under 16 h light (ca. 2,500 lx). On modified MS medium + NAA (3 mg/l) + KN (1 mg/l), callus was initiated in 5 weeks. The best growth of embryo at stage 3 was obtained when the medium contained rhizome extract. The rhizome extract serves as a superior natural medium as compared to YE and CM. The callusing and/or differentiation can be controlled by varying the concentration of rhizome extract.

A. N. Rao et al. (1982) report that it should be possible to obtain plantlets in a number of tropical fruits, by culturing cotyledonary tissue. MS medium was used with 4% sucrose and vitamins, and growth promoters NAA (5.0 mg/l) and KN (2.5–5.0 mg/l). The cultures were kept in 12 h light ($1.2–1.6 \times 200$ ft-c). Callus developed in all explants, roots differentiated in *Citrus grandis, Durio zibethinus, Mangifera indica, Nephelium lappaceum, N. malainse,* and both roots and shoots in *Averrhoa carombola, Citrus microcarpa,* and *Cucumis melo*. Wu et al. (1982) cultured shoot tip, anthers, and cotyledonary segments of *Thea sinensis* cv Chyi-Men, Pyng-Shoei, and Chin-Hsiu Oolong, on modified MS medium +2,4-D (2.0–2.8 ppm). Optimal growth of callus was obtained when IAA (2.0–4.0 ppm) and KN (2.0–4.0 ppm) were added. CM (10% v/v) proved most effective for the growth of callus. In cultures the shoot tip initiated callus only at the cut-end in contact with the medium. The callus also developed from the anther and the filament; the callus from the filament suppressed the callus from anther. Induction of callus was easier in cotyledonary segments.

In cv Chyi-Men and Pyng-Shoei the cotyledonary callus differentiated plantlets, on MS+IAA (1.0 ppm) +KN (10.0 ppm). In cv Chiu-Hsin Oolong organogenesis occurred on MS+IAA (4.0 ppm) +KN (20 ppm).

The plantlets were transferred to vermiculite, then to the nursery and, finally, to the field. These plants differed significantly from the naturally-grown plants in increase in length of main root, height and diameter of stem, length and width of leaf, leaf area, flower buds of "open" type ("close" type in naturally-grown plants), number of flowers, number of seeds/fruit, and diameter of seeds.

The in vitro-grown plants were used as female stock to cross with cv Chin-Hsin-Oolong, a high quality cv for producing Oolong tea. The average fruit-set in crosses was 32.5±15.8%. The F_1 hybrids, grown from these seeds, were transplanted from the nursery to the field. The F_1 may give better yield, and the tea may be of superior quality.

The culture of excised embryos can be advantageously exploited by the plant breeder. Very often interspecific crosses fail due to arrested development of embryo, either due to abortion of embryo at an early stage, or due to the degeneration of endosperm. Laibach (1925) obtained hybrids from a cross *Linum perenne* × *L. ostriacum*. The capsules that developed contained shrunken seeds which failed to germinate in soil. However, the excised embryos germinated when soaked with sucrose solution and placed on cotton. Subsequently, a number of hybrids have been successfully raised through embryo culture: interspecific hybrids in *Abelmoschus* (Gadwal et al. 1968), and *Secale* (Mullins 1970); intergeneric hybrids in *Hordeum vulgare* Linn. and Bowden × *Secale cereale* (Kruse 1974), *H. vulgare* Linn. emend Bowden × *Agropyron repens* (Linn.) P. Beauv., and *H. vulgare* × *Triticum* sp. (see Raghavan 1976).

Bajaj and Gosal (1982) refer to the work of hybrid embryo culture of *Lathyrus clymanum* × *L. articulatus, Phaseolus vulgaris* × *P. acutifolius, P. vulgaris* × *P. ritensis, P. mungo* × *P. calcaratus, P. aureus* × *P. angularis*, and *P. vulgaris* × *P. lunatus*. The application of a mixture of GA (100 ppm) +NAA (25 ppm) +KN (10 ppm) retained the pods on the plant for a longer duration. Plants were obtained in a hybrid *Phaseolus mungo* (♀) × *P. aureus* (♂), in cultures of developing embryo, on MS+NAA (0.5 ppm) +KN (0.2 ppm). Thus, embryo culture has been very useful in obtaining interspecific hybrids in *Phaseolus*. Hybrids of distantly related parents may be possible by adopting in vitro pollination and fertilization procedure, if the incompatibility barrier is at the fertilization level.

In the production of Triticale, when tetraploid (× *Secale cereale*) and hexaploid (× *S. cereale*) wheat varieties are used, the seeds do not develop because of the presence of two dominant genes Kr 1 and Kr 2. Through embryo culture, more than 50% embryos could be raised to mature plants (G. Rédei and G. Rédei 1955b). Intergeneric hybrids have also been raised through embryo culture in *Tripsacum* × *Zea* (Farquharson 1957).

In barley efforts have been made to produce winter-hardy and mildew-resistant plants. In crosses of strains of *Hordeum vulgare* (susceptible to *Erysiphe graminis*) × *H. bulbosum* (wild resistant strain), the embryo failed to grow beyond 15 days after pollination, but by excising and culturing the embryos on a nutrient medium, normal seedlings were obtained (Konzak et al. 1951). Davies (1960) succeeded in raising hybrids of interspecific crosses between *H. vulgare* × *H. bulbosum*, *H. brachyantherum* × *H. depressum*, and *H. jubatum* × *H. spontaneum*.

In rice the embryos are generally malformed in interspecific crosses. However, through embryo culture, many investigators have reared hybrids which are capable of withstanding unfavourable environment, and are also resistant to diseases (Iyer and Govila 1964).

The endosperm of barley has been used as a "nurse tissue" for growing hybrid embryos of *Hordeum vulgare* Linn. × *Agropyron repens* Linn., and *H. vulgare* × *Triticum aestivum*. Ziebur and Brink (1951) recorded successful growth of *Hordeum* embryos transplanted onto cellular *Hordeum* endosperm in sterile culture. Endosperm extracts also stimulate embryo growth. Kruse (1974) implanted the embryos of intergeneric hybridizations involving *Hordeum, Triticum, Agropyron*, and *Secale*, on to *Hordeum* endosperm on the surface of nutrient medium. Williams and de Lautour (1980) developed an analogous endosperm transplant technique for pasture legumes, in *Trifolium, Lotus*, and *Ornithopus*. The hybrid embryo, dissected from a normally developing, intraspecifically pollinated ovule, inserted into cellular endosperm, and the embryo and endosperm together were transferred to the surface of a nutrient agar medium.

Embryo culture has been used for recovering haploid plants by the elimination of a set of chromosomes. Kasha and Kao (1970) obtained a high frequency of haploids in barley. *Hordeum bulbosum* provides a system whereby haploid plants of *Hordeum vulgare* and *Triticum aestivum* can be obtained by selective somatic elimination of either one or more genomes.

Embryo culture can be applied to overcome self-sterility of seeds. For instance, in *Musa balbisiana* Colla., the seeds do not germinate under natural conditions. Cox et al. (1960) excised the embryos, and cultured them to obtain seedlings. Other successful examples of embryo culture in interspecific and 2n × 4n crosses are: *Trifolium* (*T. repens* × *T. ambiguum* and *T. ambiguum* + *T. hybridum*), *Melilotus alba* × *M. officinalis, Phaseolus vulgaris* × *P. acutifolius, Medicago truncatula* × *M. orbicalaris*, and *Lathyrus clymenum* × *L. articulatus* (see Raghavan 1976). Asano (1980) cultured the immature hybrid embryos (0.3–0.4 mm long) of a cross *L. langi* of *Lilium longiflorum* × *L. sugehime* and *L. shikayama* × *L. henryi*, on normal endosperm as nurse tissue.

The culture of embryos before they abort has led to the recovery of hybrid seedlings in many important crops: *Gossypium arboreum* Linn. (2n = 26) × *G. hirsutum* Linn. (4n = 52), *Corchorus capsularis* × *C. olitorius* (see Raghavan 1976), *Carica papaya* × *C. cauliflora* (Jagannathan 1979), and *Allium cepa* × *A. fistulosum* (Dolezel et al. 1980). *Solanum melongena* is seriously attacked by an insect borer (*Leucinoides orbonalis*) which causes extensive damage to shoot and fruit. A few related species of brinjal such as *S. khasianum* and *S. sisymbrifolium* are resistant to this pest. Sharma et al. (1980) obtained hybrid plants, through culture of hybrid embryos, between *S. melongena* × *S. khasianum*.

Hybrid plants have also been raised, by embryo culture, in *Malus pumila* (Nickell 1955), *Prunus persica,* and crosses of *P. persica* × *P. tomentosa,* and *Ribes nigrum* × *R. uva-crispa* (Tukey 1934).

In *Coffea* embryo culture is an important method in dealing with genetic variability. The cultivation of immature embryos obtained by interspecific crosses in the same subsection, or between subsections in the genus *Coffea,* has a great potential for obtaining new hybrids (Monacco et al. 1977).

Embryo culture, to a large extent, has benefited the orchid industry. The Singapore Botanic Garden has to-date produced 1,431 hybrids, and 142 selfed species (A.N. Rao 1977, see also Len 1982). According to A.N. Rao (1977), "Callus induction from embryo can be very useful to obtain an increased number of hybrid plants where the seed production is limited, or the percentage of germination is low".

In addition, embryo culture has been utilized for breaking seed dormancy, e.g. in *Magnolia soulangeana*. In *Iris* it has been possible to shorten the life cycle from seed to flowering, by embryo culture (see Raghavan 1976).

There are extensive investigations on the factors involved in dormancy, age, water content, and duration of soaking of seeds for the in vitro culture of the embryos of oil palm (*Elaeis guineensis*) (see Raghavan 1976).

Although the applications of embryo culture in horticulture and plant breeding are of prime importance, basically the technique has been usefully employed to investigate fundamental problems. The growth requirements of embryos, the role of hormonal substances affecting embryo growth in vitro, and the nutrition and metabolism during different phases of embryogenesis, are some of the more important aspects (see Raghavan and Srivastava 1982).

The tissue culture technique provides information about the processes involved in the transformation of the embryo into a seedling. Whenever there is no structural differentiation of the embryo, as in stem- and root-parasites, saprophytes, and even several autotrophs, the pre-germinal and germinal changes can be followed by culturing their seeds. The seeds of root-parasites germinate profusely in the absence of the host plant, if the medium contains essential organic growth supplements. The presence of a specific growth hormone determines the course of morphogenetic response, as in *Orobanche aegyptiaca, Cistanche tubulosa,* and *Santalum album* (see Raghavan and Srivastava 1982).

16.10 Polyembryony

In nature, besides the zygote, synergids, nucellus, and integumentary cells form embryos, i.e. embryo formation is restricted to ovular tissues. Experimental induction of embryos has been demonstrated in both somatic and reproductive tissues of a wide variety of plants.

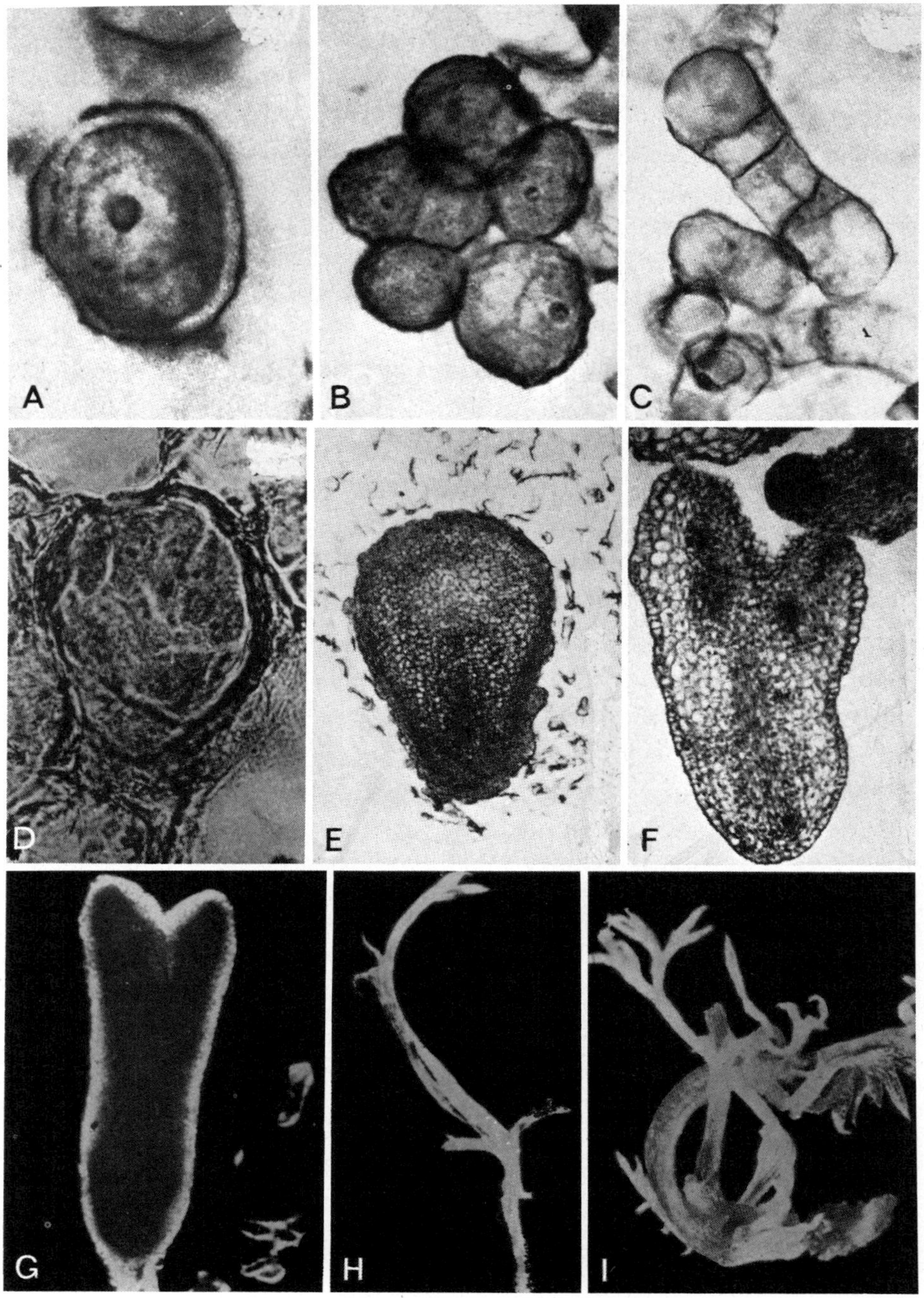

Fig. 16.8 A–I. Carrot (*Daucus carota*), organization of embryos in cell cultures. **A–C** One-, two-, four-, and six-celled embryos formed in parenchyma of carrot cultures. **D–F** Globular and heart-shaped stages. **G** Torpedo-shaped embryo. Note the suspensor-like structure at radicular pole. **H, I** Carrot plants with main root and cotyledons. (After Reinert et al. 1977)

Fig. 16.9 A–G. Sandalwood (*Santalum album* L.), somatic embryogenesis and plantlet formation. **A** Differentiation of embryos on BM + BAP (1 mg/l). **B–D** Growth of embryos in shale cultures on BM + NAA (1 mg/l), BM + IAA (1 mg/l) and BM + IAA (1 mg/l) + BAP (1 mg/l). Note the formation of roots. **E** Regeneration of shoot-buds from callus on BM + BAP (1 mg/l). **F** 4-week-old growth of callus on BM + 2,4-D (1 mg/l). **G** Established plantlets in earthern pots. (After Bapat and Rao 1979)

Somatic embryogenesis giving rise to entire plants was first demonstrated in domestic carrot. Discs of carrot in the region of cambium, with adjoining secondary phloem, and probably also xylem, were cultured on a nutrient medium containing coconut milk; the callus was capable of indefinite growth. From the callus, eventually, differentiated numerous bipolar embryos by a developmental sequence through pro-embryonal, globular, and torpedo stages (Fig. 16.8 A–G); all strikingly similar to those originating from the zygote. These embryos, ultimately, developed into entire plantlets with an organized shoot and root (Fig. 16.8 H, I).

The free cells of carrot, when provided with the proper nutrient conditions, formed an enormous number of embryos which faithfully reproduced different stages of embryogeny reminiscent of zygotic embryos. Such embryos formed in cultures developed into complete plantlets. Steward suggested two important parameters for a cell to behave like a zygote: first, the cell(s) should be freed from the restraining influence of tissue to which it belonged and, second, the presence of coconut milk (liquid endosperm of tender green coconut) in the media was absolutely essential. However, subsequent investigations on carrot revealed that none of these two preconditions are necessary (see Raghavan 1976).

Embryogenesis in vitro has been reported in cultures of vegetative tissues and organs of many other plants belonging to widely different families of both dicotyledons and monocotyledons. Included in this list are (1) floral buds of *Nigella damascena,* (2) reproductive and vegetative tissues of *Ranunculus sceleratus,* (3) pericarp of *Cucurbita pepo,* (4) nucellus of *Citrus,* (5) roots of *Atropa belladonna,* (6) leaf explants of almond, *Asparagus officinalis, Macleya cordata, Mesembryanthemum floribundum, Petunia hybrida, P. inflata, Rauwolfia serpentina, Thalictrum urbaini* (Yang and Chang 1980), *Brassica oleracea* (Pareek and Chandra 1978), *Sorghum bicolor* (Wernicke and Brettel 1980), *Coffea arabica* (Sondahl and Sharp 1977), *Kalanchoe pinnata, Apium graveolens* (Zee and Wu 1980), *Panicum maximum* (Lu and Vasil 1981), shoots of *Cymbidium,* (7) hypocotyl of *Asparagus officinalis, Anethum graveolens, Cucurbita pepo, Sinapsis alba, Nicotiana tabacum,* Chinese kale (Zee et al. 1978), *Ammi majus* (Grewal et al. 1976). *Liquidambar styraciflua* (Sommer and Brown 1980), *Santalum album* (Fig. 16.9 A–G), (8) stem segments of *Antirrhinum majus* (Sangwan and Harada 1975) and *Petunia inflata, P. hybrida, Asclepias curassavica, Asparagus officinalis, Didiscus coerulea, Euphorbia pulcherrima, Foeniculum vulgare, Nicotiana tabacum, Pergularia minor, Tylophora indica,* bulbs of garlic (Abo-El-Nil 1977) and tuberose (Narayanaswamy and Prabhudesai 1979), (9) peduncle of *Carica stipulata* (Litz and Conover 1980), and (10) petiole of *Coriandrum* (Zee 1981, for other citations see Raghavan 1976).

Embryogenesis has also been observed in the cultured embryos: *Santalum album, Coriandrum sativum, Arabidopsis thaliana, Anethum graveolens, Solanum melongena, Sium suave, Ilex aquifolium, Chichorium endivia, Corylus avellana, Theobroma cacao* (Pence et al. 1979), and in monocots such as barley, oil palm, *Bromus inermis,* and sugarcane (Nadar et al. 1978, for other citations see Raghavan 1976).

Embryogenesis has recently been reported in callus derived from immature embryos of *Lolium multiflorum* (Dale 1980), and immature embryos, inflorescences and suspension cultures of *Pennisetum americanum* and *P. americanum* × *P. purpureum* hybrid (V. Vasil and I. K. Vasil 1981 a, b), and *Panicum maximum* (Lu and Vasil 1981).

Somatic embryogenesis was reported by Butenko et al. (1968) in cultures derived from various tissues such as leaf, petiole, anthophore and stem of ginseng (*Panax ginseng*). However, complete plantlets did not regenerate. Chang and Hsing (1980) succeeded in obtaining whole-plant regeneration through embryogenesis in root-derived callus of ginseng, and have even reported in vitro flowering. Since flowering occurs in ginseng after a long juvenile phase (3 years), the precocious flowering in cultures should prove advantageous in ginseng breeding programmes.

16.11 Control of Embryogenesis

Steward's concept that isolation of cells is a necessary step to cause them to embark on embryogenic pathway, is no longer accepted. Backs-Huesmann and Reinert (1970) proved that isolated parenchymatous single cells from carrot cultures must first produce a multicellular aggregate before embryos can be initiated. Further proof came from the electron microscope studies of callus of *Ranunculus* (Thomas et al. 1972). Likewise, the role of coconut milk as an important factor for embryo induction has also been doubted. Several investigators observed that coconut milk is a superfluous component in the medium for inducing embryos and, in certain cases, its presence may even be inhibitory. According to Halperin and Wetherell (1964), vegetative parts of wild carrot form embryos in a simple medium supplemented with AD and 2,4-D, or KN and 2,4-D. 2,4-D is a potential auxin to induce embryogenesis in many other plants as well (P. S. Rao et al. 1970). In fact, Halperin (see Raghavan 1976) reported inhibition of differentiation of embryos when plated on a medium containing coconut milk.

The role of nitrogen compounds on embryogenesis has been studied by many investigators. The decisive step in this direction is the transfer of tissue from auxin-containing to auxin-free medium, and addition of nitrogen in the form of amino acids. Tazawa and Reinert (1969) observed that embryogenesis in vitro could be induced by both inorganic (KNO_3, NH_4NO_3) and organic (amino acids, amides) compounds. It appears that, for the initiation of embryogenesis, most salts must be present at a high concentration (Butenko et al. 1967), whereas compounds like ammonium and glutamine are effective at low levels. These substances are not very specific for embryogenesis in vitro; they are interchangeable. Therefore, it is not the quantity of nitrogen which determines embryogenesis, but the ratio of nitrogen to oxygen is important.

Dougall and Verma (1978), working with wild carrot (*Daucus carota*), observed that cell suspension cultures produced embryos on ammonium-ion as a sole source of nitrogen, in the absence of any exogenous Kreb's cycle acid, when the pH of the medium was controlled by continuous titration with KOH or $KHCO_3$. Fujimura and Komamine (1975) made a detailed study of the effects of various growth regulators on embryogenesis in carrot cell suspension cultures. Embryogenesis was inhibited by 2,4-D, or IAA. While zeatin promoted embryogenesis, other cytokinins inhibited it. Kessel et al. (1974) reported that there is a critical level of dissolved oxygen in cell suspension cultures of carrot tissue, and a relationship exists between dissolved oxygen concentration, ATP, and embryogenesis, both of which enhanced embryogenesis. Tisserat and Murashige (1977), working with carrot, observed that ethephon, ethylene, and 2,4-D suppressed somatic embryogenesis.

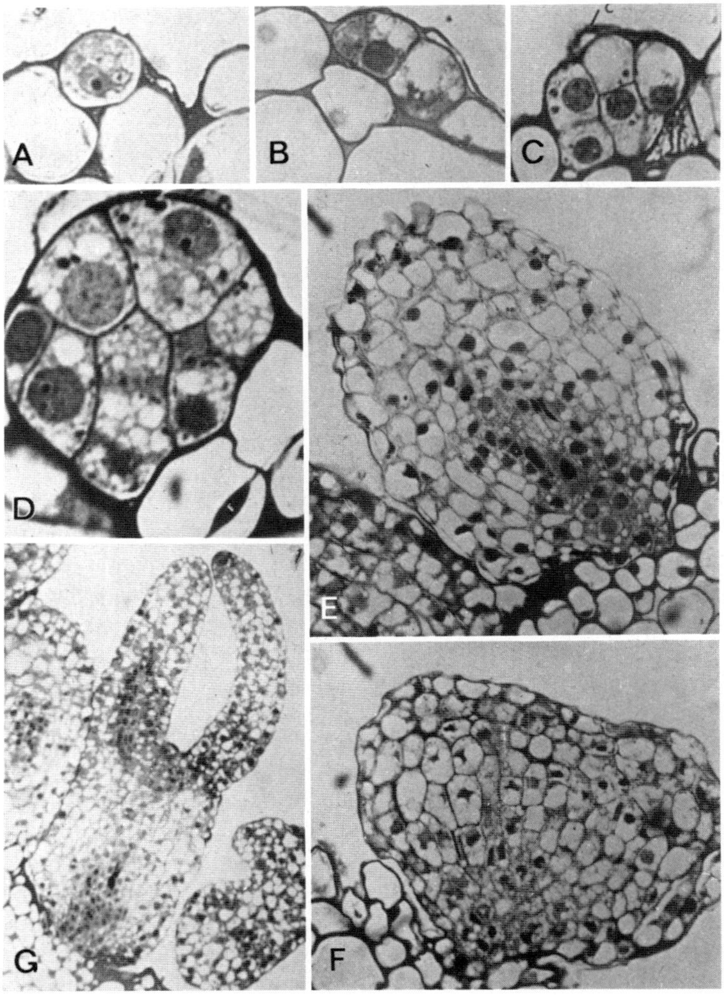

Fig. 16.10 A–G. *Ranunculus sceleratus,* ontogeny of embryos from stem. **A, B** Single and a pair of cytoplasm-rich epidermal cells. **C** Two 2-celled proembryos, and a single cytoplasm-rich epidermal cell. **D** Oblique section through a young proembryo. Note darkly-stained amyloplasts. **E** Late globular proembryo showing a central core of meristematic cells. **F** Heart-shaped embryo. **G** Mature embryo showing cotyledons, and root and shoot apices. (After Konar et al. 1972)

16.11.1 Loss of Embryogenic Potency

The loss of embryogenic potency of carrot cells, after a period of prolonged culture, has been reported by several workers (see Raghavan and Srivastava 1982). This shows a correlation between the rate of growth of cells in culture, and their capacity for embryogenesis. Generally, the slow-growing cells retain their embryogenic capacity much longer than fast-growing cells. Smith and Street (1974)

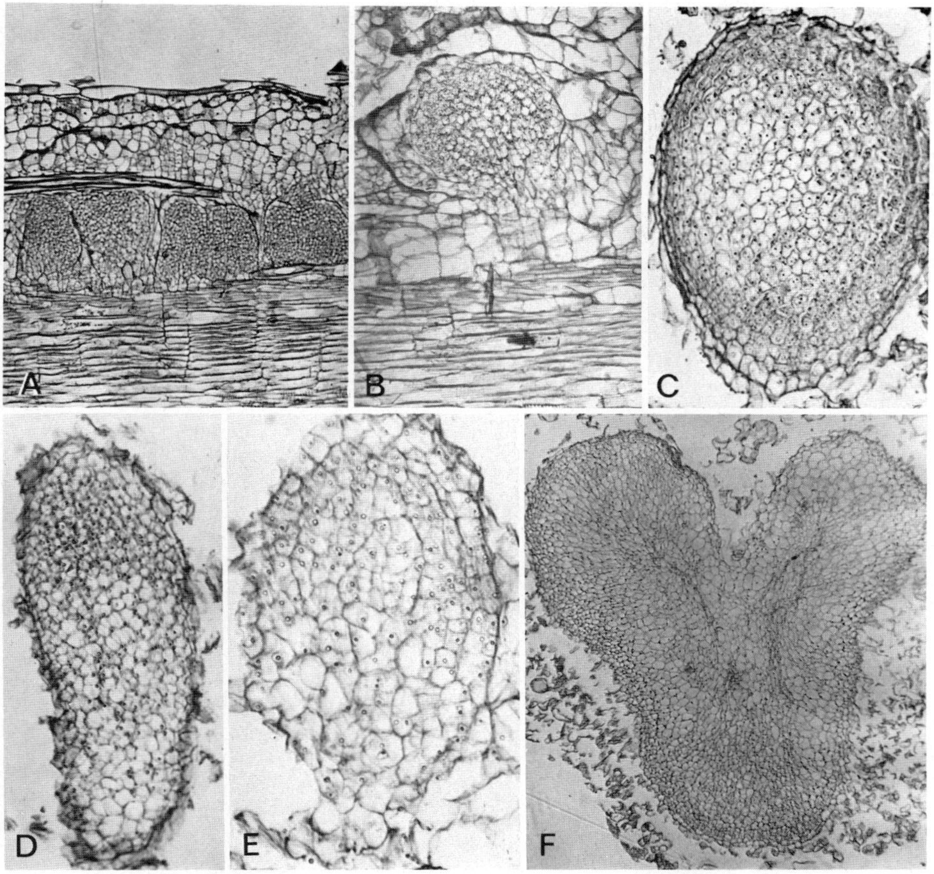

Fig. 16.11 A–F. *Petunia inflata,* ontogeny of somatic embryos. **A** Longisection through a 2-week-old culture of stem, grown on MS + 2,4-D, showing the origin of groups of embryos from cambium. **B** An embryo from a section similar to that in **A.** Note the origin of embryos from cambial cells. **C–F** Developmental stages of embryos obtained from 4-week-old stem culture, grown on MS + 2,4-D. (After Handro et al. 1973)

suggested that the loss of embryogenic potency is related to the cytological changes in cultured cells.

The recent techniques of cryopreservation might possibly be effectively used to store embryogenic strains without any appreciable loss of morphogenetic potential, as has been demonstrated in carrot (Nag and Street 1973).

16.11.2 Ontogeny of Embryos

P. Maheshwari and Baldev (1961), working with embryonal callus of *Cuscuta reflexa,* observed that epidermal cells undergo divisions and develop into globular embryos which elongate and organize the characteristic shoot apex flanked by

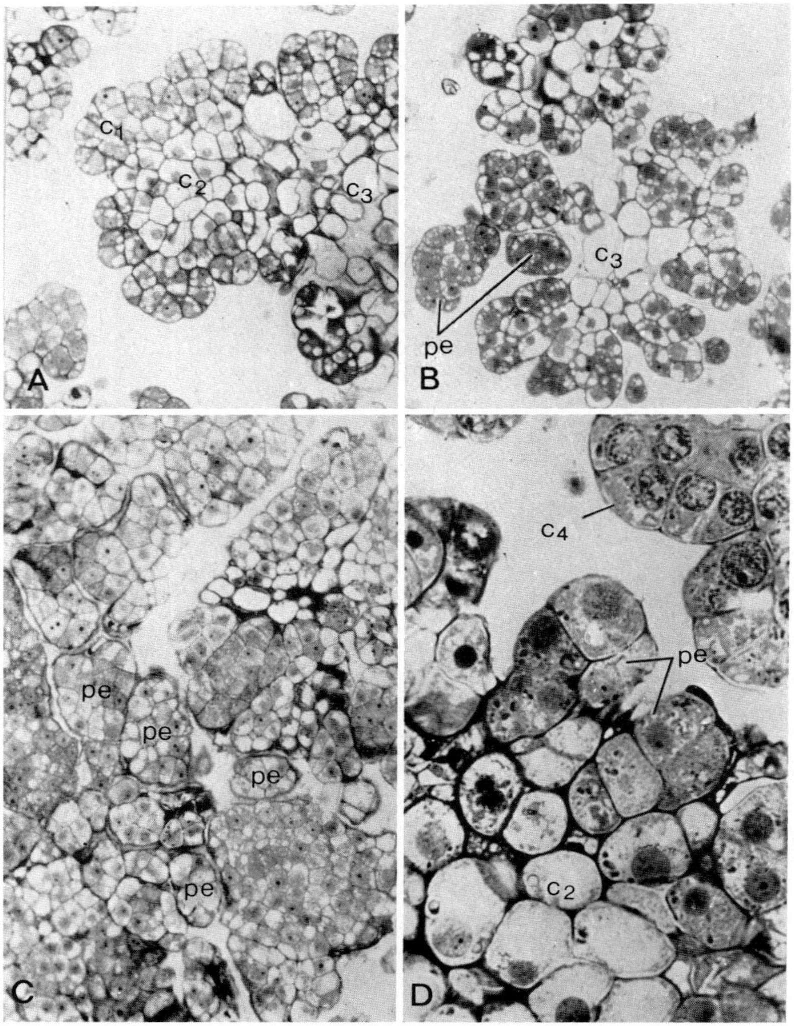

Fig. 16.12 A–D. *Ranunculus sceleratus,* histological study of embryogenic callus. **A** Section of callus showing large aggregates of superficial cytoplasm-rich cells (C_1), along with internal, more vacuolate cells (C_2), and senescent or dead cells (C_3). **B–D** Sections of callus showing presumptive proembryoids (*pe*), and superficial cells with characteristic condensed chromatin (C_4) in their nuclei. In **D** note the early stages of proembryoid (*pe*) development. (After Thomas et al. 1972)

scale leaves. In *Ranunculus sceleratus* (Fig. 16.10 A–G), and carrot (Kato and Takeuchi 1966), it has been demonstrated that epidermal cells produce embryos, whereas in *Petunia* (Fig. 16.11 A–F) embryogenesis has been traced to cortical cells.

In carrot tissue cultures the embryogenic cells divide repeatedly and form a meristematic tissue of "embryogenic masses" which are the progenitors for the production of embryos (Jones 1974).

In a continuously growing callus culture any cell within the undifferentiated tissue is potentially capable of producing an embryo. Differentiation of proembryos from callus of *Ranunculus sceleratus* is shown in Fig. 16.12A–D. Some of the characteristic features of the embryogenic cells are their dense cytoplasm, prominent nuclei, and high ribosomal density. Kohlenbach (1970) and Sussex (1972) observed intense RNA synthesis during the initial steps of embryogeny, and Wochok (1973) reported that the transition of the callus cells to an embryogenic state is accompanied by a change in the distribution of microtubules from a random orientation to linear, parallel arrays (see Raghavan 1976).

P.S. Rao and Narayanaswamy (1972) reported that in the callus cultures of *Tylophora indica,* the mode of ontogeny of embryos depended upon the relative position of the embryonal cell in callus parenchyma. In those cells which were embedded in the callus mass the segmentation pattern was reminiscent of zygotic embryogeny, whereas in the embryonal cells situated in the superficial layers the sequence of cell division was irregular. It would appear, therefore, that confinement of the embryogenic competent cells within the callus parenchyma simulate the physical environment obtaining around the zygote within the ovule.

In petiole explants of Chinese celery, embryoids are formed on a medium supplemented with 2,4-D. Zee et al. (1979) made a detailed anatomical study of the sequence of development leading to the formation of proembryoids, using SDS-gel electrophoresis. The stages of formation of proembryoids were closely linked with the changes in the types of soluble proteins. A similar correlation between structural and protein-composition changes has been noticed in Chinese kale hypocotyl explants (Zee et al. 1978). These findings strongly suggest that the synthesis and/or reconstitution of certain types of proteins in the explants are essential for organogenesis and/or embryogenesis.

Another feature of interest in in vitro developed embryos relates to the nature of suspensor. Konar et al. (1972), from their extensive investigations on the origin and structure of embryos on the stem (in vitro raised plantlets) of *Ranunculus,* observed that although the development of embryoid proceeds by the formation of an embryonal cell and a suspensor cell, a functional suspensor (as known in a zygotic embryo) is absent. Suspensor-like structures have been observed in several tissue culture systems (see Raghavan 1976). The in vitro developed embryos presumably absorb nutrients all over their surface, and the suspensor, even if present, is functionless (see Haccius 1965, Ammirato and Steward 1971).

Normally, the in vitro developed embryos grow into normal plants capable of flowering. However, embryos developed from the placental callus of *Eschscholtzia californica* (Kavathekar et al. 1977) failed to develop into plantlets, on the same medium on which they differentiated, or on subculture on a fresh medium. If the embryos were subjected to continuous chilling at 6 °C, they developed into plantlets.

The reasons for the inability of several cultured tissues to form embryos, although they may form shoot-buds/roots, remain obscure. According to Halperin (1967), embryogenic competence rests exclusively with the primary explant which contains induced and non-induced cells, and the capacity for embarking upon the embryogenic pathway depends mainly on the cytological stability of the induced cells in the primary explant, as well as the appropriate nutrient conditions. Probably, some taxa seem to be predisposed to form embryos in culture. This concept is interesting, but with further understanding of growth-promoting substances and synthesis of more exacting nutrient media it should be possible to induce embryogenesis in most somatic and reproductive tissues.

16.12 Protoplast Culture

The isolation and culture of protoplasts is a comparatively recent area of investigation. Cellulases and pectinases are used for large-scale isolation of protoplasts; there is less osmotic shrinkage, the cells are not injured, and remain intact. The yield of protoplasts depends on the physiological state of experimental plant, purity and composition of enzymes, choice of osmotic solution, and pH. After the degradation of wall, the enzymes are removed by gentle filtration on a 8 µm filter, or by low-speed centrifugation (see Gamborg et al. 1981, Constabel 1982, P. S. Rao 1982, Reinert and Yeoman 1982).

The droplets of protoplast suspension are placed in liquid media in plastic dishes. It facilitates exchange of gases, diffusion of excretory products, and addition of fresh medium. Alternatively, a 2 ml aliquot of protoplasts is suspended in liquid culture medium, poured into "falcon" plastic Petri dishes, and mixed gently with an equal volume of the same medium containing 1% agar (*T* below 45 °C). The Petri dishes are sealed with parafilm, and kept in an inverted position in a growth-room at 28 °C under continuous light (2,300 lx). The entire sequence of growth, development, and differentiation can be easily observed.

Generally, modified MS and Nagata and Takebe media are used. The media are supplemented with growth promoters according to the type and requirement of the species.

The success of protoplast culture depends on the ability of the protoplast to synthesize new cell-wall, undergo division, and develop into a new plantlet.

16.12.1 Embryogenesis and Regeneration

Embryogenesis has been observed in protoplast cultures of herbaceous plants: *Antirrhinum majus, Asparagus officinalis, Atropa belladonna, Bromus innermis,* and *Daucus carota* (see P. S. Rao 1982), *Brassica napus* (Kohlenbach et al. 1982) and *Lycopersicon peruvianum* (Zapata and Sink 1981). Citrus is the only woody species where protoplast regeneration and embryo development have been reported (Vardi et al. 1975). Embryoids formed in irradiated *Citrus* callus, while embryoids did not develop in non-irradiated protoplast colonies.

V. Vasil and I. K. Vasil (1980) report embryogenesis and plantlet formation in isolated protoplasts derived from suspension cultures of immature embryos of pearl millet (*Pennisetum americanum*). Embryoids could be induced in mesophyll protoplast of an important forage crop, alfalfa (*Medicago sativa*), and regenerated into plants (Kao and Michayluk 1980). Callus formation and occasional differentiation of shoots have been reported in mesophyll protoplasts of cassava

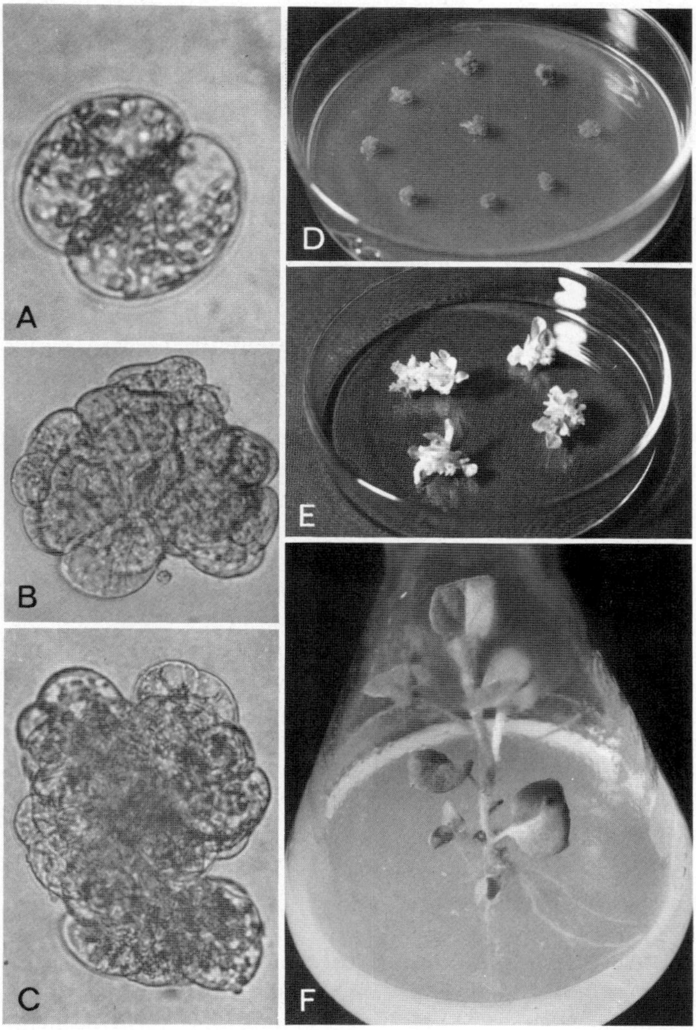

Fig. 16.13 A–F. *Nicotiana tabacum*, regeneration of plants from isolated tobacco mesophyll protoplasts. **A–C** Successive stages of colony formation by protoplasts, embedded in agar medium. **D** Colonies transferred to a plate with B_3 medium. Note callus masses with indications of shoot differentiation. **E** Shoots from callus, on B_3 medium. **F** Plantlet regenerated from single protoplast. (After Nagata and Takebe 1971)

(*Mannihot esculenta*) (Shahin and Shepard 1980), of complete plantlets in suspension-derived protoplasts of *Trifolium repens* (Gresshoff 1980) and *Salpiglossis sinuata* (Boyes and Sink 1981).

I. K. Vasil (1982) questions the misleading conclusion that "cereal protoplasts are constitutionally incapable of sustained division". He refers to protoplasts isolated from embryogenic cell suspension cultures of *Pennisetum americanum, P. purpureuns, Panicum maximum,* and *Zea mays.* These protoplasts form embryogenic callus from which embryoids and plantlets have been recovered. Further improvements in the survival of the regenerated plants to maturity are essential.

Nagata and Takabe (1971) described, for the first time, the regeneration of entire plantlets from mesophyll protoplasts of *Nicotiana tabacum* (Fig. 16.13 A–F). Gill et al. (1979), and Wernicke and Thomas (1980) also observed organogenesis in mesophyll protoplast cultures of tobacco. Next to tobacco, another solanaceous member, *Petunia,* has been extensively studied. Durand et al. (1973) obtained callus from mesophyll protoplasts, while Nagata and Takebe (1971) reported differentiation of shoots from the callus. Further work on *Petunia* protoplasts has been reported by Frearson et al. (1973), V. Vasil and I. K. Vasil (1974), and Binding (1974a, b). The parallel growth requirements for successful culture of *Petunia* and tobacco protoplasts suggest a biological compatibility between the two genera, serving a suitable system for all hybridization studies. Callus and regeneration of protoplasts have been reported in *Brassica* (Kartha et al. 1974a), *Datura* (Schieder 1975, 1977, Furner et al. 1978), *Ranunculus* (Dorion et al. 1975), *Solanum melongena* (Bhatt and Fassuliotus 1981, Saxena et al. 1981), *S. tuberosum* (Gunn and Shepard 1981), *S. dulcamara* (Binding and Nehls 1977), *Nicotiana* sp. (Passiatone and Sink 1981), *N. debreyi* (Scowcraft and Larkin 1980), and *Hyoscyamus muticus* (Wernicke and Thomas 1980).

Protoplasts isolated from haploid species of *Nicotiana tabacum* (Ohyama and Nitsch 1972) and *Petunia hybrida* (Binding 1974a, b) have been successfully regenerated into entire plantlets. A recent promising approach, in this direction, is the isolation of protoplasts from pollen tetrads and attempts to induce division and, eventually, regeneration of entire plantlets. In flowering plants each spore in the young microspore tetrad is a naked haploid cell enclosed in cellulosic wall. Pollen-tetrad protoplasts have been isolated in *Nicotiana tabacum* (Bhojwani and Cocking 1972), *Atropa, Hyoscyamus, Nicotiana*, and *Petunia* (Bajaj 1974), and *Datura metel* (Rajasekhar 1973).

16.12.2 Protoplast Fusion and Somatic Hybridization

The most promising application of protoplast research is "parasexual hybridization" which offers immense possibilities of genetic manipulation. The phenomenon of induced fusion was first envisaged by Power et al. in 1970. The inducing agent they used, $NaNO_3$ (0.25 M), had the overall effect similar to sendai virus-induced fusion in animal cells (Harris 1970). Induced fusion is influenced by an alkaline pH, and high calcium concentration (Keller and Melchers 1973). Compared to low frequency fusion by $NaNO_3$, the PEG (polyethylene glycol)-induced

Fig. 16.14 A–I. *Datura,* protoplast culture and somatic hybridization. **A** Cell colonies derived from protoplasts of *Datura innoxia* (Schieder, unpublished). **B** Cell colony derived from protoplasts of *D. innoxia,* showing organization of leaf primordia. **C** Cell colony derived from protoplasts of *D. innoxia* showing shoot formation. (**B, C** After Schieder 1975). **D** Fusion product between protoplasts derived from an albino mutant of *D. innoxia* and the wild type (green) of *D. stramonium.* (After Schieder 1978a). **E** Somatic hybrid colony (green) developed after fusion of protoplasts derived from an albino mutant of *D. innoxia* and the wild type of *D. sanguinea.* (After Schieder and Vasil 1980). **F** Grafted somatic hybrid of *D. innoxia* + *D. sanguinea.* (After Schieder 1980). **G** *D. innoxia* plant. **H** *D. stramonium* plant. **I** Somatic hybrid plant *D. innoxia* + *D. stramonium.* (**G–I** After Schieder 1978a)

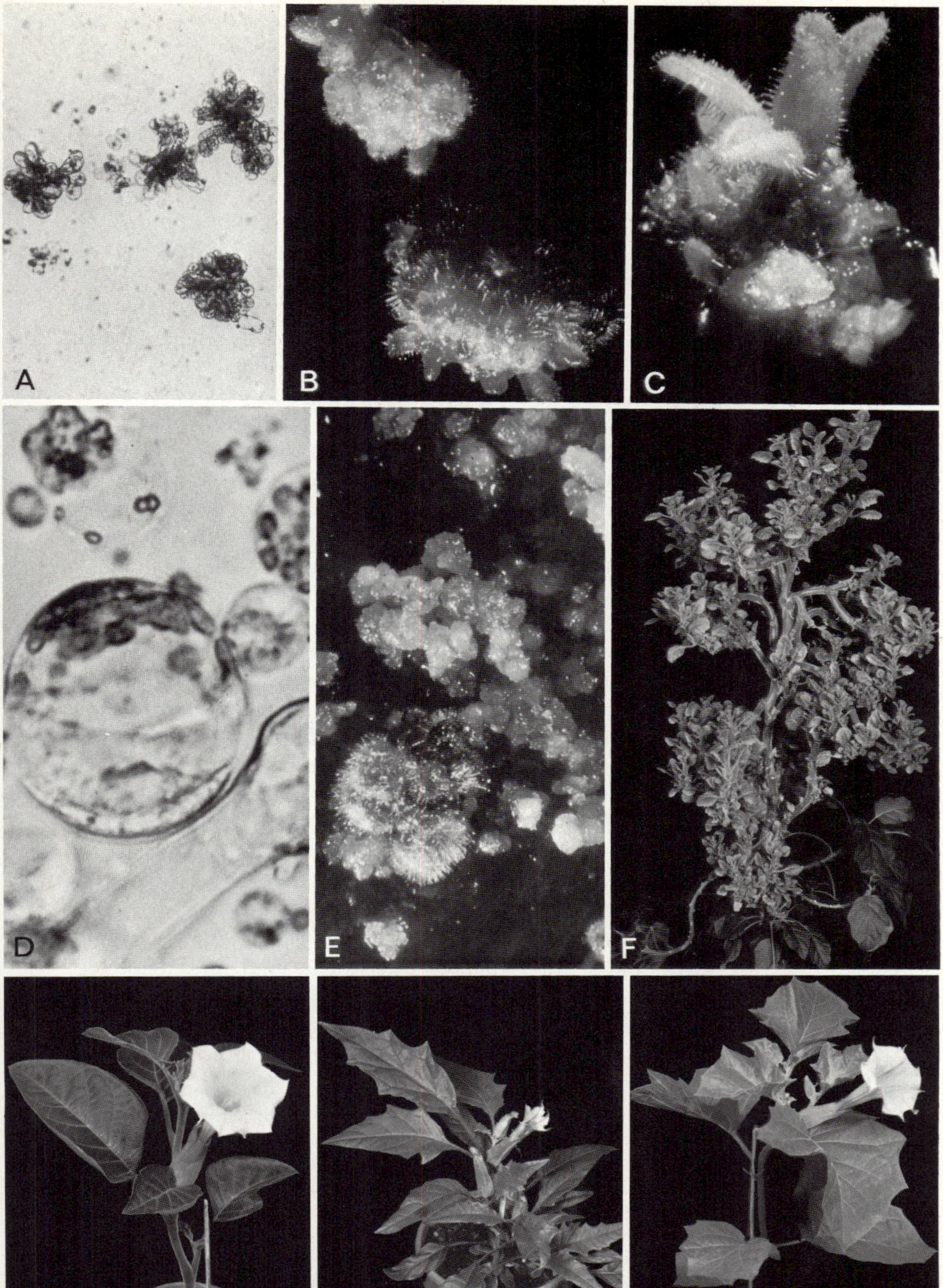

Fig. 16.14 A–I

fusion gives better results (Gamborg et al. 1974). The process of fusion is considerably enhanced if the protoplast suspension is pre-treated with a membrane-activating agent such as lysozyme (Constable and Kao 1974, Kao and Michayluk 1974). Fusion between protoplasts of the same species results in homokaryons, while from different species in heterokaryons (see Kao 1982).

Many interspecific hybrids have been obtained in carrot, *Petunia*, and *Nicotiana* (Carlson et al. 1972, Dudits et al. 1977, Maliga et al. 1977, Power et al. 1977). Examples of intergeneric hybrids are: *Glycine max* with various other genera such as *Angelica, Cicer, Hordeum, Medicago, Melilotus, Pisum,* and *Vicia* (Gamborg et al. 1974, Kao et al. 1974), and between *Nicotiana* and *Petunia* (Power and Frearson 1973).

In order to identify heteroplasmic fusion products of protoplasts, "selective markers" are necessary (see Schieder and Krumbiegel 1981). Plastid characteristics served as markers in fusion between rape seed and soybean (Kartha et al. 1974 b), tobacco and carrot, *Atropa belladonna* + *Petunia hybrida, A. belladonna* + *Nicotiana tabacum* (Gosch et al. 1975, Gosch and Reinert 1978) and in interspecific hybridization of *Torenia* (Potrykus 1971). Besides plastids, there are many other parameters which aid in somatic hybrid identification.

The first report of successful interspecific hybrid, by fusion of protoplasts, is in *Nicotiana glauca* + *N. longsdorfii* (Carlson et al. 1972). The hybrid was identical, in all aspects, with the naturally occurring amphidiploid. A similar somatic hybrid was obtained in *Petunia* (Power et al. 1976, Cocking et al. 1977). By making use of green and albino parents, Dudits et al. (1977) reported successful fusion product in *Daucus,* and Maliga et al. (1978) in tobacco. Maliga et al. (1977) also reported somatic hybrid plantlets of tobacco from a kanamycin-resistant cell line *Nicotiana sylvestris* and *N. knightiana,* both potentially deficient to produce shoots. By complementation of chlorophyll deficient mutants, protoplast heterokaryons which, eventually, gave rise to hybrid plantlets were induced in *Nicotiana* (Melchers and Labib 1974, Gleba et al. 1975), and *Datura* (Schieder 1978 a). Somatic hybrids have also been obtained between herbaceous and tree species in *Datura* (Fig. 16.14 A–I), *Datura innoxia* and *Atropa belladonna* (Krumbiegel and Schieder 1979), *Nicotiana rustica* and *N. tabacum* (Douglas et al. 1981 a–c), and *Nicotiana glauca* + *N. tabacum* (Evans et al. 1980).

The recent example of *Arabidobrassica,* from protoplast fusion product between *Arabidopsis* and *Brassica* (Gleba et al. 1978, Gleba and Hoffmann 1980) is the first report of intertribal fusion of protoplasts.

Fig. 16.15 A–E. *Solanum tuberosum* and *Lycopersicon esculentum,* somatic hybrid plants regenerated from fused protoplasts. **A** Protoplasts arisen by fusion of tomato and potato protoplasts. The cytoplasmic part of the tomato is recognizable by the chloroplasts present in the tomato mesophyll protoplasts. The cytoplasmic part of the potato is colourless as potato protoplasts were made from submerged cultured callus cells. **B, C** Potato-tomato hybrid 1b/2h/3. **B** Hybrid growing on its own roots with filled flowers (possibly because of abnormal chromosome number). Flower colour very dilute-purple (from potato?) but also weak-yellow (from tomato?). **C** Same plant dug out with filiform roots and rhizome which is slightly thickened and has produced a shoot with filled flowers. An underground flower bud is also present. **D** Fruit of the somatic hybrid line 2a/2a/36d S_2 and S_4, "S_2, S_4, S_5 morphologically" very similar, S_1, S_3 similar but different from others. Fruit obtained after pollination with *Solanum stenotomum.* **E** "Tubers" of the above hybrid (*wood label* 6.5 cm; 60% of natural size). (**A–C** After Melchers et al. 1978; **D, E** after Schiller et al. 1982; courtesy Prof. G. Melchers, Tübingen, Germany)

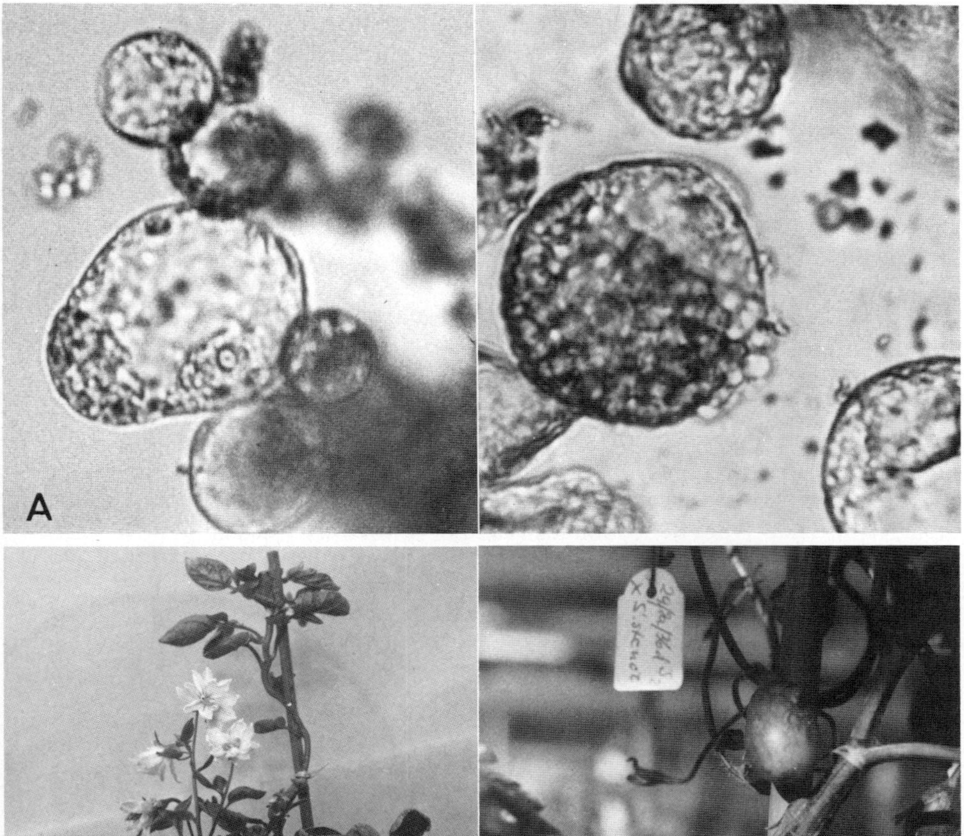

Fig. 16.15 A–E

16.12.3 "Pomatoes" and "Topatoes"

Somatic hybrids (Fig. 16.15 A–E) were obtained by fusion of mesophyll protoplasts of *Lycopersicon esculentum* Mill. var. *cerasiforme* (Dunal) Alef, mutant yellow green ygb (Rick), and protoplasts from a submersion callus culture of the originally dihaploid *Solanum tuberosum,* line HH 258 (Melchers et al. 1978, see also Melchers 1981, 1982). It was possible to grow the fusion products between the protoplasts of the two parents, up to calli. The hybrids could be clearly identified from products of the tomato line because the calli differentiated only roots (under the applied conditions).

Tomato and potato possess isoelectric variants for the small, nuclear-coded subunit, and the hybrid regenerates exhibited the components from both parents, demonstrating that they represented genuine nuclear hybrids. Only the large subunit of one parental line, either of tomato or of potato, could be detected, suggesting the existence of two hybrid categories, "pomatoes" (with plastids of potato) and "topatoes" (with plastids of tomato) (see Schiller et al. 1982). These researches have opened up vast areas of hybridization and, in due course, it should be possible to raise new types of economically important plants, especially by the combination of conventional breeding and tissue culture techniques (see Wenzel et al. 1979).

16.13 Concluding Remarks

Experimental embryology has produced significant results, and production of androgenic haploids has dominated all other areas. The raising of hybrids has been possible through embryo culture, ovule culture and/or ovary culture, control of fertilization and, more recently, through somatic hybridization. There are still many problems which need attention: (1) Can the dormant floral buds, in dioecious plants like *Salix,* after initiation, undergo a sex reversal? (2) Can a cleistogamous flower be changed into a chasmogamous flower? (3) Why does not the nucellar tissue in cucurbits undergo morphogenesis and proliferation in cultures? (4) The in vitro pollination and fertilization is somewhat tricky, and refinements in techniques and composition of nutrient media are essential before the pollen of the same or different species can bring about fertilization and development of embryo in excised ovules. This may solve the serious problem of incompatibility. (5) Even though of limited utility, at present, triploid plants can be raised from endosperm. However, such plants can be obtained only in a few taxa. Why does not the endosperm in many other plants give rise to triploids? (6) In certain angiosperms the mature embryo remains arrested at the globular stage. In not even a single taxon such an embryo has been made to differentiate radicle, cotyledon(s) and plumule (shoot tip). (7) The explants of total and semi-parasites *can* be grown in synthetic medium, without the host tissue or extract thereof. The media used, in most cases, have growth promoters like yeast extract, casein hy-
media used, in most cases, have growth promoters like yeast extract, casein hydro-
becomes "determined" at a certain stage, and its "commitment" cannot be changed.

does explain why a particular tissue must express itself in a definite pattern of growth, development and differentiation. It does not add to our knowledge as to how to "break" this determination and commitment.

References

Abo-El-Nil MM (1977) Organogenesis and embryogenesis in callus cultures of garlic (*Allium sativum* L.). Plant Sci Lett 9:259–264

Ammirato PV, Steward FC (1971) Some effects of environment on the development of embryos from cultured free cells. Bot Gaz 132:149–158

Anagnostakis SL (1974) Haploid plants from anthers of tobacco: Enhancement with charcoal. Planta (Berl) 115:281–283

Anand VV, Arekal GD, Swamy BGL (1980a) Chimeral embryoids of pollen origin in tobacco. Curr Sci (Bangalore) 49:603–604

Anand VV, Arekal GD, Swamy BGL (1980b) Androgenic chimeral plantlets in tobacco. Curr Sci (Bangalore) 49:750

Anand VV, Arekal GD, Swamy BGL (1981a) Foliar histology of androgenetic haploid "vegetative" and "generative" plants of tobacco. Curr Sci (Bangalore) 50:327–328

Anand VV, Arekal GD, Swamy BGL (1981b) Variation among flowers of androgenetic tobacco haploids. Curr Sci (Bangalore) 50:372–373

Angrish R, Nanda KK (1982a) Seasonal culture of dormant reproductive buds of *Salix tetrasperma:* Analysis of the flowering process. Plant Cell, Tissue and Organ Culture 1:181–193

Angrish R, Nanda KK (1982b) Dormancy and flowering process in reproductive buds of *Salix babylonica* cultured in vitro. Z Pflanzenphysiol 106:263–269

Asano Y (1980) Studies on the crosses between distantly related species of Lilies. 4. The culture of immature hybrid embryos 0.3–0.4 mm long. J Jpn Soc Hortic Sci 49:114–119

Backs HD, Reinert J (1970) Embryobildung durch isolierte Einzelzellen aus Gewebekulturen von *Daucus carota*. Protoplasma 70:49–60

Bajaj YPS (1964) Development of ovule of *Abelmoschus esculentus* L. var. Pusa Sawani in vitro. Proc Natl Inst Sci India 30:175–185

Bajaj YPS (1974) Isolation and culture studies on pollen tetrad and pollen mother cell protoplasts. Plant Sci Lett 3:93–99

Bajaj YPS, Gosal SS (1982) Induction of genetic variability in grain-legumes through tissue culture. In: Rao AN (ed) Proc COSTED Symp "Tissue culture of economically important plants". Natl Univ Singapore, pp 25–41

Balatkova V, Tupý J (1972) Some factors affecting the seed-set after in vitro pollination of excised placentae of *Nicotiana tabacum*. Biol Plant Acad Sci Bohemoslov 14:82–88

Bapat VA, Rao PS (1979) Somatic embryogenesis and plantlet formation in tissue cultures of sandalwood (*Santalum album* L.). Ann Bot 44:629–630

Barclay IR (1975) High frequencies of haploid production in wheat (*Triticum aestivum*) by chromosome elimination. Nature 256:410–411

Beasley CA (1971) In vitro culture of fertilized cotton ovules. Bioscience 21:906–907

Beasley CA (1973) Hormonal regulation of growth in unfertilized cotton ovules. Science (Wash, D.C.) 179:1003–1005

Berghoef J, Bruinsma J (1979a) Flower development of *Begonia franconis* Liemb. II. Effects of nutrition and growth regulating substances on the growth of flower buds in vitro. Z Pflanzenphysiol 93:345–357

Berghoef J, Bruinsma J (1979b) Flower development of *Begonia franconis* Liemb. III. Effects of growth-regulating substances on organ initiation in flower buds in vitro. Z Pflanzenphysiol 93:377–386

Berghoef J, Bruinsma J (1979c) Flower development in *Begonia franconis Liemb*. IV. Adventitious flower bud formation on excised inflorescence pedicels in vitro. Z Pflanzenphysiol 94:407–416

Bergmann L (1960) Growth and division of single cells of higher plants in vitro. J Gen Physiol 43:841–851

Bhatt DP, Fassuliotis G (1981) Plant regeneration from mesophyll protoplasts of egg plant. Z Pflanzenphysiol 104:81–89

Bhojwani SS, Cocking EC (1972) Isolation of protoplasts from pollen tetrads. Nature New Biol 239:29–30

Bhojwani SS, Razdan MK (1983) Plant tissue culture: Theory and practice. Elsevier, Amsterdam

Binding H (1974a) Cell cluster formation by leaf protoplasts from axenic cultures of haploid *Petunia hybrida* L. Plant Sci Lett 2:185–188

Binding H (1974b) Regeneration of haploid and diploid plants from protoplasts of *Petunia hybrida* L. Z Pflanzenphysiol 74:327–356

Binding H, Nehls R (1977) Regeneration of isolated protoplasts to plants in *Solanum dulcamara* L. Z Pflanzenphysiol 85:279–280

Binding H, Binding K, Straub J (1970) Selektion in Gewebekulturen mit haploiden Zellen. Naturwissenschaften 57:138–139

Bitters WP, Murashige T, Rangan TS, Nauer E (1972) Investigations on establishing virus-free *Citrus* plants through tissue culture. In: Pierce WC (ed) Proc 5th Conf Intl Organ *Citrus* Virologists. Univ Florida Press, Gainesville, pp 267–271

Bo-ji WU, Kuo-chang C (1982) Cytological and embryological studies on haploid and diploid plant production from cultured unpollinated ovaries of *Nicotiana tabacum* and *Nicotiana rustica*. In: Fujiwara A (ed) Plant tissue culture 1982. Proc 5th Intl Congr Plant Tissue and Cell Culture, Tokyo, pp 119–120

Boyes CJ, Sink KC (1981) Regeneration of plants from callus-derived protoplasts of *Salpiglossis*. J Am Soc Hortic Sci 106:42–46

Burk LG, Gwynn GR, Chaplin JF (1972) Diploidized haploids from aseptically cultured anthers of *Nicotiana tabacum:* A colchicine method applicable to plant breeding. J Hered 63:355–360

Butenko RG, Strogonov BP, Babaeva ZA (1967) Somatic embryogenesis in carrot tissue cultures under conditions of high salt concentrations in the medium. Dokl Akad Nauk SSSR 175:1179–1181

Butenko RG, Grushvitskii RV, Slepyan LI (1968) Organogenesis and somatic embryogenesis in a tissue culture of ginseng (*Panax ginseng*) and other *Panax* L. species. Bot Zh 53:906–911

Button J, Bornman CH (1971) Development of nucellar plants from unpollinated and unfertilized ovules of the Washington Navel Orange in vitro. J S Afr Bot 37:127–133

Camara A (1943) Transplantacae de embroies. Agron Lusit 5:375–386

Carlson PS, Smith HH, Dearing RD (1972) Parasexual interspecific plant hybridisation. Proc Natl Acad Sci USA 69:2292–2294

Carrel A, Burrows MT (1911) Cultivation of tissues in vitro and its technique. J Exp Med 13:387–396

Chang WC, Hsing YW (1980) In vitro flowering of embryoids from mature root callus of ginseng (*Panax ginseng*). Nature 284:341–342

Cheema GS, Mehra PN (1982) Morphogenesis in endosperm cultures. In: Fujiwara A (ed) Plant tissue culture 1982. Proc 5th Intl Congr Plant Tissue and Cell Culture, Tokyo, pp 111–112

Chen Y, Li LT (1978) Investigation and utilization of pollen-derived haploid plants in rice and wheat. In: Proc Symp Plant Tissue Culture. Science Press, Peking, pp 199–211

Chen Y, Lin MH (1976) Induction of rice plantlets from anther culture. Bot Bull Acad Sin 17:18–24

Chopra RN (1962) Effect of some growth substances and calyx on fruit and seed development of *Althaea rosea* Cav. In: Plant Embryology – A Symposium. CSIR New Delhi, pp 170–181

Chopra RN, Sabharwal PS (1963) In vitro culture of ovules of *Gynandropsis gynandra* Briq. and *Impatiens balsamina* L. In: Maheshwari P, Rangaswamy NS (eds) Plant Tissue and Organ Culture – A Symposium. Intl Soc Plant Morphologists, Univ Delhi, pp 257–264

Chuang CC, Ouyang TW, Chia H, Chou SM, Ching CK (1978) A set of potato media for wheat anther culture. In: Proc Symp Plant Tissue Culture. Science Press, Peking, pp 51–56

Clapham D (1971) In vitro development of callus from the pollen of *Lolium* and *Hordeum*. Z Pflanzenzuechtg 65:285–292

Clapham D (1977) Haploid induction in cereals. In: Reinert J, Bajaj YPS (eds) Applied and fundamental aspects of plant cell, tissue, and organ culture. Springer, Berlin Heidelberg New York, pp 279–340

Cocking EC, George D, Price-Jones MJ, Power JB (1977) Selection procedures for the production of inter-species somatic hybrids of *Petunia hybrida* and *Petunia parodii*. II. Albino complementation selection. Plant Sci Lett 10:7–12

Constabel F (1982) Isolation and culture of plant protoplasts. In: Wetter LR, Constabel F (eds) Plant tissue culture methods. Natl Res Council Canada, Saskatoon, Saskatchewan, pp 38–48

Constabel F, Kao KN (1974) Agglutination and fusion of plant protoplasts by polyethylene glycol. Can J Bot 52:1603–1606
Cox EA, Stetzky G, Goos RD (1960) In vitro culture of *Musa balbisiana* Colla embryos. Nature 185:403–404
Dale PJ (1980) Embryoids from cultured immature embryos of *Lolium multiflorum*. Z Pflanzenphysiol 100:73–77
Dale PJ, Thomas E, Brettell RIS, Wernicke W (1981) Embryogenesis from cultured immature inflorescences and nodes of *Lolium multiflorum*. Plant Cell, Tissue and Organ Culture 1:47–55
Davies RD (1960) The embryo culture of interspecific hybrids of *Hordeum*. New Phytol 59:9–14
Debergh P, Nitsch C (1973) Premiers résultats sur la culture in vitro de grains de pollen isolés chez la tomate. CR Acad Sci Paris 276:1281–1284
De Ropp RS (1955) The growth and behaviour in vitro of isolated plant cells. Proc R Soc Lond B 144:86–93
Dhindsa RS, Beasley CA, Ting IP (1976) Effects of abscisic acid on in vitro growth of cotton fiber. Planta (Berl) 130:197–201
Dodds JH, Reynolds TL (1980) A scanning electron microscope study of pollen embryogenesis in *Hyoscyamus niger*.. Z Pflanzenphysiol 97:271–276
Dodds JH, Roberts LW (1982) Experiments in plant tissue culture. Cambridge Univ Press, Cambridge
Dolezel J, Novak FJ, Luzny J (1980) Embryo development and in vitro culture of *Allium cepa* and its interspecific hybrids. Z Pflanzenzuechtg 85:177–184
Dorion N, Chupeau Y, Bourgin JP (1975) Isolation, culture, and regeneration into plants of *Ranunculus sceleratus* L. leaf protoplasts. Plant Sci Lett 5:325–332
Dougall DK, Verma DC (1978) Growth and embryo formation in wild carrot suspension cultures with ammonium-ion as a sole nitrogen source. In Vitro 14:180–182
Douglas GC, Keller WA, Setterfield G (1981 a) Somatic hybridization between *Nicotiana rustica* and *N. tabacum*. I. Isolation and culture of protoplasts and regeneration of plants from cell cultures of wild-type and chlorophyll-deficient strains. Can J Bot 59:208–219
Douglas GC, Keller WA, Setterfield G (1981 b) Somatic hybridization between *Nicotiana rustica* and *N. tabacum*. II. Protoplast fusion and selection and regeneration of hybrid plants. Can J Bot 59:220–227
Douglas GC, Wetter LR, Nakamura C, Keller WA, Setterfield G (1981 c) Somatic hybridization between *Nicotiana rustica* and *N. tabacum*. III. Biochemical, morphological, and cytological analysis of somatic hybrids. Can J Bot 59:228–237
Dudits D, Hadlaczky GY, Levi E, Fejir O, Haydu Zu, Lazar G (1977) Somatic hybridisation of *Daucus carota* and *D. capillifolius* by protoplast fusion. Theor Appl Genet 51:127–132
Dulieu H (1966) Pollination of excised ovaries and culture of ovules of *Nicotiana tabacum*. Phytomorphology 16:69–75
Durand J, Potrykus I, Doon G (1973) Plants tissues de protoplasten de *Petunia*. Z. Pflanzenphysiol 69:26–34
Earle ED, Torrey J (1963) Differentiation in colonies derived from isolated cells of *Convolvulus arvensis*. Am J Bot 50:613
Eid AAH, DeLanghe E, Waterkeyn L (1973) In vitro culture of fertilized cotton ovules. The growth of cotton embryos. Cellule 69:359–371
Engvild KC, Linde LI, Lindqvist A (1972) Anther cultures of *Datura innoxia:* Flower bud stage and embryoid level of ploidy. Hereditas 72:331–332
Evans DA, Wetter LR, Gamborg OL (1980) Somatic hybrid plants of *Nicotiana glauca* and *Nicotiana tabacum* obtained by protoplast fusion. Physiol Plant 48:225–230
Farquharson LI (1957) Hybridisation of *Tripsacum* and *Zea*. J Hered 48:295–299
Foroughi-Wehr B, Mix G (1979) In vitro response of *Hordeum vulgare* L. anthers cultured from plants grown under different environment. Environ Expt Bot 19:303–309
Frearson EM, Power JB, Cocking EC (1973) The isolation, culture, and regeneration of *Petunia* leaf protoplasts. Dev Biol 33:130–137
Fujimura T, Komamine A (1975) Effects of various growth regulators on the embryogenesis in carrot cell suspension culture. Plant Sci Lett 5:359–364
Fujiwara A (ed) (1982) Plant tissue culture 1982. Proc 5th Intl Congr Plant Tissue and Cell Culture, Tokyo
Furner IJ, King J, Gamborg OL (1978) Plant regeneration from protoplasts isolated from predominantly haploid suspension culture of *Datura innoxia* Mill. Pl Sci Lett 11:169–176

Gadwal VR, Joshi AB, Iyer RD (1968) Interspecific hybrids in *Abelmoschus* through ovule and embryo culture. Indian J Genet Plant Breed 28:269–274

Gamborg OL, Miller RA, Ojima K (1968) Nutrient requirements of suspension cultures of soybean root cells. Exp Cell Res 50:148–151

Gamborg OL, Constabel F, Fowke L, Ohyama K, Kartha KK, Kao KN, Pelcher L (1974) Protoplast and cell culture methods in somatic hybridisation in higher plants. Can J Genet Cytol 16:737–750

Gamborg OL, Shyluk JP, Shahin EA (1981) Isolation, fusion, and culture of plant protoplasts. In: Thorpe TA (ed) Plant tissue culture: Methods and applications in agriculture. Academic Press, London New York, pp 115–153

Gautheret RJ (1938) Sur le repriquage des cultures des tissue cambial de *Salix capraea*. CR Acad Sci Paris 206:125–127

Gautheret RJ (1939) Sur la possibilité de réaliser la culture indefinie des tissues de tubercles de carrotte. CR Acad Sci Paris 208:118–121

Gautheret RJ (1982) Plant tissue culture: The history. In: Fujiwara A (ed) Plant tissue culture 1982. Proc 5 th Intl Congr Plant Tissue and Cell Culture, Tokyo, pp 7–12

George L, Narayanaswamy S (1973) Haploid *Capsicum* through experimental androgenesis. Protoplasma 78:467–470

George L, Rao PS (1979) Experimental induction of triploid plants of *Physalis* through anther culture. Protoplasma 100:13–19

George L, Rao PS (1982) In vitro induction of pollen embryos and plantlets in *Brassica juncea* through anther culture. Plant Sci Lett 26:111–116

Gill R, Rashid A, Maheshwari SC (1979) Isolation of mesophyll protoplasts of *Nicotiana rustica* and their regeneration into plants flowering in vitro. Physiol Plant 47:7–10

Gleba YY, Hoffmann F (1980) "Arabidobrassica": A novel plant obtained by protoplast fusion. Planta (Berl) 149:112–117

Gleba YY, Butenko RG, Stynik KM (1975) Fusion of protoplasts and parasexual hybridisation in *Nicotiana tabacum*. Dokl Akad Nauk USSR 221:1196–1198

Gleba YY, Kohlenbach HW, Hoffmann F (1978) Root morphogenesis in somatic hybrid cell lines *Arabidopsis thaliana* + *Brassica campestris*. Naturwissenschaften 65:655

Gosch G, Reinert J (1978) Cytological identification of colony formation of intergeneric somatic hybrid cells. Protoplasma 96:23–38

Gosch G, Bajaj YPS, Reinert J (1975) Isolation, culture, and fusion studies on protoplasts from different species. Protoplasma 85:327–336

Gresshoff PM (1980) In vitro culture of white clover: Callus, suspension, protoplast culture, and plant regeneration. Bot Gaz 141:157–164

Gresshoff PM, Doy CH (1972a) Haploid *Arabidopsis thaliana* callus and plants from anther culture. Aust J Biol Sci 25:259–264

Gresshoff PM, Doy CH (1972b) Development and differentiation of haploid *Lycopersicon esculentum* (tomato). Planta (Berl) 107:161–170

Gresshoff PM, Doy CH (1974) Derivation of a haploid cell line from *Vitis vinifera* and the importance of the stage of meiotic development of anthers for haploid culture of this and other genera. Z Pflanzenphysiol 73:132–141

Grewal S, Sachdeva U, Atal CK (1976) Regeneration of plants by embryogenesis from hypocotyl cultures of *Ammi majus*. Indian J Expt Biol 14:716–717

Guha S, Johri BM (1966) In vitro development of ovary and ovule of *Allium cepa* L. Phytomorphology 16:353–364

Guha S, Maheshwari SC (1964) In vitro production of embryos from anthers of *Datura*. Nature 204:497

Guha-Mukherjee S (1973) Genotypic differences in the in vitro formation of embryoids from rice pollen. J Exp Bot 24:139–144

Gunn RE, Shepard JP (1981) Regeneration of plants from mesophyll-derived protoplasts of British potato (*Solanum tuberosum* L.) cultivars. Plant Sci Lett 22:97–101

Haberlandt G (1902) Kulturversuche mit isolierten Pflanzenzellen. SB Akad Wiss Wien Mathnatur Kul 111:69–92

Haccius B (1965) Haben „Gewebekulturen-Embryonen" einen Suspensor? Ber Dtsch Bot Ges 78:11–21

Halperin W (1967) Population density effects in embryogenesis in vitro. Science (Wash D.C.) 153:1287–1288

Halperin W, Wetherell DF (1964) Adventive embryony in tissue cultures of the wild carrot (*Daucus carota*). Am J Bot 51:274–283

Handro W, Rao PS, Harada H (1973) A histological study of the development of buds, roots, and embryos in organ cultures of *Petunia inflata* R. Fries. Ann Bot 37:817–821

Harris H (1970) Cell fusion (The Dunhan Lectures). Clarendon, Oxford

Haynes FL (1954) Potato embryo culture. Am Pot J 31:282–288

Hess D, Wagner G (1974) Induction of haploid parthenogenesis in *Mimulus luteus* by in vitro pollination with foreign pollen. Z Pflanzenphysiol 72:466–468

Heuser CW, Apps DA (1976) In vitro plantlet formation from flower petal explants of *Hemerocallis* cv. Chipper Cherry. Can J Bot 54:616–618

Hicks GS (1979) Feminised outgrowths on the stamen primordia of tobacco in vitro. Plant Sci Lett 17:81–89

Hu H, Hsi Ty, Tseng CC, Ouyang TW, Ching CK (1978) Application of anther culture to crop plants. In: Thorpe (ed) Frontiers of plant tissue culture 1978. IAPTC, Calgary, pp 123–130

Inomata N (1968) In vitro culture of ovaries of *Brassica* hybrids between 2× and 4×. I. Culture medium. Jpn J Breed 18:139–148

Iyer RD (1982) Embryo and tissue culture for crop improvement, especially of perennials, germplasm conservation and exchange – relevance to developing countries. In: Rao AN (ed) Proc COSTED Symp Tissue culture of economically important plants. Natl Univ Singapore, pp 219–230

Iyer RD, Govila OP (1964) Embryo culture of interspecific hybrids in the genus *Oryza*. Ind J Genet Plant Breed 24:116–121

Jagannathan V (1979) Application of plant tissue culture to agriculture. 11 th Philip R White Memorial Lecture. Hyderabad, India. Philip R White Mem Com, pp 1–17

Jensen CJ (1977) Monoploid production by chromosome elimination. In: Reinert J, Bajaj YPS (eds) Applied and fundamental aspects of plant cell, tissue, and organ culture. Springer, Berlin Heidelberg New York, pp 299–340

Johansson L, Eriksson T (1982) Improvement of anther culture technique: Activated charcoal bound in agar medium in combination with liquid medium and elevated CO_2 concentration. In: Fujiwara A (ed) Plant tissue culture 1982. Proc 5th Intl Congr Plant Tissue and Cell Culture, Tokyo, pp 543–544

Johri BM (ed) (1982) Experimental embryology of vascular plants. Springer, Berlin Heidelberg New York

Johri BM, Bhojwani SS (1965) Growth responses of mature endosperm in cultures. Nature 208:1345–1347

Johri BM, Guha S (1963) In vitro development of onion plants from flowers. In: Maheshwari P, Rangaswamy NS (eds) Plant tissue and organ culture – A Symposium. Intl Soc Pl Morphologists, Univ Delhi, pp 215–223

Johri BM, Sehgal CB (1966) Growth responses of ovaries of *Anethum*, *Foeniculum*, and *Trachyspermum*. Phytomorphology 16:364–378

Johri BM, Srivastava PS (1973) In vitro growth responses of mature endosperm of *Ricinus communis* L. In: Murthy YS, Johri BM, Mohan Ram HY, Varghese TM (eds) Advances in plant morphology (V. Puri Comm Vol). Sarita Prakashan, Meerut, India, pp 339–358

Johri BM, Srivastava PS, Raste AP (1980) Endosperm culture. In: Vasil IK (ed) Perspectives in plant cell and tissue culture. Intl Rev Cytol Suppl 11B. Academic Press, London New York, pp 157–182

Jones LE, Hildebrandt AC, Riker AJ, Wu JH (1960) Growth of somatic tobacco cells in microculture. Am J Bot 47:468–475

Jones LH (1974) Factors influencing embryogenesis in carrot cultures (*Daucus carota* L.). Am J Bot 38:1077–1088

Joshi PC, Johri BM (1972) In vitro growth of ovules of *Gossypium hirsutum*. Phytomorphology 22:195–209

Kanta K, Rangaswamy NS, Maheshwari P (1962) Test-tube fertilization in a flowering plant. Nature 194:1214–1217

Kao KN (1982) Plant protoplast fusion and isolation of heterokaryocytes. In: Wetter LR, Constabel F (eds) Plant tissue culture methods. Natl Res Council Canada, Saskatoon, Saskatchewan, pp 49–56

Kao KN, Michayluk MR (1974) Methods for high frequency intergeneric fusion of plant protoplasts. Planta 115:355–367

Kao KN, Michayluk MR (1980) Plant regeneration from mesophyll protoplasts of Alfalfa. Z Pflanzenphysiol 96:135–141

Kao KN, Constabel F, Michayluk MR, Gamborg OL (1974) Plant protoplast fusion and growth of intergeneric hybrid cells. Planta (Berl) 120:215–227

Kapoor M (1959) Influence of growth substances on the ovules of *Zephyranthes*. Phytomorphology 9:313–315

Kartha KK, Michayluk MR, Kao KN, Gamborg OL, Constabel F (1974a) Callus formation and plant regeneration from mesophyll protoplasts of rape plants (*Brassica napus* L. cv Zephyr). Plant Sci Lett 3:265–271

Kartha KK, Gamborg OL, Constabel F, Kao KN (1974b) Fusion of rapeseed and soybean protoplasts and subsequent division of heterokaryocytes. Can J Bot 52:2435–2436

Kasha KJ, Kao KN (1970) High frequency haploid production in barley (*Hordeum vulgare* L.). Nature 225:874–875

Kasperbauer MJ, Collins GB (1972) Reconstitution of diploids from leaf tissue of anther-derived haploids in tobacco. Crop Sci 12:98–101

Kasperbauer MJ, Buckner RC, Springer WD (1980) Haploid plants by anther panicle culture of tall fescue. Crop Sci 20:103–106

Kato H, Takeuchi M (1966) Embryogenesis from the epidermal cells of carrot hypocotyl. Sci Papers Colloq Gen Edu Univ Tokyo 16:245–254

Kavathekar AK, Ganapathy PS, Johri BM (1977) Chilling induces development of embryoids into plantlets in *Eschscholtzia california*. Biol Plant 20:98–106

Keller WA, Armstrong KC (1978) High frequency production of microspore-derived plants from *Brassica napus* anther cultures. Z Pflanzenzuechtg 80:100–108

Keller WA, Melchers G (1973) Effect of high pH and calcium on tobacco leaf protoplast fusion. Z Naturforsch 28:731–741

Keller WA, Stringham GR (1978) Production and utilization of microspore-derived haploid plants. In: Thorpe TA (ed) Frontiers of plant tissue culture 1978. IAPTC, Calgary, Canada, pp 113–122

Kessel RHJ, Goodwin C, Philip J (1974) The relationship between dissolved oxygen concentration, ATP and embryogenesis in carrot (*Daucus carota*) tissue cultures. Plant Sci Lett 10:265–274

Kimber G, Riley R (1963) Haploid angiosperms. Bot Rev 90:480–531

Knudson L (1919) Viability of detached root cap cells. Am J Bot 6:309–310

Kochba J, Spiegel-Roy P (1977) The effect of auxins, cytokinins, and inhibitors on embryogenesis in habituated ovular callus of the shomouti orange (*Citrus sinensis*). Z Pflanzenphysiol 81:283–288

Kochba J, Spiegel-Roy P, Neumann H, Saad S (1978) Stimulation of embryogenesis in *Citrus* ovular callus by ABA, ethephon, CCC, and Alar and its suppression by GA_3. Z Pflanzenphysiol 89:427–432

Kohlenbach HW (1970) Das durch Kinetin induzierte Nukleolus Kern- und Plasmawachstum isolierter Epidermiszellen von *Rhoeo spathacea*. Z Pflanzenphysiol 63:297–307

Kohlenbach HW, Wenzel G, Hoffmann F (1982) Regeneration of *Brassica napus* plantlets in cultures from isolated protoplasts of haploid stem embryos as compared with leaf protoplasts. Z Pflanzenphysiol 105:131–142

Konar RN, Thomas E, Street HE (1972) Origin and structure of embryoids arising from epidermal cells of the stem of *Ranunculus sceleratus* L. J Cell Sci 11:77–93

Konzak CF, Randolph LF, Jensen NF (1951) Embryo culture of barley species hybrids: Cytological studies of *Hordeum sativum* × *H. bulbosum*. J Hered 42:125–134

Kotte W (1922) Kulturversuche mit isolierten Wurzelspitzen. Beitr Allgem Bot 2:413–434

Krumbiegel G, Schieder O (1979) Selection of somatic hybrids after fusion of protoplasts from *Datura innoxia* Mill. and *Atropa belladonna* L. Planta 145:371–375

Kruse A (1974) An in vivo/vitro embryo culture technique. Heriditas 77:219–224

Laibach F (1925) Das Taubwerden von Bastardsamen und die künstliche Aufzucht früh absterbender Bastardembryonen. Z Bot 17:417–459

Lakshmi Sita G, Raghava Ram NV, Vaidyanathan CS (1980) Triploid plants from endosperm cultures of sandalwood by experimental embryogenesis. Plant Sci Lett 20:63–69

LaRue CD (1942) The rooting of flower in sterile culture. Bull Torrey Bot Club 69:332–341

LaRue CD (1947) Growth and regeneration of the endosperm of maize in culture. Am J Bot 34:585–586

LaRue CD (1949) Cultures of endosperm of maize. Am J Bot 36:798

Len L-HC (1982) Tissue culture of local orchid hybrids at the Singapore Botanic Gardens. In: Rao AN (ed) Proc COSTED Symp Tissue culture of economically important plants. Natl Univ Singapore, pp 295–300

Litz RE, Conover RA (1980) Somatic embryogenesis in cell cultures of *Carica stipulata*. Hortic Sci 15:733–735

Litz RE, Conover RA (1981) In vitro polyembryony in *Carcia papaya* ovules. Z Pflanzenphysiol 104:285–288

Lord EM, Mayers AM (1982) Effect of gibberellic acid on floral development in vivo and in vitro in the cleistogamous species, *Lamium amplexicaule* L. Ann Bot 50:301–307

Lu C, Vasil IK (1981) Somatic embryogenesis and plant regeneration from leaf tissues of *Panicum maximum* Jacq. Theor Appl Genet 59:275–280

Maheshwari N (1958) In vitro culture of excised ovules of *Papaver somniferum*. Science (Wash D.C.) 127:342

Maheshwari N, Lal M (1958) In vitro culture of ovaries of *Iberis amara* L. Nature 181:631–632

Maheshwari P (1950) An introduction to the embryology of angiosperms. McGraw-Hill, New York

Maheshwari P, Baldev B (1961) Artificial production of buds from the embryos of *Cuscuta reflexa* Roxb. In: Plant Embryology – A Symposium. CSIR, New Delhi, pp 129–138

Maheshwari P, Kanta K (1964) Control of fertilization. In: Linskens HF (ed) Pollen physiology and fertilization. North-Holland, Amsterdam, pp 187–194

Maheshwari SC, Rashid A, Tyagi AK (1982) Haploids from pollen grains: Retrospect and prospects. Am J Bot 69:865–879

Maliga P, Breznovitsa SZ, Marton L (1973) Streptomycin-resistant plants from callus culture of haploid tobacco. Nature New Biol 244:29–30

Maliga P, Lázár G, Joó F, Nagy AH, Menczel L (1977) Restoration of morphogenetic potential in *Nicotiana* by somatic hybridisation. Mol Gen Genet 157:291–296

Maliga P, Kiss ZR, Nagy AH, Lázár G (1978) Genetic instability in somatic hybrids of *Nicotiana tabacum* and *Nicotiana knightiana*. Mol Gen Genet 163:145–151

Massart J (1902) Sur la pollination sans fécondation. Bull Jard Bot Brux 1:89–95

Mastrangelo IA, Smith HH (1977) Selection and differentiation of aminopterin-resistant cells of *Datura innoxia*. Plant Sci Lett 10:171–179

McHughen A (1977) Development of tobacco petals in vitro. Ann Bot 41:1073–1076

Melchers G (1981) Somatic and sexual genetics combined in higher plants. Proc 14th Intl Congr Genet Moscow 2(1):208–220

Melchers G (1982) The first decennium of somatic hybridization. In: Fujiwara A (ed) Plant tissue culture 1982. Proc 5th Intl Congr Plant Tissue and Cell Culture, Tokyo, pp 13–18

Melchers G, Labib G (1974) Somatic hybridization of plants by fusion of protoplasts. I. Selection of light-resistant hybrids of "Haploid" light-sensitive varieties of tobacco. Mol Gen Genet 135:277–294

Melchers G, Sacristan MO, Holder AA (1978) Somatic hybrid plants of potato and tomato regenerated from fused protoplasts. Carlsberg Res Commun 43:203–218

Mitchell AZ, Hanson MR, Skvirsky RC, Ausubel FM (1980) Anther culture of *Petunia*. Genotypes with high frequency of callus, root, or plantlet formation. Z Pflanzenphysiol 100:131–146

Mitra GC, Chaturvedi HC (1972) Embryoids and complete plants from unpollinated ovaries and from ovules of in vivo-grown emasculated flower buds from *Citrus* spp. Bull Torrey Bot Club 9:184–189

Molliard M (1921) Sur le devéloppement des plantules fragmentées. CR Soc Biol 84:770–772

Monacco LC, Sondahl MR, Carvalho A, Crocomo OJ, Sharp WR (1977) Application of tissue culture in the improvement of coffee. In: Reinert J, Bajaj YPS (eds) Applied and fundamental aspects of plant cell, tissue, and organ culture. Springer, Berlin Heidelberg New York, pp 109–126

Mu SK, Kwei Y-L, Liu S-K, Chang FC, Lo F, Yang M-Y, Wang F-h (1977) Induction of callus in *Vitis* endosperm cultured in vitro. Acta Bot Sinica 19:93–94

Muir WH, Hildebrandt AC, Riker AJ (1954) Plant tissue cultures produced from single isolated plant cells. Science (Wash D.C.) 119:877–887

Mullins MG (1970) Tissue culture of some monocotyledonous plants. Aust J Biol Sci 23:473–477

Mullins MG, Srinivasan C (1976) Somatic embryos and plantlets from an ancient clone of the grapevine (cv. Cabernet-Sauvignon) by apomixis in vitro. J Exp Bot 27:1022–1030

Murashige T, Skoog F (1962) A revised medium for rapid growth and bioassays with tobacco tissue cultures. Physiol Plant 15:473–497

Nadar HM, Soepraptopo S, Heinz J, Ladd SL (1978) Fine structure of sugarcane (*Saccharum* sp.) callus and role of auxin in embryogenesis. Crop Sci 18:210–216

Nag KK, Street HE (1973) Carrot embryogenesis from frozen cultured cells. Nature 245:270–272

Nagata T, Takabe I (1971) Plating of isolated tobacco mesophyll protoplasts on agar medium. Planta 99:12–20

Nakajima T (1962) Physiological studies of seed development, especially embryonic growth and endosperm development. Univ Osaka Prefect Ser B 13:13–48

Nakamura A, Yamada T, Kadotani N, Itagaki R, Oka M (1974) Studies on the haploid method of breeding in tobacco. SABRAO J 6:107–131

Nakano H, Toshiro T, Maeda E (1975) Plant differentiation in callus tissue induced from immature endosperm of *Oryza sativa* L. Z Pflanzenphysiol 76:444–449

Nanda KK, Angrish R, Kaur S (1981) In vitro culture of dormant reproductive buds of *Salix tetrasperma* Linn. and *S. babylonica* Roxb.: Effect of growth-regulatory stubstances on their development during post- and pre-dormancy periods. Z Pflanzenphysiol 104:25–35

Narayanaswamy S (1963) Studies on growth of excised grass embryos in culture. In: Maheshwari P, Rangaswamy NS (eds) Plant Tissue and Organ Culture – A Symposium. Intl Soc Plant Morphologists, Univ Delhi, pp 302–313

Narayanaswamy S, Chandy L (1971) In vitro induction of haploid, diploid, and triploid androgenic embryoids and plantlets in *Datura metel* L. Ann Bot 35:535–542

Narayanaswamy S, George L (1982) Anther culture. In: Johri BM (ed) Experimental embryology of vascular plants. Springer, Berlin Heidelberg New York, pp 79–103

Narayanaswamy S, Prabhudesai VR (1979) Somatic pseudoembryogeny in tissue cultures of tuberose, *Polianthes tuberosa*. Indian J Exp Biol 17:873–875

Nickell LG (1955) Nutrition of pathological tissues caused by plant viruses. Annee Biol 31:107–121

Niimi Y (1970) In vitro fertilization in the self-incompatible plant *Petunia hybrida*. J Jpn Soc Hortic Sci 39:345–352

Niizeki H, Oono K (1968) Induction of haploid rice plant from anther culture. Proc Jpn Acad 44:554–557

Nitsch C (1974a) La Culture de pollen isolé sur milieu synthetique. CR Acad Sci Paris 278:1031–1034

Nitsch C (1974b) Pollen culture: A new technique for mass production of haploid and homozygous plants. In: Kasha KJ (ed) Haploids in higher plants – Advances and potential. Univ Guelph, Canada, pp 123–135

Nitsch C (1981) Production of isogenic lines: Basic technical aspects of androgenesis. In: Thorpe TA (ed) Plant tissue culture: Methods and applications in agriculture. Academic Press, London New York, pp 241–252

Nitsch C, Norreel B (1973) Effect d'un choc thérmique sur le pouvoir embryogène du pollen de *Datura innoxia* cultivé dans l'anthère on isole de l'anthère. CR Acad Sci Paris 276:303–306

Nitsch JP (1951) Growth and development in vitro of excised ovaries. Am J Bot 38:566–577

Nitsch JP (1963) The in vitro culture of flowers and fruits. In: Maheshwari P, Rangaswamy NS (eds) Plant tissue and organ culture – A Symposium. Intl Soc Plant Morphologists, Univ Delhi, pp 198–214

Nitsch JP (1969) Experimental androgenesis in *Nicotiana*. Phytomorphology 19:389–404

Nobécourt P (1938) Sur les proliferations spontanées de fragments de tubercules de carrotte et leur culture sur milieu synthetique. Bull Soc Bot Fr 85:1–7

Norstog K (1956) Growth of rye grass endosperm in vitro. Bot Gaz 117:253–259

Novak FJ, Havel L (1981) Shoot production from in vitro cultured flower heads of *Allium porrum* L. Biol Plant 23:266–269

Ohyama K, Nitsch JP (1972) Flowering haploid plants obtained from protoplasts of tobacco leaves. Pl Cell Physiol 13:229–236

Orton TJ (1980) Haploid barley regenerated from callus cultures of *Hordeum vulgare* × *H. jubatum*. J Hered 71:280

Overbeek J van (1942) Hormonal control of embryo and seedlings. Cold Spring Harbour Symp Quant Biol 10:126–133

Pal A, Sharma AK (1982) Cytological and developmental aspects of callus in *Costus speciosus*. In: Rao AN (ed) Proc COSTED Symp Tissue culture of economically important plants. Natl Univ Singapore, pp 85–89

Papachatzi MP, Hammer A, Hasegawa PM (1980) In vitro propagation of *Hosta plantaginea*. Hortic Sci 15:506–507

Pareek LK, Chandra N (1978) Somatic embryogenesis in leaf callus from cauliflower (*Brassica oleracea* var. *Botrytis*). Plant Sci Lett 11:311–316

Passiatore JE, Sink KC (1981) Plant regeneration from leaf mesophyll protoplasts of selected ornamental *Nicotiana* sp. J Am Soc Hortic Sci 106:799–803

Pence VC, Hasegawa PM, Janick J (1979) Asexual embryogenesis in *Theobroma cocoa* L. J Am Soc Hortic Sci 104:145–148

Potrykus I (1971) Intra- and interspecific fusion of protoplasts from petals of *Torenia baillonii* and *Torenia fournieri*. Nature New Biol 231:51–58

Power JB, Frearson EM (1973) The inter- and intraspecific fusion of plant protoplasts, subsequent development in culture, with reference to crown gall callus and tobacco and *Petunia* leaf system. In: Protoplasts et fusion de cellules somatique vegétale. Colloq Intl CNRS Paris 212:409–421

Power JB, Cummins SE, Cocking EC (1970) Fusion of isolated plant protoplasts. Nature 225:1016–1018

Power JB, Frearson EM, Hayward C, George D, Evans PK, Berry SF, Cocking EC (1976) Somatic hybridisation of *Petunia hybrida* and *P. parodii*. Nature 263:500–502

Power JB, Berry SF, Frearson EM, Cocking EC (1977) Selection procedures for the production of interspecies somatic hybrids of *Petunia hybrida* and *Petunia parodii* I. Nutrient media and drug sensitivity complementation selection. Plant Sci Lett 10:1–6

Prat S (1927) The toxicity of tissue juices for cell of the tissue. Am J Bot 14:121–125

Puri P (1963) Growth in vitro of parthenogenetic embryos of *Aerva tomentosa* Forsk. In: Maheshwari P, Rangaswamy NS (eds) Plant Tissue and Organ Culture – A Symposium. Intl Soc Plant Morphologists, Univ Delhi, pp 281–291

Radojevic L (1979) Somatic embryos and plantlets from callus cultures of *Paulownia tomentosa* Steud. Z Pflanzenphysiol 91:57–62

Raghavan V (1976) Experimental embryogenesis in vascular plants. Academic Press, London New York

Raghavan V (1978) Origin and development of pollen embryoids and pollen calluses in cultured anther segments of *Hyoscyamus niger* (henbane). Am J Bot 65:984–1002

Raghavan V (1980) Embryo culture. In: Vasil IK (ed) Perspectives in plant cell and tissue culture. Intl Rev Cytol Suppl. 11B. Academic Press, London New York, pp 209–240

Raghavan V (1982) Pollen embryogenesis in *Hyoscyamus niger:* A review. In: Rao AN (ed) Proc COSTED Symp "Tissue culture of economically important plants". Natl Univ Singapore, pp 262–268

Raghavan V, Srivastava PS (1982) Embryo culture. In: Johri BM (ed) Experimental embryology of vascular plants. Springer, Berlin Heidelberg New York, pp 195–230

Rahaman WA, Gandimathi H, Othman R, Paranjothy K (1982) Recent developments in tissue culture of *Hevea*. In: Rao AN (ed) Proc COSTED Symp on "Tissue culture of economically important plants". Natl Univ Singapore, pp 152–158

Rajasekhar EW (1973) Nuclear divisions in protoplasts isolated from pollen tetrads of *Datura metel*. Nature 246:223–224

Raman K, Greyson RI (1974) In vitro induction of embryoids in tissue culture of *Nigella damascena*. Can J Bot 52:1988–1989

Rangan TS (1982) Ovary, ovule, and nucellus culture. In: Johri BM (ed) Experimental embryology of vascular plants. Springer, Berlin Heidelberg New York, pp 105–129

Rangan TS, Murashige T, Bitters WP (1969) In vitro studies of zygotic and nucellar embryogenesis in *Citrus*. Proc 1st Intl *Citrus* Symp. Uni Calif, Riverside, pp 225–229

Rangaswamy NS (1961) Experimental studies on female reproductive structures of *Citrus microcarpa* Bunge. Phytomorphology 11:109–127

Rangaswamy NS (1977) Applications of in vitro pollination and in vitro fertilization. In: Reinert J, Bajaj YPS (eds) Applied and fundamental aspects of plant cell, tissue, and organ culture. Springer, Berlin Heidelberg New York, pp 412–425

Rangaswamy NS (1982) Nucellus as an experimental system in basic and applied tissue culture research. In: Rao AN (ed) Proc COSTED Symp on "Tissue culture of economically important plants". Natl Univ Singapore, pp 269–286

Rangaswamy NS, Rao PS (1963) Experimental studies on *Santalum album:* Establishment of tissue culture of endosperm. Phytomorphology 13:450–454

Rangaswamy NS, Shivanna KR (1967) Induction of gametic compatibility and seed formation in axenic cultures of a diploid self-incompatible species of *Petunia*. Nature 216:937–939

Rangaswamy NS, Shivanna KR (1969) Test-tube fertilization in *Dicranostigma franchetianum* (Prain) Fedde. Curr Sci (Bangalore) 38:257–259

Rao AN (1977) Tissue culture in the orchid industry. In: Reinert J, Bajaj YPS (eds) Applied and fundamental aspects of plant cell, tissue, and organ culture. Springer, Berlin Heidelberg New York, pp 44–69

Rao AN (ed) (1982) Tissue culture of economically important plants. Proc COSTED Symp on Natl Univ Singapore

Rao AN, Yeow MS, Kothagoda N, Hutchinson JF (1982) Cotyledon tissue culture of some tropical fruits. In: Rao AN (ed) Proc COSTED Symp on "Tissue culture of economically important plants". Natl Univ Singapore, pp 124–138

Rao PS (1982) Protoplast culture. In: Johri BM (ed) Experimental embryology of vascular plants. Springer, Berlin Heidelberg New York, pp 231–262

Rao PS, Narayanaswamy S (1972) Morphogenetic investigations in callus cultures of *Tylophora indica*. Physiol Plant 27:271–276

Rao PS, Raghava Ram NV (1983) Propagation of sandalwood (*Santalum album* Linn.) using tissue and organ culture technique. In: Sen SK, Giles KL (eds) Plant cell culture in crop inprovement. Plenum, New York, pp 119–124

Rao PS, Rangaswamy NS (1972) In vitro development of the pollinated pistils of *Nicotiana rustica* L. Bot Gaz 133:350–355

Rao PS, Narayanaswamy S, Benjamin BD (1970) Differentiation ex-ovulo of embryos and plantlets in stem tissue cultures of *Tylophora indica*. Physiol Plant 23:140–144

Rashid A, Reinert J (1980) Selection of embryogenic pollen from cold-treated buds of *Nicotiana tabacum* var. Badischer Burley and their development into embryos in cultures. Protoplasma 105:161–167

Rashid A, Reinert J (1981 a) Differentiation of embryogenic pollen in cold-treated buds of *Nicotiana tabacum* var. Badischer Burley and nutritional requirements of the isolated pollen to form embryos. Protoplasma 106:137–144

Rashid A, Reinert J (1981 b) In vitro differentiation of embryogenic pollen, control by cold treatment and embryo formation in ab initio pollen cultures of *Nicotiana tabacum* var. Badischer Burley. Protoplasma 109:285–294

Rashid A, Siddiqui AW, Reinert J (1981) Ultrastructure of embryogenic pollen of *Nicotiana tabacum* var. Badischer Burley. Protoplasma 107:375–385

Rechinger C (1893) Untersuchungen über die Grenzen der Teilbarkeit im Pflanzenreich. Abh Zool Bot Ges Wien 43:310–334

Rédei G, Rédei G (1955 a) Rearing wheats from various parts cultured in vitro. Acta Bot Acad Sci Hung 2:183–186

Rédei G, Rédei G (1955 b) Developing wheat embryos excised from ovaries cultured in vitro. Experientia 11:387–388

Reinert J (1973) Aspects of organisation: Organogenesis and embryogenesis. In: Street HE (ed) Plant tissue and cell culture. Blackwell, Oxford, pp 338–355

Reinert J, Yeoman MM (eds) (1982) Plant cell and tissue culture: A laboratory manual. Springer, Berlin Heidelberg New York

Reinert J, Bajaj YPS, Heberle E (1975) Induction of haploid tobacco plants from isolated pollen. Protoplasma 84:191–196

Reinert J, Bajaj YPS, Zbell B (1977) Aspects of organization – Organogenesis, embryogenesis, cytodifferentiation. In: Street HE (ed) Plant tissue and cell culture. Botanical monographs 11. Blackwell, Oxford, pp 389–427

Ringe F, Nitsch JP (1968) Conditions leading to flower formation on excised *Begonia* fragments cultured in vitro. Plant Cell Physiol 9:639–652

Robbins WJ (1922) Cultivation of excised root tips and stem tips under sterile conditions. Bot Gaz 73:376–390

Sabharwal PS (1963) In vitro culture of ovules, nucelli, and embryos of *Citrus reticulata* Blanco var. *Nagpuri*. In: Maheshwari P, Rangaswamy NS (eds) Plant tissue and organ culture – A symposium. Intl Soc Plant Morphologists, Univ Delhi, pp 265–274

Sachar RC, Guha S (1962) In vitro growth of achenes of *Ranunculus sceleratus*. In: Plant Embryology – A Symposium. CSIR New Delhi, pp 244–253

Sachar RC, Iyer RD (1959) Effect of auxin, kinetin, and gibberellin on the placental tissue of *Opuntia dillenii* Haw. cultured in vitro. Phytomorphology 9:1–3

Sachar RC, Kapoor M (1959) In vitro culture of ovules of *Zephyranthes*. Phytomorphology 9:147–156

Sanders ME, Ziebur NK (1963) Artificial culture of embryos. In: Maheshwari P (ed) Recent advances in the embryology of angiosperms. Intl Soc Plant Morphologists, Univ Delhi, pp 297–325

Sangwan RS, Harada H (1975) Chemical regulation of callus growth, organogenesis, plant regeneration, and somatic embryogenesis in *Antirrhinum majus* tissue and cell cultures. J Exp Bot 26:868–881

Sangwan RS, Norreel B (1975) Induction of plants from pollen grains of *Petunia* cultured in vitro. Nature 257:222–224

Sankhla N, Sankhla D (1967) Growth response of excised ovaries of *Reseda odorata* in sterile culture. Biol Plant 9:61–63

Sawhney VK, Greyson RI (1979) Interpretations of determination and canalization of stamen development in a tomato mutant. Can J Bot 57:2471–2477

Saxena PK, Gill R, Rashid A, Maheshwari SC (1981) Plantlet formation from isolated protoplasts of *Solanum melongena* L. Protoplasma 106:355–357

Schieder O (1975) Regeneration of haploid and diploid *Datura innoxia* Mill. mesophyll protoplasts to plants. Z Pflanzenphysiol 76:462–466

Schieder O (1977) Attempts in regeneration of mesophyll protoplasts of haploid and diploid wild type lines and those of chlorophyll-deficient strains from different Solanaceae. Z Pflanzenphysiol 84:275–281

Schieder O (1978 a) Somatic hybrids of *Datura innoxia* Mill + *D. discolor* Bernh and of *D. innoxia* Mill and *D. stramonium* L. var. *Tatula* L. Mol Gen Genet 162:113–119

Schieder O (1978 b) Haploids from *Datura innoxia* as a tool for the production of homozygous lines with high content of scopolamine and for induction of mutants. In: Alfermann AW, Reinhard E (eds) Production of natural compounds by cell culture methods. Proc Intl Symp Plant cell culture, Munich, pp 330–336

Schieder O (1980) Somatic hybrids between a herbaceous and two tree *Datura* species. Z Pflanzenphysiol 98:119–127

Schieder O, Krumbiegel G (1981) Selection of somatic hybrids in plants. Intl Cell Biol 1980–1981, pp 872–878

Schieder O, Vasil IK (1980) Protoplast fusion and somatic hyridization. In: Vasil IK (ed) Perspectives in plant cell and tissue culture. Intl Rev Cytol Suppl 11B. Academic Press, London New York, pp 21–46

Schiller B, Herrmann RG, Melchers G (1982) Restriction endonuclease analysis of plastid DNA from tomato, potato and some of their somatic hybrids. Mol Gen Genet 186:453–459

Scowcroft WR, Larkin PJ (1980) Isolation, culture and plant regeneration from protoplasts of *Nicotiana debreyi*. Aust J Plant Physiol 7:635–644

Sehgal CB (1969) Experimental studies on maize endosperm. Beitr Biol Pflanz 46:233–238

Sehgal CB (1974) Growth of barley and wheat endosperm in cultures. Curr Sci (Bangalore) 43:38–40

Shahin EA, Shepard JF (1980) Cassava mesophyll protoplasts: Isolation, proliferation and shoot formation. Plant Sci Lett 17:459–465

Sharma DR, Chowdhury JB, Ahuja V, Dhankhar BS (1980) Interspecific hybridization in genus *Solanum:* A cross between *S. melongena* and *S. khasianum* through embryo culture. Z Pflanzenzuechtg 85:248–253

Shivanna KR (1965) In vitro fertilization and seed formation in *Petunia violacea* Lindl. Phytomorphology 15:183–185

Shivanna KR (1982) Pollen-pistil interaction and control of fertilization. In: Johri BM (ed) Experimental embryology of vascular plants. Springer, Berlin Heidelberg New York, pp 131–174

Singh UP (1963) Raising nucellar seedlings of some Rutaceae in vitro. In: Maheshwari P, Rangaswamy NS (eds) Plant Tissue and Organ Culture – A symposium. Intl Soc Plant Morphologists, Univ Delhi, pp 275–277

Smith MM, Stone BA (1973) Studies on *Lolium multifolium* endosperm in tissue culture. I. Nutrition. Aust J Biol Sci 26:123–133

Smith SM, Street HE (1974) The decline of embryogenic potential as callus and suspension cultures of carrot (*Dacus carota* L.) are serially subcultured. Ann Bot 38:223–241

Sommer HE, Brown CL (1980) Embryogenesis in tissue culture of sweetgum. Forest Sci 26:257–260

Sondahl MR, Sharp WR (1977) High-frequency induction of somatic embryos in cultured leaf explants of *Coffea arabica* L. Z Pflanzenphysiol 81:395–408

Sopory SK (1977) Development of embryoids in isolated pollen culture of dihaploid *Solanum tuberosum*. Z Pflanzenphysiol 84:453–457

Sopory SK, Maheshwari SC (1976) Development of pollen embryoids in anther cultures of *Datura innoxia*. 2. Effects of growth hormones. J Exp Bot 27:58–68

Sopory SK, Jacobsen E, Wenzel G (1978) Production of monohaploid embryoids and plantlets in cultured anthers of *Solanum tuberosum*. Plant Sci Lett 12:47–54

Srivastava PS (1971) In vitro growth requirements of mature endosperm of *Ricinus communis*. Curr Sci (Bangalore) 40:337–339

Srivastava PS (1982) Endosperm culture. In: Johri BM (ed) Experimental embryology of vascular plants. Springer, Berlin Heidelberg New York, pp 175–193

Srivastava PS, Steinauer A (1981) Regeneration of birch plants from catkin tissue cultures. Plant Sci Lett 22:379–386

Steward FC, Shantz EM (1959) The chemical regulation of growth (some substances and extracts which induce growth and morphogenesis). Annu Rev Plant Physiol 10:379–404

Stingel G (1907) Experimentelle Studien über die Ernährung von pflanzlichen Embryonen. Flora 97:308–331

Straus J (1960) Maize endosperm tissue grown in vitro. III. Development of a synthetic medium. Am J Bot 47:641–647

Straus J, LaRue CD (1954) Maize endosperm tissue grown in vitro. I. Culture requirements. Am J Bot 41:687–694

Street HE, Henshaw GG (1966) Introduction and methods employed in plant tissue culture. In: Willmer EN (ed) The biology of cells and tissues in culture. Academic Press, London New York, p 459–532

Sunderland N (1978) Strategies in the improvement of yields in the anther culture. In: Proc symposium Plant tissue culture. Sci Press, Peking, pp 65–86

Sunderland N, Dunwell JM (1977) Anther and pollen culture. In: Street HE (ed) Plant tissue and cell cultures. Blackwell, Oxford, pp 223–265

Sunderland N, Roberts M (1977) New approach to pollen culture. Nature 270:236–238

Sunderland N, Wicks FM (1969) Cultivation of haploid plants from tobacco pollen. Nature 224:1227–1229

Sunderland N, Collins GB, Dunwell JM (1974) The role of nuclear fusion in pollen embryogenesis of *Datura innoxia* Mill. Planta (Berl) 117:227–241

Sussex IM (1972) Somatic embryos in long-term carrot tissue culture: Histology, cytology and development. Phytomorphology 18:339–349

Tamaoki T, Ullstrup AJ (1958) Cultivation in vitro of excised endosperm and meristem tissues of corn. Bull Torrey Bot Club 85:260–272

Tazawa M, Reinert J (1969) Extracellular and intracellular chemical environments in relation to embryogenesis in vitro. Protoplasma 68:157–173

Thomas E, Konar RN, Street HE (1972) The fine structure of the embryogenic callus of *Ranunculus sceleratus* L. J Cell Sci 11:95–109

Thorpe TA (ed) (1981) Plant tissue culture: Methods and applications in agriculture. Academic Press, London New York

Tisserat B, Murashige T (1977) Effects of ethephon, ethylene, and 2,4-dichlorophenoxyacetic acid on asexual embryogenesis in vitro. Plant Physiol 60:437–439

Torrey JG (1957) Cell division in isolated single cells in vitro. Proc Natl Acad Sci USA 43:887–891

Torrey JG, Reinert J (1961) Suspension culture of higher plant cells in synthetic medium. Plant Physiol 36:483–491

Tukey HB (1934) Artificial culture methods for isolated embryos of deciduous fruits. Am Soc Hortic Sci 32:313–322

Tyagi AK, Rashid A, Maheshwari SC (1979) High frequency production of embryos in *Datura innoxia* from isolated pollen grains by combined cold treatment and serial culture of anthers in liquid medium. Protoplasma 99:11–17

Tyagi AK, Rashid A, Maheshwari SC (1980) Enhancement of pollen embryo formation in *Datura innoxia* by charcoal. Physiol Plant 49:296–298

Usha SV (1965) In vitro pollination in *Antirrhinum majus* L. Curr Sci (Bangalore) 34:511–513

Vardi A, Spiegel-Roy P, Galun E (1975) *Citrus* cell culture: Isolation of protoplasts, plating densities, effect of mutagens and regeneration of embryos. Plant Sci Lett 4:231–236

Vasil IK (ed) (1980) Perspectives in plant cell and tissue culture. Intl Rev Cytol Suppl IIA, 11B. Academic Press, London New York

Vasil IK (1982) Somatic embryogenesis and plant regeneration in cereals and grasses. In: Fujiwara A (ed) Plant tissue culture 1982. Proc 5th Intl Congr Plant tissue and cell culture, Tokyo, pp 101–104

Vasil V, Vasil IK (1974) Regeneration of tobacco and *Petunia* plants from protoplasts and culture of corn protoplasts. In vitro 10:83–96

Vasil V, Vasil IK (1980) Isolation and culture of cereal protoplasts. II. Embryogenesis and plantlet formation from protoplasts of *Pennisetum americanum*. Theor Appl Genet 56:97–99

Vasil V, Vasil IK (1981 a) Somatic embryogenesis and plant regeneration from suspension cultures of pearl millet (*Pennisetum americanum*). Ann Bot 47:669–678

Vasil V, Vasil IK (1981 b) Somatic embryogenesis and plant regeneration from tissue cultures of *Pennisetum americanum* and *P. americanum* × *P. purpureum* hybrid. Am J Bot 68:864–872

Wagner G, Hess D (1973) In vitro Befruchtungen bei *Petunia hybrida*. Z Pflanzenphysiol 69:262–269

Wakizuka T, Nakajima T (1974) Effect of cultural condition on the in vitro development of ovules in *Petunia hybrida* Vilm. Jpn J Plant Breed 24:182–187

Walker DB (1978) Morphogenetic factors controlling differentiation and dedifferentiation of epidermal cells in the gynoecium of *Catharanthus roseus*. Planta (Berl) 142:181–186

Wenzel G (1978) Production of haploids of rape and rye. In: Alfermann AW, Reinhard E (eds) Production of natural compounds by cell culture methods. Proc Intl Symp Plant cell culture, Munich, pp 312–318

Wenzel G (1980) Anther culture and its role in plant breeding. In: Rao PS, Heble MR, Chadha MS (eds) Proc Symp Plant tissue culture: Genetic manipulation and somatic hybridization of plant cells. Bhabha Atomic Research Centre, Bombay, pp 68–78

Wenzel G, Hoffmann F, Potrykus I, Thomas E (1975) The separation of viable rye microspores from mixed populations and their development in culture. Mol Gen Genet 138:293–297

Wenzel G, Hoffman F, Thomas E (1976) Heterozygous microspore-derived plants in rye. Theor Appl Genet 48:205–208

Wenzel G, Hoffmann F, Thomas E (1977) Increased induction and chromosome doubling of androgenetic haploid rye. Theor Appl Genet 51:81–86

Wenzel G, Schieder O, Prezewozny T, Sopory SK, Melchers G (1979) Comparison of single cell culture-derived *Solanum tuberosum* L. plants and a model for their application in breeding programs. Theor Appl Genet 55:49–56

Wernicke W, Brettel R (1980) Somatic embryogenesis from *Sorghum bicolor* leaves. Nature 287:138–139

Wernicke W, Kohlenbach HW (1977) Experiments on the culture of isolated microspores in *Nicotiana* and *Hyoscyamus*. Z Pflanzenphysiol 81:330–340

Wernicke W, Thomas E (1980) Studies on morphogenesis from isolated plant protoplasts: Shoot formation from mesophyll protoplasts of *Hyoscyamus muticus* and *Nicotiana tabacum*. Plant Sci Lett 17:401–407

Wernicke W, Harms CT, Lorz H, Thomas E (1978) Selective enrichment of embryogenic microspore populations. Naturwissenschaften 65:540

Wetter LR, Constabel F (eds) (1982) Plant tissue culture methods. 2nd edn. Natl Res Council Canada, Saskatoon, Saskatchewan, Canada

White PR (1934) Potentially unlimited growth of excised tomato root tips in a liquid medium. Plant Physiol 9:585–600

White PR (1939) Potentially unlimited growth of excised plant callus in an artificial medium. Am J Bot 26:59–64

Williams EG, de Lautour G (1980) The use of embryo culture with transplanted nurse endosperm for the production of interspecific hybrids in pasture legumes. Bot Gaz 141:252–257

Wilson HM (1977) Culture of whole barley spikes stimulates high frequencies of pollen calluses in individual anthers. Plant Sci Lett 9:233–238

Wochok ZS (1973) Microtubules and multivesicular bodies in cultured tissues of wild carrot: Changes during transition from the undifferentiated to the embryonic condition. Cytobios 7:87–95

Wu C-t, Huang TK, Chen GR, Chen SY (1982) A review on the tissue culture of tea plants and on the utilization of callus-derived plantlets. In: Rao AN (ed) Proc COSTED Symp on "Tissue culture of economically important plants". Natl Univ Singapore, pp 104-106

Yang YW, Chang WC (1980) Embryoid formation and subsequent plantlet regeneration from callus culture of *Thalictrum urbaini* Hayata (Ranunculaceae). Z Pflanzenphysiol 97:19–24

Yeung EC, Thorpe TA, Jensen CJ (1981) In vitro fertilization and embryo culture. In: Thorpe TA (ed) Plant tissue culture: Methods and applications in agriculture. Academic Press, London New York, pp 253–271

Zapata FG, Sink KC (1981) Somatic embryogenesis from *Lycopersicon peruvianum* leaf mesophyll protoplasts. Theor Appl Genet 59:265–268

Zatyko JM, Simon I, Szabo CS (1975) Induction of polyembryony in cultivated ovules of red currant. Plant Sci Lett 4:281–283

Zee SY (1981) Studies on adventive embryo formation in the petiole explants of coriander (*Coriandrum sativum*). Protoplasma 107:21–26

Zee SY, Wu SC (1980) Somatic embryogenesis in the leaf explants of Chinese celery. Aust J Bot 28:429–436

Zee SY, Wu SC, Yue SB (1978) Morphogenesis of the hypocotyl explants of Chinese kale. Z Pflanzenphysiol 90:155–163

Zee SY, Wu SC, Yue SB (1979) Morphological and SDS-polyacrylamide gel electrophoretic studies of pro-embryoid formation in the petiole explants of Chinese celery. Z Pflanzenphysiol 95:397–403

Zeibur NK, Brink RA (1951) The stimulative effect of *Hordeum* endosperms on the growth of immature plant embryos in vitro. Am J Bot 38:253–256

Zenkteler M (1971) In vitro production of haploid plants from pollen grains of *Atropa belladonna* L. Experientia 27:1087

Zenkteler M (1980) Intraovarian and in vitro pollination. In: Vasil IK (ed) Perspectives in plant cell and tissue culture. Intl Rev Cytol Suppl 11B. Academic Press, London New York, pp 137–156

Zhenghua C, Changfa Q, Xuen X, Zhongtao D (1982) Anther culture techniques of rubber tree and sugarcane. In: Fujiwara A (ed) Plant tissue culture 1982. Proc 5th Intl Congr Plant tissue and cell culture, Tokyo, pp 533–534

Plant Index

Abelmoschus 757, 771
A. esculentus 756
Acacia 209, 218, 240, 244, 253
A. baileyana 244
A. conferta 244
A. farnesiana 244
A. laccata 244
A. subulata 208, 221, 240
Acacieae 253
Acalypha indica 23
A. lanceolata 704, 715–717
Acanthaceae 61, 330, 458, 594, 598, 605
Acanthus ilicifolius 458
Acer 85
A. campestris 88
A. negundo 87
A. opalus 88
A. platanoides 88
A. pseudoplatanus 88
Aceraceae 129, 217, 575, 585
Achillea millefolium 535
Aconitum napellus 534, 536
A. neomontanum 535
A. ranunculifolium 535
A. variegatum 526, 527, 535
A. vulparia 535
Actaea 425
Actinophloeus 403
Adenanthera 23, 252
Adoxa 707, 715, 724
Adoxaceae 53
Aegiceras 240
Aegilops 74, 764
Aeginetia 426
A. indica 36, 432
Aegle 594
A. marmelos 448, 761
Aerva tomentosa 756
Aesculus 573
A. hippocastanum 764
Aextoxicaceae 129
Agapanthus 94
Agavaceae 129
Agave 60, 174, 331
Agropyron 772
A. intermedium 75
A. repens 534, 771, 772
A. scabrum 479, 498, 510
Agrostemma 285
A. gracile 144
Aizoaceae 587, 667, 684
Aizoideae 667
Alangiaceae 148
Albizzia lophanta 602
Alchemilla 486–489, 502
Alectra thomsoni 61, 335, 337
Alectrolophus 337
Aleurites 728
A. triloba 23
Alfalfa (see *Medicago sativa*)
Alisma 377, 548
A. lanceolata 538, 546, 547
A. plantago 2, 377, 398
A. plantago-aquatica 546–548
Alismataceae 39, 146, 391, 538, 546
Allium 95, 161, 164, 166, 167, 274, 531, 533, 707, 715, 724, 742
A. ammophilum 529, 531, 533, 534, 537
A. amplectens 529
A. angulosum 531, 533, 534, 537
A. cepa 64, 69, 74, 75, 79, 92, 532, 533, 544, 545, 559, 741, 753, 756, 770, 772
A. fistulosum 92, 772
A. nutans 480, 531, 532, 560
A. odorum 33, 480, 532
A. porrum 742, 796
A. pulchellum 531, 533–535
A. sativum 776
A. schoenoprasum 530
A. tuberosum 109
A. ursinum 531, 532, 534, 535, 538, 541
A. vitifolia 746
Allotropa 427
Almond (see *Prunus amygdalus*)
Aloe 274
A. ciliaris 526
A. obscura 526
Alopecurus 287
Alpinia 274
Althaea rosea 741, 753
Althenia 669, 671
Alyssum 410
A. maritimum 324–327, 346, 347, 349, 350, 355, 358, 359, 361, 362, 365, 366, 368, 369
Amaranthaceae 132, 144, 391, 681
Amaryllidaceae 57, 127, 140, 148, 400, 430, 535, 540, 585, 589
Amaryllis 203
Ambrosia 222, 254
A. elatior 256
A. trifida 205
Ambrosiaceae 725
Amentiferae 572
Ammannia 168
Ammi majus 776
Ammobroma 426
Ammophila arenaria 534–536
Amphibolis 199, 216, 223–225, 242, 248, 710, 711
A. antarctica 208, 215, 222, 225, 237, 249
Amyema 15, 19, 60, 211
A. congener 18
A. gravis 15
A. miquelii 15, 16, 18, 23
A. pendula 18, 19, 215
A. preissii 18
Amylotheca dictyophleba 15
Anacardiaceae 57, 140, 148, 452, 575, 585, 605, 672, 673
Andrographis 343
A. echioides 343
A. serpyllifolia 343

Andropogoneae 484
Anemone 427
A. canadensis 746
A. coronaria 616,618
A. dichotoma 746
A. hepatica 535
A. hupehensis 746
A. nemorosa 538, 705, 709, 721, 722, 724
A. pavonia 67
A. rivularis 61
A. vitifolia 746
Anethum 754
A. graveolens 754, 776, 793
Angelica 786
Anneslea crassipes 583
Annona squamosa 604
Annonaceae 53, 148, 209, 238, 343, 572, 574, 578, 582, 584, 603, 710
Antennaria 478, 482, 502, 505, 507
A. alpina 478, 488
A. carpatica 479
A. prosildii 478
Anthurium 425
Antirrhinum 290, 292
A. majus 64, 69, 79, 526, 527, 762, 776, 782
Aphanes 486, 489, 502
Apiaceae 709, 725
Apinagia 652, 715, 716, 721, 727
Apium graveolens 363, 616–618, 626, 629, 776, 781
Apocynaceae 585, 595, 605, 705
Aponogetonaceae 339
Aporosa 590
Apple (see *Malus pumila*)
Aptenia 667
Apteria 428
Aquifoliaceae 616, 585
Aquilegia 174, 178
Arabidobrassica 786
Arabidopsis 786
A. thaliana 430, 746, 776
Arabis Holboellii 479, 488, 489, 500
A. lyallii 460, 461
Araceae 54, 57, 391, 535, 538, 581, 599
Arachis hypogaea 605, 606, 613, 620, 623
Araliaceae 132, 581, 605
Araucaria 699, 703, 704, 706
Arbutoideae 669, 670
Arceuthobium 54, 58
A. douglasii 152
A. minutissimum 334, 335
Archytaea 672
Areca 403
Argemone 60
A. mexicana 459–461, 737, 761, 762
A. ochroleuca 761
Aristolochia 58, 248
A. bracteata 461
Aristolochiaceae 238, 603
Armeria maritima 280
Artabotrys odoratissimus 343, 344
Artemisia 233, 234, 239, 240, 243, 247
A. mutellina 85
A. pynocephala 208
A. vulgaris 233
Arum 248
A. maculatum 535, 538, 541
Asarum europaeum 534
Asclepiadaceae 23, 54, 123, 148, 208, 595
Asclepias 702
A. curassavica 776
Asimina 768
Asparagus officinalis 462, 776, 782
A. setacea 599
Asperula 28
Asphodelus tenuifolius 339
Asteraceae 85, 217, 224, 390, 430, 705, 709, 725, 728
Asterales 395, 725
Asteranae 394, 395
Asteranthe 209
Atalantia racemosa 586
Atamosco (see *Zephyranthes*)
Atkinsonia 19
A. ligustrina 16
Atrichum 698, 699
Atriplex hymenelytra 715
Atropa 746, 749, 751, 784
A. belladonna 776, 782, 786
Aucuba 684
Austrobaileya 53
Avena 60, 64, 66, 79, 401, 509, 764
A. fatua 403
A. sativa 402, 403, 567, 632
Averrhoa carombola 771
Azadirachta indica 460, 461, 463
Azima 662, 664
A. tetracantha 662
Azolla 706–708, 710

Balanophora 152, 153, 426
B. elongata 152
Balanophoraceae 14, 133, 385, 391, 426
Balsaminaceae 127, 140, 581, 661
Barathranthus 19
B. axanthus 14, 16
Barleria cristata 330, 331
Barley (see *Hordeum vulgare*)
Bartsia alpina 526, 527, 538
Basella rubra 655
Basellaceae 655–657
Baselleae 655
Bauhinia 589
Bean (see *Phaseolus*)
Bean, broad (see *Vicia faba*)
Bean, lima 625
Beet (see *Beta vulgaris*)
Begonia 287, 290, 293, 569, 575, 582, 599, 601
B. epipsila 587, 588
B. franconis 742
B. leptotricha 601
B. quadriata 588
B. squamulosa 578
Begoniaceae 57, 127, 144, 147, 569, 575, 582, 599, 601, 706
Bellardia trixago 526, 527
Bellevalia 94
B. romana 534
Benincasa cerifera 322
Benthamidia 686–688
Berberidaceae 127, 144, 582
Beta 28, 60, 66, 72, 74, 233, 484, 506, 510, 725
B. vulgaris 64, 189, 510, 735
Betula 238
B. odorata 225
B. pendula 742
Betulaceae 127
Bidens 621, 625
B. cernua 620–622, 625
Bignoniaceae 61, 390, 594
Biophytum 598
Biota 722
Bixaceae 53, 148, 579
Blandfordia 682
B. nobilis 682
Blandfordieae 682

Blastania garcini 322
Blepharis 598
B. ciliaris 602
Blumenbachia hieronymi 332
B. insignis 332, 334
Boerhaavia diffusa 31, 32
Bombacaceae 129, 144, 578, 584, 585, 589
Bonnetia 672
Bonnetiaceae 671, 672
Bonnetieae 672
Boraginaceae 23, 77, 148, 274, 538, 539
Borassus flabellifer 359, 360, 538
Boschniakia 427
B. himalaica 330
B. hookeri 595
Bothriochloa 484, 491, 502, 503, 506
Bouvardia 151
Brachiaria 36
B. setigera 36, 37, 446
Brassica 199, 200, 241, 251–254, 280, 284, 287, 745, 752, 784, 786
B. chinensis 752
B. juncea 746, 747
B. napus 746, 752, 782
B. oleracea 200, 544, 776
B. pekinensis 752
Brassicaceae 390, 417, 430
Brassicoraphanus 289
Brodiaea 165
Bromeliaceae 599
Bromus 509
B. inermis 776, 782
Broussonetia 58
Bryonia 414
B. dioica 524, 538, 542
Buckleya lanceolata 61
Bulbostylis 679
Burmannia 428
B. barbata 679
B. candida 680
B. championii 680
B. coelestis 478, 487, 488, 679, 680
B. pusilla 430, 679, 680
B. tuberosa 680
Burmanniaceae 428, 595
Burseraceae 129, 578, 585, 672
Butomaceae 146
Butomopsis 39, 40, 43, 47, 319, 718
B. lanceolata 39, 40, 724
Cactaceae 132, 144, 146, 454, 574, 600, 684
Caesalpiniaceae 133, 605, 628
Caesalpinioideae 218, 589
Calamagrostis 478, 488, 489, 495, 502
Calendula 171, 174
C. suffruticosa 24
Calenduleae 133
Caliphruria 274
Calla palustris 596, 597
Calliandra 323
Callitris 720, 721
Calotis 6
C. lappulacea 6, 7
Caltha 149, 425, 427, 592, 741
C. palustris 150, 306, 535, 536
Calycanthus 57
Calyceraceae 725
Camelina sativa 523
Camellia 672
C. japonica 764
Camissonia 209
Campanula 594
C. pyramidalis 530, 538
Campanulaceae 45, 538, 599, 685, 725
Campanulales 394
Cananga odorata 55
Canellaceae 57, 579
Canna 80, 107, 235, 242, 248, 274, 602, 606
C. generalis 89, 95, 224
C. jaegeriana 580
C. maculata 589
C. paniculata 588
Cannaceae 224, 238, 582, 602
Cannabinaceae 57
Cannabis 60, 64, 79
Cansjera 150, 151
C. rheedii 332, 333
Capillipedium 484, 502
Capparales 662
Capparidaceae 23, 572, 574, 584, 585
Caprifoliaceae 526, 581, 605, 616
Capsella 298, 305, 308, 377, 379, 381, 406, 407, 410, 413, 418
C. bursa-pastoris 2, 99, 138, 174, 176, 178, 183, 185, 186, 188, 324, 325, 327, 329, 345–350, 358, 364, 368, 369, 377, 382, 390, 392, 406, 408, 411, 413, 416, 570
Capsicum 163, 233, 792
C. annuum 64, 73
C. frutescens 461, 464
Caraway (see *Carum carvi*)
Caricoideae 678
Cardamine 82
Cardiospermales 148
Carex 679
C. hirta 524
Carica cauliflora 757, 772
C. papaya 577, 757, 772
C. stipulata 776
Caricaceae 14, 23, 127, 578, 585
Carob (see *Ceratonia siliqua*)
Carrot (see *Daucus carota*)
Carthamus roseus 739
C. tinctorius 455, 461, 463
Carum carvi 613
Caryocaraceae 242, 590, 656, 659, 661
Caryophyllaceae 132, 211, 224, 391, 526, 546, 549, 762
Caryophyllales 655–657, 661, 681, 684
Caryota 405
Cassava (see *Manihot esculenta*)
Cassia 323
C. australis 526
C. laevigata 526
C. occidentalis 3
C. sativus 322
Castilleja 381
Castor bean (see *Ricinus communis*)
Casuarina 133, 706, 707
C. equisetifolia 23
Casuarinaceae 127, 128, 148
Catalpa 390
Catharanthus roseus 208, 739
Cattleya aurantiaca 425
Cauliflower (see *Brassica oleracea*)
Caulinia 222
Caytonia 40
C. nathorsti 40, 41
C. sewardi 40
C. thomasi 40, 41
Caytoniales 40, 41
Cedrus 700, 701, 703, 720
Celastraceae 54, 142, 149, 575, 584, 585, 598, 662
Celasterales 662, 664

Celastrineae 662
Celery (see *Apium graveolens*)
Celsia coromandeliana 61, 334–337, 343
Cenchrus 484, 505
C. ciliaris (see *Pennisetum ciliare*)
Centrolepidaceae 129, 664, 666
Centrolepis fascicularis 664, 666
Centrospermae 139, 146, 581, 582, 606
Centrospermales 655, 681, 684
Cephalotaxus 698, 709, 710, 712, 713, 720, 722, 723
Ceratonia 627
C. siliqua 628
Ceratophyllaceae 238
Cercidiphyllaceae 53
Cereus aethiops 601
C. jamacaru 684
Cestrum elegans 526
Centranthera hispida 337
Cereals 509
Chamaerops 403
Charophyceae 706
Cheirostylis flabella 400
Chelidonium majus 591
Chelone glabra 58, 335, 337
Chenopodiaceae 144, 391, 526, 655–657
Chenopodiales 655, 656, 681
Chenopodium rubrum 58
Chicorium endivia 776
Chimaphila 427
Chinese Kale 776, 781, 802
Chionodoxa luciliae 534, 591
Chloranthaceae 149, 426, 584
Chloranthus japanicus 426
C. serratus 426
Chlorella 233
Chlorideae 690
Chloris 484, 690
Chlorophyceae 697–699, 707, 710
Chlorophytes 697, 700, 711
Christisonia 427
Chondrilla 479, 500, 502
Chrozophora obliqua 129
Chrysanthemum 716, 717, 724
C. alpinum 535
Cicer 786
Cichorium 705, 706, 728, 729
Cimicifuga japonica 534
Cinnamomum 211, 242
Circeasteraceae 53
Cistaceae 127, 582
Cistanche 427
C. tubulosa 773
Citrullus colocynthis 31, 538
C. fistulosus 322
Citrus 66, 233, 445, 448, 449, 451, 452, 466, 508, 598, 736, 757, 760, 761, 767, 769, 776, 782
C. aurantifolia 759, 761
C. aurantium 2, 583
C. grandis 757, 770
C. limon 64, 79, 80, 95, 451, 757
C. microcarpa 448–450, 757, 771
C. paradisi 451
C. reticulata 451, 757, 761
C. sinensis 451, 757–759, 760
C. unshiu 449
Clematis 425
C. viticella 746
Cliftonia monophylla 330
Clivia 3, 203, 206, 209, 240
C. miniata 535
Clusiaceae 149, 579, 584, 672
Coccinea 761
Coccoloba uvifera 604
Cochlospermaceae 589, 595
Coconut (see *Cocos nucifera*)
Cocos 405, 413, 425
C. nucifera 323, 355, 368, 404, 405, 456, 605, 766, 770, 775–777
Codonopsis clematidea 538
Coelebogyne 445
C. ilicifolia 2
Coffea 509, 606, 613, 773
C. arabica 776
Coix 410, 510
Colchicum 3
C. autumnale 726, 728, 729
Coleochaete 697, 707, 710–712
Coleus 233
Comandra 152
C. umbellata 332
Combretaceae 61, 129, 582, 585
Combretum grandiflorum 61
Commelina 400
Commelinaceae 166, 274, 529, 540, 581, 584, 599, 683
Commelinales 683
Compositae 23–25, 54, 128, 142, 148, 171, 250, 479, 480, 484, 488, 489, 499, 500, 502, 508, 526, 535, 569, 648
Conium 164, 174
Connaraceae 571, 572, 578, 590
Conopholis 427
Convallaria majalis 537
Convolvulaceae 23, 127, 141, 148, 595, 605, 628
Convolvulus arvensis 791
Cooperia drummondii 430, 431
C. pedunculata 430, 431, 478, 489
Corchorus capsularis 772
C. olitorius 772
Cordaites 728
Coriander (see *Criandrum sativum*)
Coriandrum sativum 329, 776
Cormophytes 697, 707, 711, 719, 727
Cornaceae 585, 684
Cornus 4, 686–689
C. alternifolia 686, 688
C. amomum 686, 688
C. asperifolia 686, 688
C. florida 686–688
C. stricta 686, 688
Coronilla minima 392
Cortaderia 171, 174, 180
C. jubata 24, 448, 488, 489, 491, 493
C. selloana 24
Corydalis 387, 425
C. cava 427, 457, 535, 536
C. nobilis 535, 536
Corylaceae 140
Corylus avellana 776
Cosmos 211, 216, 233, 239, 242, 247, 284, 709
C. bipinnatus 89, 215, 224, 259
Costaceae 219, 599
Costalia 603
Costus 163, 603
C. cuspidatus 150
C. speciosus 770
Cotton (see *Gossypium*)
Cotula 171
Cotylanthera 426
Crassula aquatica 655
Crassulaceae 24, 25, 28, 132, 137, 391, 417, 652–655

Crepis 137, 165, 166, 174, 484, 487, 488, 491, 502
C. capillaris 390
C. zacyntha 525, 526
Crinum 60, 150, 151, 585, 730
Crocus 242, 243, 285, 290, 425
C. aureus 534, 535
C. suaveolens 535
C. vernus 224
Crossandra 211
C. stenostachya 211, 215
Croton 767, 769
C. bonplandianum 767
Crotonoideae 238, 584
Cruciferae 127, 140, 142, 146, 147, 250, 251, 322, 500, 546, 549, 578, 584, 598
Crowfurdia 594
Cucubalus 291
C. baccifer 546
Cucumber (see *Cucumis*)
Cucumis 757, 761, 766, 769
C. anguria 124
C. melo 74, 322, 525, 545, 771
C. sativus 54, 254, 530, 538, 542
Curburbita 30, 106
C. ficifolia 322
C. moschata 577
C. pepo 69, 322, 538, 776
Cucurbitaceae 27, 30, 31, 127, 129, 141, 148, 243, 322, 572, 578, 582, 590, 684, 761
Cuminum cyminum 329
Cunoniaceae 133
Cupressaceae 712
Cupressus 700, 709, 710
Cuscuta 701, 702, 715
C. epithymum 715
C. reflexa 368, 414, 459, 463, 779
Cuscutaceae 23, 141, 728
Cyamopsis 323, 627
Cyanastraceae 606
Cyanastrum 606
Cyanophyceae 699
Cyanotis 274
Cycads 699, 700, 704, 709, 710, 728
Cycadales 709
Cycas 703, 710
Cyclamen persicum 109
Cyclanthaceae 395
Cyclanthera pedata 538
Cymodocea 669, 671
Cymodoceaceae 222, 224, 671
Cymbidium 776
C. bicolor 28, 391, 414, 457
Cynomoriaceae 426
Cynomorium coccineum 426
Cyperaceae 148, 678, 679, 701, 702
Cyperales 666
Cyperoideae 678
Cyperus 274, 678, 679
C. alternifolius 524
Cypripedium 413
C. insigne 382
Cyrilla racemiflora 330, 331
Cyrillaceae 330
Cystisus 722
C. laburnum 414
Cytinaceae 140
Cytinus 427

Dactylorchis 218
Daphne mezereum 530, 538
Date (palm) (see *Phoenix dactylifera*)
Datiscaceae 588
Datura 127, 141, 211, 422, 736, 784–786
D. arborea 215
D. discolor 799
D. fatuosa 467
D. innoxia 558, 745, 746, 749, 784, 786
D. fusticosa 360
D. metel 784
D. sanguinea 784
D. stramonium 382, 466, 467, 784
Daucus carota 735, 737, 774–780, 782, 786
Degeneriaceae 53, 57, 584, 603
Degeneria 647, 710
Delosperma 667
Delphinium kotulae 538, 548
Dendrobium nobile 414
Dendrophthoe 19, 766, 769
D. falcata 14, 18, 19, 767
D. neelgherrensis 15, 19, 23
D. pentandra 15, 19
Desmodium 323
Desmopsis 710
Dianthus 1, 168
D. chinensis 351, 546
Diapensiales 394
Dicentra spectabilis 535
Dichanthium 484, 496, 502, 503, 506
D. aristatum 500, 501, 506
D. caricosum 500
Dichapetalaceae 142, 571, 575, 578, 581
Dicraea 715, 716, 721, 722, 727
D. elongata 653
D. stylosa 28, 414
Dicranostigma franchetianum 762
Dicranum majus 721
Dictyostegia 428
Didiscus coerulea 776
Digitaria 484
Dilleniaceae 581, 582
Dilleniidae 657
Dinizia 209, 252
Dioscorea composita 460, 461
Dioscoreaceae 594
Dipcadi 166, 167
Diplacrum caricinum 320
Diplotaxis 277, 407, 418, 421
D. erucoides 408, 417, 419
Dipsacaceae 385, 391, 526, 725
Dipsacales 395
Dipteracanthus patulus 330, 331, 458
Dipterocarpaceae 23, 148, 603
Dobera 662
Dodonaeoideae 672
Doratoxyleae 672
Dorstenia 58
Dracena 274
Drimys 209
D. winteri 107
Droseraceae 127, 144, 391, 392, 587, 595, 706
Drusa 716, 724
Durio zibethinus 770
Dysoxylon 572
Dyssapindaceae 672

Ebenaceae 581, 603, 606
Ebenales 394
Echballium elaterium 530, 538, 542
Echinocystis lobata 54, 524, 530, 538
Echinodorus tenellus 546, 547
Echium vulgare 538, 539
Elaeagnaceae 582
Elaeis 631

Elaeis guineensis 605, 631, 636, 773, 776
Elaeocarpus 594
E. mammillata 524
Elaeocarpaceae 23, 142, 148, 585
Elaeocharis 100
Elatinaceae 599, 659, 661
Elatostema 33, 488
E. acuminatum 484
E. eurhynchum 478
E. eusinuatum 33
Elymus 534
E. virginicus 534
Elytranthe 19
Elytrantheae 16
Elytraria acaulis 343
Empetrum nigrum 225
Endymion 233, 707, 715, 724
E. nonscriptus 92
Enhalus 222, 223
Entada 252
Epacridaceae 233
Ephedra 39, 703, 704, 711, 712, 720, 723
Epidendrum 174, 298, 300, 308, 379, 381
Epilobium 133, 137, 209, 218
E. angustifolium 448
E. dodonei 448
E. hirsutum 448
E. montanum 448
E. palustre 134
E. rigidum 221
Epipactis 134, 274
E. palustris 137
Episcia melittifolia 591
Equisetum arvense 699
Eragrostis curvula 484, 496, 507
Eranthis 427, 449
E. hiemalis 368, 467, 535
Eremurus 67, 69, 526
Ericaceae 142, 218, 221, 596, 669, 670, 702
Ericales 330, 394, 667
Ericanae 394
Ericoideae 669, 670
Erigeron 478–480, 500
Eriocaulaceae 428, 683
Eriocaulales 683
Eriocaulon 425, 428
E. hookerianum 683
E. robusto-brownianum 425
E. xeranthemum 425, 431
Eriochloa 484
Eruca 422
E. sativa 546, 548, 549, 559
Erycibe 344
E. griffithii 344
E. tomentosa 344
Erysiphe graminis 772
Erythraea centaurium 461
Erythrina 573
Erythronium americanum 456
E. dens-canis 456
Erythroxylaceae 142, 147, 582, 680, 681
Erythroxylum 680, 681
E. coca 680
Eschscholtzia 174
E. californica 589, 761, 762, 781
Eubrachion ambiguum 152
Eucalyptus 211, 217
E. ficifolia 215
Eucommia ulmoides 58
Eucypereae 678
Eugenia 448
E. jambos 452, 454
Eulophia 457
E. epidendraea 457
Eupatorium 478
E. glabrum 536, 537
E. riparium 482, 496
Euphorbia 174, 186, 218, 414, 592
E. dulcis 721, 722
E. pulcherrima 526, 527, 776
Euphorbiaceae 23, 55, 127, 129, 138, 148, 218, 526, 572, 578, 581, 582, 584, 585, 589, 590, 717, 728
Euphrasia minima 539
E. picta 526, 527
E. rostkoviana 526, 527, 539
Eupteleaceae 53
Eruca 555
E. sativa 559
European birch (see *Betula pendula*)
Euryale 596
Eustachys 690
E. glauca 690
E. petraea 690
Exocarpus 152, 769
E. bidwillii 152
E. cupressiformis 765, 769
E. menziesii 24, 25
E. sparteus 24, 25, 332, 333
Exospermum 245
Fabaceae 140, 417, 618, 623, 625–628
Fagaceae 575
Farinosae 682, 683
Farmeria indica 654
Fenugreek (see *Trigonella foenum-graecum*)
Ferocactus wislizeni 684
Festuca 174, 499, 509
F. arundinacea 752
Ficoidaceae 667
Ficoidea (Ficoideae) 667
Filicinae 707, 713, 714, 727
Filicium 672, 673
F. decipiens 672
Fimbristylis 679
Flacourtiaceae 144, 149, 657, 659
Flagellariaceae 664, 666
Flindersia 594
Foeniculum 754
F. vulgare 329, 754, 776
Forsythia 270
Fragaria 508, 510
Frankenia 659, 661
F. hirsuta 659
F. pulverulenta 659, 660
Frankeniaceae 659, 661
Fraxinus 425, 611, 616, 625, 664
F. excelsior 426, 616
Fritillaria 2, 12, 29, 717, 721, 724
F. meleagris 537
F. persica 12, 13
F. tenella 2, 319
Fucales 383
Fuchsia hybrida 134
Fucus 383, 384
Fuirena 679
Fumaria 425
Fumariaceae 391, 417, 535
Funkia 445
F. ovata 2
Furcraea 133

Gagea 359
G. lutea 537
Galanthus 44, 552
G. nivalis 45, 46
Galium 557
G. mollugo 25, 26
Garlic (see *Allium sativum*)
Garrya 684
G. elliptica 684
G. flavescens 684
G. veatchii 684

Garryaceae 684
Gasteria 137, 164–167, 178, 242, 274
G. verrucosa 137
Gentiana 141, 594
G. acaulis 64, 79
G. asclepiadea 588
Gentianaceae 54, 147, 426
Gentianae 394, 395
Gentianales 395, 662, 664
Geodorum densiflorum 457
Geraniaceae 127, 142, 390, 395, 539, 546, 574, 575, 582, 584, 589, 661
Geraniales 139, 581, 661, 662, 681
Geranium phaeum 531, 539, 546, 550
G. pratense 583
G. sylvaticum 588
Gerbera jamesonii 103, 208
Gerris 519, 521
G. lateralis 520
Gesneriaceae 30, 127, 142, 330
Geum rivale 463
Gherkin (see *Cucumis*)
Ginkgo 42, 43, 698, 699, 704, 708, 712, 721, 726, 728, 729
G. biloba 764
Ginseng (see *Panax ginseng*)
Gisekia 667
Gladiolus 211, 215, 240, 279
G. gandavensis 208, 224
Gloriosa 133
Glottidium 705
Glumiflorae 666
Glycine 688
G. max 135, 606, 688, 689, 786
Glycosmis 572
Gnetum 39, 43, 698, 706, 712–714, 717, 721, 723, 725–727
Gomortegaceae 238
Gonianthes 428
Goodeniaceae 208, 209, 221, 725
Gossypium 174, 176, 178, 180, 181, 185, 186, 188, 209, 278, 283, 298, 301, 379, 382, 406, 430, 595, 746, 756, 757, 761
G. arboreum 761, 772
G. barbadense 430, 432, 433, 447, 462, 465, 761
G. hirsutum 199, 348, 349, 359, 379, 380–383, 407, 410, 413, 545, 757, 761, 771, 772, 789, 791
G. davidsonii 465
Gramineae (Poaceae) 24, 57, 77, 129, 200, 211, 215, 219, 274, 286, 484, 489, 502, 508, 526, 535, 540, 569, 606, 612, 666, 753
Grape (see *Vitis vinifera*)
Grasses (see *Gramineae/ Poaceae*)
Green corn (see *Sorghum*)
Grevillea robusta 27, 324
Grossulariaceae 584
Guar (see *Cyamopsis*)
Gymnadaenia 221
Gymnosiphon 428
Gynandropsis gynandra 757

Habenaria platyphylla 457
Haemanthus 3, 44
H. katherinae 341–343, 349, 351, 353, 355, 359, 360
Haemaria 221
Haemodoraceae 677, 678, 683
Haemodorales 682
Halodule 199, 225, 228
Halophila 222, 223
H. ovata 338, 396, 399
Haloragaceae 391, 681
Haloragis micrantha 28, 414
Hamamelidaceae 582, 657
Hamamelidae 657
Hamamelidales 657
Hamamelis virginiana 457
Heckeria 427
Hedycaria 209, 245
Hedychium 248
Hedyosmum nutans 426
Heisteria cyanocarpa 152
Heleocharis 283
Helianthemum 598
Helianthus 62, 174, 176, 178, 185, 186, 188, 190, 233, 243, 247, 298, 409, 418
H. annuus 62, 75, 199, 200, 208, 240, 349, 351, 353, 364, 365, 567, 605
Helicanthes 15, 19
H. elastica 15, 16, 18, 19
Heliconia 219, 235, 242
H. nutans 236
Heliconiaceae 219
Helixanthera 19, 152
H. ligustrina 14, 16, 19
Helleborus 60, 66, 79, 82, 88, 94, 107, 233
H. foetidus 62, 64, 66, 79, 88, 100, 109, 524
H. niger 535
Helobiae 238, 710, 711, 724
Hemerocallideae 682
Hemerocallis 682, 742
Henbane (see *Hyoscyamus niger*)
Hennecartia 38, 719
Hepatica 425
H. acutiloba 67
Heracleum 616
H. sphondylium 616
Hereroa 667
Hernandia peltata 605
Hernandiaceae 238, 605
Hesperocallis 682
Heteropogon 484
Heterozostera 222
Heuchera 78
H. micrantha 11
Hevea braziliensis 748, 772
Hexalobus 209
Hibiscus 23, 216, 295, 307, 346, 382, 704
H. syriacus 1
Hieracium 488, 491, 498, 502, 503, 506
H. aurantiacum 493, 495
H. boreale 529
H. laevigatum 529
H. subg. Euhieracium 478
H. subg. Pilosella 484
Hierochloë 484, 491, 500
Himantandraceae 53
Himantoglossum nircinum 225
Homoptera 519
Honey locust 628
Hordeum 274, 298, 308, 407, 532, 752, 764, 766, 771, 772, 786
H. brachyantherum 403, 772
H. bulbosum 772
H. depressum 772
H. distichum 532, 535, 543
H. jubatum 534, 772
H. secalenum 403
H. spontaneum 772
H. vulgare 178, 200, 208, 245, 499, 500, 509, 532, 567, 613, 630–632, 636, 745, 751, 766, 771, 772, 776

Horsfieldia 710
Hosta 682
H. plantaginea 742
Houttuynia 427
Howea 403
Hyacinthus 6
H. orientalis 6, 8, 10, 11, 109, 535
Hydnora 391
H. africana 426
Hydnoraceae 140, 391, 426, 584
Hydrilla verticillata 31, 32
Hydrocharis morsus-nanae 598
Hydrocharitaceae 30, 132, 222, 224, 339, 400
Hydropteridae 707
Hydrostachyaceae 652
Hymenocallis occidentalis 589
Hyoscyamus 1, 749, 751, 784
H. muticus 784
H. niger 746, 748, 749, 751, 764
Hyparrhenia 484
Hypecoum procumbens 534
Hypericaceae 127, 539, 584, 599
Hypericum olympicum 539
H. perforatum 484, 489, 491, 495, 506
Hyphaene 702

Iberis 251, 253, 287
I. amara 324–327, 349, 350, 355, 357, 359, 362, 365, 368, 369, 523, 741
I. glandulifera 292
Icacinaceae 27, 33, 238, 331
Ilex 429
I. aquifolium 776
I. opaca 616
Illiciaceae 57
Illicium 107
Impatiens 164, 166, 282, 293, 594, 606
I. parviflora 580
I. textorii 387
I. walleriana *143*
Indotristicha 652, 654
I. ramosissima 28, 654
Inga 209
Iodina 152
I. rhombifolia 332, 333
Impomoea 69, 79, 107, 211, 418
I. coerulea 215
I. purpurea 62, 418
Iridaceae 211, 224, 274, 400, 535, 539
Iris 3, 274, 613, 773
I. bucharica 539
I. orchioides 539
I. pseudacorus 596
I. reticulata 539
Isoetes 701
Isomeris 33, 34
I. arborea 33, 34, 446
Isoplexis canariensis 337
Ivory nut (see *Phytelephas macrocarpa*)
Ixeris dentata 480

Jasione 44, 166, 169, 174, 176, 186, 188
J. montana 36, 45, 46, 382
Jasminoideae 673, 675, 677
Jasminum 673, 675, 677
J. angustifolium 675, 676
J. calophyllum *675, 676*
J. sambac 675, 676
Jatropha 766–769
J. panduraefolia 767
Jubaeopsis 405
Juglandaceae 144, 146–148, 574, 575, 589
Juglans 166
J. regia 606, 765, 767
Juncaceae 274, 390, 678, 679
Juncales 666
Juncus 390, 398
J. bufonius 679
J. prismatocarpus 339

Kadsura 107
Kalanchoe 741
K. obtusa 64, 79
K. pinnata 776
Kale (see *Brassica oleracea*)
Kibara 38, 719
Killingia triceps 524
Kingiodendron 344
K. pinnatum 604
Kleinia ficoides 535
Klugia notoniana 330
Knautia arvensis 525, 526
K. drymeia 525, 526
Kniphofia 682
K. meleagris 537
Kniphofieae 682
Korthalsella dacrydii 152
Kyllinga 678, 679
K. brevifolia 678, 679
K. melanospora 678, 679
K. triceps 678, 679

Labiatae 61, 539, 540, 594
Lactuca 411, 637
L. sativa 361, 362, 545, 613, 618, 620, 621, 623, 625–627
Lagenaria 167, 168, 180, 190, 761
Lagenostoma 703, 704, 708
Lamiaceae 390, 394, 701
Lamiales 395
Lamianae 395
Lamium amplexicaule 742
L. purpureum 735
Lamprothyrsus 489, 493
Larix 722
Lathraea squamaria 526, 527
Lathyrus 252, 556, 590
L. articulatus 771, 772
L. clymenum 771, 772
L. latifolius 556, 557
L. luteus 556
L. odoratus 94
L. silvestris 556
L. vernus 556
Lauraceae 57, 137, 148, 215, 219, 224, 238, 574, 579, 584
Laurelia 107
Lavandula 233
Lavatera 107
Lecythidaceae 142, 582
Leeaceae 578
Legume/s (see Fabaceae)
Leguminosae 23, 27, 146, 148, 252, 322, 323, 377, 574, 578, 581, 582, 589, 598, 602, 606, 612, 613, 628, 725
Leiphaimos 426
Leitneria 469
Lemnaceae 54
Lennoa 426
Lennoaceae 426
Lens 58, 60, 66, 88, 95, 100, 102, 410
L. culinare 64
Lentibulariaceae 426, 589, 595
Leontodon hispidus 506, 507
Lepeostegeres 19
L. gemmiflorus 15
Lepidophytes 699, 711
Leptomeria 5, 6, 766
L. acida 766, 767
L. billardierii 5, 6

Leschenaultea 209, 218
L. formosa 221
Lettuce (see *Lactuca sativa*)
Leucinoides orbenalis 772
Leucocrinum 682
Levenhookia dubia 392
Liliaceae 23, 57, 77, 109, 127–129, 142, 148, 224, 305, 390, 462, 526, 535, 538, 539, 581, 599, 677, 678, 682
Liliales 683
Liliflorae 606, 682
Lilium 2, 12, 29, 57, 60, 64, 70, 77, 79, 80, 83, 85, 88, 89, 92, 94, 98, 99, 107, 164, 166–168, 174, 209–211, 216, 217, 228, 231, 233, 239, 242, 244, 247, 249, 276, 277, 279, 282, 284, 287, 288, 290, 293, 298, 379, 390, 398, 462, 537, 594
L. croceum 79
L. henryi 70, 103, 462, 772
L. langi 772
L. longiflorum 72, 85, 98, 224, 293, 772
L. martagon 2, 12, 319, 460–462
L. shikayama 772
L. sugehime 772
Lily (see *Lilium*)
Limnanthes 319
Limnobium 274
Limnocharis emarginata 39
Limnophyton 410
Linaceae 142, 143, 238, 391, 578, 584, 598, 680, 681
Linaria bipartita 338
Linales 681
Linostoma pauciflorum 23
Linseed (see *Linum usitatissimum*)
Linum 174, 178, 186, 190, 298, 379, 407, 409, 410, 579, 681
L. ostriacum 771
L. perenne 771
L. usitatissimum 142, 324, 349, 363, 462, 577, 680
Liquidambar styraciflua 776
Listera ovata 462
Livistona 405
Loasa papaverifolia 541
Loasaceae 332
Lobelia 209, 381
L. dunnii 338, 349
Lobeliaceae 127, 338, 725
Lodoicea maldivica 567
Loganiaceae 595, 605, 606
Lolium 509, 543, 709, 766, 769
L. multiflorum 742, 766, 776
L. perenne 205, 208, 215, 256, 257
Lomatia polymorpha 27, 324
Loranthaceae 14, 16, 54, 152, 218, 332, 385, 391, 392, 585, 704, 727, 728
Lorantheae 16
Lotus 390, 761, 772
Lucerne (see *Medicago sativa*)
Ludwigia 209
L. uruguayensis 687
Luffa 148, 761
L. aegyptica 322
Lunaria annua 139, 140
Lupinus 546, 606
L. albus 620
L. luteus 620
L. regalis 524
Luzula 274, 592
Lychnis 210, 247
Lycopersicum 379
L. esculentum 526, 746, 786, 788, 792
Lycopersicon 66, 88, 91, 95, 165, 209, 210, 249, 252, 278, 282
L. peruvianum 64, 66, 209, 247, 278, 782
Lycopodium 697, 707, 709, 710, 735
Lycopsis 414
Lysiana 15, 19
L. exocarpi 16, 19
Lysianeae 16
Lythraceae 390, 598, 600

Machaerocarpus californicus 404
Macleya cordata 776
Macromitrium 706, 708
M. salakanum 698, 699, 706, 708
Macrosolen 15, 19
M. cochinchinensis 14, 15
Macrozanonia macrocarpa 594
Magallana 661
Magnolia 53, 61, 107, 590, 598, 647
M. kobus 106
M. liliflora 106, 107
M. soulangeana 577, 773
M. stellata 58
M. tripetala 107
Magnoliaceae 53, 57, 58, 127, 139, 141, 148, 395, 575, 578, 579, 581, 584, 589, 648, 710, 719, 728
Magnoliales 584
Magnoliideae 719
Magonia glabrata 598
Mahonia 64
M. aquifolium 62, 534
Maize (see *Zea mays*)
Malpighiaceae 129, 584
Malpighiales 681
Malus 278, 485, 486, 489, 491, 503, 507, 508
M. domestica 510
M. pumila 74, 767, 768, 773
Malva 703, 704, 722
M. alcea 241
Malvaceae 23, 35, 53, 138, 140, 147, 211, 234, 430, 574, 582, 584, 602
Malvales 578, 589
Malvaviscus 211, 216, 242
M. arboreus 215, 224
Mammillaria tenuis 454
Mandarin cleopatra (see *Citrus reticulata*)
Manettia bicolor 524
Mangifera 448, 466, 605, 759, 760
M. indica 452, 453, 508, 760, 770
M. odorata 452, 453
Mango (see *Mangifera indica*)
Mannihot esculenta 782, 799
Marantaceae 238, 592, 599
Marattia 700
Marrubium 232
Marsilea 707, 708
Marsileaceae 706
Mathiola 1
Mayacaceae 683
Maydeae 510
Medeola virginiana 129
Medicago 628, 786
M. lupulina 392
M. orbicalaris 772
M. sativa 463, 628, 782
M. truncatula 772
Melampodium divaricatum 455, 456
Melampyrum 542
M. cristatum 539
M. lineare 27, 334, 335, 337

Melampyrum nemorosum 539
M. pratense 539
M. silvaticum 539
Melandrium 763
M. album 546, 549
M. rubrum 546, 549, 586
Melastomataceae 600
Melia azedarach 586
Meliaceae 129, 571–575, 578, 584, 585, 590, 594, 598
Melianthaceae 581, 582
Melilotus 464, 786
M. alba 464, 465, 772
M. messanensis 464, 465
M. officinalis 772
M. polonica 465
Melothria 322
M. hetrophylla 322
M. maderaspatana 322
Menianthales 395
Menispermaceae 140, 395, 585, 603
Menodora 673, 675
Mentha 390
Mentzelia laevicaulis 332
Menyanthaceae 595
Menyanthes 596
M. trifoliata 597
Mercurialis annua 1
Mesembryanthamaceae 667
Mesembryanthemoideae 667
Mesembryanthemum 29, 667
M. floribundum 776
Mesembryeae 667
Microcycas 709
Mida salicifolia 331, 332
Mimosa 252, 323
Mimosaceae 209, 628
Mimoseae 252
Mimosoideae 252
Mimulus luteus 762
Mitrastemon 427
Mnium undulatum 720, 721
Mollinedieae 39
Molluginaceae 667
Mollugineae 667
Molluginoideae 667
Monandrae 721, 722
Monimiaceae 37, 140, 209, 238, 245, 579, 719
Monotropa 2, 427
M. hypopitys 1, 2, 427
M. uniflora 425, 427, 429
Monotropaceae 45, 127, 427, 595
Monotropsis 427
Montioideae 655
Moquiniella rubra 16, 18
Mora oleifera 567
Moraceae 57
Moringa 23, 469
M. oleifera 594
Moringaceae 23, 54
Morning glory (see *Ipomoea coerulea*)
Murraya 594
M. paniculata 593
Musa 208, 509
M. balbisiana 772, 791
Musaceae 132, 219, 238, 599
Muscari 29, 398
M. racemosum 534
Muscinae 706
Myosotes 391, 414
Myosurus 390, 410
Myricaceae 147, 148
Myriophyllum 391
Myristica 574
M. fragrans 23, 604
Myristicaceae 147, 148, 579, 581, 584, 589, 603, 710
Myrtaceae 140, 452, 582, 600
Myrtales 149, 681

Najadaceae 146, 224, 669, 671
Najas 222, 225, 229, 339, 341, 669
N. flexilis 224, 339, 341
N. major 459
N. marina 339, 341
Nandinaceae 584
Narcissus pseudonarcissus 534
N. rotundifolia 534
Nardus stricta 478, 489, 500
Narthecium ossifragum 597
Nechamandra alternifolia 461
Nephelium lappaceum 770
N. malainse 770
Nerium 174
Nicotiana 106, 167, 174, 178, 188, 209, 210, 381, 713, 746, 749, 751, 784, 786
N. debreyii 784
N. glauca 736, 786
N. glutinosa 464, 739
N. knightiana 786
N. langsdorfii 736, 786
N. rustica 382, 754, 762, 786
N. sylvestris 786
N. tabacum 382, 463, 464, 526, 529, 739, 740, 741, 744–746, 748–750, 752, 754, 762, 764, 776, 783, 784, 786
Nigella 705
N. damascena 55, 60, 67, 82, 139, 537, 586, 741, 767, 776
N. sativa 537
Nitella 708, 711, 712
Nothapodytes foetida 23, 27, 331
Nothoscordum 445
N. fragrans 2, 355, 534, 539
Nyctanthes 673, 675, 677
N. arbor-tristis 673, 674
Nyctanthoideae 673, 677
Nutmeg (see *Myristica fragrans*)
Nuytsia 15, 19
N. floribunda 15, 16, 23
Nuytsieae 16
Nyctaginaceae 30, 140, 144
Nymphaea 209, 596
Nymphaeaceae 53, 123, 129, 131, 149, 238, 600, 606, 684
Nymphoides 595, 596
N. peltata 593

Oat (see *Avena sativa*)
Ochna serrulata 448
Octoknema orientalis 150
Odontites rubra 526, 527
Oenone 652
Oenothera 161, 162, 174, 210, 250, 282, 287, 289, 290, 381, 706, 707, 715, 722, 730
O. erythrosepala 381
O. organensis 250
Oenotheraceae 721
Oil palm (see *Elaeis guineensis*)
Olacaceae 23, 152
Olax 60, 331
Oldenlandia 151
Olea 66, 79, 245
O. europaea 64, 67, 80, 87, 109, 242, 243
Oleaceae 61, 85, 148, 217, 426, 616, 662, 664, 673, 675, 677
Oleales 394, 662, 664
Oleoideae 673
Onagraceae 47, 133, 209, 218, 221, 388, 390, 546, 581, 584, 595, 648
Ophioglossum 700, 701
Ophiopogon 677, 678

O. intermedius 677
Ophiopogonae 677
Ophris 161
Ophrys 221, 274
Opilia 151
Opiliaceae 27, 151, 332
Opuntia dillenii 454, 757
Orchidaceae 23, 28, 45, 47, 124, 127, 132, 133, 140, 147, 200, 209, 221, 252, 274, 319, 377, 390, 417, 425, 428, 429, 454, 457, 569, 589, 595, 596, 607, 616, 728
Orchidales 682
Orchids (see Orchidaceae)
Orchis 1, 221
O. maculata 133, 462
Ornithogalum 149, 209
O. boucheanum 534
O. caudatum 705, 728
O. nutans 10
Orntihopus 772
Orobanchaceae 30, 47, 127, 330, 427, 595, 684, 685
Orobanche aegyptiaca 773
O. cernua 31, 432
Orobus angustifolius 414
Orontium 425
Orthocarpus luteus 337
Orygia 667
Oryza 403
O. sativa 362, 567, 633, 752, 767, 769
Oryzopsis 166, 190, 403
Osmunda 98
Osteospermum sinuatum 24–27
Ostrya 133
Osyris 769
Othonna crassifolia 535
Ottelia alismoides 31, 32
Oxalidaceae 142, 390, 526, 584, 598, 603, 661
Oxalis 79, 82, 580, 598
O. acetosella 526, 527
O. europaea 580, 597
O. pubescens 524
O. rosea 524
Oxybaphus nyctagineus 31
Oxychloe 60
Oxyspora paniculata 319
Oxystelma esculentum 61

Paeonia 29, 42, 43, 104, 424
P. actiflora 43, 424
P. albiflora 43, 424
P. anomala 42, 43, 424
P. delavayi 43, 424
P. lactiflora 42, 43
P. moutan 42, 424
P. suffruticosa 43, 424
P. tenuifolia 527
P. wittimanniana 42
Paeoniaceae 55, 578, 584
Palisota harteri 529
Palmae 323, 599, 603, 606, 628
Palmaceae 538
Panax ginseng 776, 777
Pancratium maritimum 596
Pandanaceae 61, 428, 581
Pandanus 163, 174, 428
P. parvius 61
Panicoideae 484, 491, 501
Panicum 484, 508
P. maximum 484, 485, 488, 491, 502, 503, 505–507, 509, 776, 783
Papaver 381, 599, 725
P. heldreichii 534
P. nudicaule 379
P. radicatum 746
P. rhoeas 396, 534–536, 761
P. setigerum 746
P. somniferum 737, 755, 761, 762
Papayeraceae 127, 391, 427, 535, 574, 581, 585, 589, 762
Papilionaceae 139, 147, 390, 539, 546, 602, 705
Parietales 139, 581, 659, 661
Parietaria 258
P. officinalis 258
Paris quadrifolia 537
Parkinsonia 233
Parnassia palustris 593
Parnassiaceae 596
Parthenium 478, 489, 491, 498, 504, 507
P. argentatum 501, 506, 508
Paspalum 164, 168, 174, 176, 178, 181, 479, 484, 489, 491, 508
P. commersonii 497, 500
P. dilatatum 496
P. hexastachyum 500
P. longifolium 497
P. notatum 507
P. orbiculare 294
P. scrobiculatum 33, 446
P. secans 489
Passiflora adenophylla 31
P. calcarata 31, 604
P. racemosa 144
P. suberosa 591
P. verspitilio 144
Passifloraceae 30, 31, 603, 684
Paulownia tomentosa 757
Pea (see *Pisum sativum*)
Peanut (see *Arachis hypogaea*)
Pearl millet (see *Pennisetum typhoides* = *P. americanum*)
Pedaliaceae 581
Pedicularis 417, 527
P. palustris 359, 530, 539
P. sylvatica 526, 527, 533
Pelargonium 165, 381
P. zonale 64, 79, 381
Peltandra 425
Penaea 387, 390, 414, 715, 716, 724
Pennisetum 484, 505, 508
P. americanum 776, 782, 783
P. ciliare 483, 491, 507, 509
P. dubium 491
P. orientale 505, 507
P. pedicellatum 496
P. purpureum 776, 783
P. setaceum 500
P. squamulatum 459–461, 507
P. typhoides (= P. americanum) 497, 505, 507, 510
Pentapetes 23
Pentaphragma 685
Pentaphragmataceae 685
Pentraphragmoideae 685
Peperomia 123, 140, 427, 715, 716, 721, 724
P. hispidula 533, 537
P. pellucida 533, 537
Peraxilla 19
P. tetrapetala 16
Pergularia daemia 100, 106
P. minor 776
Persea 246, 279, 290, 291
P. americana 219, 224, 248
P. borbonia 215
Petiveriaceae 129, 132
Petroselinum hortense 767
Petunia 73, 74, 99, 102, 174, 176–178, 181, 185, 188, 209, 210, 247, 276, 277, 279, 287–291, 294, 297, 298, 300–302,

308, 345, 346, 379, 407, 410, 532, 737, 746, 749, 762, 763, 784, 786
P. axillaris 762
P. hybrida 74, 102, 410, 411, 532, 558, 757, 762, 776, 784, 786
P. inflata 776, 777
P. violacea 762
Phacellanthus 427
Phaedranassa virdiflora 92
Phalaris tuberosa 225, 231
Pharbitis 82
Phaseolae 276, 323
Phaseolus 29, 349, 358, 365, 366, 418, 419, 421–423, 550, 551, 555, 559, 560, 624, 627, 638, 688, 771
P. acutifolius 546, 556, 771, 772
P. angularis 771
P. aureus 135, 624, 688, 689, 771
P. calcaratus 771
P. coccineus 30, 349, 358, 365, 366, 416, 418–423, 539, 546, 550–556, 559, 560
P. hysterinus 419, 546
P. lunatus 546, 771
P. multiflorus 419
P. mungo 546, 624, 625, 771
P. ritensis 771
P. silvatica 414
P. tuberosus 546
P. vulgaris 366, 418, 419, 539, 545, 556, 620–624, 771, 772
Philydraceae 53, 400, 599, 682, 683
Philydrales 682
Philydrum 682
P. lanuginosum 601
Phleum 462
P. pratense 79, 80, 462
Phlomis viscosa 539, 541
Phoenix 252, 403
P. dactylifera 1, 613
Pholisma 426
Phormiaceae 682
Phormieae 682
Phormium 682
Phthirusa pyrifolia 18
Phyllanthoideae 584
Phytelephas macrocarpa 613
Phyllis 151
Physochlaina fracalta 764
Phytolacca acinosa 537
Phytolaccaceae 132, 144, 681
Pinaceae 712–714
Pinus 698, 699, 709, 710, 712, 713, 723
P. banksiana 82, 85
P. pinea 545
Piper 140, 427
P. nigrum 427
Piperaceae 54, 123, 124, 131, 140, 385, 391, 427, 574, 584, 606, 704, 721
Piperales 727
Piptadenia 252
Pistia 596
Pisum 60, 64, 66, 88, 94, 100, 102, 165, 411, 418, 421, 611, 619, 627, 637, 786
P. arvense 544, 620, 623, 624
P. sativum 28, 64, 92, 94, 95, 349, 545, 556, 559, 570, 581, 611, 618, 620, 621, 623, 625
Pittosporaceae 147
Planera aquatica 3
Plantaginaceae 85, 143, 539, 598
Plantago 174
P. atrata 539
P. lanceolata 338, 462, 559
Platanthera chlorantha 462
Platystemma violoides 330
Pleuricospora 427
Ploiarium 671, 672
P. alternifolium 672
Plumbagella 717, 724
Plumbaginaceae 146, 480, 500, 659, 661
Plumbaginales 394
Plumbago 30, 174, 181, 183, 185, 199, 209, 283, 294, 297, 381, 659, 715, 716, 723, 724, 730, 731
P. zeylanica 44, 203, 295, 296, 303, 381
Poa 133, 489, 491, 502–504, 507, 508
P. alpina 478
P. annua 64, 79, 243
P. arctica 489
P. granitica 484
P. nervosa 478, 489
P. pratensis 484, 496, 506, 509, 730
Poaceae 140, 171, 390, 401, 403, 715–717
Podolepis jaceoides 67
Podophyllum peltatum 534
Podostemaceae 28, 47, 319, 417, 651–655, 704, 721, 725, 727
Podostemales 652
Podostemonaceae 127, 138, 584, 598, 607
Pogonia japonica 391
Polemoniaceae 143, 391
Polemoniales 394
Polemonium 391
Polyanthes tuberosa 742, 776
Polycarpicae 238
Polygonaceae 85, 129, 146, 147, 390
Polygonum 707, 715–718, 722, 724
P. divaricatum 1
P. persicaria 394
Polygala 592
P. vulgaris 139, 140, 583
Polygalaceae 572, 574, 578, 584, 589, 595, 603
Polypleurum 652, 653
Polypodiaceae 698
Polytrichum 699
Pomatoes 788
Poncirus trifoliata 449, 451, 452, 575, 761
Pontederiaceae 400, 599, 682, 683
Poplar (see *Populus*)
Populus 104, 140, 211, 240, 243, 248, 595
P. alba 224
P. deltoides 208, 215
Portulaca oleracea 1
Portulacaceae 391, 655–657, 684
Posidonia 222
P. oceaniana 222
Posidoniaceae 222
Potamogeton 274, 548, 671
P. densus 540, 546, 547
P. foliosus 671
P. nodosus 341, 359
P. pectinatus 224
Potamogetonaceae 224, 339, 400, 540, 546, 669, 671
Potato (see *Solanum tuberosum*)
Potentilla 485, 486, 491, 496, 502, 503, 505–507

P. argentea 497, 500, 503, 508
P. collina 480
P. heptaphylla 138
P. tabernaemontani 489, 491, 493
P. verna 486
Prespermaphytes 698, 699, 702–704, 706, 709, 711, 712, 726
Primula 250, 698–700, 718, 719
P. auricula 457
P. obconica 251
P. vulgaris 251
Primulaceae 142
Primulales 394
Prismatocarpus 339
Pritchardia 403
Proboscidea 174
Prosopis 252
Proteaceae 218, 322, 585, 594
Protocaryotes 697
Prunus 23, 210, 278, 722
P. amygdalus 776
P. avium 204, 251
P. persica 773
P. tomentosa 773
Pseudotsuga 700
Pseudowintera 107
P. axillaris 107
Psittacanthus 335
P. cuneifolius 19
P. schiedeanus 19
Psilophytes 727
Ptelea trifoliata 586
Pteridophytes 698, 699, 700, 702, 705, 712, 727
Pteridosperms 728
Pterospermum heyneanum 23
Pterygota alba 23
Pulmonaria mollissima 735
Punicaceae 590
Purskia 589
Putoria calabrica 25
Putranjiva 767, 768, 769
P. roxburghii 767
Pyrethrum cinerariaefolium 533, 537
P. parthenifolium 533, 537
Pyrolaceae 391, 427, 595
Pyrostegia 61
Pyrus 287
P. malus 542, 559

Quercus 168, 174, 176, 178, 298, 308, 381, 407, 410
Q. gambelii 325, 351, 365, 367, 407
Quinchamalium 171
Q. chilense 24, 26, 27

Rafflesia 427
Rafflesiaceae 140, 427, 584
Ragweed (see *Ambrosia*)
Ranales 725
Ranalian taxa 702
Ranunculaceae 57, 127, 139, 140, 149, 161, 242, 390, 395, 427, 526, 535, 538, 616
Ranunculus 427, 777, 784
R. acris 306
R. argoviensis 501
R. auricomus 480, 482, 484, 485, 489–491, 495, 500–503, 505–508
R. cassubicifolius 485, 500
R. cassubicus 484, 491, 495
R. ficaria 534
R. sceleratus 324, 363, 534, 741, 776, 778, 780, 781
Rapanea 245
Raphanobrassica 499
Raphanus 74, 275, 285, 287
Raspberry (see *Rubus idaeus*)
Rauwolfia serpentina 776
Reseda 718
R. odorata 753
Resedaceae 23, 127, 581, 585, 664, 666
Restionaceae 53
Restionales 666
Rhamnaceae 23, 129, 582
Rhinanthoideae 685
Rhizophora 730
Rhizophoraceae 54
Rhododendroideae 669, 670
Rhododendron 210
R. schlippenbachii 221
Rhoeo discolor 529
R. spathacea 72
Ribes nigrum 773
R. rubrum 757
R. uva-crispa 773
Ribesiaceae 590
Rice (see *Oryza sativa*)
Ricinus 23, 211, 574, 620, 629, 766–769
R. communis 215, 626, 629, 769
Rinanthus alectorolophus 526, 527, 540, 541
R. minor 526, 527
Rivina humilis 537
Rosa 133, 704
Rosaceae 23, 127–129, 137, 139, 140, 144, 148, 390, 486, 489, 508, 578, 582, 585, 589
Rosales 725
Rosularia 24
R. pallida 24, 138
Rubia 151
Rubiaceae 25, 61, 128, 137, 151, 391, 417, 585, 594, 605, 628, 681, 705, 707
Rubiales 394, 681
Rubioideae 681
Rubus 133, 486, 491, 498, 502, 503, 507, 706, 707
R. caesius 508
R. idaeus 510
R. nitidioides 498
R. saxatilis 497
Rudbeckia 430, 480, 492
R. bicolor 430
R. laciniata 430, 431, 433, 529
R. sullivantii 431
Ruellia 557
R. tuberosa 330
Ruppia 391
R. maritima 400
Ruppiaceae 391
Ruta 390, 414
Rutaceae 127, 140, 147, 390, 448, 582, 585, 595, 736
Rutales 581
Rye (see *Secale cereale*)
Rye grass (see *Lolium perenne*)

Saccharum 480, 509
S. officinarum 747, 767
Sagittaria graminea 463
Saintpaulia 206
Salicaceae 140, 224, 238, 595
Salix 104, 595, 788
S. babylonica 742
S. capraea 735, 742
S. tetrasperma 742
Salpiglossis sinuata 783
Salvadora 662, 664
S. persica 662
Salvadoraceae 662, 664
Salvia 233
S. coccinea 540, 541
S. mellifera 61
S. pratensis 540, 541
S. splendens 540, 541
Sambucus ebulus 526
S. nigra 525–527
S. racemosa 525–527

Sandalwood (see *Santalum album*)
Sansevieria 107
Santalaceae 14, 24, 27, 151, 171, 331, 540, 585
Santalales 152
Santalum 331, 766, 769
S. album 152, 767, 768, 773, 775, 776
Sapindaceae 148, 574, 575, 581, 582, 590, 672, 673
Sapindales 662
Sarcandra glabra 426
S. irvingbaileyi 426
Sarcodes 427
Saururaceae 427, 584
Saururus 427
Saxegothaea 703, 704
Saxifragaceae 140, 596, 652
Saxifragales 139
Scabiosa 385, 391, 392
S. kladnikiana 525, 526
Scaphium 594
Sciadopitys 722, 723
Scilla 190, 592
S. biflora 534
S. bifolia 534, 535
S. nonscripta 613
S. sibirica 534
Schisandra 107
Schisandraceae 57, 58
Schizophytes 697–699
Schoenoplectus tabernaemontanum 524
Scirpus 679
Scitamineae 123, 219, 235, 238
Scleria foliosa 321
Scrophularia 141, 335, 337, 343
S. himalensis 338
S. marylandica 335, 337
Scrophulariaceae 61, 85, 123, 127, 142, 143, 274, 330, 335, 390, 417, 525, 526, 538–540, 542, 603, 684, 685
Scrophulariales 395
Scurrula 19
S. atropurpurea 19
S. montana 15
S. pulverulenta 767
Sea Island Cotton (see *Gossypium barbadense*)
Secale 200, 210, 258, 275, 283, 630, 702, 764, 771, 772
S. cereale 208, 247, 509, 629, 632, 635, 746, 749, 752, 764, 771
Sechium edule 27, 764
Sedum acre 654, 655
S. chrysanthum 24, 137
S. fabaria 655
S. populifolium 655
S. sempervivoides 24
S. spectable 457, 458
S. ternatum 28
Selaginella 698, 703, 706, 707, 709–711
Sempervivum alpinus 654
Sequoia 43
Sericocalyx scaber 458
Sesbania aculeata 462
Setaria 484, 630
S. lutescens 361, 629, 630
S. macrostachya 502
Sicyos angulata 524, 540
Silene 60, 64, 233, 275, 284
S. pendula 79
S. nemoralis 526
S. vulgaris 224
Simaroubaceae 129, 144, 578
Sinapis 618, 620, 621
S. alba 621, 776
Sium suave 776
Smilacina stellata 537
Soja hispida 79, 80, 82
Solanaceae 77, 127, 143, 274, 391, 525, 526, 599, 702, 754
Solanum dulcamara 526, 784
S. jasminioides 526
S. khasianum 773
S. melongena 772, 773, 776, 784
S. nigrum 526
S. sisymbrifolium 773
S. stenotomum 786
S. tuberosum 462, 470, 510, 526–528, 737, 745, 746, 749, 752, 754, 784, 786, 788
Solidago canadensis 33, 446
S. virgaurea 537
Sophora flavescens 546, 550
Sorbus 486
Sorghum 66, 73, 74, 110, 223, 274, 484, 510, 632
S. bicolor 64, 79, 80, 776
S. vulgare 74, 110
Soybean (see *Glycine max*)
Sparganiaceae 400
Spergularia media 593
Spinacia 30, 169, 172–176, 178, 179, 181, 182, 185–189
S. oleracea 69, 130, 131, 276, 291, 298, 307, 345, 346, 526, 527
Spinach (see *Spinacia*)
Spiranthes 221
S. cernua 454
Spironema fragrans 519, 529
Stachyuraceae 657, 659
Stachyurus 657, 659
S. chinensis 457, 458
Stanfieldiella imperforata 600, 601
Staphyleaceae 149
Statice oleaefolia 480, 500
Stellaria 167, 168, 400, 410, 413, 418, 422
S. media 351, 353, 410, 412, 413
Sterculia colorata 23
Sterculiaceae 23, 55, 127, 148, 581, 594
Stipa 165, 174, 188, 190, 403
S. elmeri 178, 183
Stratiotes aloides 598
Strawberry (see *Fragaria*)
Striga 684, 685
S. densiflora 685
S. euphrasioides 684
S. orobanchoides 330, 684
Stripe Mosaic Virus 54
Struthanthus 19
S. vulgaris 15, 18, 19, 21, 22
Styracaceae 147
Styrax obassia 225
Sugar beet (see *Beta vulgaris*)
Sugarcane (see *Saccharum officinarum*)
Sunflower (see *Helianthus annuus*)
Sutera 343
Sweet gum (see *Liquidambar styraciflua*)
Symplocaceae 133
Syzygium caryophyllifolium 454
S. cumini 454

Tamaricaceae 595, 659, 661
Tamaricales 659
Tamarix 648, 661, 686
T. erucoides 368, 459
Tambourissa 37, 39, 43, 719
T. religiosa 37, 38, 41
Tapeinochilos 219, 235, 237
Tapinanthus 19
T. rubromarginatus 18
T. uhehensis 16, 18
Tapinostemma 19
T. acaciae 15, 18, 335
T. lagenifer 18

Tapinostemmae 16
Taraxacum 479, 491, 493, 498, 500–502, 505, 506
T. kok-saghyz 508
T. megalorrhizon 479
T. scanicum 488
T. sect. Palustria 495, 496
T. vulgare 496
Taxillus 19, 766–769
T. cuneatus 767
T. vestitus 767
Taxodiaceae 712
Taxus 698–701, 707, 709, 710, 712–718, 720, 722, 723, 725, 728, 729
Tecoma 61
Terniola 652
Tetranema mexicanum 335
Tetragonia 667
Tetragoniaceae 667
Tetragonioideae 667
Tetrastylidium janeirense 152
Teucrium 233
Thalassia 222, 223
T. hemprichii 224
Thalictrum 425
T. javanicum 329
T. urbaini 776
Thea 728
T. sinensis 764, 771
Theaceae 148, 582, 594, 657, 659, 671, 672
Theales 659, 661
Theligonaceae 681
Theligoneae 681
Theligonum 681, 682
Thelycrania 686–688
Themeda triandra 484, 485, 496
Theobroma cacao 776
Thesium 123, 152
T. alpinum 540, 541
T. limophyllum 540
T. wightianum 152
Thismia 428
T. clandestina 428
Thuja 708
Thunbergia alata 330
Thymelaeaceae 23, 55, 148, 538, 585
Tilia 414
Tiliaceae 582, 585
Tillandsia recurvata 593
Tillandsioideae 595
Tobacco (see *Nicotiana tabacum*)
Toddalia asiatica 448, 580
Tolypanthus 19
T. involucratus 15, 18
Tomato (see *Lycopersicum esculentum*)
Topatoes 786
Torenia 44, 174, 176, 298
T. fournieri 45, 46, 292, 762
Torreya 712, 725
Tozzia alpina 526, 527, 540, 542
Trachyspermum 754
T. ammi 329, 754
Tradescantia 1, 69, 74, 103, 200, 233, 359, 360, 410, 540
T. bracteata 62
T. fluminensis 529
T. reflexa 225
T. virginiana 200
T. virginica 735
Trapa natans 530, 546, 550
Trapaceae 129, 319, 417, 607
Trewia polycarpa 23
Tribulocarpus 667
Trichilia grandifolia 572, 578
Trichosanthes 761
T. anguina 27, 322
Tridax 707
Trifolium 276, 772
T. ambiguum 772
T. hybridum 772
T. repens 137, 628, 772, 783
Trigonella 628
T. foenum-graecum 392, 628, 629
Trigonocarpales 148
Trillium 89, 99, 425
T. grandiflorum 537
T. sessile 39
T. undulatum 338, 454
Tripogandra pflanzii 580
Tripsacum 772
T. dactyloides 491, 510
Triticale 55, 58, 75, 104, 365, 632, 771
Triticum 72, 74, 75, 98, 166, 283, 401, 403, 411, 463, 613, 630, 632, 635, 636, 752, 764, 766, 771, 772
T. aestivum 200, 245, 351, 352, 355, 356, 454, 462, 509, 510, 535, 567, 772
T. dicoccum 462
T. durum 534, 545, 764
T. monococcum 462
T. turgidum 764
T. vulgare 462, 764
T. zhukovsky 462
Trochodendraceae 57
Trochodendron aralioides 58
Tropaeolaceae 417
Tropaeolum 418, 421, 606, 661
T. majus 413, 421, 423, 531, 546, 550, 661
Trumpet lily (see *Datura arborea*)
Tuberose (see *Polyanthes tuberosa*)
Tulbaghia violacea 92
Tulipa 98, 248, 274, 594
T. gesneriana 456
Tunica saxifraga 546
Tupeia 19
T. antarctica 16
Turk's cap (see *Malvaviscus arboreus*)
Turnera 592
T. ulmifolia 591
Turneraceae 599
Turtle grass (see *Thalassia*)
Tylophora indica 776, 781
Typha 209
Typhaceae 209, 581, 599

Ulmaceae 57
Ulmus 648, 686
U. campestris 33, 446
U. glabra 33
Umbelliferae 127, 128, 575, 585, 616
Umbelliflorae 684
Urochloa 484
Ursinia 171
U. anthemoides 24
Urtica 698, 699
Urticaceae 146, 390, 574, 681
Utricularia 58, 426
U. coerulea 387
Uvularia grandiflora 537
U. sessilifolia 225

Vacciniaceae 61, 667, 669
Vaccinioideae 669, 670
Vaccinium 3, 667, 669
V. ashei 669
V. atrococcum 669
V. constablaei 669
V. macrocarpum 30, 337, 338
V. nummularia 668, 669
V. retusum 668, 669
V. serratum 668, 669

Vaillantia 151
Valencia organe (see *Citrus sinensis*)
Valeriaceae 526
Valeriana officinalis 525, 526
Vanda 410, 413, 425
V. sanderiana 543, 544
Ventilago 23
Vepris elliotti 586
Verbascum 343
V. densiflorum 601
Verbena 527
Verbenaceae 673, 675
Veronica 390
V. chamoedrys 337
Viburnum 616
Vicia 29, 239, 242, 287, 411
V. faba 245, 524, 525, 545, 559, 613, 619–621, 623, 624, 746, 758, 760
Victoria 596
Vigna 620
V. capensis 546
V. unguiculata 619
Villarsia 595
Vinca rosea (see Catharanthus roseus)
Vincetoxicum nigrum 456
V. officinale 456
Viola 414, 592, 718, 719
V. odorata 591
Violaceae 127, 149, 584, 585
Violales 657, 659
Viscaceae 54, 152
Vitaceae 129, 141, 148, 149, 572, 574, 578, 581, 582, 584, 590, 603
Vitis 766
V. vinifera 466, 758, 760
Vochysia 595
Vochysiaceae 128
Voyria 426
V. spruciana 597
Voyriella 426

Walnut (see *Juglans regia*)
Welwitschia 706, 712, 723
W. mirabilis 631
Wheat (see *Triticum aestivum*)
Widdringtonia 712
Wilkiea 719
W. huegeliana 38
W. macrophylla 38
Winteraceae 57, 58, 127, 209, 216, 245, 579
Wolffia 58

Xyridaceae 428, 683
Xyridales 683
Xyris 428

Yucca 613, 620, 631, 637

Zannichellia 222, 669, 671
Z. aschersoniana 669
Z. palustris 669, 671
Zannichelliaceae 391, 669, 671
Zanthoxylum simulans 577, 578
Zea 74, 110, 166, 174, 185, 188, 200, 233, 252, 274, 290, 298, 346, 364, 381, 400, 411
Z. mays 55, 73, 79, 100, 106, 379, 402, 510, 526, 527, 529, 540–542, 545, 559, 567, 605, 625, 630, 632, 637, 724, 728, 765, 766, 768, 769, 772, 783
Zephyranthes 167–169, 180, 190, 355, 430, 753, 755
Z. ajax 355
Z. grandiflora 355
Z. lancasteri 355, 359, 540
Z. macrosiphon 430
Z. texana 478, 489, 492
Zeuxine longilabris 58
Z. sulcata 454
Zeylanidium 652, 653
Zingiberaceae 60, 127, 129, 131, 147–149, 166, 238, 599
Zingiberales 235, 581, 584, 606
Zostera 222, 223, 671, 710
Zosteraceae 222, 710
Zygogynum 107
Zygophyllaceae 142, 584
Zygophyllum fabago 458

Subject Index

Acetolysis 206, 242
Acid phosphatase 635
Acytokinetic mitosis 520, 525, 526, 535, 536
Adventitious embryony 475
Agamic complex 501–503
Agamospecies 501, 502
Agamospermy 475
Aleurone cell 633–636
  grains 363, 606, 628, 629, 632–634, 637
  layer 363, 629, 632, 633, 635, 636
    (see also protein bodies)
Alternation of generations (Life cycle) 475, 504
Alveoli (in endosperm) 353
Amphimixis 475
α-Amylase 625, 633, 634, 636
β-Amylase 636
Amylopectin 636
Amylose 636
Androgenesis 476, 493
Androgenetic embryos 750
Androgenetic haploids 746
Aneuspory (see Taraxacum type, Ixeris type of diplospory)
Angiocarpy 39
  hyperstigma 37, 38
  stigmatic drop 39–41
Antennaria type (see diplospory)
Anther 53, 523, 650, 656, 664, 666, 668–670, 672
  bisporangiate 53, 666
  connective 53, 523
  microsporangium 53, 669, 675, 679
  multisporangiate 54
  organogenesis 55
  specific cells 55
  tetrasporangiate 54
Anther culture 736
  activated charcoal 746, 770
  age of pollen 746
  albinos 752
  cold treatment 750
  donor plants 746
  float cultures 747
  genotype 746
  liquid medium 744
  microspore 746
  physical factors 748
  potato extract 746
  preculture 749
  procedure 743
Anther wall 55, 523
  basic type 55
  dicot type 55
  endothecium 58, 650, 656, 669–671, 673
    bilayered 58
    α-cellulose 58
    fibrous thickenings; annular, bulbous 58
    multilayered 58
    pectin and lignin 60
    reticulate 58
  epidermis 58
    exothecium 58
    fibrous bands 58
  middle layer(s) 60
    ephemeral 60
    persist until dehiscence 60
  monocot type 55
  reduced type 55
Anthesis 742
  selection of embryogenic pollen grains 750
  somatic chromosome elimination 751
  ultrastructure of embryogenic pollen 748
Apical meristem (culture) 735
Apogamogony (see apomixis, gametophytic)
Apogamy (= apogamety) 477, 487
Apomeiosis 475, 478, 486
Apomicts, diploid 497, 500, 501, 756
Apomixis 445, 475, 651
  consequences 497–503
  heterozygosity 497, 498, 504, 506, 510
  recurrent/non-recurrent 476
  terminology 475–477
  variability 498
Apomixis, gametophytic 475–518
  autonomous 476, 487–489, 491, 495
  (and) breeding 508–510
  causes 498, 499
  components 476, 498, 499, 504, 508, 509
  degree of 494–497, 504, 505
  embryology 477–493
  environmental influences 495, 496, 507
  facultative (partial) 476, 495
  genetics 503–508
  male gametes 489, 490, 492–494
  male parent 490, 491, 495, 505, 509

Apomixis, meiosis 481, 482, 485, 493, 494, 499
obligate (total) 476, 495, 498, 505, 508, 509
polyploidy 499–503
pseudogamous 476, 488–491, 493, 494
sexual potentials 476, 482, 485, 494–497, 501, 503, 505
Apospory 475, 477, 482–486, 496, 499, 506, 507
degree 494–496
generative (see diplospory)
gonial = Antennaria type of diplospory
Hieracium type (bipolar) 482–484, 532
Panicum type (monopolar, four-nucleate) 484, 487, 490, 532
Arabinoxylan 613, 635, 636
Archesporium (male) 55, 650, 664, 675
multicelled 55
regenerated initials 55
single-celled 55
sterilization 55
Artificial induction (of polyembryony) 466, 467
(by) chemicals 468, 469
hormones 467
Automixis 498
Autoradiography 560
Autosegregation 498
Autotriploids 447
Autotrophs 426–428, 767
Auxin 555, 556, 635

B_{II}, B_{III} hybrids 476, 477
Breeding gametophytic apomicts 508–510
Buds (floral, in vivo differentiation) 739
bud primordia (culture) 741
female catkins 742
male catkins 742
(development of) pre-dormant and post-dormant buds 742

Calcium oxalate 613, 626, 628
Callus tissues 742
Cambial tissues (in cultures) 735
Carbohydrate (in seeds) 613, 626–628, 632
Carpel primordia (in cultures) 741
Cell destination (during embryogeny) 397
lineages 378, 397
Cell division 625, 630
elongation 625, 630
wall 626–628, 630, 635, 636
middle lamella 630, 636
Cell lines resistant to
aminoprotein 752
5-bromodeoxyuridine 752
streptomycin 752
Cellular arrangement (in seeds) 587
Chalazosperm 606
Chilling 746
Chimeral embryoids 748
embryos 430, 431
Chimeral plantlets 748
plants 432
tissues 430
Chromosomes
polyteny 523
stickiness 520, 525
Chromosomes, types (see separate entries)
Cleistogamous flower 742
Coconut milk 323, 755, 775
Colchicine 751
Coleoptile 629
Coleorhiza 629, 630
Cotyledon(s) 390, 392, 394, 398–401, 404, 409, 411, 422, 423, 425, 544, 545, 611, 618–624, 631
Cotyledonary locus (loci) 397, 398, 431
primordium 405
region 390, 430
Cotyledonary tissue 770
Cotyledonoids 426
Crown gall tissue 738
Cryopreservation 779
Cytochemistry 559, 560
Cytokinin 556, 635

Degenerate endosperm cell 336
Development (in cultures)
(of) embryo 736, 755
endosperm 736, 755
Dictyosomes 616, 621, 625, 629, 630, 635
Differentiation (in cultures)
embryoids 741
epidermal cells 739
female buds 742
somatic embryos 757
Dihaploid 476, 501–503
Diplochromosomes 526, 528, 529
Diplospory 475, 478–482, 486, 497, 507
Allium nutans type 480
Antennaria type 475, 478, 481, 482, 493, 499, 507, 531
degree of 494–496
Ixeris type 475, 480, 481, 499, 500, 531
mitotic = Antennaria type
restitutional = Taraxacum type/Ixeris type
Taraxacum type 475, 479–482, 493, 499, 500, 531
types 475
DNA 324–326, 338, 341, 359, 365, 628, 638
amplification 549, 550, 553–555
cytophotometry 522, 534, 544, 549
hybridization of DNA with 552, 553, 560
hybridization of RNA with 552, 553, 556, 560

nuclear content 556
puffs 553–555
satellite 554
underreplication 553

Embryo 612, 616–618, 625, 629–631, 675
antipodal 32, 33
coenocytic 42
development (see embryogenesis)
differentiation 397, 398
dicot embryo 398–400, 616–624
embryonal primordia 424
endospermal embryo 33–37
epicotyl 425, 545
epicotylary locus 397, 431
region 390, 403
epiphysis 387, 390, 391, 398
epiphysis-like cell 404
foliaceous cotyledons 672
green embryo 410
haustorium (cotyledon) 631
heart-shaped embryo 387, 411, 412, 416, 419, 422, 423
histogenesis 377, 397
internal cells (in proembryo) 395, 396
hypocotyl 394, 409, 425
hypocotyledonary region 390, 391, 395
hypophysis 387, 397, 407–409, 411
mature embryo 387, 401
dermatogen 679
periblem 679
plerome 679
monocot embryo 398–400, 629–631
grass embryo 400–403
palm embryo 403–405
organless embryo 425–429
after-ripening 425
intraseminal growth 425, 426, 467, 616–618
polycotyledonous embryo 19, 452, 454
precocious embryony 491, 492, 494, 496
pseudomonocotyledonous embryo 19
reduced embryo 425
rudimentary 616

Embryo culture 769
activated charcoal 770
callus 770
differentiation 770
embryogenesis 737
coconut milk 777
growth regulators 777
loss of embryogenic potency 778
nitrogen compounds 777
ontogeny 779
protein changes 781
RNA synthesis 781
suspensor-like structures 781
hybrid embryo 771
mature embryo 770
orchid embryo 773

Embryo sac (Female gametophyte, Megagametogenesis) 126, 135, 377, 531, 543
aberrations 161
(in) anther 5, 6–9, 494
distribution of cytoplasm 166
haustoria 24, 25
quantitative analysis 687
vacuolation 163, 166
Antipodal cells 188, 487, 520, 534, 535, 651, 656, 657, 668–671, 673, 675, 684
haustoria 24–26
histochemistry 188, 190
persistence 188
proliferation 188
syncytium formation 190
ultrastructure 188, 190
wall projections (transfer cells) 188
Cellularization 170, 171
Coenocytic stage 165–167
Central cell 185–188
cell wall 186
histochemistry 186, 188
organization 186
polar nuclei 656
ultrastructure 186, 188
wall ingrowths (transfer cells) 186
Egg (cell) 183–185, 381–383, 429–431, 433, 531, 561
cell wall 183
filiform apparatus (in *Plumbago*) 183
genetic influence 163, 164
germinal vesicle 1, 377
histochemistry 185
neoteny 649
polarity 162, 163, 171, 183
relation to nucellar tissue 168–170
seasonal influence 163
supernumerary 487
ultrastructure 185
Synergid 174–183, 415, 532, 651, 656, 673, 684
cell wall 174
chemotropic activity 178, 181
degeneration 178, 179
embryo 487
enzyme production 181
filiform apparatus 174, 175, 180
haustoria 24, 26, 180
histochemistry 174, 180
pollen tube penetration 178, 181, 183
role in apomicts 488, 492
wall ingrowths (transfer cells) 174, 175,180

Embryo sac
Typology (embryo sacs) 159–161
Aposporous 446, 447, 482–484
Allium nutans type (see diplospory)
Allium type 480
Antennaria type (see diplospory)
competition 485, 494, 496, 505, 507
diplosporous 478–480
Drusa type 480
Fritillaria type 480
Hieracium type (see apospory)
Ixeris type (see diplospory)
monosporic 480
Oenothera type 484, 533, 537
Panicum type (see apospory)
Polygonum type 480, 490, 533, 537
Taraxacum type (see diplospory)
tetrasporic 480
unreduced (mature) 475, 477–490, 499, 507, 531
Bisporic 46, 132, 133, 159–161, 163, 650, 655, 670, 671, 673, 675
Allium type 648, 650, 655, 656, 659, 671, 672, 677
Apinagia type 652, 653
Endymion type 648, 659, 671, 673, 677
Podostemon type 653
Polypleurum type 653
Scilla type 533, 537, 655
Monosporic 46, 159–163, 650, 659, 661, 671, 675
Oenothera type 648
Polygonum type 646, 647–649, 656, 657, 659, 666, 669, 670–673, 677, 678, 680, 681
Tetrasporic 132, 133, 159–161, 163, 650, 661
Adoxa type 12, 46, 533, 537, 648, 650, 656
Chrysanthemum cinerariaefolium type 648
Drusa type 46, 648
Fritillaria type 12, 13, 46, 533, 537, 648
Penaea type 46, 533, 537, 648
Peperomia type 648
Plumbagella type 46, 537
Plumbago type 46, 533, 537, 648
Embryo-endosperm relationship 364
Embryogenesis 377, 378, 384, 404, 424, 430, 431
cell patterns 384
cytochemical aspects 405, 406, 408, 410–414, 434
deviation from usual pattern 424
ultrastructural aspects 405, 406–410, 425
Embryogenic classification 388, 393
(by) Johansen 388–392
others 393–397
Schnarf 388–398
Souèges 391–393
Embryogenic law(s) 393
law of correspondence of forms 392
Embryogenic types 378, 385, 388, 390, 392
Asterad 389, 390, 394, 395, 405, 651, 662, 664, 666, 672, 678
Caryophyllad 389, 391, 394, 651, 656, 657, 661, 671, 678
Chenopodiad 389, 391, 394, 651, 656, 657, 678
Geraniad 395
Graminad 403
Irregular 394, 395
Onagrad 388–390, 394, 395, 398, 403, 407, 651, 666, 675, 678, 679, 684
Piperad 388, 389, 391, 392, 651
Polygonad 394
Solanad 389, 390, 394, 395, 656, 657, 661, 664, 669, 673, 675, 677
Embryonic number 396
formulae 390
Embryony, laws of 378
Embryos in apomicts 491–493
dependence on endosperm 491
precocious development 491, 492, 494, 496
twin 487
Endo-β-mannanase 627, 628
Endo-β-manosidase 628
Endopeptidase 624
Endoplasmic reticulum 616, 621, 624, 625, 629, 630, 635
Endoreduplication 521, 522, 524–526, 530–532, 535, 536
Endosperm 2, 13, 18, 319, 421, 429, 432, 434, 446, 537–540, 545, 548–550, 570, 607, 651, 656, 673–675, 677, 680, 684, 685
breakdown gradients 616, 617, 626, 628, 629, 633, 637, 681
Cellular type 18, 19, 330, 350, 351, 369, 370, 537, 538, 651, 656, 670, 671, 673, 675, 677
central cell 308, 345, 378
covering 349
cytoplasm 346
mitochondria 346
nucleus 349
plastids 346
polysomes 346
composite 1, 18, 19, 335
cytology 355
C-mitosis 360
restitution nucleus 360
ribosomal precursors 359
(in) dicots 611, 612, 616, 626–628

haustoria 27, 321–323, 327, 330–332, 335, 338, 414, 432, 538–542, 651, 661, 668, 669, 675, 684, 685
Helobial type 47, 332, 338, 339, 350, 351, 369, 370, 537, 538, 651, 671, 677, 678, 680, 682
incorporation of
integumentary cells 368
nucellar cells 368
(in) monocots 629, 632–637, 650
nodules 327
Nuclear type 47, 319, 350, 351, 369, 370, 520, 537, 538, 651, 656, 657, 661, 662, 664, 666, 669–673, 677, 678, 681, 683
osmotic gradient 366
osmotic pressure 366
polyploidy 360, 537–543, 545, 559
ruminate 342–344, 370
vermiform appendage 324
wall formation 322, 350, 353, 355
microtubules 352
open-ended cylinders 353
PAS-reaction 350, 355
phragmoplast 353
telophase 353
total proteins 324, 350
Endosperm, in apomicts 476, 487–491
autonomous development 476, 487–489, 491, 495
cytology 488, 489
fertilization 488, 495
pseudogamous development 476, 488–491, 493, 494
Endosperm culture 737
cellular endosperm 755
differentiation of embryos 768
differentiation of tracheids 769
embryo factor 767
endosperm extract 772
free-nuclear endosperm 755
histological studies 769
immature endosperm 766
mature endosperm 766
nutritional requirements 766
physical factors 768
starchy endosperm 766
sugary endosperm 766
triploid plantlets 768
Enyzmes, hydrolytic 624, 625, 628, 632, 633, 639

Fertilization 2, 203, 249, 250, 260, 302, 305, 306, 337, 446, 447, 651
role of pollen 252, 253
syngamy 2, 304, 377
triple fusion 2, 304
Floral bud culture 741
callus 741, 742
differentiation 742
embryoids 741
flowers 742
Floral organ culture 742, 743
petals 740
stamens 741
Funicle 569
Funicular bundle 571

Galactomannan 627, 628
Galactose 627
α-Galactosidase 628
Gametophyte in apomicts
female 477–487
male 493, 494
Gametophytic apomixis 475–518
Gaseous exchange (in cultures) 738
Germination 567, 614, 616
Gibberellin 617, 627, 629, 632, 635, 636
β-Glucanase 632
β-Glucans 613, 632, 636
Gluconeogenesis 623, 629
Glucose 627
Glutamine 635
Glyoxysomes 621, 629, 630
Grafting of style, tandem 765
Gymnosperms 631

Haploids (anther culture) 736, 748
doubled haploids 752
multihaploids 736
natural haploids 743
Hemigamy (Semigamy) 430–433, 492, 493
Heterochromatin 550
Heterosis 491, 498, 510
Hieracium type (embryo sac, see apospory)
Hilum 569
Homologies
Embryogenesis
(in) angiosperms
embryoids 702
endoscopy 727
structure 722, 726
zygote 721, 726
(in) gymnosperms 726
(in) prespermaphytes 726
Endosperm
(in) angiosperms
interpretation 724–726
paternal heredity 724, 725
ploidy 721, 724
primary endosperm nucleus 722, 724
structure 722
(in) gymnosperms 714, 725
(in) prespermaphytes 728

Homologies
  Female gametophyte
    (in) angiosperms 714–716, 718, 722
      egg cell 698, 712–714, 716–718, 720, 722, 723
      synergid 713, 714, 716, 718, 722, 723
    archegonia in cormophytes 697, 698, 711–713
    (missed) archegonium in *Nitella* 711, 712
  Fertilization
    anisogamy 698, 699
    apogamy 722, 724, 727
    cleistogamy 719
    double fertilization 697, 719–723, 726, 728, 729, 731
    isogamy 698, 699
    oogamy 698, 699
    single fertilization 719–722
    siphonogamy 709, 710
  Male gametophyte
    (in) angiosperms
      reception of pollen by female organs 709, 717, 719
      sperms 698, 710, 731
      structure 710, 711, 729
      supernumerary sperm 720, 721
    (in) other archegoniates 698, 710, 711
  Megasporangia (syn. macrosporangia)
    compared to microsporangia 708, 709
    content 706, 707, 730
    ontogeny 705
    perisperm 703, 722, 725, 728
    pollen tubes 718
    structure 703–705
  Microsporangia
    compared to megasporangia 708, 709
    number and position 698, 702, 703
    structure 700–702
  Phylogeny
    evolutionary levels 697, 698
    monophyly 697, 698
    neotenic processes 701–703, 705, 707, 708, 710, 712, 713, 717, 718, 725
    polyphyly 697, 730
  Sex separation
    dioecy 698–700
    general 698, 699
    monoecy with self-fertilization barriers 700
Homozygosity 751
Homozygous diploids 751
Hybrids 463, 464
Hydration (pollen) 211, 212, 245, 247–249, 253
Hypocotyl 544, 545
Hypocotyl tissue (culture) 776

Induction of female flowers (in cultures) 742
  male flowers 742
  mutants 752
  parthenogenesis 762
Integument(s) 138, 421, 530
  ategmic 151, 152
  bitegmic 153, 650, 652, 654, 656, 657, 659, 661, 664, 666, 672, 677, 678, 680
  endothelium (tapetum) 46, 142–144, 530, 650, 657, 673, 674, 680
  initiation 124
    dermal 153
    subdermal 153
  integumentary shifting 141
  operculum 150
  micropylar collar 150, 599
  micropyle 123, 146, 151, 650, 656, 657, 664, 666, 672, 673, 683
    endostome 125, 569
    exostome 125, 569
  unitegmic 123, 141, 152, 650, 664, 669, 670, 673

Lipid (see oleosomes) 613, 625, 629–632
Lipolysis 629

Male gametophyte (see also pollen grains) 650
β-Mannanase 628
Mannans 613, 627, 628
Mannose 627
β-Mannosidase 628
Maternal offspring 475, 476
Megasporogenesis 135, 159, 162, 650
  archesporial cells (archesporium) 128, 152, 486, 650, 652, 659, 661, 662, 664, 673, 676, 677
    parietal cell(s) 128, 650, 656, 659, 661, 664, 677, 678
    sporogenous cells 650
  megasporocyte 478, 486, 650, 652, 659, 662, 664, 673, 674
  meiosis 159
    coenomegaspore 133, 480
    dyad 650, 652, 674
  megaspore(s) 132, 135, 159, 659, 662, 668, 678
    callose 133, 135, 164
    haustoria 24, 137, 138
    tetrad 650, 668, 674, 676–678
  organization 164, 166
  polarity 137, 164
  ultrastructure 164–166
Meiosis (in microspore mother cells) 97
  (special) callose wall 100
    callose activity 102
    deposition 100
    digestion of callose 102
    functions 103
    molecular filter 103

premature dissolution 104
protects from dehydration 104
supplies carbon compounds 106
cytokinesis 106
cell plates 106
constriction furrows 106
simultaneous 106
successive 106
DNA synthesis 98
duration of meiosis 99
mitotic to meiotic change-over 97
Z-DNA 98
Meiosis in aposporous apomicts 485
in diplosporous apomicts 481, 482
restitutional 479–483, 498, 499
Mesocotyl 544, 545
Microbodies (see also glyoxysomes, peroxisomes) 616, 621, 625, 635
Microsporogenesis 1, 2, 6, 88
microgametogenesis (see pollen grains) 762
microspore 237–240, 242, 253, 254, 650
development 198, 199, 204, 230–232, 235, 242
microsporocytes 664, 668
tetrad 650, 656, 668, 673
Mitochondria 616, 621, 624, 629, 630, 635, 637
Mucilage 211, 248, 249, 252

Necrohormone 11
Němec phenomenon 109
female potency 11
male potency 11
pollen embryo sac 6–10, 109
Nucellus 123, 126, 128, 146, 147, 150, 151, 421, 530, 650, 675, 682
crassinucellate 128, 650, 656, 657, 664, 672, 677, 678, 680
epistase 126, 149, 602, 650
hyponucellus 653, 655
hypostase 16, 126, 148, 149, 602, 650, 672, 673
nucellar beak 129, 656, 673
nucellar cap 129, 682
nucellar pad 29
nucellar plasmodium 652
nucellar tracheids 23, 148
podium 149
postament 149
pseudocrassinucellate 650
tenuinucellate 128, 650, 652, 654, 664, 666, 669, 670, 672, 681
Nucellus culture 757
callus 760
cucurbitacins 761
embryoid differentiation 760
factors governing embryogenesis 760
gossypol 761
monoembryonic *Citrus* 757
nucellar polyembryony 760, 773
polyembryonic *Citrus* 760
virus-free plants 760
zygotic embryos 760, 773
Nucleus 623, 630, 638
endomitosis 480, 488, 490, 520–522, 526, 528
Nurse culture technique 738
Nutrient media 739

Oleosomes (see lipid) 618, 620, 625, 626, 628–630, 632, 635, 637
Orbicular wall (see tapetal membrane)
Orbicules (see also Ubisch bodies) 218, 239, 258
Ovary culture 752, 753
effect of growth substances 753
"hull" factor 752
polyembryony 754
role of calyx 753
unpollinated ovaries 753
Ovary-ovule complex 14–16
placenta (mamelon) 14
Ovule 530, 650, 656, 677, 678, 686, 687
aril 126, 149, 150
arilloid 126
funicle 125
initiation 124, 126, 127, 138
obturator 149, 650, 672, 677, 678
wall ingrowths (transfer cells) 149
pendulous 667, 672
primordium 124, 126, 128, 151
2-zonnate 126
3-zonnate 126
raphe 125, 126, 143
reduction 140, 141, 151
types
amphitropous 124, 126, 131, 144, 146, 147, 650, 656, 657
anatropous 124–126, 144, 146, 147, 151, 650, 652, 654, 656, 659, 664, 666, 669, 670–673, 677, 678
atropous (orthotropous) 124, 144, 146, 151, 647, 650, 664, 666, 671, 678
campylotropous 124, 144, 146, 147, 650, 656, 666, 672
circinotropous 146, 661
hemianatropous 146, 147, 650, 656, 678, 682
vasculature 147, 148
Ovule culture 754
callus 757
differentiation of somatic embryos 757
effect of growth substances 755, 756
fibre and seed development 757
role of placenta 757
unfertilized ovules 757

Ovule and embryo sac
  nutritive pathway 169, 170

Palynology 685
Panicum type (see apospory)
Parasitic angiosperms 331, 429
  leafless 426
  total 426
  total, root 426, 427
Parthenocarpic fruits 467
Parthenogenesis 447, 464, 466, 475, 492, 495, 499, 505, 507, 508
  diploid = unreduced 476, 487
  haploid = reduced 476, 500, 501
Pentosans 632, 635, 636
Perisperm 612, 613, 626, 632, 637, 656, 661, 681
Perisperm culture 756
Peroxisomes 621
Phloem explants 775
Phytin 613, 632
Pistil culture 762
Placenta 421
Placental bundle 571
Plastids 616, 621, 624, 625, 629
Ploidy determination 522
Polar nuclei, fertilization in apomicts 488, 490
  fusion in apomicts 488–490
Pollen
  cytology 199–204
  development 225–232
  dispersal, adaptations for 210–225
  mutants 252, 261
  proteins and glycoproteins 206, 208, 217, 223, 240, 242–244, 247, 250–260
  quality 245–247, 261
    testing for 246
  shape 199, 217, 219, 222, 223, 253
  sterility 231, 245
  storage 246, 261
  structure and cytochemistry 198–225
Pollen allergens 245–258
Pollen apertures 199, 204, 205, 211, 219, 225, 230, 242, 243, 248, 249, 255, 256, 258
  colpate 237, 659, 661, 675, 677–679
  colporate 211
  formation 237
  furrow 237
  germinal apertures 199, 204, 237
  inaperturate grains 237, 238, 248
  links with intine 242
  omniaperturate grains 216, 237
  operculum 204, 237
  polyporate 211
  porate 211, 611
  pore 237
  sulcate 237
  triaperturate 211
  triporate 237
  types and arrangement of apertures 237, 238
Pollen-coat (surface coating) 217–219, 225, 240, 250, 259, 260
Pollen colour 218, 219
Pollen culture (see also anther culture) 735, 736, 749
  bulbosum technique 751
  chromosome elimination 751
  dimorphism 750
  diploidization 751
  embryogenic pollen 749, 750
  embryoids 749
  endoduplication 751
  gametophytic pollen 750
  nullisomics 736
  physiology 762
  plantlets 748
  triploids 751
Pollen grains(s) 1, 5, 529, 650, 656, 661, 662, 664, 666, 669–671, 673, 675, 679–681, 683
  (in) apomicts 488, 489, 493, 494
  binucleate, trinucleate (2-celled, 3-celled) 274
    generative cell (wall) 199, 200, 209, 229, 381
      sperm cells 199–203, 209, 210, 230, 260
    vegetative cell 197, 199, 229, 230
      cytoplasm 199
      hydrated cell 230
      nucleus 209
      ultrastructure 199
  callose 199, 200, 207, 209, 210, 225, 230, 232, 233, 242, 244, 245, 249, 251
  composite (or compound) pollen grains 206, 208, 218, 220, 221, 252, 253
    cohesion 209, 244, 245
    development 198
    massulae 208, 218, 221
    pollinia 198, 208, 218, 221, 252, 260
    viscin threads 218, 220, 221
  exine 204, 211, 216, 224, 225, 230, 231, 250, 251, 253–255,. 257–260
    bridges 220, 221, 244
    chemistry 204
    cytochemical tests 206, 239
    deposition 239
    differentiation and maturation 238–242
    formation 235
    genetic control 210, 245
    layers (exine)
      baculum 204, 238, 661
      cavea 204
      columella 204
      ectexine 204
      endexine (see nexine-2)
      foot layer (see nexine)

nexine 204, 216, 238, 239, 243
nexine-2 204, 216, 221, 238, 239
pili 204
sexine 204, 239
tectate 219
tectum 204, 233, 238
germination 208, 212–215, 245, 247–250, 253, 259, 279, 280, 285
hydration 280
sticking 279, 280
incompatibility
gametophytic 286
interspecific 210
self 250–252
sporophytic 286
intine 209, 219, 221, 223, 224, 236, 239, 244, 245, 247, 248, 250, 254–257, 259, 260
synthesis and deposition 242, 243
pilate 204
proteins 217, 251, 253, 254, 259
reduction 216, 219, 223, 237
sporophytic control 210
surface 240
Pollenkitt and tryphine 85
distinction between 88
lipid and carotenoids 85
role of 85
Pollen-pistil interactions 240, 245–253, 255, 260, 284
activation 285
permeability 284, 288
plasma membrane 284
plasma streaming 284
recognition 284–287
acceptance 284, 285
rejection 284, 285, 288
signal 288
ultrastructure 281, 282, 287
Pollen tube 198, 208–210, 237, 245, 246, 249–252, 259–261, 281, 377, 382, 432
callose 383
chemotropic attraction 290, 291, 292, 294
development 247–249
directed growth 285, 289
in stigmatic tissue 285
entry in embryo sac 296, 297
entry in synergid
growth through filiform apparatus 297, 299
histochemistry 298
degeneration of one synergid 298
composition of pollen tube contents 303
discharge of pollen tube contents 301
termination of pollen tube growth 235
transport 302
ultrastructure 281, 282, 287, 306
x-bodies 302
persistent tubes 30–32
Pollen vectors, relationships with air current 219–222, 254–256, 259
animals 210, 216–219, 225, 259
water 210, 222–225
Pollen wall cytochemistry 206–208, 217, 223, 253–257
development (see also primexine) 199, 234, 236, 253
structure 198, 204, 210–216, 219–225, 259, 260
terminology 204–206, 211
Pollination 736
intraovarian 761
in vitro 762
placental 737, 762
Pollination, diagnostic landmarks 249–252
Pollinia culture 762
Polyad 198, 208, 218, 240, 244
Polyembryony 447, 647, 651, 657, 666, 673, 680
development 457
autonomous 456–458
induced 468, 469
sporadic 457, 458
gametophytic 446
antipodal 446
extrasaccal 446
intrasaccal 446
multiple 446
genetic marker 449
isoenzyme patterns 451, 452
esterase 451
peroxidase 41
sporophytic 446–454
adventive 446–454
endothelial 446, 455, 456
integumentary 446, 447, 454, 455
nucellar 447–454, 466
suspensor 457, 458, 466–469
segmentation and proliferations 464, 466–469
synergid 32, 446, 458–461
reduced (haploid) 446, 447, 459, 460, 465
unreduced (diploid) 446, 447, 459, 460
triplets
diploid-diploid-diploid 461, 463, 464
diploid-triploid-diploid 463
haploid-diploid-haploid 463
twins
diploid-diploid 447, 458, 461–464
diploid-triploid 463
haploid-diploid 447, 461, 462, 464, 465
haploid-haploid 462, 464
haploid-triploid 447, 462
u-hybrids 447
zygotic (cleavage or budding) 445–447, 452, 454, 456–458, 461, 469
determinate 469
indeterminate 469
proembryonal polyembryony 424

Polyembryony (in vitro) 760, 773
Polyhaploid 476
Polymorphism 501, 502
Polyploidy 519, 558, 559
  physiological role 558
Polytene chromosomes 521, 532, 534, 536, 544, 554, 566
Primexine 232–237, 244
Proembryo 377, 408
  globular 387, 398, 404, 407–409, 419, 420, 432
  octant 385–387, 398, 416
  quadrant 385, 386, 390, 392
  terminal cell (see apical cell)
  tetrad 385, 386, 390, 392
  three-celled 406
  two-celled 384, 390, 392, 406, 424, 429, 431, 433
    apical cell 384–386, 388, 390, 392, 399, 406–408, 424, 431, 433
    basal cell 384, 385, 387, 390–392, 406, 407, 424, 431, 433
Proendospermous cell 533
Protease 624
Protein 613, 624, 629, 631, 632, 636
  bodies (see aleurone grain) 324, 338, 349, 618–620, 624–626, 629–631
  crystaloid inclusions 363
  globoid inclusions 363
  subaleurone layer 362
  synthesis 346, 545
Protoplast culture 782
  callus formation 782
  differentiation 782
    plantlet 784
    shoot 782
  embryo/embryoids 784
  embryogenesis 782
  isolation of protoplast 782
    enzymatic 782
    mechanical 782
  liquid drop method 782
  nutrient media 782
  osmotic solution 782
  protoplast fusion 784
    heterokaryons 786
    heteroplasmic fusion 786
    homokaryons 786
    hybrid (parasexual) 788
    intergeneric 786
    interspecific 786
    intertribal 786
  protoplast isolation from
    cell suspensions 782
    haploid species 784
    mesophyll 784
    pollen tetrads 784
  selective markers 786
    albino 786
    auxotrophic 786
    chlorophyll-deficient 786
Pseudogamous heterosis 491
Pseudogamy 476, 488–491, 493, 494
Pseudohomeotypic division 479 (foot-note)

Radicle 425, 611–613, 616, 618, 625–627, 629, 630, 637
Raphal bundle 571
Raphe 569
Regeneration (polyembryony) 468, 469
Regression of (in embryogenic pollen)
  cytoplasm 750
  mitochondria 750
Reproductive calendar 20–22
Restitution nucleus 479–483, 498, 499, 520, 521, 556
Restitutional mitosis 520, 525–529, 535, 536, 538–540, 558
Reversion 500
R-hybrids 447
RNA 324–326, 338, 341, 349, 359, 365, 623, 638
RNA synthesis 551
Ribosomes 616, 621, 624, 629, 630
Root cap 387, 390, 395
  cortex 388, 390, 394, 395
Root tip 621, 625, 626, 629, 630, 638
Rumination 603

Saprophytes 426–429
Scutellum 544, 629–633
Seed(s)
  aerenchymatous tissue 578
  albuminous 606, 607, 612, 651, 661, 681
  aril 590–592, 647, 650, 667, 682
  balloon 595
  bitegmic 569
  campylotropous 574
  caruncle 650
  colour 598
  cork cells 579, 581
  crystal cells 579, 581
  cuticles 579
  development 569, 578
  dispersal 606
  dormancy 611, 616, 625
  dust seeds 595
  elaisome 592
  endotegmic seed-coat 584
  endotestal seed-coat 584
  epidermis 578
  exalbuminous 570, 606, 612, 651, 661
  exotegmic 575
  exotegmic seed-coat 584
  exotesta 575
  exotestal seed-coat 582

germination (monocots) 613, 629–637
habit 567
hairs 573
hilar 573
hydrochoric 596
hypodermis 578
imbibition lids 602
incompatibility 490, 491, 502
labyrinth seeds 603
light line (in testa) 580, 581
macrosclereids 581
Malpighian cells 580, 581
mechanical cells 580, 581
mechanical layer 575
mesotegmic seed-coat 584
mesotestal seed-coat 582
monoembryonate 449, 467
mucilage cells 581
mucilage hairs 602
mucilage seeds 596
multilayered endotesta 578
multiplicative integuments 578
multiplicative middle layers 578
myxospermy 596
obcampylotropous 574
opercula 599
pachychalazal seed 574
perichalazal 574
polyembryonate 445, 449, 454
primary sculpture 587
raphe 672, 680
rupture layer 599
sarcotesta 578, 590
sclerenchyma cells 579
sclerenchyma fibres 581
secondary sculpture 587
secretory tissue of seed coats 579
seed-coat 429, 432, 651, 661, 662, 664, 675, 680, 681
chlorophyllous 589
seed-coat-less seeds 585
shedding 426, 427, 430, 432
stomata in seed coats 589
surface 585
tegmen 569
tertiary sculpture 587
testa 569
tracheidal cells 582
types 572
undifferentiated seed coats 585
unspecialized inner integument 578
wings 594
Seed ageing 637–639
endospermic 612
hydration 615, 616, 618, 619, 621, 627, 630–632, 637
morphology 612–614
non-endospermic 612
storage 637
storage compounds (see also calcium oxalate, carbohydrate, lipid, phytin, protein, starch) 611, 613
Seedlings (zygotic) 449, 451
Self-electrophoresis 384
Self-incompatibility 250–261
Shoots, multiple 449, 452, 464
Single cell culture 738
conditioned medium 738
Maximov double cover slip method 738
microchambers 738
mineral oil 739
Somatic hybrids of
Arabidobrassica 786
Datura 786
tobacco 786
Sperm cell(s) 44, 379–381, 429–431, 433
composition 303
gametic fusion 302, 304, 305, 387
transport 284, 302
ultrastructure 306, 307
Spindle fusion 520, 525, 526, 535, 536, 538, 558
Sporogenous tissue 88
before meiosis 88
multimembrane inclusions 88
restoration of ribosomes 89
substantial fall in ribosomes 88
cytomictic channels 95
microspore mother cell
persistent pectocellulosic wall of 107, 109
nuclear pores 92
nuclear vacuoles 91
nucleolar cap 92
nucleolar cycle 92
nucleolar organizer 92
plastids and mitochondria 94
differentiation and redifferentiation 95
Sporopollenin 69, 204, 206, 221, 232–242, 244, 248–258, 260
degradation 82
deposition 70
polymerization 70
precursors 69
primexine template 69
control for pattern determination 69
Stamen 53
filament 53
laminar 53
Starch (grains) 327, 613, 618, 621, 622, 625, 629, 630–632, 636
grains 327
Stem tip 378, 390, 393, 394, 398–401, 404, 405
Stigma 273
dry 273
morphology 274, 275
pellicle 275

Stigma wet 273
composition 276, 277
exudate 274, 275
stigmatoid tissue 275
Storage cotyledons 605
Storage tissue 605
Style 274
hollow 274
solid 274
transmitting tissue 277
Subsexual reproduction 498
Suspension cultures 738
plating 739
Suspensor 378, 385, 387, 390, 392, 395, 404–406, 408, 409, 411, 414–423, 425, 493, 546–551, 589, 657, 661, 671, 685
autolysis 421
basal cell 407, 408, 416
coenocytic 556, 557
endopolyploidy 419
enzymes 412, 413, 421
fresh weight increase 418
growth rate 418
growth regulators 422, 423, 429
haustoria 28, 34, 35, 415, 421, 424, 651, 655, 661
in vitro culture 422
ultrastructure 418–421
variation in organization 414, 415, 417
wall ingrowths (transfer cells) 419–421

Tannin cells 579
Tapetum 60, 61, 217, 239, 523, 650, 656, 670
activity of callase 74
biseriate 61
cytology of 67
endomitosis 67
inhibited mitosis 69
mitotic cycle 67
multinucleate 525–527
polyploidy 67
sympathetic phenomenon 67
uninucleate 523, 524
dimorphic 61
dual origin 61
external 525, 527
internal 525, 527
functions 72
abortion of pollen 73
glandular (secretory) tapetum 64, 650, 656, 657, 666, 669, 670, 673, 677, 678, 681
cell wall autolysis 67
convective cyclosis 67
microtubules 66
plasmodesmatal connections 64
polyvesicular bodies 66
pro-Ubisch bodies 66
Ubisch bodies 77
by-product 82
changes in size 77
degradation 82
esterases in sporopollenin, origin and development 79
hypertrophied 73
low proline content 74
male fertile lines
high alanine, aspartic acid and glycine content 74
male sterile lines 74
multiseriate 61
periplasmodial (amoeboid) tapetum 6, 68, 524, 529, 650, 661, 671, 678
multivesiculate bodies 62
population of organelles 62
raphides 62
reorganization 62
single layer 61
tapetal membrane 83
functions 85
persistence 83
Taraxacum type (see diplospory)
Techniques
clearing fluids 4, 5
fluorescence microscopy 3
microcinematography 44, 45
micro-well 4
Nomarski contrast optics 3, 5, 45
phase contrast microscopy 3, 4
polarizing optics 3
scanning electron microscopy 3
Test tube fertilization 737
Tonoplast 619
Transfer cells 28–30, 421, 571
Twin embryo 487

Vacuoles 616, 619–621, 625, 629, 630, 634, 635, 639
Vascular differentiation (seeds) 624, 630, 631
Vascular supply (seeds) 571
Vascular tissue 618, 630, 631
Virus (in pollen) 245
Volume ratios of ovule, endosperm and embryo 570

Water permeability 582

Zygote 308, 378, 380, 381, 391, 395, 407, 424, 433, 547, 561
PAS-positive wall 381, 382
plastid inheritance 379–381
polarity 379, 383–385
size adjustments 382
structure and composition 379–382
Zwischenskörper 206, 221, 243, 249

D. de Nettancourt

Incompatibility in Angiosperms

1977. 45 figures, 18 tables. XIII, 230 pages
(Monographs on Theoretical and Applied Genetics, Volume 3)
ISBN 3-540-08112-7

Contents: The Basic Features of Self-Incompatibility. – The Genetic Basis of Self-Incompatibility. – Morphology and Biochemistry of the Self-Incompatibility Reaction. – Modifications of the Self-Incompatibility Character and S-Mutations. – Interspecific Incompatibility. – Conclusions: Incompatibility in Basic and Applied Researches.

This book classifies and discusses the enormous amount of isolated data accumulated on the physiology, biochemistry, ultrastructure, cytology, genetics, and evolution of self- incompatibility and interspecific incompatibility in angiosperms. It outlays the probable evolution of self-incompatibility as a primitive outbreeding device and of interspecific incompatibility and self-compatibility as derived conditions. The various mechanisms of incompatibility known to operate in nature are presented in detail together with a critical examination of their importance for research in fundamental biology and in plant breeding. The numerous models and hypotheses elaborated to explain certain poorly understood features of incompatibility systems are reviewed. At the same time, and with special emphasis on the use of mutagens, chemicals and in vitro culture techniques, the author makes a detailed presentation of the arsenal of methods now available for bypassing incompatibility barriers and transforming the breeding behavior of higher plants.

Springer-Verlag
Berlin
Heidelberg
New York
Tokyo